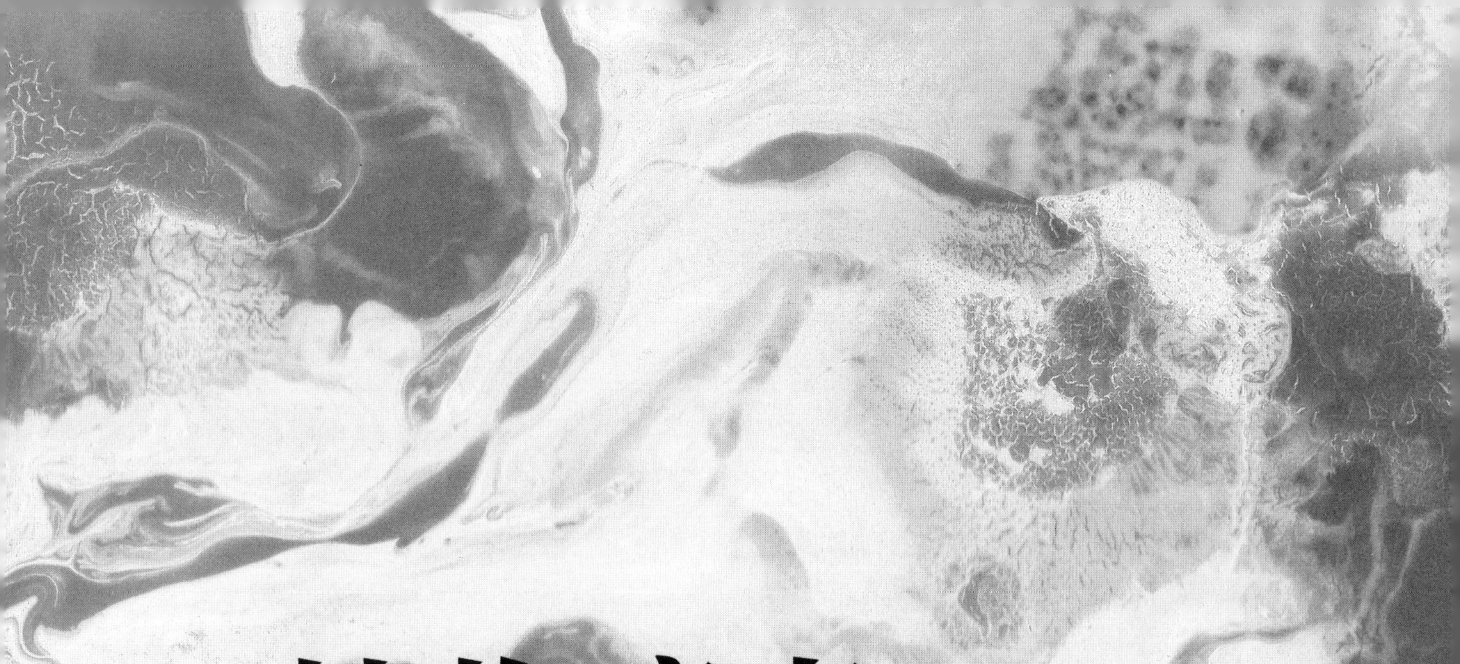

植物病毒：致病机制与病害调控

◎谢联辉　吴祖建　魏太云◎著

海峡出版发行集团 | 福建科学技术出版社
THE STRAITS PUBLISHING & DISTRIBUTING GROUP | FUJIAN SCIENCE & TECHNOLOGY PUBLISHING HOUSE

图书在版编目（CIP）数据

植物病毒：致病机制与病害调控 / 谢联辉，吴祖建，魏太云著. —福州：福建科学技术出版社，2022.4
ISBN 978-7-5335-6600-5

Ⅰ. ①植… Ⅱ. ①谢… ②吴… ③魏… Ⅲ. ①植物病毒－研究 Ⅳ. ①S432.4

中国版本图书馆CIP数据核字（2021）第258064号

书　　名　植物病毒：致病机制与病害调控
著　　者　谢联辉　吴祖建　魏太云
出版发行　福建科学技术出版社
社　　址　福州市东水路76号（邮编350001）
网　　址　www.fjstp.com
经　　销　福建新华发行（集团）有限责任公司
印　　刷　福建新华联合印务集团有限公司
开　　本　890毫米×1240毫米　1/16
印　　张　48
图　　文　768码
版　　次　2022年4月第1版
印　　次　2022年4月第1次印刷
书　　号　ISBN 978-7-5335-6600-5
定　　价　198.00元

前 言

福建农林大学植物病毒研究所，其前身为福建农学院植物病毒研究室（1979—1994）、福建农业大学植物病毒研究所（1994—2000），2000年改为现名。

本所自1979年成立以来，在大家的努力下，先后获得植物病理学科的硕士学位授予点（1984）、博士学位授予点（1990）、博士后科研流动站（1994）、福建省211重点学科（1995）、农业部重点学科（1999）和国家重点学科（2001），获准建设福建省植物病毒学重点实验室（1993）、福建省植物病毒工程研究中心（2003）、教育部生物农药与化学生物学重点实验室（2004）、财政部植物病原学特色专业实验室（2007）、农业部亚热带农业生物灾害与治理重点实验室（2008）和国家有害生物生态防控重点实验室（2016）。

42年来，本所以一个中心（培养“一高三超”英龙人才）、两个过硬（思想、业务）、三位一体（教学、科研、服务）、四个推动（科技进步、经济发展、社会文明和生态文明）、五种意识（健康、责任、团队、创新、忧患）为宗旨，以“献身、创新、求实、协作”为所训，以“敬业乐群、达士通人”为所风，主要从事以水稻为主的植物病毒和病毒病害研究，其间随着学科发展和生产实际的需求，拓展了植物病害和天然产物、植病经济与生态调控等的研究，先后主持和参加这些研究的骨干人员有谢联辉、林奇英、吴祖建、魏太云、吴建国、何敦春等，参与研究并已出站的博士后有蒋继宏等14位，已获博士学位的研究生有周仲驹等94位，发表学术论文540多篇。为了及时总结、便于查阅，特选汇这些论文，陆续出版。本集在前四集（第一集于2001年由福建科学技术出版社出版，第二至四集2009年由科学出版社出版）的基础上，从2009年至今发表的130多篇论文中选出66篇，分病毒侵染与致病机制、病毒遗传与进化分析、检测鉴定与流行调控、寄主抗性与天然产物四个部分，汇编而成。

考虑到全书格式的一致性，将原文中的作者简介和通讯作者标注予以删除。在此集编辑出版过程中，本所何敦春、林文武、高芳銮、陈启建、庄军、丁新伦、郑璐平、谭庆伟、杜振国、郑蓉蓉等博士做了大量工作，并得到福建科学技术出版社刘宜学编审、谢娟梅编辑的指导和支持，谨此致以衷心的感谢！

谢联辉

2021.03.18

前言

目 录

I 病毒侵染与致病机制

II 病毒遗传与进化分析

Ⅲ 检测鉴定与流行调控

Ⅳ 寄主抗性与天然产物

附录

I　病毒侵染与致病机制

这部分着重研究了我国较为重要的几种水稻病毒编码蛋白的功能特性，以及病毒在植物寄主和昆虫介体内侵染过程中的调控与致病机制。主要内容包括：（1）水稻病毒编码蛋白在病毒侵染过程中的功能鉴定；（2）水稻病毒编码蛋白与寄主因子互作的分子机制；（3）水稻病毒胁迫下植物寄主或昆虫介体在转录水平、表达水平以及激素水平上的变化。另外，对其他几种植物病毒的侵染与致病机制也开展了部分研究。

水稻条纹病毒与水稻互作中的生长素调控

杨金广，王文婷，丁新伦，郭利娟，方振兴，谢荔岩，
林奇英，吴祖建，谢联辉

（福建省植物病毒学重点实验室，福建农林大学植物病毒研究所，福州 350002）

摘要： 利用 real-time RT-PCR 和高效液相色谱技术对水稻条纹病毒（*Rice stripe virus*, RSV）侵染水稻（*Oryza sativa* subsp. *japonica*）植株和水稻悬浮细胞内的生长素合成酶基因 *YUCAAI* 表达量和内源生长素含量的变化分别进行了测定。结果表明，在细胞水平，RSV 侵染后的 16~64h 内能显著引起 *YUCAAI* mRNA 表达量的上调和内源生长素含量的升高。与同一生长阶段的健康水稻相比，水稻植株接种后 4~8d 内也可导致 *YUCAAI* mRNA 表达量的上调和内源生长素含量的上升，而在接种后 12d 和 16d 时，病株内的 *YUCAAI* mRNA 的表达量和内源生长素的含量均下降。这表明在 RSV 侵染水稻后的发病过程中，RSV 能够调控寄主植物内源生长素的合成。同时，利用 KPSC 缓冲液处理病株来消除其内源生长素，能够引起 RSVCP 基因表达上调近 2.9 倍；另外用 30μmol/L IAA 溶液处理病株可使其体内的 RSVCP 基因表达下调 45%，表明水稻体内生长素含量的变化能够影响 RSV 在寄主体内的复制。

关键词： 生长素；水稻；水稻条纹病毒；实时定量 RT-PCR

中图分类号： S188　**文献标识码：** A　**文章编号：** 1006-1304(2008)04-0628-07

Auxin Regulation in the Interaction between *Rice stripe virus* and Rice

Jinguang Yang, Wenting Wang, Xinlun Ding, Lijuan Guo, Zhenxing Fang, Liyan Xie, Qiying Lin, Zujian Wu, Lianhui Xie

(Key Laboratory of Plant Virology of Fujian Province, Institute of Plant Virology, Fujian Agriculture and Forestry University, Fuzhou 350002)

Abstract: The expression of *YUCAAI* gene and the amount of endogenous IAA in rice (*Oryza sativa* subsp. *japonica*) plants and rice suspension cells infected by *Rice stripe virus* (RSV) were investigated by Real-time RT-PCR and high performance liquid chromatography, respectively. And the results showed that the expression of *YUCAAI* gene and the amount of endogenous IAA increased at various times (16, 32, 48 and 64h) after post-infection by RSV in rice suspension cells. In rice plants infected by RSV, the expression of *YUCAAI* gene and the amount of endogenous IAA increased at 4-8d after post-infection as comparison with that of healthy rice plants, and decreased at 12 and 16d. These results indicated that RSV infection could regulate auxin biosynthesis in

农业生物技术学报 . 2008,16(4):628-634
收稿日期：2007-11-05；接受日期 :2007-12-10
基金项目：国家重点基础研究规划 (973) 项目（No. 2006CB100203）、教育部博士点专项科研基金 (No. 20040389002; Na 20050389006)、国家自然科学基金 (No. 30671357) 和福建省教育厅科技三项费用 (NoX03005) 资助。

rice. Additionally, the expression of RSV *CP* increased 2.9 times in rice plants after it was treated with KPSC buffer to deplete the endogenous auxins, and decreased 45% after 30μmol /L IAA treatment. All of these results suggest that the auxin may play a role among RSV replication in rice plant.

Keywords: auxin; rice; *Rice stripe virus* (RSV); real-time RT-PCR

作为纤细病毒属（*Tenuivirus*）的代表种（*Toriyama* and *Tomaru*, 1995）水稻条纹病毒（*Rice stripe virus*, RSV）是由灰飞虱（*Laodelphax striatellus*）传播的，并具有双义编码特征的负单链 RNA 植物病毒（Zhu et al., 1991, 1992; Takahashi et al., 1993; Hamamatsu and Ishihama, 1994; Toriyama et al., 1994）。1999 年以来，该病毒引起的水稻条纹叶枯病在我国水稻种植区大面积流行发生，给当地水稻生产造成巨大损失，引起了广泛的关注（魏太云 , 2003）。2007 年，对江苏、河南、云南、山东、安徽、湖北、浙江等地的水稻条纹叶枯病进行了调查，发现该病在田间的发病率一般在 5% ～ 10%，严重的可达 20% 以上，个别田块出现“秃头田”现象，已经成为当地水稻一种主要的病害。当前，抗病品种的匮乏是造成该病害大规模暴发流行的主要因素，而深入研究 RSV 与寄主互作的分子机理是解决该问题的主要突破口之一。为明确生长素在 RSV 与水稻互作中的作用，本研究利用 Real-time RT-PCR 和高效液相色谱技术（high performance liquid chromatography, HPLC）对 RSV 侵染水稻植株和水稻悬浮细胞内的生长素合成酶基因 *YUCAAI* mRNA 表达量和内源生长素含量的变化分别进行了分析，并通过消除内源生长素和外施人工合成生长素等对病株进行处理，定量检测了水稻病株内 RSV 复制的变化，以期为深入研究 RSV 与水稻互作过程中的生长素信号传导提供基本的实验依据。

1　材料和方法

1.1　植物材料和病毒毒源

高感粳稻品种武育粳 3 号（*Oryza sativa* L subsp. *japonica* cv. Wuyujing 3）为实验植物。病毒毒源为采自江苏洪泽水稻田间呈典型水稻条纹叶枯病症状的病株，RT-PCR 鉴定后，经带毒 *Laodelphax striatellus* 传毒，保存于水稻植株（台中 1 号）中。发病病株用于 RSV 提纯，提纯方法参照 Toriyama 的方法（Toriyama, 1982）。

1.2　试剂与仪器

RNA 提取试剂盒 Trizol Reagent 为美国 Invitrogen 公司产品；TaKaRa EXScript™ RT-PCR Kit 购于大连宝生物工程有限公司；荧光定量 PCR 仪 MiniOpticon™ System 系统为美国 Bio-Rad 产品；Agilent 1100 HPLC 系统为美国 Agilent 科技公司产品。

1.3　病毒接种与样品处理

2 叶期稻苗经带毒灰飞虱（*Laodelphax striatellus*）接种后，置于 MS 液体培养基中，28°C，14h 光照和 10h 黑暗交替条件下培养。样品采集分 5 次进行，分别于接种后 0、4、8、12 和 16d 采集样品。每株样品采集后置于液氮中，研磨成粉末，一部分通过 RT-PCR 对 RSV 进行检测，剩余样品置于 –80°C 保存备用。对 RSV 检测呈阳性的样品分别进行 *YUCAAI* mRNA 表达量和内源生长素含量的测定，每批样品均以相应生长时期的健康植株为空白对照。

对发病初期整株水稻进行以下 3 种处理，分别用 30μmol/L IAA 溶液和 KPSC 缓冲液（10μmol/L 磷酸钾，pH 6.0，2% 蔗糖，50μmol/L 氯霉素）浸泡处理 16h，每 2h 更换 1 次缓冲液，KPSC 缓冲液的处理可消除植株体内的内源生长素（Jain et al., 2006a）。一部分经 KPSC 缓冲液处理后的病株重新转移于新配制的含有 30μmol/L IAA 溶液中再处理 16h。每日定时采集 3 种不同处理的样品，于 –80°C 条件下保存备用。

1.4 水稻悬浮细胞培养与 RSV 侵染

武育粳 3 号水稻悬浮细胞培养与 RSV 侵染参照（Yang et al.,1996）方法。样品采集分 5 次进行，分别于接种后 0、16、32、48 和 64h 进行样品采集，每批样品均以相应生长阶段的健康水稻悬浮细胞为空白对照。

1.5 引物设计

应用 Perl Primer 软件进行 Real-time PCR 特异性引物设计（Marshall, 2004）。为保证所设计引物特异性，所选引物均在 GenBank 数据库中的 BLAST 程序下进行比较分析，其中 RSV CP 基因的引物除要求与所有已知 RSVCP 基因同源外，还要严格避免与水稻基因组中任何序列同源，水稻生长素合成酶基因引物要求避免与水稻其他基因序列同源。RSV CP 基因保守区域特异性引物：5′ 端：5′-RTTGACAGACATACCAGCCAG-3′（R=A/G）；3′ 端：5′-CATCATTCACTCCTTCCAAATAACY-3′（Y=C/T）。水稻生长素合成酶基因 *YUCAAI* 特异性引物：5′ 端：5′-TCATCGGACGCCCTCAACGTCGC-3′；3′ 端：5′-GGCAGAGCAAGATTATCAGTC-3′。水稻真核延伸因子基因（eukaryotic elongation factor 1-alpha gene, *eEF*-la）作为内参基因：引物 5′ 端：5′-TTTCACTCTTGGTGTGAAGCAGAT-3′；3′ 端：5′-GACTTCCTTCACGATTTCATCGTAA-3′（Jain et al., 2006b）。

1.6 总 RNA 提取与 Real-time RT-PCR 检测

取植物样品 50mg，于液氮条件下研磨，然后用 1mL Trizol 试剂提取总 RNA，方法按公司提供的产品说明进行。

应用 TaKaRa 公司的 Exscript™ RT ReagentKit 试剂盒进行反转录反应，20μL 反转录反应体系中含总 RNA 1μg、3′ 端引物（10pmo/L）lμL、5×Exscript™ Buffer 4μL、Exscript™ RT Enzyme Mix I 1μL, 最后用 RNase Free ddH$_2$O 补足 20μL。反转录反应条件为：42℃反应 15min，85℃灭活 5s。取 1μL cDNA 溶液为模板，然后加入以下试剂：2×Premix Ex *Tag*™ 25μL、3′ 端引物和 5′ 端引物（10pmol/L）各 1μL 和 ddH$_2$O 22μL。Real-time PCR 在 50μL 反应体系中，于 48 孔的 MiniOpticon™ System 系统中进行以下反应。95℃预变性 10s 后，进行 45 个循环。每个循环为 95℃变性 6s，62℃退火 20s。然后进行融解曲线制作，Real-time RT-PCR 扩增产物的特异性均通过 1.5% 琼脂糖凝胶电泳和每个基因的融解曲线进行鉴定。

用 EASY Dilution 将每个基因的 cDNA 溶液按 4^0、4^1、4^2、4^3 和 4^4 梯度稀释后，各取 1μL 稀释后的 cDNA 作为模板进行 Real-time PCR 灵敏性检测和标准曲线构建。根据参比基因对所有样品进行归一化处理（初始 RNA 量校正），然后确定每个目的基因在不同样品中的相对表达量。每个样品重复检测 3 次，并至少进行 2 次生物实验重复。

1.7 内源生长素的 HPLC 测定

不同样品内源生长素提取参考（Kelen et al., 2004）方法，采用 Agilent 1100 HPLC 系统，色谱柱为 Supelcosil TM LC-18 分析柱（5.0μm, 4.6mm×250.0mm），流动相为甲醇：水：乙酸溶液（45.0 ：54.2 ：0.8, V/V）；流速为 0.5mL/min，紫外检测波长 254.0nm。外标法定量，标样为 Fluka 的 HPLC 试剂，将配制的混合标样经上述纯化分离过程测定回收率，根据回收率对样品测定结果进行校正。

2 结果

2.1 生长素合成酶基因 *YUCAAI* 和 RSV *CP* 基因引物特异性验证与 Real-time RT-PCR 精确性检验

通过 Real-time RT-PCR 对水稻生长素合成酶基因 *YUCAAI* 和 RSV *CP* 基因进行扩增，得到条带单一，大小分别为 186 和 243bp（图 1A，图 B），与预期设计均相符，其融解曲线峰值单一（图 2 A，图 B）。用一系列 4 倍稀释的 cDNA 为模板，制作了 *YUCAAI* 和 RSV *CP* 基因的标准曲线，其分别是 0.998 和 0.996（图 3 A，图 B），均大于 0.995，说明其线性系数良好，具有较高的精确性，适合对不同样品中目的基因相对定量分析。表明这些引物适合基于 SYBR Green I 方法的 Real-time RT-PCR 检测。

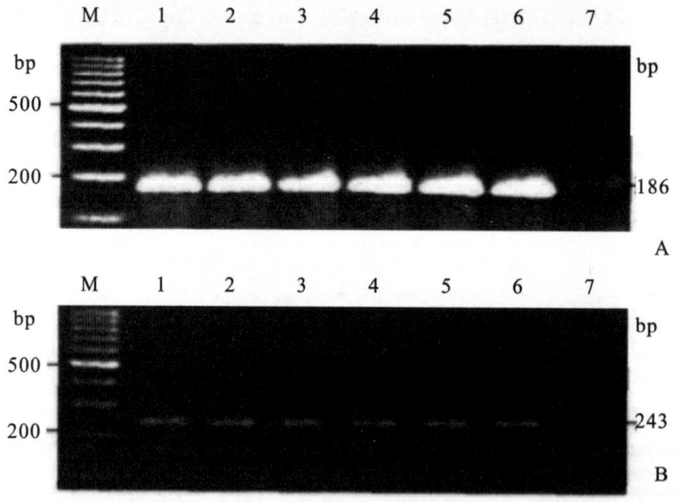

图 1 水稻生长素合成酶基因 *YUCAAI*（A）和 RSV *CP*（B）的荧光定量 RT-PCR 检测

Fig. 1 Detections of rice auxin synthase gene *YUCAAI* (A) and RSV *CP* gene (B) by Real-time RT-PCR

M,100bp maker, 1-6, PCR products of *YUCAAI* (A) and RSV *CP* gene (B);7,blank control

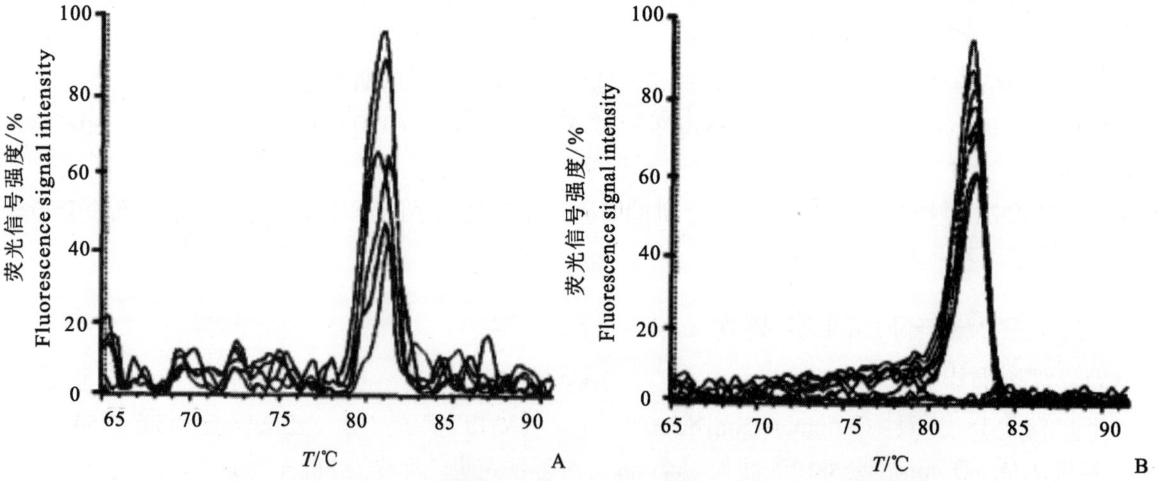

图 2 水稻生长素合成酶基因 *YUCAAI*（A）和 RSV *CP* gene（B）的融解曲线

Fig. 2 Melting curves of auxin synthase gene *YUCAAI* (A) and RSV *CP* gene (B)

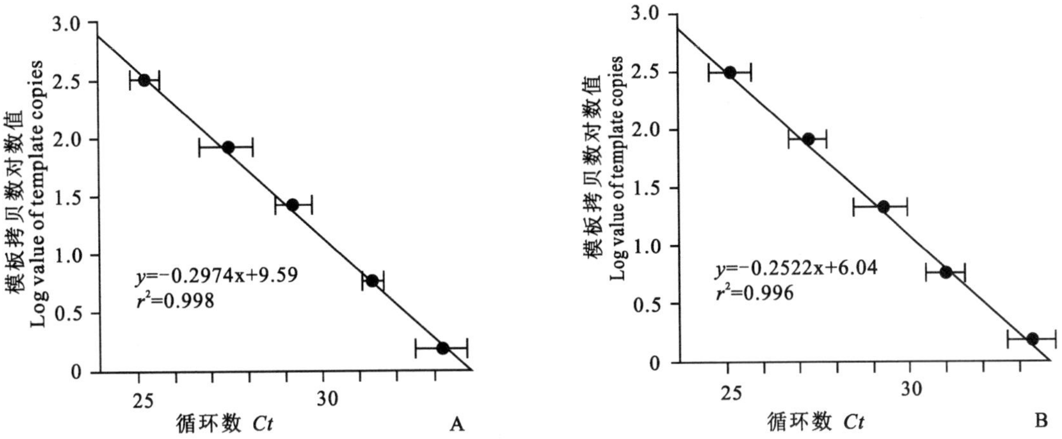

图 3 水稻 *YUCAAI* 基因（A）和 RSV *CP* 基因（B）的标准曲线

Fig. 3 Standard curves of of auxin synthase gene *YUCAAI* (A) and RSV *CP* gene (B)

Values in the figures represent the means±*SE* of three replicates

2.2 RSV 侵染对生长素合成酶基因表达的影响

根据每个样品中内参基因的表达量进行归一化处理，再对不同样品的 *YUCAAI* 的表达量与各自空白对照的 *YUCAAI* 表达量进行比较分析发现，在水稻悬浮细胞体系中，RSV 的侵染能够显著引起水稻悬浮细胞内 *YUCAAI* mRNA 表达量的上调。在 RSV 侵染后 16h，其表达量上升为健康细胞的 4.76 倍；32h 后，其表达量为健康细胞的 5.98 倍；在侵染后 48h 稍微下降，为健康细胞的 4.57 倍；在侵染后 64h，其表达量又上升至健康细胞的 6.03 倍（图 4A）。RSV 侵染水稻植株后，仅在侵染后 4d 和 8d 能够引起水稻 *YUCAAI* mRNA 表达量的上调，分别为健康植株的 5.16 倍和 1.38 倍；而在侵染后 12d 和 16d，病株内 *YUCAAI* mRNA 表达量分别为健康水稻的 20% 和 17%（图 4B），其中接种后 16d，水稻条纹叶枯病的典型褪绿条纹症状已经出现。

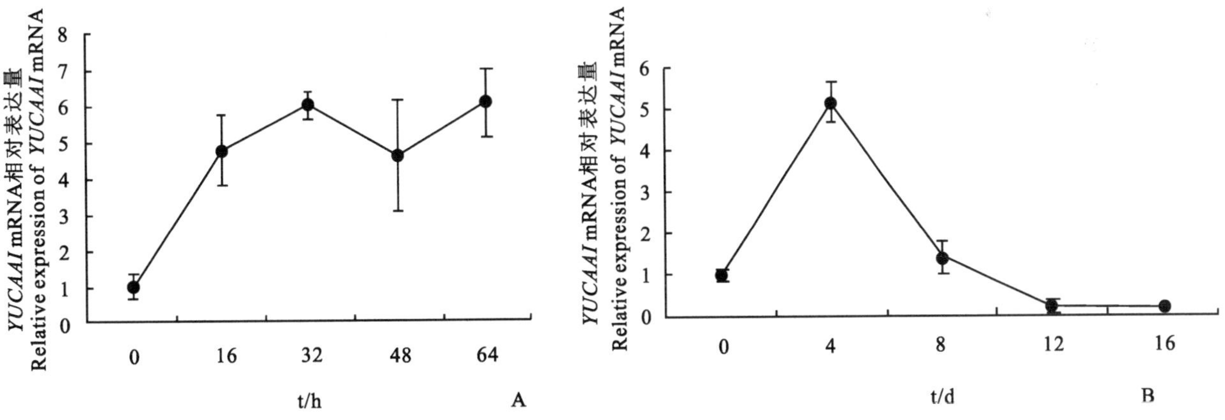

图 4 RSV 侵染水稻悬浮细胞（A）和水稻植株（B）对 *YUCAAI* mRNA 表达的影响

Fig. 4 Influences for *YUCAAI* mRNA expression in rice suspension cells (A) and rice plants (B) infected by RSV Values in the figures represent the means ± SE of three experimental replicates

2.3 RSV 侵染对水稻内源生长素的影响

图 5 A 表明，与未被侵染的水稻悬浮细胞相比，RSV 侵染后能够显著引起水稻悬浮细胞的内源生长素含量的升高。在侵染后 16 ～ 48h 细胞内的内源生长素含量维持一个较高的水平，其含量是健康细胞内的 2.34 ～ 2.59 倍。在病毒侵染后 64h，其内源生长素含量依然升高，为健康细胞的 4.43 倍。在植株水平，RSV 侵染水稻植株后第 4d 和第 8d，其内源生长素呈上升趋势，其含量分别是健康水稻 1.31 倍和 1.38 倍。而在 RSV 侵染后第 12d，病株内的内源生长素含量开始下降，是同一生长阶段健康水稻的

81%。在病毒侵染后 16d，症状已经出现，其内源生长素含量为同一生长阶段健康水稻的 73%（图 5B）。

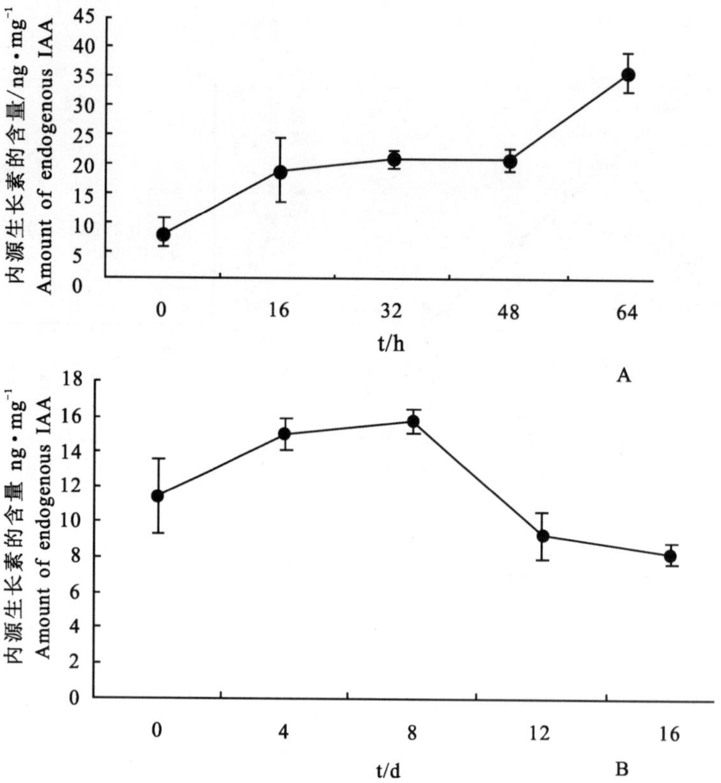

图 5 RSV 侵染水稻悬浮细胞（A）和水稻植株（B）对内源生长素的影响

Fig.5 Influences of endogenous IAA in rice suspension cells (A) and rice plants (B) infected by RSV Values in the figures represent the means ± *SE* of three experimental replicates

2.4 不同处理植株中 RSV 复制变化

为确定植株内源生长素变化对 RSV 复制的影响，本研究选取了 RSV *CP* 基因作为检测对象，应用 Real-time RT-PCR 对经 KPSC 缓冲液和 IAA 溶液处理的水稻病株与未处理的水稻病株分别进行比较分析。结果表明 RSV 病株经 KPSC 缓冲液处理消除内源生长素后，RSV*CP* 基因表达上调近 3 倍。IAA 处理能够使病株内 *CP* 基因的表达量下调 45%。KPSC 缓冲液处理后的病株再经 IAA 处理可使 RSV *CP* 基因表达量下降 55%（图 6）。

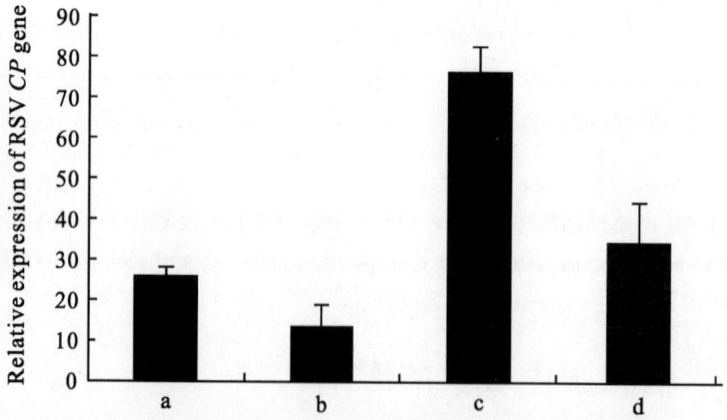

图 6 KPSC 缓冲液和 IAA 溶液对病株内 RSV 复制的影响

Fig. 6 Influences on RSV replication in diseased rice plants treated with KPSC buffer and IAA

a. 未经任何处理的病株；b. IAA 溶液处理的病株；c. KPSC 缓冲液处理的病株；d. KPSC 缓冲液处理后又经 IAA 溶液处理的病株。图中数据分别表示为平均值 ±3 次重复的标准误差

a. diseased rice plants untreated; b.diseased rice plants treated with IAA; c. diseased rice plants treated with KPSC buffer; d. diseased rice plants treated with IAA after treatment with KPSC buffer. Values in the figure represent the means ± *SE* of three experimental replicates

3 讨论

植物病毒侵染后可引起寄主植物生长发育异常，例如矮化、花叶、褪绿斑驳、黄化、顶端优势丧失，甚至死亡等表型。在植物病原物与寄主互作中，这些表型的发生并不是孤立的，众多的信号传导途径参与其内，例如水杨酸信号传导途径（Yalpani et al., 1991；Shirasu et al., 1997; Bartsch et al., 2006）和乙烯信号传导途径等（Lorenzo et al., 2003）。病原物与寄主互作的生长素信号传导已有报道。例如在 TMV 侵染拟南芥中，TMV 复制酶基因与一个生长素反应基因的调控因子 PAP1 互作，干扰 PAP1 的正常定位，导致其功能丧失，使正常的生长素信号传导发生紊乱，从而产生特异性病害症状（Padmanabhan et al., 2005）。同时生长素信号途径还参与了植物自然免疫反应，细菌鞭毛蛋白可诱导 miRNA393 对拟南芥生长素受体基因 *TIR1* 和 *ABFs* 进行负调控，从而抑制生长素信号传导，产生抗病性（Navarro et al., 2006）。通过过量表达 *OsWAKY31* 可显著提高水稻对 *Magnaporthe grisea* 侵染的抗性，并对外使人工合成生长素失去敏感性（Zhang et al., 2008）。这些研究表明，生长素信号传导途径在植物病原物发病机制 microRNA 介导的寄主植物免疫机制和寄主的抗病性等方面发挥着重要的作用。

笔者所在的研究小组已利用水稻全基因组 Affymetix 基因芯片对 RSV 与水稻互作的转录组学进行了研究，12 个生长素信号传导途径的相关基因（*Aux/IAAs, SAURs,GH3s* 和 *ARFs*）在 RSV 侵染水稻的转录谱中发生表达变化，其中有 9 个基因为生长素早期反应基因（5 个 *Aux/ZAAs* 基因，3 个 *SAURs* 基因和 1 个 *GH3s* 基因）（结果待发表）。在生长素信号传导途径中，生长素主要通过调控早期反应基因的表达来控制植物的生长发育（Abel and Theologis, 1996; Jain et al., 2006a, c, d）。为了验证是否为 RSV 侵染导致水稻内源生长素含量的变化，从而引起这些早期反应基因的表达变化，本研究选择水稻悬浮细胞和水稻植株为研究材料，对 RSV 侵染复制和发病过程中的生长素合成酶基因 *YUCAAI* 表达量与内源生长素含量的变化分别进行了比较分析，确定了 RSV 侵染对水稻内源生长素合成的影响。与健康水稻悬浮细胞相比，RSV 侵染水稻细胞 0 ～ 64h 内，可引起水稻生长素合成酶基因 *YUCAAI*；表达上调，致使水稻悬浮细胞内源生长素含量增多。而在植株水平，RSV 侵染初期，病株内的水稻 *YUCAAI* 表达上调，内源生长素含量也随之升高。但在侵染后期，特别是当症状出现后（接种后第 16d），病株内的 *YUCAAI* 表达量显著下调，其内源生长素的含量也明显低于健康水稻体内的含量。由此可见，RSV 侵染水稻后可引起水稻内源生长素合成发生变化，在侵染初期可显著提高内源生长素的合成，而在典型症状表现后病株体内内源生长素的合成受到抑制。这些结果也与本实验室已有的研究结果相一致，在 RSV 与水稻互作的转录组学的 microarray 分析中，所选择的研究材料为症状出现后 7d 的病株。从本研究结果分析表明，发病 7d 后的病株体内的内源生长素含量低于同期生长阶段的健康水稻。而内源生长素含量的降低可导致生长素信号传导途径中某些早期反应基因表达减少（Jain et al., 2006a, c, d）。其中 *OsIAA6*、*OsIAA9*、*OsIAA31* 和 *GH3-5* 和在 microarray 分析结果均表达下调，推测可能由于病株体内生长素含量的减少导致这些基因表达量的降低，并且这一结果与 Jain 等通过人工合成生长素处理的研究结论相一致（Jain et al., 2006a, c）。

由此可见，RSV 侵染后的不同阶段是造成细胞和植株水平生长素存在差异的主要因素，同时也表明生长素信号传导途径参与了 RSV 与水稻的互作，并存在一个动态过程。通常认为，侵染初期是病原物与寄主互作最激烈的阶段，也是各种信号传导最复杂时期。RSV 侵染水稻植株的过程，存在较长潜育期，需要 10 ～ 30d 才能表现症状。利用发病植株作为 RSV 与寄主互作的研究材料，从植物病原物与寄主互作角度分析，症状出现时 RSV 与水稻的互作已近末期，其初期互作过程往往被忽略。同时，水稻植株在培养过程中易受温度、光照和湿度等许多非生物环境因素的影响，造成 RSV 与水稻互作的研究结果存在较大的偏差性。因此，RSV 侵染水稻悬浮细胞体系为研究 RSV 与寄主互作的初期阶段、RSV 的侵染复制及病毒装配等提供了有利的平台。

本研究还发现，通过消除病株内源生长素可显著促进 RSV 的复制，再用人工合成生长素处理，病株内 RSV 含量又急剧下降。推测可能是由于 RSV 是专性寄生物，其病毒组分的合成与病毒粒体的装配均需要寄主细胞提供原料、能量和场所等，通过消除内源生长素和外施人工合成生长素对寄主植物的生长发育产生影响，从而间接影响了 RSV 的复制。这也表明生长素信号传导途径和 RSV 侵染水稻的过程是相互影响，相互联系的，两者之间可能存在一个 “cross-talk” 分子。当然，要深入揭示 RSV 与水稻互作的生长素信号传导途径，这些研究是远远不够的。需要根据已有的 RSV 与水稻互作的转录组谱的结果，对生长素信号传导途径下游基因进行检测和功能验证，例如生长素受体基因、相应的 *micro* RNA、生长素早期反应基因以及转录抑制因子等。

参考文献

[1] ABEL S, THEOLOGIS A. Early genes and auxin action[J]. Plant Physiology,1996,111(1):9-17.

[2] BARTSCH M, GOBBATO E, BEDNAREK P, et al. Salicylic acid-independent enhanced disease susceptibility signaling in Arabidopsis immunity and cell death is regulated by the monooxygenase FMOl and the nudix hydrolase NUDT7[J]. The Plant Cell, 2006, 18(4): 1038-1051.

[3] HAMAMATSU C, ISHIHAMA A. Ambisense *Rice stripe virus*[J]. Uirusu, 1994,44(1):19-25.

[4] KELEN M, CUBUK, DEMIRALAY E, et al. Separation of abscisic acid, indole-3-acetic acid, gibberellic acid in 99 R(*Vitisberlandieri* x vitisrupestris) and rose oil(*Rosa damascena* Mill.) by reversed phase liquid chromatography[J]. Turkish Journal of Chemistry, 2014, 28: 603-610.

[5] JAIN M, KAUR N, TYAGI A K, KHURANA J P. The auxin-respon-sive GH3 gene family in rice (*Oryza sativa*)[J]. FunctIntegr Genomics, 2006a, 6(1): 36-46.

[6] JAIN M, NIJHAWAN A, TYAGI A K, et al. Validation of housekeeping genes as internal control for studying gene expression in rice by quantitative real-time PCRBiochem. Biophys Res Commun, 2006b, 345(2): 646-651.

[7] JAIN M, KAUR N, GARG R, et al. Structure and expression analysis of early auxin-reespon- sive Aux/IAA gene family in rice(*Oryza sativa*)[J]. Functional & Integrative Genomics, 2006c, 6(1): 47-59.

[8] JAIN M, TYAGI A K, KHURANA J P. Genome-wide analysis, evolutionary expansion, and expression of early auxin-responsive SAUR gene family in rice(*Oryza sativa*)[J]. Genomics, 2006d, 88(3):360-371.

[9] LORENZO O, PIQUERAS R, SANCHEZ-SERRANO J, et al. Eethylene response factor 1 integrates signals from ethylene and jasmonate pathways in plant defense[J]. Plant Cell, 2003, 15(1): 165-178.

[10] Marshall O J. PerlPrimer: Cross-platform, graphical primer design for standard, bisulphite and real-time PCR[J]. Bioinformatics, 2004, 20(15): 2471-2472.

[11] NAVARRO L, DUNOYER P, JAY F, et al. A plant miRNA contributes to antibacterial resistance by repressing auxin signaling[J]. SCIENCE, 2006, 312(5772): 436-439.

[12] PADMANABHAN M S, GOREGAOKER S P, GOLEM S, et al. Interaction of the Tobacco mosaic virus replicase protein with the Aux/IAA protein PAP1/IAA26 is associated with disease development[J]. JVirol. 2005, 79(4): 2549-2558.

[13] SHIRASU K, NAKAJIMA H, RAJASEKHAR V K, et al. Salicylic acid potentiates an agonist-dependent gain control that amplifies pathogen signals in the activation of defense mechanisms[J]. Plant Cell, 1997, 9(2), 261-270.

[14] TAKAHASHI M, TORIYAMA S, HAMAMATSU C, et al. Nucleotide sequence and possible ambisense coding strategy of Rice stripe virus RNA segment 2[J]. Journal of General Virology, 1993, 74(4):769-773.

[15] TORIYAMA S. Characterization of Rice stripe virus: A heave component carrying infectivity[J]. Journal of General Virology, 1982, 61: 187-195.

[16] TORIYAMA S, TAKAHASHI M, SANO Y, et al. Nucleotide sequence of RNA 1, the largest genomic segment of Rice stripe vims,

the prototype of the tenuiviruses[J]. Journal of General Virology, 1994, 75(12): 3569-3579.

[17] TORIYAMA, S, TOMARU, K.Genus Tenuivirus. Murphy, FA, Fauquet, CM, Bishop, DH L, Ghabrial, SA, Jarvis, AW, Martelli, GP, Mayo, MA, Summers, MD Virus Taxonomy Classification and Nomenclature of Viruses. Sixth Report of the International Committee on Taxonomy of Viruses, Springer, Wien, 1995.

[18] Wei T Y. Molecular population genetics of Rice stripe virus[J]. Fujian Agricultural and Forestry University. (in Chinese),2003.

[19] YALPANI N, SILVERMAN P, WILSON M A, et al. Salicylic acid is a systemic signal and an inducer of pathogenesis-related proteins in virus-infected tobacco[J]. The Plant Cell, 1991, 3(8): 809-818.

[20] Yang W D, Wang X H, Wang S Y, et al. Infection and replication of a Planthopper transmitted virus-Rice stripe virus in rice protoplasts[J]. Journal of Virological Methods, 1996, 59(1-2): 57-60.

[21] Zhang J, Peng Y, Guo Z. Constitutive expression of pathogen-inducible OsWRKY31 enhances disease resistance and affects root growth and auxin response in transgenic rice plants[J]. Cell Res,2008, 18(4): 508-521.

[22] Zhu Y, Hayakawa T, Toriyama S, et al. Complete nucleotide sequence of RNA3 of Rice stripe virus: An ambisense coding strategy[J]. Journal of General Virology, 1991, 72(4): 763-767.

[23] Zhu Y, Hayakawa T, Toriyama S. Complete nucleotide sequence of RNA4 of Rice stripe vims isolate T and comparison with another isolate and with Maize stripe virus[J]. Journal of General Virology, 1992, 73(5): 1309-1312.

Pc4, a putative movement protein of Rice stripe virus, interacts with a type I DnaJ protein and a small Hsp of rice

Lianming Lu, Zhenguo Du, Meiling Qin, Ping Wang, Hanhong Lan, Xiaoqing Niu, Dongsheng Jia, Liyan Xie, Qiying Lin, Lianhui Xie, Zujian Wu

（Key Laboratory of Biopesticide and Chemibiology of Ministryof Education, Fujian Agriculture and Forestry University,Fuzhou, Fujian 350002, China）

Abstract: *Rice stripe virus* (RSV) infects rice and causes great yield reduction in some Asian countries. In this study, rice cDNA library was screened by a Gal4-based yeast two-hybrid system using pc4, a putative movement protein of RSV, as the bait. A number of positive colonies were identified and sequence analysis revealed that they might correspond to ten independent proteins. Two of them were selected and further characterized. The two proteins were a J protein and a small Hsp, respectively. Interactions between Pc4 and the two proteins were confirmed using coimmunoprecipitation. Implications of the findings that pc4 interacted with two chaperone proteins were discussed.

Keywords: *Rice stripe virus*; DnaJ protein; Small Hsp; interaction

1 Introduction

Infection cycle of plant viruses involves a phase of movement from initially infected cell into adjacent neighboring cells via plasmodesmata (PD). This cell-to-cell movement is aided by virus-encoded proteins termed movement proteins (MPs). Sometimes, the MPs could form tubules to replace PD to facilitate passage of virions. But, more often, the MPs only transiently and reversibly dilate PD openings to mediate transport of viral nucleic acids or ribonucleic acids–protein complexes(Carrington et al., 1996; Waigmann et al., 2004; Scholthof, 2005; Lucas, 2006). In the second case, the MPs usually share many features with a set of endogenous host factors named non-cell autonomous proteins (NCAPs) in terms of cell-to-cell movement (Carrington et al., 1996; Ghoshroy et al., 1997; Lucas and Lee, 2004; Oparka, 2004; Ruiz-Medrano et al., 2004; Waigmann et al., 2004; Boevink and Oparka, 2005; Gallagher and Benfey, 2005; Kim, 2005; Kurata et al., 2005; Scholthof, 2005; Lucas, 2006). Therefore, it is widely accepted that viral MPs exploit pre-existing cellular pathways to fulfill their function. Supporting this, crosscompetition experiments have demonstrated that the viral MPs and host NCAPs likely utilize a common receptor in the pathway for cell-to-cell transport (Kragler et al., 1998). Additionally, expression of a dominant-negative mutant form of NtNCAPP1, a non-cell-autonomous pathway protein (NCAPP), abolished cell-to-cell transport of TMV MP as well as specific NCAPs such as CmPP16 (Lee et al., 2003). The plant non-cell-autonomous pathway involves a set of cellular players that work coordinately (Lucas and Lee, 2004; Oparka, 2004; Ruiz-Medrano et al., 2004; Boevink and Oparka, 2005; Gallagher and Benfey, 2005;

Virus Genes. 2009, 38:320–327

Received 20 August 2008; Accepted 22 December 2008; Published online 8 January 2009

Kim, 2005; Kurata et al., 2005). Therefore, it is envisionable that many functions of viral MPs are dependent on a chain of interactions with these host factors (Carrington et al., 1996;Ghoshroy et al., 1997; Lucas and Lee, 2004;Oparka, 2004; Ruiz-Medrano et al., 2004;Waigmann et al., 2004;Boevink and Oparka, 2005; Gallagher and Benfey, 2005; Kim, 2005;Kurata et al., 2005;Scholthof, 2005; Lucas, 2006).

It was found that phloem proteins ranging from 10 to 200kDa induced an increase in size exclusion limit (SEL) to the same extent, greater than 20 but less than 40kDa, yet they all could move from cell to cell (Balachandran et al., 1997). This suggested that protein unfolding might be an essential step in plasmodesmal trafficking. Consistent with this, chemical-crosslinked KN1 that was unable to undergo conformational changes failed to mediate its own cell-to-cell transport (Kragler et al., 1998). The involvement of a phase of protein unfolding implicated a role of chaperone proteins in the NCAP pathway. Involvement of chaperone proteins in viral cell-to-cell movement was best illustrated by beet yellow virus (BYV), a member of the *closteroviridae*. This virus encodes an hsp70 homologue in its genome. The hsp70 homologue targets PDs and has been shown to be essential for cell-to-cell movement of BVY (Agranovsky et al., 1991; Peremyslov et al., 1999; Alzhanova et al., 2001). However, most viruses do not encode chaperones in their genomes. Instead, it seems that they have adapted to use the existing host cellular chaperone network. To do this, they could manipulate the host transcriptional network to induce expressions of a particular set of chaperone proteins (Whitham et al., 2003). Alternatively, they could recruit a chaperone protein directly from the host. For example, several viral movement proteins or proteins involved in viral movement were shown to interact with a set of DnaJ proteins (Soellick et al., 2000; von Bargen et al., 2001; Haupt et al., 2005; Hofius et al., 2007). Interestingly, it was shown recently that several HSP cognate 70 (hsc70) chaperones isolated from PD-rich wall fractions and from Cucubitapholem exudates could interact with PD and modify the PD SEL. Introduction of a common motif identified in these hsc70s allowed a human hsp70 protein to modify the PD SEL and move from cell to cell (Aoki et al., 2002). This raised the possibility that hsp70s might play a more direct role in viral movement than previously expected. For example, the Hsp70s, which have intrinsic ATPase activity, could serve as motor proteins facilitating the transport of viral materials through the PD. Noteworthily, no chaperone proteins other than the hsp70s family or their intimate partners have been reported to be involved in plant viral cell-to-cell movement.

RSV is the type species of the genus *Tenuivirus*, which has not been assigned to any family. It is transmitted transovarily in a circulative manner by some planthopper species (Delphacidae family), primarily the small brown planthopper (Falk, 1994; Ramírez and Haenni, 1994; Hibino, 1996; Falk and Tsai, 1998). The genome of RSV comprises four RNAs, named RNA1 to RNA4 in the decreasing order of their molecular weight (Ramírez and Haenni, 1994; Falk and Tsai, 1998). RNA1 is of negative sense and encodes a putative protein with a molecular weight of 337kDa, which was considered to be part of the RNA dependent RNA polymerase associated with the RSV filamentous ribonucleoprotein (RNP) (Toriyama, 1986; Barbier et al., 1992; Toriyama et al., 1994). RNAs 2–4 are ambisense, each containing two ORFs, one in the 5′ half of viral RNA (vRNA, the proteins they encoded named p2–p4) and the other in the 5′ half of the viral complementary RNA (vcRNA, the proteins they encoded named pc2–pc4). Pc2 shows stretches of weak amino acid similarity with membrane protein precursor of members of the Bunyaviridae that is processed into two membrane-spanning glycoproteins; however, there is no evidence that RSV forms enveloped particles that could incorporate such glycoproteins (Takahashi et al., 1993). P3 of RSV shares 46% identity with its counterpart, gene-silencing suppressor NS3 protein of the tenuivirus rice hojablanca virus (RHBV)(Bucher et al., 2003). Pc3 is the nucleocapsid protein (CP) and p4 the major non-structural protein (NSP), whose accumulation in infected plants correlates with symptom development (Kakutani et al., 1990, 1991; Zhu et al., 1991; Espinoza et al., 1993). Pc4 shares some common structures with the viral 30k superfamily movement proteins (Melcher, 2000). Recently, Xiong et al. showed that this protein localized predominantly near or within the cell walls, could move from cell to cell and complement movement defective PVX (Xiong et al., 2008). This suggested that Pc4 might be a movement protein of RSV. However, as a negative strand RNA virus that does not seem to form intact virions, it can be envisioned that cell-to-cell movement of RSV would be a very complex process, which deserves further research.

RSV infects agriculturally important crop plants such as rice and causes significant yield losses in east Asia. However, our knowledge about RSV, especially its interactions with host factors, remains sparse, partially owning to its reluctance to traditional virological methods such as infectious cloning. To bypass such obstacles, we have used Yeast two-hybrid system to investigate all the potential interactions between RSV encoded proteins and host factors (Fields and Song, 1989). Here, we report our identification of the interactions of pc4, the putative movement protein of RSV, with a DnaJ protein and an hsp20 family protein of rice.

2 Materials and methods

2.1 Plasmid construction

Total RNA was extracted from RSV-infected rice leaves with Trizol and the RSV gene segment NSvc4 was amplified by RT-PCR with primer pairs F1 and R1. PCR products were cloned into pMD18-T and then digested with *Nde*I and *Bam*HI, followed by ligation into pGBKT7 vector. The recombinant vector containing the RSV NSvc4 segment was designated as pGBK-NSvc4, and was used as the bait plasmid for following screening by yeast two-hybrid. Two cDNA fragments identified during the yeast two-hybrid screening had sequence identity with two genes encoding an hsp20 and DnaJ protein, respectively. The full-length ORF of the hsp20 and DnaJ were amplified by RT-PCR with primer pairs F2/R2, F3/R3, which were designed according to rice cDNA sequences. The specific fragments were cloned into pMD18-T. Construct containing hsp20 ORF was digested with *Nde*I and *Bam*HI, and fragment was ligated into *NdeI/Bam*HI-linearized pGADT7 vector. Construct containing DnaJ ORF was digested with *Spe*I and *Xho*I, and fragment was ligated into *SpeI/Xho*I-linearized pGADT7-CP vector containing *Spe*I site that was previously constructed. The recombinant plasmids were designated as pGAD-hsp20 and pGAD-DnaJ, respectively. There is a c-Myc-epitope tag at the 5′ terminus of NSvc4 ORF in the pGBK-NSvc4 and a HA-epitope tag at the 5′ terminus of hsp20 ORF in the pGAD-hsp20 and DnaJ ORF in the pGAD-DnaJ, respectively. The fusion gene c-Myc-NSvc4 was amplified by PCR with primer pairs F4/R4 using pGBK-NSvc4 as template and cloned into pMD18-T.Construct containing c-Myc-NSvc4 was digested with *Eco*RI and *Bam*HI, and fragment was ligated into a cauliflower mosaic virus 35S-based pEGAD transient-expression vector linearized by *EcoR*I and *BamH*I and the recombinant plasmid was designated as pEGAD-Myc-NSvc4. The c-Myc-NSvc4 fragment digested with *Hind* III and *Sac*I was inserted into *Hind* III/*Sac*I-linearized pKYLX35S2 vector and recombinant plasmid was designated as pKYLX-Myc-NSvc4. The fusion genes HA-hsp20 and HA-DnaJ were amplified by PCR with primer pairs F5/R5 using pGAD-hsp20 and pGAD-DnaJ as templates, respectively, and cloned into pMD18-T. The HA-hsp20 restriction fragment digested with *Eco*RI and *Bam*HI was ligated into *Eco*RI/*Bam*HI-linearized pEGAD vector, and the HA-hsp20 and HA-DnaJ restriction fragments digested with *Hind* III and *Xba* I were ligated into *Hind* III/*Xba*I-line-arized pKYLX35S2 vector. The recombinant plasmids were designated as pEGAD-HA-hsp20, pKYLX-

HAhsp20, and pKYLX-HA-DnaJ, respectively. The HA-hsp20 had been not ligated into Pegad vector because there were no suitable restriction enzyme sites. In addition, the following recombinant vectors constructed previously were used in this study: pGBK-CP containing RSV CP segment, pGBK-SP containing RSV NSP segment, pEGAD-Myc-CP and pKYLX-Myc-CP containing fusion gene Myc-CP, pEGAD-Myc-SP and pKYLX-Myc-SP containing fusion gene Myc-SP.

2.2 Yeast two-hybrid assay

A rice seedling yeast two-hybrid cDNA library from rice cvWuyujing 3 was constructed with CLONTECH protocols. The titer of the library was determined after amplification and was approximately 1.0×10^{11}cfu/ml. Matchmaker Gal4 Two-Hybrid System 3 and libraries were used to screen the rice cDNA library. The bait plasmid and cDNA library plasmid were transformed into yeast AH109 cells using sequential transformation or simultaneous cotransformation protocol. Colonies were selected on SD/Leu-Trp-His-Ade-medium and then Ade+/His+ positive colonies were isolated on SD/Leu-Trp-His-Ade-/X-a-gal+ medium according to the instruction manual. Primary positive candidate plasmids containing the rice cDNAs were isolated and then co-transformed into AH109 with bait plasmid pGBKNSvc4 to repeat the two-hybrid assay. The final positive candidate plasmids were selected and determined by sequencing analysis. The sequences of positive colons were subsequently used for an advanced BLAST search within the database of GenBank.

2.3 Agrobacterium-mediated transient expression

Agrobacterium strain EHA105 carrying the gene of interest expressed from a binary vector was infiltrated into leaves of *Nicotiana benthamiana*. *Agrobacterium tumefaciens* was grown overnight at 28℃ on Luria-Bertani agar containing 50μg/μL of rifampicin and 50μlg/μL of kanamycin. Cells were resuspended in induction media (10mM MES, pH 5.6, 10mM $MgCl_2$, and 150μM acetosyringone) and incubated at room temperature for 3–5h before inoculation.

2.4 Immunoprecipitation

After agrobacterium-mediated transient expression for 24h, *N. benthamiana* leaves (approximately 0.3g) were harvested and ground to a powder in liquid nitrogen. Ground tissues were resuspended in 3.0ml of IP buffer (50mM Tris, pH 7.5, 150mM NaCl, 10% glycerol, 0.1% Nonidet P-40, 5mM dithiothreitol, and 1.53 Complete Protease Inhibitor [Roche]). The crude lysates were then spun at 20,000g for 15min at 4℃. After centrifugation, 1 ml of supernatant was incubated with 0.5μg of the indicated monoclonal antibody for each immunoprecipitation. After a 1h incubation at 4℃ , immunocomplexes were collected by the addition of 50μl of protein G Sepharose-4 fast flowbeads and incubated end over end for 3–5h at 4℃. After incubation, the immunocomplexes were washed four times with 1 ml of IP buffer and the pellet was resuspended in 1×SDS-PAGE loading buffer.

2.5 Protein separation and western blotting

Protein samples were separated by SDS-PAGE on 10% polyacrylamide gels and transferred by electroblotting to PVDF membranes. Membranes were probed with anti-HA horseradish peroxidase (Roche) or anti-Myc peroxidase (Sigma-Aldrich) to detect HA- and Myc-epitope-tagged proteins, respectively. All immunoprecipitation experiments were repeated at least three times, and the identical results were obtained.

2.6 Primers used in this study

Primers used in this study are listed in Table 1.

Table 1 Primers used in this study

Primers Name	Primers sequences	Restriction enzyme site	Constructs
F1	5′-GGAATTCCATATGGCTTTGTCTCGACTTT TGTC-3′	*Nde*I	pGBK-NSnv4
R1	5′-CGGGATCCCTACATGATGACAGAAACTT CAG-3′	BamHI	
F2	5′-GGAATTCCATATGTCGCTGATCCGCCGCA GC-3′	*Nde*I	pGAD-hsp20
R2	5′-CGGGATCCCTAGCCGGAGATCTGGATGGA C-3′	BamHI	
F3	5′-GGACTAGTATGTTTGGGCGTGTACCGAG-3′	*Spe*I	pEGAD-Myc-N Snv4
R3	5′-CGCTCGAGTTACTGTTGAGCACACTGTACTC-3′	*Xho*I	
F4	5′-GGAATTCAAGCTTATCATGGAGGAGGAGA AGCTG-3′	*Eco*RI, *Hind*III	pEGAD-Myc-NSnv4
R4	5′-GGATCCGAGCTCAGGGGTTATGCTAGTTAT G-3′		pKYLX-Myc-NSnv4
F5	5′-GGAATTCAAGCTTATGGAGTACCCATACG ACG-3′	*Bam*HI, *Sac*I	pEGAD-HA-hsp20
R5	5′-CGGGATCCTCTAGATTTCAGTATCTACGAT TCAT-3′		pKYLX-HA-DnaJ
F6	5′-AAGTTCCTCCGCAGGTTCC-3′	*Bam*HI, *Xba*I	
R6	5′-GAGCACGCCGTTCTCCAT-3′		
F7	5′-GAGGCAGTGACTTCCATAATCC-3′	For *hsp*20	
R7	5′-GCCTAGTCCTATCTGTCGCATT-3′		
F9	5′-ATCCTGACGGAGCGTGGTTA-3′	For *DnaJ*	
R9	5′-CATAGTCCAGGGCGATGTAGG-3′	For *Actin*	

3 Results

3.1 Identification of RNB8 and RNB5 that interact with RSV pc4

To identify rice proteins that interacted with RSV pc4, the rice cDNA library was screened by a Gal4-based yeasttwo-hybrid system. A number of positive colonies were identified among the approximately 3.6×10^6 clones that were screened. Sequence analysis of these colonies showed that they might correspond to ten independent proteins. Two of them, designated NB8 and NB5, shared high degree of identity with a DnaJ protein (NM_001060020) and a heat shock protein 20 (hsp20, NM_001056192) from *Oryza sativa*, respectively. The full-length ORFs corresponding to NB8 and NB5 (hereafter we use RNB8 and RNB5 to represent the genes corresponding to NB8 and NB5, respectively) were cloned from rice using primers designed according to available rice sequences in NCBI. Specific interactions between the two proteins and RSV pc4 were then confirmed using entire ORFs of the two genes by yeast two-hybrid experiments (data not shown).

Sequence analysis of the coding regions indicated that the ORF of the RNB8 gene contained 1,251 nucleotides and encoded a protein of 416 amino acids; the deduced amino acids sequence of RNB8 contained conserved cysteine-rich domains and several glycine-rich regions in addition to the typical J domain (Fig. 1A), thus it represented a member of the group I dnaJ proteins (Kelley, 1998; Walsh et al., 2004; Qiu et al., 2006). The ORF of the RNB5 gene contained 486 nucleotides and encoded a protein of 161 amino acids with conserved alpha-crystallin domain (ACD) typical of a class of small Hsps(Boston et al., 1996; Waters et al., 1996; Sun and MacRae, 2005; Nakamoto and Vigh, 2007). The cDNA fragment of the NB8 we initially retrieved from screening of the rice library encoded a polypeptide of 165 amino acid residues that located on the C-terminus of the RNB8 gene (Fig. 1A). This region might be responsible for the interactions between RNB8 and RSV pc4. The cDNA fragment of the NB5 encompasses the entire ORF of the RNB5 (Fig. 1B).

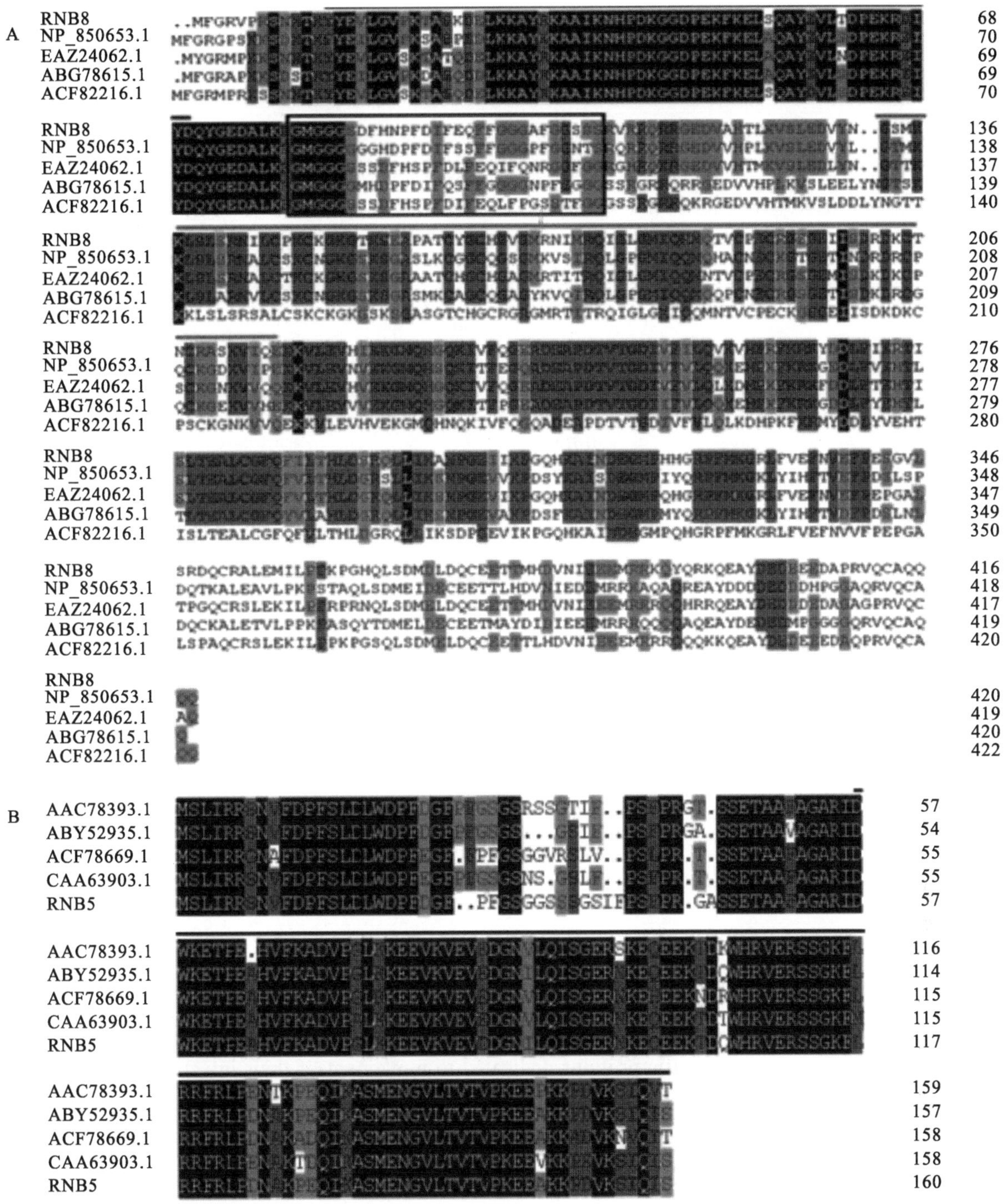

Fig. 1 Alignments of the DnaJ and hsp20 proteins from diverse species. The alignments were made using DNAMAN (4.0)
A. Alignment of DnaJ proteins from *Oryza sativa* (RNB8 and EAZ24062.1), *Arabidopsis*(NP_850653.1), *Triticuma estivum* (ABG78615.1), and *Zea mays* (ACF82216.1). The conserved J domain is marked by a *black line*. Glycine-rich regions and 4 repeats of C-X-X-C-X-G-X-G motif typical of type I J proteins are *boxed* and marked by a *red line*, respectively. Regions of identity or similarity are *colored* and gaps introduced for alignment are indicated by *dots*. **B**. Alignment of hsp20 proteins *Oryza sativa* (RNB5 and AAC78393.1), *Pennisetum glaucum* (CAA63903.1), and *Zea mays* (ACF78669.1). *Black line* indicates the conserved alpha-crystallin domain

3.2 Pc4 interacts with the rice DnaJ and hsp20 in plant cells

Specific interactions of Pc4 with rice DnaJ and hsp20 in yeast suggest functional significance. To test the interactions further, coimmunoprecipitation was used to determine whether such interactions occur in plant cells. As shown in Fig. 2, the c-Myc-epitope-tagged pc4 coimmunoprecipitated with the HA-epitope-tagged RNB8 and HA-epitope-tagged RNB5 after Agrobacterium-mediated transient expression in *Nicotiana benthamiana*. The interactions were confirmed with the reciprocal experiments, in which HA-epitope-tagged RNB8 and RNB5 were coimmunoprecipitated with the c-Myc-epitope-tagged pc4, respectively. These results provided evidence that RSV pc4 interacts with the DnaJ and hsp20 in plant cell, whereas there were no such interactions of the two rice proteins with RSV CP and SP (Fig. 2).

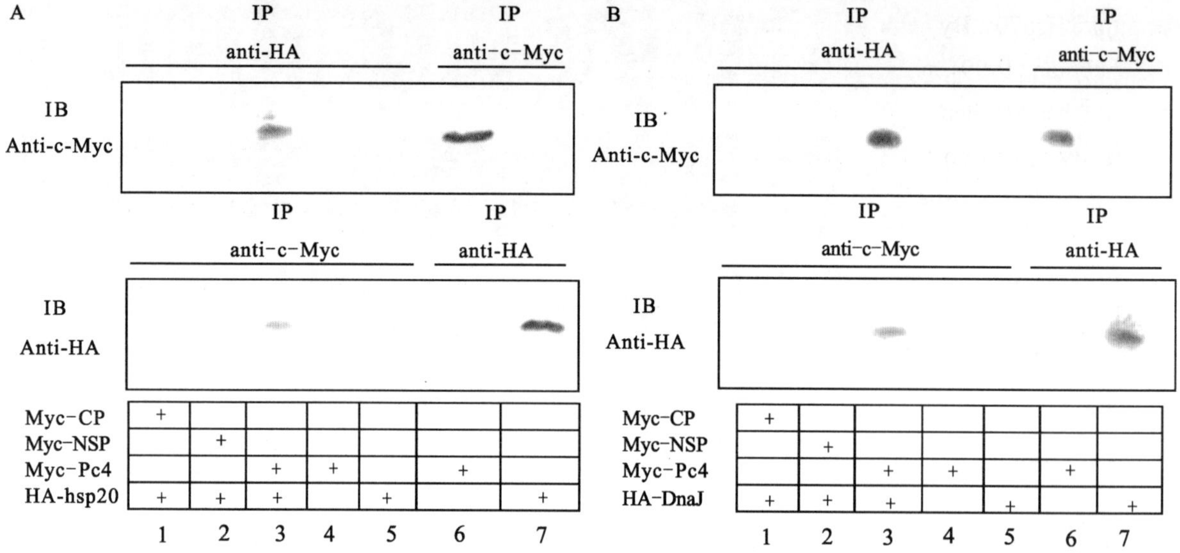

Fig. 2 RSV Pc4, but not CP and NSP, interacted with RNB5 and RNB8 proteins in plant cells

A. Immunoblot showing NSvc4 coimmunoprecipitated with the hsp20 (RNB5); for lanes 1–3, the total proteins were extracted from Agrobacterium-infiltrated Nicotiana benthamiana leaves expressing Myc-CP/HA-hsp20, (lane 1), Myc-SP/HA-hsp20 (lane 2), or Myc-pc4/HA-hsp20 (lane 3) and immunoprecipitated with anti-HA (top) or anti-c-Myc (bottom) antibodies. For lanes 4–7, the total proteins were extracted from Agrobacterium-infiltrated *Nicotiana benthamiana* leaves expressing only Myc-pc4 (lanes 4 and 6) or HA-hsp20 (lanes 5 and 7) and immunoprecipitated with anti-c-Myc (top) or anti-HA (bottom) antibodies. **B**. Immunoblot showing pc4 coimmunoprecipitated with DnaJ(RNB8)

4 Discussion

Most plant viruses encode specific proteins dedicated to movement of their infectious materials. These movement proteins exploit host NCAP pathway to fulfill their function. Identification of host factors interacting with viral MPs is essential to understand viral movement (Carrington et al., 1996; Ghoshroy et al., 1997; Lucas and Lee, 2004; Oparka, 2004; Ruiz-Medrano et al., 2004; Waigmann et al., 2004; Boevink and Oparka, 2005; Gallagher and Benfey, 2005; Kim, 2005; Kurata et al., 2005; Scholthof, 2005; Lucas, 2006). Here, using pc4, a putative movement protein of RSV, as bait, we identified two rice proteins, i.e., a J protein and a member of the hsp20 family. The two proteins interacting with pc4 did not interact with CP and NSP in yeast and in plant cell, indicating that the interactions between the two proteins and Pc4 were specific.

J proteins, featured by a 70-amino acid signature sequence through which they bind to their partner Hsp70s, are key regulators of the ATP cycle of hsp70s (Kelley, 1998; Walsh et al., 2004; Qiu et al., 2006). Three groups of J proteins have been characterized. Type I proteins are similar to E. coli DnaJ with the J domain, the Gly/Phe-rich region, and the cysteine repeats. Typ II proteins contain the J domain and the Gly/Phe-rich region, but lack the cysteine repeats. Typ III proteins do not have any of the conserved regions other than the J domain(Kelley, 1998; Walsh et al., 2004; Qiu et al., 2006).

It is believed that transport of the NCAPs to and through the PD involves a phase of conformational change of the NCAPs (Balachandran et al., 1997; Kragler et al., 1998). The need of a conformational change entails the availability of a putative chaperone protein. The most promising candidate of such a chaperone is a member of the hsp70s, a family of versatile proteins that have been implicated in protein translation, folding, unfolding, translocation, and degradation (Bukau and Horwich, 1998; Mayer and Bukau, 2005). Thus, it would be reasonable to speculate that the interaction of Pc4 with a J protein would allow the protein to locally concentrate Hsp70s. By analogy with models proposed for protein translocation into ER and mitochondria (Pilon and Schekman, 1999), unfolding of the movement protein could occur firstly at a conformationally flexible region, the hsp70s then might bind this region and promote further unfolding through trapping and sliding. For more tightly folded domains within the movement protein, the hsp70s could also provide a vigorous force to bias the equilibrium to an unfolded state (Pilon and Schekman, 1999; Sousa and Lafer, 2006). The hsp70s that were recently identified to move from cell to cell are attracting candidates to fulfill this function (Aoki et al., 2002).

Another possibility is that Hsp70s could present the viral MPs to host ubiquitin-proteosome pathway for degradation. Degradation of MP by host ubiquitin-proteosome pathway has been observed in TMV (Heinlein et al., 1998; Reichel and Beachy, 1998, 2000). It was suggested that the degradation might function to avoid extreme damage to the host and futile movement of the viral materials (Heinlein et al., 1998; Reichel and Beachy, 1998, 2000; Waigmann et al., 2007). This was consistent with the observation that Pc4 could only be detected in infected rice plants at a very early stage of infection (Qu et al., 1999; Liang et al., 2005). In this scenario, the observed larger size of Pc4 when detecting with antisera to the protein in a previous report could be the result of polyubiquination(Qu et al., 1999; Liang et al., 2005).

Previously, NSm, the movement protein of tomato spotted wilt tospovirus (TSWV), has been shown to interact with several members of J proteins from *Nicotianatabacum* and *Arabidopsis thaliana*(Soellick et al., 2000; von Bargen et al., 2001). The capsid protein (CP) of potyviruses, which is involved in movement of the virus, has been shown to interact with a set of J proteins from tobacco (Hofius et al., 2007). In the later case, transgenic plants that ectopically overexpress dominant-negative mutants of NtCPIPs showed significantly enhanced virus resistance to PVY, and the resistance was most likely due to strongly reduced cell-to-cell transport (Hofius et al., 2007). Taking these into account, the recruitment of Hsp70s through interactions with a J protein to facilitate movement protein function seems to be a widely used mechanism of plant viruses. This raises two questions; the first is why most plant viruses do not encode an Hsp70 themselves. The second is why the MPs do not interact with an Hsp70 directly. The answer to the first question is obvious. Recruiting an Hsp70 from the host is more economically reasonable than to encode one. The answer to the second question might lie in the fact that the host encodes more J proteins than Hsp70s, which implies that the functional specificity of a J protein/Hsp70 combination is determined by the J protein (Kelley, 1998; Walsh et al., 2004; Qiu et al., 2006). It is noteworthy that the J proteins interacting with TSWV NSm and PVY CP belong to type III J proteins. Yet, the J protein identified in this study was a type I DnaJ protein (Kelley, 1998; Walsh et al., 2004; Qiu et al., 2006).

This study also identified a small hsp that interacted with Pc4. sHSPs, defined by possessing a conserved alphacrystallin domain (ACD), are the most abundant and complex subset of HSPs in plants (Boston et al., 1996; Waters et al., 1996). Key function of the sHSPs is to prevent aggregation of denatured proteins. By forming a soluble complex with substrate proteins, they can create a transient reservoir of substrates for subsequent refolding by ATP-dependent chaperone systems (Boston et al., 1996; Waters et al., 1996; Nakamoto and Vigh, 2007; Sun and MacRae, 2005). It is tempting to assume that the Hsp20 forms a complex with Pc4 when the protein is partially unfolded for transport through PD. The presence of the Hsp20 keeps the denatured Pc4, and perhaps the entire viral material for movement, soluble. And when the viral material entered the neighboring cell, the presence of the Hsp20 would allow for an immediate and efficient renature of the viral material. To our knowledge, this is the first report that a plant viral MP interacts with a small Hsp. By analogy with other negative RNA viruses, the ribonucleoprotein particles (RNPs) represent the only structures responsible for transcription and replication for RSV (Elliott, 1990; Baudin et al., 1994; Klumpp et al., 1997). Thus, the infectious materials that move from cell to cell for RSV must be entire RNPs. The RSV RNPs are very complex in structure, containing at least the RdRps, CPs, and genome-length viral RNAs (Ramírez and Haenni, 1994; Falk and Tsai, 1998). This might be responsible for our results that MP of RSV needed to interact with an hsp20 in addition to a J protein.

RSV infects rice, one of the most important crop plants in the world, and poses a major threat to rice production in some Asian countries. The identification of the two host factors interacting with a putative movement protein of RSV will undoubtedly propel a step forward of our understanding of RSV. Perhaps more importantly, as has been mentioned, expression of a mutant form of a J protein that interacts with CP of PVX in tobacco dramatically increased the viral resistance of various transgenic lines (Hofius et al., 2007). Transgenic rice plants expressing RSV CP have been developed and were shown to be efficient for RSV resistance (Hayakawa et al., 1992). But the introduction of a viral gene to food crops would inevitably invoke safety concerns. It is intriguing to engineer transgenic rice plants expressing a mutant form of the J protein identified in this study and test their resistance to RSV.

References

[1] AGRANOVSKY A A, BOYKO V P, KARASEV A V, et al. Putative 65kDa protein of beet yellows closterovirus is a homologue of HSP70 heat shock proteins[J]. Journal of Molecular Biology, 1991, 217: 603-610.

[2] ALZHANOVA D V, NAPULI A J, CREAMER R, et al. Cell-to-cell movement and assembly of a plant closterovirus: roles for the capsid proteins and Hsp70 homolog[J]. Embo Journal, 2014, 20: 6997-7007.

[3] AOKI K, KRAGLER F, XOCONOSTLE-CÁZARES B, et al. A subclass of plant heat shock cognate 70 chaperones carries a motif that facilitates trafficking through plasmodesmata[J]. Proceedings of the National Academy of Sciences, 2003, 99: 16342-16347.

[4] BALACHANDRAN S, XIANG Y, SCHOBERT C, et al. Phloem sap proteins from Cucurbita maxima and Ricinus communis have the capacity to traffic cell to cell through plasmodesmata[J]. Proceedings of the National Academy of Sciences of the United States of America, 1997, 94: 14150-14155.

[5] BARBIER P, TAKAHASHI M, NAKAMURA I, et al. Solubilization and promoter analysis of RNA polymerase from rice stripe virus[J]. Journal of Virology, 1992, 66: 6171-6174.

[6] BAUDIN F, BACH C, CUSACK S, et al. Structure of influenza virus RNP I Influenza virus nucleoprotein melts secondary structure in panhandle RNA and exposes the bases to the solvent[J]. Embo Journal, 1994, 13: 3158-3165.

[7] BOEVINK P, OPARKA K J. Virus-host interactions during movement processes[J]. Plant Physiology, 2005, 138: 1815-1821.

[8] BOSTON R S, VIITANEN P V, VIERLING E. Molecular chaperones and protein folding in plants[J]. Plant Molecular Biology, 1996, 191-222.

[9] BUCHER E, SIJEN T, DE HAAN P, ET AL. Negative-strand tospoviruses and tenuiviruses carry a gene for a suppressor of gene silencing at analogous genomic positions[J]. Journal of Virology, 2003, 77: 1329-1336.

[10] BUKAU B, HORWICH A L. The Hsp70 and Hsp60 chaperone machines[J]. Cell, 1998, 92: 351-366.

[11] CARRINGTON J C, KASSCHAU K D, MAHAJAN S K, ET AL. Cell-to-cell and long-distance transport of viruses in plants[J].

The Plant Cell, 1996, 8: 1669.

[12] ELLIOTT RM. Molecular biology of the Bunyaviridae[J]. Journal of General Virology, 1990, 71: 501-522.

[13] ESPINOZA A M, PEREIRA R, MACAYA-LIZANO A V, et al. Comparative light and electron microscopic analyses of tenuivirus major noncapsid protein (NCP) inclusion bodies in infected plants, and of the NCP in vitro[J]. Virology, 1993, 195: 156-166.

[14] FALK B W. Tenuiviruses, in Encyclopedia of Virology,1994.

[15] FALK B W, TSAI, J H. Biology and molecular biology of viruses in the genus Tenuivirus[J]. Annual Review of Phytopathology, 1998, 36: 139-163.

[16] FIELDS S, SONG O K. A novel genetic system to detect protein–protein interactions[J]. Nature, 1989, 340: 245.

[17] GALLAGHER K L, BENFEY P N. Not just another hole in the wall: understanding intercellular protein trafficking[J]. Genes & Development, 2005, 19: 189-195.

[18] GHOSHROY S, LARTEY R, SHENG J, et al. Transport of proteins and nucleic acids through plasmodesmata[J]. Annual Review of Plant Physiology & Plant Molecular Biology, 1997, 48: 27-50.

[19] HAUPT S, COWAN G H, ZIEGLER A, et al. Two plant–viral movement proteins traffic in the endocytic recycling pathway[J]. The Plant Cell Online, 2005, 17: 164-181.

[20] HAYAKAWA T, ZHU Y, ITOH K, et al. Genetically engineered rice resistant to rice stripe virus, an insect-transmitted virus[J]. Proceedings of the National Academy of Sciences of the United States of America, 1992, 89: 9865-9869.

[21] HEINLEIN M, PADGETT H S, GENS J S, et al. Changing patterns of localization of the tobacco mosaic virus movement protein and replicase to the endoplasmic reticulum and microtubules during infection[J]. The Plant Cell, 1998, 10: 1107-1120.

[22] HIBINO H. Biology and epidemiology of rice viruses[J]. Annual Review of Phytopathology, 1996, 34: 249-274.

[23] HOFIUS D, MAIER A T, DIETRICH C, et al. Capsid protein-mediated recruitment of host DnaJ-like proteins is required for Potato virus Y infection in tobacco plants[J]. Journal of Virology, 2007, 81: 11870-11880.

[24] KAKUTANI T, HAYANO Y, HAYASHI T, et al. Ambisense segment 4 of rice stripe virus: possible evolutionary relationship with phleboviruses and uukuviruses (Bunyaviridae)[J]. Journal of General Virology, 1990, 71: 1427-1432.

[25] KAKUTANI T, HAYANO Y, HAYASHI T, et al. Ambisense segment 3 of rice stripe virus: the first instance of a virus containing two ambisense segments[J]. Journal of General Virology, 1991, 72: 465-468.

[26] KELLEY W L. The J-domain family and the recruitment of chaperone power[J]. Trends in Biochemical Sciences, 1998, 23: 222-227.

[27] KIM J Y. Regulation of short-distance transport of RNA and protein[J]. Current Opinion in Plant Biology, 2005, 8: 45-52.

[28] KLUMPP K, RUIGROK R W, BAUDIN F. Roles of the influenza virus polymerase and nucleoprotein in forming a functional RNP structure[J]. Embo Journal, 2014, 16: 1248-1257.

[29] KRAGLER F, MONZER J, SHASH K, et al. Cell-to-cell transport of proteins: requirement for unfolding and characterization of binding to a putative plasmodesmal receptor[J]. Plant Journal, 2010, 15: 367-381.

[30] KURATA T, OKADA K, WADA T. Intercellular movement of transcription factors[J]. Current Opinion in Plant Biology, 2005, 8: 600-605.

[31] LEE J Y, YOO B C, ROJAS M R, et al. Selective trafficking of non-cell-autonomous proteins mediated by NtNCAPP1[J]. Science, 2003, 299: 392-396.

[32] LIANG D, MA X, QU Z, et al. Nucleic acid binding property of the gene products of rice stripe virus[J]. Virus Genes, 2005, 31: 203-209.

[33] LUCAS W J. Plant viral movement proteins: agents for cell-to-cell trafficking of viral genomes[J]. Virology, 2006, 344: 169-184.

[34] LUCAS W J, LEE J Y. Plasmodesmata as a supracellular control network in plants[J]. Nature Reviews Molecular Cell Biology, 2004, 5: 712.

[35] MAYER M, BUKAU B. Hsp70 chaperones: cellular functions and molecular mechanism[J]. Cellular & Molecular Life Sciences,

2005, 62: 670.

[36] MELCHER U. The '30K'superfamily of viral movement proteins[J]. Journal of General Virology, 2000, 81: 257-266.

[37] NAKAMOTO H, VIGH L. The small heat shock proteins and their clients[J]. Cellular and Molecular Life Sciences, 2007, 64: 294-306.

[38] OPARKA K J. Getting the message across: how do plant cells exchange macromolecular complexes?[J]. Trends in Plant Science, 2004, 9: 33-41.

[39] PEREMYSLOV V V, HAGIWARA Y, DOLJA V V. HSP70 homolog functions in cell-to-cell movement of a plant virus[J]. Proceedings of the National Academy of Sciences, 1999, 96: 14771-14776.

[40] PILON M, SCHEKMAN R. Protein translocation: how Hsp70 pulls it off[J]. Cell, 1999, 97: 679-682.

[41] QIU X B, SHAO Y M, MIAO S, et al. The diversity of the DnaJ/Hsp40 family, the crucial partners for Hsp70 chaperones[J]. Cellular & Molecular Life Sciences Cmls, 2006, 63: 2560-2570.

[42] QU Z, SHEN D, XU Y, et al. Western blotting of RStV gene products in rice and insects[J]. Yi Chuan Xue Bao, 1999, 26: 512-517.

[43] RAMÍREZ B C, HAENNI A L. Molecular biology of tenuiviruses, a remarkable group of plant viruses[J]. Journal of General Virology, 1994, 75: 467-475.

[44] REICHEL C, BEACHY R N. Tobacco mosaic virus infection induces severe morphological changes of the endoplasmic reticulum[J].Proceedings of the National Academy of Sciences of the United States of America, 1998, 95: 11169-11174.

[45] REICHEL C, BEACHY R N. Degradation of tobacco mosaic virus movement protein by the 26S proteasome[J]. Journal of Virology, 2000, 74: 3330-3337.

[46] RUIZ-MEDRANO R, XOCONOSTLE-CAZARES B, KRAGLER F. The plasmodesmatal transport pathway for homeotic proteins, silencing signals and viruses[J]. Current Opinion in Plant Biology, 2004, 7: 641-650.

[47] SCHOLTHOF H B. Plant virus transport: motions of functional equivalence[J]. Trends Plant Science, 2005, 10: 376-382.

[48] SOELLICK T R, UHRIG J, BUCHER G, et al. The movement protein NSm of tomato spotted wilt tospovirus (TSWV): RNA binding, interaction with the TSWV N protein, and identification of interacting plant proteins[J]. Proceedings of the National Academy of Sciences of the United States of America, 2000, 97: 2373-2378.

[49] SOUSA R, LAFER E M. Keep the traffic moving: mechanism of the Hsp70 motor[J]. Traffic, 2010, 7: 1596-1603.

[50] SUN Y, MACRAE T H. Small heat shock proteins: molecular structure and chaperone function[J]. Cellular & Molecular Life Sciences, 2005, 62: 2460-2476.

[51] TAKAHASHI M, TORIYAMA S, HAMAMATSU C, et al. Nucleotide sequence and possible ambisense coding strategy of rice stripe virus RNA segment 2[J]. Journal of General Virology, 1993, 74: 769-773.

[52] TORIYAMA S. An RNA-dependent RNA polymerase associated with the filamentous nucleoproteins of rice stripe virus[J]. Journal of General Virology, 1986, 67: 1247-1255.

[53] TORIYAMA S, TAKAHASHI M, SANO Y, et al. Nucleotide sequence of RNA. 1, the largest genomic segment of rice stripe virus, the prototype of the tenuiviruses[J]. Journal of General Virology, 1994,75: 3569-3579.

[54] VON BARGEN S, SALCHERT K, PAAPE M, et al. Interactions between the tomato spotted wilt virus movement protein and plant proteins showing homologies to myosin, kinesin and DnaJ-like chaperones[J]. Plant Physiology and Biochemistry, 2001, 39: 1083-1093.

[55] WAIGMANN E, CURIN M, HEINLEIN M. Tobacco mosaic virus–a model for macromolecular cell-to-cell spread[J]. Springer Berlin Heidelberg, 2007, 29-62.

[56] WAIGMANN E, UEKI S, TRUTNYEVA K, et al. The ins and outs of nondestructive cell-to-cell and systemic movement of plant viruses[J]. Critical Reviews in Plant Sciences, 2004, 23: 195-250.

[57] WALSH P, BURSAĆ D, LAW Y C, et al. The J-protein family: modulating protein assembly, disassembly and translocation[J]. EMBO Journal, 2004, 5: 567-571.

[58] WATERS E R, LEE G J, VIERLING E. Evolution, structure and function of the small heat shock proteins in plants[J]. Journal of Experimental Botany, 1996, 47: 325-338.

[59] WHITHAM S A, QUAN S, CHANG H S, et al. Diverse RNA viruses elicit the expression of common sets of genes in susceptible Arabidopsis thaliana plants[J]. Plant Journal, 2010, 33: 271-283.

[60] XIONG R, WU J, ZHOU Y, et al. Identification of a movement protein of the tenuivirus rice stripe virus[J]. Journal of Virology, 2008, 82: 12304-12311.

[61] ZHU Y, HAYAKAWA T, TORIYAMA S, et al. Complete nucleotide sequence of RNA. 3 of rice stripe virus: an ambisense coding strategy[J]. Journal of General Virology, 1991, 72: 763-767.

水稻条纹病毒胁迫下灰飞虱基因的差异表达

肖冬来[1,2]，邓慧颖[1]，谢荔岩[1]，吴祖建[1]，谢联辉[1]

（1 福建农林大学植物病毒研究所，福建省植物病毒学重点实验室，福建 福州 350002；
2 福建省农业科学院食用菌研究所，福建 福州 350002）

摘要：水稻条纹病毒（*Rice stripe virus*，RSV）主要由介体昆虫灰飞虱（*Laodelphax striatellus*）以循回增殖型方式经卵传播，目前有关 RSV 与灰飞虱间的互作研究很少。为了研究 RSV 侵染对灰飞虱基因表达的影响，采用 5 条随机引物和 3 条锚定引物，利用 mRNA 差异显示（differential display RT-PCR，DDRT-PCR）技术分析了带毒和无毒灰飞虱种群基因表达差异，且利用正交实验优化了 DDRT-PCR 反应体系中的模板浓度、锚定引物浓度、随机引物浓度、dNTPs 浓度、镁离子浓度及 Taq 酶用量。结果表明：最佳 DDRT-PCR 体系（25μL）为 cDNA 3.0μg，随机引物 2.0μmol/L，锚定引物 2.5μmol/L，dNTPs 200μmol/L，Mg^{2+} 2.0μmol/L，Taq 酶 2.0U。mRNA 差异显示共获得 35 条差异片段，选取其中 6 条经 RNA 斑点杂交验证，获得了 4 条阳性差异片段。其中 3 条阳性片段为带毒灰飞虱种群特异表达，分别与 5- 羟色胺受体 1D、旋转酶 B、60S 核蛋白 L40 高度同源，无毒灰飞虱种群中特异表达的一条阳性片段在 NCBI 核酸数据库中比对无同源序列。DDRT-PCR 优化体系的建立及部分差异片段的获得为进一步研究灰飞虱与 RSV 间的互作提供了帮助。

关键词：灰飞虱；水稻条纹病毒；差异表达基因；DDRT-PCR；正交试验

中图分类号：Q965.8　**文献标识码**：A　**文章编号**：0454-6296（2010）08-0914-06

Identification of differentially expressed genes in *Laodelphaxstriatellus* (Homoptera: Delphacidae) under RSV stress

Donglai Xiao[1,2], Huiying Deng[1], Liyan Xie[1], Zujian Wu[1], Lianhui Xie[1]

(1 Key Laboratory of Plant Virology of Fujian Province, Institute of Plant Virology, Fujian Agriculture and Forestry University, Fuzhou 350002;
2 Institute of Edible Fungi, Fujian Academy of Agricultural Sciences, Fuzhou 350002)

Abstract: *Rice stripe virus* (RSV) is mainly transmitted by insect vector *Laodelphax striatellus* in circulative propagative and transovarial manners. The interaction between RSV and *L. striatellus* is largely unknown. To investigate the effects of RSV on gene expression in *L. striatellus*, viruliferous and virus-free *L. striatellus* populations were detected to reveal the differentially expressed genes with five random primers and three anchor primers by differential display RT-PCR (DDRT-PCR). Furthermore, positive-cross test was performed to find optimal conditions of DDRT-PCR by analyzing six critical parameters including cDNA

昆虫学报 . 2010, 53 (8): 914-919
收稿日期：2009-12-21
基金项目：国家“973”计划项目（2006CB100203）；国家自然科学基金项目（30671357）；公益性行业科研专项（nyhyzx07-051）

template, random primer, anchorprimer, dNTPs, Mg^{2+} and *Taq*. The results showed that the optimal conditions of DDRT-PCR (25μL) included cDNA template 3.0μg, random primer 2.0μmol/L, anchorprimer 2.5μmol/L, dNTPs 200μmol/L, Mg^{2+} 2.0μmol/L, and *Taq* 2.0U. Thirty-five differentially expressed cDNA fragments were isolated by DDRT-PCR. Six of them were verified by RNA dot blot hybridization, and four positive cDNA fragments were obtained. Three positive cDNA fragments were from viruliferous *L.striatellus* population and shared high homology with 5-hydroxytryptamine receptor 1D gene, gyrase B gene and 60S ribosomal protein L40 gene, respectively. The positive cDNA fragment from virus-free *L. striatellus* population had no similarity with sequences in NCBI nucleotide databases. The optimal DDRT-PCR system and the differentially expressed cDNA fragments obtained might provide a basis for further study on interaction between RSV and *L. striatellus*.

Keywords: *Laodelphax striatellus*; rice stripe virus(RSV); differentially expressed gene; DDRT-PCR; positive-cross test

由水稻条纹病毒（*Rice stripe virus*, RSV）引起的水稻条纹叶枯病是我国水稻生产上的重要病害之一。在我国该病害分布广泛，自从 1963 年在江苏南部发生后已在台湾、福建、浙江、上海、江苏、江西、安徽、湖北、广西、广东、云南、山东、河南、河北、北京和辽宁等地发生，对我国的水稻种植造成了重大损失（林奇英等，1990）。其中，2004 ～ 2005 年，江苏省每年发病面积达 170 万 hm^2（张恒木等，2007）。RSV 主要由介体昆虫灰飞虱 *Laodelphax striatellus* 以循回增殖型方式经卵传播（Falk and Tsai,1998），根据灰飞虱与 RSV 的亲和性可将灰飞虱分为高、中、低亲和性和非亲和性群体（刘海建等，2007）。李小力等利用 RAPD 标记技术从高亲和性灰飞虱群体中筛选出一条特异性片段（李小力等，2009），该片段是否与灰飞虱传播 RSV 有关还需进一步研究。

随着分子生物学的发展，研究基因差异表达的技术取得了很大的突破。目前筛选差异基因的主要方法有：基因芯片技术（DNA chip technique）、基因表达的系统分析（serial analysis of gene expression, SAGE）、差异显示 PCR（differential display RT-PCR, DDRT-PCR）、抑制消减杂交（suppression substractive hybridization, SSH）、代表性差异分析（representational difference analysis, RDA）等技术。张晓婷等利用基因芯片技术检测了 RSV 胁迫下水稻全基因组的表达差异（张晓婷等，2008），得到了 3517 个差异表达基因，并对差异基因进行了分类。DDRT-PCR 具有敏感性高、快速、多能性的特点，尽管该技术的假阳性率较高，分离的差异片段短（Liang et al., 1993），但是在昆虫学研究中，mRNA 差异显示技术也有着广泛的应用。刘红等通过比较野生型、七氟醚（麻醉药）敏感型和耐药型黑腹果蝇（*Drosophila melanogaster*）的基因差异得到了 3 条差异片段（刘红等，2004）。陶杰等利用该技术筛选并克隆了一些东亚飞蝗（*Locusta migratoria manilensis*）的抗药性相关基因（陶杰等，2009）。

病毒与寄主间的互作一直是病毒学研究的热点之一，目前关于 RSV 与寄主水稻和介体灰飞虱间的研究很少，尤其是 RSV 与灰飞虱在分子水平的互作研究还未见报道。灰飞虱携带 RSV 后是否会对自身生长发育产生影响还不清楚。DDRT-PCR 技术是筛选差异表达基因快速、简便的有效手段之一，本研究通过正交实验优化了适合灰飞虱的 DDRT-PCR 反应体系，并初探了灰飞虱 RSV 胁迫下基因的表达差异，以期有助于研究灰飞虱与 RSV 之间的相互关系，同时为抗病机制、病毒防治等方面的研究奠定基础。

1 材料与方法

1.1 试验材料

1.1.1 供试虫源

灰飞虱为本实验室于 2000 年从江苏洪泽田间采集并饲养于台中本地 1 号水稻幼苗上，室温保持在

25～28℃，每 4d 更换一次水稻幼苗，适时浇水。经 PCR 和传毒实验检测获得不携带 RSV 的无毒虫个体，其后代连续饲养 3 代后经检测不携带 RSV，该种群即为无毒灰飞虱种群。无毒灰飞虱个体通过饲毒、传毒筛选获得带毒虫，带毒虫后代经反复饲毒选育获得带毒灰飞虱种群，经 PCR 及传毒实验检测该种群带毒率达 90% 以上。

1.1.2 试剂

Taq DNA 聚合酶、核酸标准分子量（Lambda DNA/*Eco*R Ⅰ，*Hin*d Ⅲ marker）、M-MuLV 反转录酶、RNA 酶抑制剂（RNasin）、限制性内切酶均购自 MBI 公司；QIAEX Ⅱ GelExtractionKit 为 QIAGEN 公司产品；克隆载体 pMD18-T 为大连宝生物公司产品。Blocking Reagent、Digoxigenin-11-dUTP 为 Roch 公司产品，经典总 RNA 抽提试剂盒，锚定引物 $(T)_{13}A$、$(T)_{13}C$、$(T)_{13}G$ 和随机引物（S6，S43，S61，S306，S360）均购自上海生工生物工程有限公司。其余试剂为国产分析纯或化学纯。

1.2 总 RNA 的提取

取 3～4 龄灰飞虱若虫 0.1g 于液氮中研磨后，参照试剂盒说明书进行，稍作修改。提取的总 RNA 经 1% 琼脂糖凝胶电泳检测其完整性，并用分光光度计测定 OD_{230}、OD_{260} 和 OD_{280}，以确定 RNA 的纯度和浓度。

1.3 cDNA 第一链的合成

在 3μg 的总 RNA 中加入 2μL 的 3′ 端锚定引物 20μmol/L，于 75℃的水浴中处理 5min，然后迅速冰浴 5min。继续加入 5×M-MuLV 反转录酶缓冲液 5μL、10mmol/L 的 dNTPs 2μL、Rnasin 0.5μL、无菌去离子水 7.5μL，37℃水浴 10min，加入 M-MuLV 反转录酶 1μL，然后 37℃水浴处理 1h，95℃水浴 5min，自然冷却至室温，-20℃保存备用。

1.4 正交法优化 DDRT-PCR 的反应体系

以模板浓度、dNTPs 浓度、镁离子浓度、随机引物浓度、锚定引物浓度和 *Taq* 酶用量 6 个因素，设计 6 因素 4 水平的正交试验（表 1）。反应条件为 94℃预变性 1min, 94℃ 30s, 40℃ 2min, 72℃ 30s，循环 40 次，最后 72℃延伸 10min。反应总体积为 25μL。取 2μL 产物经 6% 变性聚丙烯酰胺凝胶电泳检测。

1.5 差异片段的回收与重扩增

回收 DDRT-PCR 反应中无毒虫或带毒虫种群特异性表达的条带放入 1.5mL 离心管中，加入 20μL TE 捣碎，75℃水浴 30min。12000g 离心 2min，取上清，加入 10μL 3mol/L NaAc、450μL 无水乙醇，混匀。于 -20℃放置 30min。4℃，12 000g 离心 10min。弃上清，沉淀用预冷的无水乙醇漂洗。4℃，12 000g 离心 10min。沉淀干燥后用 20μL 无菌去离子水溶解。取 5μL 做模板，按照 1.4 中正交法优化后的体系与反应条件进行 PCR 重扩增及电泳检测。

1.6 PCR 法标记探针与 RNA 斑点杂交

以 1.5 中的部分重扩增回收产物为模板，通过 PCR 反应合成地高辛标记的 DNA 探针，反应条件同重扩增 PCR。在 dNTPs 底物中，地高辛标记的 DIG-11-UTP 与普通 dNTPs 的比例为 1 ∶ 12。RNA 斑点杂交参照 Roche 公司的 DIG High Prime Labeling and Detection Start Kit Ⅰ操作手册进行。

表 1 DDRT-PCR 反应体系的 6 因素 4 水平正交试验方案表

Table1 Scheme of the positive-cross test of DDRT-PCR with six factors at four levels

组别 Group No.	cDNA (μg)	随机引物 Random primer (μmol/L)	dNTPs (μmoL/L)	Mg^{2+} (μmoL/L)	锚定引物 Anchor primer (μmoL/L)	Taq (U)
1	1.0	1.5	250	1.5	2.5	2.0
2	3.0	2.5	150	1.5	1.5	1.5
3	2.0	2.5	250	2.0	1.5	2.5
4	4.0	1.5	150	2.0	2.5	1.0
5	1.0	2.0	150	2.5	3.0	2.5
6	3.0	1.0	250	2.5	2.0	1.0
7	2.0	1.0	150	1.0	2.0	2.0
8	4.0	2.0	250	1.0	3.0	1.5
9	1.0	1.0	300	2.0	1.5	1. 5
10	3.0	2.0	200	2.0	2.5	2.0
11	2.0	2.0	300	1.5	2.5	1.0
12	4.0	1.0	200	1.5	1.5	2.5
13	1.0	2.5	200	1.0	2.0	1.0
14	3.0	1.5	300	1.0	3.0	2.5
15	2.0	1.5	200	2.5	3.0	1.5
16	4.0	2.5	300	2.5	2.0	2.0

1.7 差异片段序列测定和同源性分析

将 1.6 中获得的阳性差异片段连接至 pMD18-T，转化大肠杆菌 *Escherichia coli* DH5α 感受态细胞，采用蓝白斑筛选阳性克隆。挑取阳性克隆送上海博亚生物技术公司测序，测序结果用 GenBank 的 Blast 软件进行同源性比对。

1.8 数据统计与分析

通过 Gene Tools Analysis Software Version 3.03.03 分析 DDRT-PCR 正交实验各方案所能扩增出的条带数，采用直观分析法确定条带清晰、数量最多的最适反应体系。差异片段同源性比对采用的标准是在 180bp 重叠区域内序列的相似性≥ 79% 或者 E 值≤ e^{-30}。

2 结果与分析

2.1 正交法优化 DDRT-PCR 的反应体系

本试验为 6 因素、4 水平共 16 个试验组合。PCR 产物经 6% 变性聚丙烯酰胺凝胶电泳和银染显色（图 1），通过 Gene Tools Analysis Software 检测每个反应体系所产生的条带数。结果表明，反应体系 8 和 12 无扩增条带；反应体系 1~3, 5, 7, 9, 11 和 13~16 扩增条带数较少（小于 15 条）；反应体系 4，6 和 10 能够扩增出较多条带（18~20 条），而反应体系 10 能够获得最多的条带数（20 条）。所以 DDRT-PCR 的体系选用反应体系 10，即 cDNA 3.0 μg，随机引物 2.0μmol/L，锚定引物 2.5μmol/L，dNTPs 200μmol/L，Mg^{2+} 2.0μmol/L，*Taq* 酶 2.0U，反应总体积为 25μL。

2.2 差异显示分析

利用 5 条随机引物和 3 条锚定引物组合对带毒灰飞虱和无毒灰飞虱种群进行了 15 组 PCR 扩增。

从图 2 可以看出差异表达的基因来源有 3 种情况：（1）只在带毒灰飞虱种群中特异表达；（2）只在无毒灰飞虱种群中特异表达；（3）在 2 个种群中均有表达，但表达量上存在差异。本实验只回收带毒或无毒种群中特异表达的片段，共获得 35 条。其中只在带毒灰飞虱种群中特异表达的有 20 条，只在无毒灰飞虱种群中特异表达的有 15 条。

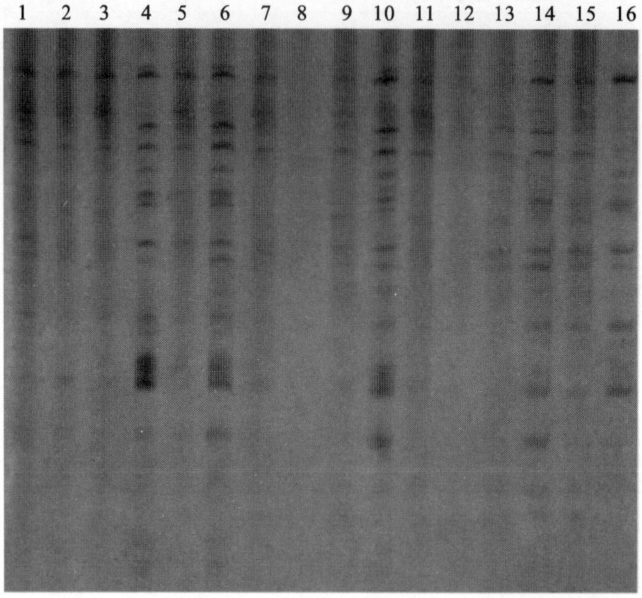

图 1　16 组不同 DDRT-PCR 反应体系正交试验银染结果

Fig.1　The positive-cross test results of sixteen different DDRT-PCR conditions by silver staining

1 ～ 16：表 1 中的 16 组不同 DDRT-PCR 反应体系

1–16: Group numbers of sixteen different DDRT-PCR conditions as in Table 1

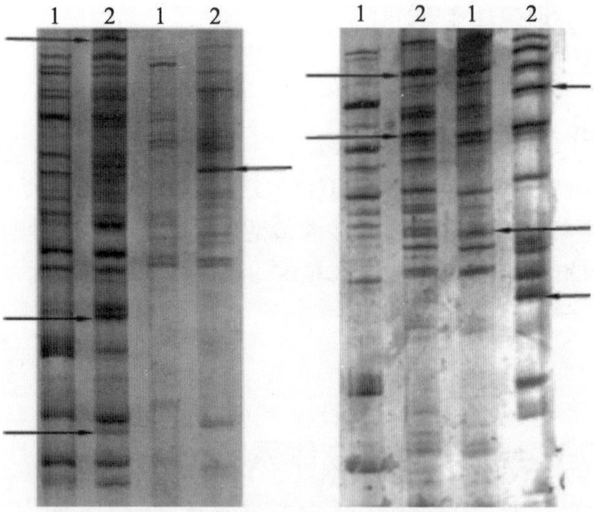

图 2　带毒与无毒灰飞虱基因的 mRNA 差异显示结果

Fig.2　Results of mRNA differential display of genes from viruliferous and virus-free *Laodelphax striatellus* populations

箭头所示为差异片段，1. 带毒灰飞虱种群，2. 无毒灰飞虱种群

Arrows indicated the differential bands, 1.Viruliferous *L. striatellus* population; 2. Virus-free *L.striatellus* population

2.3　差异片段的斑点杂交鉴定和同源性分析

在所获得的 35 条差异片段中随机选取 6 条以 PCR 法进行地高辛探针标记，分别与不同灰飞虱种群 RNA 进行斑点杂交验证。结果表明 DD2、DD3、DD4 和 DD5 片段为阳性克隆（图 3）。其中 DD4 来

自无毒灰飞虱种群，其余来自带毒灰飞虱种群。将差异基因克隆至 pMD18-T，并进行测序和同源性比对（表 2），结果显示，DD2 片段与人类 *Homosapiens*5- 羟色胺受体 1D 基因片段同源性高达 98%，DD3 片段与大肠杆菌 *E.coli* 的旋转酶 B 基因片段的同源性为 100%，DD4 片段与已知的基因序列同源性较低（重叠区域小于 30bp），推测该片段可能为一新基因。DD5 片段与白纹伊蚊（*Aedesalbopictus*）60S 核蛋白 L40 部分基因有 89% 的同源性。

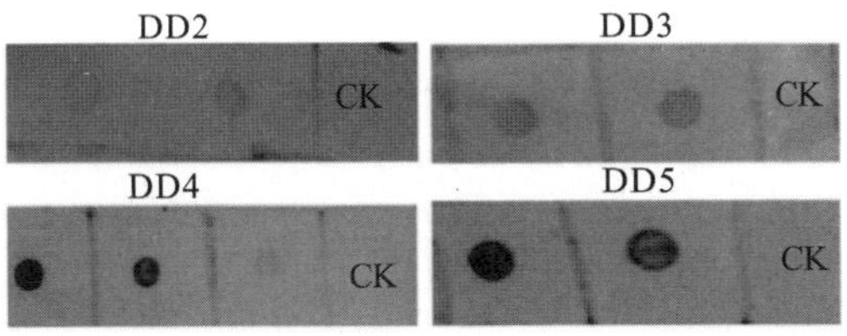

图 3 带毒与无毒灰飞虱种群中差异表达基因的 RNA 斑点杂交结果

Fig. 3 Results of RNA dot blot hybridization of differentially expressed genesin viruliferous and virus-free *Laodelphax striatellus* populations

表 2 克隆序列的同源性比较

Table2 Homology comparison of cloned cDNA fragments

片段 Sequence	片段长度 (bp) Fragment length	比对结果 (GenBank 登录号) Blast result (GenBank accession no.)
DD2	244	*Homo sapiens* 5-hydroxytryptamine receptor 1D (NM000864.3)
DD3	419	*Escherichia coli* isolate 077 gyrase B (gyrB) gene (AY832972.1)
DD4	555	None
DD5	237	*Aedes albopictus* 60S ribosomal protein L40 mRNA(AY826155.1)

3 讨论

在进行差异显示试验时，应选择发育时期、饲养条件一致或比较接近的灰飞虱作为供试材料，以减少由于选材差异造成的非相关基因的检测。在预实验中从 20 条随机引物中筛选了 8 条可以扩增出明显差异片段的引物，其中 4 条引物其差异片段回收后的重扩增效果较差，干扰了后续的验证，而另外 4 条引物重扩增效果较好，说明随机引物的不同对差异显示的结果有一定的影响，所以在进行差异显示 PCR 扩增效果不好的时候，可以考虑更换随机引物。本实验从上述 8 条随机引物中选取了 4 条重扩增效果较好及 1 条部分重扩增较好的引物进行 DDRT-PCR 分析。通过正交试验优化了 DDRT-PCR 的反应体系，明确了适合于灰飞虱的优化组合，不仅提高了实验质量，也缩短了实验时间。

病毒的侵染会带来宿主细胞转录组的变化。通常，病毒侵染后宿主基因表达量上调可能有两种作用：一是这些基因为病毒复制所必需。大多数病毒只编码有限的几个基因，为完成其复制周期，病毒需要利用一系列的宿主因子。在转录水平上提高这些基因的表达可能是病毒长期进化过程中发展出来的一种策略（Maule et al., 2002）。本研究所鉴定的 DD3 和 DD5 可能属于这一情况。DD3 与旋转酶 B 高度同源，表明它可能具有解螺旋功能域，在病毒的复制或转录中具有一定作用（Schröder, 2010）。DD5 可能编码核糖体大亚基 L40。通常，病毒都不能编码自身翻译所需的组分，而是依赖宿主的翻译机器（Kneller et al., 2006）。在长期演化过程中，病毒发展了一系列不同的机制以与宿主竞争翻译组分（Pestova et al., 2007）。病毒侵染后提高核糖体基因的表达可能是加强自身蛋白翻译的一种机制（Komarova et al., 2009）。一些转录谱实验也表明很多的 RNA 病毒在侵染植物后会带来核糖体基因表达的上调（Yang et al., 2007）。另外，在果蝇中进行的系列RNAi筛选实验表明，很多病毒对核糖体基因的表达水平十分敏感。在核糖体基因表达降低的果蝇中，这些病毒的复制严重受损，尽管宿主自身所受的影响并不大，但这些病毒均具有内部核糖体结合位点（internal ribosome entry site, IRES）（Cherry et al., 2005; Balvay et al., 2009）。目前，没有证据表明RSV具有IRES，因此对该病毒为何要提高核糖体基因表达还需要进一步的研究。宿主基因表达量上调也可能是宿主对病原物入侵的一种应答反应。比如病毒侵染后会引发寄主的某些防御反应，这些防御反应又会引起一系列基因表达的上调（Satoh et al., 2010）。二是病毒的侵染会对细胞的正常生理造成一定的影响，并偶然启动一些信号传导途径而引发某些基因表达的上调。DD2的上调可能是灰飞虱对RSV侵染的一种反应，5-羟色胺受体可能与动物的行为有关（Fineberg et al., 2010）。为了进行更为有效的传播，病毒的侵染有时会改变宿主的行为，比如促进其取食或繁殖等（James et al., 2006）。因此DD2的上调可能是RSV为其更有效传播引起的。

就病毒而言，与宿主争夺细胞中各种资源的最为有效的策略就是抑制宿主基因的表达（Arandaand Maule, 1998），这种抑制既可在转录水平上发生又可在翻译水平上体现。病毒侵染后抑制宿主基因表达的例子很多，其中与RSV具有亲缘关系的布尼亚病毒科的多种病毒已被证明会抑制宿主基因的转录，如裂谷热病毒（*Rift valley fever virus*, RVFV）S片段编码的非结构蛋白（non-structure proteins, NSs），能进入细胞核并特异地与一个与转录有关的宿主因子互作，从而抑制宿主基因的转录（May et al., 2004）。除争夺有限的资源外，病毒抑制宿主基因转录的另一种作用就是抑制宿主的防御反应。比如NSs能有效地抑制干扰素途径相关基因的转录（Billecocq et al., 2004）。因此DD4的下调很可能反映了RSV对宿主基因转录的抑制或者该基因是防御相关基因，但序列比对表明它不与任何已知的基因同源，其具体作用需进一步研究。

参考文献

[1] ARANDA M, MAULE A. Virus-induced host gene shutoff in animals and plants[J]. Virology, 1998, 243 (2): 261-267.

[2] BALVAY L, RIFO R S, RICCI E P, et al. Structural and functional diversity of viral IRESes[J]. Biochimica ETBiophysicaActa,2009,1789 (9-10): 542-557.

[3] BILLECOCQ A, SPIEGEL M, VIALAT P, et al. NSs protein of Rift Valley fever virus blocks interferon production by inhibiting host gene transcription[J]. Journal of Virology, 2004, 78(18): 9798-9806.

[4] CHERRY S, DOUKAS T, ARMKNECHT S, et al. Genome-wide RNAi screen reveals a specific sensitivity of IRES-containing RNA viruses to host translation inhibition[J]. Genes and Development, 2005,19 (4): 445-452.

[5] FALK B W, TSAI J H. Biology and molecular biology of viruses in the genus *Tenuivirus*[J]. Annual Review of Phytopathology, 1998, 36 (1): 139-163.

[6] FINEBERG N A, POTENZA M N, CHAMBERLAIN S R, et al. Probing compulsive and impulsive behaviors, from animal models to endophenotypes: a narrative review[J]. Neuropsychopharmacology, 2010, 35 (3): 591.

[7] Ng J C, Falk B W. Virus-vector interactions mediating nonpersistent and semipersistent transmission of plant viruses[J]. Annual Review of Phytopathol, 2006, 44: 183-212.

[8] KNELLER E L P, RAKOTONDRAFARA A M, MILLER W A. Cap-independent translation of plant viral RNAs[J]. Virus Research, 2006, 119 (1): 63-75.

[9] KOMAROVA A V, HAENNI A-L, RAMÍREZ B C. Virus versus host cell translation: love and hate stories[J]. Advances in Virus Research, 2009, 73: 99-170.

[10] LI X L, WEI B Q, ZHOU Y J. Preliminary study on RAPD markers dealing with compatibility of the *Laodelphax striatellus* and rice stripe virus[J]. Jiangsu Agricultural Sciences, 2009, (1): 23-25.

[11] LIANG P, AND PARDEE A. Distribution and cloning of eukaryotic mRNAs by means of differential display: refinements and

optimization[J]. Nucleic Acids Research, 1993, 21 (14): 3269-3275.

[12] LIN Q Y, XIE L H, ZHOU Z J, et al. Studies on rice stripe disease Ⅰ: Distribution and losses caused by the disease[J]. Journal of Fujian Agricultural College, 1990,19 (4): 373-379.

[13] LIU H, REN X M, CHEN L Y, et al. Cloning of the genes related to inhaled anesthetic action in *Drosophila melanogaster*[J]. Zhongguo yi xue kexue yuan xuebao. ActaAcademiaeMedicinaeSinicae, 2004, 26 (4): 385-391.

[14] LIU H J, CHENG Z B, WANG Y, et al. Preliminary study on transmission of rice stripe virus by small brown planthopper[J]. Jiangsu Journal of Agricultural Sciences, 2007, 23 (5): 385-391.

[15] MAULE A, LEH V, LEDERER C. The dialogue between viruses and hosts in compatible interactions[J]. Current Opinion in Plant Biology, 2002, 5 (4): 279-284.

[16] MAY N L, DUBAELE S, SANTIS L D, et al. TFIIH transcription factor, a target for the Rift Valley hemorrhagic fever virus[J]. Cell, 2004, 116 (4): 541-550.

[17] PESTOVA T V, LORSCH J R, HELLEN C U. The Mechanism of Translation Initiation in Eukaryotes[J]. CSHL Press, New York, 2007,48: 87-128.

[18] SATOH K, KONDOH H, SASAYA T, et al. Selective modification of rice (*Oryza sativa*) gene expression byrice stripe virus infection[J]. Journal of General Virology, 2010, 91 (1): 294-305.

[19] SCHRÖDER M. Human DEAD-box protein 3 has multiple functions in gene regulation and cell cycle control and is a prime target for viral manipulation[J]. Biochemical pharmacology, 2010, 79 (3): 297-306.

[20] TAO J, SHAN Y, JING Y J, et al. Screening and cloning of oriental migratory locust related resistance gene by mRNA differential display[J]. Acta Scientiarum Naturalium Universitatis Nankaiensis, 2009, 42 (1): 107 -111.

[21] YANG C, GUO R, JIE F, et al. Spatial analysis of *Arabidopsis thaliana* gene expression in response to turnip mosaic virus infection[J]. Molecular Plant-Microbe Interactions, 2007, 20 (4): 358-370.

[22] ZHANG H M, SUN H R, WANG H D, et al. Advances in the studies of molecular biology of rice stripe virus[J]. Acta Phytophylacica Sinica, 2007, 34 (4): 436-440.

[23] ZHANG X T, XIE L Y, LIN Q Y, et al. Transcriptional profiling in rice seedlings infected by rice stripe virus[J]. Acta Laser Biology Sinica, 2008, 17 (5): 620-629.

水稻条纹病毒 NS3 蛋白与水稻 3- 磷酸甘油醛脱氢酶(GAPDH)间的互作

肖冬来，贾东升，吴建国，杜振国，谢荔岩，吴祖建，谢联辉

(福建农林大学植物病毒研究所 福建省植物病毒学重点实验室，福建 福州 350002)

摘要： 水稻条纹病毒（*Rice stripe virus*, RSV）NS3 蛋白为病毒的沉默抑制子。利用酵母双杂交技术，从水稻 cDNA 文库中筛选出一个与 RSV NS3 蛋白互作的基因片段。推测该基因的功能是编码水稻 3- 磷酸甘油醛脱氢酶（glyceraldehyde-3-phosphate dehydrogenase, GAPDH）。双分子荧光互补实验进一步验证了 NS3 与 GAPDH 存在互作。瞬时表达实验表明，GAPDH 与 GFP 基因融合蛋白在本氏烟表皮细胞质中大量积累。此外，讨论了 GAPDH 蛋白在 RSV 侵染水稻过程中可能的功能。

关键词： 水稻条纹病毒；蛋白功能；互作；酵母双杂交；双分子荧光互补

中图分类号： S435.111 **文献标识码：** A **文章编号：** 1001-7216(2010)05-0493-04

Interaction Between Rice Stripe Virus NS3 Protein and Glyceraldehyde-3-Phosphate Dehydrogenase(GAPDH) of Rice

Donglai Xiao, Dongsheng Jia, Jianguo Wu, Zhenguo Du, Liyan Xie, Zujian Wu, Lianhui Xie

(Key Laboratory of Plant Virology of Fujian Province/Institute of Plant Virology, Fujian Agriculture and Forestry University, Fuzhou 350002, China)

Abstract: NS3 protein of *Rice stripe virus* (RSV) is an RNA silencing suppressor.Yeast two hybrid assay was used to screen the rice cDNA library with the NS3 protein as a bait. A novel rice protein interacting with RSV NS3 was obtained.The putative function of the interacting protein was glyceraldehy de-3-phosphate dehydrogenase (GAPDH). Bimolecular fluorescence complementation (BiFC) also showed that RSV NS3 could interact with the GAPDH.Cellular localization studies showed that GAPDH-GFP fusion protein accumulated predominantly in cytoplasm of tobacco (*Nicotiana benthamiana*) through transient expression assay.Possible functions of GAPDH during the infection of rice by RSV were discussed.

Keywords: *Rice stripe virus*; protein function; interaction; yeast two hybrid;bimolecular fluorescence complementation

水稻条纹病毒（*Rice stripe virus*, RSV）是纤细病毒属（*Tenuivirus*）的代表种（Haenni et al.,2005），主要由昆虫介体灰飞虱（*Laodelphax striatellus*）以持久性方式经卵传播（谢联辉等 , 2001）。RSV 可侵染 16 个属 80 多种禾本科植物，主要包括水稻、小麦等（阮义理等 , 1984）。近年来，由 RSV 引起的水

中国水稻科学 . 2010, 24(5): 493- 496

收稿日期：2010-01-22；修改稿收到日期：2010-01-27

基金项目：国家转基因生物新品种培育重大专项资助项目；国家自然科学基金资助项目；农业部农业公益性行业科研专项资助项目。

稻条纹叶枯病给江苏省的水稻生产造成了巨大损失（张恒木等 , 2007）。RSV 基因组含 4 条 RNA，除 RNA1 采用负链编码依赖 RNA 的 RNA 聚合酶（RdRp）外，RNA2、RNA3、RNA4 均采用双义编码策略共编码 6 个蛋白。RNA2 编码非结构蛋白 NS2 和 NSvc2，目前这两个蛋白的功能还不清楚。RNA3 编码 NS3 和外壳蛋白 CP，最近的研究表明 NS3 蛋白是病毒的沉默抑制子。RNA4 编码运动蛋白 NSvc4 和病害特异蛋白 SP（谢联辉等 , 2001 ；Xiong et al., 2008; Xiong et al.,2009）。目前，关于该病毒与寄主水稻间互作的研究还比较少。

酵母双杂交技术是一种研究蛋白质之间相互作用的有效手段，已被广泛应用在病毒与寄主之间的互作研究。Lu 等利用酵母双杂交体系在水稻中筛选到了与 RSVNSvc4 互作的 I 型 DnaJ 蛋白和小热激蛋白，为进一步研究 RSV 的运动机理提供了帮助（Lu et al., 2009）。本研究利用该技术从水稻 cDNA 文库中筛选出了一个与 NS3 互作的蛋白，该蛋白的基因注释为编码 3- 磷酸甘油醛脱氢酶（glyceraldehyde-3-phosphate dehydrogenase, GAPDH）。双分子荧光互补（bimolecular fluorescence complementation, BiFC）实验进一步验证了 GAPDH 与 NS3 之间的互作关系。GAPDH 是参与糖酵解过程的关键酶，除了这个重要的功能外，该酶还与膜融合、维管束形成、磷酸转移酶活性、核 RNA 输出、DNA 复制和修复、病毒致病性等功能有关（Sirover et al., 1999）。本研究对进一步研究 RSV 病毒蛋白功能、致病机制以及病害防治将具有一定意义。

1 材料与方法

1.1 实验材料

水稻酵母双杂交 cDNA 文库由北京大学李毅教授惠赠。ExTaq 聚合酶、内切酶、T4 DNA 连接酶、pMD18-T 载体、Trizol 均购自 TaKaRa 公司，反转录酶 M-M LV、λ DNA、*Eco*R Ⅰ / *Hind* Ⅲ标记购自 MBI 公司。酵母表达载体 pGADT7、pGBKT7，酵母菌株 AH109 购自 Clontech 公司。其他载体菌株均为本实验室保存。

1.2 与 NS3 互作的水稻蛋白筛选

参照 Clontech 公司 Matchmaker System Ⅲ 操作手册，以含 pGBKT7-NS3 质粒的 AH109 制备感受态，采用顺序转化法进行文库筛选。将多次验证为阳性的酵母质粒转化大肠杆菌 *DH5α*，繁殖并测序，通过水稻基因组数据库（http:// rice. plantbiology. msu.edu/）进行序列比对。

1.3 酵母双杂交系统验证 NS3 与 GAPDH 的互作

根据水稻基因组数据库中的 GAPDH 序列（LOC-Os02g38920.1）设计引物（5′-CGGAAT TCATGGGC AAGATTAAGATCG-3′，5′-CGGGATCCCTAGTTGGTGCTGTGCATG-3′），提取水稻总 RNA，RT-PCR 扩增 GAPDH 基因全长，克隆测序后构建到酵母表达载体 pGADT7 上，将构建好的 pGADT7-GAPDH 与诱饵载体 pGBKT7-NS3 共转化 AH109，再次进行互作验证。

1.4 融合蛋白 GAPDH-GFP 的定位

根据带有绿色荧光标记的瞬时表达载体 pCHF3/GFP 的多克隆位点设计引物（5′-GGTACCATGGGC AAGATTAAGATCG-3′，5′-GGATCCGTTGGTGCTGTGCATG-3′）以 pGADT7-GA PDH 为模板扩增 GAPDH，克隆测序后与 pCHF3/GFP 连接，构建好瞬时表达载体 pCHF3/GFP- GAPDH。转化农杆菌 EHA105，并注射本氏烟（*Nicotiana benthamiana*），48h 后在激光共聚焦显微镜下观察融合蛋白 GAPDH-GFP 的定位。

1.5 BiFC 验证 NS3 与 GAPDH 的互作

设计扩增引物（5′-GGTACCATGGGCAAGATTAAGATCG-3′，5′-GGATCCGTTGGTGCTG TGCA TG-3′）以 pGADT7-GAPDH 为模板扩增 GAPDH，克隆测序后与 pSAT1-nEYFP 连接，构建好表达 EYFPN 端的载体 pSAT1-nEYFPGAPDH。将 pSA T1-nEYFP-GAPDH 与表达 EYFPC 端的载体 pSAT1-cEYFP-NS3 共转化本氏烟原生质体，原生质体制备参考江力等的方法（江力等，2006）。在 100μL 原生质体（2×10^4 个）中分别加入 5μg 的 pSAT1-nEYFP-GAPDH 和 pSAT1-cEYFP-NS3，后加入 110μL 的 PEG4000 溶液，诱导 15min，室温下加入 400μL W5 溶液，轻柔混匀，100g 下离心 2min，弃上清，用 1mL W5 溶液清洗一遍，沉淀重悬于 500μL W5 溶液中，27℃下诱导 24h，在激光共聚焦显微镜下观察黄色荧光的表达。

2 结果与分析

2.1 筛选与 NS3 互作的水稻蛋白

以含 pGBKT7-NS3 质粒的 AH109 制备酵母感受态，转化水稻文库质粒，转化产物涂布 SD/ -Trp/-Leu/-His 平板，得到部分克隆，将这些克隆划线于 SD/-Trp/-Leu/-His/-Ade/X-α-gal 平板，共得到 3 个阳性克隆。将阳性克隆提取质粒并转化 DH5α 后送天根生化科技有限公司测序，测序结果与水稻基因组数据库进行同源比较，基因注释分别为包含 UBA/TS-N 域的蛋白（UBA/TS-N domain containing protein）、GAPDH、o-甲基转移酶（o-methyltransferase）。

2.2 全长 GAPDH 与 NS3 存在互作

从水稻文库中筛选到的互作片段大多不具有全长的开放阅读框。为进一步研究 GAPDH 与 NS3 的互作需要构建表达完整 GAPDH 的酵母表达载体 pGADT7-GA PDH。将 pGADT7- GAPDH 与 pGBKT7-NS3 共转化 AH109，转化产物可以在 SD/-Trp/-Leu/-His/-Ade/X-α-gal 培养基上生长且菌落变蓝（图 1）。

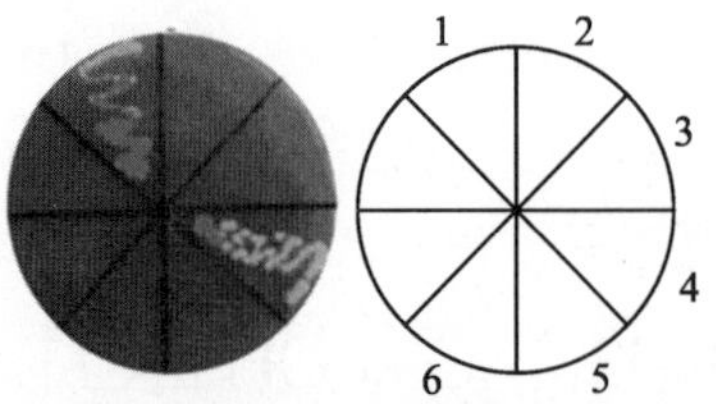

图 1 NS3 与 GAPDH 在酵母中的互作检测

Fig.1 Detection of interaction between NS3 and GAPDH by yeast two hybrid assays

1. pGADT7-GAPDH 与 pGBKT7-NS3 共转化产物生长状况；2. pGADT7-GAPDH 与 pGBKT7 共转化产物生长状况；3. pGBKT7-NS3 与 pGADT7 共转化产物生长状况；4. 阳性对照 pGADT7-T 和 pGBKT7-53 共转化产物生长状况；5. 阴性对照 pGADT7-T 和 pGBKT7-Lam 共转化产物生长状况；6. pGADT7 与 pGBKT7 共转化产物生长状况

1. Growth of co-transformants of 1-pGADT7-GAPDH and pGBKT7-NS3; 2. Growth of co-transformants of pGADT7-GAPDH and pGBKT7; 3. Growth of co-transformants of pGBKT7-NS3 and pGADT7; 4. Growth of co-transformants of pGADT7-T and pGBKT7-53; 5. Growth of co-transformants of pGADT7-T and pGBKT7-Lam (negative control); 6. Growth of co-transformants of pGADT7 and pGBKT7

2.3 融合蛋白 GAPDH-GFP 的定位

通过烟草瞬时表达体系表达了 GAPDH 和 GFP 的融合蛋白，激光共聚焦显微镜下观察到绿色荧光主要分布在细胞质中，空对照 pCHF3/GFP 的绿色荧光则分布于细胞质和细胞核中（图 2）。

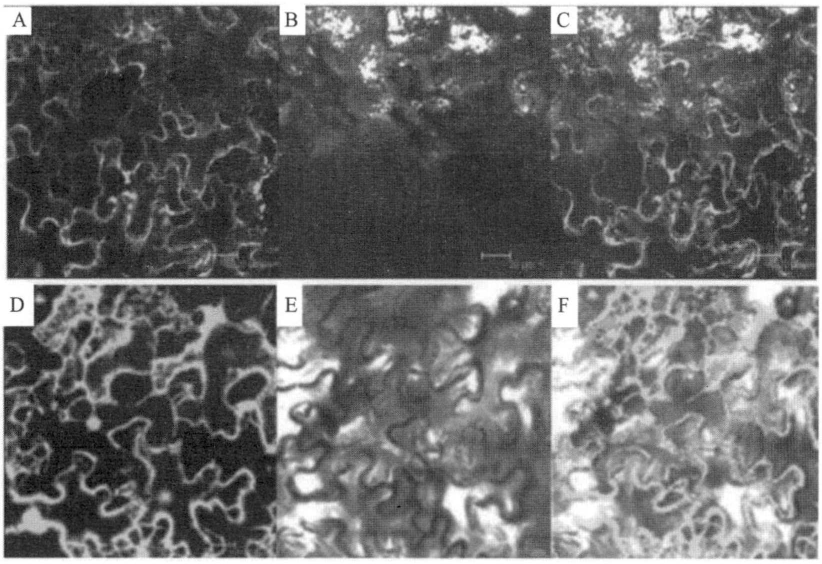

图 2 GAPDH 蛋白在烟草表皮细胞中的定位

Fig.2 Cellular localization of GAPDH protein in mesophyll cells of tobacco leaf

A.GAPDH-GFP 分布在细胞质中；B 和 E. 透射光下观察；C. A 和 B 叠加图；D. 阴性对照 pCHF/GFP;F.D 和 E 叠加图

A. GAPDH-GFP in cytoplasm; B and E. Phase image; C. Merged A and B; D. Negative control (pCHF/GFP); F. Merged D and E

2.4 NS3 与 GAPDH 在烟草原生体内存在互作

在双分子荧光互补实验中，黄色荧光蛋白 EYFP 被分为 N 端和 C 端两部分，分别与 GAPDH 和 NS3 融合，当 NS3 与 GAPDH 发生互作后，EYFP 的 N 端和 C 端相互接近，重新产生黄色荧光。将 pSAT1-nEYFP-GAPDH、pSAT1-cEYFP-NS3 质粒共同转化烟草原生质体，24h 后在激光共聚焦显微镜下进行观察，发现融合有 EYFPC 端的 NS3 和融合有 EYFPN 端的 GAPDH 存在互作，可以检测到黄色荧光信号，黄色信号可能主要集中于细胞核（图 3）。具体的共定位部位还需进一步研究。

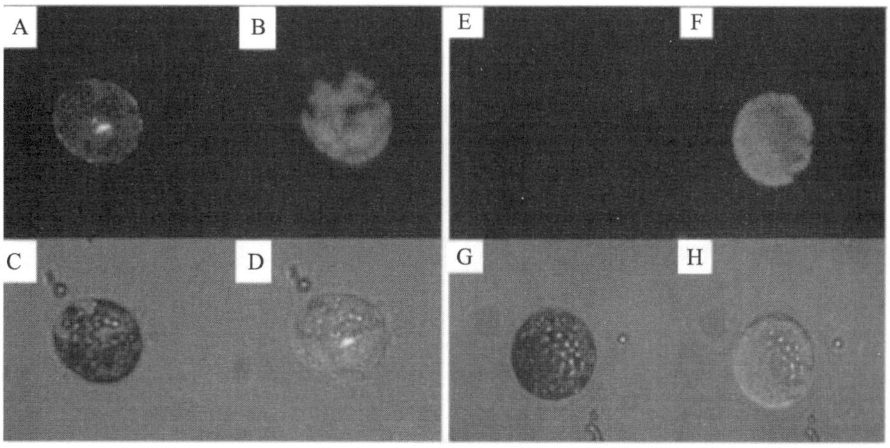

图 3 双分子荧光互补验证 NS3 与 GAPDH 互作

Fig. 3 Visualization of the interaction between NS3 and GAPDH by using BiFC

A. 重构的黄色荧光信号；B 和 F. 原生质体自发荧光；C 和 G. 透射图；D. A、B、C 叠加图；E. 阴性对照黄色荧光信号检测；H. E、F、G 叠加图

A. The reconstructed EYFP signals; B and F. Plastid autofluorescence; C and G. Phase image; D and H. Merged YFP and plastid autofluorescence signals; E.EYFP signals of pSAT1-cEYFP and pSAT1-nEYFP

3 讨论

水稻条纹病毒（RSV）近年在我国南方多个稻区持续暴发，给水稻生产带来严重危害。然而，目前我们对RSV所知甚少，这为我们制定有效措施来防控RSV带来不少困难。近来，本实验室采用酵母双杂交技术研究了RSV各蛋白与水稻蛋白间的互作。本文报道了我们利用RSV的NS3蛋白为诱饵筛选水稻cDNA文库的初步结果。我们发现NS3可能会与多个宿主因子互作，并选取了其中一个阳性克隆做了进一步验证。序列分析表明，该阳性克隆对应水稻的GAPDH。酵母双杂交及BiFC实验表明NS3与水稻全长GAPDH存在特异性互作。GAPDH是糖酵解途径中的一个关键酶，因此，NS3与该蛋白的互作很可能会干扰细胞的能量代谢。通常，GAPDH以四聚体存在，完整的四聚体是其发挥作用所必需的。因此，NS3有可能通过干扰GAPDH四聚体的形成来干扰其作用的发挥。进一步对这方面开展研究将是非常有意义的。另外，如前言所述，许多研究表明GAPDH有着多种除糖酵解外的其他作用。一般地，GAPDH通过不同的细胞定位及与不同的细胞因子相互作用来发挥不同的功能（Sirover et al., 1999; Sirover et al., 2005）。本文中的BiFC实验表明NS3与GAPDH的互作很大一部分可能发生在细胞核。在细胞核中，GAPDH可以发挥多种作用，如调节一些基因的转录，介导一些RNA的核质间运输，与端粒结合等（Sirover et al., 1999; Zheng et al., 2003; Sirover et al., 2005; Colell et al., 2007）。因此，NS3与GAPDH的互作可能与这些过程有联系。RSV与布尼亚病毒科病毒具有较近的亲缘关系（Falk et al., 1998），与NS3相对应的裂谷热病毒NSs蛋白可以抑制基因尤其是干扰素相关基因的转录（Billecocq et al.,2004）。因此，NS3与GAPDH的互作有可能是RSV特异干扰某些基因转录的一种机制。在这方面，有意思的是NS3已被证明是RSV的一个沉默抑制子，其沉默活性的发挥很大程度上依赖于其细胞核定位（Xiong et al., 2009）。已有报道表明一些沉默抑制子会通过改变某些基因的转录来发挥作用（Trink et al., 2005; Li et al., 2006）。深入研究NS3与GAPDH的互作与其抑制子活性的关系将是十分有意义的。

参考文献

[1] 江力，孙小卫，吴晓杰，等. 烟草原生质体的分离纯化 [J]. 安徽大学学报（自科版）, 2006, 30(6): 91-94.

[2] 阮义理，金登迪，许如银. 水稻条纹叶枯病毒的寄主植物 [J]. 植物保护, 1984,(3): 13.

[3] 谢联辉，魏太云，林含新，等. 水稻条纹病毒的分子生物学 [J]. 福建农林大学学报（自然版）, 2001, 30(3): 269-279.

[4] 张恒木，孙焕然，王华弟，等. 水稻条纹病毒分子生物学研究进展 [J]. 植物保护学报, 2007, 34(4): 436-440.

[5] BILLECOCQ A, SPIEGEL M, VIALAT P, et al. NSs protein of Rift valley fever virus blocks interferon production by inhibiting host gene transcription[J]. Journal of Virology, 2004, 78:9798-9806.

[6] COLELL A, RICCI JE, TAIT S, et al. Gapdh and autophagy preserve survival after apoptotic cytochrome crelease in the absence of caspase activation[J]. Cell, 2007, 129:983-997.

[7] FALK BW, TSAI JH. Biology and molecular biology of viruses in the genus Tenuivirus[J]. Annual Review of Phytopathology, 1998, 36:139-163.

[8] FAUQUET CM, MAYO MA, MANILOFF J, et al. Virus Taxonomy. Eighth Report of the International Committee on Taxonomy of Viruses[J]. Viruses, 2005,717-723.

[9] LI F, DING SW. Virus counterdefense: Diverse strategies for evading the RNA-silencing immunity[J]. Annual Review of Microbiology, 2006, 60:503-531.

[10] LU L, DU Z, QIN M, et al. Pc4, a putative movement protein of *Rice stripe virus*, interacts with a type I DnaJ protein and a small Hsp of rice[J]. Virus Genes, 2009, 38:320-327.

[11] SIROVER MA. New insights into an old protein: The functional diversity of mammalian glyceraldehyde-3-phosphate

dehydrogenase[J]. Biochimica ET Biophysica Bcta, 1999, 1432(2):159-184

[12] SIROVER MA. New nuclear functions of the glycolytic protein, glyceraldehyde-3-phosphate dehydrogenase, in mammalian cells[J]. Journal of Cellular Biochemistry, 2010, 95:45-52.

[13] TRINK D, RAJESWARAN R, SHIVAPRASAD PV, et al. Suppression of RNA silencing by a Geminivirus nuclear protein, AC2, correlates with transactivation of host genes[J]. Journal of Virology, 2005,79(4):2517-2527

[14] XIONG RY, WU JX, ZHOU YJ, et al. Characterization and subcellular localization of an RNA silencing suppressor encoded by Ricestripetenuivirus[J]. Virology, 2009, 387:29-40.

[15] XIONG RY, WU JX, ZHOU YJ, et al. Identification of a movement protein of the *tenuivirus Rice stripe virus*[J]. Journal of Virology, 2008, 82:12304-12311.

[16] ZHENG L, ROEDER RG, LUO YS. Phase activation of the histone H2B promoter by OCA-S, a coactivator complex that contains GAPDH as a key component[J]. Cell, 2003, 114:255-266.

水稻矮缩病毒对 3 种内源激素含量及代谢相关基因转录水平的影响

吴建国，王萍，谢荔岩，林奇英，吴祖建，谢联辉

（福建农林大学植物病毒研究所，福建省植物病毒学重点实验室，福州 350002）

摘要： 本文以抗病品种宜香 2292（*Oryza sativa* L. ssp. *indica* cv. Yixiang 2292）为材料，采用高效液相色谱技术（HPLC）和Real-time PCR技术研究了水稻矮缩病毒（*Rice dwarf virus*, RDV）胁迫下水稻内源赤霉素（GA_3）、生长素（IAA）和脱落酸（ABA）的动态变化。结果表明，受 RDV 侵染后，病株体内 GA_3 含量显著低于健株，在显症后的第 1d 和第 10d 最为明显，分别较健株低 6.28 和 5.92 倍；IAA 含量呈现波动变化，但病株体内的 IAA 含量始终较健株低，在显症后的 10d 最为明显，比健株低 3.58 倍；与 GA_3 和 IAA 相反，病株体内 ABA 的含量始终高于健株，显症后的第 1d 和第 13d 最为明显，分别较健株高 2.29 和 2.84 倍。Real-time PCR 定量检测了植物内源激素相关基因 mRNA 的表达，结果显示，GA_3 代谢相关的氧化还原酶基因表现为下调，而 IAA 和 ABA 代谢相关的 Cullin-1 和 P-glycoprotein1 基因表现出不同程度的上调。以上结果表明，水稻矮缩病的症状表现可能与病株体内的植物内源激素失调有关。

关键词： 水稻矮缩病毒；植物内源激素；赤霉素；生长素；脱落酸

中图分类号： S432. 41　**文献标识码：** A　**文章编号：** 0412 -0914(2010)02 -0151 -08

Affection of *Rice dwarf virus* on three phytohormones and transcriptional level of related genes in infected rice

Jianguo Wu, Ping Wang, Liyan Xie, Qiying Lin, Zujian Wu, Lianhui Xie

(Institute of Plant Virology, Fujian Agriculture and Forestry University, Key Laboratory of Plant Virology of Fujian Province, Fuzhou 350002, China)

Abstract: HPLC and Real -time PCR experiments were conducted for investigating the dynamics of GA_3, IAA and ABA in response to *Rice dwarf virus* (RDV)infection on rice variety Yixiang 2292 (*Oryza sativa* L. ssp. *indica cv.* Yixiang 2292). The results showed that GA_3 content was significantly lower in RDV infected rice than that in healthy plants, being 6.28 and 5.92 times lower at 1d and 10d after symptom appearance, respectively. The content of IAA in RDV infected rice plants was fluctuated, but always lower than that in healthy one. The lowest point was at 10th day after symptom appearance, which was 3.58 times lower than that in healthy plants. In contrast to GA_3 and IAA, the level of ABA in infected rice plants was always higher, the most significant increase was at 1 and 13d after symptom appearance, 2.29 and 2.84 times more in the diseased plants than in control. Real-time PCR experiments were conducted to investigate expression change of hormone regulated genes. The results showed that oxidoreductase, a GA_3 related gene decreased after RDV infection. Cullin-1 and P-glycoprotein 1 genes involved in the metabolism of IAA and ABA, were up-regulated to various extents. All these indicated that the symptom induced by RDV infection might be the result of phytohormones disruption.

Keywords: *Rice dwarf virus*; phytohormones; gibberellic (GA_3); auxin (IAA); abscisic acid (ABA)

植物病理学报 . 2010, 40(2): 151158

收稿日期：2009-04-24；修回日期：2010- 12- 30

基金项目：农业部农业公益性行业科研专项 (nyhyzx 07-051)；教育部博士点专项科研基金 (20050389006)；福建省科技厅项目资助 (K03005)。

病毒的侵染能引起植物一系列生理代谢过程的变化（Fraser, 1987），进而导致各种症状（Fraser et al., 1982）。目前有两种理论模式试图解释病健植物代谢差异的形成机理：一种是竞争模式，另一种是非竞争模式（Fraser, 1987; Whenham et al.,1980）。“竞争模式”认为病毒通过竞争利用核苷酸和氨基酸来抑制寄主核酸和蛋白质的合成。“非竞争模式”则认为病毒侵染引发的细胞病理效应是引起寄主代谢变化的主要原因，这种效应可能会引起植物内源激素代谢的变化（Whenham and Fraser, 1980），而植物激素的变化可能与病害症状形成有关（Jameson and Clarke, 2002）。矮缩病是水稻主要病害之一，其病原为水稻矮缩病毒（*Rice dwarf virus*，RDV）。在自然条件下，该病毒由黑尾叶蝉（*Nephotettix cincticeps*）、二点黑尾叶蝉（*N.virescens*）或电光叶蝉（*Recilia dorsasil*）传播，能在叶蝉体内复制，经卵传给子代（Xie et al., 1981; Omura et al., 1999）。该病毒广泛分布于中国、日本、朝鲜和尼泊尔等国的水稻种植区，引起植株矮缩僵硬、色泽浓绿，分蘖增多、叶片出现白色虚线状条点等症状，造成水稻大面积减产和重大经济损失。RDV 属于呼肠孤病毒科（*Reoviridae*）植物呼肠孤病毒属（*Phytoreovirus*），完整的病毒粒体为双重衣壳，球形正二十面体结构，直径 70nm 左右（Nakagawa et al., 2003; Hagiwara et al., 2004; Miyazaki et al., 2005）。RDV 全基因组由 12 条 dsRNA 组成，编码 5 个结构蛋白和 7 个非结构蛋白（Shimizu et al., 2009）。在 RDV 侵染过程中，非结构蛋白在介体昆虫细胞及水稻细胞中的功能（Wei et al., 2006; Wei et al., 2006; Wei et al., 2006c），以及 RDV 侵染水稻后的全基因组表达谱已被揭示（Shimizu et al., 2007）。然而，国内外对 RDV 与植物内源激素的互作还缺乏深入了解。我们的水稻全基因组芯片检测数据（待发表）则显示，水稻品种宜香 2292 受 RDV 胁迫后与植物内源激素合成相关的基因发生了差异表达。本文旨在研究 RDV 侵染水稻后引起的生长素（IAA）、赤霉素（GA_3）和脱落酸（ABA）的动态变化，以阐明 RDV 与激素互作关系，试图从病生理的角度揭示 RDV 对水稻生长的影响，为研究和防治水稻矮缩病提供理论依据。

1 材料与方法

1.1 试验材料

1.1.1 水稻品种

宜香 2292（*Oryza sativa* L. ssp. *indica* cv. Yixiang 2292）（四川省宜宾市农业科学研究所林纲老师惠赠），经生物学鉴定该品种对 RDV 具有中度抗性，发病潜育期为 15d 左右。

1.1.2 虫源和毒源

1）虫源：本实验室长期饲育的无毒虫经多次人工饲毒后获得的高带毒率的黑尾叶蝉种群。

2）毒源：水稻矮缩病毒由本实验室分离纯化，并保存在水稻感病品种台中 1 号上。

1.2 试验方法

1.2.1 水稻样品的培育及取样

采用单管单苗法，获毒黑尾叶蝉接种二叶一心期水稻幼苗，接种 3d 后移栽到温室栽培盆中培养（温度 28 ～ 30℃ ；空气相对湿度 50% ～ 70%，自然光照），并罩好防虫网，待幼苗表现出水稻矮缩病症状后开始取样，即发病第 1d 取幼苗地上部分 3 ～ 4g，然后每隔 3d 随机取样一次共 6 次。每次取样均以同期的健株作对照，重复 3 次，将样品做好标记保存于 -70℃冰箱中。

1.2.2 水稻内源激素（GA_3、IAA 和 ABA）的提取

植物内源激素提取参考 Arteca 等（Arteca et al., 1980）和 Kelen 等（Kelen et al., 2004）方法略有改变，取保存于 -70℃冰箱中的水稻叶片 2g，在弱光下加液氮研磨至粉末状，然后加 20mL 80% 的甲醇

再次研磨后转入 100mL 的锥形瓶中，4℃浸提 16 ～ 18 h，真空抽滤收集滤液，再将残渣反复浸提 3 次，每次加 80% 甲醇 30mL 浸提 3 ～ 5h，收集并合并其滤液。将合并的滤液转入 500mL 的旋转蒸发瓶中，在 35 ～ 40℃、弱光下减压蒸干甲醇相，收集水相并转入新的锥形瓶中。水相加入等体积的石油醚脱色至无色。用 1mol/L 的醋酸调节水相 pH 值至 2.5 ～ 2.8 后，用等体积的乙酸乙酯萃取 3 次，留酯相弃水相，在 35 ～ 40℃、弱光下减压蒸干乙酸乙酯相，用色谱纯甲醇定溶至 1mL，将定溶后的样品过 Seep-Pak-ODS-C18 小柱，小柱使用前用 2mL 色谱纯甲醇活化。样品过完小柱后再用 10mL 色谱纯甲醇冲洗，并收集滤液转入 25mL 旋转蒸发瓶中再次蒸干后用色谱纯甲醇定溶至 1mL 经 0.22μm 的微孔滤膜过滤后，待测。GA_3、IAA 和 ABA 的提取回收率分别为 90.48±12.42、102.99±19.97 和 77.94±8.37，符合 HPLC 分析的要求。

1.2.3 高效液相色谱法测定

色谱条件：Agi-lent1 100 HPLC 系统，色谱柱 :Supelcosil^{TM} LC-18 Cat #:58298 ; Col:81014-03（5μm，4.6×250mm），检测 IAA 的流动相为甲醇：水：乙酸溶液（45：54.2：0.8，V/V）；检测波长 254nm。检测 GA_3 和 ABA 的流动相为甲醇：乙腈：水（20：20：60，V/V），ABA 和 GA_3 检测波长分别为 254nm 和 208nm。柱温 35℃，柱压 70~71 bar，流速为 0.6mL/min，进样量 20μL，每次技术性重复 3 次，采用外标法定量测定。

1.2.4 Real-Time PCR 检测

选取发病后 7 d 的水稻幼苗，以健康植株为对照，采用 Trizol Reagent（Invitrogen）提取植物总 RNA，然后用 RNA free DNase Ⅰ处理（TaKaRa）。实验体系为：37℃反应 30min，然后用苯酚 / 氯仿处理，添加 1/10 体积的 3mol/L 醋酸钠，用 2.5 倍体积的冷无水乙醇沉淀 RNA，以 70% 乙醇清洗，干燥后溶于适量 DEPC 水中。然后用 1.5% 琼脂糖凝胶电泳检测经过 DNase Ⅰ处理后 RNA 的质量。以上述经过 DNase Ⅰ消化的 RNA 为模板，PCR 检测 genomic DNA 的消化情况。取 2.5μg 总 RNA 反转录形成 1st-cD-NA（Superscript Ⅱ，Invitrogen）。以同样的模板和优化后的实验体系，使用 Roche 公司 Light Cycler PCR 仪，以 SYBR green Ⅰ（Roche）为染料，并以水稻 UBQ11 基因为内参，进行 real time-PCR 扩增。实验体系为：镁离子浓度 3mmol/L，引物浓度 0.25μmol/L，退火温度 60℃，95℃变性 10min；95℃ 10s，60℃ 5s，72℃ 15s，重复 40 个循环；75℃至 95℃绘制融解曲线。

2 结果与分析

2.1 RDV 与 GA_3、IAA 和 ABA 的互作

植物体内激素含量甚微，但它们却参与了植物从生长到衰老的几乎所有的生理过程，起着十分重要的调节作用。鉴于水稻矮缩病的症状特点，研究水稻感染 RDV 后内源激素的变化，对于揭示 RDV 致病机制具有重要意义。本试验利用高效液相色谱技术（high performance liquid chromatography，HPLC）测定了水稻感染 RDV 后的 1 ～ 16d 内的生长素 IAA、GA_3 和 ABA 内源激素的动态变化（图 1）。

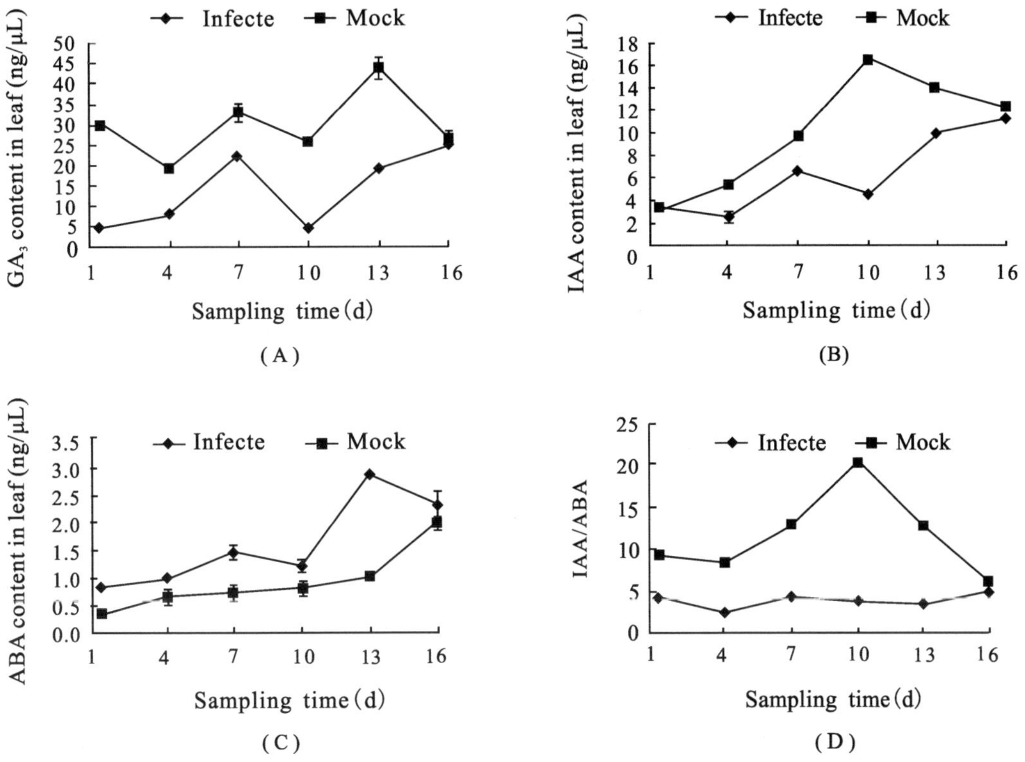

图 1　RDV 侵染水稻后 GA_3，IAA，ABA 及 IAA/ABA 的变化

Fig. 1 Variation of GA_3，IAA，ABA and IAA/ABA after RDV infecting in rice

A. GA_3 变异；B. IAA 变异；C. ABA 变异；D. IAA/ABA 变异

A. The variation of GA_3；B. The variation of IAA；C. The variation of ABA；D. The variation of IAA/ABA

图 1-A 揭示，受 RDV 侵染后，病株体内 GA_3 含量显著低于健株，在显症后的第 1d 和第 10d 最为明显，分别较对照低 6.28 和 5.92 倍；图 1-B 则表明，受 RDV 侵染后，IAA 含量呈现波动变化，但病株体内的 IAA 含量始终较健株低，在显症后的 10d 最为明显，比对照低 3.58 倍；与 GA_3 和 IAA 相反，病株体内 ABA 的含量始终高于健株，显症后的 1d 和 13d 最为明显，分别较对照高 2.29 和 2.84 倍（图 1-C）。

2.2　RDV 胁迫下 IAA/ABA 比值变化

植物生长发育过程受多种内源激素的平衡调节，单一激素的含量变化则很难揭示问题的本质。IAA/ABA 的比值通常用来表示植物内源激素对生长的作用是促进还是抑制（Wang et al., 2004）。RDV 侵染水稻后，IAA/ABA（2.55 ～ 4.89）比值明显小于对照 IAA/ABA（6.12 ～ 20.23）（图 1-D），发病后的 1 ～ 13d 比值差异明显。试验揭示了 RDV 胁迫能够改变植株体内 IAA/ABA 的平衡，当 IAA/ABA 平衡失去或打破时，其生长便受到影响。由此推测 RDV 胁迫下植物内源激素可能影响了水稻的抗病性和水稻的形态发育，参与或调控了水稻矮缩病症状的形成。

2.3　RDV 与水稻互作影响了植物激素的代谢

HPLC 试验已表明，在 RDV 与水稻互作过程中 3 种重要的植物内源激素的平衡被打破。为了进一步揭示 RDV 与植物内源激素的互作，本研究参考 Affymetrix 水稻全基因组芯片试验数据，结合 Real-Time PCR 技术，揭示了植物内源激素代谢相关基因的表达情况（表 1）。试验数据显示赤霉素生物合成和代谢相关的氧化还原酶基因或 2OG-Fe 氧化酶蛋白基因（GenBank BA000029）表现为下调，而反应对生长素和脱落酸等敏感或代谢相关的 Cullin-1（GenBank.AK073477）、P-glycoprotein1（GenBank

AK121506）和 AAA-type ATPase（GenBank AK070731）基因则表现出不同程度的上调，其中 ABA 代谢相关的 AAA-type ATPase 基因上调了 64 倍多，结果则进一步揭示 GA_3、IAA 和 ABA 参与了 RDV 与水稻互作过程；转录组水平上内源激素相关基因的差异表达可能会影响代谢组水平上的激素合成，从而影响植物内源激素的平衡。

表 1 实时聚合酶链反应验证 GA_3、IAA 和 ABA 代谢相关基因的差异表达

Table 1 Real-Time PCR to verify differential expression of GA_3, IAA and ABA pathway related genes

GenBank accession No.	Primer	Sequence	$Log_2^{RationA}$	P-Value^A	$Log_2^{RationB}$	P-Value^B	Pathway	Putative function
AK070731	Forword5′-3′	CCAACCACGTCAAGAAGCTG					Response to abscisic acid stimulus	AAA-type ATPase putative
	Reverse 5′-3′	TCACTAATGCTGCGAGGCAC	2.1667	0.0459	6.4539	0.1517		
AK073477	Forword5′-3′	TTGGGAAACTGCGTCGTCA					Response to auxin and cytokinin stimulus	Cullin-l, putative
	Reverse5′-3′	CCTTCGCATCCTATCGGTAAA	1.2667	0.3344	NA	—		
AK121506	Forword5′-3′	CTATCGTCGTAGAAGCGGTGAC					Response to auxin and cytokinin stimulus	P-glycoprotein 1, putative
	Reverse5′-3′	GCTCGTTCCGTGCTTGACTTA	1.2333	0.0828	1.0802	0.4280		
BA000029	Forword5′-3′	TCGGACTTCTCCAAGACCATC					Gibberellic acid biosynthesis, mediated signaling, and response to gibberellic acid stimulus	Oxidoreductase, 2OG-Fe oxygenase family protein
	Reverse5′-3′	GTGCTTCAGCTCAACCTCCC	-1.1667	0.0505	0.4158*	—		
UBQ11	Forword5′-3′	CTGCTGCTGTTCTTGGGTTCA						
	Reverse5′-3′	TCATTATAGTTCTTCCATGCTGCTC						

3 讨论与结论

在植物病毒与寄主互作过程中对内源激素水平的研究已有不少。1939 年 Jahnel 和 Lucas 首次报道马铃薯卷叶病毒（*Potato leaf roll virus*）侵染马铃薯后降低了生长素的活性（Pennazio et al., 1996）。此后大量研究结果显示病毒侵染寄主后能够引起生长素的活性和浓度的变化（Smith et al., 1968; Lockhart et al., 1970; Rao et al., 1974; Rajagopal, 1977）。玉米 MRC 病毒（*Mal deriocuarto virus*）侵染玉米后在离轴叶片表皮形成瘿瘤，Abdala 等（Abdala et al., 1999）研究揭示在瘿瘤形成过程中 IAA 含量和 IAA 氧化酶活性提高，随着瘿瘤的发育成熟这种差异则消失。Sheng 等揭示芜菁脉明病毒（*Turnip vein cleaning virus*）侵染拟南芥病毒诱导矮化突变体 vid1 后，生长素代谢途径被干扰会导致植株的矮化或顶端生长优势的丧失（Sheng et al., 1998）。Meenu 等证实烟草花叶病毒（*Tobacco mosaic virus*）复制酶蛋白与 Aux/IAA 蛋白 PAP1/IAA26 互作能够诱导花叶症状的产生（Meenu et al., 2005）。本试验揭示 RDV 侵染水稻后 IAA 水平呈现下降趋势，在发病后 7～13d 最为明显，采用 0.01g/L 的吲哚-3- 乙酸（Indole3-acetic acid）喷施发病植株表明，外源生长素不能使病株恢复正常株高（尚未发表的数据），进而揭示 RDV 的矮缩症状可能不是由单个生长素代谢失衡所引起的。

病毒与寄主互作过程中 GA_3 水平的变化而引起的寄主矮化已有相关报道。Russell 等的研究揭示受大麦黄矮病毒（*Barley yellow dwarf virus*）侵染后，植株体内的赤霉素含量降低，从而导致细胞的变短是大麦矮化的主要原因（Russell et al., 1971）。水稻东格鲁球状病毒（*Rice tungro spherical virus*）侵染寄主后自由态和束缚态 GA_3 类似物含量也明显降低，这可能是诱导植株矮化的一个主要原因（Sridhar et al., 1987）。Zhu 等利用酵母双杂交系统研究并揭示 RDVP2 蛋白与水稻贝克杉烯氧化酶蛋白互作，从而导致赤霉素的合成能力大大下降，通过外源赤霉素诱导感染 RDV 的水稻，发现能够使矮缩症状明显减轻，因此他们认为 RDV 通过降低赤霉素的合成而产生矮缩症状（Zhu et al., 2005）。这与本文的实验结果相类似，植物内源激素在 RDV 与水稻互作过程无论在转录组和代谢组水平都扮演一个重要角色，转录组水平上赤霉素相关基因存在差异表达，代谢组水平上同样揭示受 RDV 侵染后赤霉素含量呈现明显下降趋势，0.01g/L 的 GA_3 能够诱导发病植株基本恢复到正常株高（尚未发表的数据），这也进一步证实 GA_3 参与了水稻矮缩症状的形成。

ABA 作为胁迫激素在生物及非生物胁迫中研究的最多，植物在逆境胁迫下启动 ABA 生物合成系统，合成大量 ABA 以调控植物的正常生理代谢（Mauch-Mani et al., 2005），但脱落酸水平的积累则往往是植株叶片黄化、停止生长和老化的前兆。在很多系统感染表现矮化的病害中，寄主的脱落酸含量都明显提高（Zhang et al., 1997），而表现增生、肿瘤等症状的病害则细胞分裂素或生长素类有异常增加（Latham et al., 1997; Abdala et al., 1999; Jameson, 2000; Manes et al., 2001）。Zhang 等通过 ELISA 的方法研究香蕉束顶病毒（*Banana bunchy top virus*，BBTV）侵染香蕉植物后 ABA、GA_3 和玉米素（iPAs）含量的变化，结果表明 ABA 在 BBTV 侵染后被大量诱导并不断积累，在接种后第 35d 测定含量最高，同时认为香蕉束顶病的症状表现主要与病株中内源激素的失调有关，而与 BBTV 在体内的运转并不直接相关（Zhang et al., 1997）。本试验揭示，水稻受 RDV 侵染后 ABA 含量明显增加，在发病后的第 13d 差异最显著，推测 ABA 可能参与水稻对 RDV 的抗病反应；植物体内 IAA/ABA 平衡的破坏干扰了植物的正常生长发育。

众所周知，病毒与寄主的互作过程，水杨酸（SA）、茉莉酸（JA）和乙烯（ET）在植物抗病性方面起重要作用（Robert-Seilaniantz et al., 2007），然而其他激素如 ABA、IAA、GA、CK 和芸薹素（BL）在植物防御机制研究中了解甚少。由于相关突变材料难以获得，人们对多数植物病毒与寄主互作过程中植物内源激素的角色还缺乏深入研究。本研究揭示 RDV 侵染水稻后 ABA、GA_3 和 IAA 等均有变化，然而这种变化究竟是由病毒侵染直接所造成，还是影响了水稻同化作用或病毒诱导细胞损伤所造成，仍有待进一步研究。此外，这些内源激素代谢的变化与 SA、JA 和 ET 等之间的关系，仍需进一步揭示。从生物学以及分子生物学水平，人们对 RDV 引起植株矮缩症状的形成原因（Zhu et al., 2005; Shimizu et al., 2007），已有初步了解，然而，针对 RDV 怎样诱导产生水稻叶片的白色虚线状条点症状尚未见报道，是否由植物内源激素所调控的细胞程序性死亡或过敏性坏死反应所引起，尚需研究。

研究表明，植物激素及其类似物对某些植物病毒病具有一定的防治效果（Tomlinson et al., 1976; Jameson et al., 2002）。例如在感染不孕病毒（*Tomato aspermy virus*）的番茄上施用 GA_3，可使番茄的矮化症状有所恢复，其主要作用机制是使细胞伸长，但对细胞的有丝分裂作用不大（Bailiss, 1968）。低浓度的 JA（0.25 ～ 25nmol/L）能够抑制三叶草花叶病毒（*White clover mosaic virus*）亚基因组 dsRNA 的复制，并阻止病毒的系统性扩散，外源 SA 通过降低病毒浓度从而诱导植物的抗病毒能力，并延迟病害症状的产生（Clarke et al., 2000; Clarke et al., 1998; Murphy et al., 1999）。RDV 的症状表现可能与其内源激素失调相关，通过研究 RDV 与植物内源激素的互作，将有利于进一步揭示 RDV 的致病机制，为有效防治水稻矮缩病提供科学依据。

参考文献

[1] ABDALA G, MILRAD S, VIGLIOCCO A, et al. Hyperauxinity in diseased leaves affected by *Mal dericuarto virus* (MRCV)[J]. BIOCELL, 1999, 23:13-18.

[2] ARTECA R N, POOVAIAH B W, SMITH O E. Use of high performance liquid chromatography for the determination of endogenous hormone levels in *Solwumt uberosum* L subjected to carbon dioxide enrichment of the root zone. Journal of Plant Physiology & Pathology[J]. 1980，65:1216-1219.

[3] BAILISS K W. Gibberellins and the early disease syndrome of aspermy virus in tomato (*Lycopersicon esculentum* Mill.)[J]. Ann

Bot, 1968,32:543-552.

[4] CLARKE S F, BURRITT DJ, JAMESON PE, et al.Effects of plant hormone on white clover mosaic potexvirus double-stranded RNA[J]. Plant Pathology, 2000, 49:428-434.

[5] CLARKE S F, BURRITT D J, JAMESON P E,et al. Influence of plant hormones on virus replication and pathogenesis-related proteins in Phaseolus vulgaris L infected with white clover mosaic potexvirus[J]. Physiological & Molecular Plant Pathology, 1998, 53:195-207.

[6] FRASER R S S, WHENHAM R J. Plant growth regulators and virus infection:a critical review[J]. Plant Growth Regulation, 1982, 1:37-59.

[7] FRASER R S S. Biochemistry of virus-infected plants[J]. Letchworth, Hartfordshire, UK:Research Studies Press, 1987，641.

[8] HAGIWARA K, HIGASHI T, MIYAZAKI N, et al. The amino-terminal region of major capsid protein P3 is essential for self-assembly of single-shelled core-like particles of rice dwarf virus[J]. Journal of Virology, 2004, 78:3145-3148.

[9] JAMESON P E, CLARKE S F. Hormone-virus Interactions in plants[J]. Critical Reviews in Plant Sciences, 2002,21(3):205-228.

[10] JAMESON P. Cytokinins and auxins in plant-pathogen interactions-an overvie[J]. Plant Growth Regulation, 2000, 32:369-380.

[11] KELEN M, DEMIRALAY E C, SEN S, et al. Separation of abscisic acid, indole-3-acetic acid, gibberellic acid in 99 R (Vitis berlandieri x Vitis rupestris) and rose oil (Rosa damascena Mill.) by reversed phase liquid chromatography[J]. Turkish Journal of Chemistry, 2014, 28:603-610.

[12] LATHAM J R, SAUDERS K, PINNER M S, et al. Induction of plant cell division by beet curly top virus gene C4[J]. The Plant Journal, 1997, 11:1273-1283.

[13] LOCKHART B E, SEMANCIK J S. Growth inhibition, peroxidase and 3-indoleacetic acid oxidase activity and ethylene production in cowpea mosaic virus infected cowpea seedlings[J]. Phytopathology, 1970, 60:553-554.

[14] MANES C O, VAN MONTAGU M, PRINSEN, E, et al. De novo cortical cell division triggered by the phytopathogen Rhodococcus fascians in tobacco[J]. Molecular Plant-Microbe Interactions Journal, 2001, 14:189-195.

[15] MAUCH-MANI B, MAUCH F. The role of abscisic acid in plant-pathogen interactions[J]. Current Opinion in Plant Biology, 2005, 8:409-414.

[16] MEENU S P, SAMEER P G, SHEETAL G, et al. Interaction of the tobacco mosaic virus replicase protein with the Aux/IAA protein PAP1/IAA26 is associated with disease development[J]. JournalVirol, 2005, 79(4):2549-2558.

[17] MIYAZAKI N, HAGIWARA K, NAITOW H, et al. Transcapsidation and the conserved interactions of two major structural proteins of a pair of phytoreoviruses confirm the mechanism of assembly of the outer capsid layer[J]. Journal of Molecular Biology, 2005, 345:229-237.

[18] MURPHY A M, CHIVASA S, SINGH D P, et al. Salicylic acid-induced resistance to viruses and other pathogens: a parting of the ways[J]. Trends Plant Science, 1999, 4:155-160.

[19] NAKAGAWA A, MIYAZAKI N, TAKA J, et al. The atomic structure of rice dwarf virus reveals the self-assembly mechanism of component proteins[J]. Structure, 2003, 11:1227-1238.

[20] OMURA T, YAN J. Role of outer capsid proteins in transmission of Phytoreovirus by insect vectors[J]. Advances in Virus Research, 1999, 54:15-43.

[21] PENNAZIO S, ROGGERO P. Plant hormones and plant virus diseases-the auxin[J]. Microbiol, 1996, 19:369-378.

[22] RAJAGOPAL R. Effect of tobacco mosaic virus infection on the endogenous levels of indoleacetic, phenylacetic and abscisic acids of tobacco leaves in various stages of development[J]. Zeitschrift Für Pflanzenphysiologie, 1977, 83:403-409.

[23] RAO M R K, NARASIMHAM B. Endogenous auxin levels as influenced by ‘infectious variegation’ in Lisbon lemon leaves[J]. South Indian Horticulture, 1974, 22:138-139.

[24] ROBERT-SEILANIANTZ A, NAVARR, L, BARI R, et al. Pathological hormone imbalances[J]. Current Opinion in Plant Biology, 2007, 10:372-379.

[25] RUSSELL S L, KIMMINS W C. Growth regulators and the effect of BYDV on barley (Hordeum vulgare L)[J]. Annals of Botany,

1971, 35:1037-1043.

[26] SHENG J, LARTEY R, GHOSHROY S, et al. An Arabidopsis thaliana mutant with virus-inducible phenotype[J]. Virology, 1998, 249:119-128.

[27] SHIMIZU T, SATOH K, KIKUCHI S, et al. The repression of cell wall-and plastid-related genes and the induction of defense-related genes in rice plants infected with rice dwarf virus[J]. Molecular Plant-Microbe Interactions Journal, 2007, 20:247-254.

[28] SHIMIZU T, YOSHII M, WEI T, et al. Silencing by RNAi of the gene for Pns12, a viroplasm matrix protein of rice dwarf virus, results in strong resistance of transgenic rice plants to the virus[J]. Plant Biotechnology Journal, 2010, 7:24-32.

[29] SMITH S H, MCCALL S R, HARRIS J H. Auxin transport in curly top virus-infected tomato[J]. Pathol, 1968, 58:1669-1670.

[30] SRIDHAR R, MOHANTY S K, MOHANTY S K. Physiology of rice tungro virus disease: gibberellins in the disease syndrome[J]. International Journal of Tropical Plant Diseases, 1987, 4:85-92.

[31] TOMLINSON J A, FAITHFULL E M, WARD C M. Chemical suppression of the symptoms of two virus diseases[J]. Annals of Applied Biology, 1976, 84(1):31-41.

[32] WANG B C, SHAO J P, BIAO L, et al. Soundware stimulation triggers the content change of the endogenous hormone of the chrysanthemum mature callus[J]. Colloids and Surfaces B: Biointerfaces, 2004, 37(3-4):107-112.

[33] WEI T, KIKUCHI A, MORIYASU Y, et al. The spread of rice dwarf virus among cells of its insect vector exploits virus-induced tubular structures[J]. Journal of Virology, 2006, 80:8593-8602.

[34] WEI T, KIKUCHI A, SUZUKI N, et al. Pns4 of rice dwarf virus is a phosphoprotein, is localized around the viroplasm matrix, and forms minitubules[J]. Archives of Virology, 2006, 15:1701-1712.

[35] WEI T, SHIMIZU T, HAGIWARA K, et al. Pns12 protein of rice dwarf virus is essential for formation of viroplasms and nucleation of viral-assembly complexes[J]. Journal of General Virology, 2006c, 87:429-438.

[36] WHENHAM R J, FRASER R S S. Stimulation by abscisic acid of RNA synthesis in discs from healthy and tobacco mosaic virus-infected tobacco leaves[J]. Planta, 1980, 150:349-353.

[37] XIE L H, LIN Q Y, GUO J R. A new insect vector of rice dwarf virus[J]. International Rice Research Notes, 1981, 6(5):14.

[38] ZHANG H, ZHU X, LIU H. Effect of Banana bunchy top virus (BBTV) on endogenous hormone of banana plant (in Chinese)[J]. Acta Phytopathol Sin (植物病理学报), 1997, 27:79-83.

[39] ZHU S, GAO F, CAO X, et al. The rice dwarf virus P2 protein interacts with ent-kaurene oxidases in vivo, leading to reduced biosynthesis of gibberellins and rice dwarf symptoms[J]. Plant Physiol, 2005, 139: 1935-1945.

Rice ragged stunt virus segment S6-encoded nonstructural protein Pns6 complements cell-to-cell movement of *Tobacco mosaic virus*-based chimeric virus

Zujian Wu[1], Jianguo Wu[1], Scott Adkins[2], Lianhui Xie[1], Weimin Li[1]

(1 Institute of Plant Virology, Fujian Agriculture and Forestry University, Fuzhou, Fujian 350002,PR China; 2 United States Department of Agriculture, Agricultural Research Service, 2001 South Rock Road, Fort Pierce, FL 34945, USA)

Abstract:The protein(s) that support intercellular movement of *Rice ragged stunt virus* (RRSV) have not yet been identified. In this study, the role of three nonstructural proteins Pns6, Pns7 and Pns10 in cell-to-cell movement were determined with a movement-deficient *Tobacco mosaic virus* (TMV) vector. The results showed that only the Pns6 could complement the cell-to-cell movement of the movement-deficient TMV in *Nicotiana tabacum* Xanthi-nc and *N. benthamiana* plants, and both N- and C-terminal 50 amino acids of Pns6 were essential for the cell-to-cell movement. Transient expression in epidermal cells from *N. benthamiana* showed that the Pns6–eGFP fusion protein was present predominantly along the cell wall as well as a few punctate sites perhaps indicating plasmodesmata. Taken together with previous finding that the Pns6 has nucleic acid-binding activity (Shao et al., 2004), the possible role of Pns6 in cell-to-cell movement of RRSV were discussed.

Keywords:*Rice ragged stunt virus*; Pns6; Cell-to-cell movement

Rice ragged stunt virus (RRSV), genus *Oryzavirus*, family *Reoviridae* is the causative agent of rice ragged stunt disease, which was first discovered in 1976–1977 in Indonesia and Philippines (Ling et al., 1978) and occurs in most rice-growing countries in southeastern and far-eastern Asia (Milne et al., 1982). RRSVis transmitted in a persistent manner by brown plant hoppers after proliferation in the vector insect (Boccardo and Milne, 2008) and only infects plants in the family *Graminae* including *Oryza sativa*, *O. latifolia* and *O. nivara*, causing typical symptom such as stunting, twisted and ragged leaves, and pale spindle-shaped enations on abaxial surfaces of leaves and leaf sheaths (Ling et al., 1978).

The RRSV genome consists of 10 double-stranded RNA (dsRNA) segments with molecular weight ranging from 1.2 to 3.9kb, denoted as S1–S10, respectively (Kawano, 1984). With the exception of S4, which encodes two putative proteins (P4a and P4b), each of the other genome segments encodes one protein, termed as P1, P2, P3, P5, Pns6, Pns7, P8, P9 and Pns10, respectively. Since P8 is subjected to self-cleavage to yield P8a and P8b (Upadhyaya et al., 1996), the RRSV produces at least seven structural proteins (P1, P2, P3, P5, P8, P8b and P9), and three nonstructural proteins (Pns6, Pns7 and Pns10) (Omura et al., 1983; Hagiwara et al., 1986; Uyeda et al., 1995; Li et al., 1996; Upadhyaya et al., 1995, 1997, 1998). The 10dsRNAs as well as copies of P4a (RNA-dependent RNA polymerase, RdRP) are encapsidated with structural proteins to form an icosahedral particle about 65–70nm in diameter (Hagiwara et al., 1986; Miyazaki et al., 2008). The complexity of virus particle and genome organization has hindered the development of an infectious clone for RRSV. Lack of a reverse genetic system makes it impossible to perform standard mutagenesis for functional studies of viral

Virus research.2014, 180: 97-101

Received 26 August 2013; Accepted 18 November 2013; Available online 27 November 2013.

genes or sequence elements within the context of RRSV genomes, leading to the biological functions of RRSV being poorly linked with specific viral proteins, except for vector-transmission, which has been clearly assigned to Pns9, a spike protein (Upadhyaya et al., 1995; Zhou et al., 1999; Shao et al., 2003).

Cell-to-cell movement is a primary requirement for systemic spread of a plant virus through a compatible host. Insights into the cell-to-cell movement reveal that this process is controlled by the viral-encoded movement protein (MP), a general feature of plant viruses (Maule, 1991; Lucas and Gilbertson, 1994; Carrington et al., 1996). Growing evidence indicates that the MP is specifically evolved to circumvent the restriction of the plant cell wall and enable transport of infectious viral material from the initially infected cell to the adjacent healthy cells (Reviewed by Scholthof, 2005; Lucas, 2006; Taliansky et al., 2008). Most of the recognized MPs can be grouped as following: a "30K" superfamily, named after the 30-kDa *Tobacco mosaic virus* (TMV) MP; the products of a triple gene block, encoded by potexviruses and related viruses; the tymovirus MPs; a series of small polypeptides, less than 10kDa, encoded by carmo-like viruses and some geminiviruses (Melcher, 2000); and the hsp70-like proteins of closteroviruses (Agranovsky et al., 1998).

The RRSV-encoded protein(s) necessary for virus cell-to-cell movement has not yet been identified. The nonstructural protein Pns6 encoded by segment S6 of *Rice dwarf virus* (RDV), genus *Phytoreovirus*, family *Reoviridae* has been recognized as the virus MP (Li et al., 2004). The RDV MP currently is the only recognized MP in the family *Reoviridae*; however, computer analysis showed that it shared limited (~ 10%) sequence similarity with any of the RRSV-encoded proteins, no indirect evidence, therefore, could be generated to predict the potential RRSV MP. Given that movement proteins are nonstructural proteins encoded by many, if not all, plant viruses to enable their movement from one infected cell to neighbouring cells, in this study, we investigated roles of the three RRSV nonstructural proteins (Pns6, Pns7 and Pns10) in cell-to-cell movement with a MP-deficient *Tobacco mosaic virus* (TMV) vector, which has been successfully used to characterize the *Tomato spotted wilt virus* (TSWV) MP, NSm (Lewandowski and Adkins, 2005; Li et al., 2009).

Three TMV–RRSV hybrids, pTMV–S6, pTMV–S7 and pTMV–S10 were made as follows. Total RNA was first isolated from RRSV-infected rice leaves, the genes of Pns6, Pns7 and Pns10 were then amplified by RT-PCR using the pairs of primers, 5′S6-*Eco*RV/3′S6-*Xho*I,5′S7-*Eco*RV/3′S7-*Xho*I and 5′S10-*Eco*RV/3′S10-*Xho*I, respectively (Supplementary Table 1). The primers were designed according to the sequence (Genbank No. AF020337, U66713 and U66712 for Pns6, Pns7 and Pns10, respectively) at the 5′ and 3′ ends of the open reading frame (ORF) with *Eco*RV and *Xho*I restriction sites placed, respectively, upstream or downstream of the start or stop codons. After digestion with *Eco*RV and *Xho*I, the PCR products were ligated into the *Eco*RV/*Xho*I-treated pTMVcpGFP (Fig. 1; Lewandowski and Adkins, 2005) to create pTMV–S6, pTMV–S7 and pTMV–S10, respectively (Fig. 1). Each RRSV gene was cloned behind the TMVCP subgenomic (sg) promoter and was supposed to express its native protein from the TMV-based chimeric virus.

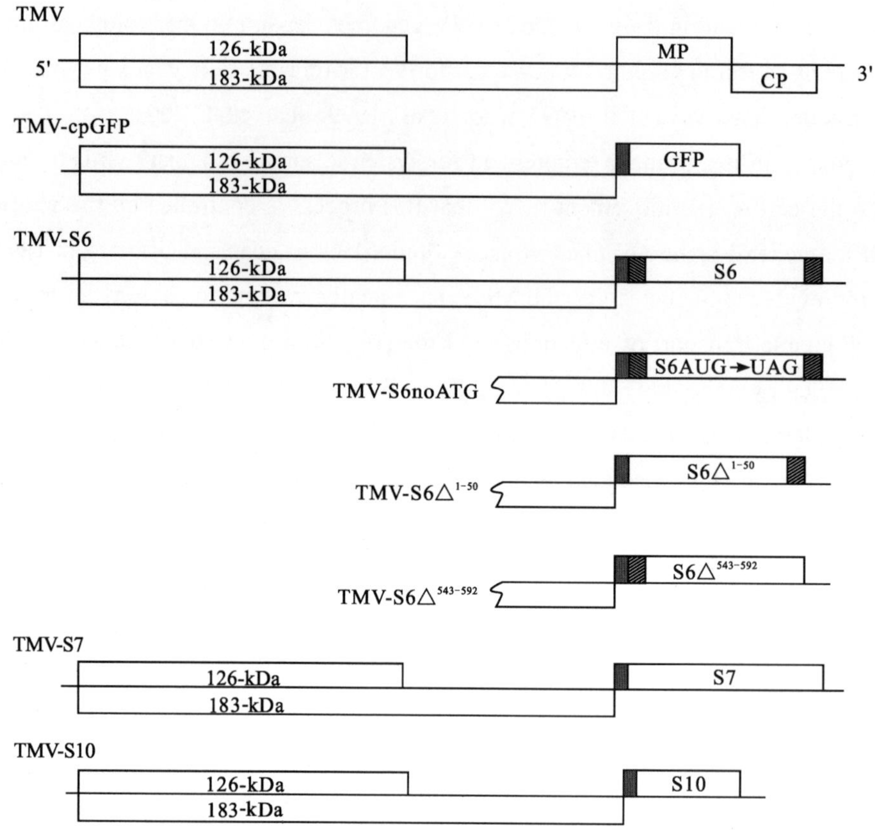

Fig. 1 Schematic diagram of the TMV genome, TMVcpGFP (Lewandowski and Adkins, 2005) and TMV-based chimeric viruses incorporating Pns6, Pns7 or Pns 10ORFs

The open boxes represent ORFs with encoded protein products, and the nameof each chimeric virus is listed, respectively ▩ TMVCP sg promoter; ▧ N-terminal 50 amino acids of Pns6; ▨ C-terminal 50 amino acids of Pns6. Expression of the RRSV-encoded proteins in the TMV–RRSV hybrids were driven by the TMV CP sg promoter

To test whether the three nonstructural proteins of RRSV could complement the cell-to-cell movement of TMV-based chimeric viruses, pTMV–S6, pTMV–S7 and pTMV–S10 were linearized with *Kpn*I and RNA transcripts were produced with T7 RNA polymerase (Lewandowski and Dawson, 1998). As a positive control, tobacco (*Nicotiana tabacum*) plants homozygous for the TMV resistance gene N and the TMV MP [NN-MP(+), Lewandowski and Adkins, 2005] were first inoculated with transcripts as described (Dawson et al., 1986). All three TMV-based chimeric viruses induced local lesions on inoculated leaves of NN-MP(+) plants 3–4 days postinoculation (dpi; Fig. 2), indicating that the RNA transcripts were viable in plants. The transcripts were then used to inoculate *N. tabacum* Xanthi nc plants. Transcripts of pTMV–S7 and pTMV–S10 did not induce any visible lesions on inoculated leaves by 10dpi, whereas tiny local lesions formed on the leaves inoculated with transcripts of pTMV–S6 at 5–7dpi (Fig. 2). The cell-to-cell movement of the TMV-based chimeric viruses was further verified by inoculation of *N. benthamiana* plants. Since none of these three TMV-based chimeric viruses induced symptoms on the inoculated leaves even after 10dpi, total RNA was extracted from the asymptomatic inoculated leaves (at 10dpi), and analyzed by Northern blotting using a TMV 3UTR-specific probe (Lewandowski and Dawson, 1998). The results showed that TMV–S6 multiplied significantly in the inoculated leaves (Fig. 3A, lane 1), whereas both TMV–S7 and TMV–S10 were undetectable (Fig. 3A, lanes 2 and 3). These data indicated that the TMV–S6, but not TMV–S7 or TMV–S10, was able to move from cell-to-cell in both Xanthi-nc tobacco and *N. benthamiana* plants.

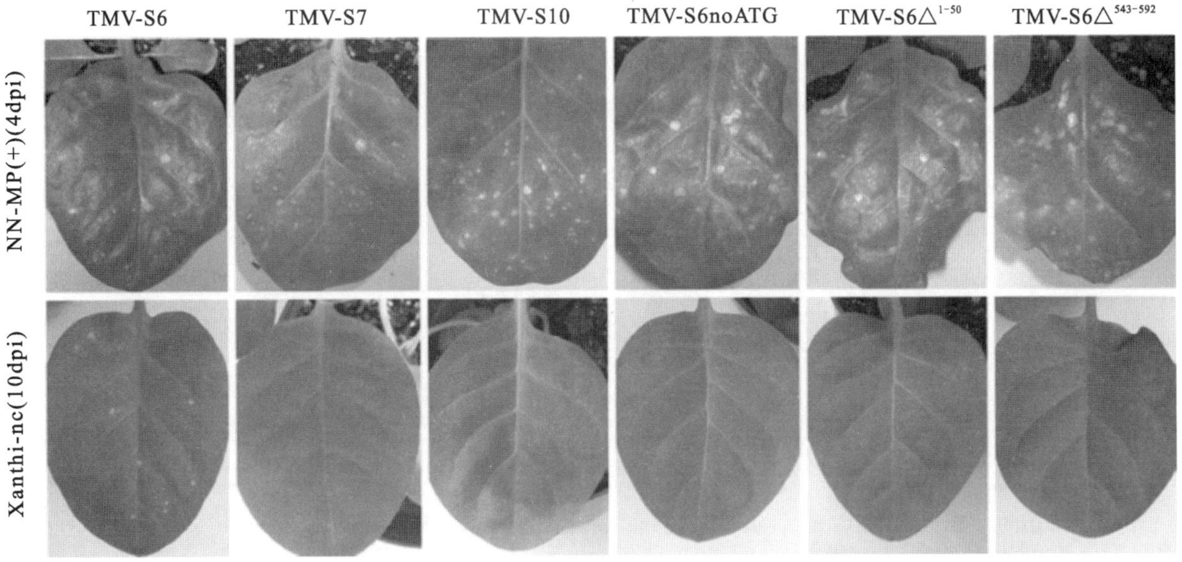

Fig. 2 The RRSV Pns6 complements TMV cell-to-cell movement in *N. tabacum* Xanthi,nc plants

Expanded leaves of NN-MP(+) and Xanthi-nc tobacco plants were inoculated with *in vitro* transcripts of the TMV–S6, TMV–S7, TMV–S10, TMV–S6noATG, TMV–S6 △ $^{1-50}$, or TMV-S6^{543592}. The photographs were taken 4dpi for NN-MP (+) and 10dpi for Xanthi-nc tobacco plants, respectively

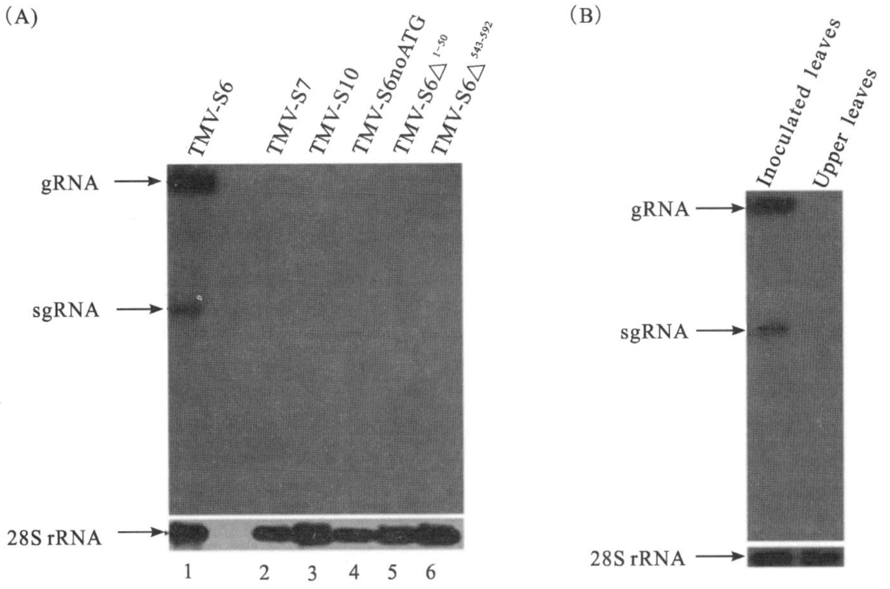

Fig. 3 The RRSV Pns6 supports TMV cell-to-cell movement but not long distance movement in *N. benthamiana plants*

(A) Northern-blot analysis of total RNA extracted from inoculated leaves (10dpi) of *N. benthamiana* plants inoculated with transcripts of TMV–S6, TMV–S7, TMV–S10, TMV–S6noATG, TMV–S6 △ $^{1-50}$, or TMV-S6$^{543-592}$. (B) Northern-blot analysis of total RNA extracted from inoculated leaves (10dpi) and upper noninoculated leaves (25dpi) of *N. benthamiana* plants inoculated with transcripts of TMV–S6. Bands corresponding to gRNA, sgRNA and 28s rRNA are indicated

The Pns6 protein is composed of 592 amino acids with molecular weight ~ 65kDa, and is able to bind single- or double-stranded nucleic acids cooperatively but with no sequence specificity (Shao et al., 2004). To clarify the role of Pns6 in cell-to-cell movement of the TMV-based chimeric virus, pTMV–S6noATG (Fig. 1) was constructed with the same strategy producing pTMV–S6 except that the forward primer was altered to 5′S6noATG (Supplementary Table 1). The pTMV–S6noATG carried the same S6 gene sequence as pTMV–S6, but was predicted to eliminate an expression of the Pns6 protein because the initial translation codon AUG was substituted with the stop codon UAG. When applied on plants, transcripts of pTMV–S6noATG induced local lesions on inoculated leaves of NNMP(+) plants 3–4dpi (Fig. 2), similar to TMV–S6, but were incapable of either forming visible lesions on *N. tabacum* (Fig. 2) or accumulating in *N. benthamiana* at 10dpi (Fig. 3A, lane 4). The failure of TMV–S6noATG to spread from cell-to-cell confirmed that the RRSV Pns6 protein, rather than the RNA sequence itself, was responsible for the cell-to-cell movement of TMV–S6.

Previous reports showed that both TMV MP and TSWV NSm can support the TMV coat protein (CP) deletion mutants move long distance in *N. benthamiana* (Knapp et al., 2001; Lewandowski and Adkins, 2005; Li et al., 2009). To test whether Pns6 functioned equivalent in this manner, the upper noninoculated leaves of *N. benthamiana* inoculated with transcripts of TMV–S6 were collected at 25dpi. Total RNA was extracted, and subsequent Northern blotting analysis showed that the TMV–S6 was undetectable in the upper noninoculated leaves (Fig. 3B). These data indicated that Pns6, unlike the TMV MP and TSWV NSm, two members of "30K" superfamily (Melcher, 2000), has inability to support long-distance movement of TMV mutant lacking the TMV CP in *N. benthamiana*.

In an attempt to more finely determine the function of Pns6 in cell-to-cell movement, sequences encoding the N-terminal 50 and C-terminal 50 amino acids of the S6 were deleted by PCR amplification using the pairs of primers 5′S6-N51/3′S6-XhoI and 5′S6-EcoRV/3′S6-C542, resulting in TMV–S6 $\triangle^{1\text{–}50}$ and TMV–S6 $\triangle^{543\text{–}592}$, respectively (Fig.1). After viability was confirmed with the NN-MP (+) plants (Fig.2), the RNA transcripts were then used to inoculate both Xanthi-nc tobacco and *N. benthamiana* plants. Neither TMV–S6$^{1\text{–}50}$ nor TMV–S6$^{543\text{–}592}$ induced lesions on Xanthi-nc tobacco (Fig.2) or accumulated in *N. benthamiana* (Fig.3A, lanes 5 and 6), indicating that removal of 50 amino acids from either the N- or C-terminus abolished the cell-to-cellmovement function of Pns6 in both plant species. The inability of both two Pns6 terminus-deletion mutants (Pns6 $\triangle^{1\text{–}50}$ and Pns6 $\triangle^{543\text{–}592}$) might be due to the incorrect folding or targeting, either of which could have prevented cell-to-cell movement. However, a defect in nucleic acid-binding activity could likely be ruled out, since the unique RNA-binding domain of Pns6 ranging from amino acid 201 to 273 was still active in two larger N- and C-terminal deletion mutants T2 (Pns6 $\triangle^{1\text{-}201}$) and T4 (Pns6 $\triangle^{273\text{–}592}$) *in vitro* (Shao et al., 2004).

To determine the subcellular location of Pns6 in plant cells, a construct p1301–S6–eGFP (Fig.4A), in which the Pns6–eGFP fusion ORF was expressed through the activity of an enhanced Cauliflower mosaic virus (CaMV) 35S promoter, was generated as described below. First, the S6 gene lacking the stop codon was amplified using a forward primer 5′S6-*Nco*I and a reverse primer 3′S6-*Nco*I (Supplementary Table 1). The PCR products were treated with *Nco*I, and inserted into the *Nco*I-digested pRTL2–eGFP (Li et al., 2004) to result in pRTL2S6–eGFP. After cutting with *Hind*III, the expression cassette harboring S6–eGFP fusion gene was inserted into the *Hind*III-treated pCAMBIA1301 (Cambia, Australia) to create p1301–S6–eGFP. Meanwhile, the p1301–eGFP expressing the free eGFP protein was constructed as a control (Fig.4A). Both constructs were finally introduced into Agrobacterium tumefaciens strain EHA105 and used to infiltrate the leaves of 4-week old *N. benthamiana* plants as described (Van der Hoorn et al., 2000). At 2 days post-infiltration, the infiltrated leaves were collected and visualized using a confocal microscope (Leica TCS STED, Germany) set at 488 nm. Fluorescence from free eGFP was generally in the cytoplasm and nucleus of epidermal cells, but the cell wall-associated punctate spots were not observed (Fig. 4B, upper pannel). In contrast, expression of the Pns6–eGFP fusion protein yielded green fluorescence mainly close to the cell wall as well as a few punctate sites perhaps indicating plasmodesmata (PD) (Fig.4B, lower pannel), suggesting the possible subcellular localization of Pns6 in infected original host, rice cells.

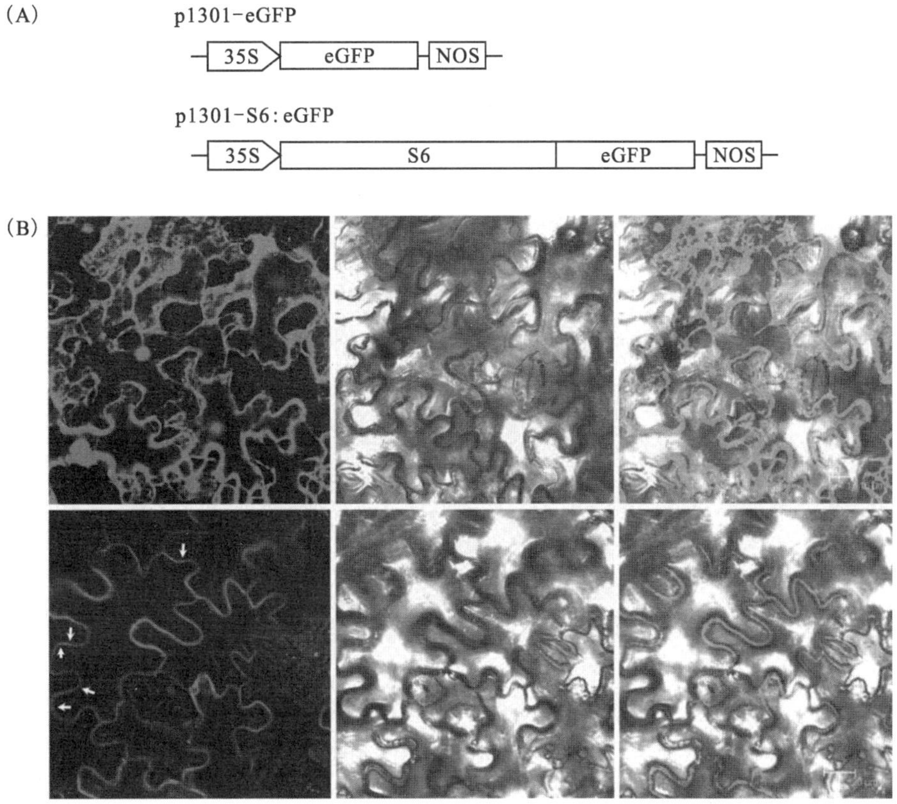

Fig. 4 Transient expression of Pns6–eGFP in epidermal cells of *N. benthamiana* leaves

(A) Schematic diagram of constructs used for agroinfiltration. The 35S indicates CaMV 35S RNA promoter and NOS represents the nopaline synthase terminator. (B) Localization of free eGFP and Pns6–eGFP fusion proteins. Four-week old leaves infiltrated with *A. tumefaciens* cells harboring p1301–S6: eGFP or p1301–eGFP were collected at 48h post-infiltration. The fluorescent and bright-field images were viewed under a confocal laser scanning microscope and Nomarski illumination, respectively, and the merged images confirmed that the Pns6–eGFP was present along the cell wall, in contrast with the free eGFP, which was localized in both nucleus and cytoplasm. The arrows showed the intense fluorescing punctate foci near the walls, indicating a possible associationof Pns6 with the PDs

In summary, our results clearly demonstrated that the RRSV Pns6 protein could complement cell-to-cell movement of the MP-deficient TMV vector in both *N. tabacum* Xanthi nc and *N. benthamiana* plants, and both the N- and C-terminal 50 amino acids of Pns6 were essential for this function. Taken together with the possible subcellular localization of Pns6 as well as the previous finding that the Pns6 is a nucleic acid-binding protein (Shao et al., 2004), the data collectively suggest that the Pns6 is responsible for cell-to-cell movement of RRSV.

Acknowledgements

We thank Dr. Dennis Lewandowski, Department of Plant Pathology, The Ohio State University, Columbus, OH, USA, for providing the TMVcpGFP and NN-MP (+) plants. This research was supported by the Major Project of Chinese National Programs for Fundamental Research and Development (grant no. 2010CB126203), the National Natural Science Foundation of China (grant no. 30970135), the Fujian Province Education Department (grant no. JB08078). We are also grateful for the support of the Fujian Human Resource Bureau to W. Li.

References

[1] AGRANOVSKY A A, FOLIMONOV A S, FOLIMONOVA S Y, et al. Beet yellows closterovirus HSP70-like protein mediates the cell-to-cell movement of a potexvirus transport-deficient mutant and a hordeivirus-based chimeric virus[J]. Journal of General Virology, 1998, 79: 889-895.

[2] BOCCARDO G, MILNE R G. Plantreovirusgroup[J]. Cmi/aab Description of Plant Viruses, 1984, 294: 4.

[3] CARRINGTON J C, KASSCHAU K D, MAHAJAN S K, et al. Cell-to-cell and long-distance transport of viruses in plants[J]. The Plant Cell, 1996, 8: 1669-1681.

[4] DAWSON W O, BECK D L, KNORR D A A, et al. Cdna cloning of the complete genome of tobacco mosaic-virus and production of infectious transcripts[J]. Proceedings of the National Academy of Sciences, 1986, 83: 1832-1836.

[5] HAGIWARA K, MINOBE Y, NOZU Y, et al. Component proteins and structures of rice ragged stunt virus[J]. Journal of General Virology, 1986, 67: 1711-1715.

[6] KAWANO S. Particle structure and double stranded RNA of rice ragged stunt virus[J]. Journal of the Faculty of Agriculture, Hokkaido University,1984, 61: 408-418.

[7] KNAPP E, DAWSON W O, LEWANDOWSKI D J. Conundrum of the lack of defective RNAs (dRNAs) associated with tobamovirus infections: dRNAs that can move are not replicated by the wild-type virus; dRNAs that are replicated by the wild-type virus do not move[J]. Journal of Virology, 2001, 75: 5518-5525.

[8] LEWANDOWSKI D J, DAWSON W O. Deletion of internal sequences results in tobacco mosaic virus defective RNAs that accumulate to high levels without interfering with replication of the helper virus[J]. Virology, 1998, 251: 427-437.

[9] LEWANDOWSKI D J, ADKINS S. The tubule-forming NSm protein from Tomato spotted wilt virus complements cell-to-cell and long-distance movement of Tobacco mosaic virus hybrids[J]. Virology, 2005, 342: 26-37.

[10] LI W, LEWANDOWSKI D J, HILF M E, et al. Identification of domains of the Tomato spotted wilt virus NSm protein involved in tubule formation, movement and symptomatology[J]. Virology New York, 2009, 390: 110-121.

[11] LI Y, BAO Y M, WEI C H, et al. Rice dwarf phytoreovirus segment S6-encoded nonstructural protein has a cell-to-cell movement function[J]. Journal of Virology, 2004, 78: 5382-5389.

[12] LI Z Y, UPADHYAYA N M, KOSITRATANA W, et al. Genome segment 5 of rice ragged stunt virus encodes a virion protein[J]. Journal of General Virology, 1996, 77: 3155-3160.

[13] LING K C, TIONGCO E R, AGUICRO V M. Rice ragged stunt, a new virus disease[J]. Plant Disease Report, 1978, 62: 701-705.

[14] LUCAS W J. Plant viral movement proteins: Agents for cell-to-cell trafficking of viral genomes[J]. Virology, 2006, 344: 169-184.

[15] LUCAS W J, GILBERTSON, R L. Plasmodesmata in relation to viral movement within leaf tissue[J]. Annual Review of Phytopathology, 1994, 32: 387-411.

[16] MAULE A J. Virus movement in infected plants[J]. Critical Reviews in Plant Sciences, 1991, 9: 457-473.

[17] MELCHER U. The '30K' superfamily of viral movement proteins[J]. Journal of General Virology, 2000, 81: 257-266.

[18] MILNE R G, BOCCASDO G, LING K C. Rice ragged stunt virus[J]. Characterization Diagnosis & Management of Plant Viruses, 2008, 16: 248.

[19] MIYAZAKI N, UEHARA-ICHIKI T, XING L, et al. Structural Evolution of Reoviridae Revealed by Oryzavirus in Acquiring the Second Capsid Shell[J]. Journal of Virology, 2008, 82: 11344-11353.

[20] OMURA T, MINOBE Y, KIMURA I, et al. Improved purification procedure and RNA segments of Rice ragged stunt virus[J]. Japanese Journal of Phytopathology, 1983, 49: 670-675.

[21] SCHOLTHOF H B. Plant virus transport: motions of functional equivalence[J]. Trends Plant Science, 2005, 10: 376-382.

[22] SHAO C G, WU J H, ZHOU G Y, et al. Ectopic expression of the spike protein of Rice ragged stunt Oryzavirus in transgenic rice plants inhibits transmission of the virus to insects[J]. Molecular Breeding, 2003, 11: 295-301.

[23] SHAO H B, WANG J M, WANG X Y,et al. Anodic dissolution of aluminum in KOH ethanol solutions[J]. Electrochemistry Communications, 2004, 6: 6-9.

[24] TALIANSKY M, TORRANCE L, KALININA N O. Role of plant virus movement proteins. In Plant Virology Protocols: From Viral Sequence to Protein Function[J]. Methods in Molecular Biology, 2008, 33-54.

[25] UPADHYAYA N M, YANG M, KOSITRATANA W,et al. Molecular analysis of Rice ragged stunt Oryzavirus segment 9 and sequence conservation among isolates from thailand and india[J]. Archives of Virology, 1995, 140: 1945-1956.

[26] UPADHYAYA N M, ZINKOWSKY E, LI Z, KOSITRATANA W, et al. The M(r) 43K major capsid protein of Rice ragged stunt Oryzavirus is a post-translationally processed product of a M(r) 67,348 polypeptide encoded by genome segment 8[J]. Archives of Virology, 1996, 141: 1689-1701.

[27] UPADHYAYA N M, RAMM K, GELLATLY J A, et al. *Rice ragged stunt Oryzavirus* genome segments S7 and S10 encode non-structural proteins of M-r 68 025 (Pns7)and M-r 32 364 (Pns10)[J]. Archives of Virology, 1997, 142: 1719-1726.

[28] UPADHYAYA N M, RAMM K, GELLATLY J A, et al.Rice ragged stunt Oryzavirus genome segment S4 could encode an RNA dependent RNA polymerase and a second protein of unknown function[J]. Archives of Virology, 1998, 143: 1815-1822.

[29] UYEDA I, SUGA H, LEE S Y, et al. Rice ragged stunt Oryzavirus genome segment-9 encodes a 38600-m(r) structural protein[J]. Journal of General Virology, 1995, 76: 975-978.

[30] VAN DER HOORN R A L, LAURENT F, ROTH R, et al. Agroinfiltration is a versatile tool that facilitates comparative analyses of Avr9/Cf-9-induced and Avr4/Cf-4-induced necrosis[J]. Molecular Plant-Microbe Interactions, 2000, 13: 439-446.

[31] ZHOU G Y, LU X B, LU H J, et al. Rice ragged stunt Oryzavirus: role of the viral spike protein in transmission by the insect vector[J]. Annals of Applied Biology, 1999, 135: 573-578.

Identification of Pns6, a putative movement protein of RRSV, as a silencing suppressor

Jianguo Wu[1,2], Zhenguo Du[1], Chunzheng Wang[2], Lijun Cai[1], Meiqun Hu[1], Qiying Lin, Zujian Wu[1], Yi Li[2], Lianhui Xie[1]

(1 Institute of Plant Virology, Fujian Agriculture and Forestry University , Key Laboratory of Plant Virology of Fujian Province, Fuzhou, Fujian, China 2Peking-Yale Joint Center for Plant Molecular Genetics and Agrobiotechnology, The National Laboratory of Protein Engineering and Plant Genetic Engineering, College of Life Sciences, Peking University, Beijing China)

Abstract: RNA silencing is a potent antiviral response in plants. As a counterdefense, most plant and some animal viruses encode RNA silencing suppressors. In this study, we showed that Pns6, a putative movement protein of *Rice ragged stunt virus* (RRSV), exhibited silencing suppressor activity in coinfiltration assays with the reporter green fluorescent protein (GFP) in transgenic *Nicotiana benthamiana* line 16c. Pns6 of RRSV suppressed local silencing induced by sense RNA but had no effect on that induced by dsRNA. Deletion of a region involved in RNA binding abolished the silencing suppressor activity of Pns6. Further, expression of Pns6 enhanced *Potato virus X* pathogenicity in *N. benthamiana*. Collectively, these results suggested that RRSV Pns6 functions as a virus suppressor of RNA silencing that targets an upstream step of the dsRNA formation in the RNA silencing pathway. This is the first silencing suppressor to be identified from the genus *Oryzavirus*.

Plant infecting reoviruses are grouped into three genera, namely *Phytoreovirus*, *Fijivirus* and *Oryzavirus*(Mertens et al., 2005). RRSV belongs to the genus *Oryzavirus*. It infects plants in the family *Graminae* and is transmitted in a persistent manner by brown plant hoppers. The disease caused by this virus was first discovered in 1976–1977 in Indonesia and Philippines. Then the disease became prevalent in most rice-growing countries in south-eastern and far-eastern Asia, causing great yield losses to rice production (Ling et al., 1978). RRSV virion has an icosahedral particle which consists of a polyhedral core surrounded by flat spikes about 20nm wide and 10 nm high (Hagiwara et al., 1986). The RRSV genome comprises 10 double stranded RNAs with molecular weights ranging from 1.2 to 3.9kb (Omura et al., 1983). The complete nucleotide sequences of all genomic segments, denoted as S1-S10, have been determined. It is now known that proteins encoded by S6, S7 and S10, with *Mrs* of about 71kD, 68kD and 32kD, respectively, are non-structural proteins, whereas proteins specified by other RNA segments have been shown to be present in or are supposed to take part in the assembly of RRSV virions (Miyazaki et al., 2008;Uyeda et al., 1995; Li et al., 1996; Upadhyaya et al., 1996; Upadhyaya et al., 1997, 1998; Hagiwara et al., 2002; Shao et al., 2004;Upadhyaya et al., 1998). However, our knowledge on the roles played by RRSV encoded proteins in virus-host interaction remains poor.

RNA silencing is now a general term that refers to a set of related processes in which small RNAs ranging from 21 to 30nt in length are used to direct sequence specific modulation of gene expression (Brodersen and Voinnet, 2006). In plants as well as in some animals, one of the firmly established roles of RNA silencing is

Virology Journal. 2010, 7:335

Received 16 March 2010 ;Accepted 22 November 2010; Published 22 November 2010.

antiviral defense (Ding and Voinnet, 2007; Ding, 2010). The antiviral RNA silencing begins with the cleavage of viral dsRNAs by members of the RNase III family enzymes called Dicer or Dicer-like (DCL) in plants, which results in the production of viral small interfering RNAs (VsiRNAs). These vsiRNAs are incorporated into an effector complex named RNA induced silencing complex (RISC) and then direct the complex to destroy viral RNAs (Ding and Voinnet, 2007; Ding, 2010). VsiRNAs also provide sequence specificity for cellular RDRs to copy viral RNAs into dsRNAs, which can be a secondary source for vsiRNAs production (Baulcombe, 2004; Voinnet, 2008). As a counterdefence, many viruses have evolved to encode one or multiple proteins to suppress RNA silencing(Lakatos et al., 2006; Ding and Voinnet, 2007). The molecular mechanisms by which these virus encoded silencing suppressors (VSRs) interfere with the RNA silencing machinery are poorly understood at present. VSRs from diverse virusesposses RNA binding activities. This led to the proposition that RNA might be a common target of VSRs (Lakatos et al., 2006; Meral et al., 2006). Some VSRs such as p19, HC-Pro and p21 efficiently form complexes with 21-nt ds-sRNA but fail to bind long dsRNA. These VSRs were believed to prevent the formation of functional RISCs by sequestering vsiRNAs. Some VSRs including CP of *Turnip crinkle virus* (TCV) bind dsRNAs without size selection. These VSRs may function to protect viral dsRNAs from being cleaved by plant DCLs (Lakatos et al., 2006; Merai et al., 2006). Many recent studies demonstrated that VSRs could also target protein components of RNA silencing for a review see (Wu et al., 2010). Protein components of the RNA silencing that have been found to be targets of VSRs include DRB4, which is an auxiliary factor of the antiviral sensor DCL4; AGO1, which forms the core of RISC; and SGS3, which is a cofactor of RDR6 functioning in the amplification step of RNA silencing (Wu et al., 2010a). Regardless of all these possibilities, it is believed that VSRs play important roles in promoting viral replication as well as in viral pathogenesis (Diaz-Pendon and Ding, 2008).

The VSRs for members of the other two genera of Phytopathogenicreoviruses have been identified (Cao et al., 2005; Zhang et al., 2005; Liu et al., 2008). However, no VSR for *Oryzavirus* has been reported, nor can it be predicted because of the low level of sequence similarity between proteins of reoviruses across genera. Given the importance of VSRs in virus-host interaction, we conducted experiments to identify the VSR of RRSV.

It was reasonable to presume that the silencing suppressor function of RRSV was encoded by a non-structural protein. Therefore, the ORFs for the three non-structural proteins, pns6, pns7 and pns10, were individually cloned into the binary vector pPZP212 (See Additional file 1 and 2 for Materials and Methods). The resultant plasmids were named 35S-S6, 35S-S7, and 35S-S10 respectively (Fig.1). Transformed agrobacterial strain carrying each of these constructs was mixed with a strain that carried 35S-GFP with a ratio of 3:1 and infiltrated into leaves from *Nicotiana benthamiana* line 16c as described previously (Hamilton et al., 2002). Then the GFP fluorescence was monitored using a handheld long wavelength UV light source. Agrobacteria harboring only the GFP gene or the 2b gene of *Tomato aspermy cucumovirus* (TAV) were used as negative and positive controls, respectively (Li et al., 1999).

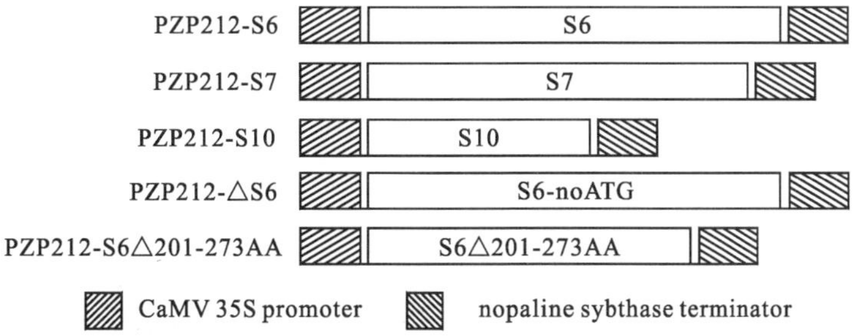

Fig.1 Schematic representation of plasmids used in this study

All infiltrated leaf patches showed bright green fluorescence at 2d post infiltration (dpi.) (Data not shown). In leaf patches infiltrated with 35S-GFP plus 35S-S6 or plus 35S-2b, the fluorescence intensity remained strong until 7dpi (Fig.2A, C). However, in leaf patches expressing GFP or GFP plus either S7 or S10, the fluorescence intensity began to decline at 3dpi and became hardly detectable at 7dpi (Fig.2D,E). Similar patterns of fluorescence decline were observed in leaf patches infiltrated with 35S-GFP plus 35S-ΔS6 (Fig. 2B). The 35S-ΔS6 carried the same S6 gene sequence as 35S-S6, but was supposed to be unable to express a functional Pns6 protein because of the deletion of the nucleotide A from the translation start codon AUG (Fig.1).

To test whether silencing suppression was responsible for the above observations, Northern blot analyses were conducted to detect steady-state levels of GFP mRNA and GFP-specific siRNAs. As shown in Fig.2G, the accumulation levels of GFP mRNA were much higher in tissues expressing 35S-GFP plus 35S-TAV2b or 35S-GFP plus 35S-S6 than in tissues expressing GFP alone or in combination with any other genes at 7dpi. This indicated that expression of S6 resulted in the stabilization of GFP mRNA and consequently higher GFP fluorescence, as that of TAV2b. In all treatments, the accumulation levels of GFP mRNA were negatively correlated with those of GFP-specific siRNAs (Fig.2G). This confirmed that Pns6 is a silencing suppressor.

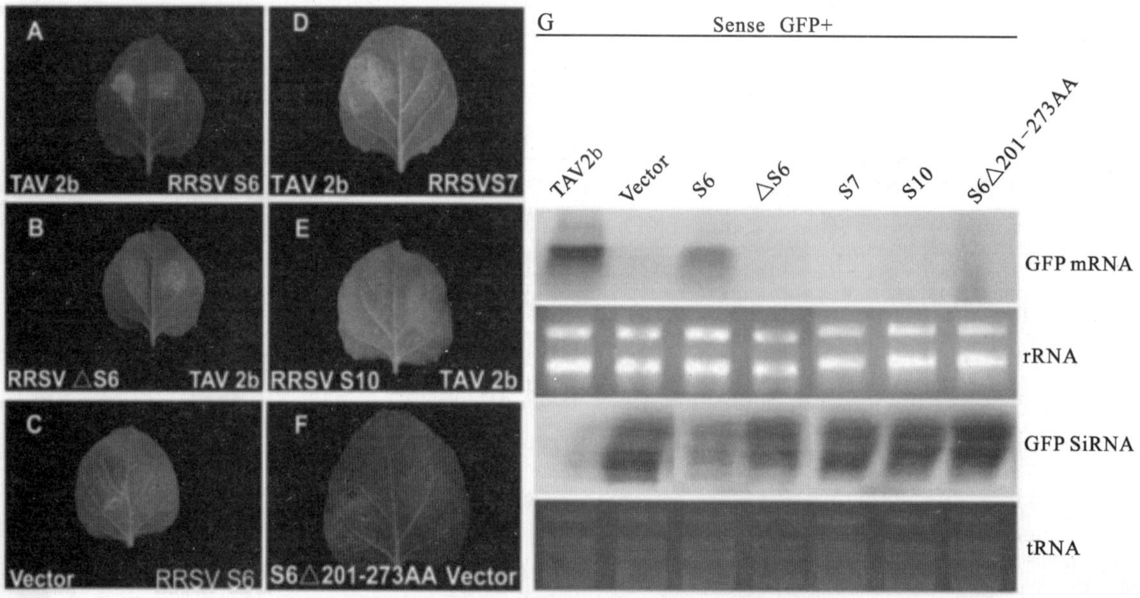

Fig.2 Suppression of local GFP silencing by RRSV Pns6

(A-F) *N. benthamiana* line 16c plants were coinfiltrated with *Agrobacterium* spp. (Agro.) mixtures carrying 35S-GFP and the individual constructs indicated in each image. GFP fluorescence was viewed under long wavelength UV light at 7d postinfiltration (dpi). (G) Northern blot analysis of the steady-state levels of GFP mRNA and siRNA extracted from different infiltrated patches shown in panel A to F. 28 S rRNA and t RNA were used as loading controls for detection of GFP mRNA and GFP siRNA respectively

As mentioned above, RNA silencing is a multi-step process (Baulcombe, 2004). To determine in which step Pns6 targets RNA silencing, we tested the effect of Pns6 on GFP dsRNA-triggered silencing. To do this, leaves of transgenic *N. benthamiana* plant line 16c were infiltrated with *Agrobacterium tumefaciens* harboring 35S-ssGFP (sense GFP RNA), 35S-dsGFP (IR-GFP), and a binary vector containing S6, ΔS6, or TAV2b under the control of the 35 S promoter. As shown in Fig.3, leaf patches infiltrated with 35S-ssGFP plus 35S-dsGFP or with 35S-ssGFP plus 35S-dsGFP plus 35S-S6 (or 35S-ΔS6) lost GFP fluorescence at 7dpi, indicating strong GFP RNA silencing. Consistently, the accumulation of GFP mRNA was hardly detectable in these leaf patches.

This suggested that Pns6 could not suppress silencing induced by dsRNA. As expected, TAV2b suppressed GFP RNA silencing triggered by dsRNA, as indicated by the bright green fluorescence and high levels of accumulation of GFP mRNA in leaves infiltrated with 35S-ssGFP plus 35S-dsGFP plus 35S-TAV2b (Fig. 3).

These data showed that pns6 of RRSV had silencing suppressor activities. It targeted an initial step of RNA silencing upstream of dsRNA production. This was consistent with a previous report which demonstrated that Pns6 of RRSV had nucleic acid binding activities and preferentially bound ssRNAs (Hagiwara et al., 2002). Nucleic acids binding is a common feature of many VSRs. It is possible that Pns6 binds ssRNAs and prevents them from being copied by cellular RDRs. In a primary attempt to exploit this possibility, we generated ΔS6201-273, in which the animo acids from 201 to 273 were deleted (Fig.1). This region of amino acids has been shown to be essential for RNA binding of RRSV pns6 (Hagiwara et al., 2002). Indeed, the S6 mutant lost silencing suppressor activity: when co-expressed with GFP in leaves of *N. benthamiana* line 16c, it could not maintain strong GFP fluorescence in infiltrated leaf patches. The GFP mRNA accumulation level was very low at 7dpi., whereas that of GFP siRNAs was markedly high (Fig. 2F).

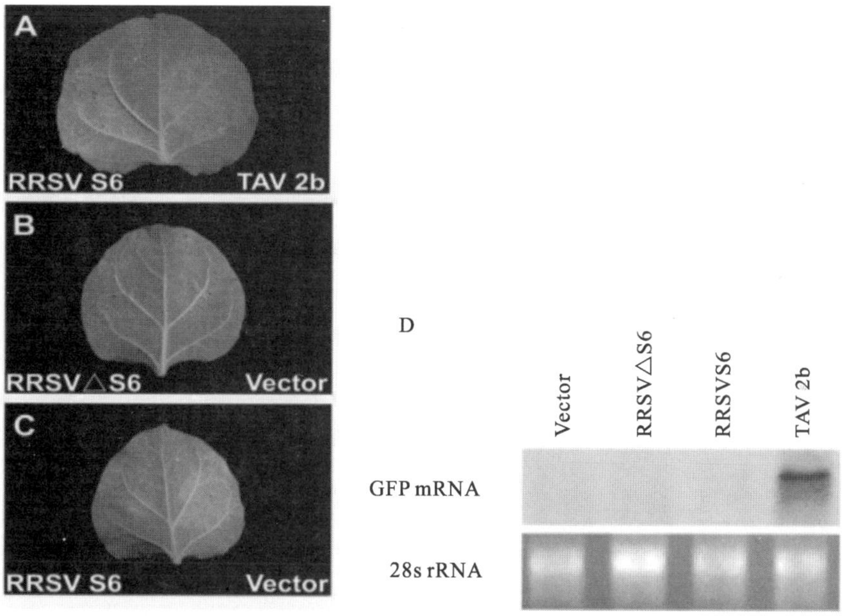

Fig.3 RRSV Pns6 can not inhibit local silencing induced by dsRNA

(A-C) *N. benthamiana* line 16c plants were coinfiltrated with *Agrobacterium* spp. (Agro.) mixtures carrying 35S-dsGFP and the individual constructs indicated in each image. GFP fluorescence was viewed under long wavelength UV light at 7d postinfiltration (dpi). (D) Northern blot analysis of the steady-state levels of GFP mRNA extracted from different infiltrated patches shown in panel A to C. The bottom gel shows rRNA with ethidium bromide staining as a loading control

Results from the above experiments indicated that RRSV Pns6 was a silencing suppressor but offered no clues with regards to whether this function would have biological implications for viral infection. The role of Pns6 in RRSV infection can not be tested directly because of the lack of an infectious clone for this virus. Therefore, we utilized pGR107, a PVX vector to express S6 in a heterologous virus. Seedlings of *N. benthamiana* plants (four-to six-leaf stage) were inoculated with PVX, PVX-S6 or PVX-ΔS6, respectively.

Variations in symptom severity were observed in individual infections. However, a general trend was that PVX-S6 elicited more severe symptoms than PVX or PVX-ΔS6 did. This was especially obvious after 9dpi. At this time, only mild chlorotic spots could be observed on some leaves from *N. benthamiana* infected by

PVX or PVX-ΔS6. However, most plants infected by PVX-S6 exhibited very severe symptoms, with some newly developed leaves being abnormal in shape (Fig.4A). The symptoms induced by PVX-S6 sustained throughout the life of the plants, whereas only a small proportion of plants infected by PVX or PVX-ΔS6 were symptomatic during later stages of observation. These results showed that RRSV S6 can accentuate symptoms when expressed by a heterologous virus. To correlate this function of RRSV S6 with silencing suppression, the accumulation levels of PVX in systemically infected *N. benthamiana* plants were detected. Northern blot analysis revealed that the concentration of PVX in plants infected by PVX-S6 was very high at 18dpi. In contrast, the presence of PVX RNA was hardly detectable in newly developed leaves from plants infected by PVX or PVX-ΔS6 at this time (Fig.4B). This indicated that RRSV S6 markedly enhanced the replication of PVX, presumably through suppression of RNA silencing. However, it was interesting to note that no differences in the accumulation of PVX RNAs were detected between plants infected by PVX-S6 and those infected by PVX or PVX-ΔS6 at 9dpi (Fig.4B). As plants infected by PVX-S6 showed more severe symptoms than those infected by PVX or PVX-ΔS6 at this time (Fig.4A), this implicated that Pns6 might have direct effects on the normal physiology of the infected plants independent of its ability to enhance viral multiplication.

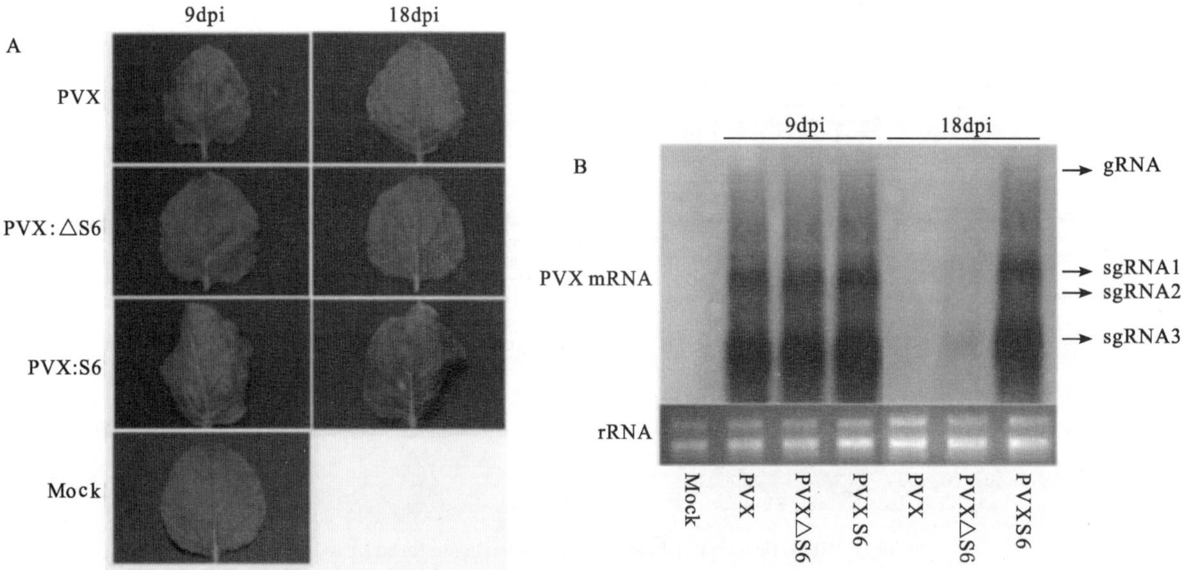

Fig.4 RRSV Pns6 enhances pathogenicity of chimeric PVX

(A) plants infected by PVXΔS6 or PVX-S6 show mild disease symptoms as a few scattered chlorotic speckles, whereas leaves infected with PVX-S6 show more severe symptoms. (B) RNA gel blot analysis of accumulation of PVX genomic (gRNA) and subgenomic mRNAs (sgRNA1 to sgRNA3) at 9 and 18dpi. The bottom gel shows rRNA with ethidium bromide staining as a loading control

Taken together, we identified pns6, which is composed of 592 amino acids with a molecular weight of about 65kDa, as a silencing suppressor of RRSV. This was the first VSR to be identified from an oryzavirus. We showed that the RNA binding activities of Pns6 might be important for its silencing suppressor function. Similar studies on other VSRs of plant infecting reoviruses have yet to be done (Cao et al., 2005; Zhang et al., 2005; Liu et al., 2008). Interestingly, we have recently reported that pns6 might be a cell-to-cell movement protein of RRSV. Pns6 could complement the cell-to-cell movement of the movement-deficient TMV in *N. tabacumXanthi nc* and *N. benthamiana* plants. When transiently expressed in epidermal cells from *N. benthamiana*, the Pns6-eGFP fusion protein was present predominantly along the cell wall (Wu et al., 2010b).

This was consistent with the notion that virus movement and RNA silencing are intimately related processes. Some well-known VSRs including p19, 2b, Hc-Pro had been implicated in virus long-range movement before the recognition of their silencing suppressor activities (Klein et al., 1994; Cronin et al., 1995; Ding et al., 1995; Scholthof et al., 1995). For Tobacco Etch PotyvirusHcPro protein, a correlation of silencing suppression and the ability to mediate long-distance virus movement has been demonstrated (Kasschau and Carrington, 2001). Through random mutagenesis of the P25 gene, Bayne et al. showed that suppression of silencing is necessary for cell-to-cell movement of PVX through plasmodesmata (Bayne et al., 2005). In addition, the movement-deficiency phenotype of a TCV CP deletion mutant could be complemented by a series of silencing suppressors in *Trans* (Powers et al., 2008). Thus, our findings may serve as an example in which a protein encoded by a rice virus functions in both RNA silencing and viral movement.

Acknowledgements

We thank Dr.Jason G. Powers for providing plasmid pPZP212 vector and Prof. David Baulcombe for providing 16c seeds and PVX vectors. This work was supported by the Major Project of Chinese National Programs for Fundamental Research and Development (grant no.2010CB126203), the National Natural Science Foundation of China (grant no.30970135), the Fujian Province Education Department (grant no.JB08078), the National Science Foundation of China (30970135) and Specialized Research Fund for the ministry of agriculture (nyhyzx 07-051).

References

[1] BAULCOMBE D. RNA silencing in plants[J]. Nature, 2004, 431: 356-363.

[2] BAYNE E H, RAKITINA D V, MOROZOV S Y, et al. Cell-to-cell movement of Potato Potexvirus X is dependent on suppression of RNA silencing[J]. The Plant Journal, 2005, 44: 471-482.

[3] BRODERSEN P, VOINNET O. The diversity of RNA silencing pathways in plants[J]. Trends in Genetics Tig, 2006, 22: 268-280.

[4] CAO X S, ZHOU P, ZHANG X M, et al. Identification of an RNA silencing suppressor from a plant double-stranded RNA virus[J]. Journal of Virology, 2005, 79: 13018-13027.

[5] Ding S W, Li W X, Symons R H. A novel naturally occurring hybrid gene encoded by a plant RNA virus facilitates long distance virus movement[J]. EMBO Journal, 1995, 14:5762-5772.

[6] CRONIN S, VERCHOT J, HALDEMAN-CAHILL R, et al. Long distance movement factor: a transport function of the potyvirus helper component proteinase[J]. The Plant Cell Online, 1995, 7:549-559.

[7] DING S W, VOINNET O. Antiviral immunity directed by small RNAs[J]. Cell, 2007, 130: 413-426.

[8] DIAZ-PENDON J A, DING S W. Direct and indirect roles of viral suppressors of RNA silencing in pathogenesis[J]. Annual Review of Phytopathology, 2008, 303-326.

[9] DING S W. RNA-based antiviral immunity[J]. Nature Reviews Immunology, 2010, 10: 632-644.

[10] HAGIWARA K, MINOBE Y, NOZU Y, et al. Component proteins and structures of rice ragged stunt virus[J]. Journal of General Virology, 1986,67:1711-1715.

[11] HAGIWARA K, RAO S J, SCOTT S W, et al. Nucleotide sequences of segments 1, 3 and 4 of the genome of Bombyx mori cypovirus 1 encoding putative capsid proteins VP1, VP3 and VP4, respectively[J]. Journal of General Virology, 2002, 83: 1477-1482.

[12] HAMILTON A, VOINNET O, CHAPPELL L, et al. Two classes of short interfering RNA in RNA silencing[J]. The EMBO Journal,

2002, 21: 4671-4679.

[13] KASSCHAU K D, CARRINGTON J C. Long-distance movement and replication maintenance functions correlate with silencing suppression activity of potyviral HC-Pro[J]. Virology, 2001, 285: 71-81.

[14] KLEIN P G, KLEIN R R, RODRIGUEZCEREZO E, et al. Mutational analysis of the tobacco vein mottling virus genome[J]. Virology, 1994, 204: 759-769.

[15] LAKATOS L, CSORBA T, PANTALEO V, et al. Small RNA binding is a common strategy to suppress RNA silencing by several viral suppressors[J]. EMBO Journal, 2014, 25: 2768-2780.

[16] LING K C, TIONGCO E R, AGUIERO V M. Rice ragged stunt, a new virus disease[J].Plant Disease Report, 1978, 62:701-705.

[17] LI Z Y, UPADHYAYA N M, KOSITRATANA W, et al. Genome segment 5 of rice ragged stunt virus encodes a virion protein[J]. Journal of General Virology, 1996, 77: 3155-3160.

[18] LI H W, LUCY A P, GUO H S, et al. Strong host resistance targeted against a viral suppressor of the plant gene silencing defence mechanism[J]. The EMBO Journal, 1999, 18: 2683-2691.

[19] LI F, DING S W. Virus counterdefense: diverse strategies for evading the RNA-silencing immunity[J]. Annual Review of Microbiology, 2006, 60:503-531.

[20] LIU F, ZHAO Q, RUAN X, et al. Suppressor of RNA silencing encoded by Rice gall dwarf virus genome segment 11[J]. Chinese Science Bull,2008, 53:96-103.

[21] MERTENS P P C, ATTOUI H, DUNCAN R, et al. Virus taxonomy. Eighth report of the International Committee on Taxonomy of Viruses[J]. Viruses, 2005, 447-454.

[22] MERAI Z, KERENYI Z, KERTESZ S, et al. Double-stranded RNA binding may be a general plant RNA viral strategy to suppress RNA silencing[J]. Journal of Virology, 2006, 80: 5747-5756.

[23] MIYAZAKI N, UEHARA-ICHIKI T, XING L, et al. Structural Evolution of Reoviridae Revealed by Oryzavirus in Acquiring the Second Capsid Shell[J]. Journal of Virology, 2008, 82: 11344-11353.

[24] OMURA T, MINOBE Y, KIMURA I, et al. Improved purification procedure and RNA segments of Rice ragged stunt virus[J]. Japanese Journal of Phytopathology, 1983, 49:670-675.

[25] POWERS J G, SIT T L, QU F, et al. A versatile assay for the identification of RNA silencing suppressors based on complementation of viral movement[J]. Molecular plant-microbe interactions : MPMI, 2008, 21: 879-890.

[26] SHAO C G, LU H J, WU J H, et al. Nucleic acid binding activity of Pns6 encoded by genome segment 6 of rice ragged stunt oryzavirus[J]. Acta Biochim Biophys Sin (Shanghai), 2004, 36: 457-466.

[27] SCHOLTHOF H B, SCHOLTHOF K B, KIKKERT M, et al. Tomato bushy stunt virus spread is regulated by two nested genes that function in cell-to-cell movement and host-dependent systemic invasion[J]. Virology, 1995,213:425-438.

[28] UPADHYAYA N M, RAMM K, GELLATLY J A, et al. Rice ragged stunt oryzavirus genome segment S4 could encode an RNA dependent RNA polymerase and a second protein of unknown function[J]. Archives of Virology, 1998, 143: 1815-1822.

[29] UPADHYAYA N M, RAMM K, GELLATLY J A, et al. Rice ragged stunt oryzavirus genome segments S7 and S10 encode non-structural proteins of M-r 68 025 (Pns7)and M-r 32 364 (Pns10)[J]. Archives of Virology, 1997, 142: 1719-1726.

[30] UPADHYAYA N M, ZINKOWSKY E, LI Z, et al. The M(r) 43K major capsid protein of rice ragged stunt oryzavirus is a post-translationally processed product of a M(r) 67,348 polypeptide encoded by genome segment 8[J]. Archives of Virology, 1996, 141: 1689-1701.

[31] UYEDA I, SUGA H, LEE S Y, et al. Rice ragged stunt Oryzavirus genome segment 9 encodes a 38,600 Mr structural protein[J]. Journal of General Virology, 1995, 76:975-978.

[32] UPADHYAYA N M, RAMM K, GELLATLY J A, et al. Rice ragged stunt oryzavirus genome segment S4 could encode an RNA dependent RNA polymerase and a second protein of unknown function[J]. Archives of Virology, 1998, 143: 1815-1822.

[33] VOINNET O. Use, tolerance and avoidance of amplified RNA silencing by plants[J]. Trends in Plant Science, 2008, 13: 317-328.

[34] WU Q, WANG X, DING S W. Viral Suppressors of RNA-Based Viral Immunity: Host Targets[J]. Cell host & microbe, 2010a, 8: 12-15.

[35] WU Z, WU J, ADKINS S, et al. Rice ragged stunt virus segment S6-encoded nonstructural protein Pns6 complements cell-to-cell movement of Tobacco mosaic virus-based chimeric virus[J]. Virus Research, 2010b, 152: 176-179.

[36] ZHANG L D, WANG Z H, WANG X B, et al. Two virus-encoded RNA silencing suppressors, P14 of Beet necrotic yellow vein virus and S6 of Rice black streak dwarf virus[J]. Chinese ence Bulletin, 2005, 50: 305-310.

水稻条纹病毒外壳蛋白叶绿体离体跨膜运输研究

程兆榜 [1,2]，於春 [1]，任春梅 [1]，周益军 [1]，范永坚 [1]，谢联辉 [2]

（1 江苏省农业科学院植物保护研究所，南京 210014；
2 福建农林大学病毒研究所，福州 350002）

摘要：明确离体条件下水稻条纹病毒（RSV）外壳蛋白（CP）能否进入叶绿体，为研究 RSV 的致病机制提供依据和方法。参考有关文献的方法提取获得水稻和小麦叶绿体，在离体条件下进行 RSV-CP 的叶绿体跨膜运输试验，同时研究孵育时间、病毒浓度等对 RSV-CP 在水稻叶绿体中离体跨膜运输的影响。在离体条件下 5min，RSV-CP 即可进入 RSV 寄主植物水稻、小麦的叶绿体内，其离体跨膜运输的基本条件为 RSV 浓度 58.1μg/mL、孵育时间 15min。随着孵育时间的延长，进入叶绿体中的 CP 量有所增加；反应体系中 RSV 浓度加大，进入叶绿体中的 RSV-CP 量随之增加。离体条件下 RSV-CP 可进入寄主植物水稻、小麦的叶绿体中，病毒外壳蛋白进入叶绿体可能是其诱发花叶症状的主要原因之一。

关键词：水稻条纹病毒；外壳蛋白；叶绿体；跨膜运输

中图分类号：S432.41 **文献标志码**：A **文章编号**：2095-1191（2011）12-1476-05

Studies on in vitro rice stripe virus coat protein import intochloroplasts

Zhaobang Cheng[1, 2], Chun Yu[1], Chunmei Ren[1], Yijun Zhou[1], Yongjian Fan[1], Lianhui Xie[2]

(1 Institute of Plant Protection, Jiangsu Academy of Agricultural Sciences, Nanjing 210014, China; 2 Institute of Virology, Fujian Agriculture and Forestry University, Fuzhou 350002)

Abstract: The present experiment was aimed to find out the possibilities of rice stripe virus coat protein (RSV-CP) import into chloroplasts in vitro in order to provide references for studying the pathogenesis of rice stripe virus. Rice and wheat chloroplast was obtained by using leaf-cut method. The system of RSV-CP import into chloroplasts in vitro, and the effects of the incubation time and the concentration of RSV on import efficiency were also studied. RSV-CP was found able to import into isolated chloroplasts of RSV host plants including rice and wheat in 5min. The lowest concentration of RSV in the import system was 58.1μg/mL and the best incubation time was 15min in the current studies. By extending the incubation time or increasing RSV concentration, the amount of RSV-CP import into the chloroplasts was increased. The study reveals that RSV-CP import into rice and wheat chloroplast in vitro is possible. This may be one of the leading causes of mosaic symtoms.

Key words: rice stripe virus; coat protein; chloroplast; transmembrane transport

水稻条纹病毒（*Rice stripe virus*, RSV）主要分布于中国、日本、朝鲜、乌克兰等地的粳稻种植区

南方农业学报 . 2011, 42(12): 1476-1480
收稿日期：2011-10-08
基金项目：国家公益性（农业）行业专项项目（201003031）；江苏省自然科学基金项目（BK2010018）

（Kisimoto and Yamada, 1991），侵染水稻引发水稻条纹叶枯病，目前该病害在我国的江苏、浙江、上海、安徽、山东、云南、河南、辽宁等省市的水稻流行为害，其中 2004 年在江苏省危害最为严重，发生面积达 157 万 hm^2，并造成数万公顷绝收，是当前粳稻生产上的主要病害之一（林奇英等 , 1990；程兆榜等 , 2002；高苓昌等 , 2006；弓利英等 , 2006；蒋耀培等 , 2005；张国鸣等 , 2006；林付根等 , 2006；史洪中等 , 2009）。因此，研究 RSV 对水稻的致病机制，对于正确理解病害流行成灾原因、科学制定病害防控策略具有重要指导意义。RSV 是纤细病毒属（*Tenuivirus*）的典型成员，其基因组由 4 条 ssRNA 组成，其中 RNA3 毒义链（vcRNA3）5′ 端的 ORF 编码外壳蛋白（Coat protein, CP，分子量约为 35.134 kDa），RNA4 的正义链（vRNA4）的 ORF 编码病害特异性蛋白（Disease- specific protein, SP）（Hayano et al., 1990），这两种蛋白均可在病叶中大量积累（林奇田等 , 1998）。RSV 侵染水稻主要症状为褪绿花叶，幼苗期感染可进一步发展形成假枯心。植物病毒病的花叶症状表现可能与 CP 蛋白进入叶绿体有关，这在烟草花叶病毒（TMV）和黄瓜花叶病毒（CMV）上已得到验证，并在离体条件下可以实现 TMV-CP 和 CMV-CP 的叶绿体跨膜运输（Banerjee and Zaitlin, 1992；梁德林等 , 1998）。RSV 的病理学研究发现，在受感染的水稻病叶细胞中观察到叶绿体的结构被破坏和淀粉粒大量积累；免疫胶体金标记发现在叶绿体和细胞质中均有 CP 蛋白存在（刘利华等 , 2000）；CP 蛋白和 SP 蛋白在病叶中的含量与病害症状有密切关系（林奇田等 , 1998）。以上研究结果暗示，CP 蛋白进入叶绿体可能是 RSV 导致水稻寄主产生花叶症状的主要原因。进行 RSV-CP 与叶绿体在离体条件下跨膜运输的尝试，并探索反应时间和 RSV 浓度对跨膜运输效率的影响，以期为探索 RSV-CP 与叶绿体的互作关系，深入探讨 RSV 侵染水稻引起花叶的机理提供研究方法和科学依据。

1　材料与方法

1.1　试验材料

RSV 毒源采自江苏姜堰、洪泽等地大田自然发生的有典型花叶症状的水稻条纹叶枯病病叶。在温室防虫罩内培育水稻（武育粳 3 号）和小麦（扬麦 158），健康苗用于叶绿体提取。RSV-CP 的兔抗血清 Sv21 自行制备，效价为 1 ∶ 16000。胰蛋白酶、胰蛋白酶抑制剂和 AP 标记的羊抗兔 IgG 均购自 Sigma 公司，底物显色液 BCIP/NBT 购自 Promega 公司，其他生化试剂购自上海生工生物工程技术服务有限公司。

1.2　病毒提纯

参考谢联辉等（谢联辉等 , 1991）的方法进行水稻条纹叶枯病病毒提纯，所得病毒浓度为 3.49mg/mL。

1.3　完整游离叶绿体的制备

参考朱水方和 Francki（朱水方和 Francki, 1992）的方法，采用切叶法制备叶绿体。叶绿体的完整度及纯度用相差显微镜（Nikon ECLIPSE 50i）检测，完整度大于 70% 者供下一步试验使用。检测 663nm 和 645nm 处的光吸收值确定叶绿素含量（斯佩克特等 , 1998）。

1.4　叶绿体蛋白酶的处理

根据 Cline 等（1984）的方法用胰蛋白酶去掉叶绿体外膜粘着蛋白，40% Percoll 垫（内含 5mmol/L EDTA）2500g 离心 5min 收集完整叶绿体。

1.5　叶绿体蛋白抽提

在酶处理过的健康和染病叶片叶绿体中分别加入抽提缓冲液（50 mmol/L Tris- HCl，pH 6.8，4.5%

SDS，9mol/L 尿素和 7.5% β-巯基乙醇）悬浮，等体积氯仿抽提除去色素，2～3 倍体积丙酮沉淀蛋白质，沉淀悬浮液在 0.1% SDS 抽提缓冲液中沸水煮 1～2min，1500g 离心 5min 除去不溶物，-20℃保存备用。

1.6 RSV- CP 的叶绿体离体跨膜运输

按照 Banerjee 和 Zaitlin（Banerjee and Zaitlin, 1992）的方法并略作修改。在水稻、小麦的叶绿体跨膜运输反应体系中 RSV 的浓度为 174.4μg/mL，叶绿体浓度为 298.5μg 叶绿素，反应体系中加入 10μL RNaseA（10mg/mL），以降解 RNA。23℃水浴保温 15min，10μL 胰蛋白酶（10mg/mL）25℃、30min 消化未进入叶绿体的 CP；水浴后加入 10μL 胰蛋白酶抑制剂（10mg/mL），4000g 离心 1min 收集叶绿体。同法进行不同温育时间梯度（1、5、15、30 和 60min）和不同 RSV 浓度梯度（19.37、21.80、58.10、174.40 和 348.70μg/mL）在水稻叶绿体中离体跨膜运输试验。

1.7 免疫印迹试验（Western blotting）

1.7.1 聚丙烯酰胺凝胶电泳（SDS- PAGE）

按 Sambrook 和 Russell（Sambrook and Russell, 2001）的方法进行。蛋白样品与 2×SDS 上样缓冲液混匀，煮沸 5min 后上样。先恒压 60V30min，随后恒压 100V2h。电泳完成后，取出凝胶，银染或转膜。

1.7.2 银染

步骤：固定液固定 10min，双蒸水漂洗 10min，固定敏化 5min，40% 乙醇漂洗 20min，双蒸水漂洗 20min，0.2g/L 硫代硫酸钠敏化 1min，双蒸水漂洗两次（1min/ 次），0.1% 硝酸银银染 20min，再用双蒸水漂洗 1min，加显色液室温显色 5～8min，然后加 5% 醋酸终止反应 30min，最后用双蒸水漂洗、拍照。

1.7.3 电转移—半干转

SDS-PAGE 电泳结束后，取下凝胶，切除浓缩胶，按照 Trans-Blot SD Semi-Dry Electrophoretic transfer cell 仪的说明书要求恒压 20V30min 进行电转移。

1.7.4 Western blotting

按照 Promega 公司 Proto blotwestern blot Ap systems 产品说明书进行。一抗用自行制备的 RSV 兔抗血清 Sv21（1 ∶ 5000），二抗用碱性磷酸酶标记羊抗兔 IgG（1 ∶ 10000），加入底物后 37℃下显色 5～10min 至特异性条带明晰后中止显色反应。

2 结果与分析

2.1 水稻条纹叶枯病病株叶绿体中 RSV–CP 蛋白的检测

水稻条纹病毒侵染水稻后可在叶肉细胞和表皮细胞中大量增殖，病毒增殖到一定程度后叶片出现花叶症状，这种症状的形成可能与病毒的外壳蛋白进入叶绿体并干扰或破坏叶绿体正常功能有关（刘利华等，2000）。利用免疫胶体金技术在叶绿体中发现 RSV-CP 的存在，为了验证这一点，本研究提取水稻条纹叶枯病病株和水稻健康植株叶绿体，经 SDS-PAGE 凝胶电泳后银染检测，结果表明，水稻健康植株叶片叶绿体中未出现而病株叶绿体中则检测到与 RSV-CP 分子量相同的蛋白条带。经 Western blotting 分析，该条带与 RSV 单克隆抗体可起清晰的血清学反应（图 1），证明在水稻条纹叶枯病病株叶绿体中有 RSV 的 CP 蛋白存在。

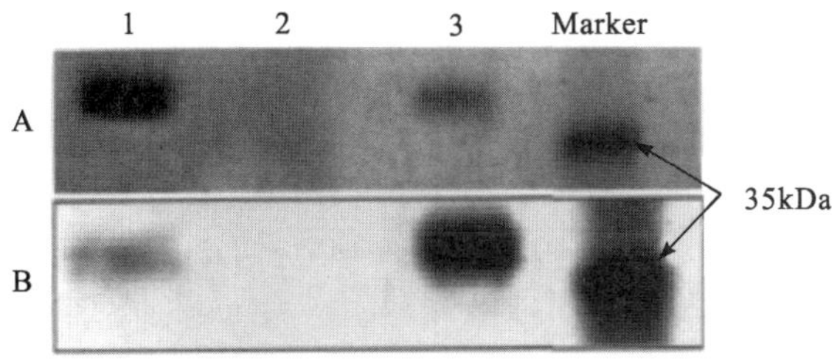

图 1 病株叶绿体中 RSV-CP 的检测结果

Fig.1 Detection of RSV- CP in the chloroplasts from diseased rice plants

A: 银染；B: 印迹转移；1: 水稻条纹叶枯病病株叶绿体；2: 健康水稻叶绿体；3: 提纯 RSV

A: Silver staining; B: Western blotting; 1: Chloroplasts in infected rice plants with RSV; 2: Chloroplasts in health rice plants; 3: Purified RSV

2.2 RSV- CP 在水稻、小麦叶片叶绿体中的离体跨膜运输

RSV- CP 如何进入水稻或其他植物的叶绿体目前尚不明确，为弄清这一问题必须建立一个简单的、可操作的技术体系，本研究进行了离体条件下 RSV-CP 对叶绿体跨膜运输的尝试。Western blotting 分析结果表明，离体状况下无需 ATP、SP、RNA 及水稻（或小麦等）寄主蛋白成分的协助，RSV-CP 可以跨膜运输进入其寄主植物水稻或小麦的完整游离叶绿体中（图 2），进入叶绿体前后 RSV-CP 分子量大小不变，其结果与感染了 RSV 病株叶绿体中的检测结果一致。

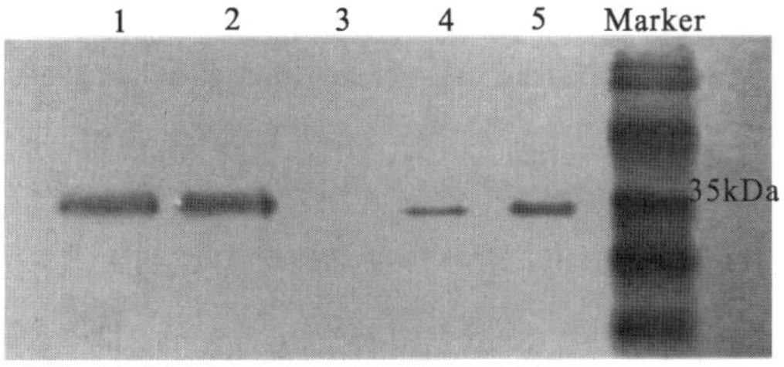

图 2 离体条件下不同孵育时间 RSV-CP 在水稻叶绿体中的跨膜运输效果

Fig. 2 Detection of RSV- CP in the intact chloroplasts by Western blotting after incubating with RSV

1: 水稻叶绿体与 RSV- CP; 2: 小麦叶绿体与 RSV-CP; 3: 健康水稻叶绿体；4: 健康小麦叶绿体；5: 提纯 RSV

1: Rice chloroplasts and RSV- CP; 2: Wheat chloroplasts and RSV-CP; 3: Health rice chloroplasts; 4: Health wheat chloroplasts; 5: Purified RSV

2.3 RSV-CP 在水稻叶绿体中的离体跨膜运输与水

水浴保温时间和 RSV 浓度的相关性为进一步完善 RSV-CP 离体叶绿体跨膜运输体系，同上法提取水稻叶绿体并与 RSV 进行孵育保温时间和 RSV 浓度对跨膜运输效率的影响试验。结果表明，水浴保温 5min 后 RSV- CP 即可进入水稻叶绿体中，15min 时条带即比较清晰。随着水浴保温时间的延长，进入叶绿体中的 RSV- CP 量略有增加（图 3）。

用 5 个浓度梯度的 RSV 与水稻叶绿体体外孵育试验结果表明，RSV- CP 的水稻叶绿体离体跨膜运输有量的基本要求。当反应体系中 RSV 浓度为 21.80μg/mL 以下时，跨膜运输后叶绿体中检测不到 RSV-CP；当反应体系中 RSV 浓度为 58.10μg/mL 时，则能在叶绿体中检测到 RSV- CP。随着反应体系中 RSV 浓度的增加，检测出的进入叶绿体的 RSV-CP 量随之明显增多。

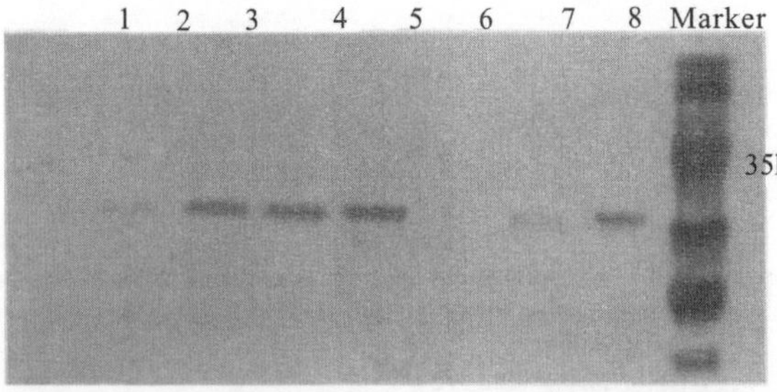

图 3 离体条件下不同孵育时间 RSV-CP 在水稻叶绿体中的跨膜运输效果

Fig. 3 Detection of RSV- CP in rice chloroplasts by Western blotting after incubating with RSV in different times

1–5：孵育时间分别为 1, 5, 15, 30, 60min; 6: 健康水稻叶绿体 ; 7: 病株水稻叶绿体 ; 8: 提纯 RSV

1–5: Incubating RSV- CP in healthy rice chloroplasts for 1, 5, 15, 30min and 1h; 6: Healthy rice chloroplasts; 7: Diseased rice chloroplasts; 8: Purified RSV

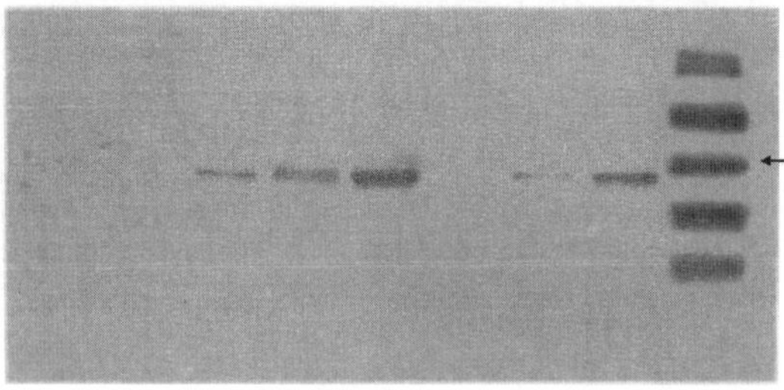

图 4 离体条件下不同病毒浓度 RSV-CP 水稻叶绿体的跨膜运输效率

Fig. 4 Detection of RSV- CP from rice chloroplasts by Western blotting after incubating with different RSV concentrations

1–5: 孵育体系中 RSV 浓度分别为 19.37, 21.80, 58.10, 174.40, 348.70μg/mL; 6: 健康水稻叶绿体 ; 7: 病株水稻叶绿体 ; 8: 提纯 RSV

1–5: 19.37, 21.80, 58.10, 174.40 and 348.70 RSV in the incubation system; 6: Healthy chloroplasts; 7: Diseased chloroplasts; 8: Purified RSV

3 讨论

植物病毒中有一大类在其寄主植物的叶片上能产生褪绿或黄色花叶症状，如 TMV、CMV 等，自然条件下 TMV 感染烟草后，需要一定的时间花叶症状才会显现并逐渐加重。（Hodgson 等，1989）研究认为这是烟草叶绿体中 TMV-CP 逐渐累积的缘故，RSV 自然侵染显症的过程与此类似。已有研究结果显示 RSV 的外壳蛋白 CP 与褪绿花叶症状的严重度密切相关，且 CP 蛋白可在 RSV 侵染的水稻中大量积累，并在寄主细胞内形成形态不一的内含体。在受感染的水稻病叶细胞中，观察到叶绿体的结构被破坏和淀粉粒大量积累（林奇田等，1998）。免疫胶体金标记发现在叶绿体和细胞质中均有 CP 蛋白存在（刘利华等，2000）。本研究也利用 Western blotting 在染病水稻叶绿体中检测到 RSV-CP。RSV-CP 在叶绿体中积累并破坏叶绿体的光合作用是否为 RSV 致害和产生花叶症状的主要原因有待进一步深入探讨，对 RSV-CP 如何进入叶绿体及其所需基本条件的探索可能会获得干扰 CP 进入叶绿体的方法，从而找到治疗该病的方法。

物质的跨膜运输有自由扩散、协助扩散、主动运输等方式，目前对植物病毒外壳蛋白的叶绿体跨膜运输机制尚不明确。Banerjee 和 Zaitlin、梁德林等研究结果表明，TMV-CP 和 CMV-CP 能进入相应寄主植物完整叶绿体内（梁德林等，1998; Banerjee and Zaitlin, 1992），跨膜运输前后蛋白分子量不变，在病毒外壳蛋白的氨基酸序列上并未发现有类似于 Rubisco SSU 等跨膜蛋白所具有的引导肽序列，在跨膜运输体系中加入或不加入 ATP 并未影响运输效率，本研究对 RSV-CP 的叶绿体跨膜运输具有类似的结果。说明 TMV、CMV 和 RSV 外壳蛋白的叶绿体跨膜运输的机制不同于 Rubisco SSU 等的主动运输，至于其确切的运输机制有待后续研究。

4 结论

本研究结果表明，RSV-CP 在离体条件下可以快速进入水稻和小麦的叶绿体；通过对孵育时间和病毒浓度对跨膜运输效率试验，初步建立了 RSV-CP 叶绿体跨膜运输的体系为 58.10μg/mL RSV、孵育时间 15min。以上研究结果为进一步探索 RSV 的致病机理打下基础。

参考文献

[1] 程兆榜，杨荣明，周益军，等．江苏稻区水稻条纹叶枯病发生新规律 [J]. 江苏农业科学，2002, (1): 39-41.

[2] 高苓昌，宋克勤，张洪瑞，等．黄淮稻区水稻条纹叶枯病的发病特点与综合防治 [J]. 山东农业科学，2006, (3): 66-67.

[3] 弓利英，夏立，马丽，等．沿黄稻区水稻条纹叶枯病的发生及防治技术 [J]. 河南农业科学，2006, (7): 64-65.

[4] 蒋耀培，李建刚，谭秀芳，等．上海地区水稻条纹叶枯病的发生与防治技术初探 [J]. 上海农业科技，2005, (5): 37-38.

[5] 梁德林，叶寅，施定基，等．黄瓜花叶病毒卫星 RNA 致弱辅助病毒的机理 [J]. 中国科学 C 辑：生命科学，1998, 28(3): 251-256.

[6] 林付根，成长庚，赵阳．水稻条纹叶枯病治虫防病策略的实践体会 [J]. 中国植保导刊，2006,(1): 13-15.

[7] 林奇田，林含新，吴祖建，等．水稻条纹病毒外壳蛋白和病害特异蛋白在寄主体内的积累 [J]. 福建农业大学学报，1998, 27(3): 322-326.

[8] 林奇英，谢联辉，周仲驹，等．水稻条纹叶枯病的研究 :I 病害的分布和损失 [J]. 福建农学院学报，1990,19(4): 421-425.

[9] 刘利华，吴祖建，林奇英，等．水稻条纹叶枯病细胞病理变化的观察 [J]. 植物病理学报，2000, 30(4): 306-311.

[10] 史洪中，郭世保，陈俊华，等．水稻条纹叶枯病成灾原因及其综合防治技术 [J]. 江西农业学报，2009, 21(4)：63-65.

[11] 斯佩克特 D L, 戈德曼 R D, 莱因万德 L A. 细胞实验指南 [M]. 北京：科学出版社，2001.

[12] 谢联辉，周仲驹，林奇英，等．水稻条纹叶枯病的研究Ⅲ．病害的病原性质 [J]. 福建农学院学报，1991, 20(2): 144-149.

[13] 张国鸣，王华弟，戴德江．浙江省水稻条纹叶枯病发生发展态势与防控对策措施 [J]. 中国植保导刊，2006, (7): 2021.

[14] 朱水方，R.I.B. FRANCKI. 黄瓜花叶病毒衣壳蛋白存在于被其侵染的烟草叶绿体中 [J]. 中国病毒学，1992, 7(3): 328-333.

[15] BANERJEE N, ZAITLIN M. Import of tobacco mosaic virus coat protein into intaet chloroplast in vitro[J]. MPMI, 1992, 5(6): 466-471.

[16] CLINE K, WERNER-WASHBURNE M, ANDREWS J, et al. Theromlys in is a suitable protease for probing the surface of intact pea chloroplasts[J]. Plant Physiology, 1984, 75(3): 675-678.

[17] HAYANO Y, KAKUTANI T, HAYASHI T, et al. Coding strategy of rice stripe virus：Major nonstructural protein is encoded in viral RNA segment 4 and coat protein in RNA complementary to segment 3[J]. Virology, 1990, 177: 372-374.

[18] HODGSON R A J, BEACHY R N, PAKRASI H B. Selective inhibition of photosystem II in spinach by tobacco mosaic virus：An effect of the viral coat protein[J]. FEBS Letters, 1989, 245(1-2): 267-270.

[19] KISIMOTO R, YAMADA Y. Present status of controlling rice stripe virus. HadidiA,KhetarpalRK,Koganezawa H Plant Virus Disease Control[J]. St. Paul Minnesota,USA, 1991, APS Press: 470-483.

[20] SAMBROOK J, RUSSELL D. Molecular Cloning:A Laboratory Manual[J]. Analytical Biochemistry, 2001, 186(1):182-183.

水稻条纹病毒胁迫下的水稻蛋白质组学

张晓婷[1,2]，谢荔岩[1]，林奇英[1]，吴祖建[1]，谢联辉[1]

(1 福建省植物病毒学重点实验室，福建农林大学植物病毒研究所，福州 350002；
2 河南农业大学植物保护学院，郑州 450002)

摘要：采用双向电泳联用 MOLDI-TOF-TOF 质谱对水稻感病品种武育粳 3 号和抗病品种 KT95-418 感染水稻条纹病毒（*Rice stripe virus*, RSV）前后的叶片进行蛋白质组学分析。结果显示，RSV 基因组编码的病害特异蛋白（disease specific protein, SP）在武育粳 3 号中的积累量明显高于 KT95-418。其他 25 个蛋白经质谱成功鉴定，包括 RSV NS2 蛋白，寄主中与光合作用、细胞氧化还原状态和离子平衡状态及蛋白的合成、转运与翻译后修饰等相关的蛋白。对这些差异表达的蛋白与水稻感、抗病的作用进行了分析。

关键词：水稻条纹病毒；水稻品种；蛋白质组学

中图分类号：S435.11　**文献标识码**：A　**文章编号**：0412-0914(2011)03-0253-09

Rice proteomics under the infection of *Rice stripe virus*

Xiaoting Zhang[1,2], Liyan Xie[1], Qiying Lin[1], Zujian Wu[1], Lianhui Xie[1]

(1 Key Laboratory of Plant Virology of Fujian Province, Institute of Plant Virology, Fujian Agriculture and Forestry University, Fuzhou 350002, China;
2 College of Plant Protection, Henan Agricultural University, Zhengzhou 450002)

Abstract: Two dimensional electrophoresis (2-DE) and MOLDI-TOF-TOFMS were used to analyse the proteomic changes in leaves of the susceptible cultivarWuYu3 and resistant cultivar KT95-418 rices infected with Rice strip virus (RSV). The results showed that disease specific protein (SP) of RSV accumulated more in susceptible cultivarWuYu3 than that in resistantcultivar KT95-418. The other 25 proteins were identified successfully by MS, including RSV NS2, host proteins related to photosynthesis, dynamic balance of cellular redox state or ions and protein translation/translocation or modification. The possible functions of these host proteins in the susceptible and resistant cultivars were discussed.

Keywords: Rice stripe virus; rice cultivars; proteomics

水稻条纹病毒（*Rice stripe virus*, RSV）是纤细病毒属（*Tenuivirus*）的代表种，由介体灰飞虱（*Laodelphax striatellus*）以持久性方式传播，寄主范围广泛。其所致的水稻条纹叶枯病在 20 世纪中期传入我国，现已成为我国水稻生产上的重要病害之一。水稻在感染 RSV 后出现展叶型条纹症状或叶片褪绿、捻转下垂的卷叶型“假枯心”症状。细胞病理学研究表明，水稻感染 RSV 后，叶肉细胞中线粒

植物病理学报 .2011,41(3):253-261

收稿日期：2010-08-20

基金项目：国家转基因生物新品种培育重大专项重点课题 (2009ZX08009-044B)；国家自然科学基金资助项目 (30770090)；农业部农业公益性行业科研专项 (nyhyzx07-051)

体增多，细胞核变大，叶绿体结构受到破坏，淀粉粒大量积累，并形成多种形态的病毒内含体（Zhou et al., 1992; Liu et al., 2000; 明艳林 , 2001; Takahashi et al., 2003; Liang et al., 2005）。作为严格的细胞内寄生生物，RSV 侵染寄主后干扰其正常的生理代谢，引起细胞形态的明显改变。同时 , RSV 还与寄主蛋白发生互作关系（Hayakawa et al., 1992; Lu et al., 2009; Daniel and Nagy, 2010），利用寄主蛋白进行复制、增殖和运动，导致其基因和蛋白表达谱发生显著的变化（Shimizu et al., 2007; Catoni et al., 2009; Brault et al., 2010; Pineda et al., 2010）。2004 年水稻条纹叶枯病大流行期间，江苏省主栽品种武育粳 3 号表现高度感病，而由其田间自然变异单株选育出来的 KT95-418 却显示高度抗病。鉴于此，针对这 2 个水稻品种的基因转录谱进行了分析，结果有 3517 个基因出现差异表达，其中 2002 个基因表达上调 , 1515 个基因表达下调（Zhang et al., 2008）。

随着蛋白分离和鉴定技术的发展，蛋白质组学已成为植物病理学研究的重要手段。病原物侵染后的蛋白质组学变化是揭示病害发生机制的一个重要方面。采用双向电泳联用质谱分析技术，稻瘟病菌（*Magnaporthe grisea*）、水稻白叶枯病菌（*Xanthomonas oryzae* pv oryszae, Xoo）和水稻黄斑驳病毒（*Rice yellow mottle virus*, RYMV）侵染后水稻的应答蛋白质组学研究已经开展（Kim et al., 2004; Ventelon-Debout et al., 2004; Chen et al., 2007; Mahmood et al., 2007），但由昆虫传播的水稻病毒，尚无蛋白质组学方面的数据。本研究以生产上发生严重的昆虫传病毒病水稻条纹叶枯病为研究对象，采用双向电泳和质谱鉴定技术对抗、感表现差异明显的武育粳 3 号和 KT95-418 水稻品种（系）进行了蛋白表达谱分析，为进一步揭示 RSV 的致病机制提供参考。

1 材料与方法

1.1 试验材料

1.1.1 水稻品种

供试水稻品种选用对水稻条纹叶枯病表现感病的粳稻品种“武育粳 3 号”及其对 RSV 表现抗病的 KT95-418。

1.1.2 介体昆虫及病毒毒源

无毒灰飞虱是 2004 年采自江苏洪泽田间的无毒灰飞虱后代，经由福建农林大学病毒所室内饲养确证不带毒的群体；病毒毒源（RSV）由采自江苏洪泽田间的病株经无毒灰飞虱饲毒接种分离纯化后，保存于粳稻品种合系 28 上。

1.2 试验方法

1.2.1 供试水稻幼苗的栽培

武育粳 3 号和 KT95-418 水稻种子经 70% 酒精表面消毒后泡在水中催芽，待白色幼芽长出时播种于盛有湿润营养土（草炭土∶沙土 =4 ∶ 1）的塑料瓶中，每瓶育苗 20 株。幼苗生长至二叶一心期时，一部分用于接种病毒，作为试验处理组（T）；另一部分在同样条件下生长但不做任何接种处理，作为试验对照组（C）。

从育苗到取样，水稻幼苗均生长于 25 ～ 28℃，光照周期为 16h 光 /8h 暗的玻璃温室中。

1.2.2 病毒接种

取 2 ～ 4 龄的无毒灰飞虱若虫 100 头在合系 28 RSV 病株上获毒取食 48h，之后将其移至健康稻苗上饲养让其渡过循回期。

以每瓶稻苗（20 株）100 头带毒灰飞虱的接种压力在武育粳 3 号和 KT95-418 健康稻苗上强迫接种取食 48h，接种后的幼苗移栽于 25℃，16h 光 /8h 暗光照周期的防虫网室中，记录发病情况直至取样。

1.2.3 样品的获取

在处理组样品中选取 3 株（显症时间相差不超过 24h）水稻条纹病株，于显症后第 7d 分别剪取其症状明显的病叶部分，做好标记，迅速冻存于液氮中。取自每株稻株的材料作为一个生物学重复。

同时，从对照组样品中任选 3 株，各自与处理组 3 个样品对应，剪取其相应叶位部分，迅速冻存于液氮中。

1.2.4 蛋白样品的制备与含量测定

采用三氯乙酸（TCA）/ 丙酮（acetone）沉淀法提取水稻蛋白。称取 0.2g 水稻叶片放入预冷的研钵，在液氮中研磨成精细的粉末，然后加入 2mL 预冷的 TCA- 丙酮蛋白提取液，在冰上研磨至匀浆。匀浆液转入 10mL 离心管中，4℃下 25 700g 离心 15min，弃上清。沉淀用 2mL 预冷的丙酮（含 10mmol/L 巯基乙醇）重悬，-20℃下沉降 3 h 后于 4℃下 25 700 g 离心 15min，弃上清。用 2mL 预冷的丙酮洗沉淀，-20℃下沉降 1.5h 后于 4℃下 25 700g 离心 15min，弃上清。再次用丙酮清洗沉淀后，-20℃下沉降 0.5h，4℃下 25 700g 离心 15min，尽量弃去上清，沉淀置 -20℃保存使丙酮挥发，即得蛋白干粉。

称取蛋白干粉 20mg，真空干燥后加入 1mL 样品裂解液（9mol/L 尿素、4% CHAPS、50mmol/LDTT、0.2% Bio-Lyte 载体两性电解质 pH 3 ~10），超声处理 10 ×1.5min，4℃下 18 000g 离心 15min，取上清。用 Bradford 法测定蛋白含量后，分装，于 -80℃保存。

1.2.5 IPG-SDS-PAGE 双向电泳

第一向固相 pH 梯度等电聚焦电泳（IPG）采用 pH 4 ～ 7，17 cm 线性 IPG 预制胶条，上样量为 408μg，上样体积 350μL。水化和聚焦过程在 20℃下自动进行。程序设置为 :50V 聚焦 12h; 250V 聚焦 1h，1 000V 聚焦 2h，8 000V 聚焦 2h; 8 000V 聚焦 60 000Vh。聚焦完毕后，取出胶条，在 5mL 平衡液 I（0.375mol/L Tris-HCl pH 8.8，6mol/L 尿素，20% 甘油，2% SDS，2% DTT）和 5mL 平衡液 II（2.5% 碘乙酰胺代替 2% DTT，其余组分同平衡液 I）中先后振荡平衡 15min。然后将经过 2 次平衡后的胶条转移至 13% SDS-PAGE 胶上，每根胶条 10mA 预电泳至样品完全走出 IPG 胶条，浓缩成一条线。加大电流以 35mA 电泳至溴酚蓝指示剂达到底部边缘时停止电泳。电泳缓冲液为 25mmol/LTris，192mmol/L 甘氨酸，0.1% SDS，pH 8.3。

1.2.6 硝酸银染色

电泳后的凝胶经超纯水清洗 3 遍后，进行银染，方法参考（夏其昌和曾嵘 , 2004）。

1.2.7 双向电泳图谱分析

用 EPSONPERFECTION1270 扫描仪扫描凝胶，图像保存为 TIFF 格式，分辨率设置为 400dpi，放大倍数为 100%，图像深度 48 比特。扫描后的图像用 Image Master 2D Platinum 6.0（GE healthcare Inc.）进行分析，蛋白点检定参数设置为：Smooth-3，Saliency-7，MinArea-60。以目标蛋白 3 个重复具有相同趋势，相邻位点相对一致，至少 2 个重复有 2 倍（蛋白点体积比）以上差异作为判定差异表达蛋白质的标准。

1.2.8 差异表达蛋白质点的质谱鉴定

选取差异蛋白质点切下，送复旦大学蛋白质组学研究中心进行 MALDI-TOF-TOF/MS 鉴定。

2 结果与分析

2.1 双向电泳图谱分析

用 2D Image Master 2D Platinum 6.0 分析“武育粳 3 号”和“KT95-418”RSV 病叶和健康叶片的电泳图谱，分别检定到 359 个（武育粳 3 号对照组），315 个（武育粳 3 号处理组），514 个（KT95-418 对照组），446 个（KT95-418 处理组）高清晰且重复性强的蛋白质点。对不同处理下蛋白质点的表达丰度进行差异检测，共得到了44个差异表达的蛋白质点（图1）。在等电点为4.5～5.5，分子量为20～25kDa的位置，2个水稻品种的试验处理组图谱与对照组图谱相比都有明显的蛋白质积累，标记为区域A（图1）。

在44个差异表达的蛋白质点中，武育粳3号处理组与对照组相比较有33个，其中25个表达量减少，8个表达量增加；KT95-418处理组与对照组相比较有15个，其中12个表达量减少，3个表达量增加（图1）。编号为4、5、7、8的4个蛋白质点在2个水稻品种处理组与对照组的比较中表达量均减少（图1）。

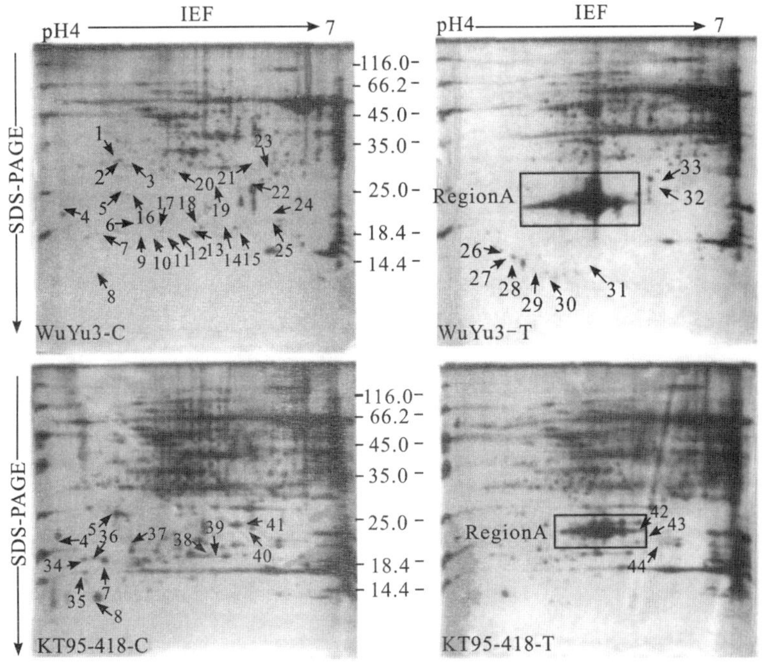

图1 WuYu3和KT95-418中RSV感染和未感染叶片蛋白质的二维电泳图

Fig.1 Two dimensional electrophoresis maps of proteins from RSV-infected and non-infected leaves in WuYu3 and KT95-418

窗口显示RSV感染样本（A区）中显著积累的蛋白；箭头和数字表示蛋白质的点和它的数量

Panes show the significantly accumulated protein in RSV-infected samples (Region A); arrows and numbers show interested spots and their numbers

2.2 差异表达蛋白质的质谱鉴定

在区域A中，分别从武育粳3号和KT95-418的病叶图谱中连续切取5个和3个蛋白质点进行了MALDI-TOF-TOF质谱鉴定。所得质谱数据提交到Mascot网站选用NCBInr数据库进行搜索，结果表明，这些蛋白质点都是RSV的病害特异蛋白（disease specific protein，SP）。

另外44个蛋白质点中，编号为4、5、7、8的4个蛋白质点在2个水稻品种的处理组中均减少，分别从2个水稻品种的对照组凝胶上切取进行质谱鉴定，其他差异蛋白质点依照图1中的显示在不同凝胶上切取，进行酶解和质谱分析，共25个差异蛋白质得到鉴定，鉴定结果见表1。

这些得到鉴定的蛋白质分别是22个基因编码的产物（18和25、19和22、40和41号点分别被鉴定为相同的蛋白质），其中包括一个RSV编码的NS2蛋白。其他蛋白质按照功能分为4个类群。

2.2.1 参与光合作用蛋白

共有10个，包括4个核酮糖1，5-二磷酸羧化酶/加氧酶（Rubisco）亚基（编号为12、13、32、35）、质体蓝素（plastocyanin，编号为8）、类囊体内腔17.4kDa蛋白（thylakoid luminal 17.4kDa protein，chloroplast precursor P17.4，编号为10）、细胞色素b6-f复合体Fe-S亚基（Cytochrome B6-F complex Fe-S subunit，编号为18和25）和光系统II放氧复合体蛋白2前体（photosystem II oxygen-evolving complex protein2 precursor, 编号为 19 和 22）。

2.2.2 氧化还原酶类

共有 5 个，包括定位于叶绿体的 Cu-Zn 超氧化物歧化酶前体（Cu-ZnSOD, 编号为 14）、Cu-Zn 超氧化物歧化酶（Cu-ZnSOD, 编号为 24）、M 型硫氧还蛋白前体（Trx, 编号为 9）和 2 个功能没有明确的蛋白质 Os02g0328300（蛋白质点编号为 1）和 Os02g0192700（蛋白质点编号为 6），它们分别与一种具有氧化还原酶活性的果实蛋白和硫氧还蛋白过氧化物酶具有较高的同源性。

2.2.3 核糖体蛋白

共有 2 个，包括一个酸性磷酸化蛋白（Os08g0116500，编号为 34）和一个定位于叶绿体的 50S 核糖体蛋白 L12 前体 [50S ribosomal proteinL12, chloroplast precursor（CL12），编号为 37]。

2.2.4 其他功能蛋白

共有 5 个，包括代谢酶类如腺苷二磷酸葡萄糖焦磷酸化酶（adenosine diphosphate glucose pyrophosphatase，编号为 40、41）、推定的核糖 -5- 磷酸异构酶（putative ribose-5-phosphate isomerase，编号为 3）、甘氨酸剪切系统蛋白 H（glycine cleavage system protein H，编号为 7）和一个推定的小泛素相关修饰物（small ubiquitin like modifier，SUMO，编号为 17）。

表 1 通过 MALDI- TOF-TOF/MS 质谱鉴定出 25 个蛋白质点

Table 1 25 Protein spots identified by MALDI-TOF-TOF/MS

Protein spot No.	Accession No.	Description	Species	Protein score	Mr(kDa)/pI
Proteins repressed by RSV in WuYu3					
Photosynthetic proteins					
8	gil115465862	Os06g0101600(plastocyanin, chloroplast precursor)	*Oryza sativa*	97	15.5/5.61
9	gil11135471	Thioredoxin M-type, chloroplast precursor (TRX-M)	*O. sativa*	174	18.5/8.16
10	gil115482792	Os10g0502000(thylakoid lumenal 17.4 kDa protein, chloroplast precursor (P17.4)	*O. sativa*	211	24.5/7.49
12	gil90968348	ribulose-1,5-bisphosphate carboxylase/oxygenase large subunit	*Bentinckia nricobarica*	108	49.1/6.8
13	gil47604672	RuBisCO large subunit	*Ypsilandra thiberica*	155	49.6/6.13
14	gil42408425	putative superoxide dismutase [Cu-Zn], chloroplast precursor	*O. sativa*	573	20.5/5.79
18、25	gil115472727	Os07g0556200(Cytochrome B6-F complex Fe-S subunit)	*O. sativa*	111、269	23.8/8.54
19、22	gil115470529	23 kDa polypeptide of photosystem Ⅱ oxygen-evolving complex protein precursor	*O. sativa*	181、318	26.9/8.66
Proteins with other functions					
1	gil115445869	Os02g0328300(Oxidoreductase NAD-binding domain. Xanthine dehydrogenases, that also bind FAD/NAD)	*O. sativa*	386	30.6/5.44
3	gil34393836	putative ribose-5-phosphate isomerase	*O. sativa*	107	26.9/4.91
5	gill1548i540	Os10g0330000 (Conserved hypothetical protein)	*O. sativa*	101	16.5/4.7
6	gil115444771	O502g0192700(Thioredoxin peroxidase)	*O. sativa*	185	23.1/6.15
7	gil115482934	Os10g0516100(glycine cleavage system protein H)	*O. sativa*	114	17.3/4.92
11	gil115469830	Os06g07051000(Conserved hypothetical protein)	*O. sativa*	315	24.3/8.74
17	gil115441855	Os01g0918300(putative SUMO protein)	*O. sativa*	138	10.9/4.95
24	gil401108	Superoxide dismutase [Cu-Zn]	*Pisum sativum*	135	15.3/5.6
Proteins induced by RSV in WuYu3					
32	gil22770406	ribulose 1 ,5-bisphosphate carboxylase large subunit	*Utricularia pubescens*	241	20.8/5.15
33	gil62363147	NS2 protein	*Rice stripe virus*	123	22.8/6.03
Proteins repressed by RSV in KT95-418					
34	gil115474515	Os08g0116500(60s Acidic ribosomal protein)	*O. sativa*	81	11/4.43
37	gil109940141	50S ribosomal protein L12, chloroplast precursor (CL12)	*O. sativa*	268	18.5/5.36
40、41	gil13160411	adenosine diphosphate glucose pyrophosphatase	*Hordeum vulgare* subsp. *vulgare*	86、83	21.8/5.68
35	gil56966763	Chain S, crystal structure of activated rice Rubisco complexed with 2-carboxyarabinitol-1 ,5-bisphosphate	*O. sativa*	109	14.9/5.89

3 讨论

3.1 病毒蛋白

已有研究表明，RSVSP 蛋白在 RSV 侵染后不同时期的水稻病株和带毒昆虫的卵巢、肠腔等部位大量存在，在病株中的表达水平比在昆虫中高 50 倍（Qu et al., 1999）。Lin 等发现 SP 蛋白在病叶中的积累量与褪绿花叶症状的严重度密切相关，且在不同品种中其达到积累高峰的时间也有所不同（Lin et al., 1998）。Liang 等用免疫胶体金标记抗体只在显症的叶片中才检测到 SP 蛋白（Liang et al., 2005）。本研究采用双向电泳技术成功分离了 RSV 侵染后武育粳 3 号和 KT95-418 的叶片总蛋白，在双向电泳图谱上发现 RSV SP 蛋白的分子量约为 20 ～ 25kDa，等电点 4.5 ～ 5.5，该蛋白质在 2 个水稻品种的病叶中都有积累，在症状较重的武育粳 3 号中的积累量明显高于 KT95-418 中，积累量与寄主的症状严重程度呈正相关。该结果进一步证实 RSV SP 与病毒在不同抗病水稻品种上的致病力相关。另外，在武育粳 3 号的 RSV 病叶蛋白双向电泳图谱中分子量约 25kDa，等电点约 6.0 的位置上，我们鉴定得到了 RSV NS2 蛋白。Liang 等采用 Western 印迹法在水稻病叶蛋白提取物的上清中检测到了该蛋白的存在，但其表达量很少，且只在新侵染叶片的叶肉细胞和表皮细胞中表达，有些内含体中也能够检测到 NS2 蛋白，分子量为 22.8kDa，与本研究中发现的 NS2 蛋白分子量位置比较一致，等电点也与预期值相符（Liang et al., 2005）。从图谱上可以看出，即使在武育粳 3 号病叶中，NS2 蛋白的表达量也很低，不易被检测出来。在 KT95-418 病叶蛋白双向电泳图谱的相应位置上则没有 NS2 蛋白质点的出现。这一研究结果表明，在不同的抗性品种中 NS2 蛋白的表达量存在差异，它与病害的发生严重度有关。

其他病毒蛋白在本研究中未得到鉴定，其原因可能有三：一是由于这些蛋白的分子量和等电点超出了本研究的检测范围（分子量小于 30kDa，等电点 4 ～ 7），因而未能得到鉴定，如 RSV RNA1 编码的 RNA 聚合酶，RNA2 和 RNA4 的负链编码产物 NSvs2 和 NSvs4，RNA3 编码的病毒外壳蛋白（coatprotein, CP）等；二是由于某些蛋白如 NS3 蛋白，其分子量约 24kDa，与 SP 的分子量比较接近，在双向电泳检测中被高丰度表达的 SP 或其他寄主蛋白所掩盖；三是由于这些蛋白的表达量相对较低，不能够在双向电泳图谱上有效显示。在进一步的研究中，可以通过层析法去除高丰度蛋白，或采用蛋白分级提取技术或窄范围 IPG 胶条电泳后拼接技术拓宽双向电泳的检测能力，深入挖掘 RSV 侵染后病毒蛋白在寄主体内的表达变化。

3.2 光合作用相关蛋白

3.2.1 核酮糖 -1，5- 二磷酸羧化酶 / 加氧酶（Rubisco）

本研究中鉴定得到 Rubisco 大亚基肽段与 3 个蛋白质点对应，其中 2 个分子量较小的肽段（蛋白质点编号为 12、13）在感染 RSV 后的武育粳 3 号叶片中表达量降低，分子量为 25kDa 左右的肽段（蛋白质点编号为 32）在感染 RSV 后的武育粳 3 号叶片中表达量增加。这反映了 Rubisco 大亚基的片段化作用，即在一定的条件下 Rubisco 大亚基发生了体内降解，不同的降解片段表达情况不同。

类似的 Rubisco 大亚基片段化作用在前人的研究中也有报道。Agrawal 等对臭氧胁迫下水稻蛋白表达谱分析表明，11 个 Rubisco 大亚基的片段，分子量介于 15kDa 和 49kDa 之间（Agrawal et al., 2002）。Zhao 等采用 MALDI-TOF/MS 鉴定了水稻双向电泳图谱上不同生长时期表达差异的 49 个蛋白质点，发现其中 32 个与 Rubisco 大亚基同源，完整的 Rubisco 大亚基等电点范围很广，分子量集中于 55kDa，片段化的 Rubisco 大亚基多数位于等电点 6.5 左右，分子量 30 ～ 40kDa 的位置（Zhao et al., 2005）。该研究所获得的双向电泳图谱相应位置，如 55kDa 和等电点 6.5 左右，分子量 30 ～ 40kDa 的位置，本研究所得图谱上都有相应的蛋白质点显示。

3.2.2 光合电子传递载体

光合电子传递是光合作用的重要过程。本研究中鉴定得到了与重要的光合电子传递载体质体蓝素和细胞色素 b6-f 复合体 Fe-S 亚基同源的蛋白，它们的表达量在感染 RSV 后的水稻叶片蛋白双向电泳图谱中相对健康叶片蛋白图谱中较少，在感染 RSV 后的武育粳 3 号水稻叶片蛋白双向电泳图谱中检测不到。

其中，与细胞色素 b6-f 复合体 Fe-S 亚基同源的蛋白质点有 2 个（蛋白质点编号为 18 和 25），它们的分子量基本一致，图谱显示约为 20kDa，等电点有所不同，分别约为 5.0 和 6.0。在本研究中有 2 个分子量相近等电点不同的蛋白质点与细胞色素 b6-f 复合体的 Fe-S 亚基对应，说明在感染 RSV 后，该蛋白亚基可能发生了翻译后修饰，或者可能产生了不同的结合方式。

质体蓝素（Plasto cyanin, Pc）的氧化还原状态的平衡及其与细胞色素 b6-f 复合体和光系统 I 的相互作用是影响光合过程中电子转运效率的重要因素。在感染 RSV 后的水稻中，质体蓝素前体的表达量下降，说明水稻寄主在感染病毒后，光合作用的电子转运效率降低，同化作用减弱，提前进入了衰老阶段。这与叶片上出现的褪绿斑驳症状相符合。比较 2 个供试水稻品种处理组和对照组的叶片双向电泳图谱可以看出，在感染 RSV 以后，武育粳 3 号中质体蓝素的表达量相对 KT95-418 中减少得更多，与 2 个水稻品种叶片上症状的严重程度一致。

3.2.3 PS Ⅱ放氧复合体

PS Ⅱ的生理功能是吸收光能，进行光化学反应，同时产生强的氧化剂，发生光水解作用（photohydrolysis），释放出氧气，并把水中的电子传递给质体醌。已有研究表明，PS Ⅱ对干旱、盐胁迫、土壤低温等胁迫环境非常敏感。Shimizu 等对水稻感染 RDV 后的基因转录谱进行分析表明，一个 33kDa PS Ⅱ外周蛋白基因在水稻感染 RDV 后转录量减少（Shimizu et al., 2007）。本研究中 23kDa PS Ⅱ外周蛋白在感染 RSV 后表达量减少，但其作用机制和与病害症状发生之间的关系还需要进一步的研究证实。

3.3 氧化还原酶类

伴随着光合成过程不可避免地会产生一些活性氧分子（reactive oxygen species, ROS）。ROS 控制植物代谢过程中的多个途径，同时对光合作用的结构产生直接破坏。ROS 平衡的调节网络非常复杂。拟南芥中参与 ROS 调控的基因至少有 152 个之多，包括多种 ROS 的生成酶类或清除酶类，是一个非常冗余和动态的调控网络（Mittler et al., 2004）。

本研究发现 5 个氧化还原酶在感染 RSV 后的 2 个水稻品种叶片中表达降低。这一现象反映了水稻在感染 RSV 以后细胞内的氧化还原平衡状态的改变。Allan 等研究表明，在抗病品种和感病品种中，TMV 或离体的 TMVCP 都能够刺激胞内 ROS 的产生，产生这一刺激过程的最初识别反应发生在寄主细胞的质膜外，并且必须有 TMVCP 的参与（Allan et al., 2001）。RSV 对寄主细胞内 ROS 的诱导作用是否也需要其外壳蛋白的参与，特异性识别病毒蛋白的寄主组分有哪些，都是值得进一步研究的问题。

另外，ROS 的过量生成可能直接破坏 PS Ⅱ或 PSI 等重要光合作用结构，导致光合成状态的氧化损伤，使光合效率降低或发生光抑制作用。同时，活性氧分子控制植物代谢中的多个途径，环境改变导致光化学能量吸收和植物代谢中能量利用的失衡将进一步加大其对光合成状态的损害。本研究证实，许多光合作用相关蛋白的表达在感染 RSV 后的水稻中发生了改变，这些改变是否与细胞的氧化还原失衡有关，它们之间的协调关系如何，是后续研究需要回答的重要问题。

3.4 核糖体蛋白

核糖体是细胞生长和分化中多肽链合成的场所，由 40S 和 60S 两个核糖核蛋白体亚基组成。本研究中鉴定得到的 2 个核糖体蛋白都属于核糖体大亚基 12kDa 蛋白家族（ribosomal protein L12P family）。

34 号蛋白质点所对应的 Os08g0116500 与玉米 60S 酸性核糖体蛋白 P1（P52855）类似，是一个酸性磷酸化蛋白（acidic ribosomal protein, ARP）, 分子量约 11.4 ～ 12.2kDa，等电点为 4.1 ～ 4.3，在体内发生磷酸化作用（Szick et al., 1998）。P1 蛋白的磷酸化修饰是植物面临胁迫环境时的一种适应方式。本研究中测得 34 号蛋白质点的分子量约为 18kDa，与 Chang 等对拟南芥 80S 核糖体进行双向电泳的图谱中 L12 蛋白的位置相近（Chang et al., 2005）。37 号蛋白质点是定位于叶绿体的 50S 核糖体蛋白 L12 前体，分子量约为 20kDa，等电点约 5，在高等植物中负责与各种翻译因子结合并催化 GTP 水解。在 RDV 侵染后的基因表达谱分析中，12 个核糖体蛋白基因的表达被抑制（Shimizu et al., 2007）。双向电泳图谱显示，在感染 RSV 后这 2 个蛋白的表达量降低，P1 蛋白可能发生了磷酸化修饰。这一结果暗示了病毒侵染对寄主基因表达调控的另一种方式，即可能通过调节核糖体蛋白的表达、互作和翻译后修饰对其他 mRNA 的翻译产生选择性作用。这一过程在玉米的发芽和淹水胁迫中已较为明确，但在寄主对病毒的响应中如何作用，目前尚没有充分的实验证据。如果该途径在病毒与寄主的互作中的确存在，是否还有其他的核糖体蛋白参与了这个调节，它们的调节机制是怎样的？这些都是很有研究价值的新课题。

3.5 其他蛋白

除前述几类蛋白外，本研究还鉴定得到了几种代谢酶类和一个推定的小泛素相关修饰物（small ubiquitin-like modifier, SUMO）。SUMO 是 20 世纪 90 年代发现的一类泛素相关的修饰蛋白，可以与各种胞内靶标蛋白结合，调节蛋白的相互作用、定位、活性和稳定性。在植物开花时间的调控、耐热性、依赖水杨酸的病害防御反应和植物对冷害、干旱、ABA 胁迫的响应等过程中都涉及 SUMO 化作用的调节。酵母双杂交实验表明，SUMO 可能通过乙烯产生的信号诱导植物防御反应（Hanania et al., 2010）。

由于 SUMO 的作用底物非常广泛，SUMO 化途径能够调控多种基因的表达，对细胞代谢影响巨大。本研究中对应的 17 号蛋白质点与区域 A 比较接近，其在水稻感染 RSV 后的信号传导途径中如何作用尚不清楚。研究表明，多种病毒如细胞巨化病毒（Cytomegalovirus）、单纯疱疹病毒（Herpes simplex virus, HSV）和腺病毒（Adenoviruses）等在与寄主的互作中涉及 SUMO 化途径（Müller and Dejean, 1999; Parkinson and Everett, 2000; Ledl et al., 2005），但其作用机制尚不明确。CELO 腺病毒 Gam1 蛋白的体内表达能够使 SUMO 化作用中的连接酶 E1、E2 失活，阻断寄主的 SUMO 化途径（Boggio et al., 2004）。RSV 病毒是否通过类似的作用途径干扰寄主的 SUMO 化途径尚需进一步验证。

4 结论

本试验比较了 2 个粳稻品种在感染 RSV 后的蛋白表达谱差异。在双向电泳图谱上显示了 RSVSP 和 RSVNS2 蛋白的相对分子质量、等电点及表达情况。用相关软件对双向电泳图谱进行分析后，经质谱鉴定获得了 25 个目标蛋白质点的注释信息。通过分析这些蛋白在感染 RSV 后的水稻中的可能作用，我们认为感染 RSV 后水稻寄主的代谢变化主要体现在光合作用、细胞氧化还原状态和离子平衡状态的改变及蛋白的合成与翻译后修饰、加工、转运等方面。在后续的研究中，我们将对这些差异表达的蛋白做进一步的功能分析，明确其在水稻条纹叶枯病症状出现过程中的具体作用，结合已开展的基因转录谱研究，最终构建水稻在感染 RSV 后的系统生物学模型，分析水稻对 RSV 的响应机制。

参考文献

[1] 明艳林 . 水稻条纹病毒在水稻原生质体内的复制与表达 [D]. 福州 : 福建农林大学 , 2001.

[2] 夏其昌 , 曾嵘 . 蛋白质化学与蛋白质组学 [M]. 北京 : 科学出版社 , 2004.

[3] AGRAWAL G K, RAKWAL R, YONEKURA M, et al. Proteome analysis of differentially displayed proteins as a tool for

investigating ozone stress in rice (*Oryza sativa* L) seedlings[J]. PROTEOMICS, 2002, 2: 947-959.

[4] ALLAN A C, LAPIDOT M, CULVER J N, et al. An early tobacco mosaic virus-induced oxidative burst in tobacco indicates extracellular perception of the virus coat protein[J]. Plant Physiol, 2002, 126: 97-108.

[5] BOGGIO R, COLOMBO R, HAY R T, et al. A mechanism for inhibiting the SUMO pathway[J]. Molecular Cell, 2004, 16: 549-561.

[6] BRAULT V, TANGUY S, REINBOLD C, et al. Transcriptomic analysis of intestinal genes following acquisition of pea enation mosaic virus by the pea aphid Acyrthosiphon pisum[J]. Journal of General Virology, 2010, 91: 802-808.

[7] CATONI M, MIOZZI L, FIORILLI V, et al. Comparative analysis of expression profiles in shoots and roots of tomato systemically infected by Tomato spotted wilt virus reveals organ-specific transcriptional responses[J]. Molecular Plant Microbe Interactions, 2009, 22: 1504-1513.

[8] CHANG F, SZICK-MIRANDA K, PAN S, et al. Proteomic characterization of evolutionarily conserved and variable proteins of Arabidopsis cytosolic ribosomes[J]. Plant Physiology, 2005, 137: 848-862.

[9] CHEN F, YUAN Y, LI Q, et al. Proteomic analysis of rice plasma membrane reveals proteins involved in early defense response to bacterial blight[J]. PROTEOMICS, 2007, 7: 1529-1539.

[10] DANIEL B, NAGY P D. Ubiquitination of tombusvirus p33 replication protein plays a role in virus replication and binding to the host Vps23p ESCRT protein[J]. Virology, 2010, 397: 358-368.

[11] LIANG D, QU Z, MA X, et al. Detection and localization of Rice stripe virus gene products in vivo[J]. Virus Genes, 2005, 31: 211-221.

[12] HANANIA U, FURMAN-MATARASSO N, RON M, et al. Isolation of a novel SUMO protein from tomato that suppresses EIX-induced cell death[J]. Plant Journal, 2010, 19: 533-541.

[13] HAYAKAWA T, ZHU Y, ITOH K, et al. Genetically engineered rice resistant to rice stripe virus, an insect-transmitted virus[J]. Proceedings of the National Academy of Sciences of the United States of America, 1992, 89: 9865-9869.

[14] KIM S T, KIM S G, HWANG D H, et al. Proteomic analysis of pathogen-responsive proteins from rice leaves induced by rice blast fungus, Magnaporthe grisea[J]. Proteomics, 2004, 4: 3569-3578.

[15] LEDL A, SCHMIDT D, MÜLLER S. Viral oncoproteins E1A and E7 and cellular LxCxE proteins repress SUMO modification of the retinoblastoma tumor suppressor[J]. Oncogene, 2005, 24: 3810.

[16] LIANMING L, ZHENGUO D, MEILING Q, et al. Pc4, a putative movement protein of Rice stripe virus, interacts with a type I DnaJ protein and a small Hsp of rice[J]. Virus Genes, 2009, 38: 320-327.

[17] LIN Q, LIN H, WU Z, et al. Accumulations of coat protein and disease-specific protein of rice stripe virus in its host[J]. Journal of Fujian Agricultural University, 1998, 27: 322-326.

[18] LIU L H, WU Z J, LIN Q Y, et al. Cytopathological observation of rice stripe[J]. Acta Phytopathologica Sinica, 2000.

[19] MAHMOOD T, KAKISHIMA M, KOMATSU S. Proteomic analysis of jasmonic acid-regulated proteins in rice leaf blades[J]. Protein & Peptide Letters, 2007, 14: 311-319.

[20] MITTLER R, VANDERAUWERA S, GOLLERY M, et al. Reactive oxygen gene network of plants[J]. Trends in Plant Science, 2004, 9: 490-498.

[21] MÜLLER S, DEJEAN A. Viral immediate-early proteins abrogate the modification by SUMO-1 of PML and Sp100 proteins, correlating with nuclear body disruption[J]. Journal of Virology, 1999, 73: 5137-5143.

[22] PARKINSON J, EVERETT R D. Alphaherpesvirus proteins related to herpes simplex virus type 1 ICP0 affect cellular structures and proteins[J]. Journal of Virology, 2000, 74: 10006-10017.

[23] PINEDA M, SAJNANI C, BARÓN M. Changes induced by the Pepper mild mottle tobamovirus on the chloroplast proteome of *Nicotiana benthamiana*[J]. Photosynthesis Research, 2010, 103: 31.

[24] QU Z, SHEN D, XU Y, et al. Western blotting of RStV gene products in rice and insects[J]. Yi Chuan Xue Bao, 1999, 26: 512-517.

[25] SHIMIZU T, SATOH K, KIKUCHI S, et al. The repression of cell wall-and plastid-related genes and the induction of defense-

related genes in rice plants infected with Rice dwarf virus[J].Mol Plant Microbe In, 2007, 20: 247-254.

[26] SZICK K, SPRINGER M, BAILEY-SERRES J. Evolutionary analyses of the 12-kDa acidic ribosomal P-proteins reveal a distinct protein of higher plant ribosomes[J]. Proceedings of the National Academy of Sciences of the United States of America, 1998, 95: 2378-2383.

[27] TAKAHASHI M, GOTO C, ISHIKAWA K, et al. Rice stripe virus 23.9 K protein aggregates and forms inclusion bodies in cultured insect cells and virus-infected plant cells[J]. Archives of Virology, 2003, 148: 2167-2179.

[28] VENTELON-DEBOUT M, DELALANDE F, BRIZARD J P, et al. Proteome analysis of cultivar-specific deregulations of Oryza sativa indica and O sativa japonica cellular suspensions undergoing rice yellow mottle virus infection[J]. Proteomics, 2004, 4: 216-225.

[29] ZHANG X, XIE L-Y, LIN Q, et al. Transcriptional profiling in rice seedlings infected by rice stripe virus[J]. Acta Laser Biology Sinica, 2008, 5: 620-629.

[30] ZHAO C, WANG J, CAO M, et al. Proteomic changes in rice leaves during development of field-grown rice plants[J]. Proteomics, 2005, 5: 961-972.

[31] ZHOU Z J, LIN Q Y, XIE L H, et al. Studies on rice stripe in China. IV Pathological changes of rice leaf cell[J]. Journal of Fujian Agriculture & Forestry University, 1992, 157-162.

p2 of *Rice stripe virus* (RSV) interacts with OsSGS3 and is a silencing suppressor

Zhenguo Du, Donglai Xiao, Jianguo Wu, Dongsheng Jia, Zhengjie Yuan, Ying Liu, Liuyang Hu, Zhao Han, Taiyun Wei, Qiying Lin, Zujian Wu, Lianhui Xie

（Institute of Plant Virology, Fujian Agriculture and Forestry University, Key Laboratory of Plant Virology of Fujian Province, Fuzhou, Fujian, 350002, China）

Abstract: A rice cDNA library was screened by a galactosidase 4 (Gal4)-based yeast two-hybrid system with *Rice stripe virus* (RSV) p2 as bait. The results revealed that RSV p2 interacted with a rice protein exhibiting a high degree of identity with *Arabidopsis thaliana* suppressor of gene silencing 3 (AtSGS3). The interaction was confirmed by bimolecular fluorescence complementation assay. SGS3 has been shown to be involved in sense transgene-induced RNA silencing and in the biogenesis of trans-acting small interfering RNAs (ta-siRNAs), possibly functioning as a cofactor of RNA-dependent RNA polymerase 6(RDR6).Giventhe intimate relationships between virus and RNA silencing, further experiments were conducted to show that p2 was a silencing suppressor. In addition, p2 enhanced the accumulation and pathogenicity of *Potato virus X* in *Nicotiana benthamiana*. Five genes that have been demonstrated to be targets of TAS3-derived ta-siRNAs were up-regulated in RSV-infected rice. The implications of these findings are discussed.

RNA silencing is a gene regulation mechanism common to virtually all eukaryotes.In RNA silencing,Dicer or Dicer-like ribonucleases (DCLs) recognize and cleave double-stranded RNAs (dsRNAs) into small interfering RNAs (siRNAs). Then, the siRNAs are recruited by an ARGONATE (AGO) protein to form an effector complex called the RNA-induced silencing complex(RISC),which can regulate gene expression at the transcriptional, posttranscriptional and translational levels using the incorporated siRNAs as guides(Carthew and Sontheimer, 2009).Many eukaryotes, including plants, have evolved to use RNA silencing as a defensive mechanism against viral infection (Ding, 2010; Ding and Voinnet, 2007). To counter this defence, viruses encode specific proteins called silencing suppressors (VSRs; Ding and Voinnet, 2007; Li and Ding, 2006). The most common target of VSRs seems to be virus-derived siRNA (Ding and Voinnet, 2007; Li and Ding,2006).As exemplified by p19 of Tombuviruses,many VSRs selectively bind siRNA duplexes.By doing this,they sequester virus-specific siRNAs (vsiRNAs) and prevent the formation of functional RISCs (Lakatos et al., 2006). However, VSRs can also use protein components of RNA silencing as targets. For example, diverse VSRs have been shown to interfere with the function or reduce the accumulation of antiviral AGOs through physical interactions (for a review, see Wu et al., 2010). Regardless of all of these possibilities,VSRs are believed to be important in viral accumulation, spread and pathogenicity (Díaz-Pendón and Ding, 2008).

In some instances,RNA silencing can be amplified in plants by cellular enzymes called RNA-dependent RNA polymerases (RDRs; Wassenegger and Krczal, 2006). In silencing amplification, RDRs use the targets of siRNAs as templates and convert them into dsRNAs, the cleavage of which by specific DCLs gives rise to

Molecular Plant Pathology. 2011, 12(8):808–814

secondary siRNAs (Wassenegger and Krczal, 2006). Although the mechanism by which RDRs recognize viral RNAs remains unclear, the importance of RDRs in plant antiviral silencing has been definitively demonstrated recently by two groups using VSR-deficient *Turnip mosaic virus* (TuMV) and *Cucumber mosaic virus* (CMV), respectively (Garcia-Ruiz et al., 2010;Wang et al., 2010). Not surprisingly, VSRs can strongly or even selectively block secondary siRNA production (Csorba et al., 2007; Mlotshwa et al., 2008; Moissiard et al., 2007; Shivaprasad et al., 2008). However, the molecular mechanisms underlying the inhibition of secondary siRNA production by plant viruses remain elusive.

Rice stripe virus (RSV) is the type species of the genus *Tenuivirus*, which has not been assigned to any family. It infects rice and causes large yield reductions in rice production in some countries of East Asia. The genome of RSV comprises four RNAs, named RNA1–RNA4, in decreasing order of their molecular weight (Falk and Tsai, 1998; Ramírez and Haenni, 1994). RNA1 is of negative sense and encodes a putative protein with a molecular weight of 337 kDa, which was considered to be part of the RDR associated with the RSV filamentous ribonucleoprotein (RNP) (Toriyama et al., 1994). RNA2–4 are ambisense, each containing two open reading frames (ORFs), one in the 5′ half of viral RNA (vRNA, the proteins they encode named p2–p4) and the other in the 5′ half of the viral complementary RNA (vcRNA, the proteins they encode named pc2–pc4; Kakutani et al., 1990, 1991; Takahashi et al., 1993; Zhu et al., 1991). Owing to its reluctance to traditional virological methods, such as infectious cloning, we know little about the functions of the seven proteins encoded by RSV at present. To obtain some insight into the functions of RSV-encoded proteins, we used the yeast two-hybrid system to investigate all the potential interactions between RSV-encoded proteins and host factors. In this article, we report our identification of the interaction between RSV p2 and the rice homologue of suppressor of gene silencing 3 (SGS3), a possible cofactor of RDRs (Dalmay et al., 2000; Mourrain et al., 2000).On the basis of this finding, we carried out further experiments to identify p2 as a silencing suppressor of RSV.

A rice cDNA library was screened by a galactosidase 4 (Gal4)-based yeast two-hybrid system using RSV p2 as bait.A number of positive colonies were identified among the approximately 3.6×10^6 colonies screened. Sequencing and BLAST analysis showed that the cDNA insert of one of the positive colonies corresponded to a rice protein exhibiting a high degree of identity with *Arabidopsis thaliana* SGS3 (AtSGS3) (Fig.1). The full-length ORF of the rice protein,designated OsSGS3 (Os12g09580),was then cloned into the vector pGBKT7, creating pGBKT7-OsSGS3 (for experimental procedures see Appendix S1).As shown in Fig. 2A, yeast cells co-transformed with pGBKT7-OsSGS3 and pGADT7-p2 grew on SD medium lacking adenine (Ade), histidine (His), leucine (Leu) and tryptophan (Trp) (SD/–Ade/–His/–Leu/–Trp), as did those co-transformed with pGADT7-T and pGBKT7-53,which were used as positive controls. By contrast, the yeast co-transformed with pGBKT7/pGADT7, pGBKT7-OsSGS3/pGADT7 or pGBKT7/ pGADT7-p2 failed to grow on SD/–Ade/–His/–Leu/–Trp, although they grew well on SD medium lacking His and Trp (Fig.2A and data not shown).

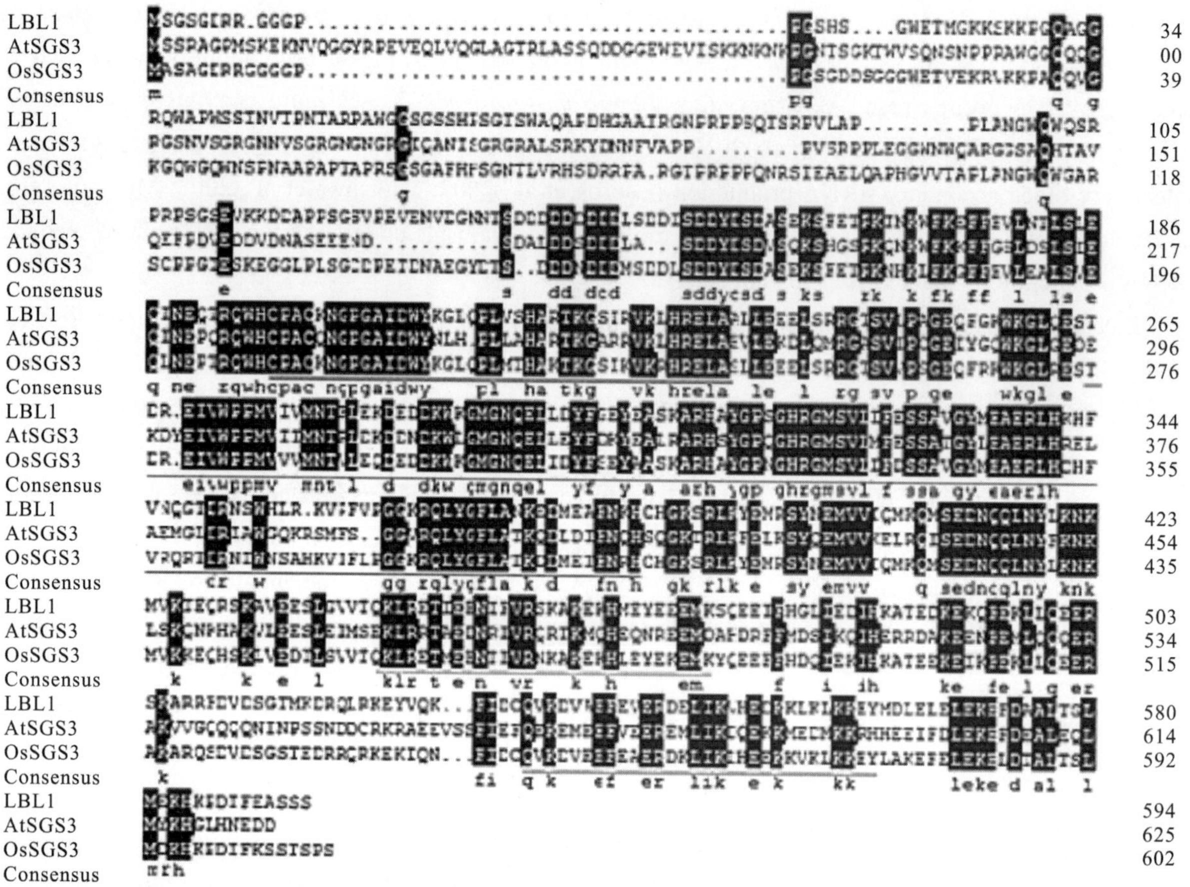

Fig.1 Alignment of rice and Arabidopsis suppressor of gene silencing 3 (OsSGS3 and AtSGS3) and maize LEAFBLADELESS1 (LBL1). Regions of identity and similarity are indicated by black and blue, respectively; gaps introduced for alignment are indicated by dots. The zinc finger XS domain, true XS domain and two coiled-coil domains are denoted by dark blue, red and green horizontal bars, respectively, under their sequences. Alignment was performed using the C_{LUSTAL}W algorithm

A bimolecular fluorescence complementation (BiFC) assay was carried out to confirm the interaction of p2 and OsSGS3 in living plant cells (Walter et al., 2010).To do this, the full-length ORF of p2 was cloned into the vector pSPYCE and that of OsSGS3 into pSPYNE. Transformed *Agrobacterium tumefaciens* EHA105 carrying each of these constructs were mixed and infiltrated into leaves of *Nicotiana benthamiana*. Strong yellow fluorescent protein (YFP) fluorescence in *N. benthamiana* leaf epidermal cells was detected as early as 2d post-infiltration (dpi) (Fig.3A–C). Similar results were obtained when p2 was fused with the N-terminal fragment of YFP and OsSGS3 was fused with the C-terminal fragment of YFP. By contrast, fluorescence was not detected in leaf cells co-infiltrated with pSPYNE/pSPYCE, pSPYCE-p2/pSPYNE or pSPYCE/pSPYNE-OsSGS3 (data not shown). As shown in Fig. 3, the interaction between p2 and OsSGS3 occurred in both the cytoplasm and the nucleus. In the cytoplasm, the p2–OsSGS3 complex formed distinct granules, which was consistent with the cellular localization patterns of AtSGS3 and OsSGS3 (Kumakura et al., 2009 and Fig.3H).

Three deletion mutants of OsSGS3 were constructed: OsSGS3-1 (amino acid residues 1–250), OsSGS3-2 (amino acid residues 250–405) and OsSGS3-3 (amino acid residues 405–609). Each of the three deletion mutants contained one type of the three conserved protein domains present in OsSGS3: a zinc finger XS domain,a true XS domain and two coiled-coil domains, respectively (Fig.1).Yeast two-hybrid experiments revealed that all three OsSGS3 fragments could interact with RSV p2 (Fig. 2B). This indicated that there were multiple p2-interacting sites on OsSGS3.

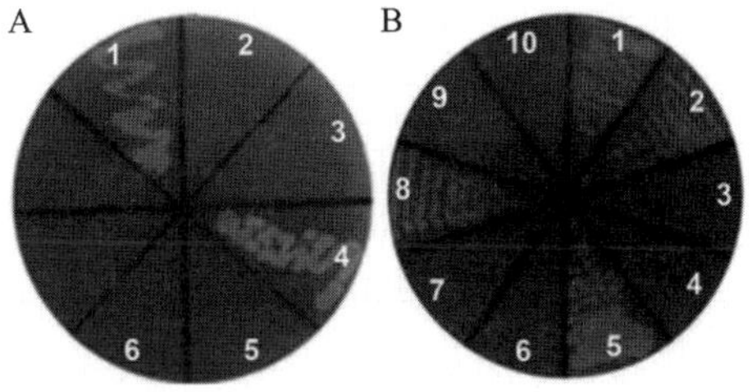

Fig.2 (A) Interaction between full-length rice suppressor of gene silencing 3 (OsSGS3) and p2

1. pGADT7-p2/ pGBKT7-OsSGS3; 2. pGADT7-p2/pGBKT7; 3. pGBKT7-OsSGS3/pGADT7; 4. pGADT7-T/pGBKT7-53 (positive control); 5. pGADT7-T/ pGBKT7-Lam (negative control); 6. pGADT7/pGBKT7.

(B) Interaction between three segments of OsSGS3 and p2

1. pGADT7-OsSGS3-3/ pGBKT7-p2; 2. pGADT7-OsSGS3-2/ pGBKT7-p2; 3. pGADT7-OsSGS3-2/ pGBKT7; 4. pGADT7-T /pGBKT7-Lam (negative control); 5. pGADT7-T/ pGBKT7-53 (positive control); 6. pGADT7/ pGBKT7; 7. pGADT7/ pGBKT7-p2; 8. pGADT7-OsSGS3-1/ pGBKT7-p2; 9. pGADT7-OsSGS3-1/ pGBKT7; 10. pGADT7-OsSGS3-3/ pGBKT7

AtSGS3 is a cofactor of RDR6 and has been implicated in antiviral silencing (Dalmay et al., 2000; Fukunaga and Doudna, 2009; Glick et al., 2007; Mourrain et al., 2000; Muangsan et al., 2010). BLAST analysis showed that OsSGS3 was the rice protein that exhibited the highest homology to AtSGS3. Given the intimate relationships between virus and RNA silencing, we speculated that RSV p2 might be a silencing suppressor (Ding, 2010; Ding and Voinnet, 2007). The finding that there were multiple p2-interacting sites on OsSGS3 and the fact that RSV can infect *N. benthamiana* encouraged us to test this possibility using a popular method for VSR identification: *Agrobacterium* infiltration assay.To this end, p2 and Dp2 were cloned into the binary vector pPZP212 and the resulting plasmids, which were named 35S-p2 and 35S-Δp2, respectively, were introduced into *Agrobacterium*. In Dp2, the nucleotide A from the translation start codon AUG was deleted.Each of the bacterial strains was mixed with a strain containing the green fluorescent protein (GFP) transgene (35S-GFP) at the ratio of 3:1 and then co-infiltrated into leaves of *N. benthamiana* line 16c as described previously (Hamilton et al., 2002). In 21 of the 24 leaf patches infiltrated with 35S-GFP plus 35S-p2, obvious fluorescence persisted for at least 5d, although the fluorescence intensity was much weaker in these leaf patches than in those co-infiltrated with 35S-GFP plus 35S-2b, which were used as positive controls, at 5dpidpi (Fig. 4, left). By contrast, in leaf patches infiltrated with 35S-GFP plus 35S-Δp2 or 35S-GFP alone, strong fluorescence first appeared at 2dpi, but decreased at 3dpi, and becamehardly detectable at 5dpi (Fig. 4, left). These data indicate that p2 may be a silencing suppressor. To confirm this, Northern blotting was conducted to detect steady-state levels of GFP mRNA and GFP-specific siRNAs in infiltrated leaf patches at 5dpi. GFP mRNA was readily detectable in leaf patches infiltrated with 35S-GFP plus 35S-p2,as well as those infiltrated with 35S-GFP plus 35S-2b. However, the accumulation levels of GFP mRNA in leaf patches infiltrated with 35S–GFP plus 35S-Δp2 or 35S–GFP alone were very low or hardly detectable. In each treatment, the concentration of GFP-specific siRNAs was inversely correlated with that of GFP mRNA (Fig.4, right).

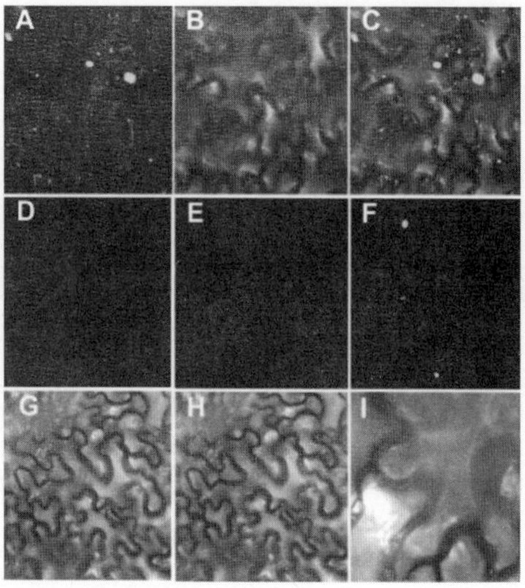

Fig.3 Bimolecular fluorescence complementation (BiFC) assay showing the interaction between p2 and rice suppressor of gene silencing 3 (OsSGS3) and the cellular localization of OsSGS3

Leaves of *Nicotiana benthamiana* were infiltrated by *Agrobacterium* tumefaciens EHA105 harbouring pSPYCE-p2/pSPYNE-OsSGS3 (A–C), pSPYCE-p2 (D, G), pSPYNE-OsSGS3 (E, H) or OsSGS3-YFP (F, I). (A, D, E, H) Yellow fluorescent protein (YFP) fluorescence. (B, F, G, I) Bright field. (C) Overlay of (A) and (B)

Many VSRs have been shown to be able to enhance the virulence of heterologous viruses (for an example, see Xiong et al., 2009). To test whether p2 had such an ability, we cloned p2 and Dp2 into a Potato virus X (PVX) vector, creating PVX-p2 and PVX-Δp2, respectively. Symptoms on newly emerged leaves of *N. benthamiana* infected with these PVX vectors were first observed at 6dpi. At this time, no symptom differences were observed between plants inoculated with PVX, PVX-p2 or PVX-Δp2 (data not shown). However, plants infected with PVX-p2 developed more severe symptoms at 9dpi, which were manifested as more severe mosaics and curling of the leaves(Fig. 5A). At later times, the symptoms on the upper leaves of plants inoculated with PVX or PVX-Δp2 became very light or hardly detectable.However,the symptoms caused by PVX-p2 were sustained throughout the life of the plants (Fig. 5A). Northern blot analysis revealed that the accumulation levels of PVX RNA were much higher in plants infected with PVX-p2 than in those infected with PVX or PVX-Δp2 at 22dpi (Fig. 5B). Thus, p2 enhanced the accumulation and pathogenicity of PVX, possibly as a result of a synergistic reaction.

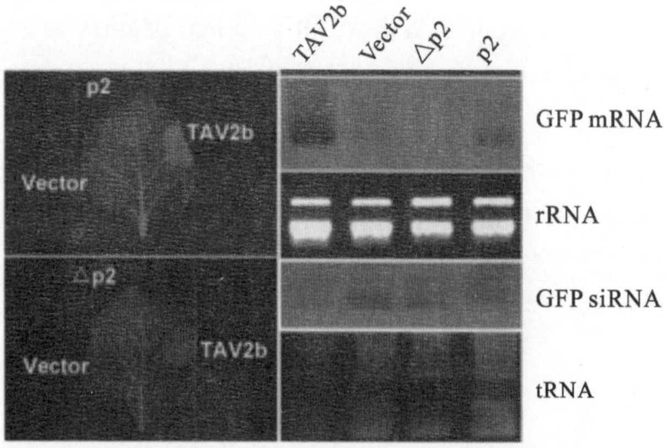

Fig.4 p2 is a silencing suppressor of *Rice stripe virus* (RSV)

Left: *Nicotiana benthamiana line* 16c plants were co-infiltrated with *Agrobacterium* mixtures carrying 35S-green fluorescent protein (35S-GFP) and the individual constructs indicated in each image. GFP fluorescence was viewed under long-wavelength UV light at 5d post-infiltration (dpi). Right: Northern blot analysis of the steady-state levels of GFP mRNA and small interfering RNA (siRNA) extracted from the different infiltrated patches shown in (A). 28S rRNA and tRNA were used as loading controls for the detection of GFP mRNA and GFP siRNA, respectively

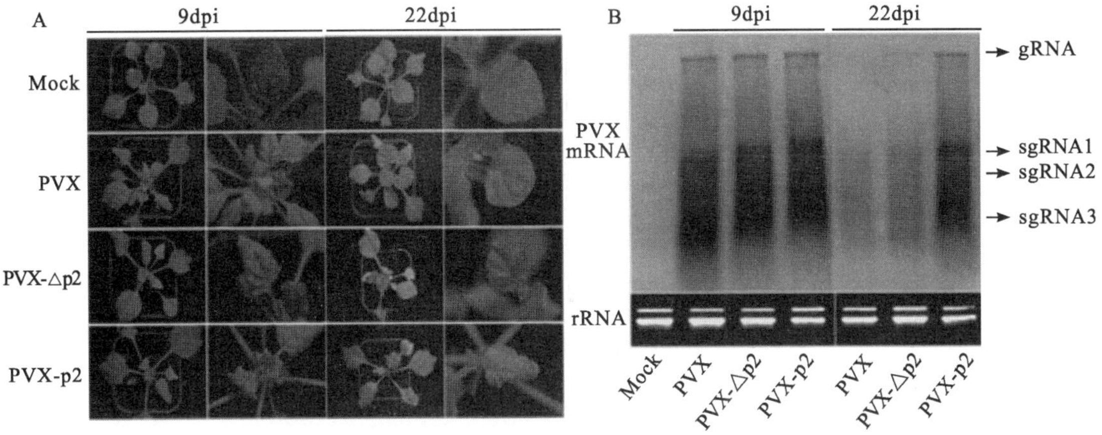

Fig.5 (A) p2 enhances the pathogenicity of chimeric *Potato virus X* (PVX). (B) RNA gel blot analysis of the accumulation of PVX genomic (gRNA) and subgenomic (sgRNA1 to sgRNA3) mRNAs at 9 and 22d posy-infiltration (dpi)

The bottom gel shows rRNA with ethidium bromide staining as a loading control

It should be mentioned that,in a previous study identifying p3 of RSV as a VSR, Xiong et al. (2009) reported that p2 had no silencing suppressor activity. Currently, we cannot explain this disparity.However,the most likely possibility is that the silencing suppressor activity of p2 is much weaker than that of p3. Moreover,in that study,equal amounts of *Agrobacterium* suspensions expressing p2 or a GFP transgene were used in the co-infiltration assay.

In addition to antiviral defence, SGS3 is also involved in the RDR6-mediated biogenesis of *trans-acting* siRNAs (ta-siRNAs), which play an important role in plant development (Peragine et al., 2004). In a primary attempt to explore the possible biological implications of the interaction between p2 and OsSGS3, we detected the expression patterns of five genes targeted by ta-siRNAs derived from *TAS3* (Liu et al., 2007). All five genes encode auxin-responsive factors (ARFs). Four have been shown to be up-regulated in transgenic rice expressing *Rice yellow mottle virus* P1, possibly resulting from the reduced accumulation of *TAS3*-derived ta-siRNAs (Lacombe et al., 2010). The results indicated that the overall accumulation levels of the five genes increased by 50%–500% in RSV-infected rice (Fig.6). In addition, the expression of OsSGS3 itself also increased. However, the expression of *SHL2* and *SHO2*, which are orthologues of Arabidopsis RDR6 and AGO7, respectively (Nagasaki et al., 2007), was not altered (data not shown). The upregulation of genes encoding ARFs might not be a general response of rice to RSV infection, because two other *ARFs*, *Os02g04810* and *06g48950*, showed decreased expression in RSV-infected rice.

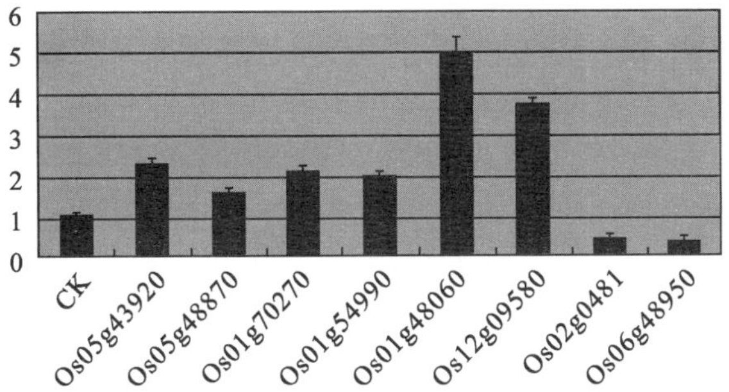

Fig.6 Quantitative reverse transcription-polymerase chain reaction (RT-PCR) of the expression of five putative target genes of rice TAS3 trans-acting siRNA (ta-siRNA) in *Rice stripe virus* (RSV)-infected rice

The determination of the expression levels of the assayed genes was carried out in triplicate and normalized according to the value of an 18S RNA. The expression of the wild-type was set as unity. Error bars indicate standard deviations. The *y* axis denotes the expression level of a particular gene in RSV-infected rice compared with that in healthy rice

Many viruses have been shown to encode more than one VSR (Fabozzi et al., 2011; Lu et al., 2004; Vanitharani et al., 2004). RSV is a virus that can multiply both in its insect vectors and in its plant hosts. As mentioned above, p3 of RSV has been shown to be a VSR (Xiong et al., 2009). Most probably, p3 suppresses RNA silencing by targeting long and/or small dsRNAs (Shen et al., 2010). In this study, we found that RSV p2 interacted with OsSGS3, and identified p2 as a silencing suppressor. We hypothesize that p2, which is a small protein having no RNA-binding activities, suppresses RNA silencing by inactivating the RDR– SGS3 pathway (Liang et al., 2005). If this is true, it would be interesting to note that RSV has two VSRs, one targeting RNA, the most conserved element in RNA silencing across kingdoms, and the other targeting a protein involved in silencing amplification, which is specific to plants. In addition to the emerging notion that RDRs are important in antiviral RNA silencing, our finding is consistent with two recent reports regarding RSV (Qu, 2010; Shimizu et al., 2011; Yan et al., 2010). In rice expressing a p2-specific RNAi construct, the appearance of symptoms caused by RSV was significantly delayed, suggesting that the p2 protein might be necessary for the proliferation of RSV and might play a role at later stages of viral infection (Shimizu et al., 2011).In the same report, the authors found that transgenic rice plants that harboured IR constructs specific for the gene for p3 only showed partial resistance to RSV, which was unexpected, to some extent, if p3 was the only VSR of RSV. In addition, (Yan et al, 2010) found that the proportion of RSV-derived siRNA within the total siRNA reads from RSV-infected rice was very low. This can be partially explained by our hypothesis that RSV encodes a protein targeting silencing amplification. Interestingly, V2 of *Tomato yellow leaf curl geminivirus* (TYLCV) has been shown to interact with SGS3 of tomato (Glick et al., 2007).This suggests that SGS3 may be a common target of diverse plant viruses.

The alteration of the ta-siRNA pathway has been reported in many plant–virus combinations and has been suggested to be involved in symptom development (Lacombe et al., 2010; Meng et al., 2008; Shivaprasad et al., 2008). In this study, we found that the ta-siRNA pathway may be affected in RSV-infected rice. This provides indirect evidence suggesting that the ta-siRNA pathway is affected by RSV infection.Transgenic rice expressing RSV p2 is being produced in our laboratory to investigate whether p2 is attributable to the alteration and to investigate the role of p2 in pathogenicity.

Acknowledgements

We thank Professor David Baulcombe (Sainsbury Laboratory, Norwich, UK) for providing 16c seeds and PVX vectors, K. Harter (Botanisches Institut, Universitätzu Köln, Germany) and J. Kudla (InstitutfürBotanik und Botanischer Garten, Molekulare Entwichlungsbiologie der Pflanzen, Universität Münster, Germany) for the vectors used in the BiFC assay,and Dr Jason G. Powers (North Carolina State University, Raleigh, NC, USA) for providing plasmid pPZP212 vector. This work was supported by the Major Project of Chinese National Programs for Fundamental Research and Development (grant no. 2010CB126203), theNational Natural Science Foundation of China (grant no. 30770090), the National Transgenic Major Program (grant no. 2009ZX08009-044B, 2009ZX08001-018B) and the Specialized Research Fund for the Ministry of Agriculture (nyhyzx 07-051).

References

[1] CARTHEW R W, SONTHEIMER E J. Origins and mechanisms of miRNAs and siRNAs[J]. Cell, 2009, 136: 642-655.

[2] CSORBA T, BOVI A, DALMAY T, et al. The p122 subunit of *Tobacco mosaic virus* replicase is a potent silencing suppressor and compromises both small interfering RNA- and microRNA-mediated pathways[J]. Journal of Virology, 2007, 81: 11768-11780.

[3] DALMAY T, HAMILTON A, RUDD S, et al. An RNA-dependent RNA polymerase gene in Arabidopsis is required for posttranscriptional gene silencing mediated by a transgene but not by a virus[J]. Cell, 2000, 101: 543-553.

[4] DÍAZ-PENDÓN J A, DING S W. Direct and indirect roles of viral suppressors of RNA silencing in pathogenesis[J]. Annual Review of Phytopathology, 2008, 46: 303-326.

[5] DING S W. RNA-based antiviral immunity[J]. Nature Reviews Immunology, 2010, 10: 632-644.

[6] DING S W, VOINNET O. Antiviral immunity directed by small RNAs[J]. Cell, 2007, 130: 413-426.

[7] FABOZZI G, NABEL C S, DOLAN M A, et al. Ebolavirus proteins suppress siRNA effects by direct interaction with the mammalian RNAi pathway[J]. Journal of Virology, 2011, 85: 2512-2523.

[8] FALK B W, TSAI J H. Biology and molecular biology of viruses in the genus *Tenuivirus*[J]. Annual Review of Phytopathology, 1998, 36: 139-163.

[9] FUKUNAGA R, DOUDNA J A. dsRNA with 5′ overhangs contributes to endogenous and antiviral RNA silencing pathways in plants[J]. The EMBO Journal, 2009, 28: 545-555.

[10] GARCIA-RUIZ H, TAKEDA A, CHAPMAN E J, et al. Arabidopsis RNA-dependent RNA polymerases and Dicer-like proteins in antiviral defense and small interfering RNA biogenesis during *Turnip mosaic virus* infection[J]. Plant Cell, 2010, 22: 481-496.

[11] GLICK E, ZRACHYA A, LEVY Y, et al. Interaction with host SGS3 is required for suppression of RNA silencing by *tomato yellow leaf curl virus* V2 protein[J]. Proceedings of the National Academy of Sciences, 2007, 105: 157-161.

[12] HAMILTON A, VOINNET O, CHAPPELL L, et al. Two classes of short interfering RNA in RNA silencing[J]. The EMBO Journal, 2002, 21: 4671-4679.

[13] KAKUTANI T, HAYANO Y, HAYASHI T, et al. Ambisense segment 4 of rice stripe virus: possible evolutionary relationship with *phleboviruses* and *uukuviruses*(Bunyaviridae)[J]. Journal of General Virology, 1990, 71: 1427-1432.

[14] KAKUTANI T, HAYANO Y, HAYASHI T, et al. Ambisense segment 3 of *rice stripe virus*: the first case of a virus containing two ambisense segments[J]. Journal of General Virology, 1991, 72: 465-468.

[15] KUMAKURA N, TAKEDA A, FUJIOKA Y, et al. SGS3 and RDR6 interact and colocalize in cytoplasmic SGS3/RDR6-bodies[J]. FEBS Lett,2009, 583: 1261-1266.

[16] LACOMBE S, BANGRATZ M, VIGNOLS F, et al. The *rice yellow mottle virus* P1 protein exhibits dual functions to suppress and activate gene silencing[J]. Plant Journal, 2010, 61: 371-382.

[17] LAKATOS L, CSORBA T, PANTALEO V, et al. Small RA binding is a common strategy to suppress RNA silencing by several viral suppressors[J]. EMBO Journal, 2006, 25: 2768-2780.

[18] LI F, DING S W. Virus counter defense: diverse strategies for evading the RNA-silencing immunity[J]. Annual Review of Microbiology, 2006, 60: 503-531.

[19] LIANG D, MA X, QU Z et al. Nucleic acid binding property of the gene products of rice stripe virus[J]. Virus Genes, 2005, 31, 203-209.

[20] LIU B, CHEN Z, SONG X, et al. Oryza sativa dicer-like4 reveals a key role for small interfering RNA silencing in plant development[J]. Plant Cell, 2007, 19: 2705-2718.

[21] LU R, FOLIMONOV A, SHINTAKU M, et al. Three distinct suppressors of RNA silencing en-coded by a 20 kb viral RNA genome[J]. Proceedings of The National Academy of Sciences of The United States of America, 2004, 101: 15742-15747.

[22] MENG C, CHEN J, DING S W, et al. Hibiscus chlorotic ringspot virus coat protein inhibits trans-acting small interfering RNA biogenesis in Arabidopsis[J]. Journal of General Virology, 2008, 89: 2349-2358.

[23] MLOTSHWA S, PRUSS G J, PERAGINE A, et al. DICER-LIKE2 plays a primary role in transitive silencing of transgenes in Arabidopsis[J]. PLOS ONE, 2008,3: e1755.

[24] MOISSIARD G, PARIZOTTO E A, HIMBER C, et al. Transitivity in Arabidopsis can be primed, requires the redundant action of the antiviral dicer-like 4 and dicer-like 2,and is compromised by viral-encoded suppressor proteins[J]. RNA, 2007, 13: 1268-1278.

[25] MOURRAIN P, BECLIN C, ELMAYAN T, et al. Arabidopsis SGS2 and SGS3 genes are required for post transcriptional gene silencing and natural virus resistance[J]. Cell, 2000, 101: 533-542.

[26] MUANGSAN N, BECLIN C, VAUCHERET H, et al. *Geminivirus* VIGS of endogenous genes requires SGS2/SDE1 and SGS3 and defines a new branch in the genetic pathway for silencing in plants[J]. Plant Journal, 2010, 38: 1004-1014.

[27] NAGASAKI H, ITOH J, HAYASHI K, et al. The small interfering RNA production path way is required for shoot meristem initiation in rice[J]. Proceedings of the National Academy of Sciences of the United States of America, 2007, 104: 14867-14871.

[28] PERAGINE A, YOSHIKAW, M, WU G, et al. SGS3 and SGS2/SDE1/RDR6 are required for juvenile development and the production of trans-acting siRNAs in Arabidopsis[J]. Genes, Development, 2004, 18: 2368-2379.

[29] QU F. Antiviral role of plant-encoded RNA-dependent RNA polymerases revisited with deep sequencing of small interfering RNAs of virus origin[J]. Molecular plant-microbe interactions:MPMI,2010, 23: 1248-1252.

[30] RAMÍREZ B C, HAENNI A L. Molecular biology of *tenuiviruses*, a remarkable group of plant viruses[J]. Journal of General Virology, 1994, 75: 467-475.

[31] SHEN M, XU Y, JIA R, et al. Size-independent and noncooperative recognition of dsRNA by the rice stripe virus RNA silencing suppressor NS3[J]. Journal of Molecular Biology, 2010, 404: 665-679.

[32] SHIMIZU T, NAKAZONO-NAGAOKA E, UEHARA-ICHIKI T, et al. Targeting specific genes for RNA interference is crucial to the development of strong resistance to *Rice stripe virus*[J]. Plant Biotechnology Journal, 2011, 9: 503-512.

[33] SHIVAPRASAD P V, RAJESWARAN R, BLEVINS T, et al. The CaMV transactivator/viroplasmin interferes with RDR6-dependent trans-acting and secondary siRNA pathways in Arabidopsis[J]. Nucleic Acids Research, 2008, 36: 5896-5909.

[34] TAKAHASHI M, TORIYAMA S, HAMAMATSU C, et al. Nucleotide sequence and possible ambisense coding strategy of rice stripe virus RNAsegment 2[J]. Journal of General Virology, 1993, 74: 769-773.

[35] TORIYAMA S, TAKAHASHI M, SANO Y, et al. Nucleotide sequence of RNA. 1, the largest genomic segment of rice stripe virus, the prototype of the *tenuiviruses*[J]. Journal of General Virology, 1994, 75: 3569-3579.

[36] VANITHARANI R, CHELLAPPAN P, PITA J S, et al. Differential roles of AC2 and AC4 of cassava *geminiviruses* in mediating synergism and suppression of posttranscriptional gene silencing[J]. Journal of Virology, 2004, 78: 9487-9498.

[37] WALTER M, CHABAN C, SCHÜTZE K, et al. Visualization of protein interactions in living plant cells using bimolecular fluorescence complementation[J]. Plant Journal, 2010, 40: 428-438.

[38] WANG X-B, WU Q, ITO T, et al. RNAi-mediated viral immunity requires amplification of virus-derived siRNAs in *Arabidopsis thaliana*[J]. Proceedings of the National Academy of Sciences of the United States of America, 2010, 107: 484-489.

[39] WASSENEGGER M, KRCZAL G. Nomenclature and functions of RNA-directed RNA polymerases[J]. Trends in Plant Science, 2006, 11: 142-151.

[40] WU Q F, WANG X B, DIN S W. Viral suppressors of RNA-based viral immunity: host targets[J]. Cell host & microbe, 2010, 8: 12-15.

[41] XIONG R, WU J, ZHOU Y, et al. Characterization and subcellular localization of an RNA silencing suppressor encoded by *Rice stripe tenuivirus*[J]. Virology, 2009, 387: 29-40.

[42] YAN F, ZHANG H, ADAMS M J, et al. Characterization of siRNAs derived from rice stripe virusin infected rice plants by deep sequencing[J]. Archives of Virology, 2010, 155: 935-940.

[43] ZHU Y, HAYAKAWA T, TORIYAMA S, et al. Complete nucleotide sequence of RNA3 of rice stripe virus: an ambisense coding strategy[J]. Journal of General Virology, 1991, 72: 763-767.

Movement Protein Pns6 of *Rice dwarf phytoreovirus* Has Both ATPase and RNA Binding Activities

Xu Ji[1], Dan Qian[1], Chunhong Wei[1], Gongyin Ye[2], Zhongkai Zhang[3], Zujian Wu[4], Lianhui Xie[4], Yi Li[1]

(1 State Key Laboratory of Protein and Plant Gene Research, Peking-Yale Joint Center for Plant Molecular Genetics and Agrobiotechnology, College of Life Sciences, Peking University, Beijing,People's Republic of China; 2 State Key Laboratory of Rice Biology, Institute of Insect Sciences, Zhejiang University, Hangzhou, People's Republic of China; 3 Biotechnology and Genetic Resources Institute, Yunnan Academy of Agricultural Sciences, Kunming, People's Republic of China; 4 Key Laboratory of Plant Virology of Fujian Province, Institute of Plant Virology, Fujian Agriculture and Forestry University, Fuzhou, People's Republic of China)

Abstract: Cell-to-cell movement is essential for plant viruses to systemically infect host plants. Plant viruses encode movement proteins (MP) to facilitate such movement. Unlike the well-characterized MPs of DNA viruses and single-stranded RNA (ssRNA) viruses, knowledge of the functional mechanisms of MPs encoded by double-stranded RNA (dsRNA) viruses is very limited. In particular, many studied MPs of DNA and ssRNA viruses bind non-specifically ssRNAs, leading to models in which ribonucleoprotein complexes (RNPs) move from cell to cell. Thus, it will be of special interest to determine whether MPs of dsRNA viruses interact with genomic dsRNAs or their derivative sRNAs. To this end, we studied the biochemical functions of MP Pns6 of *Rice dwarf phytoreo virus* (RDV), a member of *Phytoreovirus* that contains a 12-segmented dsRNA genome. We report here that Pns6 binds both dsRNAs and ssRNAs. Intriguingly, Pns6 exhibits non-sequence specificity for dsRNA but shows preference for ssRNA sequences derived from the conserved genomic 5′- and 3′- terminal consensus sequences of RDV. Furthermore, Pns6 exhibits magnesium-dependent ATPase activities. Mutagenesis identified the RNA binding and ATPase activity sites of Pns6 at the N- and C-termini, respectively. Our results uncovered the novel property of a viral MP in differentially recognizing dsRNA and ssRNA and establish a biochemical basis to enable further studies on the mechanisms of dsRNA viral MP functions.

1 Introduction

Cell-to-cell movement is required for both local and systemic infection by plant viruses. Viruses encode movement proteins (MP) to facilitate such movement. The specific movement mechanisms vary among viruses. Some viruses, such as *Tobacco mosaic virus*, encode MPs that can alter the size exclusion limit (SEL) of plasmodesmata (Wolf et al., 1989) and bind RNAs(Citovsky et al., 1990), and may thus move as ribonucleo protein complexes (Lucas, 2006). Other viruses, including *Grapevine fanleaf virus*, move through tubular structures formed inside modified plasmodesmata that are induced by the viral MPs (Laporte et al., 2003). Many MPs are multifunctional. The 25 K MP (TGBp1, a triple gene block component) encoded by *Potato virus X* (PVX), has RNA binding, RNA helicase and Mg^{2+}-dependent ATPase activities in vitro(Kalinina et al., 1996; Kalinina et al., 2002). It can increase the SEL of plasmodesmata in trichome cells of *Nicotiana clevelandii* (Angell et al.,

PLoSONE.2011, 6(9): e24986.
Received 18May 2011; Accepted 20August 2011; Published 20 September 2011.

1996). It also functions as an RNA silencing suppressor that is important for PVX spreading within a host plant (Voinnet et al., 2000; Bayne et al., 2005). A recent study indicated that the 25 K protein interacts with Argonaute proteins (AGO 1–4) and mediates their degradation through the proteasome pathway(Chiu et al., 2010).

Rice dwarf phytoreovirus (RDV) is a member of the genus *Phytoreovirus* within the family *Reoviridae* that replicates in both host plants and insect vectors. The genome of RDV is composed of 12 double-stranded RNAs (dsRNAs)(G Boccardo, 1984). The sense strand RNAs from all genome segments of RDV contain a 5′ terminal consensus sequence (5′GGCAAA--- or 5′ GGUAAA---)and a 3′ terminal consensus sequence (---UGAU 3′ or ---CGAU 3′) (Kudo et al., 1991). Many other reoviruses have similar sequences at the ends of their genomic dsRNAs(Joklik, 1999; Hull, 2002). Although such sequence conservation suggests its functional significance, the biological role of these 5′- and 3′-terminal conserved sequences for RDV and other reoviruses remain unclear. In rotavirus, another member of the family Reoviridae, the terminal sequences of mRNAs form the minimal cis-acting signal required for minus-strand synthesis (Wentz et al., 1996; Chen and Patton, 1998).

The RDV virion has an outer shell composed of structural proteins P2, P8 and P9, and a core composed of structural proteins P1, P3, P5 and P7 as well as the genomic dsRNAs. The functions of nonstructural proteins Pns4, Pns6, Pns10, Pns11 and Pns12 have previously been characterized. Pns4 can be phosphorylated in vivo and is localized at the periphery of viroplasms(Wei et al., 2006a). Pns6 is involved in cell-to-cell movement(Li et al., 2004). Pns10 is an RNA silencing suppressor in plants (Cao et al., 2005; Ren et al., 2010; Zhou et al., 2010) and is involved in the formation of tubular structures between neighboring insect cells (Wei et al., 2006c). Pns11 is a nucleic acid-binding protein (Xu et al., 1998). Pns12 is essential for the formation of cytoplasmic inclusions and is involved in virion assembly (Wei et al., 2006b). Utilizing a complementation approach, Li et al. (2004) determined that Pns6, encoded by segment S6, is responsible for the movement of RDV between cells and can restore the cell-to-cell movement ability of a PVX 25 K deletion mutant (Wei et al., 2006b). In addition, Pns6 was shown to localize to plasmodesmata in epidermal cells of both *Nicotiana tabacum* bombarded and RDV-infected rice leaves. However, Pns6 did not suppress RNA silencing in cells(Cao et al., 2005).

To advance our understanding of the cell-to-cell movement mechanisms of dsRNA viruses, we studied the biochemical functions of RDV Pns6. We report here that Pns6 binds both dsRNAs and ssRNAs. Intriguingly, Pns6 exhibits non-sequence specificity for dsRNA but shows preference for ssRNA sequences derived from the conserved genomic 5′- and 3′- terminal consensus sequences of RDV. Furthermore, Pns6 exhibits magnesium-dependent ATPase activities. Mutagenesis identified the Pns6 RNA binding and ATPase activity sites at the N- and C-termini, respectively. Our results uncovered the novel property of a viral MP in differentially recognizing dsRNA and ssRNA and establish a biochemical basis to enable further studies on the mechanisms of dsRNA viral MP functions.

2 Materials and Methods

2.1 Plasmid construction

Plasmids pGEM-S6, pGEM-S7 and pGEM-S9 were digested with *Bam*HI and *Sal* I, and the resulting S6/S7/S9 segments were inserted into the prokaryotic expression vector pET28a to yield pET28a-S6, pET28a-S7, pET28a-S9. Primers for constructing the S6 mutant segments are listed in Table 2. Segments M13 and GKS were generated using overlap extension PCR (Higuchi et al., 1988). All resulting PCR fragments were inserted into pBluescript II KS at the *Eco*RV site to generate pBS-mutant plasmids. After digesting the pBS-mutant plasmids with *Bam*HI, the resulting mutant fragments were ligated into the prokaryotic expression vector pGEX-4T-1 at the *Bam*H Isite to generate pGEX-mutant plasmids. The *Bam*HI/*Xho* I fragment released from the pGEX-mutant was inserted into pET28a at the *Bam*HI/*Xho* I site to produce pET28a-mutant plasmids.

2.2 Protein preparation

Escherichia coli BL21 cells (TaKaRa) containing pET28a-S6, pET28a-S7, pET28a-S9 or pET28a-S6 mutant were used to produce His-tagged Pns6, P7, P9 and mutant proteins, and BL21 cells carrying pGEX-mutant constructions were used to produce GST-fused S6 mutant proteins. The cells were cultured in Luria-Bertani medium until they reached an OD_{600} of 0.6, and then the recombinant proteins were induced by adding isopropyl β-D-thiogalactoside (0.5mM) for 2hours at 37℃ or overnight at 18℃. The soluble recombinant proteins were purified using affinity chromatography with a Ni^{2+} -chelating column or a Glutathione Sepharose 4B column. The His-tagged proteins were purified in a buffer containing 20mM TrisCl and 500mM NaCl, pH 7.0, and the GST fused proteins were purified in a buffer containing 40mM Tris and 50mM NaCl, pH 8.0. The resulting proteins were used for ATPase analysis and electrophoresis mobility shift assays (EMSA).

Proteins in inclusion bodies were isolated and recovered for northwestern blotting analysis. Cells were harvested by centrifugation at 5,000g at 4℃ for 15min and resuspended in lysis buffer (50mM Tris • HCl pH 7.0, 100mM NaCl and 1mM EDTA, 1% Triton X-100) at 4℃ . After sonication and centrifugation at 20,000g at 4℃ for 15min, the resulting pellet was washed once with the same lysis buffer. The washed inclusion bodies were resuspended in 6M guanidine hydrochloride, 100mM TrisHCl pH 7.0, 100mM dithiothreitol, 1mM EDTA. The solubilized proteins were separated by SDS-PAGE. The gel was stained with cold 0.5M KCl. The respective band was then cut out, crushed, and mixed with the SDS-PAGE loading buffer. After 10min incubation at 80℃, the sample was centrifuged for 15min at 13,000g and the supernatant was collected for northwestern blotting.

2.3 Probe preparation

The sequences of the synthesized RNA probes and the templates used for the in vitro transcription are presented in Table 1. Short ssRNAs and dsRNAs were synthesized and purified via HPLC by TAKARA Biotechnology (DALIAN) Co. The MA-XIscript Kit (Ambion) was used to transcribe long RNAs. The ssA-Dig probe was digoxin-labeled with the DIG RNA Labeling Kit (SP6/T7) (Roche). The radioactive probes were end-labeled with γ-^{32}P-ATP using T4 polynucleotide kinase (NEB).

2.4 Northwestern blotting and electrophoresis mobility shift assay (EMSA)

Northwestern blotting was used to detect the nucleic acid binding activity of Pns6 and its mutants as described previously(Gramstat et al., 1990). The recovered proteins were separated by SDS-PAGE and then transferred to a nitrocellulose membrane. The membrane was soaked in buffer A (20mM Tris • HCl, pH 7.5, 50mM NaCl, 0.1% Triton X-100, 1mM DTT, 1mM EDTA, pH 8.0 and 0.02% Ficoll 400) overnight at 4℃ to renature the proteins. The digoxin-labeled RNA probes or the radioactive RNA probes were then added to the buffer. After 30min of incubation, the membranes were washed three times in buffer A for 30min at 4℃ followed by UV crosslinking. The RNA probes were detected using an anti-digoxin antibody (AP conjugated, Roche) or via autoradiography with a PE Cyclone phosphor screen scanner system.

For EMSA, RNA probes were incubated with various concentrations of proteins for 20min on ice in a 10μL reaction system containing 50mM NaCl, 50mM Tris • HCl, pH 7.5, and 10% glycerol. Samples were separated by native PAGE in a 9% gel in 0.5×TBE buffer for 80 minutes at 180V, and then the gel was dried for autoradiography. In the competition assays, unlabeled RNAs were added to compete with the labeled RNA probes. The competition ability of the RNAs is related to the signal from the free RNA probes, which was quantified with ImageJ (version 1.4) software.

2.5 Thin layer chromatography (TLC) and colorimetric malachite green assay

In the TLC ATPase analysis, we used a 10μL reaction system containing 50mM Tris, 5mM $MgCl_2$, 1mM DTT, 0.5mM ATP and the respective proteins. α -^{32}P-ATP was added to the reactions at 0.05μCi/μL. After a 30min incubation at 37℃ , 0.5μL of the reaction product was dotted onto a PEI thin layer chromatography plate (Merck), and the plate was developed in 0.15M LiCl and 0.15M formic acid as described previously(Kalinina et al., 1996). Images were obtained using a PE Cyclone phosphor screen scanner system.

The colorimetric malachite green assay was performed as described previously (Baykov et al., 1988). We used a 100μL reaction system containing 50mM Tris, 5mM $MgCl_2$, 1mM DTT, 1mM ATP and the respective protein. A 96-well microplate was used to analyze multiple samples. After 60min of incubation at 37℃ , 45μL of the colorimetric mixture (0.07% malachite green, 3.7% ammonium sulfate, 2.27% ammonium molybdate tetrahydrate and 0.134% Tween-20, prepared 2 hours prior to the assay) was added to the reactions. To develop the reaction, 45μL of 15% sodium citrate was added. The OD_{570} of the samples was read by a TECAN SUNRISE basic scanner.

3 Results

3.1 Sequence analysis of Pns6 reveals potential RNA acid binding region and ATPase/helicase motif

In our previous study, Pns6 was shown to restore the cell-to-cell movement activity of a PVX 25 K deletion mutant (Li et al., 2004). The 25 K protein of PVX is a multifunctional protein that contains different functional domains (Morozov et al., 1999). We predicted the potential functional domains of Pns6 based on amino acid sequence. As Fig. 1A shows, the N-terminal region of Pns6 is rich in basic amino acids, potentially containing an RNA-binding site. In the internal region of Pns6, two transmembrane helicase domains were predicted at amino acid positions 207–228 and 254–271. A putative GKS motif, potentially related to ATPase/helicase activity, was present at amino acid positions 125–127 (Fig. 1B).

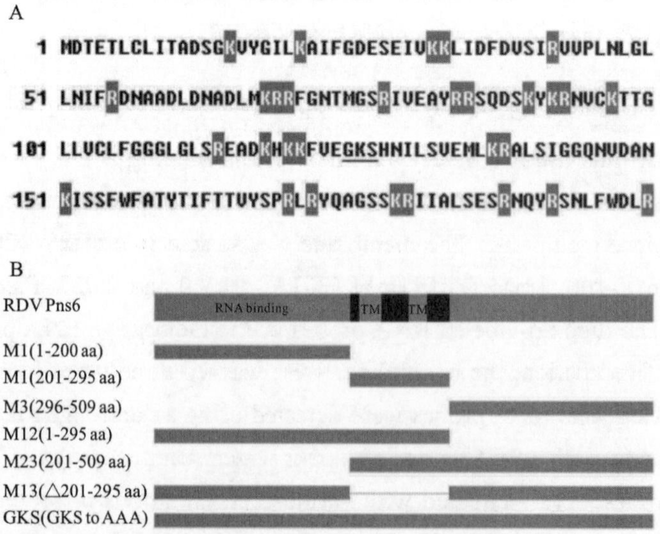

Fig. 1 Analysis of Pns6 sequence and construction of mutants

(A) The N terminal region of Pns6 is rich in basic amino acids. Lysines (K) and arginines (R) are indicated in white font on a gray background, and the GKS motif is underlined. (B) The N-terminus of Pns6 is predicted to contain the RNA binding region and its middle region is predicted to contain transmembrane domains (TM). M1-M13 indicate deletion mutants with amino acid sequences shown in parentheses. In the GKS mutant, the amino acids GKS at positions125–127 are replaced with AAA

3.2 Pns6 binds both single-stranded and double-stranded RNA

During RDV replication, both dsRNAs and ssRNAs are produced. Therefore, both forms of RNA (Table1) were used to test the RNA binding activity of Pns6. We first tested RNA probes of non-RDV sequences, including a human α-actin sequence (ssA) and a random dsRNA sequence (dsR), in northwestern blotting assays. Fig. 2 shows that Pns6 could bind both probes. We then performed electrophoresis mobility shift assay (EMSA) to further confirm the RNA binding activity of Pns6. As shown in Fig. 3A, Pns6 bound dsR as well as an RDV-specific sequence (S3-5) with equal efficiency. Pns6 also bound ssRNAs (Fig. 3B). However, binding to a random ssRNA sequence (ssR) was weaker than binding to RDV-specific ssRNA sequences (S3-5s, S3-5a, S3-3s and S3-3a). These data establish that Pns6 has RNA binding activities and suggest that it appears to have sequence preference for ssRNAs. This was further tested as described below.

Table 1 Sequences of dsRNA and ssRNA probes

Type of probes	Name of probes	Sequence(5′-3′)	Description
dsRNA	S3-5	Sense: GGCAAAAUCGAGCGAACAAU Anti-sense: AUUGUUCGCUCGAUUUUGCC	Conserved S3 5′-end 20 bp sequence
	S3-3	Sense: UGUUUUGCUUGGUUCCUGAU Anti-sense: AUCAGGAACCAAGCAAAACA	Conserved S3 3′-end 20 bp sequence
	S3m	Sense: CAAUCAAUAU GCUCGCCCCA Anti-sense: UGGGGCGAGC AUAUUGAUUG	Non-conserved S3 sequence of 1541st-1560th bp of S3 ORF
	dsR	Sense: UUCUCCGAACGUGUCACGUUU Anti-sense: ACGUGACACGUUCGGAGAAUU	Random 21 bp dsRNA sequence
ssRNA	S3-5s	GGCAAAAUCGAGCGAACAAU	Conserved S3 5′-end 20 nt sense sequence
	S3-5a	AUUGUUCGCUCGAUUUUGCC	Conserved S3 5′-end 20 nt antisense sequence
	S3-3s	UGUUUUGCUUGGUUCCUGAU	Conserved S3 3′-end 20 nt sense sequence
	S3-3a	AUCAGGAACCAAGCAAAACA	Conserved S3 3′-end 20 nt antisense sequence
	S3mss		Denatured from S3m, 20 nt non-conserved S3 sequence
	ssR	UUGUACUACACAAAAGUACUG	Random 21 nt ssRNA sequence
	ssA	In vitro transcribed from pTripleScript-actin ™ (linearized with *Sal* I)	~270 nt, Homo sapiens a-actin sequence
	ssA-Dig	In vitro transcribed from pTripleScript-actin ™ (linearized with *Sal* I)	ssA sequence with digoxin label
	ssB	In vitro transcribed from vector pSPT18 (linearized with *Hind* III)	~50 nt, sequence of EcoR I, Sac I, BamH I, Xba I, Sal I, and PstI
	M2s	In vitro transcribed from pBS M2-T7 insertion (linearized with *Hind* III)	~290 nt, the sense strand of mutant M2 of Pns6
	M2a	In vitro transcribed from pBS M2-T3 insertion (linearized with *Hind* III)	~290 nt, the antisense strand of mutant M2 of Pns6

Many MP-RNA complexes were disrupted at high salt concentrations (Citovsky et al., 1991; Carvalho et al., 2004). As shown in Fig.3C&D, Pns6 bound both dsRNA (S3-5) and ssRNA (S3-5a) in the range of 50–200mM NaCl. The Pns6-RNA complexes underwent gradual dissociation with increasing salt concentrations, as expected of the influence of salt concentrations on protein-RNA interactions. In one control, BSA did not bind these RNAs. In another control, the RDV P9 protein, prepared in the same manner as Pns6 from recombinant *E. coli,* showed no interaction with the RNAs. These controls ruled out the possibility that the observed Pns6-RNA interaction was due to (i) non-specific electrostatic interactions or (ii) contaminating proteins from *E. coli* that bound the RNAs. Rather, they indicate that Pns6 interacted with the RNAs via specific molecular recognition. This specificity is further supported by the following experimental results showing selectivity of Pns6 for some RNA sequences and identifying the Pns6 domain for RNA binding.

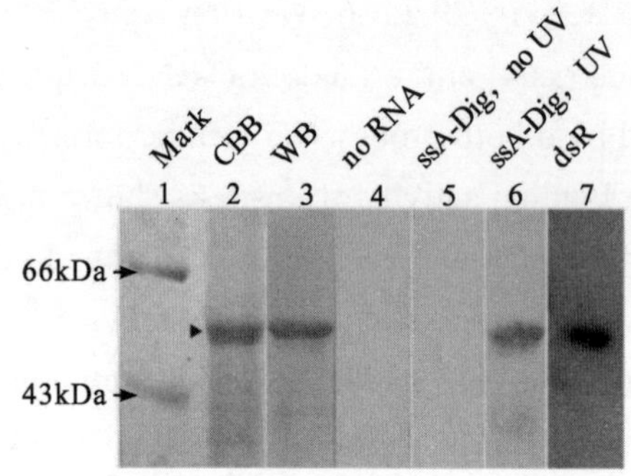

Fig. 2 Northwestern blotting showing Pns6 binding to ssRNA and dsRNA

Lanes 1 and 2: Coomassie Brilliant Blue staining. "Mark" indicates the protein molecular standards, the arrowheads indicate 66kDa and 43kDa bands. The black triangle indicates the HisPns6 band. Lane 3: His-Pns6 was detected by western blotting (WB) with anti-His antibody. Lane 6: after incubation and UV crosslinking, the digoxin-labeled single-stranded RNA probe ssA-Dig bound to His-Pns6 on the membrane was detected with anti-digoxin (AP-conjugated) antibody. Lanes 4 & 5: negative controls. Lane 7: the radiolabeled dsRNA probe dsR bound to His-Pns6 was detected after 8 hours of exposure

3.3 Pns6 binds preferentially ssRNAs derived from the terminal consensus sequences of RDV genome

Genomic RNAs of RDV consist of 12 dsRNA segments containing consensus sequences at their 5′ and 3′ ends. As described above, Pns6 showed stronger binding to ssRNA probes containing these conserved sequences than to a random ssRNA (Fig. 3B). We conducted additional experiments to determine whether Pns6 has higher affinity for the terminal consensus sequences in double-stranded as well as single-stranded forms. Four unlabeled dsRNAs were used to compete with the radiolabeled conserved dsRNA S3-5 derived from RDV, including a dsRNA of random sequence (dsR), a nonconserved RDV dsRNA (S3m) and conserved dsRNAs from the 5′ and 3′ termini of RDV S3 segments (S3-5 and S3-3), respectively. The relative amounts of free labeled dsRNAs were used to determine the binding abilities of the competitors. Fig. 4A shows that the four competitors had equivalent competitive abilities. The data further establish that Pns6 binds dsRNA in a sequence-non-specific manner.

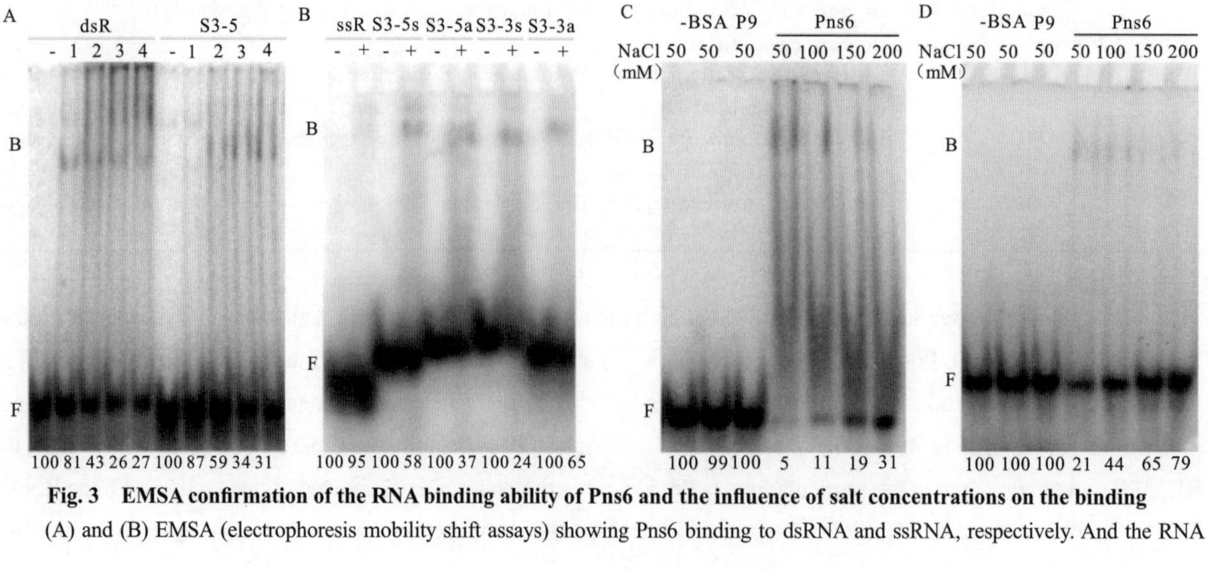

Fig. 3 EMSA confirmation of the RNA binding ability of Pns6 and the influence of salt concentrations on the binding

(A) and (B) EMSA (electrophoresis mobility shift assays) showing Pns6 binding to dsRNA and ssRNA, respectively. And the RNA binding activities of Pns6 was affected by salt concentration. Free and bound RNA probes are indicated by 'F' and 'B', respectively. Free probes were quantified with ImageJ software version 1.4. The relative values of free probe are presented below each of the lanes. In (A) and (B) 30 nanogram (ng) of BSA was added in lanes labeled '2' as a negative control. In (A), the random dsRNA (dsR) and conserved dsRNA (S3-5) were used at 5ng, and lanes 1, 2, 3, and 4 represent His-Pns6 used at 10, 20, 30 or 40ng, respectively. In (B), Pns6 was added at 30ng in lanes labeled '+'. The random ssRNA (ssR) and conserved RDV ssRNA (S3-5s, S3-5a, S3-3s and S3-3a) were used at 2.5ng. (C) and (D) Effects of salt concentrations on Pns6-RNA interaction. In (C) and (D), no protein was added in lanes labeled '2'. BSA and His-P9 served as negative controls. Proteins were added at 50ng in each lane. In (C), the dsRNA (S3-5) was used at 2.5ng. In (D), the ssRNA (S3-5a) was used at 2.5ng

To test sequence preference of ssRNA for Pns6 binding, three ssRNA competitors with random sequences and different lengths (ssR, ssB and ssA) were used to compete with a conserved ssRNA sequence (S3-5a). Figures 4B and 4C show that short random ssRNAs (21-nt ssR and 50-nt nucleotide ssB) had extremely weak competing abilities. The binding ability of the longer random ssRNA (270-nt ssA) was between that of the short random ssRNAs and that of the conserved ssRNA. Three nonconserved RDV ssRNA sequences of different lengths (S3mss, M2s and M2a) were then used to compete with S3-5a. Fig.4D shows that the binding efficiency of these competitors was similar to that of the random ssRNAs. Four ssRNAs of the RDV conserved sequences (S3-5a, S3-5s, S3-3a and S3-3s) were finally used as competitors, and they had high binding abilities (Fig. 4E). S3-5s and S3-3s had higher competition efficiency than S3-5a and S3-3a (Fig. 4E), suggesting that Pns6 has stronger affinity for sense-stranded RDV ssRNAs than for anti-sense ssRNAs. Furthermore, the higher binding affinity of Pns6 with S3-5a and S3-5s suggest that Pns6 had a preference for the 5′- end sense-stranded RDV RNAs. All together, these data indicate that Pns6 has preference for ssRNA derived from the terminal consensus sequences of the RDV genome.

3.4 The RNA binding site of Pns6 is located at the N-terminal region

The richness of basic amino acids in the N-terminal region of Pns6 suggests a role of this region in RNA binding. To test this, we generated a series of Pns6 mutants to test for RNA binding. As shown in Fig.1B, constructs M1, M2, M3, M12, M23 and M13 are deletion mutants, and the GKS construct contains a mutation at amino acids 125-127 (GKS to AAA). As shown in Fig.5A, northwestern blotting assays showed that all mutants containing the M1 region (i.e., M1, M12 and M13) as well as M3 bound ssRNA, M2 and M23 failed to bind ssRNA, and substitution of GKS with AAA had minimal effect on ssRNA binding. Furthermore, while M1 bound dsRNA of conserved RDV sequence, M3 did not exhibit clear binding activity (Fig. 5B). These results indicate that the N-terminal region of Pns6 is responsible for binding ssRNAs as well as dsRNAs. Pns6 is a magnesium-dependent ATPase sequence, M3 did not exhibit clear binding activity (Fig. 5B).These results indicate that the N-terminal region of Pns6 is responsible for binding ssRNAs as well as dsRNAs.

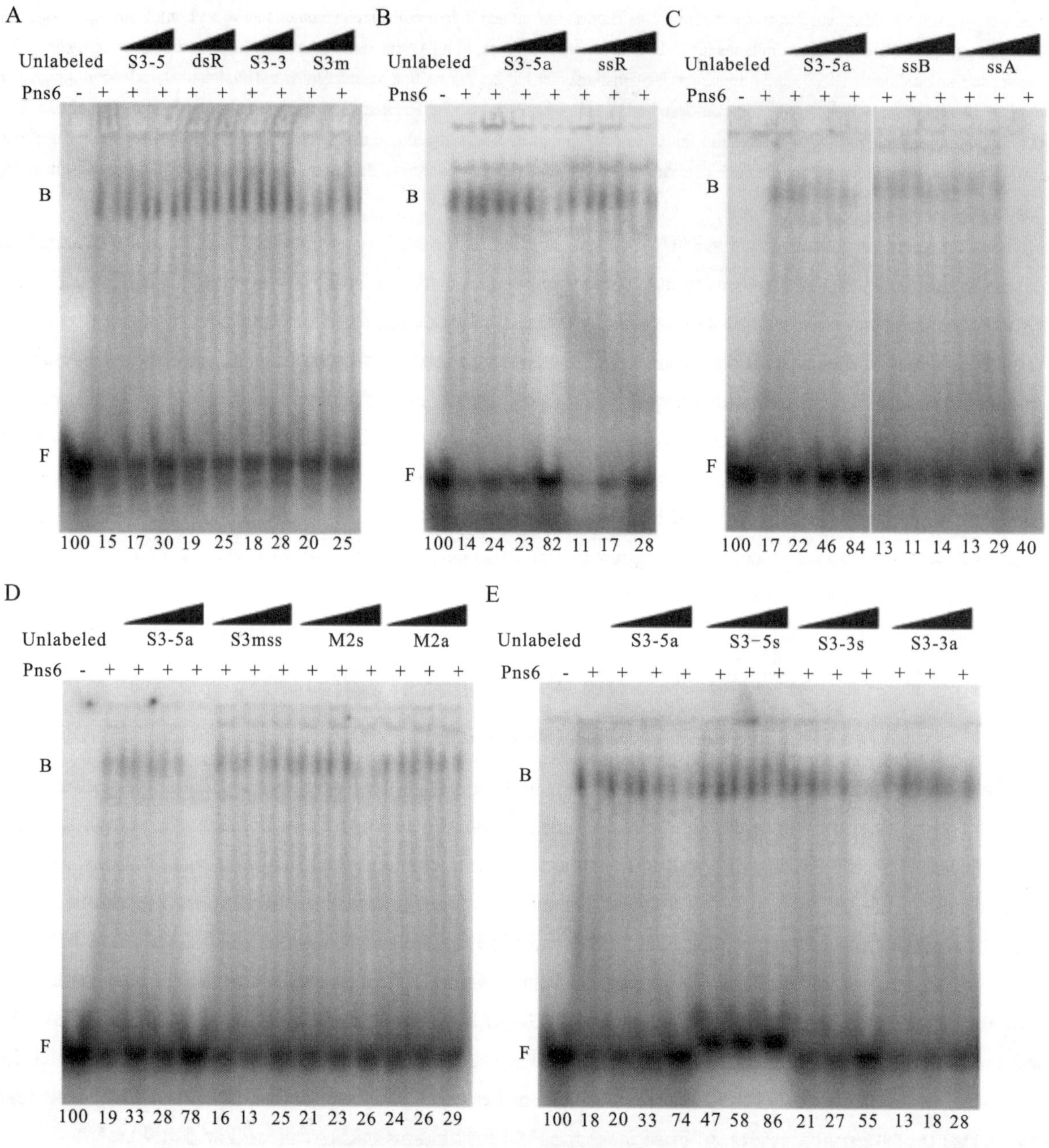

Fig. 4 RNA binding preference for Pns6

EMSA shows Pns6 binding to dsRNA in a sequence-non-specific manner (A) and binding to ssRNA with preference for the terminal consensus sequence of the RDV genome (B–E). The labeled, conserved probes were incubated with various unlabeled competitors. The binding abilities of the competitive RNAs were measured as relative levels of free labeled probes, which were quantified with ImageJ software version 1.4. Relative values are presented below each lane. Lane ‘2’ : no Pns6 and 30ng BSA was added and all labeled probes were unbound. Lane ‘+’ : 30ng Pns6 was added. (A) Labeled, conserved dsRNA (S3-5) was added at 5ng in each lane. Triangles indicate increasing amounts of unlabeled competitor (half or one-fold of S3-5). The four competitors had approximately equal competition abilities. (B–E) Each lane contains 2.5ng of labeled conserved ssRNA (S3-5a). Triangles indicate increasing amounts of unlabeled competitors (one-, two- or 20-fold of S3-5a). (B) The random ssRNA (ssR) shows weaker binding to Pns6 than the S3-5a. (C) The longer random ssRNAs (ssB, about 50 nt; ssA, about 270 nt) also shows weaker binding to Pns6. (D) Three ssRNA competitors with RDV nonconserved sequences of different lengths (S3mss, M2s and M2a) show weaker binding to Pns6. (E) Among the four conserved ssRNAs (S3-5a, S3-5s, S3-3a and S3-3s), the 59-end sense strand RDV RNAs (S3-5s) show the highest binding abilities

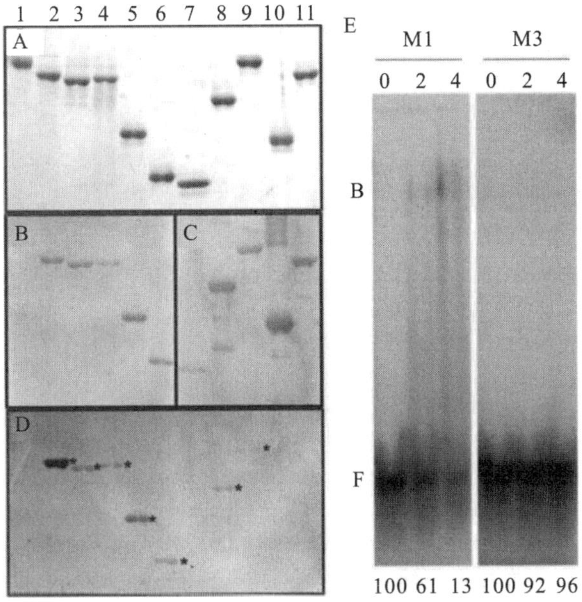

Fig. 5 N-terminal region of Pns6 is responsible for RNA binding

(A)–(D) Northwestern blotting of mutant proteins showed that mutant M3 and all mutants containing M1 had the ability to bind ssRNA. Lane 1: BSA (negative control); 2: His-P7 (positive control); 3: HisPns6; 4: His-GKS; 5: His-M12; 6: His-M3; 7: GST; 8: GST::M1; 9: GST::M13; 10: GST::M2; 11: GST::M23. (A) Coomassie Brilliant Blue staining; (B) western blot with anti-His antibody; (C) western blot with anti-GST antibody; (D) Northwestern blot detecting digoxin-labeled probe ssADig. (E) EMSA showing dsRNA-binding activity of HisM1 and lack of such activity by HisM3. Labeled S3-5 probe was used at 5ng in each lane. Lanes 0, 2 and 4 represent the proteins used at 0, 20 and 40ng. Relative values of free labeled probes are presented below each lane

3.5 Pns6 is a magnesium-dependent ATPase

The putative GKS motif at amino acid positions 125 to 127 suggests that Pns6 may be an ATPase. We performed a thin layer chromatography (TLC) assay with purified His-tagged Pns6. Fig. 6A shows that His-Pns6 can hydrolyze ATP into ADP and that the released ADP increased with increasing amounts of His- Pns6. This indicated that Pns6 has ATPase activity. BSA and P9 exhibited no ATPase activity, ruling out the possibility that the observed Pns6 ATPase activity was due to contaminating proteins from *E. coli*. It was previously reported that divalent cations play an important role in ATP hydrolysis reactions(Sriram et al., 1997; Frick et al., 2007). To examine whether these ions play a role in ATP hydrolysis by Pns6, magnesium or calcium was added to the TLC assays. As shown in Fig. 6B, the ATPase activity of Pns6 increased with increasing concentrations of magnesium, reaching the highest level at 5mM of magnesium. Increasing concentrations of calcium had little effects on the ATPase activity of Pns6 (Fig. 6C), indicating calcium-independence of this activity. Furthermore, when EDTA was added to the reaction system, the ATPase activity of Pns6 was efficiently inhibited, particularly when the concentration of EDTA was equal to or higher than that of magnesium (Fig. 6D). These data indicate that Pns6 has ATPase activities that depend on magnesium but not calcium.

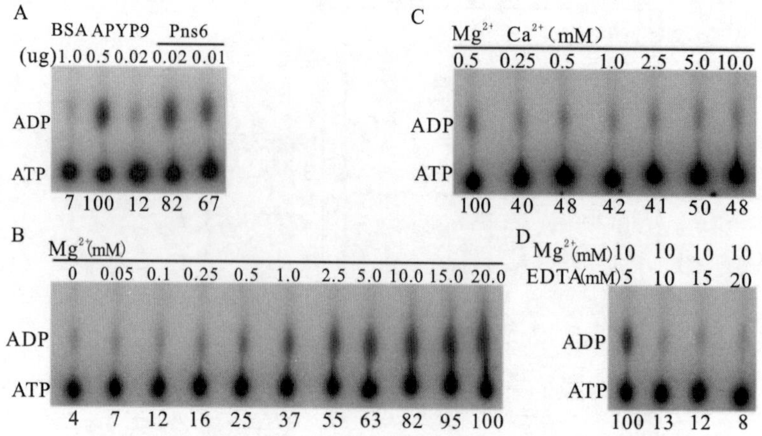

Fig. 6 Thin layer chromatography showing Pns6 as a magnesium-dependent ATPase

(A) Hydrolysis of ATP by His-Pns6. Apyrase (APY) was used as a positive control. BSA and purified His-P9 were used as negative controls. (B) The ATPase activity of Pns6 increased with increasing concentrations of magnesium. (C) The ATPase activity of Pns6 is not dependent on calcium. (D) The ATPase activity of Pns6 was inhibited when EDTA was added at a concentration equal to or greater than that of magnesium (10 mM). In (B)–(D) His-Pns6 was used at 20ng in each reaction. Relative values of released ADP are presented below each lane

3.6 Conserved motifs and active site of Pns6 ATPase

We performed TLC and colorimetric malachite green assays with purified mutant Pns6 proteins to locate the potential ATPase activity site. Mutants M1, M3, M13 and M23 showed ATPase activity equal to or higher than that of wild-type Pns6 in TLC assays (Fig. 7A). Mutants M2, M12 and GKS/AAA showed a significant reduction in ATPase activity. Consistent results with obtained from the colorimetric malachite green assays (Fig. 7B).These results suggest that the ATPase activity of Pns6 resides in the C-terminal region and that GKS is a conserved ATPase motif.

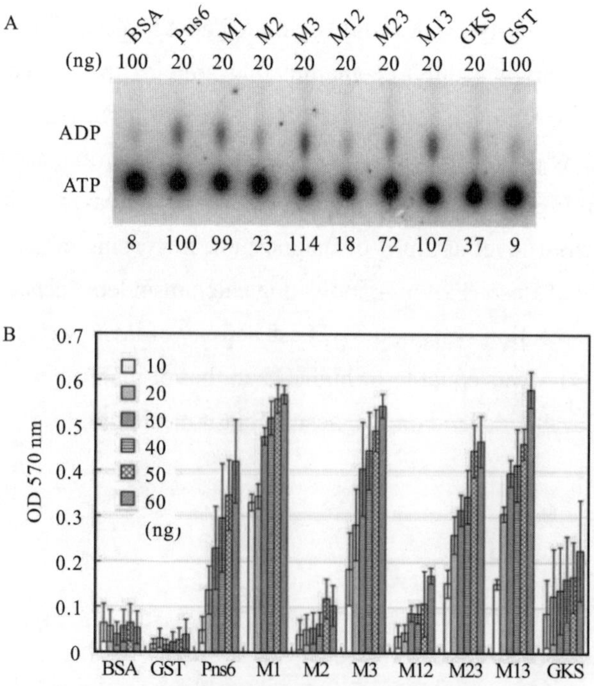

Fig. 7 Conserved motifs and the active site of Pns6 ATPase

(A) Thin layer chromatography shows that M1 and the mutants containing M3 (M3, M23 and M13) have the strongest ATPase activity. The GKS mutant shows weaker ATPase activity than Pns6, while M2 and M12 show very little activity. BSA and GST were used as negative controls. Relative values of released ADP are presented below each lane. (B) Colorimetric malachite green assay showing the ATPase activities of the mutant proteins. All proteins were used at a range of 10ng to 60ng. Error bars represent standard deviations of three biological replicates

4 Discussion

Although our knowledge of the detailed RDV replication cycle is still very limited, it has been proposed that the this virus may have a similar lifecycle to animal reoviruses(Hull, 2002). Early after infection, reoviruses become partially uncoated to form subviral particles. During this process, the viral mRNA, identical to the sense strand of the genomic dsRNA, is released from the subviral particles while the genomic dsRNA remains inside the particles. Once the viral genomic ssRNAs have accumulated to a high level and are packaged into virions, dsRNAs can be synthesized by viral RdRP (RNA-dependent RNA polymerase) inside the core. The virions can be observed in the cytoplasm at the sites where viral replication or translation have occurred (Joklik, 1999).

Virions or viral ribonucleoprotein complexes (vRNPs) can be transported through plasmodesmata into adjacent cells. In the case of RDV, it is unclear whether the virus is transported in the form of virions or vRNPs. RDV virions spread among cells of insect vectors through tubular structures composed of RDV Pns10(Wei et al., 2006c). These tubular structures are not found in the host plant cells. In our previous report, Pns6 but not RDV virions was localized to plasmodesmata in cell walls (Li et al., 2004). It is possible that the diameter of the RDV virion (about 70nm)prevents it from moving through plasmodesmata. Our previous and current data support the hypothesis that RDV may move between cells in the form of vRNPs.

Pns6 binds dsRNA in a non-sequence-specific manner. Many known MPs, such as the 30 K protein of TMV, bind ssRNA cooperatively and sequence-non-specifically (Citovsky et al., 1992). Interestingly, Pns6 shows sequence preference when it binds ssRNA. Specifically, it binds preferentially to the conserved 5′- and 3′- terminal consensus sequences of the RDV genome and exhibits a stronger binding affinity for the RDV 5′- end sense strand sequence than with the corresponding antisense strand. This is a novel property unreported for other known plant viral MPs. However, a specific property of reoviruses is that their genomic RNAs contain terminal consensus sequences. In rotavirus, the 3′- terminal consensus sequence is recognized by NSP3 and is related to the translation of viral mRNA (Poncet et al., 1994). while the 5′- and 3′- sequences are also recognized by viral RdRP for efficient dsRNA synthesis (Tortorici et al., 2003; TORTORICI et al., 2006). Thus, our data also suggest the intriguing possibility that Pns6 has additional roles in viral replication and translation, an important issue to be addressed in future studies.

The ATPase activity of Pns6 may be important for RDV movement. Numerous plant virus MPs have ATPase activities (Peremyslov et al., 1999). In potexviruses, the ATPase activity of PVX 25 K may be required to provide the driving force to traffic viral RNA through plasmodesmata or to suppress silencing (Howard et al., 2004; Bayne et al., 2005). The conserved GKS sequence is required for the NTP binding function of many proteins (Davenport and Baulcombe, 1997). In the viral RNA helicases, which exhibit NTPase activity and have been divided into three superfamilies, the GKS motif is shared by the conserved Walker A site of SF3 helicases and the conserved segment I of SF2 and SF3 helicases(Kadaré and Haenni, 1997). This motif is present in many proteins encoded by reoviruses (e.g., VP6 of the bluetongue virus) and possesses NTPase activity (Ramadevi and Roy, 1998; Kar and Roy, 2003; Nibert and Kim, 2004). The importance of GKS motif for Pns6 ATPase activity is consistent with findings from other viral proteins. When the GKS motif is changed to GAA in Bamboo mosaic virus ORF1, to GAS in the Hepatitis E virus helicase motif I, or to GES in the Hepatitis C virus NS3, the ATPase activities of these proteins decrease by 70% (Chang et al., 2000; Li et al., 2001; Karpe and Lole, 2010) .

In summary, our analyses of RDV Pns6 uncovered the novel property of a viral MP in differentially recognizing dsRNA and ssRNA. Such property, together with the identification of the RNA binding sites and ATPase activity site in the protein, establishes a biochemical basis to enable further studies on the mechanisms of dsRNA viral MP functions in movement and perhaps also other aspects of the viral life cycle.

References

[1] ANGELL S M, DAVIESC,BAULCOMBE D C, et al. Cell-to-Cell Movement of Potato Virus X Is Associated with a Change in the Size-Exclusion Limit of Plasmodesmata in Trichome Cells of *Nicotiana clevelandii*[J]. Virology, 1996, 216(1): 197-201.

[2] BAYKOV A A, EVTUSHENKO O A, AVAEVA S M. A malachite green procedure for orthophosphate determination and its use in alkaline phosphatase-based enzyme immunoassay[J]. Analytical Biochemistry, 1988, 171(2): 266-270.

[3] BAYNE E H, RAKITINA D V, MOROZOV S Y, et al. Cell-to-cell movement of Potato Potexvirus X is dependent on suppression of RNA silencing[J]. The Plant Journal, 2005, 44(3): 471-482.

[4] CAO X, ZHOU P, ZHANG X, et al. Identification of an RNA silencing suppressor from a plant double-stranded RNA virus[J]. Journal of Virology, 2005, 79(20): 13018-13027.

[5] CARVALHO C, POUWELS J, VAN LENT J, et al. The movement protein of cowpea mosaic virus binds GTP and single-stranded nucleic acid in vitro[J]. Journal of Virology, 2004, 78(3): 1591-1594.

[6] CHANG S C, CHENG J C, KOU Y H, et al. Roles of the AX4GKS and Arginine-Rich Motifs of Hepatitis C Virus RNA Helicase in ATP- and Viral RNA-Binding Activity[J]. Journal of Virology, 2000, 74(20): 9732-9737.

[7] CHEN D, PATTON J T. Rotavirus RNA replication requires a single-stranded 3′ end for efficient minus-strand synthesis[J]. Journal of Virology, 1998, 72(9): 7387-7396.

[8] CHIU M H, CHEN I H, BAULCOMBE D C, et al. The silencing suppressor P25 of Potato virus X interacts with Argonaute1 and mediates its degradation through the proteasome pathway[J]. Molecular Plant Pathology, 2010, 11(5): 641-649.

[9] CITOVSKY V, ZAMBRYSKI K P, GENE I. a potential cell-to-cell movement locus of cauliflower mosaic virus, encodes an RNA-binding protein[J]. Proceedings of the National Academy of Sciences, 1991, 88(6): 2476-2480.

[10] CITOVSKY V, KNORR D, SCHUSTER G, et al. The P30 movement protein of tobacco mosaic virus is a single-strand nucleic acid binding protein[J]. Cell, 1990, 60(4): 637-647.

[11] Citovsky V, Wong M L, Shaw A L, et al. Visualization and characterization of tobacco mosaic virus movement protein binding to single-stranded nucleic acids. Plant Cell, 1992, 4: 397-411.

[12] DAVENPORT G F, BAULCOMBE D C. Mutation of the GKS motif of the RNA-dependent RNA polymerase from *potato virus X* disables or eliminates virus replication[J]. Journal of General Virology, 1997, 78(6): 1247-1251.

[13] FRICK D N, BANIK S, RYPMA R S. Role of divalent metal cations in ATP hydrolysis catalyzed by the hepatitis C virus NS3 helicase: Magnesium provides a bridge for ATP to fuel unwinding[J]. Journal of Molecular Biology, 2007, 365(4): 1017-1032.

[14] BOCCARDO G, MILNE R G. Old Woking, England: Commonwealth Mycological Institute and Association of Applied Biologists, Unwin Brothers Ltd[J].The Gresham Press. Plant reovirus group. In: Morant AF, Harrison BD, eds. CM/AAB descriptions of plant viruses, 1984,294.

[15] GRAMSTAT A, COURTPOZANIS A, ROHDE W. The 12 kDa protein of potato virus M displays properties of a nucleic acid-binding regulatory protein[J]. Febs Letters, 1990, 276: 34-38.

[16] RUSSELL H, BARBARA K, RANDALL S. A general method of in vitro preparation and specific mutagenesis of DNA fragments: study of protein and DNA interactions[J]. Nucleic Acids Research, 1988, 16(15): 7351-7367.

[17] HOWARD A R, HEPPLER M L, JU H J, et al. Potato virus X TGBp1 induces plasmodesmata gating and moves between cells in several host species whereas CP moves only in N. benthamiana leaves[J]. Virology, 2004, 328(2): 185-197.

[18] HULL R. Matthews' Plant Virology 4th ed. San Diego: Academic Press, 2002, 336-339. .

[19] JOKLIK W K. Reoviruses (Reoviridae): molecular biology. In: Granoff A,Webster RG, eds. Encyclopedia of Virology. 2nd ed. San Diego: AcademicPress,1999, 1464-1471

[20] KADARÉ, G, HAENNI, A-L. Virus-encoded RNA helicases. Journal of Virology, 1997, 71: 2583.

[21] KALININA N O, RAKITINA D V, SOLOVYEV A G, et al. RNA helicase activity of the plant virus movement proteins encoded by the first gene of the triple gene block[J]. Virology, 2002, 296(2): 321-329.

[22] KALININA N O, FEDORKIN O, SAMUILOVA O, et al. Expression and biochemical analyses of the recombinant *potato virus X.* 25K movement protein[J]. FEBS Letters, 1996, 397(1): 75-78.

[23] KAR A K, ROY P. Defining the Structure-Function Relationships of Bluetongue Virus Helicase Protein VP6[J]. Journal of Virology, 2003, 77(21): 11347-11356.

[24] KARPE Y A, LOLE K S. NTPase and 5′ to 3′ RNA duplex-unwinding activities of the hepatitis E virus helicase domain [J]. Journal of Virology, 2010, 84(7):3595-3602.

[25] KUDO H, UYEDA I, SHIKATA E. Viruses in the phytoreovirus genus of the Reoviridae family have the same conserved terminal sequences[J]. Journal of General Virology, 1991, 72 (12): 2857-2866.

[26] LAPORTE C, VETTER G, LOUDES A-M, et al. Involvement of the secretory pathway and the cytoskeleton in intracellular targeting and tubule assembly of Grapevine fanleaf virus movement protein in tobacco BY-2 cells[J]. Plant Cell, 2003, 15(9): 2058-2075.

[27] LI Y I, SHIH T W, HSU Y H, et al. The helicase-like domain of plant potexvirus replicase participates in formation of RNA 5' cap structure by exhibiting RNA 5'-triphosphatase activity[J]. Journal of Virology, 2001, 75(24):12114-12120.

[28] LI Y, BAO Y M, WEI C H, et al. Rice Dwarf Phytoreovirus Segment S6-Encoded Nonstructural Protein Has a Cell-to-Cell Movement Function[J]. Journal of Virology, 2004, 78(10): 5382-5389.

[29] LUCAS W J. Plant viral movement proteins: agents for cell-to-cell trafficking of viral genomes[J]. Virology, 2006, 344(1): 169-184.

[30] MOROZOV S Y, SOLOVYEV A G, KALININA N O, et al. Evidence for Two Nonoverlapping Functional Domains in the Potato Virus X 25K Movement Protein[J]. Virology, 1999, 260(1): 55-63.

[31] NIBERT M L, KIM J. Conserved sequence motifs for nucleoside triphosphate binding unique to turreted Reoviridae members and coltiviruses[J]. Journal of Virology, 2004, 78(10): 5528-5530.

[32] PEREMYSLOV V V, HAGIWARA Y, DOLJA V V. HSP70 homolog functions in cell-to-cell movement of a plant virus[J]. Proceedings of the National Academy of Sciences, 1999, 96(26): 14771-14776.

[33] PONCET D, LAURENT S, COHEN J. Four nucleotides are the minimal requirement for RNA recognition by rotavirus non-structural protein NSP3[J]. EMBO Journal, 1994, 13(17): 4165-4173.

[34] RAMADEVI N, ROY P. Bluetongue virus core protein VP4 has nucleoside triphosphate phosphohydrolase activity[J]. Journal of General Virology, 1998, 79 (10): 2475-2480.

[35] REN B, GUO Y Y, GAO F, et al. Multiple functions of Rice dwarf phytoreovirus Pns10 in suppressing systemic RNA silencing[J]. Journal of Virology, 2010, 84(24): 12914-12923.

[36] SRIRAM M, OSIPIUK J, FREEMAN B C, et al. Human Hsp70 molecular chaperone binds two calcium ions within the ATPase domain[J]. Structure, 1997, 5(3): 403-414.

[37] TORTORICI M A, SHAPIRO B A, PATTON J T. A base-specific recognition signal in the 5′consensus sequence of rotavirus plus-strand RNAs promotes replication of the double-stranded RNA genome segments[J]. RNA, 2006, 12(1):133-146.

[38] TORTORICI M A, BROERING T J, NIBERT M L, et al. Template Recognition and Formation of Initiation Complexes by the Replicase of a Segmented Double-stranded RNA Virus[J]. Journal of Biological Chemistry, 2003, 278(35): 32673-32682.

[39] VOINNET O, LEDERER C, BAULCOMBE D C. A Viral Movement Protein Prevents Spread of the Gene Silencing Signal in *Nicotiana benthamiana*[J]. Cell, 2000, 103(1): 157-167.

[40] WEI T, KIKUCHI A, SUZUKI N, et al. Pns4 of rice dwarf virus is a phosphoprotein, is localized around the viroplasm matrix, and forms minitubules[J]. Archives of Virology, 2006a, 151(9): 1701-1712.

[41] WEI T, SHIMIZU T, HAGIWARA K, et al. Pns12 protein of Rice dwarf virus is essential for formation of viroplasms and nucleation of viral-assembly complexes[J]. Journal of General Virology, 2006b, 87: 429-438.

[42] WEI T, KIKUCHI A, MORIYASU Y, et al. The spread of Rice dwarf virus among cells of its insect vector exploits virus-induced tubular structures[J]. Journal of Virology, 2006c, 80(17): 8593-8602.

[43] WENTZ M J, PATTON J T, RAMIG R F. The 3'-terminal consensus sequence of rotavirus mRNA is the minimal promoter of negative-strand RNA synthesis[J]. Journal of Virology, 1996, 70(11): 7833-7841.

[44] WOLF S, DEOME CM, BEACHY RN, et al. Movement protein of tobacco mosaic virus modifies plasmodesmatal size exclusion limit[J]. Science, 1989, 246(4928): 377-379.

[45] XU H, LI Y, MAO Z, et al. Rice Dwarf Phytoreovirus Segment S11 Encodes a Nucleic Acid Binding Protein[J]. Virology, 1998, 240(2): 267-272.

[46] ZHOU P, REN B, ZHANG X, et al. Stable expression of rice dwarf virus Pns10 suppresses the post-transcriptional gene silencing in transgenic Nicotiana benthamiana plants[J]. ActaVirologica, 2010, 54(2): 99-104.

The P7-1 protein of southern rice black-streaked dwarf virus, a fijivirus, induces the formation of tubular structures in insect cells

Ying Liu,Dongsheng Jia,Hongyan Chen,Qian Chen,Lianhui Xie, Zujian Wu,Taiyun Wei

(Institute of Plant Virology, Fujian Province Key Laboratoryof Plant Virology, Fujian Agricultural and Forestry University,Jinshan, Fuzhou 350002, Fujian, People's Republic of China)

Abstract: Southern rice black-streaked dwarf virus (SRBSDV), an insect and plant-infecting reovirus of the genus *Fijivirus*, induced the formation of virus-containing tubules in infected plant and insect vector cells. Expression of the nonstructural protein P7-1 of SRBSDV in insect cells by a rccombinant baculovirus resulted in the formation of tubules with dimensions and appearance similar to those found in SRBSDV-infected cells. These tubules protruded from the cell surface, supporting the hypothesis that the P7-1 protein contains two putative transmembrane domains that are necessary for the formation of tubules. Furthermore, the self-interaction of SRBSDV P7-1 protein indicates that this protein has the capacity to form homodimers or oligomers to assemble the proposed helical symmetry structure of tubules. Taken together, our results indicate that SRBSDV P7-1 has the intrinsic ability of self-interaction to form tubules growing from the cell surface in the absence of other viral proteins.

1 Introduction

Plant-infecting reoviruses are found in the genera *Phytoreovirus*, *Fijivirus* and *Oryzavirus* of the family *Reoviridae* (Milne et al., 2005). They are transmitted by leafhopper or planthopper vectors in a persistent-propagative manner. The formation of tubules with viral particles inside is a common feature of plant and insect cells infected with plant-infecting reoviruses (Shikata et al., 1969; Hibino, 1996). Rice black-streaked dwarf virus (RBSDV) is a fijivirus that replicates both in plants and in an invertebrate insect vector (Milne et al., 2005). The viral genome consists of 10 segmented, double-stranded RNAs (dsRNAs), designated S1 through S10, in the order they appear in the electrophoretic migration pattern on SDS-polyacrylamide gels. The nonstructural protein P7-1 of RBSDV is localized to virus-containing tubules (Zhou et al., 2008), suggesting that RBSDV P7-1 is one of the constituents of the tubules. Recently, a tentative species in the genus *Fijivirus*, named Southern rice black-streaked dwarf virus (SRBSDV), has been proposed by two groups (Zhou et al., 2008; Zhang et al., 2008). In the past several years, SRBSDV has spread rapidly throughout southern China and northern Vietnam and has caused severe damage to rice in some regions (Earley et al., 2006). SRBSDV is transmitted by white-backed planthoppers, *Sogatella furcifera* (*Hemiptera: Delphacidae*), in a persistent-propagative manner. SRBSDV is most closely related to RBSDV, since both viruses have about 81% amino acid identity in P7-1, which is encoded by genomic segment S7 (Zhou et al., 2008; Zhang et al., 2008; Wang et al., 2010). Furthermore, SRBSDV infection also induces the formation of tubules in rice plants and insect

Archive of Virology. 2011, 156:1729–1736.
Received 21 March 2011; Accepted 25 May 2011;Published online 14 June 2011.

vector cells（Zhou et al., 2008; Fig. 1A）. However, the significance and origin of the tubules remain unknown. In the present study, we found that P7-1 of SRBSDV has the intrinsic ability to form tubules growing from the cell surface in the absence of other viral proteins.

2 Materials and methods

2.1 Cloning of the P7-1 gene and plasmid construction

Total RNAs were extracted from rice tissues infected with the isolate of SRBSDV from Fujian Province, China, using an RNeasy Plant Mini Kit (QIAGEN) following manufacturer's instructions. Reverse transcription was then performed using Superscript III reverse transcriptase (Invitrogen) and the appropriate reverse primers. The primer sequences used for cloning and mutagenesis in this study are listed in Supplemental Table 1. The reverse primer for SRBSDV P7-1 was fused in frame with the sequence of Strep-tag II. Gene sequences were amplified by PCR using Phusion DNA polymerase (NEB). The resulting cDNA fragments were purified and transferred by recombination into the entry vector pDONR221 (Invitrogen) using BP clonase II (Invitrogen), following the manufacturer's protocol. The insert of the resulting pDONR clone was verified by sequencing. The pDONR221 vector containing P7-1-Strep-tag II was transferred by recombination into the Gateway baculovirus vector pDESTTM8 to generate plasmid pDEST-P7-1-Strep-tag II. The pDONR221 vector containing SRBSDV P7-1 was recombined into binary destination vector pEarleygate101 (Earley et al., 2006; Wei et al., 2010) to generate plasmid P7-1-yellow fluorescent protein (P7-1-YFP). The vectors used in bimolecular fluorescence complementation (BiFC) assay were described previously (Wei et al., 2010). The pDONR221 vector containing SRBSDV P7-1 was recombined into BiFC vectors YN and YC to generate plasmids P7-1-YN and P7-1-YC. The yeast two-hybrid system used in this study was obtained from the DUAL membrane system (Dualsystems Biotech) and is suitable for identification of the interactions of transmembrane proteins. The fragment containing SRBSDV P7-1 was amplified by PCR and inserted into the DUAL membrane system bait vector pBT3-STE and the prey vector pPR3-N using the *Sfi* I restriction site to make the constructs SRBSDV-STE-P7-1 and SRBSDV-N-P7-1, respectively. The insert of the resulting DUALmembrane system clone was verified by sequencing.

Protein secondary structures were predicted based on their primary amino acid sequences using the CUBIC Predict Protein server (http://cubic.bioc.columbia.edu/pp/). Transmembrane helices (TM) in P7-1 were predicted using the programs PHDhtm (Rost et al., 1996), Tmpred (Hofmann et al., 1993) and TMHMM (Sonnhammer et al., 1998). The P7-1 mutants in which the putative TM1 and TM2 were deleted individually were amplified by the fusion PCR protocol, as described previously (Szewczyk et al., 2006; Atanassov et al., 2009). The PCR primer sequences and fusion PCR protocol are provided in Supplementary Table 1 and Fig. S1. The P7-1 mutant with the C-terminal 10 residues truncated was generated by PCR with the primers described in Supplementary Table 1. The resulting P7-1 fragments were recombined into plasmid pDONR221, and then into the Gateway baculovirus vector pDEST^{TM}8 to generate plasmids ΔTM1-Strep-tag II, ΔTM2-Strep-tag II and 1-348- Strep-tag II.

2.2 Expression of the P7-1 protein of SRBSDV by baculovirus

The recombinant baculovirus vectors P7-1-Strep-tag II, ΔTM1-Strep-tag II, ΔTM2-Strep-tag II and 1-348-Streptag II were introduced into DH10Bac (Invitrogen) for transposition into the bacmid. The recombinant bacmid was isolated and used to transfect *Spodopteafrugiperda* (Sf9) cells in the presence of

Cell FECTIN^{TM} (Invitrogen). For cytological observations, Sf9 cells on coverslips were infected with the recombinant baculoviruses. The cells were further incubated for 72h at 27℃ before being processed for immunofluorescence or electron microscopy.

2.3 Purification of tubules

The purification of tubules from baculovirus-infected Sf9 cells was performed basically as described previously (Monastyrskaya et al., 1994). Briefly, 72 hours post-inoculation (p.i.), Sf9 cells were harvested, washed with phosphate-buffered saline (PBS), and resuspended in STE buffer (150mM NaCl, 1mM EDTA, 10mM Tris-HCl [pH 7.5]) containing 0.5% Triton X-100. Cells were disrupted by homogenization, and the nuclei and cell debris were pelleted by centrifuging for 5min at 1,500g. The supernatant was loaded over a cushion of 3ml of 40% sucrose in STE buffer and centrifuged for 2h at 200,000g. The resulting pellet was resuspended in a small volume of STE buffer, loaded onto a 10% to 40% gradient of sucrose in STE buffer, and then centrifuged for 1h at 141,000g. The pellet and 1.5ml of each fraction of the gradient were collected, stained with 2% uranyl acetate, and then observed under an H-7650 electron microscope (Hitachi).

2.4 Immunofluorescence microscopy

At different times after inoculation, Sf9 cells on coverslips were fixed for 30min at room temperature in 2% paraformaldehyde (Sigma) diluted in water. Cells were then washed with PBS and permeabilized in PBS containing 0.1% Triton X-100. Then, the cells were incubated with a monoclonal anti-Strep-tag II antibody (IBA), diluted in PBS containing 0.5% bovine serum albumin (BSA), for 60min at 37℃. After three washes in PBS, fluorescein isothiocyanate (FITC) anti-mouse IgG (Sigma) diluted in PBS containing 0.5% BSA was added and incubated for 60min at 37℃. Cells were visualized using a LEICA TCSSP5 fluorescence microscope. Mock-infected cells were included in each experiment as a negative control and were processed in the same manner as the infected cells.

2.5 Immunoelectron microscopy

For electron microscopy, Sf9 cells on coverslips or salivary glands of viruliferous white-backed planthoppers were fixed, dehydrated, and embedded as described previously (Wei et al., 2006). For immunoelectron electron microscopy, sections on the grid were probed with monoclonal anti-Strep-tag II antibody followed by a 15-nm protein A-gold conjugate (Sigma) as described previously (Wei et al., 2006). All samples were observed under an electron microscope (H-7650; Hitachi). Transient expression in *Nicotiana benthamiana*.

Four-week-old *N. benthamiana* plants grown in a greenhouse at 22℃ to 24℃ were used for *Agrobacterium tumefaciens* (strain GV3101)-mediated transient expression of P7-1-YN, P7-1-YC and P7-1-YFP fusion proteins, as described previously (Sparkes et al., 2006). The relevant binary vectors for expressing P7-1-YN, P7-1-YC and P7-1-YFP fusion proteins were introduced into GV3101 by electroporation and infiltrated into leaf tissues using a 1-ml syringe by gentle pressure through the stomata on the lower epidermal surface. For leaf infiltration, agrobacteria harboring the relevant binary vectors were grown overnight in LB plus the appropriate antibiotics, collected by low-speed centrifugation, and then resuspended in 1ml of 10mM $MgCl_2$ containing 100μM acetosyringone.

2.6 Yeast two-hybrid assay

Yeast transformation was performed using the lithiumacetate-based protocol for preparing and transforming yeast competent cells following the instructions of the DUAL membrane starter kit user manual (Dualsystems Biotech). Briefly, yeast strain NMY51 was co-transformed with the bait vector SRBSDV-STE-P7-1 and the prey vector SRBSDV-N-P7-1. Simultaneously, NMY51 was co-transformed with the plasmid pTSU2-APP and pNubGFe65 or pPR3-N as a positive and a negative control, respectively. All transformants were grown on SD-trp-leu-his-ade plates for 3–4 d at 30℃. In addition, authentic transformants were confirmed by their color in the HTX galactosidase assay, as described in the DUAL membrane starter kit user manual.

3 Results

3.1 P7-1 induces tubule formation in vivo

Ultrathin sections of salivary glands of viruliferous whitebacked planthoppers showed tubules of about 85 nm in diameter with viral particles inside in the cytoplasm of infected cells (Fig.1A). To investigate whether P7-1 has an inherent ability to form these tubules, Sf9 cells were infected with a recombinant baculovirus expressing P7-1- Strep-tag II and incubated for different periods of time. By immunoblot analysis, P7-1-Strep-tag II was first detected at 24h p.i., and its expression reached maximum level at 72h p.i. (data not shown). By immunofluorescence microscopy, P7-1-Strep-tag II was shown to aggregate to form numerous tubule-like structures at 72h p.i. in all of the cells that were tested (Fig. 1B). These tubule-like structures extended from the cell surface (Fig. 1B). No specific fluorescence was detected in uninfected cells (data not shown).

Electron microscopy revealed that tubules of about 85nm in diameter were located in the cytoplasm or protruded from the Sf9 cell surface (Fig. 1C). These tubules had an appearance (e.g., diameter and length) similar to the tubules produced by wild-type virus (Fig. 1A). Furthermore, immunogold electron microscopy showed that P7-1 was localized in these tubules (Fig. 1D). Obviously, the results described above indicated that expression of P7-1 alone, in the absence of the virus replication process, is sufficient for tubule formation in Sf9 cells. Therefore, it is reasonable to conclude that P7-1 can self-aggregate to form tubules in SRBSDV infected host cells. No reaction with cellular structures was observed by incubation of uninfected cells with monoclonal anti-Strep-tag II antibody, or with viral structures, by incubation of infected cells with preimmune rabbit serum (data not shown).

To purify the tubules, infected cell extracts were prepared and subjected to sucrose density gradient centrifugation. A portion of each fraction was analyzed by immunoblotting (data not shown). The peak fractions containing P7-1 protein were pooled and pelleted. Electron microscopy confirmed that the preparation contained tubules of about 85 nm in diameter (Fig. 1E).

3.2 The putative transmembrane domains of P7-1 is required for tubuleformation in vivo

Based on secondary structure prediction using the CUBIC PredictProtein server (http://cubic.bioc.columbia.edu/pp/), P7-1 is composed of three regions, including a large N-terminal region that is predominantly composed of a-helices that are separated by short *β-strand* structures (Part I, residues 1–316), a region containing only *β-strand* (Part II, residues 323–340), and a small C-terminal tail (residues 334-358) with no apparent *α-helical* or *β-strand* structures (Fig. 2).

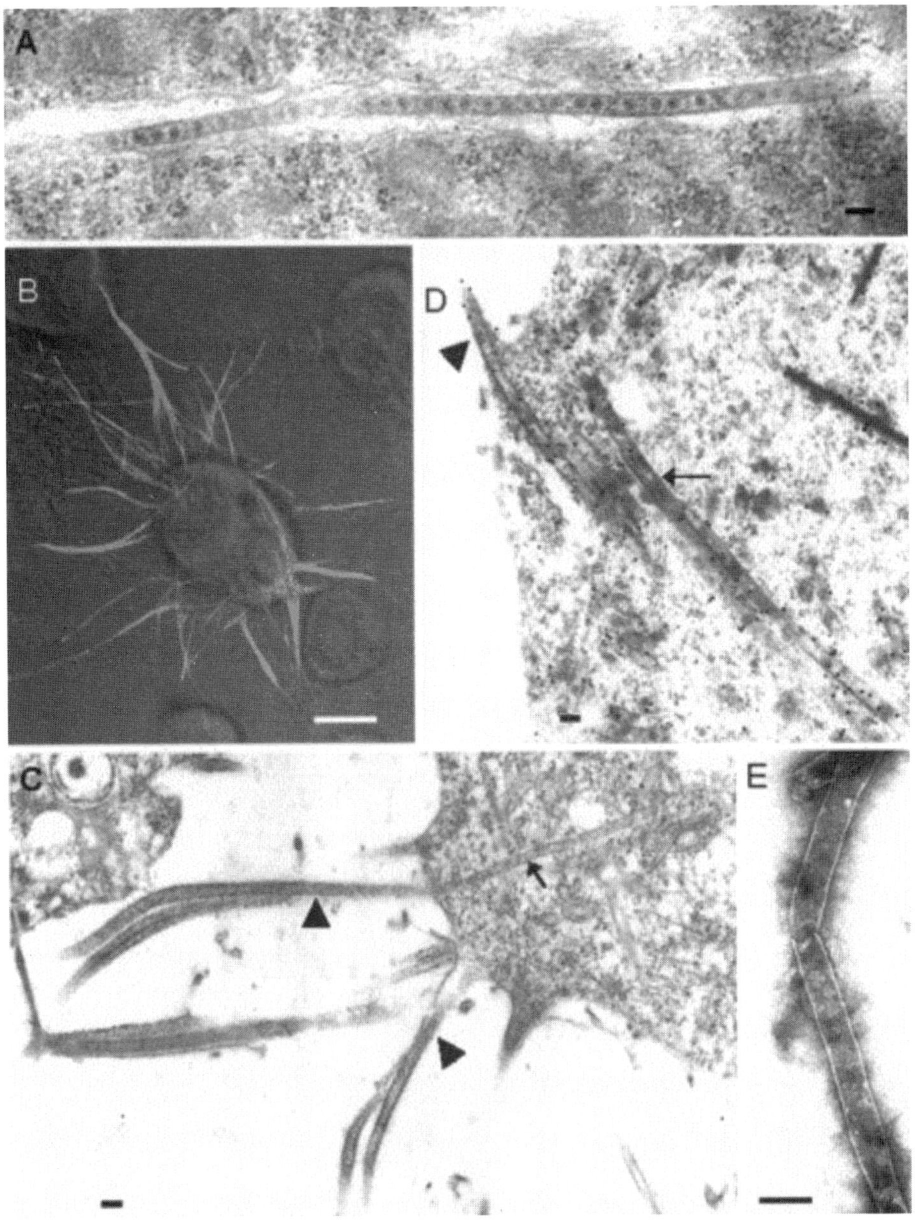

Fig. 1 Subcellular localization of P7-1 in baculovirus-infected Sf 9 cells at 72hpi

(A) Ultrathin section of salivary glands of viruliferous insect vector showing tubules of about 85nm in diameter with viral particles inside in the cytoplasm of infected cells. Bar, 100nm. (B) Immunofluorescence staining of P7-1 showing aggregation of numerous tubule-like structures in the cytoplasm or protruding from the cell surface. Bar, 10μm. (C) Electron micrograph showing the presence of tubules (of about 85nm in diameter) located in the cytoplasm (arrow) or protruding from Sf9 cell surface (arrowheads). Bar, 100nm. (D) Immunogold electron microscopy showing that the tubules specifically react with a monoclonal anti- Strep-tag II antibody in the cytoplasm (arrow) or protruding from the Sf9 cell surface (arrowhead). Bar, 100nm. (E) Electron micrograph of a purified tubule. Bar, 100nm. The image in B combines the images of P7-1-Strep-tag II fluorescence and that of differential interference contrast (DIC)

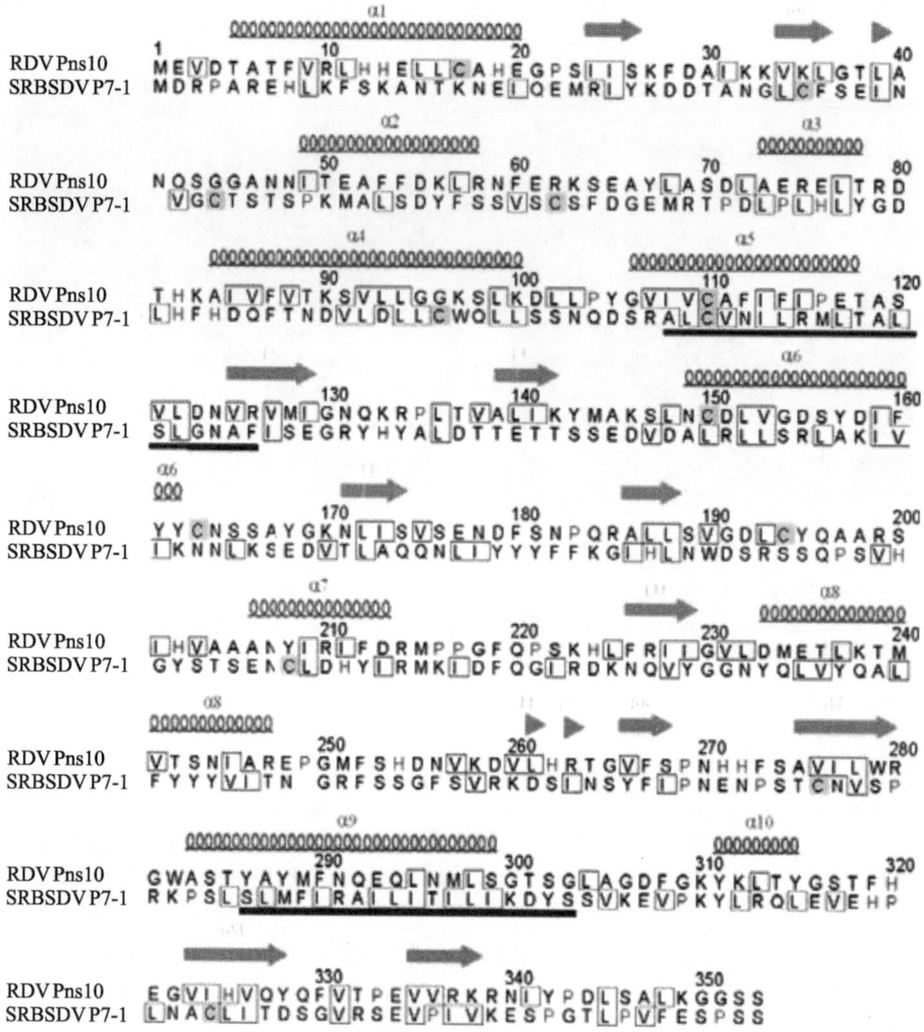

Fig.2 The amino acid sequences of the SRBSDV P7-1 and RDV Pns10 [17] proteins were submitted to the CUBIC Predict Protein server (http://cubic.bioc. columbia. edu/pp/) for secondary structure prediction

Secondary structure elements (e.g. a-helix 1–10 [a1–10] and b-strand A–N [bA-N]) are labeled above the sequence in red and blue. The two putative transmembrane domains (TM1 and TM2) of P7-1 are underlined in blue under the sequences. The hydrophobic residues Leu (L), Ile (I) and Val (V) are boxed

Computer analysis of the P7-1 amino acid sequence further revealed the presence of two putative a-helical transmembrane domains of 18 and 17 amino acid residues, respectively. The two predicted transmembrane segments were designated TM1 and TM2 (Fig. 3A). We generated two P7-1 deletion mutants (DTM1 and DTM2) in which the TM1 and TM2 were deleted, respectively, and found that both deletion mutants were diffusely distributed in the cytoplasm, different from the distribution of P7-1-Strp-tag II in the tubules (Fig. 3A, B). Thus, deletion of either the TM1 or TM2 domain of P7-1 abolished the ability to form tubules. However, another P7-1 mutant with the C-terminal 10 residues deleted retained the ability to form thetubule-like structures growing from the cell surface (Fig. 3A, B). These data indicate that the putative transmembrane domains are necessary for P7-1 to form tubules in Sf9 cells.

3.3 P7-1 has the capacity of self-assembly to form tubules

To determine whether there is direct interaction between P7-1 molecules to form homodimers or oligomers, a BiFC assay was carried out. P7-1-YN and P7-1-YC were coexpressed in *N. bethamiana* leaf cells, and indeed, there was strong BiFC fluorescence found in the inclusion bodies in the cytoplasm 48h post-infiltration (Fig. 4A), whereas no BiFC fluorescence was visible in the negative control (data not shown). This result suggested that there was self-interaction of P7-1 *in planta*. To investigate whether P7-1 could form tubule-like structures in a non-host plant, *N. benthamiana* was used for transient expression via agoinfiltration (Sparkes et al., 2006). We found that P7-1-YFP formed inclusion bodies rather than tubule-like structures in the cytoplasm of *N. benthamiana* leaf cells 48 h post-infiltration (Fig. 4B).

Because P7-1 is a membrane-protein, we took advantage of the DUAL membrane system, a variant of the yeast twohybrid approach, to detect P7-1 self-interaction on the surface of cellular membranes. Yeast cells co-transformed with all possible pairwise combinations of plasmids containing the NubG and Cub fusion proteins were then assayed for histidine and adenine prototrophs. As a result, only yeast expressing the homologous combination of P7-1 showed growth (Fig. 4 C, D). These results confirmed that the specific self-interaction is not an artifact of the BiFC assay, and thus SRBSDV P7-1 protein was found to interact specifically with itself.

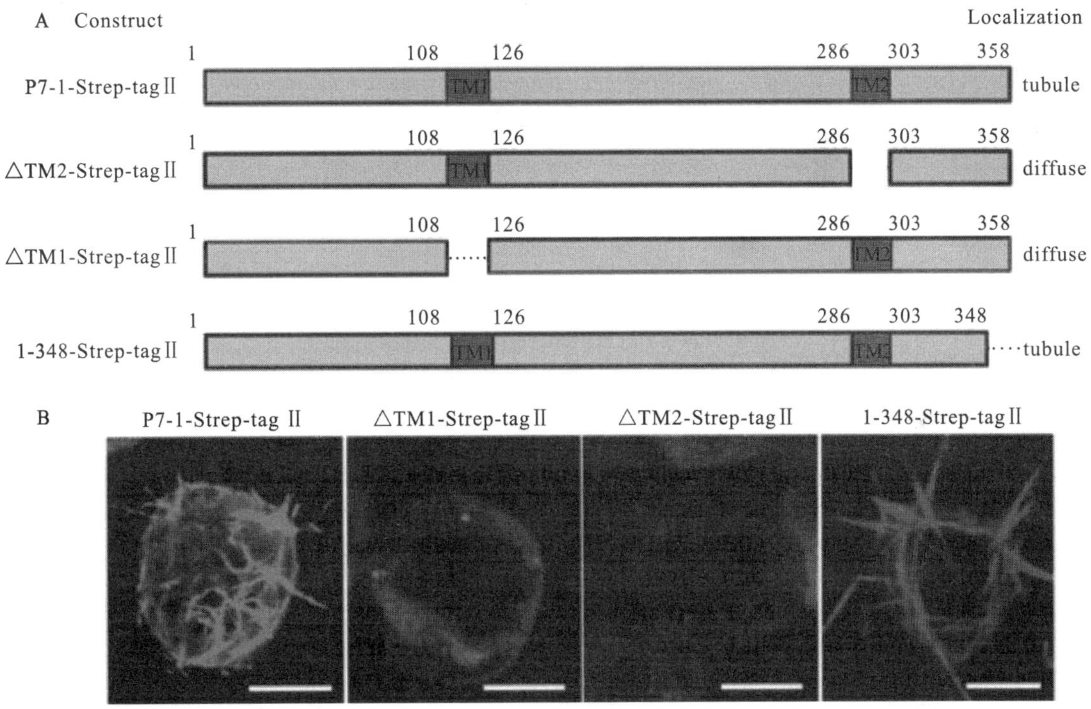

Fig. 3 (A) Schematic illustrations of SRBSDV P7-1 protein, which is 358 amino acids in length and includes two putative transmembrane (TM) domains

Portions of the putative TM1 and TM2 are colored red. Construct names were indicated at the left. Blocks represent the region of P7-1 contained in the construct. Discontinuous lines represent the deleted region. Numbers indicate the relative amino acid positions in P7-1. The subcellular distribution of protein derived from each construct is summarized at the right. (B) Subcellular localization of various deletion mutants of SRBSDV P7-1 in Sf9 cells. Sf9 cells were infected with baculovirus encoding P7-1-Strep-tag II, DTM1- Strep-tag II, DTM2-Strep-tag II or 1-348-Strep-tag II. Cells were fixed 3 h p.i., stained with monoclonal anti-Strep-tag II and FITC anti-mouse IgG as a secondary antibody, and visualized by confocal fluorescence microscopy. Bars, 10μm

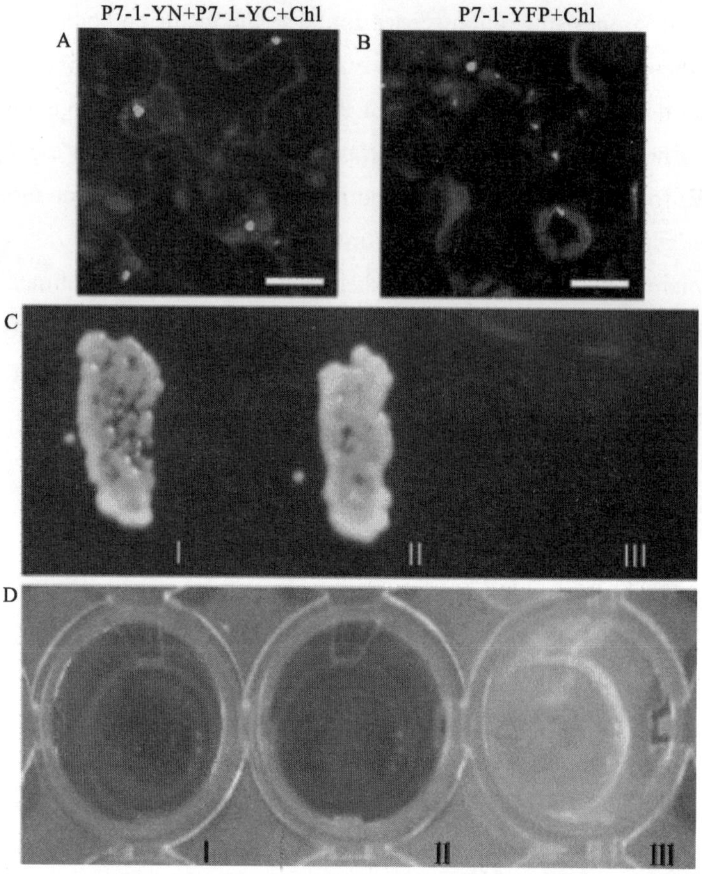

Fig. 4 The direct self-interaction of SRBSDV P7-1

(A) BiFC array showing the self-interaction of P7-1 in planta. A strong BiFC fluorescence was observed in inclusion bodies in the cytoplasm of leaf cells of *N. benthamiana* coexpressing P7-1-YN and P7-1-YC 48h post-agro-infiltration. Chl, chlorophyll autofluorescence. Bar, 10μm. (B) P7-1-YFP formed inclusion bodies in the cytoplasm in the leaf cells of *N. benthamiana* 48h postinfiltration. Bar, 10μm. (C) Yeast two-hybrid assay showing the self-interaction of P7-1 protein. Transformants were grown on SD-trp-leu-hisade plates for 3d. (D) Transformants appear colored in the HTX b-galactosidase assay. I, Positive transformant of pTSU2-APP/pNubG-Fe65. II, Transformants of pBT3-STE-P7-1/pPR3-N-P7-1. III, Negative transformant of pTSU2-APP/pPR3-N

4 Discussion

Expression of one of the SRBSDV nonstructural proteins (P7-1) fused with Strep-tag II (P7-1-Strep-tag II) in Sf9 cells can induce the formation of tubules with dimensions and appearance similar to those found in SRBSDV-infected plant and insect vector cells (Zhou et al., 2008; Fig. 1). These results demonstrate that P7-1 of SRBSDV is responsible for the formation of tubular structure and that the production of tubules is not specific to host plant or insect vector cells. The predicted secondary structure of SRBSDV P7-1 shows a high degree of similarity to the predicted secondary structure of the nonstructural protein Pns10 of rice dwarf virus (RDV), a phytoreovirus (Milne et al., 2005; Wei et al., 2006; Fig. 2). The expression of Pns10 in Sf9 cells is known to induce the formation of tubules similar to those in RDV-infected hosts (Wei et al., 2006). Therefore, SRBSDV P7-1 and RDV Pns10, both of which are major constituents of tubular structures, appear to have similar structures with a conserved assembly strategy. Katayama et al. (Katayama et al., 2007) have previously shown, using tomography and computer image processing, that the tubules of RDV Pns10 have a helical symmetry structure. We show in the current study that the purified tubules of SRBSDV P7-1 also appear to have a similar helical symmetry structure, suggesting a similar strategy for the assembly of tubules by SRBSDV and RDV. Our observations that P7-1 can interact with itself in yeast two-hybrid and BiFC assays provide evidence that this protein has the capacity of forming homodimers or oligomers (Fig. 4), which may play a role in forming the helical structure of the tubules. This type of helical structure is likely formed by end-to-end interactions between P7-1 molecules. In support of this contention, our data show that YFP fusion with the C-terminus of P7-1 (P7-1-YFP) abrogates the correct assembly of P7-1 to form tubular structures in nonhost *N. bethamiana* leaf cells, and it instead forms an inclusion body.

Our microscopic observations indicate that the tubules of SRBSDV P7-1 protrude from the plasma membrane of Sf9 cells, suggesting that P7-1 is a membrane-associated protein. Computer analysis of membrane-spanning domains within the amino acid sequence of P7-1 identifies two putative α -helical transmembrane domains of approximately 18 and 17 amino acid residues, respectively (Fig. 2, 3). Deletion of either of the two putative transmembrane domains can abolish the formation of tubules in Sf9 cells, and the deletion mutants exhibit a punctate distribution in the cytoplasm (Fig. 2, 3). By contrast, the deletion of the C-terminal 10 residues of P7-1, which did not form α -helical and β -strand structures, does not eliminate the ability to form tubules in Sf9 cells (Fig. 2, 3). These data suggest that proper membrane association is necessary for the assembly of P7-1 into tubules.

The morphology of the tubules with virions inside them in host cells infected with plant-infecting reoviruses is similar to the structures formed by the movement proteins of several other plant viruses, such as cowpea mosaic virus, which form tubules that are embedded within highly modified plasmodesmata that transport viral particles (Ritzenthaler et al., 2007). The tubules of plant-infecting reoviruses have never been found to be associated with the cell wall (Shikata 1969; Hibino et al., 1996) suggesting that these tubules are not related to the cell-to-cell movement of viral particles in plants. By using insect vector cells in monolayers, we recently determined the functional roles of the tubules induced by a phytoreovirus, RDV, and found that the virus employs the tubules formed by RDV Pns10 to move along the actin-based filopodia extending toward neighboring insect vector cells to enhance viral movement (Wei et al., 2006; Wei et al., 2008; Pu et al., 2011). The tubules containing RDV particles are found in association with the actin-based microvilli of the midgut in viruliferous vector leafhoppers (Nasu, 1965). Therefore, RDV might utilize these tubules to spread among cells of vector insects. It will be interesting to examine whether the similar tubules induced by fijiviruses play a similar role in the spread of virus among their insect vector cells. In this regard, the observation that P7-1 tubules of SRBSDV protrude from the plasma membrane of Sf9 cells suggests that the virus-containing tubules induced by fijiviruses infection can move out of the infected insect vector cells, providing a possible route for viral spread among insect vector cells.

The appearance of tubular structures during viral morphogenesis has been reported in studies of other members of the family *Reoviridae*. The infection and replication of bluetongue virus (BTV), member of the genus *Orbivirus* of the family *Reoviridae*, is accompanied by the formation of tubules of about 52nm in diameter that are composed of the nonstructural protein NS1 (Monastyrskay et al., 1994). The tubules composed of P7-1 of SRBSDV are morphologically similar to the NS1 tubules of BTV. However, SRBSDV tubules contain viral particles within them and have the ability to protrude from the plasma membrane of non-host Sf9 cells (Monastyrskay et al., 1994). By contrast, BTV tubules do not contain viral particles within them and the expression of NS1 results in formation of tubules within the cytoplasm, rather than protruding from the plasma membrane of Sf9 cells. The NS1 protein of BTV has been reported to play a direct role in the cellular pathogenesis of BTV (Owens et al., 2004). Therefore, the tubules induced by plant-infecting reoviruses and by BTV play different roles in viral infection.

Acknowledgements

This work was supported by grants from the programs for the National Basic Research Program of China (No. 2010CB126203), Specialized Research Fund for the Ministry of Agriculture (No. 201003031), New Century Excellent Talents in University (No. NCET-09-0011), the National Science Foundation of China (No. 31070130) and the Natural Science Foundation of Fujian Province (No. 2010J01075).

References

[1] ATANASSOV I I, ATANASSOV I I, ETCHELLS J P, et al. A simple, flexible and efficient PCR-fusion/Gateway cloning procedure for gene fusion, site-directed mutagenesis, short sequence insertion and domain deletions and swaps[J]. Plant Methods, 2009, 5(1): 14-14.

[2] EARLEY K W, HAAG J R, PONTES O, et al. Gateway-compatible vectors for plant functional genomics and proteomics[J]. The Plant Journal, 2010, 45(4): 616-629.

[3] HIBINO H. Biology and epidemiology of rice viruses[J]. Annual Review of Phytopathology, 1996, 34(1): 249-274.

[4] HOFMANN K, TMBASE S W. TMBASE-A database of membrane spanning protein segments[J]. Biological chemistry Hoppe-Seyler, 1993, 374(166).

[5] ISOGAI M, UYEDA I, LEE B C. Detection and assignment of proteins encoded by rice black streaked dwarf fijivirus S7, S8, S9 and S10[J]. Journal of General Virology, 1998, 79(6): 1487-1494.

[6] KATAYAMA S, WEI T, OMURA T, et al. Three-dimensional architecture of virus-packed tubule[J]. Journal of Electron Microscopy, 2007, 56(3): 77-81.

[7] Milne, RG, del Vas M, Harding R M, et al. In: Fauquet, C M, Mayo, M A, Maniloff, J, Desselberger, U, Ball, L A eds. Virus taxonomy: Classification and nomenclature of viruses, Eighth report of the international committee on taxonomy of viruses[J]. Elsevier, Academic Press; Amsterdam, Holland, 2005, 450-606.

[8] MONASTYRSKAYA K, BOOTH T, NEL L, et al. Mutation of either of two cysteine residues or deletion of the amino or carboxy terminus of nonstructural protein NS1 of bluetongue virus abrogates virus-specified tubule formation in insect cells[J]. Journal of Virology, 1994, 68(4): 2169-2178.

[9] NASU S. Electron Microscopic Studies on Transovarial Passage of Rice Dwarf Virus[J]. Japanese Journal of Applied Entomology & Zoology, 1965, 9(3): 225-237.

[10] OWENS R J, LIMN C, ROY P. Role of an arbovirus nonstructural protein in cellular pathogenesis and virus release[J]. Journal of Virology, 2004, 78(12): 6649-6656.

[11] PU Y, KIKUCHI A, MORIYASU Y, et al. Rice dwarf viruses with dysfunctional genomes generated in plants are filtered out in vector insects—implications for the virus origin[J]. Journal of Virology, 2010, 85(6): 2975-2979.

[12] RITZENTHALER C, HOFMANN C. Tubule-Guided Movement of Plant Viruses[J]. Springer Berlin Heidelberg, 1998, 7: 64-83.

[13] ROST B, CASADIO R, FARISELLI P.Refining neural network predictions for helical transmembrane proteins by dynamic programming[J].Proceedings International Conference on Intelligent Systems for Molecular Biolog,1996,4: 192-200.

[14] SHIKATA E. Electron microscopic studies on rice viruses: the virus disease of the rice plant[J]. Johns Hopkins University Press, Baltimore, 1969, 223-240.

[15] SONNHAMMER E L, VON HEIJNE G, KROGH A. A hidden markov model for predicting transmembrane helices in protein sequences[J].Proceedings International Conference on Intelligent Systems for Molecular Biological, 1998, 6: 175-182.

[16] SPARKES I A, RUNIONS J, KEARNS A, et al. Rapid, transient expression of fluorescent fusion proteins in tobacco plants and generation of stably transformed plants[J]. Nature Protocols, 2006, 1(4): 2019-2025.

[17] SZEWCZYK E, NAYAK T, OAKLEY C E, et al. Fusion PCR and gene targeting in Aspergillus nidulans[J]. Nature Protocols, 2006, 1(6): 3111-3120.

[18] WANG Q, YANG J, ZHOU G H, et al. The Complete Genome Sequence of Two Isolates of Southern rice black-streaked dwarf virus, a New Member of the Genus Fijivirus[J]. Journal of Phytopathology, 2010, 158: 733-737.

[19] WEI T, HUANG T S, MCNEIL J, et al. Sequential recruitment of the endoplasmic reticulum and chloroplasts for plant potyvirus replication[J]. Journal of Virology, 2010, 84(2): 799-809.

[20] WEI T, ZHANG C, HON, J, et al. Formation of complexes at plasmodesmata for potyvirus intercellular movement is mediated by the viral protein P3N-PIPO[J]. PLOS Pathogens, 2010, 6(6): e1000962.

[21] WEI T, KIKUCHI A, MORIYASU Y, et al. The spread of Rice dwarf virus among cells of its insect vector exploits virus-induced tubular structures[J]. Journal of Virology, 2006, 80(17): 8593-8602.

[22] WEI T, SHIMIZU T, OMURA T. Endomembranes and myosin mediate assembly into tubules of Pns10 of Rice dwarf virus and intercellular spreading of the virus in cultured insect vector cells[J]. Virology, 2008, 372(2): 349-356.

[23] ZHANG H M, YANG J, CHEN J P, et al. A black-streaked dwarf disease on rice in China is caused by a novel Fijivirus[J]. Archives of Virology, 2008, 153(10): 1893-1898.

[24] ZHOU G, WEN J, CAI D, et al. Southern rice black-streaked dwarf virus: a new proposed Fijivirus species in the family Reoviridae[J]. Chinese Science Bulletin, 2008, 53(23): 3677-3685.

Identification of Pns12 as the second silencing suppressor of *Rice gall dwarf virus*

Jianguo Wu[1,2], Chunzheng Wang[2], Zhengguo Du[1], Lijun Cai[1], Meiqun Hu[1], Zujian Wu[1], Yi Li[2], Lianhui Xie[1]

(1 Institute of Plant Virology, Fujian Agriculture and Forestry University, Key Laboratory of Plant Virology of Fujian Province, Fuzhou 350002, China; 2 Peking-Yale Joint Center for Plant Molecular Genetics and Agrobiotechnology, National Laboratory of Protein Engineering and Plant Genetic Engineering, College of Life Sciences, Peking University, Beijing 100871, China)

Abstract: RNA silencing is a conserved mechanism found ubiquitously in eukaryotic organisms. It has been used to regulate gene expression and development. In addition, RNA silencing serves as an important mechanism in plants' defense against invasive nucleic acids, such as viruses, transposons, and transgenes. As a counter-defense, most plants, and some animal viruses, encode RNA silencing suppressors to interfere at one or several points of the silencing pathway. In this study, we showed that Pns12 of RGDV (*Rice gall dwarf virus*) exhibits silencing suppressor activity on the reporter green fluorescent protein in transgenic *Nicotiana benthamiana* line 16c. Pns12 of RGDV suppressed local silencing induced by sense RNA but had no effect on that induced by dsRNA. Expression of Pns12 also enhanced *Potato virus X* pathogenicity in *N. benthamiana*. Collectively, these results suggested that RGDV Pns12 functions as a virus suppressor of RNA silencing, which might target an upstream step of dsRNA formation in the RNA silencing pathway. Furthermore, we showed that Pns12 is localized mainly in the nucleus of *N. benthamiana*leaf cells.

Key words: *Rice gall dwarf virus*; Pns12; silencing suppressor

1 Introduction

RNA silencing is a conserved mechanism used by plantsand other eukaryotes to defend themselves against virusesand transposons (Ding and Voinnet, 2007). Generally, RNA silencing is triggeredby dsRNAs. The dsRNAs are recognized by specific cellular enzymes and cleaved into small interfering RNAs(siRNAs) of 20–24nt in length. The siRNAs function as guides for a silencing complex that can block gene expression at transcriptional, post-transcriptional, and translational levels (Fire, 1999; Caplen et al., 2000; Hammond et al., 2000; Caplen et al., 2001; Elbashir et al., 2001). Most viruses produce dsRNAs during their life cycles. This means that viruses can be both triggers and targets of RNA silencing (Fire et al., 1998; Zamore, 2002). As a counter-defense, most plant viruses have evolved to encode a protein that can suppress RNA silencing (Ding, 2000; Vance and Vaucheret, 2001; Ding et al., 2004; Roth et al., 2004; Voinnet, 2005; Ding and Voinnet, 2007).

A number of studies have shown that VSRs (viral suppressors of RNA silencing) target distinct steps of RNA silencing: from the production of dsRNAs to the incorporation of siRNAs into the RNA-induced silencing

Science China Life Sciences. 2011, 54: 201–208
Received 3 April 2010; Accepted 19 July 2010.

complex (RISC)(Qi et al., 2004; Li and Ding, 2006; Ding and Voinnet, 2007). In addition, some VSRs have been shown to interfere with the movement of silencing signals. A number of VSRs have been identified from plant and animal viruses(Anandalakshmi et al., 1998; Brigneti et al., 1998; Kasschau and Carrington, 1998; Cao et al., 2005; Lecellier et al., 2005; Zhang et al., 2005; Li and Ding, 2006; Liu et al., 2008; Schnettler et al., 2008; Shi et al., 2009; Xiong et al., 2009). Although most plant viruses only have one VSR,some plant viruses encode more than one silencing suppressor, which target distinct steps of RNA silencing (Lu et al., 2004; Gopal et al., 2007).

Rice gall dwarf virus (RGDV) is a member of the genus *Phytoreovirus* under the family *Reoviridae*. Besides RGDV,viruses of the genus *Phytoreovirus* include *Rice dwarf virus*, *Wound tumor virus*, *Rice bunchy stunt virus*, *Tobacco leaf enation virus* and *Homalodisca vitripennis virus* (HVP) (Omura et al., 1982; Xie et al., 1996; Katsar et al., 2007; Picton et al., 2007). RGDV has a double-shelled icosahedral particle. Its genome consists of 12 double-stranded RNAs, which encode six structural (P1, P2, P3, P5, P6, and P8) and six non-structural (Pns4, Pns7, Pns9, Pns10, Pns11, and Pns12) proteins. The core capsid is composed of P3, the major structural protein, which encloses P1, P5, and P6. The outer layer consists of two proteins, P2 and P8 (Wei et al., 2009). RGDV is transmitted by *Recilia dorsalis* and *Nephotettix cincticepsin* a persistent and circulative manner (Francki and Boccardo, 1983). The complete sequences of RGDV Thai and Guangxi isolates have been obtained, which paved the way for molecular studies on the functions of RGDV gene products and the RGDV-rice interaction (Ichimi et al., 2002; Moriyasu et al., 2007; Zhang et al., 2007b; Zhang et al., 2008).

S12, the smallest RNA segment of RGDV, is 853bp long. It contains an ORF of 620 bp, from 31 to 650nt (Zhang et al., 2007a).To date, the function of the protein encoded by S12 and Pns12 remains unknown. A previous report showed that RGDV S11 encodes a silencing suppressor (Liu et al., 2008). In this study, we present evidence that Pns12 of RGDV also has silencing suppressor activities. Pns12 of RGDV exhibited silencing suppressor activity in coinfiltration assays with the reporter green fluorescent protein (GFP) in transgenic *Nicotiana benthamiana* line 16c. It suppressed local silencing induced by sense RNA, but had no effect on that induced by dsRNA. Expression of Pns12 also enhanced *Potato virus X*pathogenicity in *N. benthamiana*. In addition, we showed that Pns12 localizes mainly to the nucleus in *N. benthamiana* cells, which suggested that Pns12 might function in the nucleus.

2 Materials and methods

2.1 Plasmids and Agrobacterium

Rice plants infected by RGDV, which were collected from Xinyi, Guangdong Province, were used as the source of RGDV. The viruses were transmitted artificially using its insect vectors to rice seedlings and the seedlings were grown and maintained in the greenhouse of the Institute of Plant Virology, Fujian Agriculture and Forestry University.The primers for S12, 5′-CGC*AAGCTT*ATGACGAGCAACGAGGAAAAC-3′ and 5′-GAG*GAGCTC*TTACCTCGGTCTTCGTTTAC-3′, were designed according to sequencesin NCBI (GenBank accession No. EF177263; restriction enzyme cleavage sites are underlined). The fullopen reading frame of Pns12 was amplified from pMD-18T vector containing RGDV S12. The PCR products were cleaved using *Hind* III and *Sac* I and then transferred into the binary vector pPZP212 (a kind gift from Dr. Powers),creating pPZP212-S12 (named 35S:S12 for simplicity)(Powers et al., 2008).

The primer 5′-CGC*AAGCTTT*GACGAGCAACGAGGAAAAC-3′ was used to obtain ΔS12. PCR amplification using this primer results in the deletion of the first nucleicacid of the S12 ORF. Δ S12 was transferred into the same binary vector as S12, resulting in pPZP212- Δ S12 (named 35S: Δ S12 for simplicity). Deletion of the first nucleic acid of the S12 ORF leads to a construct that can be transcribed but cannot be translated into a protein. Similarly, the primer 5′-ATA*CCCGGG*TGACGAGCAACGAGGAAAAC-3′ wasused to create PVX: Δ S12 (see below).

The ORF of S12 was amplified using primers 5′-ATA*CCCGGG*ATGACGAGCAACGAGGAAAAC-3′ and 5′-ACG*GTCGAC*TTACCTCGGTCTTCGTTTAC-3′, which contains recognition sites for the enzymes *Sma* I and *Sal* I.The PCR products were ligated into the vector pGR107 (akind gift from David Baulcombe) after double digestion using *Sma* I and *Sal* I. The recombinant plasmids were named PVX:S12. PVX: Δ S12 was obtained by using the primer 5′-ATA*CCCGGG*TGACGAGCAACGAGGAAAAC-3′. The S12 ORF lacking the stop codon was obtained using the primers 5′-ATA*CCATGG*AAATGACGAGCAACGAGGAAA-3′ and 5′-ATA-*CCATG*GACCTCGGTCTTCGTTTACTTT-3′, which contained the recognition sites for the enzyme *Nco* I. The PCR products were ligated into the vector pRTL2-GFP, resulting in pRTL2-S12:GFP, which, after digestion with the enzyme *Hind* III, was transferred into the binary vector pCAMBIA1301, creating pCAMBIA1301-S12:GFP (35S:S12:GFP for simplicity).

The constructs pPZP212, 35S:S12, 35S: Δ S12, 35S:ssGFP,35S:dsGFP, 35S:TAV 2b, PVX, PVX:S12, PVX: Δ S12, and 35S:S12:GFP were sequenced to verify their correct construction and transformed into *Agrobacterium tumefaciens* GV3101 (kindly provided by David Baulcombe) by electroporation.

2.2 Agroinfiltration and GFP imaging

The *Nicotiana benthamiana* plant constitutively expressinga GFP transgene (line 16c; a gift from David Baulcombe) and the Agrobacterium infiltration operation have been described previously (Hamilton et al., 2002). The *N. benthamiana* line 16c plants were cultured in growth chambers at 22–24℃ before and after infiltration. GFP fluorescence was observed under long wave length UV light (Black Ray model B 100A; UV Products) and photographed using a Nikon D70 digital camera with a Y48 yellow filter.

2.3 RNA extraction and Northern blotting

Total RNAs were extracted from leaves with TRIzol reagent(Invitrogen) in accord with the manufacturer's instructions. Northern blotting analysis was conducted according to instructions described in "the DIG system user's guide for filter hybridization". For Northern blot analysis of siRNAs, low-molecular-weight RNAs were enriched from total RNAs by eliminating high-molecular-weight RNA using 5% polyethylene glycol (PEG 8000) plus 0.5mol • L^{-1} NaCl, separating on a 15% polyacrylamide-7mol L^{-1} urea gel, and transferring to Hybond-N membranes. The hybridization and detection of siRNAs were performed as described previously (Goto et al., 2003). The probes used in the analysis of siRNA were the same as those for Northern blots of mRNA.

2.4 Localization of Pns12

The leaves of four-week-old *N. benthamiana* plants were agroinfiltrated with *A. tumefaciens* harboring pCAMBIA1301-S12:GFP and pCAMBIA1301-GFP, respectively.By 3dpi (days postinfiltration), the GFP expression in the leaves was observed at 488nm using a confocal microscope(Leica TCS STED, Germany) and DAPI (2-(4-Amidinophenyl)-6-indolecarbamidine dihydrochloride) to counterstain the nuclei. Images were captured using LAS AF Litesoftware and converted to tagged image file format for export.

3 Results

3.1 RGDV Pns12 inhibits local silencing induced by ssGFP

To determine whether RGDV Pns12 has silencing suppressor activities, a method named the agroinfiltration bioassay was used. A transformed Agrobacterium strain carrying RGDV Pns12 was mixed with a strain that carried 35S-GFP at a ratio of 3:1 and infiltrated into leaves of *N. benthamiana*line 16c. Agrobacteria harboring only the GFP geneor the 2b gene of *Tomato aspermy cucumovirus* (TAV) was used as negative and positive controls, respectively. In each treatment, 15 plants were infiltrated and the experiments were repeated at least three times. GFP fluorescence became visible at 2dpi in localized regions. However, the green fluorescence intensity in the patches infiltrated with GFP alone or with GFP plus Δ S12, in which the first nucleic acid of the S12 ORF was deleted, became invisible after 7dpi. By contrast,the green fluorescence intensity remained strong at 7dpi in the patches coinfiltrated with 35S-GFP plus 35S-S12 and 35S-GFP plus 35S-TAV 2b, respectively (Fig. 1A–F).These results suggest that Pns12 could suppress the RNA silencing induced by sense-GFP, presumably in a similar manner to TAV 2b. In addition, the fact that Δ S12, which could not produce an intact Pns12, had no effect on RNA silencing indicated it was the protein expressed from S12,but not the RNA, that could function as a silencing suppressor.

To confirm that the above observations resulted from differential accumulation of GFP mRNA, Northern blot analyses were conducted to detect steady-state levels of GFP mRNA using digoxin labeled probes. The resultsshowed that GFP mRNA accumulated in tissues expressing 35S:GFP plus 35S:TAV 2b or 35S:GFP plus 35S:S12 and were high at 7dpi, although the accumulated levels of GFP mRNA in tissues expressing 35S:GFP plus 35S:S12 were much lower than in tissues expressing 35S:GFP plus 35S:TAV 2b. By contrast, GFP mRNA could not be detecte din tissues expressing GFP alone or 35S:GFP plus 35S: Δ S12 at the same time point (Fig. 1G).

To further test whether RNA silencing suppressor activities of Pns12 were responsible for the above observations, Northern blot analyses were conducted to detect GFP-specific siRNAs. In all treatments, higher accumulation of GFP mRNA was correlated to lower accumulation of GFP-specific siRNAs, and lower accumulation of GFP mRNA was accompanied by higher accumulation of GFP-specific siRNAs (Fig. 1G). These results indicated that transcription of the exogenously introduced sense-GFP can induceco-suppression or RNA silencing of both the exogenous and the endogenous GFP. RGDV Pns12 can suppress the RNA silencing induced by sense-GFP, in a similar manner to TAV 2b, although its silencing suppressor activities were much weaker than the latter.

3.2 Pns12 does not suppress local RNA silencing triggeredby GFP dsRNA

DsRNA, as found by Fire et al. in 1998, is the initiator of RNA silencing. RNA silencing involves three distinct stages: the generation of dsRNA, the cleavage of the dsRNAs into siRNAs 20–26 nt in length, and the incorporation of the siRNAs into an effector complex (Fire et al., 1998; Li and Ding, 2001; Roth et al., 2004). Manystudies have used dsGFP in co-infiltration assays to determine the molecular targets of a particular VSR. It was found that VSRs, including Pns10 of *Rice dwarf virus*(Cao et al., 2005), P69 of *Turnip yellow mosaic virus* (TYMV) (Chen et al., 2004), and P25 of *Potatovirus X* (PVX) (Voinnet et al., 2000) could not inhibit RNA silencing induced by dsRNAs, suggesting that these VSRs targeted astep upstream of dsRNA formation in RNA silencing. Our results indicated that Pns12 could inhibit local silencing of GFP triggered by ssGFP. However, the targets of Pns12 inthe RNA silencing were unknown.

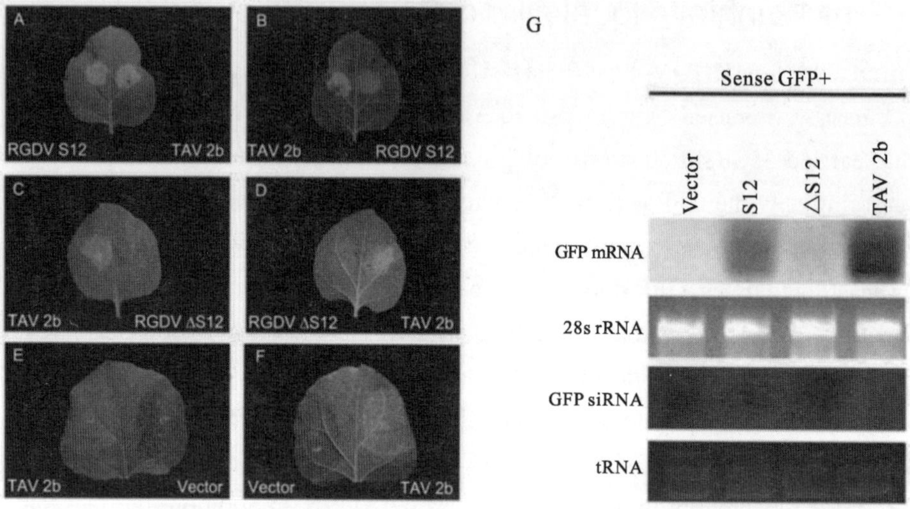

Fig. 1 Suppression of local GFP silencing by RGDV Pns12 (A–F)

N. benthamiana line 16c plants were coinfiltrated with *Agrobacterium* spp. (Agro.) mixtures carrying 35S-GFP and the individual constructs indicated in each image. GFP fluorescence was viewed under long wavelength UV light at 7d postinfiltration (dpi). G, Northern blot analysis of the steady-state levels of GFP mRNA and siRNA extracted from different infiltrated patches shown in A–F.rRNA and tRNA were used as loading controls for detection of GFP mRNA and GFP siRNA, respectively

The effect of Pns12 on dsGFP-triggered silencing wastested, using 35S:TAV 2b plus 35S:ssGFP plus 35S:dsGFP and 35S:ssGFP plus 35S:dsGFP plus 35S: Δ S12 or the empty vector as positive and negative controls, respectively. As shown in Fig. 2, leaves infiltrated with 35S-ssGFP plus 35S-dsGFP or with 35S-ssGFP plus 35S-dsGFP plus35S-S12 (or 35S- Δ S12) lost GFP fluorescence at 7dpi, indicating strong local GFP RNA silencing. This indicated that Pns12 could not suppress local silencing induced by dsRNA. As expected, strong GFP fluorescence was maintained in leaves infiltrated with 35S-ssGFP plus 35S-dsGFP plus 35S:TAV 2b, indicating that TAV 2b suppressed local GFP RNA silencing triggered by dsRNA (Fig. 2A–F).Northern blot analyses showed negligible accumulation of GFP mRNA and high accumulation of GFP-specific siRNAsin leaves infiltrated with 35S-ssGFP plus 35S-dsGFP and 35S-ssGFP plus 35S-dsGFP plus 35S:S12/35S: Δ S12 (Fig. 2G), further demonstrating that Pns12 did not suppress local RNA silencing triggered by GFP dsRNA. By contrast,leaves infiltrated with 35S:ssGFP plus 35S:dsGFP plus 35S:TAV 2b showed high accumulation of GFP mRNA and much reduced accumulation of siRNA (Fig. 2G). These results suggested that Pns12 targets a step upstream of dsRNA synthesis in the RNA silencing pathway.

3.3 Pns12 enhances PVX pathogenicity

Studies of Voinnet et al. suggested that the RNA silencing triggered by ssRNA and viruses might involve different mechanisms. P25 of PVX can inhibit ssRNA induced silencing(Voinnet et al., 2000). However, it had no apparent effects on silencing induced by viruses (Voinnet et al., 2000). Synergism refers to a phenomenonin which plants infected with two unrelated viruses displayed increased disease symptoms compared with plants infected with either of the two viruses alone (Pruss et al., 1997). Currently, it is generally accepted that synergism is the result of very effective suppression of silencing by one virus, resulting in a dramatic increase in the accumulation of the co-infectingvirus (Waterhouse and Helliwell, 2003). The above results showed that

Pns12 of RGDV can suppress local silencing induced by ssRNA when transiently expressed in *N. benthamiana*. However, the function of the silencing suppressor activity of Pns12 in natural viral infection is unknown. As a dsRNA virus with a large genome, no infectious clones for reverse genetic studies of RGDV were available, making a direct test of the role of Pns12 in RGDV infection of rice impossible.

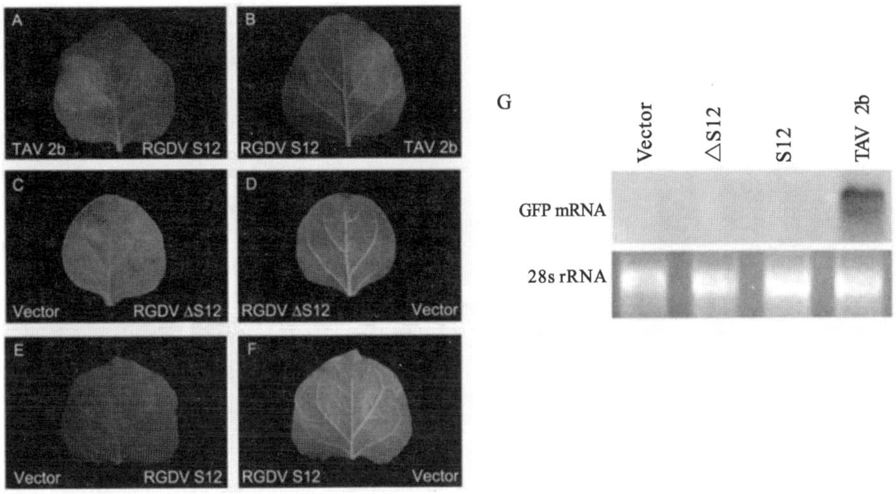

Fig. 2 RGDV Pns12 did not inhibit local silencing induced by dsRNA (A–F)

N. benthamiana line 16c plants were coinfiltrated with Agrobacterium spp.(Agro.) mixtures carrying 35S:dsGFP and the individual constructs indicated in each image. GFP fluorescence was viewed under long wavelength UV lightat 7d postinfiltration (dpi). G, Northern blot analysis of the steady-state levels of GFP mRNA extracted from different infiltrated patches shown in A–F. The bottom gel shows 28s rRNA with ethidium bromide staining as a loading control

To study the role of Pns12 in viral infection in the context of plant-virus interaction, we utilized a PVX vector toexpress S12 and an ORF frame-shift mutant (Δ S12). Seedlings of *N. benthamiana* plants (four- to six-leaf stage) were inoculated with PVX, PVX-S12, and PVX- Δ S12, respectively.All inoculated leaves were asymptomatic. However, symptoms were visible in systemically infected leaves inoculated with PVX or PVX- Δ S12 as early as 6dpi. Generally, the symptoms caused by PVX or PVX- Δ S12 developed initially as veinal chlorosis between 6 and 9dpi but subsequently (9–18dpi) as mild chlorotic spots in some leaves. In addition, some newly developed leaves became asymptomatic, a phenomenon reminiscent of recovery from viral infection. The symptoms caused by PVX-S12 developed one to two days later than those caused by PVX or PVX- Δ S12. However, the symptoms caused by PVX-S12 became indistin guishable from those caused by PVX or PVX- Δ S12 at 9dpi. At 12dpi, it became evident that thes ymptoms caused by PVX-S12 were more severe than those caused by PVX or PVX- Δ S12. This became even more obviousat 18dpi. At this time, symptoms caused by PVX-S12 were still visible, whereas only very slight symptoms could be observed in some of the leaves of PVX or PVX- Δ S12 inoculated plants (Fig. 3A).

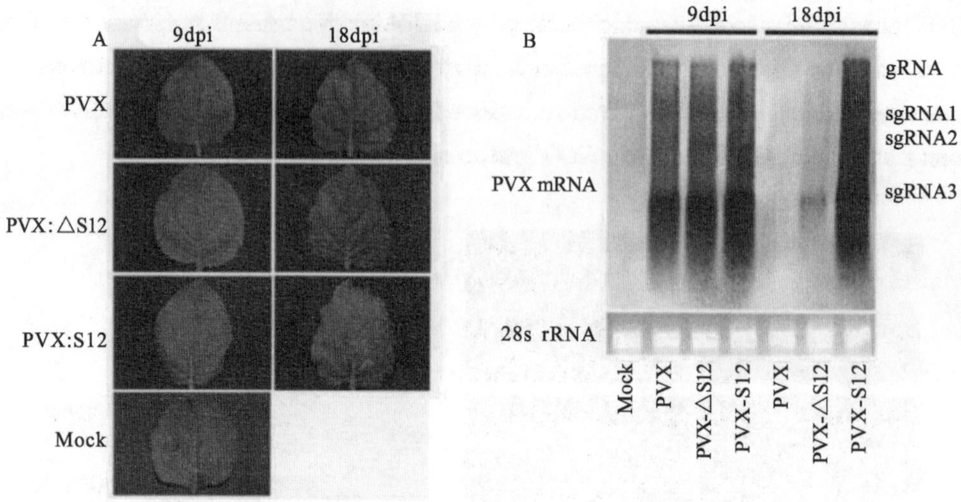

Fig. 3 RGDV Pns12 enhances pathogenicity of chimeric PVX

A, Plants infected by PVX or PVX:ΔS12 show mild disease symptoms as a few scattered chlorotic speckles, whereas leaves infected with PVX:S12 show more severe symptoms. B, RNA gel blot analysis of accumulation of PVX genomic (gRNA) and subgenomic mRNAs (sgRNA1 to sgRNA3) at 9 and 18dpi. The bottom gel shows rRNA with ethidium bromide staining as a loading control.

To investigate the mechanisms by which Pns12 caused increased disease symptoms in the heterologous system,Northern blots were carried out to detect the accumulated levels of the genomic RNAs of PVX. As shown in Fig. 3B, PVX mRNAs accumulated to high levels in systemic leaves of *N. benthamiana* infected by either PVX, PVX: Δ S12 or PVX-S12 at 6dpi. However, the accumulation of PVX mRNAs decreased significantly in plants infected by either PVX or PVX: Δ S12 at 18dpi. By contrast, the accumulated level of PVX mRNAs remained very high in plants infected by PVX-S12 at this time. Taken together, these results indicated that the presence of Pns12 resulted in increased disease symptoms in PVX infected *N. benthamiana*. The increased disease symptoms were probably a result of enhanced replication of PVX, which in turn might be caused by the suppression of RNA silencing by Pns12.

3.4 Cellular localization of RGDV Pns12

To determine the subcellular localization of Pns12, the S12 coding sequence was fused in-frame to the 5′ terminus of the GFP gene driven by a *Cauliflower mosaic virus* 35S promoter. The constructs were introduced into *N. benthamiana* leaf cells via agroinfiltration. GFP fluorescence was observed using confocal microscopy 3dpi. As shown in Fig. 4, the fluorescence occurred in the nuclei of leaves expressing the S12:GFP fusion protein. However, an examination of leaves expressing free GFP by confocal microscopy showed fluorescence distributed evenly in the cytoplasmand nuclei. The nuclear localization of Pns12 was consistent with our predications based on sequence analysis using PSORT II Prediction (http://psort.ims.u-to-kyo.ac.jp/form2.html), which showed that a region located in the Cterminus of Pns12 was rich in basic amino acids and might function as a nuclear localization signal (NLS). The relationship between the nuclear localization and the silencing suppressor activities of Pns12 warrants further investigation.

4 Discussion

Recent studies have shown that a specific plant virus can encode more than one silencing suppressors. For example, *Citrus tristeza virus* (CTV) encodes three VSRs. The three VSRs, namely p20, p23, and CP, can function independently at the intracellular or intercellular levels (Caplen et al., 2001; Lu et al., 2004). Inaddition, Cui et al. demonstrated that a Geminivirus, *Tobacco curly shoot virus* Y35 (TbCSV-Y35), could process three VSRs, namely AC2, AC4, and β C1(Cui et al., 2005). Liu et al. have reported that RGDV Pns11 was a silencing suppressor.Here, we showed that Pns12 might be another silencing suppressor of RGDV(Liu et al., 2008). Pns12 could suppress silencing inducedby sense-GFP. It can reduce, but not eliminate, the accumulation of siRNAs arising from local silencing of GFP. These results indicated that Pns12 might target aninitial step of RNA silencing, having a similar mode of action to 2b of CMV and p69 of TYMV(Li et al., 2002; Chen et al., 2004). Supporting this, we showed that Pns12 could not inhibit silencing induced by GFP dsRNA (Fig. 2). In addition, Pns12 could enhance the pathogenicity of PVX in a heterologous system. The introduction of Pns12 into the genome of PVX resultedin increased viral accumulation in infected *N. benthamiana* (Fig. 3B). This suggested that Pns12 might have an important role in RGDV infection of rice.

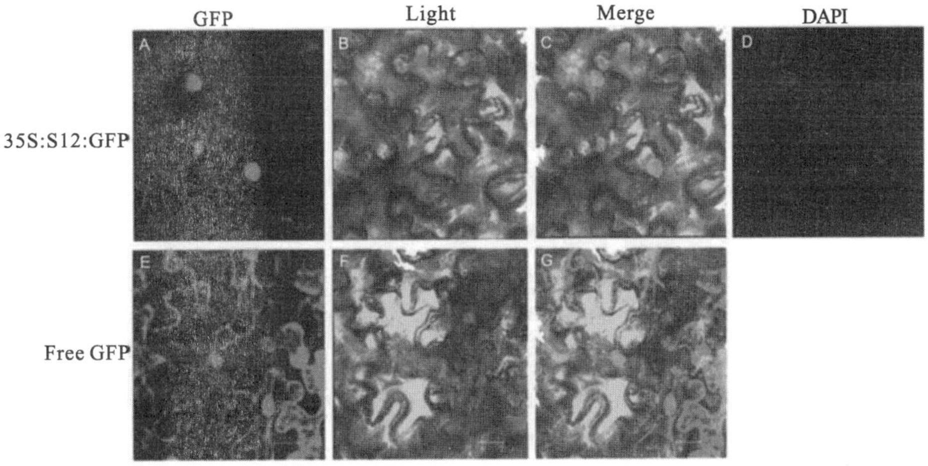

Fig. 4 Subcellular localization of 35S:S12:GFP fusion protein transiently expressed in tobacco epidermis cells

A, Fluorescence from the Pns12 tagged with GFP accumulated in the nucleus. B, The same cell as in A under bright-field illumination (C) is an overlay of images A and B. D, DAPI stained nucleus. E, Accumulation of GFP in the cytoplasm and nucleus. F, The same cell as in E under bright-field illumination. G, An overlay of bright and fluorescent illumination of E and F

Our observation indicates that the silencing suppressor activities of Pns12 are weaker than those of Pns11. Both Pns12 and Pns11 can inhibit ssRNA-induced silencing and enhance the pathogenicity of PVX in a heterologous system. However, unlike Pns11, Pns12 could not suppress systemic silencing triggered by ssRNAs (data not shown). However,the effects of Pns11 on silencing induced by dsRNAs are unknown at present. Thus, we suggcst that Pns12 and Pns11 target different steps in the RNA silencing pathway.

Transient expression of Pns12 in *N. benthamiana* showed that this protein mainly accumulated in the nucleus. Nuclear localization is essential for the silencing suppressor activities of some VSRs. For example, the nuclear localization ofthe protein 2b, a VSR of *Cucumber mosaic virus* (CMV),was closely related to its pathogenicity (Lucy et al., 2000). Studies on 2 bindicated that it could interact with AGO1, a major protein of the RNA-induced silencing complex (RISC). The interaction resulted in the suppression of the slicer activity of the RISC and thus compromised the miRNA pathway (Zhang et al., 2006).Nuclear accumulation was also required for these activities of 2b (Wang et al., 2004). Nuclear accumulation was also required for the silencing suppressor activity and induction of necrosis of *Tomato yellow leaf curl virus*-China (TYLCV-C) C2 (van Wezel et al., 2001; Dong et al., 2003).In addition, a recent study of Xiong et al. showed that NS3, the VSR of *Rice stripe virus* (RSV) localized to the nuclei(Xiong et al., 2009). Deletion of the NSL of NS3 resulted in reduced silencing suppressor activity. We hypothesize that Pns12 entersthe nucleus and inactivates certain protein components of the RNA silencing pathway through physical interactions with them. However, further studies are required to correlate the nuclear accumulation of Pns12 with its silencing suppressor activities. Additionally, it will be interesting to further characterize the role of Pns12 in RGDV infection ofrice.

Acknowledgements

This work was supported by the National Basic Research Program of China (Grant No. 2010CB126203), the National Transgenic Major Program (Grant No. 2009ZX08009-044B and 2009ZX08001-018B), the National Natural Science Foundation of China (Grant No. 30970135), the Fujian Province Education Department (Grant No. JB08078), and Specialized Research Fund for the Ministry of Agriculture (Grant No. nyhyzx07-051).

References

[1] ANANDALAKSHMI R, PRUSS G J, GE X, et al. A viral suppressor of gene silencing in plants[J]. Proceedings of the National Academy of Sciences, 1998, 95(22): 13079-13084.

[2] BRIGNETI G, VOINNET O, LI W X, et al.Viral pathogenicity determinants are suppressors of transgene silencing in nicotiana benthamiana[J]. The EMBO Journal, 1998, 17(22):6739-6746.

[3] CAO X, ZHOU P, ZHANG X, et al. Identification of an RNA silencing suppressor from a plant double-stranded RNA virus[J]. Journal of Virology, 2005, 79(20): 13018-13027.

[4] CAPLEN N J, FLEENOR J, FIRE A, et al. dsRNA-mediated gene silencing in cultured *Drosophila* cells: a tissue culture model for the analysis of RNA interference[J]. Gene, 2000, 252(1-2):95-105.

[5] CAPLEN N J, PARRISH S, IMANI F, et al. Specific inhibition of gene expression by small double-stranded RNAs in invertebrate and vertebrate systems[J]. Proceedings of the National Academy of Sciences, 2001,98:9742-9747.

[6] CHEN J, LI W X, XIE D, et al. Viral virulence protein suppresses RNA silencing–mediated defense but upregulates the role of microRNA in host gene expression[J]. The Plant Cell, 2004, 16(5):1302-1313.

[7] CUI X, LI G, WANG D, et al. A begomovirus DNA β-encoded protein binds DNA, functions as a suppressor of RNA silencing, and targets the cell nucleus[J]. Journal of Virology, 2005, 79(16):10764-10775.

[8] DING S W, VOINNET O. Antiviral immunity directed by small RNAs[J]. Cell, 2007, 130(3):413-426.

[9] DING S W, LI H, LU R, et al. RNA silencing: a conserved antiviral immunity of plants and animals[J]. Virus Research, 2004, 102(1):109-115.

[10] DING S W. RNA silencing[J]. Current Opinion in Biotechnology, 2000, 11(2):152-156.

[11] DONG X, WEZEL R V, STANLEY J, et al. Functional characterization of the nuclear localization signal for a suppressor of posttranscriptional gene silencing[J]. Journal of Virology, 2003, 77(12):7026-7033.

[12] ELBASHIR S M, LENDECKEL W, TUSCHL T. RNA interference is mediated by 21- and 22-nucleotide RNAs[J]. Genes & Development, 2001, 15(2):188-200.

[13] FIRE A. RNA-triggered gene silencing[J]. Trends in Genetics Tig, 1999, 15(9):358-363.

[14] FIRE A, XU S, MONTGOMERY M K, et al. Potent and specific genetic interference by double-stranded RNA in Caenorhabditis elegans[J].Nature, 1998,391(6669):806-8011.

[15] FRANCKI R, BOCCARDO G. The plant *Reoviridae*. In "The *Reoviridae*"[M]. Springer US, 1983.

[16] GOPAL P, KUMAR P P, SINILAL B, et al. Differential roles of C4 and βC1 in mediating suppression of post-transcriptional gene silencing: Evidence for transactivation by the C2 of *Bhendi yellow vein mosaic virus*, a monopartite begomovirus[J]. Virus Research,

2007, 123(1):9-18.

[17] GOTO K, KANAZAWA A, KUSABA M, et al. A simple and rapid method to detect plant siRNAs using nonradioactive probes[J]. Plant Molecular Biology Reporter, 2003, 21(1):51-58.

[18] HAMILTON A, VOINNET O, CHAPPELL L, et al. Two classes of short interfering RNA in RNA silencing[J]. The EMBO Journal, 2002, 21(17): 4671-4679.

[19] HAMMOND S M, BERNSTEIN E, BEACH D, et al. An RNA-directed nuclease mediates post-transcriptional gene silencing in Drosophila cells[J]. Nature, 2000, 404(6775):293.

[20] ICHIMI K, KIKUCHI A, MORIYASU Y, et al. Sequence Analysis and GTP-Binding Ability of the Minor Core Protein P5 of Rice Gall Dwarf Virus[J]. Japan Agricultural Research Quarterly, 2002, 36(2):83-87.

[21] KASSCHAU K D, CARRINGTON J C. A counterdefensive strategy of plant viruses: suppression of posttranscriptional gene silencing[J]. Cell, 1998, 95: 461-470.

[22] KATSAR C, HUNTER W, SINISTERRA X. Phytoreovirus-like sequences isolated from salivary glands of the glassy-winged sharpshooter Homalodisca vitripennis (Hemiptera: Cicadellidae) [J]. Florida Entomologist, 2007, 90(1):196-203.

[23] LECELLIER C H, DUNOYER P, ARAR K, et al. A cellular microRNA mediates antiviral defense in human cells[J]. Science, 2005, 308(5721):557-560.

[24] LI F, DING S W. Virus counterdefense: diverse strategies for evading the RNA-silencing immunity[J]. Annual Review of Microbiology, 2006, 60(1):503-531.

[25] LI H, LI W X, DING S W. Induction and suppression of RNA silencing by an animal virus. Science, 2002,296: 1319-1321.

[26] LI W X, DING S W. Viral suppressors of RNA silencing[J]. Current Opinion in Biotechnology, 2001, 12(2):150-154.

[27] LIU F, ZHAO Q, RUAN X, et al. Suppressor of RNA silencing encoded by rice gall dwarf virus genome segment 11[J]. Chinese Science Bulletin, 2008, 53: 362-369.

[28] LU R, FOLIMONOV A, SHINTAKU M, et al. Three distinct suppressors of RNA silencing encoded by a 20-kb viral RNA genome[J].Proceedings of the National Academy of Sciences of the United States of America, 2004, 101:15742-15747.

[29] LUCY A P, GUO H S, LI W X, et al. Suppression of post-transcriptional gene silencing by a plant viral protein localized in the nucleus[J]. EMBO Journal, 2000, 19(7):1672-1680.

[30] MORIYASU Y, MARUYAMA F W, KIKUCHI A, et al. Molecular analysis of the genome segments S1, S4, S6, S7 and S12 of a rice gall dwarf virus isolate from Thailand; completion of the genomic sequence[J]. Archives of Virology, 2007, 152(7):1315-1322.

[31] OMURA T, MORINAKAT, INOUE H, et al. Purification and Some Properties of Rice Gall Dwarf Virus, a New Phytoreovirus[J]. Phytopathology, 1982, 72(9): 1246-1249.

[32] ANABELA, PICTON, CHRISTIAAN, et al. Molecular analysis of six segments of Tobacco leaf enation virus, a novel phytoreovirus from tobacco[J]. Virus Genes, 2007, 35: 387-393.

[33] POWERS J G, SIT T L, F Q, et al. A versatile assay for the identification of RNA silencing suppressors based on complementation of viral movement[J]. Molecular plant-microbe interactions, 2008, 21(7):879-890.

[34] PRUSS G, GE X, SHI X M, et al.Plant viral synergism: the potyviral genome encodes a broad-range pathogenicity enhancer that transactivates replication of heterologous viruses[J]. Plant Cell, 1997, 9(6):859-868.

[35] QI Y, ZHONG X, ITAYA A, et al. Dissecting RNA silencing in protoplasts uncovers novel effects of viral suppressors on the silencing pathway at the cellular level[J]. Nucleic Acids Research, 2004, 32(22):179.

[36] ROTH B M, PRUSS G J, VANCE V B. Plant viral suppressors of RNA silencing[J]. Virus Research, 2004, 102(1):97-108.

[37] SCHNETTLER E, HEMMES H, GOLDBACH R, et al. The NS3 protein of rice hoja blanca virus suppresses RNA silencing in mammalian cells[J]. Journal of General Virology, 2008, 89(1):336-340.

[38] SHI Y, RYABOV E V, VAN WEZEL R, et al. Suppression of local RNA silencing is not sufficient to promote cell-to-cell movement of Turnip crinkle virus in Nicotiana benthamiana[J]. Plant Signaling & Behavior, 2009, 4(1):15-22.

[39] WEZEL R V, LIU H, TIEN P, et al. Gene C2 of the monopartite geminivirus tomato yellow leaf curl virus-China encodes a pathogenicity determinant that is localized in the nucleus[J]. Molecular Plant-Microbe Interactions, 2001, 14(9):1125-1128.

[40] VANCE V, VAUCHERET H. RNA silencing in plants—defense and counterdefense[J]. Science, 2001, 292(5525):2277-2280.

[41] VOINNET O. Induction and suppression of RNA silencing: insights from viral infections[J]. Nature Reviews Genetics, 2005, 6(3):206-220.

[42] VOINNET O, LEDERER C, BAULCOMBE D C. A Viral Movement Protein Prevents Spread of the Gene Silencing Signal in *Nicotiana benthamiana*[J]. Cell, 2000, 103(1):157-167.

[43] WANG Y, TZFIRA T, GABA V, et al. Functional analysis of the *Cucumber mosaic virus* 2b protein: pathogenicity and nuclear localization[J]. Journal of General Virology, 2004, 85:3135-3147.

[44] WATERHOUSE P M, HELLIWELL C A. Exploring plant genomes by RNA-induced gene silencing[J]. Nature Reviews Genetics, 2003, 4(1):29.

[45] WEI T, UEHARA-ICHIKI T, MIYAZAKI N, et al. Association of rice gall dwarf virus with microtubules is necessary for viral release from cultured insect vector cells[J]. Journal of Virology, 2009, 83(20):10830-10835.

[46] XIE L H, LIN Q Y, XIE L Y, et al. Rice Bunchy Stunt Virus: a New Member of Phytoreovirus[J]. Journal of Fujian Agricultural University, 1996, 25: 312-319.

[47] XIONG R, WU J, ZHOU Y, et al. Characterization and subcellular localization of an RNA silencing suppressor encoded by Rice stripe tenuivirus[J]. Virology, 2009, 387(1):29-40.

[48] ZAMORE P. Ancient pathways programmed by small RNAs[J]. Science, 2002, 296: 1265-1269.

[49] ZHANG H M, YANG J, XIN X, et al. Molecular characterization of the largest and smallest genome segments, S1 and S12, of *Rice gall dwarf virus*[J]. Virus Genes, 2007a, 35(3):815-823.

[50] ZHANG H M, XIN X, YANG J, et al. Completion of the sequence of rice gall dwarf virus from Guangxi, China[J]. Archives of Virology, 2008, 153(9):1737-1741.

[51] ZHANG H, YANG J, XIN X, et al. Molecular characterization of the genome segments S4, S6 and S7 of rice gall dwarf virus[J]. Archives of Virology, 2007b, 152(9):1593-1602.

[52] ZHANG L D, WANG Z H, WANG X B, et al. Two virus-encoded RNA silencing suppressors, P14 of *beet necrotic yellow vein virus* and S6 of *rice black streak dwarf virus*[J]. Chinese ence Bulletin, 2005, 50(4):305-310.

[53] ZHANG X, YUAN Y-R, PEI Y et al. Cucumber mosaic virus-encoded 2b suppressor inhibits Arabidopsis Argonaute1 cleavage activity to counter plant defense[J]. Genes, Development, 2006, 20(23): 3255-3268.

The early secretory pathway and an actin-myosin VIII motility system are required for plasmodesmatal localization of the NSvc4 protein of *Rice stripe virus*

Zhengjie Yuan[1], Hongyan Chen[1], Qian Chen[1], Toshihiro Omura[2], Lianhui Xie[1], Zujian Wu[1], Taiyun Wei[1]

(1 Institute of Plant Virology, Fujian Province Key Laboratory of Plant Virology, Fujian Agriculture and Forestry University, Jinshan, Fuzhou, Fujian 350002, PR China; 2 National Agricultural Research Center, 3-1-1 Kannondai, Tsukuba, Ibaraki 305-8666, Japan)

Abstract: Plant viruses utilize movement proteins to gain access to plasmodesmata (PD) for cell-to-cell propagation. While the NSvc4 protein of Rice stripe virus (RSV) is implicated in the passage of viruses from cell to cell, its role remains to be elucidated. We examined the mechanisms by which RSV NSvc4 is targeted to PD in cell walls. NSvc4 accumulated at PD when expressed as a fusion with yellow fluorescent protein inleaf cells of *Nicotiana benthamiana*. NSvc4 was targeted to PD via the endoplasmic reticulum-to-Golgisecretory pathway, and the actomyosin motility system was required for the delivery of NSvc4 to PD. Moreover, it appeared that NSvc4 utilized myosin VIII-1 rather than myosin XI for trafficking to PD. Taken together, our data reveal that the targeting of NSvc4 to PD exploits the early secretory pathway and the actin-myosin VIII motility system in the leaves of a non-host plant, *N. benthamiana*.

Keyword: *Rice stripe virus*; NSvc4; Plasmodesmatal localization; Early secretory pathway; Actomyosin motility system

In plants, viruses spread from cell to cell through plasmodesmata (PD), which are channels, lined with plasma membrane, that bridge the cell wall to achieve systemic continuity (Benitez-Alfonso et al., 2010). Since movement through PD is tightly regulated, viruses encode movement proteins (MPs) to overcome restrictions in the passage of materials through plasmodesmatal channel (Benitez-Alfonso et al., 2010). Recent investigations indicate that plant viruses have evolved a variety of strategies for the intracellular movement that allows them to gain access to PD via trafficking along the endoplasmic reticulum-to-Golgi (ER-to-Golgi) secretory pathway and cytoskeleton (Harries et al., 2010). In plants, the passage of proteins from the ER to the Golgi apparatus is mediated by the coat protein complex II (COPII) (daSilva et al., 2004). COPII is composed of four cytosolic components and a small GTPase known as Sar1 (daSilva et al., 2004). The mutant protein Sar1(H74L), with defective GTPase activity, specifically blocks COPII-dependent ER-to-Golgi transport (daSilva et al., 2004). Both microtubules and actin filaments, constituents the cytoskeleton, can serve as tracks for their respective motor proteins (dynein and kinesin in the case of microtubules and myosins in the case of actin filaments) for the movement of cargo (Harries et al., 2010). In plants, myosins constitute a large superfamily that can be divided into two classes (class VIII and class XI; Prokhnevsky et al.,2008; Avisar et al., 2008b; Farquharson and Staiger, 2010), and the role of myosins in the intra- and intercellular movement of plant viruses has been a focus of several recent investigations (Wright et al., 2007; Avisar et al., 2008a; Harries et al., 2009, 2010).

Rice stripe virus(RSV), the type member of the genus *Tenuivirus*, multiplies both in plants and in

Virus Research.2011,159: 62–68.

Received 16 February 2011; Accepted 23 April 2011

invertebrate insect vectors (Toriyama, 1986; Falk and Tsai, 1998). RSV is one of the most economically devastating pathogens of rice and is repeatedly epidemic in China, Japan and Korea (Jonson et al., 2009; Wei et al., 2009; Satoh et al., 2010). The genome of RSV is composed of four negative-sense single-stranded RNA segments, designated as RNA 1, 2, 3 and 4. RNA 1 has negative polarity, encoding a putative RNA-dependent RNA polymerase (Toriyama, 1986; Wei et al., 2004). The other three segments are associated with an unusual ambisense coding strategy, with both the viral-sense RNA (vRNA) and the viral complementary-sense RNA (vcRNA) having coding capacity (Toriyama, 1986;Wei et al., 2004).vcRNA4 encodes NSvc4, a protein that can move between cells and rescues the impaired intercellular spread of movement-defective Potato virus X in leaves of *Nicotiana benthamiana*, indicating that NSvc4 is a putative MP of RSV (Xiong et al., 2008). In the present study, we examined the mechanism by which NSvc4 of RSV is targeted to PD in cell walls. We found that the targeting of NSvc4 of RSV to PD required the ER-Golgi secretory pathway and the actin-myosin VIII motility system in leaves of the non-host plant *N. benthamiana*.

We used Gateway®technology (Invitrogen) to generate the expression clones used in the present study. The NSvc4 cistron of RSV and the DNA fragment encoding PDLP1a, a type I membrane plasmodesmatal protein (Thomas et al., 2008), were amplified by PCR from recombinant pGBK-NSvc4 (Lu et al., 2009) and cDNA of Arabidopsis thaliana, respectively. Destination vectors pEarleyGate101 and pEarleyGate102 were used to express a carboxy-terminal fusion protein with yellow fluorescent protein (YFP) and with cyan fluorescent protein (CFP), respectively (Earley et al., 2006). The respective DNA fragments were introduced into the entry vector pDONR221 and then into the destination vectors pEarleyGate101 and pEarleyGate102 to generate plasmids that expressed NSvc4-YFP, NSvc4-CFP, PDLP1a-CFP and PDLP1a-YFP, respectively. The plasmids for the expression of a marker of the Golgi apparatus ERD2-CFP, and of untagged Sar1 and Sar1(H74L) were provided by Dr. A. Wang (Wei and Wang, 2008). The plasmids for the respective expression of the tails of myosins XI-K, XI-2 and VIII-1 were provided by Dr. V. V. Dolja (Avisar et al., 2008a,b). Four-week- old *N. benthamiana* plants and Agrobacterium tumefaciens (strain GV3101) were used for Agrobacterium-mediated transient expression, as described previously (Agro-infiltrations; Wei and Wang, 2008). The relevant binary vectors were introduced into strain GV3101 and then infiltrated into leaf tissues. All imaging of plant tissues by confocal microscopy and related analyses were performed essentially as described previously (Wei and Wang, 2008).

To investigate the subcellular distribution of NSvc4 of RSV, we transiently expressed NSvc4-YFP in the epidermal cells of *N. benthamiana* leaves via agro-infiltration. We found that, 48 h after infiltration of leaves, NSvc4-YFP was distributed in the cytoplasm as inclusions and was localized at the cell wall as punctate spots (Fig. 1A), even after plasmolysis of infiltrated leaf tissue by treatment with 30% glycerol (Supplementary Fig. S1). The inclusions of NSvc4-YFP in the cytoplasm were associated with chloroplasts (Fig. 1A). Furthermore, paired punctate spots of NSvc4-YFP appeared to span the walls of adjoining cells, forming a pattern that resembled those of PD-rich pit fields (Fig. 1B). Next, we coexpressed the marker of PD PDLP1a-YFP with NSvc4-CFP. In this case, 48 h after infiltration of leaves, we found NSvc4-CFP and PDLP1a-YFP together as punctate spots in cell walls, confirming that NSvc4 is a PD-localized protein (Fig. 1C).

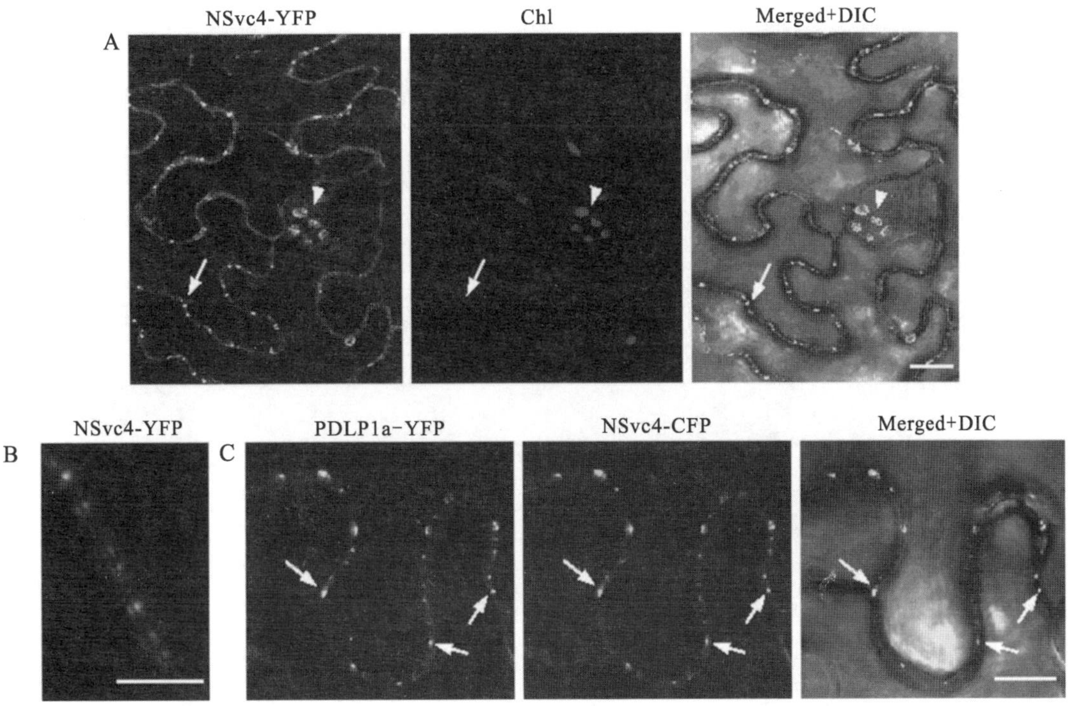

Fig. 1 The NSvc4 of RSV distributed in the chloroplasts and PD in leaves of *N. benthamiana*

(A) Confocal images of cells that expressed NSvc4-YFP, recorded 48 h after agro-infiltration. An arrow indicates punctate spots of NSvc4-YFP at the cell wall. An arrowhead indicates chloroplast-associated NSvc4-YFP in the cytoplasm. The image overlapped with NSvc4-YFP (in yellow), chlorophyll autofluorescence (Chl; in red) and differential interference contrast (DIC). (B) Confocal image of cells that expressed NSvc4-YFP, recorded 48h after agro-infiltration, showing paired punctate spots of NSvc4-YFP that appear to span the walls of adjoining cells in a pattern that resembles that of PD-rich pit fields. (C) Confocal images of cells that co-expressed the marker of PD PDLP1a-YFP and NSvc4-CFP, as recorded 48h after agro-infiltration. Arrows indicate PD at which both PDLP1a-YFP and NSvc4-CFP have accumulated. The image overlapped with PDLP1a-YFP (in yellow), NSvc4-CFP (in cyan) and DIC. Bars = 10μm

In previous studies, trafficking of certain plasmodesmatal proteins, such as PDLP1a and the MP of *Tobacco mosaic virus* (TMV), to PD has been shown to exploit the ER-to-Golgi secretory pathway (Wright., 2007; Thomas et al., 2008). We co-expressed, by agro-infiltration, untagged Sar1(H74L) with ERD2-CFP, a well characterized marker of secretion whose translocation to the Golgi apparatus in leaf cells of *N. benthamiana* depends on the presence of a functional secretory pathway (daSilva et al., 2004; Wei et al., 2010a). We found that, 48h after infiltration, the presence of Sar1(H74L) resulted in the retention of ERD2-CFP in the ER but not in the Golgi apparatus (Fig. 2A), confirming previously described results (Wei and Wang, 2008). Therefore, we used this inhibitory protein to assess the role of the ER-to-Golgi secretory pathway in the targeting of PDLP1a and NSvc4 to the PD. We co-expressed, via agro-infiltration, untagged Sar1(H74L) with the plasmodesmatal marker PDLP1a-YFP and with NSvc4-YFP, respectively, in leaf cells of *N. benthamiana*. Co-expression with Sar1(H74L) resulted in localization of PDLP1a-YFP in the ER and inhibited the trafficking of PDLP1a-YFP to PD, as visualized 48 h after infiltration (Fig. 2B). When NSvc4-YFP was co-expressed with Sar1(H74L), the trafficking of NSvc4 to PD was also inhibited and NSvc4-YFP was distributed is the cytosoli of leaf cells 48 h after infiltration (Fig. 2C). Among 100 cells observed by confocal microscopy, the numbers of PD-located punctate spots of PDLP1-YFP and NSvc4-YFP after co-expression with Sar1(H74L) were approximately 80% lower than those in the control (untreated) leaves (Fig. 4A, B). We concluded, therefore, that the ER-to-Golgi secretory pathway was involved in the delivery of NSvc4 to PD.

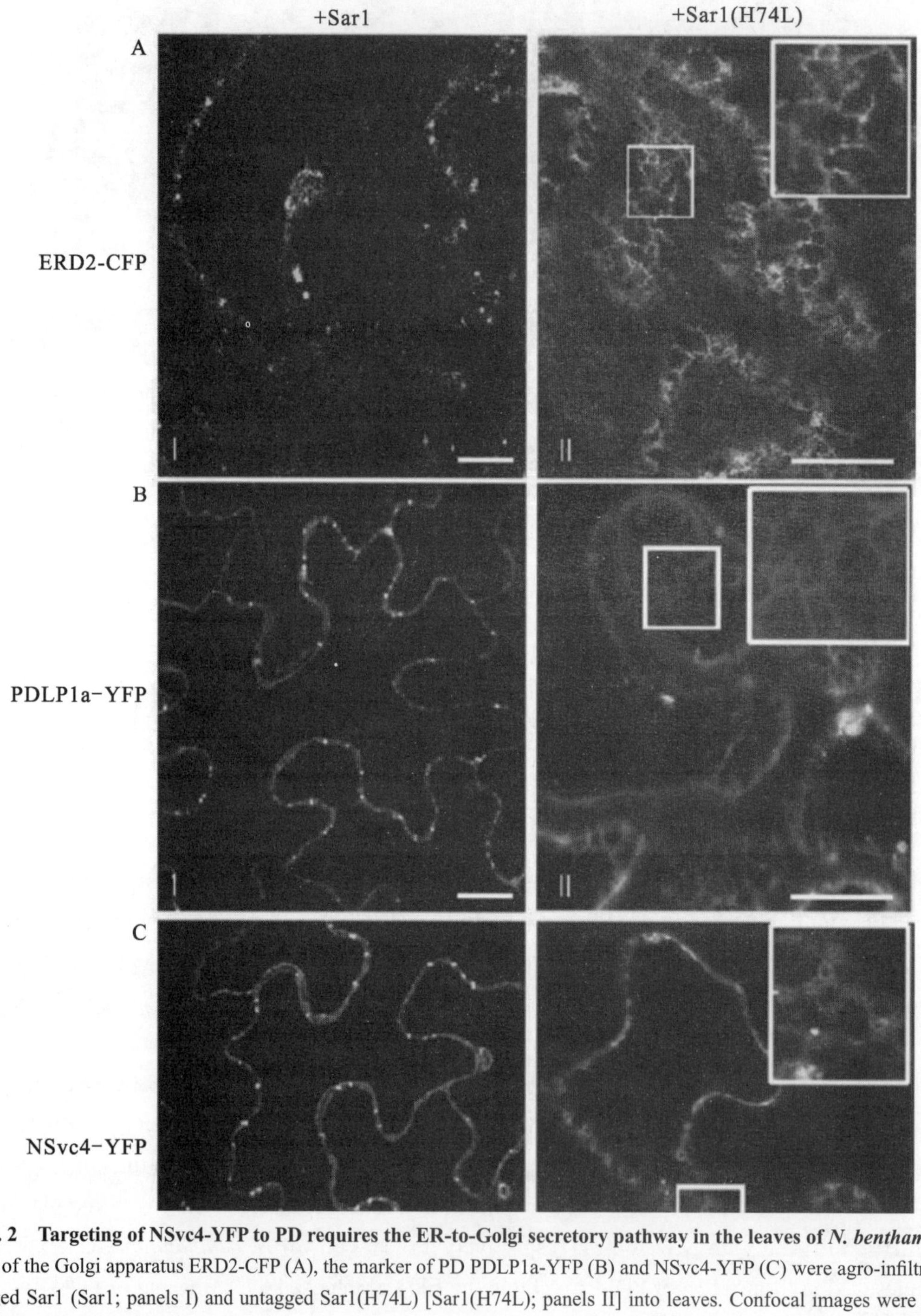

Fig. 2 Targeting of NSvc4-YFP to PD requires the ER-to-Golgi secretory pathway in the leaves of *N. benthamiana*

The marker of the Golgi apparatus ERD2-CFP (A), the marker of PD PDLP1a-YFP (B) and NSvc4-YFP (C) were agro-infiltrated together with untagged Sar1 (Sar1; panels I) and untagged Sar1(H74L) [Sar1(H74L); panels II] into leaves. Confocal images were recorded 48h after agro-infiltration. The upper-right insets in panels II show enlarged images of respective smaller rectangles and demonstrate the retention in the ER of ERD2-CFP (A, panel II) and PDLP1a-YFP (B, panel II), as well as the retention in the cytosol of NSvc4-YFP (C, panel II) in the presence of Sar1(H74L). Bars, 8μm

Our results also yielded important new insights into the targeting of PDLP1a and NSvc4 to PD. The PDLP1a protein has structural features that are typical of secreted proteins, having an aminoterminal signal peptide, a single transmembrane domain (TMD),and a short carboxy-terminal tail (Thomas et al., 2008). The targeting of PDLP1a to PD was mediated by the TMD, and disruption of the ER-to-Golgi secretory pathway resulted in the retention of PDLP1a in the ER (Fig. 2B; Thomas et al., 2008). Unlike PDLP1a, NSvc4 belongs to the 30K MP superfamily and it lacks a typical signal peptide and a TMD that might, otherwise, direct it to PD (Melcher, 2000; An et al., 2003). Furthermore, disruption of the ER-to-Golgi secretory pathway resulted in diffusion of NSvc4 into the cytosol but not into the ER (Fig. 2C). These observations suggest that PDLP1a might leave the ER and be trafficked to the Golgi apparatus or even to the PD. By contrast, transport of NSvc4 to PD might involve the interaction of NSvc4 with other cellular proteins, which are transported by Golgi-derived vesicles and eventually anchored to PD. This hypothesis is supported by previously reported evidence of the interaction of NSvc4 with two chaperone proteins, namely, a type I DnaJ and a small Hsp of rice (Lu et al., 2009). It was proposed that these two chaperones might mediate access by host proteins to PD (Scholthof, 2005). Therefore, it appears that RSV has developed the ability to recruit essential host chaperones for the cell-to-cell movement of viral particles during its infection of plant cells.

Several MPs of plant viruses are trafficked along actin filaments (Harries et al., 2010). To examine the involvement of actin filaments in the targeting of NSvc4 to PD, we disrupted the actin filaments in leaves by application of 25μM latrunculin B (Lat B; Sigma–Aldrich), an inhibitor of actin polymerization, as described previously (Wei and Wang, 2008). We confirmed the disruption of actin filaments upon exposure of leaves to 25μM Lat B for 2h by microscopic examination of actin filaments that had been labeled with the actin-binding region of mTalin fused to YFP (Fig. 2, 2008). Lat B at 25μM was infiltrated into *N. benthamiana* leaves that expressed NSvc4-YFP or PDLP1a-YFP, 46h after agro-infiltration. Two hours later, the infiltrated leaves were examined by confocal microscopy. The extent of accumulation of NSvc4-YFP in PD in the leaves of *N. benthamiana* was clearly reduced after the 2-h treatment of leaves with 25μM Lat B (Fig.3C). The accumulation of PDLP1a-YFP in the PD was also strongly inhibited by the treatment with Lat B (Fig. 3C). Among100 cells examined by confocal microscopy, numbers of PD-located punctate spots of PDLP1a-YFP and NSvc4-YFP after treatment with Lat B were approximately 75% lower than those in the control (untreated) leaves (Fig. 4A, B). We also examined the effects of oryzalin, an inhibitor of the polymerization of microtubules, on the targeting of PDLP1a-YFP and NSvc4-YFP to PD. We infiltrated 100μM oryzalin (Sigma–Aldrich) into *N. benthamiana* leaves that expressed NSvc4-YFP or PDLP1a-YFP 46h after agro-infiltration. The 2-h treatment of leaves with 100_M oryzalin did not affect the targeting of PDLP1a and NSvc4 to PD (Fig. 3B, 4A, 4B), even though, at this concentration, oryzalin disrupts the microtubules in *N. benthamiana* leaves within 2 h (data not shown). Together, our results suggest that an actin-dependent transport system was involved in the localization of the PDLP1a and NSvc4 to PD in the leaves of *N. benthamiana*.

The dependence of the targeting of NSvc4 to PD on actin filaments implies the involvement of myosin motors in the intracellular trafficking of NSvc4. Higher-plant myosins had been grouped into classes VIII and XI (Prokhnevsky et al., 2008; Avisar et al., 2008b; Farquharson and Staiger, 2010). We used the tails of myosins XI-K, XI-2 and VIII-1, which behave as dominant negative antagonists of myosin function, to examine the role of myosins in the targeting of NSvc4 to PD (Avisar et al., 2008a,b). We co-expressed PDLP1a-YFP and NSvc4-YFP, separately, with the untagged tails of myosins XI-K, XI-2 and VIII-1, respectively in leaf cells of *N. benthamiana*

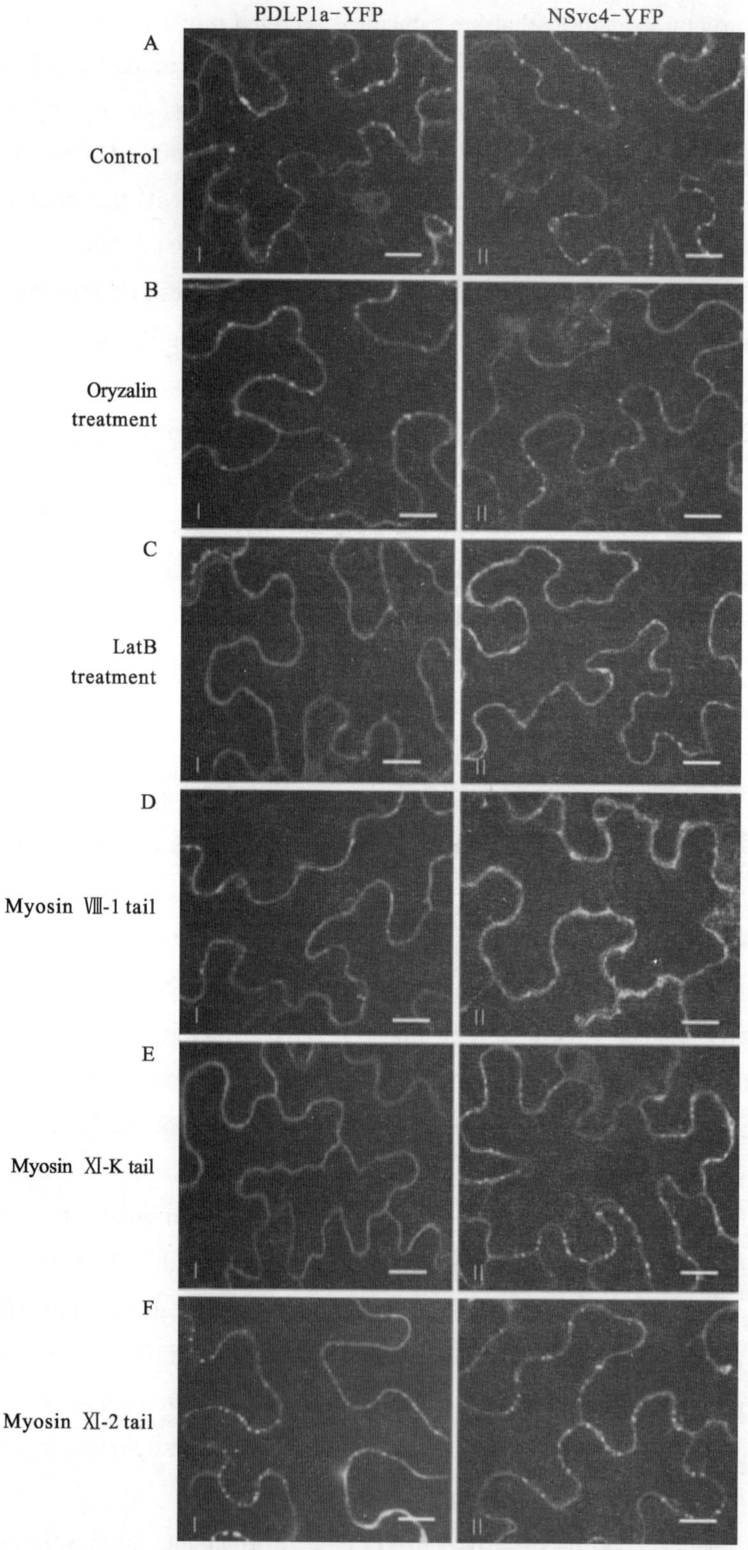

Fig. 3 The targeting of NSvc4-YFP to PD requires the actomyosin motility system in leaves of *N. benthamiana*

The marker of PD PDLP1a-YFP (panels I) and NSvc4-YFP (panels II) were transiently expressed in leaves that had been treated with DMSO (Control; A) for 2h, with 100μ Moryzalin in DMSO (Oryzalin; B) for 2h, and with 25μM latrunculin B in DMSO (Lat B; C) for 2h; or co-agro-infiltrated with the untagged tail of myosin VIII-1 (Myosin VIII-1 tail; D), the untagged tail of myosin XI-K (Myosin XI-K tail; E), and the untagged tail of myosin XI-2 (Myosin XI-2 tail; F), respectively. Confocal images were recorded 48 h after agro-infiltration. Bars, 8μm via agro-infiltration. We found that over-expression of the tails of myosins VIII-1 and XI-K, but not of the tail of myosin XI-2, blocked the targeting of PDLP1a-YFP to PD, as visualized 48h after agro-infiltration (Fig. 3D–F). Moreover, the over-expression of the tail of myosin VIII-1, but not that of the tails of myosins XI-K and XI-2, inhibited the plasmodesmatal localization of NSvc4-YFP, as visualized 48h after agro-infiltration (Fig. 3D–F). Among 100 cells observed by confocal microscopy, the number of PD-located punctate spots of PDLP1a- YFP and NSvc4-YFP after co-expression with the tail of myosin VIII-1 was approximately 80% lower than that in the control leaves (Fig. 4A, B). These results suggest that the three types of myosin function in different ways and that class VIII myosin is required for the plasmodesmatal localization of both PDLP1a and NSvc4.

Previous authors have proposed that myosin VIII-1 associates tightly with PD and is involved in the trafficking of plasmodesmatal proteins to PD (Volkmann et al., 2003). Thus, it appeared likely that both PDLP1a and NSvc4 utilized the actin-myosin VIII motility system directly to gain access to PD. By contrast, myosin XI-K is required for the rapid trafficking of plant organelles, such as the Golgi apparatus (Prokhnevsky et al., 2008; Avisar et al., 2008b; Farquharson and Staiger, 2010), and it was specifically required for the transport of PDLP1a but not of NSvc4 to PD. This scenario confirms our previous proposal that Golgi-derived vesicles might associate directly with PDLP1a but not with NSvc4 during the trafficking of these proteins to PD. Taken together, our present observations suggest that PDLPla is delivered to PD via myosin XI-K-dependent post-Golgi vesicular trafficking, whereas NSvc4 is not directly associated with Golgi vesicles but needs to reach the ER-Golgi interface for loading onto the PD-targeted actomyosin pathway. In a similar manner, the targeting to PD of the Hsp70- homologous MP of Beet yellows virus exploits class VIII but not class XI myosins (Avisar et al., 2008a). The plasmodesmatal localization of the MP of TMV, of RGP2 of A. thaliana and of P3N-PIPO of Turnip mosaic virus was unaffected by over-expression of the tail of myosin VIII-1 (Avisar et al., 2008a; Wei et al., 2010b), while, myosin XI-2 was required for the cell-to-cell movement of TMV (Harries et al., 2009). Thus, different plasmodesmatal proteins have evolved differently in terms of their requirements for myosin motors for their delivery to PD.

In conclusion, our results support the hypothesis that targeting of NSvc4 to PD requires both the ER-to-Golgi secretory pathway and the actin-myosin VIII motility system in leaves of the non-host plant *N. benthamiana* (Fig. 4C). It will be of interest to examine whether the association of NSvc4 with chloroplasts also plays a role in the targeting of this protein to PD.

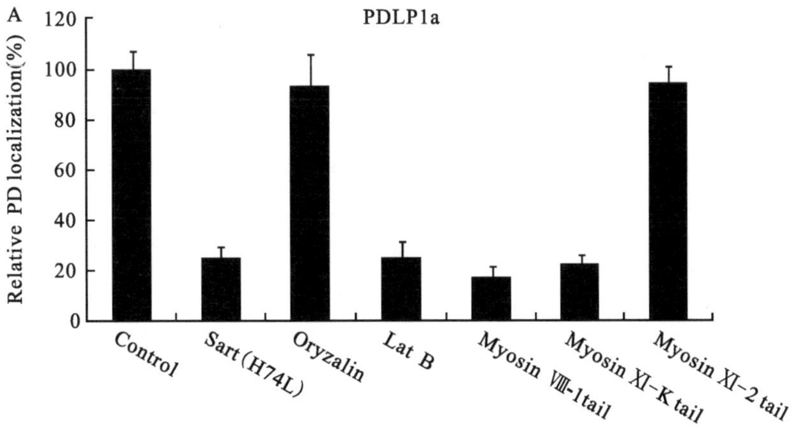

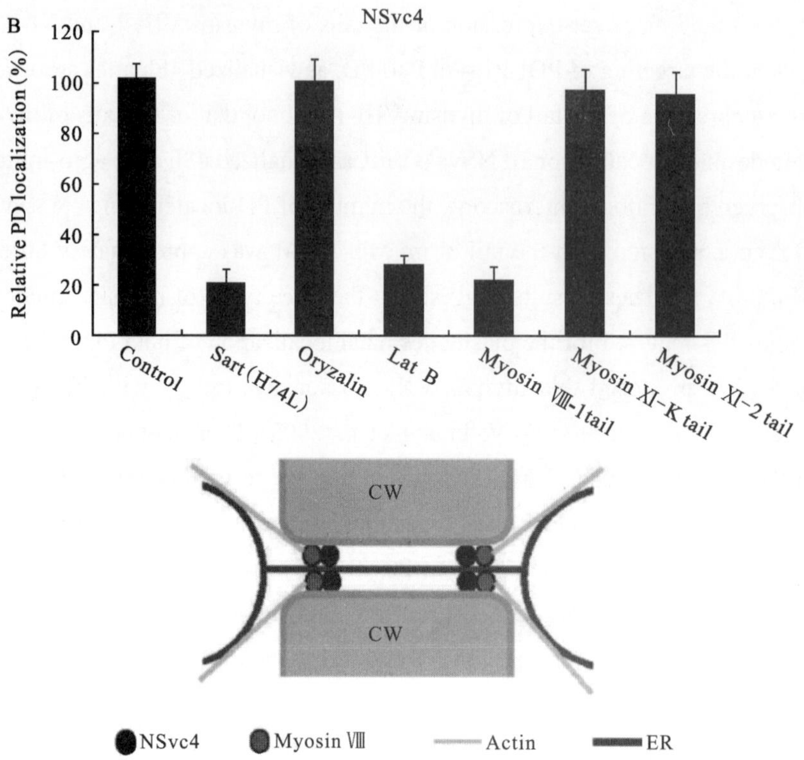

Fig. 4 Targeting of NSvc4-YFP to PD requires the ER-Golgi secretory pathway and the actomyosin motility system in *N. benthamiana*

(A and B) Quantification of the effects of oryzalin and Lat B and of the over-expression of Sar1(H74L), the tail of myosin VIII-1, the tail of myosin XI-K and the tail of myosin XI-2 on the targeting of PDLP1a-YFP (A) and NSvc4-YFP (B) to PD. Among 100 cells observed by confocal microscopy, the number of PD-located punctate spots of PDLP1-YFP and NSvc4-YFP after treatment with the inhibitory drugs or over-expression of inhibitory proteins was calculated as a percentage of that in the control leaves, 48h after agro-infiltration. Values represent mean numbers with standard errors (SE) for percentages relative to controls. (C) Hypothetical schematic presentation of the mechanism for the targeting of NSvc4 to PD. Class VIII myosin accumulates within PD and serves as an adaptor for the association between actin filaments and NSvc4. The actin-myosin VIII network is tightly aligned with the desmotubule, a modified outlet of the continuous ER network. CW, cell wall

Acknowledgements

The authors are indebted to Dr. V. Dolja (Oregon State University) for the plasmids that encoded myosin tails and to Dr. A. Wang (Agriculture and Agri-Food Canada) for the plasmids that encoded ERD2-CFP, mTalin-CFP, Sar1 and Sar1(H74L). This work was supported by grants from the programs for the New Century of Excellent Talents at Universities (no. NCET-09-0011 to T. Wei), the National Natural Science Foundation of China (no. 30770090 to Z. Wu), and the program for Promotion of Basic Research Activities for Innovative Biosciences of the Bio-oriented Technology Research Advancement Institution (BRAIN) of Japan to T. Omura.

References

[1] AN H, MELCHER U, DOSS P, et al. Evidence that the 37-kDa protein of soil-borne Wheat mosaic virus is a virus movement protein[J]. Journal of General Virology, 2003, 84:3153-3163.

[2] AVISAR D, PROKHNEVSKY A I, DOLJA V V. Class VIII myosins are required for plasmode smatal localization of a closterovirus Hsp70 homolog[J]. Journal of Virology, 2008a, 82(6): 2836-2843.

[3] AVISAR D, PROKHNEVSKY A I, MAKAROVA K S, et al. Myosin XI-K is required for rapid trafficking of Golgi stacks, peroxi somes, and mitochondria in leaf cells of Nicotiana benthamiana[J]. Plant Physiology, 2008b, 146(3):1098-1108.

[4] BENITEZA Y, FAULKNER C, RITZENTHALER C, et al. Plasmodesmata: gateways to local and systemic virus infection. molecular plant-microbe interactions, 2010, 23: 1403-1412.

[5] DASILVA L, SNAPP E L, DENECKE J,et al. Endoplasmic Reticulum Export Sites and Golgi Bodies Behave as Single Mobile Secretory Units in Plant Cells[J]. Plant Cell, 2004, 16(7):1753-1771.

[6] EARLEY K W, HAAG J R, PONTES O, et al. Gateway® -compatible vectors for plant functional genomics and proteomics[J]. Plant Journal, 2006, 45: 616-629.

[7] FALK B W, TSAI J H. Biology and molecular biology of viruses in the genus Tenuivirus[J]. Annual Review of Phytopathology, 1998, 36(1):139-163.

[8] FARQUHARSON K L, STAIGER C J. Dissecting the functions of class XI myosins in moss and Arabidopsis[J]. Plant Cell, 2010, 22(6):1649-1649.

[9] HARRIES P A, PARK J W, SASAKI N, et al. Differing requirements for actin and myosin by plant viruses for sustained intercellular movement[J]. Proceedings of the National Academy of Sciences of the United States of America, 2009,106: 17594-17599.

[10] HARRIES P A, SCHOELZ J E, NELSON R S. Intracellular transport of viruses and their components: utilizing the cytoskeleton and membrane highways[J]. Molecular Plant-Microbe Interactions, 2010, 23(11):1381-1393.

[11] JONSON M G, CHOI H S, KIM J S, et al. Sequence and phylogenetic analysis of the RNA1 and RNA2 segments of Korean *Rice stripe virus* isolates and comparison with those of China and Japan[J]. Archives of Virology, 2009, 154(10):1705-1708.

[12] LU L, DU Z, QIN M, et al. Pc4, a putative movement protein of *Rice stripe virus*, interacts with a type I DnaJ protein and a small Hsp of rice[J]. Virus Genes, 2009, 38(2):320-327.

[13] MELCHER U. The '30K' superfamily of viral movement proteins[J]. Journal of General Virology, 2000, 81: 257-266.

[14] PROKHNEVSKY A I, PEREMYSLOV V V, DOLJA V V. Overlapping functions of the four class XI myosins in Arabidopsis growth, root hair elongation, and organelle motility[J]. Proceedings of the National Academy of Sciences of the United States of America, 2008, 105: 19744-19749.

[15] SATOH K, KONDOH H, SASAYA T, et al. Selective modification of rice (Oryza sativa) gene expression by Rice stripe virus infection[J]. Journal of General Virology, 2010, 91(1):294-305.

[16] SCHOLTHOF HB. Plant virus transport: motions of functional equivalence[J]. Trends in Plant Science, 2005, 10(8):376-382.

[17] THOMAS C L, BAYER E M, RITZENTHALER C, et al. Specific Targeting of a Plasmodesmal Protein Affecting Cell-to-Cell Communication[J]. PLOS Biology, 2008, 6(1): e7.

[18] TORIYAMA S.Rice stripe virus:prototype of a new group of viruses that replicate in plants and insects[J]. Microbiological Sciences, 1986, 3(11):347-351.

[19] VOLKMANN D, MORI T, TIRLAPUR U K, et al. Unconventional myosins of the plant-specific class VIII: endocytosis, cytokinesis, plasmodesmata/pit-fields, and cell-to-cell coupling[J]. Cell Biology International, 2013, 27(3):289-291.

[20] WEI T, HUANG T S, MCNEIL J, et al. Sequential recruitment of the endoplasmic reticulum and chloroplasts for plant potyvirus replication[J]. Journal of Virology, 2010a, 84(2):799-809.

[21] WEI T, WANG A. Biogenesis of cytoplasmic membranous vesicles for plant potyvirus replication occurs at the endoplasmic reticulum exit sites in a COPI-and COPII-dependent manner[J]. Journal of virology,2008, 82(24):12252-12264.

[22] WEI T, ZHANG C, HONG J, et al. Formation of complexes at plasmodesmata for potyvirus intercellular movement is mediated by the viral protein P3N-PIPO[J]. PLOS Pathogens, 2010b, 6(6):e1000962.

[23] WEI T Y, LIN H X, WU Z J, et al. Analysis of the complete nucleotide sequences of two isolates of *Rice stripe virus* representing different subgroups in China[J]. Scientia Agricultura Sinica, 2004, 37(6):846-850.

[24] WEI T Y, YANG J G, LIAO F R, et al. Genetic diversity and population structure of Rice stripe virus in China[J]. Journal of

General Virology, 2009, 90(4):1025-1034.

[25] WRIGHT K M, WOOD N T, ROBERTS A G, et al. Targeting of TMV movement protein to plasmodesmata requires the actin/ER network: evidence from FRAP[J]. Traffic, 2007, 8(3):21-31.

[26] XIONG R, WU J, ZHOU Y, et al. Identification of a movement protein of the tenuivirus Rice stripe virus[J]. Journal of Virology, 2008, 82(24): 12304-12311.

Tubular structure induced by a plant virus facilitates viral spread in its vector insect

Qian Chen[1], Hongyan Chen[1], Qianzhuo Mao[1], Qifei Liu[1], Takumi Shimizu[2], Tamaki Uehara-Ichiki[2],Zujian Wu[1], Lianhui Xie[1], Toshihiro Omura[2], Taiyun Wei[1,2]

(1 Fujian Province Key Laboratory of Plant Virology, Institute of Plant Virology, Fujian Agriculture and Forestry University, Fuzhou, Fujian, PR China;

2 National Agricultural Research Center, Tsukuba, Ibaraki, Japan)

Abstract: Rice dwarf virus (RDV) replicates in and is transmitted by a leafhopper vector in a persistent-propagative manner. Previous cytopathologic and genetic data revealed that tubular structures, constructed by the nonstructural viral protein Pns10, contain viral particles and are directly involved in the intercellular spread of RDV among cultured leafhopper cells. Here, we demonstrated that RDV exploited these virus-containing tubules to move along actin-based microvilli of the epithelial cells and muscle fibers of visceral muscle tissues in the alimentary canal, facilitating the spread of virus in the body of its insect vector leafhoppers. In cultured leafhopper cells, the knockdown of Pns10 expression due to RNA interference (RNAi) induced by synthesized dsRNA from Pns10 gene strongly inhibited tubule formation and prevented the spread of virus among insect vector cells. RNAi induced after ingestion of dsRNA from Pns10 gene strongly inhibited formation of tubules, preventing intercellular spread and transmission of the virus by the leafhopper. All these results, for the first time, show that a persistent-propagative virus exploits virus-containing tubules composed of a nonstructural viral protein to traffic along actin-based cellular protrusions, facilitating the intercellular spread of the virus in the vector insect. The RNAi strategy and the insect vector cell culture provide useful tools to investigate the molecular mechanisms enabling efficient transmission of persistent-propagative plant viruses by vector insects.

1 Introduction

Numerous plant viruses that cause serious losses to agricultural production are transmitted by vector insects, classified according to the type of transmission: nonpersistent, semipersistent and persistent (Hogenhout et al., 2008). Viruses transmitted in a persistent manner are further separated into two groups: propagative and nonpropagative(Hogenhout et al., 2008). In persistent-propagative transmission, viruses multiply in the vector insectduring a latent period and can be transmitted to host plants until the death of the insects. Therefore, detailed analyses of viral propagation in the vector insects in persistent-propagative transmission would help disclose a mechanism that may lead to new strategies to control the transmission of the viruses by vector insects.

Plant viruses such as tospoviruses, tenuiviruses, plant rhabdoviruses and plant reoviruses, are transmitted by their respective insect vectors in a persistent-propagative manner, and thus they are designated as persistent-propagative plant viruses (de Assis Filho et al., 2002; Hogenhout et al., 2008; Ammar et al., 2009; Stafford et al., 2012). After their ingestion by insects during feeding on diseased plants, these viruses must enter and

PlosPathogens. 2012, 8(11): e1003032.
Received 9 February 2012; Accepted 2 October 2012; Available online 15 November 2012.

replicate in the epithelial cells of the alimentary canal of vector insects, then exit and move to the salivary glands tobe transmitted to healthy plants (Hogenhout et al., 2008; Ammar et al., 2009). While the replication sites and tissue tropism of these plant viruses in their respective insect vectors have been intensively studied, much less is known about how they spread from initially infected cells to adjacent cells or organs. Acquiring a better understanding of the intercellular spread of plant viruses in insects would lead to better strategies to disrupt the efficient transmission of plant viruses by insect vectors.

Rice dwarf virus(RDV), a member of the genus *Phytoreovirus* in the family *Reoviridae* (Boccardo and Milne, 1984), is transmitted by the green rice leafhopper *Nephotettixcincticeps* in a persistent-propagative manner and is transovarially transmitted (Boccardo and Milne, 1984). The alimentary canal of leafhoppers consists of the esophagus, filter chamber, midgut and hindgut (Fig. 1).The midgut is further divided into the anterior, middle and posterior regions (Fig. 1). As in other types of leafhoppers (Ammar, 1985; Cheung and Purcell, 1993; Tsai and Perrier, 1996; Chen et al., 2011), the alimentary canal of the leafhopper *N. cincticeps* is composed of a single layer of epithelial cells, with microvilli on the lumen side and basallamina on the outer side, covered with muscle fibers (Fig. 2). In epithelial tissues, cells are joined by intercellular junctional complexes(Fig. 2), which may act as the physical barrier to prevent infection by microbes (Oda and Takeichi, 2011). RDV encounters multiple membrane barriers in its path from the alimentary canal to the salivary glands in leafhopper vectors (Chen et al., 2011). After ingestion of RDV particles by leafhoppers, virions first accumulate in the epithelial cells of the filter chamber, suggesting that the microvillar membrane of the filter chamber may contain cellular receptors for viral attachment and entry. Thus, the microvillar membrane of the filter chamber may form the first membrane barrier for successful infection of RDV in the body of leafhopper. RDV spreads to adjacent cells or organs such as the anterior midgut following the initial replication and accumulation of viruses in the epithelial cells of the filter chamber (Chen et al., 2011). The molecular mechanisms by which RDV spreads from cell to cell among epithelial tissues are poorly understood.These cells or organs presumably constitute the second membrane barrier that must be penetrated by RDV to spread further in its insect host. Infection of the epithelial cells is followed by virus invasion of the visceral muscles lining the anterior midgut (Chen et al., 2011). This observation suggests that RDV crossed the basement membrane of the anterior midgut, the third membrane barrier encountered by the virus, to infect the visceral muscles, possibly through the nerves associated with the muscle tissues (Chen et al., 2011). The molecular mechanisms of cell-to-cell spread of RDV among the visceral muscle cells lining the midgut are unknown,but RDV might directly disseminate from the midgut-associated visceral muscles into the hemolymph, then into the salivary gland(Chen et al., 2011).

Studies of RDV–leafhopper interactions have provided evidence that the minor outer-capsid protein P2 and nonstructural protein Pns10 are required for vector transmissibility. Nonsense mutations in the P2 or Pns10 genes abolish transmission of the virus by insects (Pu et al., 2011). However, the absence of expression of these two proteins does not compromise the ability of the virus to replicate in plants(Pu et al., 2011). The leafhopper vector cells in monolayers (VCMs) and the intact vector insects exhibited similar productive noncy to pathic response to RDV infection (Omura et al., 1998; Wei et al., 2006; Wei et al., 2007; Zhou et al., 2007; Wei et al., 2008; Pu et al., 2011). Using VCMs, our previous studies indicated that RDV entered leafhopper cells through receptor-mediated,clathrin-dependent endocytos is using P2 as the viral attachment molecules (Wei et al., 2007; Zhou et al., 2007). Furthermore, RDV had been proposed to exploit tubular structures composed of the nonstructural protein Pns10 (Pns10 tubules) to move along actin-based filopodia extending toward neighboring cells, thus enhancing intercellular viral propagation among cultured leafhopper cells (Wei et al., 2006; Wei et al., 2008). Previously, such tubules, containing RDV particles, were observed with electron microscopy in association with actin-based microvilli and seemed to pass through the microvilli of epithelial cells of the midgut in viruliferous leafhoppers(Nasu, 1965). Therefore, a defect in this machinery might be associated with the inability of mutant RDV to be transmitted by insect vectors. All these reports suggest that RDV might have evolved novel strategies to exploit virus-containing Pns10 tubules to cross through the epithelial cell of the alimentary canal in vector insects so that it can spread intercellularly.

In the present study, by combining an immuno fluorescence technique and a feeding-based RNA interference (RNAi) technique to target Pns10 genes, we determined that RDV has evolved novel strategies to use virus-containing Pns10 tubules to move along actin-based microvilli of the epithelial cells and muscle fibers of visceral muscle tissues to facilitate intercellular spread in the intact insect vector.

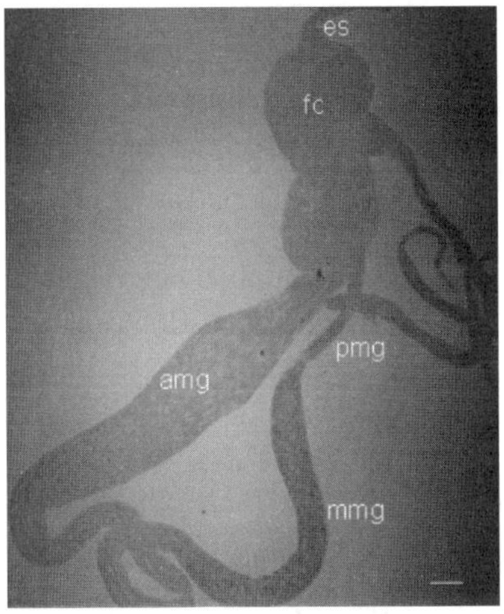

Fig. 1 Transmitted light micrograph of partially dissected alimentary canal of leafhopper vector *N. cincticeps*

The alimentary canal of leafhopper consists of the esophagus (es), filter chamber (fc), anterior midgut (amg), middle midgut (mmg), posterior midgut (pmg) and hindgut (hg). Bar, 100μm

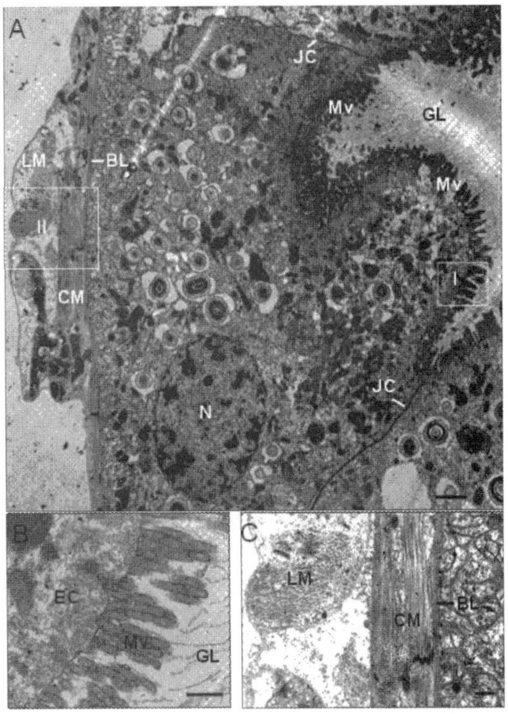

Fig. 2 Transmission electron micrographs of the posterior midgut of leafhopper vector *N. cincticeps*

(A) Midgut epithelium, showing microvilli on lumen side and a basal lamina covered with muscle fibers. Panels B and C are enlarged images of the boxed areas I and II in panel A to show the microvilli and the visceral muscle tissues, respectively. The visceral muscle tissues consist of external longitudinal and internal circular muscle fibers. (D) Diagram of midugt epithelium. BL, basal lamina. CM, circular muscle. EC, epithelial cell. GL, gut lumen. LM, longitudinal muscle. Mv, microvilli. N, nucleus. JC, junctional complex. Bars, 2μm (A) and 500nm (B–C)

2 Results

2.1 Pns10 tubules cross actin-based microvilli of the epithelial cell in viruliferous leafhoppers

To study the development and distribution of Pns10 tubules in viruliferous leafhoppers over time, we used immuno fluorescence microscopy to visualize the tubules in the epithelial cells of the alimentary canal during infection by RDV. At 2-day or 3-day post-firstaccess to diseased plants (padp), internal organs dissected from leafhoppers were immunolabeled with Pns10-specific IgG conjugated to rhodamine (Pns10-rhodamine) and viral particle-specific IgG conjugated to FITC (virus-FITC), and then examined by fluorescence microscopy. At 2-day padp, small infection foci of virus antigen simmunolabeled with virus-FITC were observed in a particular corner of the filter chamber (Fig. 3A). At 3-day padp, the number of infected cells increased, forming larger infection foci (Fig. 3B). Pns10 tubules were distributed at the periphery of infected epithelial cells and protruded from the cell surface, even at the edge of infection foci(Fig. 3C). All these results suggested that RDV might spread from the initially infected epithelial cells to adjacent cells to form infection foci and that Pns10 might be responsible for the extension of tubules from the epithelial cell surface.

In the alimentary canal of insects, bundles of parallel actin filaments form the core of a microvillus (DeRosier and Tilney, 2000; Terra et al., 2006). Our previous data showed that Pns10 tubules could traffic along actin-based filopodia in VCMs through a specific but relatively low-affinity interaction between Pns10 and actin(Wei et al., 2006). These data suggested that Pns10 tubules might pass through the bundles of actin filaments of microvilli. To examine this possibility, we dissected internal organs from leafhoppers at 3-day padp and immunolabeled them with Pns10-rhodamine and the actin dye FITC-phalloidin.This double labeling revealed that actin-based microvilli along the surface of the epithelial cells of the filter chamber were filled with Pns10 tubules (Fig. 3D, E), suggesting that Pns10 tubules could pass through the actin-based microvilli from inside of infected cells into the lumen of the filter chamber.

After replication and accumulation of progeny virions in the epithelial cells of filter chamber, RDV virions exit and move into the adjacent midgut (Chen et al., 2011). To determine whether virus-containing Pns10 tubules can pass through the microvilli of the midgut, at 6-day padp,internal organs dissected from leafhoppers were immunolabeled with Pns10-rhodamine, viral particles-specific IgG conjugated to Alexa Fluor 647 carboxylic acid (virus-Alexa Fluor 647) and actin dye FITC-phalloidin6-day padp. In the anterior midgut, abundant Pns10 tubules extended from or passed through the actin-based microvilli of infected epithelial cells (Fig. 4A). In the middle and posterior midgut,abundant Pns10 tubules extended from the actin-based microvilli along the surface of infected epithelial cells of the midgut (Fig. 4B–D),consistent with previous electron microscopy observations that tubules containing RDV particles were associated with the microvilli of the midgut in viruliferous leafhoppers(Nasu, 1965). A few Pns10 tubules had successfully crossed into the gut lumen (Fig. 4B–D). Furthermore, some Pns10 tubules were associated with the actin-based microvilli of neighboring cells (Fig. 4C), suggesting that RDV might exploit these tubules to facilitate viral cell-to-cell spread among epithelial cells in the midgut of the leafhopper.

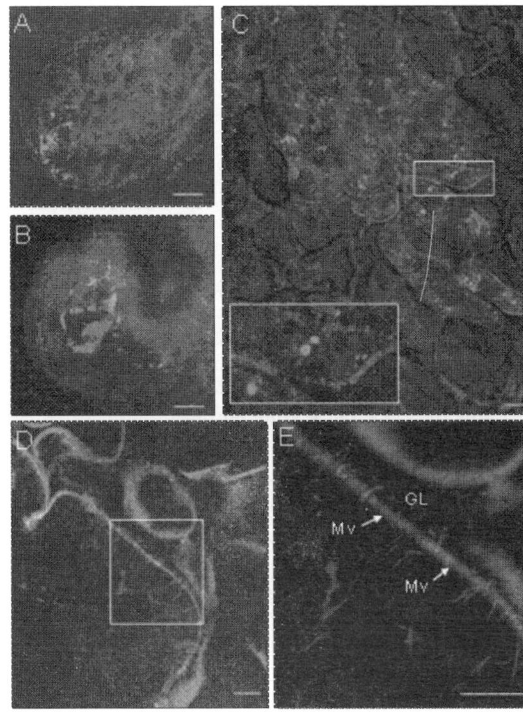

Fig. 3 Pns10 tubules on actin-based microvilli of filter chamber in viruliferous leafhoppers

At 2-day (A) or 3-day (B–E) padp, leafhopper organs were immunolabeled for Pns10 tubules with Pns10-rhodamine (red), for RDV virions with virus-FITC (green), and for actin-based microvilli with FITC-phalloidin (green), then examined by confocal microscopy. (A, B) Fluorescence micrograph of filter chamber showing green fluorescence (virus antigens) with background visualized by transmitted light. (C) Image of filter chamber merged with images with green fluorescence (virus antigens), red fluorescence (Pns10 tubules) and background visualized by transmitted light. Inset indicates an enlarged image of the boxed area. (D) Image of filter chamber merged with image of green fluorescence (actin) and of red fluorescence (Pns10 tubules). (E) Enlarged image of boxed area in panel D. GL, gut lumen. Mv, microvilli. Images are representative of multiple experiments with multiple preparations. Bars, 350μm (A–B) and 10μm (C–E)

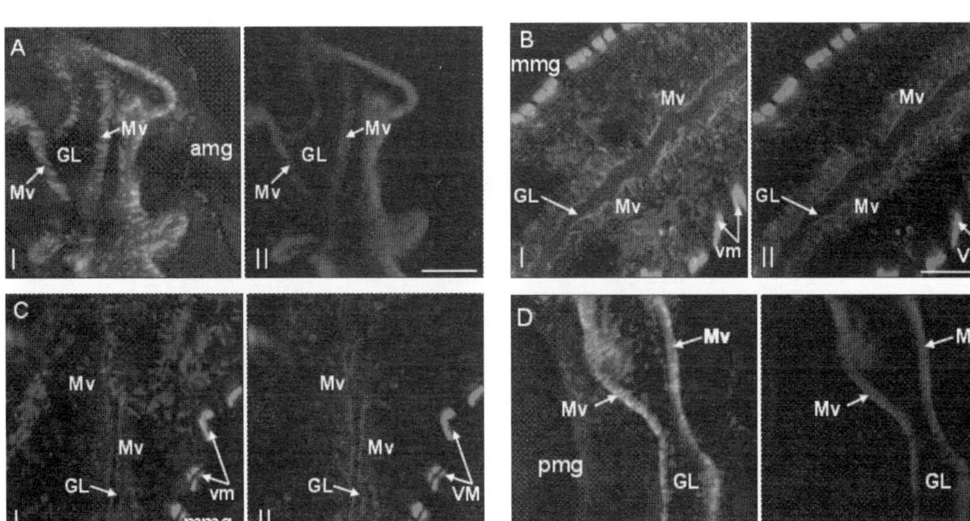

Fig. 4 Pns10 tubules on actin-based microvilli of midgut in viruliferous leafhoppers

At 6-day padp, leafhopper organs were immunolabeled for Pns10 tubules with Pns10-rhodamine (red), for RDV virions with virus-Alexa Fluor 647 (blue), and for actin-based microvilli with FITC-phalloidin (green), then examined with confocal microscopy. Images of anterior midgut (A), middle midgut (B, C) and posterior midgut (D) are shown. Images I in B and C were merged with red fluorescence (Pns10 tubules), blue fluorescence (virus antigens) and green fluorescence (actin). Images I in A and D were merged with red fluorescence (Pns10 tubules), blue fluorescence (virus antigens) and green fluorescence (actin) with background visualized by transmitted light. Images II were single green fluorescence (actin) to show the actin-based microvilli. GL, gut lumen. Mv, microvilli. VM, visceral muscles. Images are representative of multiple experiments with multiple preparations. Bars, 10μm

To confirm our observations, we further used electron microscopy to visualize the tubules in the epithelial cells of the alimentary canal during infection by RDV. The alimentary canal of leafhopper is formed by epithelial cells that lie on basal lamina surrounded by external longitudinal and internal circular muscle fibers (Fig. 2). Numerous brush border microvilli on the surface of epithelial cells extend into the lumen (Fig. 2B, Fig. 5, Fig. S1). RDV infection induced the formation of tubules approximately 85 nm in diameter, with viral particles inside, in the epithelial cells of the filter chamber and midgut (Fig. 5, Fig. S1).These tubules appeared to be passing through the microvilli from inside the infected cells into the lumen (Fig. 5, Fig. S1), consistent with the observations with immuno fluorescence microscopy (Fig. 4).

A careful analysis of electron micrographs revealed the possible process of the crossing of tubules through the microvilli. The closed-end of the tubules first was enclosed inside the microvillus (Fig. 5A), and then entire tubules extended into the microvillus (Fig. 5B). We observed that the closed-end of the tubule made contact with the inner side of the distal end of the microvillus (Fig. 5B), which may drive the elongation of the microvillus to form a membrane protrusion towards the lumen (Fig. 5C). Finally, the tubules were released in the lumen from these microvilli (Fig. 5D). Free virions were absent in the microvilli (Fig. 5, S1). Our previous electron tomographic microscopy indicated that RDV particles were fixed on the inner surface of the tubules; thus they did not freely diffuse within the tubules (Katayama et al., 2007). All these observations indicated that RDV particles were accompanied by Pns10 tubules to pass though the microvilli into the lumen.

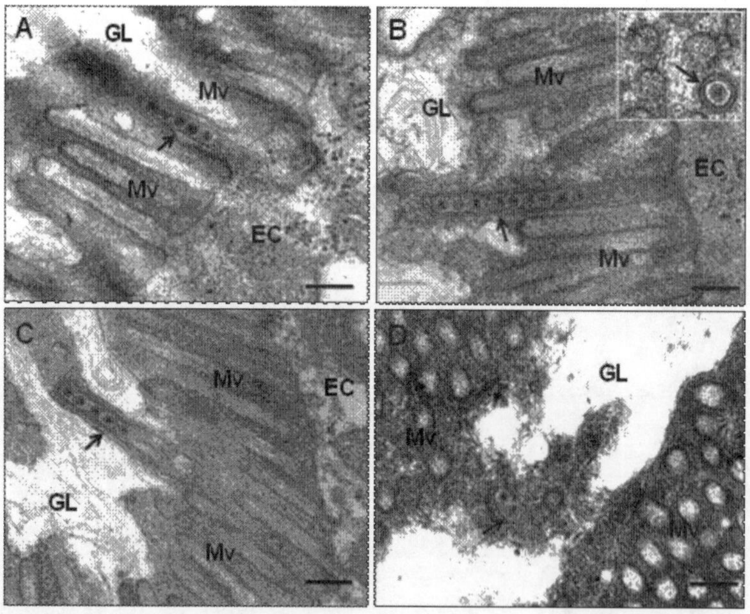

Fig. 5 Transmission electron micrographs showing the association of virus-containing tubules with microvilli of anterior midgut in viruliferous leafhoppers

(A) Closed-end of tubule (arrow) inserted into a microvillus. (B) Closed-end tubule (arrow) in contact with the inner side of distal end of microvillus. Inset, transverse section of microvillus of about 100 nm in diameter, with a virus-containing tubule (arrow) inside. Arrowheads indicate actin filaments within the microvillus. (C) Elongated tubule-associated microvillus (arrow) has formed membrane protrusion toward the lumen. (D) Tubule (arrow) in the lumen. EC, epithelial cell. GL, gut lumen. Mv, microvilli. Bars, 200nm

2.2 Pns10 tubules colocalize with actin-based muscle fibers of visceral muscle tissues surrounding the midgut of viruliferous leafhoppers

Our previous observations indicated that infection of RDV in the epithelial cells was followed by virus invasion of the visceral muscle tissues lining the anterior midgut (Chen et al., 2011). To determine whether Pns10 tubules were associated with the visceral muscle lining the anterior midgut, at 12-day padp, we dissected the internal organs from leafhoppers and immunolabeled them with Pns10-rhodamine, virus-Alexa Fluor 647 and FITC-phalloidin. As shown in Fig. 6A, the visceral muscles formed by actin-based longitudinal muscle fibers and circular muscle fibers were visible as a fluorescent lattice pattern. Short Pns10 tubules were specifically associated with single actin-based longitudinal muscle fiber that connected with circular muscle fiber bundles (Fig. 6A). In contrast, free virions were absent in these longitudinal muscle fibers (Fig. 6A). Electron microscopic observations confirmed that virus-containing tubules were associated with the visceral muscle tissues lining the anterior midgut during infection by RDV(Fig. 6B). These observations implied that RDV might exploit Pns10 tubules to traffic along actin-based muscle fibers to facilitate the lateral spread of RDV.

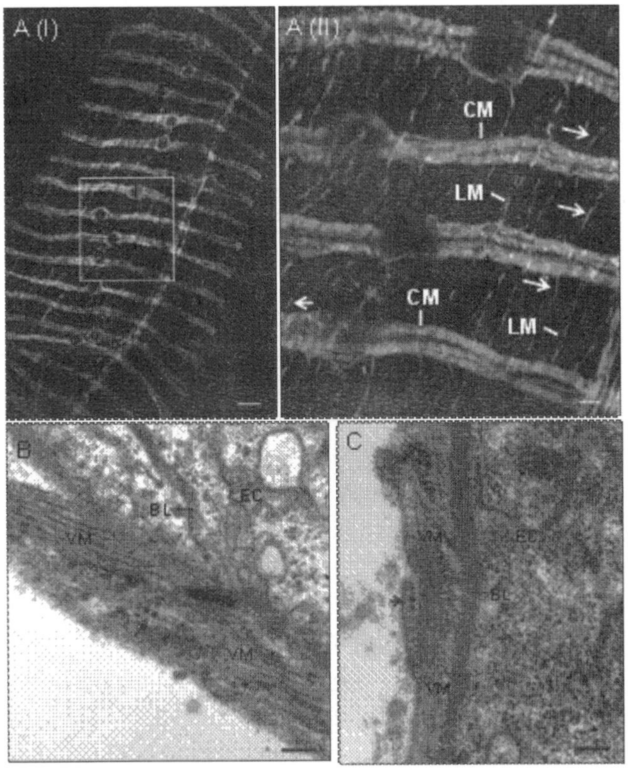

Fig. 6 Pns10 tubules localized along actin-based muscle fibers that surround the anterior midgut in viruliferous leafhopper

(A) Confocal micrograph showing the association of Pns10 tubules with actin-based longitudinal muscle fibers. At 12-day padp, leafhopper organjs were immunolabeled for Pns10 tubules with Pns10-rhodamine (red), for RDV virions with virus-Alexa Fluor 647 (blue), and for actin-based microvilli with FITC-phalloidin (green), then examined by confocal microscopy. Images show the blue fluorescence (virus antigens) and red fluorescence (Pns10 tubules) overlapped with the green fluorescence (actin). Image II is a higher magnification view of boxed area in image I to show that Pns10 tubules (red; arrows), rather than virus antigens (blue), co-localize with actin-based longitudinal muscle fibers (green) that seem to bridge two adjacent circular muscle cells. Images are representative of multiple experiments with multiple preparations. (B) Transmission electron micrograph showing the association of virus-containing tubule (arrow) with visceral muscle tissues lining anterior midgut. BL, basal lamina. CM, circular muscle. EC, epithelial cell. LM, longitudinal muscle. VM, visceral muscle. Bars, 20μm (image I in A), 80μm (image II in A) and 200nm (B)

2.3 RNAi induced by dsPns10 inhibits the assembly of Pns10 tubules and secondary infection of neighboring cells by RDV in VCMs

We initially used an in vitro system of VCMs to examine RNAi activity caused by dsRNA specific for

Pns10 gene. VCMs on a coverslip were transfected with dsRNAs specific for Pns10 gene of RDV (dsPns10) or segment encoding for yellow fluorescence protein (YFP) (dsYFP) via Cellfection-based transfection. Viability tests showed no obvious differences among transfected cells, demonstrating absence of toxicity of the transfection reagent and dsRNAs to VCMs (data not shown). RNAi is demonstrated by the appearance of small interfering RNAs (siRNAs) corresponding toan mRNA target sequence(Tomoyasu et al., 2008). Therefore, the presence of siRNAs specific for Pns10 or YFP genes was analyzed by northern blots at 72h after transfection of synthesized dsRNAs. Small RNAs approximately 21 nt long were detected from RNAs extracted from dsPns10- or dsYFP-transfected cells (Fig. 7A), showing that RNAi was induced in the VCMs after transfection with dsRNAs. Then, 24h after transfection, the VCMs were inoculated with RDV at a low MOI of 0.001 and then cultured in the presence of virus-neutralizing antibodies. At this low MOI, viral infection rate was below 1%, and secondary infection was clearly visible, as described previously (Wei et al., 2006; Wei et al., 2008). Cells were fixed 3d post-inoculation (p.i.) and immunolabeled with virus-FITC and Pns10-rhodamine. In VCMs transfected with dsYFP or treated with Cellfection alone (i.e., control), small foci of 4 to 10 infected cells were visible in the presence of virus-neutralizing antibodies, consistent with the spread of RDV from an initially infected cell to adjacent cells (Fig. 7B, C, panels I and II).Furthermore, abundant Pns10 tubules protruded from such infected cell surfaces and were scattered outside the cells(Fig. 7C, panels I, II). On the other hand, in VCMs transfected with dsPns10, RDV was restricted to the initially infected cells in the presence of virus-neutralizing antibodies (Fig. 7B, C, panelsIII). Furthermore, the formation of Pns10 tubules had been blocked (Fig. 7C, panel III), indicating that the expression of Pns10 had been knocked down by RNAi induced by dsPns10.

Our previous findings indicated that the blockage of Pns10 tubule formation by chemical inhibitors did not significantly affect viral multiplication in VCMs (Wei et al., 2006; Wei et al., 2008). To further examine whether the treatment with dsPns10 had any effects on the multiplication of RDV in VCMs, 24h after transfection with dsRNAs, VCMs were inoculated with RDV at an MOI of 10 to guarantee viral infection rate was 100%, and then cell lysates were collected at 72h. As shown in Fig. 8A, dsPns10 reduced the titer of cell-associated viruses by 7%–11%, suggesting that RNAi induced by dsPns10 had no substantial effect on the accumulation of RDV in VCMs.

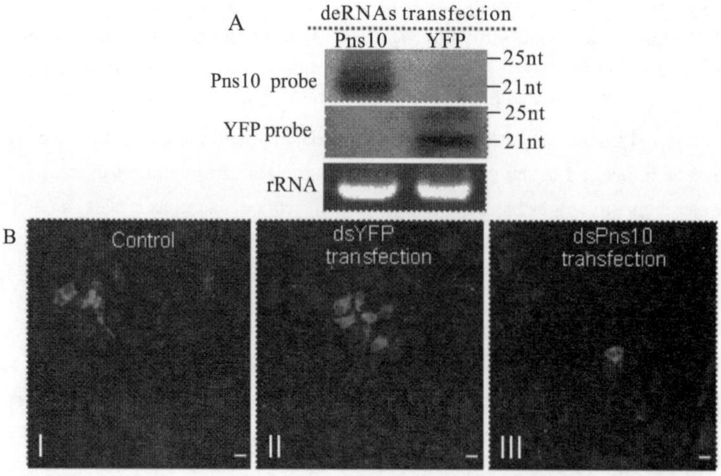

C
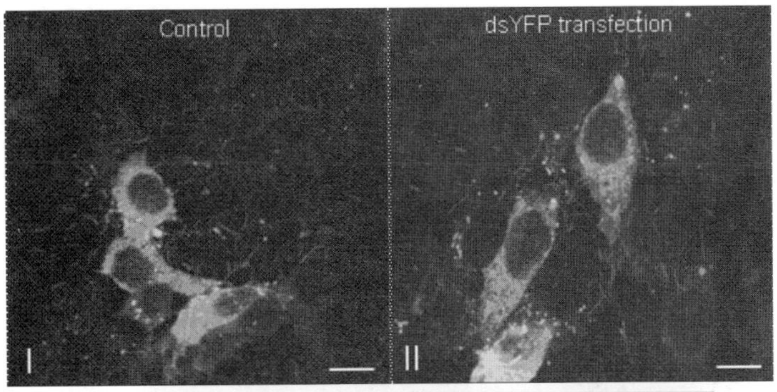

C
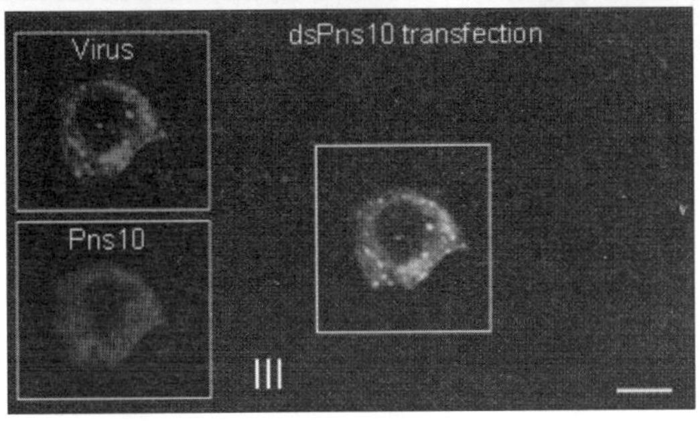

Ⅲ

Fig. 7 RNAi induced by dsPns10 inhibiting the assembly of Pns10 tubule and secondary infection of neighboring cells by RDV in VCMs
(A) Detection of siRNAs in VCMs transfected with synthesized dsPns10 or dsYFP at 72h after transfection. Approximately 5mg of total RNA was probed with DIG-labeled transfected RNA. Lower panel: detection of 5.8S rRNA as a control to confirm loading of equal amounts of RNA in each lane. (B, C) Intercellular spread of RDV was inhibited by RNAi induced by dsPns10 in VCMs. At 24h after transfection with cellfection transfection reagent (control; images I), dsYFP (images II) or dsPns10 (images III), VCMs were inoculated with RDV at a low MOI (0.001) and cultured in the presence of virus-neutralizing antibodies, which prevent free virus infection. At 3d p.i., cells were immunolabeled for RDV virions with virus-FITC (green) or for Pns10 tubules with Pns10-rhodamine (red), and then examined by confocal microscopy. Images in B show green fluorescence from virus antigens in widefield view. Images in C are composites of those for green fluorescence (virus antigens) and red fluorescence (Pns10 tubules).Insets in image III of panel C show green fluorescence (virus antigens) and red fluorescence (Pns10 antigens) of the merged images in the boxed area. Images are representative of multiple experiments with multiple preparations. Bars, 20μm

The effects of RNAi induced by dsPns10 on the synthesis of viral plus-strand RNAs were analyzed with Northern blots. As shown in Fig. 8B, transfection with dsPns10 resulted in a marked reduction of the synthesis of plus-strand RNA of Pns10 gene, but without significant effect on the synthesis of plus-strand RNA of viral major outer capsid protein P8 gene. The effects of RNAi induced by dsPns10 on the synthesis of viral plus-strand RNAs were further confirmed by quantitative real-time RT-PCR(RT-qPCR) assay. RT-qPCR assay showed that the treatment of dsPns10 caused about 80% reduction in the level of plus-strand RNA of Pns10 gene, relative to a constant amount of plus-strand RNA of P8 gene (Fig. 8C). Thus, RNAi induced by dsPns10 could specifically knock down the expression of Pns10 gene, but caused little reduction of the expression of other viral genes, a finding consistent with its failure to effectively inhibit viral multiplication in VCMs. All these results suggested that the inhibition of the assembly of Pns10 tubules, due to RNAi induced by dsPns10 resulted in the failure of Pns10 tubules to protrude beyond the cell surface and in the subsequent lack of infection of neighboring cells by RDV.

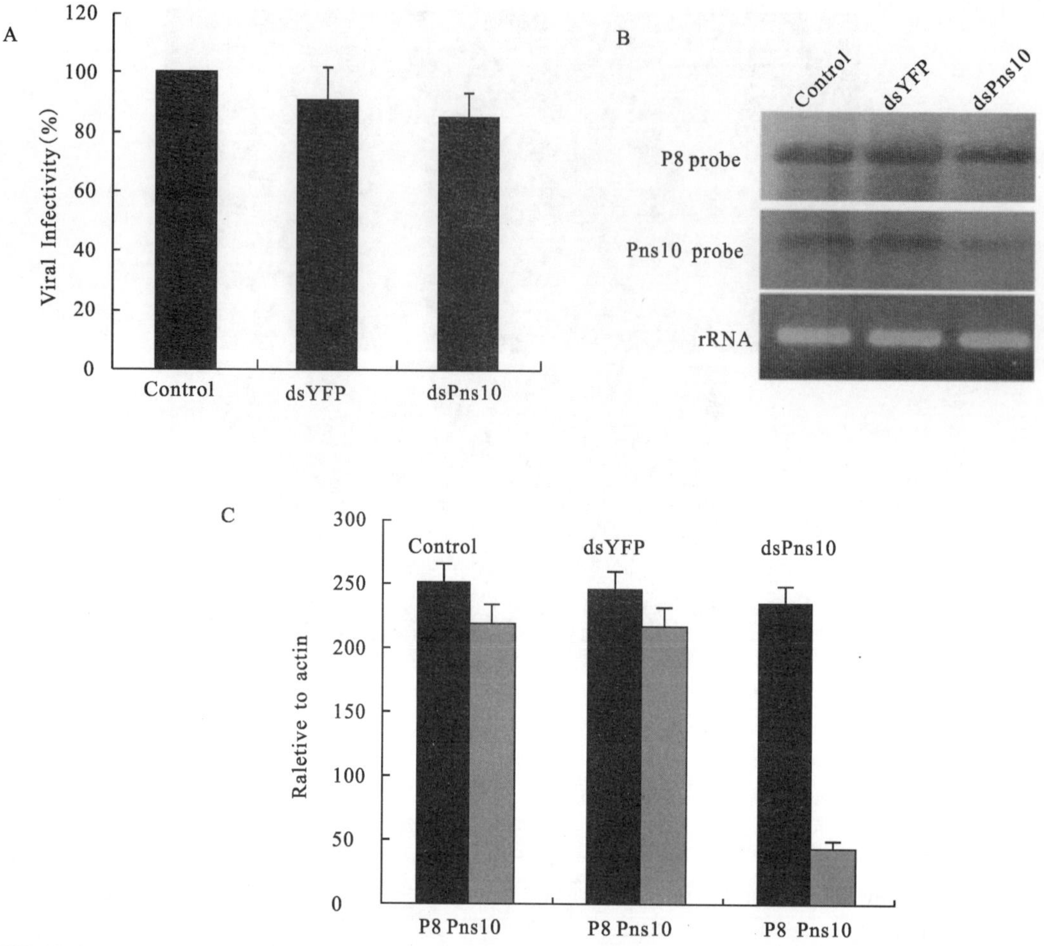

Fig. 8 RNAi induced by dsPns10 knockdown the expression of Pns10 without significantly inhibiting virus multiplication in VCMs

(A) Effects of the treatment of dsRNAs on multiplication of cell-associated RDV in VCMs. Viral titers were determined in duplicate by the fluorescent focus assay (see text for details). Error bars indicate standard deviations from three independent experiments. (B) Transfection of dsPns10 in VCMs results in a significant reduction in level of plus-strand RNA of Pns10 gene, without greatly inhibiting synthesis of plus-strand RNA of P8 gene, as revealed by northern blot. VCMs were transfected with transfection reagent (control), dsYFP or dsPns10, inoculated with RDV at an MOI of 10, then harvested 72h later. Approximately 5mg of total RNAs were probed with DIG-labeled negative-sense RNA transcripts of Pns10 or P8 genes. Lower panel: detection of 5.8S rRNA as a control to confirm loading of equal amounts of RNA in each lane. Image is representative of multiple experiments with multiple preparations. (C) Transfection of dsPns10 in VCMs caused about 80% reduction in level of plus-strand RNA of Pns10 gene, without greatly inhibiting synthesis of plus-strand RNA of P8 gene, as revealed by RT-qPCR assay. VCMs were transfected with transfection reagent (control), dsYFP or dsPns10, inoculated with RDV at an MOI of 10, then harvested 72h later. Approximately 5mg of total RNAs was extracted with TRIzol Reagent. The results of RT-qPCRs were normalized to the level of leafhopper actin gene. Error bars indicate standard deviations from three independent PCRs

2.4 Ingestion of dsPns10 knocks down expression of Pns10protein and slows RDV spread in the intact insect

To address the potential function of Pns10 in viral transmissionby insect vectors, we tested whether RNAi induced by ingestion of dsPns10 could in hibit viral infection in intact leafhoppers. Inpreliminary experiments, ingestion of dsRNA by the leafhoppersresulted in no phenotypic abnormalities (data not shown). Thepercentage of viruliferous insects after ingestion of dsRNAs was analyzed by RT-PCR of the genes for the nonstructural protein Pns10 and the major outer capsid protein P8 of virus. When one of the genes was detected, the other gene was also detected in allcases. At 12-day padp, about 55% (n = 100, 3 repetitions) of leafhoppers that received dsYFP had the Pns10 and P8 genes(Table 1). In contrast, about 12% (n = 100, 3 repetitions) of leafhoppers that received dsPns10 had the Pns10 and P8 genes(Table 1). The number of positive samples did not differsignificantly between leafhoppers that received dsYFP and asucrose diet control (Table 1). The effect of dsRNA treatment onthe synthesis of viral plus-strand RNAs from the Pns10 and P8genes was further analyzed with Northern blots. In agreement withthe RT-PCR results, accumulation of viral plus-strand RNAs ofthe Pns10 and P8 genes was greatly reduced in RNA fractionsisolated from leafhoppers receiving dsPns10 as compared with leafhoppers receiving dsYFP or the sucrose diet control (Fig. 9A).These results confirmed that the expression of Pns10 could beknocked down by RNAi induced by synthesized dsRNA that werefed to the insects using a membrane-feeding method.

To determine whether blocking the formation of virus-containingPns10 tubules would inhibit viral spread among leafhopper tissues, we dissected the internal organs from leafhoppers at 12-day padp, and then immunolabeled them with Pns10-rhodamine, virus-Alexa Fluor 647 and actin dye FITC-phalloidin.In about 36% of leafhoppers that received dsPns10, virus antigenswere restricted to a small area on the corner of the filter chamber,the initial entry site of RDV, and the formation of Pns10 tubules was mostly inhibited in these regions (Fig. 9B, Table 1). Bycontrast, both viral antigens and Pns10 tubules were detected inthe filter chamber, midgut and salivary glands in about 56% leafhoppers receiving dsYFP, but only in about 16% of thosereceiving dsPns10 (Fig. 9C, Table 1), corresponding to theproportion of positive samples checked by RT-PCR, as shownalready. All these results suggested that RNAi induced by ingestionof dsPns10 is activated in about 70% of RDV-positive leafhoppers,which may largely reflect limitations in uptake efficiency (about 70%) of dsPns10 into the intestine tissues of the leafhopper. Taken together, the formation of Pns10 tubules was heavily impaired dueto the knockdown of Pns10 expression in leafhoppers that receivedds Pns10, which would lead to significant inhibition of the efficientspread of RDV in the body of the leafhoppers. As expected, theingestion of dsPns10 by leafhoppers significantly suppressed vectortransmission of the virus (Table 1).

Table 1 Ingestion of dsPns10 via membrane feeding strongly inhibited viral spread and transmission by insect vectors

Insects	No. of positive insects with Pns10 and P8 genes detected by RT-PCR at 12-day padp (n = 100)			No. of positive insects with virus antigens and Pns10 tubules in different tissues at 12-day padp[a] (n = 30)					No. of positive insects that transmitted viruses to rice seedlings at 22-day padp (n = 60)		
	I	II	III	Exp. no.	Fc[b] (limited)	fc[b] (extensive)	mg	sg	I	II	III
dsPns10	13	10	12	I	12	3	3	2	1	2	1
				II	11	5	5	2			
				III	9	6	5	3			
dsYFP	52	58	54	I	0	18	14	9	7	8	10
				II	0	17	17	11			
				III	0	15	15	7			
diet control	58	60	63	I	0	16	16	10	8	10	10
				II	0	17	13	11			
				III	0	16	12	8			

[a], The filter chamber (fc), midgut (mg) and salivary gland (sg) of leafhoppers were examined for immunofluorescence of virus antigens and Pns10 tubules.

[b], Distribution of virus antigens and Pns10 tubules was assessed as either in a limited area or in an extensive area of filter chamber (fc).

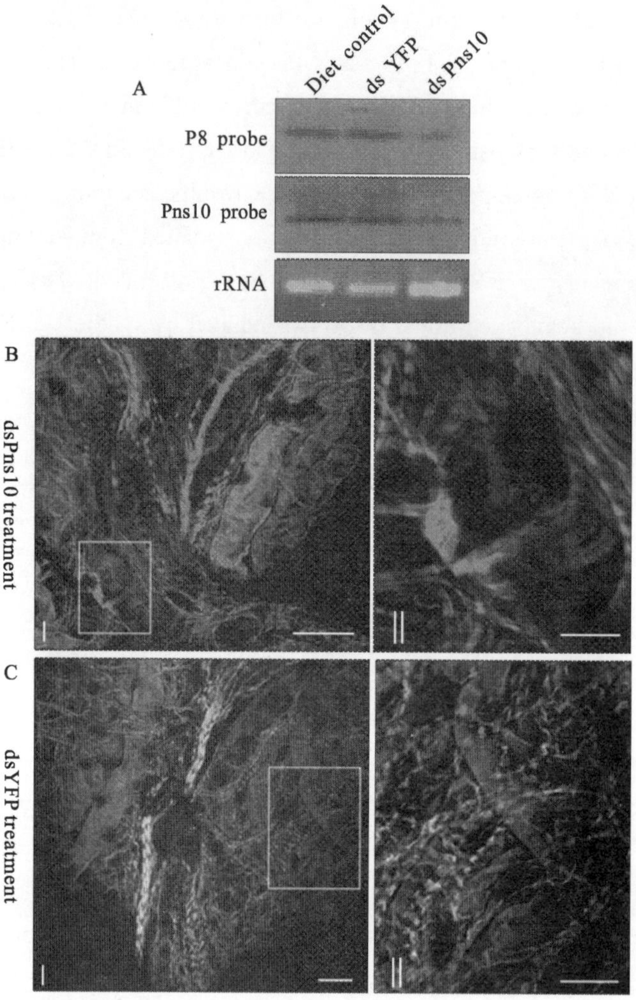

Fig. 9 Ingestion of dsPns10 knocks down expression of Pns10 proteins and slows RDV spread in the intact insect

(A) RNAi induced by dsPns10 reduced the synthesis of viral plus-strand RNAs from Pns10 and P8 genes in leafhoppers. Second-instar nymphs of leafhopper were fed with dsRNAs or sucrose diet (control) via membrane-feeding. At 12-day padp, approximately 5mg of total RNAs extracted from leafhoppers organs were probed with DIG-labeled negative-sense RNA transcripts of Pns10 or P8 genes. Lower panel: detection of 5.8S rRNA as a control to confirm loading of equal amounts of RNA in each lane. Image is representative of multiple experiments with multiple preparations. (B, C) Ingestion of dsPns10 via membrane-feeding blocking the formation of Pns10 tubules and RDV spread in the filter chamber of leafhopper vectors. The secondinstar nymphs of leafhopper were fed with dsPns10 (B) or dsYFP (C) via membrane-feeding. At 12-day padp, leafhopper organs were immunolabeled for Pns10 tubules with Pns10-rhodamine (red), immunolabeled for RDV virions with virus-Alexa Fluor 647 (blue), and immunolabeled for actin with FITC-phalloidin (green). Images II are higher magnification views of boxed areas in images I in panels A and B, respectively, to show that the assembly of Pns10 tubules (red) and the spread of RDV (blue) was blocked in the filter chamber of leafhoppers that received dsPns10. Images are representative of multiple experiments with multiple preparations. Bars, 100μm (images I in A and B), 350μm (images II in A and B)

3 Discussion

RDV particles have been observed in tubules in microvilli inultrathin sections of viruliferous vector insects(Nasu, 1965). This tubule,about 85nm in diameter, is now regarded as the Pns10 tubule and plays an important role in viral spread and infection ofneighboring uninfected cells (Wei et al., 2006; Katayama et al., 2007). Recently, we studied RDV and its sequential infection of the internal organs of its vectorinsect over time by analyzing the spread of virus antigens afteringestion of the virus by the vector insect(Chen et al., 2011). The virus firstaccumulated in epithelial cells of the filter chamber, progressed to the anterior midgut, and then spread to visceral musclessurrounding the anterior midgut (Chen et al., 2011). In the present study, wefocused on the distribution of Pns10 tubules in viruliferous vectorinsects over time and found that Pns10 tubules, with viral particlesinside, crossed actin-based microvilli of epithelial cells (Fig. 3–5), a finding for the first time in virus research, suggesting the passage of the virus through the microvillus membrane. Pns10protein, with its ability to bind actin via a specific but relativelylow-affinity interaction (Wei et al., 2006), may permit the Pns10 tubules to traffic along the actin-based microvilli of the epithelial cells ofmidgut. Results from electron microscopy revealed that virus the containing tubule extended along the actin filaments in themicrovillus and finally arrived at the distal end of the microvillus(Fig. 5, Fig. S1). The continuous contact of the tubule with thedistal end of the microvillus apparently promoted the elongation of the microvillus to form a membrane protrusion towards the lumen(Fig. 5), consistent with a mechanism underlying the outgrowth of cellular protrusions driven by actin-polymerization(Cameron et al., 1999; Baluška et al., 2004). Thetubules seemed to be released from the broken microvilli in thelumen (Fig. 5D). This scenario of RDV spread also occurredafter the virus propagated in the visceral muscle tissues surroundingthe anterior midgut, where lateral spread of the Pns10 tubulesalong actin-based muscle fiber might facilitate viral cell-to-cellmovement through the muscle tissue (Fig. 6). All these resultssuggest that movement of virus-containing Pns10 tubules alongactin-based cellular protrusions would enable RDV to efficientlyspread in various organs in viruliferous insects.

We previously provided indirect evidence that Pns10 tubuleplayed an important role in RDV spread based on data showingthat inhibition of the extension of actin-based cellular protrusionresulted in the failure of Pns10 tubule extension, which eventuallyincapacitated viral spread among insect vector cells (Wei et al., 2006; Wei et al., 2008). Togain direct evidence that Pns10 tubule plays a critical role in the intercellular spread of virus in inset vector cells, here we used anRNAi technique to interrupt the formation of Pns10 tubules, thenanalyzed its effect on viral spread. VCM is an in vitro experimentalsystem which enables us to clarify the molecular entitiesresponsible for the biological events during viral replication (Wei et al., 2006; Wei et al., 2008). In VCMs in the presence of neutralizing antibodies to avoidinfection by free virions, virus antigens spread to severalneighboring cells via Pns10 tubules (Fig. 7B–C; Control and dsYFP transfection), as reported earlier (Wei et al., 2006). In contrast, after the knockdown of Pns10 expression by RNAi induced by dsPns10,only one cell was infected even at 3d after inoculation(Fig. 7B–C, dsPns10 transfection). These results, together with the fact that dsPns10 specifically knocked down the expression of Pns10, but did not significantly affect viral multiplication (Fig. 8)demonstrate that Pns10 tubule itself plays an important role inviral spread in VCMs. Because the use of a cultured monolayersystem enables a synchronous and total cell response to treatment with dsRNAs, our RNAi system coupled with the use of VCMs should enable us to disclose the biological activities of each of thephytoreovirus-encoded proteins.

The number of RDV-positive insects detected by RT-PCR wassignificantly reduced in insect populations treated with dsPns10 ascompared to those treated with dsYFP or diet control (Table 1).When analyzed in more detail, the knockdown of Pns10 expression did not interfere with the initial infection of virus intothe filter chamber (Table 1, Fig. 9B), the primary site for viralattachment and entry(Chen et al., 2011) corresponding with a previoushypothesis that RDV uses the minor capsid protein P2 as a viralattachment molecule to bind to the cellular receptor on the microvillar membrane of the filter chamber to enable viral uptakeinto the epithelial cells(Wei et al., 2007; Zhou et al., 2007; Chen et al., 2011). In contrast, the knockdown of Pns10 expression strongly inhibited the extensive spread of RDVin the filter chamber and efficient intercellular spread of the virusto other organs such as the midgut and salivary gland in the bodyof its vector insects (Table 1). As a persistent-propagative plantvirus, after moving into the epithelial cells of the filter chamber,RDV must replicate and assemble progeny virions, which wouldthen spread to neighboring cells or other organs such as themidgut and salivary glands. Because the knockdown of Pns10 expression by treatment with dsPns10 or the blockage of Pns10 tubule formation by chemical inhibitors (Wei et al., 2008) did not significantlyaffect viral multiplication in leafhopper cells (Fig. 8), wedetermined that the slow spread of RDV in the body of the dsPns10-treated insects was directly caused by a significant loss inthe functioning of Pns10 tubules in viral intercellular spread frominitially infected epithelial cells in the filter chamber. These resultsindicated that the interference of Pns10 tubules with viral spreadin vector insect cells (Fig. 7) significantly inhibited viralproliferation in intact vector insects, which would eventually significantly reduce viral transmissibility by the vector insects(Table 1). Results mentioned earlier, together with the directevidence showing that Pns10 tubule is responsible for viral spreadin vitro (Fig.7) and the observation that virus-containing Pns10 tubules are located along actin-based cellular machinery as theyspread into various organs in the viruliferous insect (Fig. 3–6),demonstrate that Pns10 tubules containing viral particles, play acritical role in viral spread in vector insects, enabling to accomplisha latent period for the virus, and subsequent ability to transmit thevirus to plant hosts. Additional indirect evidence for theinvolvement of Pns10 tubules as viral determinants for virustransmission is that a Pns10-protein-deficient isolate of RDV failedto be transmitted by insect vectors (Pu et al., 2011). However, free RDVparticles may also exploit receptors on the intercellular junctionalcomplex in the epithelial tissues to enable viral intercellular spread,as in the case of several animal viruses (Bergelson, 2009); this issue should befurther investigated. The novel model for viral intercellular spreadpresented here is thought to be advantageous over infection by acell-free virus because the virus is protected from host immuneresponses.

Our results showing that RNAi was induced in leafhopper *N.cincticeps* by treatment with dsRNA targeting a viral gene,confirmed a recent finding that RNAi can be induced inleafhopper *Homalodiscavitripennis* (glassy winged sharpshooter) by treatment with dsRNA targeting an insect gene (Rosa et al., 2010; Rosa et al., 2012). Althoughthe mechanisms for dsRNA uptake from the gut lumen into theepithelia cells of insects through feeding are poorly understood(Price and Gatehouse, 2008; Whyard et al., 2009; Huvenne and Smagghe, 2010; Rocha et al., 2011), the silencing signal, siRNA induced by dsRNA, seems tospread from the initial entry sites via cells and tissues (Price and Gatehouse, 2008; Whyard et al., 2009; Huvenne and Smagghe, 2010; Rocha et al., 2011).Thus, siRNA is transient and not synchronous in intestine tissue of insects (Huvenne and Smagghe, 2010). However, cultured insect cells showed RNAi whensoaked in medium with dsRNA, and the dsRNA uptake into cellswas synchronous (Rosa et al., 2010; Rocha et al., 2011; Jia et al., 2012b). It may be for this reason that siRNA could be easily detected in VCMs transfected with dsRNA, but notin intact leafhoppers after ingestion of dsRNA. It is apparent thatinsects lack an RNA-dependent RNA polymerase (RdRp) toamplify siRNA induced by dsRNA (Price and Gatehouse, 2008). Our results support theearlier conclusion that insects may have alternative RdRP-likemechanisms that result in systemic RNAi (Price and Gatehouse, 2008; Whyard et al., 2009; Rosa et al., 2010; Rosa et al., 2012).

RNAi has been extensively used to investigate the functionalroles of viral proteins of reoviruses. For example, our recentfindings indicated that treatment with dsRNAs against the viralgene for P9-1 of Southern rice black-streaked dwarf virus, also aplant reovirus, inhibited viral replication in its vector insect (Jia et al., 2012b).Similarly, RNAi mediated by short-interfering RNAs against viralgenes has been extensively used for functional analyses ofrespective genome segments of animal reoviruses, which areclosely related to plant reoviruses (Zambrano et al., 2008; Ayala-Breton et al., 2009). Furthermore, rearrangementsand reassortments of genome segments of reoviruseshave long been helpful in elucidating the functions of reovirusproteins (Desselberger, 1996; Ramig, 1997; Sun and Suzuki, 2008; Eusebio-Cope et al., 2010). These together with

RNA silencing assay haveserved as alternatives to plasmid-based reverse genetics.

Thus, development of RNAi induced by synthesized dsRNA,together with the system of the rearrangement or reassortment ofgenome segments, may help overcome the lack of a reversegenetics system for plant reoviruses and provide useful tools toinvestigate the molecular mechanisms enabling efficient transmissionof viruses by vector insects.

Passage of persistent-propagative viruses through different organs in their vector insects requires specific interactions between virus and vector components to overcome different transmission barriers (de Assis Filho et al., 2002; Hogenhout et al., 2008; Ammar et al., 2009). For example, the glycoproteins of tospoviruses and rhabdoviruses and the minor capisd proteins of phytoreoviruses serve as viral ligands to mediate attachment of virions to receptors on the epithelial cells of the alimentary canal of the vector insect, a necessary step for virions to overcome midgut infection barriers of vector insect (Omura et al., 1998; Sin et al., 2005; Ullman et al., 2005; Wei et al., 2007; Zhou et al., 2007; Whitfield et al., 2008; Ammar et al., 2009). In addition to functional viral proteins that enable the early stage interaction with host cells, persistent-propagative viruses induce the formation of various cytopathological structures involved in viral replication or cell tocell movement in their hosts to facilitate viral propagation. For example, tospoviruses exploit virus-containing tubules composed of a nonstructural movement protein NSm to modify plasmodesmata, allowing the transport of entire viral particles (Li et al., 2009). In the present study, for the first time, we have provided experimental evidence to show that a virus transmitted by its insect vector in a persistent-propagative manner can also use virus-containing tubules composed of a nonstructural protein to traffic along actin-based cellular machinery, allowing efficient cell-to-cell spread of the virus in vector insect. In this respect, the transport of virus-containing tubules along actin-based cellular machinery in vector insects resembles the extension of the tubules composed of movement proteins of viruses through plasmodesmata in host plants, suggesting that viruses evolved conserved strategies for viral intercelluair spread in vector insects or host plants.

4 Materials and Methods

4.1 Cells, viruses and antibodies

The NC-24 line of *Nephotettixcincticeps* (leafhopper) cells was maintained in monolayer culture in LBM growth medium (Kimura and Omura, 1988). RDV was purified from infected rice plants without the use of CCl_4, as described by (Zhong et al., 2003). The antibodies against Pns10 and against intact viral particles (virus antigens) were described in an earlier report (Wei et al., 2006).

4.2 Immunofluorescence staining of internal organs ofleafhopper after acquisition of virus

The second-instar nymphs of *N. cincticeps* were allowed a 2-day acquisition access period (AAP) on rice plants infected with RDV. At different days after the AAP, internal organs from leafhoppers were dissected, fixed in 4% paraformaldehyde and processed for analysis of immunofluorescence as described previously (Wei et al., 2006; Chen et al., 2011). Pns10 tubules were immunolabeled with Pns10-rhodamine; viral particles were immunolabeled with virus-Alexa Fluor 647 or virus-FITC; actin was immunolabeled with FITC-phalloidin (Sigma). As controls, dissected organs from leafhoppers that fed on healthy plants were immunolabeled exactly in the same way. Samples were examined with a Leica TCS SP5 inverted confocal microscope essentially as described previously (Chen et al., 2011).

4.3 dsRNA production

DNA fragment spanning a 1062-bp segment of the Pns10 gene of RDV was amplified by PCRusing a forward primer (5′ATTCTCTA GAAGCTTAATACGACTCACTATAGGGGAAGTAGACACTGCTA CGTTTGTTCG 3′) and a reverse primer (5′ ATTCTCTAGAAGCT TAATACGACTCACTATAGGGGGAA CCGCCGCCTTTAAG3′), both possessing a T7 promoter (italics) at the 5′ end. The DNA fragment spanning a 717-bp segment of YFP was amplified by PCR using a forward primer (5′ ATTCTCTAGAAGCTTAATACG ACTCACTATAGGGGTGAGCAAGGGCGAGGAGCT 3′) and a reverse primer (5′ ATTCTCTAGAAGCTT AATACGACTCACTATAGGGCTT GTACAGCTCGTCCATGC 3′), both possessing a T7 promoter (italics) at the 5′ end. The PCR products were used for dsRNA synthesis according to the protocol of the T7 RiboMA Express RNAi System kit (Promega). The dsRNAs were purified according to the manufacturer's instructions and checked for quality by agarose gel electrophoresis and quantified by using a spectrophotometer.

4.4 Examination of in vitro spread of RDV in the presence of synthesized dsRNAs

VCMs were grown in LBM medium supplemented with 10% fetal bovine serum (FBS; Invitrogen). VCMs were transfected using lipid cellfectin using an adaptation of the manufacturer's protocol, as reported by (Rosa et al., 2010). Briefly, VCMs on a coverslip (15mm diameter) were seeded and allowed to settle for 3d to maintain exponential growth, and then 1 mg dsRNA was mixed with 8ml cellfectin transfection reagent (Invitrogen) in LBM medium without FBS supplementation. The complex was incubated at room temperature for 20min and then added to the VCMs from which normal growth medium had been removed. After a 6-h incubation, the inoculum was removed, and the coverslip was covered with LBM medium plus 10% FBS.

To gauge the effects of the dsRNAs on the direct cell-to-cell spread of RDV, after a 24-h treatment with the dsRNAs, the VCMs were inoculated with RDV at a low multiplicity of infection (MOI) of 0.001, and from 2h p.i. onward, virus-neutralizing antibodies (30mg/mL of medium) were added to the culture medium to inhibit infection by RDV particles that had been released into or were present in the culture medium, as described previously (Wei et al., 2006). VCMs were fixed 3d after viral inoculation, immunolabeled with virus-FITC and Pns10-rhodamine, and visualized by fluorescence microscopy. A minimum of four fields was examined for the foci of infection formed in three or more independent experiments.

To examine the effects of the synthesized dsRNAs on the multiplication of RDV, at 24-h after the treatment with dsRNAs, we inoculated VCMs with RDV at an MOI of 10, and cells were grown further for 72h. After harvest, cells were subjected to several cycles of freezing and thawing to release viral particles, and lysates were stored at 270uC prior to analysis. The titer of cellassociated viruses was determined in duplicate using the fluorescent focus assay (Kimura, 1986; Wei et al., 2008). Endpoint titers were calculated as means with standard deviations.

4.5 Examination of RDV spread in the intact insect in the presence of synthesized dsRNAs

A membrane-feeding method for delivering dsRNAs to leafhoppers was used. Briefly, the second-instar nymphs of leafhopper were fed with 0.5mg/ml dsRNAs diluted in 5% sucrose in water, which was held between two layers of stretched parafilm covering one open end of the tube. The nymphs of leafhoppers were maintained with this mixed diet for 1 day, allowed a 2-day AAP on RDV-infected rice plants, and then fed on healthy rice seedling. Total RNAs were extracted by TRIzol Reagent (Invitrogen). The effects of dsRNAs on the accumulation of the nonstructural protein Pns10 and major outer capsid protein P8 of virus were determined by RT-PCR. To determine whether blocking the formation of virus-containing Pns10 tubules would inhibit viral spread among leafhopper tissues, internal organs from leafhoppers that had received dsRNAs were fixed and processed for immunofluorescence microscopy as described earlier. To determine whether the ingestion of dsRNAs would inhibit viral transmission, we allowed the dsRNAs-treated leafhoppers a 2-day AAP on RDV-infected rice plants, then allowed them to feed on healthy rice seedling for 20d as described previously (Honda et al., 2007). Individual adult viruliferous insects that matured during this 20-day period were then exposed to healthy rice seedlings in individual test tubes for a 2-day inoculation access feeding. The developing leaves were scored for the first visible symptoms daily until harvest.

4.6 Electron microscopy

For transmission electron microscopy, the internal organs from RDV-infected or healthy leafhoppers were fixed and examined as described previously (Wei et al., 2006; Wei et al., 2007).

4.7 siRNA detection and viral plus-strand RNAs detection by Northern blot analysis

The basic procedure to detect siRNA was that of (Shimizu et al., 2009). Briefly, at 72 h after the treatment with dsRNAs, VCMs were harvested, and total RNA was extracted with TRIzol Reagent (Invitrogen). DIG-labeled negative-sense RNA transcripts of Pns10 or YFP genes were generated *in vitro* with T7 RNA polymerase using a DIG RNA Labeling kit (Roche), then used as a probe. Northern blots were produced using a DIG Northern starter kit (Roche) and standard protocols.

Viral plus-strand RNAs were analyzed by northern blot hybridization using a DIG Northern starter kit (Roche) according to standard protocols. VCMs transfected with dsRNAs were inoculated with RDV at an MOI of 10, and cells were further cultured for 72h. Second-instar nymphs of leafhopper were fed with dsRNAs for 1 day, allowed a 2-day AAP on RDV-infected rice plants, then fed on healthy rice seedling for 10d. Total RNAs from VCMs or intact leafhoppers were extracted with TRIzol Reagent (Invitrogen). DIG-labeled negative-sense RNA transcripts of Pns10 or P8 genes were generated in vitro with T7 RNA polymerase using a DIG RNA Labeling kit (Roche), then used as a probe. Northern blots were produced using a DIG Northern starter kit (Roche) and standard protocols.

4.8 RT-qPCR assay

The effects of RNAi induced by dsPns10 on the synthesis of viral plus-strand RNAs were further analyzed by RT-qPCR assay, essentially as described previously (Maroniche et al., 2011; Jia et al., 2012a). Briefly, VCMs transfected with dsRNAs were inoculated with RDV at an MOI of 10, and cells were further cultured for 72h. Total RNAs from VCMs were extracted with TRIzol Reagent (Invitrogen). RT-qPCR primers from the sequences of P8 and Pns10 genes of RDV were designed and tested for efficiency and specificity. RT-qPCR assay was carried out in a Mastercycler realplex4 real-time PCR system (Eppendorf) using the SYBR Green PCR Master Mix kit (QIAGEN) according to standard protocol. The level of leafhopper actin gene was used as the internal control for each RT-qPCR assay. Relative to actin gene was used for quantitative analysis using the Microsoft Excel based tools.

References

[1] AMMAR E D, NAULT L R, RODRIGUEZ J G. Internal morphology and ultrastructure of leafhoppers and planthoppers[J]. Leafhoppers & Planthoppers, 1985.

[2] AMMAR E D, TSAI C W, WHITFIELD A E, et al. Cellular and molecular aspects of rhabdovirus interactions with insect and plant hosts[J]. Annual Review of Entomology, 2009, 54: 447-468.

[3] AYALA B C, ARIAS M, ESPINOSA R, et al. Analysis of the Kinetics of Transcription and Replication of the Rotavirus Genome by RNA Interference[J]. Journal of Virology, 2009, 83(17):8819-8831.

[4] BALUŠKA F, HLAVACKA A, VOLKMANN D, et al. Getting connected: actin-based cell-to-cell channels in plants and animals[J]. Trends in Cell Biology, 2004, 14(8):404-408.

[5] BERGELSON J M. Intercellular Junctional Proteins as Receptors and Barriers to Virus Infection and Spread[J]. Cell host & microbe, 2009, 5(6):517-521.

[6] BOCCARDO G, MILNE R G. Plant reovirus group. In: Morant AF, Harrison BD, eds. CM/AAB descriptions of plant viruses[J]. Old Woking: The Gresham Press, 1984, 1–7.

[7] CAMERON L A, FOOTER M J, VAN OUDENAARDEN A, et al. Motility of ActA protein-coated microspheres driven by actin polymerization[J]. Proceedings of the National Academy of Sciences of the United States of America, 1999, 96(9):4908-4908.

[8] CHEUNG W W K, PURCELL A H. Ultrastructure of the digestive system of the Le fhopperEuscelidiusvariegatusKirshbaum (Homoptera: Cicadellidae), with and without congenital bacterial infections[J]. International Journal of Insect Morphology and Embryology, 1993, 22: 49-61.

[9] CHEN H, CHEN Q, OMURA T, et al. Sequential infection of *Rice dwarf virus* in the internal organs of its insect vector after ingestion of virus[J]. Virus Research, 2011,160(1-2): 389-394.

[10] DESSELBERGER U. Genome Rearrangements of Rotaviruses[J]. Advances in Virus Research, 1996, 46:69-95.

[11] DEROSIER D J, TILNEY L G. F-actin bundles are derivatives of microvilli: What does this tell us about how bundles might form?[J]. The Journal of Cell Biology, 2000, 148(1):1-6.

[12] FILHO F, NAIDU R A, DEOM C M, et al. Dynamics of Tomato spotted wilt virus Replication in the Alimentary Canal of Two Thrips Species[J]. Phytopathology, 2002, 92(7):729-733.

[13] EUSEBIO C A, SUN L, HILLMAN B I,et al. Mycoreovirus 1 S4-coded protein is dispensable for viral replication but necessary for efficient vertical transmission and normal symptom induction[J]. Virology, 2010,397(2):399-408.

[14] HOGENHOUT S A, AMMAR E-D, WHITFIELD A E,et al. Insect vector interactions with persistently transmitted viruses[J]. Annual Review of Phytopathology, 2008, 46:327-359.

[15] HONDA K, WEI T, HAGIWARA K, et al. Retention of Rice dwarf virus by Descendants of Pairs of Viruliferous Vector Insects After Rearing for 6 Years[J]. Phytopathology, 2007, 97(6):712-716.

[16] HUVENNE H, SMAGGHE G. Mechanisms of dsRNA uptake in insects and potential of RNAi for pest control: a review[J]. Journal of Insect Physiology, 2010, 56(3):227-235.

[17] JIA D, CHEN H, MAO Q, et al. Restriction of viral dissemination from the midgut determines incompetence of small brown planthopper as a vector of Southern rice black-streaked dwarf virus[J]. Virus Research, 2012a, 167(2):404-408.

[18] JIA D, CHEN H, ZHENG A, et al. Development of an insect vector cell culture and RNA interference system to investigate the functional role of fijivirus replication protein[J].Journal of Virology, 2012b, 86(10):5800-5807.

[19] KATAYAMA S, WEI T, OMURA T, et al. Three-Dimensional Architecture of Virus-Packed Tubule[J]. Journal of Electron Microscopy,2007, 56 (3):77-81.

[20] KIMURA I.A Study of Rice Dwarf Virus in Vector Cell Monolayers by Fluorescent Antibody Focus Counting[J]. Journal of General

Virology, 1986, 67(10):2119-2124.

[21] KIMURA I, OMURA T. Leafhopper cell cultures as a means for phytoreovirus research. Adv Dis Vector Res, 1988, 5: 111-135.

[22] LI W, LEWANDOWSKI D J, HILF M E,et al. Identification of domains of the Tomato spotted wilt virus NSmprotein involved in tubule formation, movement and symptomatology[J].Virology, 2009, 390(1):110-121.

[23] MARONICHE G A, SAGADÍN M, MONGELLI V C, et al. Reference gene selection for gene expression studies using RT-qPCR in virus-infected planthoppers[J]. Virology Journal, 2011, 8(1):308.

[24] NASU S.Electron Microscopic Studies on Transovarial Passage of Rice Dwarf Virus[J]. Japanese Journal of Applied Entomology & Zoology, 1965, 9(3):225-237.

[25] ODA H, TAKEICHI M. Structural and functional diversity of cadherin at the adherens junction[J]. Journal of Cell Biology, 2011, 193(7):1137-1146.

[26] OMURA T, YAN J, ZHONG B,et al. The P2 protein of rice dwarf phytoreovirus is required for adsorption of the virus to cells of the insect vector[J]. Journal of Virology, 1998, 72(11): 9370-9373.

[27] PRICE D R, GATEHOUSE J A. RNAi-mediated crop protection against insects[J]. Trends in Biotechnology, 2008, 26(7):393-400.

[28] PU Y, KIKUCHI A, MORIYASU Y, et al. Rice dwarf viruses with dysfunctional genomes generated in plants are filtered out in vector insects: implications for the origin of the virus[J]. Journal of Virology, 2010, 85(6):2975-2979.

[29] RAMIG R F. Genetics of the rotaviruses[J]. Annual Review of Microbiology, 1997, 185(1):225-255.

[30] ROSA C, KAMITA S G, DEQUINE H, et al. RNAi effects on actin mRNAs in Homalodiscavitripennis cells[J]. Journal of Rnai & Gene Silencing An International Journal of Rna & Gene Targeting Research, 2010, 6(1):361-366.

[31] ROSA C, KAMITA S G, FALK B W. RNA interference is induced in the glassy winged sharpshooter Homalodiscavitripennis by actin dsRNA[J]. Pest Management Science, 2012, 68(7):995-1002.

[32] ROCHA J J, KOROLCHUK V I, ROBINSON I M, et al. A phagocytic route for uptake of double-stranded RNA in RNAi. PLOSONE, 2011, 6(4):e19087.

[33] SIN S H, MCNULTY B C, KENNEDY G G, et al. Viral genetic determinants for thrips transmission of Tomato spotted wilt virus[J]. Proceedings of the National Academy of Sciences of the United States of America, 2005, 102(14): 5168-5173.

[34] SHIMIZU T, YOSHII M, WEI T, et al. Silencing by RNAi of the gene for Pns12, a viroplasm matrix protein of Rice dwarf virus, results in strong resistance of transgenic rice plants to the virus[J]. Plant Biotechnology Journal, 2009, 7(1):24-32.

[35] STAFFORD C A, WALKER G P, ULLMAN D E. Hitching a ride: Vector feeding and virus transmission[J]. Communicative & Integrative Biology, 2012, 5(1):43-49.

[36] SUN L, SUZUKI N. Intragenic rearrangements of a mycoreovirus induced by the multifunctional protein p29 encoded by the prototypic hypovirus CHV1-EP713[J]. RNA, 2008, 14(12): 2557-2571.

[37] TERRA W R, COSTA R H, FERREIRA C. et al.Plasma membranes from insect midgut cells[J]. Anais Da Academia Brasileira De Ciências, 2006, 78: 255-269.

[38] TOMOYASU Y, MILLER S C, TOMITA S, et al. Exploring systemic RNA interference in insects: a genome-wide survey for RNAi genes in Tribolium[J]. Genome Biology, 2008, 9(1):R10-R10.

[39] TSAI J, PERRIER J L. Morphology of the digestive and reproductive systems of *Dalbulusmaidis* and *Graminellanigrifrons* (Homoptera: Cicadellidae)[J]. Florida Entomologist, 1996, 79(4):563-578.

[40] ULLMAN D E, WHITFIELD A E, GERMAN T L. Thrips and tospoviruses come of age: mapping determinants of insect transmission[J]. Proceedings of the National Academy of Sciences of the United States of America, 2005, 102(14):4931-4932.

[41] WEI T, KIKUCHI A, MORIYASU Y, et al. The spread of Rice dwarf virus among cells of its insect vector exploits virus-induced tubular structures[J]. Journal of Virology, 2006, 80(17):8593-8602.

[42] WEI T, CHEN H, ICHIKI-UEHARA T, et al. Entry of Rice dwarf virus into cultured cells of its insect vector involves clathrin-mediated endocytosis[J]. Journal of Virology, 2007, 81(14):7811-7815.

[43] WEI T, SHIMIZU T, OMURA T. Endomembranes and myosin mediate assembly into tubules of Pns10 of *Rice dwarf virus* and intercellular spreading of the virus in cultured insect vector cells[J]. Virology, 2008, 372(2):349-356.

[44] WHITFIELD A, KUMAR N, ROTENBERG D, et al. A soluble form of the Tomato spotted wilt virus (TSWV) glycoprotein GN (GN-S) inhibits transmission of TSWV by Frankliniellaoccidentalis[J]. Phytopathology, 2008, 98(1):45-50.

[45] WHYARD S, SINGH A D, WONG S. Ingested double-stranded RNAs can act as species-specific insecticides[J]. Insect Biochemistry and Molecular Biology, 2009, 39(11):824-832.

[46] ZAMBRANO J L, DÍAZ Y, PEÑA F, et al. Silencing of rotavirus NSP4 or VP7 expression reduces alterations in Ca^{2+} homeostasis induced by infection of cultured cells[J]. Journal of Virology, 2008, 82(12):5815-5824.

[47] ZHONG B, KIKUCHI A, MORIYASU Y, et al. A minor outer capsid protein, P9, of Rice dwarf virus[J]. Archives of Virology, 2003, 148(11):2275-2280.

[48] ZHOU F, PU Y, WEI T, et al. The P2 capsid protein of the nonenveloped rice dwarf phytoreovirus induces membrane fusion in insect host cells[J]. Proceedings of the National Academy of Sciences of the United States of America, 2007, 104(49):19547-19552.

Assembly of the viroplasm by viral non-structural protein Pns10 is essential for persistent infection of *rice ragged stunt virus* in its insect vector

Dongsheng Jia[1], Nianmei Guo[1], Hongyan Chen[1], Fusamichi Akita[2], Lianhui Xie[1], Toshihiro Omura[2], Taiyun Wei[1]

(1 Fujian Province Key Laboratory of Plant Virology, Institute of Plant Virology, Fujian Agriculture and Forestry University, Fuzhou, Fujian 350002, PR China; 2 National Agricultural Research Center, 3-1-1 Kannondai, Tsukuba, Ibaraki 305-8666, Japan)

Abstract: Rice ragged stunt virus (RRSV), an oryzavirus, is transmitted by brown planthopper in a persistent propagative manner. In this study, sequential infection of RRSV in the internal organs of its insect vector after ingestion of virus was investigated by immunofluorescence microscopy. RRSV was first detected in the epithelial cells of the midgut, from where it proceeded to the visceral muscles surrounding the midgut, then throughout the visceral muscles of the midgut and hindgut, and finally into the salivary glands. Viroplasms, the sites of virus replication and assembly of progeny virions, were formed in the midgut epithelium, visceral muscles and salivary glands of infected insects and contained the non-structural protein Pns10 of RRSV, which appeared to be the major constituent of the viroplasms. Viroplasm-like structures formed in nonhost insect cells following expression of Pns10 in a baculovirus system, suggesting that the viroplasms observed in RRSV-infected cells were composed basically of Pns10. RNA interference induced by ingestion of dsRNA from the *Pns10* gene of RRSV strongly inhibited such viroplasm formation, preventing efficient virus infection and spread in its insect vectors. These results show that Pns10 of RRSV is essential for viroplasm formation and virus replication in the vector insect.

1 Introduction

Plant reoviruses are found in the genera *Phytoreovirus*, *Fijivirus* and *Oryzavirus* in the family *Reoviridae* (Boccardo and Milne, 1984). Rice ragged stunt virus (RRSV), an oryzavirus (Hibino, 1996; Hibino et al., 1977), has spread rapidly throughout southern China and Vietnam and causes severe damage to rice (Hoang et al., 2011). For example, since 2006, RRSV has spread widely in Fujian, Hainan, Yunnan, Guangxi and Guangdong provinces in China. RRSV has an icosahedral capsid, ~70nm in diameter, to which spikes are attached (Miyazaki et al., 2008). The RRSV genome consists of ten dsRNA segments that encode at least seven structural proteins, P1, P2, P3, P4A, P5, P8B and P9, and three non-structural proteins, Pns6, Pns7 and Pns10 (Boccardo and Milne, 1984; Hagiwara et al., 1986; Upadhyaya et al., 1996, 1997, 1998). Among the structural proteins encoded by RRSV, P2, P3, P4A and P5 are a putative guanylyltransferase, capsid shell protein, putative RNA-dependent RNA polymerase and capping enzyme, respectively (Boccardo and Milne, 1984; Hagiwara et al., 1986; Supyani et al., 2007; Upadhyaya et al., 1998), P8 is a major outer-capsid protein (Hagiwara et al., 1986) and P9 is a spike protein involved in transmission via the insect vector (Zhou et al., 1999). Among

Journal of General Virology. 2012, 93: 2299–2309
Received 25 February 2012; Accepted 16 July 2012

the non-structural proteins encoded by RRSV, Pns6 functions as a viral RNA-silencing suppressor and a viral movement protein (MP) (Wu et al., 2010a, b), whilst Pns7 has been identified as an NTP-binding protein (Spear et al., 2012; Upadhyaya et al., 1997). The functions of the remaining proteins are unknown.

RRSV is transmitted by the brown planthopper (BPH), *Nilaparvatalugens* (Stål), in a persistent propagative manner (Hibino et al., 1977, 1979). As a persistent propagative plant virus that is transmitted by an insect vector following ingestion during feeding on diseased plants, RRSV must enter the epithelial cells of the alimentary canal in its insect vector, replicate and assemble progeny virions to move into the salivary glands from which RRSV can be introduced into a plant host during feeding (Hogenhout et al., 2008). Thus, RRSV must enter insect vector cells to establish persistent infection, that is, virus must replicate and accumulate progeny virions in the body of BPHs. During RRSV infection in BPHs, cytoplasmic inclusion bodies, known as viroplasms, form in the salivary glands, gut and muscles in infected BPHs, as observed by electron microscopy (Hibino et al., 1979). Virus replication and assembly of progeny virions have been proposed to occur in viroplasms for plant reoviruses (Boccardo and Milne, 1984). Thus, the formation of a viroplasm for virus replication and assembly of progeny virions may play a crucial role in the propagation of RRSV in BPHs. Viral non-structural proteins are essential for formation of the viroplasm matrix of plant reoviruses, such as P9-1 of three fijiviruses [rice black-streaked dwarf virus (RBSDV), Southern rice black-streaked dwarf virus (SRBSDV) and Mal de Rı´o Cuarto virus (MRCV)], Pns12 of the phytoreovirus rice dwarf virus (RDV) and Pns9 of rice gall dwarf virus (RGDV), also a phytoreovirus (Akita et al., 2011, 2012; Jia et al., 2012; Maroniche et al., 2010; Wei et al., 2006b). However, viral non-structural proteins involved in formation of the viroplasms induced by oryzaviruses are unknown.

RNA interference (RNAi), a conserved sequence-specific gene-silencing mechanism induced by dsRNA (Fire et al., 1998), has been developed into an important tool to investigate the functional role of fijivirus replication proteins in insect vectors (Jia et al., 2012). Ingestion of dsRNA via membrane feeding is an effective method of inducing RNAi in BPHs to knock down specific insect genes (Chen et al., 2010; Li et al., 2011). Thus, dsRNA-mediated gene silencing offers an opportunity for us to investigate the functional roles of viral non-structural proteins in the infection cycle of RRSV in BPHs.

In this study, using immunofluorescence microscopy and an RNAi strategy, the functional roles were determined for RRSV Pns10 in formation of the viroplasm and virus infection in BPHs. Our results suggested that Pns10 of RRSV is responsible for formation of the viroplasm matrix in which the assembly of progeny virions occurs in BPHs. RNAi induced by ingestion of dsRNA of the *Pns10* gene strongly inhibited such viroplasm formation, preventing efficient virus infection and spread in BPHs. Our results indicate that assembly of the viroplasm by Pns10 is essential for persistent infection of RRSV in BPHs.

2 Results

2.1 Infection route of RRSV in BPHs revealed by confocal microscopy

To trace the infection route of RRSV within infected BPHs, immunofluorescence microscopy was used to elucidate the distribution of viral antigens in the body of BPHs after ingestion of RRSV from diseased plants. In preliminary tests, ~40% of BPHs became infected after a latent period of 9d (Table 1). At 1, 3, 4, 6 and 9d post-first access to diseased plants (p.a.d.p.), internal organs from 50 BPHs were dissected and processed for immunofluorescence microscopy. The actin-specific dye phalloidin–rhodamine was first used to stain the alimentary canal of BPHs. As in other types of planthopper (Tsai and Perrier, 1996), the alimentary canal of BPHs consists of the oesophagus, anterior diverticulum, midgut and hindgut (Fig. 1a). The midgut of BPHs consists of a single layer of epithelial cells, with extensive microvilli on the lumenal side and basal lamina on the outer side, surrounded by visceral muscle tissues (Fig. 1a). Viral antigens were observed at 1 day p.a.d.p. in the midgut lumen in ~50% of BPHs examined (Fig. 1b), suggesting that RRSV had travelled through the oesophagus into the midgut lumen. At 3d p.a.d.p., RRSV was mainly restricted to a few epithelial cells of the midgut in ~30% of the insects examined (Fig. 1c, Table 1). At 4d p.a.d.p., RRSV in the epithelial cells had traversed the basal lamina and infected the visceral muscle tissues encircling the midgut epithelium in ~28% of BPHs examined (Fig. 1d, Table 1). By 6d p.a.d.p., RRSV had spread to the oesophagus, anterior diverticulum, midgut, hindgut and salivary glands in a higher proportion of BPHs (~26%; Fig. 1e, Table 1); RRSV was present in the visceral muscle tissues of the entire midgut and hindgut but absent in the epithelium (Fig. 1f–h). By 9d p.a.d.p., the presence of RRSV was extensive in the oesophagus, anterior diverticulum, muscles of the midgut and hindgut, and salivary glands in a high proportion of BPHs tested (~ 40%) (Fig. 1i, j, Table 1). At this time, RRSV was still absent in the epithelium of the midgut and hindgut (Table 1). Taken together, these results indicated that RRSV first accumulates in the midgut epithelium, procedes to the visceral muscles surrounding the midgut, spreads throughout the visceral muscles of the midgut and hindgut, and finally spreads into the salivary glands.

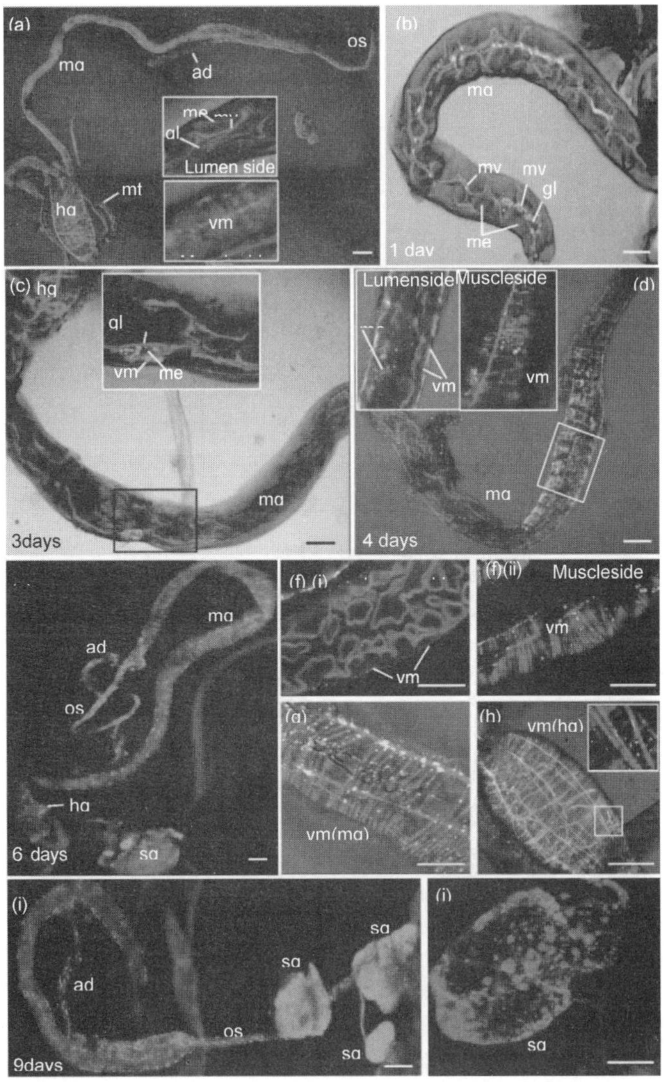

Fig. 1 Infection route of RRSV in the insect vector. The internal organs of BPHs were stained for viral antigens with virus–FITC (green) and for actin with phalloidin–rhodamine (red) and examined by confocal microscopy

(a) The alimentary canal of BPH. Insets: single optical section of the lumen side (upper panel) and muscle side (lower panel) of the midgut. (b) At 1day p.a.d.p., viral antigens accumulated in the midgut lumen. (c) By 3d p.a.d.p., viral antigens were detected in a few epithelial cells of the midgut. Inset: enlarged image of the boxed area. (d) At 4d p.a.d.p., viral antigens accumulated in the epithelium and visceral muscle of the midgut. The image is a projection of 12 optical sections taken at 0.5μM intervals. Insets: single optical sections of the lumen side (left panel) and muscle side (right panel) of the midgut. (e) By 6d p.a.d.p., viral antigens had accumulated throughout the digestive system. (f) Single optical sections of the lumen side (panel i) and muscle side (panel ii) of the midgut at 6d p.a.d.p. (g, h) Viral antigens were detected in the visceral muscle of the midgut (g) and hindgut (h) at 6d p.a.d.p. The inset in (h) is an enlargement of the boxed area. (i, j) At 9d p.a.d.p., viral antigens accumulated throughout the digestive system. ad, Anterior diverticulum; mg, midgut; hg, hindgut; mt, Malpighian tubules; os, oesophagus; sg, salivary gland; gl, gutlumen; mv, microvilli; me, midgut epithelium; vm, visceral muscle. Bars, 70μm

Table 1 Occurrence of RRSV antigens and viroplasms of Pns10 in various organs/tissues of BPHs as detected by immunofluorescence microscopy

Organ/tissue examined	No. positive insects with viral antigens and viroplasms of Pns10 in different tissues (n550)			
	3 days p.a.d.p.	4 days p.a.d.p.	6 days p.a.d.p.	9 days p.a.d.p.
Midgut epithelium	15	18	0	0
Visceral muscle (midgut)	0	14	21	20
Visceral muscle (hindgut)	0	0	14	20
Oesophagus	0	7	19	20
Anterior diverticulum	0	5	19	20
Salivary gland	0	0	13	18

2.2 RRSV Pns10 is sufficient to induce the formation of viroplasm-like structures in non-host insect cells

During infection of BPHs by RRSV, viroplasms, the putative sites for virus replication and assembly of progeny virions, are formed in the alimentary canal and salivary glands of BPHs, as revealed by electron microscopy (Boccardo and Milne, 1984; Hibino et al., 1977, 1979). Viral non-structural proteins are essential for formation of the viroplasm matrix of plant reoviruses (Akita et al., 2011, 2012; Maroniche et al., 2010; Wei et al., 2006b). To identify which non-structural protein encoded by RRSV had an inherent ability to form the viroplasm matrix, a baculovirus system was used to express each of the three non-structural proteins, Pns6, Pns7 and Pns10, fused to a 6×His tag (Pns6–His and Pns7–His) or Strep Tag II (Pns10–Strep). As seen by immunofluorescence microscopy, Pns6–His was associated exclusively with the plasma membrane (Fig. 2a, panel i), corresponding to previous evidence that Pns6 is a viral MP (Wu et al., 2010b). Pns7–His formed filament-like structures in the cytoplasm or protruding from the plasma membrane (Fig. 2a, panel ii), whilst Pns10–Strep aggregated to form punctate inclusions in the cytoplasm (Fig. 2a, panel iii), resembling the viroplasm matrix in virus-infected cells (Hibino et al., 1979). Our results indicated clearly that, among the three nonstructural proteins of RRSV, Pns10 alone was sufficient for the formation of viroplasm-like structures in *Spodopterafrugiperda*(Sf9) cells. Taken together, our results suggested that Pns10 might self-aggregate to form the viroplasm matrix in RRSV-infected host cells.

To determine whether Pns6 or Pns7 could be recruited into viroplasm-like structures formed by Pns10, recombinant baculoviruses that expressed Pns6–His or Pns7–His were co-infected with recombinant baculoviruses expressing Pns10–Strep in Sf9 cells. Co-infection led to redistribution of Pns6–His into the viroplasm-like structures formed by Pns10–Strep (Fig. 2b). By contrast, Pns7–His was not observed in association with the viroplasm-like structures formed by Pns10–Strep (data not shown). Rabbit anti-6×His tag polyclonal antibody and anti-Strep Tag II mAb in non-infected cells were not observed to react with cellular structures (data not shown). Thus, our findings suggested that Pns6 might be recruited to the viroplasm through the association of Pns6 with Pns10.

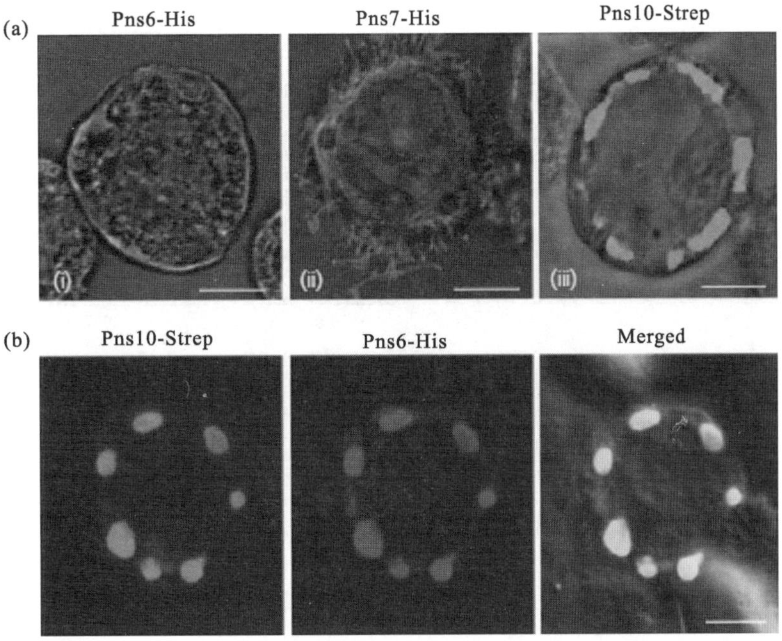

Fig. 2 RRSV Pns10 aggregates to form viroplasm-like inclusions in the absence of virus infection. Sf9 cells infected with recombinant baculoviruses containing Pns6–His, Pns7–His or Pns10–Strep were fixed 3d after infection and prepared for immunofluorescence microscopy as described in Methods

(a) Pns6–His was associated with the plasma membrane (i), Pns7–His formed filament-like structures (ii) and Pns10–Strep formed punctate structures (iii). (b) Pns6–His was associated with the punctate structures formed by Pns10–Strep when co-infected with recombinant baculoviruses containing Pns6–His and Pns10–Strep. Bars, 5μm

2.3 RRSV Pns10 is the constituent of the viroplasm matrix in the body of infected BPHs

To localize Pns10 in virus-infected cells, antibodies against this protein were prepared. Western blot analysis using anti-Pns10 antibodies showed a 33kDa protein (in accordance with the expected size of the RRSV segment 10-encoded protein) present in protein extracts from infected rice plants (Fig. S1, available in JGV Online). No reaction was observed with proteins from uninfected plants, confirming that Pns10-specific antibodies were able to specifically recognize the protein produced by RRSV infection.

To determine whether Pns10 plays a key role in formation of the viroplasm, subcellular localization of Pns10 in the body of infected BPHs was examined by immunofluorescence microscopy. Double labelling showed that viral inclusions that stained with Pns10-specific IgG directly conjugated to rhodamine (Pns10–rhodamine) co-localized with punctate inclusions that stained with viral antigen-specific IgG directly conjugated to FITC (virus–FITC) in the midgut epithelium at 3d p.a.d.p., in the visceral muscle tissues surrounding the midgut at 6d p.a.d.p. and in the salivary gland at 9d p.a.d.p. (Fig. 3a, b, Table 1). These results indicated that RRSV Pns10 is a constituent of viral inclusions where viral antigens accumulated.

To confirm whether the viral inclusions of Pns10 observed by immunofluorescence microscopy were the viroplasm where viral particles accumulate, immunoelectron microscopy was used to localize Pns10 in the salivary glands of infected BPHs. As shown in Fig. 4a, Pns10 antibodies reacted specifically with the viroplasm matrix in salivary glands, and the labelling was consistent with the results of the immunofluorescence microscopy (Fig. 3c). Careful analysis of the electron micrographs revealed that core-like particles of ~50nm in diameter were distributed within the viroplasm matrix, whereas intact, double-layered viral particles of ~70nm in diameter accumulated at the periphery of the viroplasm matrix (Fig. 4b). Moreover, RRSV particles aggregated to form paracrystalline arrays at the periphery of the viroplasm matrix (Fig. 4c). These results confirmed that Pns10 is the constituent of the viroplasm matrix induced by RRSV infection. Furthermore, our results clearly indicated that the viroplasm is the putative site of virus assembly in the body of infected BPHs.

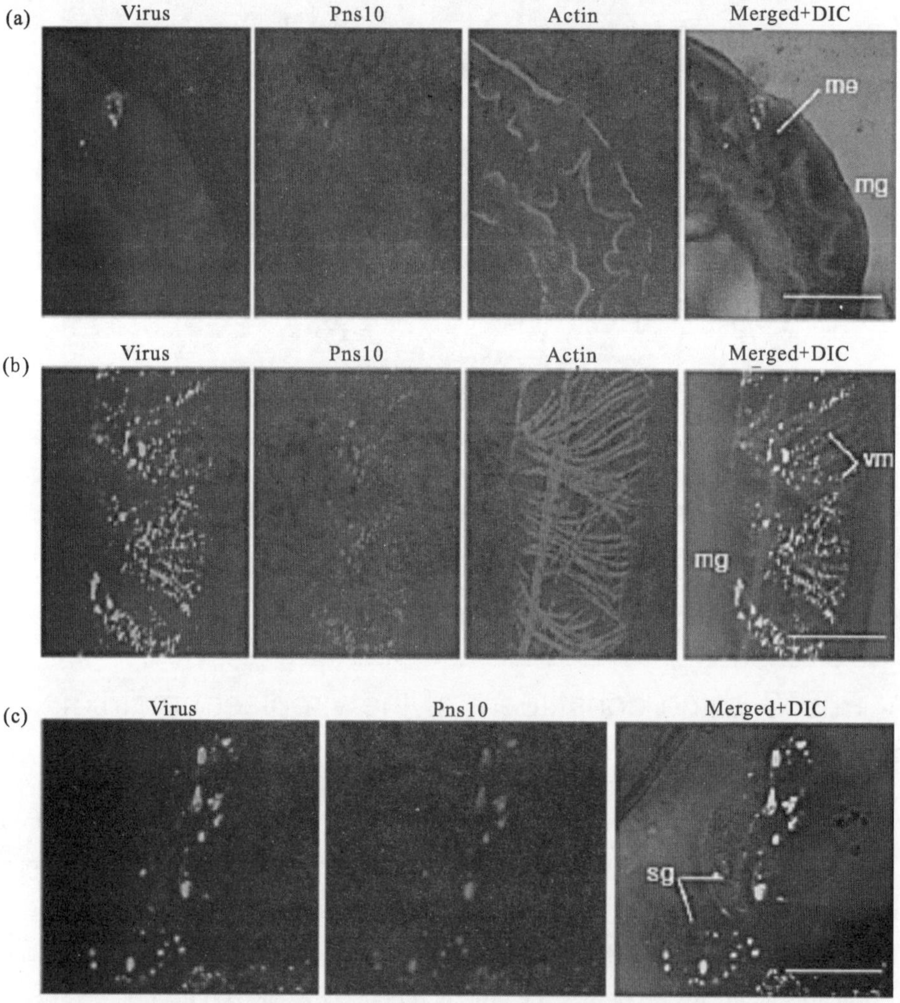

Fig. 3 Viroplasms containing Pns10 antigens co-localize with viral antigens in infected BPHs

The internal organs of BPHs were stained for viral antigens with virus–FITC (green), for viroplasms with Pns10–rhodamine (red) or for actin with phalloidin–Alexa Fluor 647 carboxylic acid (blue) and then examined by confocal microscopy. Viroplasms containing Pns10 antigens (red) and viral antigens (green) were distributed in the midgut epithelium at 3d p.a.d.p. (a), in the visceral muscle tissues surrounding the midgut at 6d p.a.d.p. (b) and in the salivary glands at 9d p.a.d.p. (c). DIC, Differential interference contrast; me, midgut epithelium; mg, midgut, sg, salivary gland; vm, visceral muscle. Bars, 70μm

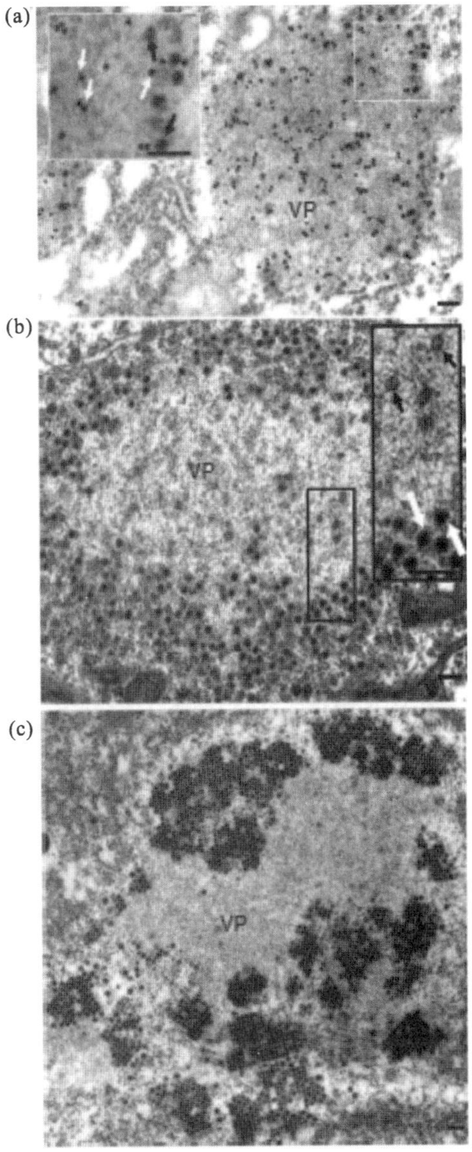

Fig. 4 RRSV Pns10 is the component of the viroplasm matrix

(a) Immunogold labelling of RRSV Pns10 in the viroplasm matrix ininfected salivary glands. Salivary glands were immunostained using Pns10-specific antibodies as the primary antibody, followed by treatment with goat anti-rabbit antibodies conjugated to 15nm gold particles as the secondary antibody. The inset shows an enlargement of the boxed area. Black arrows indicate virus particles, whilst white arrows indicate gold particles. (b, c) Morphogenesis of RRSV particles associated with the viroplasm matrix in infected salivary glands. The inset in (b) shows an enlargement of the boxed area. Black arrows indicate core-like particles, whilst white arrows indicate intact viral particles. VP, Viroplasm. Bars, 200nm

2.4 Ingestion of dsRNA of the *Pns10* gene strongly inhibits virus infection in BPHs

An RNAi strategy was used next to investigate the functional role of RRSV Pns10 in the virus replication cycle in BPHs. Second-instar nymphs of BPHs were fed 0.5μg dsRNA μl^{-1} in 10% sucrose by membrane feeding for 1 day, allowed a 2-day acquisition on RRSV-infected rice plants and then fed on fresh rice seedlings. RT-PCR was used to determine the effects of dsRNA treatment on the transcript levels of viral genes of the non-structural protein Pns10 and the major outer-capsid protein P8 in infected insects that received either dsRNA or sucrose diet alone at 9d p.a.d.p. (Upadhyaya et al., 1996, 1997). Our results showed that ~40% of BPHs (n=100, three repetitions) that received dsRNA of the *gfp* gene (dsGFP) contained transcripts for the *Pns10* and *P8* genes (Table 2). By contrast, ~12% (n=100, three repetitions) of BPHs that received dsRNA of the *Pns10* gene (dsPns10) contained transcripts for the Pns10 and P8 genes (Table 2). The number of positive samples found in BPHs that received dsGFP did not differ significantly from the controls that received the sucrose diet only (Table 2). These results indicated that ingestion of dsPns10 using the membrane feeding method can efficiently induce RNAi and inhibit virus infection in BPHs.

To analyse in more detail the inhibition of virus infection caused by ingestion of dsPns10 in the body of BPHs, internal organs from 50 BPHs receiving dsRNA or sucrose diet alone were dissected at 3, 6 and 9d p.a.d.p. and processed for immunofluorescence microscopy. Virus infection was revealed by double labelling of internal organs with virus–FITC and Pns10–rhodamine. At 3d p.a.d.p., Pns10 and viral antigens were restricted to a limited number of epithelial cells of the midgut in 30% of BPHs receiving dsGFP but only in 10% of those receiving dsPns10 (Table 2). These results suggested that RNAi induced by dsPns10 could significantly inhibit early RRSV infection in the cells of the midgut of insect vectors. At 6d p.a.d.p., Pns10 and viral antigens were detected in both visceral muscle tissues and salivary glands in 20% of BPHs receiving dsGFP but only in 6% of BPHs receiving dsPns10 (Table 2). Furthermore, RRSV was absent from the midgut epithelium in BPHs receiving dsGFP, but virus could be detected in 12% of those receiving dsPns10 (Table 2). At 9d p.a.d.p., Pns10 and viral antigens were detected in the visceral muscles of 40% of BPHs receiving dsGFP and in the salivary glands of 36% of BPHs receiving dsGFP (Table 2). However, Pns10 and viral antigens were only seen in ~12% of the visceral muscle tissues and 6% of the salivary glands in BPHs receiving dsPns10 (Table 2). No significant difference in the number of positive samples was found between BPHs that received dsGFP and diet alone (Table 2). Therefore, RNAi induced by dsPns10 inhibited efficient virus infection and spread in the body of insect vectors.

Table 2 Ingestion of dsRNA of the Pns10 gene strongly inhibits virus infection, as revealed by the accumulation of Pns10 and viral antigens of RRSV in the body of BPHs

Treatment*	No. of insects positive for *Ps10* and *P8* at 9 days p.a.d.p.(n=100)			No. of insects positive for virus and Pns10 antigens in different tissues (n=50)				
	I	II	III	Days p.a.d.p	me	vm-mg	vm-hg	sg
dsPns10	10	13	12	3	5	0	0	0
				6	6	6	4	3
				9	0	7	6	3
dsGFP	37	40	38	3	15	0	0	0
				6	0	18	13	10
				9	0	20	20	18
Diet control	40	38	41	3	16	0	0	0
				6	0	21	14	11
				9	0	19	19	17

*Second-instar nymphs of BPHs were fed with dsPns10 or dsGFP (0.5μg μl^{-1}) or diet alone for 1 day, allowed a 2-day acquisition access period (AAP) on virus-infected rice plants and then fed on uninfected rice seedling. DVirus antigens and viroplasms of Pns10 were detected in the midgut epithelium (me), visceral muscle of the midgut (vm-mg), visceral muscle of the hindgut (vm-hg) and salivary glands (sg) of BPHs by immunofluorescence microscopy.

3 Discussion

3.1 RRSV Pns10 is responsible for formation of the viroplasm matrix in virus-infected cells

Immunoelectron and immunofluorescence microscopy of RRSV-infected cells showed that the non-structural protein Pns10 of RRSV was a component of the viroplasm matrix where core-like particles and viral particles accumulated (Fig. 3, Fig. 4), confirming that viroplasms are the site of virus assembly for plant reoviruses. Among the three nonstructural proteins encoded by RRSV, only expression of Pns10 in Sf9 cells, a non-host of RRSV, resulted in formation of viroplasm-like structures, whereas neither of the other two non-structural proteins, Pns6 and Pns7, appeared to form viroplasm-like structures in Sf9 cells (Fig. 2a). These results suggested that Pns10 was the minimal viral factor required for viroplasm formation during RRSV proliferation and that formation of the viroplasm matrix was not specific to host plant or insect vector cells and did not require host-specific components. Our studies also showed that Pns10 could recruit Pns6 into the viroplasm-like structures formed by Pns10 during co-expression of these two proteins in Sf9 cells (Fig. 2b). Thus, Pns6 might be recruited to the viroplasm through the association of Pns6 with Pns10 during virus infection. Our results suggested that RRSV Pns6, which functions as a viral MP in plant hosts (Wu et al., 2010b), is probably involved in virus replication. Furthermore, the filament-like structures protruding from the plasma membrane formed by RRSV Pns7 strongly resembled those formed by P7-1 of SRBSDV and Pns10 of RDV (Liu et al., 2011;Wei et al., 2006a). The efficient spread of RDV among insect vector cells is dependent on the formation of tubular structures induced by Pns10 (Wei et al., 2006a). It will be interesting to examine whether RRSV Pns7 also forms the same kind of structures and plays a similar role in the spread of virus among insect vector cells.

Non-structural proteins essential for formation of the viroplasm matrix have common characteristics among viruses in the family *Reoviridae*. RRSV Pns10 contains the amino acid sequence motifs typical of an ATPase protein (Z.-X. Gong, unpublished data; http://www.doc88.com/p33746844520.html) and RNA-binding protein (Upadhyaya et al., 1997), suggesting that Pns10 might have ATPase and RNA-binding abilities. The viroplasm matrix protein P9-1 of RBSDV and MRCV can bind RNA (Akita et al., 2012; Maroniche et al., 2010). In addition, P9-1 of MRCV has ATPase activity (Maroniche et al., 2010). The RNA-binding and NTPase activities of the viroplasm matrix protein NSP2 of rotavirus in the family *Reoviridae* have been studied in detail (Kumar et al., 2007; Vasquez-Del Carpio et al., 2006). All these proteins are able to form viroplasm-like structures in non-host cells (Akita et al., 2012; Maroniche et al., 2010; Fabbretti et al., 1999). The parallels among Pns10 of RRSV, NSP2 of rotavirus and P9-1 of MRCV and RBSDV suggest that these proteins might play similar roles in formation of the viroplasm matrix during virus replication cycles. It is interesting to note that viroplasm matrix proteins such as rotavirus NSP2, RBSDV P9-1 and RGDV Pns9 can form octameric structures (Akita et al., 2011, 2012; Jiang et al., 2006); thus, RRSV Pns10 may have a similar structure.

3.2 Assembly of the viroplasm by Pns10 accompanies the sequential infection by RRSV of the internal organs of BPHs

In our study of the sequential infection of RRSV in the internal organs of BPHs, the accumulation of virus and the formation of viroplasms were analysed by double labelling the internal organs with virus–

FITC and Pns10–rhodamine. As early as 1 day after the virus was ingested, masses of viral particles had accumulated in the lumen or were attached to the microvillar membrane of the midgut (Fig. 1b). However, only a limited number of viral particles in the lumen had successfully crossed the microvilli into the epithelial cells of the midgut (Fig. 1c, Table 1). The fact that not all insects feeding on RRSV-infected rice plants became infected could be due to the infection barrier posed by the midgut. At an early stage of virus infection in the epithelial cells of the midgut, even in a single virus-infected cell, RRSV could initiate the formation of nascent viroplasms (Fig. 3a), which serve as the sites for assembly of progeny virions. It is interesting that most of the progeny RRSV virions traversed the basal lamina of the midgut to infect the visceral muscle tissues bordering the infected region, rather than spreading extensively into the adjacent epithelial cells of the midgut (Fig. 1d, Table 1). Subsequently, progeny virions spread to the visceral muscle tissues surrounding the midgut and hindgut (Fig. 1f–h and 3b, Table 1). RRSV might disseminate directly from the visceral muscle tissues into the haemolymph and then into the salivary glands. After a latency period of ~9d, the salivary glands were heavily infected, as shown by almost complete distribution of virus particles and viroplasms throughout the glands, which were proposed to support the assembly of a large number of progeny virions (Fig. 1i, j and 3c). Based on the preceding discussion, formation of the viroplasm for assembly of progeny virions is probably essential for the establishment of a persistent infection of RRSV in insect vectors.

Our recent findings revealed a significantly different infection route for RDV, a phytoreovirus, in its leafhopper vector (Chen et al., 2011). RDV initially infected the epithelial cells of the leafhopper filter chamber. Following accumulation of progeny virions in these cells, most RDV particles spread to the epithelial cells of other organs such as the midgut. Furthermore, RDV may exploit nerves to spread into the visceral muscle tissues surrounding the anterior midgut, although we did not observe particles associated with the neural tissues of BPHs (data not shown).

3.3 RRSV Pns10 is essential for virus infection in BPHs

Our results showed that RNAi induced by ingestion of dsPns10 strongly inhibited efficient virus infection and spread in the body of BPHs. Treatment with dsPns10 reduced the number of positive BPHs with early virus infection in the epithelial cells of the midgut by ~67% (Table 2). In RRSV-positive BPHs that received dsPns10, RRSV still could infect a limited number of the epithelial cells of the midgut, but the subsequent spread of RRSV from the initial infection site to additional tissues was strongly inhibited (Table 2). Thus, RNAi induced by dsPns10 treatment remained activated, even in RRSV-positive BPHs, which may have caused the slower virus spread in the body of BPHs. RNAi induced by dsRNA is a conserved sequence-specific gene-silencing mechanism (Fire et al., 1998). Because the sequences of *Pns10* and other viral genes of RRSV showed no homology (data not shown), we deduced that the RNAi response induced by dsPns10 most probably inhibited expression of the *Pns10* gene in BPHs. Due to the essential role of RRSV Pns10 in biogenesis of the viroplasm matrix, we thus determined that knockdown of *Pns10* expression would block formation of the viroplasm necessary for virus replication and assembly of progeny virions in the epithelial cells of the midgut, preventing efficient virus infection and spread in the body of BPHs. This conclusion agrees with our recent finding for SRBSDV, a fijivirus, on the inhibition of viroplasm assembly and virus replication in the body of white-backed planthoppers by RNAi induced by dsRNA targeting the viral gene for the viroplasm matrix protein P9-1 (Jia et al., 2012). Similarly, short-interfering RNAs against a viral gene for NSP2, the viroplasm matrix protein of rotavirus, inhibited viroplasm formation, genome replication, virion assembly and synthesis of the other viral proteins (Silvestri et al., 2004). Recently, transgenic rice plants, in which the expression of viroplasm matrix proteins of plant reoviruses, including RBSDV, RDV and RGDV, was silenced by RNAi, were shown to be strongly resistant to virus infection (Shimizu et al., 2009, 2011, 2012). These results all support the hypothesis that the viroplasmplays a pivotal role in replication of viruses in the family *Reoviridae*. In this context, identification of RRSV Pns10 as the driving force for viroplasm formation is a step towards identifying suitable targets for pathogen-derived resistance strategies to control disease. In this study, RNAi induced by synthesized dsRNA could overcome the lack of a reverse genetics system in RRSV and opens up new opportunities to understand the biological activities of viral proteins in replicative cycles *in vivo*.

4 Methods

4.1 Antibody preparation

The preparation of RRSV antigen-specific IgGs has been described previously (Takahashi et al., 1991). Rabbit polyclonal antiserum against RRSV Pns10 was prepared as described by Akita et al. (2012). IgG was isolated from specific polyclonal antiserum using a protein A–Sepharose affinity column (Pierce). RRSV antigen-specific IgGs were conjugated directly to FITC (virus– FITC) and Pns10-specific IgGs were conjugated directly to rhodamine (Pns10–rhodamine) according to the manufacturer's instructions (Invitrogen).

To determine the specificity of anti-Pns10 antibodies, total proteins were extracted from 1g RRSV-infected and healthy rice plants. The proteins were separated by SDS-PAGE, transferred to a PVDF membrane and detected on immunoblots using prepared antibodies, as described previously (Spinelli et al., 2006).

4.2 Baculovirus expression of the non-structural proteins of RRSV

A baculovirus system was used to express the three non-structural proteins of RRSV, Pns6, Pns7 and Pns10, as described previously (Wei et al., 2006b). Recombinant baculovirus vectors containing Pns6 and Pns7 fused to a 6×His tag (Pns6–His and Pns7–His) and Pns10 fused to a Strep Tag II (Pns10–Strep) were introduced into DH10Bac (Invitrogen) for transposition into the bacmid. Recombinant bacmids were transfected into Sf9 cells in the presence of Cellfectin (Invitrogen) according to the manufacturer's instructions. Sf9 cells infected with recombinant bacmids were incubated for 72h, fixed in 4% paraformaldehyde and processed for analysis by immunofluorescence microscopy, as described previously (Wei et al., 2006a, b). Cells infected with recombinant baculoviruses containing Pns6–His or Pns7–His were stained with rabbit anti-6×His tag polyclonal antibody (Abcam) and rhodamine-conjugated anti-rabbit IgG antibody (Sigma) as the secondary antibody. Cells infected with recombinant baculoviruses containing Pns10–Strep were stained with anti-Strep Tag II mAb (IBA) and FITC-conjugated anti-mouse IgG (Sigma) as the secondary antibody. Samples were examined under a Leica TCS SP5 inverted confocal microscope, as described previously (Wei et al., 2006a, b).

4.3 Immunofluorescence staining of the internal organs of BPHs after acquisition of virus

Second-instar nymphs of BPHs were allowed a 2-day AAP on rice plants infected with RRSV. On different days after the AAP, internal organs from BPHs were removed, fixed in 4% paraformaldehyde and processed for immunofluorescence microscopy, as described previously (Wei et al., 2006a, b). RRSV particles were stained with virus–FITC. Viroplasms were stained with Pns10–rhodamine. Actin was stained with phalloidin–rhodamine or phalloidin–Alexa Fluor 647 carboxylic acid (Invitrogen). The particles were then imaged by a

Leica TCS SP5 inverted confocal microscope, as described previously (Wei et al., 2006a, b).

4.4 Electron microscopy

The salivary glands of infected BPHs were fixed, dehydrated and embedded, as described previously (Wei et al., 2006a, b). Cell sections were then incubated with antibodies against Pns10 and immunogold labelled with goat antibodies against rabbit IgG that had been conjugated to 15nm gold particles (Sigma) (Wei et al., 2006a, b).

4.5 dsRNA production

A DNA fragment spanning a 793 bp segment of the *Pns10* gene of RRSV was amplified by PCR using forward primer 5′- ATTCTCTAGAAGCTTAATACGACCACTATAGGGCGTGCAATTCCCGAACTTGT-3′ and reverse primer 5′- ATTCTCTAGAAGCTTAATAC- GACTCACTATAGGGACCAGACCAATGTCGCT TGAC-3′, both possessing a T7 promoter (shown in italic) at the 5′ end. A DNA fragment spanning a 717bp segment of the *gfp* gene was amplified by PCR using forward primer 5′-ATTCTCTAGAAGCTTAATACGACTC ACTATAGGGATGAGTAAAGGAGAAGAACTT-3′ and reverse primer 5′-ATTCTCTAGAAGCTTAATACG-ACTCACTATAGGGTTATTTGTATAGTTCATCCATG-3′, also with a T7 promoter (shown in italic) at the 5′ end. PCR products were used for dsRNA synthesis according to the protocol for the T7 RiboMAX Express RNAi System kit (Promega).

4.6 Examination of the effect of dsRNAs on virus infection in the insect vectors

dsRNAs targeting viral genes were delivered to BPHs using a membrane feeding approach (Chen et al., 2010). Briefly, second-instar nymphs of BPHs were fed a diet of 0.5μg dsRNA μl^{-1} diluted in 10% sucrose on a membrane for 1day, allowed a 2-day AAP on RRSV-infected rice plants and then fed on uninfected rice seedlings. The transcript levels of viral genes for the targeted nonstructural protein Pns10 and major outer-capsid protein P8 were determined by RT-PCR (Upadhyaya et al., 1996, 1997). In addition, internal organs from BPHs receiving dsRNAs or diet alone were fixed and processed for immunofluorescence microscopy, as described elsewhere (Wei et al., 2006a, b).

Acknowledgements

This research was supported by the National Basic Research Program of China (no. 2010CB126203), projects from the Ministry of Education of China (nos 211082 and 20103515120007), the key project of Department of Education of Fujian Province, China (no. JA10097), the National Natural Science Foundation of China (nos 31130044, 31070130 and 30970135), the New Century of Excellent Talents at Universities (no. NCET-09-0011) and the program for Promotion of Basic Research Activities for Innovative Biosciences of the Bio-oriented Technology Research Advancement Institution (BRAIN) of Japan (to T. O.).

References

[1] AKITA F, MIYAZAKI N, HIBINO H, et al. Viroplasm matrix protein Pns9 from rice gall dwarf virus forms an octameric cylindrical structure[J]. Journal of General Virology, 2011, 92(9):2214-2221.

[2] AKITA F, HIGASHIURA A, SHIMIZU T, et al. Crystallographic analysis reveals octamerization of viroplasm matrix protein P9-1 of Rice black streaked dwarf virus[J]. Journal of Virology, 2012, 86(2):746-756.

[3] BOCCARDO G, MILNE R G. Plant Reovirus Group[J]. CMI/AAB Descriptions of Plant Viruses, 1984, 294.

[4] CHEN J, ZHANG D, YAO Q,et al. Feeding-based RNA interference of a *trehalose phosphate synthase* gene in the brown planthopper, *Nilaparvatalugens*[J]. Insect Molecular Biology, 2010, 19: 777-786.

[5] CHEN H, CHEN Q, OMURA T, et al. Sequential infection of*Rice dwarf virus* in the internal organs of its insect vector after ingestion of virus[J]. Virus Research, 2011,160(1-2)160: 389-394.

[6] FABBRETTI E, AFRIKANOVA I, VASCOTTO F, et al. Two non-structural rotavirus proteins, NSP2 and NSP5, form viroplasm-like structures in vivo[J]. Journal of General Virology, 1999, 80(2):333-339.

[7] FIRE A, XU S, MONTGOMERY M K, et al. Potent and specific genetic interference by double-stranded RNA inCaenorhabditis elegans[J].Nature, 1998, 391(6669): 806-811.

[8] HAGIWARA K, MINOBE Y, NOZU Y, et al. Component proteins and structures of rice ragged stunt virus[J]. Journal of General Virology,1986, 67: 1711-1715.

[9] HIBINO H. Biology and epidemiology of rice viruses[J]. Annual Review of Phytopathology, 1996, 34(1):249-274.

[10] HIBINO H, ROECHAN M, SUOARISMAN S, et al. A virus disease of rice (kerdilhampa) transmitted by brown planthopper, NilaparvatalugensStâl, in Indonesia[J].contributions central research institute for agriculture, 1977, 35: 1-15.

[11] HIBINO H, SALEH N, ROECHAN M. Reovirus-like particles associated with rice ragged stunt diseased rice and insect vector cells[J]. Japanese Journal of Phytopathology, 1979, 45(2):228-239.

[12] HOANG A T, ZHANG H M, YANG J,et al. Identification, characterization, and distribution of Southern rice black-streaked dwarf virus in Vietnam[J]. Plant Disease, 2011, 95(9):1063-1069.

[13] HOGENHOUT S A, AMMAR D, WHITFIELD A E,et al. Insect vector interactions with persistently transmitted viruses[J]. Annual Review of Phytopathology, 2008, 46:327-359.

[14] JIA D, CHEN H, ZHENG A, et al.Development of an insect vector cell culture and RNA interference system to investigate the functional role of fijivirus replication protein[J]. Journal of Virology, 2012,86(10):5800-5807.

[15] JIANG X F, JAYARAM H, KUMAR M,et al. Cryoelectron microscopy structures of rotavirus NSP2-NSP5 and NSP2-RNA complexes: implications for genome replication[J]. Journal of Virology, 2006, 80: 10829-10835.

[16] KUMAR M, JAYARAM H, VASQUEZ-DEL CARPIO R, et al. Crystallographic and biochemical analysis of rotavirus NSP2 with nucleotides reveals a nucleoside diphosphate kinase-like activity[J]. Journal of Virology, 2007, 81(22):12272-12284.

[17] LI J, CHEN Q, LIN Y,et al. 2011. RNA interference in Nilaparvatalugens (Homoptera: Delphacidae) based on dsRNA ingestion[J]. Pest Management Science, 2011, 67(7):852-859.

[18] LIU Y, JIA D, CHEN H, et al. The P7-1 protein of southern rice black-streaked dwarf virus, a fijivirus, induces the formation of tubular structures in insect cells[J]. Archives of Virology, 2011, 156(10):1729-1736.

[19] MARONICHE G A, MONGELLI V C, PERALTA A V, et al. Functional and biochemical properties of Mal de Rio Cuarto virus (Fijivirus, Reoviridae*)* P9-1 viroplasm protein show further similarities to animal reovirus counterparts[J]. Virus Research, 2010, 152: 96-103.

[20] MIYAZAKI N, UEHARA-ICHIKI T, XING L, et al. Structural evolution of Reoviridae revealed by Oryzavirus in acquiring the second capsid shell[J]. Journal of Virology, 2008, 82(22):11344-11353.

[21] SHIMIZU T, YOSHII M, WEI T,et al. Silencing by RNAi of the gene for Pns12, a viroplasm matrix protein of Rice dwarf virus, results in strong resistance of transgenic rice plants to the virus[J]. Plant Biotechnology Journal, 2010, 7(1):24-32.

[22] SHIMIZU T, NAKAZONO-NAGAOKA E, AKITA F, et al. Immunity to *Rice black streaked dwarf virus*, a plant reovirus, can be achieved in rice plants by RNA silencing against the gene for the viroplasm component protein[J]. Virus Research, 2011, 160(1-2):400-403.

[23] SHIMIZU T, NAKAZONO N E, AKITA F, et al. Hairpin RNA derived from the gene for Pns9, a viroplasm matrix protein of *Rice gall dwarf virus*, confers strong resistance to virus infection in transgenic rice plants[J]. Journal of Biotechnology, 2012, 157(3):421-427.

[24] SILVESTRI L S, TARAPOREWALA Z F, PATTON J T. Rotavirus replication: plus-sense templates fordouble-stranded RNA synthesis are made in viroplasms[J]. Journal of Virology, 2004, 78(14):7763-7774.

[25] SPEAR A, SISTERSON M S, STENGER D C. Reovirus genomes from plant-feeding insects represent a newly discovered lineage within the family Reoviridae[J]. Virus Research, 2012, 163(2):503-511.

[26] SPINELLI S, CAMPANACCI V, BLANGY S, et al. Modular Structure of the Receptor Binding Proteins of Lactococcus lactis Phages: THE RBP STRUCTURE OF THE TEMPERATE PHAGE TP901-1[J]. Journal of Biological Chemistry, 2006, 281: 14256-14262.

[27] SUPYANI S, HILLMAN B I, SUZUKI N. Baculovirus expression of the 11 mycoreovirus-1 genome segments and identification of the guanylyltransferase-encoding segment[J]. Journal of General Virology,2007, 88(1):342-350.

[28] TAKAHASHI Y, OMURA T, SHOHARA K,et al. Comparison of four serological methods for practical detection of ten viruses of rice in plants and insects[J]. Plant Disease, 1991, 75(5):458-461.

[29] TSAI J, PERRIER J L. Morphology of the Digestive and Reproductive Systems of Dalbulusmaidis and Graminellanigrifrons (Homoptera: Cicadellidae)[J]. Florida Entomologist, 1996, 79(4):563-578.

[30] UPADHYAYA N M, ZINKOWSKY E, LI Z,et al. TheM_r 43K major capsid protein of rice ragged stunt oryzavirus is a post-translationally processed product of a M_r67,348 polypeptide encoded by genome segment 8[J]. Archives of Virology, 1996, 141(9):1689-1701.

[31] UPADHYAYA N M, RAMM K, GELLATLY J A, et al. Rice ragged stunt oryzavirus genome segments S7 and S10 encode non-structural proteins of M(r)68,025(Pns7) andM(r)32,364(Pns10)[J]. Archives of Virology, 1997, 142(8):1719-1726.

[32] UPADHYAYA N M, RAMM K, GELLATLY J A, et al.Rice ragged stunt oryzavirus genome segment S4 could encode an RNA dependent RNA polymerase and a second protein of unknown function[J]. Archives of Virology, 1998, 143(9):1815-1822.

[33] VASQUEZ-DEL CARPIO R, GONZALEZ-NILO F D, RIADI G,et al. Histidine Triad-like Motif of the Rotavirus NSP2 Octamer Mediates both RTPase and NTPase Activities[J]. Journal of Molecular Biology, 2006, 362(3):539-554.

[34] WEI T, KIKUCHI A, MORIYASU Y, et al. The spread of Rice dwarf virus among cells of its insect vector exploits virus-induced tubular structures[J]. Journal of Virology, 2006a, 80(17):8593-8602.

[35] WEI T, SHIMIZU T, HAGIWARA K, et al. Pns12 protein of Rice dwarf virusis essential for formation of viroplasms and nucleation of viral-assembly complexes[J]. Journal of General Virology, 2006b, 87(2):429-438.

[36] WU J, DU Z, WANG C, et al. Identification of Pns6, a putative movement protein of RRSV, as a silencing suppressor[J]. Virology Journal, 2010a, 7(1):335.

[37] WU Z, WU J, ADKINS S, et al. Rice ragged stunt virus segment S6-encoded nonstructural protein Pns6 complements cell-to-cell movement of Tobacco mosaic virus-based chimeric virus[J]. Virus Research, 2010b, 152(1-2):176-179.

[38] ZHOU G Y, LU X B, LU H J, et al. Rice Ragged StuntOryzavirus: role of the viral spike protein in transmission by the insect vector[J]. Annals of Applied Biology, 1999, 135: 573-578.

Development of an insect vector cell culture and RNA interference system to investigate the functional role of fijivirus replication protein

Dongsheng Jia, Hongyan Chen, Ailing Zheng, Qian Chen, Qifei Liu, Lianhui Xie, Zujian Wu, Taiyun Wei

(Institute of Plant Virology, Fujian Province Key Laboratory of Plant Virology, Fujian Agriculture and Forestry University, Fuzhou, Fujian, People's Republic of China)

Abstact: An *in vitro* culture system of primary cells from white-backed planthopper, an insect vector of Southern rice black-streaked dwarf virus (SRBSDV), a fijivirus, was established to study replication of the virus. Viroplasms, putative sites of viral replication, contained the nonstructural viral protein P9-1, viral RNA, outer-capsid proteins, and viral particles in virus-infected cultured insect vector cells, as revealed by transmission electron and confocal microscopy. Formation of viroplasm-like structures in nonhost insect cells upon expression of P9-1 suggested that the matrix of viroplasms observed in virus-infected cells was composed basically of P9-1. In cultured insect vector cells, knockdown of P9-1 expression due to RNA interference (RNAi) induced by synthesized double-stranded RNA (dsRNA) from the *P9-1* gene strongly inhibited viroplasm formation and viral infection. RNAi induced by ingestion of dsRNA strongly abolished viroplasm formation, preventing efficient viral spread in the body of intact vector insects. All these results demonstrated that P9-1 was essential for viroplasm formation and viral replication. This system, combining insect vector cell culture and RNA interference, can further advance our understanding of the biological activities of fijivirus replication proteins.

1 Introduction

Plant reoviruses, comprising the genera *Phytoreovirus, Fijivirus,* and *Oryzavirus*, cause diseases of numerous important crops and are transmitted by insect vectors in a persistently propagative manner (Milne et al., 2005; Omura et al., 2005; Upadhyaya et al., 2005). For control of such viruses, understanding their mode of transmission and replication is critical. For a persistently propagative plant virus to be transmitted by its vector, the virus ingested by insect feeding on infected plants must enter the epithelial cells of the alimentary canal and then replicate and assemble progeny virions to move into the salivary glands, from which the virus can be transmitted to more plants during feeding (Hogenhout et al., 2008). Viral replication and assembly of progeny virions, critical for the propagation of plant reoviruses in their insect vectors, are thought to occur in viroplasms (Milne et al., 2005; Omura et al., 2005; Upadhyaya et al., 2005), which contain viral proteins, viral particles, and viral RNAs.

Southern rice black-streaked dwarf virus (SRBSDV), a tentatively identified species in the genus *Fijivirus*, which is transmitted by the white-backed planthopper (WBPH; *Sogatella furcifera* Horváth) has spread rapidly

American Society for Microbiology. 2012, 10:1128-07121-11
Received 15 December 2011; Accepted 27 February 2012; Published ahead of print 7 March 2012

throughout southern China and northern Vietnam and can severely damage rice (Liu et al., 2011; Wang et al., 2010; Zhang et al., 2001; Zhou et al., 2008).The icosahedral, double-layered particles of SRBSDV are ca. 70nm in diameter and contain 10 segments of double-stranded RNA (dsRNA) (Wang et al., 2010; Zhou et al., 2008). Phylogenetic analyses showed that SRBSDV, the first WBPH-borne reovirus to be identified, is most closely related to but distinct from Rice black-streaked dwarf virus (RBSDV), also a fijivirus (Wang et al., 2010; Zhou et al., 2008). Comparing the different genomic segments of SRBSDV to their counterparts in RBSDV suggests that SRBSDV encodes at least six putative structural proteins (P1, P2, P3, P4, P8, and P10) and five putative nonstructural proteins (P6, P7-1, P7-2, P9-1, and P9-2) (Wang et al., 2010).

Among the putative structural proteins encoded by SRBSDV, P1, P2, and P4 are a putative RNA-dependent RNA polymerase, a core protein, and an outer-shell B-spike protein, respectively (Wang et al., 2010; Zhang et al., 2008); P3 is a putative capping enzyme (Wang et al., 2010; Zhang et al., 2008); and P8 and P10 are putative core and major outer capsid proteins, respectively (Isogai et al., 1998; Wang et al., 2010). Among the putative nonstructural proteins encoded by SRBSDV, P6 is a viral RNA-silencing suppressor (Lu et al., 2011) P7-1 is the major constitute of the tubules and has the intrinsic ability to self-interact to form tubules in non-host insect cells (Liu et al., 2011); and P9-1 of SRBSDV has about 77% amino acid identity with its counterpart, P9-1 of RBSDV (Wang et al., 2010). In RBSDV, P9-1 forms an octameric, cylindrical structure *in vitro* and accumulates in the matrix of viroplasms in virus-infected cells (Akita et al., 2012; Isogai and Uyeda, 1998). As a major constitute of the viroplasm, P9-1 is thus likely to play an important role in the formation of viroplasm (Akita et al., 2012). Therefore, P9-1 of SRBSDV may also be essential for viroplasm formation during viral infection in the host plant and insect vector. However, the precise function(s) of the proteins in viroplasm formation and viral replication of plant reoviruses is poorly understood due in part to the lack of a reverse genetics system and useful culture systems for their respective insect vectors.

Insect vector cells in monolayer (VCM) is an *in vitro* experimental system with notable advantages over the use of whole intact insects for investigating plant viruses (Creamer, 1993; Omura et al., 1994). This is due to its capability of obtaining a uniform viral infection, which enables us to follow synchronous viral multiplication (Omura et al., 1994). The VCM also provides a very sensitive bioassay system for tracing the fate of viral infectivity under different conditions (Omura et al., 1994). We have already used VCMs derived from the leafhopper that transmits Rice dwarf virus (RDV), another phytoreovirus, to clarify that the Pns12 nonstructural protein of RDV plays a key role in the formation of viroplasms and in recruiting viral assembly complexes to the viroplasms in VCMs (Wei et al., 2006). We thus adapted the VCM system for WBPH, the vector of SRBSDV, to trace the infection and multiplication process of virus.

To further investigate the functional roles of viral proteins in the infection cycles of plant reoviruses in insect vectors, here we used RNA interference (RNAi), a conserved sequence-specific gene silencing mechanism that is induced by dsRNAs (Fire et al., 1998) By exploitation of its ability to efficiently silence gene expression, RNAi has been used in mammalian, insect, and plant cell studies to characterize the function of numerous genes (Cheng et al., 2010; Cherry et al., 2011; Huvenne et al., 2010; McGinnis et al., 2010). It has also been used to interfere with the replication of animal reoviruses (Forzan et al., 2007; Kobayashi et al., 2006; López et al., 2005), which are closely related to plant reoviruses. We thus introduced dsRNA from the *P9-1* gene of SRBSDV into VCMs or the intact insect to knock down the expression of the *P9-1* gene and examine the subsequent effect on viroplasm formation and viral replication.

In this study, by growing a primary cell culture of WBPH in a monolayer (VCM) and using the RNAi strategy, we could elucidate that the P9-1 nonstructural protein of SRBSDV functions in the assembly of viroplasm and viral replication. The P9-1 nonstructural protein appeared to be the major constituent of the matrix of viroplasms where viral RNA, major outer-capsid protein P10, and viral particles accumulated in virus-infected VCMs. RNAi induced by dsRNA from the P9-1 gene in VCMs or the intact insect strongly inhibited such viroplasm formation, preventing efficient viral replication *in vitro* and *in vivo*. Development of RNAi and VCMs can thus overcome the lack of a reverse-genetics system for persistently propagative plant viruses and advance our understanding of the biological activities of viral proteins in the replication cycle in insect vectors.

2 Materials and methods

2.1 Virus and antibodies

Rabbit polyclonal antisera against the P9-1 nonstructural protein and the P10 major outer capsid protein of SRBSDV (Wang et al., 2010; Zhang et al., 2008; Zhou et al., 2008) were prepared. The *P9-1* and *P10* genes from an SRBSDV isolate from Hunan Province, China, were amplified by reverse transcription PCR (RT-PCR), and the products were purified and engineered into Gateway vector pDEST17 (Invitrogen). The resulting pDEST17-P9-1 and pDEST17-P10 plasmids were then used to transform Escherichia coli strain Rosetta and expressed by adding isopropyl-β-α-thiogalactopyranoside (IPTG) (Sigma) (1 mmol/liter). Cells were harvested and sonicated. The final suspension containing P9-1 or P10 protein was purified with nickel-nitrilotriacetic acid (Ni-NTA) resin (Qiagen), and rabbits were immunized with the purified proteins, as described previously (Spinelli et al., 2006). IgG was isolated using specific polyclonal antiserum and a protein A-Sepharose affinity column. Eluted IgG was dialyzed exhaustively against phosphate-buffered saline (PBS). The IgG was conjugated directly to fluorescein-5-isothiocyanate (FITC) or rhodamine according to the manufacturer's instructions (Invitrogen).

To detect the specificity of P9-1 and P10 antibodies of SRBSDV, we extracted total plant proteins from 1g of either SRBSDV-infected or healthy rice plants. SRBSDV was crudely purified from infected rice plants as previously described by Miyazaki et al. (Miyazaki et al., 2008). The proteins and solution of purified viruses were separated by sodium dodecyl sulfate-polyacrylamide gel electrophoresis (SDS-PAGE), transferred onto a polyvinylidene difluoride (PVDF) membrane, and detected on immunoblots using prepared antibodies as described previously (Spinelli et al., 2006).

2.2 Establishment of primary cell cultures derived from WBPH

Primary cell cultures derived from WBPH were established by adapting the protocol described by Kimura and Omura (Kimura et al., 1988). In a preliminary test, WBPH embryos at the blastokinetic stage were found to be a suitable source for primary cell cultures. This stage can be recognized by the appearance of red eye spots on the eggs on day 8 after oviposition. Embryos at this stage were sterilized with 70% ethanol, washed with Tyrode's solution (Kimura, 1986; Kimura and Omura, 1988), and then crushed with a sterilized pestle into tissue fragments. The tissue fragments were treated with 0.25% trypsin in Tyrode's solution and then incubated with Kimura's insect medium at 25℃ (Kimura, 1986; Kimura and Omura, 1988).

2.3 Examination of viral infection in primary cell cultures derived from WBPH by immunofluorescence microscopy

SRBSDV inocula for infecting VCMs derived from WBPHs were prepared from infected plants, essentially as described previously (Kimura, 1986; Kimura and Omura, 1988). VCMs growing on a coverslip and infected with SRBSDV were fixed in 4% paraformaldehyde 48 h postinoculation (hpi.), immunostained with P10-specific IgG conjugated to FITC (P10-FITC) and P9-1-specific IgG conjugated to rhodamine (P9-1-rhodamine), and then examined with a Leica TCS SP5 inverted confocal microscope, as described previously (Wei et al., 2006).

2.4 Immunofluorescence detection of newly made viral RNA in primary cell cultures derived from WBPH

For determining whether the viroplasms are the sites of viral RNA synthesis, VCMs on a coverslip were either mock infected or infected with SRBSDV and then treated with actinomycin D (Sigma) for 1 h at 45h p.i. and incubated with bromouridine 5′-triphosphate (BrUTP) for 2h via the use of Cellfectin (Invitrogen). The samples were fixed 48 hpi., stained with monoclonal anti-bromodeoxyuridine (anti-BrdU; Sigma), and then treated with antimouse IgG conjugated to FITC (Invitrogen) and P9-1–rhodamine for imaging with confocal microscopy, as described previously (Wei et al., 2006).

2.5 Baculovirus expression of nonstructural proteins of SRBSDV

DNA fragments encoding full-length P9-1 and a carboxy-terminal (amino acids 346 though 368) deletion mutant (P9-1C) were amplified by PCR. For the control, a DNA fragment of another nonstructural protein of SRBSDV, P7-1 (Liu et al., 2011), fused in frame with the sequence of Strep-tag II (P7-1–Strep-tag II), also was amplified by PCR. The products were purified and engineered into Gateway vector pDEST8 (Invitrogen) to generate plasmids pDEST8-P9-1, pDEST8-P9-1C, and pDEST8-P7-1–Strep-tag II. The recombinant baculovirus vectors were introduced into *E.coli* DH10Bac (Invitrogen) for transposition into the bacmid. The recombinant bacmids were used to transfect *Spodoptera frugiperda* (Sf9) cells in the presence of Cellfectin. Sf9 cells, growing on a coverslip and infected with recombinant bacmids, were incubated for 72h, fixed, and stained with the monoclonal anti-Strep-tag II (IBA) followed by Rhodamine antimouse IgG (Sigma) and P9-1-FITC for imaging with confocal microscopy as described previously (Wei et al., 2006).

2.6 Effect of synthesized dsRNAs on viral infection in primary cell cultures derived from WBPH

We designed primers for PCR amplification of a 968-bp segment of the *P9-1* gene and a 717-bp segment of a green fluorescence protein (GFP)-encoding gene as a control. The PCR products were used for dsRNA synthesis according to the protocol of a T7 RiboMAX Express RNAi System kit (Promega). VCMs were transfected with dsRNAs via Cellfectin, as reported by Rosa et al. (Rosa et al., 2010). Briefly, VCMs were transfected with dsRNA (0.5g/l) from the *P9-1* gene (dsP9-1) or from the *GFP* gene (dsGFP) via the use of Cellfectin (Invitrogen) for 12h and grown further in growth medium. The basic procedure of small interfering RNA (siRNA) detection was carried out essentially as described previously (Shimizu et al., 2009). Briefly, after a 72-h treatment with dsRNAs, total RNA was extracted from harvested VCMs. Digoxigenin (DIG)-labeled riboprobes corresponding to the negative-sense RNA of the *P9-1* or *GFP* gene were generated with T7 RNA polymerase (Roche). Northern blotting was performed with a DIG Northern starter kit protocol (Roche).

For examining the effects of the dsRNAs on viral infection, after a 12-h treatment with dsRNAs, the VCMs were inoculated with SRBSDV and were fixed 3d later with 4% paraformaldehyde and stained with P10-FITC and P9-1–rhodamine for visualization with confocal microscopy (Wei et al., 2006).

2.7 Effect of synthesized dsRNAs on viral infection in insect vectors

A membrane-feeding approach for delivering dsRNAs to WBPHs was used. Second-instar nymphs of WBPH were fed with dsRNAs (0.5g/l) diluted in an artificial diet, D-97, suitable for rearing planthoppers (Fu et al., 2001). This mixed diet was held between two layers of stretched parafilm covering one open end of the chamber employed. The nymphs were maintained on the diets for 1day, allowed a 2-day period of acquisition on SRBSDV-infected rice plants, and then placed on healthy rice seedlings. Total RNAs from WBPHs receiving the dsRNA diet or the artificial diet alone were extracted by the use of TRIzol reagent (Invitrogen). The effects of dsRNAs on the transcript levels of virus *P9-1* and *P10* genes were determined by RT-PCR.

To determine whether the treatment of dsP9-1 inhibited viral replication in intact insects, at 2, 6, and 12d post-first access to diseased plants (padp), we dissected internal organs from 50 WBPHs that received dsRNAs. The organs were then fixed, stained with P10-FITC, P9-1– rhodamine, and the actin dye phalloidin-Alexa Fluor 647 carboxylic acid (Invitrogen), and processed for immunofluorescence as described previously (Chen et al., 2011).

2.8 Immunoelectron microscopy

VCMs or Sf9 cells on coverslips were fixed, dehydrated, embedded, and cut as described previously (Wei et al., 2006). Cell sections were then incubated with antibodies against P9-1 or P10 and subjected to immunogold labeling with goat antibodies against rabbit IgG that had been conjugated with 15-nm-diameter gold particles (Sigma) (Wei et al., 2006).

3 Results

3.1 Establishment of primary cell cultures derived from WBPH

To establish primary cell cultures derived from WBPH, embryos at the blastokinetic stage were sterilized, washed with Tyrode's solution, and crushed into tissue fragments, which were then treated with trypsin and incubated with growth medium. Migration of fibroblast-like cells from the tissue fragments of embryos at the blastokinetic stage were observed as early as 36h after setting up the cultures. Our observations suggested that tissue fragments of embryos at this stage could be used as explants for the migration of cells. Within 7d of explanting, fibroblast-like cells had become the dominant cell type (Fig. 1A). Monolayers of epithelium-like cells started to form as cells migrated from the original explants at 9 to 12d (Fig. 1B). They continued growing to form large epithelium-like cell sheets after 40d (Fig. 1C). Our results indicated that the epithelium-like cells of WBPHs were able to selfpropagate in controlled growth environments and formed monolayers on the surface of culture flasks. Such monolayers of primary cell cultures of WBPH, VCMs, were used to study the functional role of P9-1 in the replication of SRBSDV. Viral RNA and particles of SRBSDV accumulated in the viroplasm in virus-infected insect cell cultures. To study the localization of SRBSDV P9-1 and P10 proteins in virus-infected cells, we prepared the antibodies against these two proteins. In a Western blot analysis, the SRBSDV P9-1 and P10 antibodies recognized a 39.9-kDa and a 63-kDa protein, respectively (in accordance with the expected SRBSDV P9-1 and P10 proteins) present in protein extracts from infected rice plants (Fig. 2). Furthermore, P10 antibodies, rather than P9-1 antibodies, reacted with proteins from purified viral particles, confirming that P10 is the major outer capsid protein and P9-1 is the nonstructural protein of SRBSDV (Fig.

2). No reaction was observed with proteins from healthy plants (Fig. 2), confirming that the SRBSDV P9-1 and P10 antibodies were able to specifically recognize the proteins produced after SRBSDV infection.

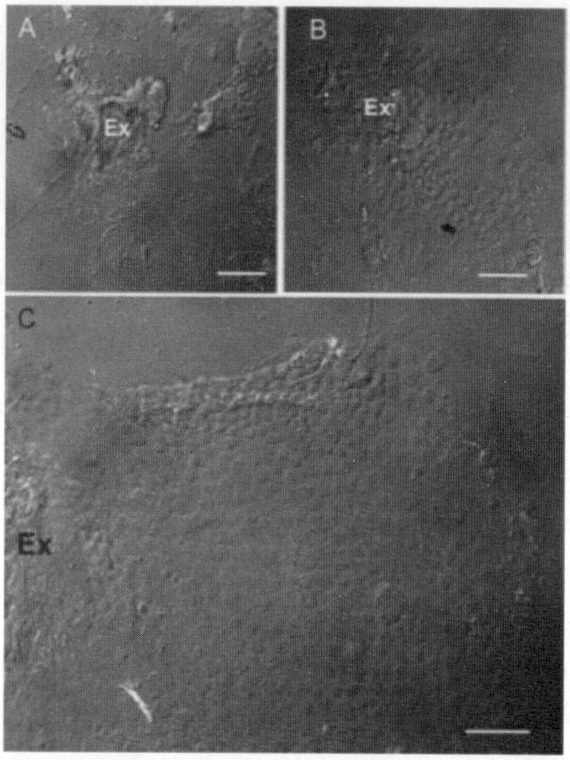

Fig. 1 Light micrograph of the primary cell cultures derived from WBPH *in vitro*

(A) Migration of fibroblast-like cells from explanted embryo fragments (Ex) of WBPH 7d after preparation. (B) Migration of the monolayers of epithelium-like cells from the explanted embryo fragments (Ex) of WBPH 12d after preparation. (C) A large epithelium-like cell sheet formed around explanted embryonic fragments (Ex) of WBPH 40d after preparation. Bars, 100μm (phase contrast)

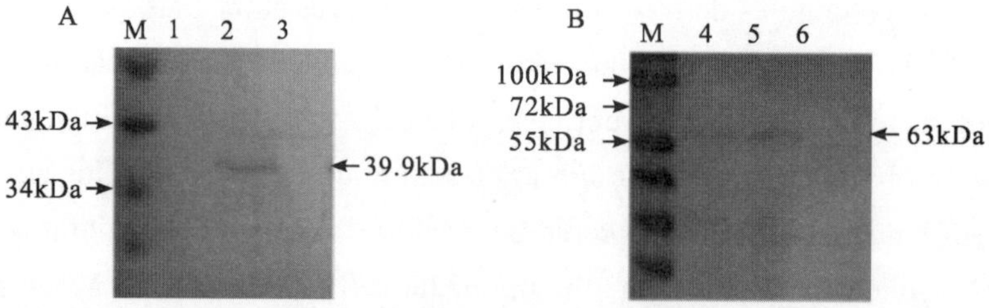

Fig. 2 Western blot analyses of P9-1 and P10 proteins

Samples were separated by SDS-PAGE and detected with P9-1-specific (A) or P10-specific (B) antibodies. Lanes M, protein marker; lanes 1 and 4, purified SRBSDV particles; lanes 2 and 5, protein extracts from rice plants infected with SRBSDV; lanes 3 and 6, protein extracts from healthy rice plants

To study the replication of SRBSDV, we used immunofluorescence microscopy to observe the localization of P9-1 in virus infected VCMs derived from WBPHs. VCMs on coverslip infected with SRBSDV were fixed 48h p.i., stained with P10-FITC and P9-1–rhodamine, and then examined with confocal microscopy. In infected cells, P9-1 localized to punctate inclusions that colocalized with punctate inclusions of P10 in the cytoplasm (Fig. 3A). P10 also formed additional punctate inclusions (Fig. 3A). Our results suggested that the viral inclusions of P9-1 accumulated outer capsid proteins of SRBSDV during viral replication.

The detection of newly made RNA in virus-infected VCMs by immunofluorescence showed that the BrUTP-labeled RNA was distributed in punctate inclusions that colocalized with viral inclusions of P9-1 (Fig.

3B). No reaction with special structures was observed in noninfected cells (Fig. 3B). All these results suggested that viral inclusions of P9-1 were the sites of viral RNA synthesis.

To confirm our observations, we fixed infected VCMs at 48h p.i. and examined them by immunoelectron microscopy using P9-1 and P10 antibodies. P9-1 antibodies specifically reacted with the matrix of the viroplasm (Fig. 3C, D), corresponding to the viral inclusions of P9-1 seen with immunofluorescence (Fig. 3A). This result suggested that SRBSDV P9-1 was a constituent of the matrix of viroplasm in virus-infected cells. Furthermore, viral particles that were distributed within the matrix of the viroplasm were specifically labeled by P10 antibodies (Fig. 3C, E), suggesting that SRBSDV particles were assembled within the matrix during viral replication.

3.2 SRBSDV P9-1 was sufficient to induce formation of viroplasm-like structures in non-host insect cells

Immunofluorescencestaining of P9-1 in Sf9 cells revealed punctate inclusions within the cells at 72h p.i. (Fig. 4A). Furthermore, P9-1 specifically reacted with the granular inclusions, whose morphology was similar to that of the viroplasm matrix, as revealed by immunogold electron microscopy (Fig. 4B). It has been shown that the carboxy-terminal portion of P9-1 of two other fijiviruses, i.e., RBSDV and *Mal de Río Cuarto virus* (MRCV), contained important domains required for the formation of viroplasm-like structures (Akita et al., 2012; Maroniche et al., 2010). To confirm a similar functional role for the carboxyterminal portion of SRBSDV in the formation of viroplasm-like structures, we generated a P9-1 mutant in which the C-terminal 20 residues were deleted (P9-1C) and found that this mutant was diffusely distributed in the cytoplasm (Fig. 4C), suggesting that the C terminus was required for the formation of viroplasm-like inclusions. Furthermore, another nonstructural protein, P7-1– Strep-tag II, formed tubule-like structures in Sf9 cells (Fig. 4D), as described previously (Lu et al. 2011), suggesting that the punctate inclusions formed by P9-1 were specific for this protein. No reaction with cellular structures after incubation with monoclonal anti-Streptag II in noninfected cells was observed (data not shown). All these results indicated that expression of P9-1 alone, in the absence of the viral multiplication process, was sufficient to induce the formation of the viroplasm-like structures in Sf9 cells, confirming an essential role of P9-1 in the biogenesis of the viroplasm in SRBSDV-infected cells.

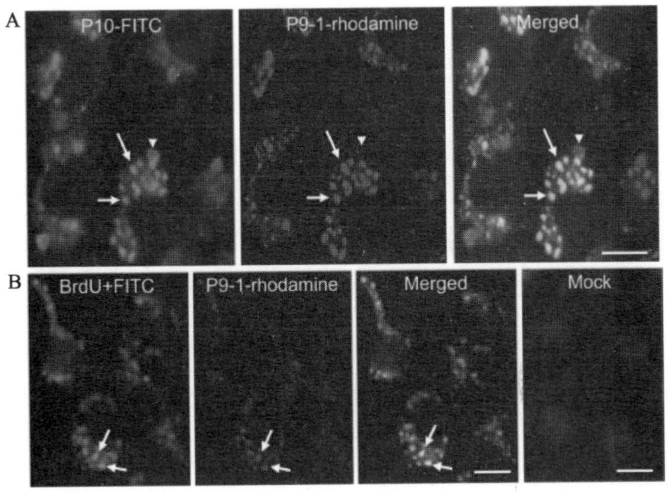

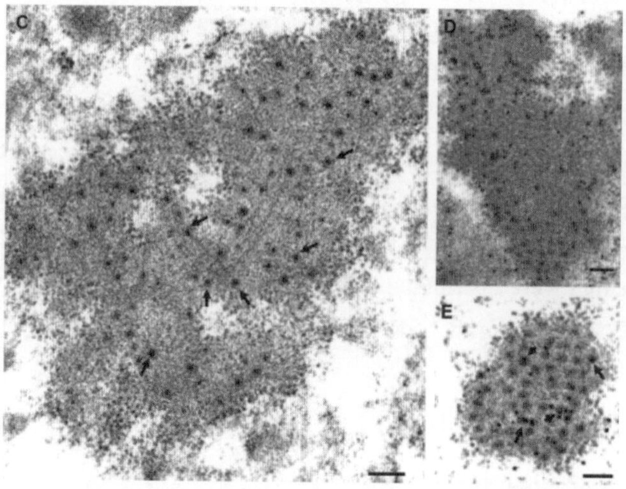

Fig. 3 Subcellular localization of P9-1 and P10 of SRBSDV in virus-infected VCMs 48h p.i.

(A) Immunstaining with P10-FITC (green) and P9-1–rhodamine (red). In merged images, colocalization of P9-1 and P10 is indicated in yellow (arrows). Arrowheads indicate indicate additional sites with P10. Bar, 10μm. (B) Intracellular sites of RNA synthesis in mock- or SRBSDV-infected VCMs. BrUTP-labeled viral RNA was stained with anti-BrdU from mouse, followed by anti-mouse IgG conjugated to FITC (green). Viroplasm is stainedwith P9-1–rhodamine (red). Colocalization of P9-1–rhodamine and BrUTP-labeled micrograph of viroplasm in virus-infected VCMs. Arrows show viral particles in the matrix of viroplasm. Bars, 250nm. (D) Immunogold labeling of P9-1 in the matrix of viroplasm. Bar, 250nm. (E) Immunogold labeling of P10with the viral particles (arrows) within the matrix of viroplasm. Bar, 250nm. Cells were immune stained for P9-1 and P10 with P9-1- and P10-specific antibodies in panels D and E, respectively, as primary antibodies, followed by treatment with goat antibodies against rabbit IgG that had been conjugated with 15-nm-diameter gold particles as secondary antibodies.

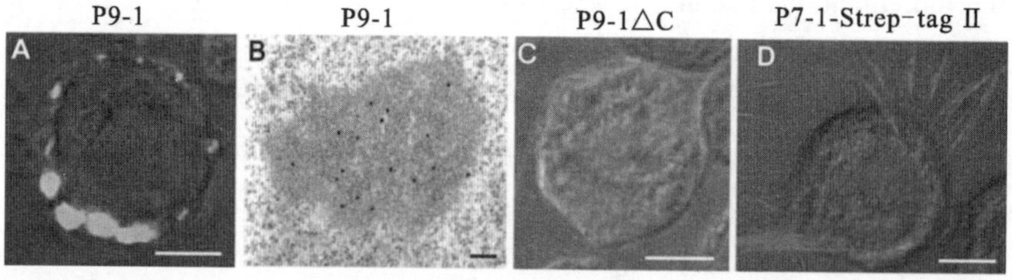

Fig. 4 Subcellular localization of P9-1 and P7-1 of SRBSDV in recombinant baculovirus-infected Sf 9 cells at 72h p.i.

(A) Immunofluorescence staining of P9-1-FITC revealed punctate inclusions of P9-1 of SRBSDV in Sf9 cells. Bar, 5μm. (B) Immunogold labeling of P9-1 associated with electron-dense inclusion. Cells were immunostained with P9-1-specific polyclonal antibodies and goat antibodies against rabbit IgG that had been conjugated with 15-nm-diameter gold particles as secondary antibodies. Bar, 100nm. (C) Immunofluorescence staining with P9-1-FITC revealed that a P9-1 mutant in which the C-terminal 20 residues had been deleted (P9-1ΔC) was diffusely distributed in the cytoplasm in Sf9 cells. Bar, 5μm. (D) Immunofluorescence staining of monoclonal anti-Strep-tag II and rhodamine anti-mouse IgG as a secondary antibody revealed that P7-1-Strep-tag II formed the tubule-like structures in Sf9 cells. Bar, 5μm

3.3 RNAi induced by dsRNAs from the *P9-1* gene inhibited the assembly of viroplasms and viral infection in insect vector cell cultures

We then used an RNAi strategy to study the functional role of P9-1 in the viral replication cycle in VCMs. RNAi is induced by dsRNAs and manifested by the appearance of small interfering RNAs (siRNAs) that correspond to the mRNA target sequence (Rosa et al., 2010). VCMs were transfected with dsP9-1 or dsGFP via Cellfectin-based transfection. Small RNAs approximately 21nt in length were detected from RNAs extracted from dsP9-1 or dsGFP-transfected cells (Fig. 5A) at 72h after transfection with the synthesized dsRNAs. These results showed that RNAi was induced by dsRNAs in VCMs.

To determine whether treatment with dsP9-1 would inhibit viroplasm formation, 24h after transfection with dsRNAs, we inoculated VCMs with SRBSDV and processed them for immunofluorescence. Immunofluorescence showed that the dsGFP treatment did not prevent formation of viroplasm staining by P9-1-rhodamine and accumulation of viral particles staining by P10- FITC (Fig. 5B, C). In contrast, the dsP9-1 treatment led to a nearly complete inhibition of viroplasm formation and viral particle accumulation (Fig. 5B-D).These results suggested that the knockdown of P9-1 expression due to RNAi induced by dsP9-1 significantly inhibited viroplasm formation and viral infection in the VCMs.

3.4 Ingestion of dsRNAs from the *P9–1* gene via membrane feeding knocked down P9–1 expression and strongly inhibited viral infection in intact WBPHs

Our results obtained using the *in vitro* VCM-based system implied that RNAi induced by dsP9-1 can be used to examine the functional role of P9-1 in intact insects *in vivo*. The preliminary experiments showed that ingestion of dsRNAs via membrane feeding resulted in no phenotypic abnormalities in WBPHs (data not shown). The virus proliferates in the WBPH vector, and the insect becomes SRBSDV infective after a latent period of from 6 to 14d (Zhou et al., 2008). The effects of dsRNAs on the transcript levels of *P9-1* and *P10* viral genes were determined by RT-PCR. At 12d post-first access to diseased plants (padp), RT-PCR analyses showed that ca. 87% (n=100; 3 experiment repetitions) of WBPHs that received dsP9-1 contained the transcripts for these two genes (Table 1). These results confirmed that the *P9-1* gene is susceptible to silencing by RNAi induced by dsP9-1 *in vivo* using a membrane-feeding approach.

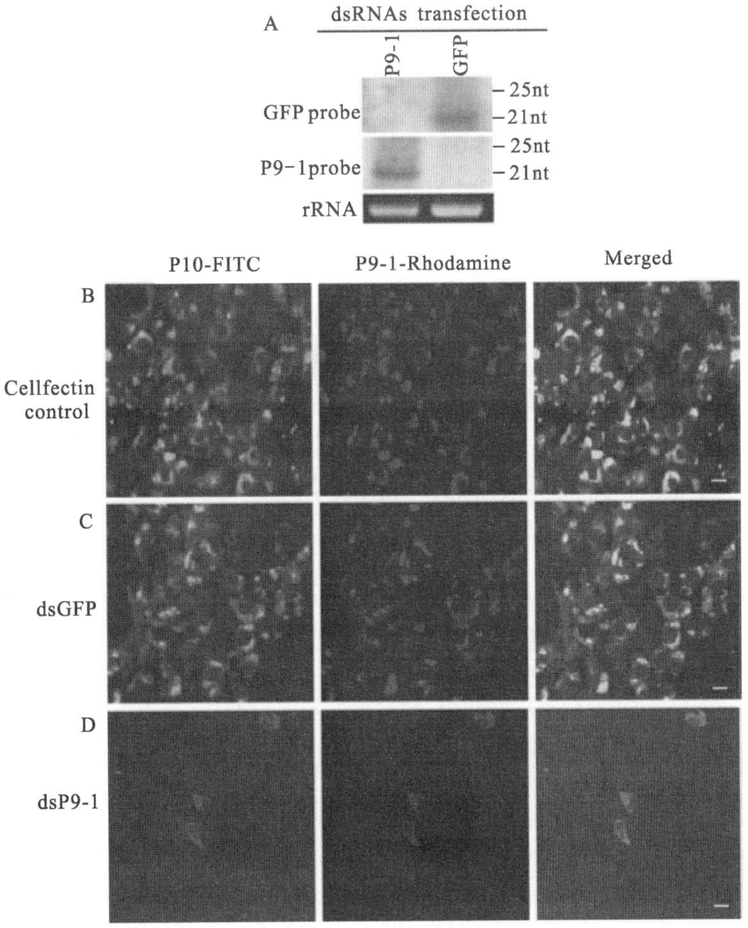

Fig. 5 RNAi induced by dsP9-1 inhibited the assembly of viroplasm and the infection of SRBSDV in VCMs

(A) Detection of siRNAs in VCMs transfected, with synthesized dsP9-1 or dsGFP at 72h after transfection. Total RNA (ca. 5μg) was probed with DIG-labeled riboprobes corresponding to the negative sense, RNA of the *P9-1* or *GFP* gene. (Lower panel) 5.8S rRNA was used as a, control to confirm loading of equal amounts of RNA in each lane. (B to D); Infection of SRBSDV was inhibited by RNAi induced by dsP9-1 in VCMs. At 24h after transfection with Cellfectin transfection reagent (B), dsGFP (C), or dsP9-1 (D), the VCMs were inoculated with SRBSDV. At 48h p.i., cells were stained with P10-FITC (green) and P9-1–rhodamine (red) and examined with confocal microscopy. Images are representative of the results of multiple experiments with multiple preparations. Bars, 10μm

Table 1 RNAi induced by dsP9-1, ingested via membrane feeding, strongly inhibited infection by SRBSDV in the body of its insect vectors, as detected by RT-PCR

Insect dieta[a]	No. of insects giving positive results (n = 100)		
	Expt I	Expt II	Expt III
dsP9-1	11	10	12
dsGFP	86	86	89
Control	88	92	90

[a] Second-instar nymphs of WBPHs were fed with dsP9-1 (0.5μg/μl), dsGFP (0.5μg/μl), or diet alone for 1day, allowed a 2-day acquisition access period on virus-infected rice plants, and then placed on healthy rice seedlings for 10d

To determine whether the treatment of dsP9-1 would inhibit viral replication in intact insects, at 2, 6, and 12d padp, we stained the internal organs from 50 WBPHs that had received dsRNAs with P10-FITC, P9-1–rhodamine, and actin dye phalloidin-Alexa Fluor 647 carboxylic acid. At 2d padp, P9-1 and P10 were restricted to a few epithelial cells of the midgut in 84% of WBPHs receiving dsGFP but in only 12% of those receiving dsP9-1 (Fig. 6A; Table 2). The results suggested that the epithelial cells of the midgut were the initial entry site of SRBSDV. Furthermore, RNAi induced by dsP9-1 could specifically knock down P9-1 expression and significantly inhibit early SRBSDV infection in the epithelial cells of the midgut of the insect vectors. At 6d padp, P9-1 and P10 were seen in an extensive area of the midgut in 83% of WBPHs receiving dsGFP but in only 6% of those receiving dsP9-1 (Fig. 6B; Table 2). However, in 12% of WBPHs that received dsP9-1, P9-1 and P10 were still restricted to a limited area of the midgut (Fig. 6B; Table 2). Therefore, the RNAi induced by dsP9-1 could inhibit efficient spread of virus from the initial infection site in the epithelial cells of the midgut. At 12d padp, in 88% of the WBPHs receiving dsGFP, P9-1 and P10 were detected in an extensive area of the midgut and in the salivary glands of 78% of the insects (Fig. 6C; Table 2). However, much lower percentages of WBPHs that received dsP9-1 had P9-1 and P10 in the midgut, and none had the proteins in the salivary glands (Fig. 6C; Table 2). No significant difference in the numbers of positive samples was found between WBPHs that received dsGFP and those that received diet alone (data not shown). These results indicated that viral infection was restricted to the alimentary canal and that spread of virus to the salivary glands had been completely blocked in insect populations treated with dsP9-1.

Table 2 Percentages of P9-1 and P10 antigens in tissues dissected from intact WBPHs after ingestion of dsP9-1 or dsGFP and acquisition feeding on virus-infected plants

Tissue[a]	% of indicated antigen					
	2 days padp		6 days padp		12 days padp	
	dsP9-1	dsGFP	dsP9-1	dsGFP	dsP9-1	dsGFP
Limited area of midgut	12	84	12	4	8	0
Extensive area of midgut	0	0	6	83	13	88
Salivary gland	0	0	0	20	0	78

[a] Second-instar nymphs of WBPHs were fed with dsP9-1 (0.5μg/μl) or dsGFP (0.5μg/μl) via membrane feeding and then maintained on a mixed diet for 1d, allowed a 2-day，acquisition access period on SRBSDV-infected rice plants, and then placed on healthy rice seedlings. Fifty individuals were tested in each group

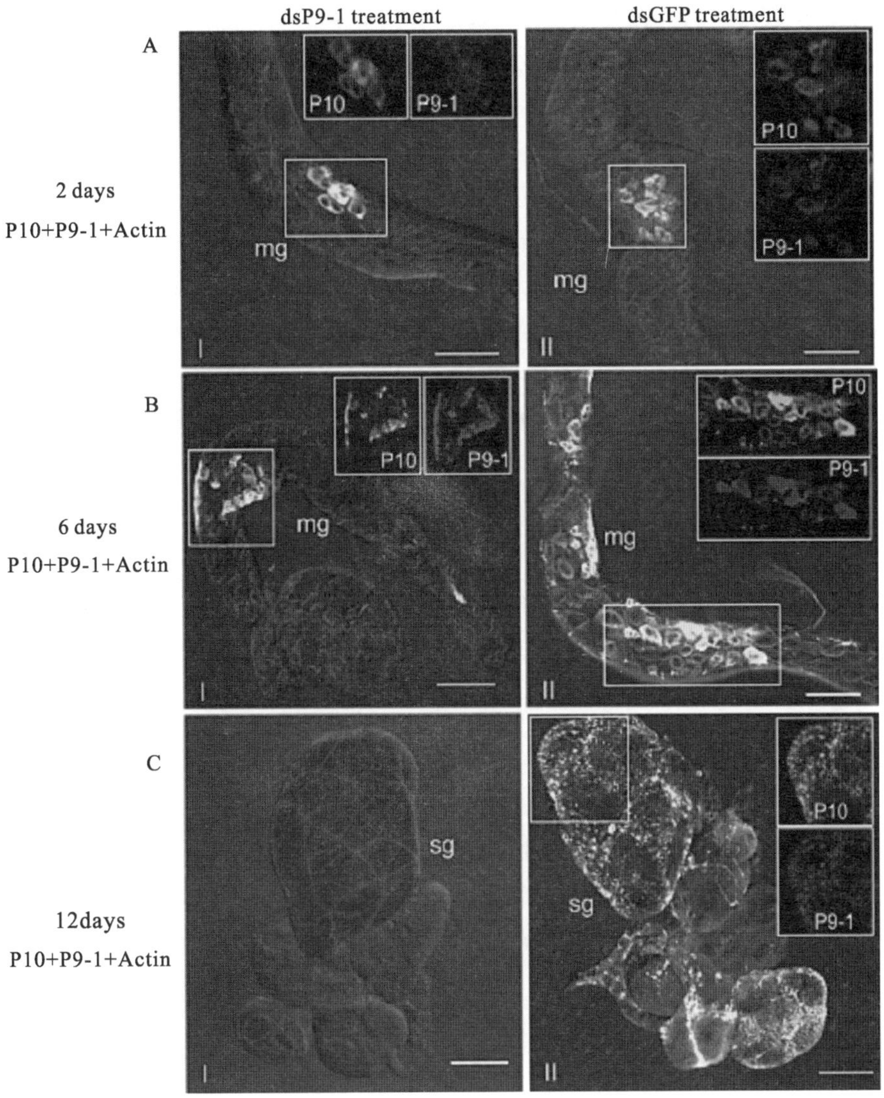

Fig. 6 Ingestion of dsP9-1 via membrane feeding suppressed viral spread *in vivo*

Second-instar nymphs of WBPHs were fed with dsP9-1 (frames I) or dsGFP, (frames II) via membrane feeding. At 2d padp (A), 6d padp (B), and 12d padp (C), internal organs of WBPHs receiving dsP9-1 or dsGFP were stained for SRBSDV virions with P10-FITC (green), stained for viroplasm with P9-1–rhodamine (red), and stained for actin with actin dye phalloidin-Alexa Fluor 647,carboxylic acid (blue). The images with green fluorescence (P10 antigens), red fluorescence (P9-1 antigens), and blue fluorescence (actin dye) were merged under a background of transmitted light. Insets show green fluorescence (P10 antigens) and red fluorescence (P9-1 antigens) of the merged images in the boxed areas in each panel. Images are representative of the results of multiple experiments with multiple preparations. mg, midgut; sg, salivary gland. Bars, 70μm

4 Discussion

VCMs enable us to study the molecular entities responsible for biological events during viral replication (Wei, et al., 2006; Wei, et al., 2006). In the present study, using VCMs derived from WBPHs, our cytopathological analysis revealed that P9-1 of SRBSDV was a component of the viroplasm matrix and, for the first time, provided direct evidence to show that newly synthesized viral RNA, outer-capsid structural proteins, and viral particles accumulated together within the matrix of viroplasms that are induced by fijivirus infection (Fig. 3), confirming the hypothesis that the viroplasm is the site for the replication and assembly of fijiviruses. Our results extend the limited information from previous electron microscopic studies that showed that P9-1 of two other fijiviruses (RBSDV and MRCV) exclusively localized in the matrix of viroplasms in virus-infected cells (Guzmán et al., 2010; Isogai et al., 1998; Maroniche et al., 2010). With the evidence indicating that P9-1 alone was necessary for the formation of viroplasm-like structures in nonhost cells (Fig. 4), our results suggest that P9-1 is the minimal viral factor required for viroplasm formation during SRBSDV infection.

To gain direct evidence that P9-1 plays a crucial role in viral replication, we used the RNAi strategy to knock down the expression of P9-1 in virus-infected VCMs and then analyzed its effect on viroplasm formation and viral infection. Because RNAi induced by dsRNA is a conserved sequence-specific gene silencing mechanism (Fire et al., 1998) and the sequences of P9-1 and other viral genes of SRBSDV had little homology (data not shown), we determined that the knockdown of P9-1 expression in VCMs was specifically caused by RNAi induced by dsP9-1. Our results indicated that the knockdown of P9-1 expression directly led to a significant loss in the function of P9-1 in the formation of viroplasms, with a significant simultaneous inhibition of the assembly of progeny virions within viroplasms in VCMs (Fig. 5). These results indicate that P9-1 was essential for the formation of the viroplasm and viral replication.

As a persistently propagative plant virus, SRBSDV, after moving into the epithelial cells of the alimentary canal, must replicate and assemble progeny virions, which eventually spread to the salivary glands in its insect vector. Our results showed that the knockdown of P9-1 expression due to RNAi induced by ingestion of dsP9-1 strongly inhibited the early formation of viroplasms and As a persistently propagative plant virus, SRBSDV, after moving into the epithelial cells of the alimentary canal, must replicate and assemble progeny virions, which eventually spread to the accumulation of viral particles in the epithelial cells of the midgut, which resulted in limited spread of the virus from the initial infection site to additional tissues, including the salivary glands in the insect (Table 2 and Fig. 6). Due to the essential role of P9-1 in the biogenesis of the viroplasm as shown above, we determined that the slow spread of SRBSDV in the body of the dsP9-1-treated insects was directly caused by a significant loss in the functioning of P9-1 in the assembly of viroplasm, leading to strong inhibition of viral replication in its insect vector. Our efforts also demonstrated for the first time that RNAi activity was inducible in WBPH.

Previously, due to the lack of insect vector cell cultures and a reverse-genetics system in *fijiviruses*, no clear function of virus-encoded proteins in the viral replicative cycle had been determined. In this study, the deployment of new biological tools, including VCMs derived from WBPH and RNAi induced by synthesized dsRNA *in vitro* and *in vivo*, helped overcome the lack of a reverse-genetics system in *fijiviruses* and advanced our understanding of SRBSDV-WBPH interactions. By combining these new approaches, the data demonstrated that P9-1 of SRBSDV is an essential protein for viral replication and suggested that this nonstructural protein may be a good target for disease control. In the viruses of family *Reoviridae*, RNAi of genes for viroplasm matrix proteins, including NS of mammalian reovirus (Kobayashi et al., 2006) and NSP5 of rotavirus (López et al., 2005), which are all orthologous to P9-1 of SRBSDV, had pleiotropic effects on the assembly of viroplasm and viral replication. Our results support the hypothesis that viroplasms play a pivotal role in replication of viruses in the family *Reoviridae*.

Acknowledgements

This work was supported by grants from the National Basic Research Program of China (2010CB126203), the Specialized Research Fund for the Ministry of Agriculture (201003031), the National Natural Science Foundation of China (31130044 and 31070130), the Natural Science Foundation of Fujian Province (2011J06008), and the Doctoral Fund of the Ministry of Education of China (20113515130002).

References

[1] AKITA F, HIGASHIURA A, SHIMIZU T, et al. Crystallographic analysis reveals octamerization of viroplasm matrix protein P9-1 of Rice black streaked dwarf virus[J]. Journal of Virology, 2012, 86(2): 746-756.

[2] CHEN H, CHEN Q, OMURA T, et al. Sequential infection of *Riced warf virus* in the internal organs of its insect vector after ingestion of virus[J]. Virus Research, 2011, 160(1-2): 389-394.

[3] CHENG G, COX J, WANG P, et al. A C-Type Lectin Collaborates with a CD45 Phosphatase Homolog to Facilitate West Nile Virus Infection of Mosquitoes[J]. Cell, 2010, 142(5): 714-725.

[4] CHERRY S. RNA is creening for host factors involved inviral infection using Drosophila cells[J]. Methods in Molecular Biology, 2011, 721(42): 375-382.

[5] CREAMER R. Invertebrate tissue culture as a tool to study insect transmission of plant viruses[J]. In Vitro Cellular and Developmental Biology, 1993, 29: 284-288.

[6] FIRE A, XU S, MONTGOMERY M K, et al. Potent and specific genetic interference by double-stranded RNA in Caenorhabditis elegans[J]. Nature, 1998, 391(6669): 806-811.

[7] FORZAN M, MARSH M, ROY P. Bluetongue virus entry into cells[J]. Journal of Virology, 2007, 81(9): 4819-4827.

[8] FU Q, ZHANG Z, HU C, et al. Achemically defined diet enables continuous rearing of the brown planthopper, *Nilaparvata lugens* (Stål) (Homoptera: Delphacidae) [J]. Applied Entomology and Zoology, 2001, 36: 111-116.

[9] GUZMÁN F A, ARNEODO J D, PONS A B S, et al. Immunodetection and subcellular localization of Cuarto virus P9-1 protein in infected plant and insect host cells[J]. Virus Genes, 2010, 41(1): 111-117.

[10] HOGENHOUT S A, AMMAR E D, WHITFIELD A E, et al. Insect vector interactions with persistently transmitted viruses[J]. Annual Review of Phytopathology, 2008, 46: 327-359.

[11] HUVENNE H, SMAGGHE G. Mechanisms of dsRNA uptake in insects and potential of RNAi for pest control: a review[J]. Journal of Insect Physiology, 2010, 56(3): 227-235.

[12] ISOGAI M, UYEDA I, LEE B C. Detection and assignment of proteins encoded by rice black streaked dwarf. fijivirus S7, S8, S9 and S10[J]. Journal of General Virology, 1998, 79(6): 1487-1494.

[13] KIMURA I. A study of rice dwarf virus in vector cell monolayers by fluorescent antibody focus counting[J]. Journal of General Virology, 1986, 67: 2119-2124.

[14] KIMURA I, OMURA T. Leafhopper cell cultures as a means for phytoreovirus research[J]. Adv Disease Vector Research, 1988, 5: 111-135.

[15] KOBAYASHI T, CHAPPELL J D, DANTHI P, et al. Gene-specific inhibition of reovirus replication by RNA interference[J]. Journal of Virology, 2006, 80(18): 9053-9063.

[16] LIU Y, JIA D, CHEN H, et al. The p7-1 protein of southern rice black-streaked dwarf virus, a fijivirus, induces the formation of tubular structures in insect cells[J]. Archives of Virology, 2011, 156(10): 1729-1736.

[17] LÓPEZ T, ROJAS M, AYALA-BRETÓN C, et al. Reduced expression of the rotavirus NSP5 gene has a pleiotropic effect on virus replication[J]. Journal of General Virology, 2005, 86(6): 1609-1617.

[18] LU Y H, ZHANG J F, XIONG R Y, et al. Identification of an RNA Silencing Suppressor Encoded by Southern rice black-streaked dwarf virus S6[J]. Scientia Agricultura Sinica, 2011, 14: 2909-2917.

[19] MARONICHE G A, MONGELLI V C, PERALTA A V, et al. Functional and biochemical properties of *Mal de Río Cuarto virus* (*Fijivirus, Reoviridae*) P9-1 viroplasm protein show further similarities to animal reovirus counterparts[J]. Virus Research, 2010, 152(1-2): 96-103.

[20] MCGINNIS K M. RNAi for functional genomics in plants[J]. Briefings in Functional Genomics, 2010(2): 111-117.

[21]MILNE R G, DEL VAS M, HARDING R M, et al. Fijivirus, In Fauquet, CM, Mayo, MA, Maniloff, J, Desselberger, U, Ball, LA (ed), Virus taxonomy: classification and nomenclature of viruses. Eighth report of the international committee on taxonomy of viruses[J]. Academic Press, Amsterdam, Holland. 2005, 534-542.

[22] MIYAZAKI N, UEHARA-ICHIKI T, XING L, et al. Structural evolution of reoviridae revealed by oryzavirus in acquiring the second capsid shell[J]. Journal of Virology, 2008, 82(22): 11344.

[23] OMURA T, MERTENS P P C. Phytoreovirus, In Fauquet CM, Mayo MA, Maniloff J, Desselberger U, Ball LA (ed), Virus taxonomy: classification and nomenclature of viruses. Eighth report of the international committee ontaxonomy of viruses[J]. Academic Press, Amsterdam, Holland, 2005, 543-549.

[24] OMURA T, KIMURA I. Leafhopper cell culture for virus research[J]. Arthropod Cell Culture Systems, 1994， 91-107.

[25] ROSA C, KAMITA S G, DEQUINE H, et al. RNAi effects on actin mRNAs in Homalodisca vitripennis cells[J]. Journal of Rnai & Gene Silencing An International Journal of Rna & Gene Targeting Research, 2010, 6(1): 361-366.

[26] SHIMIZU T, YOSHII M, WEI T, et al. Silencing by RNAi of the gene for Pns12, a viroplasm matrix protein of Riced warf virus, results in strong resistance of transgenic rice plants to the virus. Plant Biotechnol Journal, 2009, 7: 24-32.

[27] SPINELLI S, CAMPANACCI V, BLANGY S, et al. Modular Structure of the Receptor Binding Proteins of Lactococcus lactis Phages the RBP Structure of the temperate Phage TP901-1[J]. Journal of Biological Chemistry, 2006, 281: 14256-14262.

[28] UPADHYAYA N M, MERTENS P P C. Oryzavirus, In Fauquet CM, Mayo MA, Maniloff J, Desselberger U, Ball LA (ed), Virus taxonomy: classification and nomenclature of viruses. Eighth report of the international committee on taxonomy of viruses[J]. Academic Press, Amsterdam, Holland. 2005, 550-555.

[29] WANG Q, YANG J, ZHOU G H, et al. The complete genome sequence of two isolates of southern rice black-streaked dwarf virus, a new member of the genus fijivirus[J]. Journal of Phytopathology, 2010, 158: 733-737.

[30] WEI T, KIKUCHI A, MORIYASU Y, et al. The spread of rice dwarf virus among cells of its insect vector exploits virus-induced tubular structures[J]. Journal of Virology, 2006, 80(17): 8593-8602.

[31] WEI T, SHIMIZU T, HAGIWARA K, et al. Pns12 protein of rice dwarf virus is essential for formation of viroplasms and nucleation of viral-assembly complexes[J]. Journal of General Virology, 2006, 87(2): 429-438.

[32] ZHANG H, CHEN J, ADAMS M. Molecular characterisation of segments 1 to 6 of rice black-streaked dwarf virus from China provides the complete genome[J]. Archives of Virology, 2001, 146(12): 2331-2339.

[33] ZHANG H M, YANG J, CHEN J P, et al. A black-streaked dwarf disease on rice in China is caused by a novel fijivirus[J]. Archives of Virology, 2008, 153(10): 1893-1898.

[34] ZHOU G H, WEN J J, CAI D J, et al. Southern rice black-streaked dwarf virus: a new proposed *Fijivirus* species in the family *Reoviridae*[J]. Chinese Science Bulletin, 2008, 53(23): 3677-3685.

NSvc4 和 CP 蛋白与水稻条纹病毒的致病相关

袁正杰，贾东升，吴祖建，魏太云，谢联辉

（福建农林大学植物病毒研究所，福州 350002）

摘要：以水稻条纹病毒（*Rice stripe virus*, RSV）编码的 NSvc4 蛋白和 CP 蛋白的致病功能鉴定为切入点，研究 RSV 的致病机理。通过农杆菌介导在本氏烟中对 RSV 编码的 6 个蛋白进行细胞定位。采用双分子荧光互补（Bimolecular fluorescence complementation, BiFC）技术鉴定 NSvc4 蛋白与其余 5 个蛋白间的互作关系，并用酵母双杂交体系（Yeast two-hybrid, YTH）对 NSvc4 与 CP 蛋白间的互作再次验证。利用马铃薯 X 病毒（*Potato virus X*, PVX）系统在本氏烟叶片研究 NSvc4 蛋白和 CP 蛋白的致病性，采用实时荧光定量 PCR（Real-time PCR）技术验证致病性鉴定结果。RSV NSvc4 蛋白能与 CP 蛋白互作，且蛋白复合体在本氏烟细胞中能够定位到叶绿体。致病性鉴定显示，PVX-RSV-NSvc4 侵染的本氏烟表现为花叶症状；PVX-RSV-CP 能够在侵染叶上诱导花叶症状，但是系统叶却恢复健康；PVX-RSV-NSvc4 与 PVX-RSV-CP 共侵染的本氏烟和 PVX-RSV-NSvc4CP 侵染的本氏烟的侵染叶、系统叶均表现出了严重病症。Real-time PCR 结果显示，PVX 载体在各侵染本氏烟中的累积量均较低，而 RSV CP 和 NSvc4 基因的表达量与本氏烟的发病症状紧密相关。NSvc4 和 CP 蛋白均有致病性。CP 蛋白的致病力不能持久，但是与 NSvc4 蛋白互作后具有持久致病力，表明二者协同在 RSV 致病过程中发挥作用。

关键词：水稻条纹病毒；NSvc4 和 CP 蛋白；致病性；本氏烟

NSvc4 and CP Proteins Contribute to the Pathogenicity of *Rice Stripe Virus*

Zhengjie Yuan, Dongsheng Jia, Zujian Wu, Taiyun Wei, Lianhui Xie

(Institute of Plant Virology, Fujian Agriculture and Forestry University, Fuzhou 350002)

Abstract: The objective of this study is to investigate the roles of NSvc4 and CP proteins of *Rice stripe virus* (RSV) in pathogenicity to reveal the mechanisms underlying RSV pathogenesis. The subcellular localization patterns of six RSV proteins were studied by expressing them in epidermal cells of *Nicotiana benthamiana* through Agrobacterium inoculation. The interaction between NSvc4 and five other proteins of RSV was investigated using Bimolecular fluorescence complementation (BiFC), and the interaction between NSvc4 and CP was further confirmed by yeast two-hybrid experiments. The possible involvement of NSvc4 and CP in pathogenicity was exploited by expressing them individually or together in *N. benthamiana* employing the vector *Potato virus X* (PVX). NSvc4 interacted with CP and they formed protein complexes in chloroplasts of *N. benthamiana* when co-expressed, both NSvc4 and CP could induce leaf mosaic symptoms when they were expressed using PVX. However, leaf mosaic symptoms caused by CP was recovered in systemic leaves of PVX inoculated plants. When expressed together or as a

中国农业科学 . 2013,46(1): 45-53
收稿日期：2012-05-07；接受日期：2012-08-08
基金项目：国家自然科学基金重点项目（3113044）

fused protein using PVX, NSvc4 and CP induced severe symptoms in both inoculated and systemic leaves of *N. benthamiana*. Real-time PCR experiments showed that NSvc4 and CP had no effects on the accumulation of PVX, and the severity of the symptoms induced by the recombinant virus correlated with the expression levels of NSvc4 and CP. Both NSvc4 and CP have pathogenicity properties and they might have synergistical function in RSV pathogenesis.

Keywords: *Rice stripe virus*; NSvc4 and CP protein; pathogenicity; *Nicotiana benthamiana*.

近些年由纤细病毒属（*Tenuivirus*）的代表成员水稻条纹病毒（*Rice stripe virus*, RSV）引起的水稻条纹叶枯病在东亚国家大范围流行（Jonson et al., 2009a; Jonson et al., 2009b; Satoh et al., 2010），中国的水稻生产更是深受危害（王华弟等 , 2007; Wei et al., 2009; 龙亚芹等 , 2011）。近几十年来，国内外学者对RSV做了大量的研究，特别是在RSV的致病机理方面有一定的研究基础，但尚未明确病毒的致病因子。因此，进一步确定RSV编码的致病因子，揭示其致病机理对于水稻条纹叶枯病的防治有着重要的实践意义。目前，在RSV编码的7个蛋白中（Falk and Tsai, 1998），报道与致病性相关的蛋白有CP和SP（林奇田等 , 1998）。通过马铃薯X病毒（*Potato virus X*, PVX）系统在本氏烟（*Nicotiana benthamiana*）中表达具有沉默抑制子功能的NS2和NS3后，本氏烟表现致病症状（Xiong et al., 2009; Du et al., 2011），但Xiong等认为NS3不是一个致病因子，因为转基因的NS3水稻没有致病症状（Xiong et al., 2009）。对于抑制子是否为病毒编码的致病因子，目前也没有定论（Dunoyer et al., 2004; Lakatos et al.,2004 ; Ruiz-Ferrer and Voinnet, 2009; Diaz-Pendon et al., 2007）。NSvc4被鉴定为运动蛋白（Xiong et al., 2008），且Zhang等（2012）最新研究表明NSvc4能够使本氏烟产生坏死斑。笔者实验室构建了RSV编码的NS2、NSvc2-N（N端381个氨基酸）、NS3、CP、SP、NSvc4蛋白的植物定位载体，对各蛋白在本氏烟表皮细胞中定位，发现仅NSvc4单独表达可以定位到叶绿体。于是，从蛋白互作、致病性鉴定等多角度系统性研究NSvc4的致病机理，并且进一步分析在寄主体内NSvc4可能和CP互作形成蛋白复合体参与RSV的致病途径。找到RSV的致病因子，阐明致病机理，为水稻条纹叶枯病的防治提供理论依据。

1 材料与方法

试验于2011年1～10月在福建农林大学植物病毒研究所完成。

1.1 试验材料

本氏烟种植于25℃恒温光照温室。

1.2 试验方法

1.2.1 载体构建

以实验室保存的NSvc4质粒和CP质粒为模板，以构建不同目的载体的引物（表）进行PCR扩增，通过重叠PCR扩增到NSvc4CP融合蛋白的核酸序列。扩增产物经琼脂糖凝胶电泳后回收，回收产物经限制性内切酶切后再次回收并与PVX载体以及酵母双杂交系统（Yeast two-hybrid, YTH）载体pBT3-STE和pPR3-N连接。连接产物转化大肠杆菌后挑取单克隆菌落，接种于含有对应抗生素的LB培养基中培养后提取质粒。酶切验证为阳性的质粒由华大基因公司测序。测序结果正确后获得构建好的PVX-RSV-NSvc4、PVX-RSV-CP、PVX-RSV-NSvc4CP、pBT3-STE-NSvc4、pBT3-STE-CP、pPR3-N-NSvc4和pPR3-N-CP载体质粒。

使用Gateway系统载体研究蛋白的定位和互作。先用含有重组交换序列的引物扩增NSvc4和CP的目的片段，回收产物通过BP反应与pDONR221重组获得中间载体。测序正确的中间载体通过LR

反应与植物瞬时表达载体 pEarley-101 以及双分子荧光互补（Bimolecular fluorescence complementation, BiFC）载体 Yn 和 Yc 重组，最终获得 pEarley-101-NS2、pEarley-101-NSvc2-N、pEarley-101-NS3、pEarley-101-CP、pEarley-101-SP、pEarley-101-NSvc4、Yc-NSvc4、Yn-NS2、Yn-NSvc2-N、Yn-NS3、Yn-CP 和 Yn-SP 载体。

表 1 本研究中用到的引物

Table1. Primers used in this study

引物用途 Primer purpose	序列 sequence (5′-3′)	下划线部分序列 Underlined sequences
构建 PVX 载体引物 Primers for construction of PVX vectors		
PVX-RSV-Nsvc4F	CCCCCGGGATGGCTCTGTCTCGACTCTTGTCCA	*Sma* I
PVX-RSV-Nsvc4R	GCGTCGACTCACATGATGACAGAAACTTCAGAT	*Sal* I
PVX-RSV-CPF	CCCCCGGGATGGGCACCAACAAGCCAGCCACTC	*Sma* I
PVX-RSV-CPR	GCGTCGACTCAGTCATCTGCACCTTCTGCCTCG	*Sal* I
VC4CPR	GAGTGGCTGGCTTGTTGGTGCCCATCATGATGACAGAAACTTCAGAT	
VC4CPF	AAAATCTGAAGTTTCTGTCATCATGATGGGCACCAACAAGCCAGCCA	
构建酵母双杂交载体引物 Primers for construction of YTH vectors		
PBT3-STE-NSvc4F	GGCCATTACGGCCGCTCTGTCTCGACTCTTGTCCA	*Sfi* I
PBT3-STE-NSvc4R	GGCCGAGGCGGCCTTCATGATGACAGAAACTTCAGATTTT	*Sfi* I
PBT3-STE-CPF	GGCCATTACGGCCGGCACCAACAAGCCAGCCACTC	*Sfi* I
PBT3-STE-CPR	GGCCGAGGCGGCCTTGTCATCTGCACCTTCTGCCTCG	*Sfi* I
PPR3-N-NSvc4F	GGCCATTACGGCCCGGATGGCTCTGTCTCGACTCTTGTCCA	*Sfi* I
PPR3-N-NSvc4R	GGCCGAGGCGGCCGCTACATGATGACAGAAACTTCAGAT	*Sfi* I
PPR3-N-CPF	GGCCATTACGGCCCGGATGGGCACCAACAAGCCAGCCACTC	*Sfi* I
PPR3-N-CPR	GGCCGAGGCGGCCGCTAGTCATCTGCACCTTCTGCCTCG	*Sfi* I
构建 Gateway 入门载体引物 Primers for construction of pDONR221 vectors		
pDONR221- NS2F	GGGGACAAGTTTGTACAAAAAAGCAGGCTTC ATGGCATTACTCCTTTTCAATGATC	*att*L1
pDONR221- NS2R	GGGGACCACTTTGTACAAGAAAGCTGGGTC CATTAGAATAGGGCACTCATGT	*att*L2
pDONR221-NSvc2-NF	GGGGACAAGTTTGTACAAAAAAGCAGGCTTC ATGCATTTTAAATCATATTTCATCT	*att*L1
pDONR221-NSvc2-NR	GGGGACCACTTTGTACAAGAAAGCTGGGTC GGATTCTGCAGAACAAACTAATAGC	*att*L2
pDONR221- NS3F	GGGGACAAGTTTGTACAAAAAAGCAGGCTTC ATGAACGTGTTCACATCGTCTGTGG	*att*L1
pDONR221- NS3R	GGGGACCACTTTGTACAAGAAAGCTGGGTCCAGCACAGCTGGAGAGCTGCCT	*att*L2
pDONR221-CPF	GGGGACAAGTTTGTACAAAAAAGCAGGCTTCATGGGCACCAACAAGCCAGCCACTC	*att*L1
pDONR221-CPR	GGGGACCACTTTGTACAAGAAAGCTGGGTCGTCATCTGCACCTTCTGCCTCGTCC	*att*L2
pDONR221-SPF	GGGGACAAGTTTGTACAAAAAAGCAGGCTTC ATGCAAGACGTACAAAGGACAATAG	*att*L1
pDONR221-SPR	GGGGACCACTTTGTACAAGAAAGCTGGGTCTGTTTTGTGTAGAAGAGGTTGA	*att*L2
pDONR221-NSvc4F	GGGGACAAGTTTGTACAAAAAAGCAGGCTTCATGGCTCTGTCTCGACTCTTGTCCA	*att*L1
pDONR221-NSvc4R	GGGGACCACTTTGTACAAGAAAGCTGGGTCCATGATGACAGAAACTTCAGATTTT	*att*L2
实时荧光定量 PCR 引物 Primers for Real-time PCR		
RT-PVX-CPF	AACTACCTCAACCACCACAA	
RT-PVX-CPR	CCTTCCAGATAGCCTCAAT	
RT-RSV-CPF	ATGCGTTGAATTACCTGACTGC	
RT-RSV-CPR	CCGAGGACACTATCCCATACCT	
RT-NSVC4F	GGTCTGGATTGGGATAAAGGG	
RT-NSVC4R	CACGGACACCTGAAGATTATGC	
18SF	AAACGGCTACCACATCCA	
18SR	CACCAGACTTGCCCTCCA	
OligodT18	TTTTTTTTTTTTTTTTTT	

1.2.2 YTH

参照适合于研究胞膜和胞质蛋白互作的 DUAL membrane Starter Kits 系统操作手册，将 pBT3-STE-NSvc4/pPR3-N-CP、pBT3-STE-CP/pPR3-NNSvc4、阳性对照 pTSU2-APP/pNubG -Fe65 和阴性对照 pTSU2-APP/pPR3-N 等质粒组合分别转化到酵母菌 NMY51 菌株中。转化菌在 SD-trp-leu-his 三缺固体筛选培养基上画线观察能否正常生长，并进行显色反应，观察能否使显色缓冲液变蓝。

1.2.3 农杆菌转化和侵染本氏烟

分别将PVX空载体、PVX-RSV-NSvc4、PVX-RSV-CP、PVX-RSVNSvc4CP、pEarley-101-NS2、pEarley-101-NSvc2-N、pEarley-101-NS3、pEarley-101-CP、pEarley-101-SP、pEarley-101-NSvc4、Yc-NSvc4、Yn-NS2、Yn-NSvc2-N、Yn-NS3、Yn-CP和Yn-SP质粒转化到农杆菌GV3101菌株中，并注射本氏烟叶片（Wei and Wang, 2008）。单独注射含有植物瞬时表达载体的GV3101到本氏烟叶片进行细胞定位。分别共注射含有Yc-NSvc4/Yn-NS2、Yc-NSvc4/Yn-NSvc2-N、Yc-NSvc4/Yn-NS3、Yc-NSvc4/Yn-CP及Yc-NSvc4/Yn-SP的GV3101到本氏烟叶片，鉴定蛋白间的互作。48 h后在激光共聚焦显微镜下观察结果。单独注射含有PVX空载体、PVX-RSV-NSvc4、PVXRSV-CP和PVX-RSV-NSvc4CP及共注射PVX-RSVNSvc4&PVX-RSV-CP的农杆菌到本氏烟叶片，并观察致病症状。每个样品处理每次注射3～5株，试验重复3次。

1.2.4 Real-time PCR

用Trizol（Invitrogen）法分别提取PVX致病性鉴定试验中各处理的侵染叶和系统叶的总RNA，分别取1μg总RNA用M-MLV反转录酶（Promega）反转录获得cDNA，作为Real-time PCR的模板。选用稳定表达的18SrRNA基因作为内参。内参模板由OligodT18反转录，其余检测基因用各基因的R引物反转录，引物序列见表。

2 结果

2.1 NSvc4蛋白能够定位到叶绿体

RSV编码蛋白在本氏烟叶片表皮细胞中的定位结果显示，单独表达时，RSV编码的6个蛋白中仅NSvc4可以定位到叶绿体（图1A-III），其余5个蛋白均不能定位到叶绿体（结果未显示），而对照YFP蛋白在细胞各处均有分布呈弥散状（图1A-I）。NSvc4蛋白能够定位到叶绿体，暗示其可能是RSV编码的致病因子。

2.2 NSvc4蛋白能够招募CP蛋白进入叶绿体

由于RSV编码的6个蛋白中，仅NSvc4可以独立定位到叶绿体，为了验证NSvc4能否招募其他的RSV编码蛋白进入叶绿体，笔者通过BiFC分析NSvc4和其余5个蛋白间的互作关系。结果表明，NSvc4能与CP互作，且BiFC的荧光和叶绿体重叠（图1B-I）。而在本氏烟表皮细胞中，CP单独表达时，形成的颗粒状内含体分布于细胞质中（图1A-II）。以上的研究结果表明，NSvc4能够特异性地与CP互作，且能招募CP进入叶绿体。基于NSvc4是RSV的致病因子，推测NSvc4和CP能够协同致病。

同时，笔者利用YTH系统对NSvc4和CP间的互作进行了验证。结果显示，共转化NSvc4和CP的酵母菌能够在SD-leu-trp-his三缺筛选培养基上生长（图1C），并在显色试验中使显色缓冲液变蓝（图1D），这表明CP可以和NSvc4互作，与BiFC结果一致。

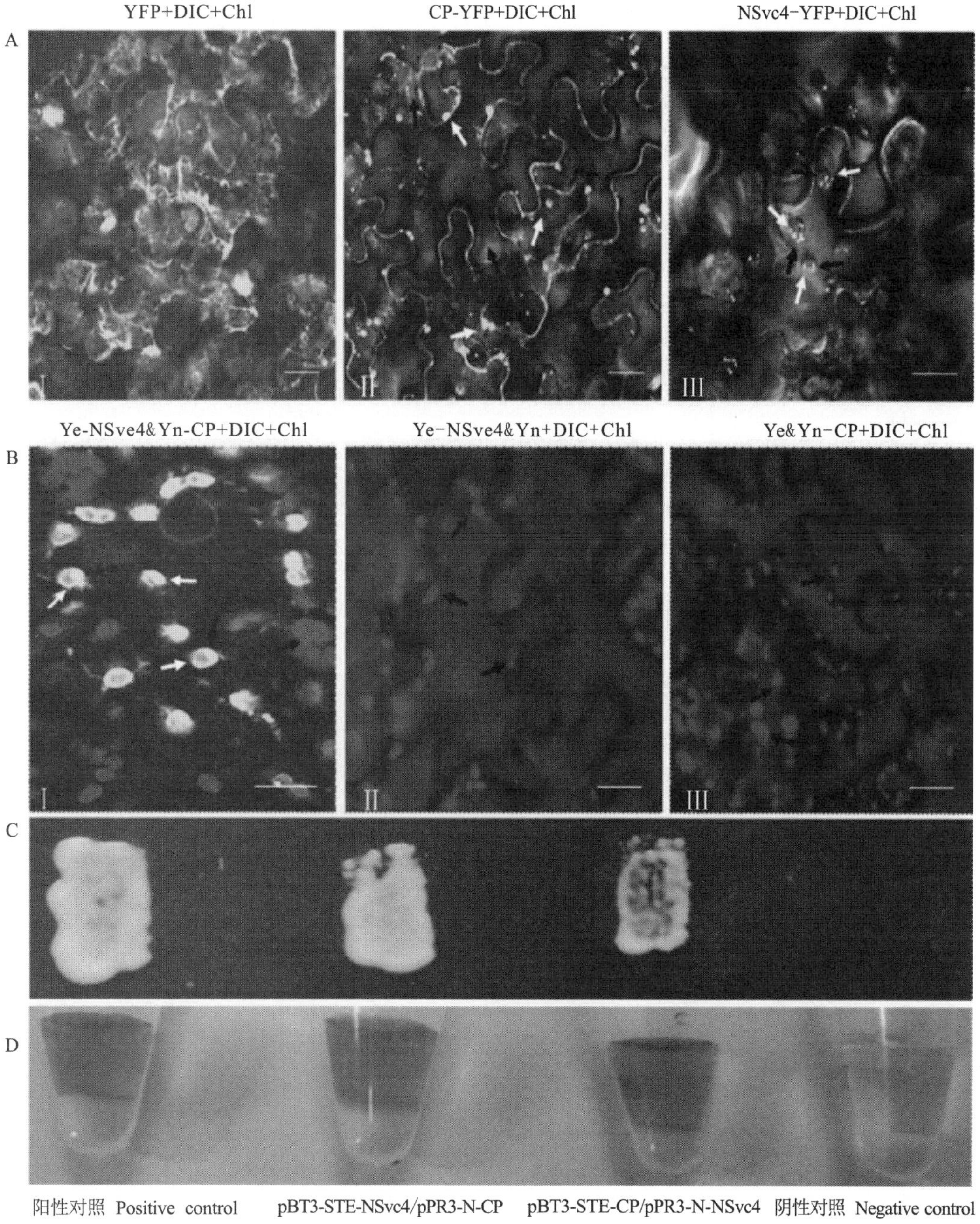

图 1　亚细胞定位、BiFC 和 YTH 试验

Fig. 1 Subcellular localization, BiFC and YTH assays

A：细胞定位，Ⅰ、Ⅱ、Ⅲ依次分别为 YFP、CP 和 NSvc4 在本氏烟表皮细胞中的定位；B：BiFC 互作检测，Ⅰ、Ⅱ、Ⅲ依次分别为 Yc-NSvc4 &Yn-CP、Yc-NSvc4 &Yn 空载体和 Yc 空载体 &Yn-CP 在本氏烟表皮细胞中共表达；C：YTH 互作检测中，共转化菌在 SD-trp-leu-his 三缺筛选平板上培养；D：YTH 互作检测中，显色实验。白色箭头指出荧光蛋白，黑色箭头指出叶绿体；Chl 为叶绿体；标尺 10μm

A: Subcellular localization assay, frames Ⅰ, Ⅱ and Ⅲ were YFP, CP and NSvc4 expressed in epidermal cells of N. benthamiana, respectively; B: BiFC assay, frames Ⅰ, Ⅱ and Ⅲ were Yc-NSvc4 &Yn-CP co-expressed, Yc-NSvc4 &Yn empty vector co-expressed and Yc empty vector &Yn-CP co-expressed in epidermal cells of *N. benthamiana*, respectively; C: YTH assay, transformed yeast NMY51 stains grew on SD-trp-leu-his plate; D: YTH, colorful assays. The white arrows indicated fluorescent viral proteins and the black arrows indicated chloroplasts; Chl: Chloroplast; Bars were 10μm

2.3 NSvc4 蛋白和 CP 蛋白的致病性鉴定

为了鉴定 NSvc4 蛋白和 CP 蛋白的致病功能，本试验构建了 NSvc4、CP 及二者融合蛋白 NSvc4CP 的 PVX 载体 PVX-RSV-NSvc4、PVX-RSV-CP 和 PVXRSV-NSvc4CP。分别单独注射 PVX-RSV-NSvc4、PVX-RSV-CP 和 PVX-RSV-NSvc4CP 及共注射 PVX-RSV-NSvc4 & PVX-RSV-CP 本氏烟，并观察发病症状。试验重复 3 次，PVX-RSV-CP 和 PVX-RSVNSvc4&PVX-RSV-CP 共侵染的本氏烟较其他样品处理早 1d 观察到致病症状，其余样品处理侵染的本氏烟致病症状表现时间较为一致，症状观察结果稳定可重复。各本氏烟发病症状如图 2 所示。侵染 9d 后表现出早期症状，PVX 空载体侵染的本氏烟表现轻微的花叶症状（图 2B- Ⅰ、Ⅱ）；PVX-RSV-NSvc4 侵染的本氏烟也表现出了花叶症状（图 2C- Ⅰ、Ⅱ）；PVX-RSV-CP、PVX-RSV-NSvc4CP 及 PVX-RSVNSvc4&PVX-RSV-CP 共侵染的本氏烟均表现出了花叶、畸形和皱缩等较严重的病症（图 2D- Ⅰ，Ⅱ，图 2E- Ⅰ，Ⅱ，图 2F- Ⅰ，Ⅱ）。侵染 18d 后表现后期症状，PVX 空载体侵染的本氏烟的系统叶症状恢复（图 2B- Ⅲ），几乎与健康本氏烟无差异（图 2A- Ⅲ）；PVX-RSV-NSvc4 侵染的本氏烟仍然能够在系统叶上保持较轻的花叶症状（图 2C- Ⅲ）；PVX-RSV-CP 侵染的本氏烟系统叶没有表现症状（图 2D- Ⅲ）；PVX-RSV-NSvc4&PVX-RSV-CP 共侵染的本氏烟的系统叶也保持有致病病症（图 2E- Ⅲ）；PVX-RSV-NSvc4CP 侵染的本氏烟的系统叶仍保持与侵染叶片一致的严重症状（图 2F- Ⅱ、Ⅲ）。

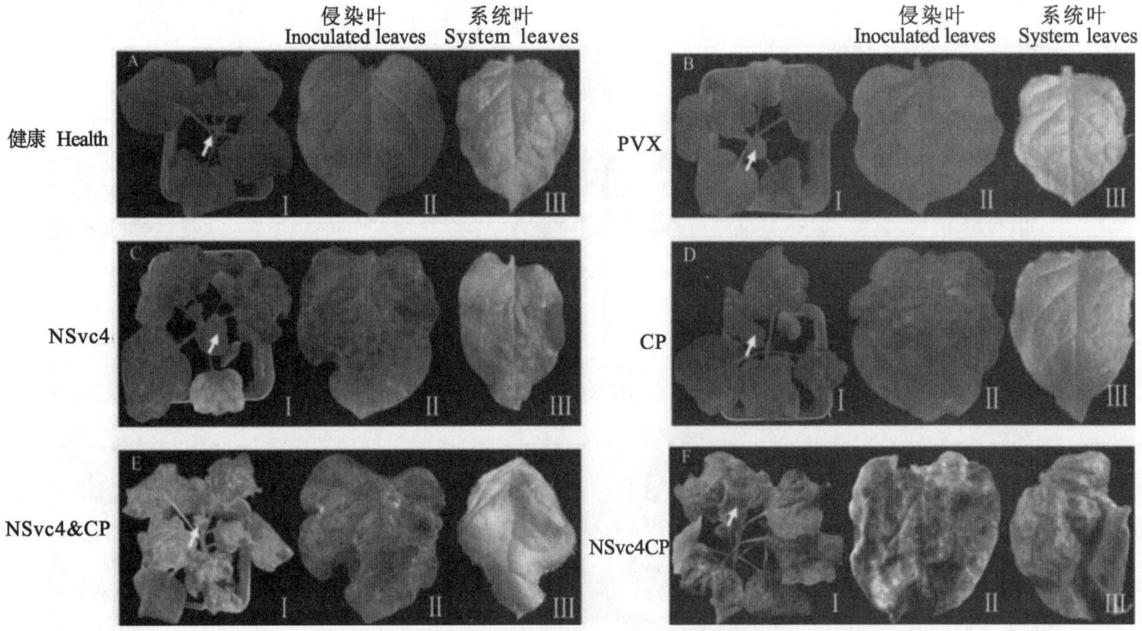

图 2　NSvc4 和 CP 的致病性鉴定

Fig. 2　Identification of the pathogenicity of NSvc4 and CP proteins of RSV

Ⅰ、Ⅱ、Ⅲ依次分别为侵染本氏烟的完整植株、侵染本氏烟的侵染叶和系统叶（红色箭头指示侵染叶，白色箭头指示系统叶）。A-F 依次分别为健康本氏烟、PVX 空载体、PVX-RSV-NSvc4、PVX-RSV-CP、PVX-RSV-NSvc4 & PVX-RSV-CP 共侵染和 PVX-RSV-NSvc4CP 侵染的本氏烟

Frames, and represented the complete plants, inoculated leaves and system leaves, respectively (red arrows: inoculated leaves; Ⅰ Ⅱ Ⅲ white arrows: system leaves). Plates A to F were health *N. benthamiana*, empty PVX vector inoculated, PVX-RSV-NSvc4 inoculated, PVX-RSV-CP inoculated, PVX-RSV-NSvc4 & PVX-RSV-CP co-inoculated and PVX-RSV-NSvc4CP inoculated *N. benthamianas*, respectively

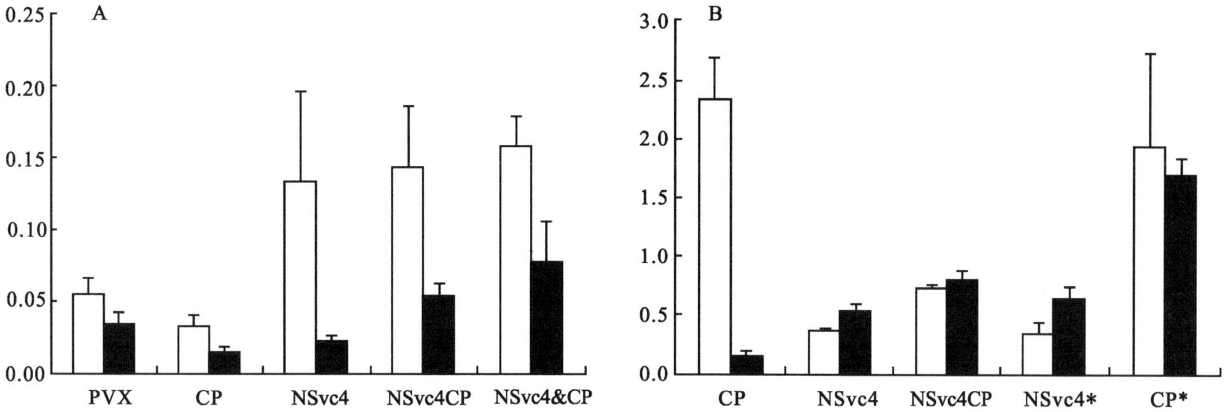

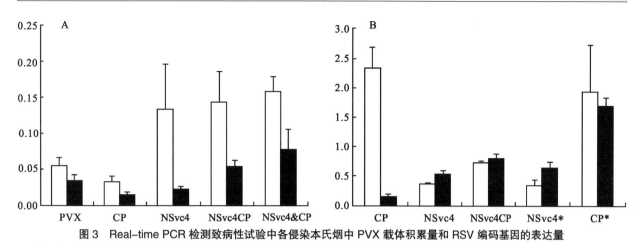

图 3 Real-time PCR 检测致病性试验中各侵染本氏烟中 PVX 载体积累量和 RSV 编码基因的表达量

Fig. 3 Real-time PCR detected the PVX vector accumulations and RSV genes expression in the leaves of *N. benthamiana*

图中白条和黑条分别代表侵染叶和系统叶。A：以 PVX 载体的 CP 基因为靶标，检测各侵染本氏烟中 PVX 载体的积累量；B：检测 RSV 的 NSvc4 或 CP 在各侵染本氏烟中的表达量。NSvc4* 和 CP* 代表这 2 个基因在 PVX-RSV-NSvc4 & PVX-RSV-CP 共侵染本氏烟中的表达量

White columns and black columns represented inoculated leaves and system leaves, respectively. A: Using PVX CP gene as targeting sequences to examine the accumulations of PVX in PVX inoculated *N.benthamiana*; B: Using RSV NSvc4 and/or CP genes as targeting sequences to examine in inoculated *N.benthamiana* except the empty P VX vector. NSvc4* and CP* were the mRNA level of the two genes in PVX-RSV-NSvc4 & PVX-RSV-CP co-inoculated *N.benthamiana*, respectively

2.4 Real-time PCR 验证致病性实验中各基因的转录水平积累程度

取 PVX-RSV-NSvc4、PVX-RSV-CP、PVX-RSVNSvc4CP、PVX-RSV-NSvc4 & PVX-RSV-CP 及 PVX 空载体侵染的本氏烟早期叶片（侵染叶）和后期叶片（系统叶），分别提取总 RNA，反转录成 cDNA 作为模板进行 Real-time PCR，检测结果如图 3 所示。

以 PVX 载体的 *CP* 基因为靶标，检测 PVX 载体在各侵染本氏烟中的积累量（图 3A）。结果显示，在各侵染的本氏烟中，PVX 载体的积累量均很低，且在系统叶中较各自的侵染叶中均有所减少，表明 PVX 载体对本氏烟的发病症状影响较小（图 2B）。RSV 编码的基因在各自侵染的本氏烟中的表达量如图 3B 所示。在 PVX-RSV-NSvc4 侵染的本氏烟中，NSvc4 在侵染叶（0.357）和系统叶（0.515）的表达量没有太大的差异，这与 NSvc4 在侵染叶和系统叶上均有花叶症状一致。与 NSvc4 表现相反，RSV *CP* 在侵染叶（2.336）和系统叶（0.139）中的表达量相差近 17 倍，差异非常显著。这与 PVX-RSV-*CP* 侵染的本氏烟所表现的症状保持一致，RSV-*CP* 在表达量高的侵染叶，表现较严重的花叶和皱缩病症，而在其表达量低的系统叶中无症状。相反，PVX 载体无论是在侵染叶（0.032）还是系统叶（0.014）都保持较低的累积量。对于 PVX-RSV-NSvc4 *CP* 侵染的本氏烟，虽然融合基因 NSvc4 *CP* 在侵染叶（0.734）和系统叶（0.788）中的表达量差异很小，但和被侵染的本氏烟的致病症状紧密相关。侵染叶和系统叶均表现花叶、畸形和皱缩的严重病症，这与 NSvc4CP 基因在本氏烟中均匀的表达相一致，同时 PVX 载体在侵染叶（0.143）和系统叶（0.054）中的积累量依然保持较低的水平。在 PVX-RSV-NSvc4&PVX-RSV-*CP* 共侵染的本氏烟中，NSvc4 在系统叶（0.633）中的表达量相对于侵染叶（0.346）有所上升，RSV *CP* 在侵染叶（1.929）和系统叶（1.679）中均保持了较高的表达量，同时 PVX 载体的积累量在侵染叶（0.158）和系统叶（0.077）中相对不高。这也与共侵染的本氏烟发病症状保持一致。

3 讨论

Xiong 等（2008）的酵母双杂交实验结果认为 NSvc4 和 CP 不互作，其原因可能是该研究使用的是

核内互作酵母系统。而通过细胞定位表明NSvc4和CP均定位在细胞质（图1A-Ⅱ、Ⅲ）。因此，笔者使用比较适合于胞质和胞膜蛋白互作的DUAL membrane公司的酵母互作系统对NSvc4和CP在细胞质环境中进行互作检测，结果更能反映真实情况。

本研究利用PVX载体对NSvc4蛋白的致病性功能进行了研究。PVX载体由马铃薯X病毒改造而来，用于研究外源基因的致病性及鉴定病毒的沉默抑制子（Chapman et al., 1992; Scholthof et al., 1995）。NS2和NS3是RSV编码的沉默抑制子（Xiong et al., 2009; Du et al., 2011），含有这2个基因的PVX载体侵染本氏烟过程中，也表现致病性。一般情况下沉默抑制子的致病机理是干扰寄主的防御系统，增加PVX载体在寄主内的累积量，从而表现出致病性。笔者认为这种类型是一种间接的致病作用，并非沉默抑制子本身的效果，并且对于病毒的沉默抑制子究竟是否就是病毒的致病因子，目前尚无定论（Dunoyer et al., 2004; Lakatos et al., 2004; Ruiz-Ferrer and Voinnet, 2009; Diaz-Pendon et al., 2007）。还有一类被认为具有直接致病的蛋白。此种情况下，外源蛋白不会增加PVX载体在寄主内的累积量，但是能够诱导寄主表现出致病症状，比如烟草花叶病毒（*Tobacco mosaic virus*，TMV）的6K蛋白和番茄黄化花叶病毒（*Tomato leaf curl New Delhivirus*，ToLCNDV）的NSP（Canto et al., 2004; Hussain et al., 2005）。本研究发现，RSV *CP*不能增加PVX载体在本氏烟中的累积量（图3A），却能够使得侵染本氏烟中高表达*CP*的侵染叶片表现出致病症状，而表达量低的系统叶片不表现病症。因此，笔者认为RSV *CP*在致病机理方面属于直接致病类型。

CP在本氏烟表皮细胞中单独表达可形成移动的颗粒分散在细胞质，当与NSvc4共同表达后，两者能发生互作形成蛋白复合体并定位于叶绿体，这与刘利华等报道的CP在发病水稻叶肉细胞的细胞核、细胞质、胞间连丝和叶绿体中均有分布保持一致（刘利华等, 2000）。一般而言，病毒的CP主要是保护病毒基因组和参与病毒在寄主内运动，目前也有一些报道表明病毒的CP具有致病功能（朱水芳等, 1992; 梁德林等, 1998; Zhao et al., 2000; Szittya and Burgyán, 2001; 曹云鹤等, 2006）。本研究中利用PVX体系也检测到了CP具有致病功能。NSvc4是RSV编码的运动蛋白，Shimizu等报道含有沉默NSvc4基因的发夹结构的转基因水稻能抵抗RSV的侵染（Shimizu et al., 2011），表明NSvc4是RSV在水稻寄主中成功侵染和复制所必不可少的。Park等也认为NSvc4对水稻条纹叶枯病症状的表现很关键（Park et al., 2011）。Zhang等(2012)的最新研究表明，NSvc4能使本氏烟产生坏死斑，是RSV的致病因子（Zhang et al., 2012）。本研究中也鉴定到了NSvc4具有致病功能，且能与CP互作加强致病力，在CP的致病力持久性保持上起到非常关键的作用。假设高浓度的CP在细胞质中自由运动分布，一部分的CP到达了可能的致病区域，诱导病症的形成。当NSvc4和CP共表达在细胞质中时，NSvc4与CP互作，并将CP带到了适当的致病区域。考虑到NSvc4单独表达可以定位到叶绿体，且BiFC试验中NSvc4和CP互作后的蛋白复合体进入叶绿体，推断可能的致病区域就是叶绿体。这个假设和推断也与林奇田等（1998）报道的发病水稻中CP在叶绿体中的积累量与水稻叶片的褪绿成正相关的结果相一致。

4 结论

RSV编码的NSvc4蛋白和CP蛋白均有致病功能。CP致病力较强，但其单独侵染的本氏烟不能持久表现致病症状。NSvc4和CP能够互作，且互作后表现出更强和持久的致病力。

参考文献

[1] 王华弟, 陈剑平, 祝增荣, 等. 浙江北部水稻条纹叶枯病的发病流行规律 [J]. 植物保护学报, 2007,34(005):487-492.

[2] 龙亚芹, 王万东, 李凡, 等. 云南水稻条纹病毒 RNA3 的分子变异及遗传多样性分析 [J]. 西南农业学报, 2011, 24(2):570-574.

[3] 朱水芳，叶寅，赵丰，等．黄瓜花叶病毒外壳蛋白质进入叶绿体与症状发生的关系 [J]. 植物病理学报 ,1992, 022(003):229-233.

[4] 刘利华，吴祖建，林奇英，等．水稻条纹叶枯病细胞病理变化的观察 [J]. 植物病理学报，2000,30(4):306-311.

[5] 林奇田，林含新，吴祖建，等．水稻条纹病毒外壳蛋白和病害特异蛋白在寄主体现人的积累 [J]. 福建农业大学学报，1998,322-326.

[6] 曹云鹤，原雪峰，王晓星，等．甜菜黑色焦枯病毒外壳蛋白与病毒致病性的关系 [J]. 生物化学与生物物理进展，2006,33:127-134.

[7] 梁德林，叶寅，施定基，等．黄瓜花叶病毒卫星 RNA 致弱辅助病毒的机理 [J]. 中国科学：生命科学 ,1998,28(3): 251-256.

[8] 林奇田，林含新，吴祖建，等．水稻条纹病毒外壳蛋白和病害特异蛋白在寄主体现人的积累 [J]. 福建农业大学学报，1998,322-326.

[9] 龙亚芹，王万东，李凡，等．云南水稻条纹病毒 RNA3 的分子变异及遗传多样性分析 [J]. 西南农业学报，2011,24:570-574.

[10] 王华弟，陈剑平，祝增荣，等．浙江北部水稻条纹叶枯病的发病流行规律 [J]. 植物保护学报，2007,34:487-492.

[11] 刘利华，吴祖建，林奇英，等．水稻条纹叶枯病细胞病理变化的观察 [J]. 植物病理学报，2000,30(4):306-311.

[12] 朱水芳，叶寅，赵丰，等．黄瓜花叶病毒外壳蛋白质进入叶绿体与症状发生的关系 [J]. 植物病理学报 ,1992，22(3)229-233.

[13] 梁德林，叶寅，施定基，等．黄瓜花叶病毒卫星 RNA 致弱辅助病毒的机理 [J]. 中国科学：生命科学 ,1998,28(3): 251-256.

[14] 曹云鹤，原雪峰，王晓星，等．甜菜黑色焦枯病毒外壳蛋白与病毒致病性的关系 [J]. 生物化学与生物物理进展，2006,33:127-134.

[15] CANTO T, MACFARLANE S A, PALUKAITIS P. ORF6 of Tobacco mosaic virus is a determinant of viral pathogenicity in Nicotiana benthamiana[J]. Journal of General Virology, 2004, 85(10):3123-3133.

[16] CHAPMAN S, KAVANAGH T, BAULCOMBE D. Potato virus X as a vector for gene expression in plants[J]. Plant Journal, 1992, 2(4):549-557.

[17] DIAZPENDON J A, LI F, LI W X, et al. Suppression of antiviral silencing by cucumber mosaic virus 2b protein in Arabidopsis is associated with drastically reduced accumulation of three classes of viral small interfering RNAs[J]. Plant Cell, 2007, 19(6):2053-2063.

[18] DU Z G, XIAO D L, WU J G, et al. p2 of rice stripe virus (RSV) interacts with OsSGS3 and is a silencing suppressor[J]. molecular plant pathology, 2011, 12: 808-814.

[19] DUNOYER P, LECELLIER C H, PARIZOTTO E A, et al. Probing the microRNA and small interfering RNA pathways with virus-encoded suppressors of RNA silencing[J]. Plant Cell, 2004, 16(5): 1235-1250.

[20] FALK B W, TSAI J H. Biology and molecular biology of viruses in the genus tenuivirus[J]. Annual Review of Phytopathology, 1998, 36(1):139-163.

[21] HUSSAIN M, MANSOOR S, IRAM S, et al. The nuclear shuttle protein of Tomato leaf curl New Delhi virus is a pathogenicity determinant[J]. Journal of Virology, 2005, 79(7):4434-4439.

[22] JONSON M G, CHOI H S, KIM J S, et al. Complete genome sequence of the RNAs 3 and 4 segments of Rice stripe virus isolates in Korea and their phylogenetic relationships with Japan and China isolates[J]. The plant pathology journal, 2009a, 25(2):142-150.

[23] JONSON M G, CHOI H S, KIM J S, et al. Sequence and phylogenetic analysis of the RNA1 and RNA2 segments of Korean Rice stripe virus isolates and comparison with those of China and Japan[J]. Archives of Virology, 2009b, 154(10):1705-1708.

[24] LAKATOS L, SZITTYA G, SILHAVY D, et al. Molecular mechanism of RNA silencing suppression mediated by p19 protein of tombusviruses[J]. Embo Journal, 2004, 23(4):876-884.

[25] PARK H M, CHOI M S, KWAK D Y, et al. Suppression of *NS3* and MP is important for the stable inheritance of RNAi-mediated *Rice Stripe Virus* (RSV) resistance obtained by targeting the fully complementary *RSV-CP* gene[J]. Molecules and Cells, 2012, 33(1):43-51.

[26] RUIZ-FERRER V, VOINNET O. Roles of plant small RNAs in biotic stress responsess[J]. Annual Review of Plant Biology, 2009, 60: 485-510.

[27] SATOH K, KONDOH H, SASAYA T, et al. Selective modification of rice (Oryza sativa) gene expression by rice stripe virus infection[J]. Journal of General Virology, 2010, 91(1):294-305.

[28] SCHOLTHOF H B, SCHOLTHOF K B, JACKSON A O. Identification of tomato bushy stunt virus host-specific symptom determinants by expression of individual genes from a potato virus X vector[J]. The Plant Cell, 1995, 7(8):1157-1172.

[29] SHIMIZU T, NAKAZONO N E, UEHARA I T, et al. Targeting specific genes for RNA interference is crucial to the development of strong resistance to rice stripe virus[J]. Plant Biotechnology Journal, 2011, 9(4):503-512.

[30] SZITTYA G, BURGYÁN J. Cymbidium ringspot tombusvirus coat protein coding sequence acts as an avirulent RNA[J]. Journal of Virology, 2001, 75(5):2411-2420.

[31] WEI T, WANG A. Biogenesis of Cytoplasmic Membranous Vesicles for Plant Potyvirus Replication Occurs at Endoplasmic Reticulum Exit Sites in a COPI- and COPII-Dependent Manner[J].Journal of Virology, 2008, 82(24):12252-12264.

[32] WEI T Y, YANG J G, LIAO F L, et al. Genetic diversity and population structure of ricestripe virus in China[J]. Journal of General Virology, 2009, 90(4):1025-1034.

[33] XIONG R Y, WU J X, ZHOU Y J, et al. Characterization and subcellular localization of an RNA silencing suppressor encoded by rice stripe tenuivirus[J]. Virology, 2009, 387(1):29-40.

[34] XIONG R Y, WU J X, ZHOU Y J, et al. Identification of movement protein of the Tenuivirus rice stripe virus[J]. Journal of Virology, 2008, 82(24):12304-12311.

[35] ZHANG C, PEI X, WANG Z, et al. The Rice stripe virus pc4 functions in movement and foliar necrosis expression in Nicotiana benthamiana[J]. Virology, 2012, 425(2):113-121.

[36] ZHAO Y, DELGROSSO L, YIGIT E, et al. The amino terminus of the coat protein of Turnip crinkle virus is the AVR factor recognized by resistant arabidopsis[J]. Molecular plant-microbe interactions:MPMI, 2000, 13(9):1015.

Identification and characterization of the interaction between viroplasm-associated proteins from two different plant-infecting reoviruses and eEF-1A of rice

Songbai Zhang[2], Zhenguo Du[3], Liang Yang[1], Zhengjie Yuan[1], Kangcheng Wu[1], Guangpu Li[4], Zujian Wu[1], Lianhui Xie[1]

(1 Institute of Plant Virology, Fujian Agriculture and Forestry University, Fuzhou 350002, Fujian, China; 2 College of Agriculture, Yangtze University, Jingzhou 434025,Hubei, China; 3 Plant Protection Research Institute, Guangdong Academy of Agricultural Sciences, Guangzhou 510640, Guangdong, China; 4 Department of Biochemistry and Molecular Biology, University of Oklahoma Health Sciences Center, Oklahoma City, Oklahoma 73104, USA)

Abstract: A rice protein homologous to eukaryotic translation elongation factor 1A (eEF-1A) was found to interact with the Pns6 of rice ragged stunt virus (RRSV), the type member of the genus *Oryzavirus*, family *Reoviridae*, in yeast two-hybrid screening. The interaction between the rice protein, designated OseEF-1A, and RRSV Pns6 was confirmed by bimolecular fluorescence complementation. Besides Pns6, OseEF-1A also interacted with the viroplasm matrix protein, Pns10, of RRSV. When expressed together, OseEF-1A co-localized with RRSV Pns10 in epidermal cells of *Nicotiana benthamiana*. Pns6 of southern rice black-streaked dwarf virus (SRBSDV), a newly reported member of the genus *Fijivirus*, family *Reoviridae*, was the only non-structural SRBSDV protein studied here that also interacted with OseEF-1A. Like Pns6 of rice black-streaked dwarf virus (RBSDV), SRBSDV Pns6 interacted with itself and co-localized with Pns9-1 in *N. benthamiana*. In the presence of Pns6, OseEF-1A co-localized with Pns9-1, the putative viroplasm matrix protein of SRBSDV.

1 Introduction

Viruses have evolved a variety of strategies to manipulate the cellular translation machinery to produce their own proteins efficiently and to suppress immune reactions by the hosts. In many cases, this involves complex interactions between viruses and their hosts. In recent years, identification and characterization of these interactions have contributed greatly to understanding viral and host translation regulation(Walsh and Mohr, 2011; Mohr and Sonenberg, 2012). Reoviruses are a group of nonenveloped, double-stranded RNA (dsRNA) viruses that replicate in discrete structures called viroplasms in the cytoplasm of their host cells. They have multi-segmented genomes enclosed in characteristic icosahedral capsids composed of one to a few concentric protein layers. Currently, the family Reoviridae includes 15 established genera, members of which infect a wide range of hosts including vertebrates, invertebrates, plants, and fungi (Attoui et al., 2012). It has long been recognized that reoviruses manipulate cellular translation systems (Katze, 1996). However, the cellular targets

Archives of virology. 2013,158:2031–2039
Received 11 December 2012; Accepted 21 March 2013; Published online 19 April 2013

involved in the manipulation have been described for only a few animal reoviruses (Lloyd and Shatkin, 1992; Piron et al., 1998; Ji et al., 2009; Chulu et al., 2010).

Plant-infecting reoviruses are classified into three genera, namely *Phytoreovirus, Oryzavirus* and *Fijivirus*. *Rice ragged stunt virus* (RRSV) is the type species of the genus *Oryzavirus*, while southern rice black-streaked dwarf virus (SRBSDV) is a newly reported member of the genus *Fijivirus* (Attoui et al., 2012;Wang et al., 2010; Zhang et al., 2008; Zhou et al., 2008).Both viruses, transmitted by different planthoppers in a circulative and propagative manner, infect rice and cause serious yield losses to rice production in some areas of Asia (Hibino, 1996; Zhang et al., 2001; Zhou et al., 2008; Wang et al., 2010). The genome of RRSV consists of 10 dsRNA segments and encodes at least 10 proteins. Among them, P1, P2, P3, P4A, P5 and P8B are structural proteins, functioning as a putative guanylyl transferase, a capsid shell protein, a putative RNA-dependent RNA polymerase, a capping enzyme, and a major outer-capsid protein, respectively (Boccardo and Milne, 1984; Hagiwara et al., 1986; Supyani et al., 2007; Upadhyaya et al., 1998). P9 is the spike protein and has been shown to be important for virus transmission by the insect vector (Zhou et al., 1999). Pns6, Pns7 and Pns10 are non-structural proteins whose functions in viral replication are not well understood (Upadhyaya et al., 1997). However, recent studies have shown that Pns6 might be a movement protein and has silencing suppressor activities when expressed in *Nicotiana benthamiana* (Wu et al., 2010a; Wu et al., 2010b). While Pns7 is an NTP-binding protein (Spear et al., 2012), Pns10 appears to be the major constituent of viroplasm (Jia et al., 2012a). Pns10 can recruit Pns6 into the viroplasm-like structures when these two proteins are co-expressed in Sf9 cells (Jia et al., 2012a). The genome of SRBSDV also has 10 segments of dsRNA. Pns7-1 encoded by segment 7self-interacts and forms tubular structures in insect cells (Liuet al., 2011). Pns9-1 encoded by segment 9 is the major constituent of the viroplasm (Jia et al., 2012b). The functions of other proteins encoded by SRBSDV remain unknown. However, based on sequence comparisons with closely related fijiviruses, P1,P2, P3, P4, P8 and P10 of SRBSDV are believed to be structural proteins, functioning as a putative RNA-dependent RNA polymerase, a core protein, a putative capping enzyme, an outer-shell B-spike protein, a putative core and a major outer capsid protein, respectively (Zhang et al., 2001; Zhang et al., 2008; Zhou et al., 2008; Wang et al., 2010).In this paper, we report the identification and characterization of the interactions between viroplasm-associated proteins of RRSV and SRBSDV and a rice protein homologous to eukaryotic translation elongation factor 1A (eEF-1A). To our knowledge, this is the first paper reporting the interaction between plant-infecting reoviruses and the translation apparatus of their hosts.

2 Materials and methods

2.1 Plasmids construction

For viral genes used in the yeast two-hybrid experiments, the complete ORFs of the genes of interest were amplified from the corresponding plasmids containing nucleotide sequences of the SRBSDV Hubei isolate (GenBank accession numbers HM585270 to HM585279) and the RRSV SX isolate (GenBank accession numbers HM125560 to HM125569), respectively, using the primers listed in Table 1. The PCR products were cloned into pMD-18T (TaKaRa). The resulting plasmids were digested with restriction enzymes for which cleavage sites were present in the forward and reverse primers and then cloned into the vector pGBK-T7 or pGAD-T7 (Clontech, see text for naming of the plasmids).

For OseEF-1A used in the yeast two hybrid experiments, total RNA was isolated from rice using TRIzol

(Invitrogen), and reverse transcription was conducted using M-MLV reverse transcriptase (Takara), with oligo(dT) as the 3′ anchor primer. The primer pair Y5 was used to amplify the entire ORF of OseEF-1A from the cDNA.

For all the genes used in Agrobacterium-mediated transient expression, a Gateway-based cloning approach (Invitrogen) was employed to construct the respective vectors. Briefly, full-length ORFs of the genes of interest were cloned into the entry vector pDONR221 (Invitrogen) using BP Clonase II (Invitrogen) following the manufacturer's protocol, and the insert was then cloned into the destination vector using LR Clonase II (Invitrogen) to generate plant expression vectors. To fuse a protein with YFP, the vector pEarleyGate101 or pEarleyGate104 was used. To fuse a protein with CFP, the vector pEarleyGate102 was used. For bimolecular fluorescence complementation (BiFC), the vectors pEarleyGate201-YC and pEarleyGate202-YN were used. All of the plasmid constructs used in this study were verified by sequencing.

2.2 Yeast two-hybrid assay

Yeast two-hybrid experiments were performed as described previously (Lu et al., 2009). A yeast two-hybrid cDNA library with a titer of approximately 1.0×10^{11}cfu/ml was constructed from healthy, complete seedlings of rice cv. Wuyujing 3 using protocols from Clontech. A Matchmaker Gal4 Two-Hybrid System 3 (Clontech) was used to screen the rice cDNA library. Prior to their use, it was confirmed that the bait plasmids did not autonomously activate the reporter genes in yeast cells. For library screening, yeast AH109 cells were transformed together with the bait plasmid and the cDNA library plasmid using a simultaneous co-transformation protocol. Colonies were selected on SD/Leu-Trp-His-medium with 15mM 3-AT, and positive colonies were then isolated on SD/Leu-Trp-His-Ade-/X-α-gal + medium according to the Yeast Protocols Handbook of Clontech. Primary positive candidate plasmids containing the rice cDNAs were isolated and introduced into AH109 together with bait plasmid pGBKT7-RRSVPns6 to repeat the two-hybrid assay. The final positive candidate plasmids were selected, and the sequences of the inserts were determined. The sequences of positive colonies were subsequently compared to those in the NCBI database using the Basic Local Alignment Search Tool (BLASTn).

Table 1 Primers used in this study

Primer name	Construct	Primer sequences Forward	Reverse	Restriction enzyme sites
Y1	pGBKT7-RRSVPns6	5′-CGGAATTCATGCAGCTCTTCATAGTCAAAC-3′	5′-ACGCGTCGACTCAATCAAGCTCCTTACATTCAGG-3′	EcoRI/SalI
Y2	pGADT7-RRSVPns7	5′-GGCATATGATGGACGAGCTAACTTTATCCATTGG-3′	5′-CCGGATCCTCCCTCGACGGGAGGCCCAAC-3′	NdeI/BamHI
Y3	pGADT7-RRSVPns10	5′-CGGAATTCATGCCTTTCGTGCAATTCCCG-3′	5′-CCGGATCCCTACTCTGCGTCATCACCAAAG-3′	EcoRI/BamHI
Y4	pGADT7-SRBSDVPns6 pGBKT7-SRBSDVPns6	5′-GGAATTCCATATGATGTCTACCAACCTCACGAACAT-3′	5′-CGGGATCCTTACTCTGAACTAAGTTGCCACAA-3′	NdeI/BamHI
Y5	pGADT7-OseEF-1A pGBKT7-OseEF-1A	5′-GGAATTCCATATGATGGGTAAGGAGAAGACGCACA-3′	5′-CGGAATTCTCACTTCTTCTTGGCGGCAGCC-3′	NdeI/EcoRI
Y6	pGBKT7-SRBSDVPns71	5′-CATATGATGGATAGACCTGCTCGAGAACA-3′	5′-CTGCAGTTAAGATGATGGAGATTCAAAAAC-3′	NdeI/PstI
Y7	pGBKT7-SRBSDVPns91	5′-GGCATATGATGGCAGACCTAGAGCGTAG-3′	5′-GGGAATTCTCAGACGTCCAATTTAAGTG-3′	NdeI/EcoRI
Y8	pGBKT7-SRBSDVPns92	5′-GGCATATGATGAACCCACAGTCTTCAGT-3′	5′-CCGGATCCTTAGTGAAACAAAGTATAAT-3′	NdeI/BamHI

2.3 Agrobacterium-mediated transient expression and microscopy

Transient protein expression was performed essentially as described (Sparkes et al., 2006). Briefly, the vectors described above were introduced into *Agrobacterium tumefaciens* strain EHA105 by transformation using the freeze-thaw method. The transformants were grown for 36–48h at 28℃ on Luria-Bertani agar containing rifampicin and kanamycin. Positive colonies were isolated, verified by PCR, and then cultivated. Agrobacterium suspensions were pelleted at 3000 rpm. The pellets were resuspended and diluted with infiltration medium to a final OD_{600} value of approximately 0.15 and incubated at room temperature for 3–5h before leaf infiltration (Hamilton et al., 2002; Sparkes et al., 2006). For co-infiltration or BiFC, equal volumes of individual Agrobacterium cultures were mixed prior to infiltration. The fluorescence was observed and imaged at room temperature using a Leica TCS SP2 inverted confocal microscope with a 63× oil immersion objective. CFP was excited at 458 nm, and the emitted light was captured at 440 to 470nm; YFP was excited at 514nm, and the emitted light was captured at 525 to 650nm. Images were captured digitally and handled using the Leica LCS software. Post-acquisition image processing was done with Adobe Photoshop 5.0 software.

3 Results

3.1 Interaction of RRSV Pns6 with eEF-1A of rice in yeast

A galactosidase 4 (Gal4)-based yeast two-hybrid system was employed to identify potential host partners of RRSV Pns6. The entire ORF of Pns6 was inserted in frame into pGBK-T7. The resulting plasmid, named pGBKT7-RRSVPns6, was introduced together with the activation plasmids containing a rice cDNA library into the yeast strain AH109 by transformation as described previously (Lu et al., 2009). Forty-five co-transformants were found to be able to grow on plates containing synthetic dropout medium lacking Trp, His and Leu (SD/Leu-Trp-His-), and more than 80% of these colonies could grow and turned blue on SD/Leu-Trp-His-Ade-/X-α-gal + (data not shown). The positive candidate plasmids containing the rice cDNAs were then isolated and sequenced. BLAST analysis showed that the sequences of six cDNA inserts were identical to the C-terminal part of the rice cDNA encoding a protein homologous to eEF-1A.

The full-length ORF of the rice protein (Os03g0177400), designated OseEF-1A in this study, was then cloned in frame into the activation vector pGAD-T7. The resulting plasmid, named pGADT7-OseEF-1A, was introduced together with pGBKT7-RRSVPns6 into the yeast strain AH109 by transformation. The co-transformants were plated on conditional media containing SD/Trp-Leu-, SD/Leu-Trp-His- and SD/ Leu-Trp-His-Ade-/X-α-gal+, respectively. As shown in Fig. 1A, co-transformants of pGBKT7-RRSVPns6 and pGADT7-OseEF-1A grew and turned blue on SD/Leu-Trp-His-Ade-/X-α-gal+, similar to those co-transformed with pGADT7-T and pGBKT7-p53, which were used as positive controls. However, the co-transformants of pGBKT7/ pGADT7, pGBKT7-OseEF-1A/pGADT7 or pGBKT7/ pGADT7-OseEF-1A failed to grow on SD/Leu-Trp-His-, although they grew well on SD/Trp-Leu- (data not shown). These results indicated that Pns6 of RRSV interacts with OseEF-1A in yeast.

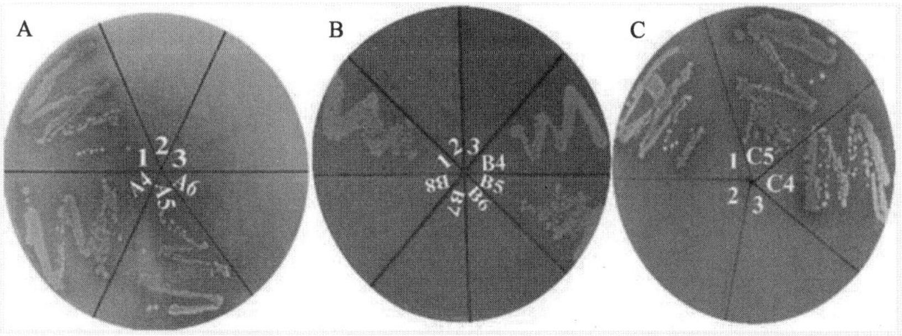

Fig.1 Yeast two-hybrid assay showing protein-protein interactions between RRSV Pns6, non-structural proteins of SRBSDV and OseEF1A
1, pGADT7-T/pGBKT7-p53 (positive control); 2,pGADT7-T/pGBKT7-Lam (negative control);3,pGADT7/pGBKT7;A4and B4, pGADT7-OseEF-1A/pGBKT7-RRSVPns6; A5, pGADT7-RRSVPns10/pGBKT7-OseEF-1A; B5, pGADT7-OseEF- 1A/pGBKT7-SRBSDVPns6; A6, pGADT7-RRSVPns7/pGBKT7-OseEF-1A;B6,pGADT7-OseEF-1A/pGBKT7-SRBSDVPns7-1;B7,pGADT7-OseEF-1A/pGBKT7-SRBSDVPns9-1;B8,pGADT7-OseEF-1A/pGBKT7-SRBSDVPns9-2;C4,pGADT7-SRBSDVPns6/ pGBKT7-SRBSDVPns6; C5, pGADT7-SRBSDVPns6/pGBKT7-SRBSDVPns9-1.Yeast co-transformants were grown on the selective medium SD/-Ade/-His/-Leu/-Trp plus X-a-Gal and incubated at 30℃ for 4d before photography

3.2 Interaction of RRSV Pns6 with eEF-1A of rice in plant cells

A bimolecular fluorescence complementation (BiFC) assay was carried out to confirm the interaction of Pns6 and eEF-1A in plant cells (Walter et al., 2004). The full-length ORFs of Pns6 and OseEF-1A were cloned separately into the GATEWAY-based recombination vectors pEarleyGate201-YC and pEarleyGate202-YN to generate plant expression vectors named as RRSVPns6-YC and OseEF-1A-YN, respectively (Tian et al., 2011). Cell cultures of the transformed *Agrobacterium tumefaciens* strain EHA105 carrying the two constructs were mixed and infiltrated into leaves of *N. benthamiana*. Yellow fluorescent protein (YFP) fluorescence was observed using a confocal microscope 2d post-infiltration(dpi). As presented in Fig. 2, strong YFP fluorescence was found in leaf epidermal cells of *N. benthamiana* infected with *A.tumefaciens* harboring RRSVPns6-YC and OseEF-1A-YN. Similar results were obtained when Pns6 and OseEF-1A were fused with the N- and C-terminal fragments of YFP, respectively (data not shown). By contrast, YFP fluorescence was never detected in leaf cells co-infiltrated with YN/YC, RRSVPns6-YC/YN or YC/OseEF-1A-YN (data not shown).

3.3 Interaction and co-localization of RRSV Pns10 and OseEF-1A

The interaction between OseEF-1A and two other non-structural proteins of RRSV, Pns7 and Pns10 was tested by yeast two-hybrid experiments similar to those described above. The results showed that Pns10, the viroplasm matrix protein of RRSV (Jia et al., 2012a), could also interact with OseEF-1A (Fig.1A). The interaction was confirmed by BiFC (Fig. 3). When expressed together, Pns10 co-localized with OseEF- 1A in epidermal cells of *N. benthamiana* (Fig. 3).

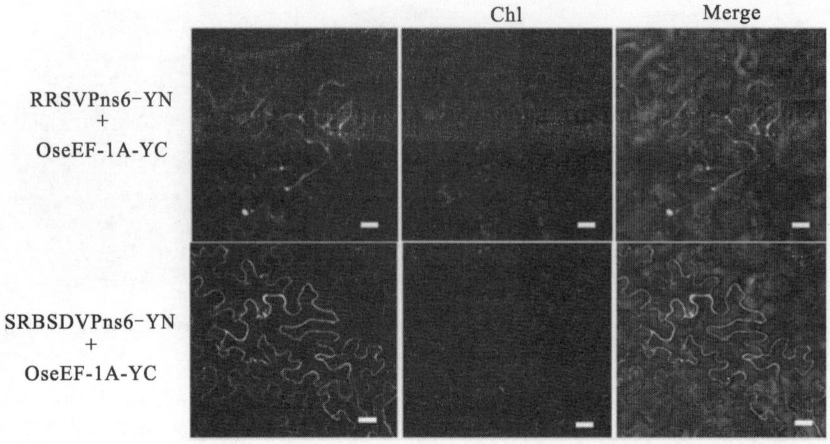

Fig. 2 Bimolecular fluorescence complementation (BiFC) showing *in planta* interactions between SRBSDV/ RRSV Pns6 and OseEF1A

Leaves of *Nicotiana benthamiana* were infiltrated with Agrobacterium EHA105 harboring the constructs indicated above. Fluorescence in epidermal cells of the infiltrated area was observed 3d after infiltration. Bars, 100μm. Chl, chloroplast autofluorescence

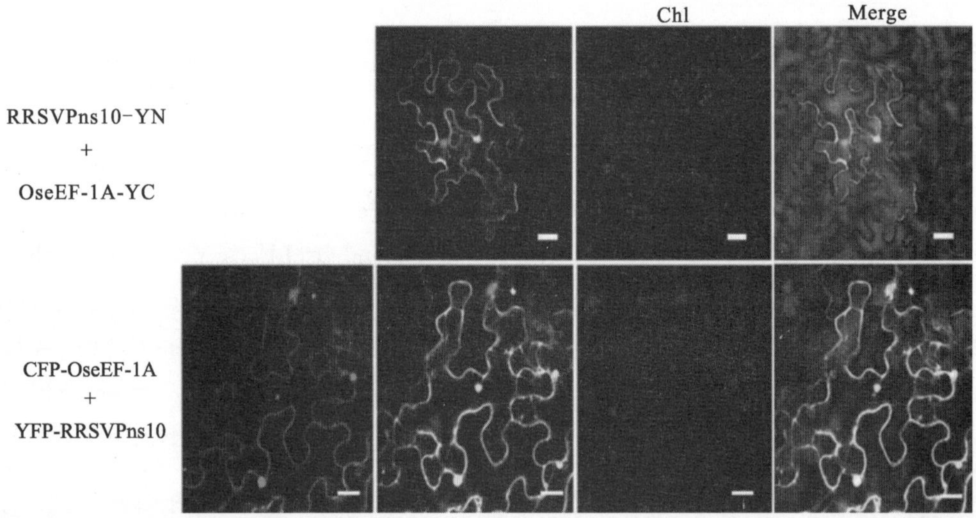

Fig.3 Bimolecular fluorescence complementation (BiFC) showing *in planta* interactions between OseEF-1A and RRSV Pns10 and cellular co-localization studies of OseEF-1A with RRSV Pns10

Leaves of *Nicotiana benthamiana* were infiltrated with *Agrobacterium* EHA105 harboring the constructs indicated to the left. Fluorescence in epidermal cells of the infiltrated area was observed 3d after infiltration. Bars, 100μm. Chl, chloroplast autofluorescence. Red arrows indicate co-localization of OseEF-1A and Pns6 or Pns10

3.4 Interaction between SRBSDV Pns6 and OseEF-1A

We tested whether the interaction with OseEF-1A is conserved in other plant-infecting reoviruses. The interaction of OseEF-1A with four putative non-structural proteins of SRBSDV, Pns6, Pns7-1, Pns9-1, and Pns9-2 (Zhang et al., 2008; Zhou et al., 2008; Wang et al., 2010) was investigated by yeast two-hybrid experiments. The results showed that Pns6 was the only non-structural SRBSDV protein that interacted with OseEF-1A (Fig. 1B). The interaction between SRBSDV Pns6 and OseEF-1A was confirmed by the BiFC experiments described above (Fig. 2). It should be noted that the Pns6 of SRBSDV shares no significant sequence similarity with that of RRSV.

3.5 Co-localization of OseEF-1A and SRBSDV Pns9 in the presence of Pns6

It has been reported that Pns6 of rice black-streaked dwarf virus (RBSDV), which shows high sequence similarity to SRBSDV Pns6, self-interacts to form punctate, viroplasm-like structures in the cytoplasm and recruits Pns9-1, the viroplasm matrix protein of RBSDV (Akita et al., 2012;Wang et al., 2011;Zhang et al., 2008;Zhang et al., 2008). If SRBSDV Pns6 shares this property with RBSDV, OseEF-1A may also associate with viroplasms of SRBSDV, although it does not interact with Pns9-1 directly. Experiments were conducted to test this hypothesis. First, we investigated whether Pns6 and Pns9-1 of SRBSDV were functionally similar to those of RBSDV. As shown in Fig.1C, Pns6 interacted with itself and with Pns9-1 in yeast. When expressed alone as a fusion protein with YFP or CFP, both Pns6 and Pns9-1 formed punctate structures in epidermal cells of *N. benthamiana* (Fig.4). When expressed together, the small punctate structures formed by the two proteins co-localized with each other. These results indicate that as in RBSDV, SRBSDV Pns6 and Pns9-1 are viroplasm-associated proteins (Wang et al., 2011).

Next, OseEF-1A was expressed together with Pns6 or and Pns9-1 of SRBSDV in epidermal cells of *N. benthamiana*. As presented in Fig. 5, OseEF-1A co-localized with Pns6, but not Pns9-1, when the two corresponding proteins were expressed together. However, when the three proteins were expressed together, co-localization of OseEF-1A and Pns9-1 was observed (Fig. 5).

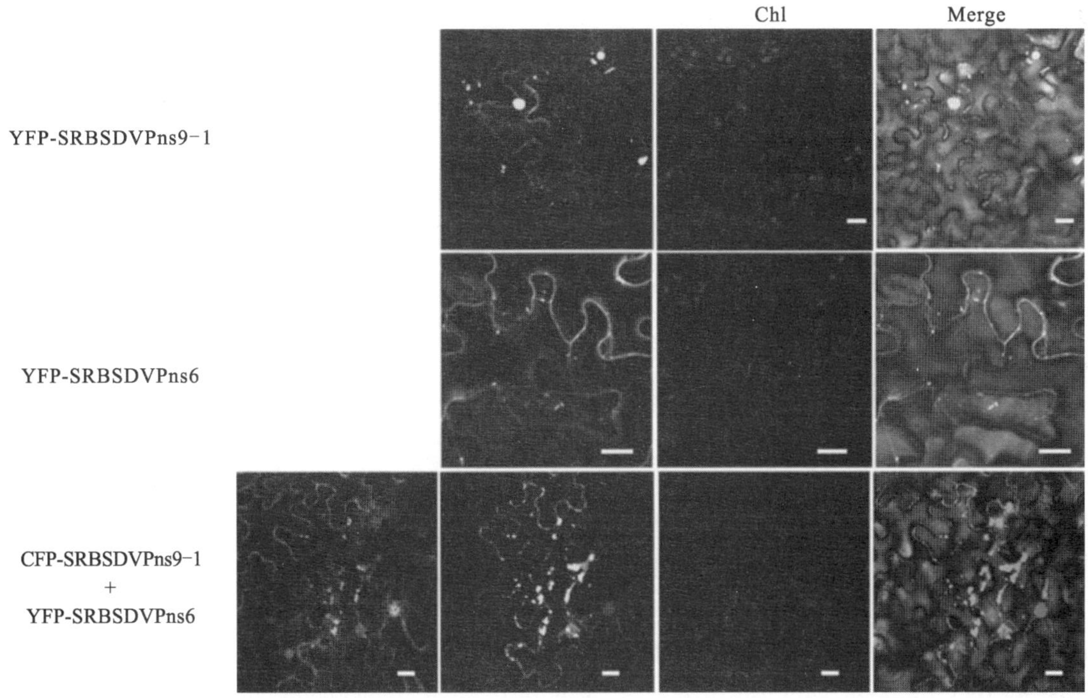

Fig.4 Cellular localization studies of SRBSDV Pns6 or/and Pns9-1: Leaves of *Nicotiana benthamiana* were infiltrated with *Agrobacterium* EHA105 harboring the constructs indicated to the left

Fluorescence in epidermal cells of the infiltrated area was observed 3d after infiltration. Bars, 100μm. Chl, chloroplast autofluorescence. Red arrows indicate co-localization of Pns6 and Pns9-1

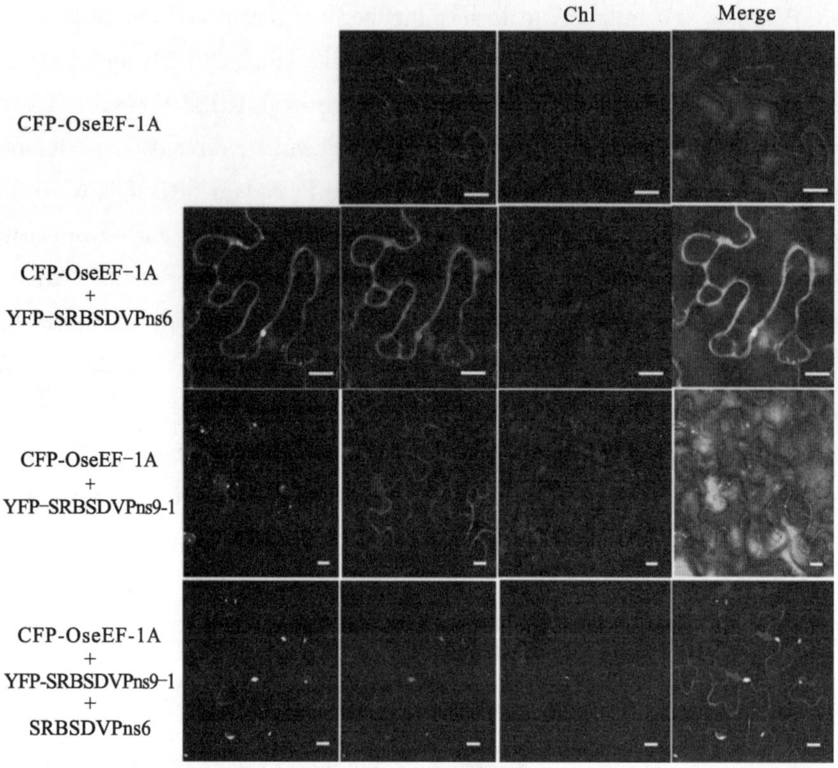

Fig.5 Cellular localization studies of OseEF-1A alone or together with SRBSDV Pns6 or/and SRBSDV Pns9-1

Leaves of *Nicotianabenthamiana* were infiltrated with *Agrobacterium* EHA105 harboring the constructs indicated to the left. Fluorescence in epidermal cells of the infiltrated area was observed 3d after infiltration. Bars, 100μm. Chl, chloroplast autofluorescence. Red arrows indicate co-localization of OseEF-1A and Pns9-1

4 Discussion

In this paper, we report the interaction between two reoviruses and a rice protein with a putative function in translation elongation (Macdonald, 2001; Gebauer and Hentze, 2004). Many viruses other than reoviruses, such as human immunodeficiency virus (HIV), bovine viral diarrhoea virus (BVDV), tobacco mosaic virus(TMV), turnip mosaic virus (TuMV), hepatitis C virus (HCV), vesicular stomatitis virus (VSV), and some herpesviruses, have been shown to encode one to several proteins that interact with host factors regulating translation elongation (Walsh and Mohr, 2011). Our observation that a translation elongation factor of rice is targeted by two different reoviruses is consistent with these findings, showing that elongation is a phase frequently targeted by viruses (Walsh and Mohr, 2011).

At present, the biological significance of the interactions identified in this study is unclear. ICP0 of herpes simplex virus 1, Tat and Gag of HIV, and NS4A of HCV, which interact with specific translation elongation factors, have been shown to inhibit translation in different in vitro systems (Kawaguchi et al., 1997; Xiao et al., 1998; Kou et al., 2006). Given the abundance of data showing that reoviruses induce shut-off of host protein synthesis (for example in ref. 13), we propose that the interaction with OseEF-1A might be a mechanism used by RRSV and SRBSDV to inhibit translation. In the case of Tat of HIV, it was found that its interaction with EF-1A may specially influence the efficiency of the translation of cellular, but not viral, mRNAs (Xiao et al., 1998). The translation inhibition by RRSV and SRBSDV may also be host specific, as this may help viral mRNAs compete for the cellular translation machinery. However, the inhibition might also be used by viruses for other purposes. For example, primary transcripts of reoviruses function as both mRNAs for protein translation and as templates for genome replication. Replication must be coupled with translation. Therefore, besides the inhibition of host protein synthesis, the inactivation of OseEF-1A might be a mechanism used by the two viruses to switch their primary transcripts from translation to replication.

It is interesting to note that both RRSV and SRBSDV use viroplasm-associated proteins to interact with OseEF-1A. Viroplasms of reoviruses can be considered as physically isolated small compartments in the cytoplasm. Therefore, the interaction of viroplasm-associated proteins with OseEF-1A may sequester the rice protein to sites not accessible by its cellular partners. This may result in inactivation of OseEF-1A. It is at present unclear why RRSV encodes one more protein than SRBSDV to interact with OsEF-1A. However, our unpublished results showed that RRSV Pns6 and Pns10 did not co-localize with each other in cells of *N. benthamiana*, although they did in insect cells (Jiaet al., 2012a). This may mean that the interaction of Pns6 and OsEF-1A alone is not enough to direct OsEF-1A to viroplasms in plant cells. This implies that the two Pns6 proteins have distinct functions, although both of them interact with OseEF-1A. Consistent with this, our studies show that thc two protcins havc different cellular distribution patterns ((Wu et al., 2010a) and this study).

It should be noted that eEF-1A has many other functions possibly unrelated to translation, such as cytoskeletal organization, apoptosis, and protein degradation (Mateyak and Kinzy, 2010). These functions can also be targeted or used by viruses. For example, translation elongation factors including homologues of eEF-1A have been shown to be components of, and play a role in, the replicase complexes of diverse viruses (Das et al., 1998; Yamaji et al., 2006; Thivierge et al., 2008; Li et al., 2010; Sasvari et al., 2011; Warren et al., 2012). It seems that the RNA binding activities of these translation elongation factors are used by these viruses (Li et al., 2010; Sasvari et al., 2011). Viroplasms are sites where reoviruses replicate and assemble themselves. It will be interesting to investigate whether OseEF-1A is recruited to function in some processes related to viral replication.

Acknowledgements

This work was supported by National Basic Research Program of China (No.2010CB126203), the National Natural Science Foundation of China (Nos. 30970135) and Key Project of the National Research Program of China (2012BAD19B03).

References

[1] AKITA F, HIGASHIURA A, SHIMIZU T, et al. Crystallographic analysis reveals octamerization of viroplasm matrix protein Pns9-1 of Rice black streaked dwarf virus[J]. Journal of Virology, 2012, 86(2):746-756.

[2] ATTOUI H, MERTENS P P C, BECNEL J, et al. Reoviridae. In: AMQ King, MJ Adams, EB Carstens, EJ Lefkowitz (eds) Virus Taxonomy: Ninth Report of the International Committee on Taxonomy of Viruses[J]. Academic Press, New York, 2012, 541-630.

[3] BCCARDO G. Plant reovirus group[J]. Cmi/aab Description of Plant Viruses, 1984, 294.

[4] CHULU J L C, HUANG W R, WANG L, et al. Avian Reovirus Nonstructural Protein p17-Induced G(2)/M Cell Cycle Arrest and Host Cellular Protein Translation Shutoff Involve Activation of p53-Dependent Pathways[J]. Journal of Virology, 2014, 88(3):1856-1856.

[5] DAS T, MATHUR M, GUPTA A K, et al. RNA Polymerase of Vesicular Stomatitis Virus Specifically Associates with Translation Elongation Factor-1 α β γ for Its Activity[J]. Proceedings of the National Academy of Sciences of the United States of America,

1998, 95(4):1449-1454.

[6] GEBAUER F, HENTZE M W. Molecular mechanisms of translational control[J]. Nature Reviews Molecular Cell Biology, 2004, 5(10):827-835.

[7] HAGIWARA K, MINOBE Y, NOZU Y, et al. Component proteins and structures of rice ragged stunt virus[J]. Journal of General Virology, 1986, 67:1711-1715.

[8] HAMILTON A, VOINNET O, CHAPPELL L, et al. Two classes of short interfering RNA in RNA silencing[J]. The EMBO Journal, 2002, 21(17):4671-4679.

[9] HIBINO H. Biology and epidemiology of rice viruses[J]. Annual Review of Phytopathology, 1996, 34(1):249-274.

[10] JI W T, WANG L, LIN R C, et al. Avian reovirus influences phosphorylation of several factors involved in host protein translation including eukaryotic translation elongation factor 2 (eEF2) in Vero cells[J]. Biochemical & Biophysical Research Communications, 2009, 384(3):301-305.

[11] JIA D, GUO N, CHEN H, et al. Assembly of the viroplasm by viral non-structural protein Pns10 is essential for persistent infection of rice ragged stunt virus in its insect vector[J]. Journal of General Virology, 2012a, 93(10):2299-2309.

[12] JIA D, CHEN H, ZHENG A, et al. Development of an Insect Vector Cell Culture and RNA Interference System To Investigate the Functional Role of Fijivirus Replication Protein[J]. Journal of Virology, 2012b, 86(10):5800-5807.

[13] KATZE M G. Translational Control in Cells Infected with Influenza Virus and Reovirus[J].Cold Spring Harb Monogr Archive, 1996, 30: 607-630.

[14] KAWAGUCHI Y, BRUNI R, ROIZMAN B. Interaction of herpes simplex virus 1 alpha regulatory protein ICP0 with elongation factor 1 delta: ICP0 affects translational machinery[J]. Journal of Virology, 1997, 71(2):1019-1024.

[15] KOU Y H, CHOU S M, WANG Y M, et al. Hepatitis C virus NS4A inhibits cap-dependent and the viral IRES-mediated translation through interacting with eukaryotic elongation factor 1A[J]. Journal of Biomedical Science, 2006, 13(6):861-874.

[16] LI Z, POGANY J, TUPMAN S, et al. Translation Elongation Factor 1A Facilitates the Assembly of the Tombusvirus Replicase and Stimulates Minus-Strand Synthesis[J]. PLOS Pathogens, 2010, 6(11):e1001175.

[17] LIU Y, JIA D, CHEN H,et al. The P7-1 protein of southern rice black-streaked dwarf virus, a fijivirus, induces the formation of tubular structures in insect cells[J]. Archives of Virology, 2011, 156(10):1729-1736.

[18] LLOYD R M, SHATKIN A J. Translational stimulation by reovirus polypeptide sigma 3: substitution for VAI RNA and inhibition of phosphorylation of the alpha subunit of eukaryotic initiation factor 2[J]. Journal of Virology, 1992, 66(12):6878-6884.

[19] LU L, DU Z, QIN M, et al. Pc4, a putative movement protein of Rice stripe virus, interacts with a type I DnaJ protein and a small Hsp of rice[J]. Virus Genes, 2009, 38(2):320-327.

[20] MACDONALD P. Diversity in translational regulation[J]. Current Opinion in Cell Biology, 2001, 13(3):326-331.

[21] MATEYAK M K, KINZY T G. eEF1A: Thinking Outside the Ribosome[J]. Journal of Biological Chemistry,2010, 285: 21209-21213.

[22] MOHR I, SONENBERG N. Host Translation at the Nexus of Infection and Immunity[J]. Cell Host & Microbe, 2012, 12(4):470-483.

[23] PIRON M, VENDE P, COHEN J, et al. Rotavirus RNA-binding protein NSP3 interacts with eIF4GI and evicts the poly(A) binding protein from eIF4F[J]. Embo Journal, 1998, 17(19):5811-5821.

[24] SASVARI Z, IZOTOVA L, KINZY T G, et al. Synergistic Roles of Eukaryotic Translation Elongation Factors 1B gamma and 1A in Stimulation of Tombusvirus Minus-Strand Synthesis[J]. PLOSPathogens, 2011, 7(12):e1002438.

[25] SPARKES I A, RUNIONS J, KEARNS A,et al. Rapid, transient expression of fluorescent fusion proteins in tobacco plants and generation of stably transformed plant[J]. Nature Protocols, 2006, 1(4):2019-2025.

[26] SPEAR A, SISTERSON M S, STENGER D C. Reovirus genomes from plant-feeding insects represent a newly discovered lineage within the family Reoviridae[J]. Virus Research, 2012, 163(2):503-511.

[27] SUPYANI S, HILLMAN B I, SUZUKI N. Baculovirus expression of the 11 mycoreovirus-1 genome segments and identification of the guanylyl transferase-encoding segment[J]. Journal of General Virology, 2007, 88(1):342-350.

[28] THIVIERGE K, COTTON S, DUFRESNE P J, et al. Eukaryotic elongation factor 1A interacts with Turnip mosaic virus RNA-dependent RNA polymerase and VPg-Pro in virus-induced vesicles[J]. Virology, 2008, 377(1):216-225.

[29] TIAN G, LU Q, ZHANG L, et al. Detection of protein interactions in plant using a gateway compatible bimo- lecular fluorescence complementation (BiFC) system[J]. Jove-journal of Visualized Experiments, 2011(55):3473.

[30] UPADHYAYA N M, RAMM K, GELLATLY J A, et al. Rice ragged stunt oryzavirus genome segments S7 and S10 encode non-structural proteins of M-r 68025 (Pns7)and M-r 32 364 (Pns10)[J]. Archives of Virology, 1997, 142(8):1719-1726.

[31] UPADHYAYA N M, RAMM K, GELLATLY J A, et al. Rice ragged stunt oryzavirus genome segment S4 could encode an RNA dependent RNA polymerase and a second protein of unknown function[J]. Archives of Virology, 1998, 143(9):1815-1822.

[32] WALSH D, MOHR I. Viral subversion of the host protein synthesis machinery[J]. Nature Reviews Microbiology, 2011, 9(12):860-875.

[33] WALTER M, CHABAN C, SCHUTZE K,et al. Visualization of protein interactions in living plant cells using bimolecular fluorescence complementation[J]. Plant Journal, 2010, 40:428-438.

[34] WANG Q, TAO T, ZHANG Y, et al. Rice black-streaked dwarf virus P6 self-interacts to form punctate, viroplasm-like structures in the cytoplasm and recruits viroplasm-associated protein P9-1[J]. Virology Journal, 2011, 8(1):24.

[35] WANG Q, YANG J, ZHOU G H, et al. The Complete Genome Sequence of Two Isolates of Southern rice black-streaked dwarf virus, a New Member of the Genus Fijivirus[J]. Journal of Phytopathology, 2010, 158:733-737.

[36] WARREN K, WEI T, LI D, et al. Eukaryotic elongation factor 1 complex subunits are critical HIV-1 reverse transcription cofactors[J]. Proceedings of the National Academy of Sciences of the United States of America, 2012, 109(24):9587-9592.

[37] WU J, DU Z, WANG C, et al. Identification of Pns6, a putative movement protein of RRSV, as a silencing suppressor[J]. Virology Journal, 2010a, 7(1):335.

[38] WU Z, WU J, ADKINS S, et al. Rice ragged stunt virus segment S6-encoded nonstructural protein Pns6 complements cell-to-cell movement of Tobacco mosaic virus-based chimeric virus[J]. Virus Research, 2010b, 152(1-2):176-179.

[39] XIAO H, NEUVEUT C, BENKIRANE M, et al. Interaction of the second coding exon of Tat with human EF-1 delta delineates a mechanism for HIV-1-mediated shut-off of host mRNA translation[J]. Biochemical and Biophysical Research Communication, 1998, 244(2):384-389.

[40] YAMAJI Y, KOBAYASHI T, HAMADA K, et al. In vivo interaction between Tobacco mosaic virus RNA-dependent RNA polymerase and host translation elongation factor 1A[J]. Virology, 2006, 347(1):100-108.

[41] ZHANG C, LIU Y, LIU L,et al. Rice black streaked dwarf virus P9-1, an a-helical protein, self-interacts and forms viroplasms in vivo[J]. Journal of General Virology, 2008, 89(7):1770-1776.

[42] ZHANG H M, YANG J, CHEN J P, et al. A black-streaked dwarf disease on rice in China is caused by a novel fijivirus[J]. Archives of Virology, 2008, 153(10):1893-1898.

[43] ZHANG H M, CHEN J P, ADAMS M J. Molecular characterisation of segments 1 to 6 of Rice black-streaked dwarf virus from China provides the complete genome e[J]. Archives of Virology, 2001, 146(12):2331-2339.

[44] ZHOU G, WEN J, CAI D, et al. Southern rice black-streaked dwarf virus: A new proposed Fijivirus species in the family Reoviridae[J]. Chinese Science Bulletin, 2008, 53(23):3677-3685.

[45] ZHOU G, LU X, LU H, et al. Rice ragged stunt oryzavirus: role of the viral spike protein in transmission by the insect vector[J]. Annals of Applied Biology, 1999, 135: 573-578.

水稻黑条矮缩病毒在灰飞虱消化系统的侵染和扩散过程

贾东升，马元元，杜雪，陈红燕，谢联辉，魏太云

（福建省植物病毒学重点实验室，福建农林大学植物病毒研究所，福州 350002）

摘要：水稻黑条矮缩病毒（*Rice black-streaked dwarf virus*，RBSDV）由介体灰飞虱（*Laodelphaxstriatellus* Fallén）以持久增殖型方式传播，其编码的 P9-1 蛋白是形成病毒复制和子代病毒粒体装配的场所——病毒原质（viroplasm）的组分之一。为了明确 RBSDV 在介体昆虫体内的侵染循回过程，本研究通过原核表达的 P9-1 蛋白免疫注射兔子制备 P9-1 抗体，应用免疫荧光标记技术研究 P9-1 在饲毒后不同时期的介体灰飞虱体内的定位。共聚焦显微镜观察到饲毒后 3d，P9-1 出现在介体中肠的少数上皮细胞内；饲毒后 6d，在中肠外表的肌肉细胞分布有 P9-1；饲毒后 10d，P9-1 分布于中肠和后肠表面的肌肉，同时在唾液腺也能观察到 P9-1 的存在。结果表明 RBSDV 在介体灰飞虱体内首先侵染中肠上皮细胞并复制，随后扩散到中肠表面的肌肉细胞，并通过环肌和纵肌扩散到中肠和后肠，最后扩散到唾液腺。本研究首次直观地阐述了 RBSDV 在灰飞虱消化系统的侵染和扩散过程，为有效阻断灰飞虱携带并传播病毒奠定基础。

关键词：水稻黑条矮缩病毒；P9-1；灰飞虱；侵染途径

中图分类号：S432.4　**文献标识码：**A　**文章编号：**0412-0914(2014)02-0188-07

Infection and spread of Rice black-streaked dwarf virus in the digestive system of its insect vector small brown planthopper

Dongsheng Jia, Yuanyuan Ma, Xue Du, Hongyan Chen, Lianhui Xie, Taiyun Wei

(Fujian Province Key Laboratory of Plant Virology,Institute of Plant Virology, Fujian Agriculture and Forestry University,Fuzhou 350002,China)

Abstract: *Rice black-streaked dwarf virus* (RBSDV), a *fijivirus*, is transmitted by small brown planthopper (SBPH) in persistent propagative manner. P9-1 encoded by segment 9 is the minimal viral factor required for viroplasm formation and virus replication during RBSDV infection. In order to elucidate the infection route of RBSDV within its insect vector SBPH after viral acquisition, the antibodies against P9-1 was prepared. Then, the distribution of P9-1 in the digestive system of SBPHs after ingestion of RBSDV was investigated by immunofluorescence microscopy with antibodies against P9-1 conjugated to FITC directly. At 3d post-acquire virus (p.a.v.), P9-1 was first detected in the epithelial cells of the midgut. At 6d p.a.v., P9-1 proceeded to the visceral muscles surrounding the midgut epithelial cells; then distributed the visceral muscles of the midgut and hindgut. At 10d p.a.v., P9-1 distributed throughout in the salivary glands. These results indicated that RBSDV first infected in the midgut epithelium, then proceeded to the visceral muscles surrounding the midgut and hindgut, and finally accumulated into the salivary

植物病理学报 .2014, 44(2): 188-194

收稿日期：2012-12-30

基金项目：国家公益性行业（农业）科研专项 (201003031)；国家自然科学基金重点项目 (3113044)；福建省杰出青年科学基金项目 (2011J06008)；教育部高等学校博士学科点专项科研基金 (20113515130002)。

glands. This is the first report about the infection and spread of RBSDV in its insect vector SBPH, which could give support to develop effective measures for blocking the transmission of RBSDV by SBPH vector.

Key words: *Rice black-streaked dwarf virus*; P9-1 protein; small brown planthopper; infection route

水稻黑条矮缩病毒（*Rice black-streaked dwarf virus*, RBSDV）属于呼肠孤病毒科（Reoviridae），与玉米粗缩病毒（*Maize rough dwarf virus*）、斐济病毒（*Fiji disease virus*）和南方水稻黑条矮缩病毒（*Southern rice black-streaked dwarf virus*, SRBSDV）等同属于斐济病毒属（*Fijivirus*）（Zhou et al., 2008; King et al., 2011）。RBSDV 可以侵染水稻、玉米和小麦等禾本科植物（Zhang et al., 2001a; Bai et al., 2002），近年来主要在我国的江苏、浙江等省发生，给农业生产带来巨大的威胁（Yang et al., 2007; Zhou et al., 2010）。RBSDV 在水稻上表现的症状为叶色浓绿、植株矮小，茎秆和叶背长有白色短条瘤状突起，瘤状突起在发病后期呈黑褐色（吕永平等, 2002）。RBSDV 主要由介体灰飞虱（*Laodelphaxstriatellus*，fallen）以持久增殖型方式传播（Chen and Zhang, 2005），其基因组由 10 条双链 RNA 组成，2001 年 Zhang 等（2001）首次完成全基因组序列测定，全基因组一共有 29410 个碱基，共编码 7 个结构蛋白和 6 个非结构蛋白。Isogai 等（1998）通过 Western blot 证明 S8 和 S10 编码的蛋白分别是病毒粒子的核心衣壳蛋白 P8 和外层衣壳蛋白 P10，通过免疫电镜观察到 S7 片段编码的 P7-1 蛋白抗体标记在包裹病毒粒子的管状结构上。Zhang 等（2005）发现 S6 编码的非结构蛋白 P6 有沉默抑制子功能，Wang 等（2011）证明 P6 可以自身互作且在植物细胞中形成类似于病毒原质的内含体，推测其是病毒原质的组分之一。

此外对 RBSDV 研究最多的是 S9 片段编码的 P9-1 蛋白的功能，早在 1998 年 Isogai 等（1998）通过电镜观察到 P9-1 蛋白是病毒原质（viroplasm）的组分；随后 Zhang 等（2008）证明非结构蛋白 P9-1 可以自身互作并在拟南芥原生质体中表达形成类似于病毒原质的包含体；Akita 等（2012）在昆虫细胞 Sf9 中也观察到类似于病毒原质的内含体，并明确了 P9-1 蛋白首先通过疏水区的互作形成二聚体，再由每个二聚体的 C 端伸向邻近的二聚体并通过疏水区互作形成四聚体。Shimizu 等（2011）应用 RNA 干扰（RNA interference, RNAi）原理获得的表达 P9-1 基因 dsRNA 转基因水稻可以有效地抑制病毒的侵染，证明 P9-1 蛋白确实与病毒的复制有关。这些研究均表明 P9-1 蛋白是 RBSDV 的病毒原质组分，参与病毒的复制和装配。但到目前为止，关于 RBSDV 在介体灰飞虱体内的复制增殖过程仍不明确，本研究通过原核表达的 P9-1 蛋白制备 P9-1 蛋白的多克隆抗体，随后应用免疫荧光标记技术系统观察了病毒侵染不同时间 P9-1 蛋白在灰飞虱体内的定位，进一步明确了 RBSDV P9-1 蛋白在病毒复制过程中的作用机理。

1 材料与方法

1.1 病毒与介体灰飞虱

RBSDV 病株采自江苏省建湖县，隔离种植于本实验室田间病毒圃。灰飞虱成虫采自福建农林大学水稻育种实验田，于温室中 25℃条件下经过无毒虫筛选和多代繁殖后饲养在水稻幼苗上。将成虫期的灰飞虱放入健康水稻幼苗上产卵 3d，12d 后水稻幼苗上孵出若虫，生长到 2 龄的若虫用于饲毒实验。

1.2 Gateway 重组技术构建原核表达载体

根据 RBSDV 第 9 条链的基因序列（GenBank No: AF540976.1），设计扩增 P9-1 基因完整开放阅读框的引物序列（正体），引物 5′ 端均含有重组序列（斜体），P9-1F:5′-GGGGACAAGTTTGTACAAAAAAGCAGGCTTCATGGCAGACCAAGAGCGGAG-3′；P9-1F:5′-GGGGACCACTTTGTACAAGAAAGCTGGGTCAACGTCCAGTTTCAAGGAGGAG-3′。以带毒灰飞虱总 RNA 为模板，用 P9-1F/R 引物进行 RT-PCR

获得目的基因片段。参照 Gateway BP Clonase Enzyme Mix 试剂盒（Invitrogen）说明，将目的基因片段重组到入门载体 pDONR221。构建好的 pDONR221-P9-1 载体经测序正确后，用于重组构建表达载体。参照 Gateway LR Clonase Enzyme Mix 试剂盒（Invitrogen）说明书将 pDONR221-P9-1 与目的表达载体 pDEST17 进行 LR 重组反应，获得原核表达载体 pDEST17-P9-1。

1.3 抗体制备

将 RBSDV-P9-1 的原核表达载体 pDEST17-P9-1 转入大肠杆菌 Rocceta 菌株，经 IPTG 诱导表达目的蛋白。通过 SDS-PAGE 凝胶电泳后回收目的蛋白条带，并用弗氏佐剂研磨，免疫注射新西兰大白兔的后退肌肉。免疫注射 5 次，动脉取血，获得 P9-1 蛋白的多克隆抗血清。应用 protein A-Sepharose affinity column（Thermo）提纯抗血清获得 immunoglobulin G（IgG），并按照 Invitrogen 说明书将 IgG 与 FITC 荧光素（Invitrogen）交联得到 P9-1-FITC 抗体。

1.4 Western blot 检测

抓取约 100 头人工饲养的 2 龄无毒灰飞虱若虫在 RBSDV 侵染的拔节期水稻病株上饲毒 2d，随后在健康水稻苗上饲养 12d，整个过程均在 25℃和 60% 湿度条件下进行。然后将饲喂 RBSDV 的带毒灰飞虱和未饲毒的无毒灰飞虱总蛋白分别用 12% 的 SDS-PAGE 凝胶电泳，电泳结束后将蛋白转印到 PVDF 膜（Whatman）上，然后用 3% 的 BSA 封闭 1h，PBST 清洗 3 次，加入制备的 P9-1 抗血清于 37℃ 孵育 1h。PBST 清洗 3 次，加入碱性磷酸酶标记的羊抗兔 IgG（Sigma）于 37℃孵育 1h。PBST 清洗 3 次，于 BCIP/NBT 显色液中进行显色反应，10min 后即可观察结果。

1.5 免疫荧光标记

抓取约 200 头 2 龄的无毒灰飞虱若虫在 RBSDV 侵染的拔节期水稻病株上饲毒 2d，随后在健康水稻苗上饲养，整个过程均在 25℃和 60% 湿度条件下进行。在饲毒后的 3、6、10、12d，分别取 30 头灰飞虱，在解剖镜下解剖昆虫的消化系统器官（包括食道、前憩室、中肠、后肠和唾液腺）。免疫荧光标记参照 Chen 等（Chen et al., 2011）方法，首先用 4% 的多聚甲醛固定消化系统器官 2h，再用 2% 的 Triton-100 渗透 30min，然后用 P9-1-FITC 抗体和肌动蛋白（Actin）染料 phalloidin-Rhodamine（Invitrogen）同时孵育消化系统器官 1h，最后通过共聚焦显微镜观察 P9-1 蛋白在介体昆虫体内的定位。

2 结果

2.1 P9-1 蛋白的原核表达

将原核表达载体 pDEST17-P9-1 转化到大肠杆菌 Rocceta 菌株，挑取阳性菌落摇菌，菌液经 0.1mmol /L IPTG 诱导后表达目的蛋白。离心收集菌体，加入上样 buffer，煮沸 10min 后通过 SDS-PAGE 凝胶电泳和考马斯亮蓝染色，可以在诱导菌液的菌体中检测到 P9-1 与 6×His 标签融合表达的大小约 43kDa 的蛋白，与预测的蛋白大小相符，而未加诱导剂的菌体没有对应的条带（图 1），表明 P9-1 蛋白特异性诱导表达。

2.2 P9-1 蛋白抗体特异性检测

分别取 30 头实验室条件下感染 RBSDV 并渡过循回期的灰飞虱成虫和人工饲养未感染 RBSDV 的灰飞虱成虫，分别提取总蛋白为样品，经 12% 的 SDS-PAGE 胶电泳后用制备的 P9-1 抗血清做 Western blot 检测，可以检测到带毒虫含有 39.9kDa 的蛋白条带，与预测的 P9-1 蛋白大小相符，而未感染 RBSDV 的灰飞虱中检测不到 P9-1 蛋白，表明所制备的抗体具有特异性，可用于免疫荧光标记实验（图 2）。

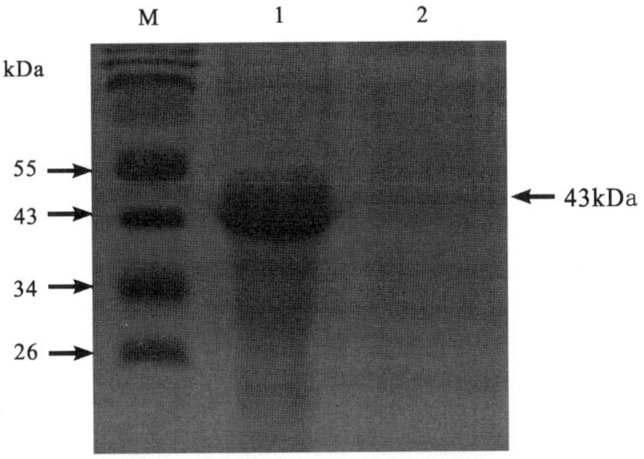

图 1 SDS-PAGE 凝胶电泳分析 RBSDV P9-1 基因在大肠杆菌中的表达情况

Fig. 1 SDS-PAGE analysis of the expression of P9-1 gene of RBSDV in *Escherichia coli*

条带 M：蛋白质标记；条带 1：IPTG 诱导的蛋白质表达；条带 2：不含 IPTG 的蛋白质表达

Lane M: Protein Marker; Lane 1: Protein expression induced by IPTG; Lane 2: Proteins expression without IPTG

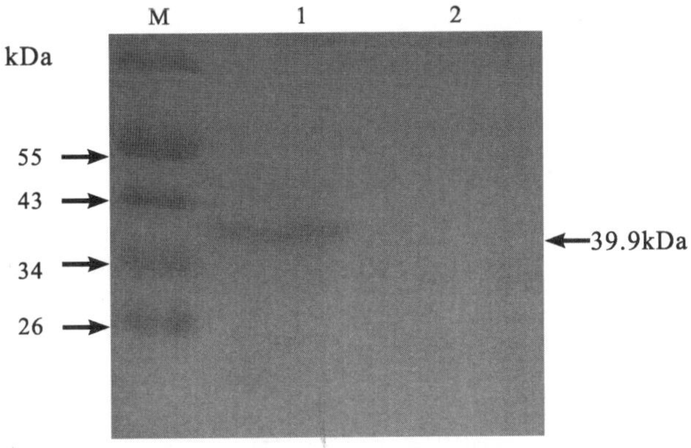

图 2 免疫印迹分析 P9- 抗体的特异性

Fig. 2 Western blot analyses of the specificity of antibodies against P9-1

条带 M：蛋白质标记；条带 1：SBPHs 感染 RBSDV 的蛋白质提取物；条带 2：SBPHs 未感染 RBSDV 的蛋白质提取物

Lane M: Protein marker; Lane 1: Protein extracts from SBPHs infected with RBSDV; Lane 2: Protein extracts from RBSDV-free SBPHs.

2.3 P9-1 蛋白在灰飞虱消化系统的定位

为了研究 RBSDV 在灰飞虱消化系统的侵染和扩散过程，本实验首先用 phalloidin-rhodamine 对灰飞虱消化系统的 Actin 进行标记（红色）。通过共聚焦显微镜可以观察到灰飞虱消化系统由食道（os）、前憩室（ad）、中肠（mg）、后肠（hg）和唾液腺（sg）组成。其中中肠从内到外分别由肠腔（gl）、单层上皮细胞（me）、基底膜（bl）和肠道表面的肌肉组织（vm）组成，在上皮细胞靠近肠腔一侧分布着大量的微绒毛（mv），而中肠外表的肌肉组织则由环肌（circular muscle）和纵肌（longitudinal muscle）组成（图 3A）。

将 2 龄灰飞虱在 25℃条件下饲毒 2d 后转到健康水稻幼苗上饲养，在饲毒后的 3、6、10 和 12d 分别解剖 30 头灰飞虱，获得的消化系统经过固定、渗透和抗体孵育，共聚焦显微镜下观察 P9-1 蛋白在灰飞虱消化系统的分布，统计结果如表 1 所示。实验中用 P9-1-FITC 标记昆虫消化系统中由于病毒侵染产生的 P9-1 蛋白（绿色），phalloidin-rhodamine 标记组成昆虫消化道的 Actin（红色）。灰飞虱饲毒后 3d，仅 33% 的灰飞虱中肠上皮细胞中可以观察到 P9-1 蛋白的表达。由于 P9-1 蛋白是形成病毒复制场所——病毒原质基质的组分，因此 P9-1 蛋白的出现表明病毒已侵入中肠上皮细胞并进行复制（图 3B）。饲毒后 6d，在 43% 的灰飞虱中肠上皮细胞和中肠表面的肌肉细胞存在 P9-1 蛋白，且 16% 的灰飞虱后肠也观察到 P9-1 蛋白，表明病毒在上皮细胞内大量复制并扩散到中肠表面的肌肉细胞（图 3C）。饲毒后 10d，47% 的灰飞虱中肠和后肠表面肌肉细胞能观察到大量的 P9-1 蛋白，约 34% 的唾液腺也有 P9-1 蛋白存在，同时上皮细胞内没有观察到 P9-1 蛋白，表明病毒已经完全扩散到中肠和后肠表面的肌肉层，且部分病毒已扩散到达唾液腺（图 3D，E）。饲毒后 12d，大量的 P9-1 蛋白分布在中肠、后肠、食道和前憩室外表的肌肉细胞和唾液腺组织，表明病毒已达到了系统性侵染（图 3F）。以上结果表明 RBSDV 在介体灰飞虱体内首先侵染中肠上皮细胞，随后扩散到中肠表面的肌肉细胞，并通过环肌和纵肌扩散到中肠和后肠，最后扩散到唾液腺。同时本研究也表明病毒可以在中肠上皮细胞，外层肌肉和唾液腺等组织或器官中复制和增殖。

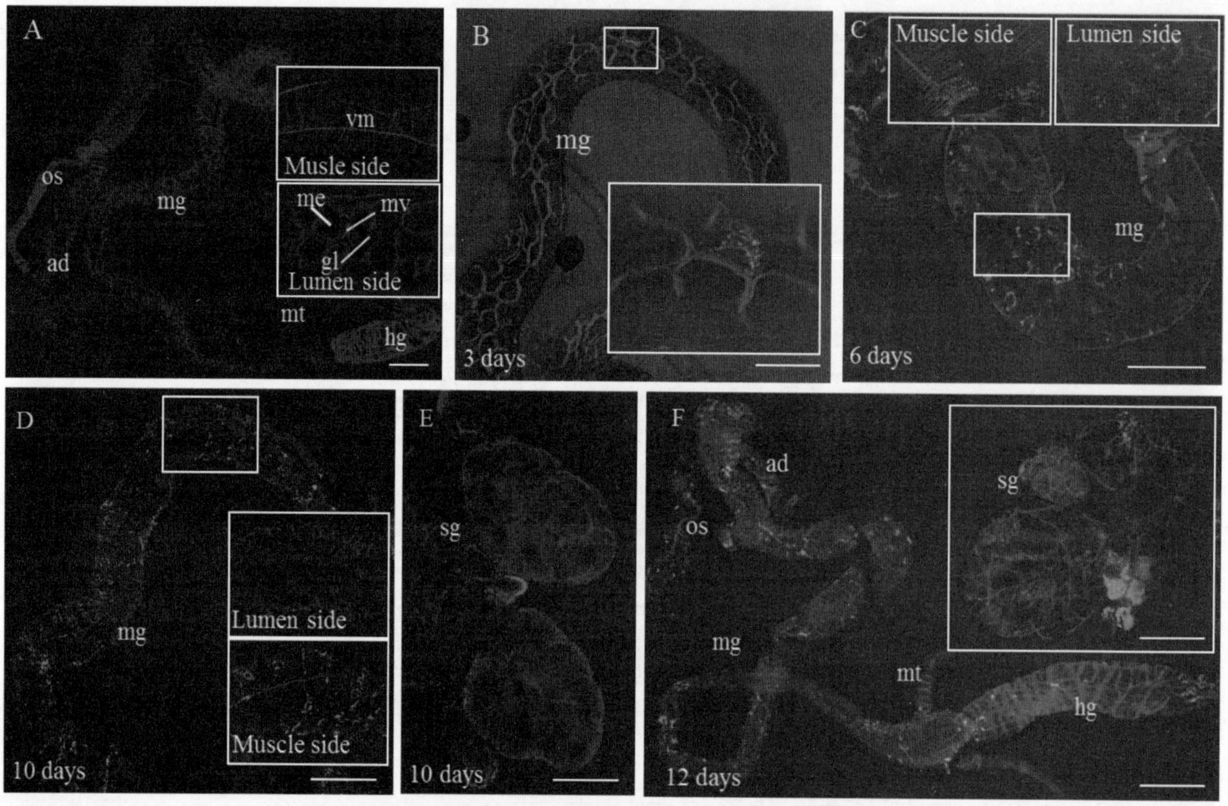

图 3 RBSDV 在昆虫载体 SBPHs 中的感染和传播

Fig. 3 Infection and spread of RBSDV in the insect vector SBPHs

用 P9-1-FITC（绿色）和 PHalloidin-rhodamine（红色）分别对 SBPH 的消化系统中 P9-1 和肌动蛋白进行染色，并用共聚焦显微镜检查。A：SBDH 的消化道。B：第 3 天，P9-1 蛋白在少数中肠上皮细胞中被检测到。C：第 6 天，在中肠上皮细胞和中肠表面的肌肉细胞存在 P9-1 蛋白。D：第 10 天，在肌肉细胞中检测到 P9-1 蛋白，而在中肠和后肠上皮细胞中检测不到 P9-1；E：第 10 天，在唾液腺中检测到 P9-1 蛋白；F：第 12 天，P9-1 蛋白在整个消化系统中积累。ad：前憩室；mg：中肠；hg：后肠；mt：马氏管；os：食道；sg：唾液腺；me：单层上皮细胞；vm：肌肉组织；gl：肠腔；比例尺条：70μm。

The digestive system of SBPHs were stained for P9-1 with P9-1-FITC (green) and for Actin with phalloidin-rhodamine (red) and examined by confocal microscopy A: The alimentary canal of SBPH. B: At 3d p.a.v., P9-1 were detectedin a few epithelial cells of the midgut. C: At 6d p.a.v., P9-1 accumulated in the epithelium and a few visceral muscle of the midgut. D: At 10d p.a.v., P9-1 were detected in the visceral muscle but without the epithelium of the midgut and hindgut. E: At 10d p.a. v., P9-1 were detected in a few salivary gland. F: At 12d p.a.v., P9-1 accumulated throughout the digestive system.ad, anterior diverticulum; mg, midgut; hg, hindgut; mt, malpighian tubules; os, oesophagus; sg, salivary gland; me, midgut epithelium; vm, visceral muscle; gl, gut lumen. Bars, 70μm.

Table 1　Occurrence of P9-1 in digestive system of SBPHs at different days post-acquire virus as detected by immunofluorescence microscopy

Days post-acquire virus/d	Number. positive insects with P9-1 in different tissues (n=30)					
	me	vm	hg	os	ad	sg
3	10	0	0	0	0	0
6	13	13	5	0	0	0
10	0	14	14	13	13	11
12	0	14	14	14	14	14

Note: me: Midgut epithelium; vm: Visceral muscle; hg: Hindgut; os: Oesophagus; ad: Anterior diverticulum; sg: Salivary gland.
注：me：中肠柱状上皮细胞；vm：脏腑肌；hg：后肠；os：食道；ad：前憩室；sg：唾腺。

3　讨论

P9-1蛋白是病毒原质的组分，参与病毒的复制过程，P9-1蛋白存在的部位即病毒侵染后复制的部位，如水稻锯齿叶矮缩病毒（*Rice ragged stuntvirus*, RRSV）的Pns10也是病毒原质的组分，其在中肠上皮细胞及唾液腺内均与病毒共定位（Jia et al., 2012b），因此通过免疫荧光技术标记P9-1蛋白在介体灰飞虱体内的分布即可明确RBSDV的复制和扩散过程。

目前关于水稻呼肠孤病毒在介体内的侵染循回已有多篇报道，如黑尾叶蝉传播的水稻矮缩病毒（*Rice dwarf virus*, RDV）、褐飞虱传播的RRSV和白背飞虱传播的SRBSDV（Chen et al., 2011; Jia et al., 2012c; Jia et al., 2012b）。这几种病毒虽然都属于介体昆虫传播的植物呼肠孤病毒，但由于基因组和传播介体的差异其在介体内的侵染循回过程也各有特点。RBSDV与SRBSDV有着非常相近的亲缘关系，因此两种病毒在介体内的侵染循回过程大体相似（Jia et al., 2012c），均表现为：病毒顺着食道进入中肠肠腔，少量病毒穿过微绒毛侵染少数的中肠上皮细胞并建立初侵染点，病毒在上皮细胞内复制增殖后穿过基底膜到达中肠表面的肌肉细胞，在肌肉细胞上进行复制并沿着环肌和纵肌扩散到中肠、后肠、前憩室和食道表面，同时病毒从中肠释放到血淋巴并扩散到唾液腺完成在介体内的整个侵染循回过程。但RBSDV在灰飞虱体内的扩散速率较SRBSDV在白背飞虱体内的扩散速率缓慢，且灰飞虱带毒率最高约为50%，低于白背飞虱80%的带毒率（Jia et al., 2012a）。与RBSDV同属呼肠孤病毒科的RDV属于植物呼肠孤病毒属，其传播介体是黑尾叶蝉，由于RBSDV与RDV属于不同病毒属且传播介体不同，因此两者的侵染循回过程也存在较大的差异（Chen et al., 2011）。首先，病毒的初侵染点不同，RBSDV首先侵染灰飞虱中肠的上皮细胞进行复制，而RDV在介体叶蝉体内首先到达滤室的上皮细胞进行初侵染复制，这可能与参与病毒侵入的受体蛋白的分布位点不同有关；其次，病毒从中肠释放的方式不同，RBSDV在上皮细胞穿过基底膜直接到达中肠表面的肌肉层，而RDV则从滤室扩散到前憩室和前中肠，病毒在未到达其他器官之前迅速扩散到与中肠和马氏管相连接的神经系统；第三，病毒在消化器官间的扩散方式不同，RBSDV沿着中肠表面的环肌和纵肌扩散到后肠和食道，而RDV不仅利用肌肉细胞，还利用神经快速扩散；第四，病毒到达唾液腺的方式不同，RBSDV通过血淋巴扩散到唾液腺，而RDV则会通过神经快速将病毒从中肠扩散到唾液腺，这种方式与玉米花叶病毒（*Maize mosaic virus*）的扩散途径相似（Ammar and Hogenhout, 2008）。

与非增殖型病毒不同，持久增殖型病毒必须在介体内复制增殖。RBSDV在介体灰飞虱体内的复制贯穿于整个侵染过程，一旦阻断病毒的复制或扩散，灰飞虱就不能有效传毒。例如：饲喂源于病毒原质组分蛋白的dsRNA可以降低介体昆虫的带毒率（Jia et al., 2012c），其原因是阻断了早期病毒在中肠上皮细胞的初侵染；SRBSDV在灰飞虱体内受中肠释放屏障的阻断，其扩散受阻而无法传毒（Jia et al.,

2012a)；此外通过转基因水稻干扰参与病毒复制的 P9-1 蛋白表达也可以成功地抵抗病毒的侵染（Shimizu et al., 2011）。因此，通过分析病毒的侵染循回过程，可以推测阻断 RBSDV 在灰飞虱体内的初侵染是防止病毒扩散的有效手段之一。

参考文献

[1] AKITA F, HIGASHIURA A, SHIMIZU T, et al. Crystallographic analysis reveals octamerization of viroplasm matrix protein P9-1 of Rice black streaked dwarf virus[J]. Journal of Virology, 2012, 86(2):746-756.

[2] AMMAR E D, HOGENHOUT S A. A neurotropic route for *Maize mosaic virus* (*Rhabdoviridae*) in its planthopper vector *Peregrinusmaidis*[J]. Virus Research, 2008, 131(1):77-85.

[3] BAI F W, YAN J, QU Z C, et al. Phylogenetic Analysis Reveals that a Dwarfing Disease on Different Cereal Crops in China is due to Rice Black Streaked Dwarf Virus (RBSDV)[J]. Virus Genes, 2002, 25(2):201-206.

[4] CHEN H, CHEN Q, OMURA T, et al. Sequential infection of *Rice dwarf virus* in the internal organs of its insect vector after ingestion of virus[J].Virus Research, 2011,160(1-2):389-394.

[5] CHEN S, ZHANG Q. Advance in researches on rice black-streaked dwarf disease and maize rough dwarf disease in China[J]. Journal of Plant Protection, 2005, 32: 97-103.

[6] ISOGAI M, UYEDA I, LEE B. Detection and assignment of proteins encoded by rice black streaked dwarf fijivirus S7, S8, S9 and S10[J]. Journal of General Virology, 1998, 79(6):1487-1494.

[7] JIA D, CHEN H, MAO Q, et al. Restriction of viral dissemination from the midgut determines incompetence of small brown planthopper as a vector of Southern rice black-streaked dwarf virus[J]. Virus Research, 2012a, 167(2):404-408.

[8] JIA D, GUO N, CHEN H, et al. Assembly of the viroplasm by viral non-structural protein Pns10 is essential for persistent infection of rice ragged stunt virus in its insect vector[J]. Journal of General Virology, 2012b, 93(10):2299-2309.

[9] JIA D, CHEN H, ZHENG A, et al. Development of an insect vector cell culture and RNA interference system to investigate the functional role of fijivirus replication protein[J]. Journal of Virology, 2012c, 86(10):5800-5807.

[10] KING A, ADAMS M J, CARSTENS E B, et al. Virus Taxonomy: Ninth Report of the International Committee on Taxonomy of Viruses[J]. archives of virology, 2012.

[11] LV Y P, LEI J L, JIN D D, et al. Detected RBSDV with RT-PCR[J]. Acta Agriculturae Zhejiang ensis, 2002, 14(2): 117-119.

[12] SHIMIZU T, NAKAZONO-NAGAOKA E, AKITA F, et al. Immunity to *Rice black streaked dwarf virus*, a plant reovirus, can be achieved in rice plants by RNA silencing against the gene for the viroplasm component protein[J]. Virus Research, 2011, 160 (1-2):400-403.

[13] WANG Q, TAO T, ZHANG Y, et al. Rice black-streaked dwarf virus P6 self-interacts to form punctate, viroplasm-like structures in the cytoplasm and recruits viroplasm-associated protein P9-1[J]. Virology Journal, 2011, 8(1):24.

[14] YANG J, ZHANG H, CHEN J, et al. Prekaryotic expression antiserum preparation and some properties of p8 protein of Rice black-streaked dwarf Fijivirus[J]. Acta Phytophylacica Sinica, 2007,34(3): 252-256.

[15] Zhang LD, Wang CH, Wang XB, et al. Function research on RNA silencing suppressors encoded by two plant viruses[J]. Chinese Science Bulletin, 2005, 50(2): 219-224.

[16] ZHANG C, LIU Y, LIU L, et al. Rice black streaked dwarf virus P9-1, an α-helical protein, self-interacts and forms viroplasms in vivo[J]. Journal of General Virology, 2008, 89(7): 1770-1776.

[17] ZHANG H, CHEN J, LEI J, et al. Sequence Analysis Shows that a Dwarfing Disease on Rice, Wheat and Maize in China is Caused by Rice Black-streaked Dwarf Virus[J]. European Journal of Plant Pathology, 2001, 107(5): 563-567.

[18] ZHANG H M, CHEN J P, ADAMS M. Molecular characterisation of segments 1 to 6 of *Rice black-streaked dwarf virus* from China provides the complete genome[J]. Archives of Virology, 2001, 146(12): 2331-2339.

[19] ZHOU G, WEN J, CAI D,et al. Southern rice black-streaked dwarf virus: a new proposed Fijivirus species in the family Reoviridae[J]. Chinese Science Bulletin, 2008, 53(23): 3677-3685.

[20] ZHOU T, WU L, WANG Y, et al. Preliminary report on the transmission of rice black-streaked dwarf virus from frozen infected leaves to rice by insect vector small brown planthopper (Laodelphax striatellus)[J]. Chinese Journal of Rice Science, 2010, 24(4):425-428.

干扰水稻瘤矮病毒（RGDV）非结构蛋白（Pns12）的表达抑制病毒在介体昆虫培养细胞内的复制

郑立敏，刘华敏，陈红燕，贾东升，谢联辉，魏太云

（福建农林大学植物病毒研究所 / 福建省植物病毒学重点实验室，福州 350002）

摘要：水稻瘤矮病毒（*Rice gall dwarf virus*, RGDV）属呼肠孤病毒科（Reoviridae）植物呼肠孤病毒属（*Phytoreovirus*），由介体电光叶蝉（*Recilia dorsalis*）以持久增殖型方式传播，其基因组第 12 条片段编码的非结构蛋白（nonstructural protein, Pns12）是提供病毒复制和子代病毒粒体装配场所——病毒原质（viroplasm）的组分之一。为了明确 Pns12 在 RGDV 侵染介体电光叶蝉培养细胞中的功能，本研究通过原核表达的 Pns12 蛋白免疫注射兔（*Oryctolagus cuniculus*），制备 Pns12 抗体，并应用免疫荧光标记和 RNA 干扰（RNA interference, RNAi）技术研究 Pns12 在介体培养细胞内的定位和参与病毒原质形成的过程。共聚焦显微镜观察到，病毒侵染的细胞中，与荧光素交联的 Pns12 抗体特异地标记在病毒原质上。干扰 Pns12 蛋白表达后，可有效地阻碍病毒原质的形成、子代病毒粒体的组装和病毒非结构蛋白 Pns12 和外壳蛋白 P8 蛋白的表达，表明 Pns12 作为病毒原质的组分参与了 RGDV 在介体培养细胞内的复制。也表明，Pns12 可作为理想的靶标用于阻断电光叶蝉携带和传播 RGDV。

关键词：水稻瘤矮病毒（RGDV）；介体昆虫培养细胞；非结构蛋白（Pns12）；RNA 干扰（RNAi）

中图分类号：S435.72　**文献标识码：**A　**文章编号：**1671-5470(2009)01-0006-05

Knockdown of nonstructural protein (Pns12) of *Rice gall dwarf virus* (RGDV) Inhibits viral replication in insect vector cells

Limin Zheng, Huamin Liu, Hongyan Chen, Dongsheng Jia, Lianhui Xie, Taiyun Wei

(Fujian Province Key Laboratory of Plant Virology, Institute of Plant Virology, Fujian Agriculture and Forestry University, Fuzhou 350002, China)

Abstract: *Rice gall dwarf virus* (RGDV), the genus *Phytoreovirus* in the family *Reoviridae*, is transmitted by the leafhopper vector (*Recilia dorsalis*) in a persistent- propagative manner. Nonstructural protein(Pns12), encoded by segment 12 of RGDV, is one of the components of viroplasm which is the site for viral replication and assembly of progeny virons during viral infection in its insect vector cells. In this study, to investigate the functional role of Pns12 in the formation of viroplasm in its insect vector cells, the polyclonal antibody against Pns12 was prepared and purified, then conjugated to fluorescein isothiocyanate. The immunoglobulin G(IgG) of P8 was purified and conjugated to rhadamine. Immunofluorescence microscopy demonstrated that Pns12 antibodies specifically distributed in the viroplasm matrix during RGDV infection in the vector cells in monolayers (VCMs),

农业生物技术学报 . 2014, 22(11): 1321-1328

收稿日期：2014-07-18；接受日期：2014-08-21

基金项目：国家自然科学基金青年基金 (No. 31200118 和 No. 31300136) 和福建省自然科学基金面上项目 (No. 2012J01087 和 No. 2014J01085)

while outer capsid protein P8 were accumulated at the periphery of the viroplasm. The viroplasm increased in size over time. Knockdown of Pns12 by RNA interference (RNAi) induced by dsRNAs, targeting Pns12 gene of RGDV, significantly inhibited the formation of viroplasm compared with dsGFP, suggesting that RGDV replication was inhibited. Western blot showed that viral Pns12 and P8 expression reduced in the VCMs treated with dsPns12. Thus, the present study indicated that Pns12 of RGDV played an important role in viroplasm formation and viral replication in its insect vector cells. RNAi induced by dsRNAs derived from the viral genes of viroplasm matrix protein may be an ideal tool for inhibiting the infection and transmission of RGDV by leafhopper vectors.

Keywords: Rice gall dwarf virus (RGDV); Continuous cell cultures of the leafhopper; Nonstructural protein (Pns12); RNA interference (RNAi)

植物呼肠孤病毒在介体细胞内可以诱导形成提供病毒复制和装配场所的球状或纤维状的电子致密内含体——病毒原质（viroplasm）（Wei et al., 2006, 2009; Akita et al., 2011; Mao et al., 2013; Chen et al., 2014）。水稻瘤矮病毒（*Rice gall dwarf virus*, RGDV）属于呼肠孤病毒科（Reoviridae）植物呼肠孤病毒属（*Phytoreovirus*），由介体电光叶蝉（*Reciliadorsalis*）以持久增殖型方式传播（Morinaka et al., 1982）。1979 年，水稻瘤矮病首次在泰国中部发生（Omura et al., 1980; Putta et al., 1980），随后在菲律宾和中国的广东、广西和福建等局部稻区均有发生（Zhang et al., 2008）。RGDV 基因组由 12 条双链 RNA 组成（Omura et al., 1982; Hibi et al., 1984），共编码 6 个结构蛋白（P1、P2、P3、P5、P6 和 P8）和 6 个非结构蛋白（Pns4、Pns7、Pns9、Pns10、Pns11 和 Pns12）（Moriyasu et al., 2000, 2007; Zhang et al.,2008）。结构蛋白 P3 为主要内层衣壳蛋白，包裹着 P1、P5 和 P6（Omura et al., 1985; Ichimi et al., 2002），P2 和 P8 分别为次要和主要外层衣壳蛋白（Omura et al., 1998; Miyazaki et al., 2005）。非结构蛋白 Pns7、Pns9 和 Pns12 是病毒原质的组分，对病毒原质的形成至关重要。Pns11 在介体培养细胞中以管状形式存在，对病毒起着运输的作用（Chen et al.,2013）。此外，利用植物瞬时表达体系发现 Pns11 和 Pns12 具有基因沉默抑制子功能（Liu et al., 2008; Wu et al., 2011）。

RGDV 非结构蛋白 Pns7、Pns9 和 Pns12 可以聚集形成病毒原质（Wei et al., 2009; Akita et al., 2011）。Akita 等（2011）利用杆状病毒表达系统，在草地贪夜蛾（*Spodopterafrugiperda*, Sf9）细胞中观察到，RGDV Pns9 单独表达能够形成类似于病毒原质的内含体结构（Akita et al., 2011）。利用 RNA 干扰（RNA interference, RNAi）技术在水稻（*Oryzasativa*）中抑制 *Pns9* 基因表达后，能有效抑制 RGDV 在水稻中的复制，抵抗病毒对水稻的侵染（Shimizu et al., 2012）。RGDV 在介体昆虫体内以持久增殖型方式进行传播，需要在介体细胞内建立初侵染点，大量复制增殖后被传播到水稻，其中，病毒在介体的复制过程至关重要，是决定介体能否传毒的关键环节，而 Pns12 是 RGDV 病毒原质组成成分之一，但其在病毒侵染过程中的功能未知。为了进一步研究 Pns12 在病毒侵染循环中的功能，本研究利用建立的电光叶蝉培养细胞体系，拟通过原核表达制备 Pns12 的多克隆抗体，并利用免疫荧光标记和 RNAi 技术研究 Pns12 在病毒侵染过程中的重要作用。

1 材料与方法

1.1 实验材料

电光叶蝉（*Recilia dorsalis*）和感染水稻瘤矮病毒（*Rice gall dwarf virus*, RGDV）水稻（*Oryza sativa*）病株采自广东省兴宁县，电光叶蝉培养细胞体系的建立参照 Kimura 和 Omura（1988）培养黑尾叶蝉（*Nephotettixcincticeps*）细胞方法。RGDV P8 抗体为日本国立农业研究中心馈赠。

1.2 实验方法

1.2.1 Pns12 多克隆抗体的制备和特异性检测

提取感染 RGDV 水稻病株总 RNA，以扩增 Pns12（EF177263.1）完整开放阅读框的引物 Pns12-F/R 进行 RT-PCR，获得目的基因片段。PCR 反应体系：去离子水 36μL，PCR 反应缓冲液 5μL，Pns12-F（10mmol/L）1μL，Pns12-R（10mmol/L）1μL，dNTP（2.5mmol/L）4μL，PCR DNA 聚合酶（2.5U/μL）1μL，cDNA 模板（0.5μg/μL）2μL。PCR 反应程序：94℃ 预变性 10min，94℃ 变性 30s，55℃ 退火 30s，72℃ 延伸 45s，30 个循环，72℃ 延伸 10min。在 BP Clonase Enzyme Mix（Invitrogen, 美国）作用下，将 Pns12 目的片段重组到入门载体 pDONR221（Invitrogen, 美国）上，经测序比对分析正确后，在 LR Clonase Enzyme Mix（Invitrogen, 美国）作用下将 Pns12 目的片段重组到原核表达载体 pDEST17（Invitrogen, 美国），获得重组质粒 pDEST17-Pns12，并转入大肠杆菌（*Escherichia coli*）表达菌 Rosetta 细胞。重组菌于 28℃，0.5mmol 异丙基硫代半乳糖苷（isopropyl β-D-1-thiogalactopyranoside, IPTG）诱导 5h，收集菌体，进行 SDS-PAGE 凝胶电泳分离。将目的蛋白条带切割回收，加入弗氏佐剂研磨乳化后，皮下多点注射新西兰兔（*Oryctolagus cuniculus*）。每隔 7d 注射 1 次，注射 5 次后动脉取血，离心收集血清即为制备的抗血清。分别提取健康和带毒电光叶蝉总蛋白，SDS-PAGE 电泳后，转至聚偏二氟乙烯膜（poly vinyli dene fluoride, PVDF）上，以制备的 Pns12 多克隆抗体为一抗，碱性磷酸酶标记的羊抗兔多克隆抗体为二抗，Westernblot 检测抗体的特异性。

1.2.2 免疫荧光标记检测

应用蛋白 A 亲和层析法 protein A-Sepharose affinity column（Thermo, 美国）提纯抗血清获得免疫球蛋白 G（immunoglobulin G, IgG），并按照 Invitrogen 说明书将 Pns12 IgGs 与异硫氰酸荧光素（fluorescein isothiocyanate, FITC）（Invitrogen, 美国）交联，结构蛋白 P8（stuctural protein, P8）IgGs 与罗丹明荧光素 rhodamine 交联得到 Pns12-FITC 和 P8-rhodamine 抗体（吴维等 , 2012）。用粗提的 RGDV 侵染液侵染培养在盖玻片上的介体单层培养细胞（vecor cells in monolayer, VCM）2h，分别在侵染后的 24、72 和 96h 用 4% 多聚甲醛固定细胞 20min，再用 2% Triton-100 渗透 10min，然后用 Pns12-FITC 和 P8-rhodamine 抗体同时孵育细胞 2h，最后通过共聚焦显微镜（Leica TCS SP5, 德国）观察 Pns12 在细胞内的定位。

1.2.3 dsRNAs 体外合成和转染细胞

设计 5′ 端均含有 T7 启动子序列的引物，扩增 Pns12 基因全长和绿色荧光蛋白基因（green fluorescent protein, GFP）（GenBank 登录号：AF324407.1），参照体外大量表达 RNAi 系统 T7 RiboMAX™ Express RNAi System 试剂盒（Promega, 美国）说明书操作，以获得的 DNA 片段为模板，用 T7 转录酶进行体外转录合成目的基因 dsPns12 和 dsGFP，经纯化和浓度测定后的 dsRNAs 用于转染细胞。按照 Cellfectin® Reagent（Invitrogen, 美国）说明书将 2μg dsRNAs 转染至盖玻片上培养的细胞 8h 后，病毒侵染 2h，分别在侵染后 24 和 72h 对细胞进行 Pns12-FITC 和 P8-rhodamine 抗体标记，共聚焦显微镜（Leica TCS SP5, 德国）下观察结果。培养瓶中干扰的细胞在侵染后 72h 提取细胞总蛋白，分别以 Pns12 和 P8 多克隆抗体为一抗，碱性磷酸酶标记的羊抗兔多克隆抗体为二抗进行 Western blot 检测。

2 结果与分析

2.1 Pns12 蛋白的原核表达和抗体制备

提取 RGDV 侵染水稻病株总 RNA，RT-PCR 扩增 Pns12 基因全长为 621bp（图 1A），经 BP 和 LR 重组酶重组后获得原核表达载体 pDEST17-Pns12。将 pDEST17-Pns12 转入大肠杆菌表达菌 Rosetta 细胞，经 0.5mmol IPTG 诱导表达蛋白，同时未加 IPTG 诱导的样品为对照。收集到的蛋白样品经 12% SDS-PAGE 分离后考马斯亮蓝染色，可以观察到大小约为 29kD 的特异性目的条带（Pns12 实际大小约为 26.5kD，His 标签蛋白大小约为 2.6kD，原核表达的蛋白为 29kD）（图 1B），与预期大小一致。表明 Pns12 蛋白在大肠杆菌中成功表达。

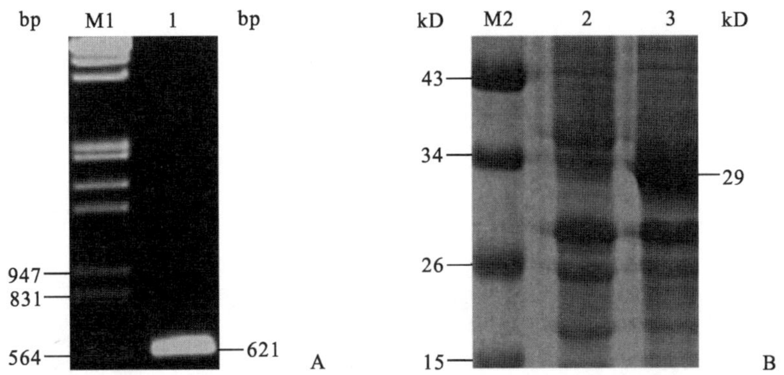

图 1 RT-PCR 扩增 RGDV *Pns12* 基因（A）和 SDS-PAGE 分析 RGDV Pns12 在大肠杆菌的原核表达（B）

Fig. 1 RT-PCR of *Pns12* gene in RGDV (A) and SDS-PAGE analysis of expression of RGDV Pns12 in *E. coli* (B)

1：*Pns12* 基因扩增片段；M2：蛋白质 marker，下同；2：未加 IPTG 诱导的蛋白样品（对照）；3：IPTG 诱导的蛋白样品

M1: Lambda DNA/*Eco*R Ⅰ+*Hin*d Ⅲ; 1: *Pns12* amplified fragment; M2: Protein marker, the same below; 2: Total proteins of *E. coli* without IPTG (control); 3: IPTG induced total proteins of *E. coli*

2.2 Pns12 蛋白抗体的特异性检测

为了检测 Pns12 多克隆抗体的特异性，本研究分别提取健康（对照）和携带 RGDV 的电光叶蝉总蛋白，经 SDS-PAGE 胶电泳后转至 PVDF 膜上，以制备的抗 Pns12 多克隆抗体为一抗，碱性磷酸酶标记的羊抗兔多克隆抗体为二抗进行 Western blot。结果表明，在携带 RGDV 的电光叶蝉体内能够检测约 26.5kD 的蛋白条带，与预测的 Pns12 蛋白大小相符，未感染 RGDV 的电光叶蝉体内检测不到 Pns12 蛋白，表明所制备的抗体具有特异性，可用于免疫荧光标记实验（图 2）。

2.3 免疫荧光检测 Pns12 在电光叶蝉培养细胞中的定位

为了研究 Pns12 在电光叶蝉培养细胞中的定位，将 Pns12 抗体和荧光素 FITC 交联，主要外层衣壳蛋白 P8 抗体和荧光素 rhodamine 交联。用粗提的 RGDV 侵染液侵染 VCMs，分别在侵染后 24、72 和 96h 进行免疫荧光标记。结果显示，Pns12 交联的多克隆抗体能够特异性标记在病毒原质上，P8 交联的多克隆抗体标记的病毒粒体分布在病毒原质周围（图 3 白色箭头所示），并且随侵染时间的延长，病毒原质形态增大，单个细胞病毒原质数量减少（图 3）。这表明，Pns12 形成的病毒原质不断地聚集增大，最终形成成熟的病毒原质。

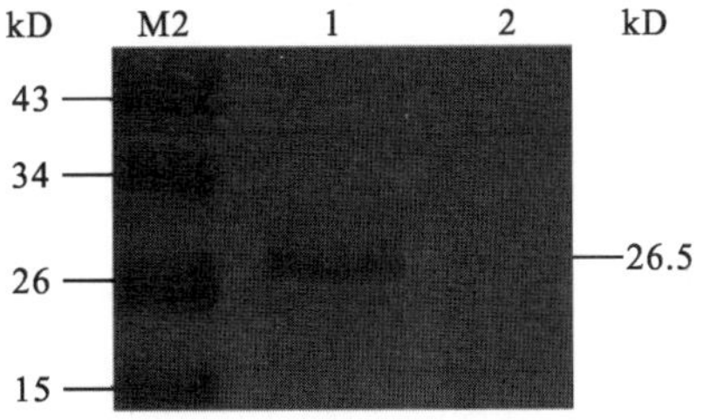

图 2 Western blot 检测 Pns12 抗体的特异性

Fig. 2 Specificity analysis of antibodies against Pns12 by Western blot

1：感染 RGDV 的电光叶蝉总蛋白；2：未感染 RGDV 的电光叶蝉总蛋白（对照）

1: Protein from RGDV infected leafhoppers; 2: Protein from RGDV-free leafhoppers (control)

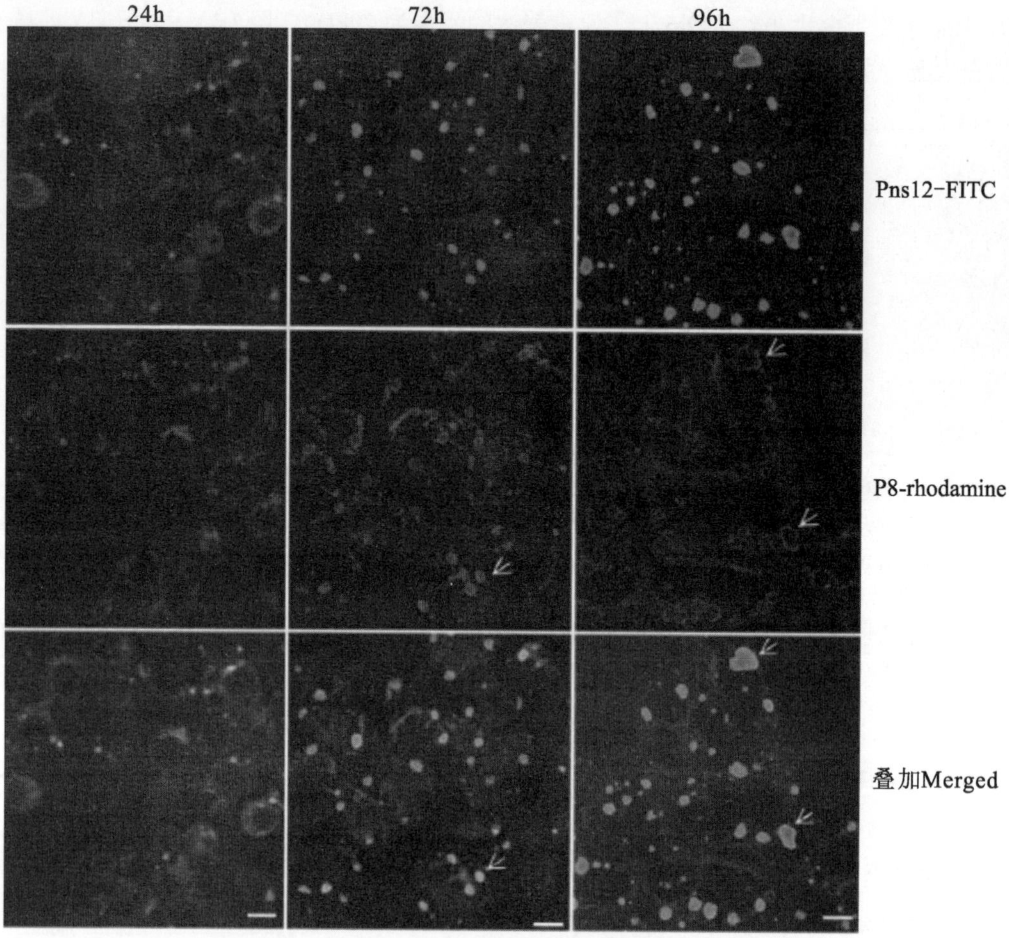

图 3　Pns12 和 P8 在电光叶蝉培养细胞中的定位 (Bar=10μm)

Fig. 3　Subcellular localization of Pns12 and P8 in VCMs.

抗体 Pns12-FITC（绿色）和 P8-rhodamine（红色）标记电光叶蝉细胞，下同；白色箭头指示 P8 标记的病毒粒体围绕在病毒原质周围；随侵染时间的延长，病毒原质形态增大，单个细胞病毒原质数量减少

VCMs were stained with Pns12-FITC (green) and P8-rhodamine (red), the same below; The arrows show ringlike profiles of P8 antigens that surrounded viral inclusion; Those viroplasms increased in size and decreased in number in the single cell over time after virus infection

2.4　dsRNAs 处理细胞对 RGDV 侵染的影响

为了进一步研究 Pns12 对 RGDV 侵染和复制的影响，本研究通过 T7 聚合酶体外合成 Pns12 基因的 dsRNA（dsPns12）和 GFP 基因的 dsRNA（ds-GFP），分别将合成的 dsPns12 和 dsGFP 与 Cellfectin® Reagent 混匀，处理培养在盖玻片上的电光叶蝉单层细胞 8h 后，RGDV 侵染细胞 2h。在病毒侵染后的 24 和 72h 对细胞进行免疫荧光标记。共聚焦显微镜观察显示，在病毒侵染后的 24h，dsFP 处理的细胞，病毒侵染率高达 80%，单个细胞病毒原质的侵染点达 5 ～ 8 个（图 4A）；dsPns12 处理的细胞，病毒原质的形成受到抑制，病毒的侵染率明显降低，仅为 8%，且单个细胞病毒原质的侵染点仅为 1、2 个（图 4B）。病毒侵染 72h，dsGFP 处理的细胞，病毒侵染率达到 100%，病毒原质形态增大，数量减少，单个细胞的侵染点为 2 ～ 5 个，可明显观察到病毒分布在病毒原质周围（白色箭头）（图 4C）；dsPns12 处理的细胞，病毒侵染率为 10%，病毒原质的侵染点仅有 1 ～ 3 个，病毒原质形态无明显变化（图 4D）。研究结果表明，由 dsPns12 诱导的 RNAi 抑制了病毒原质的形成和成熟。

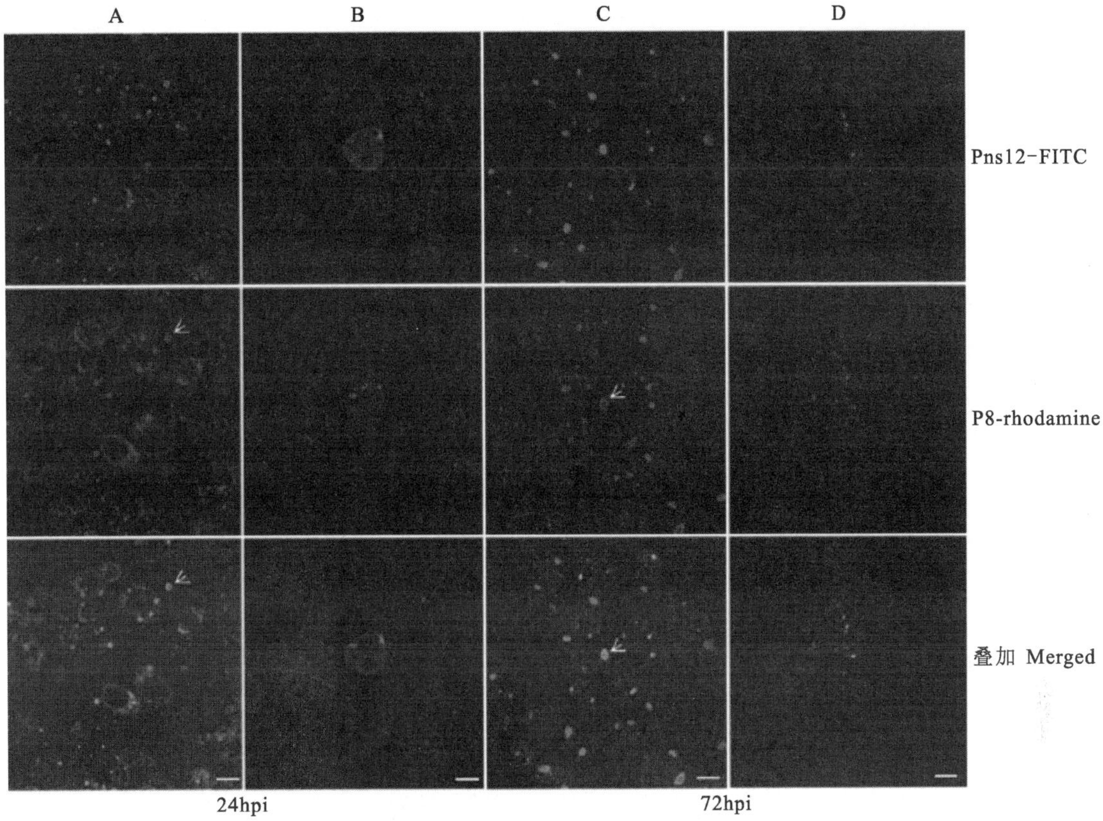

图 4 dsPns12 诱导的 RNA 干扰抑制 RGDV 原质的形成（Bar=10μm）

Fig.4 dsPns12 induced RNAi inhibited the formation of viroplasm in VCMs

A：dsGFP 转染细胞，病毒侵染后第 24h。病毒侵染率高达 80%，单个细胞病毒原质的侵染点达 5 ～ 8 个。B：dsPns12 转染细胞，病毒侵染后第 24h。病毒原质的形成受到抑制，病毒的侵染率明显降低，仅为 8%。C：dsGFP 转染细胞，病毒侵染后第 72h。病毒侵染率达到 100%，病毒原质形态增大，数量减少，单个细胞的侵染点为 2 ～ 5 个，可明显观察到病毒分布在病毒原质周围（白色箭头）。D：dsPns12 转染细胞，病毒侵染后第 72h。病毒侵染率为 10%，病毒原质的侵染点仅有 1 ～ 3 个，病毒原质形态无明显变化

A: dsGFP treated VCMs at 24h postinfection (hpi). The infection rate of virus was 80% in VCMs, the number of viroplasm reached 5–8per cell. B: dsPns12 treated VCMs at 24hpi. The formation of viroplasms was inhibited with decreased infection rate of 8% in VCMs. C: dsGFP treated VCMs at 72hpi. The infection rate of virus was 100% in VCMs, and viroplasms increased in size and decreased in number which reached 2–5per cell. D: dsPns12 treated VCMs at 72hpi, The infection rate of virus reduced to 10% in VCMs and the formation of viroplasms was still inhibited

2.5 dsRNAs 处理细胞对 RGDV 编码蛋白 Pns12 和 P8 表达水平的影响

为了分析 dsPns12 处理细胞对 RGDV 蛋白表达水平的影响，用 dsPns12 和 dsGFP 处理电光叶蝉培养细胞，病毒侵染 3d 后，分别提取细胞总蛋白进行 SDS-PAGE 电泳，以 Pns12 和 P8 抗体为一抗进行 Western blot。结果显示，在 dsGFP 处理的细胞中，非结构蛋白 Pns12 和结构蛋白 P8 都能够大量表达（图 5）；dsPns12 处理的细胞中，Pns12 和 P8 的表达量明显降低（图 5）。这表明，dsPns12 处理的细胞中，病毒蛋白的表达受到抑制。

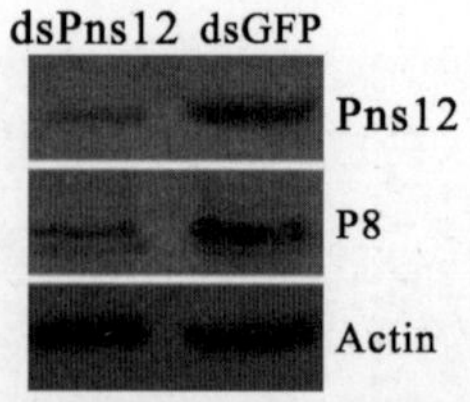

图 5　dsPns12 诱导的 RNA 干扰抑制 RGDV 蛋白的表达

Fig.5　dsPns12 induced RNAi inhibited the expression of Pns12 and P8 in VCMs

在 dsGFP 处理的细胞中，非结构蛋白 Pns12 和结构蛋白 P8 都能够大量表达；dsPns12 处理的细胞中，Pns12 和 P8 的表达量明显降低；内参蛋白：Actin

Abundant proteins of Pns12 and P8 expressed in the VCMs treated with dsGFP; While the expressions of Pns12 and P8 were reduced signifcantly in the VCMs treated with dsPns12; Internal control protein: Actin

3　讨论

植物呼肠孤病毒能够在介体昆虫培养细胞中进行增殖，需要经过侵入、复制、扩散和释放过程，其中病毒的复制过程尤为重要。近年来，利用免疫荧光标记和 RNAi 技术，已对水稻病毒在介体昆虫培养细胞中的复制功能进行了大量的研究（Jia et al., 2012; Mao et al., 2013; Chen et al., 2014）。如明确水稻矮缩病毒（*Rice dwarf virus*, RDV）非结构蛋白 Pns6、Pns11 和 Pns12 是组成球状病毒原质的组分（Wei et al., 2006），南方水稻黑条矮缩病毒（*Southern rice black-streaked dwarf virus*, SRBSDV）非结构蛋白 P5-1、P6 和 P9-1 在白背飞虱细胞内形成球状的和纤维状的病毒原质的组分（Mao et al., 2013），水稻锯齿叶矮缩病毒（*Rice ragged stunt virus*, RRSV）非结构蛋白 Pns6 和 Pns10 是组成病毒原质的组分（Jia et al., 2012; Chen et al., 2014）。这些研究表明，利用免疫荧光标记和 RNAi 技术在介体昆虫培养细胞内研究病毒编码蛋白的功能，已成为病毒蛋白功能分析的重要手段，为本研究利用该技术手段对 RGDV 编码 Pns12 蛋白的功能研究奠定了基础。

本研究首先制备了 Pns12 的多克隆抗体，Western blot 显示，抗体能够与 Pns12 特异性结合。利用免疫荧光标记技术，在电光叶蝉培养细胞内进行了 Pns12 的亚细胞定位，发现 Pns12 多克隆抗体能够特异性地标记在病毒原质上，病毒外壳蛋白 P8 分布在病毒原质周围。随着侵染时间的延长，由 Pns12 抗体标记的病毒原质的形态逐渐增大，单个细胞侵染点数量逐渐减少，表明病毒原质存在逐渐成熟和聚集的过程。与 RGDV 同属的 RDV 的病毒原质的形成过程也存在类似的现象（Wei et al.,2006）。此外，轮状病毒由非结构蛋白 NSP2、NSP5 和结构蛋白 VP2 组成的病毒原质，主要是通过细胞内的微管网络系统聚集成熟的（Fabbretti et al.,1999; Contin et al., 2010; Eichwald et al., 2012）。Pns12 作为 RGDV 病毒原质的组分之一，对病毒的复制至关重要。本研究通过 dsRNA 诱导的 RNAi 技术抑制 Pns12 在电光叶蝉培养细胞的表达，可以显著地抑制病毒原质的形成、聚集和成熟，从而抑制病毒的再侵染，表明 Pns12 直接参与了病毒的复制。同时表明，Pns12 可作为理想的靶标用于阻断电光叶蝉携带和传播 RGDV。此外，该类病毒的病毒原质由多个蛋白组成，其复制是一个复杂的过程，已知 Pns7 和 Pns9 也是病毒原质的组分，且 Pns9 参与病毒在水稻中的复制（Akita et al., 2011），但三者在病毒复制过程中的相互关系尚未弄清，有待于进一步深入研究。

已有的研究表明，RGDV 编码的 Pns12 在植物中具有基因沉默抑制子功能（Wu et al., 2011），但本研究中利用源于 Pns12 的全长 dsRNA 可成功诱导昆虫培养细胞内的 RNAi 机制，并特异性抑制 Pns12 蛋白的表达。其可能的原因是转染到细胞内的大量 dsRNA 被细胞识别后生成了数量极多的 siRNA，而病毒编码的沉默抑制子还不足以作用于所有的 siRNA，使得 RNAi 机制仍可以发挥功能而抑制 Pns12 的表达；另一方面，Pns12 是否在昆虫体内也具有沉默抑制子功能尚不清楚，有待于在昆虫细胞内进行进一步的探讨。

4 结论

本研究通过原核表达制备了 Pns12 的特异性多克隆抗体，并利用免疫荧光标记和 RNAi 技术明确了 Pns12 是病毒原质的组分，参与病毒在昆虫培养细胞中的复制，同时表明 Pns12 可作为理想的靶标蛋白用于阻断电光叶蝉携带和传播 RGDV。

参考文献

[1] 吴维，毛倩卓，陈红燕，等．应用免疫荧光技术研究水稻条纹病毒 (RSV) 侵染介体灰飞虱卵巢的过程 [J]. 农业生物技术学报 ,2012, 20(12):1457-1462.

[2] AKITA F, MIYAZAKIN, HIBINOH, et al. Viroplasm matrix protein Pns9 from rice gall dwarf virus forms an octameric cylindrical structure[J]. Journal of General Virology,2011,92(9):2214-2221.

[3] CHEN H Y, ZHENG L M, JIA D S, et al. *Rice gall dwarf virus* exploits tubules to facilitate viral spread among cultured insect vector cells derived from leafhopper *Recilia dorsalis*[J]. Frontiers in Microbiology, 2013, 4:206.

[4] CHEN H Y, ZHENG L M, MAO Q Z, et al. Development of continuous cell culture of brown plantho pper to trace the early infection process of *Oryzaviruses* in insect vector cells[J]. Journal of Virology, 2014, 88(8):4265-4274.

[5] CONTIN R, ARNOLDI F, CAMPAGNA M, et al. Rotavirus NSP5 orchestrates recruitment of viroplasmic proteins[J]. Journal of General Virology, 2010, 91(7):1782-1793.

[6] EICHWALD C, ARNOLDI F, LAIMBACHER A S, et al. Rotavirus viroplasm fusion and perinuclear localization are dynamic processes requiring stabilized microtubules[J]. PLoSOne, 2012, 7(10):e47947.

[7] FABBRETTI E, AFRIKANOVA I, VASCOTTO F, et al. Two nonstructural rotavirus proteins, NSP2 and NSP5, form viroplasm-like structures in vivo[J]. Journal of General Virology, 1999, 80(2):333-339.

[8] HIBI T, OMURA T, SAITO Y. Double- stranded RNA of *Rice gall dwarf virus*. Journal of General Virology, 1984, 65: 1585-1590.

[9] ICHIMI K, KIKUCHI A, MORIYASU Y, et al. Sequence analysis and GTP binding ability of theeminor core protein P5 of *Rice gall dwarf virus*[J]. Japan Agricultural Research Quarterly, 2002, 36(2):83-87.

[10] JIA D S, CHEN H Y, ZHENG A L, et al. Development of an insect vector cell culture and RNA interference system to investigate the functional role of fijivirus replication protein[J]. Journal of Virology, 2012, 86(10):5800-5807.

[11] KIMURA I, OMURA T. Leafhopper cell cultures as a means for phytoreovirus research[J]. Advance in Disease Vector Research, 1988, 5: 111-135.

[12] LIU F X, ZHAO Q, RUAN X L, et al. Suppressor of RNA silencing encoded by *Rice gall dwarf virus* genome segment 11[J]. Chinese Science Bulletin, 2008, 53(3): 362-369.

[13] MAO Q Z, ZHENG S L, HAN Q M, et al. New model for the genesis and maturation of viroplasms induced by fijiviruses in insect vector cells[J]. Journal of Virology, 2013, 87(12):6819-6828.

[14] MIYAZAKI N, HAGIWARA K, NAITOW H, et al. Transcapsidation and the conserved interactions of two major structural proteins of a pair of phytoreoviruses confirm the mechanism of assembly of the outer capsid layer[J]. Journal of Molecular Biology, 2005, 345(2):229-237.

[15] MORINAKA T, PUTTA M, CHETTANACHIT D, et al. Transmission of *Rice gall dwarf virus* by cicadellid leafhoppers *Recilia dorsalis* and *Nephotettix nigropictus*in Thailand[J]. Plant Disease, 1982, 66(1):703-704.

[16] MORIYASU Y, ISHIKAWA K, KIKUCHI A, et al. Sequence analysis of Pns11, a nonstructural protein of *Rice gall dwarf virus*, and its expression and detection in infected rice plants and vector insects[J]. Virus Genes, 2000, 20(3):237-241.

[17] MORIYASU Y, MARUYAMA-FUNATSUKI W, KIKUCHI A, et al. Molecular analysis of the genome segments S1, S4, S6, S7

and S12 of a *Rice gall dwarf virus* isolate from Thailand; completion of the genomic sequence[J]. Archives of Virology, 2007, 152(7):1315-1322.

[18] OMURA T, INOUE H, MORINAKA T, et al. *Rice gall dwarf*, a new virus disease[J]. Plant Disease, 1980, 64(8):795-797.

[19] OMURA T, MINOBE Y, MATSUOKA M, et al. Location of structural proteins in particles of *Rice gall dwarf virus*[J]. Journal of General Virology, 1985, 66(4):811-815.

[20] OMURA T, MORINAKA T, INOUE H, et al. Purification and some properties of *Rice gall dwarf virus*, a new phytoreovirus [Oryzasativa] [J]. Phytopathology (USA), 1982, 72(9):1246-1249.

[21] OMURA T, YAN J, ZHONG B, et al. The P2 protein of *Rice dwarf phytoreovirus* is required for adsorption of the virus to cells of the insect vector[J]. Journal of Virology, 1998, 72(11):9370-9373.

[22] PUTTA M, CHETTANACHIT D, MORINAKA T, et al. Gall dwarf a new rice virus disease in Thailand[J]. International Rice Research News, 1980, 5(8): 10-13.

[23] SHIMIZU T, NAKAZONO-NAGAOKA E, AKITA F, et al. Hairpin RNA derived from the gene for Pns9, a viroplasm matrix protein of *Rice gall dwarf virus*, confers strong resistance to virus infection in transgenic rice plants[J]. Journal of Biotechnology, 2012, 157(3):421-427.

[24] WEI T Y, SHIMIZU T, HAGIWARA K, et al. Pns12 protein of *Rice dwarf virus* is essential for formation of viroplasms and nucleation of viral-assembly complexes[J]. Journal of General Virology, 2006, 87(2):429-438.

[25] WEI T Y, UEHARA-ICHIKI T, MIYAZAKI N, et al. Association of *Rice gall dwarf virus* with microtubules is necessary for viral release from cultured insect vector cells[J]. Journal of Virology, 2009, 83(20):10830-10835.

[26] WU J G, WANG C Z, DU Z G, et al. Identification of Pns12 as the second silencing suppressor of *Rice gall dwarf virus*[J]. Science China Life Science, 2011, 54(3): 201-208.

[27] ZHANG H M, XIN X X, YANG J, et al. Completion of the genome sequence of *Rice gall dwarf virus* from Guangxi, China[J]. Archives of Virology, 2008, 153(9):1737-1741.

Transcriptome profiling confirmed correlations between symptoms and transcriptional changes in RDV infected rice and revealed nucleolus as a possible target of RDV manipulation

Liang Yang[1,2], Zhenguo Du[1,4], Feng Gao[3], Kangcheng Wu[1,2], Lianhui Xie[1,2], Yi Li[3], Zujian Wu[1,2] and Jianguo Wu[1,2,3]

(1 Key Laboratory of Plant Virology of Fujian Province, Institute of Plant Virology, Fujian Agriculture and Forestry University, Fuzhou, Fujian 350002, China; 2 Key Laboratory of Biopesticide and Chemibiology of Ministry of Education, Fujian Agriculture and Forestry University, Fuzhou, Fujian 350002, China; 3 Peking-Yale Joint Center for Plant Molecular Genetics and Agrobiotechnology, The State Key Laboratory of Protein and Plant Gene Research, College of Life Sciences, Peking University, Beijing, 100871, China; 4 Guangdong Provincial Key Laboratory of High Technology for Plant Protection, Guangzhou, 510640, China)

Abstract: *Rice dwarf virus* (RDV) is the causal agent of rice dwarf disease, which limits rice production in many areas of south East Asia. Transcriptional changes of rice in response to RDV infection have been characterized by Shimizu et al. and Satoh et al. Both studies found induction of defense related genes and correlations between transcriptional changes and symptom development in RDV-infected rice. However, the same rice cultivar, namely Nipponbare belonging to the Japonic subspecies of rice was used in both studies. In this study, gene expression changes of the indica subspecies of rice, namely *Oryza sativa* L. ssp. *indica cv* Yixiang2292 that show moderate resistance to RDV, in response to RDV infection were characterized using an Affymetrix Rice Genome Array. Differentially expressed genes (DEGs) were classified according to their Gene Ontology (GO) annotation. The effects of transient expression of Pns11 in *Nicotiana benthamiana* on the expression of nucleolar genes were studied using real-time PCR (RT-PCR). 856 genes involved in defense or other physiological processes were identified to be DEGs, most of which showed up-regulation. Ribosome and nucleolus related genes were significantly enriched in the DEGs. Representative genes related to nucleolar function exhibited altered expression in *N. benthamiana* plants transiently expressing Pns11 of RDV. Induction of defense related genes is common for rice infected with RDV. There is a co-relation between symptom severity and transcriptional alteration in RDV infected rice. Besides ribosome, RDV may also target nucleolus to manipulate the translation machinery of rice. Given the tight links between nucleolus and ribosome, it is intriguing to speculate that RDV may enhance expression of ribosomal genes by targeting nucleolus through Pns11.

Keywords: RDV; Transcriptome profiling; Pns11, Nucleolus

1 Introduction

Viruses are obligate intracellular pathogens. They hijack host functions, divert host resources and suppress host defense responses to achieve successful infection (Whitham and Wang, 2004). These involve an array of interactions with cellular factors, which, inevitably or coincidentally, often lead to host physiological disorders manifested by a variety of disease symptoms (Matthews and Hull, 2002; Pallas and García, 2011).

Virology Journal. 2014, 11:81

Received21 November 2013; Accepted 22 April 2014; Published 6 May 2014.

Understanding molecular details from infection of a virus to symptom development of the host is one major mission of plant virologists. Transcriptome profiling has been used extensively in the past decade to understand mechanisms underlying plant-virus interaction (Whitham et al., 2006; Wise et al., 2007). Transcriptional response of plants to virus infection is shown to vary depending on virus species, virus strains and the genetic backgrounds of host plants (Whitham et al., 2006; Wise et al., 2007). However, it shows a tight link with phenotypes and thus is useful to reveal how a virus colonizes a host, how a host mounts a defense response against a virus, and how a compatible virus-host interaction results in disease symptoms (Kogovšek et al., 2010; Satoh et al., 2011; Hillung et al., 2012). Also, these studies find that some genes may be commonly regulated by different viruses in different host plants (Rodrigo et al., 2012). For example, a set of ribosomal genes have been shown to be up-regulated in Arabidopsis, *Nicotiana benthamiana* and rice infected with *Turnip mosaic virus* (TuMV), Plum pox potyvirus (PPV) and *Rice stripe virus* (RSV), respectively (Dardick, 2007; Yang et al., 2007; Satoh et al., 2010). Rice, one of the main crop plants as well as a model for monocot plant research (Goff et al., 2002), is host to many viruses. Among them, *Rice dwarf virus* (RDV), a member of the genus *Phytoreovirus* in the family *Reoviridae*, is one of the most widespread and disastrous rice-infecting viruses causing great yield reduction in south East Asia (Hagiwara et al., 2003; Miyazaki et al., 2005; Wei et al., 2006c). RDV is transmitted in a propagative and circulative manner by leafhoppers (*Nephotettix spp.*) (Chen et al., 2011). Typical symptoms associated with RDV infection include severe dwarfism, increased tilling and white chlorotic specks on the infected leaves (Hibino, 1996).

RDV are icosahedral double-shelled particles of approximately 70 nm in diameter. The genome of RDV is composed of 12segments of double stranded RNAs, which are named S1-S12, respectively, according to their migration during sodium dodecyl sulfate–polyacrylamide gel electrophoresis. S1, S2, S3, S5, S7, S8, and S9 encode seven structural proteins, namely, P1, P2, P3, P5, P7, P8, and P9, respectively. P1, a putative RNA polymerase; P5, a putative guanylyltransferase; and P7, a nonspecific nucleic acid binding protein form the core of RDV together with viral dsRNAs (Suzuki, 1995). P3 and P8 are major components of the inner and outer protein shells that encapsidate the core, respectively (Omura et al., 1989; Zheng et al., 2000). P2 and P9 are minor components of the outer capsid (Yan et al., 1996; Zhong et al., 2003). The structural features and the process of assembly of RDV virions have been well studied (Zhou et al., 2001; Nakagawa et al., 2003). Besides structural proteins, RDV encodes at least five non-structural proteins, namely Pns4, Pns6, Pns10, Pns11, and Pns12, respectively. Pns6, Pns11 and Pns12 are matrix proteins of viroplasm, which is the putative site of viral replication (Wei et al., 2006b). Pns4 is a phosphoprotein and is localized around the viroplasm matrix in insect cells (Wei et al., 2006a). Several proteins of RDV have been shown to play specific roles in RDV-rice interaction. For example, Pns6 was identified as a viral movement protein and Pns10 as a RNA silencing suppressor of RDV (Li et al., 2004; Cao et al., 2005). P2 interacts with ent-kaurene oxidases of rice, which leads to reduced biosynthesis of gibberellins and rice dwarf symptoms (Zhu et al., 2005).

In this study, the transcriptome of the *indica* subspecies of rice, namely *Oryza sativa* L. ssp. indica cv Yixiang2292, in response to RDV infection was profiled using Affymetrix Gene Chips, which contains probes representing the entire genome of rice (Goff et al., 2002) (WWW.affymetrix.com). Our results further confirm the notion that induction of defense related genes is common for rice infected with RDV and there are correlations between transcriptional changes and symptom development in RDV-infected rice.

2 Results

2.1 Transcriptome profiling of RDV-infected rice

For transcriptome analysis, rice seedlings were virus- or mock- inoculated. Total RNAs were extracted at 22d post inoculation (dpi), i.e. the earliest time when infection could be confirmed by the appearance of symptoms. The GeneChip hybridization and scanning were performed at the Microarray Resource Laboratory at Beijing CapitalBio Corporation, Beijing, China, in which GeneChip microarray service was certificated by Affymetrix. The microarray data were analyzed using SAM (Significant Analysis of Microarray) software. Deferentially expressed genes (DEGs) were identified with the criteria of fold changes > 1.5 and false positive rate (q-value) < 0.058. In this way, a total of 856 genes were identified to be DEGs, in which 838genes were upregulated and 18 downregulated. A list of the genes identified is presented in Additional file 1: Table S1.

2.2 Classification of DEGs

To get an overview of the functions of the DEGs, DEGs were classified according to their function. The classification was done manually based on gene annotations (http://rice.plantbiology.msu.edu/) and literature searching. Among the 856 DEGs, 275 genes have no annotations or were simply annotated as hypothetical protein/expressed protein. These genes were not analyzed further in our study. The remaining 581 genes were classified into 14 non-redundant categories (Fig. 1, Additional file 2: Table S2). As shown in Fig.1, unclassified genes formed the largest group. They referred to genes that were difficult to be classified into groups. Of the genes that have been classified, three categories are of particular interests to us.

2.3 Defense/stress related genes

This set of genes forms the second largest group. They include PR genes, markers of defense responses; several genes encoding WRKY transcription factors, key regulators of defense responses (Eulgem et al., 2000); L-ascorbate oxidase and Peroxidase genes, important modulators of oxidative stress, among others (Arora et al., 2002).

2.4 Protein synthesis related genes

In all, 36 RDV responsive genes were classified into this category (Fig.1). The large number of this category was due mostly to ribosomal genes. These included 27 genes encoding cytosolic ribosomal subunits, 2 mitochondrial ribosomal genes and 1 gene encoding a chloroplast ribosome precursor. Other genes belonging to this category include those involved in translation initiation, termination and tRNA metabolism. All these genes were upregulated.

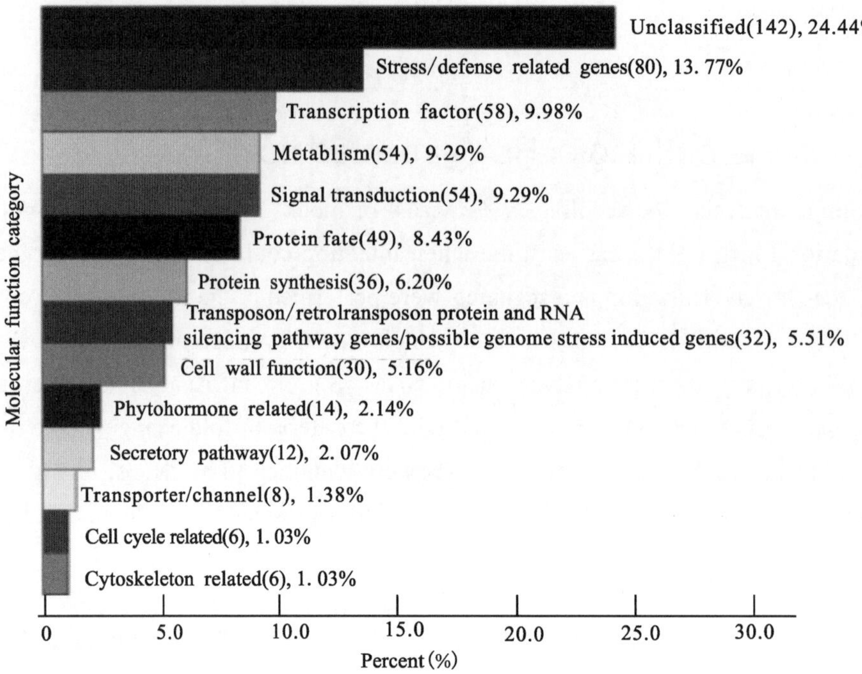

Fig. 1 An overview of the functional classification of the 581 RDV responsive genes

Number of genes and relative percentage for eachcategory were indicated. For a list of the genes in each category see Additional file 2: Table S2

2.5 Transposon/retrotransposon protein/RNA silencing pathway genes/ possible genome stress related genes

Surprisingly, a large number of genes encoding transposon/retrotransposon-related proteins were affected (Fig.1). Normally, tansposon or transposon-related genes are transcriptionally inert because of epigenetic regulations. Altered expression of these kinds of genes indicated that the rice genome was suffering a genomic stress. Indeed, genes involved in DNA recombination (AK063836, encoding a Single-strand binding protein family protein; BQ908269, encoding a RuvB-like 1 protein; AB079873, encoding a Meiotic recombination protein DMC1 homolog.), DNA repair (AK101485, encoding a DNA repair ATPase) and chromosome assembly (AK108572, encoding a complex 1 protein containing protein) were all upregulated. The RNA silencing pathway, which plays a pivotal role in epigenetic regulations, was also significantly affected. Genes functioning in this pathway such as AGOs, RDRs showed marked up-regulation (Fig. 1, Additional file 2: Table S2).

2.6 GO enrichment analysis

DEGs were also classified according to Gene Ontology (GO) cellular component, which indicates the location or suspected location of a gene in a cell (Berardini et al., 2004). As shown in Table 1, six GO cellular component terms were significantly enriched in DEGs, cell wall, nucleus, ribosome, cytosol, extracellular region, and nucleolus (p<0.01). We were interested in the concomitant enrichment of the two GO terms Ribosome and Nucleolus, because nucleolus is the site of ribosomal RNA synthesis and ribosome maturation.

Table1 Results of GO cellular component analysis with MAS 2.0 system

GO number	Cellular component	Total change genes	*p*-value	*q*-value
GO:0005618	Cell wall	112	0.0	0.0
GO:0005634	Nucleus	101	0.0	0.0
GO:0005840	Ribosome	24	0.0	0.0
GO:0005829	Cytosol	21	1.0E-6	3.0E-6
GO:0005576	Extracellular region	14	3.64E-4	5.24E-4
GO:0005730	Nucleolus	20	0.0014	0.0018
GO:0005886	Plasma membrane	23	0.0128	0.0148
GO:0016020	Membrane	103	0.0152	0.016
GO:0005740	Mitochondrial envelope	2	0.1273	0.1182
GO:0005794	Golgi apparatus	4	0.1314	0.1217
GO:0005635	Nuclear envelope	3	0.2492	0.2231
GO:0005856	Cytoskeleton	7	0.3283	0.2924
GO:0005783	Endoplasmic reticulum	2	0.3565	0.3161
GO:0016023	Cytoplasmic membrane-bound vesicle	62	0.4074	0.3457
GO:0005773	Vacuole	3	0.5308	0.4267
GO:0005654	Nucleoplasm	1	0.5373	0.4314
GO:0005777	Peroxisome	1	0.671	0.5287
GO:0009579	Thylakoid	12	0.6852	0.5375
GO:0005739	Mitochondrion	137	0.9989	0.5485
GO:0009536	Plastid	65	1.0	0.5485
GO:0005622	Intracellular	10	1.0	0.5485
GO:0005737	Cytoplasm	82	1.0	0.5485
GO:0005575	Cellular_component	3	1.0	0.5485
GO:0005623	Cell	8	1.0	0.5485

2.7 Verification of the microarray data

The accuracy of the microarray data was verified by qRT-PCR. Seventeen genes including ribosomal, nucleolar and transposon/retrotransposon related genes and genes involved in RNA silencing, auxin signal, and cell wall function were selected. The CP gene of RDV was used to as a control. As shown in Fig. 2 and Table 2, qRT-PCR results of all 17 RDV responsive genes selected were consistent with the microarray data.

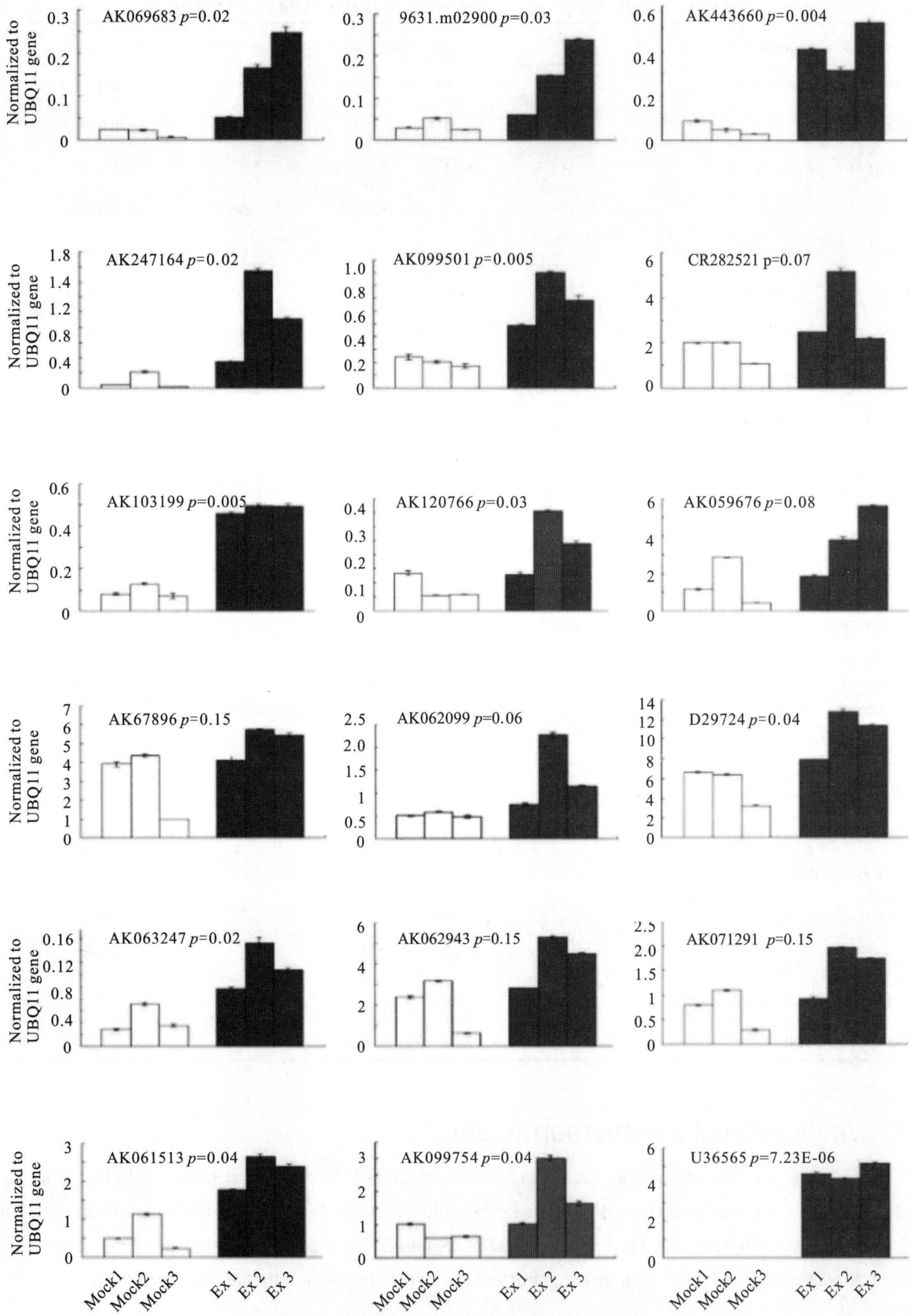

Fig. 2 Validation of microarray results using qRT-PCR

Shown are relative expression ratios to UBQ11 in inoculated (gray bars (Ex 1, 2 and 3)) and mock-inoculated (Open bars (Mock 1, 2 and 3)) rice. Means of three replicate experiments with standard deviations and p-values are shown in Table 1

Table 2 Real-time PCR to verify expression pattern of differentially expressed genes from the microarray experiment (for a list of these genes primer sequence, see Additional file 2: Table S2)

GB.accession	Fold change (Ex/Mock) ± SD		p-value[B]	p-value[A]	Description
	Microarray	qRT-PCR			
AK069685	5.4983 ± 1.581	6.6246 ± 3.6546	0	0.0204	Piwi domain containing protein, expressed
9631.m02900	2.6141 ± 0.7001	2.9988 ± 0.530	0	0.0316	Small nuclear ribonucleoprotein G, putative, expressed
AF443600	6.2009 ± 2.7810	6.5142 ± 2.1441	0.0042	0.0045	Glucan endo-1,3-beta-glucosidase GII precursor, putative, expressed
AF247164	6.1081 ± 6.8408	6.7585 ±11.4824	0.0236	0.0235	Alpha-expansin 4 precursor, putative, expressed
AK099501	2.1007 ± 0.9030	2.4711 ± 0.5329	0.0261	0.005	Ribonucleoprotein, putative, expressed
CR282531	1.7015 ± 0.2375	1.4174 ± 0.5866	0.0303	0.0694	60S acidic ribosomal protein P2A, putative, expressed
AK103199	3.2899 ± 2.5965	4.0319 ± 3.4615	0.0302	0.0057	Transposon protein, putative, CACTA, En/Spm sub-class, expressed
AK062099	1.6257 ± 0.3806	1.7461 ± 0.3890	0.0308	0.0601	Ribosomal L28e protein family protein, expressed
AK059679	1.8421 ± 0.303	1.9076 ± 0.9539	0.0308	0.0892	60S ribosomal protein L38, putative, expressed
AK067896	1.6171 ± 0.2825	1.3415 ± 0.6274	0.04342	0.1508	60S ribosomal protein L6, putative, expressed
AK120766	1.7459 ± 0.5397	2.1115 ± 0.8514	0.04342	0.0293	Piwi domain containing protein, expressed
D29724	1.529 ± 0.2032	1.51084 ±0.2562	0.04342	0.0408	Peptide chain release factor 2, putative, expressed, 60S ribosomal protein L38, putative, expressed
AK063247	2.5979 ± 1.4057	2.0925 ± 2.0497	0.04342	0.0197	Auxin-induced protein TGSAUR12, putative, expressed
AK062943	1.5919 ± 0.0932	1.5679 ± 1.0226	0.0483	0.0986	40S ribosomal protein S15a, putative, expressed
AK071291	1.6782 ± 0.3353	1.5959 ± 0.6893	0.0483	0.0831	Fibrillarin-2, putative, expressed
AK061513	2.5152 ± 0.6761	2.8184 ± 1.9757	0.0588	0.0434	Nucleoid DNA-binding protein cnd41, putative, expressed
AK099754	1.7647 ± 0.2432	1.7471 ± 0.2758	0.0588	0.0487	Nucleolar protein NOP5, putative, expressed
U36565	–	4.8787 ± 2.0832	–	7.23E-06	Rice dwarf virus coat protein mRNA, complete cds
UBQ11*	–	–	–	0.1955	

Notice: [A]p-value form microarray experiment; [B]p-value from qRT-PCR; *Rice *UBQ11* was used as a control for qRT-PCR.

2.8 The nucleoli were affected in RDV-infected rice

Transmission electron microscopy was used to determine if there are any pathologic changes related to nucleoli in RDV-infected cells. As shown in Fig. 3, two forms of nucleoli were observed in infected or control rice plants: small, round and concentrated electron-dense spheres (type 1) and big, irregular sub-cellular compartments filled with dispersive electron-dense aggregates (type 2). Statistical analysis confirmed that the number of type 2 nucleolus in RDV infected rice (61%~67%) was higher than that of type 1, whereas in control rice plants, the number of type 1 nucleolus was higher than that of type 2 (27%~33%).

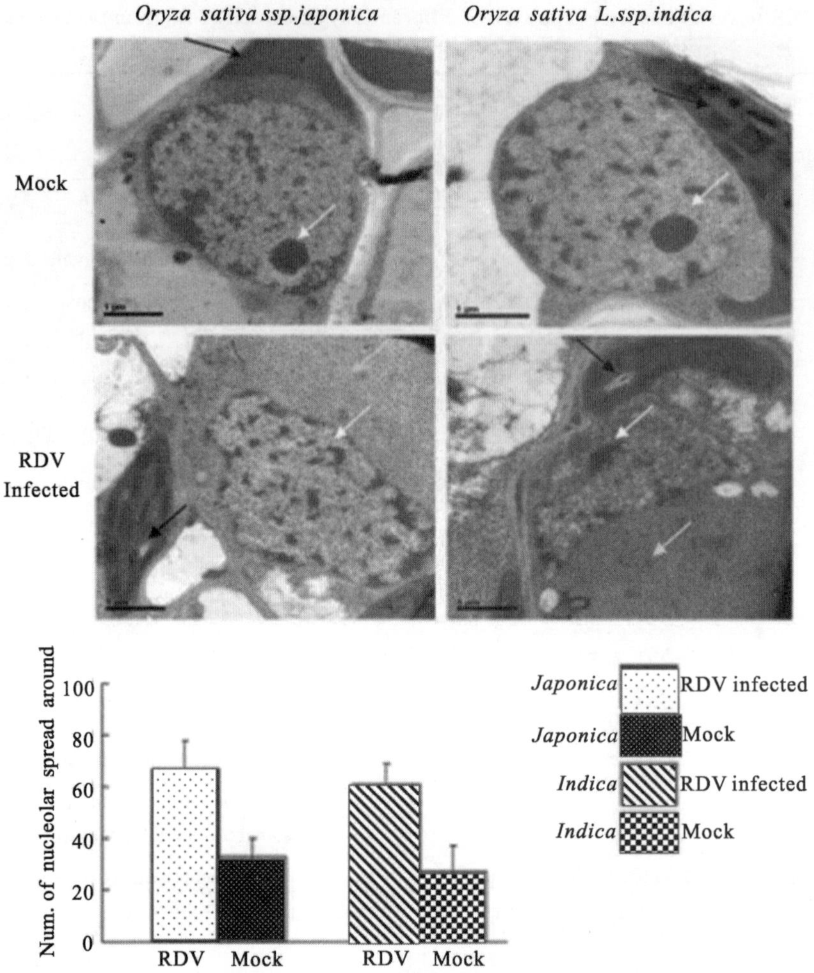

Fig. 3 The structure of nucleolus in RDV-infected rice

White arrowhead: Nucleolus; Black arrowhead: Chloroplast; Yellow arrowhead: RDV Virion. Rectangle: the biological statistics of number of nucleolar spread around

2.9 RDV Pns11 regulates the transcript levels of nucleolus-related genes in tobacco cells

The finding that many nucleolus targeting genes were de-regulated in RDV infected rice suggests that RDV may manipulate nucleolar functions. The subcellular localization of all RDV-encoded proteins was predicted by Predict NLS (http://www.biologydir.com/nls-prediction/ p1.html). Only Pns11 has a nuclear localization signal (NLS) with NLSI and NLSII domains belonging to the bipartite NLS (Abel and Theologis, 1995) (Fig. 4A). So Pns11 may be responsible for alteration of nucleolar genes in RDV infected rice. To test this possibility, the expression levels of two nucleolar genes were studied in *N. benthamiana* leaves expressing RDV Pns11. qRT-PCR results revealed that the two genes (AB207972 and AM269909 encoding fibrillarin) were upregulated significantly. As controls, two genes related to defense (Glucan endo-1,3-beta-glucosidase GII precursor, M60402 and M60403) showed reduced expression, whereas two genes functioning in RNA silencing (DQ321488 and DQ321489) remained unchanged (Fig. 4B).

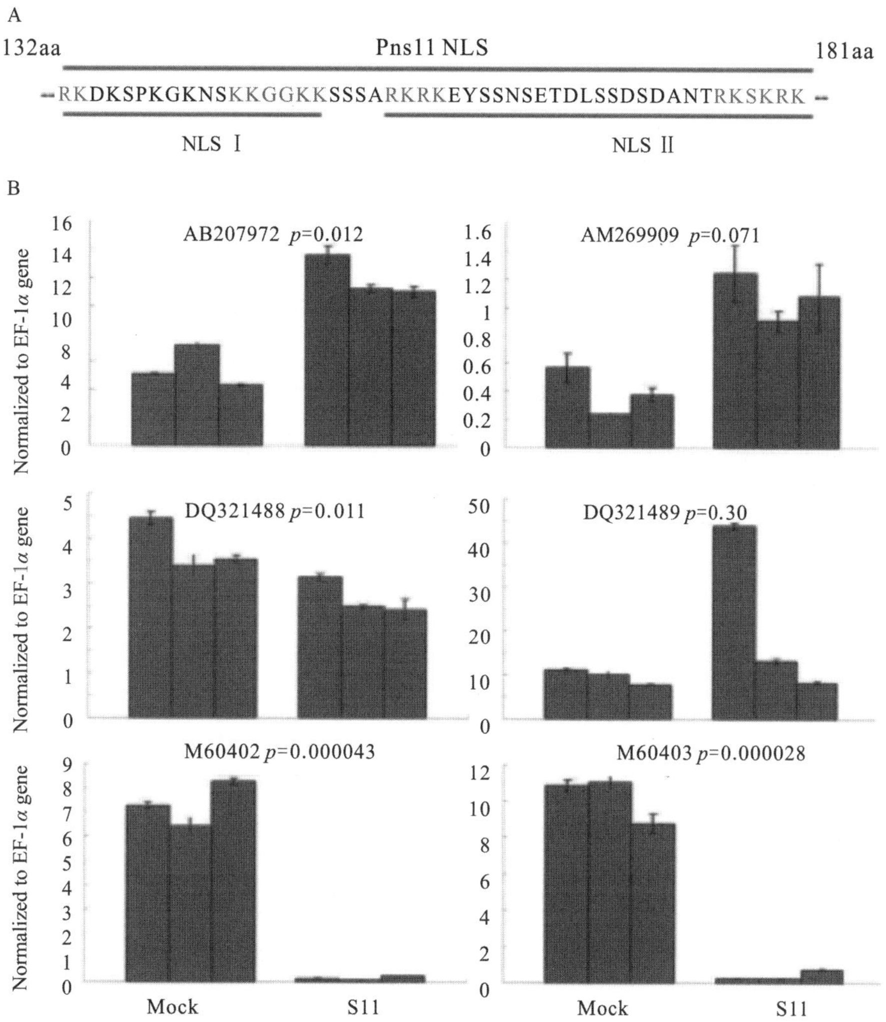

Fig. 4 RDV Pns11 regulates the transcript levels of nucleolus-related genes

(A) Schematic representation of Nuclear localization signal (NLS) of RDV Pns11. (B) qRT-PCR was used to monitor the mRNA of nucleolus-related genes. Shown are relative expression ratios to EF-1α. Three replicate experiments were done to reduce biological variation

3 Discussion

The transcriptome of RDV infected rice plants was profiled in this study. A number of genes are differentially expressed in RDV infected rice. Changes of most of these genes are consistent with previous studies carriedout using N. *benthamiana* or *Arabidopsis thaliana* (Itaya et al., 2002; Golem and Culver, 2003; Whitham et al., 2003; Ventelon-Debout et al., 2004; Whitham et al., 2006; Dardick, 2007; Babu et al., 2008). Also, we find induction of a set of defense related genes including PR genes, WRKY transcription factors and several genes functioning in RNA silencing.This is consistent with reports of Shimizu et al. (Rodrigo et al., 2012) and Satoh et al. (Satoh et al., 2011) showing that increased expression of defense related genes may be a common response of rice infected with RDV (Shimizu et al., 2007; Satoh et al., 2011). However, our results indicate that RDV induced the expression of far more genes than it suppressed. This is in sharp contrast to the report of Shimizu et al. (Shimizu et al., 2007). Multiple reasons may be responsible for the inconsistency. But the most plausible one is that transcriptome change in response to RDV infection is host-specific. In the study of Shimizu et al. (Shimizu et al., 2007), the Japonica subspecies of rice, namely *Oryza sativa L.cv. Nipponbare*, was used, whereas in this study, the *indica* subspecies of rice, namely *Oryza sativa* L. ssp. *indica cv* Yixiang 2292, was used. Yixiang 2292, the rice variety used in this study, shows moderate resistance to RDV infection. It can develop typical symptoms of RDV infection, but the symptoms are not as severe as those of more susceptible varieties. A number of recent studies have demonstrated that there is a co-relation between symptom severity and transcriptional alteration in different virus-host combinations (Kogovšek et al., 2010; Hanssen et al., 2011; Satoh et al., 2011; Hillung et al., 2012).

Many genes related to protein synthesis (Fig. 1) were found and the GO term Ribosome was significantly enriched in the DEGs (Table 1). This is consistent with several studies showing that up-regulation of ribosomal genes and a set of other genes involved in protein synthesis could be a general response of plants to many viruses (Dardick, 2007; Yang et al., 2007; Satoh et al., 2010). It has been suggested that this may be a strategy used by the virus to enhance the capacity of the cell to synthesize proteins (Dardick, 2007; Yang et al., 2007; Satoh et al., 2010).

As a two-subunit ribonucleoprotein complex comprising tens of ribosomal proteins and four species of ribosomal RNAs, the biogenesis of ribosome is one of the most energy consuming cellular processes (Green and Noller, 1997; Maguire and Zimmermann, 2001; Fatica and Tollervey, 2002). So it is anticipated that synthesis of ribosomal components should be downregulated in response to environmental cues, as it has been shown in yeast and in Arabidopsis (Gasch et al., 2000; Kojima et al., 2007). Therefore, increased expression of ribosomal genes in virus infected plants may be a result of specific virushost interaction.

Here, we show that RDV infection also causes a significant alteration of many nucleolar genes (Table 1). The fact that RDV Pns11 has a nuclear localization signal and induces the expression of nucleolus-related genes in tobacco (Fig. 4) and the observation that nucleoli seems to be affected in RDV infected rice (Fig. 3) support the notion that this alteration is specific and may be useful for RDV. Nucleolus is the site of ribosomal RNA synthesis and processing and ribosome maturation (Boisvert et al., 2007). Therefore, it is possible that, besides ribosome, RDV may also target nucleolus to manipulate the translation machinery of rice. Interestingly, there is evidence that certain nucleolar components or its overall state play a crucial role in controlling ribosomal gene expression and biogenesis (Laferté et al., 2006; Kojima et al., 2007). So it is intriguing to speculate that RDV may specifically target nucleolus to enhance expression of ribosomal genes. In this regard, it is worth noting that a number of viruses, including many RNA viruses whose primary site of replication is the cytoplasm, encode special proteins to target nucleolus (Hiscox, 2007). It would be very interesting to test the link between nucleolar targeting of these viruses and ribosome biogenesis of their hosts.

Besides ribosomal genes, malfunction of nucleolus may be responsible for altered expression of many other genes detected in this study. For example, emerging evidence suggests that nucleolus might play a role in the small interfering RNA (siRNA) pathway (Pontes et al., 2006; Boisvert et al., 2007). Therefore, many genes controlled by siRNAs may be altered because of malfunction of nucleolus in RDV infected rice. Consistent with this, we found a large number of genes encoding transposon/retrotransposon-related proteins in the DEGs (Fig. 1). It is well known that transposon or transposon-related genes are transcriptionally controlled by epigenetic modifications, in which siRNAs play an important role (Henderson and Jacobsen, 2007). To our knowledge, altered expression of transposon/retrotransposon-related genes has never been reported in virus infected plants. However, we do not favor the possibility that this is specific to RDV. Instead, DEGs were classified automatically using web-based tools in most previous studies. In this way, transposon/retrotransposon-related genes tend to be classified into "unknown" genes and be excluded for further analysis.

4 Materials and methods

4.1 Sources of virus and insects

RDV Fujian isolate, China, was maintained in "Taizhong-1" rice plants grown in greenhouses at 25 ± 3°C, 55% ± 5% RH and under natural sunlight. Insects (*Nephotettix cincticeps*) source: high infectious green rice leafhoppers cultured in our lab with five generations of artificial rearing on rice seedlings.

4.2 Plant growth and inoculation

Seeds (*Oryza sativa* L. ssp. *indicacv* Yixiang2292) were sowed and germinated on a pot (60mm in diameter and 50mm in height) that had been filled with commercial soil mixture (FAFARD SOILS, Southern Agricultural Insecticides Inc Palmetto, FL, 34221). Rice seedlings were subjected to a two-day inoculation using high infectious green rice leafhoppers or virus-free insects (for mock inoculation) by the one test tube-one-seedling method. Inoculated seedlings were transplanted to an iron dish filled with cultivation layer soil of experimental farmland. They were kept in a south-facing greenhouse at 25 ± 3°C with 55% ± 5% RH and under natural sunlight. The aerial parts of 8 entire rice plants were sampled randomly and pooled at 22dpi, i.e., 7d after appearance of the symptom (the earliest symptoms, i.e. white chlorotic specks in newly developed leaves, appeared at approximately 15dpi). The samples were flash-frozen in liquid nitrogen, and stored at −80°C for until use.

4.3 RNA preparation and microarray hybridization and scanning

Total RNA was extracted from the virus- or mock-inoculated leaves with TRIzol reagent (Invitrogen). RNA was further purified using RNeasy columns (Qiagen, Valencia, CA, USA). An aliquot of 2μg of total RNA was used to synthesize double-stranded cDNA, then produced biotin-tagged cRNA using MessageAmp™ II aRNA Amplification Kit. The resulting bio-tagged cRNA were fragmented to strands of 35 to 200bases in length according to Affymetrix's protocols. The fragmented cRNA was hybridized to Affymetrix Rice Genome Array containing 51,279 transcripts which includes approximately 48,564 japonica transcripts and 1,260 transcripts representing the *indica* cultivar (www.affymetrix.com). Hybridization was performed at 45°C with rotation for 16h (AffymetrixGeneChip Hybridization Oven 640). The GeneChip arrays were washed and then stained (streptavidin-phycoerythrin) on an Affymetrix Fluidics Station 450 followed by scanning on GeneChip Scanner 3000. We altogether used 6chips to perform the analysis of 6 RNA samples.

4.4 Microarray data analysis

Hybridization data were analyzed using GeneChip Operating software (GCOS 1.4). The scanned images were firstly assessed by visual inspection then analyzed togenerate raw data files saved as CEL files using the default setting of GCOS 1.4. A global scaling procedure was performed to normalize the arrays using dChip software. In a comparison analysis, two class unpaired method was applied in the Significant Analysis of Microarray (SAM) software to identify significantly differentially expressed genes between Test group and Control group. All differentially expressed genes were analyzed using the web-based Molecular Annotation System 3.0 (MAS 3.0, http://bioinfo.capitalbio.com/mas/). MAS 2.0 integrate three different open source pathway resources-KEGG, BioCarta and GenMAPP. In the MAS 3.0 tool, the pathways and GO were ranked with statistical significance by calculating their *P*-values based on hypergeometric distribution. The GeneChip hybridization and scanning were performed at the Microarray Resource Laboratory at Beijing CapitalBio Corporation, Beijing, China, in which GeneChip microarray service was certificated by Affymetrix.

4.5 Transient expression in leaves of *N. benthamiana*

Agro-infiltration for transient expression in leaves of *Nicotianabenthamiana*, Leuzinger was carried out as described (Leuzinger et al., 2013). Briefly, individual Agrobacterium GV3101 strains with different expression constructs (or empty vector as control) were co-infiltrated into *N. benthamiana* leaves using a syringe without needle. After 3d of transient expression, leaves were harvested for RNA extraction.

4.6 Real-time PCR assay

Total RNA used for verification of microarray data was prepared from plants that had been grown independently of those used for isolation of RNA for microarray analysis. One Step RNA PCR Kit (AMV) (TaKaRa, Japan) was used. Gene-specific primers were designed by Primer 5 (for a list of the primers used in this study, see Additional file 3:Table S3) and synthesized by Boya Company (Shanghai, China). Relative quantitation method was used. Rice UBQ11 gene and tobacco EF-1α were used as the control to normalize all data (Schmittgen and Livak, 2008) (for a list of these genes primer sequence, see Additional file 3: Table S3).

4.7 Electron microscopy

For electron microscopy experiments, RDV infected and health rice samples were fixed with 2.5% glutaraldehyde at 4°C overnight, washed in 0.1 M phosphate buffer (pH 7.0) for 3times (15 min per time), and post-fixed in phosphate-buffered 1.0% OsO_4 for 2h. Then the tissues were buffer-washed, dehydrated with ethanol (50%, 70%, 80%, 90%, 95% and 100%) and embedded in Epon-Araldite. Ultrathin sections (70–90nm) were cut with a Reichert ultramicrotome, stained with aqueous uranyl acetate and lead citrate, and examined with a Jeol JEM-1230 transmission electron microscope (Jeol, Tokyo, Japan).

Acknowledgements

We thank Guangpu Li for his critical reading of the manuscript. This work was supported by grants from the National Basic Research Program 973 (2014CB138402, 2013CBA01403 and 2010CB126203), Natural Science Foundation of China (31272018, 31201491 and 31171821), Key Project of the National Research Program of China (2012BAD19B03), Doctoral Fund of Ministry of Education of China (20113515110001, 20113515120004 and 20123515120005), Education department of Fujian Province Office Program (JA11078 and JA11080) and Natural Science Foundation of Fujian Province of China (2011 J05051 and 2013 J01089).

References

[1] ABEL S, THEOLOGIS A. A polymorphic bipartite motif signals nuclear targeting of early auxin-inducible proteins related to PS-IAA4 from pea (Pisum sativum) [J]. Plant Journal, 2010, 8(1):87-96.

[2] ARORA A, SAIRAM R, SRIVASTAVA G. Oxidative stress and antioxidative system in plants[J]. Current Science, 2002, 82(10):1227-1238.

[3] BABU M, GRIFFITHS J S, HUANG T S, et al. Altered gene expression changes in Arabidopsis leaf tissues and protoplasts in response to Plum pox virus infection[J]. BMC Genomics, 2008, 9(1):325.

[4] BERARDINI T Z, MUNDODI S, REISER L, et al. Functional annotation of the Arabidopsis genome using controlled

vocabularies[J]. Plant Physiology, 2004, 135(2): 745-755.

[5] BOISVERT F-M, VAN KONINGSBRUGGEN S, NAVASCUÉS J, et al. The multifunctional nucleolus[J]. Nature Reviews Molecular Cell Biology, 2007, 8(7): 574-585.

[6] CAO X, ZHOU P, ZHANG X, et al. Identification of an RNA silencing suppressor from a plant double-stranded RNA virus[J]. Journal of Virology, 2005, 79(20): 13018-13027.

[7] CHEN H, CHEN Q, OMURA T, et al. Sequential infection of *Rice dwarf virus* in the internal organs of its insect vector after ingestion of virus[J]. Virus Research, 2011,160(1-2):389-394.

[8] DARDICK C. Comparative expression profiling of *Nicotiana benthamiana* leaves systemically infected with three fruit tree viruses[J]. Molecular Plant-Microbe Interactions, 2007, 20(8):1004-1017.

[9] EULGEM T, RUSHTON P J, ROBATZEK S, et al. The WRKY superfamily of plant transcription factors[J]. Trends in Plant Science, 2000, 5(5):199-206.

[10] FATICA A, TOLLERVEY D. Making ribosomes[J]. Current Opinion in Cell Biology, 2002, 14(3):313-318.

[11] GASCH A P, SPELLMAN P T, KAO C M, et al. Genomic expression programs in the response of yeast cells to environmental changes[J]. Molecular Biology of the Cell, 1998, 11(12):4241-4257

[12] GOFF S A, RICKE D, LAN T-H, et al. A draft sequence of the rice genome (Oryza sativa L ssp. japonica)[J]. Science, 2002, 296(5565): 92-100.

[13] GOLEM S, CULVER J N. *Tobacco mosaic virus* induced alterations in the gene expression profile of *Arabidopsis thaliana*[J]. Molecular Plant-Microbe Interactions, 2003, 16(8):681-688.

[14] GREEN R, NOLLER H F. Ribosomes and translation[J]. Annual Review of Biochemistry, 1997, 66: 679-716.

[15] HAGIWARA K, HIGASHI T, NAMBA K, et al. Assembly of single-shelled cores and double-shelled virus-like particles after baculovirus expression of major structural proteins P3, P7 and P8 of *Rice dwarf virus*[J]. Journal of General Virology, 2003, 84(4):981-984.

[16] HANSSEN I M, VAN ESSE H P, BALLESTER A-R, et al. Differential tomato transcriptomic responses induced by pepino mosaic virus isolates with differential aggressiveness[J]. Plant Physiology, 2011, 156:301-318.

[17] HENDERSON I R, JACOBSEN S E. Epigenetic inheritance in plants[J]. Nature, 2007, 447(7143):418-424.

[18] HIBINO H. Biology and epidemiology of rice viruses[J]. Annual Review of Phytopathology, 1996, 34(1):249-274.

[19] HILLUNG J, CUEVAS J M, ELENA S F. Transcript profiling of different *Arabidopsis thaliana* ecotypes in response to *Tobacco etch potyvirus* infection[J]. Frontiers in Microbiology, 2012, 3:229.

[20] HISCOX J A. RNA viruses: hijacking the dynamic nucleolus[J]. Nature Reviews Microbiology, 2007, 5(2):119-127.

[21] ITAYA A, MATSUDA Y, GONZALES R A, et al. Potato spindle tuber viroid strains of different pathogenicity induces and suppresses expression of common and unique genes in infected tomato[J]. Molecular Plant-Microbe Interactions, 2002, 15(10):990-999.

[22] KOGOVŠEK P, POMPE-NOVAK M, BAEBLER Š, et al. Aggressive and mild *Potato virus Y* isolates trigger different specific responses in susceptible potato plants[J]. Plant Pathology, 2010, 59(6):1121-1132.

[23] KOJIMA H, SUZUKI T, KATO T, et al. Sugar-inducible expression of the nucleolin-1 gene of Arabidopsis thaliana and its role in ribosome synthesis, growth and development[J]. Plant Journal, 2007, 49(6): 1053-1063.

[24] LAFERTÉ A, FAVRY E, SENTENAC A, et al. The transcriptional activity of RNA polymerase I is a key determinant for the level of all ribosome components[J]. Genes& Development, 2006, 20(15): 2030-2040.

[25] LEUZINGER K, DENT M, HURTADO J, et al. Efficient agroinfiltration of plants for high-level transient expression of recombinant proteins[J]. Journal of Visualized ExperimentsJove, 2013,(77): e50521.

[26] LI Y, BAO Y M, WEI C H, et al. *Rice dwarf phytoreovirus* segment S6-encoded nonstructural protein has a cell-to-cell movement function[J]. Journal of Virology, 2004, 78(10):5382-5389.

[27] MAGUIRE B A, ZIMMERMANN R A. The ribosome in focus[J]. Cell, 2001, 104(6):813-816.

[28] MATTHEWS R E F. Matthews' plant virology[J]. Matthews Plant Virology, 2002.

[29] MIYAZAKI N, HAGIWARA K, NAITOW H, et al. Transcapsidation and the conserved interactions of two major structural proteins of a pair of phytoreoviruses confirm the mechanism of assembly of the outer capsid layer[J]. Journal of Molecular Biology, 2005, 345(2):229-237.

[30] NAKAGAWA A, MIYAZAKI N, TAKA J, et al. The atomic structure of rice dwarf virus reveals the self-assembly mechanism of component proteins[J]. Structure, 2003, 11(10):1227-1238.

[31] OMURA T, ISHIKAWA K, HIRANO H, et al. The outer capsid protein of rice dwarf virus is encoded by genome segment S8[J]. Journal of General Virology, 1989, 70(10):2759-2764.

[32] PALLAS V, GARCÍA J A. How do plant viruses induce disease? Interactions and interference with host components[J]. Journal of General Virology, 2011, 92(12):2691-2705.

[33] PONTES O, LI C F, NUNES P C, et al. The *Arabidopsis* chromatin-modifying nuclear siRNA pathway involves a nucleolar RNA processing center[J]. Cell, 2006, 126(1):79-92.

[34] RODRIGO G, CARRERA J, RUIZ-FERRER V, et al. A meta-analysis reveals the commonalities and differences in *Arabidopsis thaliana* response to different viral pathogens[J]. PLoS One, 2012, 7(7):e40526.

[35] SATOH K, KONDOH H, SASAYA T, et al. Selective modification of rice (Oryza sativa) gene expression by *Rice stripe virus* infection[J]. The Journal of General Virology, 2010, 91(1): 294-305.

[36] SATOH K, SHIMIZU T, KONDOH H, et al. Relationship between symptoms and gene expression induced by the infection of three strains of *Rice dwarf virus*[J]. PLoS One, 2011, 6(3): e18094.

[37] SCHMITTGEN T D, LIVAK K J. Analyzing real-time PCR data by the comparative CT method[J]. Nature Protocols, 2008, 3: 1101-1108.

[38] SHIMIZU T, SATOH K, KIKUCHI S, et al. The repression of cell wall-and plastid-related genes and the induction of defense-related genes in rice plants infected with *Rice dwarf virus*[J]. Molecular Plant-Microbe Interactions, 2007, 20(3): 247-254.

[39] SUZUKI N. Molecular analysis of the *Rice dwarf virus* genome[J]. Seminars in Virology, 1995, 6(2):89-95.

[40] VENTELON-DEBOUT M, DELALANDE F, BRIZARD J P, et al. Proteome analysis of cultivar-specific deregulations of *Oryza sativaindica* and *O sativajaponica* cellular suspensions undergoing *Rice yellow mottle virus* infection[J]. Proteomics, 2004, 4(1): 216-225.

[41] WEI T, KIKUCHI A, SUZUKI N, et al. Pns4 of *Rice dwarf virus* is a phosphoprotein, is localized around the viroplasm matrix, and forms minitubules[J]. Archives of Virology, 2006, 151(9):1701-1712.

[42] WEI T, SHIMIZU T, HAGIWARA K, et al. Pns12 protein of Rice dwarf virus is essential for formation of viroplasms and nucleation of viral-assembly complexes[J]. Journal of General Virology, 2006, 87(2):429-438.

[43] WEI T, KIKUCHI A, MORIYASU Y, et al. The spread of *Rice dwarf virus* among cells of its insect vector exploits virus-induced tubular structures[J]. Journal of Virology, 2006, 80(17):8593-8602

[44] WHITHAM S A, WANG Y. Roles for host factors in plant viral pathogenicity[J]. Current Opinion in Plant Biology, 2004, 7(4):365-371.

[45] WHITHAM S A, YANG C, GOODIN M M. Global impact: elucidating plant responses to viral infection[J]. Molecular Plant-Microbe Interactions, 2006, 19(11):1207-1215.

[46] WHITHAM S A, QUAN S, CHANG H S, et al. Diverse RNA viruses elicit the expression of common sets of genes in susceptible *Arabidopsis thaliana* plants[J]. Plant Journal, 2010, 33(2):271-283.

[47] WISE R P, MOSCOU M J, BOGDANOVE A J, et al. Transcript profiling in host–pathogen interactions[J]. Annual Review of Phytopathology, 2007, 45(1):329-369.

[48] YAN J, TOMARU M, TAKAHASHI A, et al. P2 protein encoded by genome segment S2 of *Rice dwarf phytoreovirus* is essential

for virus infection[J]. Virology, 1996, 224(2):539-541.

[49] YANG C, GUO R, JIE F, et al. Spatial analysis of Arabidopsis thaliana gene expression in response to Turnip mosaic virus infection[J]. Molecular Plant-Microbe Interactions, 2007, 20(4):358-370.

[50] ZHENG H, YU L, WEI C, et al. Assembly of double-shelled, virus-like particles in transgenic rice plants expressing two major structural proteins of *Rice dwarf virus*[J]. Journal of Virology, 2000, 74(20):9808-9810.

[51] ZHONG B, KIKUCHI A, MORIYASU Y, et al. A minor outer capsid protein, P9, of *Rice dwarf virus*[J]. Archives of Virology, 2003, 148(11):2275-2280.

[52] ZHOU Z H, BAKER M L, JIANG W, et al. Electron cryomicroscopy and bioinformatics suggest protein fold models for *Rice dwarf virus*[J]. Nature Structural Biology, 2001, 8(10):868-873.

[53] ZHU S, GAO F, CAO X, et al. The *Rice dwarf virus* P2 protein interacts with ent-kaurene oxidases in vivo, leading to reduced biosynthesis of gibberellins and rice dwarf symptoms[J]. Plant Physiology, 2005, 139(4):1935-1945.

Infection route of rice grassy stunt virus a tenuivirus in the body of its brown planthopper vector *Nilaparvata lugens* (Hemiptera: Delphacidae) after ingestion of virus

Limin Zheng, Qianzhuo Mao, Lianhui Xie, Taiyun Wei

Fujian Province Key Laboratory of Plant Virology, Institute of Plant Virology, Fujian Agriculture and Forestry University, Fuzhou, Fujian 350002, China

Abstract: *Rice grassy stunt virus* (RGSV), a tenuivirus, is transmitted by the brown planthopper (BPH), *Nilaparvata lugens* (Hemiptera, Delphacidae), in a persistent-propagative manner. In this study, immunofluorescence microscopy was used to investigate the infection route of RGSV in the internal organs of BPH after acquiring the virus by feeding on RGSV-infected rice plants. The sequential infection study revealed that RGSV initially infected the midgut epithelium, then crossed the basal lamina into the midgut visceral muscles, from where RGSV apparently spread into the hemolymph, then into the salivary glands of its BPH vector. The mechanism underlying this infection route of RGSV in its BPH vector may confer an advantage for the direct spread of RGSV from the initially infected epithelium to the salivary glands in BPH, contributing to efficient transmission of RGSV by its insect vector.

Keywords:*Rice grassy stunt virus*; Insect vector; Infection route; Midgut epithelium; Visceral muscle

Many plant viruses are transmitted by sap-sucking insects including thrips, aphids, planthoppers, leafhoppers and whiteflies in a persistent manner (Hogenhout et al., 2008; Bragard et al., 2013).After ingestion, the persistent plant viruses must move through the insect vector, from the midgut epithelium to other organs including the salivary glands, and overcome various tissue or membrane barriers within the insect (Hogenhout et al., 2008; Bragard et al.,2013). Acquiring a better understanding of this infection route of plant viruses and the barriers that they encounter in their insect vectors should lead to better strategies to block virus transmission.

Rice grassy stunt virus (RGSV), a tenuivirus, is transmitted by the brown planthopper (BPH), *Nilaparvata lugens* (Hemiptera, Delphacidae), in a persistent-propagative manner, that is, the virus also replicates in its insect vector (Hibino et al., 1985; Falk and Tsai, 1998). The virus is composed of filamentous ribonucleo protein particles (RNPs) that contain viral RNAs, nucleoprotein and RNA-dependent RNA polymerase (Hibino et al., 1985; Toriyama,1987). RGSV has caused damage to rice plants in South, Southeast, and East Asia (Hibino, 1996), and epidemics from 2006 to 2009 in southern Vietnam (Ta et al., 2013). Despite numerous studies in recent decades on the pathogenesis, molecular biology, genetic diversity, and control of RGSV (Hibino, 1996; Toriyama et al., 1998; Chomchan et al., 2003; Hiraguri et al., 2011; Shimizu et al., 2013), there are still major gaps in our knowledge about the virus, especially on the transmission mechanisms of RGSV by its BPH vector. Hence, in this study, immunofluorescence microscopy was used to investigate the infection route of RGSV in the internal organs of BPH after ingestion of virus from infected rice plants.

Virus Research.2014, 188: 170–173.

Received 20 February 2014; Accepted 12 April 2014; Available online 21 April 2014.

RGSV samples were collected in rice fields from Hainan Province in southern China. RGSV was maintained on rice plants via transmission by BPHs, as described previously (Hibino et al., 1985). The 2nd instar nymphs of BPHs were fed for 2d on rice plants infected with RGSV, then transferred to healthy rice seedlings. Because preliminary tests indicated that RGSV was transmitted after a latent period in the vector ranging from 6 to 20d, we dissected the digestive systems at 1, 3, 4, 6, 8 and 12d post-first access to diseased plants (padp) (50 insects/time point) and fixed them in 4% paraformaldehyde in phosphate-buffered saline (PBS) (137mM NaCl, 8.1mM Na_2HPO_4, 2.7mM KCl, 1.5mM KH_2PO_4, pH7.4) for 6h. After washing them twice with PBS, they were per-meabilized in 2% Triton X-100 in PBS for 4h, and then processed for immunofluorescence microscopy, as described by Chen et al.(2011). RNP of RGSV was immunolabeled with RNP-specific IgGs that had been conjugated directly to fluoresce in isothiocyanate (RNP-FITC) according to the manufacturer's instructions (Invitrogen). The RNP-specific IgGs of RGSV was isolated from polyclonal antibodies against RNPs of RGSV using a protein A-Sepharose affinity column (Pierce). The morphology of the digestive system of BPHs was visualized by immunolabeling with the actin dye phalloidin-rhodamine (Invitrogen). As controls, the digestive system dissected from BPHs which were fed on healthy rice plants was treated exactly in the same way.

As described previously, the digestive system of BPH consists of the salivary glands and alimentary canal, which includes the esophagus, anterior diverticulum, midgut and hindgut (Fig. 1A; Jia et al.,2012). The BPH midgut is composed of a single layer of epithelial cells, with microvilli lining the lumen side and basal lamina lining the outer side, surrounded by visceral muscle tissues (Fig. 1A; Jia et al., 2012). Immunofluorescence microscopy showed that, by 1day padp, RGSV had accumulated in the midgut lumen in about 74% of BPHs (Fig. 1B; Table 1). As early as 3d padp, RGSV was first observed in a limited number of the midgut epithelial cells in about 10% of BPHs (Fig.1C; Table 1), suggesting that the midgut epithelium was the initial site for RGSV entry and accumulation. Thus,the midgut microvilli seem to form a membrane barrier against RGSV and thus against persistent transmission. As a persistent-propagative virus, in the initially infected midgut epithelial cells,RGSV would replicate and assemble the progeny RNPs to further spread in the body of its insect vector (Hibino, 1996; Falk and Tsai,1998). By 4d padp, RGSV was observed to spread from the single infected epithelial cell into the visceral muscle tissues in about 6% of BPHs (Fig. 1D; Table 1). RGSV has thus apparently evolveda mechanism to directly cross the basal lamina from the initially infected midgut epithelium and then infect visceral muscle tissues. The midgut basal lamina of hemipteran insects is composed of a tightly interwoven noncellular layer (Hogenhout et al., 2008; Ammar et al., 2009), which may constitute a substantial barrier for the persistent transmission of virus by its insect vector. How RGSV crosses the midgut basal lamina of BPH is unknown.

By 6d padp, RGSV had spread into the visceral muscle tissues bordering the infected epithelium region in about 22% of the insects (Fig. 1E; Table 1). By 8d padp, RGSV had spread into the visceral muscle throughout the midgut and hindgut in about 16% of BPHs (Fig. 1G–I; Table 1). However, at this time, RGSV was apparently absent in the epithelium (Fig. 1H), suggesting that RGSV did not efficiently spread from infected epithelial cells to adjacent cells. In epithelial tissues of the BPH midgut, cells are joined by intercellular junctional complexes (Bergelson, 2009), which mayact as a physical barrier to prevent intercellular spread of RGSV.

It has been generally believed that a persistent virus is disseminated from the midgut into the hemolymph, then into the salivary glands (Hogenhout et al., 2008; Ammar et al., 2009). By 6 d padp, RGSV was first observed in the salivary glands in about 4% of BPHs (Fig. 1F; Table 1).

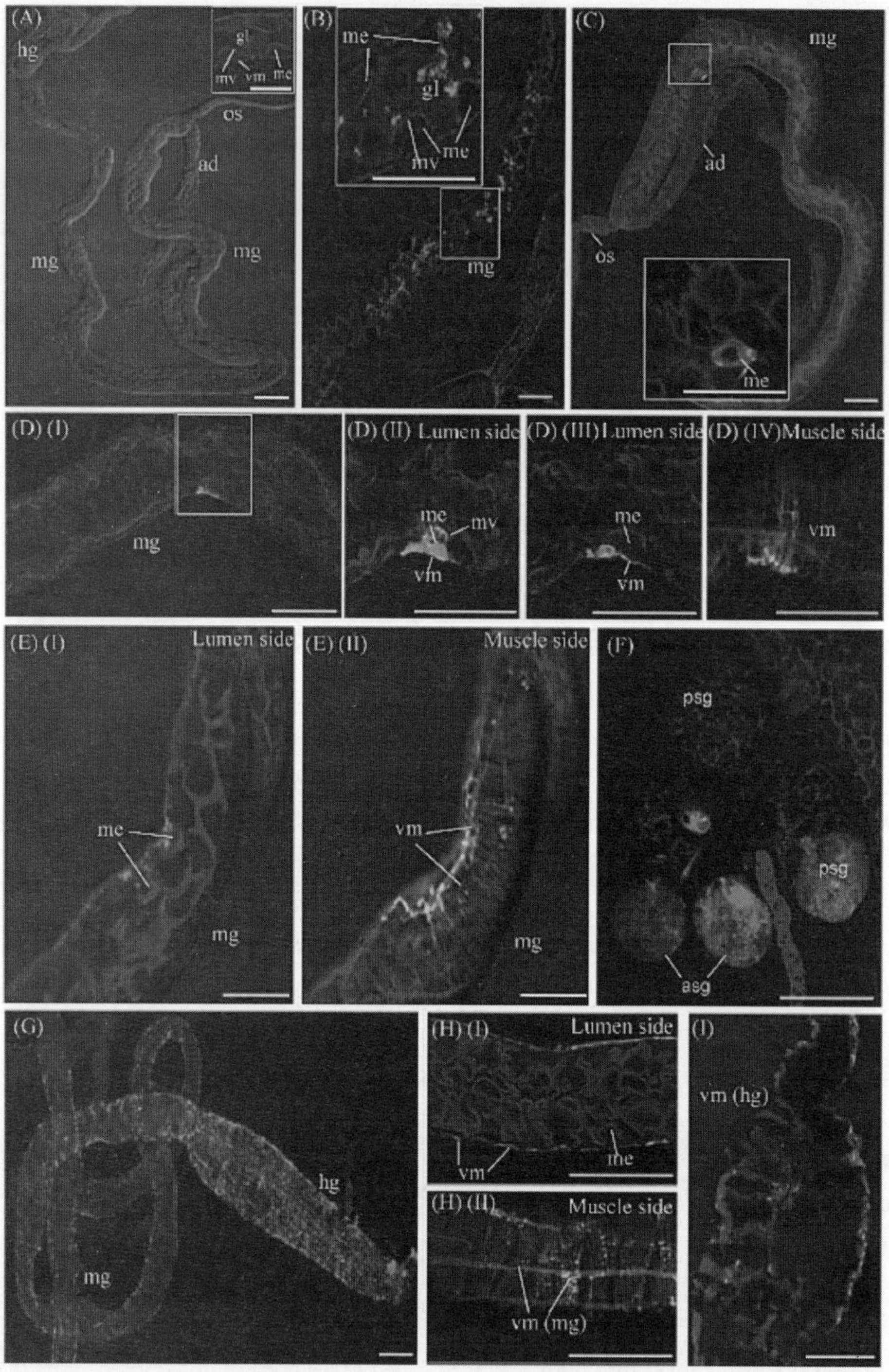

Fig. 1 Confocal micrographs of internal organs of BPH to show infection route of RGSV in its insect vector

Tissues were immunolabeled with RNP-FITC (green) and actindye phalloidin-rhodamine (red). (A) Alimentary canal of BPH. Inset: midgut epithelium, showing microvilli on lumen side and basal lamina covered with visceral muscle. (B) By 1 day post-first access to diseased plants (padp), RGSV had accumulated in the midgut. Inset: enlarged image of the boxed area. (C) By 3d padp, RGSV was foundin only one epithelial cell of the midgut. Inset: enlargement of boxed area. (D) By 4d padp, RGSV had spread from the epithelium to visceral muscle. Panels II, III, IV: enlargements of lumen (II, III) and muscle (IV) sides at boxed area in panel I. (E) By 6d padp, RGSV had spread into the visceral muscle of the midgut. Panels I, II: different sides of the midgut. (F) By 6d padp, RGSV had spread into the principal and accessory salivary glands. (G) By 8d padp, RGSV was present in the visceral muscles of the midgut and hindgut. (H) By 8d padp, RGSV spread into the visceral muscle of the midgut. Panes I, II: different sides of the midgut. (I) By 8d padp, RGSV was present in the visceral muscle of the hindgut. ad, anterior diverticulum; os, esophagus; mg, midgut; hg, hindgut; gl, gut lumen; mv, microvilli; me, midgut epithelium; vm, visceralmuscle; psg, principal salivary glands; asg, accessory salivary glands. Bars, 100μm. (For interpretation of the references to color in this figure legend, the reader is referred to the web version of this article.)

By 12d padp, RGSV had spread to the entire alimentary canal and salivary glands in about 20% of BPHs (Fig. 2; Table 1). We did not observe the association of RNP of RGSV with the neural tissues of BPHs (data not shown).Taken together, these results revealed an infection route of RGSV in the internal organs of its BPH vector after ingestion of virus. RGSV initially infected the midgut epithelium, then apparently crossed the basal lamina into the visceral muscles, from where it either spread into the hemolymph and finally into the salivary glands or had spread throughout the entire alimentary canal of its BPH vector.

Table 1 Distribution of RNP antigens of RGSV in different tissues of BPHs at different days post-first access to diseased plants, as revealed by immunofluorescence microscopy

Tissues examined[a]	Percentage of insects positive for RNPs of RGSV (%)					
	1day	3 days	4 days	6 days	8 days	12 days
Midgut lumen	74	0	0	0	0	0
Midgut epithelium	0	10	16	4	0	0
Visceral muscle (midgut)	0	0	6	22	26	26
Visceral muscle (hindgut)	0	0	0	0	16	26
Salivary glands	0	0	0	4	8	20
Esophagus	0	0	0	4	14	26
Anterior diverticulum	0	0	0	4	8	26

[a]RNP-FITC was used to indicate the presence of RGSV by green fluorescence

Rice stripe virus (RSV), the type species of the genus *Tenuivirus*, is transmitted by the small brown planthopper (SBPH), *Laodelphax striatellus*, in a persistent-propagative and transovarial manner (Toriyama, 1986; Falk and Tsai, 1998). Our recent study on the sequential infection of RSV in the internal organs of SBPH indicated that RSV could efficiently propagate in the midgut epithelialcells and visceral muscle tissues of SBPH (Wu et al., 2014). Furthermore, RSV was detected in the principal salivary glands, but notin the accessory salivary glands of SBPH. However, RGSV did not efficiently spread among midgut epithelial cells of BPH (Fig. 1D). Furthermore, RGSV could spread into the principal and accessory salivary glands of BPH (Fig. 1F). The significant difference of the infection routes for these two tenuiviruses in their insect vectors is that RSV could spread into the ovarioles of SBPH (Wu et al., 2014). *Rice ragged stunt virus* (RRSV), an oryzavirus, also is transmitted by BPH in a persistent-propagative manner (Hibino et al., 1979; Jia et al., 2012). It seemed that these two viruses did not efficiently propagate in the midgut epithelial cells, but the midgut visceral muscle tissues of BPHs were highly permissive for them (Jia et al., 2012; the present study). Most importantly, RGSV has evolved anovel mechanism to directly cross the basal lamina from the single initially infected epithelial cell toward the visceral muscle tissues in BPH midgut, but we still did not know if RRSV also evolved the similar mechanism to release from the initially infected midgut epithelial cells. The mechanism for the infection routes of RGSV in BPH vector may confer an advantage for the direct spread of virus from the initially infected epithelium to the salivary glands, on tributing to the efficient transmission of RGSV by the insectvector.

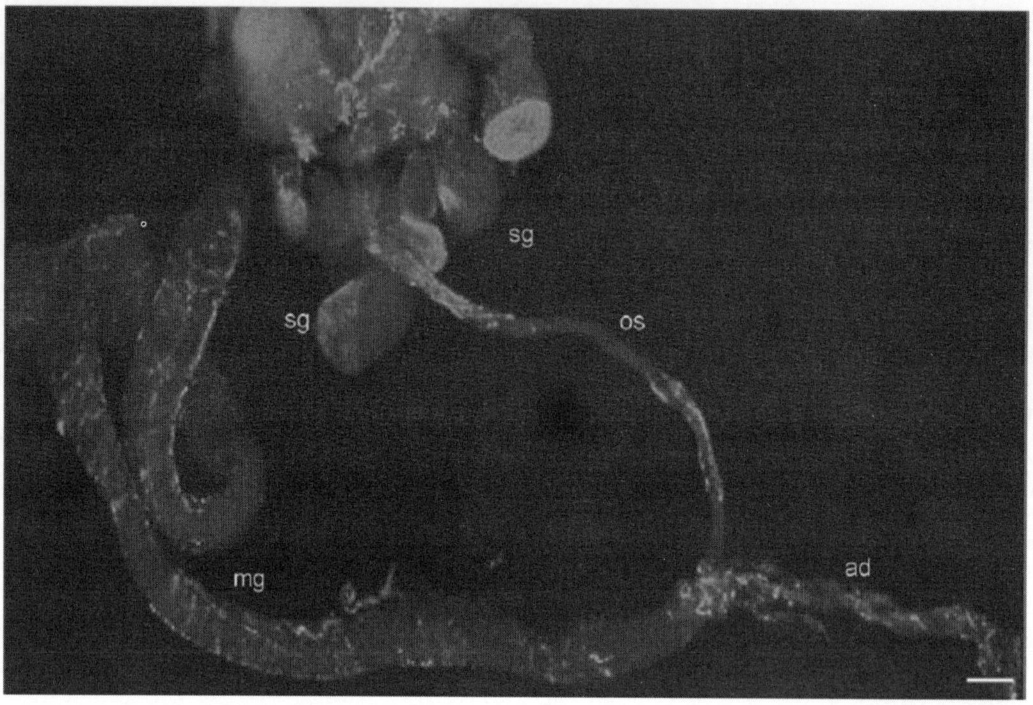

Fig. 2 Confocal micrograph of internal organs of BPH to show RGSV had spread throughout the digestive system of its BPH vector by 12d post-first access to diseased plants

The internal organs of BPHs fluoresced green, indicative of RNP-FITC. ad, anterior diverticulum; os, esophagus; mg, midgut; sg, salivary glands. Bar, 100μm. (For interpretation of the references to color in this figure legend, the reader is referred to the web version of this article.)

The tenuiviruses, rhabdovirus, phytoreoviruses, fijiviruses, oryzaviruses and tospovirusesare transmitted in a persistent-propagative manner by their respective insect vectors (Hogenhout et al., 2008; Ammar et al., 2009; Bragard et al., 2013). Through immunofluorescence microscopy, we and other groups have shown the infection routes for the tenuiviruses RGSV and RSV, the oryzavirus RRSV, and the tospovirus tomato spotted wilt virus in thrips or planthopper vectors (de Assis Filho et al., 2004; Jia et al., 2012;Wu et al., 2014; the present study). These viruses could apparently disseminate from the infected midgut epithelium into the visceral muscle tissues, from where the viruses spread into the hemolymphor other organs and finally into the salivary glands (de Assis Filho et al., 2004; Jia et al., 2012; Wu et al., 2014; the present study). However, the neurotropic route for viral dissemination from the infected vector's gut to the salivary gland also has been proposed for the phytoreovirus rice dwarf virus in the leafhopper vectoror the rhabdovirus maize mosaic virus in the planthopper vector (Ammar and Hogenhout, 2008; Chen et al., 2011). Taken together,the above analysis suggests that there are at least two viral dissemination routes from the intestinal epithelium to the salivary glands for persistent-propagative plant viruses. These two viral dissemination routes may reflect different strategies that evolved for the efficient transmission of persistent-propagative plant viruses by their respective insect vectors.

Acknowledgements

The authors are indebted to Toshihiro Omura (National Agricultural Research Center, Japan) for polyclonal antibodies against RNPs of RGSV. This research was supported by the National Basic Research Program of China (2014CB138400), the grants from theNational 863 Program of China (2012AA101505), the National Natural Science Foundation of China (31130044, 31200118), and the National Science Foundation for Outstanding Youth (31325023).

References

[1] AMMAR E-D, TSAI C W, WHITFIELD A E, et al. Cellular and molecular aspects of rhabdovirus interactions with insect and planthosts[J]. Annual Review of Entomology,2009, 54: 447-468.

[2] AMMAR E-D, HOGENHOUT S A. A neurotropic route for *Maize mosaic virus* (Rhabdoviridae) in its planthopper vector *Peregrinus maidis*[J]. Virus Research,2008, 131: 77-85.

[3] BERGELSON J M. Intercellular junctional proteins as receptors and barriers tovirus infection and spread[J]. Cell Host &Microbe,2009, 5(6): 517-521.

[4] BRAGARD C, CACIAGLI P, LEMAIRE O, et al. Status and prospects of plant virus control through interference with vector transmission[J]. Annual Review of Phytopathology, 2013, 51: 177-201.

[5] CHEN H, CHEN Q, OMURA T, et al. Sequential infection of *Rice dwarf virus* in the internal organs of its insect vector after ingestion of virus[J].Virus Research,2011, 160(1-2): 389-394.

[6] CHOMCHAN P, LI SF, SHIRAKO Y. *Rice grassy stunt tenuivirus* nonstructural protein p5 interacts with itself to form oligomeric complexes in vitro and invivo[J]. Journal of Virology,2003, 77(1): 769-775.

[7] DE ASSIS FILHO F M, DEOM C M, SHERWOOD J L. Acquisition of *Tomato spotted wilt virus* by adults of two thrips species[J]. Phytopathology, 2004, 94(4): 333-336.

[8] FALK B, TSAI J H. Biology and molecular biology of viruses in the genus *Tenuiviruses*[J]. Annual Review of Phytopathology,1998, 36(1): 139-163.

[9] HIBINO H, SALEH N, ROECHAN M. Reovirus-like particles associated with riceragged stunt diseased rice and insect vector cells[J]. Japanese Journal of Phytopathology,1979, 45(2): 228-239.

[10] HIBINO H, USUGI T, OMURA T, et al. Rice grassy stunt virus: a planthopper-borne circular filament[J]. Phytopathology,1985, 75: 894-899.

[11] HIBINO H. Biology and epidemiology of rice viruses[J]. Review of Phytopathology, 1996, 34(1): 249-274.

[12] HIRAGURI A, NETSU O, SHIMIZU T, et al. The nonstructural protein pC6 of *Rice grassy stunt virus* trans-complements the cell-to-cell spread of a movement-defective *Tomato mosaicvirus*[J]. Archives of Virology, 2011, 156(5): 911-916.

[13] HOGENHOUT S A, AMMAR E, WHITFIELD A E, et al. Insect vector interactions with persistently transmitted viruses[J]. Annual Review of Phytopathology, 2008, 46: 327-359.

[14] JIA D, GUO N, CHEN H, et al. Assembly of the viroplasm by viral non-structural protein Pns10 is essential for persistent infection of *Rice ragged stunt virus* in its insect vector[J]. Journal of General Virology, 2012, 93(10): 2299-2309.

[15] SHIMIZU T, OGAMINO T, HIRAGURI A, et al. Strong resistance against *Ricegrassy stunt virus* is induced in transgenic rice plants expressing double-stranded RNA of the viral genes for nucleocapsid or movement proteins as targets for RNA interference[J]. Phytopathology,2013, 103(5): 513-519.

[16] TA H A, NGUYEN D P, CAUSSE S, et al. Molecular diversity of *Rice grassy stunt virus* in Vietnam[J]. Virus Genes, 2013, 46(2): 383-386.

[17] TORIYAMA S. *Rice stripe virus*: prototype of a new group of viruses that replicate in plants and insects[J]. Microbiology Science,1986, 3(11): 347-351.

[18] TORIYAMA S. Ribonucleic acid polymerase activity in filamentous nucleoproteins of *Rice grassy stunt virus*[J]. Journal of General Virology, 1987, 68(3): 925-929.

[19] TORIYAMA S, KIMISHIMA T, TAKAHASHI M, et al.The complete nucleotide sequence of the *Rice grassy stunt virus* genome and genomic comparisons with viruses of the genus *Tenuivirus*[J]. Journal of General Virology, 1998, 79(8): 2051-2058.

[20] WU W, ZHENG L, CHEN H, et al. Nonstructural protein NS4 of *Rice stripe virus* plays a critical role in viral spread in the body of vector insects[J]. PLoS One, 2014, 9(2): e88636.

干扰水稻矮缩病毒（RDV）非结构蛋白 Pns11 的表达可抑制病毒在介体黑尾叶蝉内的复制

陈倩，张玲华，黄海宁，魏太云，谢联辉

（福建农林大学植物病毒研究所 / 福建省植物病毒学重点实验室，福州 350002）

摘要： 水稻矮缩病毒（*Rice dwarf virus*, RDV）主要由介体昆虫黑尾叶蝉（*Nephotettix cincticeps*）以持久增殖型方式传播。在 RDV 侵染介体黑尾叶蝉培养细胞时，病毒编码的非结构蛋白 Pns11 参与形成病毒原质，但 Pns11 的具体功能尚不知晓。本研究应用 RNA 干扰（RNA interference, RNAi）技术，将体外合成靶向 Pns11 基因的 dsRNA 转染黑尾叶蝉培养细胞，可显著降低 Pns11 的表达，并抑制病毒的积累和侵染。用显微注射技术将体外合成靶向 Pns11 基因的 dsRNA 导入到黑尾叶蝉，可显著抑制病毒在昆虫体内的侵染和扩散，最终降低介体传毒效率。结果表明，RDVPns11 是病毒在介体昆虫体内侵染和复制所必需的，是病毒复制的关键因子，可作为阻断病毒在介体内的复制和传播提供有效靶点。

关键词： 水稻矮缩病毒（RDV）；病毒原质；Pns11；复制；RNA 干扰（RNAi）；黑尾叶蝉

中图分类号： S435.72 **文献标识码：** A **文章编号：** 1671-5470(2009)01-0006-05

Interference of the expression of non-structural protein Pns11 of *Rice dwarf virus* (RDV) inhibits viral replication in leafhopper (*Nephotettix cinctic-eps*) Vector

Qian Chen, Linghua Zhang, Haining Huang, Taiyun Wei, Lianhui Xie

(Institute of Plant Virology, Key Laboratory of Plant Virology, Fujian Province, Fujian Agriculture and Forestry University,Fuzhou 350002, China)

Abstract:*Rice dwarf virus* (RDV) is mainly transmitted by leafhopper (*Nephotettix cincticeps*) inapersistent-propagative manner. The genome of RDV consists of 12 dsRNA segments, encoding at least 7 structural proteins (P1, P2, P3, P5, P7, P8 and P9) and 5 non-structural proteins (*Pns4, Pns6, Pns10, Pns11* and *Pns12*). *Pns11* of RDV is known to be a gene silencing suppressor in plant, and participates in the formation of viroplasm for viral replication and assembly of progeny virions in the cultural cells of *N. cincticeps*. However, the definite function of *Pns11* in viroplasm in insect vector is still unknown yet. In the present study, RNA interference (RNAi) was firstly applied via delivering the synthesized dsRNAs *in vitro* from *Pns11* gene into cultural cells of *N. cincticeps*. Immuno-fluorescence assay showed that when *Pns11* was knocked down,the antigens of *Pns11* were significantly reduced, and RDV was restricted within 1–2 cultural cells. The infectivity evaluated with the fluorescent antibody focus counting method showed the level of the viral infection decreased about 5 times. These results indicated that RNAI targeting on *Pns11* gene significantly reduced the expression of *Pns11*, leading the RDV accumulation and infection, was blocked in the cultural cells. Thus, *Pns11* was assumed to associate with viral replication. Then, microinjection of dsRNAs from

农业生物技术学报 . 2015, 23(11): 1401-1409.

收稿日期：2015-06-19；接受日期：2015-07-03

基金项目：国家自然科学基金（No. 31200118 和 No. 31300136）和福建省自然科学基金（No. 2014J01085）

Pns11 into the body of *N. cincticeps* was performed. qRT-PCR test showed the knockdown of *Pns11* significantly decreased the viru-liferous rate of insect more than 60%. And the RNAi effect steadily lasted for more than 12d, which was nearly a circulative transmission period. qRT-PCR assay showed that knockdown of *Pns11* caused more than 80% reduction in the relative expression levels of *P8* and *Pns11* genes. It was suggested that RNAi targeting on *Pns11* gene inhibited the viral accumulation in body of *N. cincticeps*, and decreased the number of viruliferous insect. Immuno-fluorescence assay displayed that at 6d after microinjection, RDV and *Pns11* were detected in the epithelia cells of filter chamber and midgut of *N. cincticeps* in control, while blocked in 1–2 epithelia cells of filter chamber in treatment of *Pns11* knockdown. At 12d after microinjection, RDV and *Pns11* were detected in the whole of intestines and salivary gland in control, while still blocked in 2–3 epithelia cells of filter chamber in the treatment of *Pns11* knockdown.These observations indicated that knockdown of *Pns11* inhibited the spread of RDV in insect vector via blocking viral replication. Finally, capability of viral transmission by insect was tested. The result showed that when *Pns11* was knocked down, insect vectors were unable to transmit RDV in 14d after microinjection. These results confirmed that *Pns11* was a viral replication factor, which was necessary for RDV infection and multiplication. It also verified *Pns11* was an effective target to block viral replication and transmission. This study the oretically bases for virus disease control via blocking RDV multiplication and circulation in insect vector.

Keywords: *Rice dwarf virus*(RDV); Viroplasm; *Pns11*; Replication; RNA interference (RNAi); *Nephotettix cincticeps*

水稻矮缩病毒（*Rice dwarf virus*, RDV）是引起水稻矮缩病的病原，可造成多种水稻（*Oryza sativa*）品种的矮化和减产，在东南亚地区、日本和中国南方稻区曾流行危害，引起重大的经济损失（谢联辉和林奇英，1984）。RDV 属呼肠孤病毒科（*Reoviridae*）植物呼肠孤病毒属（*Phytoreovirus*），主要由介体昆虫黑尾叶蝉（*Nephotettix cincticeps*）以循回期约 14d 的持久性增殖型方式传播。近年来，系统研究了 RDV 在介体黑尾叶蝉内的侵染循环过程，建立了黑尾叶蝉的传毒模型，病毒粒体在黑尾叶蝉取食病株时被摄入，经过食道到达滤室的内腔，随后病毒可能依赖外壳蛋白识别并结合滤室上皮细胞上的专化性受体后，以内吞作用进入细胞开始复制（Wei et al., 2007; Zhou et al.,2007）。装配成熟的子代病毒粒体随后扩散到中肠和后肠等组织（Chen et al., 2011）。同时 RDV 可利用神经系统快速穿越中肠和唾液腺的基底膜，从而进入唾液腺（Chen et al., 2011）。在黑尾叶蝉取食时，病毒随唾液释放到植物组织中，完成传毒过程。RDV 病毒粒体呈正二十面体结构，其双层衣壳包裹着 12 条 dsRNA 的基因组，分别编码 7 个结构蛋白（P1, P2, P3, P5, P7, P8 和 P9）和 5 个非结构蛋白（Pns4, Pns6, Pns10, Pns11 和 Pns12）。病毒核衣壳由 P3 蛋白组成，包裹着依赖 RNA 的 RNA 聚合酶 P1，鸟苷酸转移酶 P5 和 RNA 结合蛋白 P7（Hagiwara et al., 2004; Omura et al.,1998; Suzuki et al., 1992; 1996; Ueda and Uyeda, 1997; Zhong et al., 2005）；病毒外壳由可被介体昆虫细胞识别的 P2、主要外壳蛋白 P8 和次要外壳蛋白 P9 构成（Mao et al., 1998; 尹哲等 , 2007; Zhou et al., 2007）。RDV 的非结构蛋白 Pns4 可能具有锌指结构和 GTP-binding 活性，与病毒引起的水稻矮缩症状有关（Uyeda et al.,1990），在黑尾叶蝉培养细胞中可形成微管结构（Wei et al., 2006b）。Pns10 在植物体内起沉默抑制子功能（Cao et al., 2005; Ren et al., 2010），在黑尾叶蝉内是包裹病毒穿过消化道上皮细胞的微绒毛，并沿内脏肌肉细胞的肌肉纤维束移动的小管蛋白（Wei 2006a; Chen et al., 2012）。Pns6 具有 ATPase 和 RNA 结合活性（Wei et al., 2006c），在植物中是个运动蛋白（Li et al., 2004）；Pns12 是磷酸化蛋白（Wei et al., 2006c）；Pns11 具有结合 RNA 的能力，是 RDV 在植物中的第 2 个沉默抑制子（Xu et al., 1998）。关于非结构蛋白 Pns6、Pns11 和 Pns12 在植物中的功能有较深入的研究，但其在介体黑尾叶蝉内的功能研究较少。研究表明，在 RDV 侵染黑尾叶蝉培养细胞的早期，Pns6、Pns11 和 Pns12 首先被翻译，相互聚集成病毒原质的基质，为病毒的复制和装配提供场所（Wei et al., 2006c），推测该 3 个非结构蛋白参与了病毒在介体细胞内的复制及子代病毒粒体的装配，可能是病毒的复制因子。此前，Shimizu 等应用 RNA 干扰（RNA interference, RNAi）技术沉默 Pns12 基因，使转基因水稻的 RDV 抗性大大增强，间接证明了 Pns12 是病毒的复制蛋白（Shimizu et al., 2009）。这为研究 Pns6、Pns11 和 Pns12 在介体内的确切功能提供理论支持。

鉴此，本研究应用 RNAi 技术，在黑尾叶蝉中干扰病毒原质的组分蛋白之一 Pns11，分析病毒在介体培养细胞和黑尾叶蝉内复制和增殖受到的影响，明确 RDV 在介体内的复制关键因子，结果将有利于阐释病毒的复制机理。同时验证抑制病毒复制的有效靶点，完善叶蝉的 RNAi 体系，为虫传病害防控策略提供理论基础。

1 材料与方法

1.1 介体、培养细胞、感病水稻和抗体

无毒黑尾叶蝉（*Nephotettix cincticeps*）采集于福建福州水稻种植区。建立和维持黑尾叶蝉培养细胞的方法参照（Kimura, 1984）。

感染水稻矮缩病毒（*Rice dwarf virus*, RDV）的水稻（*Oryza sativa*）病株采自云南省寻甸县病害暴发区，用本实验室的黑尾叶蝉传毒扩繁。提纯病毒由本实验室保存。

非结构蛋白 Pns11 和 RDV 病毒多克隆抗体由日本国立农业研究中心大村敏博教授惠赠。将抗体分别与荧光素 FITC 和 rhodamine（Invitrogen 公司，美国）交联（贾东升，2013），制备成荧光抗体 RDV-FITC 和 Pns11-rhodamine。荧光抗体 actindye phalloidin-Alexa Fluor 647 carboxylic acid（actin-647）用于标记肌动蛋白 actin（Invitrogen 公司，美国）。

1.2 dsRNA 体外合成

选取绿色荧光蛋白（green fluorescent protein, GFP）基因 584 bp 和 Pns11 基因（AY375453）527 bp 长度的片段，在引物 5′ 端加上 T7 启动子序列（表 1）。以实验室保存的质粒 pDOR221-Pns11 为模板，PCR 扩增该片段。PCR 反应体系：模板 DNA 0.2μg、正反向引物各 10pmol/L、dNTP 各 200 μmol/L、TaqDNA 聚合酶 2.5U、10×buffer 10μL，无菌水补足 100μL。PCR 反应条件：94℃ 预变性 5min；94℃ 30s，55℃ 30s，72℃ 30s，共 30 个循环；72℃延伸 5min；最后 16℃保温。

根据 T7RiboMAX^{TM} Express RNAi System（Promega，美国）试剂盒说明书，以纯化后的 PCR 产物为模板，用 T7 转录酶进行 dsRNA 的体外转录。用分光光度计测定靶向 *Pns11* 和 *GFP* 基因的 dsRNA（dsPns11 和 dsGFP）浓度测定，琼脂糖凝胶电泳测定 dsRNA 纯度。

1.3 在培养细胞和黑尾叶蝉内诱导 RNAi

当黑尾叶蝉培养细胞在细胞爬片（直径 12mm）上的覆盖率达 80% 时，按 11μL 无血清无抗生素培养基分别加入 0.08μg dsRNA 和 0.06μL cellfectin Ⅱ Reagent（Invitrogen, 美国）的比例配制 dsRNA 稀释液和脂质体稀释液，而后混合两者，配成 dsRNA 转染液。将转染液孵育黑尾叶蝉培养细胞约 8h。而后，用提纯病毒 RDV（侵染复数为 10）侵染转染后的细胞 2h，最后正常培养。实验重复 3 次。

取饲喂 RDV 病株 2d 后的 2 龄黑尾叶蝉若虫，于后胸与第一腹节之间的节间膜处，显微注射浓度为 0.5μg/μL 的 dsRNA，剂量约为 0.1μL/ 头，而后转至健康水稻上饲养。

表 1 引物信息

Table 1 The primers used in this study

基因 Gene	序列 (5′~3′) Sequence	用途 Function
Pns11	Forward: ATTCTCTAGAAGCTTAATACGACTCACTATAGGGATTACCCTTGGC-TATGAC	体外转录模板 Transcriptional template *in vitro*
	Reverse: ATTCTCTAGAAGCTTAATACGACTCACTATAGGGTTACTTACGCTTT-GATTTG	
GFP	Forward: ATTCTCTAGAAGCTTAATACGACTCACTATAGGGCTTGTTGAATTAGAT-GGTGATGTT	体外转录模板 Transcriptional template *in vitro*
	Reverse: ATTCTCTAGAAGCTTAATACGACTCACTATAGGGGTTTCGAAAGGG-CAGATTGT	
actin	Forward: GGGATACAGTTTCACCACG	荧光定量 PCR
	Reverse: GACACCTGAATCGCTCGT	qRT-PCR
P8	Forward: TACAGCCATCAGCTAAGCCAAA	荧光定量 PCR
	Reverse: CCGCAACAGACCGAAACA	qRT-PCR
Pns11	Forward: TGCGGTAGCTGCCCTTTT	荧光定量 PCR
	Reverse: AGGGGACTTATCCTTTCTGTCGT	qRT-PCR

Pns11：非结构蛋白 Pns11 基因；*P8*：主要外壳蛋白基因；*GFP*：绿色荧光蛋白基因；*actin*：肌动蛋白基因；下划线：T7 启动子序列。
Pns11: Non-structural protein *Pns11* gene; *P8*: Major outer capsid protein gene; *GFP*: Green fluorescent protein gene; *actin*: Actin protein gene; Underlined: T7 promoter sequence

1.4 qRT-PCR 分析基因相对表达量

用 Primer Premier5.0 引物设计软件设计 *P8*、*Pns11* 和黑尾叶蝉 *actin* 的荧光定量 PCR 引物（表 1）。于注射后第 3、6 和 12 天收集黑尾叶蝉，用 TRIzol Reagent 提取黑尾叶蝉总 RNA。用 RevertAid Reverse Transcriptase（Thermo Scientific 公司，美国）反转录后，使用 SYBR Green PCR Master Mixkit（QIAGEN 公司，美国）试剂盒，以 *actin* 为内参基因进行 qRT-PCR。qRT-PCR 反应程序：95℃预变性 2min，95℃ 5s，60℃ 20s，72℃ 15s，40 个循环。

1.5 免疫荧光标记

免疫荧光标记方法参照 (Chen et al., 2012, 2015)，用荧光抗体 RDV-FITC、Pns11-rhodamine 和 actin-647 孵育样品。用激光共聚焦显微镜（Leica TCSSP5）观察标记好的样品，并拍照。

2 结果与分析

2.1 dsPns11 抑制 RDV 在黑尾叶蝉培养细胞内的侵染

免疫荧光观察 dsPns11 转染 RDV 侵染后的培养细胞，分析病毒侵染和 Pns11 表达受到的影响。结果显示（图 1），在对照 dsGFP 处理的培养细胞中，大部分细胞被 RDV 侵染而显示绿色荧光，Pns11 特异性地呈现出分散的点状颗粒，并且与 RDV 共定位，说明 RDV 正在复制，生成子代病毒。而在 dsPns11 处理过的细胞中，RDV 被局限在 1 ～ 2 个细胞内，Pns11 表达水平明显低于对照，甚至难以检测到。说明 dsPns11 的干扰效果较明显，显著抑制病毒在介体培养细胞内的侵染和积累。

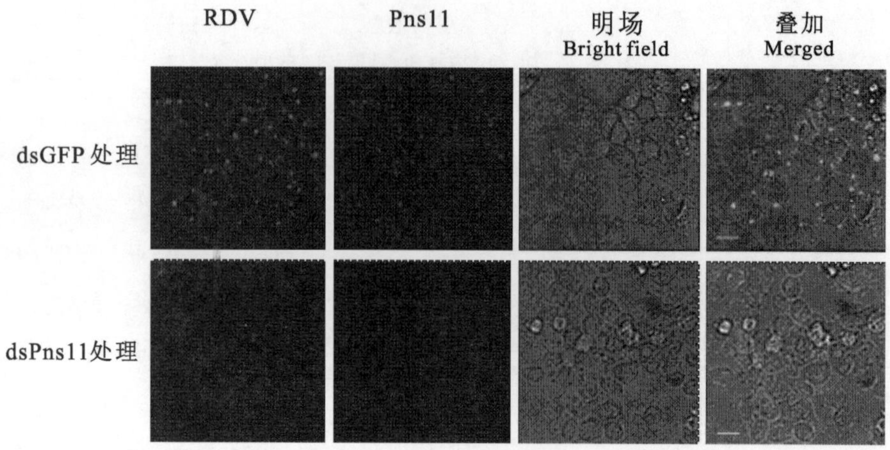

图 1 dsPns11 诱导的 RNAi 对 RDV 在介体培养细胞内侵染和积累的影响（Bar=15μm）

Fig.1 Effect of dsPns11 induced RNAi on viral infection and accumulation in cultured insect vector cells

用 RDV-FITC（绿色）标记 RDV，用 Pns11-rhodamine（红色）标记 Pns11；dsGFP 处理中，RDV 与 Pns11 共定位于大部分细胞内；dsPns11 处理中，Pns11 几乎不表达

RDV was immunolocated with RDV-FITC (green), and Pns11 antigens was immunolocated with Pns11-rhodamine (red); In dsGFP treated, RDV and Pns11 co-located in most cells, while in dsPns11treated, Pns11 was hardly detected

为定量比较 dsPns11 处理对 RDV 侵染的影响，用荧光抗体侵染点计数法（Kimura,1986）统计 RDV 在黑尾叶蝉培养细胞中的侵染率。结果显示（图 2），RDV 在对照 dsGFP 转染的细胞中平均侵染率为 93.3%，而在 dsPns11 转染的细胞中平均侵染率为 18.3% 左右，极显著降低了 5 倍左右（$P < 0.01$）。该结果说明，dsPns11 的转染显著降低了病毒积累和侵染。

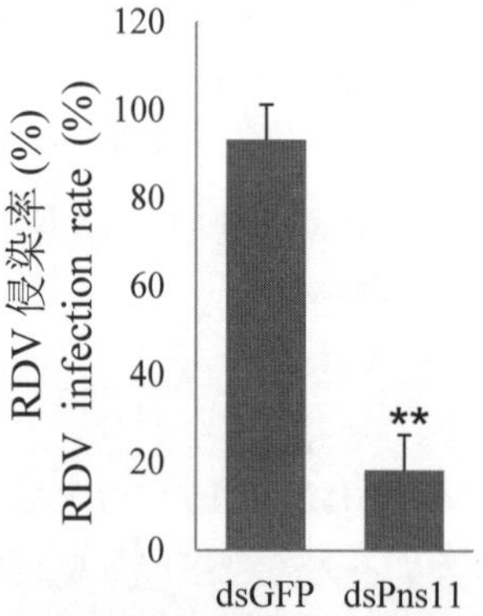

图 2 dsPns11 诱导的 RNAi 对 RDV 在介体培养细胞内侵染率的影响**：与对照（dsGFP）比差异极显著（$P < 0.01$）；n=3；下同

Fig. 2 Effect of dsPns11 induced RNAi on viral infection in cultured insect vector cells. **: Very significant differences compared with control (dsGFP) ($P < 0.01$); n=3; The same below

2.2 dsPns11 降低黑尾叶蝉的带毒率

进一步收集饲毒 2d 的黑尾叶蝉，通过显微注射将 dsPns11 导入饲毒后黑尾叶蝉体内。于注射后第 3、6 和 12d，分析黑尾叶蝉的带毒情况（> 40 头）。选择 RDV 外壳蛋白基因 *P8* 为病毒的检测基因，用 qRT-PCR 方法检测黑尾叶蝉的 P8 阳性率，统计黑尾叶蝉带毒率。

结果显示，在注射后第 3d，dsGFP 处理（对照）的黑尾叶蝉中 *P8* 基因阳性率为 52.0%，dsPns11 处理的黑尾叶蝉中 *P8* 基因阳性率为 18.5%，极显著低于对照（$P < 0.01$）（图 3）。注射后第 6 天，对照 dsGFP 处理的黑尾叶蝉 *P8* 基因阳性率为 52.5%，与注射后第 3 天的结果基本持平。而 dsPns11 处理的

黑尾叶蝉中 *P8* 基因阳性率为 19.4%，仍然低于对照。注射后第 12 天，对照 dsGFP 处理的黑尾叶蝉 *P8* 基因阳性率为 54.0%，与第 6 天基本一致，而 dsPns11 处理的黑尾叶蝉 *P8* 基因阳性率为 19.5%，与注射后第 3 和 6 天的结果基本一致，仍低于对照 60% 以上。表明以浓度为 0.5μg/μL，剂量约为 0.1μL/ 头的注射体系干扰 Pns11 的表达，至少可以稳定维持 RNAi 效果长达 12d，接近病毒在黑尾叶蝉体内的 1 个循回期。

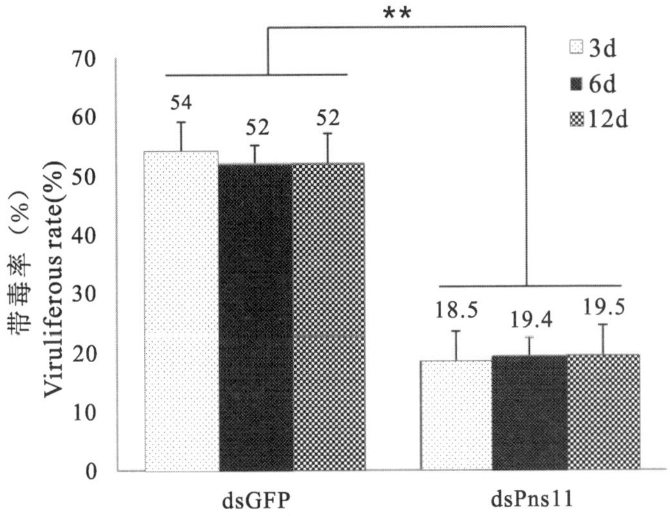

图 3 dsPns11 对黑尾叶蝉带毒率的影响

Fig. 3 Effect of dsPns11 on the viruliferous rates of leafhoppers

2.3 dsPns11 抑制 RDV Pns11 和 P8 基因在黑尾叶蝉内的表达

为解释 dsPns11 可显著降低黑尾叶蝉 RDV 带毒率的原因，注射 dsPns 1112d 后，用 qRT-PCR 验证 dsPns11 对 RDV 基因表达的影响。图 4 所示，注射 dsPns11 的带毒黑尾叶蝉中，P8 和 Pns11 基因相对表达量低于对照 80% 左右。表明 dsPns11 可特异性抑制 Pns11 基因相对表达量（$P < 0.01$），同时导致病毒其他基因的相对表达量也受到抑制，最终抑制病毒在虫体内的积累并减低黑尾叶蝉的带毒率。

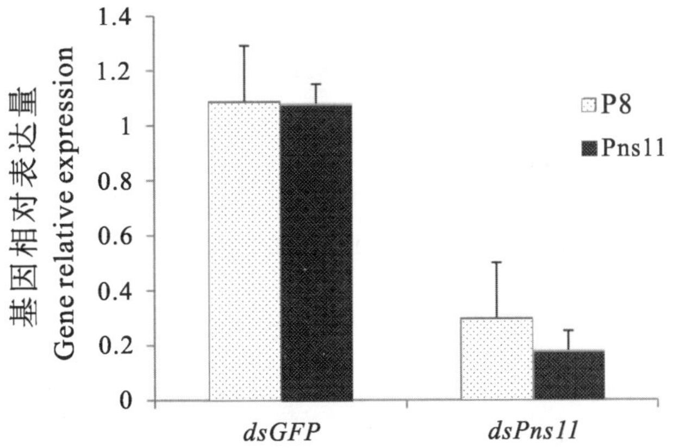

图 4 dsPns11 对黑尾叶蝉 Pns11 和 P8 基因相对表达量的影响

Fig. 4 Effect of dsPns11 on the relative expression levels of *Pns11* and *P8* genes in leafhopper vector

内参基因：actin；n=3

Reference gene: *actin*; n=3

2.4 dsPns11 减缓了 RDV 在黑尾叶蝉内的扩散

为直观地描述 dsPns11 处理对 RDV 侵染黑尾叶蝉的影响，进一步对这些 dsRNA 处理的带毒黑尾叶蝉进行解剖和免疫荧光观察。结果显示（图 5），注射后第 6 天时，在 dsGFP 处理的带毒黑尾叶蝉消化道，包括滤室、前中肠、中中肠和后中肠的上皮细胞内检测到 RDV 和 Pns11 的存在。部分黑尾叶蝉的消化道肌肉组织也可检测到 RDV 和 Pns11，表明 RDV 的侵染进入中期。而在注射了 dsPns11 的黑尾叶蝉消化道中，仅少数被检测到带毒。在这些少数的带毒滤室、前中肠或后中肠内，RDV 和 Pns11 共定位在 1 ～ 2 个上皮细胞中，类似 RDV 的初侵染阶段，说明病毒的复制和扩散受到显著抑制。对这些被 RDV 侵染的组织进行组织带毒率统计，结果显示，dsPns11 处理的黑尾叶蝉消化道中，各组织的 RDV 积累比率明显低于对照（表 2），病毒侵染还处于早期，说明 dsPns11 处理不仅抑制了 Pns11 的表达，还降低病毒在介体内的积累水平，减缓了病毒的扩散。

注射后第 12 天，在注射 dsGFP 的带毒黑尾叶蝉消化道上皮细胞、肌肉组织，以及唾液腺都检测到大量的 RDV 和 Pns11，消化道各部分的组织带毒率明显增加，说明病毒侵染处于后期。而在注射 dsPns11 的带毒黑尾叶蝉消化道中，大多数 RDV 和 Pns11 仍被局限在 2 ～ 3 个上皮细胞中，唾液腺未检测到 RDV 和 Pns11 的存在，说明病毒侵染和 Pns11 的表达状态与第 6 天相差不大，dsPns11 诱导的 RNAi 可稳定维持到第 12 天。说明 Pns11 在病毒侵染黑尾叶蝉的过程中参与了病毒的复制，干扰该基因的表达可显著减缓病毒的扩散。

2.5 dsPns11 可降低黑尾叶蝉的传毒率

分别收集注射 dsGFP 和 dsPns11 后 14d 的黑尾叶蝉，用单虫单苗传毒 5d。20d 后检测接种的水稻，统计黑尾叶蝉传毒率。用 qRT-PCR 检测 *P8* 基因，分析 *P8* 阳性率，分析 Pns11 表达被干扰后对黑尾叶蝉传毒能力的影响。结果显示，注射 dsGFP 的黑尾叶蝉接种的水稻中，*P8* 基因的平均阳性率为 30.6%，而注射 dsPns11 的黑尾叶蝉接种的水稻中未检测到 *P8* 基因，说明 dsPns11 可明显降低黑尾叶蝉的传毒效率。

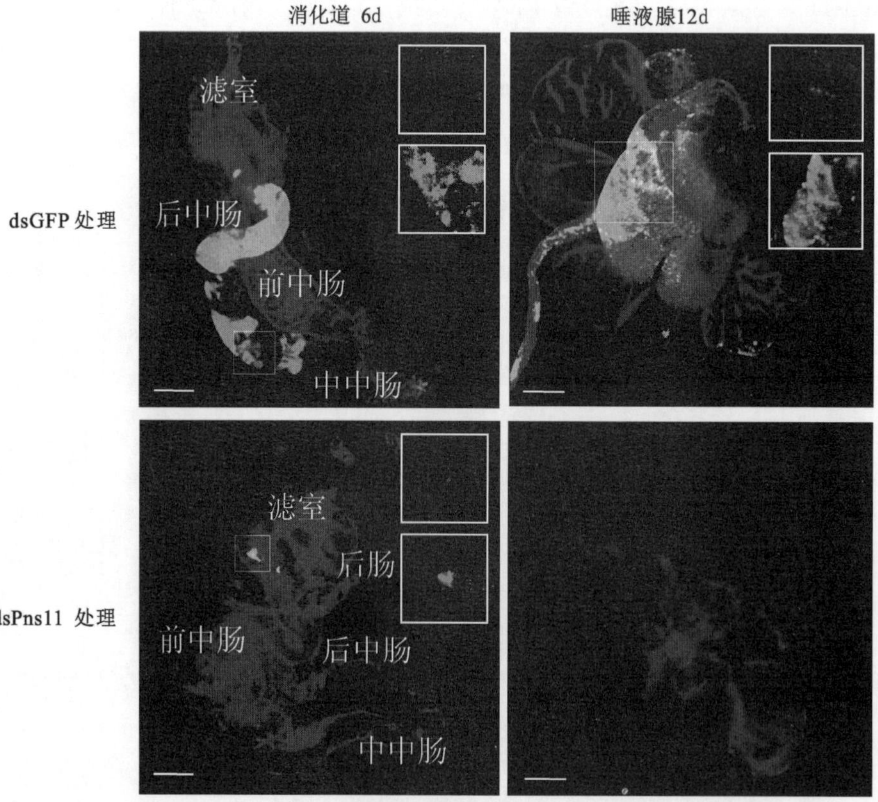

图 5　dsPns11 对 RDV 在黑尾叶蝉体内侵染的影响（Bar=100μm）

Fig. 5　Effect of dsPns11 on the RDV infection in leafhopper vector

表 2 dsPns11 对黑尾叶蝉带毒率的影响
Table 2 Effect of dsPns11 on leafhoppers viruliferous rate

处理 Treatment	注射后时间 /d Days after injection	不同组织带毒率 / % Viruliferous rate of tissues					
		滤室 Filter chamber	前中肠 Anterior midgut	中中肠 Middle midgut	后中肠 Postior midgut	后肠 Hindgut	唾液腺 Salivary gland
dsGFP	3	76.2±0.14	38.1±0.27	38.1±0.33	23.8±0.27	14.3±0.34	0
	6	82.3±0.18	41.2±0.40	35.2±0.35	41.2±0.25	41.2±0.28	0
	12	76.2±0.24	71.4±0.29	61.9±0.29	66.7±0.33	52.3±0.16	71.4±0.32
dsPns11	3	33.3±0.24	16.7±0.37	0 **	33.3±0.35	33.3±0.24	0
	6	83.3±0.19	0 **	0**	16.7±0.26	16.7±0.36	0
	12	50.0±0.27	12.5±0.26	25.0±0.38	87.5±0.24	12.5±0.22	0

3 讨论

培养细胞体系是研究病毒生活史的有效工具，将该体系与 dsRNA 或 siRNA 诱导 RNAi 技术相结合，证实了南方水稻矮缩病毒（*Southern rice black streaked dwarf virus*, SRBSDV）的病毒原质蛋白 P9-1（Jia et al., 2012），水稻瘤矮病毒（*Rice gall dwarf virus*, RGDV）的 Pns9 和 Pns12（Zheng et al., 2015; 郑立敏等，2014），轮状病毒（*Rotavirus*）病毒的 NSP5（Criglar et al., 2014）是病毒的复制蛋白，干扰了这些病毒原质蛋白的表达，病毒基因组 dsRNA 将无法正常合成，其他病毒蛋白也无法正常表达，造成病毒侵染和积累不能进行。本研究在技术手段方面，不仅将培养细胞体系与 RNAi 技术有效结合，还进一步在介体昆虫体内实现和完善 RNAi 体系，使研究数据基于介体本身，从而获得可靠的研究结果，弥补了多组分病毒无法利用反向遗传学手段研究基因功能的空缺。研究结果表明，Pns11 参与了病毒在黑尾叶蝉体内的复制，是病毒在介体内侵染和复制所必需的。干扰了该蛋白的表达，病毒在昆虫体内的侵染和扩散受抑制，介体传毒效率也被降低，表明 Pns11 是 RDV 复制的关键因子之一。该结果再次印证了病毒原质蛋白参与了病毒的复制。此前，用同样的体系也证明了 RDV 的另一个病毒原质蛋白 Pns6 是病毒复制的因子（陈倩等，2015）。

研究显示，Pns6 被干扰 12d 时，病毒积累量和介体带毒率都有明显回升（陈倩等，2015）；本研究表明，干扰了 Pns11 后，病毒积累量和介体带毒率在 dsRNA 注射 12d 内处于低水平，未见显著增长，说明病毒的 RNAi 效果至少可稳定维持近 12d，推测 Pns11 对病毒复制起到更至关重要的作用。由于 Pns11 有 RNA 结合活性（Xu et al., 1998），在病毒原质中可能聚集或挑选分拣新生成的 RNA，以便将其包裹进正在装配或已装配完毕的 RDV 内壳中。当 Pns11 被干扰后，新生成的 RNA 无法正常被包裹和装配，导致子代病毒粒体不能形成，病毒积累及其次级侵染不能顺利进行。

本研究发现，Pns11 被干扰后，介体叶蝉的传毒效率大大降低，这为抑制病毒在昆虫体内的复制提供有效靶点。由于昆虫的 RNAi 具有很高的特异性，此法已在病害防治和控制方面开展应用（Price and Gatehouse, 2008）。通过给蜜蜂（*Apis mellifera*）饲喂靶向以色列急性瘫痪病毒（*Israeli acute paralysis virus*, IAPV）两段基因的 dsRNA，可提高蜜蜂的存活率、群体数量和蜂蜜产量（Hunter et al., 2010）。基于在昆虫体内诱导 RNAi 从而实现病毒病防治的目的，在黑尾叶蝉体内，通过干扰病毒复制蛋白，抑制病毒依赖昆虫传播的效率，以防治水稻病毒病将成为可能。但在生产上，将 dsRNA 导入黑尾叶蝉群体内，饲喂法最为实用和简便。因为昆虫自行摄取 dsRNA 比显微注射法更接近自然情况。饲喂 dsRNA 的方法有多种，如将 dsRNA 与饲料混合，或者人工饲料表面包被 dsRNA，或用带有 dsRNA 的微滴饲喂，或新报道的将 dsRNA 包埋在聚合壳聚糖中形成的壳聚糖 /dsRNA 纳米颗粒饲喂，或用可表达发夹 dsRNA 的转基因植物饲喂（Yu et al., 2013）。对于咀嚼式等口器的昆虫而言，将含有 dsRNA 的饲料以叶面喷施方法，饲喂昆虫，操作简便。而黑尾叶蝉是以刺吸式方式，将口器直接伸入植物韧皮部进行取食的，似乎只能采取转基因植物饲喂的方式，把 dsRNA 导入昆虫体内。相信经过详细的对比和验证，以最佳的 dsRNA 体系将 dsRNA 导入黑尾叶蝉，诱导 RNAi，从而抑制昆虫传毒，阻断病毒病害的传播在不久的将来会实现。

4 结论

本研究应用 RNAi 技术，将体外合成的靶向病毒原质组分蛋白 Pns11 的 dsRNA 导入黑尾叶蝉培养细胞和虫体内，在培养细胞内，诱导的 RNAi 显著抑制 Pns11 的表达，并将病毒侵染率降低至 20% 左右。在虫体内，可稳定维持至少 12d 的 RNAi 通过抑制 *Pns11* 基因的表达，将病毒的积累水平降低了 80%，导致病毒在介体内的扩散明显减缓，尤其是介体带毒率降低了 60% 以上，介体传毒能力在 14d 内丧失。本研究表明，Pns11 蛋白在病毒侵染黑尾叶蝉过程中参与了病毒的复制，是 RDV 的复制因子；Pns11 是抑制病毒在昆虫体内复制的有效靶点，为阻断病毒在介体叶蝉体内的循回和增殖提供思路，为虫传病毒病的有效控制提供理论支撑。

参考文献

[1] 陈倩，张玲华，黄海宁，等．水稻矮缩病毒非结构蛋白 Pns6 是病毒的复制因子 [J]. 中国科技论文在线，2015,8(21): 11.

[2] 贾东升．SRBSDV 和 RRSV 在介体飞虱体内的侵染机理 [D]. 福州：福建农林大学，2013.

[3] 谢联辉，林奇英．我国水稻病毒病研究的进展 [J]. 中国农业科学，1984,17(6):58-65.

[4] 尹哲，吉栩，吴云锋，等．水稻矮缩病毒外壳蛋白 P9 具有体内转录激活活性 [J]. 中国农业科技导报，2007,9(3):61-65.

[5] 郑立敏，刘华敏，陈红燕，等．干扰水稻瘤矮病毒 (RGDV) 非结构蛋白 (Pns12) 的表达抑制病毒在介体昆虫培养细胞内的复制 [J]. 农业生物技术学报，2014,22(11):1321-1328.

[6] CAO X, ZHOU P, ZHANG X, et al. Identification of an RNA silencing suppressor from a plant double-stranded RNA virus[J]. Journal of Virology, 2005, 79(20): 13018-13027.

[7] CHEN H, CHEN Q, OMURA T, et al. Sequential infection of *Rice dwarf virus* in the internal organs of its insect vector after ingestion of virus[J]. Virus Research, 2011, 160(1-2): 389-394.

[8] CHEN Q, CHEN H, MAO Q, et al. Tubular structure induced by a plant virus facilitates viral spread in its vector insect[J]. PLoS Pathogens, 2012, 8(11): e1003032.

[9] CHEN Q, WANG H, REN T, et al. Interaction between nonstructural protein Pns10 of *Rice dwarf virus* and cytoplasmic actin of leafhoppers is correlated with insect vector specificity[J]. Journal of General Virology, 2015, 96(4): 933-938.

[10] CRIGLAR J M, HU L, CRAWFORD S E, et al. A novel form of rotavirus NSP2 and phosphorylation-dependent NSP2-NSP5 interactions are associated with viroplasm assembly[J]. Journal of Virology, 2014, 88(2): 786-798.

[11] HAGIWARA K, HIGASHI T, MIYAZAKI N, et al. The amino-terminal region of major capsid protein P3 is essential for self-assembly of single-shelled core-like particles of *Rice dwarf virus*. Journal of Virology, 2004, 78(6): 3145-3148.

[12] HUNTER W, ELLIS J, VANENGELSDORP D, et al. Large-scale field application of RNAi technology reducing israeli acute paralysis virus disease in honey bees (*Apis mellifera*, Hymenoptera: Apidae). PLoS Pathogens, 2010, 6(12): e1001160.

[13] JIA D, CHEN H, ZHENG A, et al. Development of an insect vector cell culture and RNA interference system to investigate the functional role of *Fijivirus* replication protein[J]. Journal of Virology, 2012, 86(10): 5800-5807.

[14] KIMURA I. Establishment of new cell lines from leafhopper vector and inoculation of its cell monolayers with *Rice dwarf virus*[J].

Proceedings of the Japan Academy Ser B Physical and Biological Sciences, 1984, 60(6): 198-201.

[15] KIMURA I. A study of *Rice dwarf virus* in vector cell monolayers by fluorescent antibody focus counting[J]. Journal of General Virology, 1986, 67(10): 2119-2124.

[16] LI Y, BAO Y M, WEI C H, et al. *Rice dwarf phytoreovirus* segment S6-encoded nonstructural protein has a cell-to-cell movement function[J]. Journal of Virology, 2004, 78 (10): 5382-5389.

[17] MAO Z J, LI Y, XU H, et al. The 42K protein of *Rice dwarf virus* is a post- translational cleavage product of the 46K outer capsid protein[J]. Archives of Virology, 1998, 143(9): 1831-1838.

[18] OMURA T, YAN J, ZHONG B, et al. The P2 protein of *Rice dwarf phytoreovirus* is required for adsorption of the virus to cells of the insect vector[J]. Journal of Virology, 1998, 72(11): 9370-9373.

[19] PRICE D R G, GATEHOUSE J A. RNAi-mediated crop protection against insects[J]. Trends in Biotechnology, 2008, 26(7): 393-400.

[20] REN B, GUO Y, GAO F, et al. Multiple functions of *Rice dwarf phytoreovirus* Pns10 in suppressing systemic RNA silencing[J]. Journal of Virology, 2010, 84(24): 12914-12923.

[21] SHIMIZU T, YOSHII M, WEI T, et al. Silencing by RNAi of the gene for Pns12, a viroplasm matrix protein of *Rice dwarf virus*, results in strong resistance of transgenic rice plants to the virus[J]. Plant Biotechnology Journal, 2009, 7(1): 24-32.

[22] SUZUKI N, KUSANO T, MATSUURA Y, et al. Novel NTP binding property of *Rice dwarf phytoreovirus* minor core protein P5[J]. Virology, 1996, 219(2): 471-474.

[23] SUZUKI N, TANIMURA M, WATANABE Y, et al. Molecular analysis of *Rice dwarf phytoreovirus* segment S1: Interviriral homology of the putative RNA-dependent RNA polymerase between plant-and animal-infecting reoviruses[J]. Virology, 1992, 190(1): 240-247.

[24] UYEDA I, Ueda s, Uyeda I. The *Rice dwarf phytoreovirus* structural protein P7 possesses non-specific nucleic acid binding activity in vitro[J]. Molecular Plant Pathology On-Line 1997.

[25] UYEDA I, KUDO H, YAMADA N, et al. Nucleotide sequence of *Rice dwarf virus* genome segment 4[J]. Journal of General Virology, 1990, 71(10): 2217-2222.

[26] WEI T, CHEN H, ICHIKI-UEHARA T, et al. Entry of *Rice dwarf virus* into cultured cells of its insect vector involves clathrin-mediated endocytosis[J]. Journal of Virology, 2007, 81(14): 7811-7815.

[27] WEI T, KIKUCHI A, MORIYASU Y, et al. The spread of *Rice dwarf virus* among cells of its insect vector exploits virus-induced tubular structures[J]. Journal of Virology, 2006a, 80(17): 8593-8602.

[28] WEI T, KIKUCHI A, SUZUKI N, et al. Pns4 of *Rice dwarf virus* is a phosphoprotein, is localized around the viroplasm matrix, and forms minitubules[J]. Archives of Virology, 2006b, 151(9): 1701-1712.

[29] WEI T, SHIMIZU T, HAGIWARA K, et al. Pns12 protein of *Rice dwarf virus* is essential for formation of viroplasms and nucleation of viral- assembly complexes[J]. Journal of General Virology, 2006c, 87(2): 429-438.

[30] XU H, LI Y, MAO Z, et al. *Rice dwarf phytoreovirus* segment S11 encodes a nucleic acid binding protein[J]. Virology, 1998, 240(2): 267-272.

[31] YU N, CHRISTIAENS O, LIU J, et al. Delivery of dsRNA for RNAi in insects: an overview and future directions[J]. Insect Science, 2013, 20(1): 4-14.

[32] ZHENG L, CHEN H, LIU H, et al. Assembly of viroplasms by viral nonstructural protein Pns9 is essential for persistent infection of *Rice gall dwarf virus* in its insect vector[J]. Virus Research, 2015, 196(22): 162-169.

[33] ZHONG B X, SHEN Y W, OMURA T. RNA-binding domain of the key structural protein P7 for the *Rice dwarf virus* particle assembly[J]. Acta Biochimica Et Biophysica Sinica, 2005, 37(1): 55-60.

[34] ZHOU F, PU Y, WEI T, et al. The P2 capsid protein of the nonenveloped *Rice dwarf phytoreovirus* induces membrane fusion in insect host cells[J]. Proceedings of the National Academy of Sciences of the United States of America, 2007, 104(49): 19547-19552.

水稻矮缩病毒非结构蛋白 Pns6 在病毒复制中的功能

陈倩[1,2]，张玲华[1,2]，黄海宁[1,2]，魏太云[1,2]，谢联辉[1,2]

（1 福建农林大学植物病毒研究所，福州 350002；2 福建省植物病毒学重点实验室，福州 350002）

摘要：水稻矮缩病毒（*Rice dwarf virus*, RDV）主要由黑尾叶蝉以持久增殖型方式传播。RDV 编码的 3 个非结构蛋白 Pns6、Pns11 和 Pns12 是病毒在介体细胞内侵染过程中形成提供病毒复制和子代病毒粒体装配的唯一场所——病毒原质（viroplasm）的主要成分。利用黑尾叶蝉培养细胞和 RNA 干扰（RNA interference, RNAi）体系，研究 Pns6 在病毒侵染介体昆虫过程中的功能。结果显示，将体外合成靶向 *Pns6* 基因的 dsRNA 转染黑尾叶蝉培养细胞后，Pns6 合成显著降低，viroplasm 的形成和病毒的侵染也受到抑制。用显微注射技术导入到黑尾叶蝉的靶向 *Pns6* 基因的 dsRNA，同样抑制 viroplasm 的形成和病毒的侵染，最终阻碍病毒在介体体内的有效扩散。这些研究明确了 Pns6 在病毒侵染黑尾叶蝉过程中的功能，Pns6 是 RDV 的复制关键因子。

关键词：水稻矮缩病毒；叶蝉；病毒原质；Pns6；复制；RNAi

中图分类号：S432.1　**文献标志码：**A　**文章编号：**2095—2783(2015)24—2840—07

Function of virus replication in non-structural protein Pns6 of *Rice dwarf virus*(RDV)

Qian Chen[1,2], Linghua Zhang[1,2], Haining Huang[1,2], Taiyun Wei[1,2], Lianhui Xie[1,2]

(1 Institute of Plant Virology, Fujian Agriculture and Forestry University, Fuzhou 350002, China；2 Fujian Province Key Laboratory of Plant Virology, Fuzhou 350002, China)

Abstract: *Rice dwarf virus* (RDV) is mainly transmitted by leafhopper *Nephotettix cincticeps* in a persistent-propagative manner. Three non-structure proteins Pns6, Pns11 and Pns12 of RDV compose the matrix of virus factory, termed as viroplasm, for viral replication and assembly of progeny virions, when RDV infects cultural cells of insect vector. In this study, the system of cultural cells derived from *N. cincticeps* and technique of RNA interference (RNAi) were used to study the role of Pns6 playing in viral infection and replication in insect vector. The results show that dsRNAs from *Pns6* gene synthesized in vitro and transfected cultural cells of *N. cincticeps* reduce the expression of Pns6. The development of viroplasm and infection of virus are also significantly inhibited. After dsRNAs from *Pns6* are introduced into *N. cincticeps* by microinjection. The formation of viroplasm and infection of virus are blocked, finally the vira1 effective spread in leafhopper is inhibited. These results confirm that Pns6 is the key factor of viral replication during RDV infection.

Keywords: *Rice dwarf virus*; leafhopper; viroplasm; Pns6; replication; RNAi

水稻矮缩病毒（*Rice dwarf virus*, RDV）是引起水稻矮缩病的病原，曾在我国的云南、福建和浙江等

中国科技论文 . 2015, (24): 2840-2846

收稿日期：2015-07-13

基金项目：国家自然科学基金资助项目 (31200118，31300136)；福建省自然科学基金资助项目 (2014J01085)

南方稻区普遍发生，引起重大经济损失。该病害引起的症状是植株明显矮缩、分蘖增多、叶片浓绿僵直，叶片上沿脉呈黄白色虚线状条点病斑（吴祖建等 , 1995; Fukushi et al., 1960; Mizuno et al., 1991; Lu et al., 1998）。RDV 为呼肠孤病毒科（*Reoviridae*）呼肠孤病毒属（*Phytoreovirus*），主要由黑尾叶蝉（*Nephotettix cincticeps*）以持久性增殖型方式传播。病毒基因组由 12 条 dsRNA 组成，分别编码 7 个结构蛋白（P1、P2、P3、P5、P7、P8 和 P9）和 5 个非结构蛋白（Pns4、Pns6、Pns10、Pns11 和 Pns12）。在病毒粒体双层衣壳结构中，P8 以三聚体形式聚合成主要外壳蛋白，与 P9 和可被介体昆虫细胞识别的 P2 组装成外壳（尹哲等 , 2007; Mao et al., 1998; Hagiwara et al., 2004; Zhou et al., 2007）。由 P3 构成的核衣壳蛋白包裹了 S1 编码的 RdRp、鸟苷酸转移酶 P5、RNA 结合蛋白 P7 及基因组 RNA（Suzuki et al., 1992; Ueda et al., 1997; Hagiwara et al., 2004; Zhong et al., 2005）。非结构蛋白的功能一直是水稻病毒研究的热点之一。RDV 的 Pns4 可能具有 GTP-binding 活性，与病毒引起的水稻矮缩症状有关（Suzuki et al., 1990; Uyeda et al., 1990），在黑尾叶蝉培养细胞中可形成微管结构（Wei et al., 2006）。Pns10 在植物体内起沉默抑制子功能（Cao et al., 2005; Ren et al., 2010），在黑尾叶蝉内可装配成包裹病毒的小管，沿 actin 纤维束构成的消化道上皮细胞的微绒毛和内脏肌肉细胞的肌肉纤维束移动，为病毒的扩散提供安全通（Wei et al., 2006; Chcn ct al., 2012）。Pns11 具有结合 RNA 的能力（Xu et al., 1998）；Pns12 是一类磷酸化蛋白（Wei et al., 2006）；Pns6 具有 ATPase 和 RNA 结合活性（Wei et al., 2006），在寄主植物中行使细胞间运动功（Li et al., 2004）。

在黑尾叶蝉培养细胞中，Pns6、Pns11 和 Pns12 参与形成病毒复制和装配的工厂——病毒原质（viroplasm）。在 RDV 侵染黑尾叶蝉培养细胞 6h 时，Pns6、Pns11 和 Pns12 首先被翻译，相互结合聚集成 viroplasm 的基质，为病毒的复制和装配提供场所（Wei et al., 2006）。Shimizu 等通过 RNA 干扰（RNA interference, RNAi）技术在水稻中沉默 *Pns12* 基因，能较强抵抗病毒的侵染，表明 Pns12 和病毒在水稻中的有效复制存在较大的相关性（Shimizu et al., 2009）。因此，RDV 编码的非结构蛋白 Pns6、Pns11 和 Pns12 与病毒在介体细胞内的复制及子代病毒粒体的装配都有一定的相关性，都有可能是病毒的复制关键因子。鉴于此，本研究利用黑尾叶蝉培养细胞和 RNAi 体系，验证 RDV 在介体细胞中侵染诱导形成的 viroplasm 组分 Pns6 在病毒侵染复制过程中的功能，鉴定病毒在介体细胞内复制的关键因子。

1 材料与方法

1.1 介体、原代细胞、感病水稻和抗体

黑尾叶蝉捕获于福建福州水稻种植区，经本实验室筛选获得无毒昆虫，并饲养于 26℃温室中。黑尾叶蝉培养细胞由本实验室建立和维持。

感染 RD 的水稻病株采自云南省寻甸县水稻种植区，用本实验室饲养的黑尾叶蝉取食传毒进行扩繁，新的感病水稻于温室中种植。提纯病毒保存于 –80℃。*Pns6* 和 RDV 病毒抗体由日本国立农业研究中心大村敏博教授馈赠。将抗体分别与购自 Invitrogen 公司的荧光素 FITC 和 rhodamine 交联，制备荧光抗体 Pns6-rhodamine 和 RDV-FITC。标记肌动蛋白 actin 的荧光抗体 actin dyephalloidin-Alexa Flu or 647 carboxylic acid（actin-647）购自 Invitrogen 公司。

1.2 dsRNA 的体外合成

分别选取 *Pns6* 基因 686bp 和绿色荧光蛋白（green fluorescent protein, GFP）基因 584bp 长度的片断，在上下游引物 5 端加上 T7 启动子序列（下划线标记），引物序列如表 1 所示。用 PCR 方法扩增该片断。

参照 T7 RiboMAX^{TM} Express RNAi System（Promega）试剂盒说明书，将得到的 PCR 产物作为 dsRNA 体外转录的模板，用 T7 转录酶进行体外转录。用琼脂糖凝胶电泳测定靶向 *Pns6* 和 GFP 基因的

dsRNA（dsPns6 和 dsGFP）纯度，分光光度计测定 dsRNA 浓度。

表 1　本研究使用的引物序列

基因	上 / 下游	引物序列 (5′-3′)	用途
Pns6	-F	ATTCTCTAGAAGCTTAATACGACTCACTATAGGGCGCGAAGCCGATAAGCACA	体外转录模板
	-R	ATTCTCTAGAAGCTTAATACGACTCACTATAGGGAGCAGTACGCCCATCCTCTAACAG	
GFP	-F	ATTCTCTAGAAGCTTAATACGACTCACTATAGGGCTTGTTGAATTAGATGGTGATGTT	体外转录模板
	-R	ATTCTCTAGAAGCTTAATACGACTCACTATAGGGGTTTCGAAAGGGCAGATTGT	
actin	-F	GGGATACAGTTTCACCACG	荧光定量 PCR
	-R	GACACCTGAATCGCTCGT	
P8	-F	TACAGCCATCAGCTAAGCCAAA	荧光定量 PCR
	-R	CCGCAACAGACCGAAACA	
Pns6	-F	TTTCGGGATAATGCTGCTGAC	荧光定量 PCR
	-R	CCGCCAAATAAGCAAACCA	

1.3　在培养细胞和黑尾叶蝉内诱导 RNAi

将黑尾叶蝉培养细胞以适当的密度种在细胞爬片（直径 12mm）上，25℃ 培养约 3d，细胞覆盖率达 80%。用无血清无抗生素培养基配制终浓度为 0.08g/L 的 dsRNA 稀释液，同时按 1μL 无血清无抗生素培养基加入 0.06μL cellfectin II Reagent 的比例配制脂质体稀释液，而后混合上述 2 种稀释液，配成 dsRNA 转染液。将该转染液于 25℃ 孵育黑尾叶蝉培养细胞 8h。接着用侵染复数（multiplicity of infection, M0I）为 10 的高浓度提纯病毒 RDV 于 25℃ 侵染转染后的细胞 2h，最后正常培养。

取 2 龄黑尾叶蝉若虫，饲喂 RDV 病株 2d 后，显微注射质量浓度为 0.5g/L 的 dsRNA，而后转至健康水稻上饲养。

1.4　RT-qPCR 分析 dsRNA 抑制基因表达水平

于注射后第 3、6 和 12 天收集黑尾叶蝉，用 TRIzol Reagent 提取黑尾叶蝉总 RNA。用 Primer Premier 5.0 引物设计软件设计 P8、Pns6 和黑尾叶蝉 actin 的荧光定量 PCR 引物，如表 1 所示。用 Revert Aid Reverse Transcriptase（Thermo Scientific 公司，美国）反转录后，使用 SYBR Green PCR Master Mix kit（QIAGEN 公司，美国）试剂盒，以 actin 为内参基因进行 RT-qPCR。RT-qPCR 反应程序：95℃ 预变性 2min，95℃ 5s，60℃ 20s，72℃ 15s，共 40 个循环。

1.5　免疫荧光标记

免疫荧光标记参照 Chen 等（Xu et al., 1998; Li et al., 2004; Wei et al., 2006; Shimizu et al., 2009; Chen et al., 2012; Chen et al., 2015）方法。

用 4% 多聚甲醛室温固定培养细胞 30min，Triton-X100/PB 渗透 10min，抗体避光孵育 45min，每步前后用 PBS（pH 值为 7.2 ～ 7.4）缓冲液清洗 3 遍。最后用甘油封片。

将解剖得到的消化道用 4% 多聚甲醛室温固定 2h，2% Triton-X100/PB 渗透 1h，37℃抗体避光孵育 2h，每步前后用 PBS（pH 值为 7.2 ～ 7.4）缓冲液清洗 3 次。最后用甘油封片。

用激光共聚焦显微镜（Leica TCS-SP5）观察标记好的样品，并拍照。

2 结果

2.1 dsPns6 诱导的 RNAi 抑制 RDV 对培养细胞的侵染

为证明 Pns6 在病毒侵染中所起的功能，首先用体外合成的靶向 Pns6 的 dsRNA 转染培养细胞，以干扰 *Pns6* 基因的表达。然后用提纯 PDV 侵染转染有 dsPns6 和 dsGFP 的细胞，从而分析 viroplasm 和病毒侵染受到的影响。

用荧光抗体 RDV-FTTC 和 Pns6-rhodamine 做免疫荧光标记，重复 3 次。共聚焦显微镜观察结果显示，在 dsGFP 处理的培养细胞中，大部分细胞被 RDV 侵染，viroplasm 也被 Pns6-rhodamine 特异性抗体标记，呈现出分散的点状颗粒（图 1）。正在装配成完整病毒粒体的子代病毒围绕在其周围，说明 RDV 正在复制（图 1I）。而在 dsPns6 处理过的细胞中，病毒和 Pns6 不易被检测到。虽然在少数的 1 ～ 2 个被 RDV 侵染的细胞簇有少量 RDV 可以被检测到，但 Pns6 仍无法观察到，说明病毒处于初侵染状态，尚未进入复制阶段（图 1）。

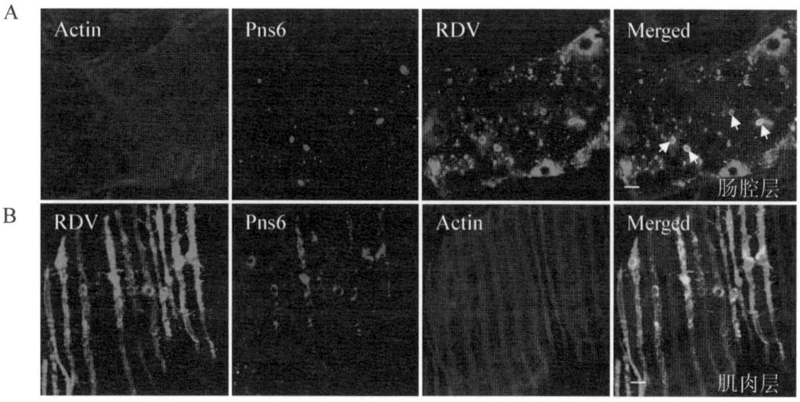

图 1 免疫荧光观察 dsPns6 处理细胞后 RDV 的侵染和 Pns6 表达情况

（插图 I 为白框部分的放大图）

为定量评估 dsPns6 诱导的 RNAi 对 RDV 侵染的影响，使用荧光抗体侵染点计数法（Kimura et al., 1986）计算 RDV 在培养细胞中的侵染率。结果如图 2（A）所示，对照 dsGFP 转染的细胞 RDV 平均侵染率为 93.3%，而 dsPns6 转染后的 RDV 平均侵染率为 20% 左右。这些结果表明 dsPns6 诱导的 RNAi 显著降低了病毒积累量和侵染率。

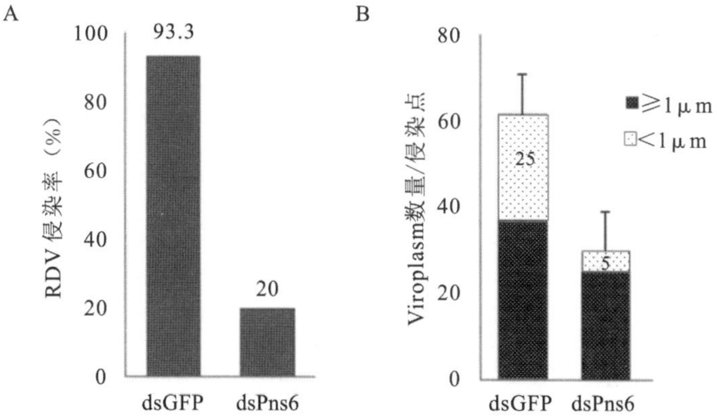

图 2 统计 dsPns6 处理细胞的 RDV 侵染率和 viroplasm 的数量

（误差线为 3 次独立实验的标准差）

2.2 dsPns6 诱导的 RNAi 阻碍 viroplasm 的动态发展

随着病毒侵染的推进，viroplasm 会呈现形成、成熟、衰老和解聚的动态过程。病毒复制进入旺盛期后，细胞内会不断有新的小 viroplasm 形成，已经形成的小 viroplasm 会聚集成大的 viroplasm，viroplasm 的数量和大小会达最高（Eichwald et al., 2012）。

为进一步描述干扰 *Pns6* 基因表达对 viroplasm 形成的影响，用激光共聚焦显微镜计算 40 倍物镜的视野下，每个侵染点的 viroplasm 数量（Eichwald et al., 2012），重复 3 次。结果发现，dsPns6 转染的细胞中，viroplasm 数量明显低于 dsGFP 处理的数量，此情况类似于病毒的早期侵染（图 2B）。为了描述 viroplasm 的大小以分析其所处的状态，利用共聚焦显微镜的标尺程序，将 viroplasm 分为直径 ≥ 1μm 和 <1μm 2 组，统计各组数量（图 2B）。可以很明显看出，dsPns6 转染的细胞中，两类大小的 viroplasm 数量较对照都有减少，说明 dsPns6 诱导的 RNAi 不仅抑制了新 viroplasm 的生成，还抑制了小 viroplasm 聚合成大 viroplasm，阻碍其进一步的成熟。

上述结果表明，由 dsRNA 诱导的 RNAi 显著抑制 *Pns6* 表达，影响了 viroplasm 的形成和病毒在介体培养细胞中的侵染。

2.3 Pns6 在黑尾叶蝉消化道参与组成 viroplasm 基质

为验证 *Pns6* 在病毒依靠介体传播中发挥的功能，首先观察 *Pns6* 在带毒黑尾叶蝉中的表达情况。取饲毒后度过 14d 循回期的黑尾叶蝉，解剖其消化道，用荧光抗体 RDV-FITC，actin-647 和 Pns6-rhrdamine 免疫标记。共聚焦显微镜观察显示，Pns6- rhodamine 标记的大量点状 viroplasms 分布在消化道上皮细胞和肌肉细胞内。病毒呈环状围绕在vircplasm 周围（图 3），与 RDV 侵染培养细胞的情况类似，从而确认 Pns6 在黑尾叶蝉体内也是 virtplasm 基质的组分。

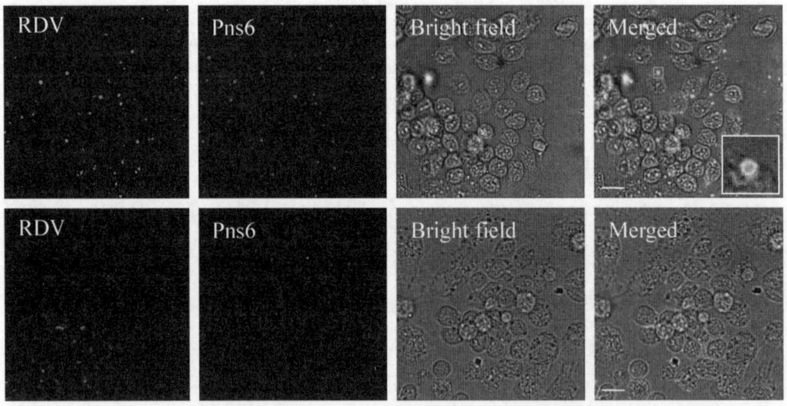

图 3 免疫荧光观察 PnW 在带毒黑尾叶蝉消化道肠腔层（a）和肌肉层（b）的表达

2.4 dsPns6 诱导的 RNAi 可降低黑尾叶蝉带毒率

在确认了 Pns6 在黑尾叶蝉体内可以参与形成 viroplasm 后，用 dsPns6 注射黑尾叶蝉，以检测 RNAi 对黑尾叶蝉获毒的影响。于注射后第 3、6 和 12 天收集黑尾叶蝉（>30 头）。用 RT-PCR 方法检测 RDV 外壳蛋白 *P8* 基因，以分析黑尾叶蝉带毒率，结果如图 4 所示。由图 4 可见：注射后第 3 天，注射 dsGFP 的黑尾叶蝉中有 50% 被检测到 *P8* 基因，注射 dsPns6 的黑尾叶蝉中，有 16% 被检测到 *P8* 基因，表明注射 dsPns6 可以诱导 RNAi，并且诱导的 RNAi 导致黑尾叶蝉带毒率显著下降；注射后第 6 天，对照 dsGFP 处理的黑尾叶蝉中，有 53% 被检测到 *P8* 基因，与注射后第 3 天结果基本保持不变。而 dsPns6 处理的黑尾叶蝉中，有 15.6% 呈 *P8* 基因阳性，仍然低于对照，说明以本实验的 dsRNA 剂量可以维持 RNAi 至少 6d。注射后第 12 天，对照 dsGFP 处理的黑尾叶蝉带毒率仍维持 51%，而 dsPns6 处理的黑尾叶蝉带毒率为 34.2%，仍显著低于对照。但相对注射后第 6 天，dsPns6 处理的带毒率有所上升，说明 RNAi 作用在逐渐变弱。

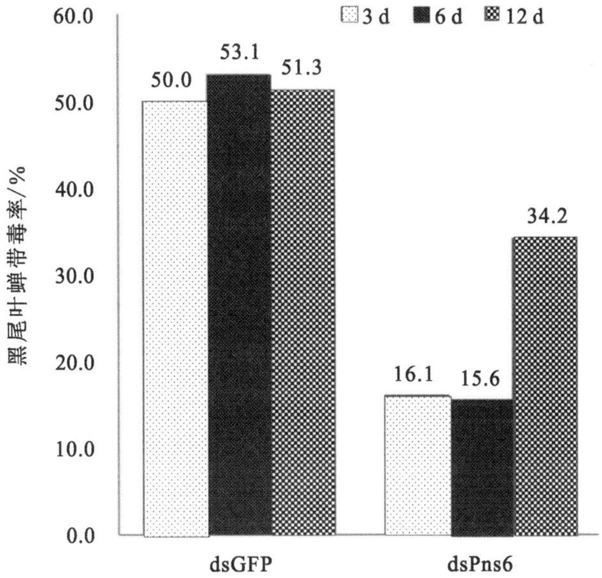

图 4　注射 dsPns6 后 3、6 和 12 天的黑尾叶蝉带毒率

2.5　dsPns6 诱导的 RNAi 同时下调 Pns6 和 P8 基因的相对表达量

用 RT-qPCR 实验验证 RNAi 对病毒基因表达的影响（3 次重复）。选择 *P8* 基因作为病毒基因相对表达量的参照，结果如图 5 所示。注射后第 6 天，注射 dsPns6 的黑尾叶蝉中，*P8* 和 *Pns6* 基因相对表达量低于对照的 40% 多。说明干扰 *Pns6* 基因，可以下调病毒基因的相对表达量，导致病毒在黑尾叶蝉体内的积累减少。

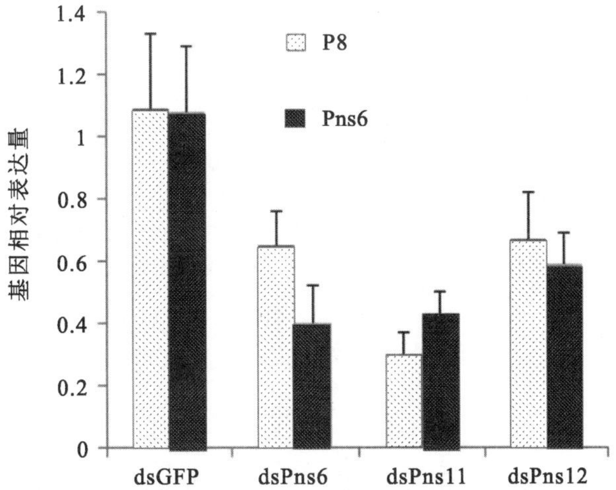

图 5　注射 dsPns6 后第 6 天的黑尾叶蝉体内 Pns6 和 P8 基因的相对表达量

（误差线为 3 次独立 PCR 的标准差）

2.6　dsPns6 诱导的 RNAi 减缓 RDV 在黑尾叶蝉内的扩散

进一步对这些 dsRNA 处理的黑尾叶蝉进行解剖，做免疫荧光标记分析。结果显示，注射后第 3 天

时，在大多数注射 dsGFP 的带毒黑尾叶蝉消化道上皮细胞内可以检测到 RDV 和 Pns6-rhodamine 荧光抗体标记的 viroplasm 的存在，包括滤室、前中肠、中中肠和后中肠。在少数黑尾叶蝉滤室和中中肠的肌肉组织还可检测到 RDV 和 viroplasm。相比之下，在注射了 dsPns6 的黑尾叶蝉中，很少有被检测到消化道被 RDV 侵染。在被检测到的带毒黑尾叶蝉中，病毒和 viroplasm 被免疫共定位在滤室、前中肠或后中肠的 1 ~ 2 个上皮细胞内，侵染情况类似 RDV 的初侵染，说明病毒的扩散被大大抑制。组织带毒率统计显示 RDV 在 dsPns6 处理的带毒黑尾叶蝉中增殖水平低下（如表 2 所示）。这些结果说明 dsPns6 诱导的 RNAi 抑制了 Pns6 的表达，影响 viroplasm 的装配，从而降低病毒在介体内的积累水平。

表 2　注射后第 3、6 和 12 天，注射 dsGFP 和 dsPns6 的带毒黑尾叶蝉不同组织的带毒率

处理	注射后时间 /d	滤室 /%	前中肠 /%	中中肠 /%	后中肠 /%	后肠 /%	唾液腺 /%
dsGFP	3	76.2	38.1	38.1	23.8	14.3	0
	6	82.3	41.2	35.2	41.2	41.2	0
	12	76.2	71.4	61.9	66.7	52.3	71.4
dsPns6	3	60	0	0	80	0	0
	6	100	20	20	80	20	0
	12	76.9	30.8	38.5	53.8	38.5	0

注射后第 6 天时，注射 dsGFP 的带毒黑尾叶蝉消化道与第 3 天相比，呈现更高的组织带毒率，并且有更多的病毒和 viroplasm 分布在上皮细胞和肌肉纤维，说明病毒在高效增殖和扩散（图 6）。在注射了 dsPns6 的带毒黑尾叶蝉中，病毒和 viroplasm 仍然分布在滤室、前中肠和后肠的 3 ~ 5 个上皮细胞侵染点中，与第 3 天相比，病毒增殖和扩散的区域有所扩大。并且组织带毒率有上升，说明病毒积累和扩散在一定程度开始恢复。到了注射后的第 12 天时，病毒和 viroplasm 遍布注射 dsGFP 的带毒黑尾叶蝉消化道上皮细胞、肌肉组织及唾液腺，处于病毒的后期侵染，消化道各部分的组织带毒率明显增加。而在 dsPns6 处理的带毒黑尾叶蝉中，大多数病毒和 viroplasm 仍然分布在消化道上皮细胞中，但侵染面积和数量都有上升，导致组织带毒率也相应提高。这些上升的数据说明病毒正在增殖和积累，也说明 RNAi 正在逐渐变弱，一旦病毒复制能力恢复，其增殖和扩散就会趋于正常。

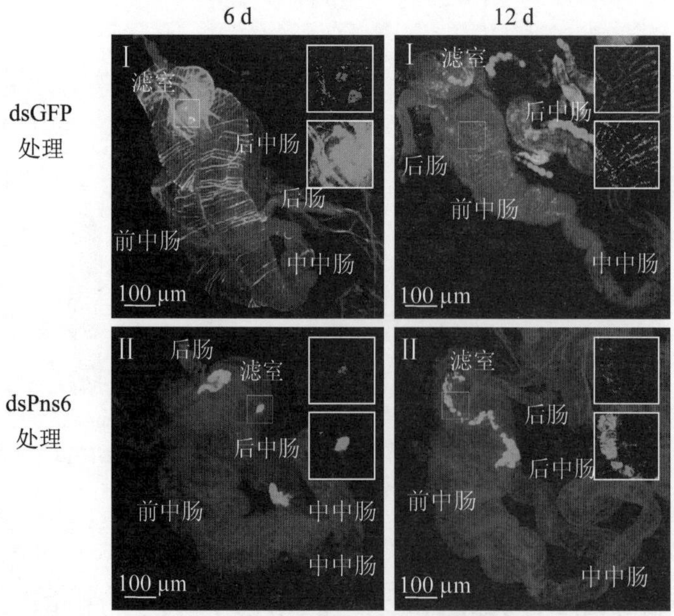

图 6　dsPns6 注射后不同时间病毒在黑尾叶蝉体内的侵染

（图片为绿色、红色和蓝色荧光的叠加，小插图为每个图中白框部分的放大分解图）

因此，上述结果都证明了 Pns6 在病毒侵染培养细胞和黑尾叶蝉过程中参与了病毒的复制，是 RDV 复制的关键因子。

3 讨论

持久性增殖型传播的病毒与非增殖型传播的病毒主要区别是病毒是否在介体内复制增殖。病毒在介体中能否有效装配 viroplasm 的基质来启动病毒复制和病毒粒体装配的程序，是病毒能否在介体内持久增殖，进而突破介体传播屏障的关键步骤。而要揭示病毒复制的机理，很大程度上要依赖于介体培养细胞。培养细胞的建立和发展，为研究病毒的生活史，包括病毒侵入、复制和扩散，提供了很好的体系。将该体系与高效的 RNAi 技术相结合，克服了由于缺乏反向遗传学技术而无法在病毒侵染的情况下研究非结构蛋白功能的弊端，弥补多组分病毒无法利用反向遗传学手段研究基因功能的空缺。利用培养细胞体系和 RNAi 技术相结合的手段，研究者们已揭示了水稻瘤矮病毒（*Rice gall dwarf virus*, RGDV）非结构蛋白 Pns12，南方水稻黑条矮缩病毒（*Southern rice black streaked dwarf virus*, SRBSDV）非结构蛋白 P9-1，以及同科的轮状病毒（*Rotavirus*）非结构蛋白 NSP2 和 NSP5 参与了病毒在介体昆虫培养细胞或宿主细胞内的复制（Jia et al., 2012; 郑立敏等，2014）。本研究将黑尾叶蝉培养细胞与 RNAi 技术相结合，观察到 Pns6 表达在被干扰的情况下，病毒侵染和积累所受到的影响，以及 viroplasm 的形成和成熟的动态变化，得到了在虫体内无法观察到的结果。此法为研究 RDV 的另外 2 个 viroplasm 基质蛋白 Pns11 和 Pns12 的功能提供良好借鉴，也为建立基于阻断病毒在介体内有效复制的调控策略提供理论依据。

呼肠孤病毒以 Rotavirus 的 viroplasm 形成机制的研究最为深入。Rotavirus 的 viroplasm 由非结构蛋白 NSP2 和 NSP5 组成。NSP5 具有多种磷酸化异构体，是病毒侵染所必需的，可与 NSP2、结构蛋白 VP1、VP2 互作，参与了多种 viroplasm 动态调控的过程（Criglar et al., 2014）。NSP2 可结合单链 RNA，并具有依赖于 ATP 的解旋特性（Taraporewala et al., 1999; Contin et al., 2010）。最新研究发现，NSP2 具有 2 种形态。在病毒早期侵染时，细胞质分散型 NSP2 以一种依赖于碟酸化方式在未成熟的 viroplasm 内与 NSP5 互作，并迅速积累。viroplasm 型 NSP2 的量随 viroplasm 的成熟和增大稳定增加，推测这种依赖于磷酸化的 NSP2-NSP5 互作方式和 NSP2 可能的构型变化是 viroplasm 形成所需的（Criglar et al., 2014）。而在 RDV 的 viroplasm 中，同样具有磷酸化活性的 Pns12、具有 ATPase 活性的 Pns6 和 RNA 结合活性的 Pns11 很可能行使和 NSP5 与 NSP2 类似的功能。但进一步的研究还需要同翅目昆虫反向遗传学技术的建立和发展。

近年，利用介体昆虫培养细胞体系，侵染植物的呼肠孤病毒 viroplasm 基质蛋白的互作机制研究方面取得了较多进展。Mao 等（Mao et al., 2013）发现 viroplasm 是由含有病毒 RNA 和非结构蛋白 P6 和 P9-1 的颗粒状区域，和含有病毒 RNA、子代内核粒体、病毒粒体、非结构蛋白 P5 和 P6 的丝状区域组成的。丝状的 viroplasm 基质是子代病毒装配的场所。因为装配的内核颗粒在丝状的 viroplasm 基质中生成病毒 RNA，推测这些病毒 RNA 可能被转运至颗粒状的 viroplasm 基质中。P5 可独立形成丝状内含体，而 P9-1 可独立形成颗粒状内含体，说明丝状和颗粒状的 viroplasm 基质主要由 P5 和 P9-1 组成。P6 通过与 P9-1 和 P5 直接互作而被招募至 viroplasm。因此，P5、P6 和 P9-1 是 viroplasm 丝状和颗粒状基质形成和成熟所必需的。Chen 等（Chen et al., 2014）利用褐飞虱继代培养细胞发现了 RRSV 的 viroplasm 由非结构蛋白 Pns6 和 Pns10 组成。大量的核衣壳蛋白 P3、内核粒体和新合成的病毒 RNA 积累于此，外壳蛋白 P8 和完整病毒粒体则围绕在 viroplasm 周围。非结构蛋白 Pns10 是 viroplasm 的主要成分，可独立形成类似 viroplasm 的内含体，说明 viroplasm 的基质主要由 Pns10 构成。Pns6 通过与 Pns10 互作而被招募至 viroplasm。核衣壳蛋白 P3 通过与 Pns6 互作而被招募至 viroplasm。Pns6 和 Pns10 的表达被干扰后，viroplasm 的形成、病毒装配、病毒蛋白的表达和病毒 dsRNA 的合成也被抑制，说明 Pns6 和 Pns10 是通过招募或保留病毒复制和装配的必要成分，从而在病毒侵染介体细胞的早期起重要作用。这些 viroplasm 蛋白的互作网络为 RDV viroplasm 的形成机制提供研究依据。先前的研究证明了非结构蛋白 Pns12 在异源细胞 Sf9 中单独表达时，可形成类似 viroplasm 的内含体，暗示了 Pns12 可能是 viroplasm 形成所需的最基本的病毒蛋白（Suzuki et al., 1990）。它可能搭建了 viroplasm 的基本构架，并招募了其他病毒蛋白，形成完整的 viroplasm，以启动病毒的复制和装配。而 Pns6 与 Pns12、Pns11 间的两两互作和三者互作研究，以及 Pns6 是如何参与病毒的复制等问题将是下一步亟需解答的。

4 结论

本研究在前期已知 Pns6、Pns11 和 Pns12 是 viroplasm 组分的基础上，将 RNAi 技术应用于黑尾叶蝉培养细胞体系与虫体，研究了 viroplasm 基质蛋白 Pns6 在病毒复制过程中起的作用。在黑尾叶蝉培养细胞中诱导耙向 *Pns6* 基因的 RNAi，可显著抑制 Pns6 的表达，影响 viroplasm 的聚合与成熟，最终病毒复制受阻，子代病毒的积累和侵染水平降低。造成这些影响的原因与 Pns6 在 viroplasm 中发挥的功能有关，推测 Pns6 参与了病毒的复制。在介体黑尾叶蝉内诱导的 RNAi，通过降低 *Pns6* 基因的相对表达量和蛋白表达水平，抑制病毒的积累水平，导致病毒在介体内的扩散明显减缓，甚至介体带毒率显著降低。这些结果确认了在病毒侵染黑尾叶蝉过程中，Pns6 蛋白参与了病毒复制，为 RDV 的复制因子，缺失了该因子，将导致病毒复制受影响。

本研究体系为揭示 RDV 的另外 2 个 viroplasm 基质蛋白 Pns11 和 Pns12 的功能提供了良好借鉴，也为建立基于阻断病毒在介体内有效复制的调控策略提供了理论依据。

参考文献

[1] 吴祖建，林奇英，林奇田．水稻矮缩病毒的提纯和抗血清制备：福建省科协第二届青年学术年会 - 中国科协第二届青年学术年会卫星会议论文集 [C]. 福州：福建科学技术出版社，1995.

[2] 尹哲，吉栩，吴云锋，等．水稻矮缩病毒外壳蛋白 P9 具有体内转录激活活性 [J]. 中国农业科技导报，2007,9(3):61-65.

[3] 郑立敏，刘华敏，陈红燕，等．干扰水稻瘤矮病毒 (RGDV) 非结构蛋白 (Pns12) 的表达抑制病毒在介体昆虫培养细胞内的复制 [J]. 农业生物技术学报，2014,22(11):1321-1328.

[4] CAO X, ZHOU P, ZHANG X, et al. Identification of an RNA silencing suppressor from a plant double-stranded RNA virus[J]. Journal of Virology, 2005, 79(20): 13018-13027.

[5] CHEN H, ZHENG L, MAO Q, et al. Development of continuous cell culture of brown planthopper to trace the early infection process of *Oryzaviruses* in insect vector cells[J]. Journal of Virology, 2014, 88(8): 4265-4274.

[6] Chen Q, Chen H, Mao Q, et al. Tubular structure induced by a plant virus facilitates vira1 spread in its vector insect. PLoS Pathogens, 8(11): e1003032.

[7] CHEN Q, WANG H, REN T, et al. Interaction between nonstructural protein Pns10 of *Rice dwarf virus* and cytoplasmic actin of leafhoppers is correlated with insect vector specificity[J]. The Journal of General Virology, 2015, 96(4): 933-938.

[8] CONTIN R, ARNOLDI F, CAMPAGNA M, et al. *Rotavirus* NSP5 orchestrates recruitment of viroplasmic proteins[J]. Journal of General Virology, 2010, 91(7): 1782-1793.

[9] CRIGLAR JM, HU L, CRAWFORD SE, et al. A novel form of *Rotavirus* NSP2 and phosphorylation-dependent NSP2-NSP5 interactions are associated with viroplasm assembly[J]. Journal of Virology, 2014, 88(2): 786-798.

[10] EICHWALD C, ARNOLDI F, LAIMIBACHER A S, et al.*Rotavirus* viroplasm fusion and perinuclear localization are dynamic processes requiring stabilized microtubules[J]. PLoS One, 2012, 7(10): e47947.

[11] FUKUSHI T, SHIKATA E, KIMURA I, et al. Electron microscopic studies on the *Rice dwarf virus*[J]. Proceedings of the Japan Academy, 2006, 36: 352-357.

[12] HAGIWARA K, HIGASHI T, MIYAZAKI N, et al. The amino-terminal region of major capsid protein P3 is essential for self-assembly of single-shelled core-like particles of *Rice dwarf virus*[J]. Journal of Virology, 2004, 78(6): 3145-3148.

[13] JIA D, CHEN H, ZHENG A, et al. Development of an insect vector cell culture and RNA interference system to investigate the functional role of *Fijivirus* replication protein[J]. Journal of Virology, 2012, 86(10): 5800-5807.

[14] KIMURA I. A study of *Rice dwarf virus* in vector cell monolayers by fluorescent antibody focus counting[J]. Journal of General Virology, 1986, 67(10): 2119-2124.

[15] LI Y, BAO YM, WEI CH, et al. *Rice dwarf phytoreovirus* segment S6-encoded nonstruetural protein has a cell-to-cell movement function[J]. Journal of Virology, 2004, 78(10): 5382-5389.

[16] LU G, ZHOU Z H, BAKER M L, et al. Structure of double-shelled *Rice dwarf virus*[J]. Virology Journal, 1998, 72(11): 8541-8549.

[17] MAO Q, ZHENG S, HAN Q, et al. New model for the genesis and maturation of viroplasms induced by *Fijiviruses* in insect vector cells[J]. Journal of Virology, 2013, 87(12): 6819-6828.

[18] MAO ZJ, LI Y, XU H, et al. 1998. The 42K protein of *Rice dwarf virus* is a post-translational cleavage product of the 46K outer capsid protein[J]. Archives of Virology, 1998, 143(9): 1831-1838.

[19] MIZUNO H, KANO H, OMURA T, et al. Crystallization and preliminary X-ray study of a double-shelled spherical virus, *Rice dwarf virus*[J]. Journal of Molecular Biology, 1991, 219(4): 665-669.

[20] REN B, GUO Y, GAO F, et al. Multiple functions of *Rice dwarf phytoreovirus* Pns10 in suppressing systemic RNA silencing[J]. Journal of Virology, 2010, 84(24): 12914-12923.

[21] SHIMIZU T, YOSHII M, WEI T, et al. Silencing by RNAi of the gene for Pns12, a viroplasm matrix protein of *Rice dwarf virus*, results in strong resistance of transgenic rice plants to the virus[J]. Plant Biotechnol J, 2009, 7(1): 24-32.

[22] SILVESTRI L S, TARAPOREWALA Z F, PATTON J T. *Rotavirus* replication: plus-sense templates for double-stranded RNA synthesis are made in viroplasms[J]. Journal of Virology, 2004, 78(14): 7763-7774.

[23] SUZUKI N, KUSANO T, MATSUURA Y, et al. Novel NTP binding property of *Rice dwarf phytoreovirus* minor core protein P5[J]. Virology, 1996, 219(2): 471-474.

[24] SUZUKI N, TANIMURA M, WATANABE Y, et al. Molecular analysis of *Rice dwarf phytoreovirus* segment S1: interviriral homology of the putative RNA-dependent RNA polymerase between plant-and animab-infecting reoviruses[J]. Virology, 1992, 190(1): 240-247.

[25] SUZUKI N, WATANABE Y, KUSANO T, et al. Sequence analysis of *Rice dwarf phytoreovirus* genome segments S4, S5 and S6: comparison with the equivalent wound tumor virus segments[J]. Virology, 1990, 179(1): 446-454.

[26] TARAPOREWALA Z, CHEN D, PATTON J T. Multimers formed by the *Rotavirus* nonstructural protein NSP2 bind to RNA and have nucleoside triphosphatase activity[J]. Journal of Virology, 1999, 73(12): 9934-9943.

[27] UYEDA I, KUDO H, YAMADA N, et al. Nucleotide sequence of *Rice dwarf virus* genome segment 4[J]. Journal of General Virology, 1990, 71 (10): 2217-2222.

[28] WEI T, KIKUCHI A, MORIYASU Y, et al. The spread of *Rice dwarf virus* among cells of its insect vector exploits virus-induced tubular structures[J]. Journal of Virology, 2006, 80(17): 8593-8602.

[29] WEI T, KIKUCHI A, SUZUKI N, et al. Pns4 of *Rice dwarf virus* is a phosphoprotein, is localized around the viroplasm matrix, and forms minitubules[J]. Archives of Virology, 2006, 151(9): 1701-1712.

[30] WEI T, SHIMIZU T, HAGIWARA K, et al. Pns12 protein of *Rice dwarf virus* is essential for formation of viroplasms and nucleation of viral-assembly complexes[J]. Journal of General Virology, 2006, 87(2): 429-438.

[31] XU H, LI Y, MAO Z, et al. *Rice dwarf phytoreovirus* segment S11 encodes a nucleic acid binding protein[J]. Virology, 1998, 240(2): 267-272.

[32] ZHOU F, PU Y, WEI T, et al. The P2 capsid protein of the nonenveloped *Rice dwarf phytoreovirus* induces membrane fusion in insect host cells[J]. Proceedings of the National Academy of Sciences of the United States of America, 2007, 104(49): 19547-19552.

[33] ZHONG BX, SHEN YW, OMURA T. RNA-binding domain of the key structural protein P7 for the *Rice dwarf virus* particle assembly[J]. Acta Biochimica Et Biophysica Sinica, 2005, 37(1): 55-60.

Nonstructural protein Pns4 of *rice dwarf virus* is essential for viral infection in its insect vector

Qian Chen, Linghua Zhang, Hongyan Chen, Lianhui Xie, Taiyun Weia

(Fujian Province Key Laboratory of Plant Virology, Institute of Plant Virology, Fujian Agriculture and Forestry University, Fuzhou, Fujian 350002, China)

Abstract: *Rice dwarf virus* (RDV), a plant reovirus, is mainly transmitted by the green rice leafhopper, *Nephotettix cincticeps*, in a persistent-propagative manner. Plant reovirusesare thought to replicate and assemble within cytoplasmic structures called viroplasms. Nonstructural protein Pns4 of RDV, a phosphoprotein, is localized around the viroplasm matrix and forms minitubules in insect vector cells. However, the functional role of Pns4 minitubules during viral infection in insect vector is still unknown yet. RNA interference (RNAi) system targeting Pns4 gene of RDV was conducted. Double-stranded RNA (dsRNA) specific for Pns4 gene was synthesized in vitro, and introduced into cultured leafhopper cells by transfection or into insect body by microinjection. The effects of the knockdown of Pns4 expression due to RNAi induced by synthesized dsRNA from Pns4 gene on viral replication and spread in cultured cells and insect vector were analyzed using immunofluorescence, western blotting or RT-PCR assays. In cultured leafhopper cells, the knockdown of Pns4 expression due to RNAi induced by synthesized dsRNA from Pns4 gene strongly inhibited the formation of minitubules, preventing the accumulation of viroplasms and efficient viral infection in insect vector cells. RNAi induced by microinjection of dsRNA from Pns4 gene significantly reduced the viruliferous rate of *N. cincticeps*. Furthermore, it also strongly inhibited the formation of minitubules and viroplasms, preventing efficient viral spread from the initially infected site in the filter chamber of intact insect vector. Pns4 of RDV is essential for viral infection and replication in insect vector. It may directly participate in the functional role of viroplasm for viral replication and assembly of progeny virions during viral infection in leafhopper vector.

Keywords: *Rice dwarf virus*, Leafhopper vector, Pns4 minitubules, Viroplasm, RNA interference

1 Introduction

Viruses in the family *Reoviridae* are thought to replicate and assemble within cytoplasmic, nonmembranous structures called viroplasms(Eichwald et al., 2012). For animal reoviruses, including reoviruses, rotaviruses and *bluetongue virus* (BTV), viral dsRNA, mRNA and proteins, and host components such as ribosomes, protein-synthesis machinery and chaperones aggregate within the viroplasm matrix (Guglielmi et al., 2010; Boyce et al., 2012; Eichwald et al., 2012; de Castro et al., 2014). Furthermore, some host components such as ribosomes, membrane components, mitochondria and microtubules have been observed to accumulate at the periphery of the viroplasm matrix (Boyce et al., 2012; Eichwald et al., 2012; de Castro et al., 2014). The host components at the periphery of the viroplasm

Virology Journal. 2015, 12:211
Received 13 August 2015; Accepted 25 November 2015

may be utilized to facilitate viral replication and assembly of progeny virions during viral infection in host cells.

Rice dwarf virus (RDV), a phytoreovirus in the family *Reoviridae*, is mainly transmitted by the leafhopper vector, *Nephotettix cincticeps*, in a persistent-propagative manner (King et al., 2011). RDV is an icosahedral and double-layered spherical virion of approximately 70nm in diameter. The RDV genome consists of 12 double-stranded RNA (dsRNA) segments (S1-S12), encoding at least seven structural proteins (P1, P2, P3, P5, P7, P8 and P9) and five nonstructural proteins (Pns4, Pns6, Pns10, Pns11 and Pns12) (Suzuki et al., 1994; Suzuki et al., 1996a). The viral core particle is composed of the inner core protein P3, which encloses P1, a putative RNA polymerase, P5, a guanylyltransferase, and P7, a protien with RNA-binding activity (Uyeda; Kano et al., 1990; Suzuki et al., 1992; Suzuki et al., 1996b). The outer capsid shell is composed of the major outer capsid protein P8, and the minor outer capsid proteins P2 and P9 (Omura et al., 1989; Yan et al., 1996; Zhong et al., 2003; Zhou et al., 2007b; Zhou et al., 2007a). The functions of nonstructural proteins in the insect vector have been determined using continuous cell cultures derived from *N. cincticeps*. Pns10 assembles into tubules that package virions to facilitate viral intercellular spread (Wei et al., 2006a; Katayama et al., 2007; Wei et al., 2008; Chen et al., 2012; Chen et al., 2015a). Pns6, Pns11 and Pns12 aggregate together to form the viroplasm matrix for viral replication and progeny virions assembly (Wei et al., 2006b). Our recent report shows that Pns12 of RDV is a principal regulator for viral replication and infection in its insect vector (Chen et al., 2015b). Pns4, a phosphoprotein, is localized around the viroplasm in continuous cell cultures of *N. cincticeps* (Suzuki et al., 1990; Uyeda et al., 1990; Wei et al., 2006a). The combination of immunofluorescence and immunoelectron microscopy has revealed that at the early stage of viral infection, Pns4 accumulates at the periphery of the viroplasm, and at later stages, it associates with novel minitubules of approximately 10nm in diameter (Wei et al., 2006a). In the viruliferous leafhopper, Pns4 assembles into bundles of minitubules that surround the viroplasm matrix (Wei et al., 2006a). However, whether the formation of the Pns4 minitubule is essential for viral replication and infection in insect vector cells is undetermined yet.

In this study, the functional role of Pns4 in the viral infection cycle was investigated using the system of RNA interference (RNAi) in cultured leafhopper cells *in vitro* and insect body *in vivo*. Our results showed that Pns4 of RDV is crucial for viral infection and replication in insect vector, which may play a role in viral replication and assembly of progeny virions in leafhopper vector.

2 Results

2.1 RNAi induced by dsPns4 inhibits the replication and infection of RDV in cultured insect vector cells

In order to investigate the functional role of Pns4 of RDV, the RNAi activity induced by dsRNA specific for the Pns4 gene (dsPns4) was examined in cultured leafhopper cells. Cultured leafhopper cells were treated with dsPns4 and dsRNAs specific for the green fluorescence protein (dsGFP) by cellfectin-base transfection. It was confirmed that cellfectin treatment had no obvious influence on cellular viability and RDV infection in cultured leafhopper cells (data not shown). It has been previously suggested that RNAi is triggered by the appearance of small interfering RNAs (siRNAs) corresponding to the mRNA target sequence (Chen et al., 2012). Northern blot assay demonstrated that siRNAs approximately 21nt in length from dsPns4 and dsGFP treatments were detected(Fig. 1A), indicating that RNAi induced by dsRNAs was active in cultured leafhopper cells.

To investigate the effects of the knockdown of Pns4 expression on viral infection, cultured leafhopper cells growing on coverslips were transfected with dsRNAs, and inoculated by purified RDV. At 36h post-inoculation (hpi), infected cells were fixed, immunolabeled with Pns4-specific IgG conjugated to fluorescein isothiocyanate (Pns4-FITC) and Pns12-specific IgG conjugated to rhodamine (Pns12-rhodamine), and observed with a confocal microscope, as described previously (Wei et al., 2006a). In cells transfected with dsGFP, viroplasm of Pns12 distributed in the cytoplasm (Fig. 1B). Abundant Pns4 appeared as minitubules which accumulated in the cytoplasm or at the periphery of the viroplam (Fig. 1B). In contrast, in cultured cells transfected with dsPns4, Pns4 was restricted in a limited number of infected cells, and the formation of viroplasm of Pns12 was almost completely blocked (Fig. 1B). These results suggested that the knockdown of Pns4 expression strongly impaired the formation of viroplasm, the machinery of viral replication, leading to the blocking of the subsequent viral infection.

For the purpose of further analyzing the effects of RNAi induced by dsPns4 on the formation of Pns4 minitubules and viroplasm, as well as viral assembly, western blotting assays were performed at 36hpi with antibodies against viral nonstructural proteins Pns4 and Pns12, or the major outer capsid protein P8, to determine the expression level of corresponding proteins. Here, the expression level of Pns12 was served to judge for viroplasm accumulation, and the expression level of P8 was served to judge for RDV accumulation. As shown in Fig.1C, the treatment of dsPns4 caused significant reduction of the accumulation of Pns4, Pns12 and P8. These results revealed that the knockdown of Pns4 expression inhibited viral replication possibly via affecting the development of the viroplasm. We deduced that Pns4 minitubules were essential for viral infection and possibly assisted the viropalsm to perform the function of replication and assembly of progeny virions.

2.2 RNAi induced by dsPns4 inhibits the replication and infection of RDV in the insect vector

The efficiency of RNAi in cultured insect vector cells suggested that it was possible to introduce dsRNA into intact insects to investigate the role of Pns4 *in vivo*. The preliminary test showed that the microinjection with dsRNAs targeting RDV Pns4 gene caused no phenotypic abnormalities in leafhoppers (data not shown). The synthesized dsPns4 was microinjected into the second-instar nymphs of leafhoppers, which were then allowed a 3-day acquisition access period (AAP) on RDV-infected rice plants. At 12d post-first access to diseased plants (padp), RT-PCR was conducted to detect the presence of the genes for Pns4 and P8 to analyze the viruliferous rate of insects. The result showed that about 50.7% (n=100, 3 repetitions) of leafhoppers microinjected with dsGFP were detected to contain Pns4 and P8 genes (Table 1). In contrast, about 15.6% (n=100, 3 repetitions) of leafhoppers microinjected with dsPns4 were detected to be positive for Pns4 and P8 genes.

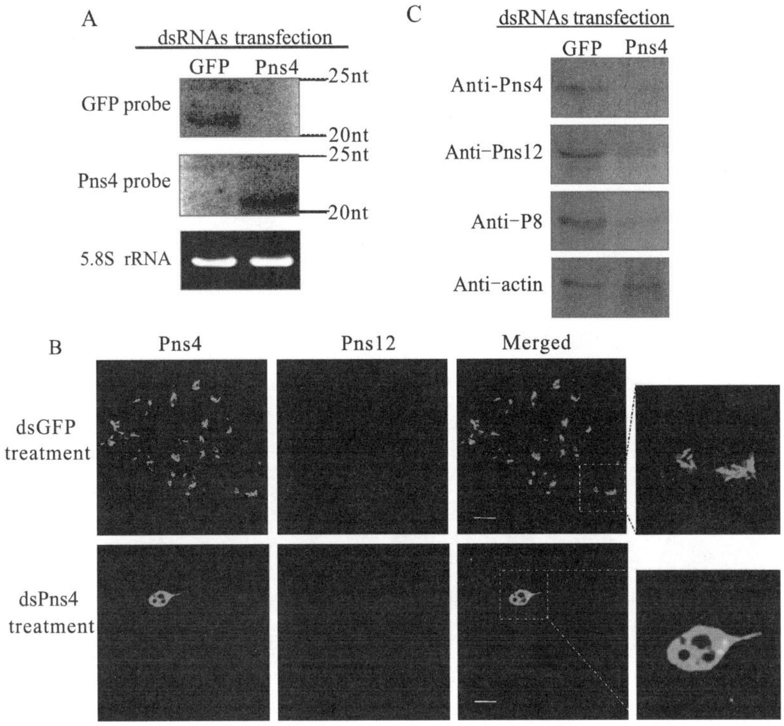

Fig. 1 RNAi induced by dsPns4 inhibited RDV infection in cultured leafhopper cells *in vitro*

A. siRNAs were active in the cultured leafhopper cellstransfected with synthesized dsGFP or dsPns4 at 72h post-transfection. 5.8S rRNA was used as a control to indicate loading of equal amounts of RNA in each lane. B. The infection of RDV was blocked by RNAi induced by dsPns4 in cultured leafhopper cells. At 36hpi, cells were immunolabeled with Pns4-FITC (green) and Pns12-rhodamine (red). The enlarged images displayed green fluorescence (Pns4-FITC) and red fluorescence (Pns12-rodanmine) of the merged images in the boxed areas in each panel, indicating that Pns4 distributed at the periphery of viroplasms. Bars, 20μm. C. The treatment of dsPns4 significantly reduced the synthesis viral proteins in cultured leafhopper cells with Western blotting assay at 36hpi. Proteins separation by DS-PAGE was performed to detect Pns4, Pns12 or P8 with Pns4-, Pns12- or P8-specific IgGs, respectively. Actin was used as a control and was detected with actin-specific IgG

Table 1 RNAi induced by dsPns4 significantly inhibited viral infection in insect vectors

Microinjection[a]	No. of insects giving positive results detected by RT-PCR (n=100)		
	Expt I	Expt II	Expt III
dsGFP	51	50	51
dsPns4	16	15	16

These results not only confirmed that Pns4 expression could be knocked down by dsPns4 microinjection, but also revealed that Pns4 contributed to efficient viral infection in leafhoppers. Then, the internal organs of leafhoppers (n= 30) microinjected with dsRNAs at 12d padp were dissected, immunolabeled with Pns12-, Pns4- or virus-specific antibodies, and examined by immunofluorescence microscopy. The results showed that in leafhoppers microinjected with dsGFP, viral antigens distributed throughout the whole intestine and salivary gland in about 54% of tested leafhoppers (Fig. 2A, C). Pns4 appeared as minitubules in the cells of intestine and salivary gland of viruliferous leafhopper (Fig. 2B, D). Some Pns4 minitubules distributed at the edge of viroplasms of Pns12 (Fig. 2B, D). In leafhoppers microinjected with dsPns4, RDV infection was restricted in a particular area of the filter chamber in the intestine in about 18% of tested leafhoppers (Fig. 2A). Pns4 distributed diffusely in the cytoplasm and some were observed at the edge of the viroplasm within the epithelial cells of the filter chamber (Fig. 2B). No salivary gland was immunolabeled by Pns12-, Pns4- or virus-specific antibodies in the dsPns4-treated leafhoppers (Fig. 2C, D). These results indicated that the dsPns4 treatment specifically knocked down Pns4 expression in the initially infected sites of the filter chamber, so that RDV infection in the filter chamber was inhibited and viral spread to the salivary glands was blocked.

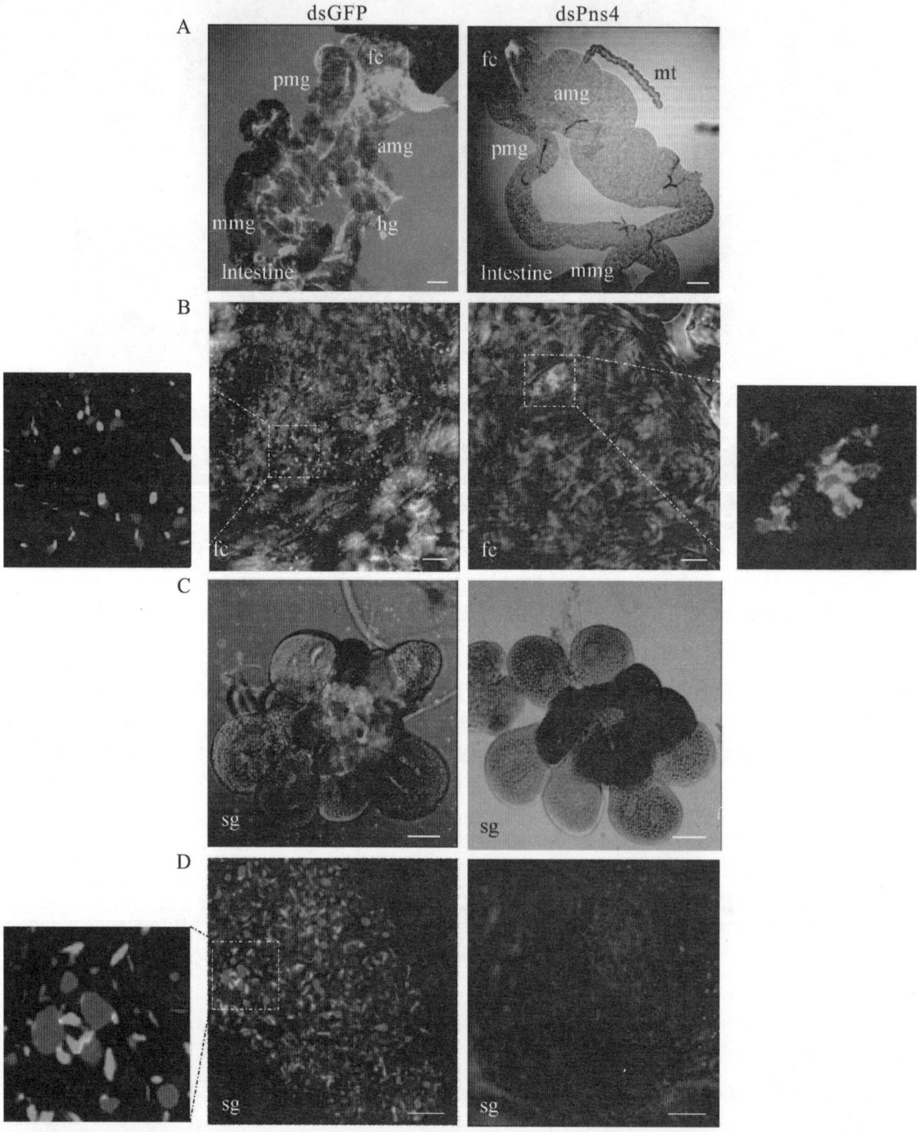

Fig. 2 Microinjection of dsPns4 inhibited RDV infection and spread in insect vectors in vivo

At 12d padp, the dissected intestines a and b and salivary glands (c and d) from leafhoppers receiving dsGFP or dsPns4 were immunolabeled with virus-FITC (green) (a and c), Pns4-FITC (green) (b and d) or Pns12-rodanmine (red) (b and d). The images with green fluorescence or red fluorescence were merged under a background of transmitted light. The enlarged images showing green fluorescence (Pns4-FITC) and red fluorescence (Pns12-rodanmine) of the merged images in the boxed areas in each panel, indicated that both of minitubules and diffusion of Pns4 distributed at the edge of viroplasms. fc, filter chamber; amg, anterior midgut; mmg, middle midgut; pmg, posterior midgut; hg, hindgut; mt, Malpighian tubules; sg, salivary gland. Bars, 100μm (a and c) and 10μm (b and d)

3 Discussion

In the present study, we have advanced the previous study that Pns4 of RDV surrounded the viroplasms in a ring-like structure at the early stage of viral infection and it formed bundles of minitubules at later stages of viral infection in cultured insect vector cells (Wei et al., 2006a). It was found that when the expression of

Pns4 was significantly reduced by RNAi induced by treatment of dsRNA from Pns4 gene, the development of viroplasms and viral replication were strongly inhibited both in cultured insect vector cells and the insect vectors (Fig. 1, 2), revealing that Pns4 of RDV is required for viral efficient infection in insect vector. Because the non-structural protein Pns10 of RDV has been confirmed to assembly tubules to package virions for viral efficient spread in insect vectors, and virions are not directly associated with the minitubules of Pns4 in insect vectors (Uyeda et al., 1990; Wei et al., 2006c; Wei et al., 2006a; Katayama et al., 2007; Wei et al., 2008; Chen et al., 2012), it seemed that Pns4 was not involved in viral spread in insect vectors. The accumulation of the minitubules of Pns4 at the periphery of viroplasm, and the essential role of Pns4 in viral infection illustrated that Pns4 may contribute to viral replication or progeny virions assembly. Viroplasms are generally thought to be devoid of translational machinery ribosomes, so the synthesis of viral proteins was thought to occur in the cytosol (Wei et al., 2006b). We have observed that ribosomes were densely accumulated on the surface of minitubules of Pns4 at the peripheral of the viroplasms in insect vectors (Wei et al., unpublished data). Thus, it was assumed that the minitubules of Pns4 was involved in viral protein synthesis during virus replication and assembly in insect vectors.

In the family *Reoviridae*, non-essential genome segments particularly are prone to rearrangements and rearranged segments are stably maintained (Matthijnssens et al., 2006; Troupin et al., 2010; Tanaka et al., 2011; Tanaka et al., 2012). This is also true for phytoreoviruses such as would tumor virus and RDV(Nuss, 1984; Murao et al., 1996; Pu et al., 2011). Despite these facts, no rearrangement on S4s of the two phytoreoviruses have been reported, further confirming that Pns4 of phytoreoviruses was essential for viral infection.

As part of their infective strategy, viruses in *Reoviridae* family exploit various tubule structures to facilitate their replication and translocation. Viral nonstructural protein P7-1 of *southern rice black-streaked dwarf virus* (SRBSDV), a *fijivirus*, has intrinsic ability to assemble into tubules that package virions to spread along the actin cytoskeleton in insect vector cells (Jia et al., 2014). The nonstructural protein NS1 of BTV, an orbivirus, is also able to constitute helically coiled ribbons to form tubules of 52.3nm in diameter and up to 100nm long (Hewat et al., 1992; Boyce et al., 2012). This shape is identical to Pns4 minitubules in morphology (Wei et al., 2006a), though it is wider than Pns4 minitubules. Some studies reveal that these tubules are involved in the translocation of virions, cellular pathogenesis and morphogenesis of BTV, but the defined role of NS1 is yet to be determined (Owens et al., 2004). In addition to viruses that are able to form tubule structures by their nonstructural proteins, there are also several viruses that exploit host microtubules to achieve their replication and trafficking of viral particles (Ohkawa et al., 2010; Dodding and Way, 2014). For example, in rotaviruses, viroplasm assembly, fusion and structural maintenance are dependent on the microtubular network of the host (Eichwald et al., 2012). Inside and adjacent to the viroplasm of reoviruses, coated microtubules are utilized to attach to mature virions and empty viral particles (de Castro et al., 2014). Further studies address that the μ2 protein of reoviruses, an essential viroplasm protein, binds microtubules and tethers viroplasms to the cytoskeleton of the host to determine the viral inclusion morphology (Parker et al., 2002; Ooms et al., 2012). These data have inspired us to presume the minitubules of Pns4 may act a role as describe above.

Unlike viruses in the genus *Orbivirus, Reovirus* or *Rotavirus*, there is no reverse-genetics system in cultured cells for RDV, and methods to purify viroplasm, composite viroplasm, and assemble viral particle in vitro are absent. Therefore, a direct evidence of Pns4 playing a role in the assembly of viroplasm was unable to be obtained here. Further studies will be conducted to investigate the mechanisms underlying the structures of Pns4 in the progress of virus infection and the assembly of virions via the interaction of Pns4 with viral proteins and host components.

4 Conclusions

The functional role of Pns4 in the viral infection cycle was investigated using the system of RNAi in cultured leafhopper cells *in vitro* and insect body *in vivo*. This study provides evidence that Pns4 is essential for viral infection and replication in insect vector. We also assume that Pns4 directly participates in executing the functional role of viroplasm for viral replication and assembly of progeny virions in insect vector.

5 Methods

5.1 Cells, viruses, vectors and antibodies

The nonviruliferous leafhopper vector *N. cincticeps*were collected in Fujian Province, China and propagated for several generations at 25±3°C in laboratory. Cultured cells, maintained in LBM growth medium, were originally developed from embryonic fragments of *N. cincticeps*(KIMURA, 1984). The rice samples infected with RDV were initially collected from Yunnan Province, China, and propagated for several generations via transmission by *N. cincticeps*. Rabbit polyclonal antisera specific for Pns4, Pns12 and virus were prepared, as previously described (Wei et al., 2006b). IgGs were purified from specific polyclonal antisera, followed by conjugation directly to FITC or rhodamine according to the manufacturer's instructions (Invitrogen). Finally, Pns4-/virus-FITC and Pns12-rhodamine were obtained for following immunofluorescence detection.

5.2 dsRNAs synthesis *in vitro*

A T7 RNA polymerase promoter with the sequence 5′- ATTCTCTAGAAGCTTAATACGACTCACTATAG-GG-3′ was added to the forward primer (5′ AGGCAGTACATCGTCACCC 3′) and reverse primer (5′ GCAAT-CACGCTCGCAAC 3′) at the 5′ terminal to amplify a region of about 815 bp of the Pns4 gene of RDV. The promoter was also added to 5′ terminal of the forward and reverse primers to amplify the full length of the GFP gene as a control. PCR products were transcribed into dsRNAs *in vitro* using the T7 RiboMAX (TM) Express RNAi System according to the manufacturer's protocol (Promega). Purified dsRNAs were examined using agarose gel electrophoresis to insure their integrity and quantified by spectroscopy.

5.3 Effects of synthesized dsRNAs on viral infection in cultured insect vector cells

8 μg dsRNA and 6μL cellfectin II Reagent (Invitrogen) were diluted individually in 50μL LBM without fetal bovine serum and antibiotics, mixed gently together at room temperature for 40min, and allowed to incubate with cultured cells for 8h. Thereafter, cultured cells were inoculated with purified RDV at a high multiplicity of infection of 10 in a solution of 0.1M histidine that contained 0.01M $MgCl_2$ (pH 6.2; His-Mg) at 25°C for 2h, and were finally recovered for complete culture (Chen et al., 2012).

At 36hpi, cultured cells were fixed in 4% paraformaldehyde in PBS for 30min, and then penetrated in 0.2% Triton-X for 10min. Cells were immunolabeled using Pns4-FITC and Pns12-rhodamine for 45min, then observed with an inverted confocal microscope (Leica TCS SP5).

5.4 Effects of synthesized dsRNAs on viral infection in insect vectors

The nonviruliferous second-instar nymphs were microinjected with 0.1μL of dsRNAs (0.5 μg/μL) at the intersegmental region of thorax, and then fed on RDV-infected rice plants for 3d. The nymphs were then kept on healthy rice seedlings for 8d. The intestines and salivary glands of the insects were dissected, fixed in 4% paraformaldehyde in PBS for 2h, penetrated with 2% Triton-X for 1h and incubated in Pns4-/virus-FITC, or Pns12-rhodamine for 2h, as described previously (Chen et al., 2012). The immunolabeled organs were observed with a confocal microscope to analyze the effect of dsRNAs on viral infection and spread.

5.5 The detection of siRNAs by Northern blot analysis

Cultured cells were harvested for total RNA extraction using the TRIzol Reagent (Invitrogen) at 72h posttreatment with dsRNA, in order to detect the presence of siRNAs specific for Pns4 or GFP genes in northern blots. Digoxigenin (DIG)-labeled ribo-probes corresponding to the negative-sense RNA of Pns4 or GFP genes were synthesized in vitro by T7 RNA polymerase using a DIG RNA Labeling kit (Roche). Approximately 5mg of total RNA was probed using a DIG Northern starter kit (Roche) following the manufacturer's protocol.

References

[1] BOYCE M, CELMA C P, ROY P. Bluetongue virus non-structural protein 1 is a positive regulator of viral protein synthesis[J]. Virology Journal, 2012, 9(1): 178-189.

[2] CHEN Q, WANG H, REN T, et al. Interaction between nonstructural protein Pns10 of *Rice dwarf virus* and cytoplasmic actin of leafhoppers is correlated with insect vector specificity[J]. The Journal of General Virology, 2015a, 96(4): 933-938.

[3] CHEN Q, CHEN H, JIA D, et al. Nonstructural protein Pns12 of *Rice dwarf virus* is a principal regulator for viral replication and infection in its insect vector[J]. Virus Research, 2015b, 210(1): 54-61.

[4] CHEN Q, CHEN H, MAO Q, et al. Tubular structure induced by a plant virus facilitates viral spread in its vector insect[J]. PLoS Pathogens, 2012, 8(11): e1003032.

[5] DE CASTRO I F, ZAMORA P F, OOMS L, et al. Reovirus forms neo-organelles for progeny particle assembly within reorganized cell membranes[J]. mBio, 2014, 5(1):e00931-13.

[6] DODDING M P, WAY M. Coupling viruses to dynein and kinesin-1[J]. Embo Journal, 2014, 30(17):3527-3539.

[7] EICHWALD C, ARNOLDI F, LAIMBACHER A S, et al. Rotavirus viroplasm fusion and perinuclear localization are dynamic processes requiring stabilized microtubules. PLoS One, 2012, 7(10): e47947.

[8] GUGLIELMI K M, MCDONALD S M, PATTON J T. Mechanism of intraparticle synthesis of the rotavirus double-stranded RNA genome[J]. Journal of Biological Chemistry, 2010, 285(24):18123-18128.

[9] HEWAT E A, BOOTH T F, WADE R H, et al. 3-D reconstruction of bluetongue virus tubules using cryoelectron microscopy[J]. Journal of Structural Biology, 1992, 108(1):35-48.

[10] JIA D, MAO Q, CHEN H, et al. Virus-induced tubule: a vehicle for rapid spread of virions through basal lamina from midgut epithelium in the insect vector[J]. Journal of Virology, 2014, 88(18), 10488.

[11] KANO H, KOIZUMI M, NODA H, et al. Nucleotide sequence of *Rice dwarf virus* (RDV) genome segment S3 coding for 114 K major core protein[J]. Nucleic Acids Resrarch, 1990, 18(22): 6700.

[12] KATAYAMA S, WEI T, OMURA T, et al. Three-dimensional architecture of virus-packed tubule[J]. Journal of Electron Microscopy, 2007, 56(3): 77-81.

[13] KIMURA, I. Establishment of new cell lines from leafhopper vector and inoculation of its cell monolayers with *Rice dwarf virus*[J]. Proceedings of the Japan Academy Ser B Physical and Biological Sciences, 1984, 60(6): 198-201.

[14] KING A M, LEFKOWITZ E, ADAMS M J, et al. Virus taxonomy: ninth report of the International Committee on Taxonomy of Viruses[M]. Elsevier, 2011.

[15] MATTHIJNSSENS J, RAHMAN M, RANST M V. Loop model: mechanism to explain partial gene duplications in segmented

dsRNA viruses[J]. Biochemical & Biophysical Research Communications, 2006, 340(1):140-144.

[16] MURAO K, UYEDA I, ANDO Y, et al. Genomic rearrangement in genome segment 12 of *Rice dwarf phytoreovirus*[J].Virology, 1996, 216(1): 238-240.

[17] NUSS D L. Molecular biology of wound tumor virus[J]. Advances in Virus Research, 1984, 29:57-93.

[18] OHKAWA T, VOLKMAN L E, WELCH M D. Actin-based motility drives baculovirus transit to the nucleus and cell surface[J]. The Journal of Cell Biology, 2010, 190(2): 187-195.

[19] OMURA T, ISHIKAWA K, HIRANO H, et al. The outer capsid protein of *Rice dwarf virus*is encoded by genome segment S8[J]. Journal of General Virology, 1989, 70(10):2759-2764.

[20] OOMS L S, JEROME W G, DERMODY T S, et al. Reovirus replication protein μ2 influences cell tropism by promoting particle assembly within viral inclusions[J]. Journal of Virology, 2012, 86(20): 10979-10987.

[21] OWENS R J, LIMN C, ROY P. Role of an arbovirus nonstructural protein in cellular pathogenesis and virus release[J]. Journal of Virology, 2004, 78(12):6649-6656.

[22] PARKER J S, BROERING T J, KIM J, et al. 2002. Reovirus core protein mu2 determines the filamentous morphology of viral inclusion bodies by interacting with and stabilizing microtubules[J]. Journal of Virology, 2002, 76(9):4483-4496.

[23] PU Y, KIKUCHI A, MORIYASU Y, et al. *Rice dwarf viruses* with dysfunctional genomes generated in plants are filtered out in vector insects: implications for the origin of the virus[J]. Journal of Virology, 2010, 85(6):2975-2979.

[24] SUZUKI N, WATANABE Y, KUSANO T, et al. Sequence analysis of *Rice dwarf phytoreovirus* genome segments S4, S5, and S6: Comparison with the equivalent wound tumor virus segments[J]. Virology, 1990, 179(1):446-454.

[25] SUZUKI N, SUGAWARA M, NUSS D L, et al. Polycistronic (tri- or bicistronic) phytoreoviral segments translatable in both plant and insect cells[J]. Journal of Virology, 1996a, 70(11):8155-8159.

[26] SUZUKI N, KUSANO T, MATSUURA Y, et al. Novel NTP binding property of *Rice dwarf phytoreovirus* minor core protein P5[J]. Virology, 1996b, 219(2):471-474.

[27] SUZUKI N, SUGAWARA M, KUSANO T, et al. Immunodetection of rice dwarf phytoreoviral proteins in both insect and plant hosts[J]. Virology, 1994, 202(1):41-48.

[28] SUZUKI N, TANIMURA M, WATANABE Y, et al. Molecular analysis of *Rice dwarf phytoreovirus* segment S1: interviral homology of the putative RNA-dependent RNA polymerase between plant- and animal-infecting reoviruses[J]. Virology, 1992, 190(1):240-247.

[29] TANAKA T, SUN L, TSUTANI K, et al. Rearrangements of mycoreovirus 1 S1, S2 and S3 induced by the multifunctional protein p29 encoded by the prototypic hypovirus *Cryphonectria* hypovirus 1 strain EP713[J]. Journal of General Virology, 2011, 92(8):1949-1959.

[30] TANAKA T, EUSEBIO-COPE A, SUN L, et al. Mycoreovirus genome alterations: similarities to and differences from rearrangements reported for other reoviruses[J]. Frontiers in Microbiology, 2012, 3: 186.

[31] TROUPIN C, DEHÉE A, SCHNURIGER A, et al. Rearranged genomic RNA segments offer a new approach to the reverse genetics of rotaviruses[J]. Journal of Virology, 2010, 84(13): 6711-6719.

[32] UYEDA I. Ueda s, Uyeda I. The *Rice dwarf phytoreovirus* structural protein P7 possesses non-specific nucleic acid binding activity in vitro[J]. Molecular Plant Pathology On-Line, 1997,0123ueda.

[33] UYEDA I, KUDO H, TAKAHASHI T, et al. Nucleotide sequence of *Rice dwarf virus* genome segment 9[J]. Nucleic Acids Research, 1990, 18(5):1297-1300.

[34] WEI T, SHIMIZU T, OMURA T. Endomembranes and myosin mediate assembly into tubules of Pns10 of *Rice dwarf virus* and intercellular spreading of the virus in cultured insect vector cells[J]. Virology, 2008, 372(2):349-356.

[35] WEI T, KIKUCHI A, SUZUKI N, et al. Pns4 of *Rice dwarf virus* is a phosphoprotein, is localized around the viroplasm matrix, and forms minitubules[J]. Archives of Virology, 2006a, 151(9):1701-1712.

[36] WEI T, SHIMIZU T, HAGIWARA K, et al. Pns12 protein of *Rice dwarf virus* is essential for formation of viroplasms and nucleation of viral-assembly complexes[J]. Journal of General Virology, 2006b, 87(2):429-438.

[37] WEI T, KIKUCHI A, MORIYASU Y, et al. The spread of *Rice dwarf virus* among cells of its insect vector exploits virus-induced tubular structures[J]. Journal of Virology, 2006c, 80(17):8593-8602.

[38] YAN J, TOMARU M, TAKAHASHI A, et al. P2 protein encoded by genome segment S2 of *Rice dwarf phytoreovirus* is essential for virus infection[J]. Virology, 1996, 224(2):539-541.

[39] ZHONG B, KIKUCHI A, MORIYASU Y, et al. A minor outer capsid protein, P9, of *Rice dwarf virus*[J]. Archives of Virology, 2003, 148(11):2275-2280.

[40] ZHOU F, WU G, DENG W, et al. Interaction of *Rice dwarf virus* outer capsid P8 protein with rice glycolate oxidase mediates relocalization of P8[J]. FEBS Letters, 2007a, 581(1):34-40.

[41] ZHOU F, PU Y, WEI T, et al. The P2 capsid protein of the nonenveloped *Rice dwarf phytoreovirus* induces membrane fusion in insect host cells[J]. Proceedings of the National Academy of Sciences of the United States of America, 2007b, 104(49):19547-19552.

Nonstructural protein Pns12 of *rice dwarf virus* is a principal regulator for viral replication and infection in its insect vector

Qian Chen, Hongyan Chen, Dongsheng Jia, Qianzhuo Mao, Lianhui Xei, Taiyun Wei

（Fujian Province Key Laboratory of Plant Virology, Institute of Plant Virology, Fujian Agriculture and Forestry University, Fuzhou, Fujian 350002, China）

Abstract: Plant reoviruses are thought to replicate and assemble within cytoplasmic structures called viroplasms. The molecular mechanisms underlying the formation of the viroplasm during infection of *rice dwarf virus* (RDV), a plant reovirus, in its leafhopper vector cells remain poorly understood. Viral nonstructural protein Pns12 forms viroplasm-like inclusions in the absence of viral infection, suggesting that the viroplasm matrix is basically composed of Pns12. Here, we demonstrated that core capsid protein P3 and nonstructural protein Pns11 were recruited in the viroplasm by direct interaction with Pns12, whereas nonstructural protein Pns6 was recruited through interaction with Pns11. The introduction of dsRNA from Pns12 gene into cultured insect vector cells or intact insect strongly inhibited such viroplasm formation, preventing efficient viral spread in the leafhopper *in vitro* and *in vivo*. Thus, nonstructural protein Pns12 of RDV is a principal regulator for viral replication and infection in its insect vector.

Keywords: *Rice dwarf virus*; Nonstructural protein Pns12; Viroplasm; Viral replication; RNA interference

1 Introduction

Plant viruses in the Reoviridae family (Plant reoviruses) comprise the genera *Phytoreovirus, Fijivirus* and *Oryzavirus* and are transmitted by cicadellid leafhoppers or planthoppers in a persistent-propagative manner (Attoui et al., 2012). Plant reoviruses are icosahedral, double-layered particles with 10 or 12 double-stranded RNA (dsRNA) segments (Attoui et al., 2012). *Rice dwarf virus* (RDV), a phytoreovirus in the family Reoviridae, is transmitted mainly by the leafhopper species *Nephotettix cincticeps* in a persistent-propagative manner (Zhong et al., 2003). The genome of RDV encodes at least seven structural proteins (P1, P2, P3, P5, P7, P8 and P9) and five nonstructural proteins (Pns4, Pns6, Pns10, Pns11 and Pns12) (Omura and Yan, 1999; Nakagawa et al., 2003). Replication of the RDV genome begins with the assortment and packaging of the 12 viral mRNA and is followed by minus strand synthesis to produce new dsRNA genome segments (Zhong et al., 2003; Miyazaki et al., 2010; Wei et al., 2006a,b). In a coupled process, core proteins P1, P3, P5 and P7 are first assembled to form core particles for the transcription of the bulk of viral mRNAs (Zhong et al., 2003; Miyazaki et al., 2010; Wei et al., 2006a,b). Subsequently, progeny core particles are thought to be coated by outer capsid proteins P2, P8 and P9 to generate mature RDV virions (Zhong et al., 2003; Miyazaki et al., 2010; Wei et al., 2006a,b). The precise timing and molecular mechanisms of these processes are largely unknown. However, the development of the cultured insect vector cell in monolayers (VCMs) derived from the rice green leafhopper (*N. cincticeps*) is starting to shed light on investigating RDV replication cycle in depth (Miyazaki et al., 2010; Wei

Virus Research. 2015, 54–61
Received 15 May 2015; Accepted 11 July 2015; Available online 19 July 2015

et al., 2006a,b, 2007; Zhou et al., 2007; Chen et al., 2011, 2012).

In our previous work using VCMs, we have demonstrated that core particles assembly takes place exclusively within the dense globular inclusions termed viroplasms in the cytoplasm of insect vector cells, while intact double-layered RDV virions are assembled at the periphery of the viroplasm (Wei et al., 2006b). Viroplasm formation is orchestrated by complex networks of interactions involving nonstructural proteins and structural proteins. The nonstructural proteins Pns6, Pns11 and Pns12 are all present within the viroplasm matrix induced by RDV infection, but only Pns12 forms viroplasm-like inclusions when expressed in non-host insect Sf9 cells, suggesting that Pns12 is a minimal viral factor required for viroplasm formation (Wei et al., 2006b). Currently, the molecular mechanisms that govern the genesis and maturation of viroplasm, the intermolecular interactions among viroplasm components, and the specific roles that viroplasms play in viral persistent infection in its insect vector all remain poorly understood.

In this work, we provided the evidence that Pns12 of RDV could directly bind to Pns11 and core capsid protein P3, whereas Pns11 could directly bind to Pns6, indicating that Pns12 may act as a scaffold for the recruitment of other viral proteins into viroplasm. To further investigate the functional roles of viroplasm in the infection cycle of RDV in insect vector, we used RNA interference (RNAi), a conserved sequence-specific gene silencing mechanism that is induced by dsRNAs (Fire et al., 1998), to knock down the expression of Pns12 gene *in vivo* and *in vitro*. We determined that the assembly of the viroplasm by viral nonstructural protein Pns12 is essential for persistent infection of RDV in its insect vector.

2 Materials and methods

2.1 Cells, viruses and antibodies

VCMs, maintained in LBM growth medium, were originally developed from embryonic fragments of *N. cincticeps* collected from Fujian Province, China (Kimura and Omura, 1988). The rice samples of RDV were initially collected from Yunnan Province, China, and propagated for several generations via transmission by *N. cincticeps*at 25 ± 3℃ in laboratory. RDV virions were purified from infected rice plants without the usage of CCl_4 as described by Zhong et al. (2003). Rabbit polyclonal antisera against the Pns6, Pns11, Pns12 and virus were prepared as described previously (Wei et al., 2006b). IgGs purified from specific polyclonal antisera were conjugated directly to fluorescein isothiocyanate (FITC) or rhodamine according to the manufacturer's instructions (Invitrogen).

2.2 Yeast two-hybrid assay

Yeast two-hybrid screening was performed using a Matchmake Gold Yeast two-hybrid system (Clontech) according to the manufacturer's protocol. Full-length cDNA of RDV P3, Pns6 and Pns12 amplified by PCR were cloned in pGBKT7 and pGADT7 vector as bait and prey plasmids, respectively. Each of these constructs was transformed into yeast strain AH109 to confirm that the baits and preys were not toxic or self-activating. For protein interaction test, yeast strain AH109 was co-transformed with the bait and the prey pairs. The transformants were cultured on SD double-dropout (DDO) medium (SD/-Leu/-Trp), SD triple-dropout(TDO) medium (SD/-His/-Leu/-Trp) and SD quadruple-dropout (QDO) medium (SD/-Ade/-His/-Leu/-Trp) to show specific interactions. Then the positive ones were picked and streaked on QDO/X plates (containing 20μg mL-1 X-α-Gal) to measuredα-Galactosidase activity. The pGBKT7-53/ pGADT7-T interaction was used as a positive control, while the vectors pGBKT7-Lam/pGADT7-T were used as a negative control.

2.3 GST pull-down assay

Vector of pGEX-3X was used to construct plasmids expressing GST fusion proteins of P3, Pns11 and Pns12 as probe proteins, and pDEST17 vector was used to construct plasmids expressing His fusion proteins of Pns6, Pns11 and Pns12 as target proteins. The constructs were transformed into Escherichia coli BL21 (DE3) pLysS (Invitrogen) and expressed GST and His fusion proteins after induction with IPTG. GST fusion proteins were immobilized on glutathione sepharose beads (glutathione-sepharose 4 Fast Flow, GE) and then incubated with lysates from *E. coli* BL21 expressing His fusion proteins with gentle shaking. Thereafter, beads were washed 6 times with washing buffer and proteins were eluted by boiling in loading buffer. Purified proteins and GST pull-down products were then analyzed by Western blotting.

2.4 Effect of synthesized dsRNAs from Pns12 gene of RDV on viral infection in the continuous cell cultures derived from leafhopper

A 600-bp segment of the RDV Pns12 gene (bases 141–740) and a 585-bp segment of the GFP gene were amplified by PCR as templates for the synthesis of dsRNAs. A T7 RiboMAX Express RNA interference System kit (Promega) was used to synthesize dsRNAs *in vitro* for Pns12 or GFP genes according to the manufacturer's instructions. VCMs were transfected with 0.5μg/μL dsRNAs in the presence of Cellfectin II reagent (Invitrogen) for 8h. At 72h after the treatment of dsRNAs, total RNA from transfected cells was extracted with TRIzol Reagent (Invitrogen). DIG-labeled negative-sense RNA transcripts of Pns12 or GFP genes were generated in vitro with T7 RNA polymerase using a DIG RNA Labeling kit (Roche). Northern blots were produced using a DIG Northern starter kit (Roche), as described previously (Chen et al., 2012).

For examining the effects of synthesized dsRNAs on viral infection, after a 8h treatment with dsRNAs, VCMs were inoculated with RDV at a multiplicity of infection (MOI) of 10 in a solution of 0.1M histidine that contained 0.01M $MgCl_2$ (pH 6.2; His-Mg) for 2h as described previously (Chen et al., 2011, 2012). At 48h post-inoculation (hpi), VCMs were fixed, immunolabeled with virus-specific IgG conjugated to FITC (virus-FITC) and Pns12- specific IgG conjugated to rhodamine (Pns12-rhodamine) and then processed for immunofluorescence microscopy as described previously (Chen et al., 2011, 2012). Furthermore, at 48hpi, total proteins were extracted from infected cells and analyzed by a Western blot assay using Pns12- and P8-specific IgGs, respectively. Insect actin was detected with actin-specific antibodies as a control to confirm loading of equal amounts of proteins in each lane.

2.5 Effect of synthesized dsRNAs from the Pns12 gene of RDV on viral infection in the leafhopper

The dsRNAs were microinjected into leafhoppers using the method reported for the brown planthopper (Liu et al., 2010). Briefly, 30 second-instar nymphs of the leafhopper were kept on RDV-infected rice for 2d and then microinjected with 64 nldsRNAs (0.5μg/μL) into the thorax using a Nanoject II Auto-Nanoliter Injector (Spring). The microinjected leafhoppers were placed on healthy rice seedlings. At different days post-first access to diseased plants (padp), the intestines and the salivary glands of leafhoppers were dissected, fixed and immunolabeled with virus-FITC, Pns12- rhodamine and actin dye phalloidin-Alexa Fluor 647 carboxylicacid (Invitrogen), as described previously (Chen et al., 2011, 2012). At 15d padp, the accumulation of Pns12 and P8 of RDV was analyzed with a Western blot assay using Pns12- and P8-specific IgGs, respectively.

3 Results

3.1 Interaction among nonstructural proteins Pns6, Pns11 and Pns12 and core capsid protein P3 of RDV

Our previous data demonstrated that the nonstructural proteins Pns6, Pns11 and Pns12 are the constituents of viroplasm matrix in virus-infected insect vector cells (Wei et al., 2006b). Furthermore, Pns12 has the ability of self-association to form the viroplasm-like inclusions in the absence of viral infection (Wei et al., 2006b). To explore how Pns6 and Pns11 are recruited in the viroplasm, the yeast two-hybrid system was used to determine whether there were interactions among Pns6, Pns11 and Pns12. Our results showed that Pns12 interacted with Pns11, but not with Pns6 (Fig. 1A). Pns6 was also found to bind to Pns11 (Fig. 1A). To confirm these interactions, we fused Pns6, Pns11 and Pns12 with a His or GST and assayed the interaction using GST pull-down methodology, as described previously (Kong et al., 2014; Huo et al., 2014). As shown in Fig.1B, positive interactions were confirmed between Pns6 and Pns11 and between Pns11 and Pns12, but not between Pns6 and Pns12. These data suggested that Pns11 was recruited in the viroplasm by direct interaction with Pns12, whereas Pns6 was recruited through interaction with Pns11. Thus, Pns6, Pns11 and Pns12 of RDV could bind to each other to form the viroplasm matrix.

Our observations show that core particles are assembled within the interior regions of viroplasm induced by RDV infection in insect vector cells (Wei et al., 2006b). To explore how the core capsid protein P3 is recruited in the viroplasm, the yeast two-hybrid and GST pull-down assays were used to determine whether proteins Pns6, Pns11 or Pns12 could specifically bind to P3. Both protein interaction methods revealed that P3 specifically interacted with Pns12, but not with Pns6 or Pns11 (Fig.1). Our results clearly showed that P3 is recruited and retained in the viroplasm by direct interaction with Pns12. Thus, Pns12 could directly bind to core particles within the viroplasm.

3.2 Transfection of dsRNAs from the Pns12 gene inhibited the assembly of viroplasmand viral infection in insect vector cells

Since we have shown the central position of Pns12 in the protein interaction network during viroplasm formation (Fig. 1) and its ability to self-associate (Wei et al., 2006b), we then explored its functional role in viroplasm formation in RDV infected insect vector cells using dsRNAs specific for Pns12 and GFP genes (dsPns12 and dsGFP)-mediated RNAi. RNAi is induced by dsRNAs and manifested by the appearance of small interfering RNAs (siRNAs) that correspond to the mRNA target sequence (Chen et al., 2011, 2012). Small RNAs approximately 21nt long were detected from RNAs extracted from dsRNA-transfected cells at 72h after transfection with the synthesized dsRNAs (Fig. 2A). These results showed that RNAi was triggered by dsRNAs in VCMs.

Eight hour after transfection of dsPns12 or dsGFP, we inoculated VCMs with RDV at a MOI of 10. At 48hpi, VCMs were fixed, immunolabeled with virus-FITC and Pns12-rhodamine, and processed for immunofluorescence, as described previously (Chen et al., 2011, 2012). In VCMs transfected with dsGFP, viral infection was observed in 100% of cells, and viroplasms containing Pns12 were abundant (Fig. 2B). However, in VCMs transfected with dsPns12, RDV infection was restricted to a limited number of the initially infected cells, and the formation of viroplasms was almost entirely blocked (Fig. 2B). The accumulation of viral proteins in the cell lysates was then analyzed using Western blotting assay with antibodies against Pns12 or outer capsid protein P8. Our results showed that the treatment of dsPns12 significantly inhibited the expression of these viral proteins (Fig. 2C). These results suggested that Pns12 of RDV was involved in viroplasm formation and viral infection. The mRNA of genomic segment S12 ofRDV.

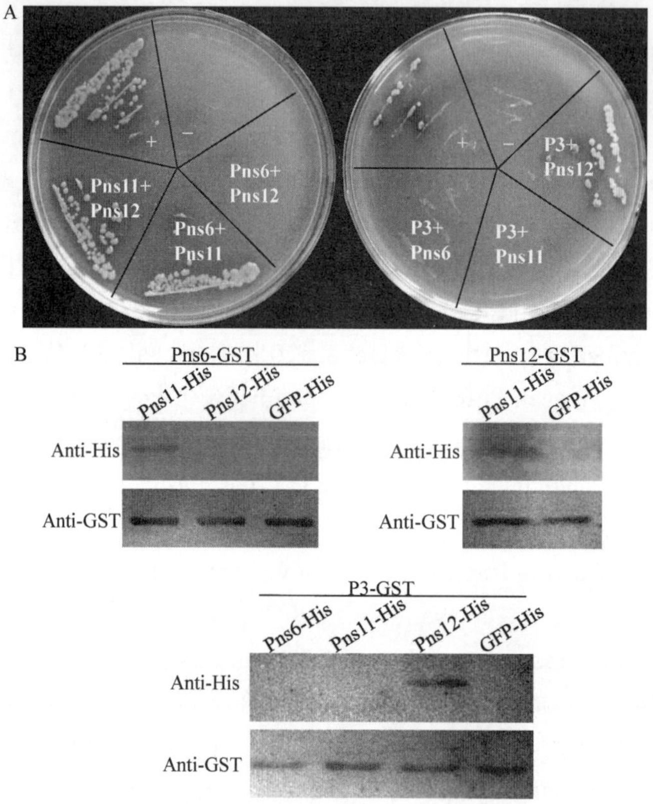

Fig. 1 Protein interactions between Pns6, Pns11, Pns12 and P3 of RDV

(A) Yeast two-hybrid assay of protein–protein interactions. Transformants on plate of SD-Trp-LeuHis-Ade medium were shown. (+) Positive control, i.e., pGBKT7-53/ pGADT7-T; (–) negative control, i.e., pGBKT7-Lam/pGADT7-T; Pns6+Pns12, pGBKT7-Pns6/ pGADT7-Pns12; Pns11+Pns12, pGADT7-Pns11/pGBKT7-Pns12; Pns6+Pns11, pGBKT7-Pns6/pGADT7-Pns11;P3+Pns6, pGBKT7-P3/ pGADT7-Pns6;P3+Pns11, pGBKT7-P3/pGADT7-Pns11; P3+Pns12, pGBKT7-P3/pGADT7-Pns12. (B) GST pull-down assay of protein–protein interactions. Pns6, Pns11 or Pns12 of RDV was fused with His to act as bait proteins; GFP was fused with His as a control. Pns6, Pns12 or P3 of RDV was fused with GST as prey proteins. GST pull-down products were analyzed by immunoblotting with antibodies against GST to detect prey proteins and with antibodies against His to detect bound proteins

3.3 Microinjection of dsRNAs from the Pns12 gene inhibits the spread of RDV from the initially infected site in the filter chamber of its insect vector

After ingested by the leafhopper vector, RDV establishes its primary infection in the epithelial cells of the filter chamber, then infects other tissues in the insect (Chen et al., 2011, 2012; Ahlquist, 2006). An RNAi strategy was used to investigate the functional role of nonstructural protein Pns12 during the early persistent infection of RDV in its leafhopper vector. Second-instar nymphs of leafhoppers were allowed a 2-day acquisition on virus-infected rice plants, and then microinjected with dsGFP or dsPns12. At 8d padp, viral antigens and viroplasms containing Pns12 were seen in the entire intestine in 60% of leafhoppers receiving dsGFP (Fig. 3A; Table 1). However, viral infection was restricted in the epithelial regions of the filter chamber and only observed in 30% of leafhoppers receiving dsPns12 (Fig. 3A; Table 1). At 15d padp, viral antigens and viroplasms containing Pns12 had accumulated in the salivary glands in 50% of leafhoppers that received dsGFP (Fig. 3; Table 1). As expected, the treatment of dsPns12 inhibited the spread of RDV in the salivary glands of its insect vector (Fig. 3B; Table 1). The decrease in the expression of Pns12 and P8 proteins of RDV in leafhoppers that received dsPns12 was confirmed using western blot (Fig. 3C). The above results indicated that the treatment of dsPns12 specifically knocked down the expression of Pns12 in the initially infected regions of the filter chamber, which led to the restriction of viral infection in the filter chamber and blocked viral spread to the salivary glands.

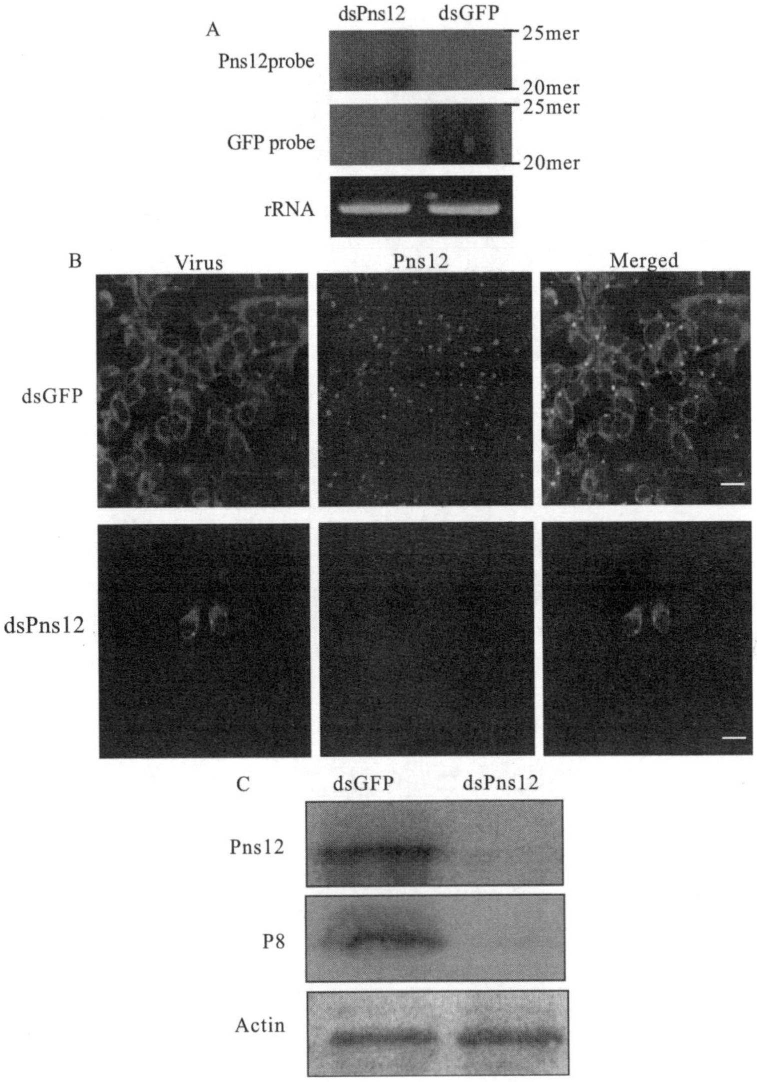

Fig. 2 The nonstructural protein Pns12 of RDV were essential for viral replication or assembly in VCMs

(A) Detection of small RNAs in VCMs transfected with dsRNAs. Total RNA was probed with DIG-labeled riboprobes corresponding to negative-sense RNA of the genes of Pns12 or GFP. The 5.8S rRNA was used as a control. (B) Eight hours after transfection with dsGFP or dsPns12, VCMs were infected with RDV. At 48hpi, VCMs were immunolabeled with virus-FITC (green) and Pns12-rhodamine (red), and then examined with confocal microscopy. (C) Inhibition of viral protein expression in VCMs transfected with dsRNAs. Proteins were analyzed by immunoblotting with antibodies against Pns12 or P8. An actin-specific antibody was used as the control (For interpretation of the references to color in this figure legend, the reader is referred to the web version of this article)

Table 1 Inhibition of viral infection and spread in the bodies of leafhoppers treated with dsPns12 *via* microinjection

	padp (days)	Limited regions of filter chamber	Entire intestine	Salivary glands
dsGFP	8	0	18	0
	15	0	25	15
dsPns12	8	9	0	0
	15	9	0	0

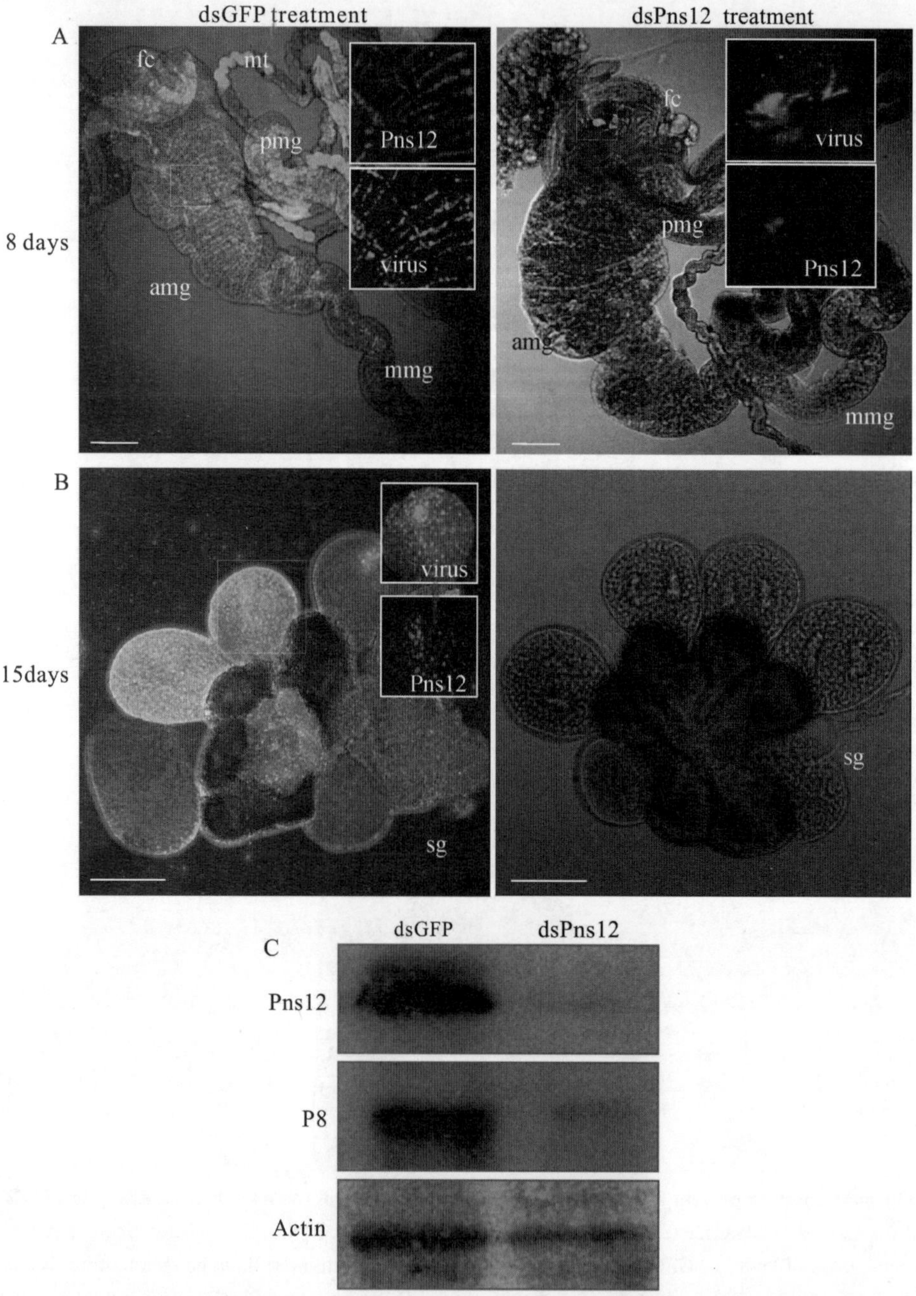

Fig. 3 RNAi induced by dsPns12 inhibited the replication and spread of RDV *in vivo*

(A and B) Second-instar nymphs of leafhoppers were microinjected with dsGFP or dsPns9. At 8 and 15d padp (panels A and B, respectively), internal organs of leafhoppers were immunolabeled with virus-FITC (green), Pns12-rhodamine (red) and actin dye phalloidin-Alexa Fluor 647 carboxylicacid (blue). The images were merged with green fluorescence (viral antigens), red fluorescence (Pns12 antigens), and blue fluorescence (actin dye). Insets show green fluorescence (viral antigens) and red fluorescence (Pns12 antigens) of the merged images in the boxed areas in each panel. (C) Treatment with dsPns12 reduced the expression of Pns12 and P8 proteins in leafhoppers as shown by Western blotting. fc, filter chamber; mg, midgut; amg, anterior midgut; pmg, posterior midgut; sg, salivary glands. Bars, 100m (For interpretation of the references to color in this figure legend, the reader is referred to the web version of this article)

4 Discussion

During RDV infection of the cultured leafhopper cells, first the mRNAs of RDV were transcribed from the viral core particles, then the nonstructural proteins Pns6, Pns11 and Pns12 were expressed by the host's cellular translational machinery and associated together to form the initial viroplasm matrix, which is the site for viral replication and assembly (Ahlquist, 2006; Akita et al., 2012; Wei et al., 2006b). In this study, we demonstrated that knockdown of Pns12 expression by dsPns12 efficiently abolished viroplasm formation and prevented viral replication in cultured leafhopper cells (Fig. 2). Moreover, microinjection of dsPns12 suppressed viroplasm formation and restricted viral infection in the initially infected epithelium of the filter chamber, thus, inhibiting the spread of RDV in the body of leafhopper vector (Fig. 3; Table 1). These data were consistent with the previous observation that Pns12 alone was able to form inclusions that are similar in morphology to viroplasm in the absence of viral infection (Wei et al., 2006b), and suggested that the formation of the viroplasm for progeny virions assembly was a prerequisite for the persistent infection of RDV in its insect vector. In our study, the segment of Pns12 gene for the synthesis of dsPns12 also contained two small out-of-phase open reading frames which encoded Pns12OPa and Pns12OPb, respectively (Suzuki et al., 1996). Thus, the treatment of dsPns12 should also specifically knock down the expression of Pns12OPa and Pns12OPb in insect vector. Whether these two small proteins are involved in the formation of viroplasm during viral infection in insect vector will be investigated in future.

Despite the critical role of RDV Pns12 in mediating viroplasm formation, the molecular mechanism underlying this process is still unclear. A group of nonstructural proteins encoded by other members in the family Reoviridae may provide some hints. For example, NSP2 of *reoviruses*, NS2 of *orbiviruses*, NSP2/NSP5 of rotaviruses, Pns9 of the *phytoreovirus* rice gall dwarf virus, P9-1 of the *fijivirus* rice black streaked dwarf virus, and Pns10 of the *oryzavirus* rice ragged stunt virus all possess similar functions as RDV Pns12 in inducing viroplasm matrix during virus replication cycles (Jia et al., 2012; Mao et al., 2013; Chen et al., 2014; Fabbretti et al., 1999; Miller et al., 2010; Theron et al., 1996; Thomas et al., 1990; Broering et al., 2004; Contin et al., 2010; Kar et al., 2007). Among them, Rotavirus NSP2, RBSDV P9-1 and RGDV Pns9 can form octameric structures that can act as a scaffold for the recruitment of viral proteins to initiate the viroplasm matrix (Akita et al., 2011, 2012; Jiang et al., 2006). The cryo-EM analysis of the native structure of the Pns12, in solution, also forms octameric ring-like structures (Akita et al., unpublished data). Such an octameric structure of Pns12 may be required to provide a suitable scaffold for the recruitment of viral proteins to initiate the viroplasm matrix for viral replication and assembly. Our finding of the interactions between Pns6, Pns11 and Pns12 (Fig. 1) supports this hypothesis and provides a potential explanation on how Pns6 and Pns11 are recruited and retained in the viraplasm matrix.

How Pns6, Pns11 and Pns12 are involved in viral replication and assembly in virus-infected insect vector cells? The nonstructural protein Pns6 is a viral movement protein and Pns11 is a viral RNA- silencing suppressor in plant hosts (Li et al., 2004; Xu et al., 1998). Moreover, both Pns6 and Pns11 have RNA binding activities (Ji et al., 2011; Li et al., 2004; Xu et al., 1998). Therefore, it is possible that Pns6 or Pns11 acts to recruit or retain viral mRNAs within the viroplasm, which is consistent with the evidence that large amounts of newly transcribed viral RNAs localize to the viroplasm in infected cells (Wei et al., 2006b). The RNA-binding activity of Pns6 and Pns11 also suggests that they might be involved in the assortment and packaging of 12 different viral genome segments of RDV into the newly assembling progeny core particles within the viroplasm.

The core capsid protein P3 was recruited to the viroplasm matrix via its direct interaction with Pns12 (Fig. 1). However, the assembly of core-like particles by P3 has been shown to occur in the absence of Pns12 (Hagiwara et al., 2003), suggesting that the direct binding of Pns12 to P3 may be required for the assembly of viral RNA- containing core particles within the viroplasms matrix. For viruses in the family Reoviridae, viral mRNAs within the viroplasm matrix are transcribed by core particles and transported to the adjacent ribosome for the synthesis of viral proteins to enlarge the nascent viroplasm (Attoui et al., 2012; Ahlquist, 2006; Mao et al., 2013; Chen et al., 2014). The addition of the outer capsid proteins of the virion to core particles would terminate viral transcription (Attoui et al., 2012; Ahlquist, 2006; Mao et al., 2013; Chen et al., 2014). Thus, the binding of Pns12 with P3 might delay the attachment of outer capsid proteins to core particles and allow core particles to continue synthesizing more viral mRNAs, which would benefit viral replication in insect vector cells.

Taken together, we propose a model for the genesis and maturation of viroplasm induced by RDV in its insect vector cells (Fig. 4). Early in the infection process, Pns12 might act as a scaffold to recruit Pns6 and Pns11 through protein–protein interaction to form the initial viroplasm matrix. During infection, core capsid protein P3 is recruited to the viroplasm through its interaction with Pns12, whereas other structural proteins are recruited to the viroplasms through interactions with as-yet-unknown factors. RDV RNAs binding with Pns6 or Pns11 form replication and assembly complexes for the production of progeny core particles within the interior regions of viroplasm. Next, Pns12 binds to core particles to enhance the production of viral RNAs by delaying outer capsid assembly on these particles. The intact viral particles are then assembled at the periphery of the viroplasm (Wei et al., 2006b). This model indicates Pns12 as a principal regulator of the viroplasm formation and viral RNAs transcription, which is supported by the fact that Pns12 gene silencing in transgenic rice plants results in complete resistance to RDV (Shimizu et al., 2009).

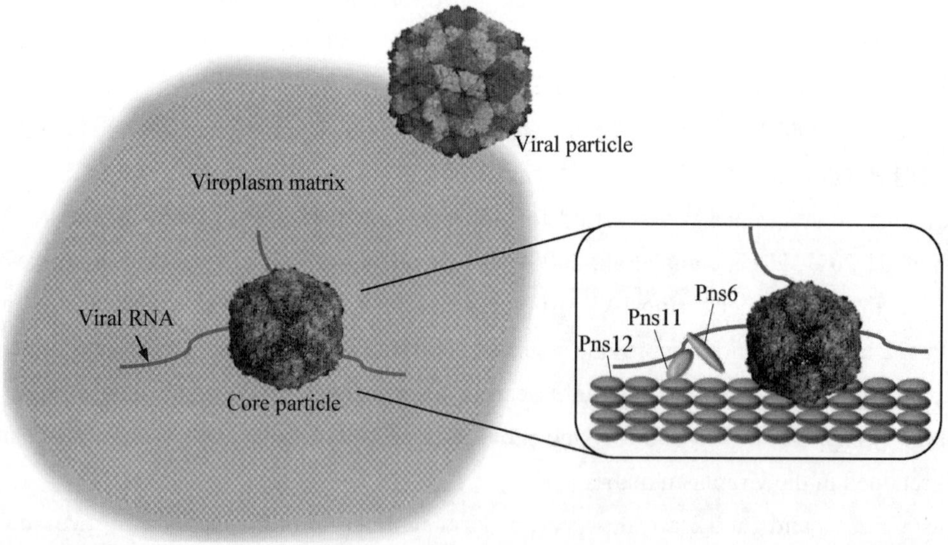

Fig. 4 Proposed model for the assembly of RDV viroplasm in infected cells of its insect vector

Pns12 acts as a scaffold for the recruitment and retaining of core capsid protein P3 and nonstructural proteins Pns6 and Pns11. RDV RNA binds with Pns6 or Pns11 to form replication and assembly complexes for the production of progeny core particles within the viroplasm. Pns12 directly binds to core particles. Intact viral particles are assembled at the periphery of the viroplasm

Acknowledgements

The authors are indebted to Toshihiro Omura (National Agricultural Research Center, Japan) for antibodies against intact viral particles, Pns6, Pns11, Pns12 and P8 of RDV. This research was supported by the National Basic Research Program of China (2014CB138400), the National Natural Science Foundation of China (31200118 and 31300136), and the Scientific Research Foundation of Graduate School of FAFU (YB2013008).

References

[1] AHLQUIST P. Parallels among positive-strand RNA viruses, reverse-transcribing viruses and double-stranded RNA viruses[J]. Nature Reviews Microbiology, 2006, 4(5):371-382.

[2] AKITA F, MIYAZAKI N, HIBINO H, et al. Viroplasm matrix protein Pns9 from *Rice gall dwarf virus* forms an octameric cylindrical structure[J]. Journal of General Virology, 2011, 92(9):2214-2221.

[3] AKITA F, HIGASHIURA A, SHIMIZU T, et al. Crystallographic analysis reveals octamerization of viroplasm matrix protein P9-1 of *Rice black streaked dwarf virus*[J]. Journal of Virology, 2012, 86(2):746-756.

[4] ATTOUI H, MERTENS PPC, BECNEL J, et al. Family reoviridae. In: King AMQ, Carstens E, Adams M, Lefkowitz E, editors[M]. Virus Taxonomy, 9th Report of the ICTV. London: Elsevier/Academic Press, 2011, pp.541-637.

[5] BROERING T J, KIM J, MILLER C L, et al. Reovirus nonstructural protein mu NS recruits viral core surface proteins and entering core particles to factory-like inclusions[J]. Journal of Virology, 2004, 78(4):1882-1892.

[6] CHEN H, CHEN Q, OMURA T, et al. Sequential infection of *Rice dwarf virus* in the internal organs of its insect vector after ingestion of virus[J]. Virus Research,2011, 160(1-2): 389-394.

[7] CHEN Q, CHEN H, MAO Q, et al. Tubular structure induced by a plant virus facilitates viral spread in its vector insect[J]. Plos Pathogens, 2012, 8(11):e1003032.

[8] CHEN H, ZHENG L, MAO Q, et al. Development of continuous cell culture of brown planthopper to trace the early infection process of oryzaviruses in insect vector cells[J]. Journal of Virology, 2014, 88(8):4265-4274.

[9] CONTIN R, ARNOLDI F, CAMPAGNA M, et al. Rotavirus NSP5 orchestrates recruitment of viroplasmic proteins[J]. Journal of General Virology, 2010, 91(7):1782-1793.

[10] FABBRETTI E, AFRIKANOVA I, VASCOTTO F, et al. Two non-structural rotavirus proteins, NSP2 and NSP5, form viroplasm-like structures in vivo[J]. Journal of General Virology, 1999, 80(2):333-339.

[11] FIRE A, XU S, MONTGOMERY M K, et al. Potent and specific genetic interference by double-stranded RNA in *Caenorhabditiselegans*[J]. Nature,1998, 391(6669):806-811.

[12] HAGIWARA K, HIGASHI T, NAMBA K, et al. Assembly of single-shelled cores and double-shelled virus-like particles after baculovirus expression of major structural proteins P3, P7 and P8 of *Rice dwarf virus*[J]. Journal of General Virology, 2003, 84(4):981-984.

[13] HUO Y, LIU W, ZHANG F, et al. Transovarial transmission of a plant virus is mediated by vitellogenin of its insect vector[J]. PLoS Pathogens, 2014, 10(3):e1003949.

[14] JI X, QIAN D, WEI C, et al. Movement protein Pns6 of *Rice dwarf phytoreovirus* has both ATPase and RNA binding activities[J]. PLoS One,2011, 6(9):e24986.

[15] JIA D, CHEN H, ZHENG A, et al. Development of an insect vector cell culture and RNA interference system to investigate the functional role of *Fijivirus* replication protein[J]. Journal of Virology, 2012, 86(10):5800-5807.

[16] JIANG X F, JAYARAM H, KUMAR M, et al. Cryoelectron microscopy structures of rotavirus NSP2-NSP5 and NSP2-RNA complexes: implications for genome replication[J]. Journal of Virology, 2006, 80(21):10829-10835.

[17] KAR A K, BHATTACHARYA B, ROY P. Bluetongue virus RNA binding protein NS2 is a modulator of viral replication and assembly[J]. BMC Molecular Biology, 2007, 8(1):4.

[18] KIMURA I, OMURA T. Leafhopper cell cultures as a means for phytoreovirus research[J]. Advancesin Disease Vector

Research,1988, 5: 111-135.

[19] KONG L, WU J, LU L, et al. Interaction between *Rice stripe virus* disease-specific protein and host PsbP enhances virus symptoms[J]. Molecular Plant, 2014, 7(4):691-708.

[20] LI Y, BAO Y M, WEI C H, et al. *Rice dwarf phytoreovirus* segment S6-encoded nonstructural protein has a cell-to-cell movement function[J]. Journal of Virology, 2004, 78(10):5382-5389.

[21] LIU S, DING Z, ZHANG C, et al. Gene knockdown by intro-thoracic injection of double-stranded RNA in the brown planthopper, *Nilaparvata lugens*[J]. Insect Biochemistry and Molecular Biology, 2010, 40(9):666-671.

[22] MAO Q, ZHENG S, HAN Q, et al. New model for the genesis and maturation of viroplasms induced by *Fijiviruses* in insect vector cells[J]. Journal of Virology, 2013, 87(12):6819-6828.

[23] MILLER C L, ARNOLD M M, BROERING T J, et al. Localization of mammalian orthoreovirus proteins to cytoplasmic factory-like structures via nonoverlapping regions of microNS[J]. Journal of Virology, 2010, 84(2):867-882.

[24] MIYAZAKI N, WU B, HAGIWARA K, et al. The functional organization of the internal components of *Rice dwarf virus*[J]. Journal of Biochemistry, 2010, 147(6):843-850.

[25] NAKAGAWA A, MIYAZAKI N, TAKA J, et al. The atomic structure of *Rice dwarf virus* reveals the self-assembly mechanism of component proteins[J]. Structure, 2003, 11(10):1227-1238.

[26] OMURA T, YAN J. Role of outer capsid proteins in transmission of phytoreovirus by insect vectors[J]. Advances in Virus Research, 1999, 54:15-43.

[27] SHIMIZU T, YOSHII M, WEI T, et al. Silencing by RNAi of the gene for Pns12, a viroplasm matrix protein of *Rice dwarf virus*, results in strong resistance of transgenic rice plants to the virus[J]. Plant Biotechnology Journal, 2010, 7(1):24-32.

[28] SUZUKI N, SUGAWARA M, NUSS D L, et al. Polycistronic (tri- or bicistronic) phytoreoviral segments translatable in both plant and insect cells[J]. Journal of Virology, 1996, 70(11):8155-8159.

[29] THERON J, HUISMANS H, NEL L H. Site-specific mutations in the NS2 protein of epizootic haemorrhagic disease virus markedly affect the formation of cytoplasmic inclusion bodies[J]. Archives of Virology, 1996, 141(6):1143-1151.

[30] THOMAS C P, BOOTH T F, ROY P. Synthesis of bluetongue virus-encoded phosphoprotein and formation of inclusion bodies by recombinant baculovirus in insect cells: it binds the single-stranded RNA species[J]. Journal of General Virology, 1990, 71(9):2073-2083.

[31] WEI T, KIKUCHI A, MORIYASU Y, et al. The spread of *Rice dwarf virus* among cells of its insect vector exploits virus-induced tubular structures[J]. Journal of Virology, 2006a, 80(17):8593-8602.

[32] WEI T, SHIMIZU T, HAGIWARA K, et al. Pns12 protein of *Rice dwarf virus* is essential for formation of viroplasms and nucleation of viral-assembly complexes[J]. Journal of General Virology, 2006b, 87(2):429-438.

[33] WEI T, CHEN H, ICHIKI-UEHARA T, et al. Entry of *Rice dwarf virus* into cultured cells of its insect vector involves clathrin-mediated endocytosis[J]. Journal of Virology, 2007, 81(14):7811-7815.

[34] XU H, LI Y, MAO Z, et al. *Rice dwarf phytoreovirus* segment S11 encodes a nucleic acid binding protein[J]. Virology, 1998, 240(2):267-272.

[35] ZHONG B, KIKUCHI A, MORIYASU Y, et al. A minor outer capsid protein, P9, of *Rice dwarf virus*[J]. Archives of Virology, 2003, 148(11):2275-2280.

[36] ZHOU F, PU Y, WEI T, et al. The P2 capsid protein of the nonenveloped *Rice dwarf phytoreovirus* induces membrane fusion in insect host cells[J]. Proceedings of the National Academy of Sciences of the United States of America, 2007, 104(49):19547-19552.

Interaction between non-structural protein Pns10 of rice dwarf virus and cytoplasmic actin of leafhoppers is correlated with insect vector specificity

Qian Chen, Haitao Wang, Tangyu Ren, Lianhui Xie and Taiyun Wei

(Fujian Province Key Laboratory of Plant Virology, Institute of Plant Virology, Fujian Agriculture and Forestry University, Fuzhou, Fujian 350002, PR China)

Abstract: Many insect-transmissible pathogens are transmitted by specific insect species and not by others, even if the insect species are closely related. The molecular mechanisms underlying such strict pathogen-insect specificity are poorly understood. Rice dwarf virus (RDV), a plant reovirus, is transmitted mainly by the leafhopper species *Nephotettix cincticeps* but is transmitted ineffectively by the leafhopper *Recilia dorsalis*. Here, we demonstrated that virus-containing tubules composed of viral non-structural protein Pns10 of RDV associated with the intestinal microvilli of *N. cincticeps* but not with those of *R. dorsalis*. Furthermore, Pns10 of RDV specifically interacted with cytoplasmic actin, the main component of microvilli of *N. cincticeps*, but not with that of *R. dorsalis*, suggesting that the interaction of Pns10 with insect cytoplasmic actin is consistent with the transmissibility of RDV by leafhoppers. All these results suggested that the interaction of Pns10 of RDV with insect cytoplasmic actin may determine pathogen-vector specificity.

The vector specificity of plant viruses is common in nature, and describes the specific relationship between virus and vector (Hogenhout et al., 2008; Ammar et al., 2009). Rice dwarf virus (RDV), a phytoreovirus in the family *Reoviridae*, is transmitted mainly by the leafhopper species *Nephotettix cincticeps* in a persistent-propagative manner (Honda et al., 2007). Our preliminary test indicated that another leafhopper species, *Recilia dorsalis*, was inefficient at transmitting RDV. Furthermore, our preliminary test also indicated that a continuous cell line of *R. dorsalis* displayed a low susceptibility to RDV infection compared with a cell line of *N. cincticeps*. The specificity of a plant virus for its vector can be explained by transmission barriers posed by different tissues in the insects (Hogenhout et al., 2008; Jia et al., 2012a; Markham et al., 1984). RDV encounters multiple barriers in its path from the intestine to the salivary gland of *N. cincticeps* (Chen et al., 2011). When ingested by *N. cincticeps* feeding on infected plants, RDV initially infects the filter chamber epithelium. Following assembly of progeny virions, RDV spreads to adjacent organs such as the anterior midgut, crosses the basal lamina of the midgut epithelium into the visceral muscles and then moves into the haemolymph and finally into the salivary glands, from which it can be introduced into rice hosts (Chen et al., 2011). Our recent study showed that RDV can exploit virus-containing tubules composed of viral nonstructural protein Pns10 to facilitate virus spread within *N. cincticeps* (Chen et al., 2012). Furthermore, a Pns10-deficient isolate of RDV failed to be transmitted by *N. cincticeps* (Pu et al., 2011). These results suggested that Pns10 of RDV may act as a viral determinant for transmission. Thus, it is reasonable to assume that the specific interaction between Pns10 of RDV and leafhopper proteins would contribute to vector transmission specificity. Here, we revealed

Journal of General Virology. 2015, 96: 933-938.
Received 27 October 2014; Accepted 28 November 2014.

that the specific interaction between Pns10 of RDV and insect cytoplasmic actins may determine insect vector specificity.

In the present study, stock cultures of *N. cincticeps* and *R. dorsalis*, collected originally from Fujian Province in eastern China, were used for laboratory experiments. RDV was maintained on rice plants via transmission by *N. cincticeps*. To compare the ability of *N. cincticeps* and *R. dorsalis* to transmit RDV, second-instar nymphs were allowed a 2-day acquisition access period (AAP) on diseased rice plants and were then placed on healthy rice seedlings. At 20d after first access to diseased plants, an individual adult insect was fed on a healthy rice seedling in test tubes for 2d. The insect intestines and salivary glands were then examined by an immunofluorescence assay for the presence of viral antigen with viral particle-specific IgG directly conjugated to Alexa Fluor 633 and with the actin-specific dye phalloidin conjugated to FITC (Invitrogen), as described previously (Wei et al., 2006; Jia et al., 2012b). Inoculated seedlings were grown to allow the development of symptoms of infection. Immunofluorescence microscopy showed that RDV extensively infected the intestine and salivary gland of *N. cincticeps* (Fig. 1a). The viruliferous *N. cincticeps* was able to transmit the virus to rice seedlings, as judged by the appearance of symptoms on rice plants (Table 1). In contrast, *R. dorsalis* was rarely detected to be viruliferous, and viral infection was restricted to the small infection regions of the filter chamber (Fig. 1b). As expected, the viruliferous *R. dorsalis* failed to transmit RDV (Table 1). Thus, the inability of *R. dorsalis* to transmit RDV may be caused by the restriction of viral spread from the initially infected intestinal epithelium of *R. dorsalis*, and thus the virus could not spread to the salivary glands for subsequent transmission.

To establish that RDV indeed cannot pass through the intestine of *R. dorsalis*, which would prevent its spread, 50μl purified virus (10μg/ml^{-1}) was microinjected into the haemocoel of third-instar nymphs of leafhoppers, as described previously (Omura et al., 1982). The insects were kept on healthy rice plants for 20d and then confined individually for 2d to inoculate the rice seedlings. Our preliminary test indicated that there was no significant difference in transmission efficiency for the second- or third-instar nymphs of leafhoppers after a 2 day AAP on diseased rice plants, but the third-instar nymphs were more suitable for microinjection experiments. The salivary glands of insects were then examined by an immunofluorescence assay for the presence of viral antigen, as described above. Immunofluorescence microscopy indicated that RDV infected the salivary glands of injected *N. cincticeps* and *R. dorsalis* (Fig. 1c). These insects were also able to transmit RDV (Table 1). The results confirmed that the movement of RDV across the intestine into the leafhopper haemolymph was a significant barrier to transmission.

Table 1 Comparison of the ability of *N. cincticeps* and *R. dorsalis* to transmit RDV after 2 day AAP on diseased rice plants or after microinjection with purified viruses (*Data are for 50 insects)

Methods	Insects	No. of viral antigen-positive insects that transmitted RDV to rice seedlings in expt no.[a]:		
		I	II	III
Feeding	*N. cincticeps*	15	18	17
	R. dorsalis	0	0	0
Microinjection	*N. cincticeps*	37	31	33
	R. dorsalis	10	9	11

We then determined whether the restriction of virus infection in the intestine was caused by the failure of spread of virus-containing Pns10 tubules in *R. dorsalis*. The intestines from viruliferous *N. cincticeps* and *R. dorsalis* were examined by an immunofluorescence assay with virus-specific IgG-Alexa Fluor 633, Pns10-

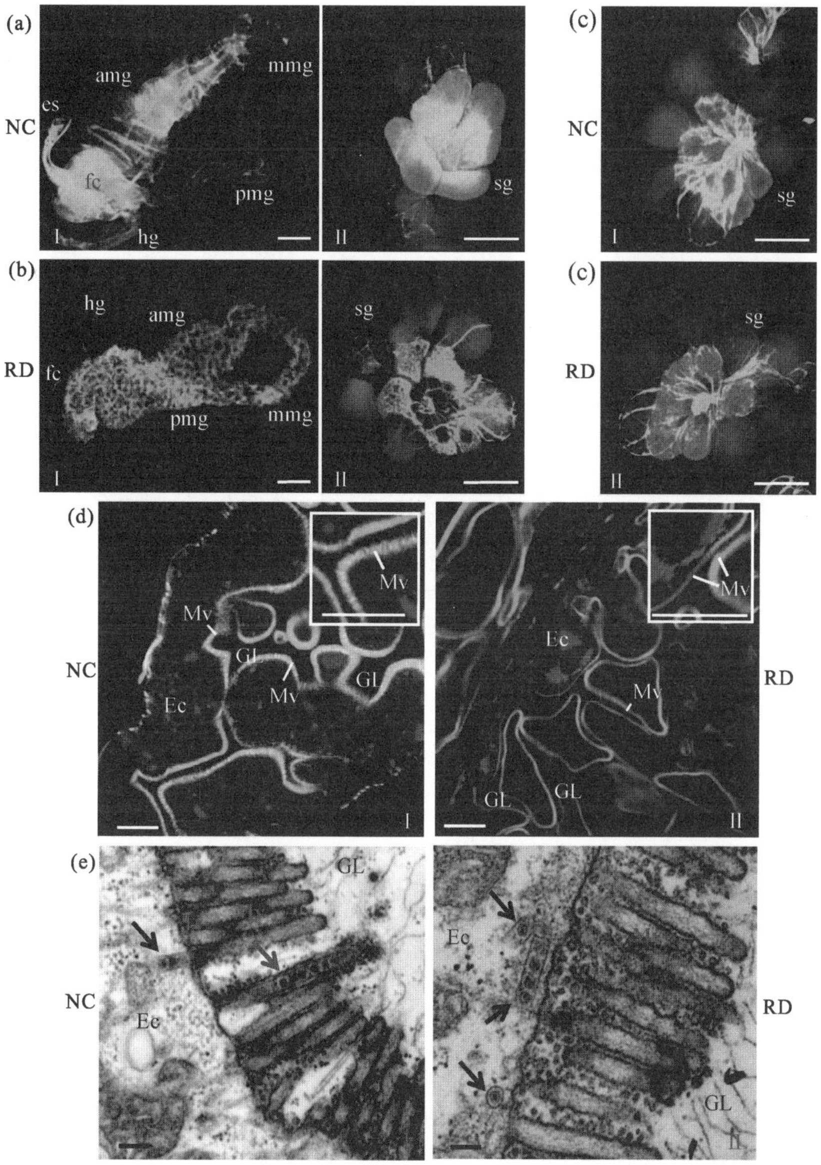

Fig. 1 Restriction of virus infection in the intestine in *R. dorsalis* is caused by the failure of spread of virus-containing Pns10 tubules

(a-d) Internal organs of *N. cincticeps* or *R. dorsalis* were immunostained for tubules with Pns10-specific IgG-rhodamine (red), for virions with virus-specific IgG-Alexa Fluor 633 (blue) and for actin filaments with phalloidin-FITC (green)and then examined by confocal microscopy. (a) RDV extensively infected the intestine (panel I) and salivary gland (panel II) of *N.cincticeps*. (b) RDV infected the small infection areas of the intestine (panel I) but failed to infect the salivary gland (panel II) of *R.dorsalis*. (c) RDV infected the salivary glands of *N. cincticeps* (panel I) and *R. dorsalis* (panel II) microinjected with purified viruses. (d) Pns10 tubules pass through microvilli of filter chamber epithelium of *N. cincticeps* (panel I) and only accumulate in the cytoplasm of filter chamber epithelium of *R. dorsalis* (panel II). (e) Electron micrographs showing that virus-containing tubules (arrows) are inserted into midgut microvilli in viruliferous *N. cincticeps* (panel I) but are only distributed in the cytoplasm of filter chamber epithelium in viruliferous *R. dorsalis* (panel II). Images in (a-c) were merged with blue fluorescence (viralantigen) and green fluorescence (actin). Images in (d) were merged with red fluorescence (Pns10 tubules), blue fluorescence(viral antigen) and green fluorescence (actin). es, Oesophagus; fc, filter chamber; amg, anterior midgut; mmg, middle midgut; pmg, posterior midgut; hg, hindgut; sg, salivary gland; Mv, microvilli; GL, gut lumen; Ec, epithelial cell; NC, *N. cincticeps*; RD, *R.dorsalis*. Bars, 100mm (a, b, c); 20mm (d); 200nm (e)

specific IgG- rhodamine and phalloidin-FITC, as described previously (Chen et al., 2012). Immunofluorescence microscopy indicated that Pns10 tubules completely passed through the actin-based epithelium microvilli of the intestine of *N. cincticeps*, as shown previously (Chen et al., 2012), but did not cross those of *R. dorsalis* (Fig. 1d). To confirm these results, the intestines from viruliferous *N. cincticeps* and *R. dorsalis* were examined by electron microscopy, as described previously (Chen et al., 2012). We observed that virus-containing tubules were associated with the midgut microvilli in the viruliferous *N. cincticeps* (Fig. 1e). The tubules were also observed in a limited number of epithelial cells of the filter chamber in the viruliferous *R. dorsalis* but were never found in association with the microvilli (Fig. 1e). All these results suggested that the failure of Pns10 tubules to associate with the actin-based intestinal microvilli may inhibit the efficient spread of RDV in *R. dorsalis*.

(a)

```
N. cincticeps   MCDDEVAALVVDNGSGMCKAGFAGDDAPRAVFPSIVGRPRHQGVMVGMGQ   50
R. dorsalis     ----------I---------------------------------------   50

N. cincticeps   KDSYVGDEAQSKRGILTLKYPIEHGIVTNWDDMEKIWHHTFYNELRVAPE  100
R. dorsalis     --------------------------------------------------  100

N. cincticeps   EHPVLLTEAPLNPKANREKMTQIMFETFNTPAMYVAIQAVLSLYASGRTT  150
R. dorsalis     --------------------------------------------------  150

N. cincticeps   GIVLDSGDGVSHTVPIYEGYALPHAILRLDLAGRDLTDYLMKILTERGYS  200
R. dorsalis     --------------------------------------------------  200

N. cincticeps   FTTTAEREIVRDIKEKLCYVALDFEQEMATAASSSSLEKSYELPDGQVIT  250
R. dorsalis     --------------------------------------------------  250

N. cincticeps   IGNERFRCPEALFQPSFLGMEACGIHETTYNSIMKCDVDIRKDLYANTVL  300
R. dorsalis     -----------M--------------------------------------  300

N. cincticeps   SGGTTMYPGIADRMQKEITALAPSTMKIKIIAPPERKYSVWIGGSILVSL  350
R. dorsalis     -----------------------------------------------A--  350

N. cincticeps   STFQQMWISKQEYDESGPSIVHRKCF                          376
R. dorsalis     --------------------------                          376
```

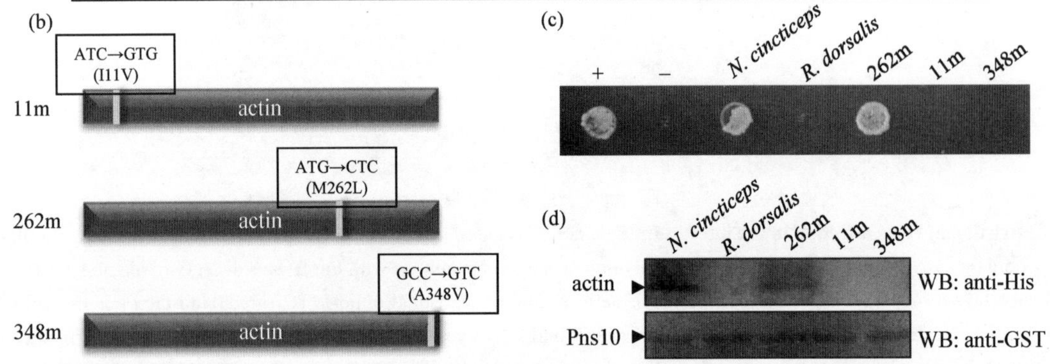

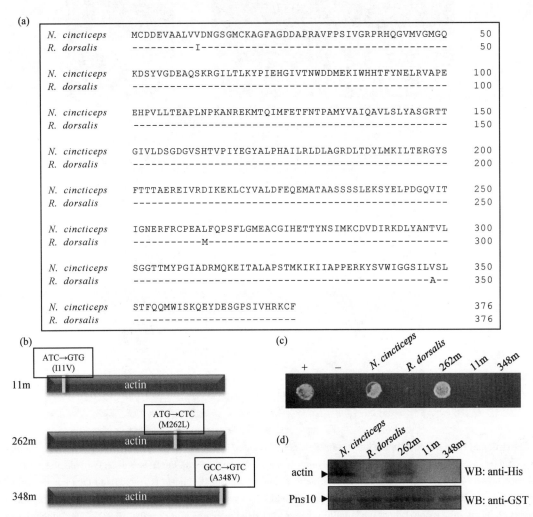

Fig. 2 Pns10 specifically interacts with cytoplasmic actin of *N. cincticeps* but not with that of *R. dorsalis*

(a) Amino acid sequence alignment of the cytoplasmic actin genes of *N. cincticeps* and *R. dorsalis*. Sequences were aligned using DNAMAN 7.0 software with default parameters. Identical residues between sequences of *R. dorsalis* and of *N. cincticeps* are indicated with dashes. (b) Schematic representations of cytoplasmic actin mutants of *R. dorsalis* generated by site-directed mutagenesis. (c)Yeast two-hybrid analysis for interactions of Pns10 with cytoplasmic actins and mutants. Transformants were grown on SD-Trp-Leu-His-Ade plates. +, Positive control (pTSU2-APP+pNubG-Fe65); –, negative control (pTSU2-APP+pRR3N); *N. cincticeps*, pBT-STE-Pns10+pPR3-N-Actin-NC; *R. dorsalis*, pBT-STE-Pns10+pPR3-N-Actin-RD; 262m, pBT-STEPns10+pPR3-N-262m; 11m, pBT-STE-Pns10+pPR3-N-11m; 348m, pBT-STE-Pns10+pPR3-N-348m. (d) GST pull-down assay to detect interactions of Pns10 with cytoplasmic actins and mutants. Lysates from *Escherichia coli* strain BL21 cells expressing GST-Pns10 were incubated with cell lysate expressing His-actin of *N. cincticeps* and *R. dorsalis* and the three mutants. GST pull-down products were analysed by Western blotting (WB); antibody to GST was used to detect Pns10 and antibody to His to detect bound proteins. *N. cincticeps*, His-actin-*N. cincticeps*; *R. dorsalis*, His-actin-*R. dorsalis*; 262m,His-262m; 11m, His-11m; 348m, His-348m

We then determined whether there was protein-protein interaction between Pns10 of RDV and actin of the leafhoppers. Because actin within the microvilli of the insect intestine is cytoplasmic actin (Popova-Butler & Dean, 2009), we firstly amplified the genes of the cytoplasmic actins of *N. cincticeps* and *R. dorsalis* by a reverse transcription-PCR method. The identity of the deduced amino acid sequences of the two cytoplasmic actin genes from *N. cincticeps* and *R. dorsalis* was 99.2%, with amino acids differing only at positions 11, 262 and 348 (Fig. 2a, b). Because Pns10 of RDV is a membrane-associated protein (Wei et al., 2008; Liu et al., 2011), we used the DUAL membrane system (Dualsystems Biotech), a split-ubiquitin membrane-based yeast two-hybrid system, to detect the interaction between Pns10 of RDV and actins of *N. cincticeps* or *R. dorsalis*. Briefly, the Pns10 gene of RDV and two actin genes of *N. cincticeps* and *R. dorsalis* were cloned into the bait vector pBT-STE and the prey vector pPR3 N, respectively. The recombinant plasmids pBT-STE-Pns10 and pPR3-N-Actin-NC, pBT3-STEPns10 and pPR3-N-Actin-RD were used to co-transform yeast strain NMY51. The transformants were confirmed on SD-Trp-Leu-His-Ade plates for 3–4d at 30℃. This yeast two-hybrid assay demonstrated that Pns10 specifically interacted with cytoplasmic actin of *N. cincticeps* but failed to interact with that of *R. dorsalis* (Fig. 2c). On the basis of these analyses, we deduced that the loss in association of Pns10 with the microvilli from *R. dorsalis* was directly caused by the lack of interaction between Pns10 and the cytoplasmic actin from *R. dorsalis*.

Because only three amino acids differed between the two cytoplasmic actin genes of *N. cincticeps* and *R. dorsalis*, we then determined which amino acid was responsible for the specific interaction of RDV Pns10 and cytoplasmic actin of leafhoppers. We substituted the amino acids at positions of 11, 262 or 348 of the cytoplasmic actin gene of *R. dorsalis* by fusion PCR according to the corresponding gene sequence of *N. cincticeps* (Fig. 2b). The primers carrying the mutations at nt 31–33 (ATC→GTG), 784–786 (ATG→CTC) or 1042–1044bp (GCC→GTC) of the DNA fragment for the cytoplasmic actin of *R. dorsalis* were designed to substitute the corresponding amino acids at positions 11 (Ile→Val), 262 (Met→Leu) or 348 (Ala→Val) (Fig. 2b). The PCR products were cloned into the prey vector pPR3 N and then used in the yeast two-hybrid assay described above. The results showed that only the mutation at position 262 (Met→Leu, 262m) led to the specific interaction, whilst the reaction remained negative with the mutations at positions 11 (Iso→Val, 11m) and 348 (Ala→Val, 348m) (Fig. 2c). To further confirm this interaction, the DNA fragments for the WT and three mutants of leafhopper cytoplasmic actins were cloned into plasmid vector pDEST17 to express His fusion proteins as the preys, and the Pns10 gene was cloned into pGEX-3X to construct a plasmid expressing glutathione S-transferase (GST) fusion protein as the bait. As described previously (Jia et al., 2014), the interaction was assayed using GST pull-down methodology, and binding between GST and His fusion proteins was detected by Western blotting. As shown in Fig. 2d, positive interactions of Pns10 with the WT of cytoplasmic actin of *N. cincticeps* and mutant 262m were confirmed, but not with the WT of cytoplasmic actin of *R. dorsalis* and the 11m or 348m mutant. These results revealed that leucine at position 262 of leafhopper cytoplasmic actin plays a key role in determining the interaction of RDV Pns10 with the cytoplasmic actin from *N. cincticeps* and lack of interaction with actin from *R. dorsalis*.

In conclusion, we demonstrated that viral non-structural protein Pns10 of RDV specifically interacts with the cytoplasmic actin of the virus-transmitting leafhopper *N. cincticeps* but not with that of the non-transmitting leafhopper *R. dorsalis*, suggesting that the interaction of RDV Pns10 with insect cytoplasmic actin is consistent with the transmissibility of RDV by the leafhoppers. Pns10 tubules, which interacted with cytoplasmic actin within the microvilli of *N. cincticeps*, are able to pass through the intestinal microvilli, facilitating virus spread in the body of its insect vector (Chen et al., 2012). In contrast, the lack of interaction of RDV Pns10 with cytoplasmic actin within the microvilli of *R. dorsalis* restricted the virus in the initially infected epithelium in the intestine. Therefore, the interaction between RDV Pns10 and insect cytoplasmic actin may determine the insect vector specificity. We further determined that a change in one amino acid in leafhopper cytoplasmic actin may be enough to alter the insect vector specificity for RDV. Similarly, the interaction of antigenic membrane protein of a phytoplasma and leafhopper actin is correlated with the phytoplasma-transmitting capability of leafhoppers (Suzuki et al., 2006). Due to the lack of reverse-genetics and transgenic expression systems for hemipteran insects, direct evidence to support the conclusion that the interaction of pathogens with insect actin determines insect vector specificity is still not available.

Plant reoviruses, plant rhabdoviruses, tospoviruses and tenuiviruses are transmitted by insect vectors in a persistent- propagative manner (Hogenhout et al., 2008). These viruses induce the formation of virus inclusions composed of viral non-structural proteins to facilitate viral propagation in insect vectors (Hogenhout et al., 2008). In this study, we revealed that the interaction between the non-structural protein of a plant virus and insect actin was correlated with the specificity of the insect vector for the transmitted virus. The ability of viruses to pass through the insect intestine is an important factor in vector determination. Actin filaments are the major component of intestinal microvilli and visceral muscle, which would constitute a substantial barrier to the persistent transmission of viruses. Thus, the specific association of viruses and insect actin suggests that viruses may directly utilize insect actin filaments to overcome the transmission barriers. All these analyses support the conclusion that the interaction of viruses and insect actin may determine insect vector specificity.

Acknowledgements

This research was supported by the National Natural Science Foundation of China (31130044, 31200118 and 31300136), the National Basic Research Program of China (2014CB138400), the National Science Foundation for Outstanding Youth (31325023), a project from the Henry Fok Education Foundation (131019), the Natural Science Foundation of Fujian Province of China (2014J01085) and the Scientific Research Foundation of Graduate School of Fujian Agriculture and Forestry University (YB2013008).

References

[1] AMMAR D, TSAI C W, WHITFIELD A E, et al. Cellular and molecular aspects of rhabdovirus interactions with insect and plant hosts[J]. Annual Review of Entomology,2009, 54: 447-468.

[2] CHEN H, CHEN Q, OMURA T, et al. Sequential infection of *Rice dwarf virus* in the internal organs of its insect vector after ingestion of virus[J]. Virus Research,2011, 160(1-2): 389-394.

[3] CHEN Q, CHEN H, MAO Q, et al. Tubular structure induced by a plant virus facilitates viral spread in its vector insect[J]. Plos Pathogens, 2012, 8(11):e1003032.

[4] HOGENHOUT S A, AMMAR D, WHITFIELD A E,et al. Insect vector interactions with persistently transmitted viruses[J]. Annual Review of Phytopathology, 2008, 46:327-359.

[5] HONDA K, WEI T, HAGIWARA K, et al. Retention of *Rice dwarf virus* by descendants of pairs of viruliferous vector insects after rearing for 6 years[J]. Phytopathology, 2007, 97(6):712-716.

[6] JIA D, CHEN H, MAO Q, et al. Restriction of viral dissemination from the midgut determines incompetence of small brown planthopper as a vector of *Southernrice black-streaked dwarf virus*[J]. Virus Research, 2012a, 167(2):404-408.

[7] JIA D, CHEN H, ZHENG A, et al. Development of an insect vector cell culture and RNA interference system to investigate the functional role of *Fijivirus* replication protein[J]. Journal of Virology, 2012b, 86(10):5800-5807.

[8] JIA D, MAO Q, CHEN H, et al. Virus-induced tubule: a vehicle for rapid spread of virions through basal lamina from midgut epithelium in the insect vector[J]. Journal of Virology, 2014, 88:10488-10500.

[9] LIU Y, JIA D, CHEN H, et al. The P7-1 protein of *Southern rice black-streaked dwarf virus*, a *Fijivirus*, induces the formation of tubular structures in insect cells[J]. Archives of Virology, 2011, 156(10):1729-1736.

[10] MARKHAM P G, PINNER M S, BOULTON M I. The transmission of maize streak virus by leafhoppers, a new look at host adaptation[J]. Bulletin of Entomological Research,1984, 57: 431-432.

[11] OMURA T, MORINAKA T, INOUE H, et al. Purification and some properties of *Rice gall dwarf virus*, a new phytoreovirus[J]. Phytopathology(USA), 1982, 72(9):1246-1249.

[12] POPOVA-BUTLER A, DEAN D H. Proteomic analysis of the mosquito *Aedes aegypti* midgut brush border membrane vesicles[J]. Journal of Insect Physiology, 2009, 55(3):264-272.

[13] PU Y, KIKUCHI A,MORIYASU Y, et al. *Rice dwarf viruses* with dysfunctional genomes generated in plants are filtered out in vector insects: implications for the origin of the virus[J]. Journal of Virology, 2010, 85(6):2975-2979.

[14] SUZUKI S, OSHIMA K, KAKIZAWA S, et al. Interaction between the membrane protein of a pathogen and insect microfilament complex determines insect-vector specificity[J]. Proceedings of the National Academy of Sciences, 2006, 103:4252-4257.

[15] WEI T, KIKUCHI A, MORIYASU Y, et al. The spread of *Rice dwarf virus* among cells of its insect vector exploits virus-induced tubular structures[J]. Journal of Virology, 2006, 80(17):8593-8602.

[16] WEI T, SHIMIZU T, OMURA T. Endomembranes and myosin mediate assembly into tubules of Pns10 of *Rice dwarf virus* and intercellular spreading of the virus in cultured insect vector cells[J]. Virology, 2008, 372(2):349-356.

Rice stripe tenuivirus p2 may recruit or manipulate nucleolar functions through an interaction with fibrillarin to promote virus systemic movement

Luping Zheng[1], Zhenguo Du[1,2], Chen Lin[1,3], Qianzhou Mao[1], Kangcheng Wu[1], Jianguo Wu[1], Taiyun Wei[1], Zujian Wu[1], andLianhui Xie[1]

(1 Institute of Plant Virology, Fujian Agriculture and Forestry University, Fuzhou 350002, China; 2 Guangdong Provincial Key Laboratory of High Technology for Plant Protection, Plant Protection Research Institute, GAAS, Guangzhou 510640, China; 3 Bayuquan Entry–Exit Inspection and Quarantine Bureau, Yingkou 115007, China)

Abstract:*Rice stripe virus* (RSV) is the type species of the genus *Tenuivirus* and represents a major viral pathogen affecting rice production in East Asia. In this study, RSV p2 was fused to yellow fluorescent protein (p2-YFP) and expressed in epidermal cells of *Nicotiana benthamiana*. p2-YFP fluorescence was found to move to the nucleolus initially, but to leave the nucleolus for the cytoplasm forming numerous distinct bright spots there at later time points. A bimolecular fluorescence complementation (BiFC) assay showed that p2 interacted with fibrillarin and that the interaction occurred in the nucleus. Both the nucleolar localization and cytoplasmic distribution of p2-YFP fluorescence were affected in fibrillarin-silenced *N. benthamiana*. Fibrillarin depletion abolished the systemic movement of RSV, but not that of Tobacco mosaic virus (TMV) and Potato virus X (PVX).A Tobacco rattle virus (TRV)-based virus-induced gene silencing (VIGS) method was used to diminish RSV NS2 (encoding p2) or NS3 (encoding p3) during RSV infection. Silencing of NS3 alleviated symptom severity and reduced RSV accumulation, but had no obvious effects on virus movement and the timing of symptom development. However, silencing of NS2 abolished the systemic movement of RSV. The possibility that RSV p2 may recruit or manipulate nucleolar functions to promote virus systemic infection is discussed.

Keywords:fibrillarin; nucleolus;*Rice stripe virus*; systemic movement.

1 Introduction

The nucleolus is a plurifunctional subnuclear structure that may harbour as many as 4500 proteins (Boisvert et al., 2007; Boulon et al., 2010; Shaw and Brown, 2012). The interaction of animal viruses with the nucleolus has been studied extensively since the beginning of the 1990s. Most, if not all, well-studied animal viruses have been found to have a nucleolar phase in their infection cycle. A link between the ability to interact with the nucleolus and the outcome of viral infection has been established in many cases (Hiscox, 2007; Salvetti and Greco, 2014). In recent years, diverse plant viruses belonging to different genera or families have also been found to encode one or multiple proteins localizing to the nucleolus (Martínez and Daròs, 2014; Rossi et al., 2014; Taliansky et al., 2010; and references therein). In the nucleolus, *Groundnut rosette virus* (GRV, an umbravirus) open reading frame 3 (ORF3) interacts with fibrillarin and recruits this nucleolar protein to form infectious ribonucleoproteins (RNPs) in the cytoplasm. This is crucial for the systemic movement of GRV

Virus Research.2014, 180: 97-101

Received 26 August 2013; Accepted 18 November 2013; Available online 27 November 2013

(Kim et al., 2007a, b). Fibrillarin is also essential for the systemic movement of *Potato leafroll virus* (PLRV, a polerovirus), but the mechanism by which PLRV uses fibrillarin remains unclear (Kim et al., 2007b). The nucleolar localization of NIa/Vpg plays an important role in Potato virus A (PVA, a potyvirus) replication, pathogenicity and systemic movement. This is partially explained by the fact that NIa/Vpg may need to enter the nucleolus to suppress RNA silencing (Rajamäki and Valkonen, 2009). NIa/Vpg also interacts with fibrillarin. However, the depletion offibrillarin reduced PVA accumulation, but had no significant effects on PVA systemic movement. For most other plant viruses, the rationale for their nucleolus targeting is currently unclear. However, a growing number of correlative studies have indicated that the nucleolus may play a role in many steps in the infection cycle of these viruses (Martínez and Daròs, 2014; Rossi et al., 2014; Taliansky et al., 2010; and references therein).

Rice stripe virus (RSV) is the type species of the genus *Tenuivirus*, which has not been assigned to any family (King et al., 2012). RSV is transmitted by the small brown planthopper (SBPH) (*Laodelphax striatellus* Fallén) in a persistent and circulative–propagative manner (Falk and Tsai, 1998). In Far East Asia, RSV represents one of the most important viralpathogens severely affecting rice production (Cho et al., 2013, Hibino, 1996; Wei et al., 2009). Recently, RSV has been reported to occur in Vietnam, a country in South-East Asia (Ren et al., 2013).

The genome of RSV comprises four single-stranded RNA (ssRNA) segments, named RNA1–RNA4, respectively, in decreasing order of size. RNA1–RNA4 are encompassed separately by nucleocapsid proteins to form distinct filamentous RNPs (Ramírez and Haenni, 1994). RNA1 is of negative sense and encodes a 337kDa protein (pc1) considered to be the RNA dependent RNA polymerase (RdRp) associated with RSV virions (Toriyama et al., 1994). RNA2–RNA4 use an ambisense coding strategy, each containing two non-overlapping ORFs, one in the 5′ half of the viral RNA (the genes they encode are named NS2–NS4 and the proteins they encode are named p2–p4, respectively) and the other in the 5′ half of the viral complementary RNA (the genes they encode are named NSVc2–NSVc4 and the proteins they encode are named pc2–pc4, respectively; Falk and Tsai, 1998; Kakutani et al., 1990, 1991; Takahashi et al., 1993; Zhu et al., 1991). pc4 (32kDa) is a movement protein which may also play a role in RSV pathogenicity (Xiong et al., 2008; Xu and Zhou, 2012; Zhang et al., 2012). p4 (20.5kDa) is the major nonstructural protein which often accumulates to very high levels in RSV-infected plants (Kong et al., 2014; Lin et al., 1998). pc3 (35kDa) is the nucleocapsid protein (Hayano et al., 1990; Kakutani et al., 1991). p3 (23.9kDa) is a silencing suppressor (Xiong et al., 2009). pc2 (94kDa) can be cleaved into two proteins, pc2-N and pc2-C, which target the Golgi apparatus and the endoplasmic reticulum (ER), respectively, in cells of *Nicotiana benthamiana* (Yao et al., 2014; Zhao et al., 2012). p2 (22.8kDa) has weak silencing suppressor activities (Du et al., 2011).

Owing to the large genome, the unusual coding strategy, the absolute insect transmission and the use of monocots as its host, RSV is resistant to genetic manipulations. As a result, studies on RSV–host interaction are difficult and our knowledge of the mechanisms by which RSV infects its host remains poor. The situation is beginning to change after the discovery that RSV can infect *N. benthamiana* and *Arabidopsis thaliana* under laboratory conditions (Sun et al., 2011 and Xiong et al., 2008). Recently, *N. benthamiana* has been shown to be a feasible model plant to investigate the roles of host factors in RSV infection (Jiang et al., 2014; Kong et al., 2014).

At least three RSV proteins, namely p3, pc3 and p4, may localize to the nucleus (Lian et al., 2014; Xiong et al., 2009). This suggests that RSV may have a nuclear phase in its replication cycle. However, this possibility has never been explored. In this article, we present our data suggesting that RSV p2 may recruit or manipulate nucleolar functions to promote virus systemic movement in *N. benthamiana*.

2 Results

2.1 RSV p2 moves to and through the nucleolus when expressed individually in epidermal cells of *N. benthamiana*.

To study the subcellular localization of RSV p2, a fusion of p2 with yellow fluorescent protein (YFP; p2-YFP) was created by cloning the p2 ORF into the plasmid vector pEarleyGate 101 (Earley et al., 2006). The recombinant plasmids were agroinfiltrated into epidermal cells of *N. benthamiana* leaves. The expression of the fusion protein was confirmed by Western blotting using commercially available antibodies against green fluorescent protein (GFP, Fig. 1G). YFP fluorescence was observed at 24, 48, 60, 72 and 84h post-infiltration (hpi) by confocal laser scanning microscopy (CLSM).

At 24hpi, YFP fluorescence was found in both the nucleus and the cytoplasm. In the cytoplasm, most of the fluorescence was diffuse. In the nucleus, however, the fluorescence accumulated and formed multiple bright spots (Fig. 1A). At 48hpi, YFP fluorescence was found to localize predominantly in the nucleus, being concentrated in structures that resembled the nucleolus and some other small subnuclear structures reminiscent of Cajal bodies (Fig. 1B; Cioce and Lamond, 2005). The nucleolar localization of p2-YFP was confirmed by its co-localization with fibrillarin (NbFib2 of *N. benthamiana*, GenBank accession: AM269909), a marker protein of the nucleolus (Fig. 1C; Barneche et al., 2000; Kim et al., 2007a, b). By 60hpi, in addition to the intense fluorescence signals in the nucleolus, many small bright spots of unknown nature were found in the cytoplasm near the plasma membrane and in perinuclear regions (Fig. 1D). By 72hpi, the numbers of small spots in the cytoplasm had increased (Fig. 1E). Fluorescence was also apparent in the nucleus. However, in most cells, the single, large fluorescent body observed at 60hpi had changed into two to four smaller ones (Fig. 1D, F). By 84hpi, numerous small spots were found in the cytoplasm near the plasma membrane. Faint fluorescence was observed from the nucleus, but, in most cases, the fluorescence did not accumulate in the nucleolus (Fig. 1E). A fusion protein with YFP at the N-terminus of p2 (YFPp2) and fusion proteins tagging p2 with cyan fluorescent protein (CFP) were also used to study the cellular localization of p2. The same results were obtained (data not shown).

2.2 p2 interacts with fibrillarin

Many nucleolus-targeting viral proteins interact with one or multiple nucleolar proteins (Hiscox, 2007; Salvetti and Greco, 2014). For plant viruses, at least five nucleolus-targeting proteins from Poasemilatent virus (PSLV; Semashko et al., 2012), Potato mop-top virus (PMTV; Wright et al., 2010), GRV (Kim et al., 2007a, b), Beet black scorch virus (BBSV; Wang et al., 2012) and PVA (Rajamäkiand Valkonen, 2009), respectively, have been shown to interact with fibrillarin. A bimolecular fluorescence complementation (BiFC) assay was carried out to investigate the interaction between p2 and NbFib2 (Walter et al., 2004). To do this, the full-length ORF of p2 was cloned into the vector pEarleyGate 201-YC and that of NbFib2 into pEarleyGate 201-YN. Transformed Agrobacterium tumefaciens EHA105 carrying each of these constructs was mixed and infiltrated into the leaves of *N. benthamiana*. Strong YFP fluorescence derived from the reconstitution of the YFP fluorophore was observed at 48hpi, indicating the interaction of p2 and NbFib2 in living plant cells (Fig.2). Similar results were obtained when p2 was fused with the N-terminal fragment of YFP and NbFib2 with the C-terminal fragment of YFP. By contrast, fluorescence was not detected in leaf cells co-infiltrated with pEarleyGate 201-

YC/pEarleyGate 201-YN, pEarleyGate 201-YC-p2/pEarleyGate 201-YN or pEarleyGate 201- YC/pEarleyGate 201-YN-NbFib2 (data not shown). As shown in Fig. 2, the interaction between p2 and NbFib2 occurred in the nucleus. Similar BiFC experiments showed that p2 also interacted with fibrillarinof rice (GenBank accession: AK103477) and Arabidopsis (GenBank accession: AAG10153; data not shown).

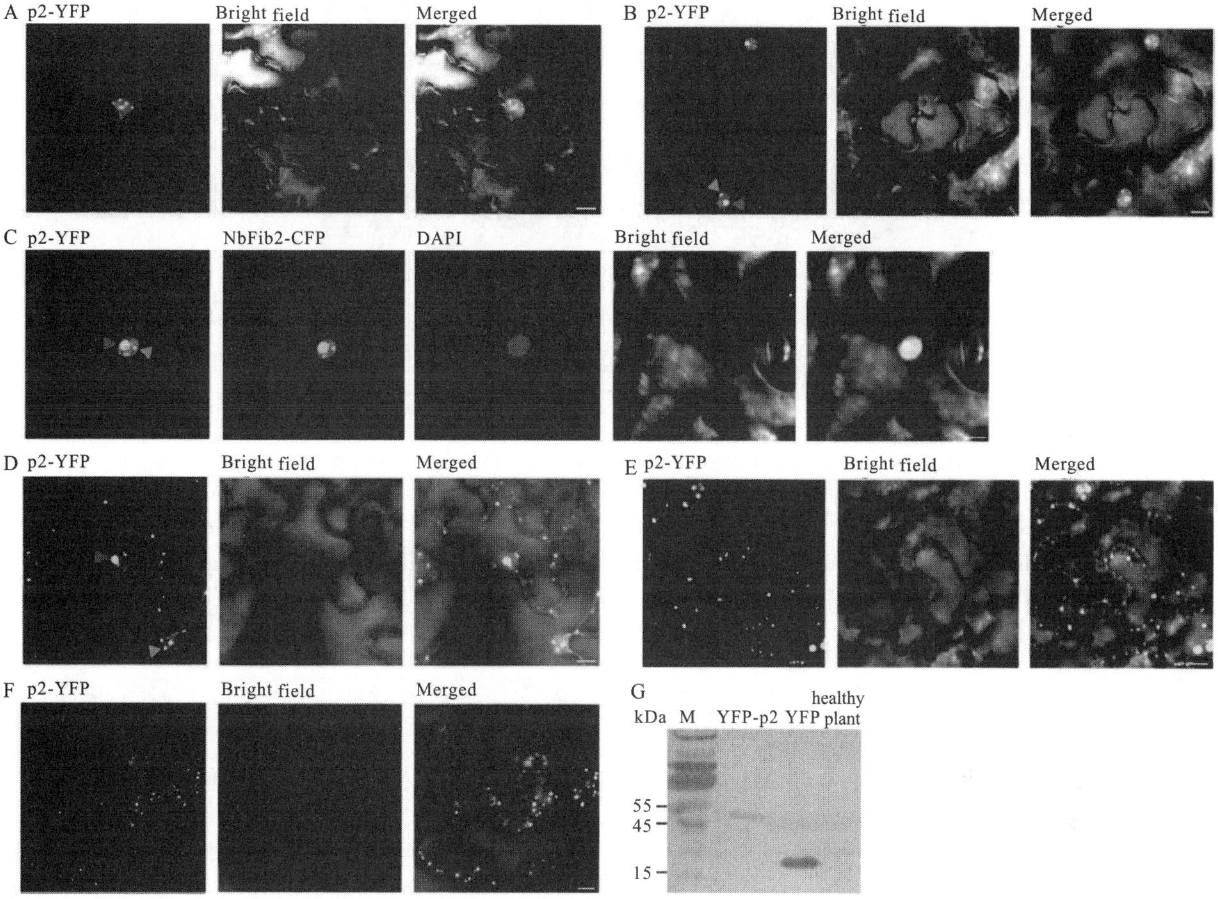

Fig. 1 Subcellular localization of *Rice stripe virus* (RSV) p2 in epidermal cells of *Nicotiana benthamiana* (A–F) and Western blotting confirming the expression of the fusion protein p2-yellow fluorescent protein (p2-YFP, G) at 48h post-infiltration (hpi)

p2-YFP was expressed individually (A, B, D, E and F) or co-expressed with NbFib2 tagged with cyan fluorescent protein (CFP, C) in leaves of *N. benthamiana* by agroinfiltration. Fluorescence was observed at 24 (A), 48 (B, C), 60 (D), 72 (E)and 84hpi (F) by confocal laser scanning microscopy. Cells co-expressing p2 and NbFib2 were stained with 4,6-diaminophenylindole (DAPI) to show the nucleus(C). Possible nucleoli or Cajal bodies described in the text are designated with red and blue arrows, respectively. Scale bars, 10μm

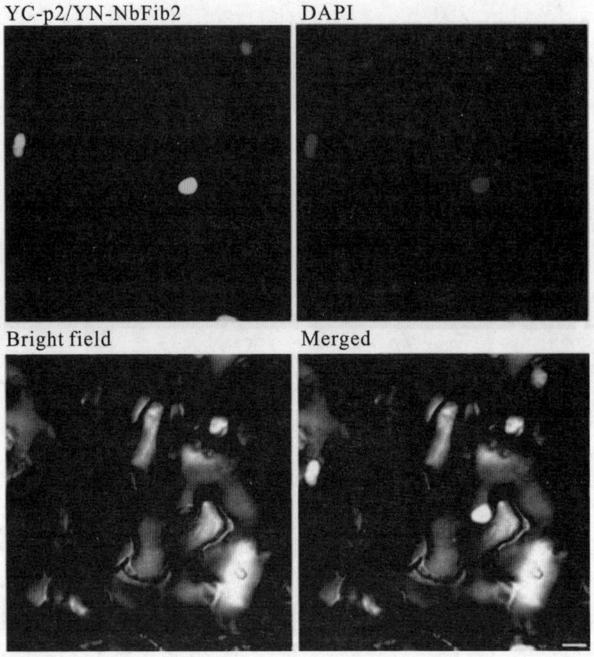

Fig. 2 Bimolecular fluorescence complementation (BiFC) assay showing interaction between p2 and NbFib2

In the figures shown here, the full-lengthopen reading frame (ORF) of p2 was cloned into the vector pEarleyGate 201-YC (YC-p2) and that of NbFib2 into pEarleyGate 201-YN (YN-NbFib2).The recombinant plasmids were introduced into epidermal cells of *Nicotiana benthamiana* by agroinfiltration. Fluorescence was observed at 48h post-infiltration by confocal laser scanning microscopy.The nucleus was stained with 4,6-diaminophenylindole (DAPI). Scale bars, 10μm

2.3 Fibrillarin depletion affects the nucleolar localization and cytoplasmic distribution of p2

The interaction with fibrillarin has been shown to be required for the nucleolar localization of GRV ORF3 (Kim et al., 2007b). To test the effects of fibrillarin on p2 localization, we down-regulated the expression of *NbFib2* in *N. benthamiana* using a *Tobacco rattle virus* (TRV)-based virus-induced gene silencing (VIGS) method in the manner described previously by Kim et al. (2007b) and Rajamäki and Valkonen (2009).

At 12d post-inoculation (dpi), plants inoculated with TRV-NbFib2 could be separated into three groups according to symptoms: Group I, severely dwarfed plants with arrested growth; Group II, plants with slightly shortened stems, mild leaf chlorosis and negligible leaf deformity; Group III, plants with very mild leaf chlorosis, but without detectable leaf deformity (Fig. 3A).The three groups of plants showed >90%, 45%–60% and <10% reduction in *NbFib2* expression, respectively, as demonstrated by real-time reverse transcription polymerase chain reaction (RT-PCR, Fig. 3B).

YFP-p2/p2-YFP was expressed in representative plants belonging to each of the three groups and in plants inoculated with TRV alone using agroinfiltration. The YFP fluorescence was observed at 2 and 3d after agroinfiltration (dai).

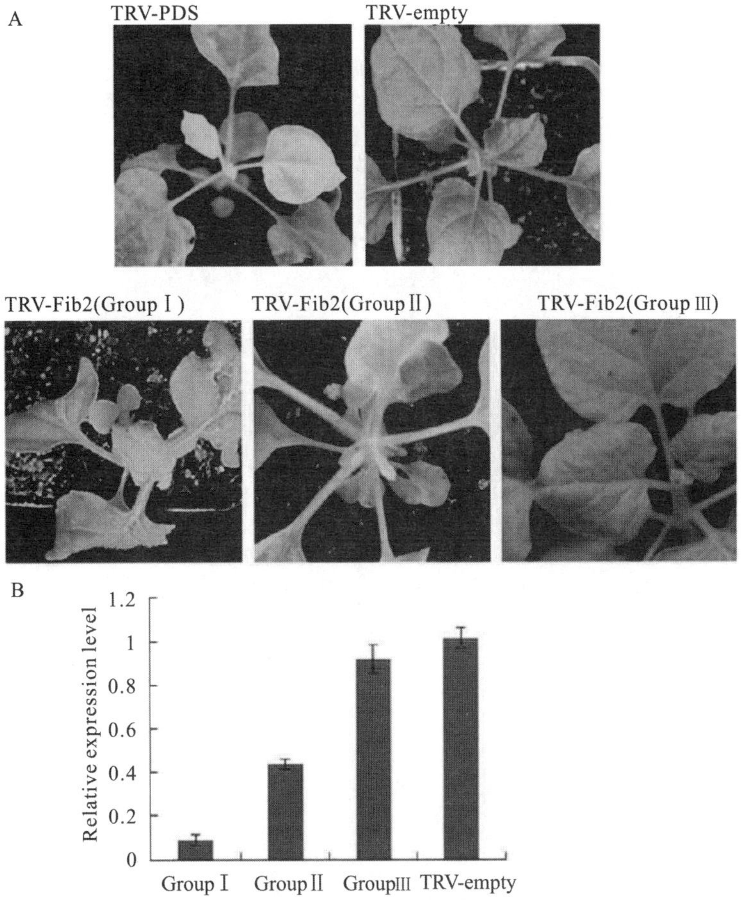

Fig. 3 Virus-induced gene silencing of NbFib2

(A) Symptoms of *Nicotiana benthamiana* infected by *Tobacco rattle virus* (TRV), TRV-PDS and TRV-Fib2 as indicated. For plantsinoculated with TRV-Fib2, representative plants belonging to each of the three groups reported in the text are shown. (B) Varying silencing efficiencies of *NbFib2* in representative plants belonging to each of the three groups shownin (A). For real-time reverse transcription-polymerase chain reaction(RT-PCR) detecting the expression of *NbFib2*, three plants belonging to each group were pooled as one sample and the experiments were performed in triplicate

As shown in Fig. 4A, B, the localization of p2-YFP in plants inoculated with TRV alone or in Group III plants was similar to that in wild-type plants, i.e. fluorescence accumulated in the nucleolus initially (2dai), but left the nucleolus for the cytoplasm and formed numerous small bright bodies there at 3dai (data not shown). In Group II plants, a nucleolar accumulation of p2-YFP was not found at 2dai. Instead, fluorescence was found in both the nucleus and the cytoplasm (Fig. 4C). The distribution of fluorescence did not show any detectable change at 3dai (data not shown). In Group I plants, p2-YFP fluorescence showed a diffuse localization in both the nucleus and the cytoplasm at 2dai (Fig. 4D). In addition, fluorescence did not show any detectable change at 3dai (data not shown). These results suggest that fibrillarin may play a role in both the nucleolar localization and the appropriate cytoplasmic distribution of p2.

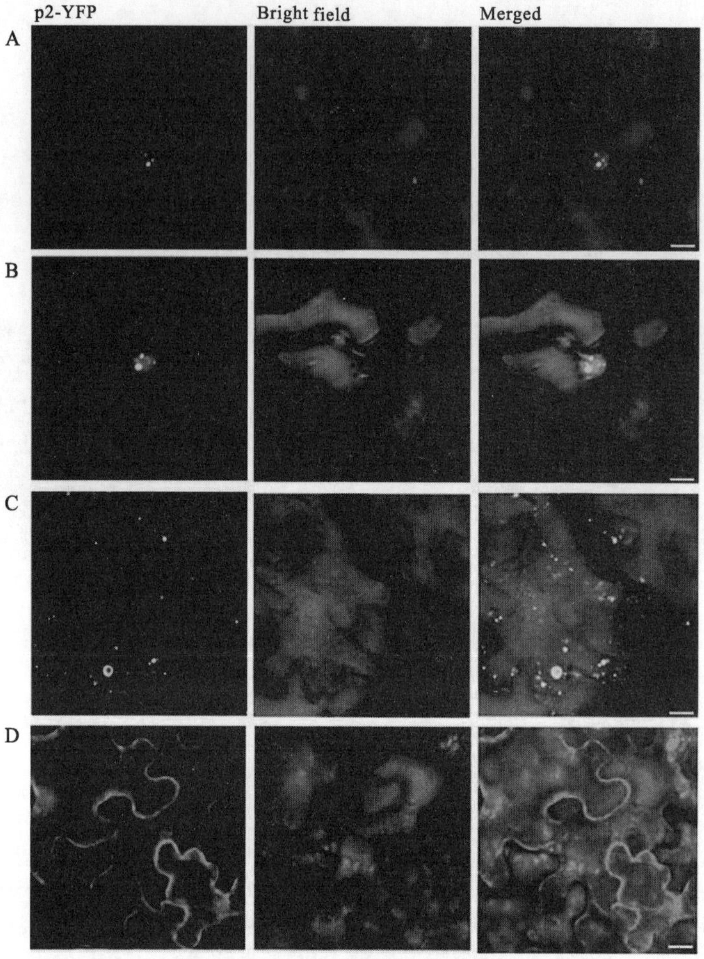

Fig. 4 Subcellular localization of *Rice stripevirus* (RSV)

p2 fused with yellow fluorescent protein (YFP) in epidermal cells of *Nicotianabenthamiana* inoculated with empty *Tobaccorattle virus* (TRV) (A) and N. benthamiana belonging to Group III (B), Group II (C) and Group I (D) as described in the text. The YFP fluorescence was observed 2d after agroinfiltration. Scale bars, 10μm

2.4 Fibrillarin depletion abolishes the systemic movement of RSV in N. benthamiana

Previously, it has been reported that the knock-down of fibrillarin abolishes the systemic movement of GRV and PLRV. Fibrillarin depletion did not prevent the systemic movement but reduced the accumulation of PVA (Kim et al., 2007b; Rajamäki and Valkonen, 2009). To investigate the role of fibrillarin in RSV infection, crude extracts of RSV-infected rice were used to mechanically inoculate Group II plants in the manner described previously (at 10d after TRV-NbFib2 inoculation; Jiang et al., 2014; Xiong et al., 2008). *Nicotiana benthamiana* plants previously infected by empty TRV were used as controls. The inoculated plants were monitored daily for symptoms associated with RSV infection. The accumulation of RSV was detected by RT-PCR and Western blotting at 22dpi of RSV using antibodies against RSV p4, the major non-structural protein of RSV, whose accumulation correlates with RSV replication and host symptom development (Kong et al., 2014; Lin et al., 1998).

In control plants, symptoms including leaf chlorosis, yellowing and downward leaf curling appeared at 12–15dpi of RSV. The symptoms persisted for months before the death of the plants (Fig. 5A). RSV could be readily detected in both inoculated and systemic leaves (Fig. 5C). This indicates that TRV does not affect RSV infection of *N. benthamiana*. A similar observation has been made recently by Jiang et al. (2014).

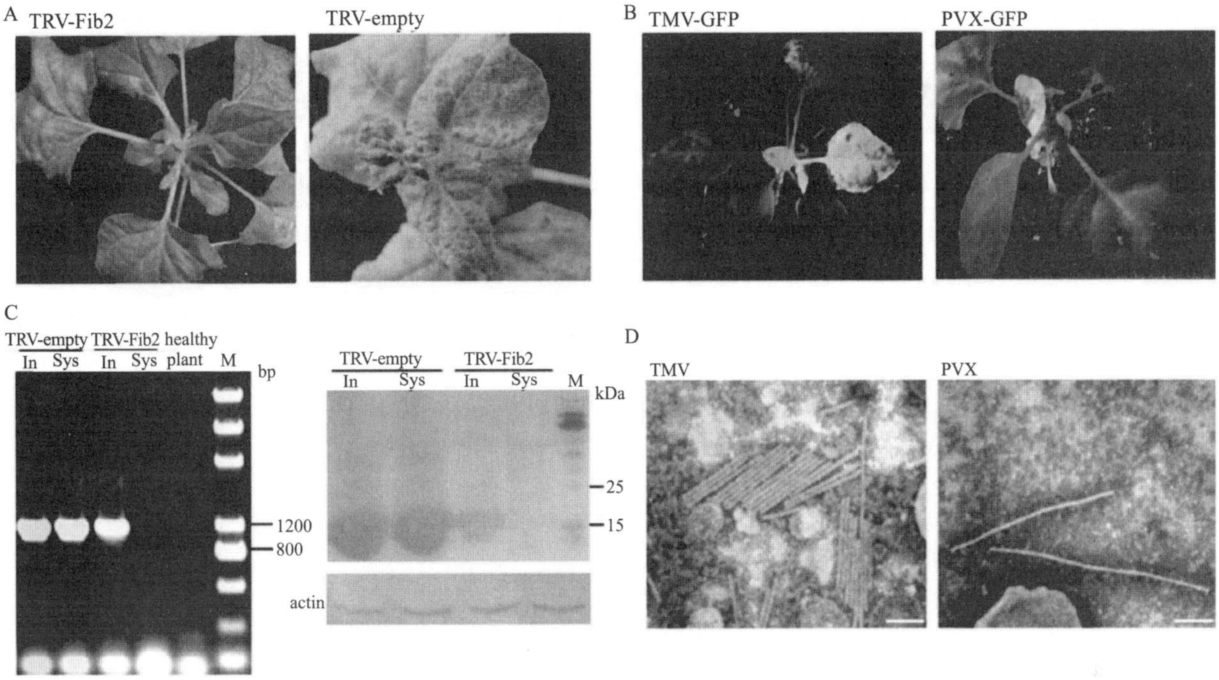

Fig. 5 Effects of *NbFib2* depletion on *Rice stripe virus* (RSV) infection

NbFib2-silenced *Nicotiana benthamiana* plants belonging to Group II were inoculated withRSV (A, left), *Potato virus* X-green fluorescent protein (PVX-GFP) (B, right) or *Tobacco mosaic virus*-GFP (TMV-GFP) (B, left), as described in the text. *Nicotiana benthamiana* previously inoculated with empty *Tobacco rattle virus* (TRV) was used as a control (A, right). *Nicotiana benthamiana* plants are photographed at 22d after virus inoculation. The accumulation of RSV 22d after virus inoculation in inoculated (In) or upper (Sys) leaves was detected by reverse transcription-polymerase chain reaction (RT-PCR, C, left) and Western blotting (C, right).The presence of PVX-GFP and TMV-GFP was visualized under UV light (B),and the presence of intact PVX and TMV virions in systemically infected leaves was demonstrated by electron microscopy (D; scale bars, 100nm)

In fibrillarin-silenced *N. benthamiana*, symptoms associated with RSV infection were not observed either at this time or later (Fig. 5A). RT-PCR and Western blotting failed to detect RSV in newly developed leaves (Fig. 5C). However, RSV accumulated in inoculated leaves (Fig. 5C). The accumulation level was comparable with, although obviously lower than, that in control plants (Fig. 5C). Fifteen *N. benthamiana* plants were used in the experiment; the experiment was carried out in triplicate and the results were reproducible. This suggests that fibrillarin depletion prevents the systemic movement of RSV in *N. benthamiana*.

Kim et al. (2007b) have demonstrated that the silencing of fibrillarin does not inhibit the normal cellular functions necessary for general virus infection. To confirm this under our experimental conditions, Group II *N. benthamiana* plants were inoculated with Tobacco mosaic virus-GFP (TMV-GFP) and Potato virus X-GFP (PVX-GFP). In plants infected by either virus, GFP fluorescence was observed as early as 9dpi of virus (Fig. 5B). TMV and PVX werereadily detectable in the leaves by RT-PCR (data not shown). Intact virions of TMV and PVX could also be observed by electron microscopy (Fig. 5D). Thus, fibrillarin silencing under our experimental conditions does not disturb the cellular functions necessary for general virus infection.

2.5 p2 is involved in systemic movement of RSV in *N. benthamiana*

The above results strongly suggest that RSV p2 may function in virus systemic movement. To test this, a TRV vector containing a partial sequence of NS2 was created and used to infect *N. benthamiana* as described above. *Nicotiana benthamiana* plants infected by empty TRV or TRV carrying a partial sequence of NS3 were used as controls. Throughout our experiments, *N. benthamiana* infected by TRV-NS2 or TRV-NS3 showed no obvious differences from those infected by TRV alone (data not shown). Thus, it seems unlikely that TRV-NS2 or TRV-NS3 may target host genes important for normal physiology. The upper leaves of these plants were inoculated with RSV at 12dpi of TRV/TRV-NS2/TRV-NS3. Symptoms associated with RSV infection were monitored daily and the accumulation of RSV was detected at 22dpi of RSV as described above.

TRV-NS3- and TRV-treated *N. benthamiana* plants were similar to each other in terms of the timing of symptom development, i.e. symptoms appeared at 12–15dpi of RSV in the top leaves and persisted for months (Fig. 6A). The accumulation of RSV was readily detectable in both inoculated and systemic leaves in TRVNS3-treated plants (Fig. 6B), although the symptoms were obviously milder and the accumulation of RSV was lower in these plants when compared with the control (TRV-treated plant; Fig. 6A, B). The accumulation of the gene NS3 in inoculated leaves of TRV-NS3-treated plants was detected by real-time RT-PCR. The results showed that the expression of NS3 was reduced by about 60%.

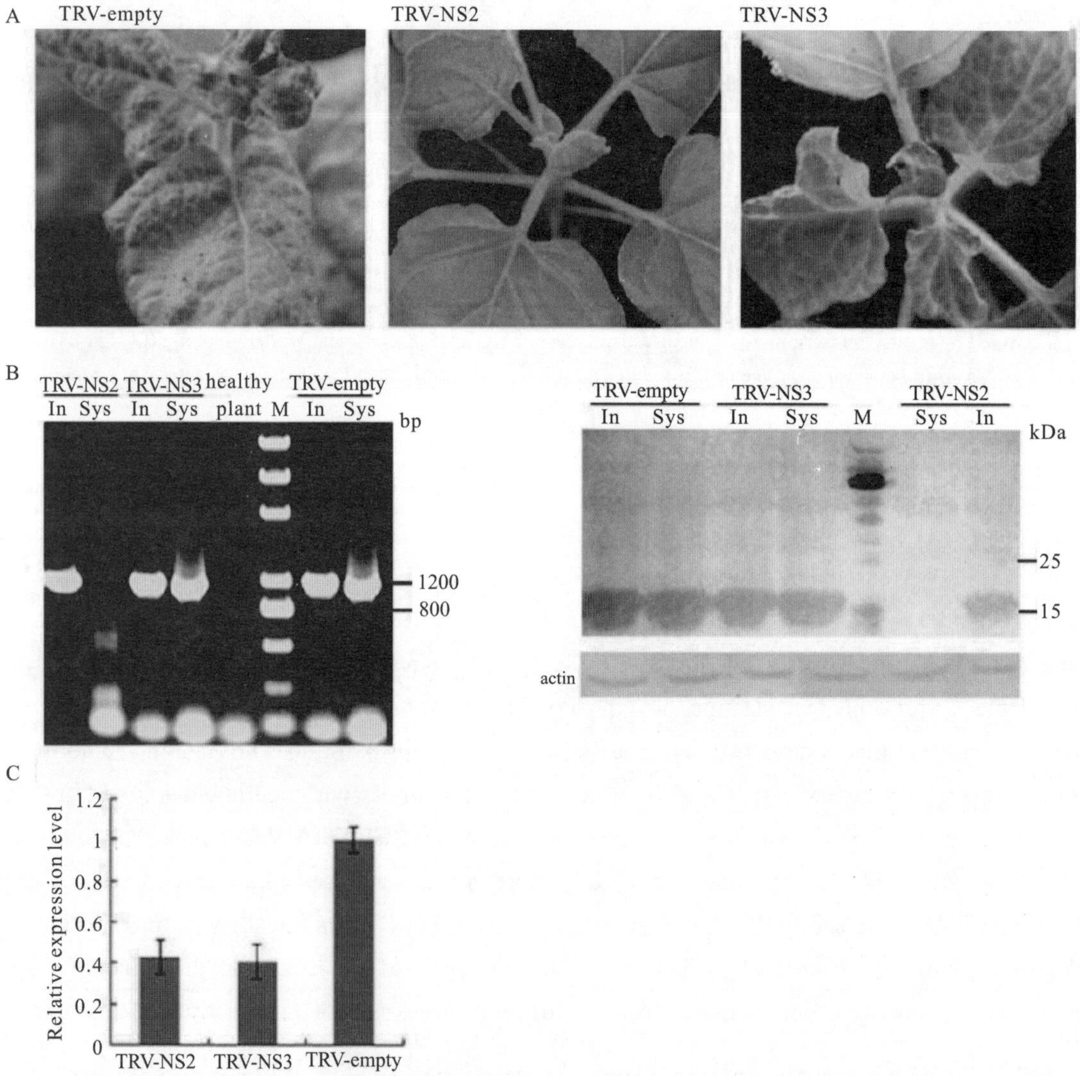

Fig. 6 Effects of NS2 silencing on *Rice stripe virus* (RSV) infection

Nicotiana benthamiana plants were inoculated with either *Tobacco rattle virus* (TRV), TRV-NS2 or TRV-NS3, as indicated, 12d before RSV inoculation.The symptoms shown in the figure (A) were photographed 22d after RSV inoculation.The accumulation of RSV in inoculated (In) and upper (Sys) leaves of TRV-NS2/TRV-NS3-treated *N. benthamiana* was detected by reverse transcription-polymerase chain reaction(RT-PCR) (B, left) and Western blotting (B,right), 24d after RSV inoculation.The accumulation of the gene NS2 in inoculated leaves of TRV-NS2-treated plants at 22d post-inoculation of RSV was detected byreal-time RT-PCR (C)

Symptoms were not observed even 40d after RSV inoculation on newly emerged leaves of plants treated with TRV-NS2 (Fig. 6A). Only four of the 15 plants developed symptoms at 60dpi of RSV. As shown in Fig. 6B, RSV was not detected in systemic leaves. However, RSV accumulated in inoculated leaves (Fig. 6B). The accumulation of the gene NS2 in inoculated leaves was detected by real-time RT-PCR. The results showed that the expression of NS2 was reduced by about 55% (Fig. 6C). Thus, p2 of RSV may indeed play a role in the systemic movement of RSV.

3 Discussion

Overall, we found that: (i) RSV p2 trafficks between the nucleolus and the cytoplasm; (ii) p2 interacts with fibrillarin and fibrillarin is important for the normal cellular distribution (and the dynamics of distribution) of p2; (iii) fibrillarin is important for the systemic movement of RSV; and (iv) p2 may play a role in RSV systemic movement. It should be noted that the localization experiments and experiments showing the interaction between p2 and fibrillarin reported here were conducted out of context of viral infection. Further experiments are needed to demonstrate the dynamics of p2 localization and the interaction between p2 and fibrillarin in the process of RSV infection.

Although many plant viruses have been shown to interact with fibrillarin, RSV is the third plant virus shown to require fibrillarin for systemic infection after GRV and PLRV (Kim et al., 2007a, b). The mechanism by which RSV uses fibrillarin is unclear at present. GRV has been shown to redirect fibrillarin to the cytoplasm and to use fibrillarin to form infectious RNPs (Kim et al., 2007a, b). It is possible that RSV may recruit a small proportion of fibrillarin to function in the cytoplasm. However, a more plausible possibility is that RSV may use fibrillarin in a different manner, as we found that the interaction between p2 and fibrillarin occurred exclusively in the nucleus (Fig. 2). In our co-localization studies of p2 and fibrillarin, fibrillarin was never found outside the nucleus. Instead, p2 was trapped in the nucleolus when NbFib2 was over expressed(Fig. 1C). Moreover, fibrillarin was not found in the cytoplasm when it was expressed as a fusion with YFP in RSV-infected *N. benthamiana* (data not shown). In this context, it is interesting to note that PVA requires fibrillarin for enhanced replication, but not for systemic movement (Rajamäki and Valkonen, 2009). This suggests that the same host protein may be used by different viruses for different purposes, or for the same purposes but in a different way.

The fact that an interaction partner of p2 is involved in RSV systemic movement implicates a role of p2 in this process. This cannot be tested directly because of the resistance of RSV to genetic manipulation. However, RSV has a segmented genome and uses independently transcribed, non-overlapping mRNAs to produce proteins (Falk and Tsai, 1998; Kormelink et al., 2011; Nguyen and Haenni, 2003; Wu et al., 2013). Because of this, we used RNA interference (RNAi) to investigate the role of p2 in RSV infection. Although RNAi has been widely used to study the gene functions of some animal viruses, especially reoviruses and some negative-strand RNA viruses, similar technologies have been used rarely in studies of plant viruses (Bitko and Barik, 2001; Kobayashi et al., 2006). One may argue that small interfering RNAs (siRNAs) derived from NS2 may target RSV genomic RNAs. Currently, we cannot rule out this possibility. However, genomic RNAs of negative-strand RNA viruses, such as RSV, are coated with nucleocapsid proteins. Nucleocapsid proteins of negative-strand RNA viruses may block siRNA-targeted recognition and degradation, as suggested by Ahlquist (2002), and demonstrated by the fact that synthetic siRNAs against Respiratory syncytial virus induce the degradation of viral mRNA, but not nucleocapsid coated genomic RNA, in cultured human cells (Bitko and Barik, 2001). Recent work by Shimizu et al. (2011) supports the feasibility of RNAi in the study of RSV gene functions.

The researchers expressed RNAi constructs targeting different RSV genes in rice. Transgenic rice showed very different levels of resistance to RSV depending on the RSV genes targeted. Importantly, the levels of resistance in plants expressing RNAi constructs targeting RSV genes expressed from the same genomic RNAs varied from being immune to no enhanced resistance. This strongly suggests that the RNAi constructs target specific genes of RSV rather than genomic RNAs. We anticipate that RNAi can be used in more sophisticated ways to study gene functions of RSV in the future.

The finding that p2 may function in the systemic movement of RSV is also consistent with observations made by Shimizu et al. (2011). In their studies, among the 80 transgenic rice plants expressing an RNAi construct targeting NS2, the appearance of symptoms was delayed by 2–35d in 48 (60%) plants. Twenty(25%) plants never exhibited symptoms prior to harvest. Also, this is consistent with our previous finding that p2 has silencing suppressor activities (Du et al., 2011). A number of silencing suppressors of plant viruses play a role in virus systemic movement (Díaz-Pendón and Ding, 2008; Hipper et al., 2013). The fact that both fibrillarin and p2 are required for the systemic movement of RSV and the finding that p2 requires fibrillarin to target the nucleolus point to a scenario in which RSV p2 may manipulate or recruit nucleolar functions to promote virus systemic infection. For example, p2 may need to enter the nucleolus to undergo necessary modifications or processing for appropriate cytoplasmic distribution. The processing or modification may be important for the normal function of p2. Alternatively, p2 may need to recruit some unknown factors localized in the nucleolus for its functions in the cytoplasm. It is also possible that p2 may enter the nucleolus to disturb certain host functions which normally restrict viral infection. However, further studies are needed to provide direct evidence showing that p2 needs to target the nucleolus to promote RSV systemic movement. Overall, the results of this study are consistent with the emerging notion that interaction with the nucleolus is a pan-virus phenomenon, even for plant viruses (Taliansky et al., 2010). In addition, the data presented here deepen our understanding of RSV–host interaction and provide potential targets for the design of new anti-viral strategies to control RSV.

4 Experimental procedures

4.1 Plasmid construction

Total RNA was extracted from *N. benthamiana* leaves and RSV-infected rice leaves using an EasyPure Plant RNA Kit manufactured by Bejing Transgen Biotech Co. Ltd. (Beijing, China). Reverse transcription was carried out using a Fast Quant RT Kit with gDNase (Tiangen Biotech Co., Ltd., Beijng, China). NbFib2 and RSV-NS2 (GenBank accession: AM269909 and EF198702) were amplified by PCR using the primers listed in Table S1 (see Supporting Information). The vectors (pEarleyGate 101-p2/NbFib2, pEarleyGate 102-p2/NbFib2, pEarleyGate 201-YC-p2/NbFib2 and pEarleyGate 201-YN-p2/NbFib2) used in cellular localization and BiFC were generated using Gateway recombination technology (Invitrogen China, Shanghai, China) by inserting RSV-NS2 and NbFib2 into the entry vector pDONR 221 (Karimi et al., 2002). For VIGS assays, RSV-NS2, RSVNS3 (GenBank accession: EF493242) and NbFib2 were amplified by the primers listed in Table S1. The PCR products were digested with *EcoR* I and *BamH* I, and ligated into the TRV vector pTRV2 digested with the same enzymes. All the plasmid constructs used in this study were confirmed by sequencing (Takara, Dalian, China).

4.2 Agrobacterium-mediated transient expression

Agrobacterium tumefaciens strains EHA105 or GV3101 carrying the genes of interest were grown separately to an optical density at 600 nm (OD_{600}) of 0.8 at 28°C on Luria–Bertani liquid medium supplemented with 50μg/μL of rifampicin and 50μg/μL of kanamycin. The cultures were centrifuged at 12000g for 1min and resuspended in induction medium [10mM 2-(N-morpholino) ethanesulfonic acid (MES), pH 5.6, 10mM $MgCl_2$ and 150μM acetosyringone]. In co-localization and BiFC assays, *A. tumefaciens* strains containing different constructs were mixed in equal volume. In VIGS assays, *A. tumefaciens* carrying pTRV2-NbPDS, pTRV2-NS2, pTRV2-NbFib2 or empty pTRV2 was mixed with an equal

volume of TRV1. The mixtures of the bacterial cultures were incubated at room temperature for 3h, and then infiltrated into fully grown upper leaves. Six-week-old *N. benthamiana* plants were used for the experiments.

4.3 Confocal imaging analysis

The fluorophores in CFP and YFP were excited at 458 and 514nm, and images were taken using BA480–495- and BA535–565-nm emission filters, respectively.

4.4 Quantitative RT-PCR

The total RNAs of the upper leaves of NbFib2- or NS2-silenced plants were extracted and RT-PCRs were performed as described previously. The primer pairs NS2-qPCRF/NS2-qPCRR, Fib2 qPCRF/Fib2-qPCRR, NS3-qPCRF/NS3- qPCRR and 18S-qPCRF/18S-qPCRR were used to detect the expression of NS2, NbFib2 and NS3. 18S RNA of *N. benthamiana* was used as a reference gene (Eppendorf China, Shanghai, China, AG 22331 Real-Time PCR System, Hamburg No. 5345 018780). The 20-μL PCR included 1.0μL of RT product, 10μL of SYBR qPCR Mix, 0.8μL of primers (Table S1) and 9.2μL of diethylpyrocarbonate (DEPC) water; the reactions were incubated in a 96-well optical plate at 95°C for 1min, followed by 40 cycles of 95°C for 15s and 60°C for 1min (TOYOBO, Shanghai, China, No. QPS-201). The Ct data were determined using default threshold settings. The threshold cycle (Ct) is defined as the fractional cycle number at which the fluorescence passes the fixed threshold.

4.5 Virus inoculation

For RSV inoculation assays, a Jiangsu isolate of RSV (LS-JSJJ03, EF198702, EF198682, EF198733, EF198702), maintained in our laboratory through transmission using *L. striatellus*, was employed in this study. RSV inoculation was performed as described previously using crude extracts of RSV-infected rice (Zhang et al., 2012).

PVX-GFP and TMV-GFP were inoculated by agroinfiltration. GFP fluorescence was observed under a hand-held, 100-W, long-wave UV lamp (UV Products, Upland, CA, USA). Then, fluorescent leaves were collected and ground. The grinding solution was dipped on the copper grid with a Formvar membrane for 1–2min, washed using double distilled water, stained with 2% uranyl acetate and observed under an H-7650 electron microscope (Hitachi, Japan).

4.6 Detection of RSV accumulation

Inoculated and upper non-inoculated leaves of *N. benthamiana* infected by RSV were collected individually. RT-PCR using a primer pair amplifying the RSV coat protein (CP) was employed to detect RSV.

For Western blotting, inoculated and non-inoculated leaves were homogenized in 2mL of extraction buffer containing 50mM phosphate (pH8.0), 10mM tris(hydroxymethyl)aminomethane (Tris) (pH 8.0), 500 mMNaCl,

0.1% Tween 20, 0.1% Nonidet P40 (NP-40), 0.1% β-mercaptoethanol, 1 mMphenylmethylsulfonylfluoride (PMSF) and onequarter of a Roche Protease inhibitor cocktail MINI tablet. The crude extracts were centrifuged at 12 000 g for 10 min. The supernatant was transferred into a new centrifuge tube and centrifuged at 12 000 g for another 15min; 10μL of supernatant were mixed with 2μL of 5 × sodium dodecylsulfate-polyacrylamide gel electrophoresis (SDS-PAGE) loading buffer. Proteins in the extracts were separated by electrophoresis by 12% SDS-PAGE at 80 V for 1 h and then at 120 V for another 40 min. Proteins in the gels were transferred onto polyvinylidene difluoride (PVDF) membranes by electrophoresis cell at 60V for 1h (Beijing WoDe Life Sciences Instrument Company, Beijing, China) and probed with RSV-P4 polyclonal antibody. The polyclonal antibody was a goat–anti-rabbit immunoglobulin G (IgG) conjugated with alkaline phosphatase (Sigma, St. Louis, MO, USA) and used at 1:10 000 (v/v) dilution. Proteins on the membrane were visualized by nitroblue tetrazolium–5-bromo-4-chloroindol-3-yl phosphate (NBT-BCIP) solution (Shanghai Promega Biological Products, Co, Ltd Shanghai, China).

Acknowledgements

We are grateful to Professor Xinzhong Cai (Zhengjiang University, Hangzhou, China) for providing the TRV vector and plasmid pTRV2-NbPDS, and to Dr Aiming Wang (AAFC-Southern Crop Protection and Food Research Centre, Canada) for providing the Gateway vectors for localization and BiFC studies. This work was supported by grants from the National Basic Research Program 973 (2014CB138402, 2014CB138403 and 2010CB126203), the Natural Science Foundation of China (31401715, 31171821 and 31272018), the Key Project of the National Research Program of China (2012BAD19B03) and the Doctoral Fund of the Ministry of Education of China (20113515110001). The authors have no conflicts of interest to declare.

References

[1] AHLQUIST P. RNA-dependent RNA polymerases, viruses, and RNA silencing[J]. Science, 2002, 296:1270-1273.

[2] BARNECHE F, STEINMETZ F, ECHEVERRI'A M. Fibrillarin genes encode both a conserved nucleolar protein and a novel small nucleolar RNA involved in ribosomal RNA methylation in *Arabidopsis thaliana*[J]. Journal of Biological Chemistry, 2000, 275(35):27212-27220.

[3] BITKO V, BARIK S. Phenotypic silencing of cytoplasmic genes using sequence-specific double-stranded short interfering RNA and its application in the reverse genetics of wild type negative-strand RNA viruses[J]. BMC Microbiology, 2001, 1(1):34-34.

[4] BOISVERTF M, VAN KONINGSBRUGGENS, NAVASCUESJ, et al. The multifunctional nucleolus. Nature Reviews Molecular Cell Biology, 2007, 8: 574-585.

[5] BOULON S, WESTMAN B J, HUTTEN S, et al. The nucleolus under stress[J]. Molecular Cell, 2010, 40(2):216-227.

[6] CHO W K, LIAN S, KIM S M, et al. Current insights into research on *Rice stripe virus*[J]. Plant Pathology Journal, 2013, 29(3):223-233.

[7] CIOCE M, LAMOND A I. Cajal bodies: a long history of discovery[J]. Annual Review of Cell and Developmental Biology, 2005, 21(1):105-131.

[8] DÍAZ-PENDÓN J A, DING S W. 2008. Direct and indirect roles of viral suppressors of RNA silencing in pathogenesis[J]. Annual Review of Phytopathology, 2008, 46(1):303-326.

[9] DU Z, XIAO D, WU J, et al. p2 of *Rice stripe virus*(RSV) interacts with OsSGS3 and is a silencing suppressor[J]. Molecular Plant Pathology, 2011, 12(8):808-814.

[10] EARLEY K W, HAAG J R, PONTES O, et al. Gateway-compatible vectors for plant functional genomics and proteomics[J]. The Plant Journal, 2010, 45(4):616-629.

[11] FALK B W, TSAI J H. Biology and molecular biology of viruses in the genus *Tenuivirus*[J]. Annual Review of Phytopathology, 1998, 36(1):139-163.

[12] HAYANO Y, KAKUTANI T, HAYASHI T, et al. Coding strategy of *Rice stripe virus*: major nonstructural protein is encoded in viral RNA segment 4 and coat protein in RNA complementary to segment 3[J]. Virology, 1990, 177(1):372-374.

[13] HIBINO H. Biology and epidemiology of rice viruses[J]. Annual Review of Phytopathology, 1996, 34(1):249-274.

[14] HIPPERC,BRAULTV,ZIEGLER-GRAFFV, et al. Viral and cellular factors involved in phloem transport of plant viruses[J]. Frontiers in Plant Science, 2013, 4:154.

[15] HISCOX J A. RNA viruses: hijacking the dynamic nucleolus[J]. Nature Reviews Microbiology, 2007, 5(2):119-127.

[16] JIANG S, LU Y, LI K, et al. Heat shock protein 70 is necessary for *Rice stripe virus* infection in plants[J]. Molecular Plant Pathology, 2015, 15(9):907-917.

[17] KAKUTANIT, HAYANOY, HAYASHIT, et al. Ambisense segment 4 of rice stripe virus: possible evolutionary relationship with *phleboviruses* and *uukuviruses* (*Bunyaviridae*)[J]. Journal of General Virology, 1990, 71(7):1427-1432.

[18] KAKUTANIT,HAYANOY,HAYASHIT, et al. Ambisense segment 3 of *Rice stripe virus*: the first case of a virus containing two ambisensesegments[J]. Journal of General Virology, 1991, 72(2):465-468.

[19] KARIMIM,INZÉD, DEPICKERA. GATEWAY vectors for *Agrobacterium*-mediated plant transformation[J]. Trends in Plant Science, 2002, 7(5):193-195.

[20] KIMS H,RYABOVE V,KALININAN O, et al. Cajal bodies and the nucleolus are required for a plant virus systemic infection[J]. The EMBO Journal, 2007a, 26(8):2169-2179.

[21] KIMS H,MACFARLANES,KALININAN O,et al. Interaction of a plant virus-encoded protein with the major nucleolar protein fibrillarin is required for systemic virus infection[J]. Proceedings of the National Academy of Sciences of the United States of America, 2007b, 104(26):11115-11120.

[22] KMATTHEWS R. Virus taxonomy: classification and nomenclature of viruses: Ninth report of the international committee on taxonomy of viruses[J]. Intervirology, 1979, 12(3-5):129-296.

[23] KOBAYASHI T, CHAPPELL J D, DANTHI P, et al. Gene-specific inhibition of reovirus replication by RNA interference[J]. Journal of Virology, 2006, 80(18):9053-9063.

[24] KONGL,WUJ,LUL,et al.Interaction between *Rice stripe virus* disease-specific protein and host PsbP enhances virus symptoms[J]. Molecular Plant, 2014, 7(4):691-708.

[25] KORMELINK R, GARCIA M L, GOODIN M, et al. Negative-strand RNA viruses: the plant-infecting counterparts[J]. Virus Research, 2011, 162(1-2):184-202.

[26] LIAN S, CHO WK, JO Y, et al. Interaction study of *Rice stripe virus* proteins reveals a region of the nucleocapsid protein NP required for NP self-interaction and nuclear localization[J]. Virus Research, 2014, 183:6-14.

[27] LIN Q T, LIN H X, WU Z J, et al. Accumulations of coat protein and disease-specific protein of *Rice stripe virus* in its host[J]. Journal of Fujian Agricultural University, 1998, 27:257-260.

[28] MARTÍNEZ F, DARÒS J A. Tobaccoetch virus protein P1 traffics to the nucleolus and associates with the host 60S ribosomal subunits during infection[J]. Journal of Virology, 2014, 88(18):10725-10737.

[29] NGUYENM, HAENNIA L. Expression strategies of ambisense viruses[J]. Virus Research, 2003, 93(2):141-150.

[30] RAJAMÄKI M L, VALKONEN J P. Control of nuclear and nucleolar localization of nuclear inclusion protein a of picorna-like potato virus A in *Nicotiana species*[J]. Plant Cell, 2009, 21(8):2485-2502.

[31] RAMÍREZB C, HAENNIA L.Molecular biology of tenuiviruses, are markable group of plant viruses[J]. Journal of General Virology, 1994, 75(3):467-475.

[32] RENC,CHENGZ,MIAOQ,et al. First report of *Rice stripe virus* in Vietnam[J]. Plant Disease, 2013, 97(8):1123-1123.

[33] ROSSI M, GENRE A, TURINA M. Genetic dissection of a putative nucleolar localization signal in the coat protein of ourmia

melon virus[J]. Archives of Virology, 2014, 159(5):1187-1192.

[34] SALVETTI A, GRECO A. Viruses and the nucleolus: the fatal attraction[J]. Biochimica et Biophysica Acta (BBA) - Molecular Basis of Disease, 2014, 1842(6):840-847.

[35] SEMASHKO M A, GONZÁLEZ I, SHAW J, et al. The extreme N-terminal domain of a hordeivirus TGB1 movement protein mediates its localization to the nucleolus and interaction with fibrillarin[J]. Biochimie, 2012, 94(5):1180-1188.

[36] SHAW P, BROWN J. Nucleoli: composition, function, and dynamics[J]. Plant physiology, 2012, 158(1):44-51.

[37] SHIMIZU T, NAKAZONO-NAGAOKA E, UEHARA-ICHIKI T, et al. Targeting specific genes for RNA interference is crucial to the development of strong resistance to *Rice stripe virus*[J]. Plant Biotechnology Journal, 2011, 9(4):503-512.

[38] SUN F, YUAN X, ZHOU T, et al. Arabidopsis is susceptible to *Rice stripe virus* infections[J]. Journal of Phytopathology, 2011, 159(11-12):767-772.

[39] TAKAHASHI M, TORIYAMA S, HAMAMATSU C, et al. Nucleotide sequence and possible ambisense coding strategy of *Rice stripe virus* RNA segment 2[J]. Journal of General Virology, 1993, 74(4):769-773.

[40] TALIANSKY M E, BROWN J W S, RAJAMÄKI M L, et al. Involvement of the plant nucleolus in virus and viroid infections: parallels with animal pathosystems[J]. Advances in Virus Research, 2010, 77:119-158.

[41] TORIYAMA S, TAKAHASHI M, SANO Y, et al. 1994. Nucleotide sequence of RNA. 1, the largest genomic segment of *rice stripe virus*, the prototype of the tenuiviruses. Journal of General Virology, 75:3569-3580.

[42] WALTER M, CHABAN C, SCHÜTZE K, et al. Visualization of protein interactions in living plant cells using bimolecular fluorescence complementation[J]. Plant Journal, 2010, 40:428-438.

[43] WANG X, ZHANG Y, XU J, et al. The R-rich motif of *Beet black scorch virus* P7a movement protein is important for the nuclear localization, nucleolar targeting and viral infectivity[J]. Virus Research, 2012, 167(2):207-218.

[44] WEI T Y, YANG J G, LIAO F L, et al. Genetic diversity and population structure of *Rice stripe virus* in China[J]. Journal of General Virology, 2009, 90(4):1025-1034.

[45] WRIGHT K M, COWAN G H, LUKHOVITSKAYA N I, et al. The N-terminal domain of PMTV TGB1 movement protein is required for nucleolar localization, microtubule association, and long-distance movement[J]. Molecular plant-microbe interactions : MPMI, 2010, 23(11):1486-1497.

[46] WU G, LU Y, ZHENG H, et al. Transcription of ORFs on RNA2 and RNA4 of *Rice stripe virus* terminate at an AUCCGGAU sequence that is conserved in the genus Tenuivirus[J]. Virus Research, 2013, 175(1):71-77.

[47] XIONG R, WU J, ZHOU Y, et al. Identification of a movement protein of the tenuivirus *Rice stripe virus*[J]. Journal of Virology, 2008, 82(24):12304-12311.

[48] XIONG R, WU J, ZHOU Y, et al. Characterization and subcellular localization of an RNA silencing suppressor encoded by *Rice stripe tenuivirus*[J]. Virology, 2009, 387(1):29-40.

[49] XU Y, ZHOU X. Role of *Rice stripe virus* NSvc4 in cell-to-cell movement and symptom development in *Nicotiana benthamiana*[J]. Frontiers in Plant Science, 2012, 3:269.

[50] YAO M, LIU X, LI S, et al. *Rice stripe tenuivirus* NSvc2 glycoproteins targeted to the Golgi body by the N-terminal transmembrane domain and adjacent cytosolic 24 amino acids via the COP I- and COP II-dependent secretion pathway[J]. Journal of Virology, 2014,88: 3223-3234.

[51] ZHANG C, PEI X, WANG Z, et al. *The Rice stripe virus* pc4 functions in movement and foliar necrosis expression in *Nicotiana benthamiana*[J]. Virology, 2012, 425(2):113-121.

[52] ZHAO S L, DAI X J, LIANG J S, et al. Surface display of *Rice stripe virus* NSvc2 and analysis of its membrane fusion activity[J]. Virologica Sinica, 2012, 27(2):100-108.

[53] ZHU Y, HAYAKAWA T, TORIYAMA S, et al. Complete nucleotide sequence of RNA3 of *Rice stripe virus*: an ambisense coding strategy[J]. Journal of General Virology, 1991, 72(4):763-767.

Assembly of viroplasms by viral nonstructural protein Pns9 is essential for persistent infection of rice gall dwarf virus in its insect vector

Limin Zheng, Hongyan Chen, Huamin Liu, Lianhui Xie, Taiyun Wei

(Fujian Province Key Laboratory of Plant Virology, Institute of Plant Virology, Fujian Agriculture and Forestry University, Fuzhou, Fujian 350002, PR China)

Abstract: Rice gall dwarf virus (RGDV), a plant reovirus, is transmitted by leafhopper vector *Recilia dorsalis* in a persistent-propagative manner. In a sequential study of RGDV infection of its insect vector, the virus initially infected the filter chamber epithelium, then directly crossed the basal lamina into the visceral muscles, from where it spread throughout the entire midgut and hindgut. Finally, RGDV spread into the salivary glands. During RGDV infection of the continuous cultured cells of *R. dorsalis*, viroplasm that was mainly comprised of viral nonstructural protein Pns9 was formed and acted as the site of viral replication and assembly of progeny virions. Knockdown of Pns9 expression in cultured insect vector cells using synthesized dsRNAs from the Pns9 gene strongly inhibited viroplasm formation and viral infection. The microinjection of dsRNAs from the Pns9 gene strongly abolished viroplasm formation in the initially infected filter chamber epithelium and prevented viral spread into leafhopper visceral muscles. These results indicated that the assembly of viroplasms was essential for the persistent infection and spread of RGDV in its insect vector.

Keywords: Rice gall dwarf virus; leafhopper vector; viral infection route; Pns9 protein; viroplasm formation; RNA interference.

1 Introduction

Transmission by insect vectors is essential for the infection cycle of numerous viruses that cause diseases in humans, animals and plants. Hemipteran insects, including leafhoppers, planthoppers, aphids and whiteflies, are global pests and vectors that transmit plant viruses in a persistent manner (Hogenhout et al., 2008). Rice gall dwarf virus (RGDV), a member of the genus *Phytoreovirus* in the family *Reoviridae*, is mainly transmitted by its leafhopper vector *Recilia dorsalis* (Hemiptera: Cicadellidae) in a persistent-propagative manner (Boccardo and Milne, 1984). RGDV is first described in 1979 in Thailand (Omura et al., 1980) and causes a severe disease of rice in China and Southeast Asia (Putta et al., 1980; Ong and Omura, 1982; Fan et al., 1983). In the past several years, RGDV has spread to Guangdong and Guangxi provinces of China (Zhang et al., 2008; Fan et al., 2010). As a persistent propagative virus, RGDV must enter the insect intestine to establish productive infection, suggesting that the virus must replicate and assemble progeny virions in insect vectors. Understanding the mechanisms on how RGDV infect, spread and propagate in its leafhopper vector should lead to better strategies to disrupt the efficient transmission of RGDV. RGDV has icosahedral and double-shelled particles approximately 65–70nm in diameter (Boccardo and Milne, 1984). The viral genome consists

Virus Research.2015, 196: 162-169

Received 2 September 2014; Accepted 23 November 2014; Available online 29 November 2014 of 12 segments (S1 through S12) of double-stranded RNAs (dsRNAs) (Omura et al., 1982; Hibi et al., 1984). S1, S2, S3, S5, S6 and S8 encode structural proteins P1, P2, P3, P5, P6 and P8, respectively (Hibi et al., 1984; Boccardo et al., 1985). The other segments of RGDV genome encode nonstructural proteins (Pns4, Pns7, Pns9, Pns10, Pns11 and Pns12) (Hibi et al., 1984; Boccardo et al., 1985; Koganezawa et al., 1990; Noda et al., 1991; Moriyasu et al., 2000, 2007). The formation of viral inclusions composed of nonstructural proteins is essential for the replication or spread of other persistent-propagative viruses in the bodies of insect vectors (Hogenhout et al., 2008). The nonstructural proteins Pns7, Pns9 and Pns12 of RGDV are the components of the viroplasm, the site for viral replication and assembly of progeny virions in insect vector cells (Wei et al., 2009; Akita et al., 2011). When Pns9 of RGDV is expressed alone, viroplasm-like inclusions can form in the non-vector insect *Spodoptera frugiperda*(Sf9) cells, suggesting that Pns9 is the main viral factor responsible for the formation of viroplasm in infected cells (Akita et al., 2011). Pns11 forms virus-containing tubules to facilitate viral spread among insect vector cells (Chen et al., 2013). The functions of Pns4 and Pns10 are unknown. The functional roles of the viral inclusions formed by viral nonstructural proteins during persistent infection of RGDV in its insect vector are poorly understood.

To further investigate the functional roles of viral proteins in the infection cycle of RGDV in insect vector, here we used RNA interference (RNAi), a conserved sequence-specific gene silencing mechanism that is induced by dsRNAs (Fire et al., 1998), to knock down the expression of viral genes *in vivo*and *in vitro*. RNAi strategies have been used to inhibit the infection of persistent-propagative viruses in insect vectors (Chen et al., 2012, 2013, 2014; Jia et al., 2012a, b, 2014; Wu et al., 2014). In this study, we determined that the assembly of the viroplasm by viral nonstructural protein Pns9 is essential for persistent infection of RGDV in its insect vector.

2 Materials and methods

2.1 Virus, cells and antibodies

RGDV samples were collected in rice fields from Guangdong Province in southern China. Continuous monolayer cultures of the vector cells (VCM) were developed from the leafhopper *R. dorsalis* and maintained on the growth medium as described previously (Chen et al., 2013). Polyclonal antibodies against the nonstructural protein Pns9, viral antigens and the major outer capsid protein P8 of RGDV were prepared as described previously (Jia et al., 2012a). IgGs were isolated from specific polyclonal antiserum using a proteinA-Sepharose affinity column (Pierce) and then directly conjugated to fluorescein isothiocyanate (FITC) or rhodamine (Invitrogen) (Jia et al., 2012a).

2.2 Immunofluorescence staining of internal organs of leafhoppers after acquisition of virus

Second-instar nymphs of leafhopper *R. dorsalis* were fed for 2d on rice plants infected with RGDV and then transferred to healthy rice seedlings. At different days after viral acquisition by leafhoppers, internal organs including intestines, hemolymphs and salivary glands from 30 leafhoppers were dissected, fixed in 4% paraformaldehyde in phosphate-buffered saline (PBS) (137mM NaCl, 8.1mM Na_2HPO_4, 2.7mM KCl, 1.5mM KH_2PO_4, pH7.4) and permeabilized at room temperature in 2% Triton X-100 in PBS. The internal organs were then treated with viral antigens-specific IgG conjugated to rhodamine (virus-rhodamine) and actin dyephalloidin-FITC (Invitrogen), as described previously (Chen et al., 2011). The samples were then imaged with a Leica TCS SP5 inverted confocal microscope, as described previously (Chen et al., 2011). As controls, internal organs were dissected from leafhopper that had fed on healthy rice plants and treated the same way.

2.3 Effect of synthesized dsRNAs from the Pns9 gene of RGDV on viral infection in the continuous cell cultures derived from leafhopper

A 972-bp segment of the Pns9 gene of RGDV and a 717-bpsegment of the GFP gene were amplified by PCR as templates for the synthesis of dsRNAs. A T7 RiboMAX Express RNA interference (RNAi) System kit (Promega) was used to synthesize dsRNAs *in vitro* for these two genes according to the manufacturer's instructions. For examining the effects of synthesized dsRNAs targeting GFP (dsGFP) or Pns9 (dsPns9) genes on viral infection, VCMs derived from *R. dorsalis*were transfected with 0.5μg/μL dsRNAs in the presence of Cellfectin (Invitrogen) for 8h and then infected with RGDV at RGDV at a multiplicity of infection (MOI) of 10 for 2h, as described previously (Chen et al., 2013). At 72h post-inoculation (hpi), VCMs were fixed, immunolabeled with virus-rhodamine and Pns9-specific IgG conjugated to FITC (Pns9-FITC) and then processed for immunofluorescence microscopy. Furthermore, at 72hpi, total proteins were extracted from infected leafhopper cells and further analyzed by immunoblotting with Pns9 or P8-specific antibodies, respectively. Insect actin was detected with actin-specific antibodies as a control to confirm loading of equal amounts of proteins in each lane.

2.4 Effect of synthesized dsRNAs from the Pns9 gene of RGDV on viral infection in the leafhopper

The dsRNAs were microinjected into leafhopper *R. dorsalis* using the method reported for the brown planthopper (Liu et al., 2010). The second-instar nymphs of leafhoppers were kept on RGDV-infected rice for 2d, and were microinjected with 0.5μg/μL dsPns9 and dsGFP into the thorax using a dissecting microscope and a CellTram Oil injector (Eppendorf). After microinjection, leafhoppers were transferred to healthy rice plants. At 4, 10 and 15d post-first access to diseased plants (padp), the internal organs from 30 leafhoppers that received dsRNAs were dissected, fixed, and then processed for immunofluorescence microscopy. At 15d padp, the accumulation of Pns9 and P8 of RGDV were analyzed with a Western blot assay using Pns9- and P8-specific IgGs, respectively.

3 Results

3.1 Infection route of RGDV in the body of its insect vector

To trace the infection route of RGDV within infected leafhoppers, we used immunofluorescence microscopy to elucidate the distribution of viral antigens in the bodies of leafhoppers that have ingested RGDV from diseased plants. The second-instar nymphs were given an acquisition access period of 2d, then transferred to healthy rice. At 2, 4, 6, 9 and 12d padp, internal organs from 30 leafhoppers were dissected and processed for immunofluorescence microscopy. The intestine of the leafhopper *R. dorsalis* consists of esophagus, filter chamber, midgut and hindgut (Fig. 1A). The anterior and posterior midgut and the anterior hindgut were enclosed in the sheathed filter chamber (Fig. 1A–E, G). In addition, the leafhopper intestine is composed of a single layer of epithelial cells with microvilli on the lumen side and basal lamina on the outer side which is covered with muscle fibers (Chen et al., 2011). As early as 2d padp, RGDV was first observed to accumulate in a particular corner of the filter chamber in 73% of leafhoppers (Fig. 1B; Table 1), suggesting that RGDV ingested from diseased rice plants has crossed the microvilli into the epithelial regions of the filter chamber. At 4d padp, RGDV had spread from the initially infected regions of the filter chamber into the visceral muscle tissues in 67% of leafhoppers (Fig. 1C, D; Table 1). At this time, RGDV was first observed in the hemolymphs in about 20% of leafhoppers (Fig. 2; Table 1), suggesting that RGDV had spread from the visceral muscle tissues into the hemolymph. At 6d padp, RGDV had spread from the infected visceral muscle tissues of the filter chamber into the adjacent tissues including the anterior and posterior midgut and the anterior hindgut in 50% of leafhoppers (Fig. 1E; Table 1). At this time, RGDV was distributed in the visceral muscle tissues of the intestine, but was undetectable in the epithelium regions (Fig. 1E), suggesting that RGDV has the ability to spread along the muscle fibers of the filter chamber through the adjacent intestinal muscle tissues. At 9d padp, RGDV was first observed in the salivary glands in about 20% of leafhoppers (Fig.1F; Table 1). At 12d padp, RGDV had spread into the visceral muscle throughout the midgut and hindgut in 83% of leafhoppers (Fig. 1G; Table 1). The virus also distributed extensively in the hemolymphs and salivary glands in about 50% of leafhoppers (Fig. 2; Table 1). Taken together, these results revealed an infection route of RGDV in the internal organs of its vector after virus ingestion. RGDV initially infected the filter chamber epithelium, then crossed the basal lamina into the visceral muscles, from where it spread into the hemolymph or throughout the entire midgut and hindgut. Finally, RGDV spread into the salivary glands.

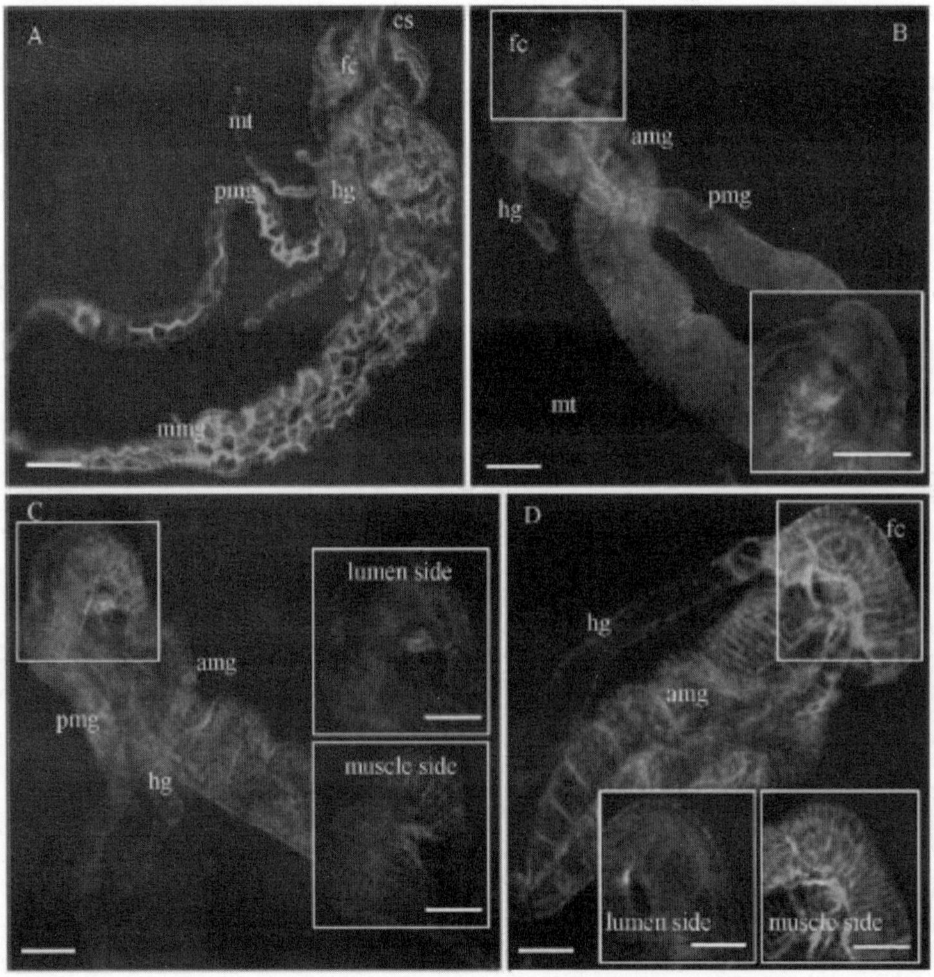

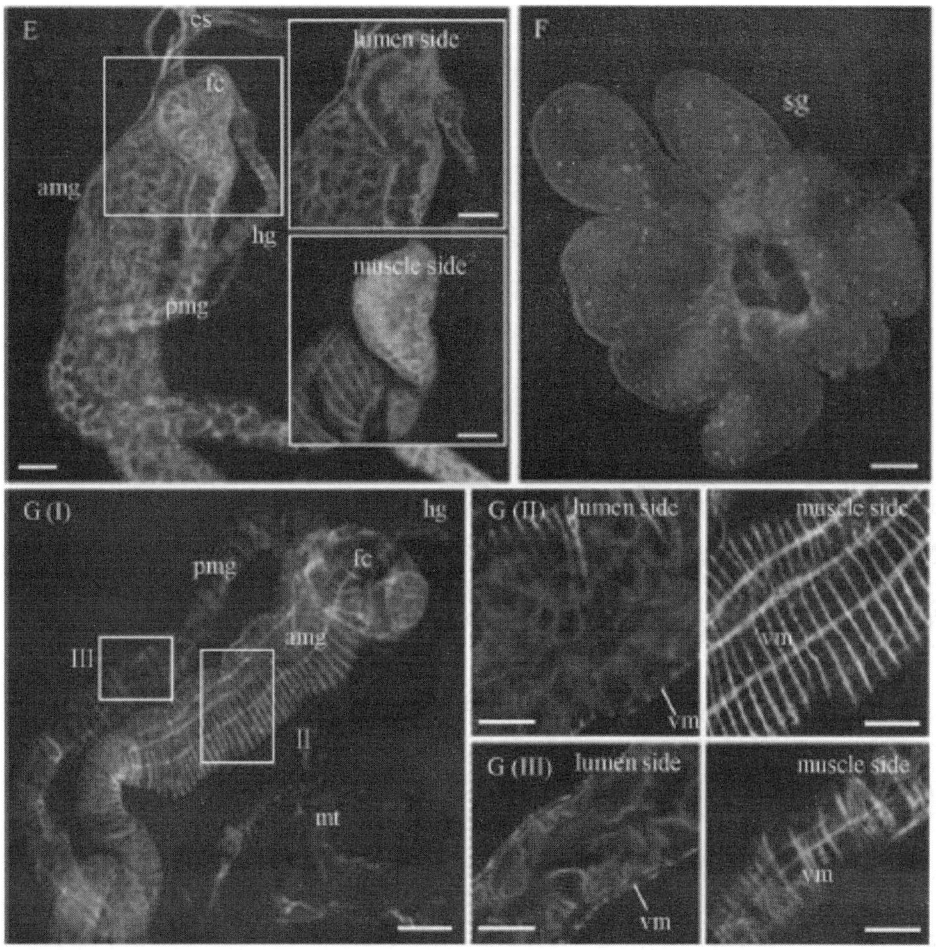

Fig. 1 Confocal micrographs of internal organs of the leafhopper *R. dorsalis* immunolabeled with virus-rhodamine (red) and actin dye phalloidin-FITC (green) to show infection route of RGDV in its insect vector

Each image was projected from 10 optical sections at 0.5mm intervals. (A) Alimentary canal of the leafhopper fed on the healthy rice plants. (B) At 2d padp, RGDV was detected in the filter chamber. Inset: enlarged image of boxed area. (C and D) By 4d padp, RGDV had spread from the epithelium to visceral muscle tissues of filter chamber. Insets: single optical section of different sides of the filter chamber. (E) By 6d padp, RGDV was detected in the esophagus, filter chamber, parts of the anterior midgut, posterior midgut and hindgut. Insets: single optical section of different sides of the boxed area. (F) By 9d padp, RGDV had spread into salivary glands. (G) At 12d padp, RGDV was detected in the esophagus, filter chamber, the whole midgut, hindgut and malpighian tubules. Panels II, III: enlargements of different sides of anterior midgut (II) and posterior midgut (III) at boxed areas in panel I. es, esophagus; fc, filter chamber; mg, midgut; amg, anterior midgut; mmg, middle midgut; pmg, posterior midgut; hg, hindgut; mt, Malpighian tubules; vm, visceral muscle; sg, salivary glands. Bars, 100μm

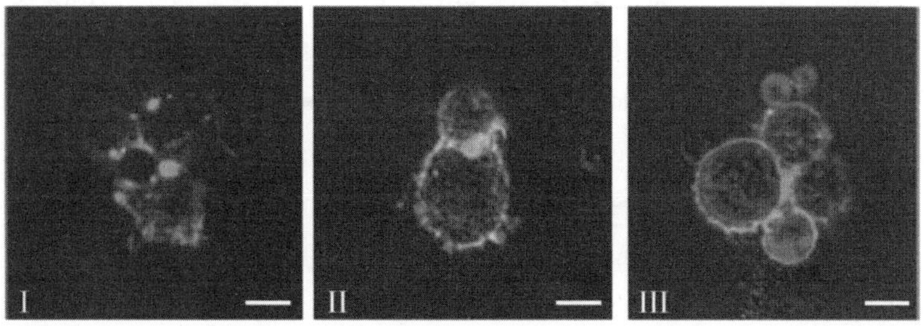

Fig. 2 Detection of viral accumulation in the leafhopper hemolymphs by immunofluorescence microscopy

At 2, 4 and 12d padp (panels I–III, respectively), the hemolymphs isolated from leafhoppers were immunolabeled with virus-rhodamine (red) and actin dye phalloidin-FITC (green). Images were merged with green fluorescence (actin) and red fluorescence (virus antigens) under background visualized by transmitted light. Images are representative of multiple experiments with multiple preparations. Bars, 10μm (For interpretation of the references to color in this figure legend, the reader is referred to the web version of this article)

3.2 RNAi induced by dsPns9 inhibited the assembly of viroplasm and viral infection in insect vector cells

In order to investigate the functional role of the nonstructural protein Pns9 of RGDV during the early process of viral infection in its leafhopper vector cells, we examined whether RNAi induced by dsPns9 could effectively inhibit viroplasm formation and viral infection in VCMs derived from *R. dorsalis*. Eight hours after dsPns9 transfection, the VCMs were inoculated with RGDV (MOI=10) and examined with immunofluorescence using virus-rhodamine and Pns9-FITC. At 72hpi, immunofluorescence microcopy showed that Pns9 formed granular inclusions throughout the cytoplasm, whereas viral antigens were localized to fibrillar structures at the margins of the granular inclusions of Pns9 in the cells transfected with dsGFP (Fig. 3A). However, the treatment with dsPns9 strongly inhibited the accumulation of viroplasms and viral antigens (Fig. 3A). As expected, the expression of the nonstructural protein Pns9 and the major outer capsid protein P8 decreased significantly in VCMs transfected with dsPns9 (Fig. 3B). These results suggested that the knockdown of Pns9 expression due to RNAi induced by dsPns9 significantly inhibited viroplasm formation and early viral infection in VCMs.

3.3 RNAi induced by dsPns9 inhibits the spread of RGDV from the initially infected site in the filter chamber of its insect vector

An RNAi strategy was used to investigate the functional role of RGDV Pns9 in the process of viral infection in the body of its leafhopper vector. Second-instar nymphs of *R. dorsalis* were microinjected with dsGFP or dsPns9, allowed a 2-day acquisition on virus-infected rice plants and then fed on healthy rice seedlings. At 4d padp, viral antigens and viroplasms containing Pns9 had accumulated in the visceral muscle tissues of filter chamber in 67% of leafhoppers receiving dsGFP (Fig. 4A; Table 2). However, viral antigens were restricted to the initially infected epithelial regions of the filter chamber in 30% of leafhoppers that received dsPns9 (Fig. 4A; Table 2). Furthermore, the treatment with dsPns9 significantly inhibited the formation of viroplasms in these initially infected regions (Fig. 4A). At 10d padp, viral antigens and viroplasms containing Pns9 were seen in the visceral muscle tissues of the entire intestine in 83% of leafhoppers receiving dsGFP (Fig. 4B; Table 2). However, in 30% of leafhoppers receiving dsPns9, viral infection was still restricted in the epithelial regions of the filter chamber, where the viroplasm formation was strongly inhibited (Fig. 4B; Table 2). At 15d padp, viral antigens and viroplasms containing Pns9 had accumulated in the salivary glands in 57% of leafhoppers that received dsGFP (Fig. 4C; Table 2). As expected, the treatment of dsPns9 inhibited the spread of RGDV in the salivary glands of its insect vector (Fig. 4C; Table 2). The decrease in the expression of Pns9 and P8 of RGDV in leafhoppers that received dsPns9 was confirmed using a Western blot assay (Fig. 4D). The above results indicated that the treatment of dsPns9 specifically knocked down the expression of Pns9 in the initially infected regions of the filter chamber, which led to the restriction of viral infection in the filter chamber and blocked viral spread to the salivary glands.

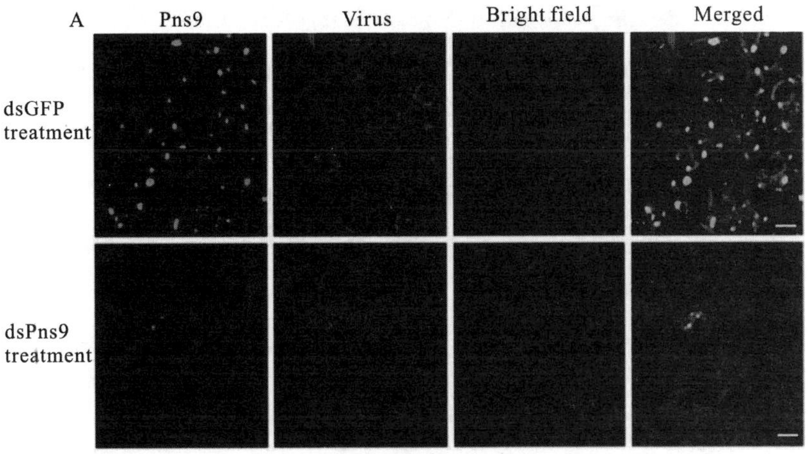

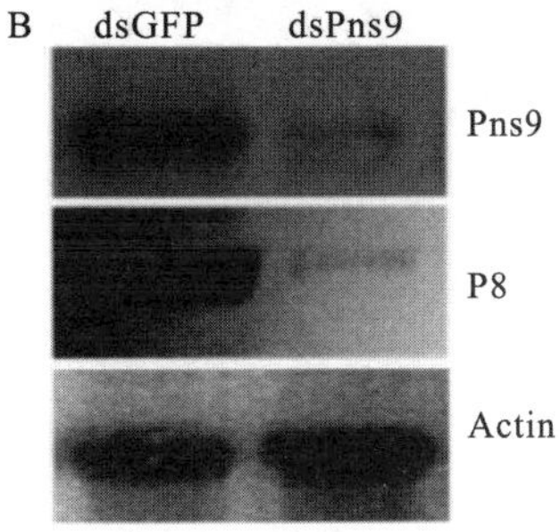

Fig. 3 RNAi induced by dsPns9 inhibiting the assembly of viroplasms and viral infection in RGDV-infected VCMs

(A) Eight hours after transfection with dsGFP or dsPns9, VCMs were infected with RGDV. At 72hpi, VCMs were immunolabeled with Pns9-FITC (green) and virus-rhodamine (red), and then examined with confocal microscopy. Images were shown with green fluorescence (Pns9 antigens) and red fluorescence (virus antigens) under background visualized by transmitted light. (B) At 72hpi, Western blotting test indicated that transfection of dsPns9 in VCMs reduced the expressions of Pns9 and P8 proteins in virus-infected VCMs. Bars, 10μm (For interpretation of the references to color in this figure legend, the reader is referred to the web version of this article)

Table 2 Treatment with dsPns9 via microinjection strongly inhibited viral infection and spread in the bodies of leafhoppers

Insects	No. of Pns9- or viral antigens-positive insects detected by immunofluorescence at 4, 10 and 15 days padp (*n*-30)				
	padp (days)	Epithelium regions of filter chamber	Visceral muscles of filter chamber	Entire intestine	Salivary glands
dsGFP	4	4	20	0	0
	10	0	25	25	8
	15	0	25	25	17
dsPns9	4	9	0	0	0
	10	9	0	0	0
	15	7	2	0	0

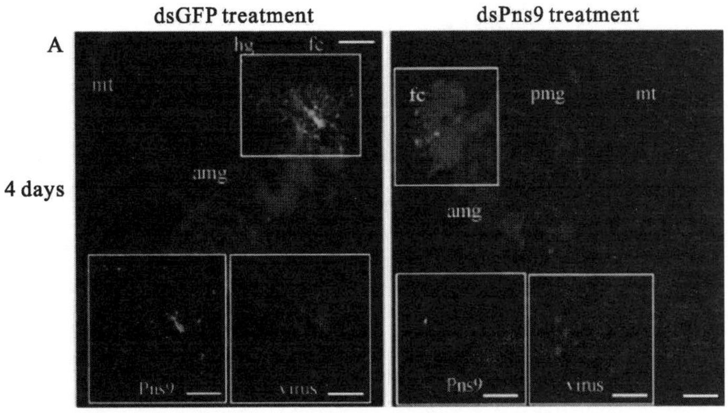

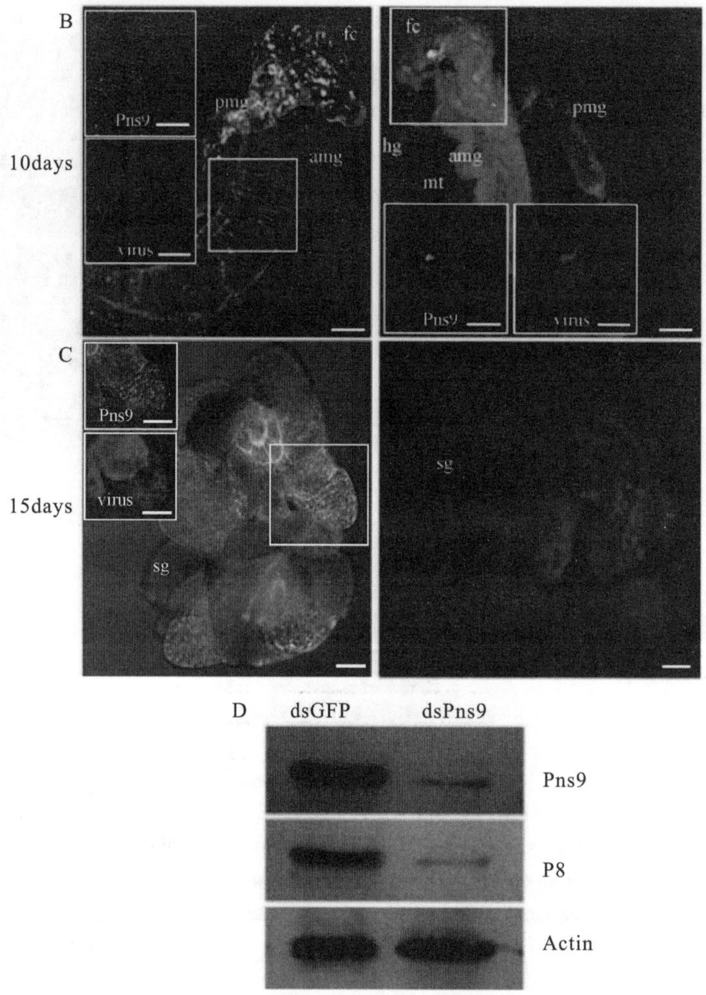

Fig. 4 RNAi induced by dsPns9 inhibiting the replication and spread of RGDV *in vivo*. (A–C) Second-instar nymphs of leafhoppers were microinjected with dsGFP or dsPns9

At 4, 10 and 15d padp (panels A–C, respectively), internal organs of leafhoppers were immunolabeled with Pns9-FITC (green), virus-rhodamine (red) and actin dyephalloidin-Alexa Fluor 647 carboxylicacid (blue). The images were merged with green fluorescence (Pns9 antigens), red fluorescence (viral antigens), and blue fluorescence (actin dye). Insets show green fluorescence (Pns9 antigens) and red fluorescence (viral antigens) of the merged images in the boxed areas in each panel. (D) Treatment with dsPns9 reduced the expression of Pns9 and P8 proteins in leafhoppers as shown by Western blotting. fc, filter chamber; mg, midgut; amg, anterior midgut; pmg, posterior midgut; hg, hindgut; mt, Malpighian tubules; sg, salivary glands. Bars, 100μm (For interpretation of the references to color in this figure legend, the reader is referred to the web version of this article)

4 Discussion

Phytoreoviruses including rice dwarf virus (RDV), RGDV and wound tumor virus are transmitted in a persistent-propagative mode by cicadellid leafhopper vectors (Hogenhout et al., 2008). The leafhopper filter chamber, where the midgut makes contact with the hindgut (Tsai and Perrier, 1996), is the initially infected site for several phytoreoviruses (Sinha, 1965; Chen et al., 2011; Fig. 1). Following initial infection of a limited number of epithelial cells of the filter chamber, RGDV crossed the basal lamina into the visceral muscle tissues around the filter chamber. It then either spread into the hemolymphs and eventually into the salivary glands, or spread throughout the visceral muscles of the midgut and hindgut of its leafhopper vector (Fig.1, Fig. 2; Table 1). Following infection of the epithelial cells of filter chamber, RDV spread from cell to cell among epithelial tissues to adjacent anterior midgut, and then infected the visceral muscles lining the anterior midgut (Chen et al., 2011). Furthermore, RDV may also exploit nerves to spread into the salivary glands from the infected vector's midgut (Chen et al., 2011), but RGDV was not reported to use the neural tissue of its leafhopper vector (data not shown). Thus, it seemed that RDV preferred to spread among the epithelial cells of insect intestine, while RGDV spread directly from the initially infected epithelial cells of insect intestine toward the visceral muscle tissues. Our recent findings revealed that the tenuivirus rice grassy stunt virus, fijivirus southern black streaked dwarf virus (SRBSDV) and oryzavirus rice ragged stunt virus (RRSV) may exploit a similar pathway to directly pass through the basal lamina from the initially infected intestine epithelium toward the visceral muscles of the insect vectors (Jia et al., 2012a, b; Zheng et al., 2014). It is interesting that SRBSDV could exploit virus-induced tubules as vehicles for viral spread through the basal lamina, facilitating viral dissemination from the initially infected midgut epithelium of its planthopper vector (Jia et al., 2014). Whether the tubules induced by RGDV infection could facilitate viral spread from the initially infected epithelium into the visceral muscles of the insect vectors is unknown and needs further investigation.

To understand the mechanism underlying the efficient infection of RGDV in the body of its insect vector, here, we first used VCMs derived from leafhopper *R. dorsalis*, an *in vitro* tissue culture system, to define the early process of viral infection in insect vector cells. During RGDV infection of VCMs, viroplasm that was mainly comprised of viral nonstructural protein Pns9 was formed and acted as the site of viral replication and assembly of progeny virions (Akita et al., 2011; Fig. 3). Because RNAi induced by dsRNA is a conserved sequence-specific gene silencing mechanism (Fire et al., 1998), we deduced that the synthesized dsRNAs targeting RGDV Pns9 gene could efficiently knock down the expression of Pns9 in virus-infected VCMs. Due to the essential role of Pns9 of RGDV in the biogenesis of viroplasm, we observed that the knockdown of Pns9 expression by dsPns9 treatment could efficiently abolish viroplasm formation, thus preventing the early viral replication and assembly of progeny virions in the insect vector cells (Fig. 3).

We then determined whether the formation of viroplasm for assembly of progeny virions is essential for the establishment of a persistent infection of RGDV *in vivo* in its insect vector.

After the initial infection of RGDV in a limited number of epithelial cells of the filter chamber, the nascent viroplasms, which served as the sites for assembly of progeny virions, were formed (Fig. 4). Thus, the formation of viroplasm for assembly of progeny virions was a consequence for the initial infection of RGDV in its insect vector. We then showed that the microinjection of dsPns9 efficiently suppressed viroplasm formation and restricted viral infection in the initially infected epithelium, thus inhibiting the spread of RGDV into the visceral muscles of the filter chamber (Fig. 4; Table 2). All these data suggested that the formation of the viroplasm for assembly of progeny virions was a prerequisite for the persistent infection of RGDV in its insect vector. We already revealed similar findings that viroplasms induced by other plant reoviruses including oryzavirus RRSV and fijivirus SRBSDV were essential for persistent infection of their respective insect vectors (Jia et al., 2012a; Chen et al., 2014). Thus, the formation of viroplasms for viral replication and assembly of progeny virions in the initially infected intestine of the insect vector was essential for the transmission of several plant reoviruses in their insect vectors.

The propagative persistent viruses such as plant reoviruses establish their primary infection in a few epithelial cells of the insect intestine when first ingested by their vectors. They then replicated and assembled progeny virions, which gradually spread to more cells of the intestine, and finally into the salivary glands of their vectors after 6–14d (Hogenhout et al., 2008; Ammar et al., 2009; Chen et al., 2011; Jia et al., 2012a,b, 2014; Blanc et al., 2014; Wu et al., 2014; Zheng et al., 2014). Only a limited number of epithelial cells of the vector intestine contained receptors for the attachment and entry of propagative persistent viruses (Chen et al., 2011; Jia et al., 2012a,b, 2014; Wu et al., 2014; Zheng et al., 2014). The nonpropagative persistent viruses such as begomoviruses and luteoviruses passed through the vector intestinal layers and exited from the salivary glands at 0.25–8h padp, much faster than the propagative persistent viruses in their vectors (Hogenhout et al.,2008; Ammar et al., 2009; Blanc et al., 2014; Wang et al., 2014). Thus, for the propagative persistent viruses, the formation of viral inclusions to facilitate viral multiplication in initially infected cells may be the essential step for the accumulation of enough number of virions to overcome the transmission barriers posed by different tissues in the insects, which may compensate for the insufficiency of invading viruses from the gut lumen. Our data suggest that during the co-evolution of persistent viruses and insect vectors, viruses exploited different mechanisms to survive and coexist within their vectors in order to be transmitted efficiently by respective insect vectors.

Acknowledgements

This research was supported by the National Natural Science Foundation of China (31130044, 31200118 and 31300136), the National Basic Research Program of China (2014CB138400), the National Science Foundation for Outstanding Youth (31325023), a project from the Henry Fok Education Foundation (131019), and the Natural Science Foundation of Fujian Province (2014J01085).

References

[1] AKITA F, MIYAZAKI N, HIBINO H, et al. Viroplasm matrixprotein Pns9 from *Rice gall dwarf virus* forms an octameric cylindrical structure[J]. Journal of General Virology, 2011, 92(9):2214-2221.

[2] AMMAR EL-D, TSAI C W, WHITFIELD A E, et al. Cellular and molecular aspects of rhabdovirus interactions with insect and planthosts[J]. Annual Review of Entomology, 2009, 54:447-468.

[3] BLANC S, DRUCKER M, UZEST M. Localizing viruses in their insect vectors[J]. Annual Review of Entomology, 2014, 52: 403-425.

[4] BCCARDO G. Plant reovirus group[J]. Cmi/aab Description of Plant Viruses, 1984, 294.

[5] BOCCARDO G, MILNE R G, DISTHAPORN S, et al. Morphology and nucleic acid of *Rice gall dwarf virus*[J]. Intervirology, 1985, 23(3):167-171.

[6] CHEN H, CHEN Q, OMURA T, et al. Sequential infection of *Rice dwarf virus* in the internal organs of its insect vector after ingestion of virus[J]. Virus Research, 2011, 160(1-2):389-394.

[7] CHEN H, ZHENG L, JIA D, et al. *Rice gall dwarf virus* exploits tubules to facilitate viral spread among cultured insect vector cells derived from leafhopper *Recilia dorsalis*[J]. Frontiers in Microbiology, 2013, 4:206-213.

[8] CHEN Q, CHEN H, MAO Q, et al. Tubular structure induced by a plant virus facilitates viral spread in its vector insect[J]. Plos Pathogens, 2012, 8(11):e1003032.

[9] CHEN H, ZHENG L, MAO Q, et al. Development of continuous cell culture of brown planthopper to trace the early infection process of oryzaviruses in insect vector cells[J]. Journal of Virology, 2014, 88(8):4265-4274.

[10] FAN G C, GAO F L, WEI T Y, et al. Expression of *Rice gall dwarf virus* outer coat protein gene (S8) in insect cells[J]. Virologica Sinica, 2010, 25(6):401-408.

[11] FAN H, ZHANG S, HE X, et al. *Rice gall dwarf*: a new virus disease epidemic in the west of Guangdong province of south China[J]. Acta Phytopathologica Sinica, 1983, 13:1-6.

[12] FIRE A, XU S, MONTGOMERY M K, et al. Potent and specific genetic interference by double-stranded RNA in *Caenorhabditis*

elegans[J]. Nature, 1998, 391(6669): 806-811.

[13] HIBI T, OMURA T, SAITO Y. Double-stranded RNA of *Rice gall dwarf virus*[J]. Journal of General Virology, 1984, 65(9):1585-1590.

[14] HOGENHOUT S A, AMMAR EL-D, WHITFIELD A E, et al. Insect vector interactions with persistently transmitted viruses[J]. Annual Review of Phytopathology, 2008, 46:327-359.

[15] JIA D, CHEN H, ZHENG A, et al. Development of an insect vector cell culture and RNA interference system to investigate the functional role of *Fijivirus* replication protein[J]. Journal of Virology, 2012a, 86(10):5800-5807.

[16] JIA D, GUO N, CHEN H, et al. Assembly of the viroplasm by viral non-structural protein Pns10 is essential for persistent infec-tion of *Rice ragged stunt virus* in its insect vector[J]. Journal of General Virology, 2012b, 93(10):2299-2309.

[17] JIA D, MAO Q, CHEN H, et al. Virus-induced tubule: a vehicle for rapid spread of virions through basal lamina from midgut epithelium in the insect vector[J]. Journal of Virology, 2014, 88:10488-10500.

[18] KOGANEZAWA H, HIBINO H, MOTOYOSHI F, et al. Nucleotide sequence of segment S9 of the genome of *Rice gall dwarf virus*[J]. Journal of General Virology, 1990, 71(8):1861-1863.

[19] LIU S, DING Z, ZHANG C, et al. Gene knockdown by intro-thoracic injection of double-stranded RNA in the brown planthopper, *Nilaparvata lugens*[J]. Insect Biochemistry and Molecular Biology, 2010, 40(9):666-671.

[20] MORIYASU Y, ISHIKAWA K, KIKUCHI A, et al. Sequence analysis of Pns11, a nonstructural protein of *Rice gall dwarf virus*, and its expression and detection in infected rice plants and vector insects[J]. Virus Genes, 2000, 20(3):237-241.

[21] MORIYASU Y, MARUYAMA-FUNATSUKI W, KIKUCHI A, et al. Molecular analysis of the genome segments S1, S4, S6, S7 and S12 of a *Rice gall dwarf virus* isolate from Thailand; completion of the genomic sequence[J]. Archives of Virology, 2007, 152(7):1315-1322.

[22] NODA H, ISHIKAWA K, HIBINO H, et al. Nucleotide sequences of genome segments S8, encoding a capsid protein, and S10, encoding a 36K protein, of *Rice gall dwarf virus*[J]. Journal of General Virology, 1991, 72(11):2837-2842.

[23] OMURA T, INOUE H, MORINAKA T, et al. *Rice gall dwarf*, a new virus disease[J]. Plant Disease, 1980, 64(8):795-797.

[24] OMURA T, MORINAKA T, INOUE H, et al. Purification and some properties of *Rice gall dwarf virus*, a new phytoreovirus[J]. Phytopathology (USA), 1982, 72(9):1246-1249.

[25]ONG C A, OMURA T. *Rice gall dwarf virus* occurrence in peninsular Malaysia[J]. International Rice Research Newsletter, 1982, 7: 7.

[26] PUTTA M, CHETTANACHIT D, MORINAKA T, et al. *Gall dwarf-new rice virus* disease in Thailand. International Rice Research Newsletter, 1980, 5:10-11.

[27] SINHA R C. Sequential infection and distribution of wound-tumor virus in the internal organs of a vector after ingestion of virus[J]. Virology, 1965, 26(4):673,681-679,686.

[28] TSAI J H, PERRIER J L. Morphology of the digestive and reproductive systems of *Dalbulus maidis* and *Graminella nigrifrons* (Homoptera: Cicadellidae)[J]. Florida Entomologist, 1996, 79(4):563-578.

[29] WANG Y, MAO Q, LIU W, et al. Localization and distribution of *Wheat dwarf virus* in its vector leafhopper, *Psammotettixalienus*[J]. Phytopathology, 2014, 104(8):897-904.

[30] WEI T, UEHARA-ICHIKI T, MIYAZAKI N, et al. Association of *Rice gall dwarf virus* with microtubules is necessary for viral release from cultured insect vector cells[J]. Journal of Virology, 2009, 83(20):10830-10835.

[31] WU W, ZHENG L, CHEN H, et al. Nonstructural protein NS4 of *Rice stripe virus* plays a critical role in viral spread in the body of vector insects[J]. PLoS One, 2014, 9(2):e88636.

[32] ZHANG H, XIN X, YANG J, et al. Completion of the genome sequence of *Rice gall dwarf virus* from Guangxi, China[J]. Archives of Virology, 2008, 153(9):1737-1741.

[33] ZHENG L, MAO Q, XIE L, et al. Infection route of *Rice grassy stunt virus*, a tenuivirus, in the body of its brown planthopper vector, *Nilaparvata lugens* (Hemiptera: Delphacidae) after ingestion of virus[J]. Virus Research, 2014, 188(1):170-173.

Characterisation of siRNAs derived from new isolates of bamboo mosaic virus and their associated satellites in infected ma bamboo (*Dendrocalamus latiflorus*)

Wenwu Lin, Wenkai Yan, Wenting Yang, Chaowei Yu, Huihuang Chen, Wen Zhang, Zujian Wu, Liang Yang, Lianhui Xie

(Fujian Key Lab of Plant Virology, Institute of Plant Virology, Fujian Agriculture and Forestry University, Fuzhou 350002, China)

Abstract: We characterised the virus-derived small interfering RNAs (vsiRNA) of bamboo mosaic virus (Ba-vsiRNAs) and its associated satellite RNA (satRNA)-derived siRNAs (satsiRNAs) in a bamboo plant (*Dendrocalamuslatiflorus*) by deep sequencing. Ba-vsiRNAs and satsiRNAs of 21–22nt in length, with both (+) and (-) polarity, predominated. The 5′-terminal base of Ba-vsiRNA was biased towards A, whereas a bias towards C/U was observed in sense satsiRNAs, and towards A in antisense satsiRNAs. A large set of bamboo genes were identified as potential targets of BavsiRNAs and satsiRNAs, revealing RNA silencing-based virus-host interactions in plants. Moreover, we isolated and characterised new isolates of bamboo mosaic virus (BaMV; 6,350nt) and BaMV-associated satRNA (satBaMV; 834nt), designated BaMV-MAZSL1 and satBaMV-MAZSL1, respectively.

RNA silencing, a gene regulatory mechanism that uses small RNAs for sequence-specific gene expression inhibition, protect eukaryotic genomes against aberrant endogenous or exogenous RNA molecules (Zamore, 2002; Baulcombe, 2004; Csorba et al., 2009; Poulsen et al., 2013). RNA silencing pathways are triggered by double-stranded RNAs (dsRNA) or single-stranded RNAs (ssRNA) with foldback structures that are processed into small interfering RNAs (siRNAs) of 21–24 nucleotides (nt) by RNase III-type DICER enzymes (Bernstein et al., 2001; Baulcombe, 2004; Meister and Tuschl, 2004). These siRNAs are recruited into the RNA-induced silencing complex (RISC) by proteins of the Argonaute (AGO) family, to facilitate the cleavage of target RNAs through various base-pairing mechanisms (Vazquez, 2006; Chapman and Carrington, 2007).

In higher plants, RNA silencing can serve as an adaptive, antiviral defence system, which is transmitted systemically in response to localised virus challenge (Voinnet, 2001). Plant viruses can activate the RNA machinery and can also be targets of RNA silencing. In infected cells, high levels of virus-derived small interfering RNAs (vsiRNAs) can be processed from either viral dsRNA replicative intermediates, or local self-complementary regions of the viral genome, or dsRNA resulting from the action of RNA-dependent RNA polymerases (RDRs) on viral RNA templates (Molnár et al., 2005; Ding and Voinnet, 2007; Qi et al., 2009; Garcia-Ruiz et al., 2010; Szittya et al., 2010; Wang et al., 2010). These vsiRNAs are key elements in guiding auto-silencing of viral RNA as part of an antiviral self-defence response in plants (Ding and Voinnet, 2007). Multiple AGO proteins, such as AGO1, AGO2, AGO4, AGO5, AGO7, AGO10 and AGO18, are involved in this process (Fang and Qi, 2016). Loading of siRNAs onto a particular AGO complex is preferentially, but not exclusively, dictated by their 5′terminal nucleotides (Brodersen et al., 2008; Mi et al., 2008; Fang and Qi,

Archive of Virology. 2017, 162: 505–510.

Received 11 September 2016; Accepted 26 September 2016; Published online 14 October 2016.

2016). Moreover, compelling evidence indicates that the biogenesis of vsiRNA of different size classes involves the same Dicer-like (DCL)-dependent pathways responsible for the formation of endogenous siRNAs (Xie et al., 2004; Blevins et al., 2006; Bouché et al., 2006). Thus, vsiRNAs share some features with host siRNA and can mediate RNA silencing. It is worth noting that some of these vsiRNAs can participate in degrading complementary cellular transcripts to create cellular conditions suitable for viral infection (Ruiz-Ferrer and Voinnet, 2009; Shimura et al., 2011; Smith et al., 2011).

Bamboo mosaic virus (BaMV) has been investigated intensively and has thus become one of the most important models for studying plant-virus interactions (Liou et al., 2015). This virus has a single-stranded, positive-sense RNA genome of 6,400nt comprising five open reading frames (ORFs) flanked by 5′- and 3′-untranslated regions (UTRs) of 94 and 142nt, respectively (Lin et al., 1994). ORF1 encodes a 155kDa replication-related protein with three functional domains: an N-terminal mRNA capping enzyme, a central RNA helicase and a C-terminal RDR (Li et al., 2001a; Li et al., 2001b; Huang et al., 2005). ORF2–4, which are overlapping and are referred to as the 'triple gene block', are required for viral movement (Liou et al., 2015). The 25-kDa coat protein encoded by ORF5 is associated with virion encapsidation, replication and both cell-to-cell and long distance viral movement (Lan et al., 2010). The satellite RNA (satRNA) of BaMV (satBaMV), the only example of satRNA associated with a potexvirus, totally depends on BaMV for replication, assembly and movement (Lin and Hsu, 1994). P20, which is encoded by satBaMV, is not required for satBaMV replication or cell-to-cell movement; however, it indeed facilitates long-distance movement of satBaMV in *Nicotiana benthamiana* co-infected with BaMV(Lin et al., 1996; Palani et al., 2006).

In recent years, deep sequencing of vsiRNAs in different host–virus systems, along with functional characterisation, has provided insights into the origin and composition of vsiRNAs and their potential role in virus–host interactions (Prabha et al., 2013). The composition of BaMV and satBaMV-derived siRNAs in infected *N. benthamiana* and *Arabidopsis thaliana* has been investigated (Lin et al., 2010). However, few studies have focused on vsiRNA profiles from bamboo, which is the natural host of BaMV; such studies are indispensable for the understanding of virus-plant interactions and, most importantly, for developing sustainable methods of controlling viral infections. Here, we report on vsiRNAs of new divergent BaMV isolates (Ba-vsiRNAs) and their associated satBaMV in a cultivated bamboo species (*Dendrocalamus latiflorus*).

Bamboo leaves of a single plant with mosaic symptoms were collected in the Fuzhou National Forest Park, Fujian, China, once a month from July to October, 2014. Total RNA was extracted with an RNeasy Plant Mini Kit (Qiagen, Hilden, Germany) according to the manufacturer's instructions. The complete genomes of BaMV and satBaMV in infected *D. latiflorus* were sequenced as described previously (Lin et al., 2016). Small RNA isolation and library construction were performed essentially as described (Lin et al., 2010). Deep sequencing was performed on the Illumina Solexa platform following the manufacturer's protocol. The 5′and 3′ adapter sequences were trimmed from the Solexa reads, and vsiRNAs and satRNA derived siRNAs (satsiRNAs) of 18- to 28-nt were identified by a BLAST search against the complete genomic sequences of the BaMV-MAZSL1 isolate and the sat-BaMV-MAZSL1 isolate (accession numbers: KU870664 and KU870665, respectively) identified in this study. Only sequences that contained no more than two-position mismatches were further analysed. Library characterization and determination of the BaMV and satBaMV siRNA profiles were performed using in-house scripts. Further statistical analyses and summaries were conducted using Microsoft Excel 2010. Potential targets of the Ba-vsiRNAs and satsiRNAs were identified by psRNA Target (http://plantgrn.noble.org/psRNATarget/), using the default parameters (Yang et al., 2014). Due to our limited knowledge of the complete genome of *D. latiflorus*, the cds_DNA of *Phyllostachys heterocycla* (Moso Bamboo) was used as the pool for putative target prediction.

To date, only eight complete genomes of wild BaMV isolates have been deposited in GenBank. The genome of BaMV-MAZSL1 (6,350nt) has the same genomic structure as that of the other isolates, except for five amino acid deletions close to the N-terminus of the coat protein gene. BaMV-MAZSL1 and satBaMV-MAZSL1 share the highest overall nucleotide sequence identity (83% and 94%, respectively) with all BaMV and satBaMV isolates available in GenBank. Detailed genomic information about the isolates is presented in Table S1. Phylogenetic analysis placed nine complete BaMV genomic sequences into two clusters; however, BaMV-MAZSL1 was clustered into a new phylogenetic sub-lineage, which is closer to isolates from Taiwan than those from Fuzhou (Fig. S1A). A similar result was obtained for satBaMV-MAZSL1, which was also grouped into a new sub-cluster despite the large number of satBaMV isolates analysed in this study (Fig. S1B).

For siRNAs analysis, a total of 21,006,558 reads with length between 18 and 28nt (after trimming the sequences) were searched against the corresponding viral genomes. As a result, 3,190,981 and 727,234 reads, accounting for 15.2% and 3.5% of 18- to 28-nt reads, were identified as BaMV-MAZSL- and satBaMV-MAZSL-specific siRNA, respectively, representing 323,031 and 67,856 unique reads(Fig. 1). For both vsiRNAs and satsiRNAs, the 21-nt class was clearly the most dominant, followed by the 22-nt class, together accounting for 81.7% and 74.6% of the total selected reads, respectively (Fig. 1). The result suggest that the bamboo homologues of DCL4 and DCL2 may be the predominant DCL ribonucleases involved in vsiRNAs and satsiRNAs biogenesis and that the 21- and 22-nt siRNAsare the predominant antiviral silencing components, in accordance with many studies of various plant viruses (Donaire et al., 2009; Pantaleo et al., 2010; Yang et al., 2014; Li et al., 2016). By contrast, we detected far fewer 23- and 24-nt vsiRNAs and satsiRNAs, suggesting that DCL3 plays only a minor role in the biogenesis of BaMV- and sat-BaMV-specific siRNAs in bamboo.

To investigate the frequency distribution of vsiRNAs and satsiRNAs in the BaMV- and satBaMV-MAZSL genomes, respectively, single-base resolution maps of all redundant BaMV- and satBaMV-derived siRNAs along the genomes were constructed using Bowtie tools (Langmead et al., 2009). The results showed that the most abundant Ba-vsiRNAs were mainly located within both positive and negative strands of the capping enzyme domain of the replicase and the 5’ UTR (Fig. 2A, B). A similar pattern was found in *A. thaliana*, whereas extremely different results were obtained in *N. benthamiana*, with the CP and 3’ UTR being the major sources of Ba-vsiRNAs(Lin et al., 2010). The highly abundant region of satsiRNAs was in the positive strand of the P20-encoding region, and much fewer satsiRNAs were matched to the negative strand (Fig. 2C, D). Moreover, a total of 18 hotspots (V1–V13 on BaMV and S1–S5 on satBaMV) of viral-specific siRNAs, with more than 10,000 reads, were identified (Fig. 2 B, D). The mapping patterns of vsiRNA and satsiRNA of different size classes (21–24nt) were clarified, revealing that 21-nt sequences, followed by 22nt sequences, predominated within these hotspots (Fig. 2A, B). In contrast to the negative-strand dominance of BavsiRNAsand satsiRNAs found in *N. benthamiana* in a previous study (Lin et al., 2010), there were slightly more vsiRNAs in the positive strand (52%) than in the negative strand (48%) of *D. latiflorus* (Fig. 2B). Moreover, there were substantially more satsiRNAs in the positive strand(88.9%) than in the negative strand (11.1%) (Fig. 2D). Our results suggest that the high or low abundance of antisense vsiRNAs compared to sense vsiRNAs is not a specific feature of a plant virus. Instead, the mechanism responsible for strand polarity might depend on other factorspossibly related to a specific virus, different hosts or the environment.

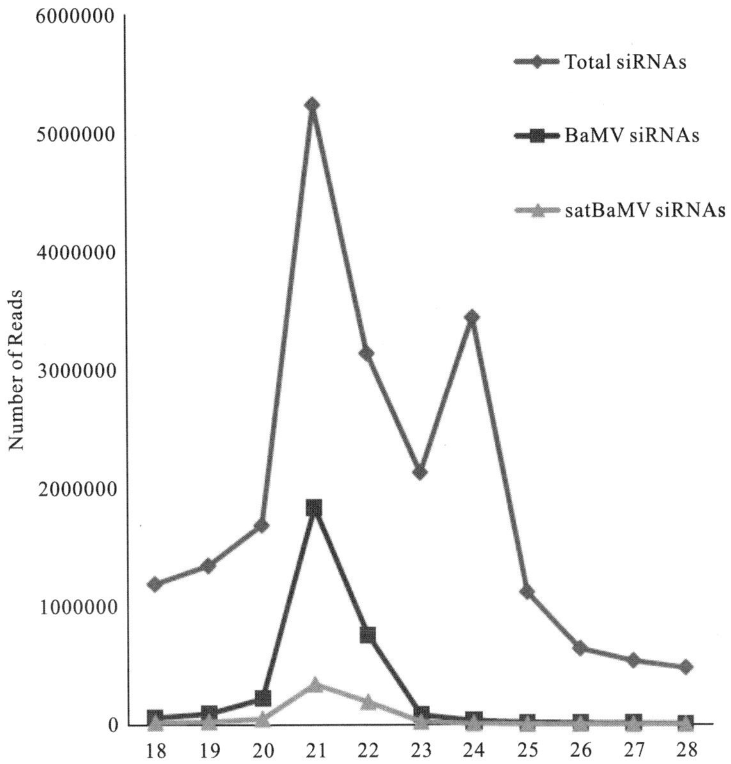

Fig. 1 Number of total siRNAs, BaMV-derived siRNAs and satBaMV-derived siRNAs of 18–28 nt in size in libraries prepared from BaMV-infected *Dendrocalamus latiflorus*

Previous studies have indicated that loading of siRNAs onto a particular AGO complex is preferentially dictated bytheir 5′-terminal nucleotides. The bioinformatics data revealed that sense Ba-vsiRNAs with an adenine (A) at their 5′-termini were the most abundant, accounting for 35.92% of all sense vsiRNAs, followed by cytosine (C), uracil (U) and guanine (G) with 32.43%, 19.24% and 12.41%, respectively (Fig. 3A; Fig. S3). Likewise, antisense Ba-vsiRNAs with an A residue at the 5′-termini were the most abundant (37.39% of all antisense vsiRNAs) (Fig. 3A; Fig. S3). However, based on the nucleotide sequence, only 30.31% and 18.55% of the sequence of genomic and complementary RNAs of BaMV-MAZSL1 contain A. This suggested that no matter what polarity these Ba-vsiRNAs were, their 5′-terminal base was biased towards A (Fig. S3). Interestingly, satsiRNAs exhibited a different pattern: a biasfor C and U was evident in sense satsiRNAs and the preference towards A was strong in antisense satsiRNAs (Fig. S3). These results are in contrast with the previous finding that there is no nucleotide preference in the generation of BaMV and satBaMV siRNAs in *N. benthamiana* and *A. thaliana*(Lin et al., 2010). To gain further insights into vsiRNA and satsiRNA biogenesis and sorting, different-sized species(21–24nt) were analysed (Fig. 3B, C). For all vsiRNAs of different size classes, a clear preference for A as the 5′- terminal nucleotide was observed, which is indicative of vsiRNAs with high binding affinity for AGO2 and AGO4 homologues; however, a strong bias for sequences beginning with a 5′-C was observed in satsiRNAs of all size classes, suggesting the high binding affinity of AGO5 homologues for satsiRNAs in *D. latiflorus*.

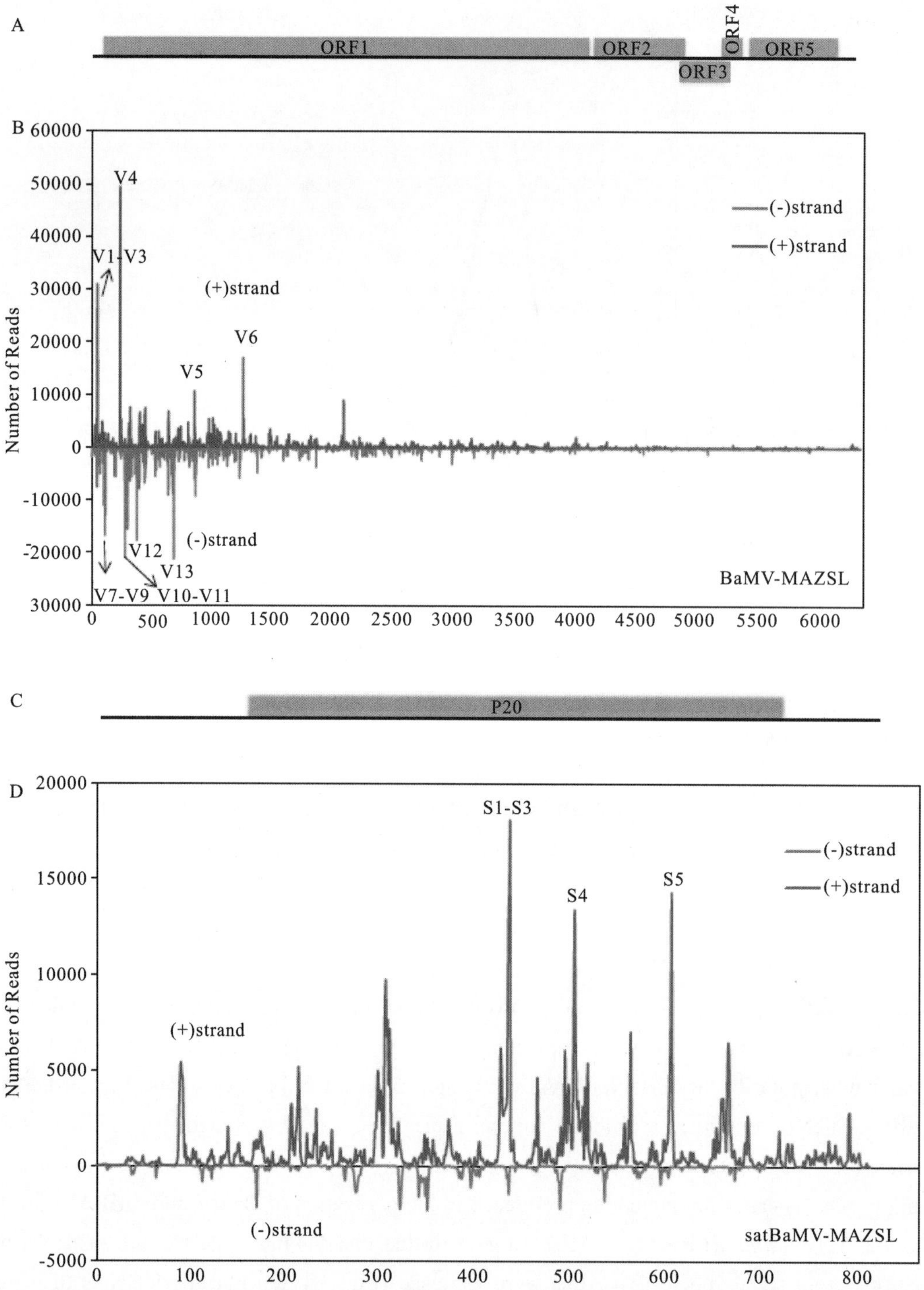

Fig. 2 The distribution of siRNAs on the BaMV (A and B) and satBaMV (C and D) genomes of *Dendrocalamus latiflorus*

The siRNAs are shown in orange above (positive strand) and in blue below (negative strand) the horizontal line. Hotspots accumulating vsiRNAs and satsiRNAs at precise positions are indicated by V1–13 and S1–5 (10,000 reads), respectively. The X axis represents the length of the genome, and the Y axis represents the number of siRNAs

A total of 1389 and 584 bamboo genes were predicted to be possible targets of Ba-vsiRNAs and satsiRNAs, respectively; these potential targets were predicted to be involved in a broad range of biological processes. Meanwhile, twelve hotspots of these siRNAs were predicted to have one or more targets. The current experimental evidence supporting a functiona linteraction between host mRNAs and virus-specific siRNAs is weak. Nevertheless, this finding suggests that many host genes and their regulatory sequences might be targeted by Ba-vsiRNA- and satsiRNA-mediated downregulation during viral function.

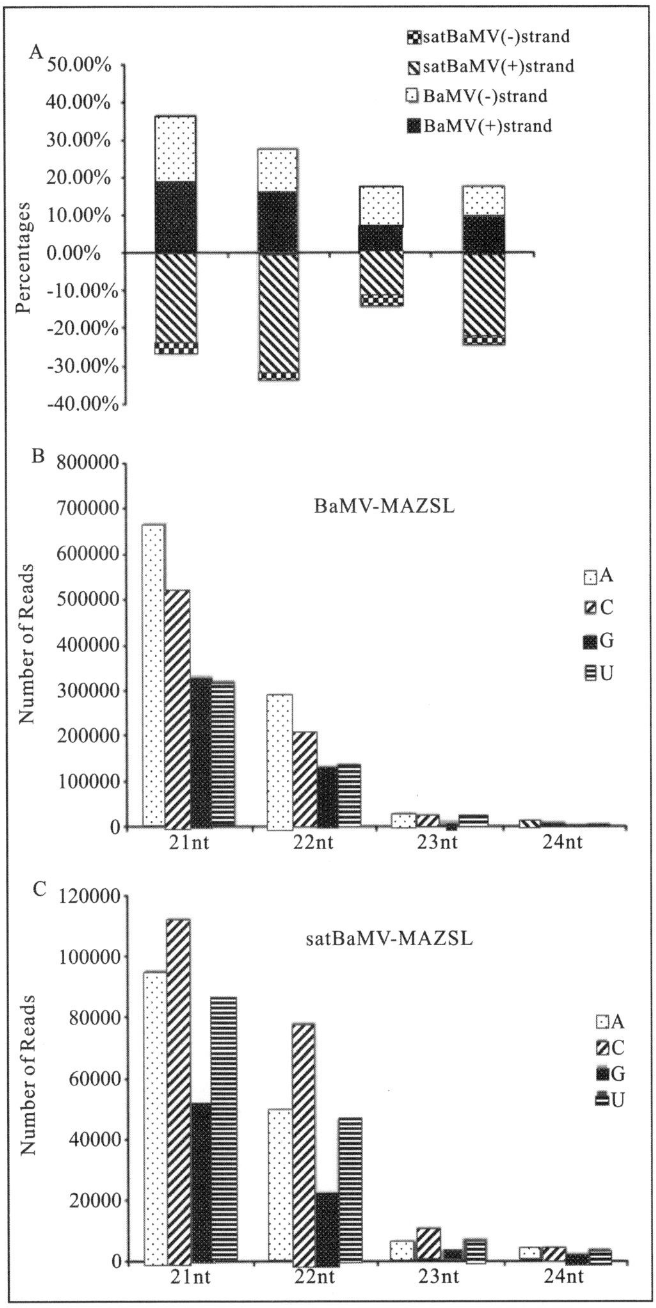

Fig. 3 The 5′-terminal nucleotides of Ba-vsiRNAs and satsiRNAs

(A): Relative frequency of the 5'-terminal nucleotides of all BavsiRNAsand satsiRNAs (sense and antisense). (B): Number of 21–24nt reads of Ba-vsiRNAs. (C): Number of 21–24nt reads of satsiRNAs

In this study, we identified and characterised new divergent BaMV and satBaMV isolates and their specific siRNA profiles, which differ somewhat from those of previous studies, supporting the view that the combined action of viruses, satRNA, DCLs and AGOs, in different host plants results in the high diversity of the vsiRNAs pool found in nature. The challenge ahead is to further determine the extent of these functional interactions between vsiRNAs and their targets from a biological perspective.

Acknowledgements

This work was supported by grants from Natural Science Foundation of Fujian Province of China (Grant No. 2014J06011), Education Department of Fujian Province Office Program (Grant No. JA13092), FAFU Science Fund for Distinguished Young Scholars (Grant No. xjq201402) and the China Scholarship Council (CSC No. 201608350081).

References

[1] BAULCOMBE D. RNA silencing in plants[J]. Nature,2004, 431: 356.

[2] BERNSTEIN E, CAUDY A A, HAMMOND S M, et al. Role for a bidentate ribonuclease in the initiation step of RNA interference[J]. Nature,2001, 409: 363.

[3] BLEVINS T, RAJESWARAN R, SHIVAPRASAD P V, et al. Four plant Dicers mediate viral small RNA biogenesis and DNA virus induced silencing[J]. Nucleic Acids Research, 2006, 34: 6233-6246.

[4] BOUCHÉ N, LAURESSERGUES D, GASCIOLLI V, et al. An antagonistic function for Arabidopsis DCL2 in development and a new function for DCL4 in generating viral siRNAs[J]. EMBO Journal,2006, 25: 3347-3356.

[5] BRODERSEN P, SAKVARELIDZE-ACHARD L, BRUUN-RASMUSSEN M, et al. Widespread translational inhibition by plant miRNAs and siRNAs[J]. Science, 2008, 320: 1185-1190.

[6] CHAPMAN E J, CARRINGTON J C. Specialization and evolution of endogenous small RNA pathways[J]. Nature Reviews Genetics, 2007, 8: 884.

[7] CSORBA T, PANTALEO V, BURGYÁN J. RNA silencing: an antiviral mechanism[J]. Advances in Virus Research, 2009, 75: 35-230.

[8] DING S-W, VOINNET O. Antiviral immunity directed by small RNAs[J]. Cell, 2007, 130: 413-426.

[9] DONAIRE L, WANG Y, GONZALEZ-IBEAS D, et al. Deep-sequencing of plant viral small RNAs reveals effective and widespread targeting of viral genomes[J]. Virology journal, 2009, 392: 203-214.

[10] FANG X, QI Y. RNAi in plants: an Argonaute-centered view[J]. Plant Cell, 2016, 28: 272-285.

[11] GARCIA-RUIZ H, TAKEDA A, CHAPMAN E J, et al. Arabidopsis RNA-dependent RNA polymerases and dicer-like proteins in antiviral defense and small interfering RNA biogenesis during Turnip Mosaic Virus infection[J]. Plant Cell, 2010, 22: 481-496.

[12] HUANG Y-L, HSU Y-H, HAN Y-T, et al. mRNA guanylation catalyzed by the S-adenosylmethionine-dependent guanylyltransferase of Bamboo mosaic virus[J]. Journal of biological chemistry,2005, 280: 13153-13162.

[13] LAN P, YEH W-B, TSAI C-W, et al. A unique glycine-rich motif at the N-terminal region of Bamboo mosaic virus coat protein is required for symptom expression[J]. Molecular Plant-Microbe Interactions,2010, 23: 903-914.

[14] LANGMEAD B, TRAPNELL C, POP M, et al. Ultrafast and memory-efficient alignment of short DNA sequences to the human genome[J]. Genome biology, 2009, 10(3): R25.

[15] LI Y-I, CHEN Y-J, HSU Y-H, et al. Characterization of the AdoMet-dependent guanylyltransferase activity that is associated with the N terminus of Bamboo mosaic virus replicase[J]. Journal of virology, 2001a, 75(2): 782-788.

[16] LI Y-I, SHIH T-W, HSU Y-H, et al. The helicase-like domain of plant potexvirusreplicase participates in formation of RNA. 5′ cap structure by exhibiting RNA. 5′-triphosphatase activity[J]. Journal of virology,2001b, 75(24): 12114-12120.

[17] LI Y, DENG C, SHANG Q, et al. Characterization of siRNAs derived from Cucumber green mottle mosaic virusin infected cucumber plants[J]. Archives of virology,2016, 161(2): 455-458.

[18] LIN K-Y, CHENG C-P, CHANG BC-H, et al. Global analyses of small interfering RNAs derived from bamboo mosaic virus and its associated satellite RNAs in different plants[J]. PLoS One,2010, 5: e11928.

[19] LIN N-S, HSU Y-H. A satellite RNA associated with bamboo mosaic potexvirus[J]. Virology,1994, 202: 707-714.

[20] LIN N-S, LEE Y-S, LIN B-Y, et al. The open reading frame of bamboo mosaic potexvirus satellite RNA is not essential for its replication and can be replaced with a bacterial gene[J]. Proceedings of the National Academy of Sciences of the United States of America, 1996, 93: 3138-3142.

[21] LIN N-S, LIN B-Y, LO N-W, et al. Nucleotide sequence of the genomic RNA of bamboo mosaic potexvirus[J]. Journal of general virology,1994, 75: 2513-2518.

[22] LIN W, GAO F, YANG W, et al. Molecular characterization and detection of a recombinant isolate of bamboo mosaic virus from China[J]. Archives of virology, 2016, 161(4): 1091-1094.

[23] LIOU M-R, HU C-C CHOU, et al. Viral elements and host cellular proteins in intercellular movement of bamboo mosaic virus[J]. Current opinion in virology,2015, 12: 99-108.

[24] MEISTER G, TUSCHL T. Mechanisms of gene silencing by double-stranded RNA[J]. Nature, 2004, 431: 343.

[25] MI S, CAI T, HU Y, et al. Sorting of small RNAs into Arabidopsis argonaute complexes is directed by the 5′ terminal nucleotide[J]. Cell, 2008, 133: 116-127.

[26] MOLNÁR A, CSORBA T, LAKATOS L, et al. Plant virus-derived small interfering RNAs originate predominantly from highly structured single-stranded viral RNAs[J]. Journal of virology,2005, 79(12): 7812-7818.

[27] PALANI P V, KASIVISWANATHAN V, CHEN JC-F, et al. The arginine-rich motif of Bamboo mosaic virus satellite RNA-encoded P20 mediates self-interaction, intracellular targeting, and cell-to-cell movement[J]. Molecular Plant-Microbe Interactions,2006, 19: 758-767.

[28] PANTALEO V, SALDARELLI P, MIOZZI L, et al. Deep sequencing analysis of viral short RNAs from an infected Pinot Noir grapevine[J]. Virology,2010, 408: 49-56.

[29] POULSEN C, VAUCHERET H, BRODERSEN P. Lessons on RNA silencing mechanisms in plants from eukaryotic argonaute structures[J]. Plant Cell, 2013, 25: 22-37.

[30] PRABHA K, BARANWAL V, JAIN R. Applications of next generation high throughput sequencing technologies in characterization, discovery and molecular interaction of plant viruses[J]. Indian Journal of Virology, 2013, 24: 157-165.

[31] QI X, BAO F S, XIE Z. Small RNA deep sequencing reveals role for Arabidopsis thaliana RNA-dependent RNA polymerases in viral siRNA biogenesis[J].PloS one,2009, 4: e4971.

[32] RUIZ-FERRER V, VOINNET O. Roles of plant small RNAs in biotic stress responses[J]. Annual Review of Plant Biology, 2009, 60: 485-510.

[33] SHIMURA H, PANTALEO V, ISHIHARA T, et al. A viral satellite RNA induces yellow symptoms on tobacco by targeting a gene involved in chlorophyll biosynthesis using the RNA silencing machinery[J]. PLoSPathogens, 2011, 7: e1002021.

[34] SMITH N A, EAMENS A L, WANG M-B. Viral small interfering RNAs target host genes to mediate disease symptoms in plants[J]. PLoSPathogens, 2011, 7: e1002022.

[35] SZITTYA G, MOXON S, PANTALEO V, et al. Structural and functional analysis of viral siRNAs[J]. PLoSP athogens, 2010, 6: e1000838.

[36] VAZQUEZ F. Arabidopsis endogenous small RNAs: highways and byways[J]. Trends Plant Science, 2006, 11: 460-468.

[37] VOINNET O. RNA silencing as a plant immune system against viruses[J]. Trends in Genetics, 2001, 17: 449-459.

[38] WANG X-B, WU Q, ITO T, et al. RNAi-mediated viral immunity requires amplification of virus-derived siRNAs in *Arabidopsis thaliana*[J]. Proceedings of the National Academy of Sciences of the United States of America, 2010, 107: 484-489.

[39] XIE Z, JOHANSEN L K, GUSTAFSON A M, et al. Genetic and functional diversification of small RNA pathways in plants[J]. PLoS Biology,2004, 2: e104.

[40] YANG J, ZHENG S-L, ZHANG H-M, et al. Analysis of small RNAs derived from Chinese wheat mosaic virus[J]. Archives of virology, 2014, 159(11): 3077-3082.

[41] ZAMORE P D. Ancient pathways programmed by small RNAs[J]. Science, 2002, 296: 1265-1269.

Translation initiation factor eIF4E and eIFiso4E are both required for peanut stripe virus infection in peanut (*Arachis hypogaea* L.)

Manlin Xu[1,2], Hongfeng Xie[1], Juxiang Wu[1], Lianhui Xie[2], Jinguang Yang[3] and Yucheng Chi[1]

(1 Shandong Peanut Research Institute, Qingdao, China; 2 Fujian Agriculture and Forestry University, Fuzhou, China; 3 OpenProject Program of Key Laboratory of Tobacco Pest Monitoring Controlling and Integrated Management, Tobacco ResearchInstitute of Chinese Academy of Agricultural Sciences, Qingdao, China)

Abstract: *Peanut stripe virus*(PStV) belongs to the genus *Potyvirus* and is the most important viral pathogen of cultivated peanut (*Arachis hypogaea*L.). The eukaryotic translation initiation factor, eIF4E, and its isoform, eIF(iso)4E, play key roles during virus infection inplants, particularly *Potyvirus*. In the present study, we cloned the *eIF4E* and *eIF(iso)4E* homologs in peanut and named these as *PeaeIF4E* and *PeaeIF(iso)4E*, respectively. Quantitative real-time PCR (qRT-PCR) analysis showed that these two genes wereexpressed during all growth periods and in all peanut organs, but were especially abundant in young leaves and roots. These also had similar expression levels. Yeast two-hybrid analysis showed that PStV multifunctional helper component proteinase(HC-Pro) and viral protein genome-linked (VPg) both interacted with PeaeIF4E and PeaeIF(iso)4E. Bimolecular fluorescence complementation assay showed that there was an interaction between HC-Pro and PeaeIF4E/PeaeIF(iso)4E in the cytoplasm and between VPg and PeaeIF4E/PeaeIF(iso)4E in the nucleus. Silencing either *PeaeIF4E*or *PeaeIF(iso)4E*using a virus-induced gene silencing system did not significantly affect PStV accumulation. However, silencing both *PeaeIF4E* and *PeaeIF(iso)4E* genes significantly weakened PStV accumulation. The findings of the present study suggest that PeaeIF4E and PeaeIF(iso)4E play important roles in the PStV infection cycle and may potentially contribute to PStV resistance.

Keywords: peanut; Peanut stripe virus; translation initiation factor 4E; protein–protein interaction; gene silencing

1 Introduction

Peanut is one of the most important oil crops and food legumes in the world. In China, peanutsare grown on 3.5 million hectares of land each year (Xu, 2008). *Peanut stripe virus* (PStV; genus *Potyvirus*, family *Potyviridae*) is one of the most widely distributed peanut viruses constraining peanut production. PStV has been detected in various countries, including China, the US, the Philippines, Thailand, Indonesia, Malaysia, and Korea (Xu et al., 1983; Demski and Lovell, 1985;Saleh et al., 1989; Choi et al., 2001, 2006). Recently, PStV has been reported in India and some African countries, which were possibly caused by exchanges in peanut seed resources (Xu, 2008).

In China, PStV is a very serious infectious disease that afflicts peanut, particularly those grown in northern China. Thein fection incidence has reached 50% and recently, even 100% in some fields. PStV has also

Frontiers in microbiology. 2017, 8: 338.
Received8 December 2016; Accepted 17 February 2017;Published 10 March 2017

infected various other crops, including soybean (*Glycine max*), sesame (*Sesamum indicum*), cowpea (*Vigna unguiculata*), hyacinth bean (*Dolichos lablab*), white lupin (*Lupinus albus*), and patchouli (*Pogostemon cablin*; Xu, 2008;Singh et al., 2009). To date, no effective method for controlling this virus has been established. Peanut stripe virus is a member of the genus *Potyvirus*, an economically significant and one of the largest groups of viruses that infect plants. These viruses are about 10kb in length, carrya single positive-strand RNA, and contain a 350-kD polyprotein that is translated by a single open reading frame (ORF). The polyprotein is cleaved by three virus-encoded proteases into 10mature proteins and an additional protein called PIPO, which isembedded in the P3 cistron (Urcuqui-Inchima et al., 2001; Wei et al., 2010).

One of the three virus-encoded proteases is the multifunctional helper component proteinase (HC-Pro), which consists of C-proximal, central, and N-proximal domains.The C-proximal domain separates HC-Pro from the polyprotein precursor via proteolysis (Carrington and Herndon, 1992). HC-Pro contributes to various essential steps that are related to viral replication and infection cycles. HC-Pro is in volved in some processes, including virus transmission by aphids (Govier et al., 1977) and virus movement from cell-to-cell (Rojas et al.,1997) to long-distance migration (Saenz et al., 2002). In addition,HC-Pro facilitates the development of virulence and symptom amplification (Atreya et al., 1992; Redondo et al., 2001); it is also a regulator of gene silencing suppression (Llave et al., 2000). HC-Pro interacts with numerous host proteins and some virus proteins, as well as mediates the function of host proteins andother viral proteins (Jin et al., 2007; Ala-Poikela et al., 2011).

Another important region of the potyviral protein is VPg, which is translated into the polyprotein, NIa, which is also known as VPg-Pro. During the potyvirus infection, VPg participates in replication and proteolysis and is composed of N-terminal and C-terminal protease domains (Revers et al., 1999). VPgis a multifunctional protein that plays a crucial role in race-specific replication (translation and RNA synthesis), as well asin cell-to-cell and long-distance movement; it also interacts with host proteins as well as various recessive potyvirus resistancegenes in different host species (Lellis et al., 2002; Rajamaki andValkonen, 2002).

Because viruses have relatively small genomes and a limited number of proteins, they rely on the host-cell environment to complete their infection cycle. The characterization of host proteins, membranes, and nucleic acids, using a model host system, functional genomics, and modern molecular biology methods, help in the understanding of plant–virus interactions (Whitham and Wang, 2004). For example, positive-sense ssRNA viruses replicate in association with host endomembranes (Mackenzie, 2005) and different host factors (Ahlquist et al.,2003; Whitham and Wang, 2004). One of the most important genes is the translation initiation factor, *eIF4E*, which initiates the translation of mRNA and regulates protein synthesis (Sonenberg et al., 1978; Jackson et al., 2010). *eIF4E* also interacts with the 50-terminal cap of mRNA and was initially named the'cap-binding protein'. Moreover, *eIF4E* and its isoform *eIF(iso)4E* are functionally redundant and one or both of them interact with HCpro and VPg, which are in dispensable for viruses to complete their infection cycle; therefore, abolishing this interaction may prevent the viral infection (Lellis et al., 2002; Browning, 2004;Kang et al., 2005; Jin et al., 2007; Charron et al., 2008; Ala-Poikela et al., 2011; Wang and Krishnaswamy, 2012; Sanfacon,2015). Based on this concept, silencing or incurring mutation in the gene may disrupt infection. *pvr2* is a two-nucleotide substitution of the amino acid of pepper eIF4E and is resistant to PVY (Ruffel et al., 2002). A small number of amino acidsubstitutions in the tomato eIF4E *pot-1* confer resistance against PVY and Tobacco etch virus (TEV) in tomato. Barley *rym4* and *rym5* are also amino acid substitutions that confer eIF4E resistance to *Barley yellow mosaic virus* (BaYMV) and *Barley mild mosaic virus* (BaMMV) in Barley (Kanyuka et al., 2005; Stein et al., 2005). Pepper *pvr1(2)* contains an eIF4E mutation and *pvr6* is an eIF(iso)4E mutation;

simultaneous mutations in eIF4E and eIF(iso)4E confer resistance to *Chilli veinal mottle virus*(ChiVMV) in pepper, and silencing eIF4E and eIF(iso)4E reduces the ChiVMV accumulation (Ruffel et al., 2006; Hwang et al.,2009). In plum, the silencing of eIF(iso)4E results in resistance to *Plum pox virus* (PPV; Wang et al., 2013; Cui and Wang, 2016). Thus, the dependence of potyviruses on eIF4E and/or eIF(iso)4E varies with each virus–host interaction.

To date, no effective way of controlling PStV such asusing genetically resistant varieties has been established, mainly because no resistance genes have been identified. We hypothesize that peanut eIF4E/eIF(iso)4E controls the effect of PStV inpeanut. To test this hypothesis, we investigated the effects of silencing the translation initiation factor, eIF4E/eIF(iso)4E, to confer PStV resistance in peanut. Moreover, we examined the interaction between HC-Pro and VPg of PStV with eIF4E/eIF(iso)4E using Y2H and BiFC. We also detected the expression of eIF4E/eIF(iso)4E in different peanut tissues.

2 Materials and methods

2.1 Cloning and sequencing of PeaeIF4E and PeaeIF(iso)4E genes

Total RNA was extracted from peanut (*Arachis hypogaea*) leaves using TRIzol (Invitrogen, Carlsbad, CA, USA), and cDNA was synthesized using an M-MLV RTase cDNA synthesis kit (Takara,Dalian, China), following the manufacturer's recommendations.To design primers for cloning the *eIF4E* and *eIF(iso)4E* genes of peanut, we compared and downloaded the *eIF4E* sequences of *Medicago truncatula* (XM_003593785), *Pisum sativum* (AY423375), *Pisum sativum* (DQ641471), *Phaseolus vulgaris* (EF571276), *Phaseolus vulgaris* (EF571275), *M. tornata*(HQ735878), and *M. truncatula* (HQ735877). The conserved sequences were used in designing the primer pairs PeaeIF4E2-R and PeaeIF4E2-F (Supplementary Table S1) to amplify the peanut *eIF4E* gene. The PCR products showing the expected lengths were sequenced and compared. The product with the correct sequence was then used in designing primers (Supplementary Table S1) for 5′amplification of cDNA ends (5′-RACE) and 3′-RACE to obtain the full-length cDNA of *PeaeIF4E*. A 5′-RACE kit (Invitrogen) was used according to the manufacturer's instructions to obtain the 5′ terminus of the *PeaeIF4E* gene. The 5′-RACE *eIF4E* outer and inner primers and 3′-RACE *eIF4E* outer and inner primer(Supplementary Table S1) were used to obtain the full-length *PeaeIF4E* cDNA. The *PeaeIF(iso)4E* gene was amplified using the primers listed in Supplementary Table S1. Phusion high-fidelity DNA polymerase (Takara, Dalian, China) was used to perform all the PCRs. A gel extraction kit (TianGen, Beijing, China) was used to purify the PCR products, which were then cloned into a pMD-18T easy vector (Takara) for sequencing. DNAMAN 6.0 wasused for multiple sequence alignment to homologous proteins of different plant species. MEGA5 with the Equal input model was used for phylogenetic analyses by using the neighbor-joining (NJ) method, and confidence was estimated by using 1,000 bootstrap replicates (Tamura et al., 2011).

2.2 Cloning and sequencing of PStVVPg and HC–Pro

Total RNA extraction and cDNA synthesis of the PStVVPg and HC-Pro genes were similar to the method used in cloning *PeaeIF4E* and *PeaeIF(iso)4E*. Based on the reported cDNA sequence of PStV (GenBank accession nos. KF439722, U05771, and U34972), we designed primers for the amplification of segments that corresponded to PStVVPg and PStVHCPro (Supplementary Table S1). Phusion high-fidelity DNA polymerase (Takara) was used for all PCRs. A gel extraction kit (TianGen) was used to purify the PCR products. Then the purified PCR products were cloned into a pMD-18T easy vector (Takara) for sequencing.

2.3 qRT-PCR analysis

Total RNA samples were extracted from the roots, stems,leaves, flower buds, leaf buds of "Huayu 20" and cDNAs were synthesized using the same method employed in cloning the *PeaeIF4E* and *PeaeIF(iso)4E* genes. All peanut tissueswere sampled from three different peanut plants as biological replicates. For the analysis of gene silencing in peanut, total RNA was extracted from the leaves of eIF4E-silenced, eIF(iso)4E-silenced, or eIF4E-eIF(iso)4E-double silenced peanuts and used in RT-PCR as previously described. Quantitative real-time PCR(qRT-PCR) reactions were conducted by using a SYBR Premix Ex Taq PCR kit (Takara) on an ABI7500 real-time PCR system(ABI, Foster, CA, USA). The primer pairs YG4E-R/YG4E-F and YG4IE-R/YG4IE-F were used to detect the expression of *PeaeIF4E* and *PeaeIF(iso)4E*. The primer pair YGPStVR/YGPStV-F was used to detect the accumulation of PStV after inoculation. The primer pair actin-R/actin-F was used to amplify the actin gene of *A.hypogaea*, which was used as a reference. The PCR reaction system consisted of a total volume of 20μL, which included 2μL of the RT product, 10μL of Ex *Taq*, 0.8μL of the primers (Supplementary Table S1), and 7.2μL of DEPC-water. All the reactions were performed in a 96-well optical plate. The PCR conditions were as follows: 94℃ for 15s, 94℃ for 6s, and 60℃ for 30s for a total of 40 cycles. Data analysis was performed by using an ABI7500 real-time PCR system, and standard curveswere also constructed.

2.4 Subcellular localization of PStVVPg,PStV HC-Pro, and eIF4E/eIF(iso)4E

The ORF of the target gene without its stop codon was amplified using the Phusion high-fidelity DNA polymerase (Takara) using the corresponding primer pairs, ORF4E-R/ORF4E-F, 4E(isoORF)F/4E(isoORF) R, PV-F/PV-R, and PH-R/PH-F, and then cloned into a pMD-18T easy vector (Takara) for sequencing. After confirmation of the correct clone from pMD-18T, these were then introduced into an entry vector pGWCm by TA cloning, and finally, via LR gateway recombination reaction (Invitrogen), was transferred to the plant expression vector, pHZM03. Plasmid DNA with green fluorescent protein (GFP) was transiently introduced into *Arabidopsis* protoplasts (Meng, 2012). After incubating for 12–16h in the dark, GFP expression was visualized using a confocal laser microscope (Leica SP5, Mannheim, Germany).

2.5 Yeast two-hybrid assay

Yeast two-hybrid screening was conducted using a Matchmaker Gold Yeast two-hybrid system (Clontech, Mountain View, CA, USA). The coding sequences of PStVVPg, PStV HC-Pro, and eIF4E/eIF(iso)4E were PCR amplified by using Phusion high-fidelity DNA polymerase (Takara) with the primer pair listed in Supplementary Table S1. PStVVPg, PStV HC-Pro were cloned into the prey vector, pGADT7, and eIF4E/eIF(iso)4E were cloned into the bait vector, pGBKT7. Confirmed correct clones were transformed into *Escherichia coli* DH5a cells for subsequent DNA sequencing. Both the confirmed correct prey and bait vectors were then co-transformed into AH109 yeast cells. SD/-Leu- Trp, SD/-Leu-Trp-His, SD/-Leu-Trp-His-Ade, and SD/-Leu-Trp-His-AdeCX-a-gal (Clontech) were used as selective media to detect any interactions. Yeast that contained both empty pGADT7 and pGBKT7 were used as negative controls, and yeast containing both pGBK-p53 and pGAD-RecT were used as positive controls.

2.6 Bimolecular fluorescence complementation (BiFC)

The Gateway compatible BiFC vectors pEarleyGate202- NYFP and pEarleyGate202-CYFP were used.

DNA fragments corresponding to PStVVPg, PStV HC-Pro, and eIF4E/eIF(iso)4E were introduced individually into the entry vector pGWCm as previously described. pGWCm-VPg and pGWCm-HC-pro were transferred to the pEarleyGate202-CYFP vector, whereas pGWCm-eIF4E and pGWCm-eIF(iso)4E were transferred to the pEarleyGate202-NYFP vector via the LR Gateway recombination reaction (Invitrogen). Plasmid DNA with YFP fusion was introduced into Arabidopsis protoplasts for transient expression (Meng, 2012). After incubation for 12–16h in the dark, confocal laser-scanning microscopy (Leica SP5) was performed to evaluate YFP expression.

2.7 Silencing of peanut using the virus-induced gene silencing (VIGS) vector system

For VIGS assays, PeaeIF4E and PeaeIF(iso)4E were amplified by using the primers listed in Supplementary Table S1. The PCR products were digested with *Xho*I and *BamH*I (New England Biolabs) and ligated into vector ALSV-RNA2, which was digested with the corresponding enzymes. The plasmid constructs were then sequenced (Takara) to confirm that we obtained the correct inserts. Apple latent spherical virus (ALSV)-based VIGS method was used. The constructed ALSVRNA2 and ALSV-RNA1 were transformed into *E. coli* DH5a cells for sequencing; then the confirmed correct plasmid was cultured in *E. coli* DH5a cells using TransGEN plasmid Maxi Kit (TransGEN) for purification. Approximately 1mg/mL of modified ALSV-RNA2 and ALSV-RNA1 were mechanically inoculated into *Chenopodium quinoa* plants. Two to three weeks later, symptomatic leaves were collected, homogenized in three volumes of extraction buffer (0.1MTris-HCl, pH7.8, 0.1MNaCl, 5mM $MgCl_2$), and re-inoculated in *C. quinoa* plants. Then the total RNA of symptomatic leaves were extracted and diluted to about 0.02 mg/mL to be mechanically inoculated into peanuts (2 weeks after sprouting; Igarashi et al., 2009).

2.8 Virus inoculation

In the present study, the PStV isolate was from Laixi, Qingdao city, Shandong province, China. It was cultured in our laboratory via mechanical inoculation for maintenance. PStV-infected plants were maintained at 25℃ with 8h photoperiod. Two weeks after silenced peanut, PStV was mechanically inoculated into peanut. After infection, the presence of PStV was tested by real-time RT-PCR.

3 Results

3.1 Cloning and sequencing of *PeaeIF4E* and *PeaeIF(iso)4E* genes

The sequence of the peanut *eIF4E* gene was amplified by using 5′-RACE and 3′-RACE. The full-length cDNA sequence of the eIF4E gene was deposited in GenBank (accession no. HE985069). The *eIF4E* gene was 764-bp long, with a 5′ 39-bp untranslated region, a 696-bp ORF, and a 29-bp 3′ untranslated region, and encoded a putative 231-amino acid polypeptide. Moreover, it had 92% nucleotide sequence identity to its homolog in *Pisum sativum* cultivar (GenBank accession no. AY423375.2) and 70% nucleotide identity to its homolog in *M. truncatula* (GenBank Accession No. BT134162.1).

The peanut *eIF(iso)4E* gene was isolated as described above. The obtained peanut eIF(iso)4E ORF was 612bp (GenBank accession no. KF956378) and was predicted to encode a 203-amino acid polypeptide. The cloned peanut eIF4E and eIF(iso)4E genes were designated as *peaeIF4E* and *peaeIF(iso)4E*, respectively. The identity of the *peaeIF4E* and *peaeIF(iso)4E* genes at the nucleotide sequence level was 53.55%, whereas that at the amino acid sequence level was 42.67%.

The eIF4E and eIF(iso)4E protein sequences of other plant species were aligned for phylogenetic reconstruction. In the phylogenetic tree, eIF4E and eIF(iso)4E formed two distinct branches (Fig.1). Furthermore, peaeIF4E was clustered with the eIF4E subgroup, whereas peaeIF(iso)4E was classified into the eIF(iso)4E subgroup. Analysis using the conserved domain search service of NCBI confirmed that the two proteins contained the eIF4E family conserved sequence.

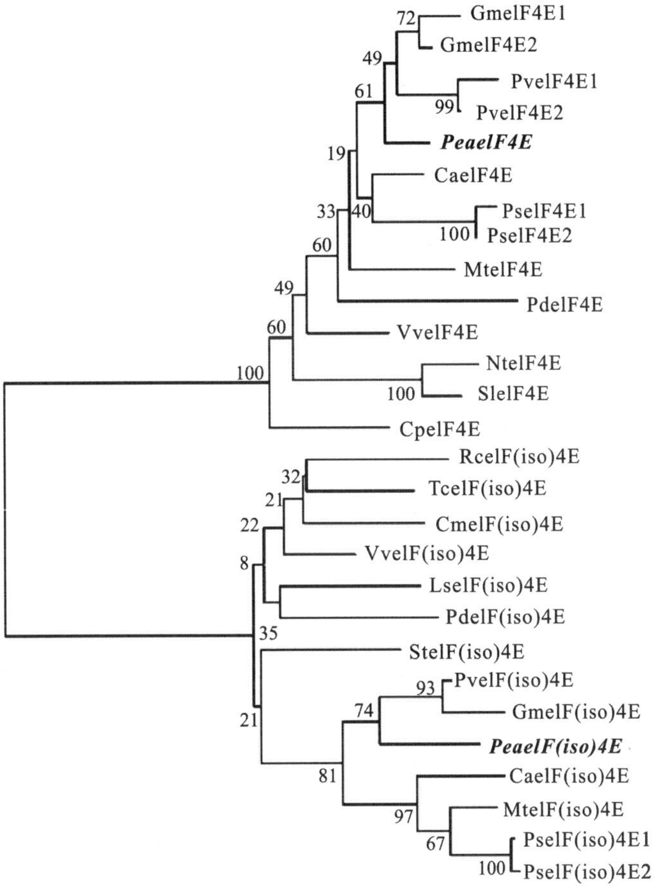

Fig.1 Phylogenetic analysis of eIF4E and eIF(iso)4E sequences of different plant species

The phylogenetic tree was constructed using ClustalW(http://www.ebi.ac.uk/clustalw/). The GenBank accession numbers of the amino acid sequences used are listed in Supplementary Table S2. The two peanut sequences are highlighted in bold and italics

3.2 Expression profiles of *PeaeIF4E*/*PeaeIF(iso)4E* in different tissues of peanut

To detect the expression levels of peanut *PeaeIF4E* and *PeaeIF(iso)4E* in various tissues, quantitative real-time PCR was performed. RNA was isolated from various tissues such as the roots, stems, leaves, flower buds, and leaf buds of "Huayu 20". The expression of the peanut actin gene is constant under different conditions and in different tissues (Chi et al., 2012), and was thus selected as a reference. The mRNA transcript levels showed significant differences in different tissues from peanut plants. The expression patterns and the mRNA transcript levels of *PeaeIF4E* and *PeaeIF(iso)4E* were similar in all tissues (Fig. 2A). The highest transcript levels, for both genes, were observed in the leaf bud and the lowest in flowers (Fig. 2A).

3.3 Subcellular localization of PeaeIF4E and PeaeIF(iso)4E in *Arabidopsis*

To test the subcellular localization of PeaeIF4E and PeaeIF(iso)4E, PeaeIF4E and PeaeIF(iso)4E were fused to the GFP by cloning of the ORFs of PeaeIF4E and PeaeIF(iso)4E into the entry vector pGWCm. Recombinant plasmids that expressed the PeaeIF4E-GFP and PeaeIF(iso)4E-GFP fusion proteins were introduced into Arabidopsis protoplasts cell. The *Arabidopsis* protoplasts cell were cultured in the dark at 23℃ for about 12–16h and observed by the confocal laser scanning microscopy (Leica SP5). The results suggested that the PeaeIF(iso)4E and PeaeIF4E fusion proteins were present in both the nucleus and the cytoplasm (Fig. 2B).

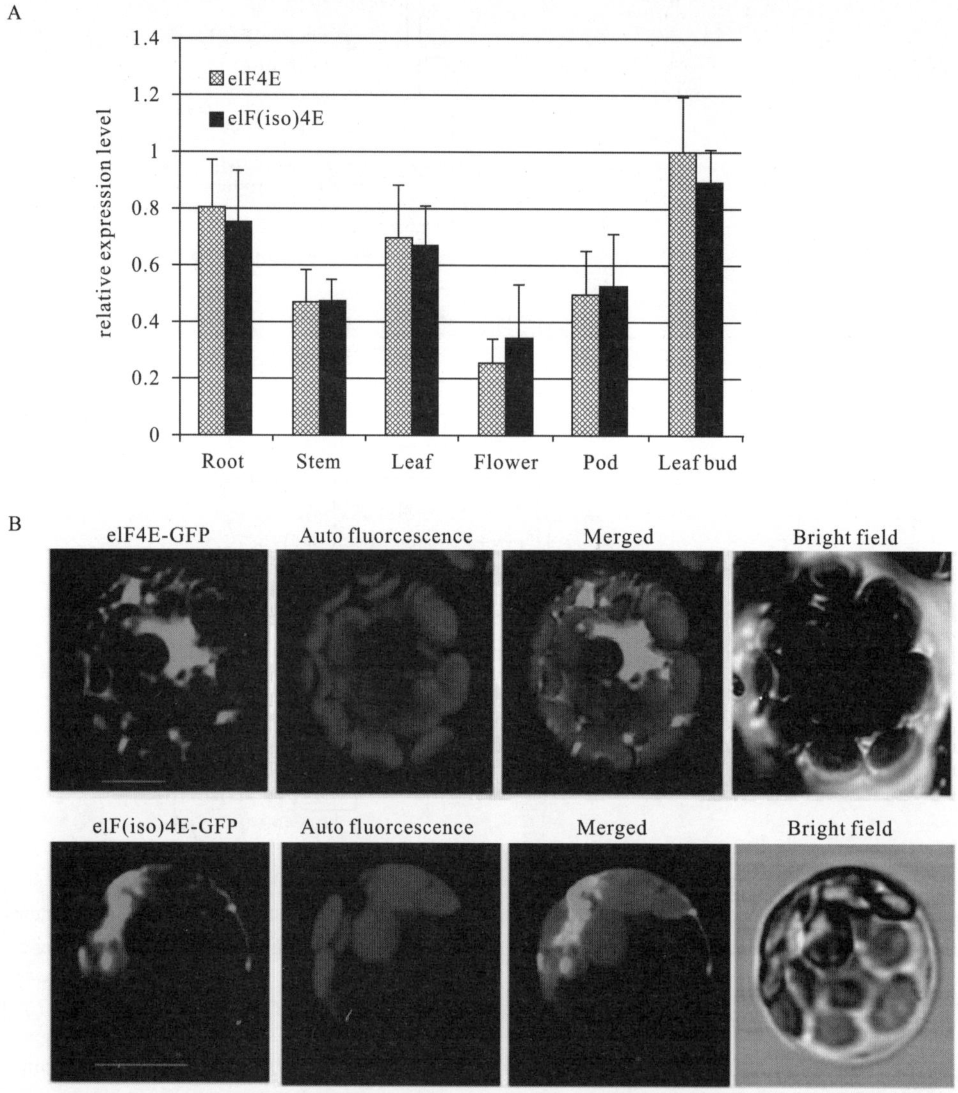

Fig. 2 mRNA transcript levels (A) of PeaeIF4E and PeaeIF(iso)4E and their subcellular localization (B)

Relative mRNA expression levels of *peaeIF4E* and *peaeIF(iso)4E* were determined by real-time reverse transcript PCR (RT-PCR). The values represent means of three biological repeats and the value of each biological repeat is the mean of three technical repeats. All values were normalized to the reference gene peanut actin. PeaeIF4E and PeaeIF(iso)4E were fused with green fluorescent protein (GFP) are delivered into protoplasts of Arabidopsis. The GFP fluorescence was observed 12–16h after transfection. Scale bars=10μm

3.4 Subcellular Localization of VPg and HC-Pro in Arabidopsis

We obtained the VPg and HC-Pro protein cDNAs from PStV by RT-PCR, and their ORFs were fused to GFP as above. The fusion proteins were expressed in *Arabidopsis* protoplasts as above. The results suggested that the VPg fusion protein was present in the nucleus, and HC-Pro fusion protein was observed in the cytoplasm (Fig. 3).

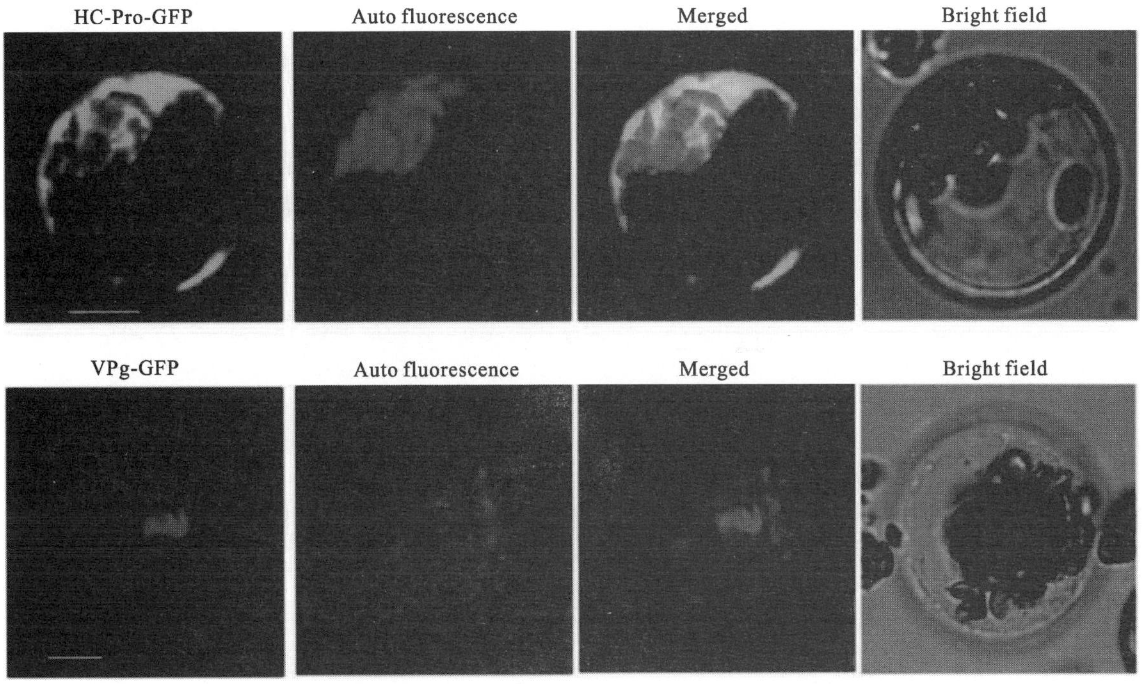

Fig.3 Subcellular localization of PStV HC-Pro and PStV VPg

PStV HC-Pro and PStVVPg fused with GFP were transfected into protoplasts of Arabidopsis. The GFP fluorescence was observed 12–16h after transfection. Scale bars=10μm

3.5 Interaction analysis between eIF4E/eIF(iso)4E and PStV HC-Pro/PStVVpg

Yeast two-hybrid analysis was used to determine whether there was an interaction between viral proteins and peanut proteins. Yeast two-hybridization showed interactions between VPg and PeaeIF4E/PeaeIF(iso)4E, and between HC-Pro and PeaeIF4E/PeaeIF(iso)4E (Fig. 4). The interactions were further confirmed by using BiFC. In this system, the YFP was split into N-terminal and C-terminal fragments, and the PeaeIF4E and PeaeIF(iso)4E were attached to the N-terminal fragment of YFP (eIF4E-NY and eIF(iso)4E-NY). VPg and HC-pro were fused to the C-terminal fragment of YFP (VPg-CY and HCpro-CY). The eIF(iso)4E-NYCVPg-CY, eIF4E-NYCVPg-CY, eIF4E-NYCHC-pro-CY, and eIF(iso)4E-NYCHC-pro-CY plasmids were then transformed into Arabidopsis protoplasts cells. A nuclear fluorescence signal was, respectively, observed in eIF(iso)4E-NYCVPg-CY and eIF4E-NYCVPg-CY combination and the signal was observed throughout the nucleus (Fig. 5C). Cytoplasmic fluorescence signals were observed in the eIF4E-NYCHC-pro-CY and eIF (iso) 4E-NYCHC-pro-CY combinations, and the signals were observed throughout the cytoplasm (Fig. 5A). As expected, the negative controls, i.e., the combinations of eIF4E-NYCCY and NYCHC-Pro-CY did not emit fluorescence signals (Fig. 5B). Taken together, these results show that PeaeIF4E/PeaeIF(iso)4E interacts with VPg in the nucleus, whereas PeaeIF4E/PeaeIF(iso)4E interacts with HC-Pro in the cytoplasm.

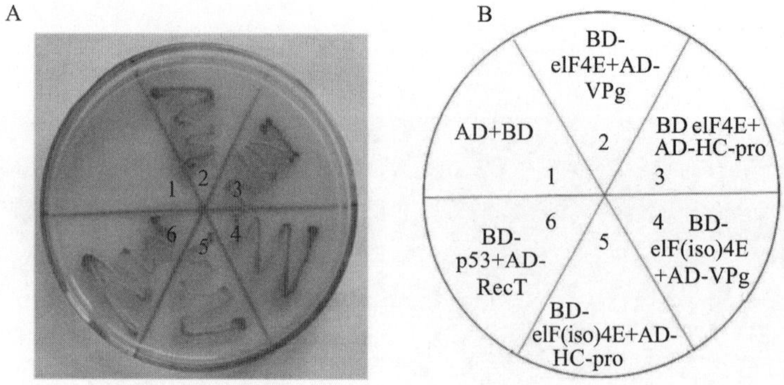

Fig.4 Yeast two-hybrid assay of protein–protein interaction between the PeaeIF4E/PeaeIF(iso)4E from peanut and PStV-HC-pro/PStV-VPg

Yeast co-transformants were grown on selective medium SD/-Leu-Trp-His-Ade plus X-a-Gal and incubated for 4d at 30℃ (A). Frame (B) corresponds to the clones left (A)

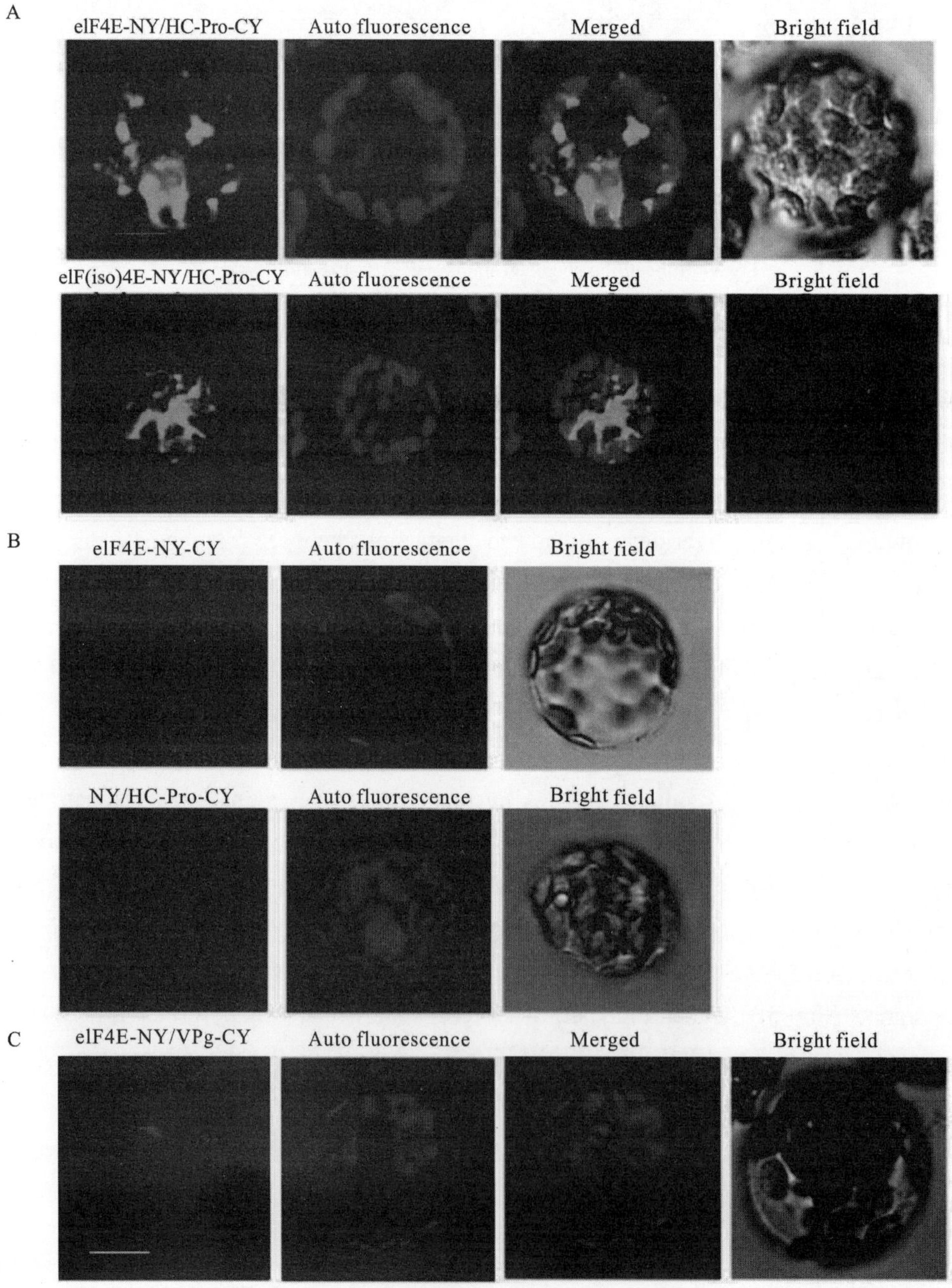

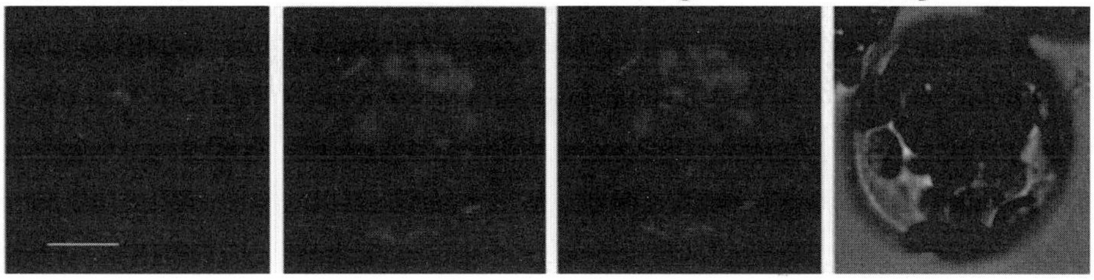

Fig.5 Bimolecular fluorescence complementation (BiFC) assay showing interaction between PeaeIF4E/PeaeIF(iso)4E and PStV HC-Pro/PStV VPg

The full-length open reading frame (ORF) of PeaeIF4E/PeaeIF(iso)4E was cloned into the vector pEarleyGate202-NYFP [eIF4E-NY, eIF(iso)4E-NY] and that of PStV HC-Pro/PStV VPg into pEarleyGate202-CYFP (HC-Pro-CY, VPg-CY). The recombinant plasmids were transfected into protoplasts of Arabidopsis. Fluorescence was observed at 14–16h post-transfection by confocal laser-scanning microscopy. Scale bars = 10μm. (A) BiFC analysis of PeaeIF4E/PeaeIF(iso)4E and PStV HC-Pro. (B) A range of negative controls. (C) BiFC analysis of PeaeIF4E/PeaeIF(iso)4E and PStV VPg

3.6 Silencing of *PeaeIF4E* and *PeaeIF(iso)4E* genes confers resistance against PStV in peanut

To confirm the role of the PeaeIF4E and PeaeIF(iso)4E genesin PStV infection, gene silencing was performed. The *PeaeIF4E*(355-nt) and *PeaeIF(iso)4E* (326-nt) fragments were inserted into the ALSV-RNA2 vector. The recombinant viruses (ALSV-eIF4E and ALSV-eIF(iso)4E) were then inoculated into a peanut.Two weeks after inoculation, real-time RT-PCR analysis wasperformed, which demonstrated that the expression levels of *PeaeIF4E* and *PeaeIF(iso)4E* were significantly lower in the inoculated plants as compared with control (Fig.6A), although no significant phenotypic alterations were observed in the transgenic plants (Fig. 6C) . The expression level of PeaeIF4E decreased by 60% (upon inoculation with ALSVeIF4E) while that of *PeaeIF(iso)4E* decreased by 65% (ALSVeIF(iso)4E inoculation) as compared with control. When inoculated with ALSV-eIF4ECALSV-eIF(iso)4E, the expression levels of *PeaeIF4E* and *PeaeIF(iso)4E* decreased by 53% and 57%, respectively, as compared with control (Fig. 6A). Peanut plants where either the *PeaeIF4E* or *PeaeIF(iso)4E* were silenced, showed mosaic symptoms of infection at about 10–14d after inoculation with PStV. On the other hand, silencing of both *PeaeIF4E* or *PeaeIF(iso)4E* caused the symptoms to appear later at about 18–20d after inoculation with PStV, and the symptoms were milder as compared with plants with only one gene silenced and the control (Fig. 6C). Real-time RT-PCR analysis indicated that silencing both *PeaeIF4E* or *PeaeIF(iso)4E* reduced PStV accumulation by 70% compared to control plants. No significant differences were observed between plants in which either PeaeIF4E or PeaeIF(iso)4E was silenced as compared with controls (Fig. 6B) suggesting that the two isoforms play overlapping or redundant roles in the virus multiplication cycle.

4 Discussion

The cap-binding protein eIF4E/eIF(iso)4E confers resistance to some RNA viruses in specific plant species (Nicaisen et al., 2003; Nieto et al., 2006; Ruffel et al., 2006; Hwang et al., 2009). In the present study, we cloned the peanut *eIF4E*/*eIF(iso)4E* genes and analyzed their protein sequences. Phylogenetic analyses of these sequences demonstrated that *PeaeIF4E* and *PeaeIF(iso)4E* showed high homologies with orthologs from related plant species. *PeaeIF4E* and *PeaeIF(iso)4E* were closely related totheir homologs from soybean (G. max) and kidney bean (*Phaseolus vulgaris*). The expression levels of *PeaeIF4E* and *PeaeIF(iso)4E* were similar in different peanut tissues with both genes being upregulated in leaf buds and roots and down regulated in flowers (Fig. 2). Previous studies have also shown that these two genes are upregulated in young tissues and down regulated in mature tissues of *Arabidopsis* and plum, which corroborated our results (Rodriguez et al., 1998; Wang et al.,2013).

A

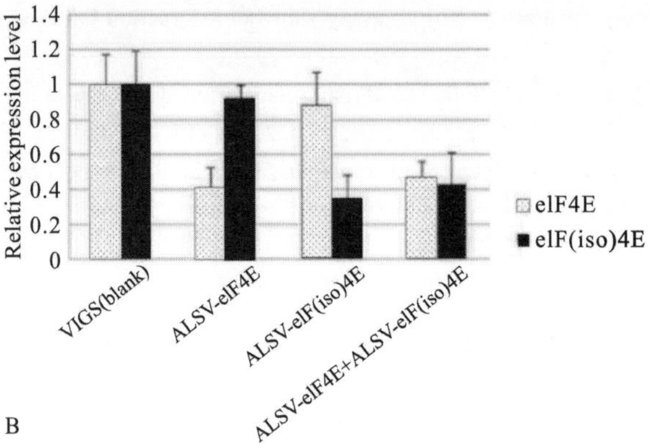

B

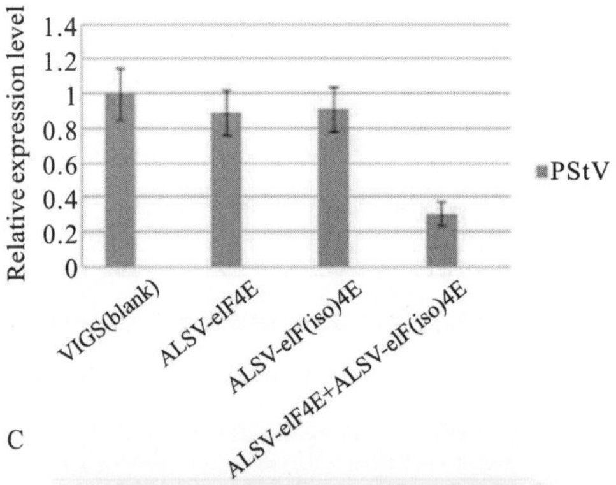

C

Fig.6 Real-time PCR analysis for target gene expression in peanut (A), accumulation of PStV RNA (B) and the growth condition and symptoms of peanut after inoculated PStV (C)

Virus-induced gene silencing of *PeaeIF4E*/*PeaeIF(iso)4E* in representative plants belonging to each of the four different treatments. For qRT-PCR detection of the expression of PeaeIF4E/PeaeIF(iso)4E, three plants from each group were pooled as one sample and the experiments were performed in triplicate (A). Effects of silencing of *PeaeIF4E*/*PeaeIF(iso)4E* on PStV infection. The accumulation of PStV RNA in inoculated peanut plants was detected by RT-PCR 15d after PStV inoculation (B). The growth condition and symptoms of peanut after inoculated PStV (C). Four different treatments peanuts were mechanically infected by PStV. Three plants from each group were pooled as one sample and the experiments were performed in triplicate, 10–14d later, PStV disease symptom began to appeared. The peanut growth condition after inoculated with PStV, except the different symptoms on peanut leaves, different treatments peanuts growth condition showed no significant differences compared with control. 1, control group; 2, silencing *PeaeIF4E*; 3, silencing *PeaeIF(iso)4E*; 4, silencing both *PeaeIF4E* and *PeaeIF(iso)4E*

Confocal microscopy showed that PeaeIF4E and PeaeIF(iso)4E were both localized in the nucleus and the cytoplasm of Arabidopsis cells (Fig.2B). In Chrysanthemum morifolium, eIF(iso)4E was also localized in the nucleus, cytoplasm, and cytomembrane (Song et al., 2013). In Arabidopsis, in quiescent cells, eIF4E was localized in the nucleus, whereas inproliferating cells, this was detected in the cytoplasm. Both inquiescent and proliferating cells, eIF(iso)4E has been observed in the cytoplasm and nucleus (Bush et al., 2009). In mature Arabidopsis cells, PeaeIF4E and PeaeIF(iso)4E were localized to both the nucleus and the cytoplasm, but whether the two proteins have different cellular locations during different growth stages needs further investigation. In animals, eIF4E has beendetected in both cytoplasm and nucleus, and about 68% ofthe eIF4E was detected in mammalian nuclei. eIF4E plays different roles depending on its subcellular location; when itis localized in cytoplasm, it functions in translation initiation.When it is localized in the nucleus, it participates in the export of mRNAs that contain 4E-sensitive elements (SE; Iborra et al.,2001; Culjkovic et al., 2007, 2008). The two peanut proteins that localized in different places may play different roles that require further investigation.

Confocal microscopy showed that the VPg fusion protein was localized to the nucleus, whereas the HC-Pro fusion protein was observed in the cytoplasm. In the case of MDMV (*Maize dwarfmosaic virus*) and TuMV (*Turnip mosaic virus*), HC-Pro was detected in the cytoplasm (Li et al., 2001; Zheng, 2011). In the case of Potato virus Y, HC-Pro was localized throughout the cytoplasm, whereas it displays different subcellular localization patterns depending on the cellular environment (del Toro et al.,2014). HC-Pro was also distributed throughout the cytoplasmin CABMV (*Cowpea aphid-borne mosaic virus*) infected plants(Mlotshwa et al., 2002). In WYMV (*Wheat mosaic virus*) infected plants, VPg occurred in two forms in the nucleus; one gathered into one or several irregular shape inclusions, whereas the other was evenly distributed across the entirenucleus (Bian, 2013). These results were consistent with the findings of our study as well as with the results of BiF Canalysis of HC-Pro and PeaeIF4E/PeaeIF(iso)4E, and that of VPg and PeaeIF4E/PeaeIF(iso)4E. We also observed interactions between VPg and PeaeIF4E/PeaeIF(iso)4E and between HCProand PeaeIF4E/PeaeIF(iso)4E. The interaction between VPg and PeaeIF4E/PeaeIF(iso)4E interactions was observed in the nucleus. These results coincided with the findings on viral protein location in our study. The interaction between VPg and PeaeIF4E/PeaeIF(iso)4E in the nucleus provides additional evidence that both proteins are localized in the nucleus. In potyvirus, the interaction between VPg and eIF4E/eIF(iso)4E plays a major role in cellular transport and localization of RNA (Lellis et al., 2002). HC-Pro and PeaeIF4E/PeaeIF(iso)4E interactions were detected in the cytoplasm but not in the nucleus, which supported the absence of HC-Pro in the nuclei of infected plant cells (Rajamaki and Valkonen, 2003).

The interaction between translation initiation factors andviral proteins is essential for viral replication and infection(Dreher and Miller, 2006; Robaglia and Caranta, 2006). The interaction between eIF4E/eIF(iso)4E and HC-Pro/VPg maybe necessary for potyvirus infection and amplification. The silencing of both *PeaeIF4E* and *PeaeIF(iso)4E* conferred moderate resistance against PStV in peanut, as evidenced by symptom delay and reduced virus accumulation. This findings strongly suggest that PeaeIF4E and PeaeIF(iso)4E play important rolesto facilitate virus infection and that they are functionally interchangeable. The silencing of *PeaeIF4E* and *PeaeIF(iso)4E* hindered the interaction between the host and the virus,which in turn prevented infection and viral replication inthe host. Furthermore, viral accumulation was lower in gene-silencedpeanut plants. Silenced plants with decreased expression of PeaeIF4E and/or PeaeIF(iso)4E did not phenotypically differ from control plants. In tobacco, antisense depletion of either eIF4E and two eIF(iso)4E isoforms displayed normal development, but antisense depletion of both eIF4E and eIF(iso)4E resulted in semi-dwarf phenotype

(Combe et al.,2005). It is possible that the remaining low levels of expression of PeaeIF4E and PeaeIF(iso)4E in the silenced plants were sufficient to sustain peanut growth. Alternatively, it is possible that the twogenes are dispensable for peanut growth and that some other genes possess complementary functions. In other virus-host compositions, simultaneous mutations in the eIF4E and eIF(iso)4E genes result in a decrease inviral resistance, such as resistance to Pepper veinal mottle virus (PVMV) and ChiVMV in pepper (Ruffel et al., 2006; Hwang et al., 2009). Furthermore, knocking down the eIF(iso)4E in peach plants results in peach resistance to PPV (Cui and Wang, 2016).Silencing of the eIF(iso)4E gene in plum confers resistance to PPV(Wang et al., 2013) and silencing the eIF4E gene in melon plantsconfers upon it broad-spectrum viral resistance (Rodriguez-Hernandez et al., 2012). These reports suggest that these viruses probably utilize one or two translation initiation factors during infection. PStV may use either translation initiation factors of PeaeIF4E and PeaeIF(iso)4E during its infection because silencing only one gene does not confer resistance to PStV in peanut. Our study suggests that the eIF4E/eIF(iso)4E gene may be utilized in increasing PStV resistance in peanut by gene silencing, gene mutation, or the TILLING strategy (Ruffel et al., 2002; Kanyuka et al., 2005; Stein et al., 2005; Piron et al., 2010). We have a variety of peanut cultivars that could be employed in TILLING to detect allelic variants of a target gene.

Therefore, the findings of the present study suggest that eIF4E/eIF(iso)4E plays important roles in the PStV infection cycle and may serve as a novel method for increasing the PStV resistance in economically important peanut cultivars. The two genes may also be used as genetic resources for improving PStV resistance in peanut breeding programs.

Acknowledgements

This research was funded by The Youth Scientific Research Foundation of Shandong Academy of Agricultural Sciences (2014QNM01); Natural Science Foundation of Shandong Province (ZR2015YL065, ZR2014CQ025, 2015GNC111029).

We would like to thank Dr. Chengming Fan (The Chinese Academy of Sciences, Beijing, China) for providing Gatewayvectors for our localization and BiFC studies, Dr. Qijun Chen(China Agricultural University, Beijing, China) for providing pGWCm, and Dr.Na Chen (Shandong Peanut Research Institute,Qingdao, China) for providing vectors for the Y2H assays.

References

[1] AHLQUIST P, NOUEIRY A O, LEE W M, et al. Host factors in positive-strand RNA virus genome replication[J]. Journal of virology,2003, 77: 8181-8186.

[2] ALA-POIKELA M, GOYTIA E, HAIKONEN T, et al. Helper component proteinase of the genus Potyvirus is an interaction partner of translation initiation factors eIF(iso)4E and eIF4E and contains a 4E binding motif[J]. Journal of virology, 2011,85: 6784-6794.

[3] ATREYA C D, ATREYA P L, THORNBURY D W, et al. Sitedirected mutations in the potyvirus HC-Pro gene affect helper component activity, virus accumulation, and symptom expression in infected tobacco plants[J]. Virology,1992, 191, 106-111.

[4] BIAN L. Structural and Subcellular Distribution of VPg Protein Encode by Wheat Yellow Masaic Virus. Masterthesis[D]. Zhejiang Normal University, 2013.

[5] BROWNING K S. Plant translation initiation factors: it is not easy to be green[J]. Biochemical Society Transactions, 2004, 32(4): 589-591.

[6] BUSH M S, HUTCHINS A P, JONES A M, et al. Selective recruitment of proteins to 5' cap complexes during the growth cycle in

Arabidopsis[J]. The Plant Journal,2009, 59(3): 400-412.

[7] CARRINGTON J C, HERNDON K L. Characterization of the potyviral HC-pro autoproteolytic cleavage site[J]. Virology,1992, 187: 308-315.

[8] CHARRON C, NICOLAI M, GALLOIS J L, et al. Natural variation and functional analyses provide evidence for co-evolution between plant eIF4E and potyviral VPg[J]. The Plant Journal,2008, 54: 56-68.

[9] CHI X, HU R, YANG Q, et al. Validation of reference genes for gene expression studies in peanut by quantitative real-time RT-PCR[J]. Molecular genetics and genomics,2012, 287(2): 167-176.

[10] CHOI H S, KIM J S, CHEON J U, et al. First report of Peanut stripe virus (Family Potyviridae) in South Korea[J]. Plant Disease,2001, 85:679.

[11] CHOI H S, KIM M, PARK J W, et al. Occurrence of Bean common mosaic virus (BCMV) infecting Peanut in Korea[J]. Plant Pathology Journal,2006, 22: 97-102.

[12] COMBE J P, PETRACEK M E, VAN ELDIK G, et al. Translation initiation factors eIF4E and eIFiso4E are required for polysome formation and regulate plant growth in tobacco[J]. Plant molecular biology, 2005, 57(5): 749-760.

[13] CUI H, WANG A. An efficient viral vector for functional genomic studies of Prunus fruit trees and its induced resistance to Plum pox virus via silencing of a host factor gene[J]. Plant Biotechnol Journal, 2016, 15: 344-356.

[14] CULJKOVIC B, TAN K, OROLICKI S, et al. The eIF4E RNA regulon promotes the Akt signaling pathway[J]. The Journal of cell biology, 2008, 181(1): 51-63.

[15] CULJKOVIC B, TOPISIROVIC I, BORDEN K L. Controlling gene expression through RNA regulons: the role of the eukaryotic translation initiation factor eIF4E[J]. Cell Cycle,2007, 6(1): 65-69.

[16] DEL TORO F, FERNANDEZ F T, TILSNER J, et al. Potato virus Y HCPro localization at distinct, dynamically related and environment-influenced structures in the cell cytoplasm[J]. Molecular plant-microbe interactions, 2014, 27(12): 1331-1343.

[17] DEMSKI J W, LOVELL G R. Peanut stripe virus and the distribution of peanut seed[J]. Plant Disease, 1985, 69: 734-738.

[18] DREHER T W, MILLER W A. Translational control in positive strand RNA plant viruses[J]. Virology, 2006, 344: 185-197.

[19] GOVIER D A, KASSANIS B, PIRONE T P. Partial purification and characterization of the potato virus Y helper component[J]. Virology, 1977, 78: 306-314.

[20] HWANG J, LI J, LIU W Y, et al. Double mutations in eIF4E and eIFiso4E confer recessive resistance to Chilli veinal mottle virus in pepper[J]. Molecules and cells, 2009, 27: 329-336.

[21] IBORRA F J, JACKSON D A, COOK P R. Coupled transcription and translation within nuclei of mammalian cells[J]. Science,2001, 293: 1139-1142.

[22] IGARASHI A, YAMAGATA K, SUGAI T, et al. Apple latent spherical virus vectors for reliable and effective virus-induced gene silencing among a broad range of plants including tobacco, tomato, Arabidopsis thaliana, cucurbits, and legumes[J]. Virology,2009, 386: 407-416.

[23] JACKSON R J, HELLEN C U, PESTOVA T V. The mechanism of eukaryotic translation initiation and principles of its regulation[J]. Nature reviews. Molecular cell biology, 2010, 11(2): 113-127.

[24] JIN Y, MA D, DONG J, et al. The HC-pro protein of potato virus Y interacts with NtMinD of tobacco[J]. Molecular plant-microbe interactions, 2007, 20(12): 1505-1511.

[25] KANG B C, YEAM I, FRANTZ J D, et al. The pvr1 locus in Capsicum encodes a translation initiation factor eIF4E that interacts with Tobacco etch virus VPg[J]. The Plant journal,2005, 42(3): 392-405.

[26] KANYUKA K, DRUKA A, CALDWELL D G, et al. Evidence that the recessive by movirus resistance locus rym4 in barley corresponds to the eukaryotic translation initiation factor 4E gene[J]. Molecular plant pathology,2005, 6(4):449-458.

[27] LELLIS A D, KASSCHAU K D, WHITHAM S A, et al. Lossof- susceptibility mutants of Arabidopsis thaliana reveal an essential role for eIF(iso)4E during potyvirus infection[J]. Current biology,2002, 12(12): 1046-1051.

[28] LI X D, FAN Z F, LI H F, et al. Accumulation and immunolocalization of Maize dwarf mosaic virus HC-Pro in infected maize leaves[J]. Acta Phytopathologica Sinica,2001, 31: 310-314.

[29] LLAVE C, KASSCHAU K D, CARRINGTON J C. Virus-encoded suppressor of posttran scriptional gene silencing targets a maintenance step in the silencing pathway[J]. Proceedings of the National Academy of Sciences of the United States of America, 2000, 97(24): 13401-13406.

[30] MACKENZIE J. Wrapping things up about virus RNA replication[J]. Traffic, 2005, 6(11): 967-977.

[31] MENG Y Y. Functional Analysis of CRYs and CIB3 in Flowering and Senescence Regulation in Soybean (Glycine max)[D]. Ph.D thesis, Graduate School of Chinese Academy of Agricultural Sciences, 2012.

[32] MLOTSHWA S, VERVER J, SITHOLE-NIANG I, et al. Subcellular location of the helper component proteinase of Cowpea aphid-borne mosaic virus[J]. Virus Genes,2002, 25(2): 207-216.

[33] NICAISE V, GERMAN-RETANA S, SANJUAN R, et al. The eukaryotic translation initiation factor 4E controls lettuce susceptibility to the Potyvirus Lettuce mosaic virus[J].Plant Physiology, 2003, 132: 1272-1282.

[34] NIETO C, MORALES M, ORJEDA G, et al. An eIF4E allele confers resistance to an uncapped and non-polyadenylated RNA virus in melon[J].The Plant Journal,2006, 48, 452-462.

[35] PIRON F, NICOLAI M, MINOIA S, et al. An induced mutation in tomato eIF4E leads to immunity to two potyviruses[J]. PLoS ONE,2010, 5:e11313.

[36] RAJAMAKI M L,VALKONEN J P. Viral genome-linked protein (VPg) controls accumulation and phloem-loading of a potyvirus in inoculated potato leaves[J]. Molecular plant-microbe interactions,2002, 15: 138-149.

[37] RAJAMAKI M L, VALKONEN J P. Localization of a potyvirus and the viral genome-linked protein in wild potato leaves at an early stage of systemic infection[J]. Molecular plant-microbe interactions,2003, 16: 25-34.

[38] REDONDO E, KRAUSE-SAKATE R, YANG S J, et al. Lettuce mosaic virus pathogenicity determinants in susceptible and tolerant lettuce cultivars map to different regions of the viral genome[J]. Molecular plant-microbe interactions,2001, 14: 804-810.

[39] REVERS F, VAN DER VLUGT R A, SOUCHE S, et al. Nucleotide sequence of the 3′ terminal region of the genome of four lettuce mosaic virus isolates from Greece and Yemen[J]. Archives of virology,1999, 144: 1619-1626.

[40] ROBAGLIA C, CARANTA C. Translation initiation factors: a weak link in plant RNA virus infection[J]. Trends Plant Science,2006, 11: 40-45.

[41] RODRIGUEZ C M, FREIRE M A, CAMILLERI C, et al. The Arabidopsis thaliana cDNAs coding for eIF4E and eIF(iso)4E are not functionally equivalent for yeast complementation and are differentially expressed during plant development[J].The Plant Journal,1998,13: 465-473.

[42] RODRIGUEZ-HERNANDEZ A M, GOSALVEZ B, SEMPERE R N, et al. Melon RNA interference (RNAi) lines silenced for Cm-eIF4E show broad virus resistance[J]. Molecular plant pathology,2012, 13(7): 755-763.

[43] ROJAS M R, ZERBINI F M, ALLISON R F, et al. Capsid protein and helper component-proteinase function as potyvirus cell-to-cell movement proteins[J]. Virology,1997, 237:283-295.

[44] RUFFEL S, DUSSAULT M H, PALLOIX A, et al. A natural recessive resistance gene against potato virus Y in pepper corresponds to the eukaryotic initiation factor 4E (eIF4E) [J].The Plant Journal,2002, 32:1067-1075.

[45] RUFFEL S, GALLOIS J L, MOURY B, et al. Simultaneous mutations in translation initiation factors eIF4E and eIF(iso)4E are required to prevent pepper veinal mottle virus infection of pepper[J]. The Journal of general virology,2006, 87(7): 2089-2098.

[46] SAENZ P, SALVADOR B, SIMON-MATEO C, et al. Host-specific involvement of the HC protein in the long distance movement of potyviruses[J]. Journal of virology,2002, 76(4): 1922-1931.

[47] SALEH N, HORN N M, REDDY D V R, et al. Peanut stripe virus in Indonesia[J]. European journal of plant pathology,1989, 99:123-127.

[48] SANFACON H. Plant translation factors and virus resistance[J]. Viruses, 2015, 7: 3392-3419.

[49] SINGH M K, CHANDEL V, HALLAN V, et al. Occurrence of Peanut stripe virus on patchouli and raising of virus-free patchouli plants by meristem tip culture[J]. Journal of Plant Diseases and Protection,2009, 116(1): 2-6.

[50] SONENBERG N, MORGAN M A, MERRICK W C, et al. A polypeptide in eukaryotic initiation factors that crosslinks specifically to the 5′-terminal cap in mRNA[J]. Proceedings of the National Academy of Sciences of the United States of America,1978, 75(10):4843-4847.

[51] SONG A, LOU W, JIANG J, et al. An isoform of eukaryotic initiation factor 4E from Chrysanthemum morifolium interacts with Chrysanthemum virus B coat protein[J]. PLoS ONE,2013, 8:e57229.

[52] STEIN N, PEROVIC D, KUMLEHN J, et al. The eukaryotic translation initiation factor 4E confers multiallelic recessive Bymovirus resistance in Hordeum vulgare (L) [J]. The Plant journal,2005,42(6): 912-922.

[53] TAMURA K, PETERSON D, PETERSON N, et al. MEGA5: molecular evolutionary genetics analysis using maximum likelihood, evolutionary distance, and maximum parsimony methods[J].Molecular biology and evolution,2011, 28(10): 2731-2739.

[54] URCUQUI-INCHIMA S, HAENNI A L, BERNARDI F. Potyvirus proteins: a wealth of functions[J]. Virus research,2001, 74: 157-175.

[55] WANG A, KRISHNASWAMY S. Eukaryotic translation initiation factor 4E-mediated recessive resistance to plant viruses and its utility in crop improvement[J]. Molecular plant pathology, 2012, 13(7): 795-803.

[56] WANG X, KOHALMI S E, SVIRCEV A, et al. Silencing of the host factor eIF(iso)4E gene confers plum pox virus resistance in plum[J]. PLoS ONE,2013, 8:e50627.

[57] WEI T, ZHANG C, HONG J, et al. Formation of complexes at plasmodesmata for potyvirus intercellular movement is mediated by the viral protein P3N-PIPO[J]. PLoS Pathogens,2010, 6(6):e1000962.

[58] WHITHAM S A, WANG Y. Roles for host factors in plant viral pathogenicity[J]. Current opinion in plant biology,2004, 7(4): 365-371.

[59] XU Z Y. Viruses and Viral Diseases of Oil Crops[M]. Beijing: Chemical Industry Press,2008.

[60] XU Z Y, YU Z, LIU J L, et al. A virus causing peanut mild mottle in Hubei province, China[J]. Plant Disease,1983, 67: 1029-1032.

[61] ZHENG H Y. The Self Interaction of TuMV HC-Pro and its Interaction with Rieske Fe/S Protein Encode by Arabidopsis thaliana[D]. Ph.D dissertation, Huazhong Agricultural University, Wuhan, 2011.

Rice stripe virus NS3 protein regulates primary miRNA processing through association with the miRNA biogenesis factor OsDRB1 and facilitates virus infection in rice

Lijia Zheng[1,2], Chao Zhang[1], Chaonan Shi[1], Zhirui Yang[2], Yu Wang[2], Tong Zhou[3],Feng Sun[3], Hong Wang[1], Shanshan Zhao[2], Qingqing Qin[2], Rui Qiao[2], Zuomei Ding[1],Chunhong Wei[2], Lianhui Xie[1], Jianguo Wu[1],Yi Li[2]

(1 State Key Laboratory of Ecological Pest Control for Fujian and Taiwan Crops, Fujian Province Key Laboratory of Plant Virology, Institute of Plant Virology, Fujian Agriculture and Forestry University, Fuzhou, China; 2 The State Key Laboratory of Protein and Plant Gene Research, College of Life Sciences, Peking University, Beijing, China; 3 Institute of Plant Protection, Jiangsu Academy of Agricultural Sciences, Nanjing, China)

Abstract: MicroRNAs (miRNAs) are small regulatory RNAs processed from primary miRNA transcripts, and plant miRNAs play important roles in plant growth, development, and response to infection by microbes. Microbial infections broadly alter miRNA biogenesis, but the underlying mechanisms remain poorly understood. In this study, we report that the *Rice stripevirus*(RSV)-encoded nonstructural protein 3 (NS3) interacts with OsDRB1, an indispensable component of the rice (*Oryza sativa*) miRNA-processing complex. Moreover, the NS3-OsDRB1 interaction occurs at the sites required for OsDRB1 self-interaction, which is essential for miRNA biogenesis. Further analysis revealed that NS3 acts as a scaffold between OsDRB1 and pri-miRNAs to regulate their association and aids *in vivo* processing of pri-miRNAs. Genetic evidence in *Arabidopsis* showed that NS3 can partially substitute for the function of double-stranded RNA binding domain (dsRBD) of AtDRB1/AtHYL1 during miRNA biogenesis. As a result, NS3 induces the accumulation of several miRNAs, most of which target pivotal genes associated with development or pathogen resistance. In contrast, a mutant version of NS3 (mNS3), which still associated with OsDRB1 but has defects in pri-miRNA binding, reduces accumulation of these miRNAs. Transgenic rice lines expressing *NS3* exhibited significantly higher susceptibility to RSV infection compared with non-trans-genic wild-type plants, whereas the transgenic lines expressing *mNS3* showed a less-sensitive response. Our findings revealed a previously unknown mechanism in which a viral protein hijacks OsDRB1, a key component of the processing complex, for miRNA biogenesis and enhances viral infection and pathogenesis in rice.

1 Introduction

MicroRNAs (miRNAs), a class of endogenous small RNAs processed from their primary transcripts (pri-miRNAs), are crucial for plant development and responses to abiotic and biotic stresses (Baulcombe, 2004; Chen, 2009; Iwakawa and Tomari, 2015). Invading pathogens can manipulate the biogenesis and stability of many miRNAs to promote infection, or affect plant defense. For example, the accumulation of miR168 is

PLoS Pathogens. 2017, 13(10): e1006662
Received 3 April 2017; Accepted 22 September 2017

elevated by infections with *Cymbidum ringspot virus* (CymRSV), *crucifer-infecting Tobacco mosaic virus* (crTMV), *Potato virus X* (PVX), and *Tobacco etch virus* (TEV) in *Nicotianabenthamiana* and by *Rice stripe virus* (RSV) and *Rice dwarf virus* (RDV) in rice (*Oryzasativa*). Indeed, this induced accumulation was found to promote the infection process of these viruses (Ding and Voinnet, 2007; Várallyay et al., 2010; Duet al., 2011a; Pumplin and Voinnet, 2013; Wu et al., 2015; Zhang et al., 2015). *Rice ragged stunt virus* (RRSV) and *Rice black streaked dwarf virus*(RBSDV) infections in rice increase the level of miR319, which suppresses jasmonic acid mediated antiviral defense in rice (Zhang et al., 2016). Over expression of miR528 in transgenic rice plants reduces the accumulation level of reactive oxygen species (ROS) compared with that of wild type (WT), and these plants are more sensitive to RSV infection (Wu et al., 2017). *Arabidopsis* miR393 suppresses auxin signaling in response to bacterial infection (Navarro et al., 2006), while miR398b is involved in defense against fungal pathogens (Li et al., 2014). In addition, some miRNAs that function in basal metabolism are regulated by pathogen invasion. For instance, miR395 and miR399, which are involved in the regulation of sulfur assimilation (Takahashi et al., 2000; Maruyama-Nakashita et al., 2003) and response to phosphorus starvation (Fujii et al., 2005; Chiou et al., 2006; Hu et al., 2015), respectively, are up-regulated by RSV infection in rice (Lian et al., 2016).

In plants, the miRNA biogenesis pathway is regulated by the cooperation of several host factors (Francisco-Mangilet et al., 2015). Work in *Arabidopsis* showed that miRNAs are processed from primary transcripts that contain partially complementary fold-back regions of variable lengths (pri-miRNAs) by a processing complex consisting of the RNAse III enzyme DICER-LIKE 1 (AtDCL), the double stranded RNA (dsRNA) binding protein HYPONASTIC LEAVES1 (AtDRB1/AtHYL1), and the zinc finger protein SERRATE (AtSE) (Fang and Spector, 2007; Fujioka et al., 2007; Dong et al., 2008). At HYL1 and AtSE are essential for the accurate and efficient cleavage of pri-miRNAs by AtDCL1 (Dong et al., 2008). Homodimerization/self-interaction of AtHYL1 (orthologous to the OsDRB1) ensures the correct selection of cleavage sites on pri-miRNAs (Yang et al., 2014). After processing, the miRNA/miRNA* duplex is methylated by HUA ENHANCER 1 (AtHEN1) at the 3′-terminus to ensure the stability of mature miRNA (Li et al., 2005; Yu et al., 2005). The miRNA strand is recruited by diverse ARGONAUTE (AtAGO) proteins to form miRNA-induced silencing complexes to mediate post-transcription gene silencing or translation repression (Djuranovic et al., 2012; Cui et al., 2016; Fang and Qi, 2016). Although the miRNA biogenesis pathway in plants has been well documented and many studies have indicated that miRNAs are involved in host–virus interactions, little is known about how pathogens regulate miRNA processing and accumulation.

Rice stripe virus (RSV), the type member of the genus *Tenuivirus*, causes severe disease and yield losses in many Asian rice cultivars. The RSV genome comprises four negative-sense, single-stranded RNA segments, RNA1, 2, 3, and 4. RNA1 uses a negative sense coding strategy while RNA2, 3, and 4 use ambisense coding strategy (Zhu et al., 1991; Takahashi et al., 1993; Jiang et al., 2012; Kong et al., 2014). RNA1 encodes RNA-dependent RNA polymerase (RdRp, 337kDa) (Toriyama et al., 1994). RNA2 encodes NS2 (22.8kDa), a weak suppressor of RNA silencing (Toriyama et al., 1994) and NSvc2 (94.2kDa), a glycoprotein that targets the Golgi body and the endoplasmic reticulum (ER) (Yao et al., 2014). RNA3 encodes nonstructural protein 3 (NS3, 23.9kDa) and the coat protein (CP, 35.1kDa) (Xiong et al., 2009; Shen et al., 2010); NS3 was identified as a viral-encoded RNA silencing suppressor (VSR) which suppresses post-transcriptional gene silencing (PTGS) in *N. benthamiana* and binds single- or double-stranded RNA without sequence preference (Xiong et al., 2009; Shen et al., 2010). RNA4 encodes disease specific protein (SP, 20.5kDa) which interferes with photosynthesis by interaction with an oxygen-evolving complex protein (Kong et al., 2014) and NSvc4 (32.4kDa), a cell-to-cell movement protein (Xiong et al., 2008). Several studies have demonstrated that RSV infection perturbs miRNA accumulation, for example, 38 miRNAs including miR167, miR168, miR395, miR399 etc., were induced upon RSV infection between 7 to 15dpi (Takahashi et al., 2000; Maruyama-Nakashita et al., 2003; Fujii et al., 2005; Chiou et al., 2006; Navarro et al., 2006; Du et al., 2011a; Li et al., 2014; Hu et al., 2015; Lian et al., 2016; Zhang et al., 2016; Wu et al., 2017), but the underlying mechanism for this is unclear. In this study, we found that RSV-encoded NS3 is responsible for the over-accumulation of several miRNAs, many of them known to regulate biotic or abiotic stress response genes, in a dsRNA-binding activity-dependent manner both in rice and *Arbidopsis*. *In vivo* experiments demonstrated that NS3 enhanced pri-miRNA processing through its dsRNA binding domain. Additionally, NS3 interacts specifically with the second dsRNA-binding domain (dsRDB2) of OsDRB1. Importantly, the NS3-interacting sites in OsDRB1 are also required for the homodimerization of the OsDRB1 and NS3 acts as a scaffold to regulate the association of OsDRB1 and pri-miRNA. Genetic analysis showed that NS3-ΔdsRBD-AtHYL1 fusion protein could partially rescue the phenotype of *hyl1-2*, but NS3 or ΔD1-AtHYL1 could not do so. *NS3*-over expressing transgenic rice lines enhanced RSV pathogenicity compared with the control transgenic lines expressing a mutant form of *NS3* (mNS3) and the WT plants. Our data revealed that the RSV NS3 protein regulates the association between OsDRB1 and pri-miRNAs, induces accumulation of a number of miRNAs, and enhances viral pathogenicity in rice.

2 Results

2.1 NS3 is responsible for the RSV-induced over-accumulation of a set of miRNAs

RSV-infected rice plants show stunting, rolled-leaf and chlorotic mottling in leaves symptom compared to mock-infected rice plants at early stage (Fig. 1A, C). In a previous study, we found that RSV infection resulted in an increased accumulation of several rice miRNAs, including miR168, miR395, miR398, miR399 and miR528 (20-nt and 21-nt forms) (Du et al., 2011a; Wu et al., 2015; Lian et al., 2016; Wu et al., 2017). To confirm the RSV-induced increase in the accumulation of these miRNAs in a different set of plants, we analyzed the differential expression of these rice miRNAs in mock-inoculated and RSV-infected rice plants by northern blotting. All tested miRNAs, with the exception of miR528 (21-nt), were up-regulated by RSV infection (Fig. 1D), which was consistent with previous reports (Du et al., 2011b; Wu et al., 2015; Yang et al., 2016; Wu et al., 2017).

To determine which RSV-encoded protein triggers the up-regulation of miRNAs in the RSV-infected rice plants, we measured the expression of these miRNAs in rice plants over expressing various RSV-encoded proteins, driven by the *ACTIN 1* promoter, with a 4×Myc epitope tag at the N-terminus. Western blot assays confirmed the expression of myc-NS2,myc-NS3, myc-SP, and myc-NSvc4 in the corresponding transgenic rice lines (Fig. 1E) (bottom two panels). As shown in Fig.1E, (top three panels), the levels of miR168 and miR395, measured by northern blot hybridizations, were strongly elevated by RSV infection and transgenic expression of NS3 (Fig. 1E), but not other RSV proteins. Also, RSV infection and NS3 transgene expression is produced similar elevations of miRNA levels (Fig. 1E). Therefore, expression of NS3 alone fully recapitulates the perturbation of miRNA levels caused by RSV infections.

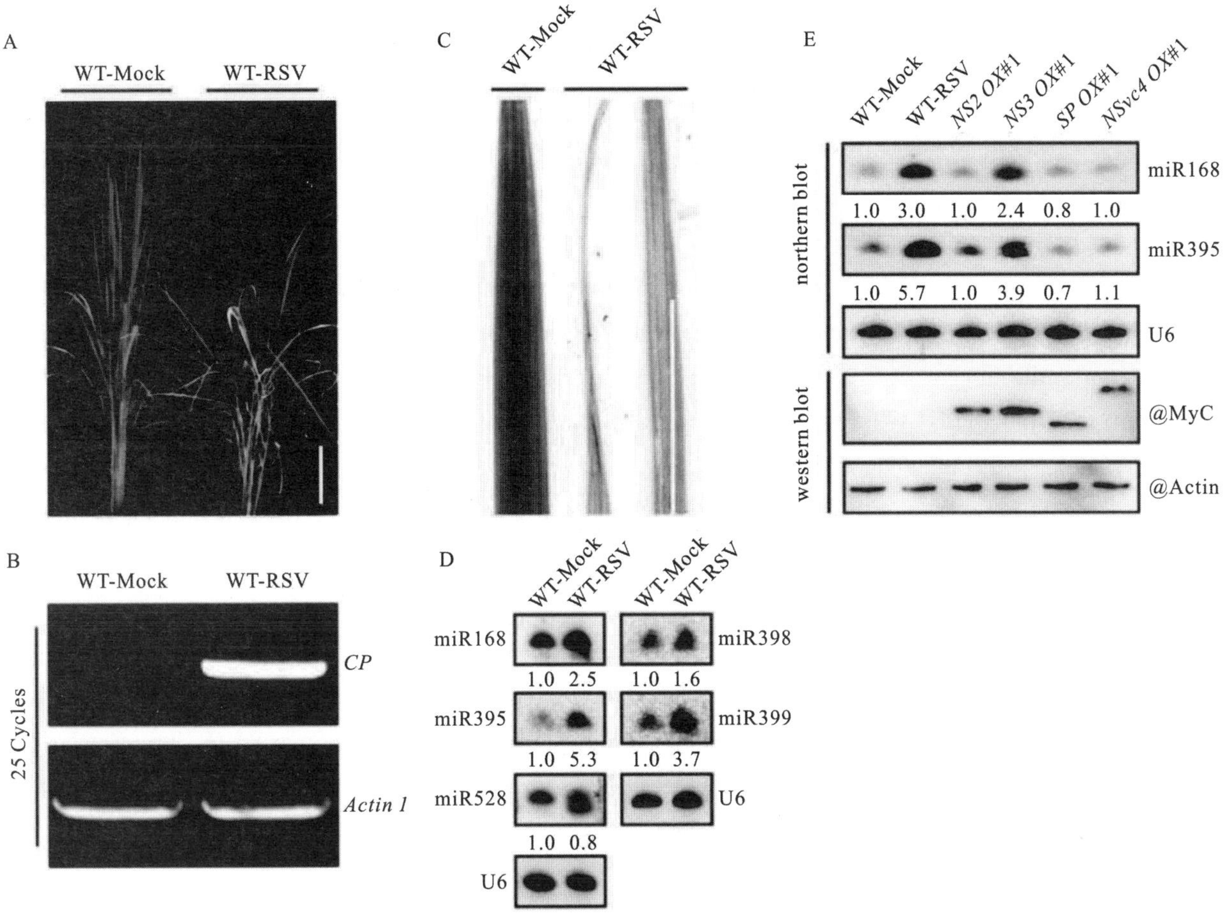

Fig. 1 RSV symptoms and miRNAs induced by NS3 in rice

(A) Images of whole plants exhibiting stunted phenotypes. Scale bar = 15cm. (B) Detection of the RSV *CP* gene by RT-PCR. (C) Images of RSV-infected leaves exhibiting chlorotic mottling and rolled-leaf phenotypes. Scale bar = 5cm. (D) Detection of miRNA (miR168, 395, 398, 399, and 528) accumulations in healthy (mock-infected) and RSV-infected rice plants by northern blotting. Redstar, 21-nt miR528. (E) Northern and western blot assay detection of miRNA (miR168 and miR395) accumulations and protein expression in mock-infected, RSV-infected (RSV), *NS2* OX#1, *NS3* OX#1, *SP* OX#1, and *NSvc4* OX#1 rice plants. In (D and E), U6 served as a loading control, the expression levels in the WT-Mock plants are set to a value of 1.0 and the expression levels in the other plants are relative to this reference value

2.2 NS3 triggers the accumulation of several miRNAs through its dsRNA binding domain

The dsRNA binding domain of NS3 is important for its activity (Shen et al., 2010). To determine whether NS3-promoted miRNA accumulation depends on its dsRNA binding activity, we constructed a transgenic rice line overexpressing Myc-tagged *mNS3*,in which the dsRNA binding domain was disrupted by the replacement of a 173K174K175R motif with 173E174D175E (Fig. 2A). We then conducted de novo sequencing of small RNAs in the WT, and the *NS3* overexpression (OX) #1 and *mNS3*OX#1 rice lines. The results showed that miR168, miR395, miR398, miR399, and miR528 were all up-regulated in the *NS3*-overexpressing rice line, but down-regulated or unchanged in the *mNS3*-overexpressing rice line compared with the WT (See S1 Table for more details) (Fig. 2B). To verify the *denovo* sequencing results, we carried out northern blot and western blot assays in two independent overexpression lines each for *NS3* and *mNS3*(*NS3*OX#1 and #7 and *mNS3*OX#1 and #4). As shown in Fig. 2C, most miRNAs accumulated at higher levels in the *NS3* overexpression lines compared with the mock-inoculated WT, whereas lower levels were detected in the *mNS3*overexpression lines. These results indicated that NS3 plays a critical role in the positive regulation of miRNA accumulations in rice and this regulation is dependent on its dsRNA binding activity.

To verify that the increased accumulation of these miRNAs further reduced the expression levels of their target genes, we examined the expression levels of their target mRNAs by quantitative RT-PCR (RT-qPCR). As shown in Fig. 2D, mRNA levels of *OsAGO1a* (miR168), *OsSULTR2;1* (miR395), *OsCDS1*(miR398), *Os08g45000* (miR399) and *OsRFPH2-10* (miR528) were all reduced in the *NS3* overexpression lines compared with the mock-inoculated WT plants, whereas expression levels in the *mNS3* overexpression lines were relatively unchanged compared to the WT.

To further test whether NS3 triggers accumulation of miRNAs through its dsRNA binding domain in dicots, we constructed *NS3*and *mNS3* overexpression transgenic *Arabidopsis* plants. Northern blot showed that the levels of miR168 and miR395 were up-regulated by RSV infection and *NS3* overexpression in these *Arabidopsis* plants (Fig. S1A,B). Although overexpression of *NS3* in rice did not result in disease-like symptoms, *NS3*-overexpressing *Arabidopsis* plants exhibited a severely stunted phenotype that phenocopied the disease symptoms of the RSV infected *Arabidopsis* (Fig. S1C,D). In contrast, overexpression of *mNS3* had no influence on the growth or the levels of miR168 and miR395 in the *Arabidopsis* plants (Fig. S1B, D). These results indicate that induction of miRNA accumulations by *NS3* overexpression is conserved in dicots and monocots.

To determine whether RSV infection and NS3 overexpression increases the accumulation of these miRNAs through promoting primary miRNA (pri-miRNA) processing, we measured the primary transcript levels of miR168, miR395, miR398, miR399, and miR528 by RT-qPCR. The results demonstrated that NS3, but not mNS3, down-regulated all of the pri-miRNAs, except for pri-miR399d, which suggests that NS3 is involved in pri-miRNA processing in a dsRNA-binding activity-dependent manner (Fig. 2E).

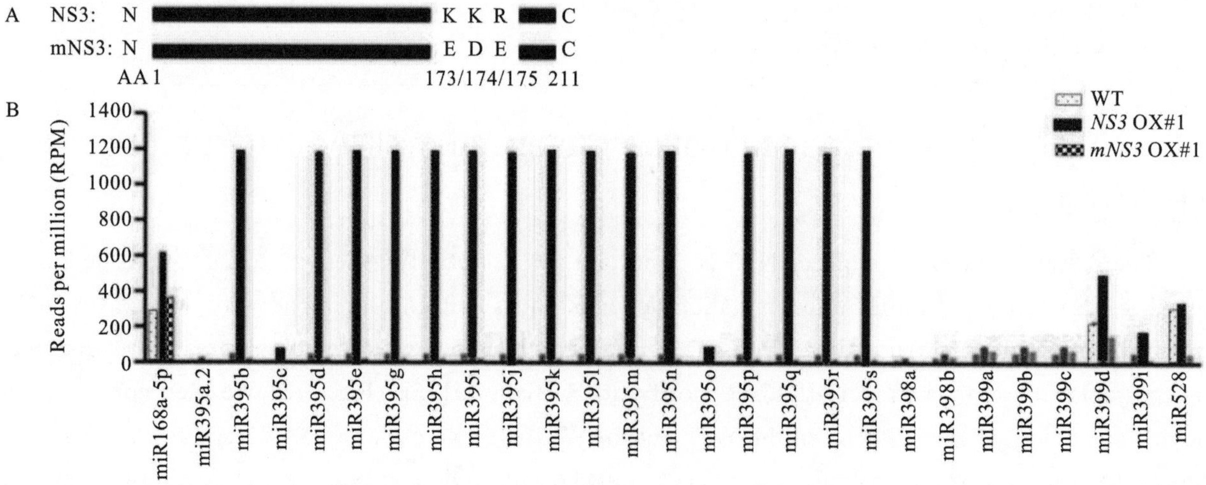

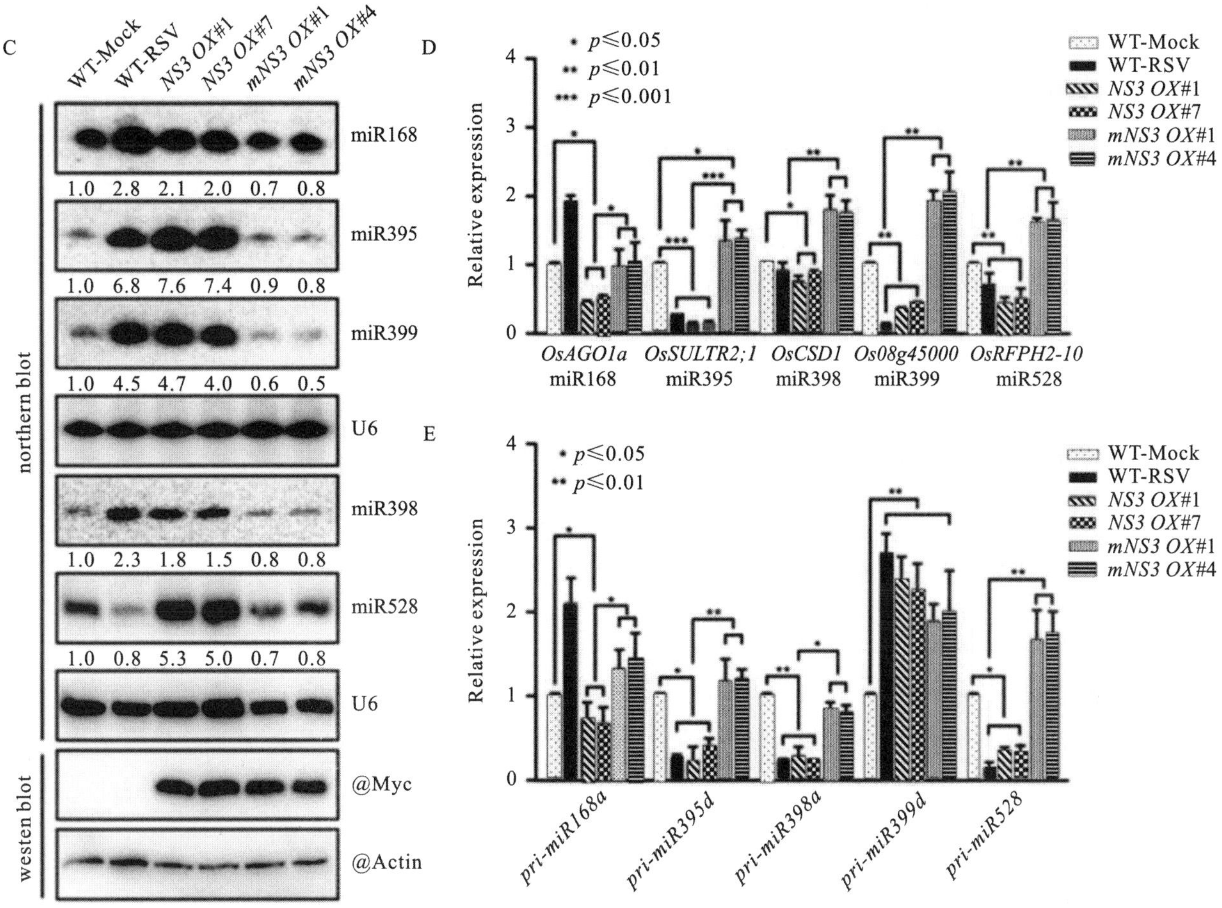

Fig. 2 Expressionof NS3 promotes the accumulation of several miRNAs and reduces the expression of their targets

(A) Mutation site of the mutant NS3 (mNS3). (B) Measurement of miRNA (miR168, 395, 398, 399, and 528) accumulation in the WT, *NS3*OX#1, and *mNS3*OX#1 rice lines by small RNA sequencing. (C) Detection of miRNA (miR168,395, 398, 399, and 528) accumulations in mock-infected, RSV-infected (RSV), *NS3*OX#1, *NS3*OX#7, *mNS3*OX#1, and *mNS3*OX#4 rice plants by northern blotting. U6 served as a loading control, the expression levels in the WT-Mock plants are set to a value of 1.0 and the expression levels of in the other plants are relative to this reference value. (D) Relative expression levels of the target genes of the miRNAs (miR168, 395, 398, 399, and 528), including *OsAGO1a*, *OsSULTR2;1*, *OsCSD1*, *Os08g45000*, and *OsRFPH2-10* in the mock-infected, RSV-infected (RSV), *NS3*OX#1, *NS3*OX#7, *mNS3*OX#1, and *mNS3*OX#4 rice plants. (E) Relative expression levels of the miRNA (miR168, 395, 398, 399, and 528) precursors, including *pri-miR168a*, *pri-miR395d*, *pri-miR398a*, *pri-miR399b*, and *pri-miR528* in the mock-infected, RSV-infected (RSV), *NS3* OX#1, *NS3* OX#7, *mNS3*OX#1, and *mNS3* OX#4 rice plants. Average (± SD) values based on RT-qPCR analysis of three biological replicates are shown. ***,$P \leq 0.001$; **,$P \leq 0.01$; *, $P \leq 0.05$

2.3 NS3 aids *in vivo* processing of pri-miRNA

Given that NS3 reduces the accumulation of a set of pri-miRNAs but induces the accumulation of corresponding mature miRNAs, we suggested that NS3 may promote the recruitment of pri-miRNAs by affecting the miRNA-processing complex. Since the thermo stability of the end of the miRNA/miRNA* duplex is important for mature miRNA accumulation (Eamens et al., 2009), we designed an experiment in which a pri-miRNA can be recognized by NS3, but not by the processing complex, which would indicate that NS3 directly assists in the association of the processing complex with pri-miRNAs when NS3is co-expressed with the pri-miRNA. We constructed a 35S promoter-driven artificial *primary miR528* (*apri-miR528*) and mutant artificial *primary miR528* (*mapri-miR528*) with three additional C/G pairs at the end of the miRNA/ miRNA* duplex using an *Arabidopsis primary miR159a* backbone (Fig. 3A). miR528 is only expressed in monocot plants; therefore, we carried out an *in vivo* transient expression assay toco-express apri-miR528 and *vector/NS3/mNS3* and also *mapri-miR528* and *vector/NS3/mNS3* in the leaves of the dicots *N. benthamiana* and measured mature artificial miR528 (amiR528) levels in each group by northern blotting at 3d post-infiltration (dpi). We found that expression of *NS3* had little influence on the accumulation of amiR528 in the *apri-miR528* group, but expression of *mNS3* reduced the levels of miR528 relative to the vector control. As expected, no miR528 accumulation was observed in the negative control. Additionally, no mature miR528 accumulation was detected in the mapri-miR528 group in the absence of NS3 expression. The accumulation of mature miRNA was only detected in the leaves co-expressing mapri-miR528 with NS3 (Fig. 3B). To test if NS3 interacts with apri-miR528 and *mapri-miR528*, an *in vitro* microscale thermophoresis assay with GST-mNS3 serving as the negative control was used to reveal that both apri-miR528 and mapri-miR528 were recognized by GST-NS3 (Fig. 3C). These results provide further evidence for the role of NS3 in mature miRNA biogenesis.

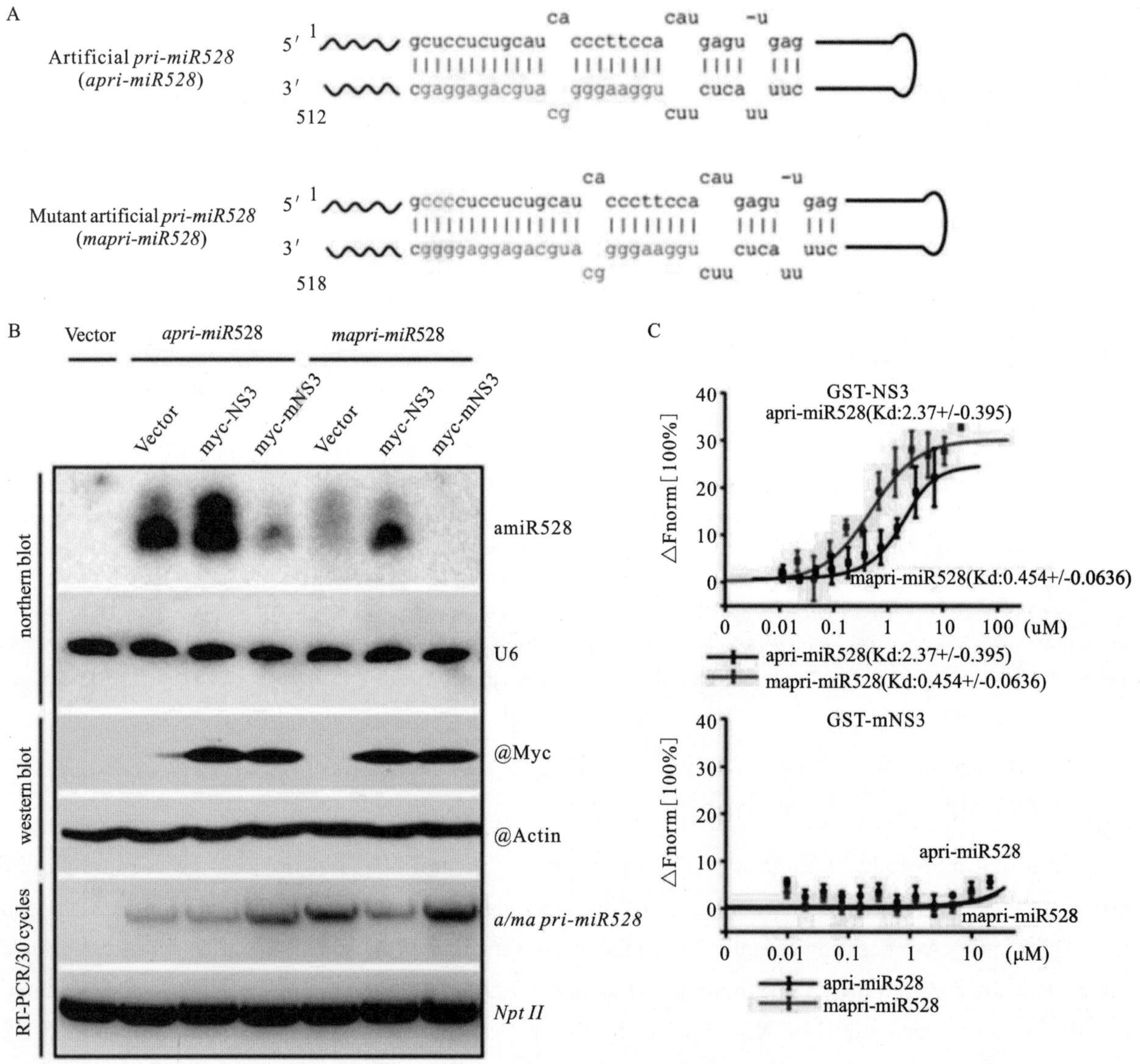

Fig. 3 NS3 promotes pri-miRNA processing

(A) Structures of artificial *pri-miR528* (*apri-miR528*) and mutated artificial *pri-miR528* (*mapri-miR528*). (B) Northern blot, western blot, and RT-PCR detection of the products of transiently co-expressed pri-miRNAs (apri-miR528 or mapri-miR528) and proteins (empty vector, NS3, or mNS3) in *N. benthamiana*. (C) Results of a microscale thermophoresis assay shows the interactions between pri-miRNA (apri-miR528 or mapri-miR528) and protein (GST-NS3 or GST-mNS3)

2.4 Association of NS3 with the dsRBD2 domain of OsDRB1

NS3 does not contain an RNase III domain; therefore, it cannot promote miRNA accumulation by itself. We hypothesized that NS3 may function in miRNA processing through its association with components of the Dicing body (D-body) to promote the recruitment of the pri-miRNA by the processing complex. To test this hypothesis, we performed bimolecular fluorescence complementation (BiFC) assays to co-express *Arabidopsis DCL1*(*AtDCL1*), *SE* (*AtSE*), *HYL1* (*AtHYL1*), or *CBP20*(*AtCBP20*) with NS3 in *N. benthamiana* leaves. NS3 associated with AtHYL1 and AtSE but not with AtDCL1 or AtCPB20. In addition, we found that AtHYL1, but not AtSE, specifically interacted with NS3 in the D-body, and the interaction between NS3 and AtHYL1 may involve in miRNA maturation (Fig. S2A). Using the basic local alignment search tool (BLAST) and the UniProt protein database (uniprot.org), we found six AtHYL1 homologs in rice, OsDRB1a, OsDRB1b, OsDRB1c, OsDRB2, OsDRB3 and OsDRB4 (Fig. S2B).To confirm that OsDRB1 has the same function as AtHYL1, we measured the levels of miR164, miR166, and miR168 by small-RNA RT-qPCR in an OsDRB1-knockdown rice line and found that these miRNAs were down-regulated (Fig. S2C). We also analyzed OsDRB1 protein levels, and found that the level of OsDRB1 was reduced in the OsDRB1-knockdown line (Fig. S2D). OsDRB1a contains all the domains of the other two OsDRB1s, as well as a unique C-terminus, so we chose OsDRB1a to test the interaction between NS3 and OsDRB1, BiFC and co-immunoprecipitation (CoIP) assays demonstrated that OsDRB1a does interact with NS3 (Fig. 4A, B). We also tested interactions between NS3 and the other OsDRBs by BiFC assay, and found that only OsDRB2 has a weak interaction with NS3 in cytoplasm (Fig. S2E). Previous studies have shown that homodimerization of AtHYL1 is required for AtDCLs to locate the correct cleavage sites in pri-miRNAs, while the G147 and L165 residues of AtHYL1 are critical for homodimer formation (Yang et al., 2014). The G162 and L180 residues of OsDRB1a correspond to the G147 and L165 residues of AtHYL1 and may be essential for homodimer formation according to amino acid alignment (Fig. S2F). To test our hypothesis, we constructed an OsDRB1 amutant (*mOsDRB1a*) by replacing the G162 and L180 residues with E162 and E180, respectively (Fig. 4C). In a subsequent BiFC assay, we found that mOsDRB1a could not interact with itself (Fig. 4D), as expected. We further confirmed that wild-type OsDRB1a, like AtHYL1, formed a homodimer (Fig. 4E) (Yang et al., 2010; Yang et al., 2014). Also, mOsDRB1a could not interact with NS3 or mNS3 (Fig. 4F), but NS3 interacts with itself, mNS3 has weak interaction with NS3, and OsDRB1a has weak interaction with mOsDRB1a too (Fig. S2G). We summarize the interaction of each pair between NS3, mNS3, OsDRB1a and mOsDRB1a in Table 1. These results demonstrated that the association between NS3 and OsDRB1a depends on the dsRBD2 domain of OsDRB1a and is essential for miRNA processing.

Table 1 Interactions of each two proteins as shown below

	NS3	mNS3	OsDRB1a	mOsDRB1a
NS3	√ √	√	√ √	×
mNS3	√	×	√ √	×
OsDRB1a	√ √	√ √	√ √	√
mOsDRB1a	×	×	√	×

√√: strong interaction;√: weak interaction; ×: no interaction.

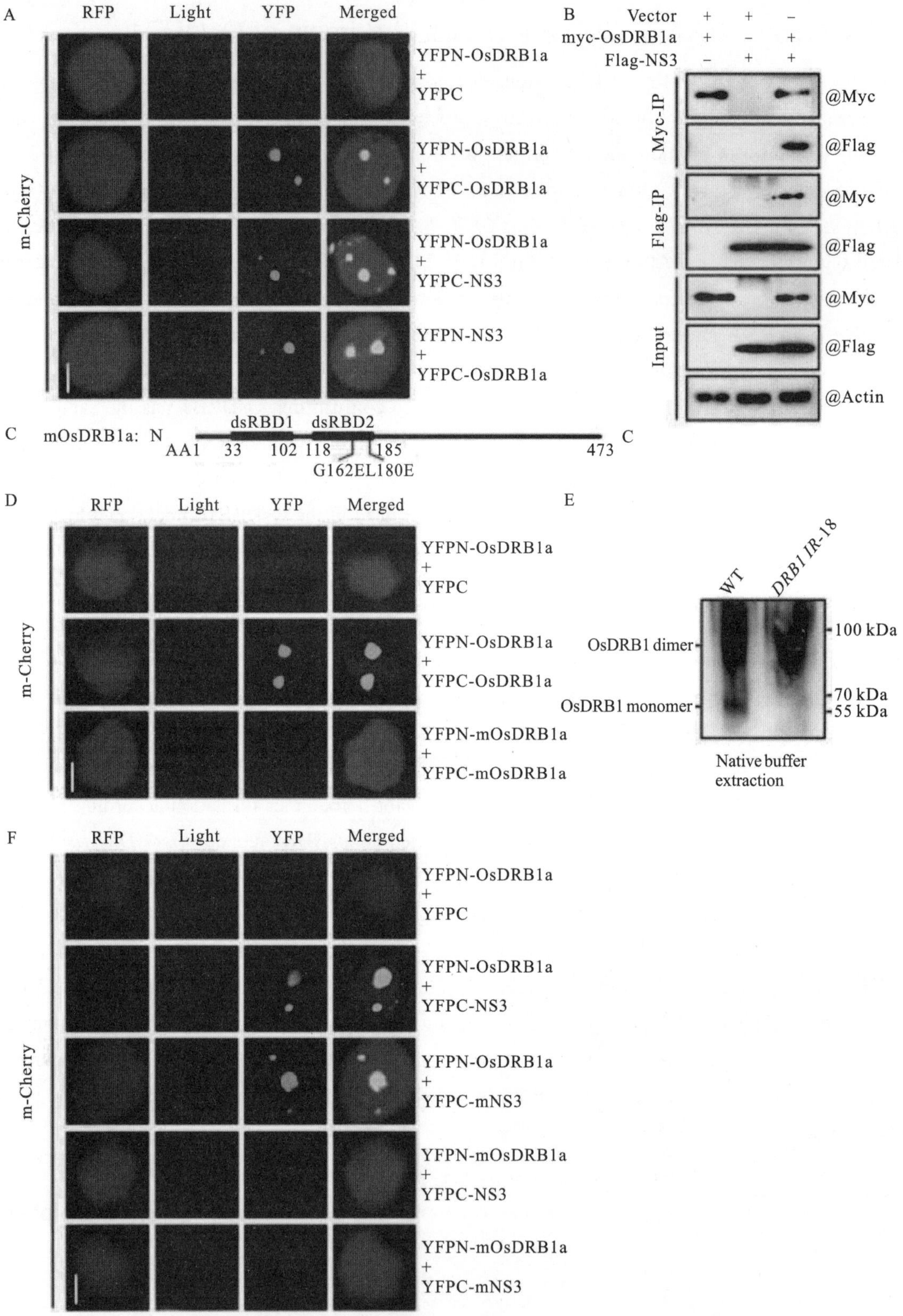

Fig. 4 NS3 interacts with OsDRB1, a pri-miRNA processing factor

The BiFC assay was conducted in *N. benthamiana* epidermal cells, mCherry is a nuclear localization marker fused with red florescent protein (RFP). (A) Results of a BiFC assay showing the interaction between NS3 and OsDRB1a. Scale bar = 0.1μm. (B) Results of a co-immunoprecipitation analysis showing the interaction between NS3 and OsDRB1a. (C) The accuracy of mutation sites of mOsDRB1a. (D) Results of a BiFC assay showing the accuracy of interaction sites between OsDRB1a. Scale bar = 0.1μm. (E) Formation of OsDRB1a dimers *in vivo*. Total rice protein extractsfrom the WT and OsDRB1- knockdown lines were treated with "native" buffer and detected using anti-HYL1antibodies. (F) Results of a BiFC assay showing the accuracy of interaction sites between NS3 and DRB1a. Scale bar = 0.1μm

2.5 NS3 mimics the dsRBD domain of DRB1 in miRNA processing

Given that the dsRNA-binding activity of NS3 is important for the processing of pri-miRNAs and the accumulation of miRNA, and NS3 interacts with DRB1, we deduced that NS3, rather than DRB1, recognizes pri-miRNAs during NS3–DRB1 interactions. To test this hypothesis, we transiently co-expressed *OsDRB1a/mOsDRB1a*, *apri-miR528/mapri-miR528*, or *empty vector/NS3/mNS3* in *N. benthamiana* and detected the CoIP products of OsDRB1a/mOsDRB1a by RT-PCR. We found that both OsDRB1a and mOsDRB1a associated with apri-miR528, but neither of them recognized mapri-miR528, and NS3 and mNS3 associated with OsDRB1a instead of mOsDRB1a. With the expression of NS3, both OsDRB1a and mOsDRB1a associated with apri-miR528. However, with the expression of mNS3, only mOsDRB1a associated with apri-miR528. Additionally, only co-expression with NS3 resulted in an OsDRB1a interaction with mapri-miR528 (Fig. 5A), see Table 2 for more details. Using a microscale thermophoresis assay, we also found that OsDRB1a associated with apri-miR528 but not mapri-miR528 (Fig. 5B). We also found that NS3, but not mNS3, could bind with the endogenous miRNA precursors (pre-miR168a, pre-miR395d, pre-miR398a, pre-miR399d, and pre-miR528) in an electrophoretic mobility shift assay (EMSA) (Fig. S3A). These results indicated that by interacting with DRB1, NS3 replaced the dsRNA-binding activity of DRB1. To test this hypothesis, we overexpressed AtHYL1, ΔD1-HYL1, a double-stranded RNA binding domain 1 deletion form of AtHYL, NS3-ΔD1D2-HYL1,a fusion protein of NS3 and ΔD1D2-HYL1 and *NS3* (Fig. 5C) in *Arabidopsishyl1-2* mutant background, the transgenes were all driven by a 35S promoter with the proteins tagged with a myc-epitope tag at the N-terminus, we found that the fusion protein NS3-ΔD1D2-AtHYL1 could ameliorated the phenotype (Fig. 5D) and miRNA (miR156, miR164, miR168 and miR395) levels (Fig. 5E) of *hyl1-2* mutant as HYL1 did but NS3 or ΔD1-HYL1 could not (Fig. 5D, E), we test these transgenic *Arabidopsis* plants by western blotting (Fig. 5F), and this results indicated that NS3 could substitute for the dsRBD domain of AtHYL1 in miRNA processing.

Table 2 Interaction of apri–miR528 or mapri–miR528 with single protein or protein complex as shown below

	NS3	mNS3	OsDRB1a	OsDRB1a+NS3	OsDRB1a+NS3	mOsDRB1a	mOsDRB1a+NS3	mOsDRB1a+NS3
apri-miR528	√	×	√	√	×	√	√	√
mapri-miR528	√	×	×	×	×	×	√	×

√:Yes;×: No

2.6 NS3 enhances virus pathogenicity in rice

Because NS3 induced miRNA accumulation along with a decreased antiviral defense responsein rice, we speculated that NS3 may play a role in regulating viral pathogenicity. We usedvirus-free (mock) and viruliferous (RSV) planthoppers (*Laodelphax striatellus*) to inoculate the WT, and *NS3*-and *mNS3*-overexpression rice lines (*NS3*OX#1 and OX#7 and *mNS3*OX#1and OX#4) and found that the *NS3*OX#1 and OX#7 lines, but not the *mNS3*OX#1 and OX#4lines, were hypersensitive to RSV infection compared with WT plants, with most serious stunted and chlorisis phenotypes (Fig. 6A). To examine whether the increased susceptibility of the NS3OX lines was due to the increased accumulation of RSV, we used RT-qPCR to measure the transcript levels of the RSV CP gene. We found that the expression of CP mRNA in the*NS3*OX#1 and #7 lines was much higher than that in the WT plants, with no obvious changes detected in the *mNS3*OX#1 and #4 lines when compared with the WT (Fig. 6B). We also monitored differences in RSV infection rates among the WT and the *NS3*OX#1, *NS3*OX#7, *mNS3*OX#1, and *mNS3*OX#4 lines every 3d until 21dpi. These observations indicated that NS3

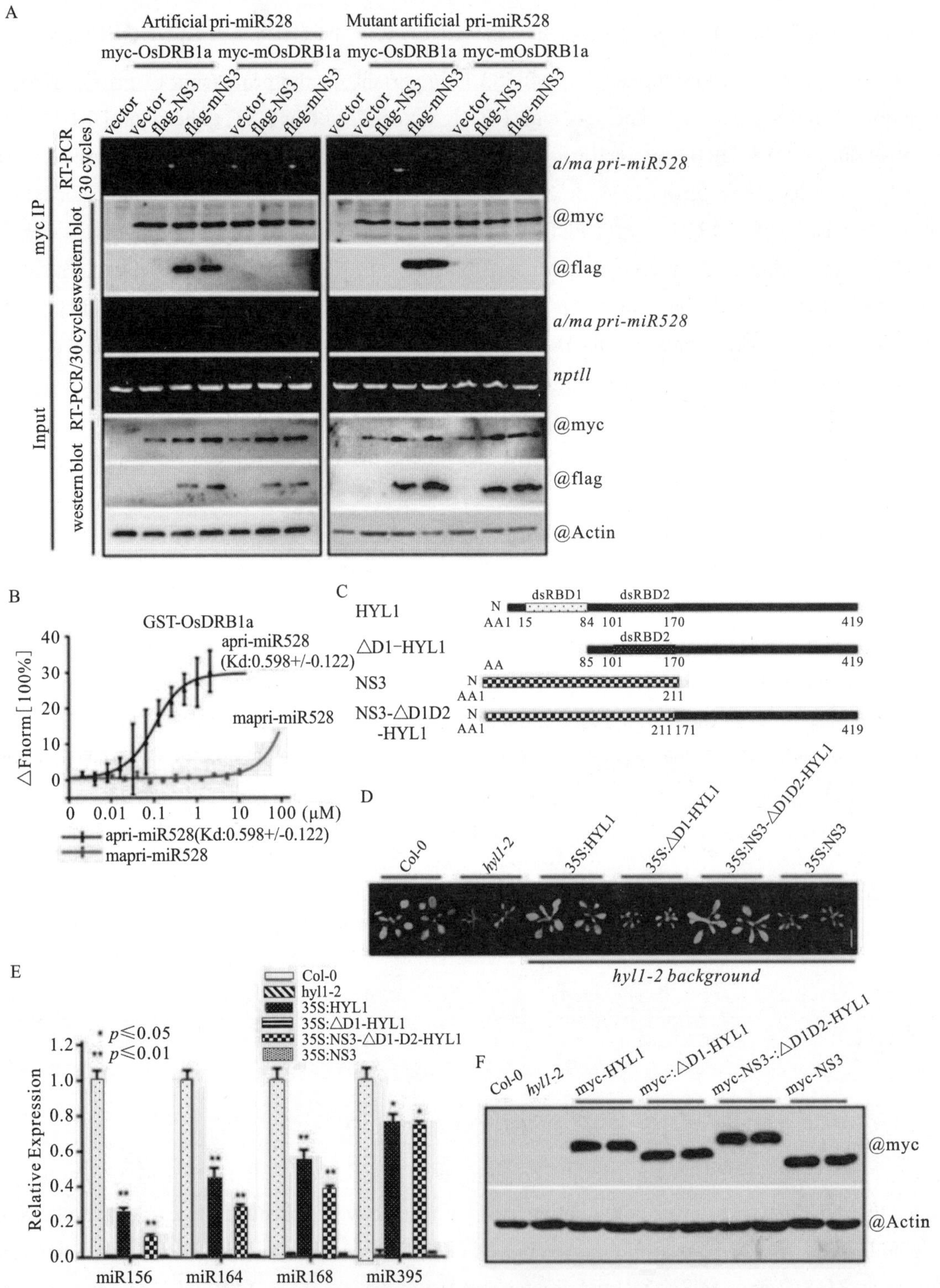

Fig. 5 NS3 acts as a scaffold between DRB1 and pri-miRNA

(A) RT-PCR and western blot detection of co-immunoprecipitated and input products of transiently co-expressed protein (DRB1a or mDRB1a), pri-miRNA (apri-miR528 or mapri-miR528), and protein (empty vector, NS3, or mNS3) in *N. benthamiana*.(B) Results of a microscale thermophoresis assay shows the interactions between pri-miRNA (apri-miR528 or mapri-miR528) and GST-OsDRB1a. (C) Gene structure of *HYL1*, ΔD1-HYL1, *NS3* and *NS3-ΔD1D2-HYL1*. (D) Phenotype of Col-0, *hyl1-2* and transgenic *Arabidopsis* plants overexpressed *AtHYL1*, ΔD1-HYL1, *NS3-ΔD1D2-HYL1* and *NS3* with the 35S promoterin the *hyl1-2* mutant background. (E) miRNA (miR156, miR164, miR168 and miR395) levels in *Col-0*, *hyl1-2* and transgenic *Arabidopsis* plants overexpressing *AtHYL1*, ΔD1-HYL1, *NS3-ΔD1D2-HYL1* and *NS3* with the 35S promoterin the hyl1-2 mutant background. (F) Western blot of AtHYL1, ΔD1-HYL1, NS3-ΔD1D2-HYL1 and NS3 transgenic *Arabidopsis* plants. Fnorm, normalized fluorescence increased RSV pathogenicity in rice . To test if NS3 OX plants are alsomore sensitive to infection with other viruses, we used *Rice ragged stunt virus* (RRSV), a member of the genus *Oryzavirus*, to infect the WT, and the *NS3*OX#1, *NS3*OX#7, *mNS3*OX#1, and *mNS3*OX#4 lines. The results showed that NS3 OX plant lines displayed a stunted phenotype (Fig. S4A) and accumulated more RRSV CP genes (Fig. S4B) compared with other riceplant lines.

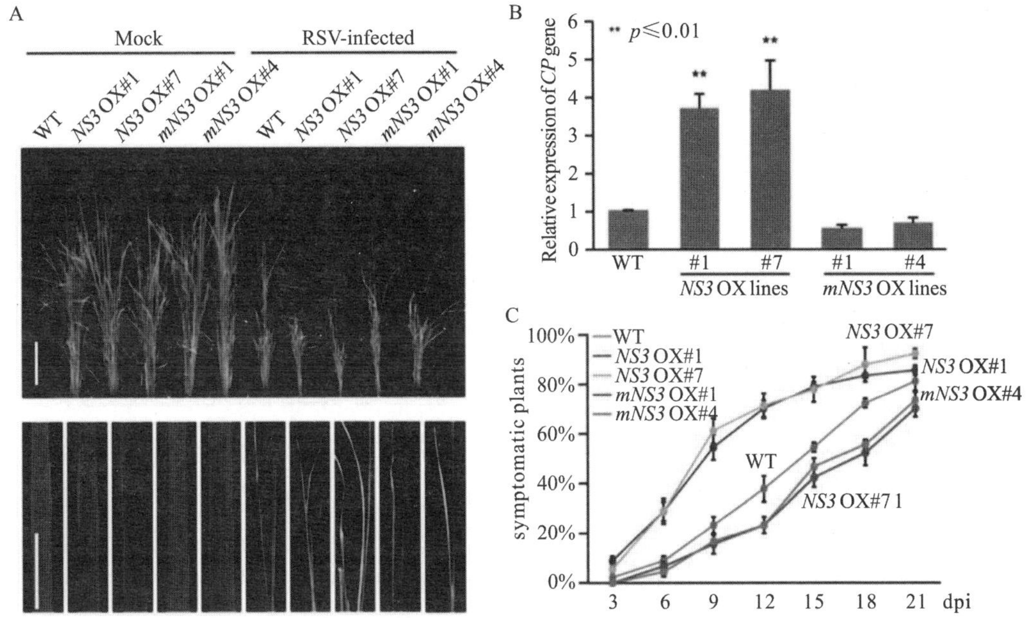

Fig. 6 Disease symptoms induced by RSV infection in WT, *NS3* OX, and *mNS3* OX rice lines

(A) Images of whole plants and details showing stunted or folded-leaf phenotypes of the wild-type, *NS3* OX#1, *NS3* OX#7, *mNS3* OX#1 and *mNS3* OX#4 rice plants. Scale bars = 15cm (upper panel) and 5cm (lower panel). (B) Detection of the RSV *CP* gene by RT-PCR. (C) Time course of RSV symptom development in the WT and *NS3* OX or *mNS3* OX transgenic plants. Values represent the percentage of RSV-infected plants at various days post inoculation (dpi). Thirty plants were used for each treatment. **, $P \leqslant 0.01$. Average (± SD) values from three biological replicates are shown

3 Discussion

RNA interference (RNAi) is a conserved and effective antiviral mechanism in plants and insects (Ding, 2010). To counter the host's antiviral defense mechanisms, viruses encode VSR (s) to interfere with the host's RNAi. Most reported VSRs act to suppress the host's RNA silencing system, such as by inhibiting viral RNA recognition, blocking dicing, suppressing assembly of the RNA-induced silencing complex (RISC), and preventing siRNA amplification (Li and Ding, 2001; Chapman et al., 2004; Cao et al., 2005; Li and Ding, 2006; Ding, 2010; Ren et al., 2010; Duan et al., 2012). RSV NS3, a reported RNA silencing suppressor, suppresses post-transcriptional gene silencing (PTGS) in *N. benthamiana* through its dsRNA binding ability (Xiong et al., 2009). Our previously obtained small RNA sequencing data of changes associated with RSV infection revealed that many miRNAs are induced by RSV infection (Du et al., 2011a; Lian et al., 2016; Yang et al., 2016). Up to now, only a few studies have focused on how a virus hijacks the RNA silencing pathway to regulate miRNA accumulation and advance its own pathogenicity, with very limited reports focusing on how virus regulates miRNA processing. In the present study, we revealed that NS3 exploits OsDRB1, a key component of the D-body, to promote the processing of pri-miRNA along with the regulation of miRNA target gene expression

(Fig. 7). NS3 showed a weak interaction with OsDRB2 in the cytoplasm. Since miRNA processing is occurred in the nucleus, we deduced that the association of NS3 and OsDRB2 would not affect NS3-mediated miRNA processing.

Previous studies showed that NS3 suppression of the PTGS pathway may depend on its siRNA and long dsRNA binding activity (Xiong et al., 2009; Shen et al., 2010), which is different from the results of this study showing that NS3 enhanced the miRNA pathway through its function of bridging pri-miRNA and OsDRB1, a key player in miRNA processing complex. These two different functions may act in parallel. These results widen our knowledge of the molecular mechanisms underlying viral-host interactions.

Maturation of miRNAs is a complex process. miRNA-coding genes (MIRs) are transcribed by DNA-dependent RNA polymerase II (Pol II) and subjected to splicing and addition of a 5′ 7-methyguanosine cap and a 3′ polyadenylated tail (Xie et al., 2005). Besides Pol II, many other PolII-associated factors such as RNA-binding proteins PLEIOTROPIC REGULATORY LOCUS 1 and DAWDLE, CAP BINDING PROTEIN 20 and 80, SE, CAM33/XAP CIRCADIAN TIME- KEEPER and TOUGH have been reported to regulate miRNA transcription (Wang et al., 2009; Wu et al., 2009; Yamasaki et al., 2009; Kim et al., 2011; Baek et al., 2013; Wang et al., 2013; Zhang et al., 2013; Fang et al., 2015). The structures of pri-miRNAs also affect pri-miRNA processing (Cuperus et al., 2010; Kim et al., 2016). RSV infection results in the differential expression of pri-miRNAs in rice. For example, pri-miR168a and pri-miR399d were up-regulated, while other pri-miRNAs were down-regulated (Fig. 2E). We revealed that NS3 mainly decreases the accumulation of pri-miRNAs, but the mechanism by which NS3 influences pri-miRNA expression levels remains unclear.

Denovo small RNA sequencing showed that NS3 and mNS3 have a strong influence on the accumulation of miRNAs, but not on other types of small RNAs. The total number of miRNA reads in the *NS3*OX lines was higher than that in the WT, and the total number of miRNAreads in the *mNS3*OX lines was lower than that in the WT (Fig. S3B), suggesting that NS3 may directly promote pri-miRNA recruitment by the processing complex. This hypothesis was supported by the finding that NS3 aids in *in vivo*pri-miRNA processing (Fig. 3B). Further investigations showed that NS3 associated with OsDRB1a at sites required for OsDRB1a self-interaction (Fig. 4F) and promoted the association of OsDRB1a with pri-miRNAs (Fig. 5A). Small RNA sequencing showed that the lengths and sequences of miRNAs were not changed by NS3 or mNS3 overexpression in rice (S1 Table). These results suggested that NS3 promotes recruitment of pri-miRNAs by the processor complex through an interaction with DRB1. The observation that NS3 overexpressing rice plants showed no notable developmental abnormities at vegetative stages under normal growth conditions may relate to the presence of target genes of miRNAs, regulated by NS3, in a complex genome, and a similar phenomenon was previously described (Henderson et al., 2006; Chen, 2009; Wang et al., 2013). A study in *Arabidopsis* showed that a set of miRNAs is regulated by DRB1 (Szarzynska et al., 2009). NS3 up-regulates several miRNAs in a DRB1 dependent manner (Fig. 5D). These results may explain why NS3 could not enhance some miRNA which are DRB1-independent. But it's still unknown how the preferential regulation of DRB1-dependent miRNA by NS3, we did not find similarities of sequences or secondary structure between precursors of DRB1-dependent miRNA.

Understanding the effects of NS3 of miRNA biogenesis will widen our understanding of the functions of virus-encoded proteins in regulating miRNA metabolism. Our findings of the association of NS3 with miRNA processing and accumulation provides insights into RSV pathogenicity and helps identify important targets for RSV infection, which can offer new strategies for the breeding or genetic engineering of RSV-resistant rice.

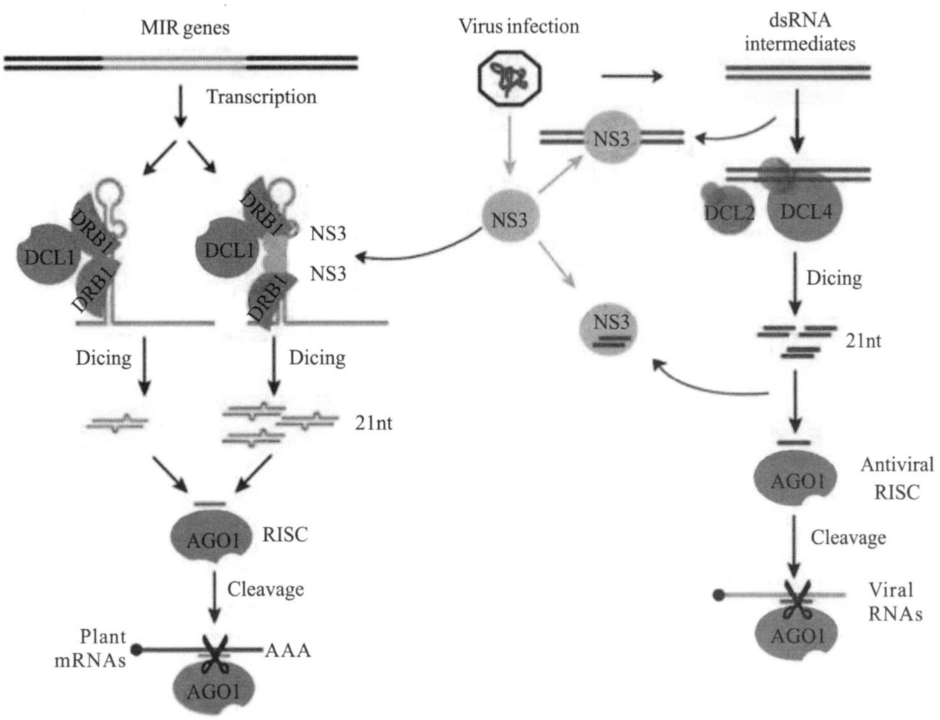

Fig. 7 Proposed model. NS3 represses the PTGS pathway

Viral-derived long double-stranded RNAs (dsRNA) are processed into 21-nt siRNAs, the 21-nt siRNAs are loaded into AGO protein complex for cleavage of viral RNA genome RNAs. However, NS3 binds viral long dsRNA or siRNAs and inhibits siRNA production or the loading of siRNAs into AGO complex. Therefore, the PTGS pathway is blocked by NS3. NS3 enhances the miRNA biogenesis pathway. In mock-infected rice plants, a given pri-miRNA has a low probability of being recognized by dimeric DRB1 and then being processed into a mature miRNA by the processor complex. Consequently, the mature miRNA is insufficient to repress its target gene, which plays a crucial role in antiviral activity or development. Under RSV infection, pri-miRNA-bound NS3 interacts with DRB1 in sites required for DRB1 self-interaction and acts as a scaffold to regulate the association between DRB1 and the pri-miRNA. This increases the chance that this pri-miRNA will be processed into a mature miRNA. In addition, this compromises the expression of the target gene and enhances RSV pathogenicity

4 Materials and methods

4.1 Plant growth and virus inoculation

Rice (*Oryzasativa spp. japonica*) seedlings were grown in a greenhouse at 28–30°C and 60% ± 5% relative humidity under natural sunlight for 4 weeks. *Arabidopsis thaliana* plants were grown at 22°C under long-day conditions (16-h light/8-h dark). When the seedlings were about 10d old, they were inoculated using viruliferous (RSV) or virus-free (mock) planthoppers (*L.striatellus*) at a ratio of four insects per plant for 72h. After removal of the insects, the inoculated plants were returned to the greenhouse and monitored daily for the appearance of viral symptoms. The number of symptomatic rice plants of each line was recorded (Fig. 6C).

4.2 Vector construction and plant transformation

The Gateway system (Invitrogen) and the enzyme digestion connection method were used to make binary constructs. Binary gateway vectors pSAT4A-DEST-N(1–174)EYFP-N1, pSA- T5A-DEST-C(175-END)EYFP-N1, pEarleyGate202, and pEarleyGate203 (Karimi et al., 2007) were used for transient expression in *N. benthamiana* and stable transformation of *Arabidopsis*, while binary vectors pGEX-4T-1 and pCam2300:Actin1:OCS were used for Escherichia coli and stable rice transformations. Most cDNA and miRNA genes were cloned into pENTR/D and pEASY Blunt Zero vectors, and pENTR/D-mNS3 and mOsDRB1a clones were prepared using a Quik Change site-directed mutagenesis kit (Stratagene, La Jolla, CA, USA). After confirmation by sequencing, all clones were transferred to the appropriate destination vector by recombination using the Gateway LR Clonase II Enzyme mix (Invitrogen) or T4 DNA ligase (TransGen Biotech, Beijing, China). *Agrobacterium tumefaciens*-mediated rice transformation was carried out at Weiming Kaituo Co., Ltd. (Beijing, China). Transgenic *Arabidopsis* plants were selected by their resistance to Basta on soil.

4.3 RNA extraction, quantitative RT–PCR, RT–PCR, and northern blotanalysis

Total RNAs were extracted from plants using Trizol (Invitrogen). The extracted RNAs were treated with RNase-free DNase I (Promega, Madison, WI, USA) to remove DNA contamination and then reverse-transcribed with SuperScript III reverse transcriptase (Invitrogen) using oligo $_{(dT)}$ primers and a Mir-X miRNA first-strand synthesis kit (Clontech Laboratories). The resulting cDNAs were then used as templates for RT-qPCR and RT-PCR. RT-qPCR was performed using SYBR Green Real-time PCR Master Mix (Toyobo, Osaka, Japan). The rice OsEF-1a gene was detected in parallel and used as an internal control. Northern blot analyses were performed as previously described (Qi et al., 2005). Briefly, RNA samples were separated by 15% denaturing gel electrophoresis and transferred onto Hybond-N+ membranes (Amersham, Fairfield,CT, USA). Membranes were UV cross-linked and hybridized to ^{32}P end-labeled oligonucleotide probes. Sequences of primers used in RT–qPCR and northern blot assays are listed in S2 Table.

4.4 Microscale thermophoresis assay

The Microscale Thermophoresis assays were performed as previously described (Wienken et al., 2010; Jin et al., 2016). GST-NS3, GST-OsDRB1a, and GST-mNS3 proteins were individually labeled with NHS red fluorescent dye according to the instructions of the RED-NHS Monolith NT Protein Labeling kit (NanoTemper Technologies GmbH, München, Germany). In protein and RNA interaction assays, the concentration of NHS-labeled protein was maintained at 100nM, whereas RNA concentrations were gradient-diluted (20,000nM, 10,000nM, 5,000nM, and then 2-fold dilutions until 10nM). The RNA was denatured and annealed to form dsRNA, followed by addition of RNase inhibitor (1 U per group). After a short incubation, the samples were loaded into MST standard-treated glass capillaries. Measurements were performed at 25°C and LED and MST powers of 20% in buffer containing 20mM Tris (pH 8.0) and 150mM NaCl. The assays were repeated two times for each affinity measurement. Data analyses were performed using Nanotemper Analysis and OriginPro 8.0 software provided by the manufacturer.

4.5 Electrophoretic mobility shift assays

The pGEX-4T-1-NS3, pGEX-4T-1-mNS3, and pGEX-4T-1-OsDRB1a constructs, as well as empty pGEX-4T-1 vectors, were individually transformed into *E. coli* Transetta (DE3) (Trans-gene, Beijing, China) with protein expression induced by IPTG. The soluble GST fusion proteins were extracted and immobilized onto glutathione sepharose beads (Amersham). RNAs were labeled with ^{32}P–UTP using a T7 RNA Production system (Promega) and purified with a MEGAclear kit (Ambion, Waltham, MA, USA). Approximately 2,000cpm of radioactive RNA was denatured and then annealed to form dsRNA. The dsRNA was incubated with individual GST fusion proteins in the presence of RNase inhibitor (1 U per group) for 30min on ice. The mixtures were then run on an RNase-free native PAGE gel for 1 h at 4°C.

4.6 Protein preparation and western blot analysis

The procedures used for protein extraction and western blotting have been described previously (Wu et al., 2015). The following antibodies were purchased from commercial sources: anti-flag-peroxidase (Sigma, St. Louis, MO, USA), anti-Myc-peroxidase (Sigma), anti-Myc (Sigma), anti-FLAG (Sigma), anti-Actin (Easybio, Beijing, China), and anti-HYL1 (Agrisera AB Box 57, SE-911 21 Vännäs, Sweden).

4.7 Small RNA sequencing

Small RNA cloning for Illumina sequencing was carried out at Bainuodacheng Co., Ltd (Bei-jing, China). Rice miRNA annotations were obtained from miRBase (http://microrna.sanger. ac.uk/sequences, Release 14). Statistical analysis of the small RNA data sets was performed using in-house Perl scripts (Wu et al., 2015).

4.8 Bimolecular fluorescence complementation

BiFC assays were performed as previously described (Fang and Spector, 2007). Briefly, we transformed the recombinant constructs into *A. tumefaciens* strain EHA105 and injected the *Agrobacterium* cultures into *N.benthamiana* leaves. After 3d, we observed the tobacco epidermal cells with a confocal laser scanning microscope (LSM 710 NLO and Duo Scan System, Zeiss).

4.9 RNA immunoprecipitation

RNA immunoprecipitation was carried out essentially as previously described (Wierzbicki et al., 2008) using 1% formaldehyde-treated leaves, transiently expressed protein, and RNA for 72h. RIP was carried out with the following modifications: nuclei were isolated, resuspended in high salt nuclear lysis buffer (20mM Tris-HCl, pH 7.5, 500mM NaCl, 4mM $MgCl_2$, 0.2% NP-40). The chromatin supernatant was diluted five times with dilution buffer (20mM Tris-HCl, pH 7.5, 4mM $MgCl_2$, 0.2% NP-40) and immunoprecipitated using antibody. Pri-miRNAs were detected in the immunoprecipitates by RT-PCR using the artificial primers miR528-5p and -3p.

4.10 Accession numbers

Sequence data generated in this study can be found in GenBank (https://www.ncbi.nlm.nih. gov/genbank/), the Rice Genome Database (http://rice.plantbiology.msu.edu/) and the *Arabidopsis* Information Resource (http://www.*Arabidopsis*.org/) under the following accession numbers: *NS3*(AY284945.1), *AGO1a* (LOC_Os02g45070), *SULTR2;1* (LOC_Os03g09930), *CDS1*(LOC_Os07g46990), *LOC_Os08g45000*,*OsRFPH2-10*(LOC_Os06g06050), *OsDRB1a* (LOC_Os11g01869), *OsDRB1b*(LOC_Os12g01916), *OsDRB1c*(LOC_Os05g24160),*OsDRB2* (LOC_Os10g33970), *OsDRB3*(LOC_Os09g33460), *OsDRB4*(LOC_Os01g56520), *AtHYL1* (At1g09700),*AtDCL1* (At1g01040),*AtSE* (At2g27100), and *AtCBP20* (At5g44200), *AtDRB2* (At2g28380), *AtDRB3* (At3g26932), *AtDRB4* (At3g62800), *AtDRB5* (At5g41070).

References

[1] BAEK D, KIM M C, CHUN H J, et al. Regulation of miR399f transcription by AtMYB2 affects phosphate starvation responses in Arabidopsis[J]. Plant physiology, 2013, 161(1): 362-373.

[2] BAULCOMBE D. RNA silencing in plants. Nature, 2004, 431(7006), 356–363

[3] CAO X, ZHOU P, ZHANG X, et al. Identification of an RNA silencing suppressor from a plant double-stranded RNA virus[J]. Journal of Virology,2005, 79(20): 13018-13027.

[4] CHAPMAN E J, PROKHNEVSKY A I, GOPINATH K, et al. Viral RNA silencing suppressors inhibit the microRNA pathway at an intermediate step[J]. Genes Development,2004, 18(10): 1179-1186.

[5] CHEN X. Small RNAs and their roles in plant development[J]. Annual Review of Cell and Developmental Biology, 2009, 25: 21-44.

[6] CHIOU T-J, AUNG K, LIN S-I, et al. Regulation of phosphate homeostasis by microRNA in Arabidopsis[J]. Plant Cell,2006, 18: 412-421.

[7] CUI Y, FANG X, QI Y. TRANSPORTIN1 Promotes the Association of MicroRNA with ARGONAUTE1 in Arabidopsis[J]. Plant Cell,2016, 28: 2576-2585.

[8] CUPERUS J T, MONTGOMERY T A, FAHLGREN N, et al. Identification of MIR390a precurs or processing-defective mutants in *Arabidopsis* by direct genome sequencing[J]. Proceedings of the National Academy of Sciences of the United States of America,2010, 107(1): 466-471.

[9] DING S-W. RNA-based antiviral immunity[J]. Nature reviews immunology,2010, 10: 632.

[10] DING S-W, VOINNET O. Antiviral immunity directed by small RNAs[J]. Cell,2007, 130: 413-426.

[11] DJURANOVIC S, NAHVI A, GREEN R. miRNA-mediated gene silencing by translational repression followed by mRNA deadenylation and decay[J]. Science,2012, 336: 237-240.

[12] DONG Z, HAN M-H, FEDOROFF N. The RNA-binding proteins HYL1 and SE promote accurate *in vitro* processing of pri-miRNA by DCL1[J]. Proceedings of the National Academy of Sciences of the United States of America, 2008, 105(29): 9970-9975.

[13] DU P, WU J, ZHANG J, et al. Viral infection induces expression of novel phased microRNAs from conserved cellular microRNA precursors[J]. PLoS Pathogens,2011a, 7: e1002176.

[14] DU Z, XIAO D, WU J, et al. P2 of rice stripe virus (RSV) interacts with OsSGS3 and is a silencing suppressor. Molecular plant pathology,2011b, 12(8): 808-814.

[15] DUAN C-G, FANG Y-Y, ZHOU B-J, et al. Suppression of *Arabidopsis* ARGONAUTE1-mediated slicing, transgene-induced RNA silencing, and DNA methylation by distinct domains of the Cucumber mosaic virus 2b protein[J]. Plant Cell, 2012, 24: 259-274.

[16] EAMENS A L, SMITH N A, CURTIN S J, et al. The *Arabidopsis thaliana* double-stranded RNA binding protein DRB1 directs guide strand selection from microRNA duplexes[J]. RNA,2009, 15(12): 2219-2235.

[17] FANG X, QI Y. RNAi in plants: an Argonaute-centered view[J]. Plant Cell, 2016, 28: 272-285.

[18] FANG X, CUI Y, LI Y, et al. Transcription and processing of primary microRNAs are coupled by Elongator complex in *Arabidopsis*[J]. Nature Plants, 2015, 1: 15075.

[19] FANG Y, SPECTOR D L. Identification of nuclear dicing bodies containing proteins for microRNA biogenesis in living *Arabidopsis* plants[J]. Current biology, 2007, 17(9): 818-823.

[20] FRANCISCO-MANGILET A G, KARLSSON P, KIM M H, et al. THO_2, a core member of the THO/TREX complex, is required for micro RNA production in Arabidopsis[J].The Plant Journal, 2015, 82(6): 1018-1029.

[21] FUJII H, CHIOU T-J, LIN S-I, et al. A miRNA involved in phosphate-starvation response in *Arabidopsis*[J]. Current biology, 2005, 15(22): 2038-2043.

[22] FUJIOKA Y, UTSUMI M, OHBA Y, et al. Location of a possible miRNA processing site in SmD3/SmB nuclear bodies in *Arabidopsis*[J]. Plant Cell Physiology,2007, 48: 1243-1253.

[23] HENDERSON I R, ZHANG X, LU C, et al. Dissecting *Arabidopsis thaliana* DICER function in small RNA processing, gene silencing and DNA methylation patterning[J]. Nature genetics, 2006, 38(6): 721.

[24] HU B, WANG W, DENG K, et al. MicroRNA399 is involved in multiple nutrient starvation responses in rice[J]. Frontiers in plant science,2015, 6: 188.

[25] IWAKAWA H-O, TOMARI Y. The functions of microRNAs: mRNA decay and translational repression[J]. Trends Cell Biology,

2015, 25: 651-665.

[26] JIANG L, QIAN D, ZHENG H, et al. RNA-dependent RNA polymerase 6 of rice (*Oryza sativa*) plays role in host defense against negative-strand RNA virus, Rice stripe virus[J]. Virus Research, 2012, 163(2): 512-519.

[27] JIN L, QIN Q, WANG Y, et al. Rice dwarf virus P2 protein hijacks auxin signaling by directly targeting the rice OsIAA10 protein, enhancing viral infection and disease development[J]. PLoS Pathog,2016, 12: e1005847.

[28] KARIMI M, DEPICKER A, HILSON P. Recombinational cloning with plant gateway vectors[J]. Plant physiology, 2007, 145(4): 1144-1154.

[29] KIM W, KIM H-E, JUN A R, et al. Structural determinants of miR156a precursor processing in temperature-responsive flowering in Arabidopsis[J]. Journal of experimental botany, 2016, 67(15): 4659-4670.

[30] KIM Y J, ZHENG B, YU Y, et al. The role of Mediator in small and long noncoding RNA production in Arabidopsis thaliana[J]. The EMBO journal, 2011, 30(5): 814-822.

[31] KONG L, WU J, LU L, et al. Interaction between Rice stripe virus disease-specific protein and host PsbP enhances virus symptoms[J]. Molecular plant, 2014, 7(4): 691-708.

[32] LI F, DING S-W. Virus counterdefense: diverse strategies for evading the RNA-silencing immunity[J]. Annual review of microbiology, 2006, 60: 503-531.

[33] LI J, YANG Z, YU B, et al. 2005. Methylation protects miRNAs and siRNAs from a 3′-end uridylation activity in Arabidopsis[J]. Cancer Biology, 15(16): 1501-1507.

[34] BURGYÁN J, HAVELDA Z. Viral suppressors of RNA silencing[J]. Trends in plant science, 2011, 16(5): 265–272.

[35] LI Y, LU Y-G, SHI Y, et al. Multiple rice microRNAs are involved in immunity against the blast fungus *Magnaporthe oryzae*[J]. Plant Physiology, 2014, 164(2): 1077-1092.

[36] LIAN S, CHO W K, KIM S-M, et al. Time-course small RNA profiling reveals rice miRNAs and their target genes in response to rice stripe virus infection[J]. PloS one, 2016, 11: e0162319.

[37] MARUYAMA-NAKASHITA A, INOUE E, WATANABE-TAKAHASHI A, et al. Transcriptome profiling of sulfur-responsive genes in *Arabidopsis* reveals global effects of sulfur nutrition on multiple metabolic pathways[J]. Plant Physiology, 2003, 132: 597-605.

[38] NAVARRO L, DUNOYER P, JAY F, et al. A plant miRNA contributes to antibacterial resistance by repressing auxin signaling[J]. Science, 2006, 312: 436-439.

[39] PUMPLIN N, VOINNET O. RNA silencing suppression by plant pathogens: defence, counter-defence and counter-counter-defence[J]. Nature reviews Microbiology, 2013, 11(11): 745.

[40] QI Y, DENLI A M, HANNON G J. Biochemical specialization within *Arabidopsis* RNA silencing pathways[J]. Molecular Cell, 2005, 19(3): 421-428.

[41] REN B, GUO Y, GAO F, et al. Multiple functions of Rice dwarf phytoreovirus Pns10 in suppressing systemic RNA silencing[J]. Journal of virology, 2010, 84(24): 12914-12923.

[42] SHEN M, XU Y, JIA R, et al. Size-independent and noncooperative recognition of dsRNA by the Rice stripe virus RNA silencing suppressor NS3[J]. Journal of virology, 2010, 404: 665-679.

[43] SZARZYNSKA B, SOBKOWIAK L, PANT B D, et al. Gene structures and processing of *Arabidopsis thaliana* HYL1-dependent pri-miRNAs[J]. Nucleic acids research, 2009, 37(9): 3083-3093.

[44] TAKAHASHI H, WATANABE-TAKAHASHI A, SMITH F W, et al. The roles of three functional sulphate transporters involved in uptake and translocation of sulphate in *Arabidopsis thaliana*[J]. The Plant journal, 2000, 23(2): 171-182.

[45] TAKAHASHI M, TORIYAMA S, HAMAMATSU C, et al. Nucleotide sequence and possible ambisense coding strategy of rice stripe virus RNA segment 2[J]. The Journal of general virology, 1993, 74: 769-773.

[46] TORIYAMA S, TAKAHASHI M, SANO Y, et al. Nucleotide sequence of RNA. 1, the largest genomic segment of rice stripe virus, the prototype of the tenuiviruses[J]. The Journal of general virology, 1994, 75: 3569-3579.

[47] VÁRALLYAY É, VÁLÓCZI A, ÁGYI Á, et al. Plant virus-mediated induction of miR168 is associated with repression of ARGONAUTE1 accumulation[J]. The EMBO journal, 2010, 29: 3507-3519.

[48] WANG J-W, CZECH B, WEIGEL D. miR156-regulated SPL transcription factors define an endogenous flowering pathway in *Arabidopsis thaliana*[J]. Cell, 2009, 138: 738-749.

[49] WANG L, SONG X, GU L, et al. NOT2 proteins promote polymerase II–dependent transcription and interact with multiple microRNA biogenesis factors in *Arabidopsis*[J]. Plant Cell, 2013, 25: 715-727.

[50] WIENKEN C J, BAASKE P, ROTHBAUER U, et al. Protein-binding assays in biological liquids using microscale thermophoresis[J]. Nature communications, 2010, 1: 100.

[51] WIERZBICKI A T, HAAG J R, PIKAARD C S. Noncoding transcription by RNA polymerase Pol IVb/Pol V mediates transcriptional silencing of overlapping and adjacent genes[J]. Cell,2008, 135: 635-648.

[52] WU G, PARK M Y, CONWAY S R, et al. The sequential action of miR156 and miR172 regulates developmental timing in *Arabidopsis*[J]. Cell, 2009, 138: 750-759.

[53] WU J, YANG Z, WANG Y, et al. Viral-inducible Argonaute18 confers broad-spectrum virus resistance in rice by sequestering a host microRNA[J]. Elife, 2015, 4: e05733.

[54] WU J, YANG R, YANG Z, et al. ROS accumulation and antiviral defence control by microRNA528 in rice[J]. Nature plants, 2017, 3: 16203.

[55] XIE Z, ALLEN E, FAHLGREN N, et al. Expression of *Arabidopsis* MIRNA genes[J]. Plant Physiology, 2005, 138: 2145-2154.

[56] XIONG R, WU J, ZHOU Y, et al. Identification of a movement protein of the tenuivirus rice stripe virus[J]. Journal of virology, 2008, 82(24): 12304-12311.

[57] XIONG R, WU J, ZHOU Y, et al. Characterization and subcellular localization of an RNA silencing suppressor encoded by Rice stripe tenuivirus[J]. Virology, 2009, 387(1): 29-40.

[58] YAMASAKI H, HAYASHI M, FUKAZAWA M, et al. SQUAMOSA promoter binding protein–like7 is a central regulator for copper homeostasis in *Arabidopsis*[J]. Plant Cell, 2009, 21: 347-361.

[59] YANG J, ZHANG F, LI J, et al. Integrative analysis of the microRNAome and transcriptome illuminates the response of susceptible rice plants to rice stripe virus[J]. PloS one, 2016, 11: e0146946.

[60] YANG S W, CHEN H-Y, YANG J, et al. Structure of *Arabidopsis* HYPONASTIC LEAVES1 and its molecular implications for miRNA processing[J]. Structure, 2010, 18(5): 594-605.

[61] YANG X, REN W, ZHAO Q, et al. Homodimerization of HYL1 ensures the correct selection of cleavage sites in primary miRNA[J]. Nucleic acids research, 2014, 42(19): 12224-12236.

[62] YAO M, LIU X, LI S, et al. Rice stripe tenuivirus NSvc2 glycoproteins targeted to the golgi body by the N-terminal transmembrane domain and adjacent cytosolic 24 amino acids via the COP I-and COP II-dependent secretion pathway[J].Journal of virology, 2014, 88(6): 3223-3234.

[63] YU B, YANG Z, LI J, et al. Methylation as a crucial step in plant microRNA biogenesis[J]. Science,2005, 307: 932-935.

[64] ZHANG C, WU Z, LI Y, et al. Biogenesis, function, and applications of virus-derived small RNAs in plants[J]. Frontiers in microbiology, 2015, 6: 1237.

[65] ZHANG C, DING Z, WU K, et al. Suppression of jasmonic acid-mediated defense by viral-inducible microRNA319 facilitates virus infection in rice[J]. Molecular plant, 2016, 9(9): 1302-1314.

[66] ZHANG S, XIE M, REN G, et al. CDC5, a DNA binding protein, positively regulates posttranscriptional processing and/or transcription of primary microRNA transcripts[J]. Proceedings of the National Academy of Sciences of the United States of America, 2013, 110: 17588-17593.

[67] ZHU Y, HAYAKAWA T, TORIYAMA S, et al. Complete nucleotide sequence of RNA. 3 of rice stripe virus: an ambisense coding strategy[J]. The Journal of general virology, 1991, 72: 763-767.

Cleavage of the babuvirus movement protein B4 into functional peptides capable of host factor conjugation is required for virulence

Jun Zhuang[1,2], Wenwu Lin[1,2], Christopher J. Coates[3], Pengxiang Shang[2], Taiyun Wei[1,2], Zujian Wu[1,2], Lianhui Xie[1,2]

(1 State Key Laboratory of Ecological Pest Control for Fujian and Taiwan Crops, Fujian Agriculture and Forestry University, Fuzhou 350002, China; 2 Institute of Plant Virology, Fujian Agriculture and Forestry University, Fuzhou 350002, China; 3 Department of Biosciences, College of Science, Swansea University, Swansea SA2 8PP, Wales, UK)

Abstract: *Banana bunchy top virus* (BBTV) poses a serious danger to banana crops world wide. BBTV-encoded protein B4 is a determinant of pathogenicity. However, the relevant molecular mechanisms underlying its effects remain unknown. In this study, we found that a functional peptide could be liberated from protein B4, likely via proteolytic processing. Site-directed mutagenesis indicated that the functional processing of protein B4 is required for its pathogenic effects, including dwarfism and sterility, in plants. The released protein fragment targets host proteins, such as the large subunit of RuBisCO (RbcL) and elongation factor 2 (EF2), involved in protein synthesis. Therefore, the peptide released from B4 (also a precursor) may act as a non-canonical modifier to influence host-pathogen interactions involving BBTV and plants.

Keywords: *Banana bunchy top virus* (BBTV); Movement protein B4; Functional peptide; Pathogenicity

1 Introduction

Single-stranded (ss) DNA viruses (such as geminiviruses) are among the most successful groups of plant pathogens (Hanley-Bowdoin et al., 2013). Mastreviruses and begomoviruses in *Geminiviridae* are transmitted by leafhoppers and whiteflies, respectively, and can infect monocots and dicots (Hanley-Bowdoin et al., 2013). Aphid-borne *Banana bunchy top virus* (BBTV), an ssDNA virus belonging to the genus *Babuvirus* in the family *Nanoviridae*, naturally infects only species in the *Musaceae* family and is a severe threat to commercial banana production (Wu and Su, 1990).

The BBTV genome is made up of at least six circular ssDNA components, all of which exhibit rolling-circle replication (Burns et al., 1995). BBTV genomic DNA components are displayed as a single open reading frame (ORF) and a 5′ intergenic region including the origin of rolling-circle replication, similar to geminiviral components, exemplified by DNA component 4 (Burns et al., 1995). The BBTV DNA 4 protein can functionally restore the movement of the CMV-Fny-△MP mutant (Cucumber mosaic virus-Fny isolate with the deletion of *MP*) (Sun et al., 2002). Protein B4 is capable of redirecting DNA 6 gene products, which preferentially translocate into the nucleus, to the cell periphery (Wanitchakorn et al., 2000). Therefore, BBTV DNA 4 and 6 gene products are considered analogous to movement protein (MP) and nuclear shuttle protein

Virologica Sinica. 2019, 1-11.
Received 15 October 2018;Published 13 March 2019

(NSP), respectively (herein referred to as proteins B4 and B6). As previously reported, geminivirus NSP and MP orchestrate the ingress/egress of viral DNAs into and out of the nucleus and mobility between cells (Rojas et al., 2001; Fondong 2013; Zorzatto et al., 2015). However, BBTV MP and NSP do not display detectable sequence identity/similarity to bipartite geminiviral counterparts.

Uniquely, BBTV movement protein B4 contains a single transmembrane (TM) motif adjacent to the N-terminal domain, thereby facilitating membrane insertion (Zhuang et al., 2016). The single N-terminal TM motif along with lateral charged residues is likely equivalent in function to a signal peptide sequence used to guide proteins through the secretory pathway. Thus, the functional cleavage of protein B4, *i.e.*, the removal of the N-terminal TM domain by proteases, may be required for its activation. In the present study, we confirmed the membrane-bound subcellular localization of protein B4, which coexists in the vasculature with DNA 4 components. We further identified a B4-derived cryptic peptide whose functional release is associated with BBTV pathogenicity. The cryptic peptide appears to act as a non-canonical modifier by host (target) protein conjugation.

2 Materials and methods

2.1 Plant materials and growth conditions

Nicotiana benthamiana and its GFP-transgenic 16c line were used. The seeds were plated on nutrient soil and imbibed for 1 week at 25℃ under a 16-h-light/8-h-dark cycle. The seedlings were transferred individually into vials and kept at 25℃ under a continuous 16-h light/8-h dark cycle until reaching the 5-leaf stage (2-week-old *N. benthamiana* seedlings) for agrobacterial inoculation. In addition, BBTV-infected bananas maintained under natural conditions (20~30℃ and natural lighting) were harvested from the banana field at Fujian Agriculture and Forestry University.

2.2 Plasmid construction and viral inoculations

The PVX:B4 and PVX:B4GFP constructs were assembled by cloning protein B4 and B4GFP fused/ inserted between *Sal* I and *Cla*I sites of the PVX vector, respectively. The B4/B4GFP mutants were amplified using primers listed in Supplementary Table S1 and the corresponding PVX:B4 and PVX:B4GFP derivatives were obtained using a Fast Mutagenesis Kit V2 (Vazyme Biotech, Nanjing, China). All constructs were introduced into *Agrobacterium tumefaciens* strain GV3101 and cultured in Luria–Bertani (LB) broth overnight at 28℃ . The inocula were transferred into fresh medium (1:50 dilution) and incubated for 6–8h. Bacteria were pelleted and re-suspended in infiltration buffer (5g/L glucose, 10mmol/L $MgCl_2$, 50mmol/L MES-KOH, pH 5.7; including 100μmol/L acetosyringone) until an OD 600 value of 1.0 was obtained. Cultures of agrobacteria were maintained at room temperature for 1–2h prior to infiltration into the lower epidermis of 14-day-old *N. benthamiana* leaves using 1-mL syringes without needles. *Agrobacterium* with PVX alone and infiltration buffer alone were used as negative controls. Seven-to-ten days post-infiltration, leaf tissues were harvested for total protein extraction. DNA extracted from BBTV-infected banana petioles was inoculated into the stem of using a syringe, using DNA from healthy bananas as a control.

2.3 Fluorescence in *situ* hybridization (FISH) detection of BBTV

BBTV-infected banana petioles with (symptomatic) dark green streaks were sliced into thin sections. The samples were permeabilized in PBS (pH 7.4) containing 2% (v/v) Triton X-100 and 5% (v/v) β-mercaptoethanol for 4h at 20℃ .

They were then fixed in PBS containing 1% Triton X-100 and 4% paraformaldehyde for 4h. Post-fixation, samples were washed with hybridization solution three times, placed into a hybridization solution containing 1% β-mercaptoethanol and 301mol/L DNA fluorescent probe for 5h, washed in 1×SSC three times, and washed twice in PBS (pH 7.4). Confocal microscopy was used to evaluate the samples for the presence of BBTV. The following FAM-labeled probe was used to visualize DNA4: B4 FAM-RC-Probe, 5′-FAM__CAGAAACCATTCGAAGAATAGTTTCACC-3′.

2.4 Protoplast preparation of *N.benthamiana* mesophyll cells infected by PVX:B4 -GFP

Approximately 7d post-inoculation, newly developed leaves infected by chimeric PVX were harvested. The lower epidermis of tobacco leaves was separated/peeled using fine tweezers and tobacco mesophyll cells were placed into an enzyme solution containing 1% cellulase R-10 (Yakult Honsha, Tokyo, Japan), 0.25% macerozyme (Yakult Honsha), 0.4mol/L mannitol, 10mmol/L $CaCl_2$, 20mmol/L KCl, 0.1% bovine serum albumin, and 20mmol/L MES (pH 5.8) for 1–2h at room temperature (with constant agitation at~30rpm). The lysate was filtered using fiber sheets. The filtrate was centrifuged at low speed (~100×g) for 5min. The green pellet was rinsed three times using W5 buffer (150mmol/L NaCl, 125mmol/L $CaCl_2$, 5mmol/L KCl, 5mmol/L glucose, 2mmol/L MES pH 5.7) and re-suspended in W5 buffer. Approximately 20μL of protoplast suspension was removed and placed onto sterile glass slides for inspection by microscopy (Nikon, Tokyo, Japan).

2.5 Preparation of thylakoid components from BBTV petioles

The peeled BBTV petioles (BBTV-infected and healthy samples) were individually harvested and cut into small (1–2cm) pieces. Cooled samples were macerated for several minutes in extracting buffer [0.15mol/L Tris–Cl pH 8.0, 0.5% (w/v) $NaSO_3$, 0.2% (w/v) PVP-K30], ground on ice for several minutes, and filtered using 100 meshes. Chloroplasts were isolated from filtrates using an extraction kit (Sigma-Aldrich, St. Louis, MO, USA), following the supplier's guidelines. Briefly, the macerates were centrifuged for 3min at 200×g to remove cellular debris. The supernatant was retained and centrifuged at 1000×g for 7–10min. This time, the supernatant was discarded and the green pellet was re-suspended in 1×CIB. The chloroplast extracts were loaded onto a discontinuous 40%–80% Percoll gradient. After centrifugation, the green layers containing chloroplasts were harvested to isolate thylakoid membranes.

To purify the thylakoid membrane, the chloroplast suspensions were subjected to ultrasonic treatment at 4℃ . The homogenate was centrifuged at 10,000×g for 10–20min. The green supernatants were loaded onto 15%–40% sucrose gradients and centrifuged in a swinging bucket rotor (type 32, Beckman) at 4℃ for 4h at 110,000g. The separated fractions were retained for SDS- PAGE and western blot analyses.

2.6 Immuno–EM assay

Tobacco stems were cut into 0.5cm^2 sections and fixed for 2h in PBS (pH 7.2) containing 4% paraformaldehyde and 0.1% glutaraldehyde at room temperature. After fixation, the samples were washed three times using PBS. The samples were dehydrated through an ethanol gradient of 50%, 70%, 90%, and 100% for 1h at -20℃ . Postdehydration, mixtures of the embedding agent LR-Gold and ethanol at 1 ∶ 1 and 2 ∶ 1 ratios were used to permeabilize the samples for 1h at -20℃ .The samples were embedded incomplete LR-Gold for 1h at -20℃ .Samples were then transferred to LR-Gold containing 0.1% BENZIL for 3h at -20℃ . Finally, the samples were placed into thin-walled tubes with LR-Gold containing BENZIL and polymerization was achieved under UV light at 20℃ over a 7-day period. For immuno-electron microscopy, ultra-thin sections were blocked using 2% BSA for 30min prior to labeling with the anti-GFP antibody and 10-nm gold particles conjugated to goat antibodies against rabbit IgG (GAR10; British Bifocals International, Cardiff, UK). The specificity of labeling was monitored by incubating noninfected tobacco samples with anti-GFP IgG.

2.7 Imaging and microscopy

B4GFP fluorescence was detected using a confocal laser microscope Zeiss LSM 710 (Carl Zeiss, Oberkochen, Germany) equipped with a 40×C-Apochromat objective. GFP expressed in tobacco leaves was detected by long wave UV excitation at 488nm and emission at 509nm. Ultrathin sections of banana petioles and immuno-gold particles were observed by transmission electron microscopy (Hitachi H-7650).

The banana samples were harvested and cut into thin ($0.2cm^2$) slices and fixed in PBS (pH 7.2) containing 4% paraformaldehyde, 0.1% glutaraldehyde, 0.1% Triton X-100, and 2% β-mercaptoethanol at 26℃ for several hours. After fixation, the samples were dehydrated using a gradient of ethanol and then coated in embedding agent at 26℃ for several hours. The polymerizing patches were prepared for ultrathin slicing and stained in 1% uranyl acetate prior to observation by TEM.

2.8 Total protein extraction from BBTV-infected stems

Approximately 20g of ‘green streak’ sections from BBTV-infected banana petioles were peeled, flash frozen in liquid nitrogen, and ground into a fine powder. PBS pH 7.4 [containing 0.5% (w/v) $NaSO_3$, 0.2% (w/v) PVP-30, and complete protease inhibitor (Roche, Basel, Switzerland)] was added to the sample prior to vortexing and homogenization in an ice bath. The sample was passed through a 100-mesh filter and subsequently centrifuged once at 10,000×g for 30 min and again at 40,000×g for 20min at 4℃ . The remaining supernatant was centrifuged at 150,000 ×g for 4.5h at 4℃ . The pellet was re-suspended in 500μL of 50mmol/L PBS, pH 7.4. The suspension was aliquoted (50μL per vial) for co-immunoprecipitation (CoIP), and stored at -70℃ . Total proteins from healthy bananas were extracted following the above procedure.

2.9 Western blotting

Approximately 0.5g of PVX:B4GFP-infected *N. benthamiana* (GFP-expressing *N. benthamiana* as a control) leaves were homogenized using lysis buffer [50mmol/L Tris–HCl pH 7.5, 5mmol/L $MgCl_2$, 150mmol/L NaCl, 0.1% Tween 20, 5% (v/v) β-mercaptoethanol, and a protease inhibitor cocktail (Roche)] on ice. Lysates were centrifuged twice at 30,000×g at 4℃ for 20min. Approximately 40μL of supernatant was transferred to a 0.5mL tube and 10μL of 5×SDS sample buffer (CW Biotech) was added. The remaining supernatant was incubated with 20μg of anti-GFP antibody at 16℃ for 0.5h and then transferred to another column pre-filled with protein A/G agarose beads (Abmart). Next, the beads were rinsed thoroughly using a sixfold column volume of the lysis buffer and 50μL of 5×SDS sample buffer was used to suspend the beads after the washing step. Subsequently, all samples were boiled at 98℃ for 5min to disrupt disulfide bonds and linearize the peptides. The treated protein samples were centrifuged at 14,000 ×g for 5–10min, separated by 12%–15% SDS-PAGE, and transferred onto a PVDF membrane. Immunoblotting (IB) was performed using anti-B4 and anti-GFP antibodies (Genscript). Specifically, 1×TBST (50mmol/L Tris–Cl pH 8.0, 150mmol/L NaCl, 0.05% Tween 20) supplemented with 5mmol/L $MgCl_2$ was used for anti-B4 immunoblotting at 16℃ .

2.10 Co-immunoprecipitation (CoIP) assay

Approximately 0.5g of PVX:B4GFP-infected *N. benthamiana* (GFP-expressing *N. benthamiana* as a control) leaves was homogenized using lysis buffer [50 mmol/L Tris–HCl pH 7.5, 5mmol/L $MgCl_2$, 150mmol/L NaCl, 0.1% Tween 20, 5% (v/v) β-mercaptoethanol, and a protease inhibitor cocktail (Roche)] on ice. Lysates were centrifuged twice at 30,000×g at 4℃ for 20min. Approximately 40μL of supernatant was transferred to a 0.5mL tube and 10μL of 5×SDS sample buffer (CW Biotech, Beijing, China) was added. The remaining supernatant was incubated with 20μg of anti-GFP antibody at 16℃ for 0.5h and then transferred to another column pre-filled with protein A/G agarose beads (Abmart, Shanghai, China). Next, the beads were rinsed thoroughly using a sixfold column volume of the lysis buffer. Approximately 50μL of 5×SDS sample buffer was used to suspend the beads after the washing step. Subsequently, all samples were boiled at 98℃ for 5min to disrupt disulfide bonds and linearize the peptides. The treated protein samples were centrifuged at 14,000×g for 5–10min, separated by 12%–15% SDS-PAGE, and transferred onto a PVDF membrane. Immunoblotting (IB) experiments were respectively performed using anti-B4 and anti-GFP antibodies (GenScript, Piscataway, NJ, USA). Specifically, 1×TBST (50mmol/L Tris–Cl pH 8.0, 150mmol/L NaCl, 0.05% Tween 20) supplemented with 5mmol/L $MgCl_2$ was used for anti-B4 immunoblotting at 16℃ .

To capture the target protein fragments associated with the B4-derived cryptic peptide in infected banana, the total banana protein extracts separated by centrifugation at 40,000×g and the supernatant were used for CoIP at 16℃ . Similarly, CoIP using anti-B4 IgG required the addition of appropriate amounts of Mg^{2+} into protein suspensions to a final concentration of 1.0mmol/L. Next, 20–30μg of anti-B4 IgG was sequentially added and incubated at 16℃ . The mixture was aspirated into columns pre-filled with protein A/G agarose beads (Abmart). The beads were gently rinsed by running buffer through the column and re-suspended in 5×SDS sample buffer.

2.11 Mass spectrometry detection

Protein bands were stained using Coomassie brilliant blue (CBB) R-250 staining and excised from gels for further analysis. The Coomassie-stained gel slices of target proteins were excised and applied to a 96-well plate, destained twice in 200μL of 15mmol/L potassium ferricyanide and 50mmol/L sodium thiosulfate (1:1) and then dried twice with 200μL of acetonitrile. The dried gels were incubated in a pre-chilled digestion solution (trypsin 12.5ng/μL and 20mmol/L NH_4HCO_3) for 20min and incubated overnight at 37℃ . Finally, the fragmented peptides were pelleted using an extraction solution (5% formic acid in 50% acetonitrile) and desiccated under a stream of N_2.

The dried peptides were dissolved in solvent A (5% acetonitrile, 0.1% formic acid) and analyzed using a Triple-TOF 5600 system (AB SCIEX, Framingham, MA, USA). Briefly, peptides were separated on a reverse-phase column (ZORBAX 300SB-C18 column, 5μm, 300 Å , 0.1×15mm; MicroMass, Cary, MA, USA) using an Eksigent 1D PLUS system (AB SCIEX) at an analytical flow rate of 300nL/min. The peptides were separated with a linear gradient from 5% to 40% of solvent B (0.1% formic acid/90% acetonitrile) over 2h. Survey scans were acquired from 800 to 2500Da with up to 15 precursors selected for MS/MS and dynamic exclusion for 20s. For protein identification, MS/MS data were processed using MASCOT version 2.3.02 (Matrix Science, London, UK) and the *Nicotiana* subset of the NCBI sequence databases. Peptides with significance scores over the “identity threshold” were recorded and the sequence was confirmed by an artificial analysis according the b and y ions.

3 Results

3.1 BBTV DNA4 components localize in vascular tissue and neighboring chlorophyllou scompanion cells

BBTV infection usually causes typical dark green streaks in banana petioles (Fig. 1A), suggesting the gross aggregation of chloroplasts in BBTV-infected vasculature, relative to the evenly distributed light green coloration in healthy petioles (Fig. 1B). To further evaluate the association of the vasculature and chloroplasts with BBTV, the BBTV-infected petiole veins were harvested for in situ DNA hybridization. BBTV DNA4 component was detected within plant vascular tissues and adjacent chlorophyllous companion cells of the phloem during viremia (Fig. 1C) and was absent from the same tissues of healthy bananas (Fig. 1D).

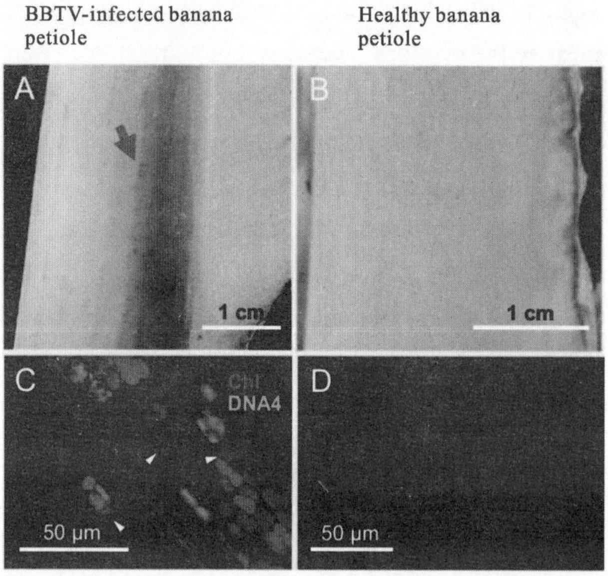

Fig. 1　Banana bunchy top virus DNA4 component exists in the vasculature and companion cells

Dark green streaks (marked by a red arrow) of BBTV-infected banana petioles are clearly visible (A), whereas light green coloration is distributed evenly in the petiole of a healthy banana (B). DNA hybridization signals produced from fluorescent probes indicated that BBTV DNA4 component accumulates in the vasculature and chlorophyllous companion cells (denoted by white arrow heads) (C). No hybridization signals were detected in healthy banana cells; only chlorophyll was detected (D)

3.2　Subcellular localization of BBTV movement protein B4

To determine the molecular characteristics of B4 protein, we first traced its definitive subcellular localization using a functionally intact B4GFP conjugate. We selected protein B4 from the BBTV Hainan isolate, which consists of 117 amino acids ordered into three domains: an N-terminal region (1–13 aa), a TM motif (position 14 to position 36), and a C-terminal tail (Fig. 2A). Based on PVX:B4GFP construct expression in *N. benthamiana* via the *Agrobacterium* infiltration method, the green fluorescence of B4GFP after 7d post inoculation (d.p.i.) was distributed throughout the stem, particularly in the vasculature of *N. benthamiana* under blue-light excitation (Fig. S1A). B4GFP was found in the cytoplasmic membrane and appeared to travel via a symplastic route to the neighboring cells/vasculature in leaf veins (Fig. S1B). Furthermore, B4GFP accumulation in foci in the apoplastic region of the stem was identified by immunogold labelling (Fig. S1C, S1D).

After B4GFP entered the foliage along with vascular tissues, expression states in different mesophyll cells were visualized. First, B4GFP circled the periphery of mesophyll cells prior to entry into the cytoplasm (Fig. 2B). Once inside the cell, it mainly aggregated into granule-like structures with distinct green fluorescence; a small portion overlapped with chloroplasts (containing chlorophylls), identified by red fluorescence (Fig.

2C). To further examine the ultrastructural location of B4GFP, ultrathin sections were assayed by immunoelectron microscopy. The granular compartment composed of filaments was specifically marked by colloidal gold particles after anti-GFP immunogold labelling (Fig. 2D). B4GFP deposited in the chloroplastic thylakoid membranes and matrix from PVX:B4GFP-infected leaves was also identified by immunolabelling (Fig. 2E).

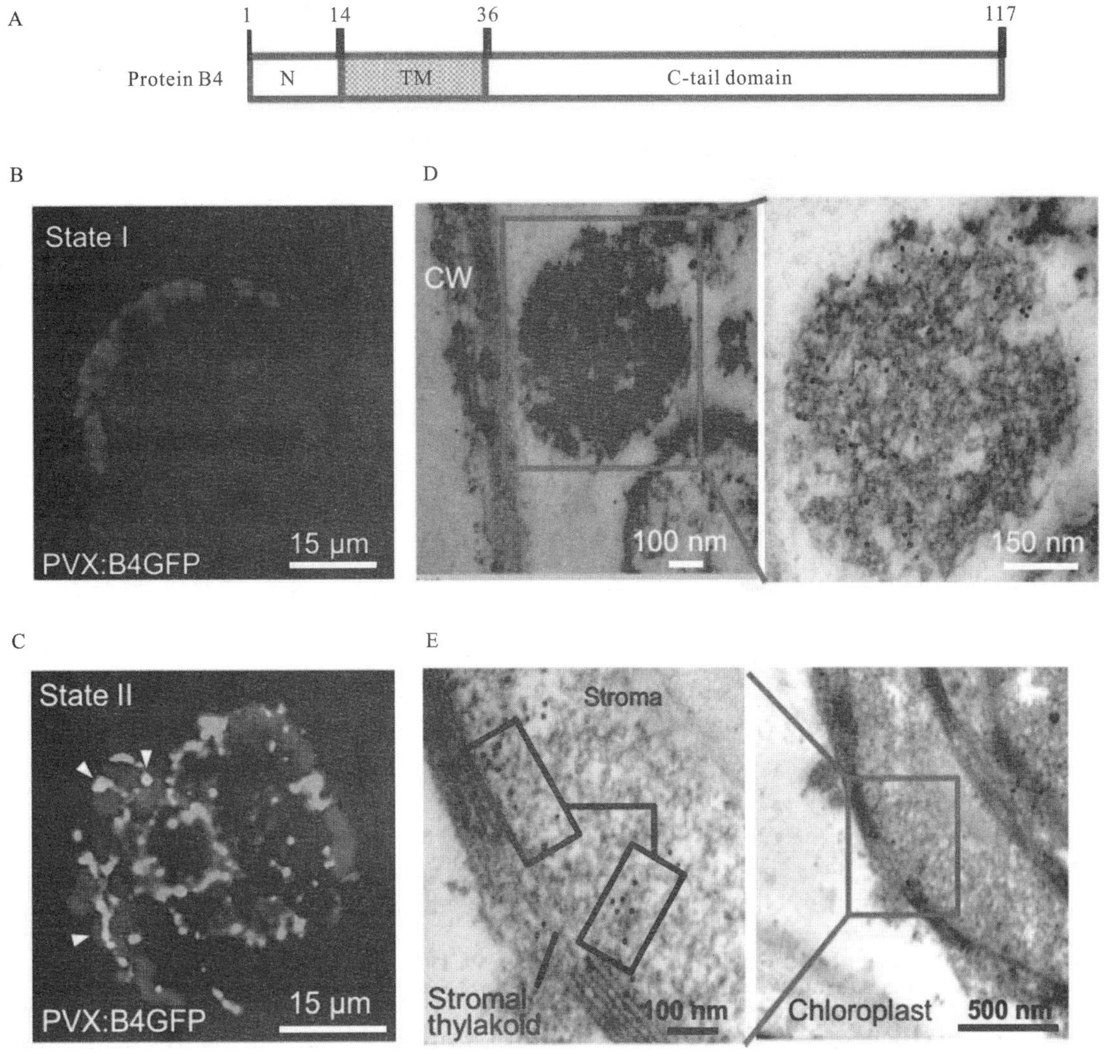

Fig. 2 Subcellular localization of Banana bunchy top virus protein B4 fused to GFP

A schematic representation of protein B4 domains from the BBTV Hainan isolate. Positions 1–13 make-up the N-terminal domain; positions 14–36 form the transmembrane domain (TM), and the remaining 37–117 C-tail domain. B. B4GFP was located in the periphery of protoplasts prepared from PVX:B4GFP infected *Nicotiana benthamiana* in expression state I. C.B4GFP moves into the intracellular reticula and chloroplasts in granular form in expression state II. D. The granular compartment adjacent to the cell wall (CW) of infected leaves was labeled immunologically using 10-nm gold particles (denoted by yellow arrowheads) for B4GFP. E. Immuno-labeled gold particles localized at the stroma and thylakoid membrane of B4GFP entering the chloroplast. Images represent experiments performed on at least two independent occasions

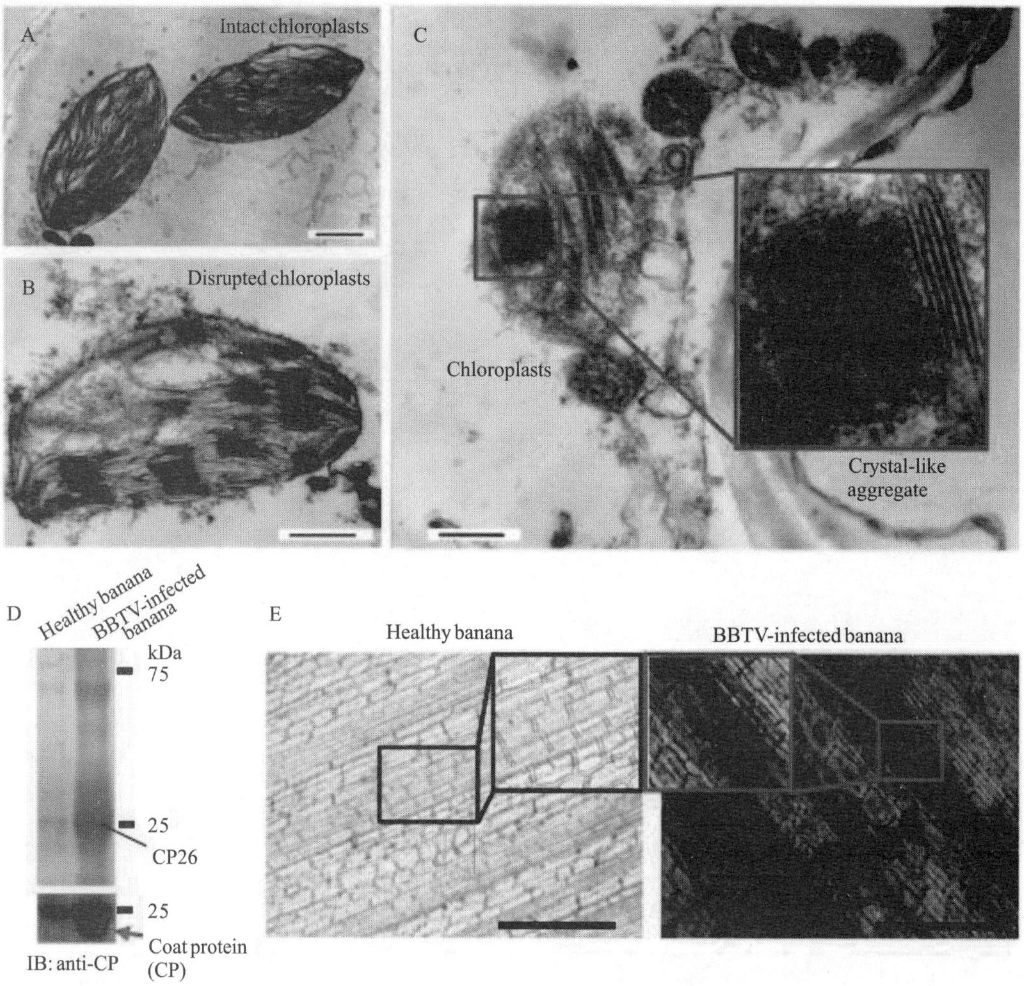

Fig. 3 The disruption of chloroplasts upon BBTV infection accompanies the crystalline aggregation of virus-like particles enveloped by the thylakoid membrane and a substantial increase of ROS

A,B disruption of the outer membrane and vacuolization of the broken chloroplast from healthy banana, in contrast to the intact envelope of chloroplasts from healthy banana. C. The thylakoid membrane fused to vesicle/organelles encompassed the crystalline-like particles. D. BBTV coat protein (CP) was specifically detected in the thylakoid pellet from BBTV-infected banana. E. Significantly deeper NBT staining in sieve-tube plate between neighboring cells in infected petioles indicated that ROS were robustly elevated relative to levels in healthy banana. Scale bars 1μm (A, B), 500nm (C), and 60μm (E)

3.3 Chloroplast disruption and ROS elevation in BBTV-infected banana petioles

BBTV colonizes phloem tissues to establish infection in bananas, leading to the formation of dark green streaks (longitudinal in direction) along the vasculature of the petioles/pseudostems. Relative to the normal morphology of chloroplasts in healthy banana (Fig. 3A), the disruption of the outer membranes of chloroplasts is induced by BBTV infection (Fig. 3B). Surprisingly, virus-like particles accumulate into crystalline assemblages that are surrounded by thylakoid membranes in BBTV-infected banana petioles (Fig. 3C). Accordingly, the coat protein of BBTV was detected in thylakoid components using an anti-CP antibody (Fig. 3D).

A dramatic increase in reactive oxygen species (ROS) levels in sections of BBTV-infected petioles was indicated by dark NBT staining compared to light dyeing in corresponding tissues of healthy bananas, especially in sievetube plates of vascular cells of infected bananas (Fig. 3E). Collectively, our data suggest a close relationship between chloroplasts and BBTV infectivity, replication, and dissemination.

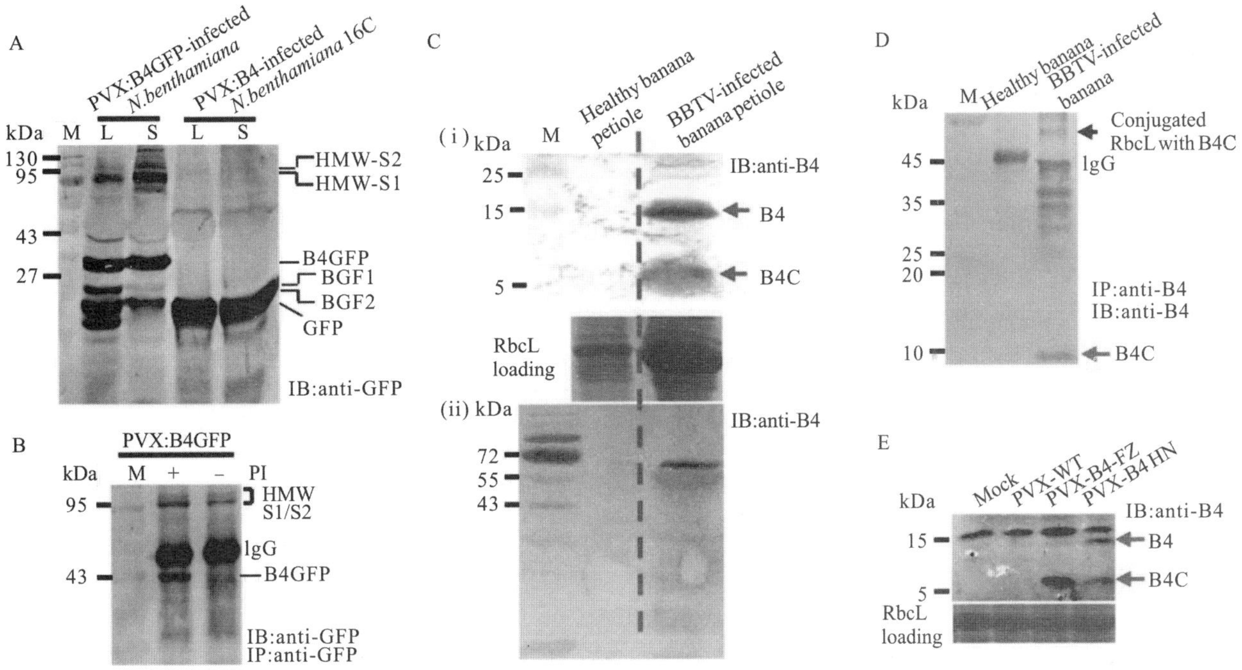

Fig. 4 The discrete B4-derived functional peptide and its conjugated proteins were identified by CoIP and immunoblotting

A B4GFP and BGF1/BGF2 fragments were detected by anti-GFP immunoblotting. Higher molecular weight (HMW) species were denoted HMW-S1 and HMW-S2. B HMW species were resistant to proteolytic degradation in the absence of a protease inhibitor (PI). C Protein B4 and B4-derived cryptide (denoted B4C) were confirmed in BBTV-infected banana by anti-B4 immunoblotting. The shifted bands (denoted by an arrow) above RbcL were probed in total protein extracts of BBTV-fected banana petioles. There were no positive signals in healthy banana petioles. D The discrete B4C and B4C-conjugated RbcL species were immuno-precipitated by anti-B4 Ab and were sequentially subjected to SDS-PAGE and immunoblotting by an anti-B4 Ab. E Protein B4 and B4C also were detected in PVX:B4HN/B4FZ (from BBTV HN and BBTV FZ isolates, respectively) inocula

3.4 A functional peptide is released from protein B4 and covalently targets host proteins

To characterize protein B4, the total protein contents of agro-infiltrated *N. benthamiana* were initially extracted in the presence of a protease inhibitor cocktail. Anti-GFP immuno-blotting revealed that protein B4GFP was cleaved into two smaller fragments (BGF1 and BGF2) by a putative peptidase (Fig. 4A). BGF1 is ~ 35kDa, highly similar to the predicted size of the C-terminal domain of protein B4 (i.e., B4C) with intact GFP (26.9kDa). Intriguingly, immuno-blots showed several anti-GFP-positive bands above 95kDa after B4GFP challenge. Prior to SDS-PAGE, the total proteins resolved in loading buffer containing SDS underwent disulfide bond disruption and polymer disassembly by the addition of reducing agent dithiothreitol (DTT) and heat treatment. Therefore, it is highly unlikely that these protein bands are B4GFP self-aggregates. Concurrent with the PVX:B4GFP distribution in nascent tissues, the abundance of modified protein species of higher molecular weight (HMW) continued to increase from the original site of inoculation (Fig. 4A). Conversely, the abundance of BGF1 fragments from systemic (distal) leaves was lower than that in inoculated leaves (Fig. 4A), in dicative of HMW species as a result of host factor conjugation with BGF1. In addition, B4-derived peptides inPVX:B4-infected 16C (GFP-transgenic *N. benthamiana* line) do not target GFP to form the HMW species (Fig. 4A), implying that protein B4 targets specific host factors, rather than random substrates (such as GFP).

We utilized CoIP to further investigate the target(s) of protein B4. The immuno-precipitates from

PVX:B4GFP-infected *N. benthamiana* revealed a prominent HMW GFP-positive signal, which seemed to be resistant to degradation in the absence of protease inhibitors and perhaps resistant to proteasomal cleavage (Fig.4B). The immunoprecipitated HMW species were excised from an SDS-PAGE gel, processed by trypsin, and evaluated by liquid chromatography mass spectrometry (LC–MS). Peptide mass spectra corresponded to ribulose bisphosphate carboxylase oxygenase large subunit (Rubisco LS, or RbcL), elongation factor 2 (EF2), and BGF1 fragment (Fig.S2, Fig.S3). The migration patterns of the modified proteins HMW-S1 (~90kDa) and HMW-S2 (~132kDa) were consistent with those of RbcL (~55kDa) and EF2 (~95kDa) in complexes with a single BGF1 (~34kDa, B4C+GFP).

The full-length protein B4 and its cleaved B4C fragment were specifically probed in the protein extract from natural host banana petioles infected by BBTV (Fig. 4C-i). Similarly, an HMW anti-B4-positive signal above RbcL was specifically probed in the BBTV-infected banana protein extract by western blotting (Fig. 4C-ii). We co-immunoprecipitated the HMW species in the protein extract of BBTV-infected banana using anti-B4 antibodies. The HMW species above RbcL was identified as the conjugate of RbcL and B4C by matrix-assisted laser desorption ionization time-of-flight tandem mass spectrometry (MALDITOF–MS/MS) (Fig.S4). In parallel, the immuno-precipitate was subjected to anti-B4 immunoblotting. This analysis showed that these HMW species contained the RbcL–B4C conjugate, other presumptive targets, and discrete B4C peptide (Fig. 4D). B4C could be detected in the PVX:B4 inocula as well (B4HN and B4FZ individually representing protein B4 from Hainan and Fuzhou isolates) (Fig. 4E). These results provide clear evidence that the functional cleavage of protein B4 is conserved in various BBTV isolates and different plant taxa.

3.5 Proteolytic cleavage of protein B4 is a prerequisite for switching-on protein targeting activity

The tangible scission of protein B4 is likely executed by proteases. To identify if highly conserved arginine residues are involved in the release of the functional fragment B4C and the covalent linkage with host factors, a series of invariant positively charged arginine residues were replaced with alanine residues using site-directed mutagenesis (Fig. 5A). In contrast to the infertility induced by the PVX:B4-inoculum, PVX:B4Mut1 and PVX:B4Mut2 inocula restored flowering and seeding. All other B4 mutants retained the functionality of protein B4, leading to dwarfism and sterility (Fig. 5B). Importantly, the B4-derived peptide (B4C) was not detected in the PVX:B4GFPMut2 ($RRR_{55-57}\rightarrow$ AAA)-inoculated plants (Fig. 5C), consistent with the absence of BGF1 from the B4GFPMut2 inoculum (Fig. 5D). Additionally, the HMW species were almost completely lacking in B4GFPMut1 ($K_{40}K_{43}R_{47}\rightarrow$AAA) and B4GFPMut6/7 ($R_{99-100}\rightarrow$AA) inocula, despite the presence of BGF1 (Fig. 5D). Presumably, RRR_{55-57} is necessary for the release of the B4-derived cryptide and other arginine residues could be responsible for conjugation to host factors. Because the mutation of protein B4 at RRR_{55-57} prevented cryptide release from full-length protein B4 as well as B4C (~8kDa) and BGF1 (~34kDa), the BGF1 cleavage site (named the P1 position) was located at RRR_{55-57} or adjacent residues. However, the definitive cleavage site at protein B4 needs to be confirmed, in addition to the identification of host/plant endopeptidases involved in this proteolytic processing.

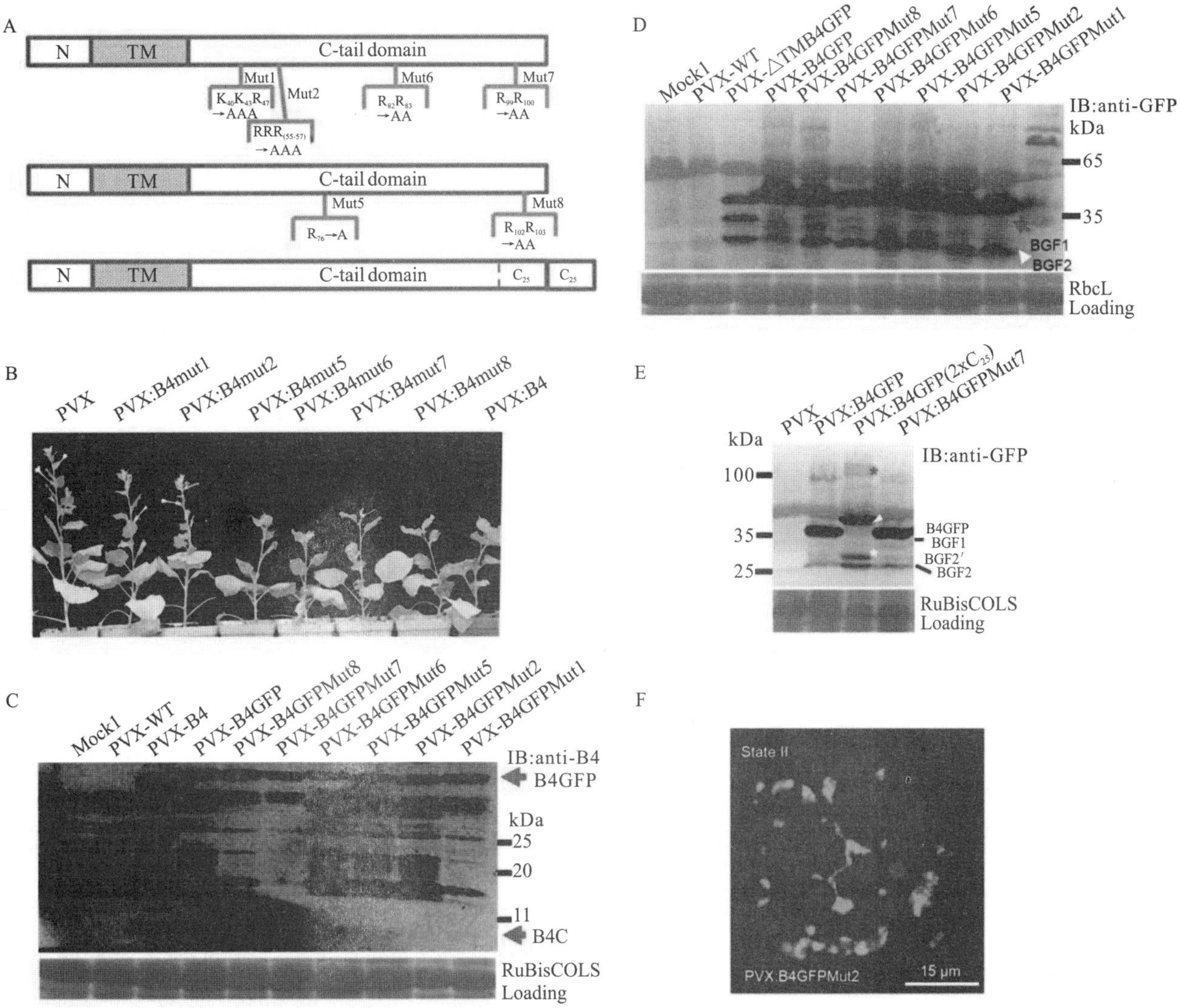

Fig. 5 Crucial arginine (Arg) residues are required for the release of the B4-derived cryptide

A schematic diagrams of mutation positions in protein B4. B. BBTV B4Mut1 and B4Mut2 were not pathogenic when exposed to *Nicotiana benthamiana*. However, B4 mutants 5–8 displayed similar wide-type symptoms. C. The B4-derived functional peptide is absent in the PVX:B4GFPMut2 inoculum. D. BGF1 can be detected in PVX:B4GFP and its mutant inocula, other than the PVX:B4GFPMut2 inoculum. The anti-GFP-positive HMW species are lacking in the Mut2 inoculum, concomitant with the absence of BGF1. BGF2 and BGF1 are denoted by red and white arrowheads, respectively. E. The C_{25} duplicate contains the two P2 cleavage sites; BGF2' fragment (denoted by a white asterisk) resulted from the incomplete proteolysis of the C_{25} duplicate. BGF1′ is marked by a white arrowhead. The up-shifted migration of HMW species (denoted by a red asterisk) might indicate a conjugate with BGF1′ containing another C_{25} duplicate. F. GFP fluorescence surrounds the periphery of chloroplasts but does not appear inside the organelles in plants treated with B4GFP Mut2

It is likely that the P2 position in the C_{25} region contributes to the generation of BGF2, because BGF2′ and B4GF2 fragments were both generated from B4GFP (2×C_{25}) fusion proteins containing two C_{25} repeats (Fig. 5E). BGF2′ may be a product of incomplete cleavage at the P2 position. Notably, B4GFP (2×C_{25}) also can be processed into BGF1′ at the P1 position, which contains two C_{25} repeats and is larger than B4GF1. The upward shift of HMW species (denoted by a red asterisk in Fig. 5D) from the B4GFP (2×C_{25}) inoculum was easily identifiable, unlike the changes of HMW species in B4GFP and B4GFPMut7 inocula. These observations imply that HMW species formation is mediated by conjugation with BGF1 rather than BGF2. Unlike the intact protein B4GFP, B4GFPMut2 did not overlap (co-localize) with chloroplasts but accumulated around the periphery of the organelle (Fig. 5F), thereby providing subcellular evidence for noncovalent linking to chloroplast RbcL or other proteins. Even though RbcL and EF2 are likely targets of this B4- derived modifier, these complexes does not account for the infertility of plants under B4 challenge; therefore, we predict the presence of other molecular targets or functional contribution by the liberated N-terminal fragment. It is possible that the combined activities of protein B4 and its released cryptides have indirect consequences that compromise plant-BBTV homeostasis during viral infection.

4 Discussion

Virus-infected plants display a common set of symptoms, such as chlorosis and mottling of the leaves, indicating that these organelles are deconstructed and damaged (Mandadi and Scholthof, 2013). BBTV DNA4 components localize in vascular tissues and co-occur in companion cells where they enter chloroplasts (indicated by the merged image of the hybrid signal of the DNA 4 probe and auto-fluorescence of chlorophyll in Fig. 1D). Protein B4 enters the chloroplast as well (Fig. 2C, E). Furthermore, the virus-like particles are enveloped by the thylakoid membrane in BBTV-infected banana petioles (Fig. 3C). This suggests that BBTV exploits the chloroplast platform for replication and colonization. It is plausible that BBTV could directly usurp the energy produced in chloroplasts by photophosphorylation for self-advantage. Chloroplasts play pivotal roles in early immune responses, such as the production of antimicrobial ROS (de Torres Zabala et al., 2015). A homolog of bacterial respiratory Complex I, NDH-1, is involved in cyclic electron transfer around photosystem I within chloroplasts (Endo et al., 2008). NDH-1 catalyzes the substantial turnover of NADH into NAD^+ accompanying the production of ROS. Therefore, a vast amount of ROS likely results from a significant elevation of NDH catalytic activity upon BBTV infection. The chloroplast essentially acts as a battling arena between host and BBTV.

BBTV movement protein B4 is a pathogenic determinant and is associated with chloroplasts. Nevertheless, intact protein B4 acts as a precursor and is processed proteolytically into the B4C fragment, which subsequently targets RbcL. Although the biological function of the conjugate of RbcL and B4-derived peptide (i.e., B4C) remains unclear, EF2, a ribosomal component that facilitates and orchestrates translational elongation, is also conjugated to B4C in PVX:B4GFP *N. benthamiana* inocula. Similar to the circumvention of antiviral translational suppression used by geminiviruses in plant hosts, BBTV likely intercepts cellular factors or hijacks the translational apparatus for the synthesis of viral proteins in plant organelles, such as chloroplasts.

The conjugation of the B4C fragment to other proteins seems to be similar to that of ubiquitin (Ub) in eukaryotes. Ub and Ub-like factors are regulatory proteins, functioning via their covalent attachment to substrates/targets. Usually, Ub is also produced from precursors by deubiquitinase and is further conjugated to substrates via the E1, E2, and E3 cascades (Komander and Rape, 2012). Viral pathogens directly recruit the ubiquitination system by the mimicry of editing proteins or eukaryotic Ub itself (Randow and Lehner, 2009; Alcaide-Loridan and Jupin, 2012). Host Ubs also exist in certain mammalian (e.g., poxvirus, herpes simplex virus, parainfluenza virus) and insect viruses in phospholipid-modified forms, and viral Ub-like genes in the genomes of *Baculoviridae* (group I nucleopolyhe-drovirus) enhance pathogenicity (Guarino et al., 1995). Although sequence composition differs between the B4-derived peptide and Ub, they have similar molecular weights and are capable of conjugation with protein targets.

In addition to canonical ATP-dependent ubiquitination, NAD-dependent ubiquitin transfer to host proteins is mediated by a single bacterial effector SdeA (Bhogaraju et al., 2016; Qiu et al., 2016; Kotewicz et al., 2017).

This non-canonical serine ubiquitination of target proteins via a phosphoribosyl moiety linkage with Arg42 in ubiquitin involves ADP-ribosyltransferase and phosphohydrolase activities with the aid of an NAD^+ cofactor (Bhogaraju et al., 2016). The Sde-mediated phosphoribosylation of ubiquitin blocks the classical ubiquitination cascade involved in protein degradation. Similarly, protein B4 is full of conserved arginine residues and the target species conjugated with the B4-derived cryptide was resistant to proteolytic degradation. Whether the ATP-independent ligase-like bacterial effector SdeA in plants is responsible for the covalent linkage between the B4-derived peptide and plant proteins warrants further study. Additionally, the definitive linkage mode of the B4-derived modifier to host factors remains to be determined. Our data provide strong evidence that a novel modifier encoded by a plant viral protein can modulate the interactions between ssDNA viruses and plant hosts.

Acknowledgements

This work was supported by grants from the Natural Science Foundation of China (No. 31301641 to J.Z.) and the Program for Qualified Personnel of Taiwan Strait West Coast (No. K8812007 to L.H.X.). We are grateful to Haitao Wang for assistance with TEM, Dr. Zhixin Liu for providing the anti-B3 (coat protein) antibody, and Dr. Zhenguo Du for valuable discussions.

References

[1] ALCAIDE-LORIDAN C J, JUPIN I. Ubiquitin and plant viruses, let's play together[J]. Plant Physiology, 2012, 160(1): 72-82.

[2] BHOGARAJU S, KALAYIL S, LIU Y, et al. Phosphoribosylation of ubiquitin promotes serine ubiquitination and impairs conventional ubiquitination[J]. Cell, 2016, 167: 1636-1649.

[3] BURNS T M, HARDING R M, DALE J L. The genome organization of banana bunchy top virus: analysis of six ssDNA components[J]. The Journal of general virology, 1995, 76: 1471-1482.

[4] DE TORRES, ZABALA M, LITTLEJOHN G, et al. Chloroplasts play a central role in plant defence and are targeted by pathogen effectors[J]. Nature Plants, 2015, 1:15074.

[5] ENDO T, ISHIDA S, ISHIKAWA N, et al. Chloroplastic NADPH dehydrogenase complex and cyclic electron transport around photosystem I[J]. Molecules and cells, 2008, 25(2):158-162.

[6] FONDONG V N. Geminivirus protein structure and function[J]. Molecules Plant Pathology, 2013, 14:635-649.

[7] GUARINO L A, SMITH G, DONG W. Ubiquitin is attached to membranes of baculovirus particles by a novel type of phospholipid anchor[J]. Cell, 1995, 80: 301-309.

[8] HANLEY-BOWDOIN L, BEJARANO E R, ROBERTSON D, et al. Geminiviruses: masters at redirecting and reprogramming plant processes[J]. Nature reviews Microbiology, 2013, 11(11): 777-788.

[9] KOMANDER D, RAPE M. The ubiquitin code[J]. Annual review of biochemistry, 2012, 81: 203-229.

[10] KOTEWICZ K M, RAMABHADRAN V, SJOBLOM N, et al. A single legionella effector catalyzes a multistep ubiquitination pathway to rearrange tubular endoplasmic reticulum for replication[J]. Cell Host Microbe,2017, 21(2): 169-181.

[11] MANDADI K K, SCHOLTHOF K B. Plant immune responses against viruses: how does a virus cause disease?[J]. Plant Cell, 2013, 25: 1489-1505.

[12] QIU J, SHEEDLO M J, YU K, et al. Ubiquitination independent of E1 and E2 enzymes by bacterial effectors[J]. Nature, 2016, 533: 120-124.

[13] RANDOW F, LEHNER P J. Viral avoidance and exploitation of the ubiquitin system[J]. Nature cell biology, 2009, 11(5): 527-534.

[14] ROJAS M R, JIANG H, SALATI R, et al. Functional analysis of proteins involved in movement of the monopartite begomovirus, tomato yellow leaf curl virus[J]. Virology, 2001, 291: 110-125.

[15] SUN D J, SUN H, WEI H Y, et al. Functional analysis of DNA. 4 coding region from a Chinese Zhangzhou isolate of banana bunchy top virus[J]. Progress in Natural Science, 2002, 12: 426-430.

[16] WANITCHAKORN R, HAFNER G J, HARDING R M, et al. Functional analysis of proteins encoded by banana bunchy top virus DNA-4 to -6[J]. The Journal of general virology, 2000, 81: 299-306.

[17] WU R Y, SU H J. Purification and characterization of banana bunchy top virus[J]. The Journal of general virology, 1990, 128: 153-160.

[18] ZHUANG J, COATES C J, MAO Q, et al. The antagonistic effect of Banana bunchy top virus multifunctional protein B4 against Fusarium oxysporum[J]. Molecular plant pathology, 2016, 17(5): 669-679.

[19] ZORZATTO C, MACHADO J P, LOPES K V, et al. NIK1-mediated translation suppression functions as a plant antiviral immunity mechanism[J]. Nature,2015, 520: 679-682.

Co-opting the fermentation pathway for tombusvirus replication: Compartmentalization of cellular metabolic pathways for rapid ATP generation

Wenwu Lin[1,2], Yuyan Liu[2], Melissa Molho[2], Shengjie Zhang[1], Longshen Wang[1], Lianhui Xie[1], Peter D. Nagy[2]

(1 State Key Laboratory of Ecological Pest Control for Fujian and Taiwan Crops, Fujian Agriculture and Forestry University, Fuzhou, China; 2 Department of Plant Pathology, University of Kentucky, Lexington, Kentucky, United States of America)

Abstract: The viral replication proteins of plus-stranded RNA viruses orchestrate the biogenesis of the large viral replication compartments, including the numerous viral replicase complexes, which represent the sites of viral RNA replication. The formation and operation of these virus-driven structures require subversion of numerous cellular proteins, membrane deformation, membrane proliferation, changes in lipid composition of the hijacked cellular membranes and intensive viral RNA synthesis. These virus-driven processes require plentiful ATP and molecular building blocks produced at the sites of replication or delivered there. Toobtain the necessary resources from the infected cells, tomato bushy stunt virus (TBSV) rewires cellular metabolic pathways by co-opting aerobic glycolytic enzymes to produce ATP molecules within the replication compartment and enhance virus production. However, aerobic glycolysis requires the replenishing of the NAD^+ pool. In this paper, we demonstrate the efficient recruitment of pyruvate decarboxylase (Pdc1) and alcohol dehydrogenase (Adh1) fermentation enzymes into the viral replication compartment. Depletion of Pdc1 in combination with deletion of the homologous *PDC5* in yeast or knockdown of Pdc1 and Adh1 in plants reduced the efficiency of tombusvirus replication. Complementation approach revealed that the enzymatically functional Pdc1 is required to support tombusvirus replication. Measurements with an ATP biosensor revealed that both Pdc1 and Adh1 enzymes are required for efficient generation of ATP within the viral replication compartment. *In vitro* reconstitution experiments with the viral replicase show the pro-viral function of Pdc1 during the assembly of the viral replicase and the activation of the viral p92 RdRp, both of which require the co-opted ATP-driven Hsp70 protein chaperone. We propose that compartmentalization of the co-opted fermentation pathway in the tombusviral replication compartment benefits the virus by allowing for the rapid production of ATP locally, including replenishing of the regulatory NAD^+ pool by the fermentation pathway. The compartmentalized production of NAD^+ and ATP facilitates their efficient use by the co-opted ATP-dependent host factors to support robust tombusvirus replication. We propose that compartmentalization of the fermentation pathway gives an evolutionary advantage for tombusviruses to replicate rapidly to speed ahead of antiviral responses of the hosts and to outcompete other pathogenic viruses. We also show the dependence of turnip crinkle virus, bamboo mosaic virus, tobacco mosaic virus and the insect-infecting Flock House virus on the fermentation pathway, suggesting that a broad range of viruses might induce this pathway to support rapid replication.

1 Introduction

Similar to other positive-strand RNA viruses, the plant-infecting tombusviruses cause major structural

PLoS pathogens.2019, 15(10): e1008092.
Received 29 March 2019; Accepted 18 September 2019; Published 24 October 2019.

rearrangements and metabolic changes in infected cells. The changes include the subversion of pro-viral host factors to support their replication and the induction of lipid synthesis, lipid transfer, membrane proliferation and alteration of vesicular trafficking. The major outcome of all these virus-driven processes is the biogenesis of the unique and extensive viral replication compartments and the formation of numerous viral replicase complexes (VRCs) on subverted subcellular membrane surfaces (den Boon and Ahlquist, 2010; Heaton et al., 2010; Wang et al., 2011; Belov and van Kuppeveld, 2012; Nagy and Pogany, 2012; de Castro et al., 2013; Wang, 2015; van der Schaar et al., 2016; Altan-Bonnet, 2017). All these cellular changes serve several purposes, including supporting robust viral RNA replication, and protection of the viral RNA, including the dsRNA replication intermediate, from recognition by the cellular innateimmune system or from elimination by the host RNAi machinery, which is also called post transcriptional gene silencing in plants (Andino, 2003; Carbonell and Carrington, 2015; Shulla and Randall, 2016; Kovalev et al., 2017). Also, sequestration of viral and co-opted hostproteins together with the viral (+)RNA into the replication compartment results in high local concentrations and efficient macromolecular assembly needed for the optimal formation of VRCs (Diamond et al., 2010; Heaton and Randall, 2010; Hsu et al., 2010; Syed et al., 2010; Perera et al., 2012; Schoggins and Randall, 2013; Altan-Bonnet, 2017). Our increasing knowledge of the roles of various lipids/membranes and coopted host factors in RNA virus replication will be useful to control RNA viruses.

Tombusviruses, such as tomato bushy stunt virus (TBSV) and carnation Italian ringspot virus (CIRV), are small (+)RNA viruses, which can replicate in the surrogate model host yeast (*Saccharomyces cerevisiae*) (Nagy et al., 2014; Nagy, 2016, 2017). Intensive genome-wide and proteome-wide research withthe TBSV–yeast system has led to a catalog of host factors co-opted for viral RNA replication (Xu and Nagy, 2014; Nagy, 2016). Induction of global phospholipid biosynthesis and redistribution of sterol and alteration of vesicular trafficking has been revealed to play major roles in the formation of VRCs and the activation of the viral-coded p92 RdRp (Pogany and Nagy, 2015; Xu and Nagy, 2015, 2016, 2017). All these subcellular changes areguided by the p33 replication protein, which is the master regulator of VRC assembly and viral (+) RNA recruitment into the VRCs (Pogany et al., 2005; Xu and Nagy, 2017). Additional characteristic alterations caused by TBSV include the subversion of the actin network, the induction of subcellular membrane proliferation, peroxisome aggregation and the formation of membrane contact sites to support efficient virus replication (Nagy, 2016; Nagy et al., 2016; de Castro et al., 2017).

Most of these viral-induced processes require ATP-based energy and the production of new metabolites in infected cells. Accordingly, we have previously discovered that several glycolyticenzymes, such as glyceraldehyde-3-phosphate dehydrogenase (GAPDH, Tdh2/3 in yeast), phosphoglycerate kinase (Pgk1) and pyruvate kinase (PK, Cdc19 in yeast) are recruited into the viral replication compartment (Wang and Nagy, 2008; Huang and Nagy, 2011; Chuang et al., 2017; Prasanth et al., 2017) and Eno2 phosphopyruvate hydratase binds top92^{pol}replication protein (Pogany and Nagy, 2015). These findings suggest that tombusviruses co-opt the aerobicglycolytic pathway, which leads to the production of plentiful ATP within the viral replicationcompartment (Chuang et al., 2017; Prasanth et al., 2017). Our studies also revealed that the locally produced ATP is used up tofuel the co-opted Hsp70 proteins and DEAD-box helicases, and possibly the ESCRT-associated Vps4 AAA ATPase to promote viral replication within the viral replication compartment (Chuang et al., 2017; Prasanth et al., 2017)

Sustaining aerobic glycolysis pathway, however, requires the replenishing of NAD^+, whichis a critical regulatory compound in glycolysis (Vander Heiden et al., 2009; Lunt and Vander Heiden, 2011). Because a previous proteomic-basedscreen indicated that p92^{pol} replication protein binds to Pdc1 pyruvate decarboxylase fermentation protein (Pogany and Nagy, 2015), in this work we studied the role of the fermentation pathway in tombusvirus replication.

Aerobic glycolysis occurs in many fast-growing microbes, and in cancer cells, some embryonic cells and immune cells, such as fibroblasts and lymphocytes (Vander Heiden et al., 2009; Lunt and Vander Heiden, 2011; Olson et al., 2016). Aerobic glycolysis ishijacked by the malaria parasite (Lunt and Vander Heiden, 2011). It is a process that regulates the balance between fast ATP production and biosynthesis of ribonucleotides, lipids and several amino acids. However, aerobic glycolysis requires the replenishing of NAD^+ compound, which is produced by fermentation in eukaryotic cells. NAD^+ is converted to NADH during glycolysis by GAPDH, which is required for many biosynthetic processes (Vander Heiden et al., 2009; Lunt and Vander Heiden, 2011; Olson et al., 2016).

Our surprising discovery is that tombusviruses co-opt the host Pdc1 and the alcohol dehydrogenase (Adh1) fermentation enzymes by re-localizing them from the cytosol into the large viral replication compartment through direct interaction with the tombusvirus replication proteins. The subversion of the fermentation enzymes is critical to support VRC assembly, the activation of p92 RdRp protein and viral RNA synthesis. Most important, however, is the role of the co-opted fermentation enzymes in the maintenance of ATP synthesis within the viral replication compartment. Altogether, we discovered that tombusviruses compartmentalize entire cellular metabolic pathways to promote intensive viral replication within the viral replication compartment at the expense of the infected host cells. Based on these and previous findings, we propose that compartmentalization of the fermentation pathway gives anevolutionary advantage for tombusviruses to replicate quickly to speed ahead of antiviral responses of the hosts.

2 Results

2.1 Pdc1 fermentation enzyme is required for tombusvirus replication

Pdc1 was previously identified in a co-purification assay with the TBSV $p92^{pol}$ replication protein (Pogany and Nagy, 2015). To test the relevance of *PDC1* in tombusvirus replication, first we created a double mutant, which allowed for the depletion of Pdc1p in the absence of Pdc5p paralog in yeast (GAL::PDC1 pdc5Δ). Then, we used this double mutant yeast to launch TBSV replication viaco-expressing the p33 and $p92^{pol}$ replication proteins and the replicon repRNA. Northern blot analysis revealed a ~10-fold decrease in TBSV repRNA accumulation when Pdc1p expressionwas suppressed versus induced (Fig. 1, lanes 13–15 versus 16–18). These data suggest that Pdc1p and Pdc5p are critical for TBSV replication. Expression of Pdc1p from a plasmid inpdc1Δ yeast increased TBSV repRNA accumulation by ~2-fold (Fig. 1B). We observed similar~2-fold enhanced replication of the closely-related CIRV, which replicates on the outer membranes of mitochondria (Fig. 1C). Depletion of Pdc1p in double mutant yeast (GAL::PDC1pdc5Δ) also inhibited the replication of the unrelated *Flock House virus* (FHV), which is aninsect-infecting RNA virus (S1 Fig, lanes 17–20 versus 1–24). FHV replicates on the outer membranes of mitochondria (Kopek et al., 2007). These data suggest that Pdc1p has a pro-viral role in different subcellular microenvironments.

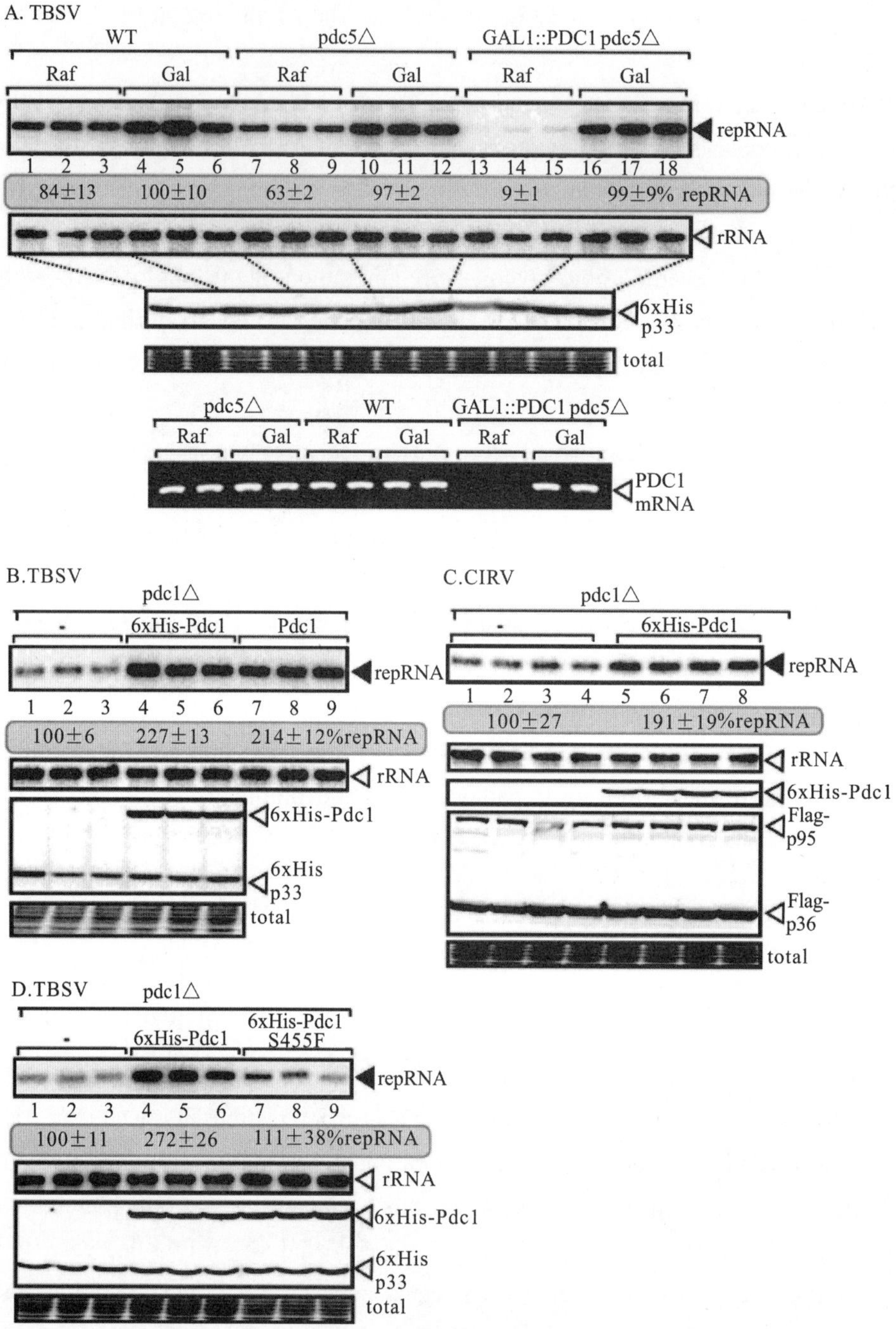

Fig. 1 Pdc1 fermentation enzyme is an essential host factor for tombusvirus replication in yeast

(A) Depletion of pyruvate decarboxylase (Pdc1p) in combination with deletion of the homologous *PDC5* inhibits TBSV replicon (rep) RNA replication in yeast. Top panels: northern blot analyses of TBSV repRNA using a 3′ end specific probe demonstrates reduced accumulation of repRNA in GAL::PDC1 pdc5Δ yeast strain with depleted Pdc1p (raffinose-containing media) in comparison with the WT yeast strain or GAL::PDC1 pdc5Δ yeast strain with induced Pdc1p (galactose-containing media). Viral proteins His_6-p33 and His_6-p92 of TBSV were expressed from plasmids from the copper-inducible *CUP1* promoter, while DI-72(+) repRNA was expressed from the constitutive *TET1* promoter. Second panel: northern blot with an 18S ribosomal RNA specific probe was used as a loading control. Bottom images: western blot analysis of the level of His6-tagged p33 protein with anti-His antibody. Coomassie bluestained SDS-PAGE was used for protein loading control. The down-regulation of Pdc1 mRNA was confirmed with RT-PCR. Each experiment was repeated three times. (B-D) Expression of Pdc1p from a plasmid increases tombusvirus replication in pdc1Δ yeast strain. His6-p33 and His6-p92 were expressed from the *GAL1* promoter, whereas (+)repRNA was expressed from the *GAL10* promoter. For panel C, Flag-p36 and Flag-p95 were expressed from the *CUP1* promoter, whereas the repRNA from the *GAL10* promoter. The untagged or His6-tagged Pdc1 were expressed from the *TET* promoter in all these experiments. See further details in panel A above

To test if the canonical enzymatic function of Pdc1p is needed for TBSV replication, we expressed $Pdc1^{S455F}$ mutant, which has a reduced pyruvate decarboxylase activity (Eberhardt et al., 1999), inpdc1Δ yeast replicating TBSV repRNA. Unlike the WT Pdc1p, $Pdc1^{S455F}$ mutant was unable to enhance the replication of TBSV repRNA (Fig. 1D, lanes 7–9 versus 4–6). Therefore, we suggest that the canonical role of Pdc1p in the fermentation pathway is required for efficient TBSV replication in yeast.

To obtain additional evidence for the pro-viral role of Pdc1p, we replaced the original promoter of *PDC2* gene with the regulatable *GAL1* promoter in the haploid yeast chromosome. Pdc2p is a transcription factor regulating the transcription of both *PDC1* and *PDC5* genes inyeast (Hohmann, 1993). Depletion of Pdc2p in the above yeast (GAL1::HA-PDC2) resulted in ~3-fold reduction in TBSV repRNA accumulation (S2 Fig, lanes 13–16 versus 17–24). Therefore, these findings confirm the pro-viral role of Pdc1/5p in TBSV replication in yeast.

2.2 Pdc1 protein interacts with the tombusvirus replication proteins

To test if Pdc1p could interact with the tombusvirus replication proteins, we used the membrane yeast two-hybrid assay (MYTH), which is based on the split-ubiquitin strategy (Snider et al., 2010). We found that the yeast Pdc1p and the homologous *Arabidopsis* AtPdc1 proteins interacted with the TBSV p33 replication protein (Fig. 2A).

We then purified the TBSV replicase from yeast membrane fraction through detergent-solubilization and Flag-affinity purification. Interestingly, the yeast Pdc1p was co-purified with the Flag-p33/Flag-p92 replication proteins (Fig. 2B). We also found that the enzymatically inactive $Pdc1^{S455F}$ mutant was co-purified with the TBSV replicase from yeast (Fig. 2C, lane 3). The mitochondrial CIRV Flag-p36/Flag-p95 showed a comparable co-purification profile with the WT Pdc1p and $Pdc1^{S455F}$ mutant to that observed with the TBSV replication proteins (Fig. 2C). The homologous AtPdc1 was also co-purified with either TBSV Flag-p33/Flag-p92 or the CIRV Flag-p36/Flag-p95 replication proteins from the membrane-fraction of yeast (Fig. 2D and 2E). Similar co-purification experiments with the Flag-p33 replication protein from detergent-solubilized fraction of *Nicotiana benthamiana* also confirmed the interaction of the replication proteins with AtPdc1 protein (Fig. 2F). These data suggest that the interaction between Pdc1/AtPdc1 and p33 replication protein occurs in both yeast and plant cells.

To confirm direct interactions between the TBSV p33 and Pdc1p, we applied a pull-down assay with MBP-tagged Pdc1p or MBP-AtPdc1 and GST-His_6-tagged p33C (the C-terminal, soluble portion) proteins from *E. coli* (Fig. 2G). Both MBP-Pdc1p and MBP-AtPdc1 captured the GST-His_6-p33C protein on the maltose-column, indicating direct interaction between these host and viral proteins. This conclusion was confirmed using the TBSV MBP-p33C or the CIRV MBP-p36C proteins, which captured the GST-His_6-AtPdc1 in the second pull-down assay (Fig. 2H). In the pull-down assay, we used truncated TBSV p33 and CIRV p36 replication proteins missing their membrane-binding regions to aid their solubility in *E. coli* (Fig. 2G, H). Altogether, these data suggest that the direct interactions between the replication proteins of TBSV and CIRV and Pdc1p/AtPdc1 host proteins occur within the viral protein C-terminal domain facing the cytosolic compartment.

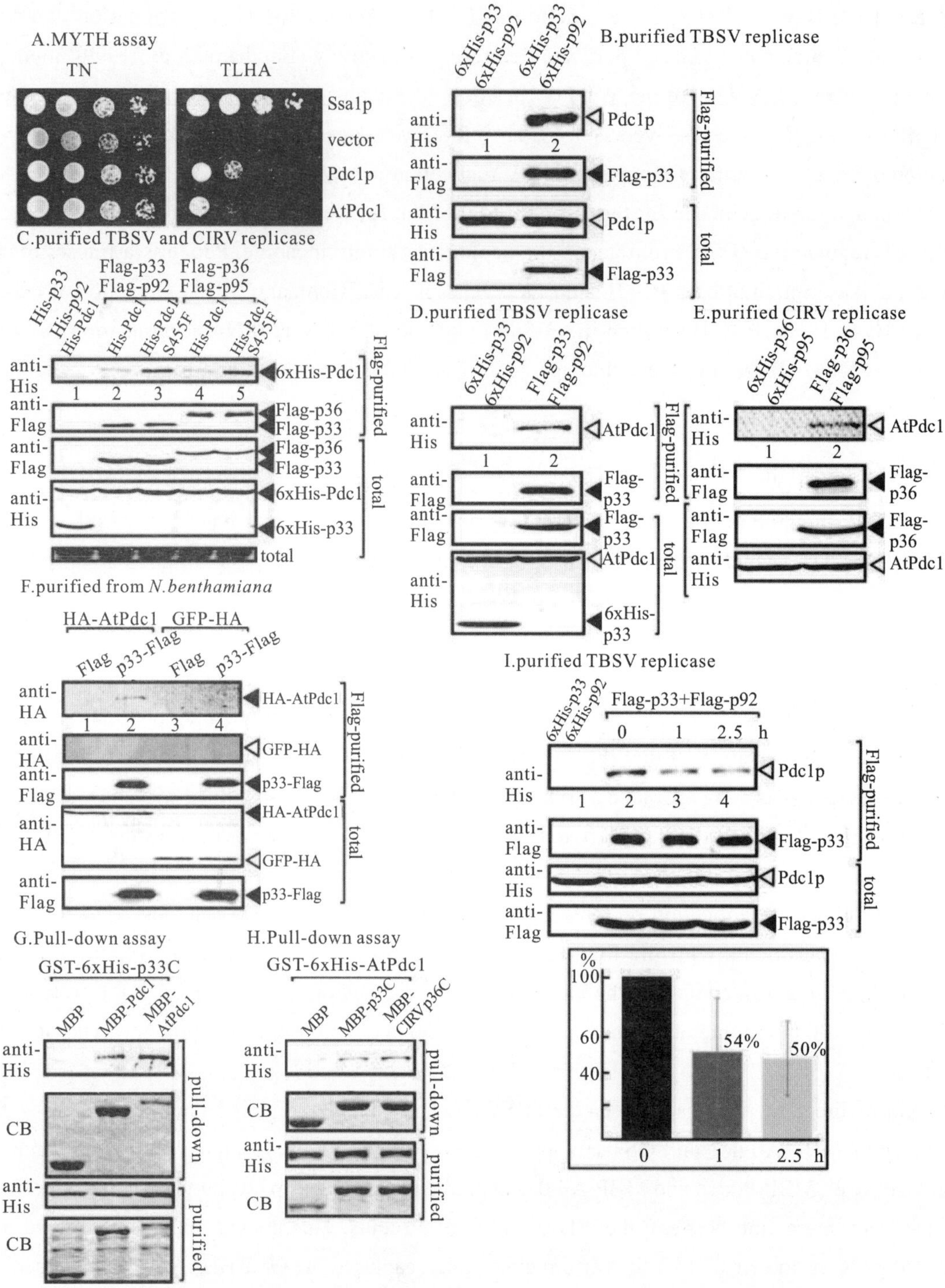

Fig.2 Interaction between tombusvirus replication proteins and Pdc1 fermentation enzyme

(A) The split ubiquitin-based MYTH assay was used to test binding between the TBSV p33 and the yeast Pdc1p and *Arabidopsis* Pdc1 proteins in yeast. The bait p33 was co-expressed with the shown prey proteins. The Ssa1p heat shock protein 70 (Hsp70) and the empty prey vector (NubG) were used as the positive and the negative controls, respectively. The right panel shows p33: Pdc1 interactions, the left panel demonstrates that comparable amounts of yeasts were used for these experiments. (B-C) Co-purification of the yeast His6-Pdc1p and His6-Pdc1^{S455F} mutant with TBSV Flag-p33 and Flag-p92pol or the carnation Italian ring spot virus (CIRV) Flag-p36/Flag-p95 replication proteins from subcellular membranes. Top two panels: western blot analysis of the co-purified WT His_6-Pdc1p and His6-Pdc1^{S455F} mutant with the Flag-affinity purified replication proteins. The His_6-tagged proteins were detected with an anti-His antibody, while Flag-p33 and Flag-p36 were detected with an anti-Flag antibody. The negative control was from yeast expressing His_6-p33 and His_6-p92^{pol} purified using a Flag-affinity column (lane 1). Samples were cross-linked with formaldehyde in intact yeast cells. Bottom two panels: western blot of total His_6-Pdc1p and Flag-p33 and Flag-p36 in the total yeast extracts. (D-E) Co-purification of the Arabidopsis His_6-Pdc1 with the TBSV Flag-p33 and Flag-p92^{pol} or the CIRV Flag-p36/Flag-p95 replication proteins from subcellular membranes of yeast. Top two panels: western blot analysis of the co-purified His$_6$-AtPdc1 with Flag-affinity purified replication proteins. The His$_6$-tagged proteins were detected with anti-His antibody, while the Flag-p33 and Flag-p36 were detected with an anti-Flag antibody. The negative control was from yeast expressing the His$_6$-tagged replication proteins purified using a Flag-affinity column (lane 1). Samples were cross-linked with formaldehyde. Bottom two panels: western blot of the total Flag-p33 or Flag-p36 and the His6-AtPdc1 and in the total yeast extracts (F) Co-purification of HA-AtPdc1 with the TBSV p33-Flag replication protein from N. benthamiana. Top two panels: western blot analysis of the co-purified HA-tagged AtPdc1 (lane 2) with the Flag-affinity purified Flag-p33. HA-Pdc1 was detected with an anti-HA antibody, while the p33-Flag was detected with an anti-Flag antibody as shown. Bottom two panels: western blot of the total plant extracts. (G) Pull-down assay including the GST-His$_6$-p33 replication protein and the MBP-tagged yeast Pdc1p or the MBP-AtPdc1. Note that we used the soluble C-terminal region of the TBSV p33 replication protein, which lacked the N-terminal sequence, including the trans-membrane TM domain. Top panel: western blot analysis of the captured His6-p33 with the MBP affinity purified MBP-Pdc1 was performed with an anti-His antibody. The negative control was MBP (lane 1). Middle panel: Coomassie-blue stained SDS-PAGE of the captured yeast MBP-Pdc1p or MBP-AtPdc1 and MBP. Bottom panels: western blot analysis of the His6-p33 in the total extracts. Coomassie-blue stained SDS-PAGE of the MBP-Pdc1p or MBP-AtPdc1 and MBP in the total extracts. Each experiment was repeated three times. (H) Pull-down assay including the GST-His6-AtPdc1 and the MBP-tagged p33 or the CIRV MBP-p36 replication proteins. Please see further details in panel G. (I) Decreasing level of co-purification of His$_6$-Pdc1p with the Flag-tagged viral replicase after blocking new VRC assembly. The yeast samples were collected at the shown time points after the addition of cycloheximide (blocks cellular translation, thus new VRC formation) to the yeast culture. Note that samples were from yeasts replicating TBSV repRNA. Top panel: western blot analysis of the co-purified His$_6$-Pdc1 with the Flag-affinity purified Flag-p33 and Flag-p92pol from membrane fraction of yeast. The His$_6$-Pdc1p was detected with an anti-His antibody. The negative control was the His6-p33 and His$_6$-p92^{pol} purified from yeast extracts using a Flag-affinity column. Middle panel: western blot of the purified Flag-p33 detected with an anti-Flag antibody. Bottom panels: western blots of His$_6$-Pdc1p and Flag-p33 proteins in the total yeast extracts using an anti-His and an anti-Flag antibodies. The graph shows the % of co-purified His$_6$-Pdc1p with the tombusviral replication proteins with standard deviation. Each experiment was repeated three times

To examine if Pdc1p was co-opted as a permanent or temporary component of the tombusvirus replicase, first, we stopped the formation of new tombusvirus replicase complexes byblocking ribosomal translation via adding cycloheximide to the yeast growth media (Barajas et al., 2014b). Second, we performed Flag-affinity-purification of the tombusvirus replicase from the membrane fraction of yeast at various time-points. Interestingly, the amount of the co-purified Pdc1p wasdecreased by ~50% in the purified replicase preparations at the 2.5h time point (Fig. 2I, lanes3–4 versus 2). The reduction of Pdc1p amount suggests that Pdc1p is likely released from the replicase. The release of Pdc1p likely occurs before the final assembly of the viral replicase,which ultimately forms a rather closed vesicle-like structure during replication (Barajas et al., 2014b; Kovalev et al., 2014; de Castro et al., 2017). Based on this observation, we suggest that the function of Pdc1p is temporary with the replication proteins, which likely takes place during the early steps of tombusvirus replication.

2.3 The Adh1 family of fermentation enzymes is co-opted for tombusvirus replication

The fermentation pathway consists of two different sets of enzymes, pyruvate decarboxylase(Pdc1/5 in yeast) and alcohol dehydrogenase (Adh1-5 in yeast). The end result of the pathway is ethanol, howcver, the critical product is NAD^+ from NADH (Lunt and Vander Heiden, 2011; Olson et al., 2016). Importantly, NAD^+ is required to replenish the glycolytic pathway via providing the regulatory compound of Glyceraldehyde-3-phosphate dehydrogenase (GAPDH, coded by *TDH2/3* in yeast). Because the subversion of the catalytically active Pdc1 enzyme is needed to support TBSV replication (seeabove), we tested if the NAD^+ producing Adh family members are also co-opted by tombusviruses.

To test if Adh1-5p could interact with the tombusvirus replication proteins, we used the MYTH assay (Snider et al., 2010), which revealed that the five members of the yeast Adh family as well as the homologous *Arabidopsis* AtAdh1 protein interacted with the TBSV p33 replication protein (Fig. 3A). To confirm this unexpected finding, we purified the TBSV replicase from yeast membrane fraction through detergent-solubilization and Flag-affinity purification. We found that the yeast Adh1p, Adh2p and Adh3p were all co-purified with the Flag-p33/p92 replication proteins (Fig. 3B). Adh1p was also co-purified with the mitochondrial CIRV Flag-p36/p95 replication proteins (Fig. 3C), suggesting that different tombusviruses recruit Adh proteins into the membrane fraction of yeast. Similar co-purification experiments with the TBSV Flag-p33 orthe CIRV Flag-p36 replication proteins from detergent-solubilized membranous fraction of yeast confirmed the interactions of the TBSV and CIRV replication proteins with AtAdh1 protein (Fig. 3D). These data suggest that the interactions between Adh1/AtAdh1 and the tombusviral replication proteins occur in yeast cells.

To confirm direct interactions between TBSV p33 and Adh1p, we applied a pull-downassay using the TBSV MBP-p33C or the CIRV MBP-p36C proteins, which captured the GST-His6-Adh1p in the pull-down assay (Fig. 3E). Similar pull-down experiment also confirmed the direct interaction between GST-His_6-AtAdh1 and the viral replication proteins (Fig. 3F). As above, we used the truncated TBSV p33 and CIRV p36 replication proteins missing their membrane-binding regions to increase their solubility in *E. coli* in the pull-down assay(Fig. 3E–F). Altogether, these data suggest that the direct interactions between the replication proteins of TBSV and CIRV and the Adh1 host protein occur within the viral protein domain facing the cytosol.

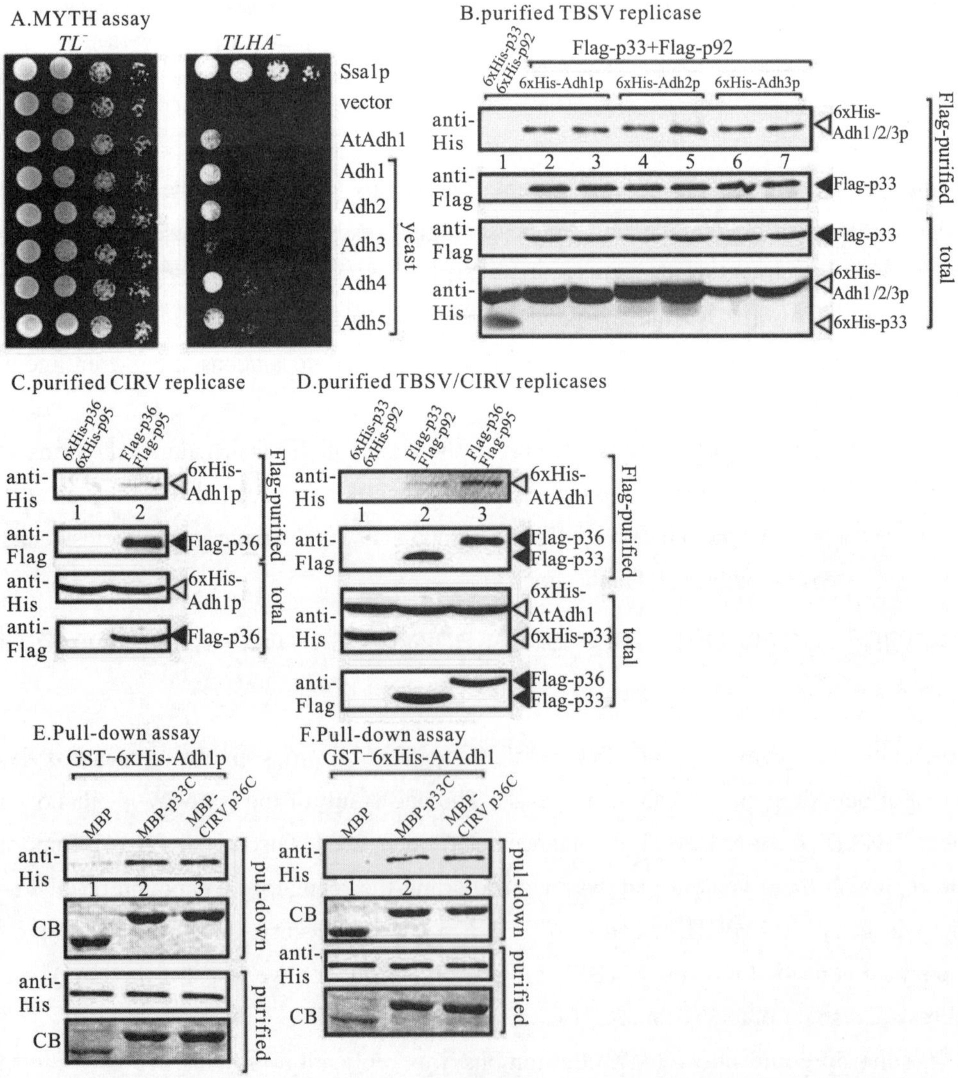

Fig. 3 Interaction between the tombusvirus replication proteins and Adh1 fermentation enzyme

(A) The split ubiquitin-based MYTH assay was used to test binding between the TBSV p33 and the yeast alcohol dehydrogenase Adh1-5p and the *Arabidopsis* Adh1 proteins in yeast. The bait p33 was co-expressed with the shown prey proteins. The Ssa1p Hsp70 and the empty prey vector (NubG) were used as the positive and the negative controls, respectively. The right panel shows p33: Adh1-5 interactions, the left panel demonstrates that comparable amounts of yeasts were used for these experiments. (B) Co-purification of the yeast His6-Adh1, 2, 3p with the TBSV Flag-p33 and Flag-p92pol replication proteins from subcellular membranes. Top two panels: western blot analysis of the co-purified His6-Adh1-3p with the Flag-affinity purified replication proteins. His6-tagged proteins were detected with an anti-His antibody, while Flag33 was detected with an anti-Flag antibody. The negative control was from yeast expressing His6-p33 and His6-p92pol purified using a Flag-affinity column (lane 1). Samples were cross-linked with formaldehyde. Bottom two panels: western blot of the total His6-Adh1-3p and Flag-p33 in the total yeast extracts. (C) Co-purification of the yeast His6-Adh1p with the CIRV Flag-p36 and Flag-p95pol replication proteins from subcellular membranes. See further details inpanel B. (D) Co-purification of His6-AtAdh1 with either the TBSV or the CIRV replicase from yeast subcellular membranes. Top two panels: western blot analysis of the co-purified His6-Adh1p (lanes 2–3) with the Flag-affinity purified Flag-p33 or CIRV Flag-p36. His6-Adh1p was detected with an anti-His antibody, while the Flag-p33 or CIRV Flag-p36 replication proteins were detected with an anti-Flag antibody as shown. Bottom two panels: western blot of the total plant protein extracts. (E-F) Pull-down assay including GST-His6-Adh1p or GST-His6-AtAdh1 with the TBSV MBP-p33C or the CIRV MBP-p36C replication proteins and MBP. See further details in panel B. Each experiment was repeated three times

2.4 Both Pdc1 and Adh1 have pro-viral functions in plants

The homologous *PDC1* and *ADH1* genes are present in plants, but they are expressed at a verylow level in plant cells under normal growth conditions (Ismond et al., 2003; Mithran et al., 2014). Therefore, tombusviruses likely need to induce the expression of Pdc1 and Adh1 mRNAs in order to exploit Pdc1 and Adh1 for pro-viral functions during plant infections. Indeed, RT-PCR analysis of Pdc1 mRNA levels in TBSV-infected versus mock-treated *N. benthamiana* leaves revealed robust up-regulation of Pdc1 mRNA level in the TBSV inoculated leaves (Fig. 4A) as well as the leaves expressing only the p33 replication protein (Fig. 4C, lanes 1–3 versus 4–6). We observed a comparable up-regulation of Pdc1 mRNA level in CIRV infected *N. benthamiana* leaves or the leaves only expressing the CIRV p36 replication protein (Fig. 4B, C). Based on these observations, we propose that TBSV and CIRV replication induces a high level of Pdc1 expression.

Similarly, RT-PCR analysis of Adh1 mRNA level in TBSV-infected versus mock-treated *N.benthamiana*leaves revealed an up-regulation of Adh1 mRNA level in the TBSV and CNV inoculated leaves (Fig. 4D–F) and the systemically-infected leaves (Fig. 4E). Based on these observations, we propose that TBSV and CNV replication induces a high level of Adh1 expression in plant leaves.

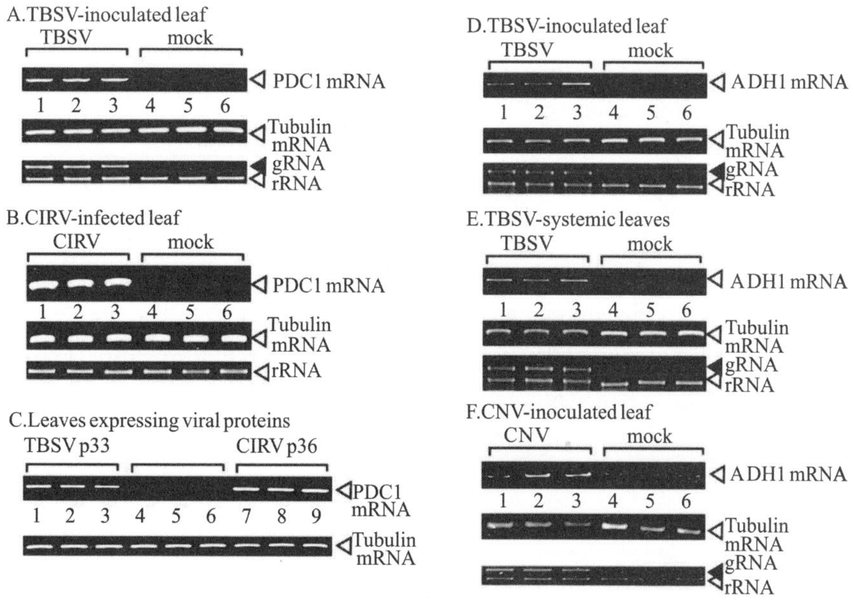

Fig. 4 Tombusvirus infection induces the expression of Pdc1 and Adh1 mRNAs in *N. benthamiana*

(A) Top panels: semi-quantitative RT-PCR analysis of the NbPdc1 mRNA level at 1.5dpi in *N. benthamiana* leaves infected with either

TBSV or mock-infected. (B) Semi-quantitative RT-PCR analysis of the NbPdc1 mRNA level at 3dpi in *N.benthamiana* leaves infected with either CIRV or mock-inoculated. The samples were taken 3d after tombusvirus inoculation. Second panel: RT-PCR analysis of the tubulin mRNA level in the same plants. Each experiment was repeated three times. Bottom panels: Ethidium-bromide-stained agarose gels show the comparable amounts of RNA loading, as shown for the ribosomal RNA. (C) Top panel: semi-quantitative RT-PCR analysis of the NbPdc1 mRNA level at 3d post agro-infiltration in *N. benthamiana* leaves agro-infiltrated to express TBSV p33, CIRV p36 or no-expression control. See further details in panels A-B. (D-F) Semi-quantitative RT-PCR analysis of the NbAdh1 mRNA level at 1.5dpi in *N. benthamiana* leaves infected with either TBSV or mock-infected or CNV- or mock-infected at 3dpi. See further details in panels A-B

To study if tombusviruses depend on the Pdc1 function in plants, we knocked-down Pdc1 expression via virus-induced gene-silencing (VIGS) in *N. benthamiana* plants. Knockdown of Pdc1 in *N. benthamiana* resulted in a ~3-fold reduction of TBSV RNAs in the inoculated leaves (Fig. 5A). Knockdown of Pdc1 level did not cause an obvious phenotype in *N. benthamiana* (Fig. 5A). To test the effect of Pdc1 depletion on virus accumulation in the absence of cell-to-cell spread, we also tested TBSV replication in Pdc1 knockdown protoplasts. Interestingly,TBSV RNA accumulation was reduced by ~7-fold in Pdc1 knockdown protoplasts in comparison with control protoplasts (Fig. 5B), suggesting that Pdc1 affects the viral replication process.

Similar experiments with CIRV in Pdc1 knockdown *N. benthamiana* plants also revealed a ~3-fold reduced level of tombusvirus accumulation (Fig. 5C). These data confirmed the proviral function of Pdc1 is also exploited by a more distantly-related carmovirus, we measured the accumulation of *turnip crinkle virus* (TCV) in Pdc1 knockdown *N. benthamiana* plants. The accumulation of TCV RNAs decreased by ~3-fold in Pdc1 knockdown plants (Fig. 5D). It seems that tombusviruses and a carmovirus can exploit Pdc1 functions to support viral replication.

Knockdown of Adh1 in *N. benthamiana* resulted in a ~2-fold reduction of TBSV RNAs in the inoculated leaves (Fig. 6A). Knockdown of Adh1 in *N. benthamiana* also reduced the accumulation of the peroxisomal-replicating CNV and the mitochondrial-replicating CIRV by ~2-fold (Fig. 6B, C). These data confirmed the pro-viral role of Adh1 in supporting tombusvirus replication in plants.

2.5 Both Pdc1 and Adh1 proteins are recruited into the tombusvirus replication compartment in plants

To determine if Pdc1 is recruited by TBSV into the extensive viral replication compartment,we co-expressed the BFP-tagged TBSV p33 replication protein and the RFP-tagged AtPdc1 with the GFP-SKL peroxisome matrix protein in *N. benthamiana* leaves, followed by confocal imaging. These experiments revealed a high level of co-localization of the TBSV p33 replication protein and the RFP-AtPdc1 within the replication compartments consisting of aggregated peroxisomes, even in the absence of TBSV replication (Fig. 7A). We observed a similarre-distribution of the RFP-AtPdc1 in the presence of CIRV p36-BFP within the replication compartments consisting of aggregated mitochondria (Fig. 7B). Therefore, we suggest that the TBSV p33 and the CIRV p36 replication proteins alone are enough to recruit Pdc1 to the replication compartment to a similar extent as the actively replicating TBSV or CIRV (Fig. 7A, B). In the absence of viral components, AtPdc1 is localized in the cytosol (Fig. 7C). Based on these experiments, we propose that Pdc1 is efficiently recruited by the tombusvirus replication proteins to the extensive tombusvirus replication compartments in plants.

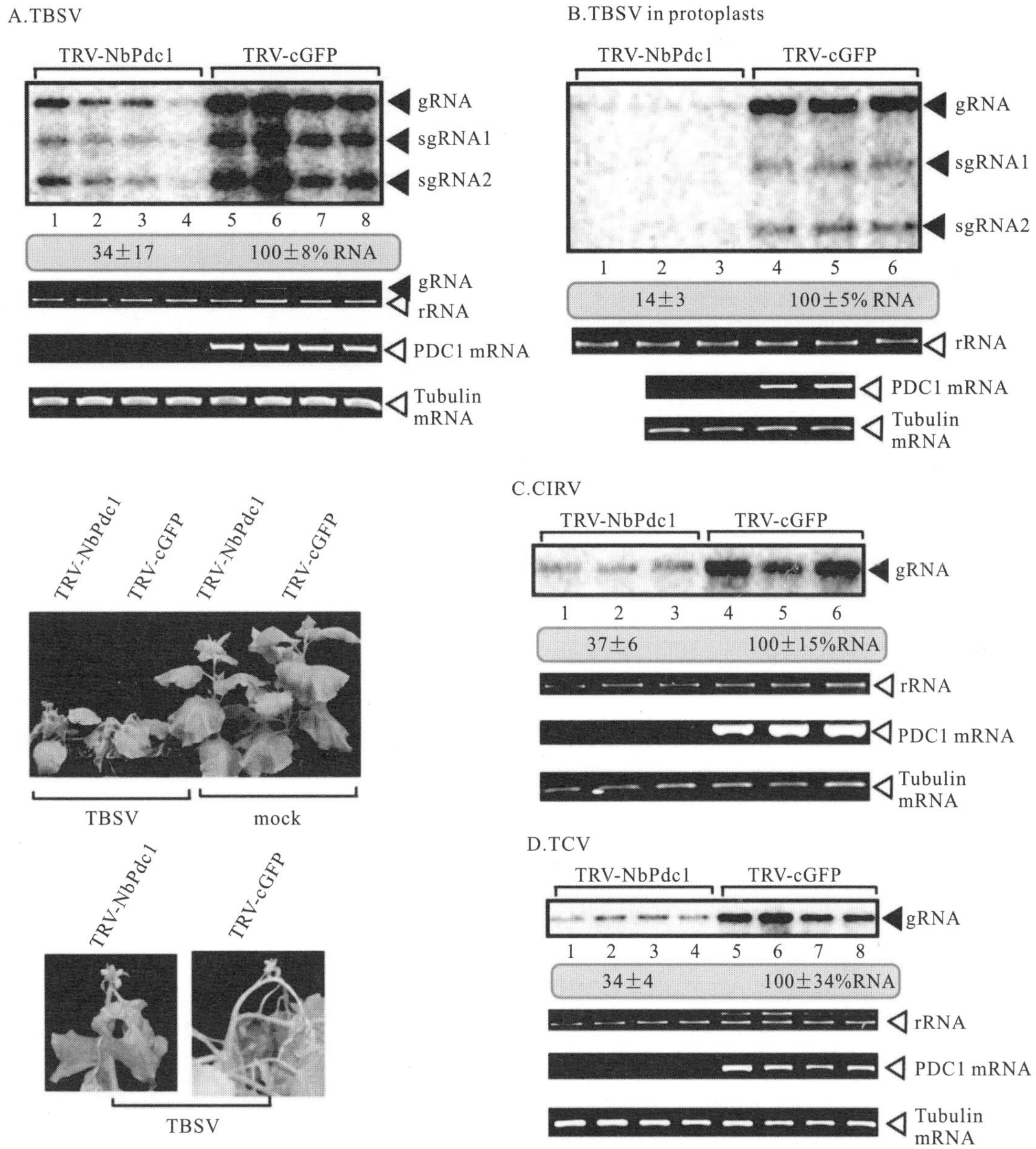

Fig. 5 Knockdown of Pdc1 mRNA level inhibits tombusvirus replication in *N. benthamiana plants*

(A) Top panel: Accumulation of the TBSV genomic (g)RNA in *Pdc1*-silenced *N. benthamiana* plants 1.5d post-inoculation (dpi) in the inoculated leaves was measured by northern blot analysis. Inoculation of TBSV gRNA was done 12d after silencing of Pdc1 expression. Agroinfiltration of *tobacco rattle virus*(TRV) vector carrying NbPdc1 or 3′-terminal GFP (as a control) sequences was used to achieve virus-induced gene silencing (VIGS). Secondpanel: Ribosomal RNA is shown as a loading control in an ethidium-bromide stained agarose gel. Third panel: RT-PCR analysis of NbPdc1mRNA level in the silenced and control plants. Fourth panel: RT-PCR analysis of tubulin mRNA level in the silenced and control plants. Each experiment was repeated three times. Delayed development of TBSV-induced symptoms is observed in the Pdc1-silenced *N. benthamiana* plants as compared with the control plants. Note the lack of phenotype in the Pdc1-silenced and mock-inoculated *N. benthamiana* plants. Note the severe wilting and beginning stage of necrosis in the control TBSV-infected plant versus the lack of those symptoms in the Pdc1-silenced *N.benthamiana* plants. The pictures were taken at 8dpi. (B) Top panel: Accumulation of the TBSV gRNA in protoplasts isolated from Pdc1-silenced *N. benthamiana* was measured by northern blot analysis 16 hours after virus transfection. Protoplasts were isolated 12d after silencing of Pdc1 expression. Agro-infiltration of TRV-NbPdc1 or TRV-cGFP (as a control) was used to induce VIGS. Second panel: Ribosomal RNA is shown as a loading control in an ethidium-bromide stained agarose gel. Third panel: RT-PCR analysis of NbPdc1 mRNA level in the silenced and the control protoplasts. Fourth panel: RT-PCR analysis of tubulin mRNA level in the silenced and the control protoplasts. Each experiment was repeated three times. (C) Accumulation of the CIRV gRNA in the Pdc1-silenced *N. benthamiana* plants 3dpi in the inoculated leaves and at 5dpi in the systemically-infected leaves was measured by northern blot analysis. See further details in panel A. (D) Accumulation of the TCV gRNA in the Pdc1-silenced *N. benthamiana* plants 6dpi in the inoculated leaves was measured by northern blot analysis. See further details in panel A

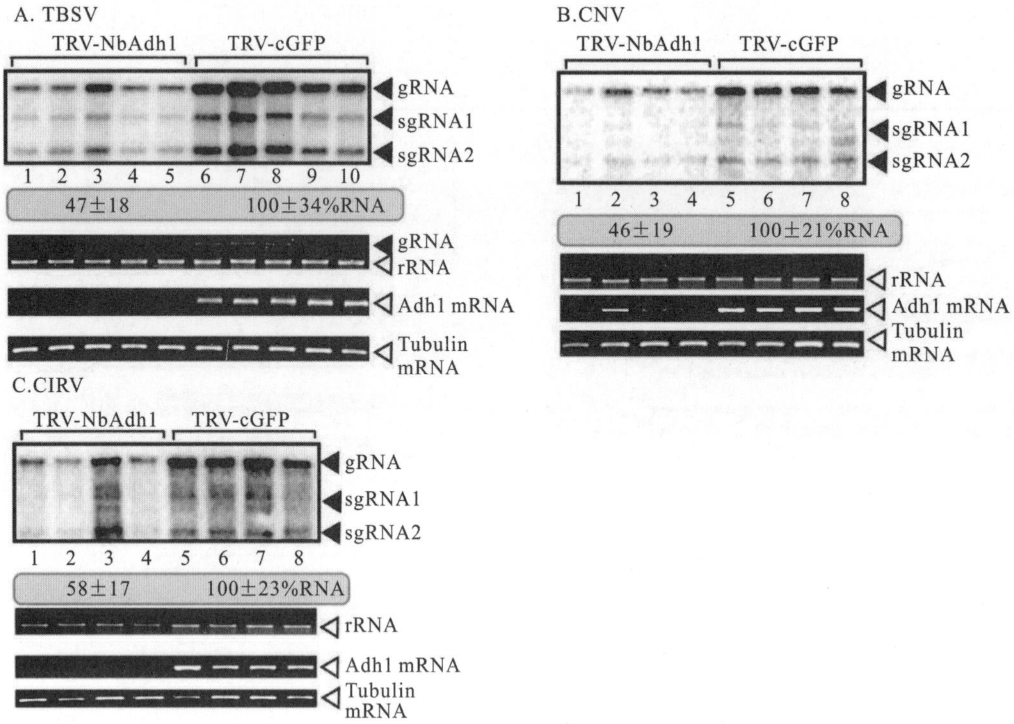

Fig. 6 Knockdown of Adh1 mRNA level inhibits tombusvirus replication in *N. benthamiana*plants

(A-C) Accumulation of the TBSV, CNV and CIRV gRNA in the Adh1-silenced *N. benthamiana* plants. The experimental data are presented as in Fig. 5

Similar co-localization experiments revealed a high level of co-localization of the TBSVp33-RFP replication protein and the BFP-AtAdh1 within the replication compartments consisting of aggregated peroxisomes in the absence or presence of TBSV replication (Fig. 7D). We also observed re-distribution of the BFP-AtAdh1 in the presence of CIRV p36-RFP within the replication compartments consisting of aggregated mitochondria (Fig.7E). Based on these observations, we suggest that the TBSV p33 and the CIRV p36 replication proteins alone are capable of recruiting AtAdh1 to the replication compartment (Fig. 7). In the absence of viral components, AtAdh1 is localized in the cytosol (Fig. 7F). These experiments support a model that Adh1 is efficiently recruited by the tombusvirus replication proteins to the extensive tombusvirus replication compartments in plants.

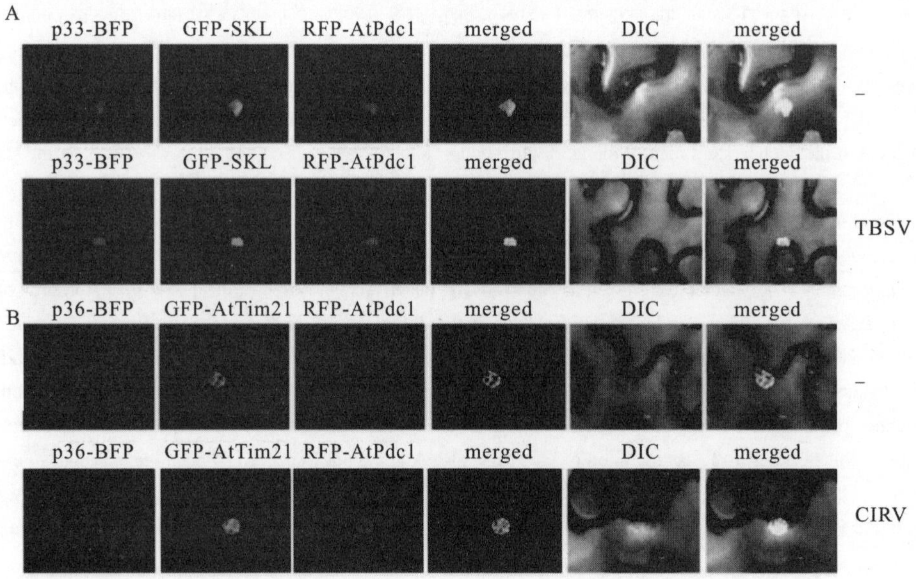

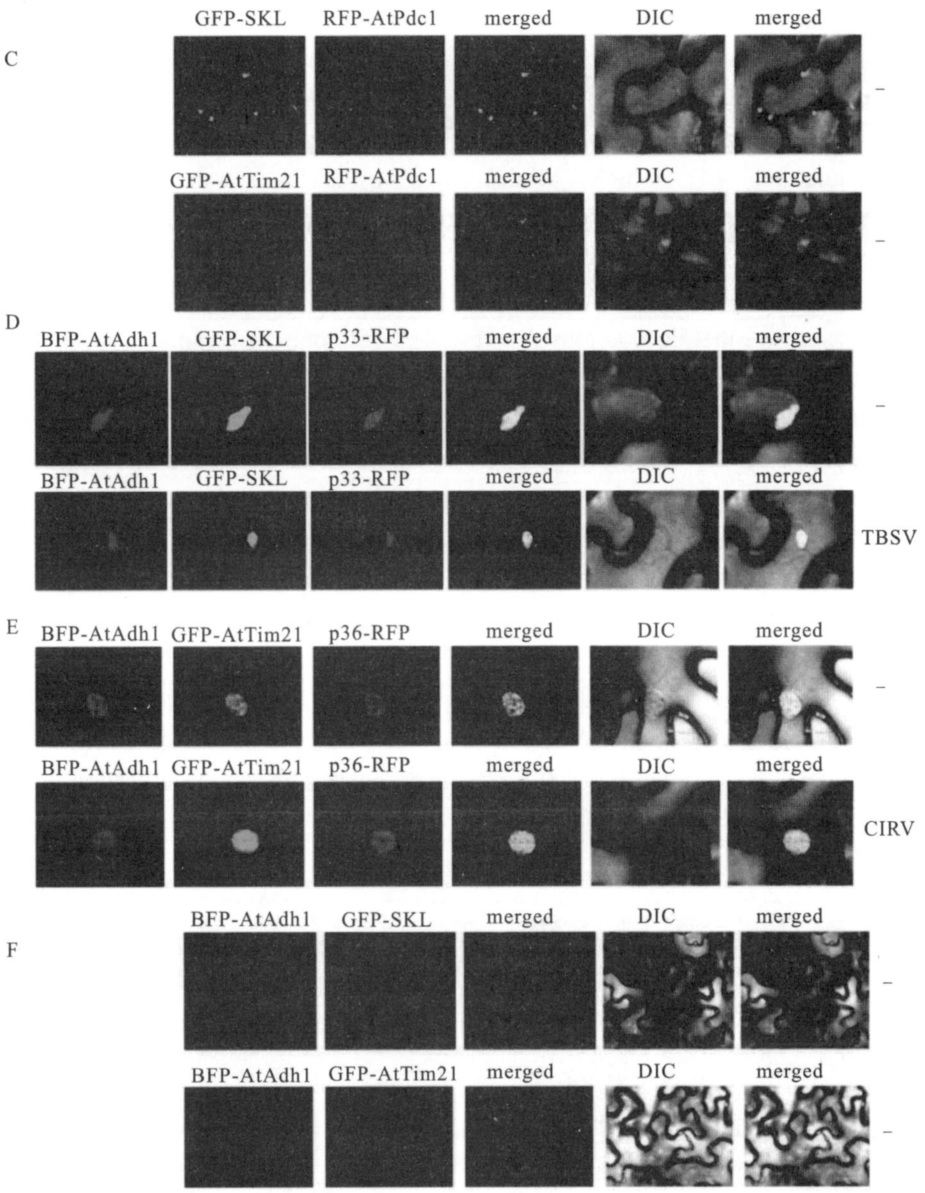

Fig. 7 Recruitment of Pdc1 and Adh1 fermentation enzymes by the tombusvirus replication protein into the viral replication compartment in *N. benthamiana*

(A) Confocal microscopy images show efficient co-localization of the TBSV p33-BFP replication protein and the RFP-AtPdc1 within the viral replication compartment, marked by GFP-SKL peroxisomal luminal marker *in N. benthamiana* leaves. Expression of these proteins from the 35S promoter was done after co-agroinfiltration into *N. benthamiana* leaves. The plant leaves were either TBSV-infected or mock-inoculated as shown. The images were taken 1.5d after TBSV inoculation of plant leaves. Scale bars represent 10μm. (B) Recruitment of Pdc1 by the CIRV p36 replication protein into the mitochondria-derived viral replication compartment in *N. benthamiana*. Confocal microscopy images show efficient co-localization of CIRV p36-BFP replication protein and the RFP-AtPdc1 within the viral replication compartment, marked by GFP-AtTim21mitochondrial marker *in N. benthamiana* leaves. The images were taken 1.5d after agro-infiltration of plant leaves. See further details in panel A. (C) Confocal microscopy imaging shows the cytosolic localization of RFP-AtPdc1 in the absence of viral components. See further details in panel A. (D) Confocal microscopy images show efficient colocalization of TBSV p33-RFP replication protein and the BFP-AtAdh1 within the viral replication compartment,marked by GFP-SKL peroxisomal luminal marker *in N. benthamiana* leaves. See further details in panel A. (E) Recruitment of Adh1 by the CIRV p36 replication protein into the mitochondria-derived viral replication compartment in *N. benthamiana*. Confocal microscopy images show efficient co-localization of CIRV p36-RFP replication protein and the BFP-AtAdh1 within the viral replication compartment, marked by GFP-AtTim21 mitochondrial marker *in N. benthamiana* leaves. See further details in panel A. (F) Confocal microscopy imaging shows the cytosolic localization of BFP-AtAdh1 in the absence of viral components. See further details in panel A

To demonstrate whether AtPdc1 is recruited into the TBSV replication compartment,which actively replicates the viral RNAs, we utilized a modified repRNA carrying an ssRNAsensor (Panavas et al., 2005). This sensor consists of six repeats of a hairpin RNA from MS2 bacteriophage,which is specifically recognized by the MS2 coat protein (MS2-CP) (Bertrand et al., 1998). Co-expression of the TBSV p33-BFP with the GFP-tagged AtPdc1 and the RFP-tagged MS2-CP revealed the relocalization of AtPdc1 to the active TBSV replication compartment containing the new (+) repRNA product (Fig. 8A; Fig. S3A) or the (-) repRNA, which is part of the replication intermediate (Fig. 8C). In the control experiments, in the presence of only the TBSV repRNA and p33-BFP (no replication due to the absence of $p92^{pol}$ replication protein), AtPdc1 was still localized in the viral replication compartment with p33, whereas RFP-MS2-CP was located in the nucleus (Fig. 8B, 8D, Fig. S3B). Therefore, we conclude that Pdc1 is present at the sites of tombusvirus replication and Pdc1 likely plays a role in the formation of the tombusvirus replication compartments.

Similar experiments with AtAdh1 revealed the re-localization of AtAdh1 to the active TBSV replication compartment containing the new (+) repRNA product (Fig. 8E) or the (-) repRNA, which is part of the replication intermediate (Fig. 8G, Fig. S3C). In the control experiments, when only the TBSV repRNA and p33-BFP were expressed without the $p92^{pol}$ replication protein, then AtAdh1 was still re-localized into the viral replication compartment with p33, but RFP-MS2-CP was located in the nucleus (Fig. 8F, H, Fig. S3D). Therefore, we conclude that AtAdh1, similar to AtPdc1, is present at the sites of active tombusvirus replication.

To provide additional evidence that the AtPdc1 is recruited into the viral replication compartments through interacting with the TBSV p33 or CIRV p36 replication proteins, we have conducted bimolecular fluorescence complementation (BiFC) experiments in *N. benthamiana* leaves. The BiFC experiments revealed robust interactions between AtPdc1 and either the TBSV $p33/p92^{pol}$ or the CIRV p36 replication proteins within the replication compartment (Fig. 9A, see also Fig.S4 for the negative control BiFC experiments).

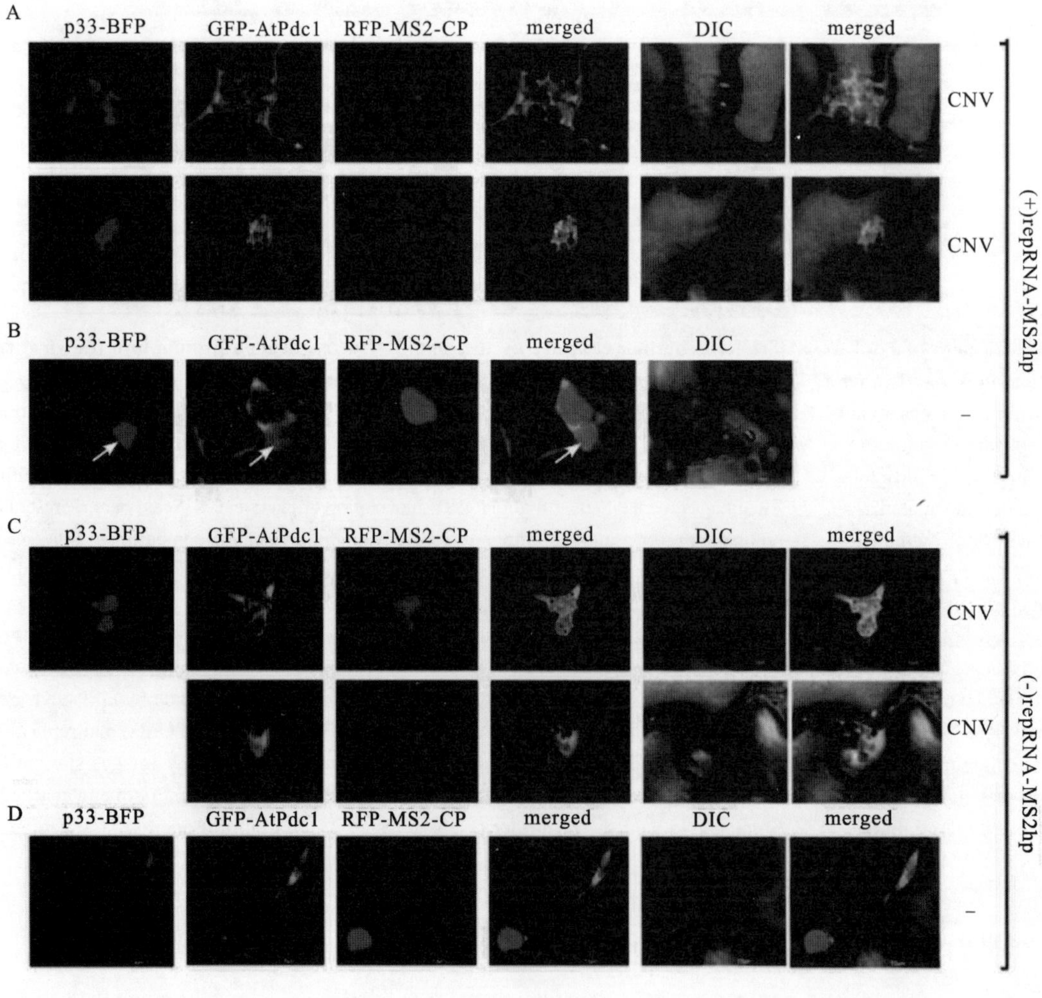

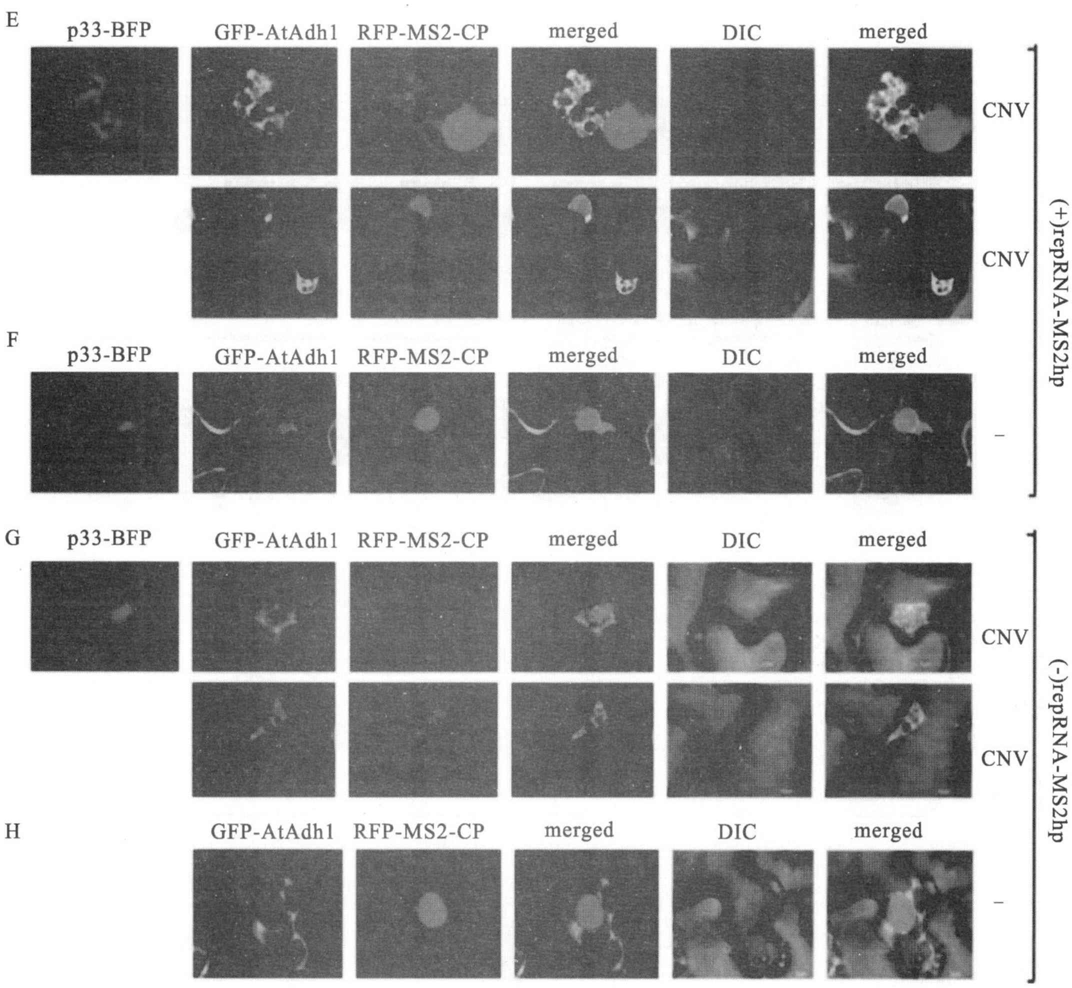

Fig. 8 Confocal microscopy shows co-localization of the co-opted fermention enzymes with the viral repRNAs in whole plants infected with CNV

(A-B) Most of GFP-AtPdc1 is re-targeted into the replication compartment where the viral RNA synthesis takes place. The viral (+) repRNA carried six copies of the MS2 bacteriophage RNA hairpin (MS2hp), which is recognized by the MS2 coat protein (RFP-MS2-CP). The replication compartment was marked by the BFP-tagged p33 replication protein in *N. benthamiana*. Panel B shows images from plants mock-inoculated (no viral RNA replication). Note that RFP-MS2-CP contains a week nuclear localization signal, therefore this protein ends up in the nucleus in the absence of replicating (+) repRNA-MS2hp in the cytosol. Expression of the above proteins from the 35S promoter was done after co-agroinfiltration into *N. benthamiana* leaves. The images were taken 3.5d after agro-infiltration of plant leaves. Scale bars represent 10μm. Each experiment was repeated three times. (C-D) Similar experimental set-up as in panel A-B, except the six MS2hps form the suitable structures on the viral (-) repRNA-MS2hp, which is recognized by RFP-MS2-CP. See further details in Panel A. (E-H) Most of GFP-AtAdh1 is re-targeted into the replication compartment where the viral RNA synthesis takes place. See further details in Panel A and C

Similar BiFC experiments in *N. benthamiana* leaves revealed interactions between AtAdh1 and the TBSV p33/p92^{pol} or the CIRV p36 replication proteins within the replication compartment (Fig. 9B). These data confirmed the replication protein-driven re-localization of AtPdc1 and AtAdh1 into the viral replication compartment.

2.6 Pdc1 is required for efficient tombusvirus replication *in vitro*

To obtain direct evidence of the role of Pdc1 in TBSV replication, we used an *in vitro* replicase reconstitution assay based on a cell-free extract (CFE) from GAL::PDC1 pdc5Δ yeast strain with depleted Pdc1p level. Programming the CFE with the (+)repRNA and purified replication proteins led to ~3-to-4-fold reduced replication, including the production of both dsRNA replication intermediate and the (+)repRNA progeny when compared with CFE prepared from the same yeast strain with induced Pdc1p expression (Fig. 10A). Based on these data, we suggest that Pdc1p is required for robust replication and likely during the replicase assembly step since all TBSV repRNA products were reduced when Pdc1p was depleted.

We also performed another approach to test the efficiency of replicase assembly in yeast, which is based on the purification of the tombusvirus replicase from yeast, followed by *in vitro* RdRp assay with added template RNA. The purified replicase prepared from GAL::PDC1 pdc5Δ yeast strain with depleted Pdc1p level had a reduced activity in comparison with the replicase obtained from the same yeast strain with induced Pdc1p expression on both (-) and (+) RNA templates (Fig. 10B). Because the replicase has to pre-assemble in yeast in this approach, the reduced activity of replicase with depleted Pdc1p is likely due to defect in the replicase assembly.

Interestingly, the replicase assembly involves the activation of the p92 RdRp, which depends on several viral- and host factors, most notably the co-opted Hsp70 protein chaperone (Pogany and Nagy, 2015). We have tested the p92 RdRp activation in a simplified *in vitro* assay, based on a purified N-terminally-truncated p92 RdRp and the soluble fraction of yeast CFEs, which should provide the needed host components (Pogany and Nagy, 2012, 2015). We observed a ~40% reduction in p92 RdRp activation when the CFE was derived from the double-mutant GAL::PDC1 pdc5Δ yeast strain with depleted Pdc1p level versus the CFE obtained from the same yeast strain with induced Pdc1p expression (Fig. 10C). This reduction might indicate a low-level activity for the ATP-dependent Hsp70 in CFE, which could be due to reduced ATP production by glycolysis in yeast with a depletedPdc1p level (see Discussion). Overall, all *in vitro* assays suggest the direct involvement of Pdc1 in tombusvirus replication, which is likely due to reduced ATP production by glycolysis.

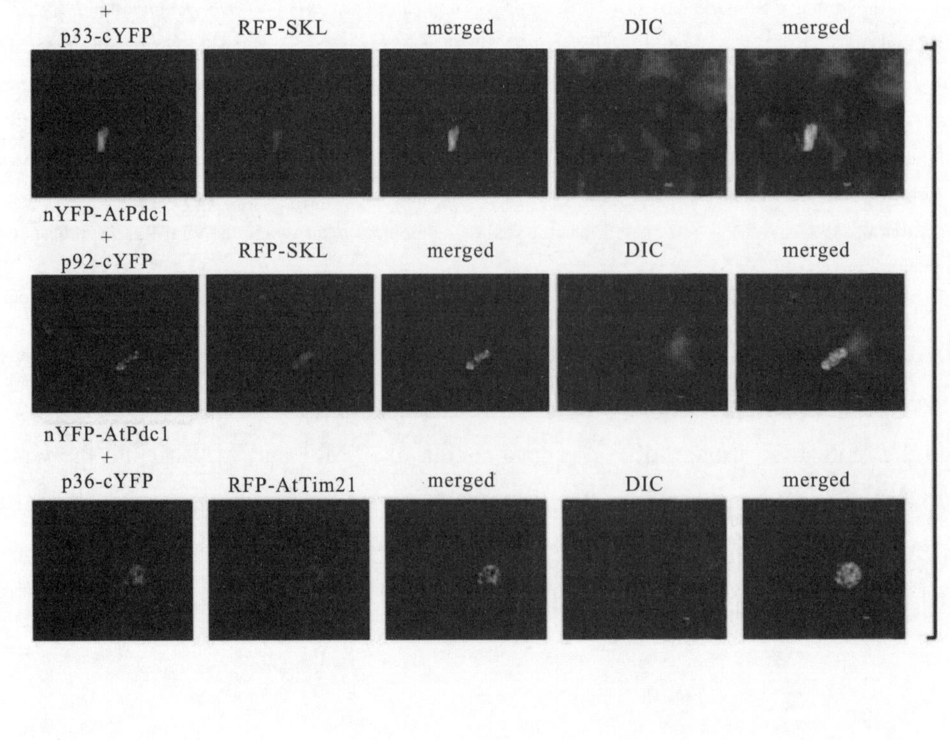

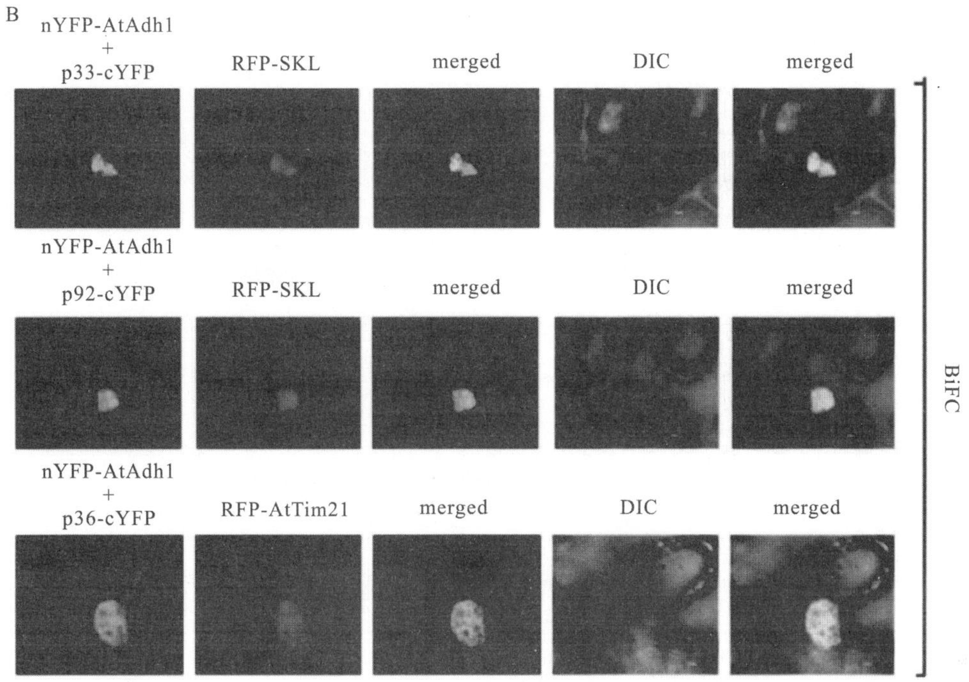

Fig. 9 Interactions between TBSV p33/p92 or CIRV p36 replication proteins and the AtPdc1 or AtAdh1 proteins were detected by BiFC

The TBSV p33-cYFP or p92-cYFP or CIRV p36-cYFP replication proteins and the nYFP-AtPdc1 (panel A) or nYFP-AtAdh1 (panel B) proteins and the marker proteins were expressed via agroinfiltration. The merged images show the efficient co-localization of the peroxisomal RFP-SKL or the mitochondrial RFP-AtTim21 with the bimolecular fluorescence complementation (BiFC) signals, indicating that the interactions between the tombusvirus replication proteins and the co-opted AtPdc1 or AtAdh1 proteins occur in the large viral replication compartments, which consist of either aggregated peroxisomes or aggregated mitochondria. Scale bars represent 10μm

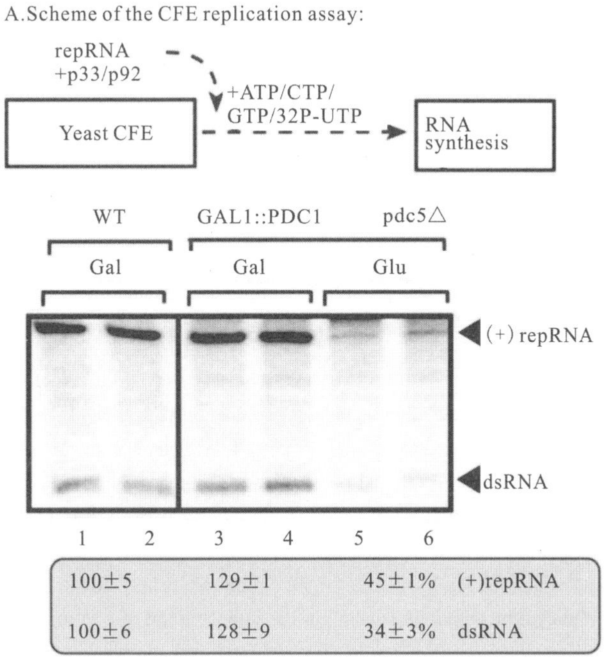

B. purified replicase assay:

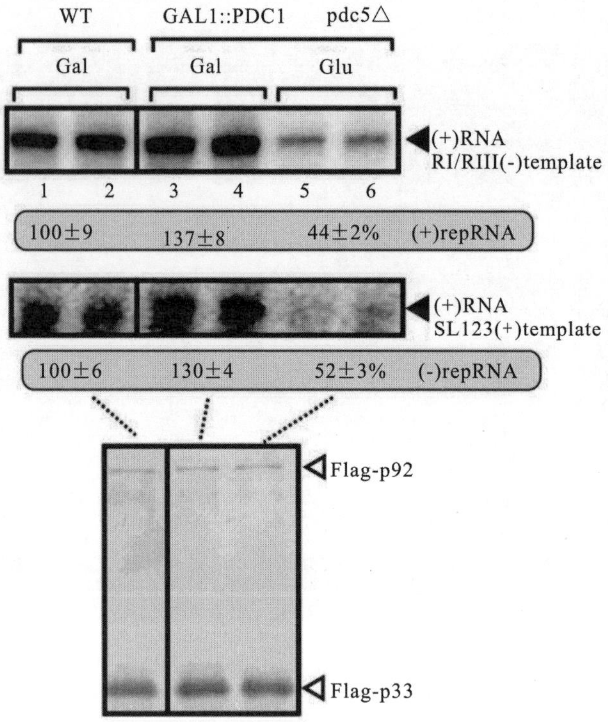

C. RdRp activation assay:

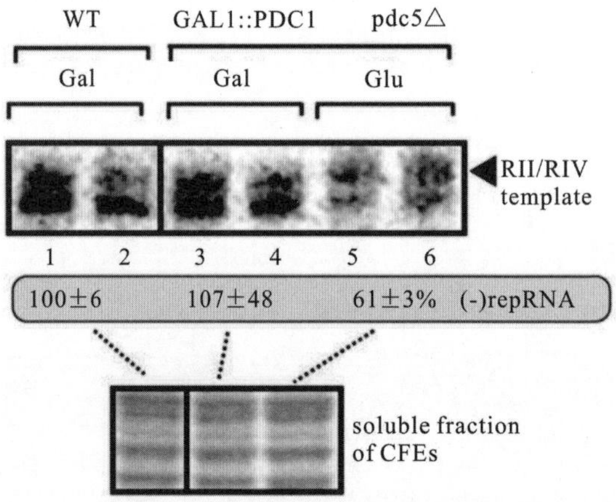

Fig. 10 Dependence of TBSV repRNA accumulation on Pdc1/5 in an *in vitro* replicase reconstitution assay based on CFE obtained from yeast with depleted Pdc1/5

(A) Top: A scheme of the *in vitro* replicase reconstitution assay based on yeast cell-free extracts (CFEs). The purified recombinant TBSV p33 and $p92^{pol}$ replication proteins from *E. coli* were added in combination with the (+) repRNA template to program the *in vitro* tombusvirus replication assay. The CFEs were prepared from yeast strains cultured in the shown media prior to CFE preparation. Bottom: Non-denaturing PAGE shows the accumulation of ^{32}P-labeled (+) repRNAs and the dsRNA replication intermediate products made by the reconstituted replicases in the shown CFE preparations. All the samples shown were loaded on the same PAGE gel. Each experiment was repeated. (B) RdRp assay with Flag-affinity purified tombusvirus replicase preparations. The shown yeast strains expressing Flag-p33 and Flag-$p92^{pol}$ from the *CUP1* promoter and (+) repRNA from the *TET* promoter were cultured in the shown media prior to preparation of the purified replicase preps. The replicase preparations containing the same amount of p33 replication protein were programmed with the shown (+) or (-) RNA templates. The denaturing PAGE gels show the produced complementary RNA products by the given replicase preparations. All the samples shown were loaded on the same PAGE gel. Each experiment was repeated. (C) The *in vitro* RdRp activation assay is based on (+) repRNA and p92-Δ167N RdRp protein in the presence of the soluble fraction of yeast CFE. The CFEs were prepared from yeast strains cultured in the shown media prior to CFE preparation. Denaturing PAGE analysis of the ^{32}P-labeled RNA products obtained in an *in vitro* assay with recombinant p92-Δ167N RdRp. Each experiment was repeated three times

2.7 Robust generation of ATP by glycolysis in the tombusvirus replication compartment is dependent on the co-opted Pdc1 and Adh1 in yeast and plants

Because several pro-viral co-opted host proteins, such as Hsp70, the ESCRT-associated Vps4AAA ATPase and DEAD-box helicases, require plentiful ATP within the replication compartment to fuel robust viral replication (Wang et al., 2009a; Wang et al., 2009b; Kovalev et al., 2012; Barajas et al., 2014b; Chuang et al., 2015), it is possible that the co-opted Pdc1 and Adh1 are needed within the replication compartment to rapidly supply the NAD^+ substrate. NAD^+ is critical to replenish the glycolytic pathway, which is dependent on reducing NAD^+ to NADH via the co-opted GAPDH (Tdh2/2p in yeast, GAPC in plants) (Lunt and Vander Heiden, 2011; Olson et al., 2016). The facts that both fermentation enzymes are recruited to the sites of virus replication and the catalytic activity of Pdc1 is required for its pro-viral function (Fig. 1D), also support this hypothesis.

To estimate the ATP level within the tombusviral replication compartment, we used aFRET-based biosensor (Imamura et al., 2009), which was previously adapted to estimate ATP levels (Chuang et al., 2017; Prasanth et al., 2017).Briefly, ATeam- $p92^{pol}$ can measure ATP level due to the conformational change in theenhanced ε subunit of the bacterial F_0F_1-ATP synthase upon ATP binding (Chuang et al., 2017; Prasanth et al., 2017). The ε subunit bound to ATP draws the CFP and YFP fluorescent tags in close vicinity, increasing the FRET signal in confocal laser microscopy (Fig. 11A). On the contrary, the ε subunit in the ATP-free form is present in an extended conformation, which places CFP and YFP tags in adistal position, thus reducing the FRET signal (Fig. 11A) (Imamura et al., 2009). We found previously (Chuang et al., 2017; Prasanth et al., 2017) that the ATeam-tagged $p92^{pol}$ is a fully functional RdRp, which localizes to the viral replication compartment representing aggregated peroxisomes. Since these experiments are best performed in the presence of glucose in yeast media (Chuang et al., 2017), we used pdc1Δ yeast strain expressing Pdc1p from a plasmid. We found that the ATP level was ~4-fold higher in pdc1Δ yeast strain expressing WT Pdc1p than in the control lacking *PDC1* or expressing $Pdc1^{S455F}$ mutant (Fig. 11B). Expression of the low-sensitive variant of ATeam ($ATeam^{RK}$-p92) (Imamura et al., 2009; Chuang et al., 2017) in pdc1Δ yeast strain expressing Pdc1p showed low FRET values, confirming that the FRET data is derived from the ATP biosensor in this assay. Overall, the obtained data support the model that Pdc1pis recruited into the tombusvirus replication compartment in yeast to facilitate the generation of ATP locally for viral RNA synthesis.

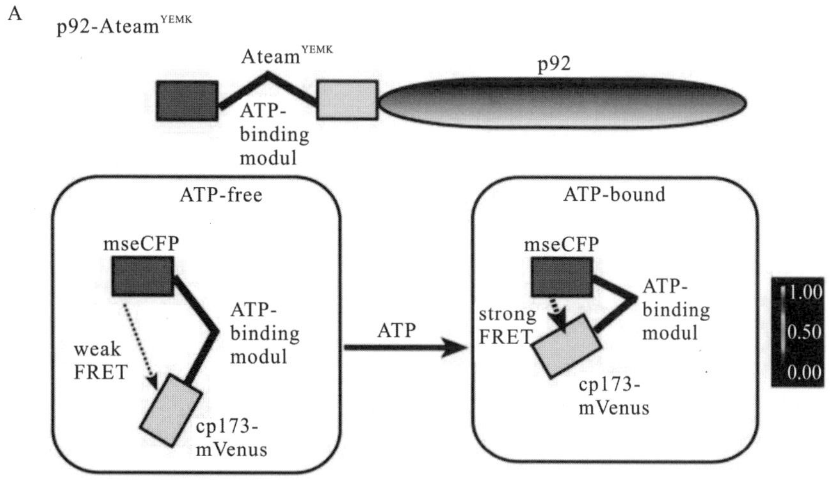

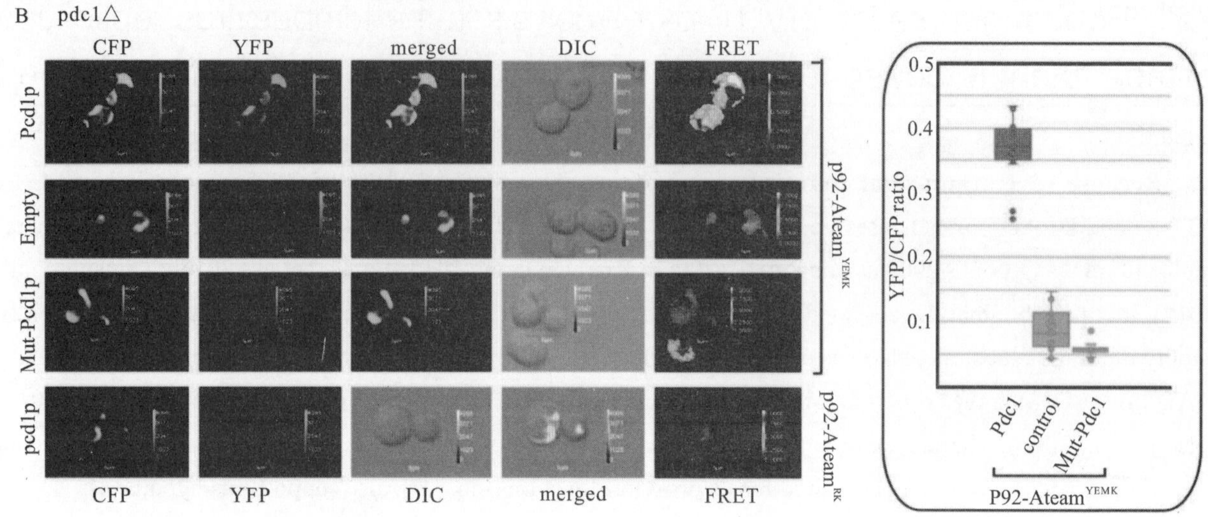

Fig. 11 The co-opted cytosolic Pdc1 fermentation enzyme affects ATP accumulation within the tombusvirus replication compartment in yeast
(A) A scheme of the FRET-based detection of ATP within the tombusvirus replication compartment. The enhanced ATP biosensor, $ATeam^{YEMK}$ was fused to TBSV $p92^{pol}$ replication protein. See further details in the main text. (B) Comparison of the ATP level produced within the tombusvirus replication compartment in pdc1Δ yeast strain expressing WT Pdc1p, $Pdc1^{S455F}$ mutant or without Pdc1 expression using $ATeam^{YEMK}$ -$p92^{pol}$. The more intense FRET signals are white and red (between 0.5 to 1.0 ratio), whereas the low FRET signals (0.1 and below) are light blue and dark blue. We show the quantitative FRET values (obtained with ImageJ) for a number of samples in the graph. Note that we also used a reduced ATP-sensitive version of $ATeam^{RK}$-p92 (bottom panel) to demonstrate that the FRET signal is due to ATP-sensing. Scale bars represent 5μm. Each experiment was repeated three-four times

To confirm that tombusviruses co-opt Pdc1 into the viral replication compartment to support efficient ATP generation in plants, we expressed p33-ATeam replication protein in *N. benthamiana* leaves, which were either silenced for Pdc1 expression or not (Fig. 12A). The obtained data showed up to a ~4-fold reduction in ATP production within the viral replication compartment in the Pdc1 knockdown plants versus the control plants (Fig. 12B). Similar experiments with *N. benthamiana* infected with TBSV showed a ~3-fold reduction in ATP level within the viral replication compartment in the knockdown plants versus the control plants (Fig. 12C). Intensive TBSV replication likely uses up some of the produced ATP in the latter experiments as we observed previously (Chuang et al., 2017). Applying the same approach showed that similar pictures on ATP production within the viral replication compartment in the Pdc1-knockdown plants versus the control plants exist during the peroxisomal CNV (Fig. 12D) and the mitochondrial CIRV (Fig. 13) infections of *N. benthamiana* leaves. However, we did observe arange in ATP production [between ~3-fold (Fig. 13A) and ~2-fold (S5 Fig)] in the absence of CIRV replication within the viral replication compartment in the Pdc1-knockdown plants versus the control plants in different experiments. It is possible that glucose concentrations and/or the efficiency of fermentation within the viral replication compartments in leaves are influenced by several physiological processes in plants. Nevertheless, the emerging picture is that subversion of Pdc1 into the viral replication compartment is required to support efficient ATP generation locally.

Because Pdc1 works together with Adh1 in the fermentation pathway, we were curious if tombusviruses also co-opt Adh1 into the viral replication compartment to support efficient ATP generation locally in plants. Therefore, we expressed the p33-ATeam replication proteinin *N. benthamiana* leaves, which were either silenced for Adh1 expression or not. We observeda ~2-fold reduction in ATP production within the viral replication compartment in the Adh1knockdown plants versus the control plants (Fig. 14A). Similar experiments with *N. benthamiana* infected with TBSV or CIRV also showed a ~2-fold reduction in ATP-level within the viral replication compartment in the knockdown plants versus the control plants (Fig. 14). Based on these findings, we propose that subversion of both Pdc1 and Adh1 fermentation enzymes by tombusviruses facilitates the glycolytic process to produce plentiful ATP locally within the replication compartment in *N. benthamiana* leaves.

2.8 Dependence of bamboo mosaic virus and tobacco mosaic virus replicationon the fermentation pathway in plants

To investigate if additional plant viruses also depend on the fermentation pathway for robust replication, we chose bamboo mosaic virus (BaMV), a potexvirus, and tobacco mosaic virus(TMV), a tobamovirus, which are unrelated to TBSV.

We found that BaMV and TMV replication led to the efficient induction of both Pdc1mRNA and Adh1 mRNA expression in the inoculated as well as the systemically-infected *N.benthamiana* leaves (Fig. 15A and 15B and S6A and S6B Fig). VIGS-based silencing of Pdc1level in *N. benthamiana* leaves resulted in a ~60% reduction in the accumulation of both BaMV and TMV gRNAs (Fig. 15C and S6C Fig). Similarly, knocking down Adh1 level in *N.benthamiana* leaves reduced the accumulation of BaMV and TMV gRNAs by 70% and 60%, respectively (Fig. 15C and S6C Fig). To test if BaMV can recruit the fermentation proteins directly through protein-protein interactions, we used a BiFC approach. Co-expression of either the capping enzyme domain or the helicase domain of the BaMV replicase with Pdc1 in *N. benthamiana* leaves resulted in punctate and cytosolic signals, respectively (Fig. 15D). Incontrast, co-expression of the RdRp domain of the BaMV replicase with Pdc1 did not produce signals, suggesting the lack of interaction. Interestingly, we also observed interaction between the capping enzyme domain or the helicase domain, but not the RdRp domain of the BaMV replicase with Adh1 in *N. benthamiana* leaves (Fig. 15E). Therefore, it is possible that BaMV also exploits the fermentation pathway via interaction between the viral replicase and the fermentation enzymes. Based on these observations, we suggest that similar to tombusviruses, other unrelated and rapidly replicating plant viruses also depend on the fermentation pathway in plants. Further experiments will be needed on the mechanistic details on the role of the fermentation pathway in the replication of BaMV and TMV.

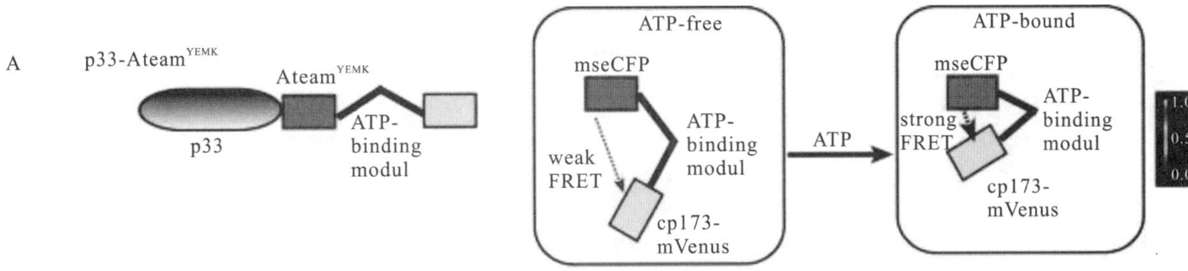

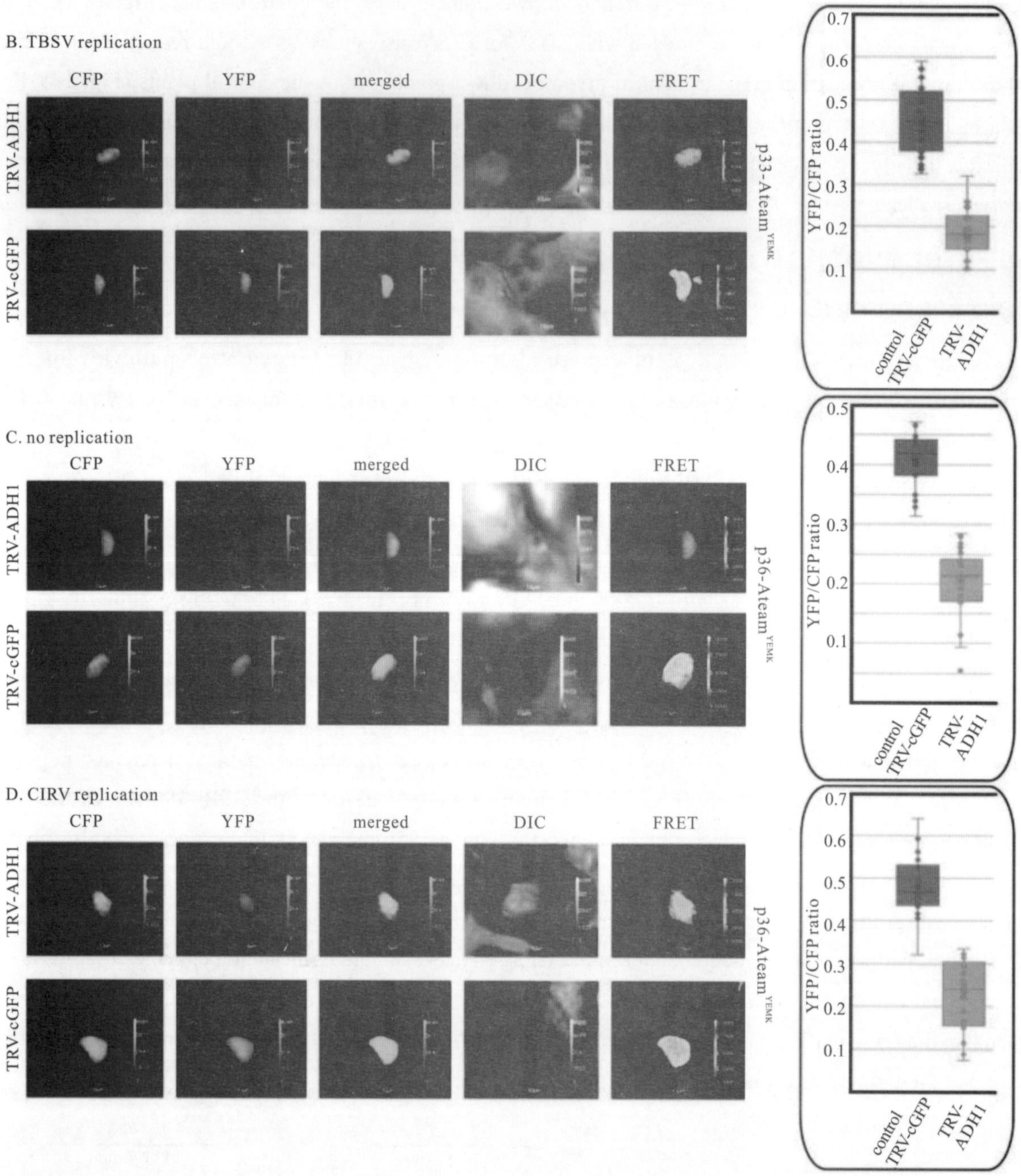

Fig.12 Knockdown of the cellular Pdc1 fermentation enzyme inhibits ATP accumulation within the tombusvirus replication compartment in *N. benthamiana*

(A) A scheme of the FRET-based detection of cellular ATP within the replication compartment. The enhanced ATP biosensor, ATeam ^{YEMK} was fused to TBSV p33 replication protein. (B) Knock-down of Pdc1 mRNA level by VIGS in *N. benthamiana* was done using a TRV vector. Twelve days later, expression of p33-ATeam ^{YEMK} was done in upper *N. benthamiana* leaves by agroinfiltration. The YFP signal was generated by mVenus in p33-ATeam^{YEMK} via FRET 1.5d after agro-infiltration. The FRET signal ratio is shown in the right panels. The more intense FRET signals are white and red (between 0.5 to 1.0 ratio), whereas the low FRET signals (0.1 and below) are light blue and dark blue. We also show the average quantitative FRET values (obtained with ImageJ) for 10–20 samples on the graph. Note that *N. benthamiana* plants were mock-inoculated. (C-D) Comparable experiments with p33-ATeam^{YEMK} in the Pdc1 knockdown *N. benthamiana* plants infected with the peroxisomal TBSV and CNV tombusviruses. See further details in panel B

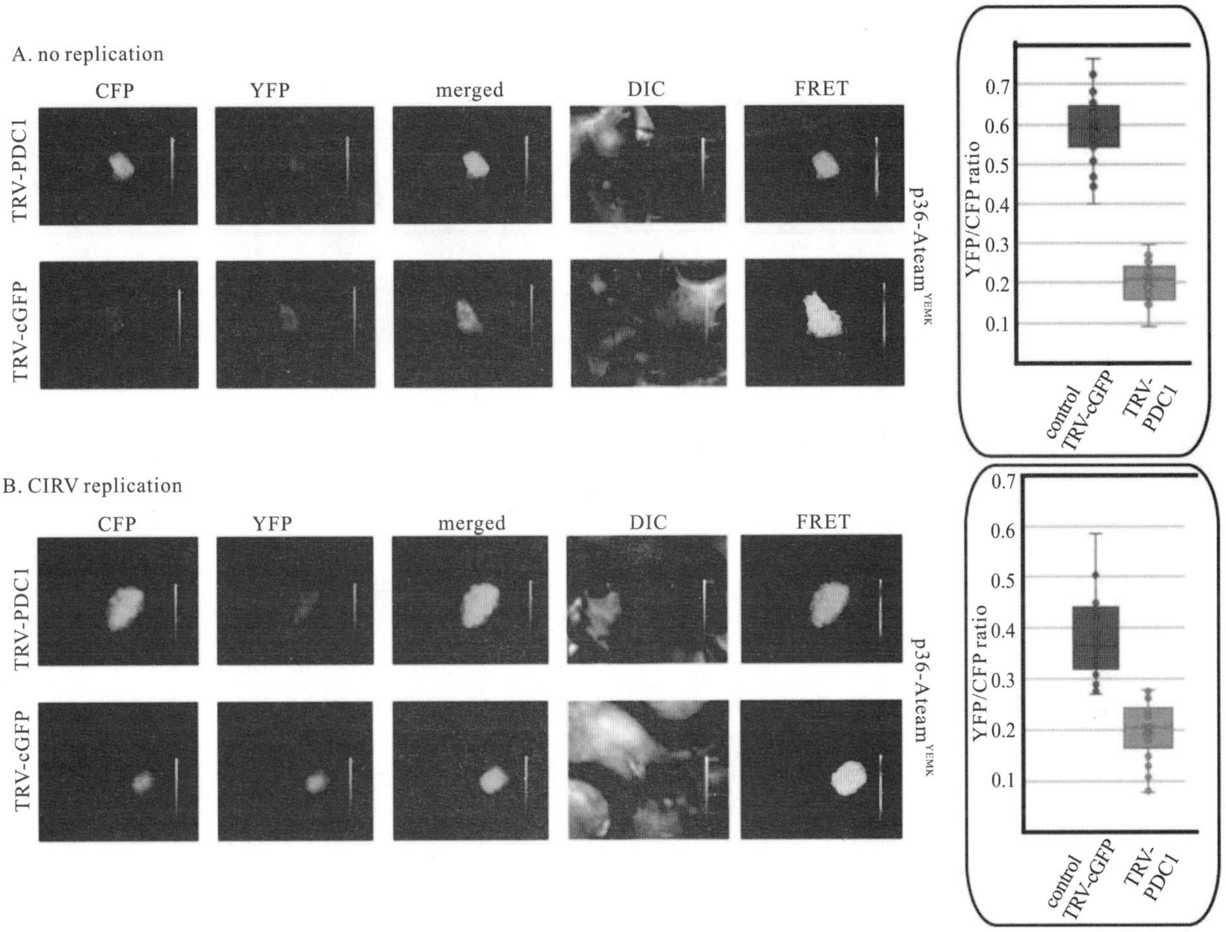

Fig. 13 The co-opted cellular Pdc1 fermentation enzyme affects ATP accumulation within the CIRV replication compartment in *N. benthamiana*

(A) Knock-down of Pdc1 mRNA level by VIGS in *N. benthamiana* was done using a TRV vector. Twelve days later, expression of the CIRV p36-ATeam^{YEMK} was done in upper *N. benthamiana* leaves by agroinfiltration. The FRET-based confocal microscopy analysis was performed 1.5d after agro-infiltration. The FRET signal ratio is shown in the right panels. We show the average quantitative FRET values for 10–20 samples on the graph. Note that *N. benthamiana* plants were mock-inoculated. (B) Comparable experiments to those in panel A, except CIRV supported repRNA replication in the cells. See further details in panel A

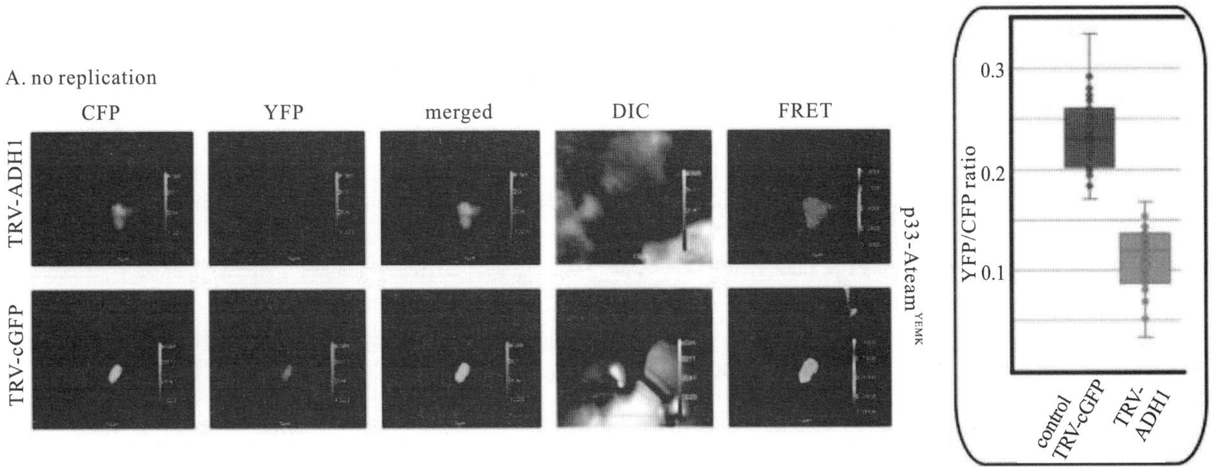

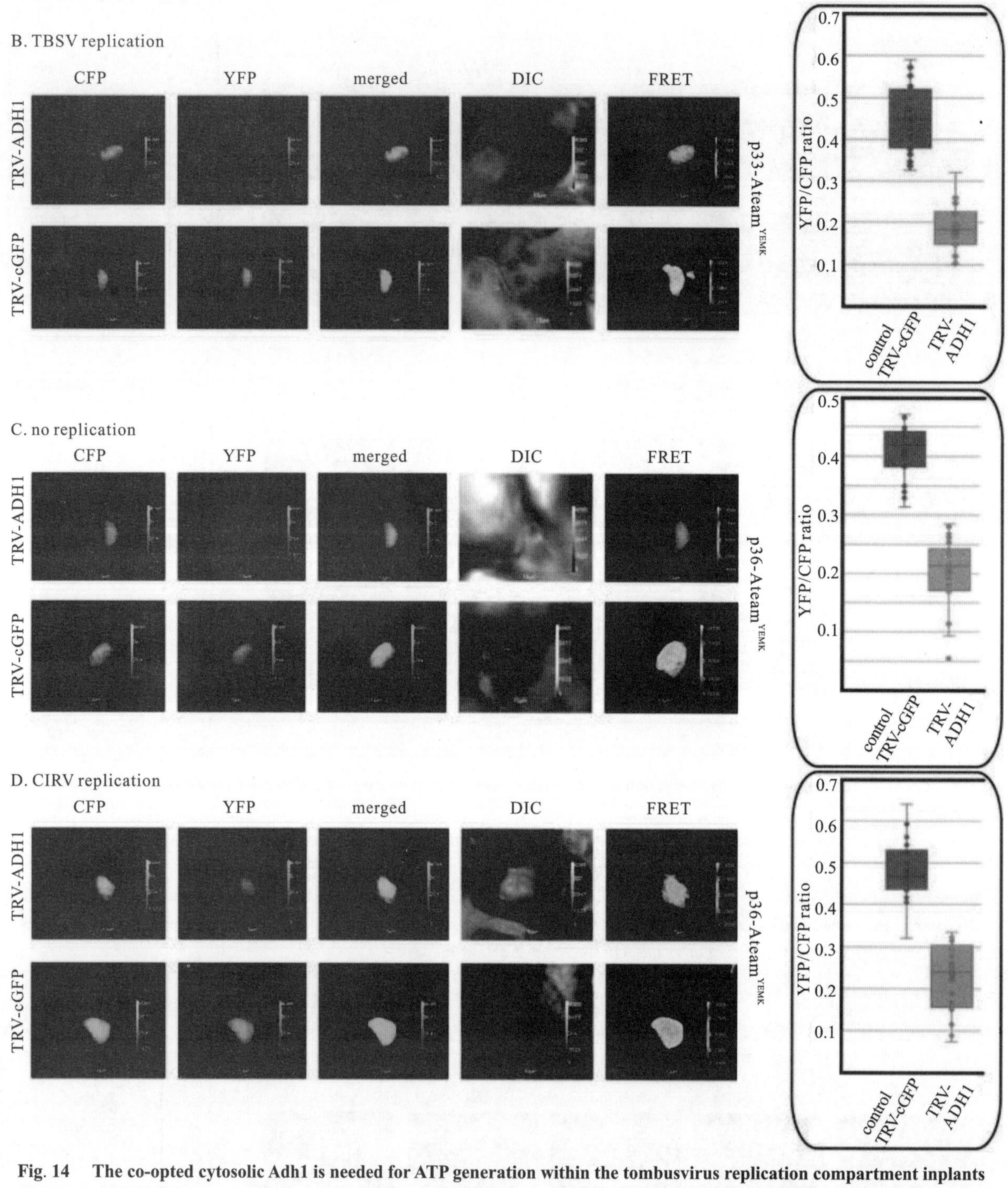

Fig. 14 The co-opted cytosolic Adh1 is needed for ATP generation within the tombusvirus replication compartment inplants
(A-D) Knock-down of Adh1 mRNA level by VIGS in *N. benthamiana* was done using a TRV vector. Twelve days later, expression of TBSV p33-ATeamYEMK (panels A-B) or the CIRV p36-ATeamYEMK (panels C-D) was done in upper *N. benthamiana* leaves by agroinfiltration. The YFP signal was generated via FRET 1.5d after agro-infiltration. The FRET signal ratio is shown in the right panels. We show the average quantitative FRET values for 10–20 samples on the graph. Note that *N. benthamiana* plants were mock-inoculated or the plants supported TBSV and CIRV repRNA replication as shown. See further details in Fig. 12B. Scale bars represent 10μm. Each experiment was repeated three or four times

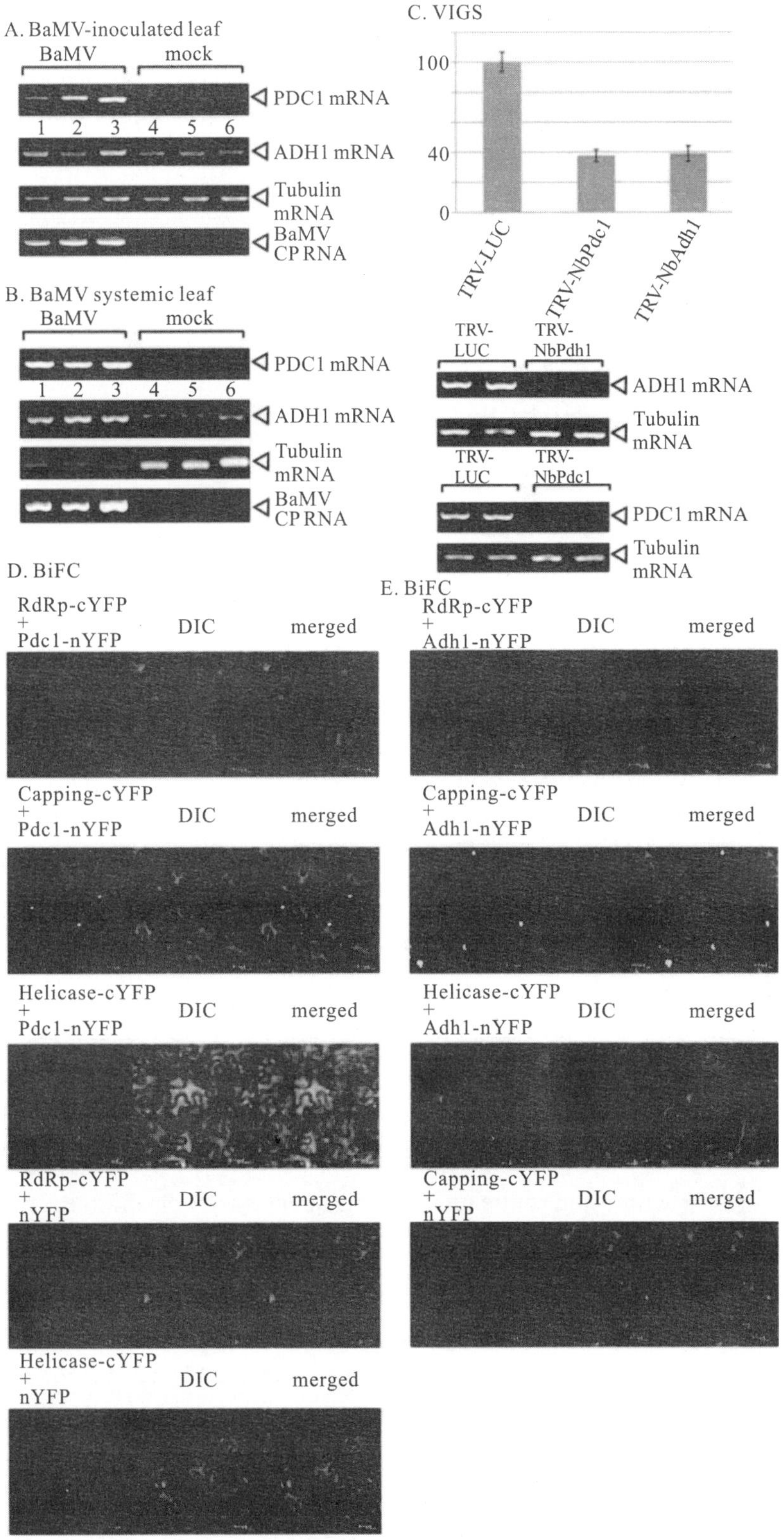

Fig. 15 Dependence of BaMV replication on Pdc1 and Adh1 proteins in *N. benthamiana*

(A) Top panels: semi-quantitative RT-PCR analysis of NbPdc1 and NbAdh1 mRNA levels at 3dpi in *N. benthamiana* leaves infected with *bamboo mosaic virus* (BaMV) or mock-inoculated. Third panel: RT-PCR analysis of tubulin mRNA level in the same plants. Bottom panel: RT-PCR detection of the BaMV CP subgenomic RNA. (B) Semi-quantitative RT-PCR analysis of NbPdc1 and NbAdh1 mRNA levels at 7dpi in *N. benthamiana* leaves infected with BaMV or mock-inoculated. Third panel: RT-PCR analysis of tubulin mRNA level in the same plants. Bottom panel: RT-PCR detection of the BaMV CP subgenomic RNA. Each experiment was repeated three times. (C) Knockdown of Pdc1 or Adh1 mRNA levels inhibits BaMV replication in *N. benthamiana* plants. Top panel: Accumulation of the BaMV genomic (g) RNA in the Pdc1-silenced *N. benthamiana* plants 2.5d post-inoculation (dpi) in the inoculated leaves was measured by quantitative RT-PCR. Inoculation of BaMV gRNA was done 12d after silencing of Pdc1 or Adh1 expression. Agroinfiltration of *tobacco rattle virus* (TRV) vector carrying NbPdc1 or NbAdh1 or luciferase (LUC, as a control) sequences was used to induce VIGS. Second panel: RT-PCR analysis of tubulin mRNA level in the silenced and control plants. Each experiment was repeated three times. (D-E) The BaMV ORF1-capping-cYFP or ORF1-helicase-cYFP or ORF1-RdRp-cYFP domains of the replicase protein and the AtPdc1-nYFP (panel D) or AtAdh1-nYFP (panel E) proteins were expressed via agroinfiltration. The DIC and the merged images are also shown. The BiFC signals were detected via confocal microscopy 2d after agroinfiltration to *N. benthamiana* plants. Scale bars represent 25μm

3 Discussion

3.1 Compartmentalization of the co-opted fermentation pathway in the tombusviral replication compartment to support tombusvirus replication

The viral replication proteins orchestrate the biogenesis of the large tombusviral replication compartment, including the numerous spherules/VRCs, which represent the sites of viral RNA replication (Nagy, 2015, 2016; de Castro et al., 2017). The formation and operation of these virus-driven structures require subversion of numerous cellular proteins, membrane deformation, membrane proliferation, changes in lipid composition of the hijacked cellular membranes and intensive viral RNA synthesis. To obtain the necessary resources from the infected cells, tombusviruses haveto rewire cellular pathways to fuel the biogenesis of the replication compartment. These robust processes require plentiful ATPs and molecular building blocks produced at the sites of replication or delivered there. The emerging picture with tombusviruses is that by co-opting theaerobic glycolysis, the ATP molecules are produced and utilized within the replication compartment (Chuang et al., 2017; Prasanth et al., 2017). However, the aerobic glycolysis requires the replenishing of the NAD^+ pool, which is used by the glycolytic GAPDH to produce NADH. NAD^+ is efficiently generated by the fermentation pathway, which also utilizes pyruvate, the end product of the glycolytic pathway (Lunt and Vander Heiden, 2011; Olson et al., 2016). Accordingly, in the current work we show the efficient recruitment of Pdc1 and Adh1 fermentation enzymes into the viral replication compartment. Depletion of Pdc1 combined with deletion of the homologous *PDC5* in yeast or knockdown of Pdc1 and Adh1 inplants reduced the efficiency of tombusvirus replication. A complementation approach revealed that the enzymatically functional Pdc1p is required to support tombusvirus replication. We provide evidence that both Pdc1 and Adh1 enzymes are required for efficient generation of ATP within the replication compartment based on the measurements with an ATP biosensor inside the viral replication compartment (Fig. 11–14). Moreover, *in vitro* works show the pro-viral function of Pdc1 during the assembly of the viral replicase and the activation of the p92 RdRp, both of which require the co-opted ATP-driven Hsp70 protein chaperone.

Is the co-opted fermentation pathway only required for facilitating ATP production within the tombusviral replication compartment? Albeit not studied in this work, it is very likely that the co-opted aerobic glycolysis in combination with the subverted fermentation pathway also provide plentiful metabolic precursors, which could be utilized by the cell to make molecular building blocks, such as ribonucleotides, lipids and amino acids (Vander Heiden et al., 2009; Lunt and Vander Heiden, 2011). These newly made molecular building blocks are likely exploited by tombusviruses to build the viral replication compartment and support intensive viral RNA synthesis. Accordingly, high glucose concentration stimulates TBSV replication in yeast, whereas blocking the aerobic glycolysis with 2DG compound strongly inhibited TBSV accumulation in yeast and plants (Inaba and Nagy, 2018).

Why are the relatively inefficient aerobic glycolytic and fermentation pathways co-opted by tombusviruses?

These metabolic pathways are present in the cytosol, thus easily accessible for subversion by the cytosolic tombusviruses. Moreover, the ATP generation by the aerobic glycolytic and fermentation pathways is fast if plentiful glucose is present in the cells. Plants produce plentiful glucose based on chloroplasts, thus glucose is not expected to be rate limiting for the aerobic glycolytic and fermentation pathways in the infected plant cells. Moreover, these metabolic pathways do not require free oxygen, which could be an advantage for tombusviruses that also replicate efficiently in plant roots. Moreover, tombusviruses require the synthesis of new phospholipids and ribonucleotides (Barajas et al., 2014a; Xu and Nagy, 2015). The nexus point of the metabolic pathways, which is pyruvate, the end-product of glycolysis, has to be re-routed into the fast fermentation pathway. This then leads to the rapid regeneration of NAD^+ to replenish the glycolytic pathway. NAD^+ is also necessary for the biosynthesis of nucleotides and amino acids, and the fermentation pathway supports fast glucose flux through glycolysis. Thus, the rapid regeneration of NAD^+ allows fast incorporation of glucose into metabolites (Vander Heiden et al., 2009; Lunt and Vander Heiden, 2011; Olson et al., 2016). Altogether, by providing plentiful precursor compounds in the cytosol, the aerobic glycolytic and fermentation pathways are far more efficient to facilitate the production of molecular building blocks than the oxidative phosphorylation pathway (Vander Heiden et al., 2009; Lunt and Vander Heiden, 2011; Olson et al., 2016). Then, the generated new metabolites canbe exploited by tombusviruses to build extensive replication.

Why is compartmentalization of the aerobic glycolytic and fermentation pathways in the replication compartment advantageous for tombusviruses? The combined subversion of the aerobic glycolytic and fermentation pathways allows for the rapid production of ATP locally, including replenishing of the regulatory NAD^+ pool by the fermentation pathway. Then, the locally produced ATP could be used efficiently by the co-opted ATP-dependent host factors,such as the Hsp70 protein chaperone, the ESCRT-associated Vps4 AAA ATPase and the proviral DEAD-box helicases (Chuang et al., 2017; Prasanth et al., 2017). These co-opted host factors are required for pro-viral processes, including VRC assembly, the activation of p92 RdRp, and the utilization of both ssRNA templates and dsRNA replication intermediates for viral RNA synthesis (Nagy, 2016; Chuang et al., 2017; Prasanth et al., 2017). By producing the ATP locally within the replication compartment, tombusviruses do not need to compete with cellular processes for the common ATP pool and all the molecular processes could be accelerated by the high local concentration of ATP within the replication compartment. Itis also possible that the feedback regulation of these metabolic processes by the cell is less efficient when compartmentalized in the viral replication compartment. Overall, there is an evolutionary pressure for tombusviruses to replicate fast and speed ahead of antiviral responses ofthe hosts and to outcompete other pathogenic viruses. Therefore, there are numerous advantages for tombusviruses to subvert the cellular aerobic glycolytic and fermentation pathways to support the infection process.

Aerobic glycolysis is induced during cancer and other diseases as well, including type 2 diabetes, amyloid-based brain diseases, wound repair and oncogenic virus infections (Vaishnavi et al., 2010; Vlassenko et al., 2010; Palm and Thompson, 2017; Yu et al., 2018). Switching to the aerobic glycolytic metabolism can also occur with healthy cells, for example, during Endothelial cell differentiation, monocytes-based trained immunity, in rapidly dividing cells during embryogenesis, during T cell differentiation and motor adaptation learning in human brain (Jones and Bianchi, 2015; Shannon et al., 2016; Palm and Thompson, 2017). The fetal heart primarily produces ATP via glycolytic metabolism (Olson et al., 2016). All these cells/tissues utilize aerobic glycolysis as a metabolic compromise to provide ATP and produce enough new metabolic compounds to perform their functions.

In summary, we show evidence that TBSV exploit the fermentation pathway to support rapid virus replication. The dependence on the fermentation pathway is also shown for several other related and unrelated plant viruses. These viruses induce the fermentation pathway, thus indicating that a broad range of viruses takes advantage of the rapid cytosolic generation of ATP and numerous metabolic precursors. It will also be interesting to learn if other (+)RNA viruses exploit the aerobic glycolytic and fermentation pathways for their replication. Becauseall plant, animal and human (+)RNA viruses require the biogenesis of the membranous viral replication compartment/ organelle, thus they likely use plenty ATP and they might depend on the production of new metabolic precursors, it is possible that hijacking the aerobic glycolytic and fermentation pathways occurs in other virus-host interactions as well. This could open upnew common antiviral strategies targeting the fermentation pathway.

4 Materials and methods

4.1 Plant materials, yeast strain and plasmids

Wild type *N. benthamiana* plants were potted in soil and placed in growth room at 25°C undera 16-h-light/8-h-dark cycle. *S. cerevisiae* strain BY4741 (MATa his3Δ1 leu2Δ0 met15Δ0ura3Δ0) was purchased from Open Biosystems. Yeast strains pdc1Δ was from the YKO library (Openbiosystems). To create pdc5Δ yeast strain, the hygromycin resistance gene hphNTI was PCR-amplified from vector pFA6a–hphNT1 (Euroscarf) (Janke et al., 2004) with primers #7504 and primers#7505 and the PCR product was transformed into BY4741. To generate GAL1::PDC1 pdc5Δ and GAL1::HA-PDC2 yeast strains, the transformants with GAL1 promoter along with the nourseothricin resistance gene were PCR-amplified from plasmid pYM-N23 with primers#7508 and #7509 or from pYM-N24 with #7475 and #7476 and then transformed into pdc5Δ and BY4741 yeast strains, respectively. Yeast strain NMY51 was obtained from Dualsystems. Plasmids and their constructions are listed in S1 and S2 Tables and primers used are described in S3 Table.

4.2 Analysis of virus replication in yeast

To determine the effect of Pdc1 on the replication of TBSV in yeast, BY4741, pdc5Δ and GAL1::PDC1 pdc5Δ strains were transformed with HpGBK-CUP1-Flagp33, LpGAD-CUP1-Flag92 and UpCM189-Tet-DI72. TBSV replication was induced by growing cells at 23°C in SC-ULH- (synthetic complete medium without uracil, leucine and histidine) medium supplemented with 2% galactose or 2% raffinose for 16h. Then, yeast cultures were resuspended in SC-ULH- medium supplemented with 50μM $CuSO_4$ and 2% galactose or 2% raffinose, and grown for 24h at 23°C.

To complement TBSV or CIRV replication with Pdc1 in pdc1Δ yeast strain, plasmids HpGBK-CUP1-Hisp33/Gal-DI72 and LpGAD-CUP1-Hisp92 or HpESC-CUP1-Flagp36/GalDI72 and LpESC-CUP1-Flagp95, respectively, were co-transformed with UpCM189-Tetempty or UpCM189-Tet-HisPdc1 or UpCM189-Tet-Pdc1 into yeast strain. To test if the enzymatic function of Pdc1 is required for TBSV replication, plasmids HpGBK-CUP1-Hisp33/Gal-DI72, LpGAD-CUP1-Hisp92 and UpCM189-Tet-Pdc1S455F were co-transformed intopdc1Δ yeast strain. Transformed yeast cells were pre-grown in 2ml SC-ULH- medium supplemented with 2% galactose and 100μM BCS for 16h at 23°C. Then, yeast cultures were re-suspended in SC-ULH$^-$ medium supplemented with 2% galactose and 50μM $CuSO_4$ and grown for 24h at 23°C.

4.3 Co–purification assay

To understand the dynamics of Pdc1 association with the viral replicase, transformed yeast cells were pre-grown in SC-ULH- medium supplemented with 2% glucose and 100μM BCS at 29°C for 16h. Then yeast cultures were centrifuged and the pellets were resuspended in SC-ULH- medium supplemented with 2% galactose and 100μM BCS and grown at 23°C for 24h, followed by culturing yeast cells in SC-ULH- medium supplemented with 2% galactose and 50μM CuSO4 at 23°C for 6h. Next, the yeast cells were shifted to SC-ULH- medium supplemented with 2% glucose and cycloheximide (100μg/mL) and samples were taken at 0, 1hand 2.5h time points. Yeast cultures were treated with formaldehyde and glycine and performed Flag-immunoaffinity purification as described below.

Co-purification assay from plants was performed by slight modifications of a previously described method (Jaag and Nagy, 2009). Briefly, *N. benthamiana* leaves were co-infiltrated with agrobacterium carrying pGD-HA-AtPdc1, pGD-GFP-HA, pGD-T33-Flag, pGD-p19 and pGD-empty. Then, samples were harvested at 2.5d post agroinfiltration and ground in cooled mortar in PPEB buffer (10% [v/v] glycerol, 25mM Tris-HCl, pH 7.5, 1mM EDTA, 150mM NaCl, 10mM DTT, 0.5% [v/v] Triton X-100 and protease inhibitor cocktail). The supernatant was incubated with anti-FLAG M2 affinity agarose (Sigma-Aldrich) using Bio-spin chromatography columns (Bio-rad) for 2h at 4°C on a rotator, followed by washing with the CP buffer (10% [v/v] glycerol, 25mM Tris-HCl, pH 7.5, 1mM EDTA, 150mM NaCl, 1mM DTT and 0.1% [v/v] TritonX-100). Elutions of the purified proteins were as described in the co-purification assay in yeast (Li et al., 2008).

4.4 Knockdown of NbPdc1 and NbAdh1 in *N. benthamiana* plants by VIGS

The VIGS-based knockdown of host genes in *N. benthamiana* was performed as described previously (Jaag and Nagy, 2009). To generate VIGS constructs TRV2-NbPdc1 and TRV2-NbAdh1, cDNA fragments were PCR-amplified with primers #5847/#5848 and #7911/#7912 from *N. benthamiana*c DNA preparations and inserted into the plasmid pTRV2 (Bachan and Dinesh-Kumar, 2012). At 12d after VIGS treatment of *N. benthamiana* (pTRV1 together with pTRV2-NbPdc1 or pTRV2-NbAdh1 orpTRV2-cGFP or TRV-LUC), the levels of *N. benthamiana* NbPdc1 and NbAdh1 mRNAs were determined by semi-quantitative RT-PCR. Then, the silenced leaves were either sap inoculated with TBSV, CIRV or TCV inocula or agroinfiltrated with pGD-$CNV^{20KSTOP}$ or pGD-CIRV, to launch virus replication. At different time points, samples from the inoculated and systemically-infected leaves were collected, followed by total RNA extraction and northern blot analysis as described previously (Jaag and Nagy, 2009). In case of BaMV, the VIGS-silenced leaves were sap-inoculated with BaMV inocula to launch viral replication. Then, the inoculated leaves werecollected at 2.5dpi, followed by total RNA extraction and quantitative real-time RT-PCR analysis as described previously (Lin et al., 2017). The *N. benthamiana* EF1α gene was used as an internal control to normalize the level of viral gene expression.

The VIGS-silenced NbPdc1 and NbAdh1 leaves of *N. benthamiana* were agroinfiltrated with pJL-36 vector carrying TMV cDNA and plant samples were collected 2d after infection from the inoculated leaves. The TMV RNA levels were measured by northern blot analysis.

4.5 Plant protoplasts preparation and viral RNA transfection

Protoplasts preparation from plant leaves was performed by some modifications of a previously described method (Yoo et al., 2007). Briefly, the NbPdc1-silenced or the mock-treated leaves were harvested at 12d post VIGS silencing. Then, the leaves were sliced into 0.5–1mm strips,digested with an enzyme solution containing 1.2% [w/v] Cellulase, 0.16% [w/v] Macerozyme,0.12% [w/v] BSA and 0.5 M mannitol. To improve the isolation of protoplasts, leaf strips were vacuum infiltrated for 20 min in the dark using Vacufuge Plus (Eppendorf) and furtherdigested in the dark for at least 3h at room temperature. The protoplasts preparations were passed through a sieve set (Scienceware Mini-Sieve Microsieve Set from Fisher cat# 14-306A)and collected by centrifugation at 900rpm for 2min, followed by washing once with the W5 solution (154mM NaCl, 125mM $CaCl_2$, 5mM KCl, 2mM MES pH 5.7) and re-suspending in the W5 solution. Then, 0.6 M sucrose was layered under the W5 solution with protoplasts and centrifuged at 900 rpm for 3 min. Protoplasts were transferred from the interface between the W5 solution and 0.6 M sucrose layers in the same amount of the W5 solution, followed by washing once with the W5 solution and re-suspending at $2 \times 10^5 ml^{-1}$ in the MMG solution (4mM MES pH5.7, 0.4M mannitol and 15mM $MgCl_2$). For RNA transfection, protoplasts were incubated with the PEG-calcium transfection solution containing 40% PEG 4000, 0.2M mannitol and 100mM $CaCl_2$ and either viral RNA transcripts or total RNA extracts obtained from virus-infected plants at room temperature for up to 15min. The transfection mixtures werediluted with the W5 solution and centrifuged at 100g for 2min at room temperature and incubated in the WI solution (4mM MES, pH 5.7, 0.5M mannitol and 20mM KCl). The protoplasts were harvested at 16h or 24h post-transfection and subjected to RNA extraction and northern blot analysis as mentioned above.

4.6 Visualization and measurement of ATP levels in yeast and plants

To visualize ATP production within the TBSV replication compartments in yeast, the previously adapted ATeam-based biosensor LpGAD-ADH-ATeam^{YEMK}-p92 (high sensitivity) and LpGAD-ADH-ATeam^{RK}-p92 (low sensitivity) were utilized (Imamura et al., 2009; Chuang et al., 2017). pdc1Δ yeast strain wasco-transformed with HpGBK-CUP1-Hisp33 and UpCM189-Tet-HisPdc1 or UpCM189-TetHisPdc1^{S455F} or UpCM189-Tet. The transformed yeast cells were pre-grown in 2ml SC-ULH$^-$medium supplemented with 2% glucose and 100μM BCS for 16h at 23°C. Then, the yeast cultures were re-suspended in SC-ULH- medium supplemented with 2% glucose and 50μM $CuSO_4$ and grown for 3h at 23°C. Then samples were collected for confocal laser microscopy analysis. FRET values (YFP/CFP ratio) were obtained based on the quantification of CFP and Venus images using ImageJ software and calculation using Microsoft Excel as described (Chuang et al., 2017).

To measure the ATP level within the tombusvirus replication compartments in the NbPdc1- or the NbAdh1-silenced *N. benthamiana* leaves, the previously adapted ATeam based biosensor pGD-p33-ATeam^{YEMK} or pGD-p36-ATeam^{YEMK}(Chuang et al., 2017) were transformed into agrobacterium strain C58C1. In case of TBSV, the silenced or the control leaves of *N.benthamiana* (at 12d after VIGS treatment using pTRV1 in combination with pTRV2-NbPdc1 or pTRV2-NbAdh1 or pTRV2-cGFP control) were co-agroinfiltrated with plasmids pGD-p33-ATeam^{YEMK} with or without pGD-p92 and pGD-DI72. In case of CNV, the silenced or the control leaves were co-agroinfiltrated with plasmids pGD-p33-ATeam^{YEMK} with or without pGD-CNV^{20KSTOP}. In case of CIRV, the silenced or control leaves were coagroinfiltrated with plasmids pGD-p36-ATeam^{YEMK} with or without pGD-p95 and pGD-DI72. Then samples were harvested at 1.5d after agroinfiltration for the confocal microscopy analysis. FRET values (YFP/CFP ratio) were obtained based on the quantification of CFP and Venus images using ImageJ software and calculation using Microsoft Excel (Chuang et al., 2017).

4.7 Yeast cell free extract (CFE)–based *in vitro* replication assay

CFEs from BY4741, pdc5Δ and GAL1::PDC1 pdc5Δ yeast strains were prepared as described earlier (Pogany and Nagy, 2008; Pogany et al., 2008). These yeast stains were pre-grown in YPD or YPG media at 29°C for 16h. Then, the yeast cultures were diluted (to 0.4 OD_{600}) with fresh YPD or YPG media and grown at 29°C for 5h, followed by 37°C treatment for 30 min. The individual CFE preparations were made following the published protocol (Pogany and Nagy, 2008; Pogany et al., 2008) and adjusted to contain comparable amounts of total proteins. The *in vitro* CFE assay was performed in 20μl total volume containing 2μl of adjusted CFE, 0.5 μg DI-72 (+)

RNA transcripts, 0.5μg affinity-purified MBP-p33, 0.5μg affinity-purified MBP-p92 (both recombinant proteins were obtained from E. coli) (Rajendran and Nagy, 2003, 2006), 30 mM HEPES-KOH, pH 7.4, 150mM potassium acetate, 5mM magnesium acetate, 0.13M sorbitol, 0.2μl actinomycin D (5mg/ml), 2μl of 150mM creatine phosphate, 0.2μL of 10mg/ml creatine kinase, 0.2μl of RNase inhibitor, 0.2μl of 1 M dithiothreitol (DTT), 2μl of 10mM ATP, CTP, and GTP and 0.1mM UTP and 0.2μl of ^{32}P-UTP. Reaction mixtures were incubated for 3h at 25°C, followed by phenol/chloroform extraction and isopropanol/ammonium acetate(10:1) precipitation. The ^{32}P-UTP-labeled RNA products were analyzed in 5% acrylamide/8 Murea gels (Pogany and Nagy, 2008; Pogany et al., 2008). Additional methods used are described in S1 text.

Acknowledgements

The authors thank Dr. Chingkai Chuang and Judit Pogany and Ms. Paulina Alatriste for valuable suggestions. The BaMV-S isolate was kindly provided by Prof. Yau-Heiu Hsu from National Chung Hsing University, Taichung, Taiwan.

References

[1] ALTAN-BONNET N. Lipid tales of viral replication and transmission[J]. Trends in cell biology, 2017, 27(3): 201-213.

[2] ANDINO R. RNAi puts a lid on virus replication[J]. Nature biotechnology, 2003, 21(6), 629-630.

[3] BACHAN S, DINESH-KUMAR S P. Tobacco rattle virus (TRV)-based virus-induced gene silencing[J]. Methods in Molecular Biology, 2012, 894(894):83-92.

[4] BARAJAS D, XU K, SHARMA M, et al. Tombusviruses upregulate phospholipid biosynthesis via interaction between p33 replication protein and yeast lipid sensor proteins during virus replication in yeast[J]. Virology,2014a, 471: 72-80.

[5] BARAJAS D, DE CASTRO MARTIN I F, POGANY J, et al. Noncanonical role for the host Vps4 AAA+ ATPase ESCRT protein in the formation of Tomato bushy stunt virus replicase[J]. PLoS Pathog,2014b, 10: e1004087.

[6] BELOV G, VAN KUPPEVELD F. (+) RNA viruses rewire cellular pathways to build replication[J]. Current Opinion in Virology,2012, 2(6): 1.

[7] BERTRAND E, CHARTRAND P, SCHAEFER M, et al. Localization of ASH1 mRNA particles in living yeast[J]. Molecular Cell,1998, 2(4): 437-445.

[8] CARBONELL A, CARRINGTON J C. Antiviral roles of plant ARGONAUTES[J]. Current Opinion in Plant Biology,2015, 27: 111-117.

[9] CHUANG C, PRASANTH K R, NAGY P D. Coordinated function of cellular DEAD-box helicases in suppression of viral RNA recombination and maintenance of viral genome integrity[J]. PLoS Pathogens,2015, 11: e1004680.

[10] CHUANG C, PRASANTH K R, NAGY P D. The glycolytic pyruvate kinase is recruited directly into the viral replicase complex to generate ATP for RNA synthesis[J]. Cell Host Microbe,2017, 22: 639-652.

[11] DE CASTRO I F, VOLONTÉ L, RISCO C. Virus factories: biogenesis and structural design[J]. Cell Microbiology,2013, 15: 24-34.

[12] DE CASTRO I F, FERNÁNDEZ J J, BARAJAS D, et al. Three-dimensional imaging of the intracellular assembly of a functional viral RNA replicase complex[J]. Journal of Cell Science,2017, 130: 260-268.

[13] DEN BOON J A, AHLQUIST P. Organelle-like membrane compartmentalization of positive-strand RNA virus replication factories[J]. Annual Review of Microbiology,2010, 64(1): 241-256.

[14] DIAMOND D L, SYDER A J, JACOBS J M, et al. Temporal proteome and lipidome profiles reveal hepatitis C virus-associated reprogramming of hepatocellular metabolism and bioenergetics[J]. PLoS Pathogens,2010, 6: e1000719.

[15] EBERHARDT I, CEDERBERG H, LI H, et al. Autoregulation of yeast pyruvate decarboxylase gene expression requires the

enzyme but not its catalytic activity[J]. European Journal of Biochemistry,1999, 262(1): 191-201.

[16] HEATON N S, RANDALL G. Dengue virus-induced autophagy regulates lipid metabolism[J].Cell host microbe,2010, 8: 422-432.

[17] HEATON N S, PERERA R, BERGER K L, et al. Dengue virus nonstructural protein 3 redistributes fatty acid synthase to sites of viral replication and increases cellular fatty acid synthesis[J].Proceedings of the National Academy of Sciences of the United States of America, 2010, 107(40):17345-17350.

[18] HOHMANN S. Characterisation of PDC2, a gene necessary for high level expression of pyruvate decarboxylase structural genes in Saccharomyces cerevlsiae[J]. Molecular and general genetics,1993, 241: 657-666.

[19] HSU N-Y, ILNYTSKA O, BELOV G, et al. Viral reorganization of the secretory pathway generates distinct organelles for RNA replication[J]. Cell,2010, 141: 799-811.

[20] HUANG T-S, NAGY P D. Direct inhibition of tombusvirus plus-strand RNA synthesis by a dominant negative mutant of a host metabolic enzyme, glyceraldehyde-3-phosphate dehydrogenase, in yeast and plants[J]. Journal of virology,2011, 85: 9090-9102.

[21] IMAMURA H, NHAT K P H, TOGAWA H, et al. Visualization of ATP levels inside single living cells with fluorescence resonance energy transfer-based genetically encoded indicators[J]. Proceedings of the National Academy of Sciences of the United States of America, 2009, 106: 15651-15656.

[22] INABA J-I, NAGY P D. Tombusvirus RNA replication depends on the TOR pathway in yeast and plants[J]. Virology,2018, 519: 207-222.

[23] ISMOND K P, DOLFERUS R, DE PAUW M, et al. Enhanced low oxygen survival in Arabidopsis through increased metabolic flux in the fermentative pathway[J]. Plant Physiology, 2003, 132: 1292-1302.

[24] JAAG H M, NAGY P D. Silencing of Nicotiana benthamiana Xrn4p exoribonuclease promotes tombusvirus RNA accumulation and recombination[J]. Virology,2009, 386: 344-352.

[25] JANKE C, MAGIERA M M, RATHFELDER N, et al. A versatile toolbox for PCR-based tagging of yeast genes: new fluorescent proteins, more markers and promoter substitution cassettes[J]. Yeast,2004, 21(11): 947-962.

[26] JONES W, BIANCHI K. Aerobic glycolysis: beyond proliferation[J]. Frontiers in Immunology,2015, 6: 227.

[27] KOPEK B G, PERKINS G, MILLER D J, et al. Three-dimensional analysis of a viral RNA replication complex reveals a virus-induced mini-organelle[J]. PLoS biology, 2007, 5(9): e220.

[28] KOVALEV N, POGANY J, NAGY P D. A co-opted DEAD-box RNA helicase enhances tombusvirus plus-strand synthesis[J]. PLoS Pathogens,2012, 8: e1002537.

[29] Kovalev N, Pogany J, Nagy P D. Template role of double-stranded RNA in tombusvirus replication[J].Journal of Virology, 2014, 88(10):5638-5651.

[30] KOVALEV N, INABA J-I, LI Z, et al. The role of co-opted ESCRT proteins and lipid factors in protection of tombusviral double-stranded RNA replication intermediate against reconstituted RNAi in yeast[J]. PLoS Pathogens,2017, 13: e1006520.

[31] LI Z, BARAJAS D, PANAVAS T, et al. Cdc34p ubiquitin-conjugating enzyme is a component of the tombusvirus replicase complex and ubiquitinates p33 replication protein[J].Journal of Virology, 2008, 82(14):6911-6926.

[32] LIN W, WANG L, YAN W, et al. Identification and characterization of Bamboo mosaic virus isolates from a naturally occurring coinfection in Bambusa xiashanensis[J].Archives of Virology, 2017, 162(5):1335-1339.

[33] LUNT S Y, VANDER HEIDEN M G. Aerobic glycolysis: meeting the metabolic requirements of cell proliferation[J].Annual Review of Cell and Developmental Biology, 2011, 27(1):441-464.

[34] MITHRAN M, PAPARELLI E, NOVI G, et al. Analysis of the role of the pyruvate decarboxylase gene family in A rabidopsis thaliana under low-oxygen conditions[J]. Plant Biology,2014, 16: 28-34.

[35] NAGY P D. Viral sensing of the subcellular environment regulates the assembly of new viral replicase complexes during the course of infection[J].Journal of Virology, 2015, 89(10): 5196-5199.

[36] NAGY P D. Tombusvirus-host interactions: co-opted evolutionarily conserved host factors take center court[J]. Annual Review of

Virology,2016, 3: 491-515.

[37] NAGY P D. Exploitation of a surrogate host Saccharomyces cerevisiae, to identify cellular targets and develop novel antiviral approaches[J].Current Opinion in Virology, 2017, 26:132-140.

[38] NAGY P D, POGANY J. The dependence of viral RNA replication on co-opted host factors[J].Nature Reviews Microbiology,2012, 10: 137.

[39] NAGY P D, POGANY J, LIN J-Y. How yeast can be used as a genetic platform to explore virus–host interactions: from 'omics' to functional studies[J].Trends in Microbiology, 2014, 22(6):309-316.

[40] NAGY P D, STRATING J R, VAN KUPPEVELD F J. Building viral replication organelles: close encounters of the membrane types[J].PLoS Pathogens, 2016, 12(10):e1005912.

[41] OLSON K A, SCHELL J C, RUTTER J. Pyruvate and metabolic flexibility: illuminating a path toward selective cancer therapies[J]. Trends in Biochemical sciences,2016, 41: 219-230.

[42] PALM W, THOMPSON C B. Nutrient acquisition strategies of mammalian cells[J]. Nature,2017, 546: 234-242.

[43] PANAVAS T, HAWKINS C M, PANAVIENE Z, et al. The role of the p33: p33/p92 interaction domain in RNA replication and intracellular localization of p33 and p92 proteins of Cucumber necrosis tombusvirus[J]. Virology,2005, 338: 81-95.

[44] PERERA R, RILEY C, ISAAC G, et al. Dengue virus infection perturbs lipid homeostasis in infected mosquito cells[J]. PLoS Pathogens,2012, 8: e1002584.

[45] POGANY J, NAGY P D. Authentic replication and recombination of Tomato bushy stunt virus RNA in a cell-free extract from yeast[J].Journal of Virology, 2008, 82(12):5967-5980.

[46] POGANY J, NAGY P D. p33-independent activation of a truncated p92 RNA-dependent RNA polymerase of tomato bushy stunt virus in yeast cell-free extract[J].Journal of Virology, 2012, 86(22):12025.

[47] POGANY J, NAGY P D. Activation of Tomato bushy stunt virus RNA-dependent RNA polymerase by cellular heat shock protein 70 is enhanced by phospholipids in vitro[J].Journal of Virology, 2015, 89(10): 5714-5723.

[48] POGANY J, WHITE K A, NAGY P D. Specific binding of tombusvirus replication protein p33 to an internal replication element in the viral RNA is essential for replication[J].Journal of Virology, 2005, 79(8):4859-4869.

[49] POGANY J, STORK J, LI Z, et al. In vitro assembly of the Tomato bushy stunt virus replicase requires the host Heat shock protein 70[J].Proceedings of the National Academy of Sciences of the United States of America, 2008, 105(50): 19956-19961.

[50] PRASANTH K R, CHUANG C, NAGY P D. Co-opting ATP-generating glycolytic enzyme PGK1 phosphoglycerate kinase facilitates the assembly of viral replicase complexes[J]. PLoS Pathogens,2017, 13: e1006689.

[51] RAJENDRAN K, NAGY P D. Characterization of the RNA-binding domains in the replicase proteins of tomato bushy stunt virus[J].Journal of Virology, 2003, 77(17):9244-9258.

[52] RAJENDRAN K, NAGY P D. Kinetics and functional studies on interaction between the replicase proteins of Tomato Bushy Stunt Virus: requirement of p33: p92 interaction for replicase assembly[J]. Virology,2006, 345(1): 270-279.

[53] SCHOGGINS J W, RANDALL G. Lipids in innate antiviral defense[J].Cell Host Microbe, 2013, 14(4):379-385.

[54] SHANNON B J, VAISHNAVI S N, VLASSENKO A G, et al. Brain aerobic glycolysis and motor adaptation learning[J]. Proceedings of the National Academy of Sciences of the United States of America,2016. 113(26): E3782-E3791.

[55] SHULLA A, RANDALL G. (+) RNA virus replication compartments: a safe home for (most) viral replication[J]. Current Opinion in Microbiology, 2016, 32:82-88.

[56] SNIDER J, KITTANAKOM S, CURAK J, et al. Split-ubiquitin based membrane yeast two-hybrid (MYTH) system: a powerful tool for identifying protein-protein interactions[J].Journal of Visualized Experiments, 2010, 36(36): e1698.

[57] SYED G H, AMAKO Y, SIDDIQUI A. Hepatitis C virus hijacks host lipid metabolism[J]. Trends Endocrinol Metab, 2010, 21(1):33-40.

[58] VAISHNAVI S N, VLASSENKO A G, RUNDLE M M, et al. Regional aerobic glycolysis in the human brain[J]. Proceedings of the National Academy of Sciences of the United States of America,2010, 107: 17757-17762.

[59] VAN DER SCHAAR H M, DOROBANTU C M, ALBULESCU L, et al. Fat (al) attraction: picornaviruses usurp lipid transfer at membrane contact sites to create replication organelles[J].Trends in Microbiology, 2016, 24(7):535-546.

[60] VANDER HEIDEN M G, CANTLEY L C, THOMPSON C B. Understanding the Warburg effect: the metabolic requirements of cell proliferation[J]. Science,2009, 324: 1029-1033.

[61] VLASSENKO A G, VAISHNAVI S N, COUTURE L, et al. Spatial correlation between brain aerobic glycolysis and amyloid-β (Aβ) deposition[J]. Proceedings of the National Academy of Sciences of the United States of America,2010, 107: 17763-17767.

[62] WANG A. Dissecting the molecular network of virus-plant interactions: the complex roles of host factors[J].Annual Review of Phytopathology, 2015, 53(1):150504162158003.

[63] WANG RY-L, NAGY P D. Tomato bushy stunt virus co-opts the RNA-binding function of a host metabolic enzyme for viral genomic RNA synthesis[J].Cell host microbe, 2008, 3(3):178-187.

[64] WANG RY-L, STORK J, NAGY P D. A key role for heat shock protein 70 in the localization and insertion of tombusvirus replication proteins to intracellular membranes[J].Journal of Virology, 2009, 83(7):3276.

[65] WANG RY-L, STORK J, POGANY J, et al. A temperature sensitive mutant of heat shock protein 70 reveals an essential role during the early steps of tombusvirus replication[J]. Virology,2009b, 394: 28-38.

[66] WANG X, DIAZ A, HAO L, et al. Intersection of the multivesicular body pathway and lipid homeostasis in RNA replication by a positive-strand RNA virus[J].Journal of Virology, 2011, 85(11):5494-5503.

[67] XU K, NAGY P D. Expanding use of multi-origin subcellular membranes by positive-strand RNA viruses during replication[J]. Current Opinion in Virology,2014, 9: 119-126.

[68] XU K, NAGY P D. RNA virus replication depends on enrichment of phosphatidylethanolamine at replication sites in subcellular membranes[J].Proceedings of the National Academy of Sciences of the United States of America, 2015, 112(14):1782-91.

[69] XU K, NAGY P D. Enrichment of phosphatidylethanolamine in viral replication compartments via co-opting the endosomal Rab5 small GTPase by a positive-strand RNA virus[J]. PLoS Biology,2016, 14: e2000128.

[70] XU K, NAGY P D. Sterol binding by the tombusviral replication proteins is essential for replication in yeast and plants[J].Journal of Virology, 2017, 91(7):JVI.01984-16.

[71] YOO S-D, CHO Y-H, SHEEN J. Arabidopsis mesophyll protoplasts: a versatile cell system for transient gene expression analysis[J].Nature Protocols, 2007, 2(7):1565-1572.

[72] YU L, CHEN X, WANG L, et al. Oncogenic virus-induced aerobic glycolysis and tumorigenesis[J].Journal of Cancer, 2018, 9(20):3699-3706.

An engineered mutant of a host phospholipid synthesis gene inhibits viral replication without compromising host fitness

Guijuan He[1, 2], Zhenlu Zhang[1, 2, 3], Preethi Sathanantham[2], Xin Zhang[2], Zujian Wu[1], Lianhui Xie[1], Xiaofeng Wang[2]

(1 Fujian Province Key Laboratory of Plant Virology, Institute of Plant Virology, Fujian Agriculture and Forestry University, Fuzhou, Fujian 350002, P. R. China; 2 School of Plant and Environmental Sciences, Virginia Tech, Blacksburg, VA 24061, USA; 3 National Key Laboratory of Crop Biology, National Research Center for Apple Engineering and Technology, College of Horticulture Science and Engineering, Shandong Agricultural University, Tai-An, Shandong 271018, P.R. China)

Abstract:Viral infections universally rely on numerous hijacked host factors to be successful. It is therefore possible to control viral infections by manipulating host factors that are critical for viral replication. Given that host genes may play essential roles in certain cellular processes, any successful manipulations for virus control should cause no or mild effects on host fitness. We previously showed that a group of positive-strand RNA viruses enrich phosphatidylcholine (PC) at the sites of viral replication. Specifically, brome mosaic virus (BMV) replication protein 1a interacts with and recruits a PC synthesis enzyme phosphatidylethanolamine methyltransferase, Cho2p, to the viral replication sites that are assembled on the perinuclear ER membrane. Deletion of the *CHO2* gene inhibited BMV replication by 5-fold, however, slowed down cell growth as well. Here, we show that an engineered Cho2p mutant supports general PC synthesis and normal cell growth but blocks BMV replication. This mutant interacts and co-localizes with BMV 1a but prevents BMV 1a from localizing to the perinuclear ER membrane. The mislocalized BMV 1a fails to induce the formation of viral replication complexes. Our study demonstrates an effective antiviral strategy in which a host lipid synthesis gene is engineered to control viral replication without comprising host growth.

Keywords:Viroid; *Coleus blumei*; Infectious clone; Host

1 Introduction

Plant viruses are a major threat to stable agronomic production and cause multibillion-dollar losses each year. The planting of virus-resistant cultivars and application of insecticides to control virus-transmitting insects are common practices for controlling viral diseases in the field. However, pesticides cause pollution and leave chemical residue and become ineffective when insects develop resistance (Pimentel, 2005). Breeding for crop cultivars with broad-spectrum and stable viral resistance is, therefore, crucial. Conventional breeding has incorporated many available resistance (*R*) genes into elite cultivars to control viral diseases (Dangl and Jones, 2001; Kang et al., 2005; Maule et al., 2007; Galvez et al., 2014). *R* genes can be dominant or recessive. Dominant resistance is governed by a specific interaction between an *R* gene and a corresponding avirulence (*Avr*) gene from a pathogen to elicit a hypersensitive response (Kang et al., 2005; Martin et al., 2003).

Journal of Biological Chemistry. 2019, 118: 7-51

Received 6 December 2018; Accepted 30 July 2019; Available online 13 August 2019

However, there are several issues related to dominant resistance. One is that high mutation rates during viral infection can generate *Avr* gene derivatives that escape the recognition and thus, overcoming the resistance. This is particularly true for RNA viruses whose RNA-dependent RNA polymerases (RdRp) lack proofreading activity. Another issue is strain specificity, although *R* gene products can recognize *Avr* gene products from several virus strains, they are unable to control many other strains (Galvez et al., 2014).

Recessive resistance is usually conferred by a mutated form of a host gene whose wild-type (wt) form encodes a protein that is critical for viral infection (Hashimoto et al., 2016; Truniger and Aranda, 2009). Mutants of these genes do not support viral infection but cause no, or only mild, growth phenotypes in virus-resistant plants as the mutants retain their ability to support normal cellular processes. This resistance mechanism is also likely to work in a wide-range of crops. For instance, mutated forms of either eIF4E (eukaryotic translation initiation factor 4E) or eIF4(iso)E have been reported as a recessive *R* gene in multiple crops, including barley, lettuce, melon, pepper, pea and tomato [reviewed in (Wang and Krishnaswamy, 2012)]. Additionally, recessive *R* gene-mediated resistance is durable because the mutation rate of host DNA-dependent DNA polymerase is much lower than viral RdRp (Galvez et al., 2014).

Being the largest among seven viral classes, positive-strand RNA viruses [(+)RNA viruses] include numerous pathogens that infect and cause severe consequences in humans, animals, and plants. Human illnesses-causing viruses include severe acute respiratory syndrome (SARS) coronavirus, Zika virus, Dengue virus (DENV), and foot-and-mouth disease virus, among many others. Importantly, the vast majority of plant viruses are (+) RNA viruses. All (+) RNA viruses have a small genome, encode a limited number of viral proteins, and rely heavily on host factors to achieve a successful infection (Wang, 2015; Nagy, 2016; Diaz and Wang, 2014; Belov and van Kuppeveld, 2012; Belov, 2014). Numerous host factors involved in the replication of hepatitis C virus (HCV) (Belov and van Kuppeveld, 2012; Randall et al., 2007; Paul et al., 2014), picornavirus (Belov, 2014), brome mosaic virus (BMV) (Diaz and Wang, 2014; Kushner et al., 2003), cucumber mosaic virus (Guo et al., 2017), turnip mosaic virus (Wang, 2015; Li et al., 2018; Jiang et al., 2015), and tomato bushy stunt virus (TBSV) have been identified and characterized (Nagy, 2016). In theory, loss or mutation of host genes required for viral replication should provide recessive resistance to (+) RNA viruses. It is thus, crucial to identify, characterize, and manipulate host genes required for viral replication to develop virus-resistant hosts.

BMV, which serves as a model system to study (+) RNA viruses, is the type member of the *Bromoviridae* family, and a representative member of the alphavirus-like superfamily (Wang and Ahlquist, 2008). The genome of BMV is composed of RNA 1, 2, and 3. RNA1 and 2 encode replication proteins 1a and 2a polymerase ($2a^{pol}$), respectively (Wang and Ahlquist, 2008). RNA3 encodes the movement protein 3a and the coat protein (CP). BMV 1a and $2a^{pol}$ are necessary and sufficient for its replication in barley, its natural host, and in an alternative host, the baker's yeast *Saccharomyces cerevisiae* (Wang and Ahlquist, 2008). BMV replication in yeast cells duplicates nearly all major features of its replication in plant cells (Wang and Ahlquist, 2008), and enables systematic screenings of host genes that are involved in viral infection (Kushner et al., 2003). About 100 host genes have been identified to be involved in BMV replication (Kushner et al., 2003), including a group of genes involved in lipid metabolism (Zhang et al., 2012; Lee et al., 2001; Lee and Ahlquist, 2003; Zhang et al., 2016; Zhang et al., 2018). This is consistent with the fact that (+)RNA viruses assemble their viral replication complexes (VRCs) in a tight association with remodeled host intracellular membranes. For example, deleting the *ACB1* gene, which encodes acyl-CoA binding protein that is involved in the trafficking of long chain acyl-CoA esters among organelles (Du et al., 2016; Neess et al., 2015), affects host lipid homeostasis and results in a more than 10-fold decrease in BMV genome replication (Zhang et al., 2012). Additionally, a single point mutation in *OLE1,* which encodes a cellular Δ9 fatty acid desaturase that converts saturated FA (SFA) to unsaturated FA (UFA), causes a mild reduction in host UFA levels but inhibits BMV RNA replication by 20-fold (Lee et al., 2001; Lee and Ahlquist, 2003). On the contrary, deleting the *PAH1* gene, which encodes for Pah1p (phosphatidic acid phosphohydrolase) and is the sole yeast ortholog of human *LIPIN* genes, results in a 3-fold increase of BMV replication (Zhang et al., 2018). The requirement of and high sensitivity of viral replication to host lipid composition is a common feature shared by numerous other (+)RNA viruses, including HCV (Hsu et al., 2010; Berger et al., 2009), West Nile virus (WNV) (Martín-Acebes et al., 2011), red clover necrotic virus (RCNV) (Hyodo et al., 2015), and TBSV (Xu and Nagy, 2016; Sharma et al., 2011; Barajas et al., 2014). This highly conserved feature of viral genome replication raises the possibility of developing a novel, durable, and broad-spectrum strategy to control (+) RNA viruses by manipulating host genes involved in lipid metabolism.

Phospholipids are major components of cellular membranes. Phosphatidylcholine (PC) in particular is important as it accounts up to 50% of the total phospholipids in cells (Li and Vance, 2008). A significant increase in PC levels is associated with the replication of multiple (+) RNA viruses, DENV (Perera et al., 2012), flock house virus (Castorena et al., 2010), and poliovirus among others (Vance et al., 1980). BMV also significantly increases PC content in yeast and barley cells (Zhang et al., 2016). It was further shown that PC was specifically enriched at the sites of BMV replication (Zhang et al., 2016). PC enrichment at the sites of viral replication was also present in HCV- and poliovirus-infected cells, indicating a conserved feature among a group of (+) RNA viruses (Zhang et al., 2016; Banerjee et al., 2018). In yeast cells, Cho2p (choline requiring 2), a phosphatidylethanolamine (PE) methyltransferase (PEMT) that converts PE to mono-methyl PE (PMME) (Henry et al., 2012), is recruited to viral replication sites by an interaction with BMV 1a (Zhang et al., 2016). The increase in PC levels and the recruitment of Cho2p to the viral replication sites suggest that the enhanced PC content is synthesized at the sites of BMV replication. Moreover, disrupting PC synthesis by deleting the *CHO2* gene results in a decrease of up to ~80% of BMV RNA synthesis (Zhang et al., 2016), highlighting the critical role of PC in viral replication. However, deleting *CHO2* also disrupts general PC synthesis and thus, affects host cell growth (Summers et al., 1988), preventing the deletion of *CHO2* from being a promising antiviral strategy.

We now report the engineered manipulation of *CHO2* that leads to an inhibition of BMV RNA replication without affecting cell fitness. Specifically, we introduced substitutions in *CHO2* to make the *Cho2p-aia* mutant, in which glycine102 and glycine104 were replaced by an alanine residue. The *Cho2p-aia* functions as well as wt Cho2p in supporting host cell growth and general PC synthesis but does not support BMV RNA replication. We further demonstrate that the *Cho2p-aia* mutant disrupts the proper localization of BMV 1a and the formation of VRCs. Our data provide a proof of concept for a novel antiviral strategy by manipulating a host lipid metabolism gene to mistarget the viral replication protein away from the proper viral replication sites and prevent the formation of VRCs.

2 Results

2.1 The *Cho2p-aia* mutant complements the defect of yeast cell growth in *cho2*Δ cells

The biosynthesis of PC in yeast cells in the absence of exogenous free choline proceeds via the CDP-DAG (Cytidine diphosphodiacylglycerol) pathway. Cho2p is a key enzyme in the CDP-DAG pathway where it catalyzes the addition of a methyl moiety to PE to produce PMME (Henry et al., 2012) (Fig. 1A). PMME is further converted to PC by Opi3p (overproducer of inositol 3) (Fig. 1A), also known as phospholipid methyltransferase (Henry et al., 2012). Even though deleting *CHO2* inhibits BMV RNA synthesis substantially (Zhang et al., 2016), it also affects host cell growth (Zhang et al., 2016; Summers et al., 1988). We set out to identify *CHO2* mutants that disrupt viral replication without affecting host cell growth. To this end, we made various deletions based on structural analysis of Cho2p by using Pfam and Protein Homology/analogy recognition engine V 2.0 (PHYRE2) programs (El-Gebali et al., 2018; Kelley et al., 2015). Cho2p has two PEMT domains and a SKIP carboxy homology (SKICH) domain (Fig. 1B). We found that the N-terminal 466 amino acids or 365 amino acids of Cho2p (F1 & F2 fragments, Fig. 1B), without the second PEMT domain and SKITCH domain, was sufficient to support wt level of cell growth. Deleting the first 108 amino acids, however, abolished the ability of F1 to support wt-level cell growth (F3 fragment, Fig. 1B). Searching the first 108 amino acids, a putative GXG motif (GIG, amino acids 102-104) was identified. The GXG motif was reported to play a critical role in binding the methyl donor, Nadenosylmethionine, to PEMT in human cells (Shields et al., 2003). We constructed a Cho2p mutant with two alanine substitutions for glycine 102 and 104, referred as the *Cho2p-aia* mutant (*Cho2p-aia*, Fig. 1C).

In synthetic defined medium containing galactose as the carbon source without supplemented choline, we found that the doubling time of wt cells was ~4.7h/generation (Fig. 1D). The doubling time of *cho2Δ* cells with the expression of a negative control, an ER-resident protein Dpm1p (dolichol phosphate mannose synthase), was 6.5h/per generation, which was significantly slower than that of wt cells (Fig. 1D, labeled as NC). The presence of HA-tagged wt Cho2p, nevetherless, improved cell growth and led to a doublting time of 5.3h/generation (Fig. 1D). The presence of *Cho2p-aia*-HA improved cell growth of *cho2Δ* cells. The doubling time became ~4.8h/generation, which was similar to that of wt cells. We also confirmed that both HA-tagged wt Cho2p and *Cho2p-aia* accumulated to similar levels as determined via Western blotting (Fig. 1D). Pgk1p (phosphoglycerate kinase) was used as a loading control to show the equal loading of total proteins. We concluded that *Cho2p-aia* functions as well as wt Cho2p in supporting cell growth (Fig. 1D).

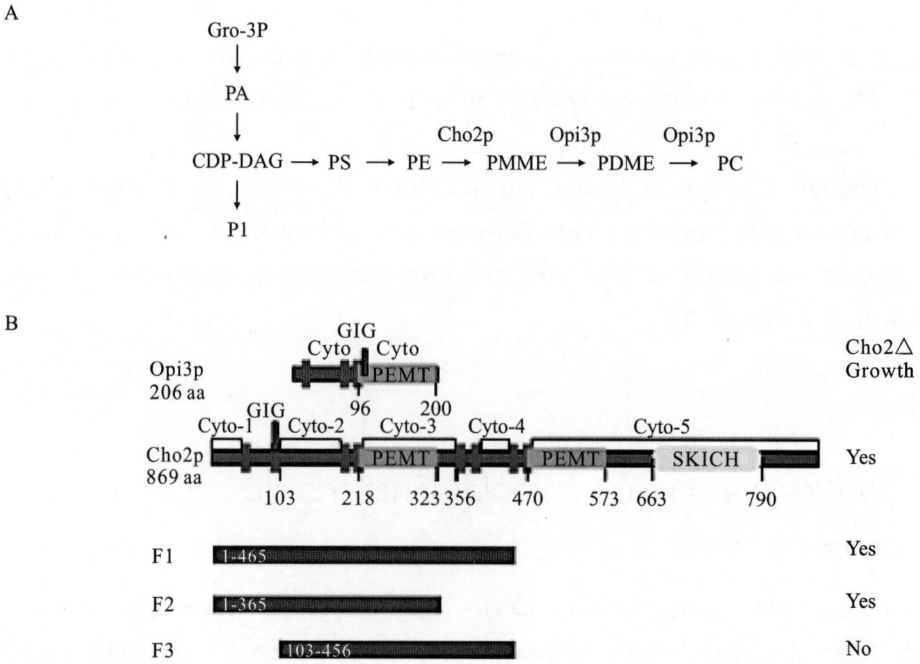

C

	AAT	TTA	GGT	ATT	GGA	CTG	GTT
Cho2p	N	L	G	I	G	L	V
	AAT	TTA	GCT	ATT	GCA	CTG	GTT
Cho2p-aia	N	L	A	I	A	L	V
			102		104		

D

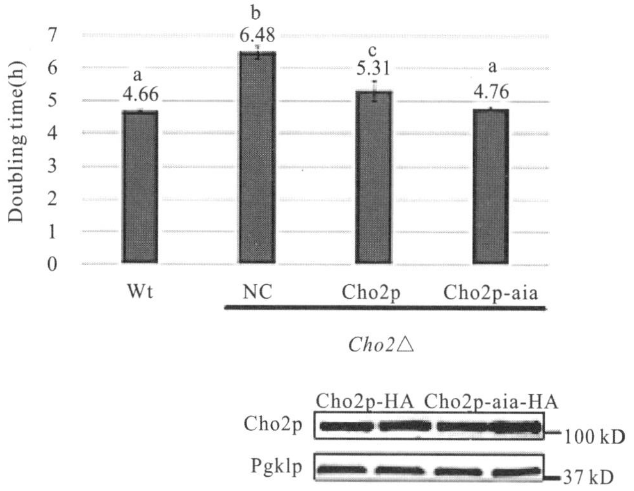

Fig. 1 The *Cho2p-aia* mutant complements the cell growth defect in *cho2Δ* cells

(A) The CDP-DAP pathway for PC biosynthesis in yeast. Cho2p, a PE methyltrasferase, catalyzes the addition of a methyl moiety to PE to produce PMME. PMME is subsequently methylated by Opi3p, a phospholipid N-methyltransferase, to produce PC. (B) The schematic of Opi3p, Cho2p, and various Cho2p deletion mutants. PEMT: PE *N*-methyltransferase. SKICH: skeletal muscle and kidney-enriched inositol phosphatase carboxyl homology. (C) Nucleic acid and amino acid sequences of aa 100-106 of wt Cho2p and *Cho2p-aia* mutant. Mutated sites are highlighted in yellow. (D) The doubling time (hours/generation) of wt cells, or *cho2Δ* cells with the expressed Dpm1p, a negative control (NC), the wt Cho2p, or the *Cho2p-aia* mutant. All cells were grown in synthetic defined medium with galactose as the carbon source in the absence of free choline. According to the one-way ANOVA test, means do not differ significantly ($P>0.05$) if they are indicated with the same letter, and error bars represent standard deviation. Western blot showing the accumulation of HA-tagged wt Cho2p or *Cho2p-aia* in *cho2Δ* cells. Pgk1p served as a loading control

2.2 The *Cho2p-aia* mutant largely restores the PC synthesis in *cho2Δ* cells

The restored growth rate of *cho2Δ* cells in the presence of the *Cho2p-aia* mutant suggests that *Cho2p-aia* may function as well as wt Cho2p in supporting PC synthesis. To confirm this assumption, total lipids were extracted from wt and mutant cells, and subjected to mass spectrometry. We quantified phospholipid species by comparing to a defined amount of phospholipid standards and reported the mol percentage of each of phospholipid classes in Fig. 2. In *cho2Δ* cells with the expressed negative control Dpm1p (labeled as NC in Fig. 2), there was a two-fold decrease in PC levels compared to wt cells (27% compared to 55%). Expressing *CHO2-aia* in *cho2Δ* cells (*cho2Δ* + *Cho2p-aia*) increased the percentage of PC levels to 46%, which was significantly higher than that of negative control ($P<0.05$) and not statistically different from that in wt cells. The levels (21%) of phosphatidylinositol (PI) in *cho2Δ* cells were siginificantly higher than those of wt (16%). The expressed *Cho2p-aia* largely restored the PI levels to 17% that is not significantly different from the wt levels (Fig. 2). For PE, which is the substrate of PC, deleting *CHO2* had a dramatic effect as PE levels increased to 42% of total PLs from the 16% in wt cells. Expressing *CHO2-aia* partially restored PE levels because it decreased PE levels to 27%, which was significantly lower from that measured in cells lacking *CHO2*, even though it was still higher than wt levels (Fig. 2, comparing NC and wt). Similarly, decreased levels of phosphatidylserine (PS) was not restored to wt levels when *Cho2p-aia* was expressed in *cho2*Δ cells. Moreover, there was a significant decrease in phosphatidic acid (PA) levels in the presence of *Cho2p-aia* over the wt levels. Reasons for the decrease of PA levels and the partial restoration of alterations in PE and PS levels by *Cho2p-aia* are not clear. Nevertheless, our data suggest that *Cho2p-aia* is able to largely restore general PC synthesis for cell growth. Our data also indicate that cell growth is not noticeably affected by the moderate differences in phospholipid composition (Fig. 1-2).

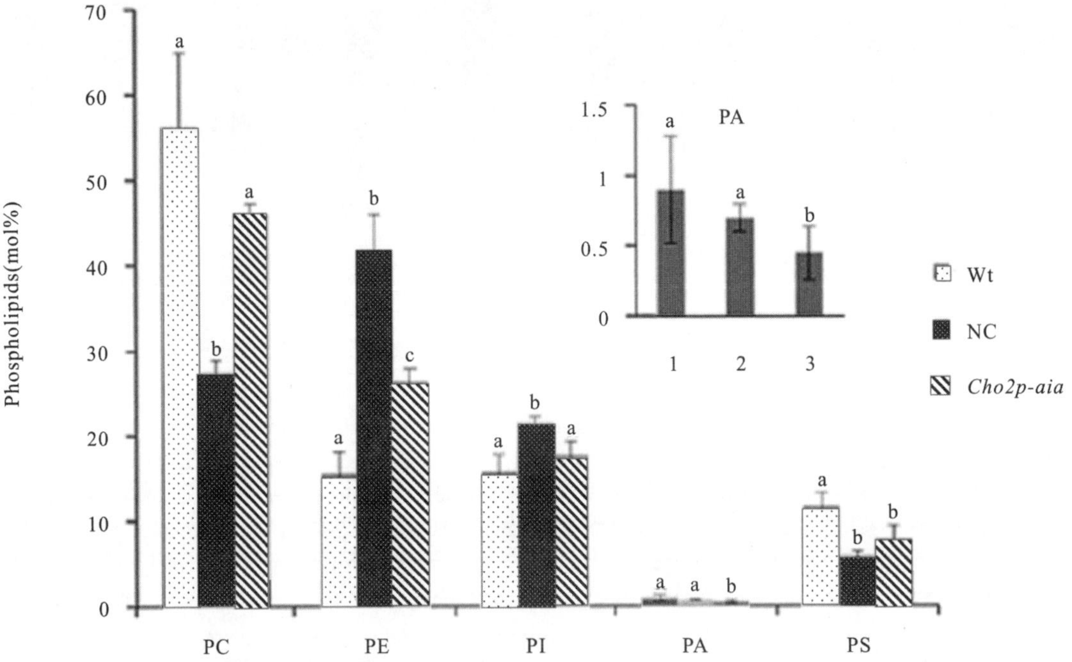

Fig. 2 The *Cho2p-aia* mutant largely restores PC content in *cho2*Δ cells

Phospholipid compositions are expressed as mol percentage (mol%) in wt, or *cho2*Δ cells expressing *CHO2-aia* or the negative control (NC) *DPM1*. According to the one-way ANOVA test, means do not differ significantly ($P>0.05$) if they are indicated with the same letter, and error bars represent standard deviation

2.3 The *Cho2p-aia* mutant fails to rescue the BMV replication defect in *cho2*Δ cells

Given that the *Cho2p-aia* mutant support wt-level cell growth and largely restored PC content to wt levels in *cho2*Δ cells, we next checked whether *Cho2p-aia* is able to complement the viral RNA replication defect. In the engineered BMV-yeast system (Ishikawa et al., 1997), BMV 1a, 2a^{pol}, and RNA3 are expressed from plasmids to provide the necessary components required for BMV RNA replication (Fig. 3A). In wt yeast cells, negative-strand RNA3 and positive-strand RNA4 were detected using BMV RNA strand-specific probes, indicating full replication (Fig. 3B). In *cho2*Δ cells with the expressed Dpm1p, which is not known to be involved in BMV replication and is used as a negative control, accumulated positive-strand RNA4 and negative-strand RNA3 were about 17% and 24% of wt levels, respectively. Expressing *CHO2-aia* did not improve BMV replication as positive-strand RNA4 and negative-strand RNA3 levels were similar to those in *cho2*Δ cells expressing *DPM1* (Fig. 3B), indicating that *Cho2p-aia* failed to complement the BMV replication defect in *cho2*Δ cells.

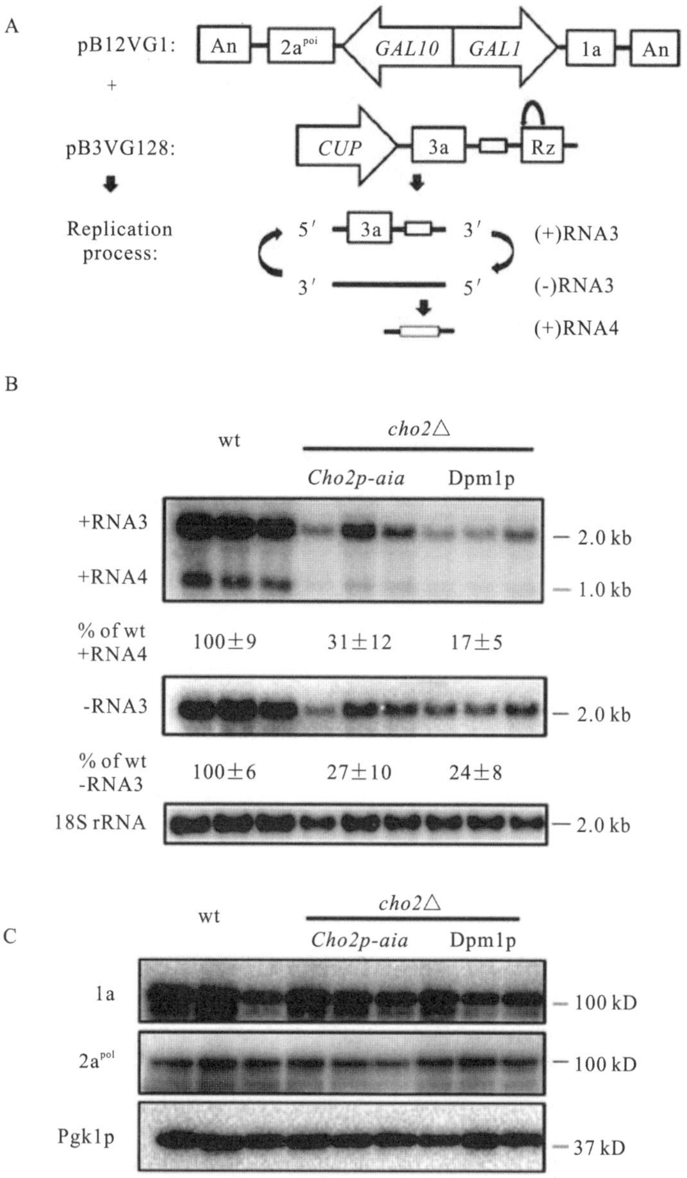

Fig. 3 The *Cho2p-aia* mutant fails to rescue the BMV replication defect in *cho2Δ* cells

(A) Diagram of BMV replication process in yeast. BMV 1a and 2apol were expressed from plasmid pB12VG1 under the control of the *GAL1* and *GAL10* promoter, respectively. BMV RNA3 was launched from plasmid pB3VG128 under the control of the copper-inducible *CUP1* promoter but no copper was added to the growth media. (B) BMV replication in wt cells, or in *cho2Δ* cells with the expressed Dpm1p or *Cho2p-aia*. BMV positive- and negative-strand RNAs were detected by 32P-labeled BMV strand-specific probes. 18S rRNA served as a loading control. The blot detecting (-) RNA3 was exposed for a longer time than that of (+) RNAs. (C) BMV 1a and 2apol accumulation in wt cells, or in *cho2Δ* cells with the expressed *Cho2p-aia* or Dpm1p. Pgk1p served as a loading control

To identify specific reasons why BMV replication defects were not complemented by *Cho2p-aia*, we first checked the accumulation of BMV 1a and $2a^{pol}$, which are necessary and sufficient for BMV replication. As shown in Fig. 3C, there was no significant difference in accumulated levels of 1a and $2a^{pol}$ in these cells, ruling out the possibility that the inhibited BMV genome replication was due to negative effects on expression and/or stability of the viral replication proteins.

2.4 The *Cho2p-aia* mutant interacts with BMV 1a

Since the *Cho2p-aia* mutant failed to complement the defective BMV RNA synthesis, we wanted to confirm if *Cho2p-aia* was stably expressed during viral replication. Western blotting determined that levels of HA-tagged *Cho2p-aia* and wt Cho2p were similar (Fig. 4A), indicating that *Cho2p-aia*-HA is stably expressed and accumulated in the absence (Fig. 1D) or presence (Fig. 4A) of BMV replication.

Cho2p interacts with and is recruited by BMV 1a to the sites of viral replication to facilitate BMV genome replication (Zhang et al., 2016). It is possible that *Cho2p-aia* fails to interact with 1a and therefore, is not recruited to the VRCs. To confirm or reject this notion, we performed a co-immunoprecipitation (co-IP) assay by co-expressing HA-tagged Cho2p, *Cho2p-aia*, or Dpm1p with His6-tagged 1a in yeast cells. Similar to Fig. 4A, Cho2p-HA and *Cho2p-aia*-HA accumulated to similar levels in the presence of BMV 1a (Fig. 4B, Total). As expected from our previous work, precipitating Cho2p-HA with anti-HA antibody pulled down 1a-His6 (Zhang et al., 2016) (Fig. 4B). On the contrary, 1a-His6 was not coprecipitated with Dpm1p-HA, which has been used as a negative control and is not expected to interact with BMV 1a (Zhang et al., 2016). To our surprise, 1a-His6 was also pulled down along with *Cho2p-aia*-HA (Fig. 4B), indicating that *Cho2p-aia*-HA is still associated or interacted with BMV 1a.

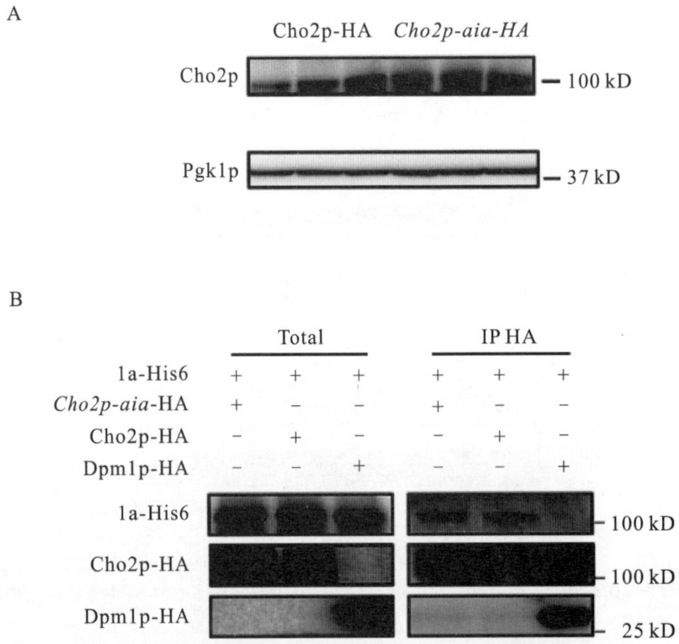

Fig. 4 The *Cho2p-aia* mutant interacts with BMV 1a

(A) Western blot showing the accumulation of HA-tagged *Cho2p-aia* or wt Cho2p in *cho2Δ* cells with the present of BMV components. Pgk1p served as a loading control. (B) Co-immunoprecipitation of *Cho2p-aia*-HA and BMV 1a-His6. Cell lysates were subjected to immunoprecipitation using an anti-HA (IP-HA) pAb, followed by Western blotting with an anti-His6 mAb or anti-HA pAb. Wt Cho2p served as a positive control and Dpm1p served as a negative control

2.5 *Cho2p-aia* prevents the localization of BMV 1aat the perinuclear ER membrane

BMV 1a is associated with the perinuclear ER (nER) membrane, where it invaginates the outer nER membrane to form VRCs (Diaz and Wang, 2014). BMV 1a recruits host proteins such as reticulon homology proteins (RHPs) (Diaz et al., 2010), sucrose non-fermenting7 (Snf7) (Diaz et al., 2015), and Cho2p to assemble functional VRCs (Zhang et al., 2016). Like the majority of lipid enzymes, Cho2p is normally localized in ER membranes where lipids are synthesized. Agreeing with our previous results and as shown in Fig. 5A (Zhang et al., 2016), Cho2p-HA was detected by immunofluorescence microscopy and appeared as a two-ring localization pattern in 43.1% of cells (n=153). The larger ring represents peripheral ER membranes and the smaller ring surrounding the DAPI-stained nucleus corresponds to the nER membrane (Fig. 5A, upper left panels). In the presence of 1a, Cho2p is depleted from peripheral ER membranes, enriched in the nER membrane and co-localized with 1a-His6 (59 of total 178 cells) (Fig. 5A, upper left panels).

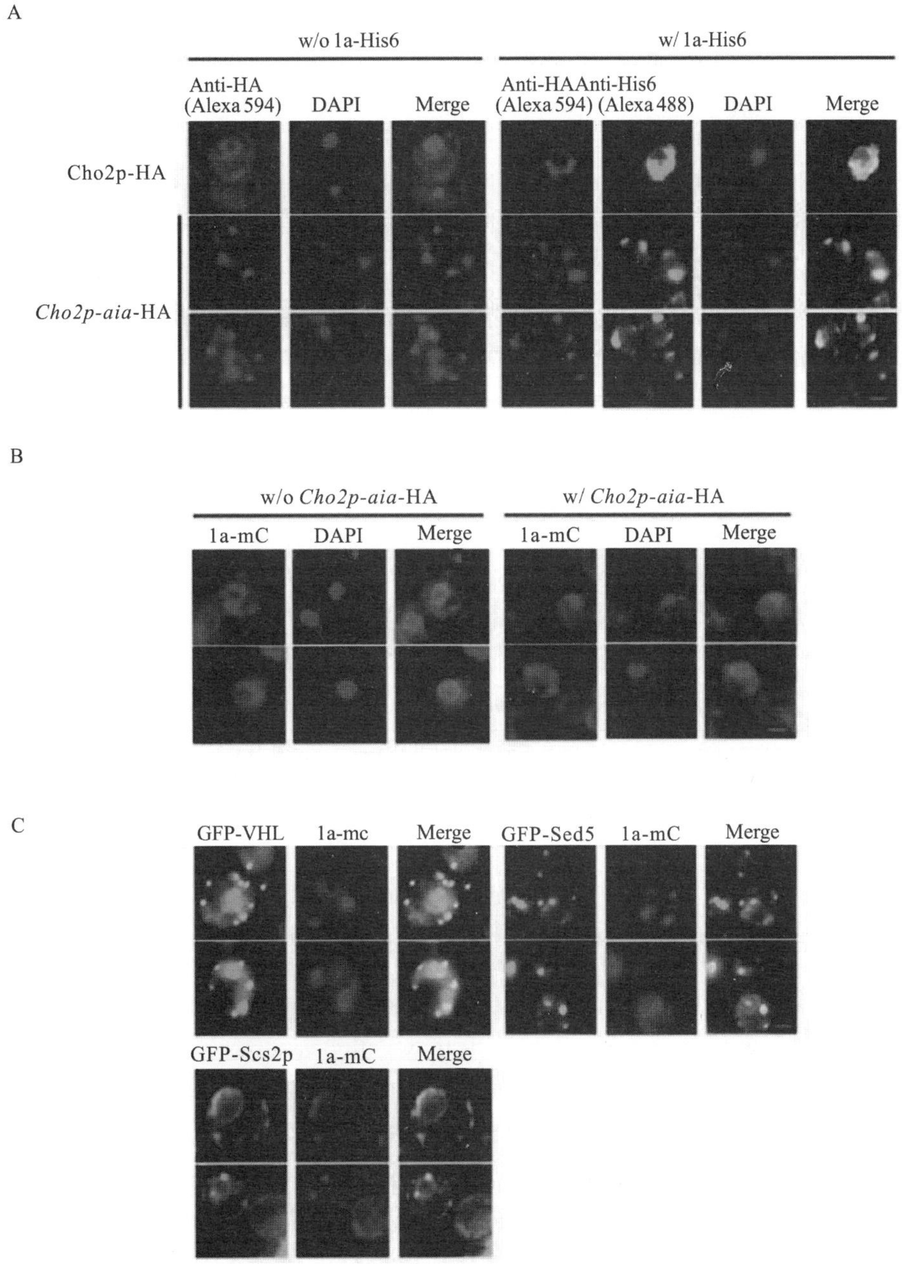

Fig. 5 BMV 1a mislocalizes in the presence of *Cho2p-aia*

(A) Microscopy images showing the distribution patterns of Cho2p-HA and *Cho2p-aia*-HA in *cho2Δ* cells in the absence or presence of BMV 1a-His6. Cho2p-HA and *Cho2p-aia*-HA were detected using an anti-HA pAb and a secondary antibody conjugated to Alexa Fluor 594. BMV 1a was detected using an anti-His6 mAb and a secondary antibody conjugated to Alexa Fluor 488. (B) Distribution patterns of BMV 1a-mCherry in *cho2Δ* cells in the absence or presence of *Cho2p-aia*-HA. (C) Distribution pattern of BMV 1a-mCherry in *cho2Δ* cells in the presence of *Cho2p-aia*-HA and GFP-tagged organelle markers for inclusion bodies (VHL), Golgi (Sed5p), or ER membrane (Scs2p). Nuclei were stained with DAPI and scale bars are 2.5μm in all figures

To determine if *Cho2p-aia* can be relocalized by 1a, we first expressed HA-tagged *Cho2p-aia* in the absence of BMV replicaiton in *cho2Δ* cells and tested its distribution. To our surprise, among all cells (n=326) that we can detect *Cho2p-aia*-HA, only approximately 6 out of the total 326 cells (1.8%) had a two-ring distribution pattern, and in sharp contrast, *Cho2p-aia*-HA was found as puncta in about 72% of cells (Fig. 5A, lower left panels). Intriguingly, when *Cho2p-aia*-HA and 1a-His6 were expressed simultaneously, the localization of 1a was altered. BMV 1a-His6 was localized to the nER membrane in only approximately 3.7% cells (12 out of 328 cells) but as puncta in 73.2% cells where we can clearly detect signals of 1a-His6, *Cho2p-aia*-HA, and DAPI signals (Fig. 5A, lower right panels). It should be noted that even though localization patterns of both *Cho2p-aia* and 1a changed, they co-localized with each other in all cells both can be detected, consistent with our co-IP results that *Cho2p-aia* is associated or interacts with 1a (Fig. 4B).

To better and easily characterize the nature of punctate structures of BMV 1a in the presence of *Cho2p-aia*, we used an mCherry-tagged 1a, which we have previously shown to have a similar localization pattern as that of 1a-His6 in wt cells (Li et al., 2016). In *cho2Δ* cells, about 35.5% of cells (n=262) expressing 1a-mCherry has a ring structure in the nER membrane (Fig. 5B, left panel), similar to that in wt cells (Zhang et al., 2016; Li et al., 2016). The distribution pattern of 1a-mCherry was also similar to that of 1a-His6 in the presence of *Cho2p-aia*, as it was associated with punctate structures in 74.6% (n=386) cells (Fig. 5A, B). To determine if BMV 1a was being redistributed to a different membrane in the presence of *Cho2p-aia*, we co-expressed 1a-mCherry, *Cho2p-aia* and GFP-tagged organelle markers. By comparing 1a-mCherry distributions with those of organelle markers, we found that the majority of puncta did not co-localize with the inclusion body marker VHL (Von Hippel-Lindau) (Kaganovich et al., 2008) or the Golgi marker Sed5p (suppressor of Erd2 deletion 5) (Hardwick and Pelham, 1992) (Fig. 5C). In contrast, these dots were co-localized with the ER marker Scs2p (suppressor of choline sensitivity 2) (Manford et al., 2012) in peripheral ER membranes in about 49% of cells that we counted (n=123) and only 2.3% of cells has 1a-mCherry localized in the nER membrane (Fig. 5C). Taken together, our results indicate that *Cho2p-aia* disrupts the normal distribution of 1a to the nER membrane, however, 1a is still primarily associated with ER membranes.

2.6 The *Cho2p–aia* mutant prevents the formation of BMV VRCs

BMV 1a invaginates the outer nER membrane into the lumen to form spherular VRCs during BMV replication (Diaz and Wang, 2014; Schwartz et al., 2002). These spherular compartments are about 60-80nm in diameter with an ~10nm neck connecting to the cytoplasm (Schwartz et al., 2002; Wang et al., 2005). The abundance and size of these VRCs can change significantly with deletion or mutation of host genes involved in membrane shaping or lipid metabolism (Zhang et al., 2012; Diaz et al., 2010; Diaz et al., 2015). We previously reported that in *cho2Δ* cells, 1a was detected at the nER membrane, and spherular VRCs were formed but they were larger than those in wt cells and were not functional in genome replication (Zhang et al., 2016). Because the expressed *Cho2p-aia* affected 1a's association with the nER membrane (Fig. 5), we predicted that the VRCs formation could be affected. To confirm this, we checked the formation of VRCs by transmission electron microscopy in wt cells, or *cho2Δ* cells expressing *CHO2-aia* in the presence of 1a, $2a^{pol}$, and RNA3. In wt cells with BMV components, the percentage of cells with spherular VRCs was approximately 16% (11 out of 67 cells). The average number of spherules was 6 (SD=1.8) with an average diameter of 65 ± 8nm (Fig. 6A). However, among the 100 cells checked, no spherical structures were observed in *cho2Δ* cells expressing the BMV replication components and *Cho2p-aia*. Of note, the inner and outer nER membranes were clearly observed in these cells (Fig. 6B). The failure to identify spherular VRCs is consistent with our results that *Cho2p-aia* inhibited the proper targeting of BMV 1a to the nER membrane in *cho2Δ* cells (Fig. 5).

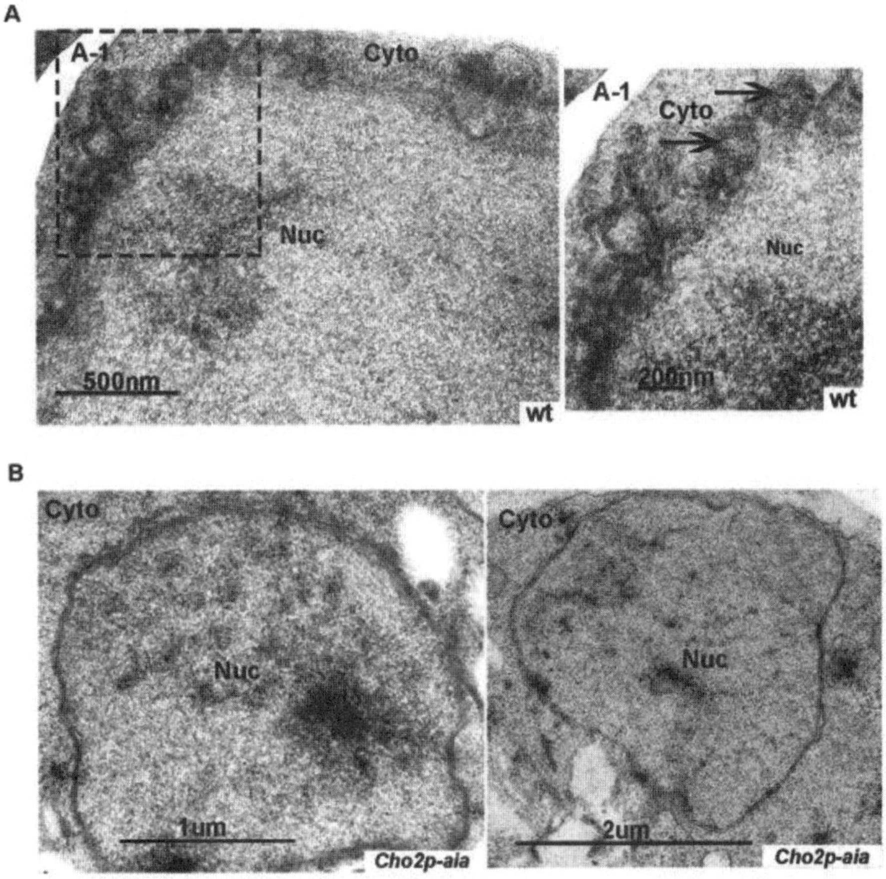

Fig. 6 The *Cho2p-aia* mutant affects the formation of BMV replication complexes

(A) Electron micrographs of VRCs formed in wt cells in the presence of all BMV replication components. Image A-1 is a higher magnification of the boxed area from the image on the left. Black arrows in A-1 point to VRCs. The spherular VRCs were observed in 11 out of 67 cells. (B) Electron micrographs of *Cho2Δ* cells with the expressed *Cho2p-aia* in the presence of all BMV replication components. 100 cells with clear nuclear membranes were recorded. Nuc: nucleus; Cyto: cytoplasm

3 Discussion

Positive-strand RNA viruses have limited protein-coding capacity and replicate in a tight association with host intracellular membranes to form the viral replication complexes. As major components of cellular membranes, lipids play critical roles in the replication of multiple (+) RNA viruses (Strating and van Kuppeveld, 2017; Stapleford and Miller, 2010; Romero-Brey and Bartenschlager, 2014; Belov, 2016). For example, PI4P plays a key role in the replication of multiple (+) RNA viruses, including HCV and enteroviruses (Hsu et al., 2010; Berger et al., 2009; Reiss et al., 2011; Borawski et al., 2009; Berger et al., 2011). Poliovirus and several enteroviruses promote an enrichment of free cholesterol to the sites of their replication (Roulin et al., 2014; Ilnytska et al., 2013). Disrupting sterol biosynthesis and composition inhibits the replication of WNV and DENV (Rothwell et al., 2009; Mackenzie et al., 2007). Among plant viruses, RCNMV promotes PA production by hijacking phospholipase D to the sites of its replication (Hyodo et al., 2015). TBSV and Carnation Italian ringspot virus transport PE to viral replication sites to build the PE-enriched VRCs for their replication (Xu and Nagy, 2016; Xu and Nagy, 2015). For BMV, its replication also requires balanced lipid homeostasis, phospholipids PC and PA are particularly important (Zhang et al., 2012; Lee et al., 2001; Lee and Ahlquist, 2003; Zhang et al., 2016; Zhang et al., 2018). Thus, dissecting the role of phospholipids in BMV replication and in turn, manipulating phospholipid metabolism may potentially provide antiviral strategies with broad-range and durable resistance. A major challenge for engineering lipid synthesis genes to control viruses and other pathogens, however, is that lipids also play crucial roles in cellular processes. Any manipulation of lipid synthesis genes will potentially affect host growth. Here, we generated a mutant, *Cho2p-aia*, which largely complemented PC synthesis and host growth defects but failed to support BMV replication (Fig. 1-3). Although the *Cho2p-aia* mutant still interacted with BMV 1a (Fig. 4), it disrupted the normal distribution of BMV 1a (Fig. 5) and thus, affected the formation of BMV VRCs (Fig. 6).

All well-studied (+)RNA viruses assemble their VRCs in association with various organelle membranes. As such, the trafficking of viral replication proteins to their destination organelles is the first and a critical step for the VRCs formation. Without transmembrane domains, BMV 1a is associated with the nER membrane via an amphipathic alpha-helix domain, helix A (Restrepo-Hartwig and Ahlquist, 1999; den Boon et al., 2001). Mutations in helix A result in two major phenotypes, 1a either no longer localizes to the nER or it induces the formation of smaller, abundant, non-functional VRCs compared to wt 1a (Liu et al., 2009). Besides 1a's helix A domain, host genes also regulate 1a's distribution. For example, the COPII (coat protein complex II) vesicle cargo receptor Erv14p (ER-vesicle protein of 14kD) is required for targeting 1a to the nER membrane (Li et al., 2016). In cells lacking *ERV14*, 1a no longer associates with the nER membrane but is distributed in punctate dots and large clusters that are located at peripheral ER membranes (Li et al., 2016). Yet, the mechanism by which Erv14p and COPII vesicles regulate 1a's localization remains unclear. Expressing *CHO2-aia* resulted in a similar phenotype as the majority of 1a localized in peripheral ER-associated punctate dots (Fig. 5). It is unclear whether the punctate structures present in *erv14*Δ cells or cells with the *Cho2p-aia* mutant are related or formed in a similar fashion.

It should be noted that 1a interacts (Fig. 4B) and completely colocalizes with the *Cho2p-aia* mutant in puncta (Fig. 5A). Because *Cho2p-aia* interacts with 1a and localizes to punctate structures when expressed alone, therefore, it is most likely that *Cho2p-aia* recruits 1a to these dot structures. If this is the case, it is contrary to wt Cho2p, which is recruited to the nER membrane-localized viral replication sites by 1a via their interaction (Fig. 5A) (Zhang et al., 2016). Alternatively, although *Cho2p-aia* largely complemented PC defects in *cho2*Δ cells, the phospholipid composition was not identical to that in wt cells: levels of PE were significantly higher and levels of PA and PS were lower than wt levels (Fig. 2). These lipid compositional changes might affect the physical properties of membranes, including thickness, intrinsic curvature, or fluidity. Such changes may affect membrane fusion and fission or the conformation of membrane-associated/ integral proteins (Renne and de Kroon, 2018; Li et al., 2006; de Kroon, 2007), including ER membrane-localized *Cho2p-aia* and 1a. It is also remotely possible that both 1a and *Cho2p-aia* are similarly affected and targeted to punctate structures independently. However, the localization of another ER resident protein, Scs2p, was not affected and a two-ring localization pattern was still observed in the presence of *Cho2p-aia* for Scs2p (Fig. 5C), arguing against a similar effect on conformation and/or localization of all ER membrane proteins. Cho2p is involved in converting PE to PC. In the presence of *Cho2p-aia*, we oberserved a close-to-wt levels of PC but PE contents were still significantly higher than wt levels (Fig. 2). PE is primarily synthesized by phosphatidylserine decarboxylase from PS in mitochondria and then transported to ER membranes, where PE is methylated first by Cho2p (Henry et al., 2012). It is possible that the mislocalization of *Cho2p-aia* from entire ER membranes to puncta may affect the access of PE by *Cho2p-aia* and as such, leads to an enhanced levels of PE compared to wt levels.

All (+)RNA viruses depend entirely on host factors such as proteins or lipids to complete their life cycles. In theory, host genes required for viral infection (also termed susceptibility genes) can be potentially targeted to develop viral resistance (Garcia-Ruiz, 2018), similar to recessive *R* genes in nature. Some advantages of engineering host genes to achieve virus resistance include: 1) viral resistance caused by the loss of or mutation(s) of a host gene is durable because host genes are less likely to mutate than viral genes under infections; 2) multiple genes can be targeted in combination to achieve a better or complete control of viral infection; and 3) the same or similar manipulation of a specific gene can be applied to several crops. However, these host genes may play critical roles in cellular processes and in turn, alterations of such host genes may lead to side effects on host survival. This is true for *CHO2*, whose deletion inhibited BMV replication more than 5-fold but also caused host cell growth defects (Zhang et al., 2016) (Figs. 1-3). However, cells expressing *CHO2-aia* grew as well as wt cells but still inhibited BMV replication by ~5-fold. This provides proof-of-principle evidence that host lipid genes can be engineered to control (+) RNA virus genome replication without compromising host fitness. We note that the deletion of or mutations in *CHO2* substantially inhibited but did not eliminate BMV replication, suggesting that BMV may acquire PC through the Kennedy pathway, which is a secondary PC synthesis pathway in yeast but the primary pathway in higher eukaryotes. In addition to manipulating *CHO2*, targeting other component(s) of the Kennedy pathway or other related processes coordinately may provide complete blockage of BMV replication.

In summary, we identified a *CHO2* mutant, *Cho2p-aia,* which supports general PC synthesis required for cell growth but failed to support BMV replication. The *Cho2p-aia* mutant prevented BMV replication protein 1a from localizing to the perinuclear ER membrane and as such, blocked the formation of VRCs and inhibited viral replication. This indicates that engineering host lipid synthesis genes can be an effective way of controlling viral replication without compromising host fitness.

4 Experimental procedures

4.1 Yeast strains and growth conditions

The *Saccharomyces cerevisiae* strain YPH500 (*MATaura3-52, lys2-801, ade2-101, trp1-Δ63, his3-Δ200, leu2-Δ1*) and YPH500-based *CHO2* deletion mutant were used in all experiments (Zhang et al., 2016). Yeast cells were grown at 30°C in synthetic defined (SD) medium containing 2% galactose as the carbon source. Histidine, leucine, uracil, or combinations of them were omitted from the medium to maintain selection for different plasmid combinations. After two passages (36 to 48h) in SD medium, cells were harvested when the optical density at 600nm (OD_{600}) reached between 0.4-1.0. The doubling time was calculated following the equation: [hours cells grown * Ln (2)] / Ln [(final OD_{600} / initial OD_{600})].

4.2 Plasmids

To launch BMV replication, plasmid pB12VG1 expressing BMV 1a and $2a^{pol}$, and plasmid pB3VG128 transcribing BMV RNA3 were used (Li et al., 2016). His6- or mCherry-tagged BMV 1a was expressed under the control of the *GAL1* promoter from pB1YT3-cH6 or pB1YT3-mC, respectively. G102A and G104A mutations were introduced into *CHO2* by an overlapping PCR-based approach using a pair of primers 5′- GT AATGCAGAACCAGTgCAATAgCTAAATTATAAGCAATTCTCCA-3′ and 5′- TGGAGAATTGCTTATAAT TTAGcTATTGcACTGGTTCTGCATTAC-3′ Note: nucleic acids in the lower case were intended to introduce mutations. Both Cho2p-HA and *Cho2p-aia*-HA were expressed under the control of the *GAL1* promoter from a centromeric plasmid. Plasmids expressing organelle markers: GFP-VHL, GFP-Scs2, and GFP-Sed5 were provided by Dr. Maya Schuldiner (Weizmann Institute of Sciences, Israel).

4.3 Lipid analysis

Ten OD_{600} units of yeast cells were harvested and total lipids were extracted as described previously (Matyash et al., 2008). Total lipids were reconstituted in chloroform: methanol: water (65:35:8, v/v/v; 500ul) with the addition of the phospholipid internal standard mixture purchased from Kansas Lipidomics Research Center. The phospholipid compositions were analyzed by a Waters I-class UPLC interfaced with a Waters Synapt G2-S mass spectrometer (Waters Corp.) using conditions as described (Castro-Perez et al., 2010). Data normalization to the internal standards for mol percentage calculations was performed as described previously (Zhou et al., 2011).

4.4 RNA extraction and Northern blotting

Yeast cells were harvested, and total RNA was extracted by a hot phenol method (Kong et al., 1999). Equal amounts of total RNA were prepared for Northern blotting analysis. P^{32}-labeled probes specific to BMV positive- or negative-strand RNAs or 18S rRNA were used for the hybridization. 18S rRNA was used as a loading control to eliminate loading variations. Radioactive signals of BMV positive-, negative-strand, or 18S rRNA were scanned using a Typhoon FLA 7000 phosphoimager and the intensity of signals was quantified by using Image Quant TL software (GE healthcare).

4.5 Protein extraction and Western blotting

Two OD_{600} units of yeast cells were harvested and total proteins were extracted as described previously (Li et al., 2016). Equal volumes of total proteins were analyzed using sodium dodecyl sulfate polyacrylamide gel electrophoresis (SDS-PAGE) and transferred to a polyvinylidene difluoride (PVDF) membrane. Rabbit anti-BMV 1a antiserum (1∶10,000 dilution, a gift from Dr. Paul Ahlquist at the University of Wisconsin-Madison), mouse anti- BMV $2a^{pol}$ (1∶3,000 dilution), rabbit anti-HA (1∶3,000 dilution, Thermo Fisher Scientific, cat #: 71-5500), mouse anti-His (1∶3,000 dilution Genscript, cat #: A00186), mouse anti-Pgk1 (1∶10,000 dilution, Thermo Fisher Scientific cat #: 459250) or mouse anti-Dpm1p (1∶3,000 dilution, Thermo Fisher Scientific cat #: A6429) was used as the primary antibody; horseradish peroxidase (HRP)-conjugated anti-rabbit or anti-mouse antibody (1∶10,000 dilution, Thermo Fisher Scientific cat #: 32460 & 32430) together with the Supersignal West Femto substrate (Thermo Scientific Thermo Fisher Scientific, cat #: 32460) were used to detect target proteins.

4.6 Co-immunoprecipitation assay

Ten OD_{600} units of yeast cells were harvested and the co-IP assay was performed as previously described (Li et al., 2016). Briefly, harvested cells were lysed in RIPA (radioimmunoprecipitation assay) buffer (50mM Tris at pH 8.0, 1% Nonidet P-40, 0.1% SDS, 150mM NaCl, 0.5% sodium deoxycholate, 5mM EDTA, 10mM NaF, 10mM NaPPi, and protease inhibitor mix). Cell debris was removed and the supernatant was mixed with Protein a Sepharose beads and anti-HA pAb overnight at 4°C. Beads were washed three times with RIPA buffer, resuspended in 1× SDS gel loading buffer and boiled for 10min. Samples were loaded onto a SDS-PAGE gel, followed by Western blotting with anti-His6 mAb and anti-HA pAb to detect target proteins.

4.7 Immunofluorescence assay

Two OD_{600} units of yeast cells were harvested and fixed with 4% (vol/vol) formaldehyde for 30 minutes at 30℃ . After removing the cell wall by lyticase, spheroplasts were permeabilized with 0.1% Triton X-100, incubated with specified primary antibodies (anti-His6 mAb or anti-HA pAb at 1∶100 dilution) overnight at 4°C, and followed by an incubation with secondary antibodies (1∶100 dilution) for 1h at room temperature.

Secondary antibodies were Alexa Fluor 488-conjugated antimouse (Thermo Fisher Scientific A11001) and Alexa Fluor 594-conjugated anti-rabbit antibodies (Jackson Immuno Research cat #: 711-545-152). The nucleus was stained with DAPI (Vector laboratories, Cat #: H-1200) for 10min. Images were captured using a Zeiss epifluorescence microscope at the Fralin microscopy facility, Virginia Tech.

4.8 Electron microscopy

Ten OD_{600} units of yeast cells were harvested. Fixation, dehydration, and embedding were performed as previously described (Zhang et al., 2016). Images were captured using a JEOL JEM 1400 transmission electron microscope at the Virginia- Maryland College of Veterinary Medicine, Virginia Tech.

4.9 Statistical analysis

One-way ANOVA analysis was used to compare phospholipids or doubling time in wt and mutant cells. Error bars represent the standard deviation.

Acknowledgements

We would like to thank Haijie Liu, Jianhui Li, Nicholas Todd, and Elizabeth Barton for general help and assistance. We appreciate the help from Ms. Kathy Lowe at the Virginia-Maryland College of Veterinary Medicine, VT for electron microscopy work, Dr. Kristi DeCourcy at the Fralin Life Science Institute, VT for fluorescence microscopy work, and Drs. Sherry Hildreth and Richard Helm at the Mass Spectrometry Incubator, VT for measuring phospholipids. We are grateful to Drs. George Carman, Arturo Diaz, and Janet Webster for critical reading of our manuscript. Guijuan He and Zhenlu Zhang were supported by a scholarship from the Chinese Scholarship Council.

References

[1] BERGER K L, COOPER J D, HEATON N S, et al. Roles for endocytic trafficking and phosphatidylinositol 4-kinase III alpha in hepatitis C virus replication[J]. Proceedings of the National Academy of Sciences of the United States of America, 2009, 106: 7577-7582.

[2] BORAWSKI J, TROKE P, PUYANG X, et al. Class III phosphatidylinositol 4-kinase alpha and beta are novel host factor regulators of hepatitis C virus replication[J].Journal of Virology, 2009, 83: 10058-10074.

[3] BERGER K L, KELLY S M, JORDAN T X, et al. Hepatitis C virus stimulates the phosphatidylinositol 4-kinase III alpha-dependent phosphatidylinositol 4-phosphate production that is essential for its replication[J].Journal of Virology, 2011, 85: 8870-8883.

[4] BELOV G A, VAN KUPPEVELD F J. (+) RNA viruses rewire cellular pathways to build replication organelles[J]. Current Opinion in Virology, 2012, 2: 740-747.

[5] BARAJAS D, XU K, SHARMA M, et al. Tombusviruses upregulate phospholipid biosynthesis via interaction between p33 replication protein and yeast lipid sensor proteins during virus replication in yeast[J]. Virology,2014, 471: 72-80.

[6] BELOV G A. Modulation of lipid synthesis and trafficking pathways by picornaviruses[J]. Current Opinion in Virology, 2014, 9: 19-23.

[7] BELOV G A. Dynamic lipid landscape of picornavirus replication organelles[J]. Current Opinion in Virology,2016, 19: 1-6.

[8] BANERJEE S, APONTE D D, YEAGER C, et al. Hijacking of multiple phospholipid biosynthetic pathways and induction of membrane biogenesis by a picornaviral 3CD protein[J].PLoS Pathogensensens, 2018, 14: e1007086.

[9] CASTORENA K M, STAPLEFORD K A, MILLER D J. Complementary transcriptomic, lipidomic, targeted functional genetic analyses in cultured Drosophila cells highlight the role of glycerophospholipid metabolism in flock house virus RNA replication[J]. BMC Genomics, 2010, 11: 183.

[10] CASTRO J M, KAMPHORST J, DEGROOT J, et al. Comprehensive LC–MS^E lipidomic analysis using a shotgun approach and its

application to biomarker detection and identification in osteoarthritis patients[J].Journal of Proteome Research,2010, 9: 2377-2389.

[11] DANGL J L, JONES J D. Plant pathogens and integrated defence responses to infection[J]. Nature, 2001, 411: 826-833.

[12] DEN BOON J A, CHEN J, AHLQUIST P. Identification of sequences in brome mosaic virusreplicase protein 1a that mediate association with endoplasmic reticulum membranes[J]. Journal of Virology, 2001, 75: 12370-12381.

[13] DE KROON A I. Metabolism of phosphatidylcholine and its implications for lipid acyl chain composition in *Saccharomyces cerevisiae*[J]. Biochimica et Biophysica Acta (BBA) - Molecular and Cell Biology of Lipids, 2007, 1771: 343-352.

[14] DIAZ A, WANG X, AHLQUIST P. Membrane-shaping host reticulon proteins play crucial roles in viral RNA replication compartment formation and function[J]. Proceedings of the National Academy of Sciences of the United States of America, 2010, 107: 16291-16296.

[15] DIAZ A, WANG X.Bromovirus-induced remodeling of host membranes during viral RNA replication[J].Current Opinion in Virology, 2014, 9: 104-110.

[16] DIAZ A, ZHANG J, OLLWERTHER A, et al. Host E SC RT proteins are required for bromovirus RNA replication compartment assembly and function[J]. PLoS Pathogens, 2015, 11: e1004742.

[17] DU Z Y, ARIAS T, MENG W, et al. Plant acyl-CoA-binding proteins: an emerging family involved in plant development and stress responses[J].Prog Lipid Research, 2016, 63: 165-181.

[18] EL G S, MISTRY J, BATEMAN A, et al. The Pfam protein families database in 2019[J]. Nuclc Acids Research,2018,47: D427-D432.

[19] GALVEZ L C, BANERJEE J, PINAR H, et al. Engineered plant virus resistance[J]. Plant Science, 2014,228: 11-25.

[20] GUO Z, LU J, WANG X, et al. Lipid flippases promote antiviral silencing and the biogenesis of viral and host siRNAs in *Arabidopsis*[J]. Proceedings of the National Academy of Sciences of the United States of America,2017, 114: 1377-1382.

[21] GARCIA-RUIZ H. Susceptibility Genes to Plant Viruses[J]. Viruses,2018, 10: 484.

[22] HARDWICK K G, PELHAM H R. S ED5 encodes a 39-kD integral membrane protein required for vesicular transport between the ER and the Golgi complex[J].The Journal of Cell Biology, 1992, 119: 513-521.

[23] HSU N Y, ILNYTSKA O, BELOV G, et al. Viral reorganization of the secretory pathway generates distinct organelles for RNA replication[J]. Cell,2010, 141: 799-811.

[24] HENRY S A, KOHLWEIN S D, CARMAN G M. Metabolism and regulation of glycerolipids in the yeast *Saccharomyces cerevisiae*[J]. Genetics,2012, 190: 317-349.

[25] HYODO K, TANIGUCHI T, MANABE Y, et al. Phosphatidic acid produced by phospholipase D promotes RNA replication of a plant RNA virus[J].PLoS Pathogens, 2015, 11: e1004909.

[26] HASHIMOTO M, NERIYA Y, YAMAJI Y, et al. Recessive resistance to plant viruses: potential resistance genes beyond translation initiation factors[J]. Frontiers in Microbiology, 2016, 7: 1695.

[27] ISHIKAWA M, JANDA M, KROL M A, et al. In vivo DNA expression of functional brome mosaic virus RNA replicons in *Saccharomyces cerevisiae*[J]. Journal of Virology, 1997, 71: 7781-7790.

[28] ILNYTSKA O, SANTIANA M, HSU N Y, et al. Enteroviruses harness the cellular endocytic machinery to remodel the host cell cholesterol landscape for effective viral replication[J]. Cell Host Microbe, 2013, 14: 281-293.

[29] JIANG J, PATARROYO C, CABANILLAS D G, et al. The vesicle-forming 6K2 protein of turnip mosaic virus interacts with the COPII coatomer Sec24a for viral systemic infection[J]. Journal of Virology, 2015, 89: 6695-6710.

[30] KONG F, SIVAKUMARAN K, KAO C. The N-terminal half of the brome mosaic virus 1a protein has R NA capping-associated activities: specificity for GTP and s-adenosylmethionine[J]. Virology, 1999, 259: 200-210.

[31] KUSHNER D B, LINDENBACH B D, GRDZELISHVILI V Z, et al. Systematic, genome-wide identification of host genes affecting replication of a positive-strand RNA virus[J]. Proceedings of the National Academy of Sciences of the United States of America, 2003, 100: 15764-15769.

[32] KANG B C, YEAM I, JAHN M M. Genetics of plant virus resistance[J]. Annual Review of Phytopathology, 2005, 43: 581-621.

[33] KAGANOVICH D, KOPITO R, FRYDMAN J. Misfolded proteins partition between two distinct quality control compartments[J].

Nature, 2008, 454: 1088.

[34] KELLEY L A, MEZULIS S, YATES C M, et al. The Phyre2 web portal for protein modeling, prediction and analysis[J]. Nature Protocols, 2015, 10: 845-858.

[35] LEE W M, ISHIKAWA M, AHLQUIST P. Mutation of host Δ9 fatty acid desaturase inhibits brome mosaic virus RNA replication between template recognition and RNA synthesis[J]. Journal of Virology, 2001, 75: 2097-2106.

[36] LEE W M, AHLQUIST P. Membrane synthesis, specific lipid requirements, localized lipid composition changes associated with a positive-strand RNA virus RNA replication protein[J]. Journal of Virology, 2003, 77: 12819-12828.

[37] LI Z, VANCE D E. Thematic review series: glycerolipids. Phosphatidylcholine and choline homeostasis[J].Journal of Lipid Research, 2008, 49(6):1187.

[38] LIU L, WESTLER W M, DEN BOON J A, et al. An amphipathic α-helix controls multiple roles of brome mosaic virus protein 1a in RNA replication complex assembly and function[J].PLoS Pathogens, 2009, 5(3):e1000351.

[39] LI J, FUCHS S, ZHANG J, et al. An unrecognized function for C OP II components in recruiting a viral replication protein to the perinuclear ER[J]. Journal of Cell Science, 2016, 129(19):3597.

[40] LI F, ZHANG C, LI Y, et al. Beclinl restricts RNA virus infection in plants through suppression and degradation of the viral polymerase[J]. Nature Communications, 2018, 9(1):1268.

[41] MARTIN G B, BOGDANOVE A J, SESSA G. Understanding the functions of plant disease resistance proteins[J]. Annual Review of Plant Biology, 2003, 54(54):23.

[42] MACKENZIE J M, KHROMYKH A A, PARTON R G. Cholesterol manipulation by West Nile virus perturbs the cellular immune response[J]. Cell Host Microbe, 2007, 2: 229-239.

[43] MAULE A J, CARANTA C, BOULTON M I. Sources of natural resistance to plant viruses: status and prospects[J]. Molecular Plant Pathology, 2010, 8(2):223-231.

[44] MATYASH V, LIEBISCH G, KURZCHALIA T V, et al. Lipid extraction by methyl-tert-butyl ether for high-throughput lipidomics[J]. Journal of Lipid Research, 2008, 49(5):1137-1146.

[45] MARTÍN-ACEBES M A, BLÁZQUEZ A B, DE OYA N J, et al. West Nile virus replication requires fatty acid synthesis but is independent on phosphatidylinositol-4-phosphate lipids[J]. PLOS ONE, 2011, 6: e24970.

[46] MANFORD A G, STEFAN C J, YUAN H L, et al. E R-to-plasma membrane tethering proteins regulate cell signaling and ER morphology[J]. Developmental Cell, 2012, 23(6):1129-1140.

[47] NEESS D, BEK S, ENGELSBY H, et al. Long-chain acyl-CoA esters in metabolism and signaling: role of acyl-CoA binding proteins[J]. Progress in Lipid Research, 2015, 59:1-25.

[48] NAGY P D. Tombusvirus-host interactions: co-opted evolutionarily conserved host factors take center court[J]. Annual Review of Virology, 2016, 3(1): 491-515.

[49] PIMENTEL D. Environmental and economic costs of the application of pesticides primarily in the United States[J]. Environment Development Sustainability, 2005, 7(2):229-252.

[50] PERERA R, RILEY C, ISAAC G, et al. Dengue virus infection perturbs lipid homeostasis in infected mosquito cells[J]. PLOS Pathogens, 2012, 8: e1002584.

[51] PAUL D, MADAN V, BARTENSCHLAGER R. Hepatitis C virus RNA replication and assembly: living on the fat of the land[J]. Cell Host Microbe, 2014, 16(5): 569-579.

[52] RESTREPO H M, AHLQUIST P. Brome mosaic virus RNA replication proteins 1a and 2a colocalize and 1a independently localizes on the yeast endoplasmic reticulum[J]. Journal of Virology, 1999, 73: 10303-10309.

[53] RANDALL G, PANIS M, COOPER J D, et al. Cellular cofactors affecting hepatitis C virus infection and replication[J]. Proceedings of the National Academy of Sciences of the United States of America, 2007, 104: 12884-12889.

[54] ROTHWELL C, LEBRETON A, NG C Y, et al. Cholesterol biosynthesis modulation regulates dengue viral replication[J]. Virology, 2009, 389(1-2): 8-19.

[55] REISS S, REBHAN I, BACKES P, et al. Recruitment and activation of a lipid kinase by hepatitis C virus NS5A is essential for integrity of the membranous replication compartment[J]. Cell Host Microbe, 2011, 9(1): 32-45.

[56] ROMERO B I, BARTENSCHLAGER R. Membranous replication factories induced by plus-strand RNA viruses[J]. Viruses, 2014, 6(7): 2826-2857.

[57] ROULIN P S, LÖTZERICH M, TORTA F, et al. Rhinovirus uses a phosphatidylinositol 4-phosphate/cholesterol counter-current for the formation of replication compartments at the ER-Golgi interface[J]. Cell Host Microbe, 2014, 16(5): 677-690.

[58] RENNE M F, DE KROON A I. The role of phospholipid molecular species in determining the physical properties of yeast membranes[J]. FEBS Letters, 2018, 592(8): 1330-1345.

[59] SUMMERS E F, LETTS V A, MCGRAW P, et al. *Saccharomyces cerevisiae* cho2 mutants are deficient in phospholipid methylation and cross-pathway regulation of inositol synthesis[J]. Genetics, 1988, 120(4):909-22.

[60] SCHWARTZ M, CHEN J, JANDA M, et al. A positive-strand RNA virus replication complex parallels form and function of retrovirus capsids[J]. Molecular Cell, 2002, 9(3):505-514.

[61] SHIELDS D J, ALTAREJOS J Y, Wang X, et al. Molecular dissection of the S-adenosylmethionine-binding site of phosphatidylethanolamine N-methyltransferase[J]. Journal of Biological Chemistry, 2003, 278(37):35826-35836.

[62] STAPLEFORD K A, MILLER D J. Role of cellular lipids in positive-sense RNA virus replication complex assembly and function[J]. Viruses, 2010, 2(5): 1055-1068.

[63] SHARMA M, SASVARI Z, NAGY P D. Inhibition of phospholipid biosynthesis decreases the activity of the tombusvirus replicase and alters the subcellular localization of replication proteins[J]. Virology, 2011, 415(2): 141-152.

[64] STRATING J R, VAN KUPPEVELD F J. Viral rewiring of cellular lipid metabolism to create membranous replication compartments[J]. Current Opinion in Cell Biology, 2017, 47: 24-33.

[65] TRUNIGER V, ARANDA M A. Recessive resistance to plant viruses[J]. Adv Virus Research, 2009, 75: 119-159.

[66] VANCE D E, TRIP E M, PADDON H B. Poliovirus increases phosphatidylcholine biosynthesis in HeLa cells by stimulation of the rate-limiting reaction catalyzed by CTP: phosphocholine cytidylyltransferase[J]. Journal of Biological Chemistry, 1980, 255(3):1064-1069.

[67] WANG X, LEE W M, WATANABE T, et al. Brome mosaic virus 1a Nucleoside Triphosphatase/Helicase Domain Plays Crucial Roles in Recruiting RNA Replication Templates[J]. Journal of Virology, 2005, 179: 13747-13758.

[68] WANG X, AHLQUIST P. *Brome mosaic virus* (*Bromoviridae*). In: Mahy BWJ, van Regenmortel MHV (eds) Encyclopedia of Virology, 3rd edn[M]. New York, 2008, 381-386.

[69] WANG A, KRISHNASWAMY S. Eukaryotic translation initiation factor 4E mediated recessive resistance to plant viruses and its utility in crop improvement[J]. Molecular plant pathology, 2012, 13(7): 795-803.

[70] WANG A. Dissecting the molecular network of virus-plant interactions: the complex roles of host factors[J]. Annual Review of Phytopathology, 2015, 53: 45-66.

[71] XU K, NAGY P D. RNA virus replication depends on enrichment of phosphatidylethanolamine at replication sites in subcellular membranes[J]. Proceedings of the National Academy of Sciences of the United States of America, 2015, 112: E1782-E1791.

[72] XU K, NAGY P D. Enrichment of phosphatidylethanolamine in viral replication compartments via co-opting the endosomal Rab5 small GTPase by a positive-strand RNA virus[J]. PLoS Biology, 2016, 14: e2000128.

[73] ZHOU Z, MAREPALLY S R, NUNE D S, et al. LipidomeDB data calculation environment: online processing of direct-infusion mass spectral data for lipid profiles[J]. Lipids, 2011, 46: 879-884.

[74] ZHANG J, DIAZ A, MAO L, et al. Host acyl-CoA binding protein regulates replication complex assembly and activity of a positive-strand RNA virus[J]. Journal of Virology, 2012, 86: 5110-5121.

[75] ZHANG J, ZHANG Z, CHUKKAPALLI V, et al. Positive-strand RNA viruses stimulate host phosphatidylcholine synthesis at viral replication sites[J]. Proceedings of the National Academy of Sciences of the United States of America, 2016, 113: E1064-E1073.

[76] ZHANG Z, HE G, HAN G S, et al. Host Pah1p phosphatidate phosphatase limits viral replication by regulating phospholipid synthesis[J]. PLoS Pathogens, 2018, 14: e1006988.

II　病毒遗传与进化分析

这部分论文着重分析了马铃薯 Y 病毒、水稻条纹病毒等几种植物病毒在我国的遗传多样性与群体结构特征，对病毒分子变异与病害流行的驱动因素作了比较深入的分析和探讨。另外，发现了不同病毒分离物间存在重组和共侵染同株寄主植物的现象。

水稻条纹病毒楚雄分离物一个重组 RNA 序列分析

程文金，吴祖建，谢联辉

（福建农林大学植物病毒研究所，福建省植物病毒学重点实验室，福州 350002）

摘要： 为检测水稻条纹病毒（*Rice stripe virus*, RSV）基因组存在的重组 RNA，采用 RT-PCR 方法扩增并克隆云南楚雄分离物（YCX07）的 RNA 重组片段 YCX07R。克隆重组片段并测定序列。序列分析表明，YCX07R 由 RSV RNA2 的 3′端序列与 RNA3 的 3′端序列重组而成，具有 RSV 基因组结构特征，采取双义编码策略，即在正、负链的 5′端分别存在一个阅读框，编码两个重组蛋白 YCX07R-5P、YCX07R-3P。分析推测重组更可能通过类似于脊髓灰质炎病毒（Poliovirus）的引物互补延伸机制发生。

关键词： RSV；重组 RNA；双义编码

中图分类号： S432.4+1 **文献标识码：** A **论文编号：** 2009-0828

A Recombinant RNA from Chuxiong Isolate of Rice Stripe Virus (RSV) was Sequenced

Wenjin Cheng, Zujian Wu, Lianhui Xie

（Key Lab of Plant Virology of Fujian Province, Institute of Plant Virology, Fujian Agriculture and Forestry University, Fuzhou, Fujian 350002）

Abstract: To detect recombinant RNA in genome of Rice stripe virus (RSV), a recombinant RNA of one RSV isolate from Chuxiong(YCX07), Yunnan Province, China, was amplified by RT-PCR. The recombinant RNA was cloned and sequenced. Analysis of the sequence showed that YCX07R was recombined from the 3′ segment nucleotide of RNA2 and RNA3 of RSV. The recombinant RNA uses ambisense coding strategy and encodes two proteins, namely YCX07R-5P and YCX07R-3P. The two proteins located on the 5′ end of the plus and minus strand, respectively. This is similar to the four RNAs of RSV It was discussed that the RNA was more likely recombined from the alignment and extension mechanism similar to Poliovirus.

Key words: RSV; recombinant RNA;ambisense

RNA 重组是一种普遍现象，存在于多种 RNA 病毒中。早在 20 世纪 60 年代就观察到脊髓灰质炎病毒（*Poliovirus*）和流感病毒（*Influenza virus*）中存在 RNA 重组（Hirst, 1962）。近年来，对病毒 RNA 重组的研究更加扩展与深入。对雀麦花叶病毒（*Brome mosaic virus*，BMV）的研究表明，其 RNA 重组涉及复制酶及基因组末端非编码序列的参与（Figlerowicz and Bujarski, 1998; Olsthoorn et al., 2002）。对丁肝病毒的研究也表明，RNA 重组在患者体内以及培养细胞中都能发生（Wang and

中国农学通报 . 2009, 25(18): 352-355

收稿日期：2009-04-20

基金项目：农业部公益性行业（农业）科研专项（nyhyzx07-051）；国家自然科学基金项目（30671357，30770090）。

Chao, 2005)。Qβ 噬菌体的 RNA 重组系统已成为研究 RNA 重组的一个有效方法，利用该系统进行的研究表明，不同的 RNA 病毒可能存在不同的重组机制（Chetverin, 1997; Chetverin et al., 2005)。RNA 重组不仅发生在病毒 RNA 之间，还发生在病毒与寄主的 RNA 之间，例如：在牛病毒性腹泻病毒（*Bovine viral diarrhea virus*, BVDV）的基因组中曾检测到重组 mRNA（Baroth et al., 2000)，编码寄主细胞蛋白 NEDD8，重组的结果使该病毒的 NS3 蛋白产生变异的 N 端。像这种寄主与病毒间的 RNA 重组发生在流感病毒与寄主 28S rRNA 之间时，会导致病毒致病力和侵染力的提高（Khatchikian et al., 1989)。

RSV 是纤细病毒属（*Tenuivirus*）的代表种，具有独特的基因组结构和编码策略。在 RSV 基因组四条 RNA 中，除 RNA1 采取负链编码策略（编码 RdRp）外，RNA2、3、4 均采取双义（ambisense）编码策略：即在 RNA 毒义链（vRNA）与毒义互补链（vcRNA）的近 5′ 端处均存在一个开放阅读框，各编码一个蛋白质（Ramírez and Haenni, 1994)。魏太云（魏太云等, 2003; 魏太云和王辉, 2003）根据对 RSV 分子变异的研究认为 RSV 可能存在重组现象，但还没有直接证据表明 RSV 存在重组，尤其是不同 RSV 分离物间的重组（重配)。笔者报道了 RSV 云南楚雄分离物（YCX07）基因组中的一个重组 RNA，这个重组 RNA 由 RNA2 与 RNA3 的两个 3′ 端序列通过一个互补序列重组形成，具有 RSV 基因组结构的特征。

1 材料与方法

1.1 材料

供试 RSV 毒源于 2007 年 7 月采自云南省楚雄田间，经灰飞虱（*Laodelphax Striatellus*）传毒后接种于水稻品种武育粳 3 号上，取症状明显的病叶冻存于 −70℃冰箱中。供试灰飞虱采自云南田间，经测定无毒后饲养于水稻品种台中 1 号上。供试水稻品种武育粳 3 号及大肠杆菌 DH5α 为福建省植物病毒学重点实验室保存。RNA 提取试剂盒 BIOZOL、RT-PCR 两步合成试剂盒、胶回收试剂盒均购自杭州博日公司；pMD18-T 连接试剂盒以及核酸内切酶 *EcoR*I 和 *Hind*III 均购自 TaKaRa 公司。

1.2 方法

1.2.1 引物设计

引物参照 RSV 云南楚雄分离物 YCX1 的 RNA2（GenBank 登录号：AY186790）与 RNA3（GenBank 登录号：AF508913）序列设计，由 TaKaRa 公司合成（表 1)。

1.2.2 RNA 提取及 RT-PCR

采用 BIOZOL 试剂提取 RNA，采用两步合成试剂盒进行 RT-PCR 扩增，操作按说明进行。

1.2.3 克隆与测序 PCR

产物经回收纯化（按试剂盒说明进行操作）后，与 pMD18-T 连接。连接产物与采用 $CaCl_2$ 法制备的大肠杆菌 DH5α 感受态细胞进行转化。将转化产物涂布于含 100μg/mL 氨苄西林的 LB 固体培养基上，于 37℃培养 20h。于平板上挑取单菌落，置于含 100μg/mL 氨苄西林的 LB 液体培养基中震荡培养。提取质粒 DNA，使用 *EcoRI* 和 *Hind*III 筛选阳性克隆。序列测定由北京百泰克生物技术有限公司完成。

1.2.4 序列分析

利用软件 DNAMAN. 6.0.3.48 及 http://blast.ncbi.nlm.nih.gov/Blast.cgi 进行序列分析。

表 1 参照 YCX1 与 JHZ 分离物基因组序列设计的引物

引物编号	序列	在基因组中的位置
PR207	5′GCAATGGAGTCTTCTACAAC3′	与 RNA2:2457-2476nt 一致
PR208	5′ACACAAAGTCTGGGTATAAC3′	与 RNA2:3495-3514nt 互补
PR305	5′GTCTAGTCATCTGCACCTTC3′	与 RNA3:1448-1467nt 一致
PR306	5′ACACAAAGTCTGGGTAATAA3′	与 RNA3:2491-2510nt 互补
PRN5	5′GGAATTCCATTTTAAATCATATTTCATC3′	与 RNA2:3460-3480nt 互补
PRCP	5′CGGGATCCATGGGTACCAACAAGCCAAC3′	与 RNA3:2398-2417nt 互补

注：限制性内切酶 *EcoRI* 与 *Bam*HI 位点在引物中以下划线表示。

2 结果与分析

2.1 RSV 重组 RNA

提取病叶总 RNA，以 PR306 为引物进行反转录，并以 PR306 及 PR305 为引物进行 PCR 反应，扩增获得片段 YCX07CP 与 YCX07R。测序结果表明，YCX07CP 是 RSV RNA3（GenBank 登录号：EU931496）3′ 端长约 1060 的多核苷酸的序列；YCX07R 是全长为 867 的多核苷酸链，这条多核苷酸链是由 RSV 的 RNA2 与 RNA3 重组而成（GenBank 登录号：FJ807502）。YCX07R 两个末端序列与 RNA3 的末端序列一致，5′ 端的 568 个碱基与 RNA2 正链的 3′ 端互补，3′ 端的 312 个碱基与 RNA3 正链 3′ 末端一致。

为进一步验证重组 RNA 的存在，设计另一对引物 PRN5 与 PRCP，并引入了酶切位点（以便可能的遗传操作），以 PR306 为引物获得的反转录产物为模板进行 PCR，获得 YCX07R 内部 754 核苷酸长的片段，命名为 YCX07RI（图 1）。回收该片段进行测序。结果表明，YCX07RI 两个克隆在两个引物之间的序列与 YCX07R 相应片段分别仅存在一个碱基的差异。

另以引物 PR207 和 PR208 作 RT-PCR，克隆并测序获得 RSV RNA2（GenBank 的登录号：EU931495）3′ 末端长约 1060 个核苷酸的片段 YCX072D。YCX07R、YCX07CP 和 YCX072D 的序列比对显示：YCX07CP 和 YCX072D 除末端序列相似外，在内部都存在一个长为 13 个碱基的反向互补序列：TGCAA(T)CACCATGA（重组点）；YCX07R 是由 YCX07CP 的 3′ 端与 YCX072D 的 3′ 端互补序列（即 RNA2 负链的 5′ 端）通过该重组点联结而成；重组的两部分片段在原基因组中的极性相反。YCX07R 的 5′ 端与 YCX072D 的 3′ 端互补部分除引物外有两个碱基不能配对；YCX07R 的 3′ 端与 YCX07CP 的 3′ 端仅在重组点有一个碱基差异，即 YCX07R 在重组点的序列与 YCX072D 一致。因而，YCX07R 是由 YCX072D（RNA2）与 YCX07CP（RNA3）通过重组形成。

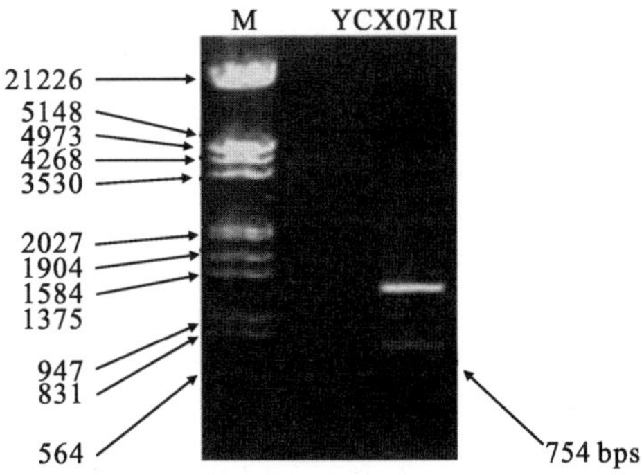

图 1 YCX07RI PCR 产物

M：λDNA/*Eco*R Ⅰ +*Hind* Ⅲ marker

2.2 重组 RNA 序列分析

使用软件 DNAMAN. 6.0.3.48 分析重组 RNA 的结果显示，YCX07R 具有 RSV 基因组结构特征，其两个末端的序列互补。在正义链及反义链上均存在一个开放阅读框，分别编码一个蛋白；两个阅读框存在重叠区域，不存在 RSV 典型的基因间隔区（图 2）。正义链编码的蛋白 YCX07R-5P 全长 228 个氨基酸残基，其中前 179 个氨基酸残基（下划线）来自 RSV NSvc2 蛋白的前 179 个氨基酸残基，后 49 个氨基酸残基由重组 RNA 编码生成；反义链编码的蛋白 YCX07R-3P 全长 98 个氨基酸残基，其中前 73 个氨基酸残基来自 RSV CP 蛋白的前 73 个氨基酸残基（下划线），后 25 个氨基酸残基由重组 RNA 编码生成。两个预测蛋白的氨基酸残基序列如下：

YCX07R-5P: MHFKSYFIYTTIFNMAWGAPIPFPDTHSWMRNREREPSEIVKVPCSARAPPCKLAYDLNGYFIENGLICYNRASVNYFETCYTGNYNYKLPLHPSFSKFGGHVYLSCDDAILQNVSLVGIQQTEYTSSPLLITTSNSERISYSNLRTGFLGMVYAVETRACIQPDQAKKPEEIINHGVAHSHILGQCVTTFVLQYANQSCSIIACILYLTSKSHNISFMSSQVIQRIF

YCX07R-3P:MGTNKPATLADLQKAINDISKDALNYLTAHKADVVTFAGQIEYAGYDAATLIGILKDKGGDTLAKDMTMCNTMIYDFFWFLSLIRLNTCSSFNSIHHS

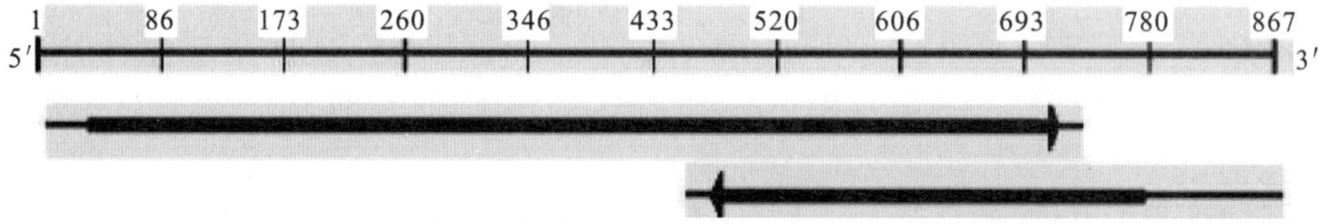

图 2 DNAMAN.6.0.3.48 分析预测的 YCX07R 两个开放阅读框

3 讨论

RSV 基因组中各 RNA 末端的序列具有高度相似性，而 RNA2 与 RNA3 的 3′ 端序列相似性极高（PR208 与 PR306）；因而，YCX07R 可能是在 PCR 扩增 YCX07CP 片段时通过 PR306 单引物扩增而得。对 YCX07R 的序列分析显示，该重组 RNA 互补的两个末端分别来自 RSV 基因组中 RNA2 的 3′ 端与 RNA3 的 3′ 端；YCX07R 是由 CX07CP 的 3′ 端与 YCX072D 的 3′ 端的互补序列（即 RNA2 负链的 5′ 端）通过重组点联结而成；重组的两部分片段在原基因组中的极性相反。研究表明（Chetverin et al., 2005），脊髓灰质炎病毒（Poliovirus）利用引物互补延伸机制（Primer Alignment and Extension Mechanism）产生同源重组，重组点的序列与原序列一致，不存在插入或缺失序列；Qβ 噬菌体不是利用该机制而更可能是通过两个重组片段的酯化反应产生重组，其重组点的序列与原序列不一致，存在不同程度的插入序列。这些插入序列能形成互补的二级结构。序列分析显示，YCX07R 在重组点的序列与原序列一致，不存在插入或缺失序列。因此，YCX07R 更可能是 RNA2 与 RNA3 通过重组点的序列一致性发生类似于脊髓灰质炎病毒的同源重组，重组过程如下：RSV 的复制酶 RdRp 以正链 RNA2 为模板进行复制，产生互补的负链 RNA2（cRNA2）。当复制进行到重组点时，复制中断，复制体系解体（即流产性复制）。新合成的负链 RNA2 仅长 568 个碱基。但复制酶并未与新链 RNA 解离，而是携带着新链 RNA 跳到新合成的负链 RNA3 上，并以负链 RNA3 为模板，以不完整的负链 RNA2 的 3′ 端为引物，通过重组点的序列互补性继续 RNA 复制，合成由 RNA2 的 3′ 端与 RNA3 的 3′ 端联结而成的重组 RNA，即 YCX07R（如图 3 中的粗箭头所示）。由于重组点自身存在两个回文结构（TCATGGTGTTGCA），可使 RNA2 与 RNA3 在重组点处形成互补的二级结构（图 4）。因此，YCX07R 也有可能通过类似于 Qβ 噬菌体的酯化反应产生，即新合成的负链 RNA2 在重组点通过末端的羟基攻击正链 RNA3，形成 YCX07R（图 3 中细箭头所示）。由于 3 个克隆的重组点序列均不存在插入序列，因此这种重组模式不大可能发生。重组机制的确定需要进行进一步的研究。

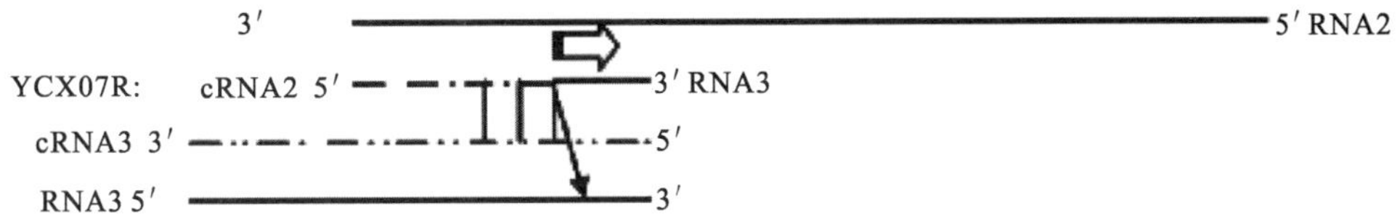

图 3　YCX07R 重组的可能机制

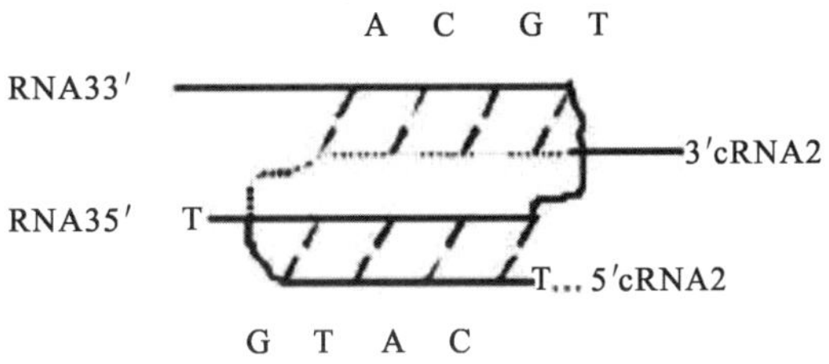

图 4　两个一致的重组点序列 TCATGGTGTTGCA 可能存在的二级结构

参考文献

[1] 魏太云 , 王辉 , 林含新 , 等 . 我国水稻条纹病毒 RNA3 片段序列分析 —— 纤细病毒属重配的又一证据 [J]. 生物化学与生物物理学报 : 英文版 , 2003, 35(01):97-103.

[2] 魏太云 , 林含新 , 吴祖建 , 等 . 水稻条纹病毒 RNA4 基因间隔区序列分析 —— 混合侵染及基因组变异证据 [J]. 微生物学报 , 2003, 43(5):577-585.

[3] BAROTH M, ORLICH M, THIEL H J, et al. Insertion of cellular NEDD8 coding sequences in a pestivirus[J]. Virology, 2000, 278(2):456-466.

[4] CHETVERIN A B. Recombination in Bacteriophage Qβ and Its Satellite RNAs: Thein Vivoandin VitroStudies[J]. Seminars in Virology, 1997, 8(2):121-129.

[5] CHETVERIN A B, KOPEIN D S, CHETVERINA H V, et al. Viral RNA-directed RNA polymerases use diverse mechanisms to promote recombination between RNA molecules[J]. Journal of Biological Chemistry, 2005, 280(10):8748-8755.

[6] FIGLEROWICZ M, BUJARSKI J J. RNA recombination in brome mosaic virus, a model plus strand RNA virus[J]. Acta Biochimica Polonica, 1998, 45(4):847-868.

[7] HIRST G K. Genetic recombination with Newcastle disease virus, polioviruses, and influenza[J]. Cold Spring Harbor Symposia on Quantitative Biology, 1962, 27(7):303-309.

[8] KHATCHIKIAN D, ORLICH M, ROTT R. Increased viral pathogenicity after insertion of a 28S ribosomal R NA sequence into the haemagglutinin gene of an influenza virus[J]. Nature, 1989, 340(6229):156-157.

[9] OLSTHOORN R, BRUYERE A, DZIANOTT A, et al. RNA recombination in brome mosaic virus: effects of strand-specific stem-loop inserts[J]. Virol. 2002, 76(24):12654-12662.

[10] RAMÍREZ B C, HAENNI A L. Molecular biology of tenuiviruses, a remarkable group of plant viruses[J]. Journal of General Virology, 1994, 75 (3):467-475.

[11] WANG T C, CHAO M. RNA recombination of hepatitis delta virus in natural mixed-genotype infection and transfected cultured cells[J]. Journal of Virology, 2005, 79(4):2221-2229.

Genetic diversity and population structure of rice stripe virus in China

Taiyun Wei[1], Jinguang Yang[1], Furong Liao[1], Fangluan Gao[1], Lianming Lu[1], Xiaoting Zhang[1], Fan Li[1,2], Zujian Wu[1], Qiyin Lin[1], Lianhui Xie[1], Hanxin Lin[1]

(1 Institute of Plant Virology, Fujian Agricultural and Forestry University,Fuzhou, Fujian 350002, PR China; 2 Key Laboratory of Agricultural Biodiversity and Pest Management of Ministry of Education, Yunnan Agricultural University, Kunming, PR China)

Abstact: Rice stripe virus(RSV) is one of the most economically important pathogens of rice and is repeatedly epidemic in China, Japan and Korea. The most recent outbreak of RSV in eastern China in 2000 caused significant losses and raised serious concerns. In this paper, we provide a genotyping profile of RSV field isolates and describe the population structure of RSV in China, based on the nucleotide sequences of isolates collected from different geographical regions during 1997–2004. RSV isolates could be divided into two or three subtypes, depending on which gene was analysed. The genetic distances between subtypes range from 0.050 to 0.067. The population from eastern China is composed only of subtype I/IB isolates. In contrast, the population from Yunnan province (southwest China) is composed mainly of subtype II isolates, but also contains a small proportion of subtype I/IB isolates and subtype IA isolates. However, subpopulations collected from different districts in eastern China or Yunnan province are not genetically differentiated and show frequent gene flow. RSV genes were found to be under strong negative selection. Our data suggest that the most recent outbreak of RSV in eastern China was not due to the invasion of new RSV subtype(s). The evolutionary processes contributing to the observed genetic diversity and population structure are discussed.

1 Introduction

Due to their error-prone RNA replication, large population size and short generation time, RNA viruses have high mutation rates (Drake and Holland, 1999). Therefore, RNA viruses exhibit high potential for genetic variation, and a large number of nucleotide variations could exist in natural populations. Analysing the polymorphic pattern of these variations will help us to understand the phylogenetic relationships, epidemiological routes, population structures and underlying evolutionary mechanisms of RNA viruses. In turn, this information will facilitate the development of effective control strategies for plant viral diseases.

Rice stripe virus (RSV) is one of the most important plant pathogens in China. The rice stripe disease induced by RSV was first recorded in Jiangsu–Zhejiang–Shanghai (JZS) district in 1963 and was later discovered in 16 provinces (Lin et al., 1990). Outbreaks of this disease were reported in JZS district in 1966, in Taiwan in 1969, in Yunnan province in 1974, in Beijing in 1975, in Shandong in 1986 and in Liaoning in the early 1990s (Lin et al., 1990; Xie et al., 2001). The disease is endemic in Yunnan, Jiangsu, Shanghai, Shandong,

Journal of General Virology. 2009,90: 1025-1034

Received 26 August 2008; Accepted 29 December 2008

Beijing and Liaoning provinces. Since 2000, the disease has circulated widely in Jiangsu province and become more severe. For instance, approximately 780000 ha rice was infected in 2002 in Jiangsu province. This increased to 957000 ha in 2003 and 1571000 ha in 2004, accounting for 80% of the rice fields and a 30%~40% yield loss. The outbreak of RSV also spread to adjacent provinces, such as Henan, Zhejiang, Anhui, Shandong, Shanghai and Hebei. However, the reasons behind this outbreak are not well understood. For example, it is unclear whether the invasion was triggered by a new RSV genotype. Outside China, RSV has been reported only in Japan, Korea and the Far East (ex-USSR) and has been epidemic in Japan and Korea since the 1960s, causing significant losses of rice yield (Hibino, 1996).

RSV is the type member of the genus *Tenuivirus*. It mainly infects rice plants, but also infects some other species in the family Poaceae, such as wheat and maize. RSV is transmitted by a small brown planthopper, *Laodelphax striatellus* Fallen (Hemiptera, Delphacidae), in circulative– propagative and transovarial manners (Falk and Tsai, 1998). The genome of RSV is composed of four negative-sense single-stranded RNA segments, designated RNA 1, 2, 3 and 4 according to decreasing size (Fig. 1) (Takahashi et al., 1993; Toriyama et al., 1994; Zhu et al., 1991, 1992). RNA 1 is of negative polarity, encoding the putative RNA dependent RNA polymerase (Toriyama et al., 1994). The other three segments adopt an unusual ambisense coding strategy, i.e. both the viral-sense RNA (vRNA) and viral complementary-sense RNA (vcRNA) possess coding capacity, but the functions of the translated proteins are unclear. What is known is that the vRNA 2 encodes a membrane-associated protein and the vcRNA 2 encodes a polyglycoprotein (Falk and Tsai, 1998;Takahashi et al.,1993). The nucleocapsid (N) protein gene has been mapped to vcRNA 3 (Kakutani et al., 1991; Zhu et al., 1991), and the NS3 protein encoded by vRNA 3 could act as an RNA-silencing suppressor based on the function of the analogous NS3 of rice hojablanca virus, another member of the genus *Tenuivirus* (Bucher et al., 2003). vRNA 4 encodes a protein known as major non-capsid protein (NCP) that accumulates in infected plants and may be involved in pathogenesis (Toriyama, 1986), whilst vcRNA 4 encodes a protein that has recently been shown to be involved in movement (Xiong et al., 2008).

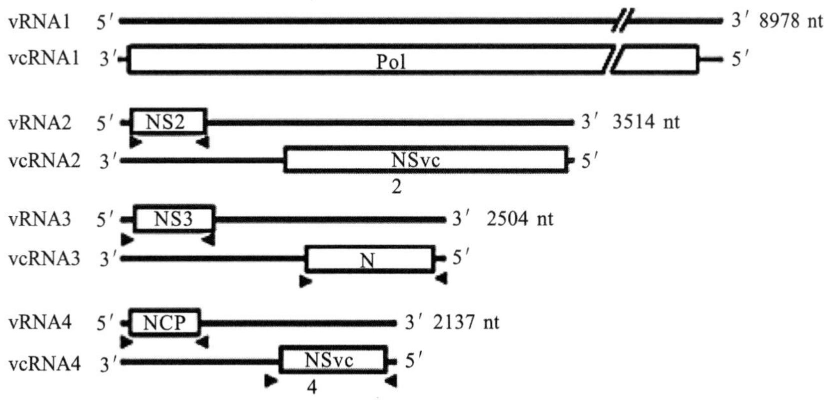

Fig. 1 Schematic representation of the RSV genome

The lines represent RNA segments and the empty boxes denote genes encoded by the vRNAs (viral-sense RNAs) or vcRNAs (viral complementary-sense RNAs). Oligonucleotide primers used for RT-PCR of RSV genes are indicated by arrows. Pol, RNA dependent RNA polymerase

Numerous studies have been performed in recent decades on RSV aetiology, pathogenesis, ecology, molecular biology, control strategies etc. (Falk and Tsai, 1998).Nevertheless, there are still major gaps, especially in our understanding of the genetic diversity and population structure of RSV. Our previous studies have shown the biological diversity and genetic variations of several RSV isolates (Lin et al., 1999, 2001, 2002; Wei et al., 2003a, b). In this report, we analysed five genes (NS2, N, NS3, NCP and NSvc4) of 136 RSV isolates collected from different areas in China during 1997–2004. Our data showed that RSV isolates could be divided into two or three subtypes, depending on which gene was analysed. The distribution of these subtypes is correlated with their geographyical locations, but not with the collection years. All of the isolates collected from eastern China and Japan belong to subtype I/IB, whereas isolates from Yunnan province are much more diverse, belonging to different subtypes with subtype II being predominant.

2 Methods

2.1 Virussamples

RSV samples were collected in rice fields from eight provinces (Fujian, Jiangsu, Shanghai, Zhejiang, Henan, Shandong, Beijing and Liaoning) in eastern China and from Yunnan province in southwest China during 1997–2004. A virus sample from an individual rice plant was considered as one isolate. Infected rice plants were either used for extraction of total RNA or stored at –80℃ for future use. Some isolates were maintained on suitable rice varieties (e.g. Hexi 28) via transmission by small brown planthoppers, *L. striatellus* Fallen. The geographical locations, collection years and rice varieties of RSV isolates used in this study are listed in Supplementary Table S1 (available in JGV Online).

2.2 RT–PCR, cloning and sequencing

Extraction of total RNA from infected rice leaves, cDNA synthesis and PCR amplification were performed as described previously (Lin et al., 2001). Forward and reverse primers were designed according to the nucleotide sequences of RNAs 2, 3 and 4 of RSV isolate T (GenBank accession numbers NC_003754,NC_003776 and NC_003753,respectively): 5′-ATGGCATTACTCCTTTTCAATG-3′ (nt81–102)and 5′-CCAAATTCACATTAGAATAGG-3′ (nt 666–686) for the NS2 gene;5′-GTTCAGTCTAGTCATCTGCAC-3′ (nt 1437–1457) and 5′-ACACAAAGTCTGGGT-3′ (nt2490–2504) for the N gene; 5′-ACACAAAGTCCTGGGT-3′ (nucleotides nt 1–15) and 5′-CTACAGCACAGCTGGAGAGCTG-3′ (nt 681–701) for the NS3 gene; 5′-ACACAAAGTCCTGGG-3′ (nt 1–15) and 5′-GGTGGAAAATGTGATATGCAAT-3′ (nt 597–616) for the NCP gene; and 5′-TGGAACTGGTTATCTCACCT-3′ (nt 1208–1228) and 5′-ACACAAAGTCATGGC-3′ (nt 2123–2137) for the NSvc4 gene.

RT-PCR products were purified by using a QIA quick PCR extraction kit (Qiagen). The purified PCR products were inserted into the pGEM-T vector (Promega) followed by transformation into *Escherichia coli* DH5α. Nucleotide sequences were determined by the Shanghai Jikang Biotech Company. For each gene of each isolate, one or two clones were sequenced. The GenBank accession numbers of these sequences are listed in Supplementary Table S1.

2.3 Phylogenetic analysis

Multiple nucleotide sequence alignments were performed by using CLUSTAL W (Thompson et al., 1994). Alignments were also adjusted manually to guarantee correct reading frames. Non-coding sequences were removed before alignment. Eight RSV isolates (T, M, O, C, Y, JS-YM, JSYD-05 and SD-JN2) that had been sequenced by other laboratories were included as references (Kakutani et al., 1990, 1991; Qu et al., 1997; Wang et al., 1992; Zhu et al., 1991, 1992). Maize stripe virus (MStV), the most closely related virus to RSV in the genus *Tenuivirus*, was used as outgroup for phylogenetic analyses. The GenBank accession numbers for RNAs 2, 3 and 4 of MStV are U53224, S40180 and AJ969410, respectively. All of the sequence alignments used in this study are available upon request. Phylogenetic trees were reconstructed by using the neighbour-joining (NJ) method with the Kimura two-parameter model implemented in PAUP* 4.0b10.0 (Swofford, 2002). Gaps were treated as a fifth character state. Evaluation of statistical confidence in nodes was based on 1000bootstrap replicates. Branches with <50% bootstrap value were collapsed.

2.4 Estimation of genetic distance and selection pressure

Genetic distances (the average number of nucleotide substitutions between two randomly selected sequences in a population) within and between subtypes were calculated by MEGA 2.1 based on the Kimura two parameter model(Kumar et al., 2001). The d_n:$d_s$ratio is used to estimate selection pressure.d_n (the average number of non-synonymous substitutions per non-synonymous site) and d_s (the average number of synonymous substitutions per synonymous site) were estimated by using the Pamilo–Bianchi–Li method implemented in MEGA 2.1. SEM was computed by MEGA 2.1 using a bootstrap with 100 replicates.

2.5 Statistical tests of genetic differentiation and measurement of gene flow

Genetic differentiation between populations was examined by three permutation-based statistical tests, Ks*, Z and Snn, which represent the most powerful sequence-based statistical tests for genetic differentiation and are recommended for use in cases of high mutation rate and small sample size (Hudson, 2000; Hudson et al., 1992). The extent of genetic differentiation or the level of gene flow between populations was measured by estimating F_{st} (the interpopulational component of genetic variation or the standardized variance in allele frequencies across populations). F_{st} ranges from 0 to 1 for undifferentiated to fully differentiated populations, respectively. Normally, an absolute value of F_{st}>0.33 suggests infrequent gene flow. The statistical tests for genetic differentiation and estimation of F_{st} were performed by DnaSP 4.0 (Rozas et al., 2003).

3 Results

3.1 Genotype profile of RSV field isolates

To provide a genotype profile of RSV field isolates, we performed phylogenetic analyses of RSV isolates collected from different provinces in eastern and southwest China during 1997–2004 (Table S1). Neighbour-joining (NJ) trees were constructed by using datasets for five RSV genes: NS2 (600bp, 55 sequences), NS3 (636bp, 44 sequences), N (969bp, 87 sequences), NCP (543bp, 65 sequences) and NSvc4 (861bp, 33 sequences). As illustrated in Fig. 2 and Fig. 3, RSV isolates fell into two monophyletic clades. As all of the isolates collected from eastern China form one of the monophyletic clades, we refer to it as subtype I and to the other as subtype II. The mean genetic distance between these two subtypes ranges from 0.054 to 0.067, and those within subtypes range from 0.008 to 0.038 (Table 1). Unlike the polytomic topology of subtype I in the NS2 and NS3 gene trees, subtype I in the NCP and NSvc4 gene trees is dichotomic (Fig. 2). Noticeably, one of the sister clades only contains isolates that were collected from Yunnan province. We thereby refer to this clade as subtype IA, and to the other as subtype IB (Fig. 2c, d). Indeed, the genetic distances between subtypes IA, IB and II fall into a range (0.050–0.060) similar to those between subtypes I and II (Table 2). Therefore, for the NCP and NSvc4 genes, we divided RSV isolates into three subtypes.

Table 1 Genetic distances within and between subtypes I and II for five genes

Gene	Genetic distance		
	Within subtype I	Within subtype II	Between subtypes I and II
NS2	0.022±0.003	0.025±0.004	0.054±0.008
NS3	0.021±0.002	0.029±0.004	0.067±0.009
N	0.017±0.002	0.023±0.002	0.057±0.009
NCP	0.036±0.004	0.008±0.001	0.055±0.007
NSvc4	0.038±0.004	0.021±0.003	0.058±0.006

Genetic distance refers to the average number of nucleotide substitutions between two randomly selected sequencesin a population and was estimated by the program MEGA2 based on the Kimura two-parameter model. SEM was calculated by using a bootstrap of 100 replicates. Subtypes are designated based on the phylogenetic analyses of RSV in Fig. 2 and Fig. 3

Table 2 Genetic distances within and between subtypes IA, IB and II of the NCP and NSvc4 genes Subtypes are designated based on the phylogenetic analyses of RSV in Fig. 2 and 3

Gene	Genetic distance					
	Within subtype IA	Within subtype IB	Within subtype II	Between IA and IB	Between IA and II	Between IB and II
NCP	0.025±0.004	0.025±0.004	0.008±0.001	0.051±0.008	0.057±0.009	0.050±0.007
NSvc4	0.012±0.002	0.024±0.003	0.021±0.003	0.055±0.006	0.060±0.007	0.056±0.006

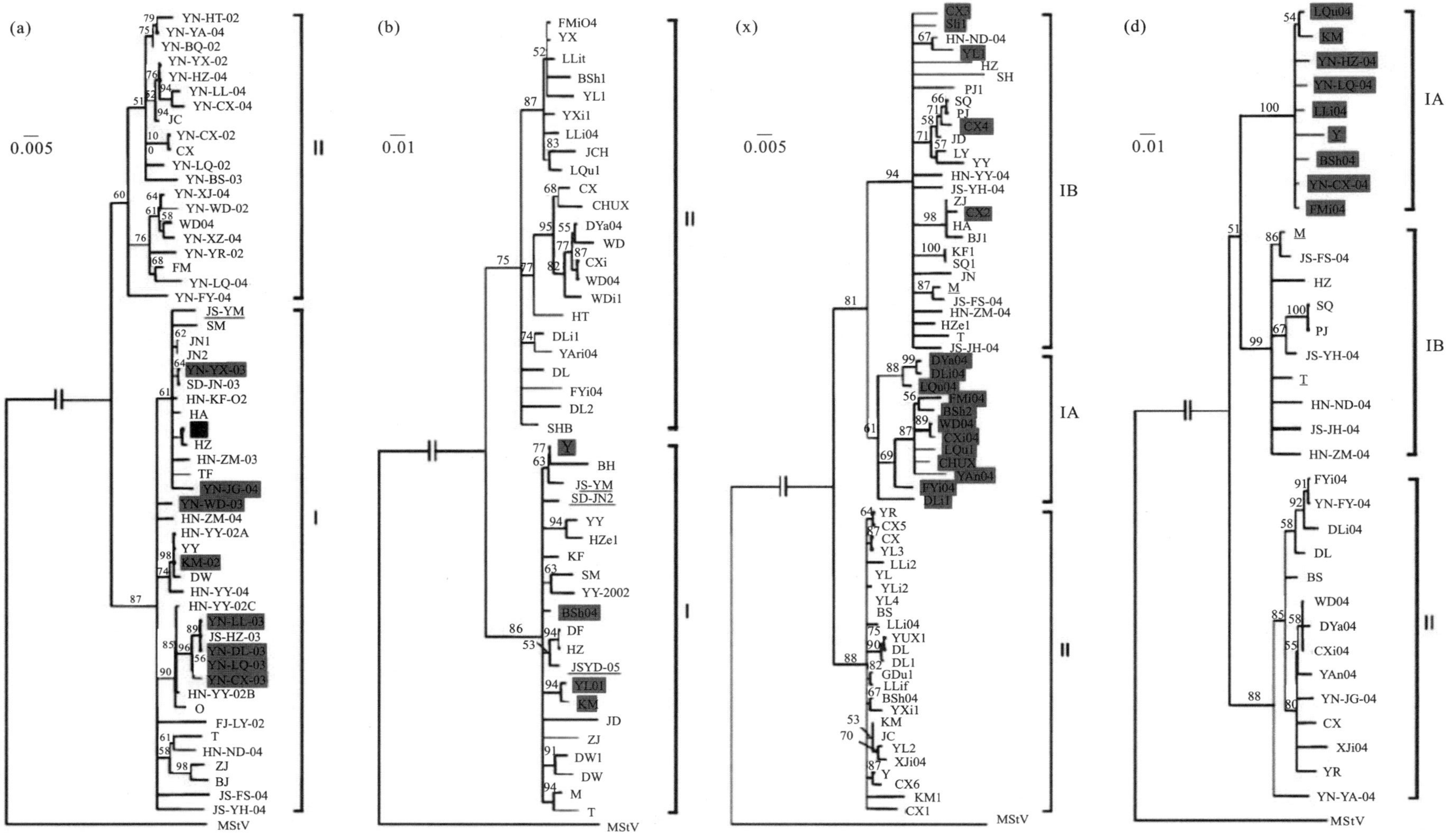

Fig. 3 Phylogenetic analysis of four RSV genes

Trees were constructed by the NJ algorithm with the Kimura two-parameter model implemented in PAUP*4.0b10.0.

Branches were collapsed when the bootstrap value was < 50%.Bootstrap values are given above the branches of each clade.The trees were rooted by using *maizestripevirus*(MStV) as the out group. Eight RSVisolates (T, M, O, C, Y, JS-YM, JSYD-05 and SD-JN2)that were sequenced by other laboratories were included as references and are underlined. Taxa on grey backgrounds indicate the subtype I, IB and IA isolates that were collected in Yunnan province. Bars, number of substitution spersite. (a)NS2gene; (b)NS3 gene; (c)NCPgene; (d)NSvc4gene

3.2 Spatial and temporal distribution of RSV isolates in nature

To better understand the spatial and temporal distribution patterns, the collection years and geographical locations of RSV isolates corresponding to taxa in the phylogenetic trees were depicted (information for the N gene tree is shown in Fig. 3). Considering the geographical separation/ proximity, the RSV population in China was first divided into two subpopulations, YN and E, containing isolates collected from Yunnan province and eastern China, respectively. RSV isolates collected from Yunnan province were further grouped into five districts: Baoshan–Shaba–Banqiao–Xinjie–Hetu–Jiaguan (BS), Dali–Xizhou–Fengyi–Weishan (DL), Chuxiong–Yongren–Dayao–Yaoan (CX), Kunming–Luquan–Wuding–Fumin–Yiliang–Shilin–Luliang (KM) and Yuxi–Jiangchun (YX). Those from eastern China were further grouped into six districts: Fujian (FJ), Jiangsu– Zhejiang–Shanghai (JZS), Henan (HN), Shandong (SD), Beijing (BJ) and Liaoning (LN) (Fig. 3).

Two interesting patterns are seen in Fig. 3. Firstly, all of the isolates collected from eastern China and Japan belong to subtype I. Secondly, all of the subtype II isolates were collected from Yunnan province. However, two isolates, Y and CX2 (shown on grey backgrounds in Fig. 3), collectedfrom Yunnan province also belong to subtype I. Therefore, isolates collected from Yunnan province are more diverse, belonging to different subtypes. A similar genetic structure was observed inthe trees for the other four genes(Fig. 2). We also examined the spatial distribution pattern of isolates collected from different districts of eastern China and Yunnan province, but failed to see any obvious pattern.

Despite examining the phylogenetic grouping of RSV isolates and their collection years closely, we are unable to find a general pattern as clear as that for spatial distribution. However, we discovered that most of the subtype IA isolates in the NCP and NSvc4 gene trees were collected in 2004 (Fig. 2c, d).

3.3 Genetic differentiation between subpopulations

To test the genetic differentiation between and within subpopulations E and YN, three statistical tests, Ks*, Z and Snn, were used (Hudson, 2000; Hudson et al., 1992). As shown in Table 3, all of the tests strongly support the hypothesis that these two subpopulations are genetically differentiated (P=0). Although, for some genes, significant genetic differentiation could also be observed within subpopulations YN and E, it was not supported by all three statistical tests. To estimate the extent of genetic differentiation, we measured the coefficient F_{st}, which is also an estimate of gene flow. Except for the NS2 gene, the values of F_{st} between subpopulations are all $>$ 0.33(Table 3), an indication of infrequent gene flow. The absolute values of F_{st} for within subpopulations are all $<$ 0.33, suggesting frequent gene flow. This was also seen when F_{st} between any two districts within YN or E was measured (data not shown).

3.4 Strong selective pressures acting on RSV genes

We also estimated the negative selection pressure acting on RSV genes. Overall, the values of the d_n:d_s ratio for five genes were markedly low (0.046–0.123; Table 4), implying that all of these genes are under strong negative selection. The value of the d_n:d_s ratio for the NS2 gene is slightly higher than that for the N and NS3 genes, but is 2.6 times higher than that of the NCP and NSvc4 genes. Therefore, the NCP and NSvc4 genes are subject to stricter selective constraints. Interest-ingly, the values of the d_n:$d_s$ratio for genes on the same RNA segment, i.e. the NS3 and N genes on RNA 3, and the NCP and NSvc4 genes on RNA 4, are almost identical (Table 4). This suggests that the single RNA segments of the divided genome may be the evolutionary unit of selection. With regard to selection pressure acting on different subtypes, there is no general pattern of which subtype is under stronger or weaker selection pressure.

Table 3 Genetic differentiation between and within subpopulations YN and E

Gene	Test	Subpopulation		
		Between YN and E	Within YN	Within E
NS2	P-value of Ks*	0.000†	0.510	0.126
	P-value of Z	0.000†	0.767	0.403
	P-value of Snn	0.000†	0.353	0.032†
	F_{st}‡	0.310	-0.048	-0.010
NS3	P-value of Ks*	0.000†	0.098	0.232
	P-value of Z	0.000†	0.189	0.066
	P-value of Snn	0.000†	0.030†	0.033†
	F_{st}	0.476	0.133	0.281
N	P-value of Ks*	0.000†	0.385	0.436
	P-value of Z	0.000†	0.197	0.344
	P-value of Snn	0.000†	0.206	0.281
	F_{st}	0.604	0.069	0.042
NCP	P-value of Ks*	0.000†	0.648	0.514
	P-value of Z	0.000†	0.863	0.686
	P-value of Snn	0.000†	0.087	0.831
	F_{st}	0.415	-0.043	-0.059
NSvc4	P-value of Ks*	0.000†	0.007†	0.318
	P-value of Z	0.000†	0.008†	0.334
	P-value of Snn	0.000†	0.090	0.858
	F_{st}	0.435	0.282	-0.095

Subpopulation division is based on the geographical sites of origin ofRSV samples

The RSV population in China was first divided into two, subpopulations, Yunnan province (YN) and eastern China (E).Subpopulation YN was further divided into five districts: Baoshan(BS), Dali (DL), Chuxiong (CX), Kunming (KM) and Yuxi (YX). Subpopulation E was further divided into six districts: Liaoning(LN), Beijing (BJ), Henan (HN), Shandong (SD),Jiangsu–Zhejiang–Shanghai (JZS) and Fujian (FJ)

†$P < 0.05$, which was considered as significantly rejecting the null hypothesis that there is no genetic differentiation between two subpopulations.‡Fst is a coefficient of the extent of genetic differentiation and provides an estimate of the extent of gene flow. An absolute value of Fst > 0.33 suggests infrequent gene flow.

Table 4 Estimation of nucleotide diversity of five RSV genes

Gene		Nucleotide diversity			
		Subtype IA*	Subtype I/IB	Subtype II	All†
NS2	d_n	NA	0.006±0.001	0.009±0.002	0.013±0.003
	d_s	NA	0.058±0.008	0.073±0.014	0.105±0.015
	$d_n : d_s$	NA	0.103	0.123	0.123
NS3	d_n	NA	0.010±0.002	0.006±0.001	0.014±0.003
	d_s	NA	0.052±0.006	0.083±0.013	0.139±0.019
	$d_n : d_s$	NA	0.192	0.072	0.100
N	d_n	NA	0.004±0.001	0.008±0.002	0.011±0.003
	d_s	NA	0.047±0.006	0.064±0.008	0.111±0.012
	$d_n : d_s$	NA	0.085	0.125	0.099
NCP	d_n	0.003±0.002	0.008±0.002	0.004±0.001	0.007±0.002
	d_s	0.090±0.017	0.065±0.012	0.021±0.006	0.151±0.025
	$d_n : d_s$	0.033	0.123	0.190	0.046
NSvc4	d_n	0.004±0.001	0.007±0.002	0.004±0.001	0.007±0.001
	d_s	0.032±0.006	0.067±0.012	0.061±0.009	0.149±0.015
	$d_n : d_s$	0.125	0.104	0.066	0.047

Nucleotide diversity, defined here as the average number of nucleotide substitutions per site, was computed separately for non-synonymous sites (d_n) and synonymous sites (d_s). The d_n: d_s ratio gives an estimate of selection pressure; if d_n:d_s < 1.0, negative selection is implied

*NA, Not applied. The subdivisions of subtype IA, IB and II were applied only to the NCP and NSvc4 genes.

†All isolates were included in the analysis.

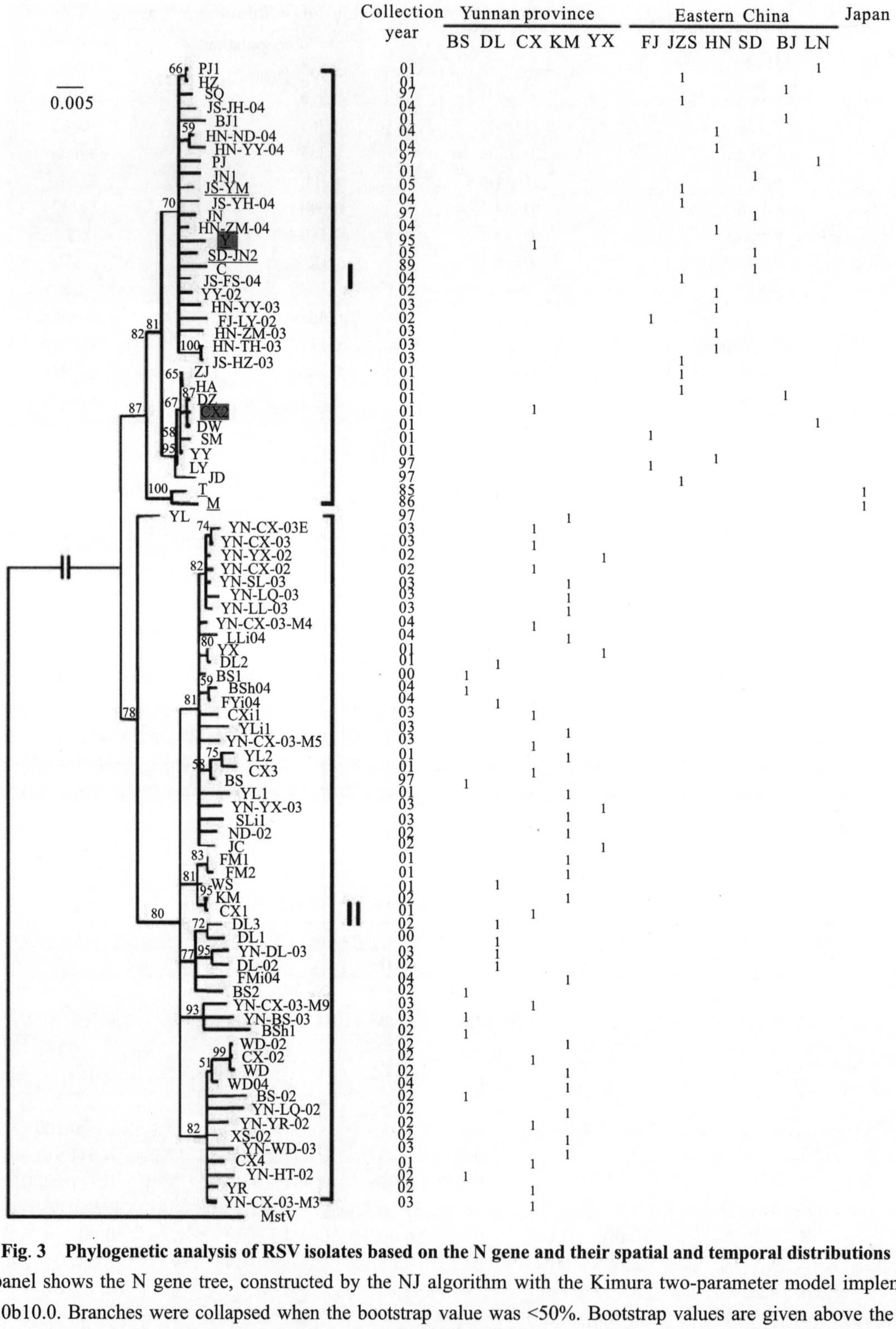

Fig. 3 Phylogenetic analysis of RSV isolates based on the N gene and their spatial and temporal distributions

The left panel shows the N gene tree, constructed by the NJ algorithm with the Kimura two-parameter model implemented in PAUP* 4.0b10.0. Branches were collapsed when the bootstrap value was <50%. Bootstrap values are given above the branches of each clade.The tree was rooted by using MStV as the outgroup. RSV isolates (T, M, C, Y, JS-YM and SD-JN2) that were sequenced by other laboratories were included as references and are underlined. Taxa on grey backgrounds indicate the subtype I, IB and IA isolates that were collected in Yunnan province. Bar, 0.005 substitutions per site. In the right panel, the collection year and sitefor each isolate are shown. Collecting sites in China fall into two geographically large categories, Yunnan (YN) province andeastern China. There are five districts [Baoshan–Shaba–Banqiao–Xinjie–Hetu–Jiaguan (BS), Dali–Xizhou–Fengyi–Weishan(DL), Chuxiong–Yongren–Dayao–Yaoan (CX), Kunming–Luquan–Wuding–Fumin–Yiliang–Shilin–Luliang (KM) and Yuxi–Jiangchun (YX)] in YN province and six districts, Fujian, Jiangsu–Zhejiang–Shanghai, Henan, Shandong and Liaoning province (FJ, JZS, HN, SD, BJ and LN, respectively), in eastern China

4 Discussion

A number of methods have been used to differentiate virus isolates in order to provide a genotyping basis for examining the genetic composition of viral populations. These include restriction fragment-length polymorphism (RFLP; ArboledaandAzzam, 2000), PCR–single-strand conformation polymorphism (PCR-SSCP; Lin et al., 2004), RNase-protection assays (Fraile et al., 1997) and molecular phylogeny (Abubakar et al., 2003). In this paper, we applied a distance-based NJ method to construct the phylogenetic trees of RSV isolates collected from different areas in China during 1997–2004. Overall, these isolates fell into two monophyletic clades, namely subtypes I and II (Fig. 2, Fig. 3). However, the subtype I clade in the NSvc4 gene tree is only poorly supported (51% bootstrap value, Fig. 2d). Furthermore, some subtype clades collapsed when the character-based maximum-parsimony method was used (data not shown). The low resolution of the phylogenetic trees of RSV isolates could be due to the low genetic diversity (0.05–0.067 between subtypes and up to 0.092 between isolates) and insufficient informative sites regarding evolution of the genome. Nevertheless, our analyses provided a genotyping profile of RSV field isolates, and this allowed us to investigate the genetic structure of RSV populations in China.

Our data show that the population collected from eastern China is composed only of subtype I or IB isolates. In contrast, the population from Yunnan province is composed of different subtypes, with subtype II predominating (Fig. 3). Subpopulations collected from different districts in eastern China or Yunnan province are not genetically differentiated and show frequent gene flow (Table 3). Particularly, subtype I/IB isolates prevailing in the outbreak sites (Jiangsu–Zhejiang–Shangha and Henan districts) show high sequence identities to isolates collected from other provinces in eastern China (Fig. 3). These data indicate that the most recent outbreak of RSV in these provinces was not due to the invasion of a new subtype(s) from Yunnan province.

Although the prevailing genotype in Yunnan province is subtype II, we noticed a recent expansion of a 'new' lineage, subtype IA. Subtype IA isolates were collected from different districts in Yunnan province and mainly in the year 2004 (Fig. 2c, d; Table S1). In fact, some of them already existed in nature before 2004. For example, isolate CHUX was collected in 2001, and the collection year of isolate Y was 1995(Qu et al., 1997). It would be interesting to see whether this subtype expands furtherin the field and to compare the biological properties of isolates of subtype IA with those of subtypes IB and II.

The population structure described here for RSV in China is distinct from that of cucumber mosaic virus (CMV), but is somewhat similar to those of two insect vector-borne rice viruses, rice tungro bacilliform virus (RTBV) and rice yellow mottle virus(RYMV). The genetic composition of 11 CMV populations collected in Spain was not correlated with their geographical locations and collection years, and was described as metapopulation with local colonization, extinction and recolonization (Fraile et al., 1997). In contrast, RTBV populations also showed a spatial distribution pattern with a greater genetic diversity in the Indonesian population than in Philippine or Vietnamese populations (Azzam et al., 2000). Similarly, the genetic diversity of the RYMV population is highest in east Africa, especially in eastern Tanzania, and decreases progressively from the east to the west of Africa (Abubakar et al., 2003).

The geographical origin of a virus can be inferred from the extent of its genetic diversity. If a viral population shows higher genetic diversity, it is normally considered more ancient (Azzam et al., 2000; Fargette et al., 2006; Gessain et al., 1992; Giri et al., 1997; Koralnik et al., 1994; Moya and Garcia, 1995). The historical record of disease may or may not be consistent with the extent of genetic diversity. For example, the higher genetic diversity of the RTBV Indonesian population is consistent with historic records of rice tungro disease in 1840 in Indonesia, but this disease was not reported in the Philippines until 1940 and has been described in Vietnam only recently (Azzam et al., 2000). Likewise, eastern Tanzania is believed to be the centre of origin for RYMV, as the most divergent isolates were found in this area (Abubakar et al., 2003; Fargette et al., 2006). The virus was first reported in Kenya; however, Kenya is located directly north-east of Tanzania (Fargette et al., 2006). According to the population structure described in our report, Yunnan province could be the geographical origin of RSV in China. Although the report of rice stripe disease in this province occurred almost a decade after its discovery in Jiangsu–Zhejiang– Shanghai district (Lin et al., 1990; Xie et al., 2001), the disease could have been unnoticed in Yunnan province for many years. This is likely considering the fact that plant virology research in Yunnan province greatly lagged behind that in eastern China. The scenario of a Yunnan origin is reinforced by the fact that this province is well known as the 'kingdom of plants and animals' that may harbour primary indigenous hosts and efficient insect vectors of RSV. Analogously, Tanzania is also reputed to be a biodiversity hotspot for plants and animals, and this rich biodiversity has been proposed to account for the origin of RYMV as well as another insect-borne plant virus, cassava mosaic virus, in Africa (Fargette et al., 2006). Despite this evidence, we cannot exclude the possibility that different subtypes might have been present in eastern China a long time ago, giving rise to subtype I/IB only in recent years. Interestingly, although the occurrence of rice stripe disease dates back to the early 1990s in Japan (Hibino, 1996), the only three Japanese isolates (T, M and O) with available nucleotide sequences belong to subtype I/IB (Fig. 2, Fig. 3). Apparently, more sequences, especially for Japanese and Korean isolates, are needed to fully disclose the secrets of the origin of RSV.

Why is the RSV Yunnan population composed of different subtypes, whereas the eastern China population is only composed of a single subtype? This can be explained simply by the founder effect, which is often invoked to explain the low genetic diversity of certain populations of various plant viruses, including the RYMV population in western Africa (Fargette et al., 2006; Garcia et al., 2001). However, we believe that the interplay between RSV and its insect vector, L. striatellus, could explain the observed population structure of RSV in China more specifically. Unlike CMV or RYMV, which can be transmitted by contact and several different vectors (Fargette et al., 2006; Palukaitis et al., 1992), RSV is strictly transmitted by L. striatellus in persistent propagative and transovarial manners (Toriyama, 1986). Therefore, the RSV–vector interaction probably played a critical role in shaping the population structure. It is generally believed that *L. striatellus* lacks the ability to migrate long distances unless helped by a monsoon and therefore has evolved to adapt to the local climate (Hoshizaki, 1997). Therefore, it is possible that *L. striatellus* populations in Yunnan and eastern China are genetically differentiated and have different affinities for RSV subtypes. The vector for the Yunnan population may be favourable for the transmission of the predominant subtype II isolates, whereas the vector for the eastern China population is favourable for the subtype I/IB isolates. Such biased transmissibility has been reported for other plant virus–vector systems. For example, whitefly *Bemisia tabaci* populations from various geographical locations had different transmission efficiencies with different isolates of tomato leaf curl geminivirus or other geminiviruses (Bedford et al., 1994; McGrath and Harrison, 1995). This scenario is supported by our unpublished data indicating that four L. *striatellus* populations collected from Yunnan province grouped together with similar random-amplified polymorphic DNA (RAPD) patterns, and were distinct from another group containing nine insect populations collected from eastern China. The specific interaction between RSV subtypes and insect vectors is being investigated further in our laboratory.

The strict manner of insect-vector transmission, together with the narrow host range of RSV, may explain the observed low genetic diversity. Compared with the highly variable CMV isolates (genetic distance up to 0.4 between subgroup I and II isolates), RSV isolates show very low genetic diversity (0.05–0.067 between subtypes and up to 0.092 between isolates). Similarly, compared with a list of plant viruses (Chare and Holmes, 2004; Garcia-Arenal et al., 2001), RSV genes are subjected to very strong negative selection (Table 4). It has been shown that the genetic diversities of several plant viruses were correlated with their host ranges and controlled by virus–host inter-actions (Schneider and Roossinck, 2000, 2001). In addition, plant virus variability is believed to be constrained more by the specificity of virus–vector interaction than by the specificity of virus–host plant interaction (Chare and Holmes, 2004; Power, 2000). Considering the facts that (i) RSV only infects rice and some species in the family Poaceae, whereas CMV is able to infect more than 1000 species in 85 plant families (Palukaitis et al., 1992), and (ii) RSV is transmitted by *L. striatellus* in a persistent propagative manner, whereas CMV can be transmitted by sap and different species of aphids in a non-persistent manner, it is understandable that RSV has low genetic diversity and is under strong negative selection.

Acknowledgements

We are grateful to Bashan Huang, Jun Zheng, KexianZong, Yushan Wang, Weimin Li, Dechun Fang, Lingeng Gong, Xiaomin Li,Wenwu Shi, Yexiu Huang, Dunfan Fang, Yun Xu, Yunkun He, Yijun Zhou, Zhaoban Chen, Maosheng Li, Xuelian Lai, Honglian Li, Dalin Shen, Chongguan Pan and Guanghe Zhou for their help in collecting RSV samples. We also thank Dr Bryce Falk and Dr Dustin Johnson for their help with the writing. This work was supported by projects from the National Natural Scientific and Technological Foundation of China (items 39900091 and 30000002 to L.-H.X. and H.-X.L., respectively), Natural Scientific and Technological Foundation of Fujian province (item B0110014 to H.-X.L.), the Ministry of Education of China (item 020401 to H.-X.L.) and the Natural Scientific and Technological Foundation of Yunnan province (item 2003C0042M to F.L.).

References

[1] ABUBAKAR Z, ALI F, PINEL A, et al. Phylogeography of rice yellow mottle virus in Africa[J]. Journal of General Virology, 2003, 84(3):733-743.

[2] ARBOLEDA M, AZZAM O. Inter- and intra-site genetic diversity of natural field populations of rice tungro bacilliform virus in the Philippines[J]. Archives of Virology, 2000, 145(2):275-289.

[3] AZZAM O, ARBOLEDA M, UMADHAY K M, et al. Genetic composition and complexity of virus populations at tungro-endemic and outbreak rice sites[J]. Archives of Virology, 2000, 145(12):2643-2657.

[4] BEDFORD I D, BRIDDON R W, ROSELL R, et al. Geminivirus transmission and biological characterization of *Bemisia tabaci* (Gennadius) biotypes from different geographic regions[J]. Annals of Applied Biology, 1994,125(2): 311-325.

[5] BUCHER E, SIJEN T, DE HAAN P, et al. Negative-strand tospoviruses and tenuiviruses carry a gene for a suppressor of gene silencing at analogous genomic positions[J]. Journal of Virology, 2003, 77(2):1329-1336.

[6] CHARE E R, HOLMES E C. Selection pressures in the capsid genes of plant RNA viruses reflect mode of transmission[J]. Journal of General Virology, 2004, 85(10):3149-3159.

[7] DRAKE J W, HOLLAND J J. Mutation rates among RNA viruses[J]. Proceedings of the National Academy of Sciences, 1999, 96(24):13910-13913.

[8] FALK B W, TSAI J H. Biology and molecular biology of viruses in the genus tenuivirus[J]. Annual Review of Phytopathology, 1998, 36(1):139-163.

[9] FARGETTE D, KONATE G, FAUQUET C,et al. Molecular ecology and emergence of tropical plant viruses[J]. Annual Review of Phytopathology, 2006, 44(1):235-260.

[10] FRAILE A, ALONSO-PRADOS J L, Aranda M A, et al. Genetic `exchange by recombination or reassortment is infrequent in natural populations of a tripartite RNA plant virus[J]. Journal of Virology, 1997, 71(2):934-940.

[11] GARCIA A F, FRAILE A, MALPICA J M. Variability and genetic structure of plant virus populations[M]. Annual Review of Phytopathology,2001, 39: 157-186.

[12] GESSAIN A, GALLO R C, FRANCHINI G. Low degree of human T-cell lymphotropic virus type I genetic drift in vivo as a means of monitoring viral transmission and movement of ancient human populations[J]. Journal of Virology, 1992, 66(4):2288-2295.

[13] GIRI A, SLATTERY J P, HENEINE W, et al. The tax gene sequences form two divergent monophyletic lineages corresponding to types I and II of simian and human T-cell leukemia/lymphotropic viruses[J]. Virology, 1997, 231(1):96-104.

[14] HIBINO H. Biology and epidemiology of rice viruses[J]. Annual Review of Phytopathology, 1996, 34(1):249-274.

[15] HOSHIZAKI S. Allozyme polymorphism and geographic variation in the small brown planthopper *Laodelphax striatellus* (Homoptera: Delphacidae) [J]. Biochemical Genetics, 1997, 35(11):383-393.

[16] HUDSON R R. A new statistic for detecting genetic differentiation[J]. Genetics, 2000, 155(4):2011-2014.

[17] HUDSON R R, BOOS D D, KAPLAN N L. A statistical test for detecting geographic subdivision[J]. Molecular Biology & Evolution, 1992(1):138-151.

[18] KAKUTANI T, HAYANO Y, HAYASHI T, et al. Ambisense segment 4 of rice stripe virus: possible evolutionary relationship with phleboviruses and uukuviruses (Bunyaviridae)[J]. Journal of General Virology, 1990, 71 (7):1427-1432.

[19] KAKUTANI T, HAYANO Y, HAYASHI T,et al. Ambisense segment 3 of rice stripe virus: the first instance of a virus containing two ambisense segments[J]. Journal of General Virology, 1991, 72(2):465-468.

[20] KORALNIK I J, BOERI E, SAXINGER W C, et al. Phylogenetic associations of human and simian T-cell leukemia/lymphotropic virus type I strains: evidence for interspecies transmission[J]. Journal of Virology, 1994, 68(4):2693-2707.

[21] KUMAR S, TAMURA K, JAKOBSEN I B,et al. MEGA2: molecular evolutionary genetics analysis software[J]. Bioinformatics, 2001, 17(12):1244-1245.

[22] LIN Q Y, XIE L H, ZHOU Z J, et al. Studies on rice stripe I Distribution of and losses caused by the disease[J]. Journal of Fujian Agricultural College, 1990, 19: 421-425.

[23] Lin H, Wei T, Wu Z, et al. Molecular variability in coat protein and disease-specific protein genes among seven isolates of Rice stripe virus in China. In Abstracts of the XIth International Congress of Virology. SydneyAustralia, 9-13 August 1999, pp. 235-236. 1999,Utrecht The Netherlands: International Union of Microbiological Societies.

[24] LIN H, WEI T, WU Z, et al. Sequence analysis of RNA4 of a severe isolate of Rice stripe virus in China[J]. Acta Microbiologica Polonica,2001, 41: 25-30.

[25] LIN H, WEI T, WU Z, et al. Comparison of pathogenesis and biochemical properties of seven isolates of Rice stripe virus[J]. Journal of Fujian Agricultural University, 2002, 31(2):164-167.

[26] LIN H X, RUBIO L, SMYTHE A B, et al. Molecular population genetics of Cucumber mosaic virus in California: evidence for founder effects and reassortment[J]. Journal of Virology, 2004, 78(12):6666-6675.

[27] MCGRATH P F, HARRISON B D. Transmission of tomato leaf curl geminiviruses by Bemisiatabaci: effects of virus isolate and vector biotype[J]. Annals of Applied Biology, 1995, 126(2):307-316.

[28] MOYA A, GARCIA-ARENAL F. Molecular Basis of Virus Evolution: Population genetics of viruses: an introduction[M]. 1995: 213-223.

[29] PALUKAITIS P, ROOSSINCK M J, DIETZGEN R G, et al. Cucumber mosaic virus[J]. Advances in Virus Research, 1992, 41(1):281-348.

[30] POWER A G. Insect transmision of plant viruses: a constraint on virus variability[J]. Current Opinion in Plant Biology, 2000,

3(4):336-340.

[31] QU Z, LIANG D, HARPER G,et al. Comparison of sequences of RNAs 3 and 4 of rice stripe virus from China with those of Japanese isolates[J]. Virus Genes, 1997, 15(2):99-103.

[32] ROZAS J, SANCHEZ-DELBARRIO J C, MESSEGUER X,et al. DnaSP, DNA polymorphism analyses by the coalescent and other methods[J]. Bioinformatics, 2003, 19(18):2496-2497.

[33] SCHNEIDER W L, ROOSSINCK M J. Evolutionarily related Sindbis-like plant viruses maintain different levels of population diversity in a common host[J]. Journal of Virology, 2000, 74(7):3130-3134.

[34] SCHNEIDER W L, ROOSSINCK M J. Genetic diversity in RNA virus quasispecies is controlled byhost-virus interactions[J]. Journal of Virology, 2001, 75(14):6566-6571.

[35] SWOFFORD L, SWOFFORD D L,SWOFFORD D. PAUP: phylogenetic analysis using parsimony (and other methods) Version 4.0M0, mac version, 2002.

[36] TAKAHASHI M, TORIYAMA S, HAMAMATSU C,et al. Nucleotide sequence and possible ambisense coding strategy of rice stripe virus RNA segment 2[J]. Journal of General Virology, 1993, 74 (4):769-773.

[37] THOMPSON J D, HIGGINS D G, GIBSON T J. CLUSTAL W: improving the sensitivity of progressive multiple sequence alignment through sequence weighting, position-specific gap penalties and weight matrix choice[J]. Nucleic Acids Research, 1994, 22(22):1673-1680.

[38] TORIYAMA S. Rice stripe virus: prototype of a new group of viruses that replicate in plants and insects[J]. Microbiological Sciences, 1986, 3(11):347-351.

[39] TORIYAMA S, TAKAHASHI M, SANO Y,et al. Nucleotide sequence of RNA. 1, the largest genomic segment of rice stripe virus, the prototype of the tenuiviruses[J]. Journal of General Virology, 1994, 75 (12):3569-3579.

[40] WANG Z F, QIU B S, TIEN P. Molecular biology of rice stripe virus III Sequence analysis of coat protein gene[J]. Virologica Sinica, 1992, 7: 463-466.

[41] WEI T, LIN H, WU Z,et al. Sequence analysis of intergenic region of rice stripe virus RNA4: evidence for mixed infection and genetic variation[J]. Acta Microbiologica Sinica, 2003a, 43(5):577-585.

[42] WEI T Y, WANG H, LIN H X,et al. Sequence analysis of RNA3 of rice stripe virus isolates found in China: evidence for reassortment in Tenuivirus[J]. actabiochimica et biophysicasinica, 2003b, 35(1):97-103.

[43] XIE L H, WEI T Y, LIN H X, et al. The molecular biology of *Rice stripe virus*[J]. Journal of Fujian Agricultural University, 2001, 30: 269-279.

[44] XIONG R, WU J, ZHOU Y, et al. Identification of a movement protein of *rice stripe tenuivirus*[J]. Journal of Virology, 2008, 82(24):12304-12311.

[45] ZHU Y, HAYAKAWA T, TORIYAMA S, TAKAHASHI M. Complete nucleotide sequence of RNA. 3 of rice stripe virus: an ambisense coding strategy[J]. Journal of General Virology, 1991, 72 (4):763-767.

[46] ZHU Y, HAYAKAWA T, TORIYAMA S. Complete nucleotide sequence of RNA. 4 of rice stripe virus isolate T, and comparison with another isolate and with maize stripe virus[J]. Journal of General Virology, 1992, 73 (5):1309-1312.

水稻矮缩病毒基因组遗传多样性的初步研究

章松柏[1,2]，吴祖建[2]，段永平[2]，王盛[2]，林奇英[2]，谢联辉[2]

(1 长江大学农学院，湖北荆州 434025；2 福建农林大学农药生物化学教育部重点实验室，福建福州 350002)

摘要：田间调查发现，云南、福建两地的水稻矮缩病症状存在差异。生物学接种实验排除了水稻品种引起这种差异的可能性。通过 10% SDS-PAGE 对福建、云南两地的 7 个水稻矮缩病毒分离物基因组遗传多样性进行分析，结果表明基因组之间存在明显差异，主要体现在 S2、S3、S11、S124 个基因片段上，其中 S2、S3 片段之间的差异与症状差异有地域上的一致性，推测这 4 个基因片段中的一个或几个与病毒的致病性相关。

关键词：水稻矮缩病毒；遗传多样性；SDS- 聚丙烯酰胺凝胶电泳

中图分类号：346.5　**文献标识码：**A　**文章编号：**1673-1409(2009)04-S037-03

水稻矮缩病毒（*Rice Dwarf Virus*, RDV）是隶属于呼肠孤病毒科（*Reoviridae*）中的植物呼肠孤病毒属（*Phytorevirus*）成员（Van Regenmortel et al., 2000），主要由黑尾叶蝉（*Nephotettix cincticeps*）、电光叶蝉（*Recilia dorsalis*）和二条黑尾叶蝉（*N.nigropictus*）以持久方式且可经卵传播（Lin et al., 1981; 林奇英等 , 1982；陈声祥，1996）。该病毒引起的水稻矮缩病广泛地分布于中国、日本及菲律宾南部等水稻产区，造成水稻严重减产（谢联辉等 , 1988）。近年来在我国福建、云南、河南等水稻产区又有局部流行的趋势（乔玉昌等 , 2003）。水稻矮缩病的典型症状是在苗期至分蘖期染病后，植株矮缩，分蘖增多，叶片浓绿、僵直，生长后期病稻不能抽穗结实，始病叶以上新叶都出现点条斑。但是在福建和云南两省进行调查时发现，侵染 RDV 的水稻病株症状存在明显的差别，福建病株都呈现典型的矮化症状，高度一般只有健株的一半，分蘖增多，其病稻不能抽穗结实，而很多采自云南的病株矮化症状并不明显，能正常分蘖、抽穗和结实。植物病毒病症状差异的原因往往与植物病毒基因组遗传多样性密切相关，而植物病毒的株系分化和分子变异在自然界中广泛存在。由于病毒致病性的变异，往往导致品种抗性的丧失，因此，对病毒致病性分化和基因变异进行研究，是非常迫切的任务。日本学者首先发现 RDV 在自然条件下存在基因变异，深入研究了 RDV 的遗传多样性以及其与致病性之间的关系（Uyeda et al., 1995），从而发现 RDV 不同株系间致病力存在很大差异。为了研究田间调查过程中发现的福建、云南水稻矮缩病病株外部症状差异的原因，对两地采集的水稻矮缩病毒基因组进行了研究，以初步揭示基因组遗传差异与外部症状差异的关系。

1　材料与方法

1.1　材料

水稻矮缩病代表性病株于 2001 ～ 2003 年采自福建福州、龙岩、松溪和云南农大、西山、禄劝、武定等地，分别编号为 RDV-FZ（福州）、RDV-SX（松溪）、RDV-LY（龙岩）、RDV-ND（云南农大）、

长江大学学报 (自然科学版).2009,6(4): 37-40

收稿日期：2009-08-31

基金项目：国家自然科学基金项目 (39900091)

RDV-LQ（禄劝）、RDV-WD（武定）、RDV-XS（西山），经过接种纯化后，种植于防虫网室。介体昆虫黑尾叶蝉采自福建省农科院水稻所无病试验田。水稻矮缩病感病品种为台中 1 号水稻品种。

1.2 生物学接种

将采集的 RDV-FZ（福州）、RDV-SX（松溪）、RDV-LY（龙岩）、RDV-ND（云南农大）、RDV-LQ（禄劝）、RDV-WD（武定）、RDV-XS（西山）等 7 种分离物接种到水稻矮缩病毒感病品种台中 1 号水稻品种上，种植于防虫网室，观察发病情况。

1.3 dsRNA 病毒基因组的提取

病毒基因组（dsRNA）的提取参考周雪平（1995）等（周雪平等，1995）病株中直接提取病毒基因组的方法，略有改进。步骤如下：采集以上 7 种分离物具有典型症状的水稻病叶，液氮研磨成粉末。用 2 倍体积的 2×STE 及酚、氯仿抽提，低速离心。上清液用无水乙醇调至含量为 17%，注入用含 17% 乙醇的 1×STE 平衡过的纤维素（CF-11）柱中，然后用 80mL 含 17% 乙醇的 1×STE 洗脱，再用不含乙醇的 1×STE 洗脱，洗脱液用乙醇沉淀后真空抽干，即为 RDV-dsRNA。

1.4 SDS- 聚丙烯酰胺凝胶电泳

提取的病毒 dsRNA 基因组加等体积 2× 凝胶加样缓冲液（100mmol/L Tris-HCl、pH 8.0，2% 巯基乙醇，4% SDS，0.2% 溴酚蓝，20% 甘油）煮沸 10min，取 4μL 加样于 10% SDS- 聚丙烯酰胺凝胶上。采用 TBE 缓冲体系，165V，16h。EB 染色或银染观察结果。

2 结果与分析

2.1 生物学接种结果

将采集的 7 种分离物接种到感病品种台中 1 号上，种植于防虫网室，观察发病情况。结果如表 1。症状的差异主要由两种原因引起的：（1）病毒的基因组差异，导致致病性存在差异，从而引起自然寄主发病差异；（2）水稻品种抗病性的差异。从表 1 可得知，生物学接种结果符合田间调查发现，说明这种症状差异不是水稻品种不同所导致的，推测与病毒基因组差异相关。

表 1 RDV 7 种分离物在原水稻品种和感病品种台中 1 号的表现

Table 1 The performances of original or Taizhong No.1 rice varietie inoculated by 7 RDV isolates

	原水稻品种	台中 1 号
RDV-FZ（福州）	典型症状	典型症状
RDV-SX（松溪）	典型症状	典型症状
RDV-LY（龙岩）	典型症状	典型症状
RDV-ND（云南农大）	矮化不明显、分蘖正常	矮化不明显、分蘖正常
RDV-LQ（禄劝）	矮化不明显、分蘖正常	矮化不明显、分蘖正常
RDV-WD（武定）	矮化不明显、分蘖正常	矮化不明显、分蘖正常
RDV-XS（西山）	矮化不明显、分蘖正常	矮化不明显、分蘖正常

2.2 SDS- 聚丙烯酰胺凝胶电泳结果

RDV 福州（RDV-FZ）、松溪（RDV-SX）、龙岩（RDV-LY）、云南农大（RDV-ND）、禄劝（RDV-LQ）、

武定（RDV-WD）、西山（RDV-XS）等7个地方分离物的10% SDS-PAGE电泳图谱如图1（其中A为银染，B、C为EB染色）。图谱显示：福建、云南两地的RDV7种分离物的基因组dsRNA存在明显的遗传多样性，基因组差异体现在S2/S3、S11/S12这4个基因片段上，其中S2/S3之间的差异有地域性，即福建分离物和云南地方分离物在S2/S3上存在地理差异。而田间调查发现福建、云南两省的各分离物其引起的水稻矮缩病的症状存在较大的差异，生物学接种实验排除了水稻品种引起这种差异的可能性，推测与病毒基因组差异相关。暗示这4个基因片段中的一个或几个可能与水稻的矮化症状相关。

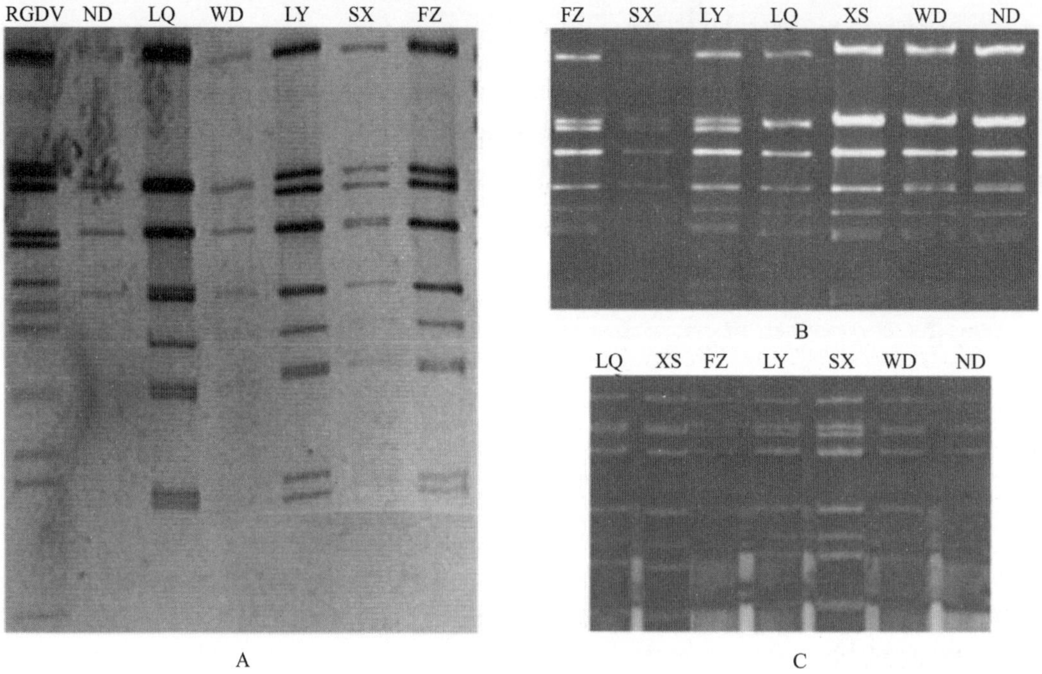

图1 RDV分离物基因组10%聚丙烯酰胺电泳图谱

Fig.1 10% PAGE showing dsRNA profiles from some RDV isolates rice dwarf gall virus

RGDV，水稻瘤矮病毒；ND、LQ、WD、LY、SX、XS、FZ：水稻矮缩病毒分离物

3 讨论

植物病毒的株系分化和分子变异在自然界中广泛存在。株系分化和分子变异为病毒分类、分化和抗病育种以及病毒生物功能和机制的比较研究提供重要材料和依据。我国幅员辽阔，各地生境千差万别，RDV分布地理跨度极大，很有可能产生新株系，引起遗传多样性和致病性变化。由此可见RDV遗传多样性研究具有重要的理论和应用意义。本研究初步分析了RDV基因组的遗传多样性。通过生物学接种和致病性测定首先排除水稻矮缩病症状差异由水稻品种导致，明确症状差异是由基因组差异引起的。在此基础上对症状上有差异的7个病毒分离物的基因组遗传多样性进行了初步分析，显示S2、S3、S11、S12存在较明显的差异。值得注意的是，S2/S3片段之间的差异与症状差异具有一致的地域性。所以推测S2/S3、S11/S12这4个基因片段中的某一个或多个片段与病毒致病性有一定的关系。这种推测与报道中的一些推测相符合（Yan et al., 1996; Tomaru et al., 1997; Chen et al., 1998）。因此，这4个基因片段中的一个或几个可能与水稻的矮化症状相关，值得进一步研究。进一步研究的重点可放在S2/S3、S11/S12这4个差异基因片段的核酸序列分析和比较上，找出可能与致病性相关的基因片段，在此基础上进行分子间互作研究。

参考文献

[1] 陈声祥 . 水稻病毒病发生现状及研究进展 [J]. 浙江农业科学 , 1996,(1):41-42.

[2] 林奇英 , 谢联辉 . 传带水稻矮缩病毒的二点黑点叶蝉 [J]. 福建农业科技 ,1982, (3):24-25.

[3] 乔玉昌 , 王阳 , 黄雅柯 , 等 . 水稻矮缩病毒病及其防治 [J]. 湖北农业科学 , 2003,(1):12.

[4] 谢联辉 , 林奇英 . 中国水稻病虫综合防治进展 [M]. 杭州 : 浙江科学技术出版社 , 1988,255-264.

[5] 周雪平 , 李德葆 . 双链 RNA 技术在植物病毒研究中的应用 [J]. 生物技术 , 1995,5(1):1-4.

[6] CHEN D Y, PATTON J T. Rotavirus RNA replication requires a single-stranded 3′end for efficient minus-strand synthesis[J]. Journal of Virology, 1998, 72(9):7387-7396.

[7] LIN Q Y, XIE L H. A new insect vector of rice dwarf virus. IRRN, 1981, 6(5):14.

[8] TOMARU M, MARUYAMA W, KIKUCHI A, et al. The loss of outer capsid protein P2 results in nontransmissibility by the insect vector of rice dwarf phytoreovirus[J]. Journal of Virology, 1997, 71(10):8019-8023.

[9] UYEDA I, ANDO Y, MURAO K, et al. High resolution genome typing and genomic reassortment events of rice dwarf phytoreovirus[J]. Virology, 1995, 212(2):724-727.

[10] REGENMORTEL M H, FAUQUET C M, BISHOP D H. Taxonomy, classification and nomenclature of viruses: Seventh report of the international committee on taxonomy of viruses[M]. Academic Press, 2000, 622-627.

[11] YAN J, TOMARU M, TAKAHASHI A, et al. P2 protein encoded by genome segment S2 of rice dwarf phytoreovirus is essential for virus infection[J]. Virology, 1996, 224(2):539-541.

灰飞虱来源的水稻条纹病毒外壳蛋白基因遗传多样性

程兆榜[1,2]，任春梅[2]，周益军[2]，季英华[2]，范永坚[2]，谢联辉[1]

(1 福建农林大学病毒研究所，福州 350002；2 江苏省农业科学院植物保护研究所，南京 210014)

摘要：采用单雌产卵法获得来自江苏、云南、山东、河北等地灰飞虱（*Laodelphax striatellus* Fallén, SBPH）来源的 21 个水稻条纹病毒（*Rice stripe virus*, RSV）分离物，提取灰飞虱总 RNA，经 RT-PCR 扩增，获得 21 个 RSV 分离物的包含外壳蛋白（*cp*）基因在内的约 1000bp 左右的 DNA 片段。测序结果显示，参试分离物的 *cp* 由 969 个核苷酸组成，编码 322 个氨基酸。采用 DNASTAR 软件进行分析，21 个灰飞虱来源的 RSV-*cp* 核苷酸序列和推导出的编码蛋白的氨基酸序列同源性分别为 95.7% ～ 100% 和 96.0% ～ 100%。与已报道水稻来源的 32 个 RSV 分离物一起进行序列同源性比较和系统进化树分析结果表明，总体而言 RSV-*cp* 较为保守，其遗传多样性首先与地缘相关，从地理位置上可以分成中国云南、中国沿海和日本 3 个地理种群；其次与寄主相关，在同一地理种群中可以划分为灰飞虱和水稻两个寄主种群。

关键词：水稻条纹病毒；外壳蛋白基因；灰飞虱；遗传多样性；地理种群；寄主种群

中图分类号：S432.45　**文献标识码**：A　**文章编号**：0412-0914(2012) 06-0585-09

Genetic diversity of coat protein gene in *Rice stripe virus* isolated from *Laodelphax striatellus*

Zhaobang Cheng[1,2], Chunmei Ren[2], Yijun Zhou[2], Yinghua Ji[2], Yongjian Fan[2], Lianhui Xie[1]

(1 Institute of Virology, Fujian Agriculture and Forestry University, Fuzhou 350002; 2 Institute of Plant Protection, Jiangsu Academy of Agricultural Sciences, Nanjing 210014)

Abstract: Twenty-one *Laodelphax striatellus* Fallén isofemale lines infected with *Rice stripe virus*(RSV) were collected from Jiangsu, Yunnan, Shandong and Hebei Province, China. The RSV coat protein(*cp*) gene was amplified using RT-PCR, generating a fragment with the size of 1 000bp. Sequence analysis revealed that the RSV-*cp* contained 969bp which encoded 322 amino acids. Further analysis with DNASTAR software showed that the identities of *cp* nucleotide and amino acid sequences among the isolates ranged from 95.7% to 100% and 96.0% to 100%, respectively. The results together with the 32 sequence data published previously suggested that RSV-*cp* was conserved and had little molecular variation. The results also suggested that the genetic diversity of RSV was strongly associated with its geographical origin and the virus could be grouped into three geographical populations, i.e. Yunnan, China and Coastal areas of China and Japan. The genetic diversity of RSV was also associated with its host origin. Within a geographical location, RSV could be grouped into *Laodelphax striatellus* Fallén and rice populations.

植物病理学报.2012,42（06）：585-593

收稿日期：2011-10-26；接受日期：2012-06-28

基金项目：国家自然科学基金重点项目（30300230）

Keywords:*rice stripe virus*; *cp*; *Laodelphax striatellus* Fallén; genetic diversity; geographical population; host population

水稻条纹病毒（*Rice stripe virus*, RSV）引起的水稻条纹叶枯病是我国水稻上一种重要的植物病毒病害（Lin et al., 1990）。自 2002 年在江苏、河南等地暴发以来，病害发生范围和危害程度逐年加大（Cheng et al., 2002）。据文献报道和田间调查，目前在浙江、上海、山东、安徽、辽宁、河北、云南等地也均有不同程度的发生（Jiang and Li, 2005; Gao et al., 2006; Li et al., 2006; Zhang et al., 2006; Wang et al., 2007; Wang et al., 2007; Xiang and Gao, 2007）成为影响当地水稻生产的主要制约因素之一。据江苏省农林厅统计，2004—2007 年江苏水稻条纹叶枯病的发生面积分别达 157 万、146 万、141 万、135 万 hm^2，每年均有一定的绝收面积。

RSV 是纤细病毒属（*Tenuivirus*）的典型成员，由灰飞虱（*Laodelphax striatellus* Fallén，SBPH）以持久性方式传播，其独特的双义编码策略（Takahashi et al., 1993）使其成为研究植物病毒变异与寄主互作及协同进化关系的模式对象之一。RSV 基因组由 4 条 ssRNA 组成（Zhu et al., 1991; Toriyama et al., 1994），其中 RNA3 的 5′ 端毒义链（viral RNA3）编码一个 35.1kD 的蛋白，即外壳蛋白（coat protein, CP），该蛋白可能与致病性和症状表达有关（Lin et al., 1998）。

RSV 在北纬 109° 的中国云南到北纬 131° 的俄罗斯海参崴，东经 27° 的中国四川至东经 41° 的日本北海道均有分布（Kismoto and Yamada, 1991），可侵染水稻、小麦、大麦、燕麦、稗草、狗尾草等 37 种禾本科植物和灰飞虱（Toriyama, 2000），其分布生态位的多样化决定了 RSV 可能具有丰富的遗传多样性。魏太云、林含新等根据来自水稻病株 RSV 分离物的 SP、CP、NS2 和 NS3 等基因遗传多样性，认为我国 RSV 存在云南和云南以外病区为代表的 2 个亚种群，各分离物的变异及亲缘关系与其地理分布位置有一定的关系（Lin et al.,2001; Wei et al., 2009）。

RSV 不仅通过灰飞虱在禾本科植物之间水平传播，也可在灰飞虱体内增殖并经卵垂直传播。因此一般认为 RSV 不仅是一种植物病毒，也是一种昆虫病毒。灰飞虱作为昆虫，其基因组结构及细胞内环境构成的对 RSV 的选择压与植物相比有较大的差别。如果将 RSV 作为昆虫病毒，其遗传多样性与将其作为植物病毒而显示的遗传多样性有何异同？围绕这一问题，本文以与致病性密切相关的 RSV-*cp* 为研究对象对 RSV 的遗传多样性进行了研究，以下报道相关结果。

1 材料与方法

1.1 病毒分离物

2003—2005 采集来自云南大理，保山，河北唐海，山东济宁，江苏高邮、洪泽、丹阳、盐都、靖江、姜堰、海安、沛县的秧田或稻田灰飞虱，采用单雌产卵法进行毒源纯化得到 21 个纯分离物（表 1），纯化方法参见文献（Garcia et al., 2001）。

1.2 试剂

提取 RNA 的 Trizol 试剂来自 Invitrogen 公司，用于 cDNA 合成的 M-MuLV 反转录酶、RNasinRNA 酶抑制剂来自 MBI 公司，*Taq plus* DNA 聚合酶、dNTP 等 PCR 试剂均购自上海申能博采有限公司。其余试剂为国产分析纯。

1.3 引物设计

根据已报道的 RSV 日本 T 分离物 RNA3 的基因序列，设计一对用于扩增 RSV-*cp* 的寡聚核苷酸引物，由上海英俊生物技术公司合成。引物序列如下：CP1：5′-AACTTACACTTAAAGAGACAAC-3′（与 RNA35′ 端 1415-1437bp 相对应）；CP2:5′-TTCGATTTTGCTTTTACATTCC-3′（与 RNA35′ 端 2419-2441bp 互补）。

表 1 本研究中检测到的水稻条纹病毒分离物

Table 1 *Rice stripe virus* isolates tested in the study

SN.	Isolate	Sampling place	Time of collection
1	LS-JSYD051	Yandu, Jiangsu	2005-10-5
2	LS-SDJN051	Jining, Shandong	2005-4-30
3	LS-JSJJ051	Jingjiang, Jiangsu	2005-9-30
4	LS-JSYD052	Yandu, Jiangsu	2005-10-5
5	LS-JSJJ052	Jingjiang, Jiangsu	2005-9-30
6	LS-JSJJ053	Jingjiang, Jiangsu	2005-9-30
7	LS-SDJN052	Jining, Shandong	2005-4-30
8	LS-JSJJ054	Jingjiang, Jiangsu	2005-9-30
9	LS-JSDY03	Danyang, Jiangsu	2003-7-27
10	LS-JSHZ03	Hongze, Jiangsu	2003-9-3
11	LS-JSHA03	Haian, Jiangsu	2003-8-29
12	LS-JSGY03	Gaoyou, Jiangsu	2003-7-21
13	LS-JSJJ03	Jingjiang, Jiangsu	2003-7-26
14	LS-JSPX03	Peixian, Jiangsu	2003-6-14
15	LS-JSYD03	Yandu, Jiangsu	2003-7-21
16	LS-JSJY03	Jiangyan, Jiangsu	2003-7-20
17	LS-HBTH05	Tanghai, Hebei	2005-7-29
18	LS-HBTH03	Tanghai, Hebei	2003-7-27
19	LS-YNBS041	Baosan, Yunnan	2004-7-16
20	LS-YNBS042	Baosan, Yunnan	2004-7-16
21	LS-YNDL04	Dali, Yunnan	2004-7-16

1.4 灰飞虱总 RNA 提取

取带毒灰飞虱 10 头，冰浴匀浆，用 Trizol 试剂盒按产品说明书提取灰飞虱总 RNA。

1.5 RT-PCR

反转录程序如下：取总 RNA 3μL，3′ 端引物（10pmol/L）1μL，ddH_2O 6μL，70℃下变性 5min，冰浴 1min，依次加入下列试剂，5×M-MuLV 反转录酶缓冲液 2μL、dNTP5μL（10mmol/L）、RNasin（40U/μL）0.5μL、M-MuLV 反转录酶（20U/μL）0.5μL、ddH_2O 1.5μL。42℃水浴 1h，70℃灭活 10min，–20℃保存备用。

PCR 扩增在 50μL 反应体系中进行，包括反转录产物 6μL，5′ 端和 3′ 端引物（10pmol/μL）各 2μL、dNTP（10mmol/L）1μL、10×buffer 8μL、ddH_2O 30μL、Taq plus DNA 聚合酶（4U/μL）1μL。扩增条件：94℃变性 2min 后，94℃变性 1min，47℃退火 2min，72℃延伸 1.5min，共 30 个循环，最后一个循环结束后，72℃保温 10min。

1.6 测序及序列分析

PCR 产物经电泳检测后直接送至上海生工采用双脱氧核苷酸终止法在 ABI3730DNA 自动测序仪上进行。在 GenBank 中利用 Blast 搜索（网址 http://www. ncbi. nlm. nih. gov/blast）RSV 已发表的中国及日本来自水稻的分离物进行比较分析。序列比对和进化树分析用 DNASTAR（5.0）软件完成，进化树分析采用 ClustalW 方法。

2 结果与分析

2.1 CP 扩增与测序

提取的病毒 RNA 经 RT-PCR 扩增后，21 个灰飞虱来源的 RSV 分离物均可得到一个长度为 1000bp 左右的扩增片段，与预期目的片段大小相符（图 1）。

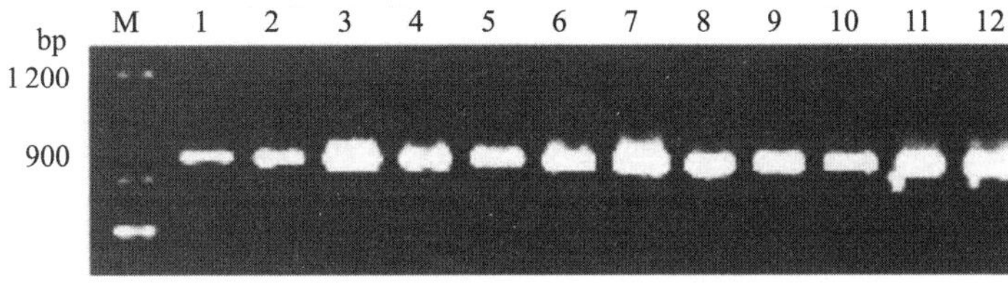

图 1 RSV-cp 扩增片段通过 RT-PCR 检测灰飞虱分离株

Fig. 1 Amplified fragment of RSV-*cp* by RT-PCR in some isolates from *Laodelphax striatellus*

M：标记；1-12：从 SBPH 得到的 RSV 分离株，另见表 1

M: Marker; 1-12: RSV isolates from SBPH, see also Table 1

PCR 产物直接测序，结果表明 21 个 RSV 分离物的 cp 均由 969nts 组成，编码 322 个氨基酸和 1 个终止密码子（GenBank 登录号为：EF198680-EF198699、EF493227）。

2.2 *cp* 基因序列分析

2.2.1 序列同源性比较

应用 DNASTAR 软件对 21 个 RSV 分离物的 *cp* 核苷酸序列和推导的编码蛋白氨基酸序列同源性进行比较分析（表 2），各分离物间核苷酸序列同源性为 95.7% ～ 100%，推导的氨基酸序列同源性为 96.0% ～ 100%。

2.2.2 RSV-cp 遗传多样性

根据水稻条纹叶枯病的发生及分布特点，选取已发表的 32 个水稻来源的 RSV 分离物（含最早报道的日本 T、M2 个分离物）的 *cp* 序列，以其地名首字母、采集年份以及登录序号命名。对以上 53 个分离物的 *cp* 进行核苷酸序列同源性聚类分析结果表明（图 2），水稻来源的 RSV 分离物，在 94.9% 的相似性水平上可以明显分为 2 组：中国云南和云南以外，这与 Wei（Wei et al., 2009）等人报道的结果一致。进一步细分，在 96.4% 的相似性水平上，云南以外的又可分为日本与中国沿海（包含沿黄淮地区）2 个亚组。灰飞虱来源的 RSV 分离物，在 94.9% 的相似性水平上与水稻来源的中国云南以外分离物同为一组，进一步细分，在 96.4% 相似性水平上，灰飞虱来源的 RSV 云南分离物分属 2 个亚组，而其余中国沿海分离物归为同一亚组。将水稻和灰飞虱来源的分离物综合分析，从地缘划分的中国云南分离物可分为 2 个独立的组群，中国沿海地区分离物中 18 个有 15 个分离物在 97.6% 相似性水平上归为同一组而与中国沿海的多数水稻分离物不在同一组。对各分离物进行氨基酸同源性分析，趋势与核苷酸同源性一致（图 3）。由以上分析可知，RSV 总体变异较小，是一种较为保守的病毒，其 RSV 的遗传多样性首先与地缘相关，从地理位置上可以分成中国云南、中国沿海和日本 3 个地理群。其次与寄主相关，在同一地理群中可以划分为灰飞虱和水稻 2 个寄主群。

表 2　利用 DNASTAR 软件对 21 株来自灰飞虱的 RSV 分离株 CP 中核苷酸（左下）和氨基酸（右上）的百分率进行分析

Table 2　Percentages of identified nucleotides (Lower left) and amino acids (upper right) in cp of 21 RSV isolates from *Laodelphax striatellus* with DNASTAR soft ware (%)

	1	2	3	4	5	6	7	8	9	10	11	12	13	14	15	16	17	18	19	20
1		100	100	100	100	99.7	99.7	100	97.8	99.4	97.5	97.5	98.8	100	99.7	99.1	100	100	98.4	98.8
2	100		100	100	100	99.7	100	100	97.8	99.4	97.5	97.5	98.8	100	99.7	99.1	100	100	98.1	98.4
3	99.0	99.0		100	100	99.7	99.7	100	97.8	99.4	97.5	97.5	98.8	100	99.7	99.1	100	100	98.4	98.8
4	99.7	99.7	99.1		100	99.7	99.7	100	97.8	99.4	97.5	97.5	98.8	100	99.7	99.1	100	100	98.4	98.8
5	99.3	99.3	98.7	99.4		99.7	99.7	100	97.8	99.4	97.5	97.5	98.8	100	99.7	99.1	100	100	98.4	98.8
6	99.2	99.2	99.0	99.5	98.9		99.4	99.7	98.1	99.1	97.8	97.8	98.4	99.7	99.4	99.4	99.7	99.7	98.1	98.4
7	99.8	99.8	98.8	99.7	99.1	99.2		99.7	97.5	99.1	97.2	97.2	98.4	99.7	99.4	98.8	99.7	99.7	100	99.7
8	99.4	99.4	99.4	99.7	99.1	99.4	99.4		97.8	99.4	97.5	97.8	98.8	100	99.7	99.4	100	100	98.4	98.8
9	99.1	99.1	98.0	99.0	98.3	98.7	99.1	98.7		98.4	97.8	97.8	97.2	97.8	97.5	98.1	97.8	97.8	96.3	98.4
10	99.5	99.5	98.9	99.8	99.2	99.3	99.5	99.5	99.2		97.5	97.5	98.1	99.4	99.1	99.1	99.4	99.4	97.8	98.1
11	98.7	98.7	99.5	99.0	98.3	98.9	98.7	99.3	98.3	99.0		98.1	96.9	97.5	97.2	97.2	97.2	97.5	96.0	96.3
12	97.5	97.5	97.1	97.8	97.8	97.7	97.5	97.5	97.8	97.8	97.0		98.8	97.5	97.8	97.2	97.5	97.5	96.6	96.6
13	98.7	98.7	98.2	99.0	98.3	98.7	98.7	98.7	98.3	98.8	97.9	98.0		98.8	99.1	99.1	98.8	98.8	97.8	98.1
14	99.6	99.6	99.0	99.7	99.3	99.6	99.4	99.4	98.7	99.5	98.7	97.5	98.7		99.7	99.1	100	100	98.4	98.8
15	99.7	99.7	98.7	99.6	99.0	99.1	99.7	99.3	99.0	99.4	98.6	97.8	98.8	99.3		98.8	99.7	99.7	98.8	99.1
16	98.0	98.0	97.8	98.3	97.7	98.2	98.0	98.2	98.1	98.3	97.7	97.8	97.7	98.0	97.9		99.1	99.1	97.5	97.8
17	99.3	99.3	98.7	99.6	99.4	99.1	99.3	99.3	98.6	99.4	98.6	98.2	98.6	99.3	99.2	97.9		100	98.4	98.8
18	99.9	99.9	98.9	99.8	99.2	99.3	99.9	99.5	99.2	99.6	98.8	97.6	98.8	99.5	99.8	98.1	99.4		98.4	98.8
19	97.3	97.3	97.4	97.6	97.0	97.3	97.3	97.8	96.6	97.4	97.3	95.7	96.8	97.3	97.4	96.2	97.2	97.4		100
20	98.0	98.0	97.9	98.1	97.5	97.8	98.0	98.3	97.3	97.9	97.8	96.2	97.3	97.8	98.1	96.7	97.7	98.1	98.9	
21	98.3	98.3	98.2	98.7	98.0	98.3	98.3	98.9	97.6	98.5	98.1	96.5	97.6	98.3	98.2	97.6	98.2	98.5	98.2	98.3

1-21: RSV isolates from *Laodelphax striatellus*, see also Table 1

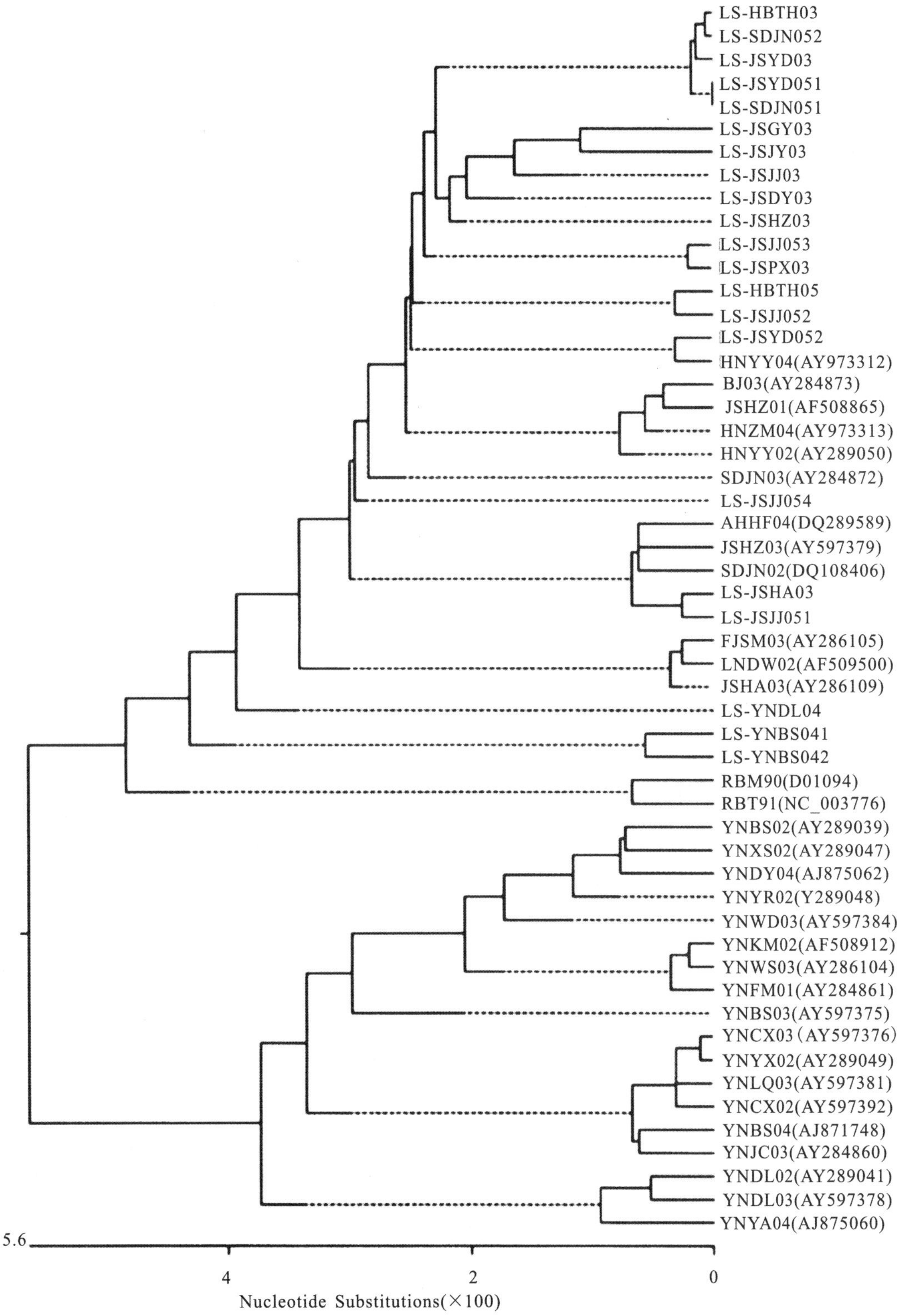

图 2 通过对 53 株 RSV 分离株 CP 核苷酸序列比对，构建了系统发育树

Fig. 2 Phylogenetic tree constructed from alignment of nucleotide sequences of the CP of 53 RSV isolates

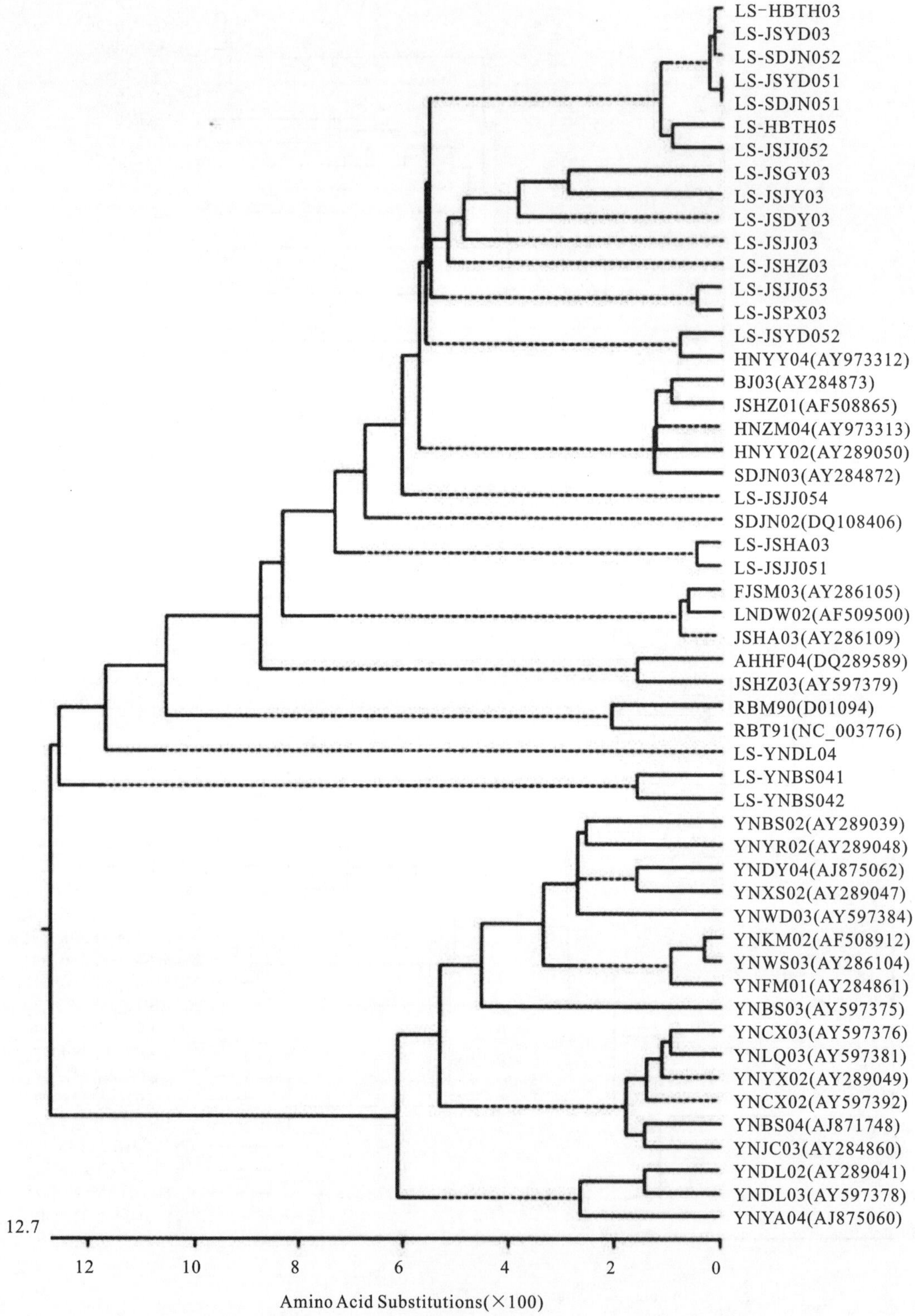

图3 通过对53株RSV分离株CP蛋白氨基酸序列比对，构建了系统发育树

Fig. 3 Phylogenetic tree constructed from alignment of amino acids sequences of the CP proteins of 53 RSV isolates

3 讨论

病毒是一个准种（quasispecies）的概念正逐渐被人们接受，即病毒并不是一个单一的序列结构，而是由一种优势序列和一些低频率的与优势序列差异微小的序列组成的一个序列群（Saenz et al., 2001）。准种的概念首先是在动物病毒上提出来的，对乙型肝炎病毒（HBV）、丙型肝炎病毒（HCV）和艾滋病毒（HIV）研究的结果表明，准种是病毒与寄主选择压相适应的一种方式，在不同个体、不同发病进程和不同治疗手段下，病毒的准种结构和优势序列是不同的（Schneider et al., 2001）。对 TMV、CMV 等病毒的研究表明，植物病毒亦符合准种结构特征，并且其准种结构与寄主有关（Cheng et al., 2008）。本研究测定灰飞虱来源的 21 个 RSV 分离物的 *cp* 并与水稻来源的 RSV 分离物序列进行了比较，发现 RSV-*cp* 遗传结构与地缘及其寄主（水稻和灰飞虱）相关，这暗示了 RSV 可能也符合病毒准种的特征。RSV 是否为准种及其准种特征有待进一步研究。

灰飞虱是 RSV 传播的介体，同时 RSV 又可在灰飞虱体内增殖并能经卵传播，因此 RSV 被认为既是一种植物病毒，也是一种昆虫病毒。本项研究显示 RSV 的遗传结构在灰飞虱和水稻上有各自的特点，且同一地区灰飞虱来源的 RSV-*cp* 同源性高于水稻来源，这可能与同一地区种植的水稻品种的异质性高于生活在这一地区的灰飞虱种群的异质性有关。这里可能暗含了 RSV 与不同类型寄主之间的相互适应性的机制，在 RSV 侵染后灰飞虱和水稻不同品种构成不同的遗传背景对 RSV 造成不同的选择压，从而导致 RSV 出现遗传多样性上的变化。RSV-*cp* 与不同类型寄主之间的相应关系有何生物学功能尚不清楚，比较两种类型分离物之间基因序列的异同及变异特点，或许有助于我们揭开 RSV 从昆虫传播到水稻的一些奥秘。

本项研究结果所显示的 RSV-*cp* 的变异首先与地缘其次与寄主相关的特性，反映了植物病毒变异的两种主要动力：地理隔离和寄主选择。这两种动力导致病毒向某种特定的方向变异，最终形成病毒的不同株系甚至成为不同的种，这在植物病毒上已有多个例证。就 RSV 而言，从地理位置和地貌特点来看，中国云南和日本均是相对孤立的，形成了一个封闭系统，而中国沿海一带虽然南北跨度极大，但由于周年的季风作用使其成为一个大的开放系统。一年之中多次的南、北季风，使灰飞虱携带 RSV 在沿海地区不同省份之间进行较为频繁的基因交流，从而使其成为一个共同的地理种群。从水稻品种来看，由于各地的气候条件的差别和种植习惯的不同，云南种植的水稻品种与中国沿海及日本有一定的区别，因此处于不同地理环境中的 RSV 长期受到不同水稻品种所构成寄主选择压力的定向选择，这可能逐渐导致三地形成了相对独特的地理种群。

通过已有文献报道和本研究结果显示，RSV 是一种相对保守的病毒，虽然可以将其分为中国云南、中国沿海和日本 3 个地理种群，同一地理种群中也可以分为水稻和灰飞虱 2 个寄主群，但不同种群之间相似度较高，这种所谓的种群划分只是相对而言。由此可以反映出，RSV 与 HIV、流感病毒以主动变异的方式来适应环境不同，RSV 应是以寻找适合其生存的生态系统来被动的适应环境。RSV 近年来在江苏等地暴发流行，从与致病性相关的 *cp* 变异较小及其不同地区分离物之间致病性差异不显著的特点来看，在其暴发流行中决定因素并不是 RSV 变异，而是与病毒传播有关的介体或环境因素（Garcia et al., 2001）。由于介体种群受环境因素的影响较大，由此推测目前 RSV 的暴发流行除了品种这一基本要素外，环境因素可能起主要作用，在当前病害流行地区的农田生态系统中出现了适应灰飞虱发生和 RSV 传播的有利因素从而促进甚至决定了 RSV 的流行。因此在控制 RSV 流行方面，从环境因素考虑并适当采取应对措施可能才是控制其危害的基础。

参考文献

[1] CHENG Z B, REN C M, ZHOU Y J, et al. Pathogenicity of Rice stripe tenuivirus isolates from different areas[J]. Acta Phytopathologica Sinica, 2008, 38(2):126-131.

[2] CHENG Z B, YANG R M, ZHOU Y J. New law of rice stripe disease occurred in Jiangsu province[J]. Agricultural Science of Jiangsu, 2002, (1): 39-41.

[3] GAO L C, SONG K Q, ZHANG H R, et al. The characters and integrated control of rice stripe disease in Huang huai area[J]. Agricultural Science of Shandong,2006, (3): 66-67.

[4] GARCIA A F, FRAILE A, MALPICA J M. Variability and genetic structure of plant virus populations[M]. annual review of phytopathology, 2001,39:157-186.

[5] JIANG Y P, LI J G. The occurrence and control techniques of rice stripe disease in Shanghai[J]. Agricultural Science and Technology of Shanghai, 2005, (5):37-38.

[6] KISIMOTO R, YAMADA Y. Present status of controlling Rice stripe virus. 1991.

[7] HADIDI A, KHETARPAL RK, KOGANEZAWA H. Plant Virus Disease Control[J]. 1988, St.Paul. Minnesota: APS Press, 470-481.

[8] LI F, YANG J G WU Z J. Cloning and sequence comparison analysis of the coat protein gene of rice stripe virus isolates in Yunnan[J]. Journal of Yunnan Agricultural University, 2006, 21(1): 48-51.

[9] LIN H X, WEI T Y, WU Z J, et al. Cloning and sequence analysis of coat protein gene and disease-specific protein gene of Rice stripe virus[J]. Journal of Fujian Agricultural University (Natural Science), 2001, 30(1): 53-58.

[10] LIN Q T, LIN H X, WU Z J, et al. Accumulations of coat protein and disease-specific protein of rice stripe virus in its host[J]. Journal of Fujian Agricultural University, 1998, 27(3): 322-326.

[11] LIN Q T, XIE L H, ZHOU Z J, et al. Studies on rice stripe disease I.Distribution and losses caused by the disease[J]. Journal of Fujian Agricultural College, 1990, 19(4): 421-425.

[12] SAENZ P, QUIOT L, QUIOT J B, et al. Pathogenicity determinants in the complex virus population of a Plum pox virus isolate[J]. molecular plant-microbe interactions, 2001, 14(3):278-287.

[13] SCHNEIDER W L, ROOSSINCK M J. Genetic diversity in RNA virus quasispecies is controlled by host-virus interactions[J]. Journal of Virology, 2001, 75(14):6566-6571.

[14] TAKAHASHI M, TORIYAMA S, HANANATSU C, et al. Nucleotide sequence and possible ambisense coding strategy of rice stripe virus RNA segment 2[J]. Journal of General Virology, 1993, 74 (4):769-773.

[15] TORIYAMA S, TAKAHASHI M, SANO Y, et al. Nucleotide sequence of R NA1, the largest genomic segment of rice stripe virus, the prototype of the tenuiviruses[J]. Journal of General Virology, 1994, 75 (12):3569-3579.

[16] TORIYAMA S. Rice stripe virus.CMI /AAB Scotland: Descriptions of plant viruses, 2000, 375.

[17] WANG Q, ZHAI J C, XU D K. The occurrence and integrated control of rice stripe disease [J]. Journal of Hebei Agricultural Sciences,2007, 11(3):49-50.

[18] WANG X D, WU H G, BO X S. The occurrence characteristics and control techniques of rice stripe disease[J]. Anhui Agricultural Science Bulletin, 2007, 13(5): 146.

[19] WEI T Y, YANG J G, LIAO F L, et al. Genetic diversity and population structure of Rice stripe virus in China[J]. Journal of General Virology, 2009, 90(4):1025-1034.

[20] XIANG Y P, GAO Y. Cause and control tactics of rice stripe disease in Liaoning [J]. Liaoning Agricultural Sciences, 2007, (3): 110-111.

[21] ZHANG G M, WANG H D, DAI D J. The occurrence and control strategy of rice stripe disease in Zhejiang area[J]. China Plant Protection, 2006, (7):20-21.

马铃薯 Y 病毒 *pipo* 基因的分子变异及结构特征分析

高芳銮[1]，沈建国[2]，史凤阳[1]，常飞[1]，谢联辉[1]，詹家绥[1]

(1 福建农林大学植物病毒研究所，福建省植物病毒学重点实验室，福州 350002;

2 福建出入境检验检疫局检验检疫技术中心，福州 350001)

摘要：为揭示马铃薯 Y 病毒（*Potato virus Y*, PVY）*pipo* 基因的分子变异和结构特征，文章根据文献报道的马铃薯 Y 病毒属（*Potyvirus*）*pipo* 基因保守区序列设计一对简并引物，从感染 PVY 的马铃薯病叶中克隆获得 *pipo* 基因的 cDNA 全长序列，分析其核苷酸序列和氨基酸序列的特征，并基于氨基酸序列使用贝叶斯法重建了 *Potyvirus* 的系统发育树。结果显示：20 个 PVY 分离物成功扩增出预期大小（约 235bp）的特异性片段，其核苷酸序列与已报道的其他 PVY 株系的 *pipo* 基因核苷酸序列一致性均在 92% 以上；5′ 端均含有典型的 $G_{1\text{-}2}\,A_{6\text{-}7}$ 基序（motif），无碱基插入 / 缺失，所有的核苷酸变异都是碱基置换，共发现 13 个多态性位点，其中 4 个简约信息位点，9 个单一变异位点，表明该基因高度保守，但不同分离物也存在一定的分子变异；PIPO 蛋白理论等电点 11.26 ～ 11.62，无信号肽和跨膜区，是可溶的亲水性蛋白；整个蛋白含有 3 个保守区，其中位于 10 ～ 59aa 的基序最为保守。该蛋白主要定位于线粒体中，可能是线粒体导肽。系统发育分析结果显示，源于 PVY 不同株系优先相聚成簇，而向日葵褪绿斑驳病毒（*Sunflower chlorotic mottle virus*, SuCMoV）与辣椒重花叶病毒（*Pepper severe mosaic virus*, PepSMV）的亲缘关系较 PVY 相比更近，与前人的结果相一致，表明 PIPO 蛋白可以作为研究 *Potyvirus* 系统发育关系的新的分子标记。

关键词：马铃薯 Y 病毒；*pipo* 基因；病毒进化

Sequence variation and protein structure of *pipo* gene in *Potato virus Y*

Fangluan Gao[1], Jianguo Shen[2], Fengyang Shi[1], Fei Chang[1], Lian-Hui Xie[1], Jiasui Zhan[1]

(1 Fujian Key Laboratory of Plant Virology, Institute of Plant Virology, Fujian Agriculture and Forestry University, Fuzhou 350002; 2 Inspection & Quarantine Technology Center, Fujian Entry-Exit Inspection and Quarantine Bureau, Fuzhou 350001)

Abstract: The objectives of this study were to understand the sequence variation and the putative protein structure of *pipo* gene in the *Potato virus Y* (PVY) collected from *Solanumtu berosum*. The *pipo* gene in PVY was cloned using a pair of degenerate primers designed from its conserved region and its sequences were used to re-construct phylogenetic tree in *Potyvirus* genera by a Bayesian inference method. An expected fragment of 235bp was amplified in all 20 samples by RT-PCR and the *pipo* genes in the 20samples assayed shared more than 92% nucleotide sequence similarity with the published sequences of PVY strains. Among the 20 *pipo* gene sequences, 13 polymorphic sites were detected, including 4 parsimony informative sites and 9 singleton variable sites. These results indicate that PVY *pipo* gene is highly conserved but some sequence variations exist. Further analyses suggest

遗传 . 2013, 35(9): 1125-1134

收稿日期：2013-04-15

基金项目：国家现代农业马铃薯产业技术体系（编号：CARS-10），福建省自然科学基金项目（编号：2013J01088），福建省教育厅科技项目（编号：JA12119）和福建出入境检验检疫局科技项目（编号：FK2011-07）资助

that the *pipo* gene encodes a hydrophilic protein without signal peptide and transmembrane region. The protein has theoretical isoelectric points (pI) ranging from 11.26 to 11.62 and contains three highly conserved regions, especially between aa 10 and 59. The protein is likely located in the mitochondria and has α-helix secondary structure. Bayesian inference of phylogenetic trees reveals that PVY isolates are clustered in the same branch with high posterior probability, while *Sunflower chlorotic mottle virus* (SoCMoV) and *Pepper severe mosaic virus* (PepSMV) are closely related, consisting with the classification of *Potyvirus* genera using other approaches. Our analyses suggest that the *pipo* gene can be a new marker for phylogenetic analysis of the genera. The results reported in this paper provide useful insights in the genetic variation and the evolution of PVY and can stimulate further research on structure and function of the PIPO protein.

Keywords: *Potato virus Y*; *pipo* gene; evolution of virus

病原物群体遗传结构及其演化机制直接影响植物病害的发生、发展及控制。在寄主、病原和环境的互作体系中，高群体遗传多样性的病原具有相对强的生存和进化优势，能更快适应新的抗病寄主或环境。病原物通过基因突变、遗传迁移和基因重组不断引进和产生新的变异，而这些变异能否在群体中生存和繁衍则由遗传漂变和自然选择所决定。在自然选择下，有害的变异会被逐渐淘汰，而有利的变异则会被不断累积和放大（祝雯和詹家绥，2012）。然而，不是所有有利突变都能在自然界中生存下来，也不是所有的有害突变都能从群体中消失。有些有益突变会通过遗传漂变和搭便车效应（Hitchhiking effect）从群体中消失，反之，那些对病原物的生存和繁衍有害的变异也会通过这两种途径在群体中保存下来（祝雯和詹家绥，2012）。因此开展病原物的分子变异及群体结构特征分析有利于了解植物病毒的发生、流行和防控，具有重要的理论和实践意义。

马铃薯 Y 病毒（*Potato virus Y*, PVY）是马铃薯生产上造成经济损失最为严重的植物病毒之一。除了马铃薯，PVY 还可以侵染烟草、番茄等经济作物。PVY 是马铃薯 Y 病毒科（*Potyviridae*）马铃薯 Y 病毒属（*Potyvirus*）成员之一，其基因组由正义单链 RNA 分子组成，大小约 9.7kb，包含一个大的开放阅读框（Open reading frame, ORF），两端含有非编码区（Untranslated region,UTR），表达时先翻译成一个单一的多聚蛋白质（Polyprotein），再通过自身编码的蛋白酶将多聚蛋白质分割加工为 P1、HC-Pro、P3 等多个成熟蛋白（King et al., 2011）。*Potyvirus* 病毒编码蛋白的功能极为复杂，一般均是多功能的，通过与其他编码蛋白及寄主蛋白的互作共同实现病毒的侵染循环（Gibbs and Ohshima, 2010）。多聚蛋白切割是 PVY 基因组通常采用的表达策略，通过多聚蛋白切割使一个单顺反子 RNA 最终产生一系列成熟且功能各异的蛋白，也使得病毒基因组中不同基因的遗传变异呈现多样化，如：突变、缺失、插入、重组等。除了多聚蛋白切割的表达策略外，最近研究表明 PVY 基因组也可以通过移码翻译（Reading frame shift）的表达策略产生新蛋白，所以高变异性是 PVY 基因组的重要特征。

pipo 基因的发现，改变了长期以来对 PVY 基因组只编码 10 个蛋白的共识（Betty Y-W et al., 2008）。新近研究发现，*pipo* 基因广泛分布于 *Potyvirus* 中，位于 P3 顺反子内部，相对于 P3 以 +2 阅读框相位编码，其核苷酸序列 5′ 端含有高度保守的 GGAA AAA A（A）基序（空格为 P3 蛋白的密码子编码相位）。Wen 和 Hajimorad（2010）通过研究发现大豆花叶病毒（*Soybean mosaic virus*, SMV）PIPO 蛋白参与了病毒的细胞间运动，首次确认 *Potyvirus pipo* 基因的功能，并首次证实 SMV *pipo* 基因 5′ 末端保守的 GA_6 基序具有影响病毒运动的生物学功能。尽管 Chung 等（2008）通过免疫印迹实验检测到芜菁花叶病毒（*Turnip mosaic virus*, TuMV）PIPO 蛋白，证实其存在性，但研究结果表明 TuMV PIPO 蛋白不单独表达而以 P3N-PIPO 融合蛋白的方式表达。Wei 等（2010）进一步研究表明 TuMV P3N-PIPO 是一个胞间连丝（Plasmodesmata, PD）定位蛋白，在植物体内可以和柱状内含体蛋白（Cylindrical inclusion protein, CI）互作。目前，国内外专家学者对 *Potyvirus* P3N-PIPO 的功能研究作了一些基础研究，但关于 PVY *pipo* 基因变异的报道仅 Cuevas 等（2012）对 68 个全球 PVY 分离物 PIPO 蛋白的氨基酸长度变异作了初步分析，结果显示不同分离物的蛋白存在长度变异，说明 PVY 的适应性进化可能为植物寄主所驱动。

PVY 在世界上广泛流行，在我国及周边国家呈上升发展趋势。随着近年来农产品贸易的全球化，PVY 变异频繁，株系分化严重（陈士华等 , 2010, 2011; 孙琦和张春庆 , 2005; Ogawa et al., 2008; Hu et al., 2011; Ogawa et al.,2000, 2012），迫切需要对其遗传变异及分子进化开展研究。PIPO 蛋白是 *Potyvirus* P3 蛋白内移码方式产生的新蛋白，使得 *pipo* 基因的相关研究成为目前国内外植物病毒学的研究热点，但对于 PVY *pipo* 基因的分子变异以及 PIPO 蛋白的结构特征的相关研究国内外鲜有报道。为此，本研究通过扩增、克隆陕西、湖南两省的 20 个 PVY 分离物 *pipo* 基因的全长 cDNA 序列，并应用生物信息学方法对其核苷酸序列、氨基酸序列的分子变异及结构特征等进行了一系列的系统分析，旨在为探索 PVY *pipo* 基因的遗传变异、分子进化以及进一步开展 PIPO 蛋白的功能研究奠定基础。

1 材料和方法

1.1 毒源采集及检测

马铃薯病株是 2011 年采用随机采样法从陕西（4 月）、湖南（8 月）两省的马铃薯主要种植区采集的，样品经过 Indirect-ELISA 检测和电镜观察以确保毒源为 PVY（高芳銮等 , 2014）。

1.2 RNA 提取及 *pipo* 基因克隆

采用 Trizol 试剂法从感染 PVY 的马铃薯病叶中提取总 RNA，提取方法参照 RNAsimple Total RNA 试剂盒说明书进行。取 2 μL 总 RNA 以 oligo（dT）为引物按照操作说明书进行反转录，获得 PVY 全长的 cDNA。根据文献（Betty Y-W et al., 2008）报道的 *pipo* 基因序列保守区，设计用于扩增 *pipo* 基因的简并引物 PIPO-F（5′-YGAGYGTYGTRCAGATYATGG-3′）和 PIPO-R（5′-TTRARTCGCTCRYTCAABCCTG-3′），预期片段大小为 235bp，引物由南京金斯瑞生物技术有限公司合成。

PCR 扩增采用 50μL 反应体系：10×Trans*Taq*^{TM}HiFi Buffer II 5 μL，dNTPs（2.5mmol/L）4μL，CP-F（10μmol/L）2μL，CP-R（10μmol/L）2μL，ddH_2O 34.5μL，Trans*Taq*^{TM}HiFi Polymerase（5 U/μL）0.5μL，cDNA 2μL。PCR 扩增条件为：94℃预变性 5min; 94℃变性 30s，50℃复性 30s，72℃延伸 1min，共 30 个循环；最后再 72℃延伸 10min。

PCR 扩增结束后，取产物 5μL 用 1% 琼脂糖凝胶电泳进行检测。PCR 产物纯化后克隆至 pEASY TM-T5 Zero 载体上，并转化到 Trans1-T1 感受态细胞中。经菌落 PCR 鉴定获得阳性重组质粒后，随机选择其中 3 ～ 6 个阳性克隆子由南京金斯瑞生物技术有限公司测序，并通过测序峰图及序列比对分析排除由 PCR 引起的突变。

1.3 *pipo* 基因的核苷酸序列分析

测序获得的序列使用 BioEdit 等软件进行处理，*pipo* 基因的核苷酸一致序列使用 MEGA 软件中的 Muscle（Codons）子程序进行多重比对，并根据对应的编码氨基酸序列手动校正。序列同源性主要使用 BLAST（http://blast.ncbi.nlm.nih.gov/）计算，分子变异通过 DnaSP（Librado and Rozas, 2009）等软件进行分析。*pipo* 基因的选择压力分别采用 FEL（Fixed effects likelihood）、IEFL（Internal branches fixed-effects likelihood）、MEME（Mixed effects model of evolution）3 种方法通过在线服务器（http:// www.datamonkey.org）进行检验。判断选择的方向，主要根据 ω 值（d_N/d_S，即非同义替换 - 同义替换速率的比值）进行，当 ω>1 且模型具显著性差异时，说明为正向选择；反之，当 1>ω>0 之间时，则说明是净化选择。

1.4 PIPO 蛋白的结构特征分析

PIPO 蛋白的理化性质使用 ExPASy 服务器（http://www.expasy.org）的 ProtParam 分析，疏水性、跨

膜区、信号肽等功能区分别使用 BioEdit、TMpred 在线程序（http://ch.embnet.org/software/TMPRED_form.html/）和 SignalP 4.1 服务器（http://www.cbs.dtu.dk/services/SignalP/）进行分析，并提交其氨基酸质序列到 TargetP 1.1 服务器（http://www.cbs.dtu.dk/services/TargetP/）进行亚细胞定位。PIPO 蛋白的保守区使用 MEME 数据库（http://meme.nbcr.net）进行分析，二级、三级结构及功能通过 I-TASSER、COFACTOR 在线服务器（http://zhanglab.ccmb.med.umich.edu）进行预测。

1.5 系统发育分析

选取已报道的 *Potyvirus* 48 个种不同病毒及 8 个 PVY 不同株系的 *pipo* 基因氨基酸序列作为参考序列，并从本研究的 20 个 PVY 分离物中选取 8 个特异的氨基酸序列用于构建系统发育树。建树前，使用 MAFFT（Kazutaka and Standley, 2013）软件对建树的 64 条蛋白质序列进行多重序列比对，利用 ProtTest 3.0（Abascal et al.,2005）选择最优化的蛋白质进化模型 --JTT（Jones, Taylor, Thornton）+F 模型，使用 Mrbayes3.04 b（Ronquistet al.,2012）软件重建 *Potyvirus* 贝叶斯系统发育树。在重建贝叶斯树过程中，建立 4 个马尔可夫链，以随机树为起始树，共运行 4 000 000 代。每 100 代抽样 1 次，舍弃 25% 老化样本后，根据剩余的样本构建一致树，并计算后验概率（Posterior probability）。

2 结果与分析

2.1 *pipo* 基因扩增与序列分析

简并引物对 PIPO-F/PIPO-R 成功地扩增到 *pipo* 基因，扩增产物的片段大小与预计的目的产物一致，约 235bp（图 1），20 个分离物以及阳性对照均扩增到目的片段，而阴性对照（健康马铃薯叶片）和空白对照均未扩增到相应的片段。

测序分析后，确定 pipo 基因核苷酸序列长度均为 231bp（GenBank 登录号：KC577425 ～ KC577444），编码 75 个氨基酸长度的 PIPO 蛋白，序列同源性分析显示 20 个 PVY 分离物 *pipo* 基因与已登录 GenBank 的 PVY 不同株系核苷酸序列一致性均在 92% 以上，其中与 PVY^{C} 株系的序列一致性最低，为 92% ～ 96%，而与其他株系的序列一致性均超过 96%（表 1），表明 PVY *pipo* 基因序列高度保守。

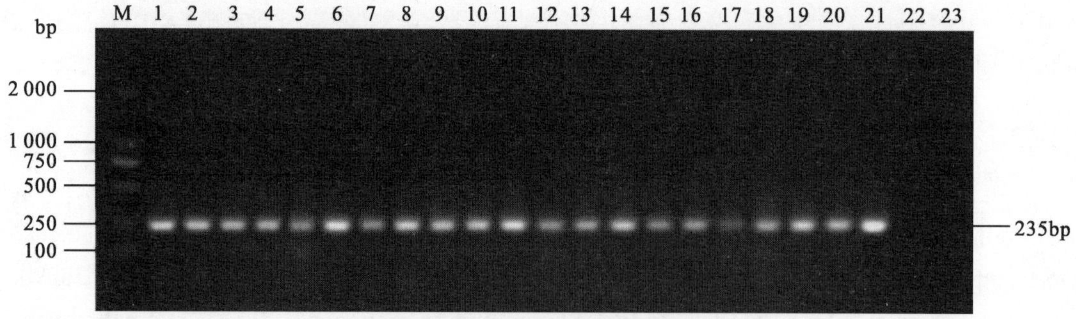

图 1 PVY *pipo* 基因扩增结果

M：DL2000 DNA 分子量标准；泳道 1 ～ 23：分离物 CS13、CS14、CS15、CS16、CS17、CS18、CS19、CS20、CS23、CS26、CS36、ShX03、ShX08、ShX12、ShX14、ShX16、ShX18、ShX19、ShX20、ShX21、阳性对照、阴性对照、空白对照

表 1 PVY 不同分离物 *pipo* 基因核苷酸序列一致性（%）

分离物	O-139	PRI-509	N-fr	12-94	PB209	N1	SYR-II-2-8
CS13	98	93	96	98	98	99	98
CS14	98	92	96	99	99	97	99
CS15	98	93	97	99	99	98	99
CS16	99	93	96	98	98	99	98
CS17	99	93	96	98	98	99	98
CS18	99	93	97	99	99	99	99
CS19	98	96	98	98	98	99	98
CS20	99	93	96	98	98	99	98
CS23	99	93	97	99	99	99	99
CS26	99	93	97	99	99	99	99
CS36	99	93	96	98	98	99	98
ShX03	99	93	96	98	98	99	98
ShX08	99	93	96	98	98	99	98
ShX12	99	93	96	98	98	99	98
ShX14	99	93	96	98	98	99	98
ShX16	99	92	96	98	98	99	98
ShX18	99	93	96	98	98	99	98
ShX19	99	93	96	98	98	99	98
ShX20	98	92	97	99	99	98	99
ShX21	98	93	97	99	99	98	99

注：O-139（PVY^{O}），PRI-509（PVY^{C}），N-fr（PVY^{N}），12-94（PVY^{NTN}），PB209（$PVY^{N:O}$），N1（$PVY^{N\text{-}W}$），SYR-II-2-8（$PVY^{NTN\text{-}NW}$）为已登录 GenBank 的 PVY 不同株系的 *pipo* 核苷酸序列，GenBank 登录号分别为 U09509、EU563512、NC_001616、AJ889866、EF026076、HQ912863、AB461451。

20 个 PVY 分离物的 *pipo* 基因均以 +2 阅读框开始（图 2A），都含有典型的 $G_{1\text{-}2}A_{6\text{-}7}$ 基序（图 2B），不存在碱基插入 / 缺失现象，所有的核苷酸变异都是碱基置换，有 13 个（占总分离位点的 6.04%）多态性位点（Polymorphic sites），其中 4 个为简约信息位点（Parsimony informative sites），9 个为单一变异位点（Singleton variable sites）。PIPO 蛋白的第一个氨基酸为赖氨酸（Lysine, K）（图 2B），整个蛋白存在 8 个变异的氨基酸位点（表 2），其中分离物 CS13、CS14、CS19、ShX20、ShX21 各有一个特异性氨基酸突变位点。在这 20 个 *pipo* 基因的 228 个位点中，除了 FEL 法检测到 2 个净化选择（$1>\omega>0$）位点外，其他均未检测到正向选择（$\omega>1$）压力位点（表 3），大部分密码子处于中性选择，说明 *pipo* 基因可能主要以中性进化为主。

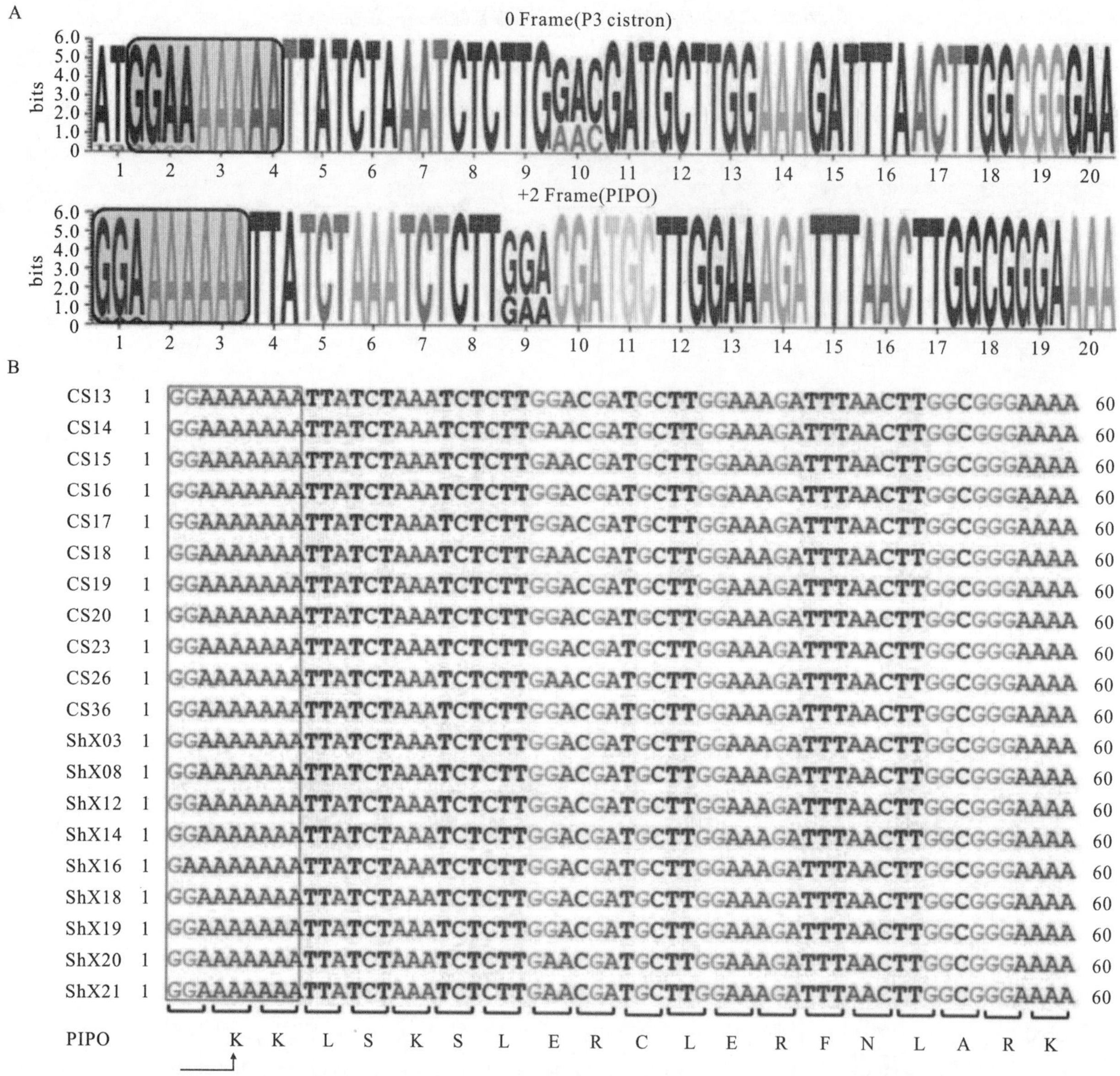

图2 PVY *pipo* 基因的序列特征

A：移码翻译；B：PVY *pipo* 基因的多重序列比对（部分）。其中加框显示的为 *pipo* 基因高度保守的 $G_{1\text{-}2}A_{6\text{-}7}$ 基序；箭头为编码PIPO蛋白的第一个氨基酸位置

表 2 20 个 PVY 分离物 *pipo* 基因氨基酸变异位点的位置

分离物	氨基酸位置							
	8	21	50	58	60	63	67	71
CS13	G	I	Q	G	K	**S**	Y	I
CS14	E	·	·	·	R	G	H	**F**
CS15	E	·	·	·	R	G	H	·
CS16	·	·	·	·	·	G	·	·
CS17	·	·	·	·	·	G	·	·
CS18	E	·	·	·	·	G	·	·
CS19	·	·	**R**	·	·	G	·	·
CS20	·	·	·	·	·	G	·	·
CS23	·	·	·	·	·	G	·	·
CS26	E	·	·	·	·	G	·	·
CS36	·	·	·	·	·	G	·	·
ShX03	·	·	·	·	·	G	·	·
ShX08	·	·	·	·	·	G	·	·
ShX12	·	·	·	·	·	G	·	·
ShX14	·	·	·	·	·	G	·	·
ShX16	·	·	·	·	·	G	·	·
ShX18	·	·	·	·	·	G	·	·
ShX19	·	·	·	·	·	G	·	·
ShX20	E	·	·	**E**	R	G	H	·
ShX21	E	**F**	·	·	·	G	H	·

注：加框灰度显示的为特异性氨基酸。

表 3 3 种方法检测 PVY *pipo* 基因的选择压力位点

方法	净化选择	中性选择	正向选择
FEL	2(Arg47*、Arg73**)	226	0
IFEL	0	228	0
MEME	0	228	0

注：*P= 0.06; **P= 0.09。

2.2 PIPO 蛋白质分析

PIPO 蛋白的分子量大约 7.4 ～ 8.9kDa，理论等电点（pI）为 11.26 ～ 11.62，溶解度为 98%，属于可溶蛋白，不稳定指数为 47.19 ～ 55.62，脂肪指数为 92.27 ～ 97.47，有两个疏水性区域（20 ～ 24 位、50 ～ 54 位氨基酸），平均疏水性值均不高（分别为 1.0 和 1.35），表现较强的亲水性；蛋白内不具有典型的跨膜区，表明可能不是跨膜蛋白；N′ 端无信号肽，推测其为非分泌型蛋白。该蛋白主要定位于线粒体中（预测分值为 0.491 ～ 0.755），可能是线粒体导肽（Mitochondrial targeting peptide, mTP）。

PIPO 蛋白含有 3 个明显的保守区，根据显著性从高到低顺序分别为 Motif 1、Motif 2、Motif 3（表 4），其中位于 PIPO 蛋白的 10 ～ 59 位氨基酸的 Motif 1 区域最保守，表明其可能在该蛋白行使功能中起着非常重要的作用；二级结构主要成分为 α 螺旋（α1-α4）（图 3A）。PyMOL 软件渲染后的 PIPO 蛋白三维结构如图 3B 所示。同时，COFACTOR 服务器预测显示 PIPO 蛋白存在两个活性位点，分别是 23 位的天冬酰胺（Asn）和 59 位的脯氨酸（Pro），各自具有固醇硫酸酯酶（Steryl-Sulfatase, STS）和核糖核酸酶 P（Ribonuclease P, RNase P）活性（图 3C）。

表 4 PIPO 蛋白的保守区

保守区	位置	E- 值	保守序列
Motif 1	10~59aa	1.6e-824	
Motif 2	61~75aa	2.7e-208	
Motif 3	1~8aa	1.9e-072	

2.3 系统发育分析

PVY 不同株系（含本研究中的 PVY 不同分离物）相聚在同一簇（Clade A，后验概率为 100%），显示出明显的物种特异性。向日葵褪绿斑驳病毒（*Sunflower chlorotic mottle virus*, SuCMoV）先与辣椒重花叶病毒（*Pepper severe mosaic virus*, PepSMV）聚成另一簇（Clade B）后，才与 PVY 进一步聚成一个大簇（后验概率为 99%）。通过 BLASTP 比对，发现 PVY 不同株系的 PIPO 蛋白与 SuCMoV、PepSMV 的氨基酸序列一致性分别为 37% ～ 51% 和 32% ～ 44%，而 SuCMoV 与 PepSMV 的序列一致性则高达 61%。系统发育分析结果说明，与 PVY 相比，SuCMoV 在进化史上与 PepSMV 的亲缘关系更近。

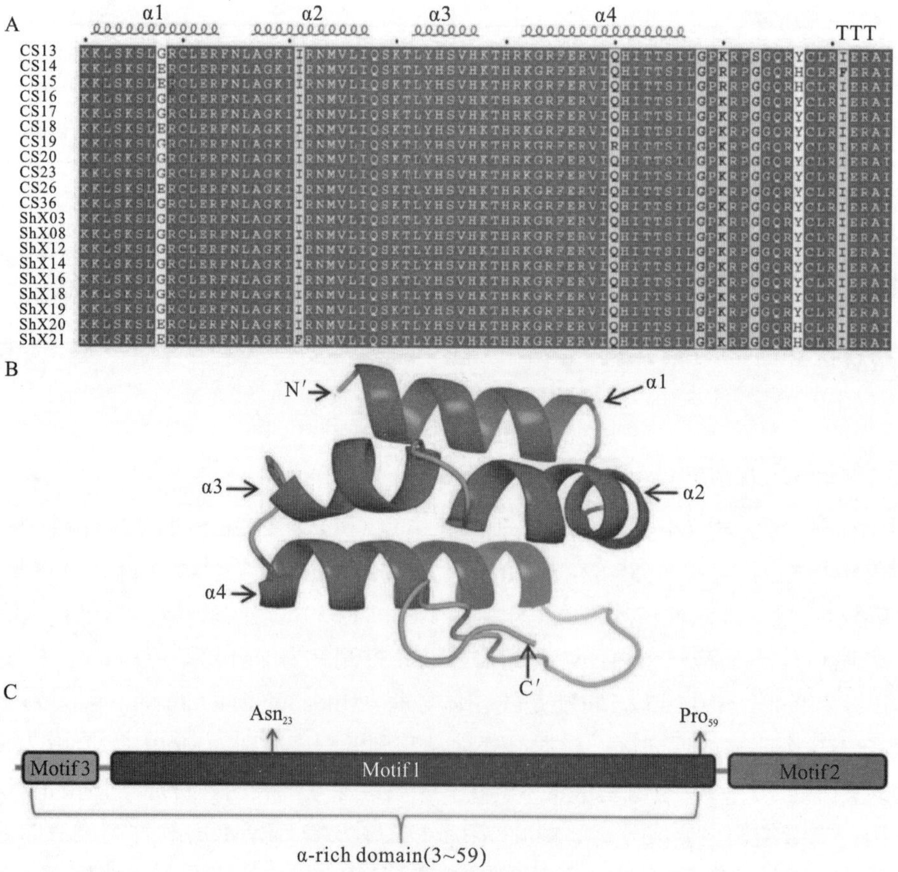

图 3 PVY PIPO 蛋白的结构特征

A：PVY PIPO 蛋白的多重序列比对及二级结构预测（α1-α4 为 α- 螺旋）；B：PVY PIPO 蛋白的三维结构预测（红色为 α - 螺旋；绿色为无规则卷曲）；C：PVY PIPO 蛋白的结构特征示意图（预测的活性位点：Asn 23：固醇硫酸酯酶；Pro 59：核糖核酸酶 P）

3 讨论

3.1 *pipo* 基因的分子变异与分子进化

PVY *pipo* 基因具有高度的保守性（表 1），20 个不同分离物 *pipo* 基因与已知的核苷酸序列一致性均在 92% 以上，说明该基因在病毒的生存和繁殖中起着十分重要的作用。虽然 PVY *pipo* 基因高度保守，但不同地区分离物也存在一定的分子变异，20 个分离物共发现有 13 个碱基多态性位点，其中单一变异位点约占 70%。这些突变体都是通过核苷酸碱基置换形成的，没有出现序列插入或缺失现象，说明点突变在 PVY 进化中起着重要的作用。有些突变体是地区特有的，比如：分离物 CS13、CS14、CS19、ShX20、ShX21 均有一个地区特异性的氨基酸位点，暗示 PVY 群体间存在一定的地理间隔。

pipo 基因整体上以中性进化为主（表 3），使用 FEL 法在 *pipo* 基因中检测到 2 个净化选择位点（Arg47、Arg73），而另外两种方法均未检测到显著的选择压力位点。究其原因，可能有两个方面：一是 *pipo* 基因可能具有古老的进化历史，但因其编码的氨基酸序列较短，可用于检测位点较少，从而导致适合性进化信号无法被检测。我们对 126 个来自不同寄主和地区的 PVY 分离物 *pipo* 基因进行进化分析和选择压力检测时，表明该基因受中性选择为主，但也存在一个显著的正向选择压力位点，且该位点在氨基酸的组成上存在寄主、地区特异性（未发表数据）；二是 *pipo* 基因可能处于突变冷点（Martincorena et al., 2012）。

3.2 PIPO 蛋白的结构与功能

组成蛋白质的 20 种氨基酸各自带有不同极性的侧链基因，疏水性氨基酸位于蛋白质内部，通过疏水的相互作用，在保持蛋白质三级结构的形成和稳定中起着重要的作用（吴祖建等 , 2010）。PIPO 蛋白理化蛋白质分析结果可知，其溶解度为 98%，且稳定指数高于 40，表明 PIPO 蛋白为不稳定的可溶蛋白（低于 40 为稳定蛋白）（Guruprasad et al., 1990）。当前蛋白质三维结构的主要预测方法包括同源模建、折叠识别和从头预测 3 种，其中同源模建是目前应用最成功的一种方法，由于已知蛋白质结构数据库中找不到 PIPO 蛋白的同源蛋白质，故采用折叠识别法预测 PIPO 蛋白三维结构，I-TASSER 服务器预测结果显示，折叠 1fc3A 与 PIPO 蛋白的匹配性非常高，其各项参数综合值在预测的前 10 个模板蛋白中最高。通过 PIPO 蛋白的保守区（表 4）及空间结构（图 3）分析结果，显示 PIPO 蛋白含有两个富含 α 螺旋的结构域：Motif 1、Motif 3，这两个结构域都为 α 螺旋所组成的重复片段结构，可以参与蛋白 - 蛋白相互作用。同时，位于 10 ～ 59 位氨基酸的 Motif 1 最为保守，且两个活性位点 Asn 23、Pro 59 也位于其内，说明该结构域在 PIPO 蛋白功能中发挥着非常重要的作用。蛋白质的结构信息在蛋白质的功能研究上起决定性作用，PIPO 蛋白的结构和功能预测分析可以为 PIPO 蛋白后期的表达纯化、功能研究等相关的实验工作提供重要的参考价值。

3.3 *Potyvirus* 的系统发育分析

在进化关系上，由于 SuCMoV 与 PVY 非常接近，以 *CP* 基因等为分子标记未能明显区别两者，故而一直以来被视为 PVY 的分组成员，归入 PVY 种内。Bejerman 等（2010）通过系统分析表明 SuCMoV 基因组具有明显区别于 PVY 的典型特征，应作为 *Potyvirus* 的一个新病毒种。从重建的 *Potyvirus* 贝叶斯树（图 4）可以看出，与 PVY 相比，SoCMoV 和 PepSMV 的亲缘关系更近。*CP* 基因由于高度保守的特点，在系统发育研究中，通常作为病毒分类的分子标记。相比 *CP* 基因而言，*pipo* 基因也具有高度保守的特点。和 Bejerman 等的研究结果相同，以 *pipo* 基因氨基酸序列构建的系统发育树也表明 SuCMoV 是 *Potyvirus* 的一个新病毒种（Guruprasad et al., 1990）。除外，*Potyvirus* 属内 49 个病毒种以较高的后验概率优先相聚成簇大致形成 10 个大簇，同一种内病毒总是以较高的后验概率优先相聚成簇，种内差异小于种间差异。尽管部分分支的后验概率不高，但基本能够反映出种间的系统发育关系。而基于 *pipo* 基因氨基酸序列构建系统发育树的拓扑结构与目前的分类地位大体一致（King et al., 2011）。综上结果说明，以 *pipo* 基因为分子标记用来研究 *Potyvirus* 的系统发育关系是完全可行的。

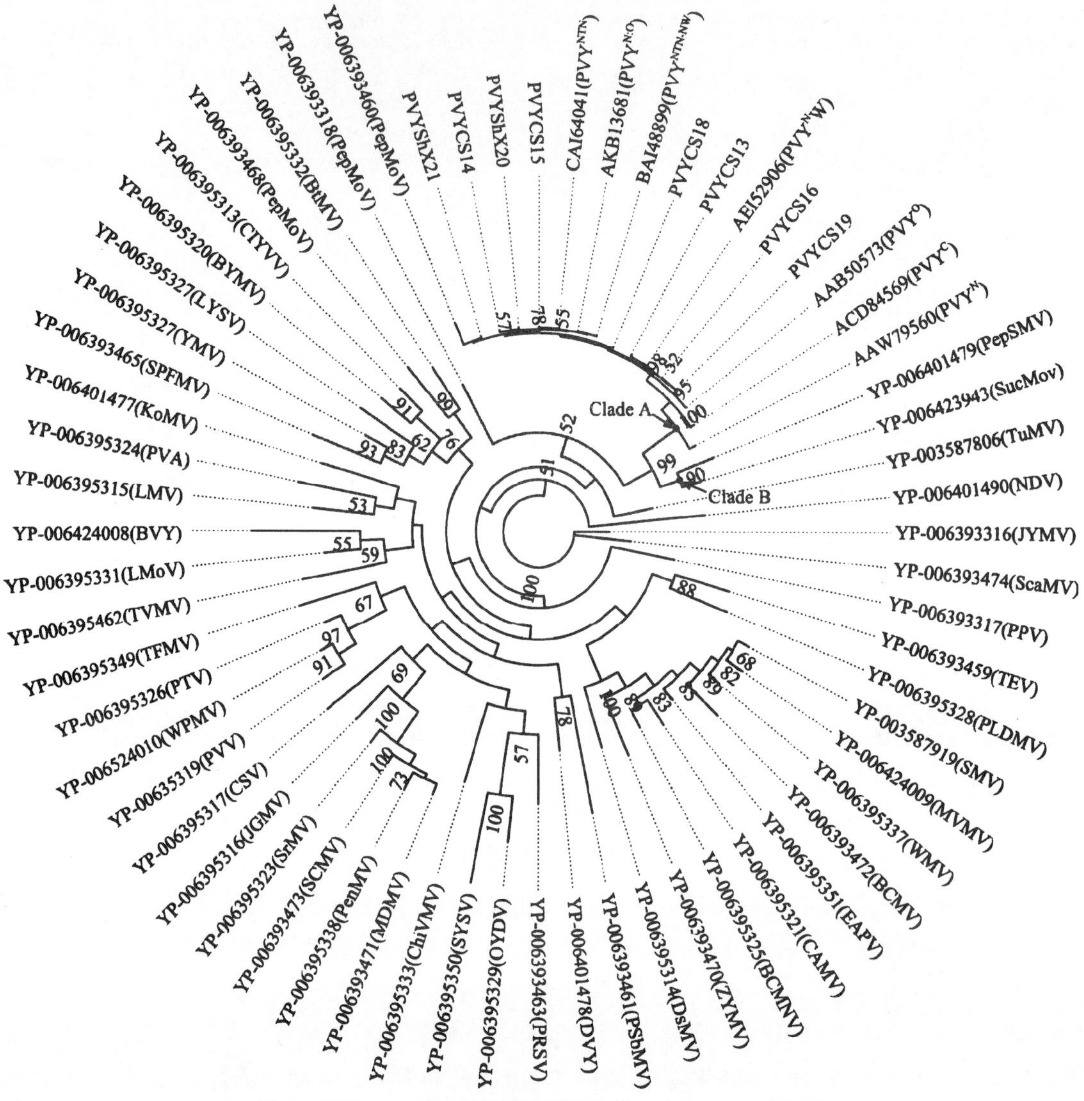

图 4　基于 *pipo* 基因氨基酸序列重建的 *Potyvirus* 贝叶斯树

分支节点上的数值为后验概率（>50%），蓝色显示的为 PVY 分离物（含本研究的 20 个分离物）

参考文献

[1] 陈士华，刘晓磊，张晓婷，等．河南 PVY 高致病性株系的发现及其分子特征研究 [J]. 河南农业大学学报，2010, 44(04):443-447.

[2] 陈士华，刘晓磊，张晓婷，等．中国部分马铃薯产区马铃薯 Y 病毒 (PVY) 的株系分化与鉴定 [J]. 河南农业大学学报，2011(05):548-551.

[3] 高芳銮，沈建国，史凤阳，等．中国马铃薯 Y 病毒的检测鉴定及 CP 基因的分子变异 [J]. 中国农业科学，2013,46(015):3125-3133.

[4] 孙琦，张春庆．PVY^{N} 与 PVY^{O} 病毒 RT-PCR 快速检测体系研究 [J]. 中国农业科学，2005(01):213-216.

[5] 吴祖建，高芳銮，沈建国．生物信息学分析实践 [M]. 北京：科学出版社，2010:107-110.

[6] 祝雯，詹家绥．植物病原物的群体遗传学 [J]. 遗传，2012(02):157-166.

[7] ABASCAL F, ZARDOYA R, POSADA D. ProtTest: selection of best-fit models of protein evolution[J]. Bioinformatics, 2005, 21(9):

2104-2105.

[8] BEJERMAN N, GIOLITTI F, DE BREUIL S, et al. Molecular characterization of *Sunflower chlorotic mottle virus*: a member of a distinct species in the genus *Potyvirus*[J]. Archives of Virology, 2010, 155(8): 1331-1335

[9] CHUNG B Y, MILLER W A, ATKINS J F, et al. An overlapping essential gene in the Potyviridae[J]. Proceedings of the National Academy of Sciences of the United States of America, 2008, 105(15): 5897-5902.

[10] CUEVAS J M, DELAUNAY A, VISSER J C,et al. Phylogeography and molecular evolution of *Potato virus Y*[J]. PLOS ONE, 2012, 7: e37853.

[11] GIBBS A, OHSHIMA K. Potyviruses and the digital revolution[J]. Annual Review of Phytopathology, 2010, 48(1): 205-223.

[12] GURUPRASAD K, REDDY B V, PANDIT M W.Correlation between stability of a protein and its dipeptide composition: a novel approach for predicting in vivo stability of a protein from its primary sequence[J]. Protein engineering, 1990, 4(2):155-161.

[13] HU X, NIE X, HE C, et al. Differential pathogenicity of two different recombinant $PVY^{N\ TN}$ isolates in Physalis floridana is likely determined by the coat protein gene[J]. Virology Journal, 2011, 8 (1): 207

[14] KAZUTAKA K, STANDLEY D M. MAFFT multiple sequence alignment software version 7: improvements in performance and usability[J]. Narnia, 2013, 30: 772-780.

[15] KING A M, LEFKOWITZ E, ADAMS M J, et al.Virustaxonomy: Ninth report of the international committee ontaxonomy of viruses[M]. Amsterdam: Elsevier Academic Press, 2011.

[16] LIBRADO P, ROZAS J. DnaS P v5: a software for comprehensive analysis of DNA polymorphism data[J]. Bioinformatics, 2009, 25: 1451-1452.

[17] MARTINCORENA I, SESHASAYEE A S, LUSCOMBE N M. Evidence of non-random mutation rates suggests an evolutionary risk management strategy[J]. Nature, 2012, 485(7396):95-98.

[18] OGAWA T, TOMITAKA Y, NAKAGAWA A, et al. Genetic structure of a population of *Potato virus Y* inducing potato tuber necrotic ringspot disease in Japan; comparison with North American and European populations[J]. Virus Research, 2008, 131: 199-212.

[19] OGAWA T, NAKAGAWA A, HATAYA T, et al. The Genetic structure of populations of *Potato virus Y* in Japan; based on the Analysis of 20 full genomic sequences[J]. Journal of Phytopathology, 2012, 160: 661-673.

[20] OHSHIMA K, SAKO K, HIRAISHI C, et al. Potato tuber necrotic ringspot disease occurring in Japan: its association with potato virus Y necrotic strain[J]. Plant Disease, 2000, 84: 1109-1115.

[21] RONQUIST F, TESLENKO M, van der Mark P, et al. MrBayes 3.2: Efficient Bayesian phylogenetic inference and model choice across a large model space[J]. Systematic Biology, 2012, 61(3): 539-542.

[22] WEI T, ZHANG C, HONG J, et al. Formation of complexes at plasmodesmata for potyvirus intercellular movement is mediated by the viral protein P3N-PIPO[J]. PLoS Pathogens, 2010, 6: e1000962.

[23] WEN R H, HAJIMORAD M R. Mutational analysis of the putative pipo of soybean mosaic virus suggests disruption of PIPO protein impedes movement[J]. Virology, 2010, 400: 1-7.

中国马铃薯 Y 病毒的检测鉴定及 CP 基因的分子变异

高芳銮[1]，沈建国[2]，史凤阳[1]，方治国[1]，谢联辉[1]，詹家绥[1]

（1 福建省植物病毒学重点实验室，福建农林大学植物病毒研究所，福州 350002；

2 福建出入境检验检疫局检验检疫技术中心，福州 350001）

摘要：查明马铃薯 Y 病毒（*Potato virus Y*, PVY）病在中国的发生情况，及时、准确地鉴定出 PVY 并对其分子变异进行分析。采用 ELISA 方法对采自中国 14 个省（直辖市）马铃薯种植区疑似受 PVY 感染的样品进行了检测，并对其中的部分材料镜检验证，然后根据 CP 基因序列设计 1 对简并引物针对随机选择感染 PVY 的 14 个省（直辖市）代表样品进行 CP 基因扩增克隆，将测序得到的序列进行分子变异分析，并使用贝叶斯法（Bayesian inference, BI）重建系统发育关系。ELISA 检测结果表明，691 份样品中有 220 个样品与 PVY 抗体呈阳性反应，其余呈阴性反应；ELISA 检测的阳性材料在透射电镜下均可观察到明显的风轮状内含体，14 个 PVY 分离物均成功扩增出预期大小（约 800bp）的特异性片段，CP 基因的核苷酸序列与已报道 PVY 不同株系的核苷酸序列一致性均在 88% 以上；在 14 个 PVY 分离物 CP 基因中共发现有 29 个多态性位点，其中 6 个简约信息位点，23 个单一变异位点。系统发育分析结果显示，14 个 PVY 分离物与 $PVY^{N:O}$ 株系相聚成簇，表明其在系统发育关系上，与 $PVY^{N:O}$ 株系的亲缘关系最近。PVY CP 基因高度保守，但不同地区分离物也存在一定的分子变异，本研究可为今后了解 PVY 病毒病流行、变异趋势及其防治提供依据。

关键词：马铃薯 Y 病毒；ELISA；外壳蛋白基因；分子变异；贝叶斯法

Detection and Molecular Variation of *Potato virus Y* CP Gene in China

Fangluan Gao[1], Jianguo Shen[2], Fengyang Shi[1], Zhiguo Fang[1], Lianhui Xie[1], Jiasui Zhan[1]

(1 Key Laboratory of Plant Virology of Fujian Province, Institute of Plant Virology, Fujian Agriculture and Forestry University,Fuzhou 350002;

2 Inspection & Quarantine Technology Center, Fujian Exit-Entry Inspection and Quarantine Bureau, Fuzhou 350001)

Abstract:The objectives of this study are to investigate the occurrence of *Potato virus Y* (PVY) disease in China,to develop fast and accurate PVY detection methods and to understand sequence variations in PVY CP gene. ELISA method was used to detect the infected leaf samples randomly collected from 14 provinces (municipalities) in China during 2011 and 2012. A subsample of the infected leaves was examined using a transmission electron microscope. CP gene of PVY was amplified by reverse-transcription polymerase chain reaction (RT-PCR) using a pair of degenerate primers designed from the conserved regions of published CP gene sequences. Sequence variation was analyzed and phylogenetic tree was re-constructed using Bayesian inference method. ELISA revealed that 220 out of the 691 samples assayed were infected with PVY. The pin-wheel inclusion bodies were visualized clearly under the transmission electron microscope in the subsamples of ELISA-positive leaves. RT-

中国农业科学 .2013,46(15):3125-3133

收稿日期：2013-01-25；接受日期：2013-05-16

基金项目：国家现代农业马铃薯产业技术体系（CARS-10）、福建省自然科学基金项目（2013J01088）、福建省教育厅科技项目（JA12119）、福建出入境检验检疫局科技项目（FK2011-07）

PCR amplifications of the 14 ELISA-positive subsamples randomly selected from different provinces (municipalities) all generated an expected fragment of ~800bp in size. CP genes in the selected 14 subsamples shared more than 88% nucleotide identity with the reported sequences in other PVY strains. Among the 14 CP gene sequences in the subsamples, 29 polymorphic sites were detected, including 6 parsimony informative and 23 singleton variable sites. Some of the sequence variations were geography-specific. Bayesian inference of phylogenetic trees revealed that 14 PVY isolates were in the same branch with reference $PVY^{N:O}$ strains suggesting that the 14 PVY isolates analyzed in this study shared high homology with $PVY^{N:O}$ strain. CP gene in PVY is highly conserved but some geographical variation in sequences exists. The results reported in this manuscript provide useful insights in understanding the evolution of PVY, the epidemiology and control of PVY disease in potato.

Key words: *Potato virus Y*; ELISA; coat protein gene (CP gene); molecular variation; Bayesian inference method

马铃薯是继水稻、小麦、玉米之后的第四大作物。中国是世界马铃薯第一生产大国，种植面积占全球种植面积约 25%，总产约占全球的 20%（Hu et al., 2009b）。马铃薯病毒病是马铃薯生产上非常重要的病害，其中马铃薯 Y 病毒（*Potato virus Y*, PVY）是目前中国感染马铃薯病毒病并造成最严重经济损失的病毒之一。PVY 可以侵染马铃薯、烟草、番茄和茄子等多种植物，侵染马铃薯后引起叶脉坏死、褐脉病、黄斑坏死等症状，可导致马铃薯种质退化、产量降低，严重时可造成高达 80% 以上的产量损失（Glais et al., 2002; Nolte et al., 2004; Whitworth et al., 2006）。由 PVY 引起的马铃薯病毒病自从 1931 年首次发现以来，在世界上广泛流行。自 20 世纪 90 年代初始，PVY 在中国及亚洲地区的中国周边国家呈上升发展趋势（Beczner et al., 1984; Ohshima et al., 2000; Sun and Zhang, 2005; Ogawa et al., 2008; Nie et al., 2012; Ogawa et al., 2012; Wang et al., 2012）。因此，快速准确检测 PVY 并研究该病毒在中国马铃薯各生产区的分布和发生情况及其分子变异情况，将有助于了解该病毒病流行和演化趋势，为其有效防控提供理论基础。目前生产上检测 PVY 的方法有多种，主要是传统生物学方法、血清学方法、分子生物学技术以及多种方法的综合应用。传统生物学主要通过指示寄主接种鉴定或电镜观察，具有准确、直观等特点，但由于存在周期长、易受环境及人为因素影响等问题，不适于 PVY 的快速及大批量检测鉴定。酶联免疫吸附测定法（Enzyme-linked immunosorbent assay, ELISA）是目前最常用的血清学方法之一，操作简便、快速灵敏、特异性强，适用于大量田间样品的检测，但受 ELISA 试剂盒质量和病毒在植株分布不均匀以及灵敏度低等的影响，在 PVY 检测中容易出现非特异性反应（Si et al., 2011）。RT-PCR 技术是近年来广泛用于检测诊断 PVY 的分子生物学方法，其最大优势在于高度的特异性和灵敏性，此法可以检测到 10^{-18}g 水平的病毒。由于序列的高度保守性以及在 PVYRNA 的衣壳化中起着重要的作用，CP 基因已成为 RT-PCR 扩增的首选基因。孙琦等（2005）针对 PVY 外壳蛋白同源区设计 1 对通用引物建立了 PVY^N 与 PVY^O 的 RT-PCR 快速检测体系。RT-PCR 检测技术虽然具有灵敏度高、耗时短、特异性强等优点，但在实际应用中对 RNA 的提取、引物的特异性要求高，不适于大批量样品的检测。鉴于所采样品大部分为田间自然发病的样品，采用单一的检测方法无法满足 PVY 的检测需求，多种检测方法的综合应用是当前 PVY 检测技术的主流趋势，如高秀妍等（2009）综合应用生物学、血清学、RT-PCR3 种方法对 PVY 黑龙江分离物 H2-1 进行株系鉴定，3 种方法复合使用，取长补短，可以进一步提高 PVY 检测的准确性。迄今为止，PVY 分子变异的研究主要集中于个别省份或地区的 PVY 株系分化及鉴定，如吴志明等（2005）对河北张家口感病马铃薯的 1 个 PVY 分离物的 CP 基因进行了克隆和序列分析，判断其可能归属于 PVY 株系。但对于多省份大范围的 PVY 分子变异报道较少，目前仅陈士华等（2011）分别克隆了黑龙江、山西、贵州等 3 个省内的 4 个马铃薯产区的 P1 和 CP 基因，并通过基因序列分析对各分离物的株系进行了鉴定，研究结果表明中国部分马铃薯产区 PVY 株系分化明显，且 $PVY^{N:O}$ 株系分布普遍。中国是一个幅员辽阔、地形复杂、生境多样的国家，且随着植物种质资源的引进和生态环境的变化，PVY 变异频繁，株系分化大，迫切需要对 PVY 在全国范围内的分布情况展开调查，密切关注变异株系的发生和分布，为 PVY 病毒病流行、变异趋势及其有效防控奠定基础。在中国南北 14 个省（直辖市）20 个地区进行了毒源采集，先通过 ELISA 方法对 PVY 疑似样品进行了初步筛查，并进一步通过电镜检测进行验证，最后设计适合于检测各株系 CP 基因的简并引物，对随机抽样的 14 个省（直辖市）PVY 分离物的 CP 基因进行 RT-PCR 扩增，同时进行分子变异分析，旨在研究 PVY 在中国的发生、分布及其遗传变异，确定 PVY 株系的变异是否具有区域性。

1 材料与方法

试验于 2011 年 8 月至 2012 年 12 月在福建农林大学植物病毒研究所和福建出入境检验检疫局检验检疫技术中心完成。

1.1 主要试剂

PVY 检测试剂为美国 Agdia 公司产品；RNAsimple TotalRNAKit、DNA Marker、Gel Purification Kit 为天根（TianGen）生化科技公司产品；Trans *Taq* DNA Polymerase HiFi、pEASY-T5 Zero Cloning Kit 为北京全式金（TransGen）公司产品。

1.2 毒源采集

2011—2012 年间从种植马铃薯的 14 个省（直辖市）采用随机采样法采集带有重度花叶、条斑坏死、点条斑坏死等明显症状的马铃薯叶片，共计 691 份（表 1），新鲜样本置于密封袋中，保存于 −85℃超低温冰箱。

1.3 ELISA 检测

具体检测方法参照 PVY 检测试剂盒说明书进行，先在预包被抗体的酶标板中加入待测样品，4℃冰箱孵育过夜，PBST 洗涤后加入酶标抗体、检测抗体，室温孵育 2h，PBST 洗涤后加入底物（pNPP）显色，于 Thermo Multiskan MK3 酶标仪 405nm 处读取 OD 值，比较判定阴、阳性。共检测 691 份样品，每样设 3 个重复，同时设置阴性、阳性及空白对照。

1.4 电镜观察

采用文献（谢联辉和林奇英 , 2011）的超薄切片法制样，将待检病毒样品置于 H-7650 透射电镜下观察植物组织病变情况，观察内含体的形状以判断病原。

1.5 RNA 提取及 CP 基因克隆

采用 Trizol 试剂法提取 PVY 马铃薯病叶中的总 RNA，提取方法参照试剂盒说明书进行。取 2μL 总 RNA 用 $oligo(dT)_{18}$ 为引物按照操作说明书进行反转录获得 cDNA。根据 Gen Bank 已公布的 PVY 常见株系 CP 基因的序列保守区，设计用于扩增 CP 基因的简并引物 CP-F（5′-GSAAAYGAYACAATYGATGC-3′）和 CP-R（5′-CACATGTTYTTVACTCCAAG-3′），预期扩增的目的片段大小约为 800bp，引物由南京金斯瑞生物技术有限公司合成。PCR 扩增采用 50μL 反应体系：10×Trans*Taq* HiFi BufferII 5μL，dNTPs（2.5mmol/L）4μL，CP-F（10μmol/L）2μL，CP-R（10μmol/L）

2μL，ddH_2O 34.5μL，Trans*Taq* HiFi Polymerase（5U/μL）0.5μL，cDNA 2μL。PCR 反应条件为：94℃预变性 5min；94℃变性 30s，55℃复性 30s，72℃延伸 1min，共 30 个循环；最后一轮循环后 72℃延伸 10min。

PCR 反应结束后，取产物 5μL 用 1% 琼脂糖凝胶电泳进行检测。PCR 产物纯化后克隆至 pEASY-T5Zero 载体上，并转化 Trans1-T1 感受态细胞中。经菌落 PCR 鉴定获得阳性重组质粒，随机选择其中 3~6 阳性克隆子委托南京金斯瑞生物技术有限公司测序，并通过测序峰图及序列比对分析排除由 PCR 扩增引起的突变。

1.6 序列分析及系统发育树重建

测序获得的序列分别采用 DNAMAN（Lynnon Bio Soft, Canada）、BLAST、DnaSP（Librado and Rozas, 2009）等软件进行序列分析。为了进一步了解 PVY 不同分离物的系统发育关系，从 GenBank 选取 24 个已知株系的 PVY CP 基因序列作为参考，进行系统发育分析。应用 Clustlax2.0（Larkin et al., 2007）多重序列比对后，利用 Gblocks0.91b（Talavera and Castresana, 2007）进行序列保守区的选择，MrModeltest2.3（Nylander, 2008）选择最优化的核苷酸替换模型并计算相关参数，Mrbayes3.04b（Ronquist et al., 2012）重建 PVY CP 基因的贝叶斯树。在重建贝叶斯树过程中，以芜菁花叶病毒（*Turnip mosaic virus*, TuMV）的 CP 基因为外群（Outgroup），建立 4 个马尔可夫链，以随机树为起始树，共运行 5000000 代，每 100 代抽样 1 次。舍弃 25% 老化样本后，根据剩余的样本构建一致树，并计算后验概率（Posterior probability）。

2 结果

2.1 ELISA 检测结果

在检测的 691 份样品中，共有 220 份样品与 PVY 抗体呈阳性反应，其余样品均呈阴性反应（表 1），不同地区样本的阳性检出率在 6.52% ～ 87.50%。其中采自福建闽侯、河南郑州及重庆石柱的样本阳性检出率均低于 10%，而采自福建长乐和霞浦的样本阳性检出率都在 80% 以上。按年份算，2011 年的平均检出率比 2012 年高，分别为 54.67% 和 27.52%。

2.2 电镜检测

ELISA 阳性反应的材料中随机选择 10 份，感病植株在 H-7650 透射电镜下均可以观察到明显的风轮状内含体（图 1），具备典型的 *Potyvirus* 细胞病理特征。

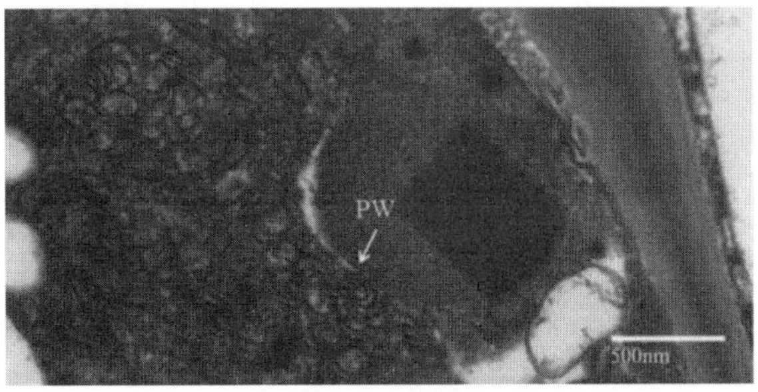

图 1 感病植株的细胞病理学特征

Fig. 1 Cyto-pathological characteristics of potato plant infected with PVY

PW：风轮状内含体

Pin-wheel inclusion body

表 1 PVY ELISA 检测结果

年份 Year		采样地点 Location	采样数 Sample sites	检测数 Tested sample	阳性数 Positive sample	检出率 Positive rate (%)
2011	陕西 Shaanxi	榆林 Yulin	24	24	9	37.50
	贵州 Guizhou	毕节 Bijie	6	6	1	16.70
	四川 Sichuan	西昌 Xichang	14	14	6	42.86
	湖南 Hunan	长沙 Changsha	41	41	25	60.98
	福建 Fujian	长乐 Changle	25	25	20	80.00
		福清 Fuqing	7	7	4	57.14
		霞浦 Xiapu	8	8	7	87.50
2012	福建 Fujian	龙岩 Longyan	40	40	9	22.50
		漳州 Zhangzhou	29	29	8	27.60
		闽候 Minhou	52	52	5	9.60
	广东 Guangdong	惠东 Huidong	16	16	5	31.25
	湖南 Hunan	常德 Changde	45	45	5	11.11
	河南 Henan	郑州 Zhengzhou	44	44	3	6.82
	湖北 Hubei	襄阳 Xiangyang	43	43	10	23.26
	重庆 Chongqing	石柱 Shizhu	46	46	3	6.52
	山东 Shandong	济南 Jinan	92	92	31	33.69
	云南 Yunnan	曲靖 Qujing	69	69	32	46.37
	河北 Hebei	张家口 Zhangjiakou	44	44	13	29.55
	内蒙古 Inner Mongolia	乌兰察布 Wulanchabu	18	18	12	66.67
	黑龙江 Heilongjiang	哈尔滨 Harbin	28	28	12	42.86

2.3 CP 基因的 PCR 扩增与序列分析

简并引物对 CP 基因的扩增产物与预计的目的片段大小一致，大小约为 0.8kb（图 2）。从 20 个地区随机抽样的 14 个省（直辖市）代表分离物以及阳性对照均可扩增到目的片段，阴性（健康马铃薯叶片）和空白对照均未扩增到相应的片段。

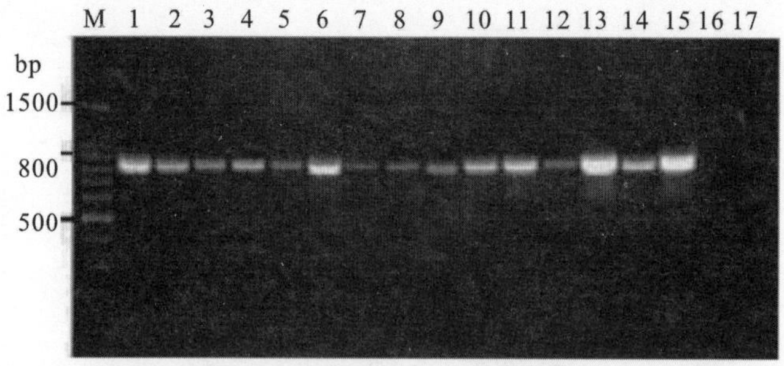

图 2 CP 基因 PCR 扩增结果

Fig. 2 RT-PCR amplification of CP gene from PVY

M：DNA 分子量标准；
1 ～ 14：陕西、贵州、四川、湖南、福建、广东、河南、湖北、重庆、山东、云南、河北、内蒙古、黑龙江的分离物；
15 ～ 17：阳性对照、阴性对照、空白对照
M: Marker DNA (100bp);
1–14: PCR products of CP gene in the PVY isolates collected from Shaanxi, Guizhou, Sichuan, Hunan, Fujian,Guangdong, Henan, Hubei, Chongqing, Shandong, Yunnan, Hebei, Inner Mongolia, Heilongjiang; 15–17: Positive control (with known PVY infected plant), negative control 1 (healthy plant) and negative control 2 (water), respectively

RT-PCR 扩增到的目的片段克隆、测序后经 DNAMAN、BLAST 等分析，确定目的基因片段大小为 801bp，为 PVY 外壳蛋白（coatprotein）的编码基因序列（GenBank 登录号见图 3）。14 个省（直辖市）PVY 分离物 CP 基因的序列与 GenBank 上已公布的 PVY 不同株系 CP 基因序列核苷酸序列一致性均在 88% 以上（表 2），其中与 $PVY^{N:O}$ 株系的序列一致性均高达 99%。根据 *Potyvirus* 病毒划分种的标准，当 CP 基因的核苷酸序列一致性高于 76% 时，应归为同一个种。显然，这 14 个不同省份的分离物均为 PVY，说明所检测样品中携带的病原为 PVY。

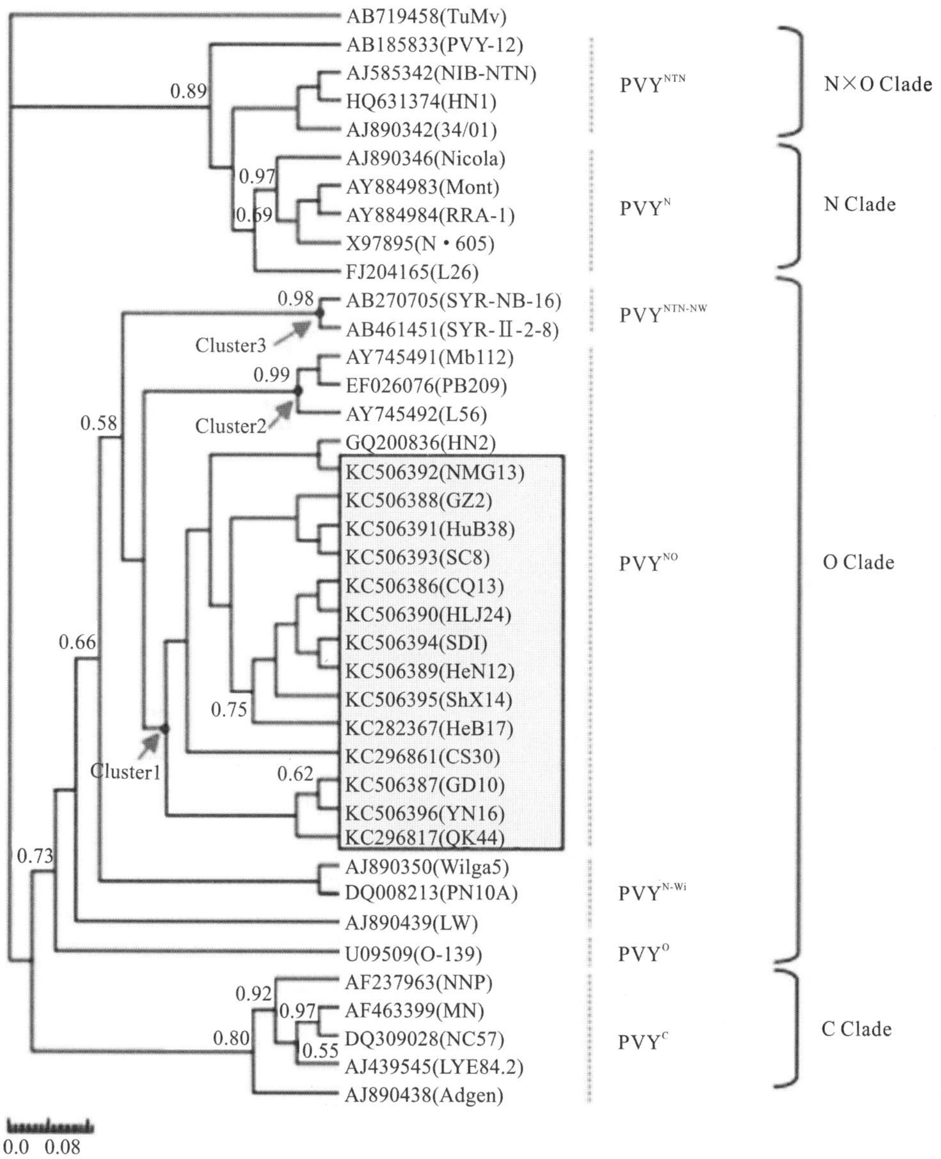

图 3 PVY CP 基因核苷序列基于贝叶斯法重建的系统发育树

Fig. 3 Phylogenetic tree of PVY CP gene constructed using Bayesian inference method

分支节点上的数值为后验概率（> 0.50），粗体加框显示的为本研究 14 个省（直辖市）分离物

The numbers on the branches were posterior probability (> 0.50). The 14 PVY isolates from current study were boxed with bold font

14 个不同省份 PVY 分离物 CP 基因的核苷酸序列均无阅读框的改变，也没有存在碱基插入 / 缺失现象，所有的核苷酸变异都是碱基置换，并找到 29 个（3.62%）多态性位点（Polymorphicsites），其中 6 个简约信息位点（Parsimony informative sites），23 个单一变异位点（Singleton variable sites）。翻译后的 CP 蛋白存在 11 个变异的氨基酸位点（表 3），其中河南分离物 HeN12、云南分离物 YuN16、河北分离

物 HeB17、内蒙古分离物 NMG13、黑龙江分离物 HLJ24 各有 1 个特异性的氨基酸突变位点，广东分离物 GD10 有 2 个特异性氨基酸突变位点，贵州分离物 GZ2 则有 3 个特异性氨基酸突变位点。

表 2　不同 PVY 分离物 CP 基因核苷酸序列一致性

Table 2　Nucleotide identity (%) in the CP gene among the PVY isolates

分离物 Isolate	Adgen	O-139	N-605	RRA-1	PB209	v942490	SYR-II-2-8
ShX14	93	97	90	90	99	92	99
GZ2	93	97	89	89	99	91	99
SC8	93	97	89	90	99	92	99
CS30	93	97	90	90	99	92	99
QK44	93	97	90	90	99	92	98
GD10	93	97	89	89	99	91	99
HeN12	93	97	89	90	99	92	99
HuB38	93	97	89	90	99	92	99
CQ13	93	97	90	90	99	92	99
SD1	93	97	89	90	99	92	99
YuN16	93	97	89	90	99	92	99
HeB17	93	97	90	90	99	92	99
NMG13	93	97	90	90	99	92	99
HLJ24	93	97	89	89	99	91	99

ShX14、GZ2、SC8、CS30、QK44、GD10、HeN12、HuB38、CQ13、SD1、YuN16、HeB17、NMG13、HLJ24 分别为陕西、贵州、四川、湖南、福建、广东、河南、湖北、重庆、山东、云南、河北、内蒙古、黑龙江分离物；Adgen（PVY^{C}）、O-139（PVY^{O}）、N-605（PVY^{N}）、RRA-1（NA-PVY^{N}）、PB209（$PVY^{N:O}$）、v942490（PVY^{NTN}）、SYR-II-2-8（$PVY^{NTN\text{-}NW}$）的 GenBank 登录号分别为 AJ890348、U09509、X97895、AY884984、EF026076、EF016294、AB461451

3　讨论

3.1　PVY 检测

病毒病的发生和流行是寄主、病毒、介体以及环境等各因子综合作用的结果，各因子相互作用、相互制约形成一个变化的植物病毒动态体系。病毒病流行中的暴发性、间歇性和迁移性（“三性”）是与病毒生态系统中各生物因子、非生物因子的作用密切相关（谢联辉，林奇英，2011）。由于病毒病的流行具有“三性”等特点，使其发生情况变得复杂，从而给病毒病的流行预测带来诸多困难。因此快速、准确检测病原，对有效防控病毒病十分重要。ELISA 是在马铃薯上应用较多的 PVY 检测方法，该方法具有快速、简便、样品检测量大等优点，可用于 PVY 种薯的大规模筛查。在 691 份材料中，本研究用 ELISA 方法只检测到 220 份呈阳性反应，约占检测总量的 1/3，造成材料低阳性检出率的原因可能有 2 种：一是有些材料看似 PVY 症状，其实是由其他一种或多种病毒感染造成的。例如在重庆的绝大部分材料中，ELISA 检测结果呈阴性反应，RT-PCR 也未扩增到任何目的条带。二是由于 ELISA 方法灵敏度低，对于 PVY 含量低的材料容易漏检。本研究的有些材料 ELISA 检测呈阴性反应，但用 RT-PCR 检测时能扩增到目的条带，例如在 28 份采自黑龙江的样本中，ELISA 只检测 12 份阳性材料，但 RT-PCR 则检测到所有的样本都携带有 PVY 病毒。在电镜下，一些 ELISA 阴性材料也能观察到风轮状内含体。在植物病毒分类上，这种专化性内含体是作为鉴定线状病毒粒体归属于 *Potyvirus* 的依据之一。这些结果进一步说明准确检测马铃薯 PVY 病毒需要多种方法结合使用。

表 3 14 个 PVY 分离物 CP 基因氨基酸变异位点的位置

Table 3 The positions of unique amino acid mutation in the CP gene of the 14 PVY isolates sampled across China

分离物	氨基酸位置 Amino acid position										
Isolate	1	12	35	84	101	127	136	153	166	240	256
ShX14	G	K	N	R	Y	N	D	K	V	S	P
GZ2	A	·	·	W	D	·	G	·	·	·	·
SC8	A	·	·	·	·	·	·	·	·	·	·
CS30	A	·	·	·	·	·	·	·	·	·	·
QK44	A	·	·	·	·	·	·	·	·	·	·
GD10	A	·	·	·	·	·	·	·	A	G	·
HeN12	·	·	·	·	·	S	·	·	·	·	·
HuB38	A	·	·	·	·	·	·	·	·	·	·
CQ13	·	·	·	·	·	·	·	·	·	·	·
SD1	·	·	·	·	·	·	·	·	·	·	·
YuN16	A	·	·	·	·	·	·	·	·	·	S
HeB17	·	E	·	·	·	·	·	·	·	·	·
NMG13	A	·	D	·	·	·	·	·	·	·	·
HLJ24	·	·	·	·	·	·	·	E	·	·	·

本研究的结果（表 1）说明，PVY 在中国大部分地区都可以检测到，但频率在地区间存在明显的差异。究其原因有 3 种：一是不同地区种植的马铃薯品种不同，不同品种对病毒病的抗性存在差异；二是不同地区的地理位置、田间栽培管理、气候条件等存在差异，因而发病率也有所不同；三是各地区使用的种薯质量和来源不同。

3.2 PVY CP 基因的多态性和系统发育树

PVY CP 基因具有高度的保守性（表 2），来自 14 个省（直辖市）PVY 分离物 CP 基因序列与 $PVY^{N:O}$ 株系的核苷酸序列一致性为 99%，说明该基因在病毒的生存和繁殖中起着十分重要的作用。前人研究表明，在 PVY 全基因组上，$PVY^{N:O}$、$PVY^{N\text{-}Wi}$ 株系和 PVYO 株系序列高度相似，只是在 P1 和 HC-Pro 基因区段重组了 PVY^{N} 株系的序列（Hu et al., 2009a），而 $PVY^{NTN\text{-}NW}$ 株系的 CP 基因序列也是 O 株系型（OClade）（Chikh et al., 2010），故而图 3 中 $PVY^{NTN\text{-}NW}$、$PVY^{N:O}$ 和 $PVY^{N\text{-}Wi}$ 三者优先相聚成簇。虽然 PVY CP 基因高度保守，但不同地区分离物也存在一定的分子变异，14 个 PVY 分离物共发现有 29 个碱基多态性位点，其中单一变异位点约占 80%。这些突变体都是通过核苷酸碱基置换形成的，没有出现序列插入或缺失现象，说明点突变在 PVY 进化中起着重要的作用。有些突变体是地区特有的，如河南分离物 HeN12、云南分离物 YuN16、河北分离物 HeB17、内蒙古分离物 NMG13、黑龙江分离物 HLJ24、广东分离物 GD10、贵州分离物 GZ2 均发现 1 ～ 3 个的地区特异性氨基酸位点，说明 PVY 群体间存在一定的地理间隔。

14 个省（直辖市）PVY 分离物与 $PVY^{N:O}$ 株系的亲缘关系最近，故而推断这 14 个省（直辖市）分离物可能归属于 $PVY^{N:O}$ 株系，说明中国的 PVY 群体可能以 $PVY^{N:O}$ 株系为主，这和陈士华等（2011）的结果相似。但由于 PVY 株系分化严重，重组频繁，单独使用 CP 基因序列为分子标记进行系统发育分析，可能无法对 PVY 全部株系种类进行准确鉴定，如图 3 中的 HN2 分离物（$PVY^{NTN\text{-}NW}$ 株系）与本研究的14个分离物（$PVY^{N:O}$株系）尽管属于不同株系，但因CP基因区段都是O株系型（OClade）而聚在一起（后验概率小于0.50）。但是，当笔者联合P1和CP基因以及多重PCR等多种方法进一步验证后，得到与本文一致的系统发育分析结果（未发表数据），14个省份PVY分离物多为$PVY^{N:O}$株系，这也是目前全球各地流行的重组株系之一（Quenouille et al., 2013）。目前分子系统发育研究中的建树方法，常见有邻接法（Neighbor-Joining, NJ）、最大似然法（Maximum Likehood, ML）、最大简约法（Maximum Parsimony, MP）和BI法等（吴祖建，等, 2010），各有优缺点。进化生物学认为，在进化模型确定的情况下，ML法是与进化事实吻合最好的建树算法（黄原, 2012）。对于构建系统发育树的准确性，Hall（2004）认为BI法最好，其次是ML法，然后是MP法。Yang等（2012）认为BI法得到的系统发育树不需要利用自举（Bootstrap）法进行检验，其后验概率直观地反映了系统发育树的可信度，既能根据分子进化的现有理论和各种模型用概率重建系统发育关系，又克服了ML法高计算强度、不适用于大数据集样本的缺陷。因此，本研究采用BI法对PVY的CP基因进行了系统发育分析，尽管部分分支的后验概率不高，但相同株系的PVY分离物以较高的后验概率优先相聚，总体上能够反映出不同株系的系统发育关系。

4 结论

通过ELISA方法对中国14个省（直辖市）691份疑似PVY样品进行检测，发现了220份阳性反应，对随机选取的14个省（直辖市）代表样品的CP基因RT-PCR扩增和镜检也验证了这些材料携带PVY病毒，说明PVY在中国马铃薯种植区普遍存在，且多为$PVY^{N:O}$株系。PVY的CP基因高度保守，但不同地区分离物也存在一定的由点突变引起的碱基差异。

参考文献

[1] 黄原. 分子系统发生学[J]. 遗传, 2012,34(11):1471.

[2] 吴祖建, 高芳銮, 沈建国. 生物信息学分析实践[M]. 北京：科学出版社, 2010:107-110.

[3] 谢联辉, 林奇英. 植物病毒学.2011.

[4] BECZNER L, HORVÁTH J, ROMHANYI I, et al. Studies on the etiology of tuber necrotic ringspot disease in potato[J]. Potato Research, 1984, 27: 339-352.

[5] CHIHK ALI M, MAOKA T, NATSUAKI K, et al. The simultaneous differentiation of *Potato virus Y* strains including the newly described strain PVY^{NTN-NW} by multiplex PCR assay[J]. Journal of Virological Methods, 2010, 165: 15-20.

[6] CHEN S, LIU X, ZHANG X, et al. Study on the PVY strain differentiation and identification of some potato producing areas in China[J]. Journal of Henan Agricultural University, 2011,45: 548-551.

[7] GAO X, GAO Y, GENG H, et al. Identification of a *potato virus Y* isolate from the Heilongjiang Province in China[J]. Chin Potato Journal, 2009, 23: 11-14.

[8] GLAIS L, TRIBODET M, KERLAN C. Genomic variability in *Potato potyvirus Y* (PVY): evidence that PVY^{NW} and PVY^{NTN} variants are single to multiple recombinants between PVY^{O} and PVY^{N}isolates[J]. Archives of Virology, 2002,147: 363-378.

[9] HALL B G. Comparison of the Accuracies of Several Phylogenetic Methods Using Protein and DNA Sequences[J]. Molecular biology and evolution,2005,22: 792-802.

[10] HU X, KARASEV A V, BROWN C J, et al. Sequence characteristics of *potato virus Y* recombinants[J]. Journal of General Virology, 2009a, 90: 3033-3041.

[11] HU X, HE C, XIONG X, et al. Molecular characterization and detection of recombinant isolates of *potato virus Y* from China[J]. Archives of Virology, 2009b, 23: 293-300.

[12] LARKIN M A, BLACKSHIELDS G, BROWN N, et al. Clustal W and Clustal X version 2.0[J]. Bioinformatics, 2007, 23: 2947-2948.

[13] LIBRADO P, ROZAS J. DnaS Pv5: a software for comprehensive analysis of DNA polymorphism data[J]. Bioinformatics, 2009, 25: 1451-1452.

[14] NIE B, SINGH M, MURPHY A, et al. Response of potato cultivars to five isolates belonging to four strains of *Potato virus Y*[J]. Plant Dis, 2012, 96: 1422-1429.

[15] NOLTE P, WHITWORTH J L, THORNTON M K, et al. Effect of seedborne *Potato virus Y* on performance of Russet Burbank Russet Norkotah, and Shepody potato[J]. Plant Disease, 2004,88: 248-252.

[16] NYLANDER J A A.Mr Model test v2.3. Program distributed by the author.http://www.abc.se/nylander/mrmodeltest2/mrmodeltest2.html, 2004.

[17] OGAWA T, TOMITAKA Y, NAKAGAWA A, et al. Genetic structure of a population of *Potato virus Y* inducing potato tuber necrotic ringspot disease in Japan; comparison with North American and European populations[J].Virus Research, 2008, 131: 199-212.

[18] OGAWA T, NAKAGAWA A, HATAYA T, et al. The genetic structure of populations of *Potato virus Y* in Japan; based on the analysis of 20 full genomic sequences[J]. Journal of Phytopathology, 2012, 160: 661-673.

[19] OHSHIMA K, SAKO K, HIRAISHI C, et al. Potato tuber necrotic ringspot disease occurring in Japan: its association with *Potato virus Y* necrotic strain[J]. Plant disease, 2000, 84: 1109-1115.

[20] QUENOUILLE J, VASSILAKOS N, MOURY B. Potato virus Y: a major crop pathogen that has provided major insights into the evolution of viral pathogenicity[J]. Molecular Plant Pathology, 2013, 14: 439-452.

[21] RONQUIST F, TESLENKO M, VAN DER MARK P, et al.MrBayes 3.2: efficient Bayesian phylogenetic inference and model choice across a large model space[J]. Systematic Biology,2012, 61: 539-542.

[22] SI X, TUYONG Y, QIMING X. Research Progress in Strains Differentiation and Detection Research of *Potato Virus Y*[J]. Chinese Agricultural Science Bulletin, 2011.

[23] SUN Q, ZHANG C. Studies on the methods of detecting PVY^N and PVY^O by RT-PCR[J]. Scientia Agricultura Sinica, 2005, 38: 213-216.

[24] TALAVERA G, CASTRESANA J. Improvement of phylogenies after removing divergent and ambiguously aligned blocks from protein sequence alignments[J]. Systematic Biology, 2007, 56: 564-577.

[25] WANG F, WU Y, GAO Z, et al. First report of *Potato virus Y* in *Kalimerisindica* in China[J]. Plant Disease, 2012, 96: 1827-1827.

[26] WHITWORTH J L, NOLTE P, MCINTOSH C, et al. Effect of *Potato virus Y* on yield of three potato cultivars grown under different nitrogen levels[J]. Plant Disease, 2006,90: 73-76.

[27] WU Z, DONG Z, LIU X, et al. Coat protein gene sequence analysis and identification of a *potato virus Y* Hebei isolate[J]. Acta Horticulturae Sinica, 2005, 32: 324-326.

[28] YANG Z, RANNALA B. Molecular phylogenetics: principles and practice[J]. Nature Reviews Genetics, 2012, 13: 303.

PVY NTN-NW 榆林分离物的全基因组序列测定与分析

高芳銮[1]，常飞[1]，沈建国[2]，谢联辉[1]，詹家绥[1]

（1 福建农林大学植物病毒研究所 / 福建省植物病毒学重点实验室，福州 350002；
2 福建出入境检验检疫局检验检疫技术中心 / 福建省检验检疫技术研究重点实验室，福州 350001）

摘要：马铃薯 Y 病毒（*Potato virus Y*, PVY）是马铃薯病毒科（Potyviridae）马铃薯 Y 病毒属（*Potyvirus*）代表成员，是马铃薯和烟草生产中分布最为广泛并造成严重经济损失的病毒之一。研究旨在揭示 PVY 榆林分离物 ShX14 的全基因组特征，并准确判断其株系分子类型。根据已经报道的 PVY 不同基因保守区设计 11 对简并引物，采用片段重叠法从感染 PVY 的马铃薯病叶中扩增、克隆获得榆林分离物 ShX14 的全长序列，并对其序列特征、重组位点、系统发育关系等进行分析。同时，应用系统发育与性状关联分析（phylogeny-trait association analysis）评估 PVY 分离物与株系的关联性。测序获得的 ShX14 分离物核苷酸序列全长为 9724nt（不含 3′- 端的多聚腺苷酸尾）。该病毒基因组含有一个 9186nt 的开放阅读框，编码 3061 个氨基酸的多聚蛋白（polyprotein）。在 P3 顺反子中 +2 相位上也发现由移码（reading frame shift）产生的 PIPO 蛋白。全基因组序列比较分析显示，ShX14 分离物与 HN2、SYR-NB-16 分离物（PVY^{NTN-NW} 株系 SYR-I 型）的核苷酸、氨基酸序列的一致性分别为 98%~99% 和 98%~100%。该分离物的 P1、HC-Pro/P3 和 CP 基因的 5′- 端均检测到显著的重组信号，重组位点分别位于 2318、5674 和 8385nt，与 PVY^{NTN-NW} 株系（SYR-I 型）的重组位点相似。系统发育分析结果显示该分离物与 HN2、SYR-NB-16 分离物相聚成簇，表明其在系统发育关系上，与 PVY^{NTN-NW} 株系（SYR-I 型）的亲缘关系最近。系统发育分析与性状关联分析结果显示该分离物与 PVY^{NTN-NW} 株系（SYR-I 型）的关联系数（association index, *AI*）、简约分值（parsimony score, *PS*）和最大单系分支（maximum monophyletic clade, *MC*）3 个统计检验均显著，表明其与 PVYNTN-NW 株系（SYR-I 型）存在显著的关联性。同时，应用多重 RT-PCR 成功扩增出约为 1000、600 和 400bp 的 3 个特异性片段，与 PVY^{NTN-NW} 株系（SYR-I 型）的特异条带大小相一致。这些结果也进一步确定该分离物属于 PVY^{NTN-NW} 株系（SYR-I 型）。ShX14 分离物为 N×O 重组分离物，属于 PVY^{NTN-NW} 株系（SYR-I 型），为后续深入开展该分离物的生物学等研究奠定了基础。

关键词：马铃薯 Y 病毒；全基因组；重组；系统发育；系统发育与性状关联分析

Complete genome analysis of a PVYNTN-NW recombinant isolate from Yulin of China

Fang-luan Gao[1], Fei Chang[1], Jianguo Shen[2], Lianhui Xie[1], Jiasui Zhan[1]

Abstract: *Potato virus Y* (PVY) is a typical member of the genus *Potyvirus* in the family Potyviridae. It is one of the most

中国农业科学 . 2015,48(2):270-279
接受日期：2014-08-10
基金项目：国家现代农业马铃薯产业技术体系（CARS-10）、福建省自然科学基金（2013J01088）、国家质检总局科技计划（2013IK094）、质检公益性行业科研专项（201410076）

destructive pathogens affecting potato and tobacco and causes significant economic losses worldwide. The objective of this study is to determine the genomic structure of an isolate of PVY from Yulin of China and its phylogenetic relationship with reported PVY strains. The complete genome of the ShX14 was amplified and sequenced from overlapping fragments using 11 pairs of primers designed from the conserved regions of the known PVY isolates. Genomic structure and recombination events of the isolate were evaluated by various bioinformatics approaches. Phylogenetic tree was reconstructed by maximum likelihood (ML) method using nucleotide sequences of coding regions. In addition, phylogeny-trait association analysis was used to evaluate the relationship between PVY isolate and the reported stains. The complete sequence of ShX14 had 9 724 nucleotides, excluding the 3′- terminal poly (A) tail. It contains a single open reading frame of 9 186 nucleotides and encodes a polyprotein of 3 061 amino acids. An additional protein, termed 'PIPO', is also translated by +2 nucleotide frame shifting within the P3 cistron. The isolate shares 98%–99% nucleotide identity and 98%-100% amino acid identity with HN2 and SYR-NB-16 (PVY^{NTN-NW} strain, SYR-I genotype), respectively. Similar to the PVY^{NTN-NW} (SYR-I) genomic structure, three recombination breakpoints were identified at nucleotide positions 2 318, 5 674 and 8 385 in the P1, HC-Pro/P3 and the 5′-terminus of CP gene respectively with high confidence. Phylogenetic analysis indicated that ShX14 was clustered together with HN2 and SYR-NB-16, suggesting that it shared high sequence homology with PVY^{NTN-NW} (SYR-I) strain. Association index (*AI*), parsimony score (*PS*) and maximum monophyletic clade (*MC*) all indicated that ShX14 was strongly associated with the PVY^{NTN-NW} (SYR-I) group. Three fragments of about 1 000, 600 and 400bp in size were also amplified from the isolate by a multiplex RT-PCR, consisting with the expected band of the PVY^{NTN-NW} (SYR-I) strain. Overall, these analyses strongly indicate that ShX14 is likely to be a PVY^{NTN-NW} strain (SYR-I). ShX14 is a N×O recombinant isolate, classified to PVY^{NTN-NW} strain (SYR-I). It will provide useful information in the further study of the biology of this pathogen.

Keywords: *Potato virus Y*; complete genome; recombination; phylogeny; phylogeny-trait association analysis

马铃薯 Y 病毒（*Potato virus Y*, PVY）是马铃薯、烟草生产中最为广泛并造成重要经济损失的病毒之一。近十年来，在日本和中国等亚洲地区呈上升发展趋势，在全球马铃薯种植区也广泛流行（Glais et al., 2002; Ogawa et al., 2008; Hu et al., 2011）。PVY 是马铃薯 Y 病毒科（*Potyviridae*）马铃薯 Y 病毒属（*Potyvirus*）的代表成员。该病毒在自然条件下，主要由蚜虫以非持久性方式传播，也可以通过汁液摩擦、嫁接等方式传播，主要侵染包括马铃薯、烟草、番茄等茄科寄主植物。PVY 侵染马铃薯后，可以引起叶脉坏死、褐脉、黄斑坏死等症状，因寄主品种和病毒株系不同，可以使马铃薯产生不同的症状并导致马铃薯种质退化、产量降低，严重时减产高达 80% 以上，给马铃薯生产造成极大损失（Glais et al., 2002; Nolte et al., 2004; Rahman and Akanda, 2009; Whitworth et al., 2006）。同时，PVY 侵染烟草等作物后，使烟草产量和质量大幅度降低，在东北烟区和黄淮烟区危害特别严重，成为阻碍烟草生产的瓶颈（罗红香等 , 2012）。PVY 的病毒粒体呈线状，每个病毒粒体各含有一套完整的基因组，基因组由正单链 RNA 组成，开放阅读框（open reading frame, ORF）编码的多聚蛋白（polyprotein）经过自身编码的 3 个蛋白酶切割后，得到 P1、HC-Pro、P3 等蛋白。同时，PVY 基因组内还有一个由移码翻译（read frame shift）产生的蛋白 -PIPO，该蛋白与 P3 蛋白的 N′ 端以 P3N-PIPO 融合形式存（Chung et al., 2008; 高芳銮等 , 2013）。PVY 基因组通过这两种方式最终生成 11 个成熟的多功能蛋白，使这些基因呈现不同程度的遗传多样性。在寄主、病原和环境的互作体系中，高遗传多样性的病原具有相对强的生存和进化优势，能更快速适应新的抗病寄主或环境。除了高度变异，基因重组在 PVY 基因组中也频繁发生，这使得 PVY 株系分化严重。根据寄主植物反应不同，已被广泛认可的 PVY 株系种类包括常见的 3 种：普通株系（ordinary strain, PVY^{O}）、点刻斜条株系（stipple streak strain, PVY^{C}）和叶脉坏死株系（necrotic strain, PVY^{N}）。但随着农产品贸易的全球化，PVY 变异频繁，不断有新的重组株系出现，并在全球的不同地区蔓延（Singh et al., 2008; Quenouille et al., 2013）。这些新分化出的株系一般被认为由 PVY^{N} 和 PVY^{O} 重组而来，主要包括 PVY^{NTN} 和 PVY^{N-Wi} 株系。由于重组位点数量和位置不同，这两个株系都具有两种不同的基因结构。PVY^{NTN}株系的第一种结构（PVY^{NTN-a}）在HC-Pro/P3、VPg和CP基因内含有3个重组位点，而另一种结构（PVY^{NTN-b}）在P1基因内还有1个重组位点（Ogawa et al., 2008）；而PVY^{N-Wi}株系的第一种结构含有两个重组位点，分别位于P1和HC-Pro/P3基因区域，另一种结构则只含有1个重组位点（位于HC-Pro/P3基因之间），北美称之为PVY$^{N:O}$（Nie et al., 2004; Hu et al., 2009a）。然而，Chikh等（2010）研究发现PVY叙利亚分离物的基因组含有3种不同的基因结构（SYR-I、SYR-II和SYR-III型），前两者具有明显的PVY^{NTN}和PVY^{N-Wi}株系的重组特征，并命名为PVY^{NTN-NW}株系。PVY病毒基因组的高度多样性，使其逐渐成为植物病毒学研究的热点之一。国内外学者对中国分离物部分功能基因的检测鉴定、遗传变异等开展相关研究（Tian et al., 2011; 罗红香等, 2012; 高芳銮等, 2013），而对于PVY全基因组的研究工作，主要是烟草分离物的序列测定，如Yang等（2013）利用PCR和RACE技术扩增到一个PVYN株系烟草分离物的全长序列，Wang等（2012）完成两个烟草分离物AQ4和FZ10的基因组序列测定，但关于PVY的初始寄主——马铃薯分离物全基因组序列的报道并不多见。迄今，仅Hu等（2009b）报道一株血清型为PVY^{O}、马铃薯上表型为PVY^{NTN}的HN2分离物。该分离物在生物学特征及基因组结构方面与PVY^{NTN-NW}株系（SYR-I型）极为相似（Lole et al., 1999）。由于PVY群体遗传多样性高，株系分化严重，且所采样品大部分为田间自然发病病株，采用单一的传统生物学或血清学方法已无法准确判断PVY的株系类型。本文在前期病理特征等研究基础上（高芳銮等, 2013），拟通过马铃薯寄主上PVY榆林分离物ShX14的全基因组序列测定，对其序列特征、重组位点、系统发育关系等进行系统分析，并应用系统发育与性状关联分析（phylogeny-trait association analysis）评估PVY分离物与株系的关联性，以期为该分离物后续功能研究等奠定基础。

1 材料与方法

试验于2013年11月至2014年2月在福建农林大学植物病毒研究所和福建出入境检验检疫局检验检疫技术中心完成。

1.1 试验材料

榆林分离物ShX14于2011年8月随机采自陕西省榆林市马铃薯主要种植区，马铃薯品种为费乌瑞它（Favorite），呈典型的花叶症状，前期经血清学检测、电镜观察及CP基因扩增等确定为PVY分离物（高芳銮等, 2013），新鲜样本置于密封袋中，保存于–80℃超低温冰箱。

1.2 主要试剂

RNA Simple Total RNA Kit、DNA Marker、Gel Purification Kit为天根（TianGen）生化科技公司产品；TransScript First-Strand cDNA Synthesis SuperMix、Trans*Taq* DNA Polymerase HiFi、pEASY-T5 Zero Cloning Kit为北京全式金（TransGen）公司产品。

1.3 试验方法

1.3.1 RNA提取及基因克隆

采用Trizol试剂法从感染PVY的马铃薯病叶中提取总RNA，提取方法参照RNA Simple Total RNA试剂盒（TianGen）说明书进行。取2μL总RNA以Oligo(dT)$_{18}$为引物按照操作说明书进行反转录，获得PVY全长的cDNA。根据GenBank已报道的PVY常见株系各基因片段的保守区，设计11对用于扩增各基因的简并引物（表1），简并引物委托南京金斯瑞生物技术有限公司合成。

PCR扩增采用25μL反应体系：10×Trans*Taq* HiFi Buffer II 2.5μL，dNTPs（2.5mmol • L^{-1}）2μL，正向引物（10μmol • L^{-1}）1μL，反向引物（10μmol • L^{-1}）1μL，ddH$_2$O 34.5μL，Trans*Taq* HiFi Polymerase（5U • μL^{-1}）0.5μL，cDNA 2μL。PCR反应条件为：94℃预变性4min；94℃变性30s，复性30s（退火温

度视基因片段而定，表 1)，72℃延伸若干分钟（延伸时间按 1 min • kb^{-1} 计)，共 30 个循环；最后一轮循环后 72℃延伸 5min。

表 1 PVY 全基因组扩增和测序的简并引物

Table 1 Degenerate primers used for amplification and sequencing of PVY genome

片段 Segment	引物 Primer	引物序列 Primer sequence (5′-3′)	预期大小 Expected size (bp)	退火温度 T_m (℃)
5′UTR	5-F	AAATTAAAACAACTCAATACAAC	250	53
	5-R	GTGAGTATGGTAGCTTGCATTCAA		
P1	P1-F	CVATGGCAAYYTACAYGTCAAC	915	53
	P1-R	AGGRTATCTCADYHGTGCCC		
HC-Pro	HC-F	GTTACYCARRGKGTTWTG	1412	52
	HC-R	GGAAYDCCACCAACTCTATA		
P3	P3-F	GGWRTTCCTRRWGCATGCCC	1095	55
	P3-R	CTGRTGTCKCACWTYATATTC		
6K1	6K1-F	CTCAGATAGTTCARTTTGCTCAAG	370	55
	6K1-R	CTGTAATGTGGAAGTGTATGTCC		
CI	CI-F	CAGTCYTTRGAYGATGTRATC	1906	55
	CI-R	CTTGRTGRTGMACRAACTGY		
6K2/VPg	KV-F	TAATATTGGTGGAGAGAYTGCTTG	991	55
	KV-R	GCTTCATGCTCYACYTCCTG		
NIa	Na-F	GTRGAGCATGAAGCYAAATC	745	55
	Na-R	TGCTCYRCRACWWCATCATG		
NIb	Nb-F	AGCARGCTAARCAYTCTGCG	1564	55
	Nb-R	CYTGRTGRTRYACTTCATAAG		
CP	CP-F	GSAAAYGAYACAATYGATGC	803	55
	CP-R	CACATGTTYTTVACTCCAAG		
3′UTR	3-F	CTTGGAGTYAAGAACATGTG	350	55
	3-R	GTCTCCTGATTGAAGTTTACAGTC		

F：正向引物；R：反向引物

F: Forward primers; R: Reverse primers

PCR 反应结束后，取产物 5μL 用 1% 琼脂糖凝胶电泳进行检测。PCR 产物经胶回收试剂盒纯化后与 pEASY-T5 Zero 克隆载体连接，并转化到 Trans1-T1 感受态细胞中。经菌落 PCR 鉴定筛选得到的阳性克隆子，随机选择其中 3 ～ 6 个委托南京金斯瑞生物技术有限公司测序，并通过测序峰图及序列比对分析排除由 PCR 扩增引起的突变。

1.3.2 *序列分析及重组检测*

测序获得的序列使用 DNAMAN 等软件进行处理，多聚蛋白的剪切位点参考在线网站（http://www.dpvweb.net/potycleavage/index.html）进行分析，序列一致性使用 BLAST（http://blast.ncbi.nlm.nih.gov/）工具在线比对。

以文献（Hu et al., 2009a）报道的 PVY 分离物 Oz（EF026074，PVYO）和 Mont（AY884983，PVYN）作为参考亲本，并以分离物 SYR-NB-16 作为备选参考（Hu et al., 2009b）。使用 SimPlot3.5 进行可能重组事件的序列相似性（similarity plot）作图分析（Lole et al., 1999）。同时，使用遗传算法重组检测法（genetic algorithm recombination detection，GARD）进行重组位点的检测及可靠性评估（Kosakovsky Pond et al., 2006）。

1.3.3 系统发育分析

为分析 PVY 的系统发育关系，从 GenBank 中选取 31 个其他 PVY 分离物的核苷酸序列作为参考，使用最大似然法（maximum likelihood，ML）基于 PVY 全基因组的核苷酸序列（不含 UTR 区）重建 PVY 的系统发育树。建树前，使用 MEGA5（Tamura et al., 2011）对建树序列基于密码子（codons）方式进行多重序列比对，利用 Mrmodeltest（Nylander, 2008）选择最优化的核苷酸替换模型，并参照 AIC（Akaike information criterion）标准设置相应参数，最后应用自举法（Bootstrap）评估各分支节点的置信值（Bootstrap confidence level）。

1.3.4 PVY 分离物与株系关联分析

采用系统发育与性状关联分析验证PVY分离物与主要株系是否存在关联。根据参考分离物可能归属的株系类别，先后不同定义了 11 种状态，分别为：C（PVY^{C}）、O（PVY^{O}）、N（PVY^{N}）、NA-N（$PVY^{NA-N/NTN}$）、NTN-a（PVY^{NTN}）、NTN-b（PVY^{NTN}）、N-Wi（PVY^{N-Wi}）、N:O（$PVY^{N:O}$）、SYR-I（PVY^{NTN-NW}）、SYR-II（PVY^{NTN-NW}）和 E（PVY^{E}）。使用 BEASTv1.8.0 基于马尔科夫蒙特卡洛法（ Markov Chain Monte Carlo，MCMC）对上节多重比对后的序列进行贝叶斯分析（Bayesian analysis）（Drummond et al., 2012）。为确保所有参数的收敛性，MCMC 共运行了 100000000 代。当有效样品大小（effective sample size，ESS）大于 200 时，弃去样本树的前 10% 老化样本，使用 BaTS 2.0（Parker et al., 2008）计算树后验分布中样本树（sample trees）的 3 个统计参数：关联系数（association index，*AI*）、简约分值（parsimony score，*PS*）和最大单系分支（maximum monophyletic clade，*MC*）。这 3 个统计参数中，*AI* 是通过计算每个节点下 PVY 分离物出现在指定株系的频率判断分离物与株系的关联性；*PS* 是通过计算整个系统发育树中发生迁移的株系类别数判断分离物与株系的关联性；而 *MC* 则是通过评估同一株系内不同分离物聚为一簇的最大值衡量分离物和株系的关联强度。分离物和株系之间关联是否显著，主要根据三者相应的 P 值判断。若 *P* 值均≤ 0.05，则判定为分离物与株系存在显著的关联性。

1.3.5 ShX14 分离物的多重 RT-PCR 验证

为进一步验证 ShX14 分离物的株系归属，采用 Chikh 等（2010）建立的多重 RT-PCR 体系对该分离物进行验证。该体系主要通过 6 对特异性引物对单一或混合侵染的 PVY 样品进行 RT-PCR 扩增，根据 PCR 产物凝胶电泳产生的条带大小及不同条带的组合判定 PVY 的主要株系，包括 PVY^{O}、PVY^{N}、PVY^{N-Wi} 等。

2 结果

2.1 ShX14 分离物的全基因组序列特征

克隆、测序获得的 ShX14 分离物基因组除去 3′- 端的 Poly(A) 外，确定全长序列大小为 9724nt（NCBI 登录号为 KJ634024），其中 190—9375nt 为多聚蛋白编码区序列，编码一个包含 3061 个氨基酸长度的多聚蛋白。该分离物基因组 5′ 端和 3′ 端的非编码区（untranslated regions, UTRs）的长度分别为 189 和 349nt。通过剪切位点分析显示，ShX14 分离物的 9 个剪切位点分别为 Q/G、G/G、Q/R、Q/S、Q/A、Q/G、E/A、Q/A 和 Q/G。ShX14 分离物不同基因的位置、编码蛋白的大小以及其 4 个相关参考分离物的序列一致性比较如表 2 所示。

序列一致性分析显示，在全基因组水平上（不含有 UTR 区），ShX14 分离物与 4 个参考分离物的核苷酸、氨基酸序列一致性分别为 89% ～ 99%、93% ～ 99%（表 2）。在 P1、HC-Pro、VPg、NIa 和 NIb 基因区段上，ShX14 分离物和 Mont 分离物（PVY^{N} 株系）的核苷酸序列一致性较高（97% ～ 99%），而与 Oz 分离物（PVY^{O} 株系）的核苷酸序列一致性值较低（73% ～ 88%）；而在 P3、PIPO、6K1、CI、6K2 和 CP 基因区段上则相反，ShX14 分离物与 Oz 分离物的序列一致性较高（97% ～ 99%），与 Mont

分离物的核苷酸序列一致性较低（83% ～ 90%）。

表 2　ShX14 分离物全基因组信息及其序列一致性比较

Table 2　Genomic information and sequence identity of ShX14 isolate with the reference PVY isolates

片段 segment	序列位置与长度 Sequence position and length			序列一致性 Sequence identity (%)*			
	位置 Position	nt	aa	Mont	Oz	HN2	SYR-NB-16
Polyprotein	190—9374	9186	3061	93/97	89/94	99/99	98/99
5′ UTR	1—189	189	-	97/-	-/-	99/-	98/-
P1	190—1014	825	275	98/99	73/72	99/99	98/99
HC-Pro	1015—2409	1395	465	99/99	82/91	99/99	99/99
P3	2410—3504	1095	365	86/92	98/98	99/99	96/97
PIPO	2922—3146	228	75	86/76	99/99	99/100	98/96
6K1	3505—3660	156	52	83/87	97/96	99/98	98/96
CI	3661—5562	1902	634	84/96	99/99	99/99	98/99
6K2	5563—5718	156	52	84/92	98/98	99/100	98/98
VPg	5719—6282	564	188	97/99	88/95	99/100	99/99
NIa	6283—7014	732	244	99/99	81/93	99/99	99/99
NIb	7015—8571	1557	519	99/99	85/95	99/100	99/99
CP	8572—9372	801	267	90/94	99/99	99/99	99/99
3′ UTR	9376—9724	349	-	86/-	99/-	99/-	99/-

* 核苷酸 / 氨基酸一致性 : Nucleotide/amino acid identity

虽然 ShX14 分离物与 Mont 和 Oz 分离物在不同基因区段上，序列一致性值差异较大，但其与分离物 HN2 和 SYR-NB-16 之间，不论核苷酸或氨基酸序列的一致性（分别为 98% ～ 99% 和 98% ～ 100%）则非常高，表明 ShX14 分离物与 HN2、SYR-NB-16 高度同源。

2.2　重组分析

通过 Simplot 软件分析（图 1A），在 ShX14 分离物中检测 3 个潜在的重组信号，进一步通过 GARD 确认这 3 个重组位点（RJ）分别位于 2318、5674 和 8385nt（图 1B），模型平均支持率分别为 93.64%、99.99% 和 99.63%，且 KH 检验均极显著（P<0.01）。这 3 个 RJ 分别位于 HC-Pro/P3、VPg 以及 NIb/CP 基因区域内，与 PVYNTN-NW 株系（SYR-I 型）的重组位点非常接近。

2.3　系统发育分析

建树序列多重比对后的序列长度为 9195bp，经 Gblock 裁去冗余序列后得到长度为 9171bp 的序列保守区。根据 MrModeltest 软件的 AIC 标准，建树序列最适合的核苷酸替换模型为 $GTR+I+G$。ML 法重建的系统发育树如图 2 所示，32 个 PVY 不同分离物共形成 11 个簇，其中相同株系的分离物以较高的置信值优先相聚，显示出明显的株系特异性。ShX14 分离物先与 HN2 以高置信值（100%）聚为一亚簇后，与 SYR-NB-16 进一步相聚成簇。系统发育分析结果说明，该分离物在系统发育关系上与 PVY^{NTN-NW} 株系（SYR-I 型）最近。

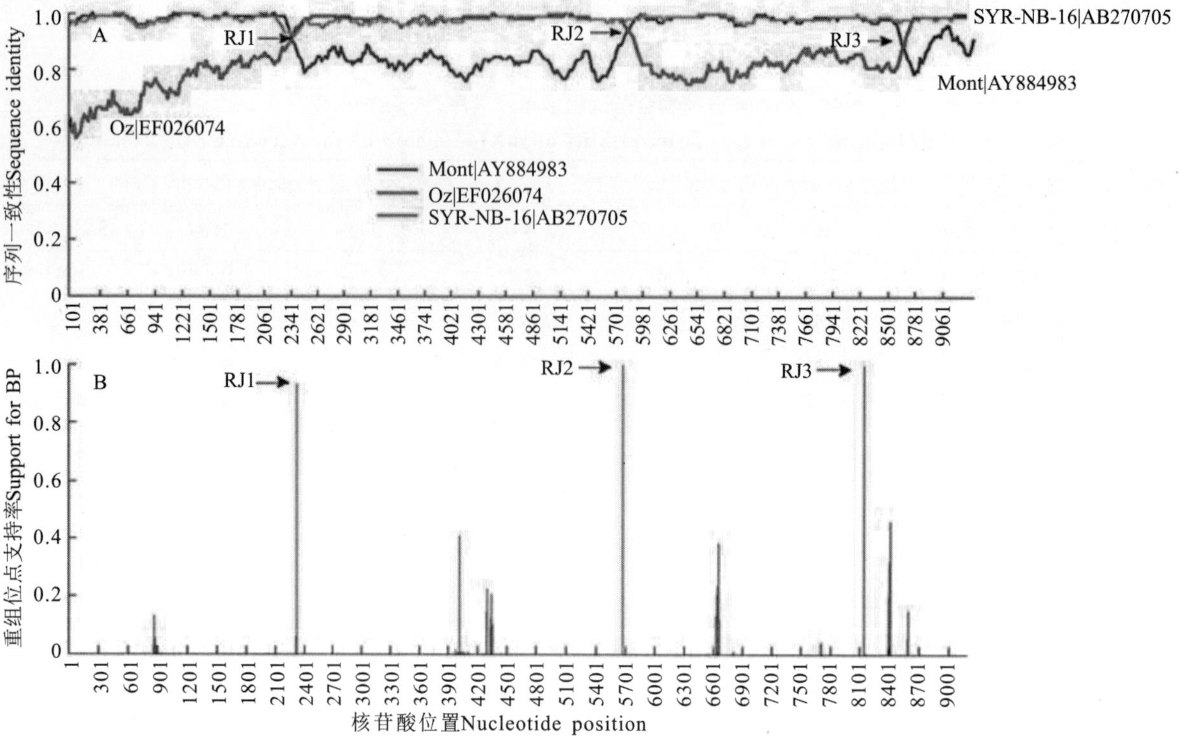

A：重组分离物的 Simplot 相似性分析；B：GARD 检测重组位点。RJ1—RJ3：潜在的重组位点

A: Similarity analysis of the putative recombinants by Simplot; B: Detection of recombination breakpoints with GARD. RJ1-RJ3: The putative recombination breakpoints

图 1　ShX14 分离物重组信号的检测及验证

Fig.1　Detection and verification of recombination events in the PVY ShX14 isolate

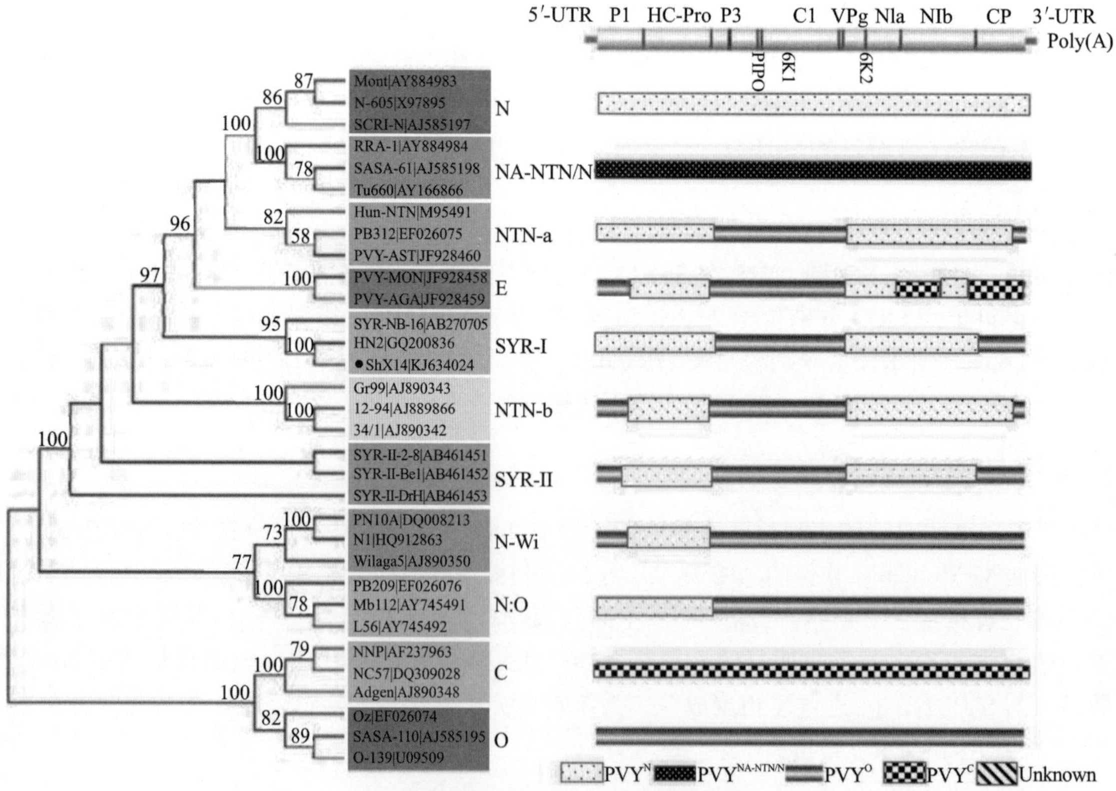

图 2　PVY 系统发育关系与基因组结构特征

Fig. 2　Phylogenetic relationships and genomic structures of PVY

2.4 PVY 分离物与株系关联分析

系统发育与性状关联分析结果显示（表 3），除了 PVY^{NTN-NW}（SYR-II）、PVY^{N} 株系的 *MC* 统计不显著外（*P*>0.05）外，其他各株系 *AI*、*PS*、*MC* 值的统计检验均呈显著水平（P ≤ 0.05），表明 PVY 各分离物与所属株系之间存在较强的关联性。ShX14 分离物所在分支 SYR-I 的 *MC* 值为 2.37（介于 2 和 3 之间），说明该分支内的 3 个分离物 SYR-NB-16、HN2、ShX14 聚为一个分支的概率极高（*P*=0.05），且 *AI*、*PS* 统计检验均显著，表明 ShX14 分离物与 PVY^{NTN-NW} 株系（SYR-I 型）密切相关。

表 3 PVY 株系的系统发育 – 性状关联检验

Table 3 Phylogeny-trait association test of PVY-strain clustering

统计 Statistic	BaTS 检验 BaTS estimate (95% *HPD CIs)	*P* 值 *P* value
AI	0.24 (0.19, 0.31)	<0.001
PS	11.00 (11, 11)	<0.001
MC (C)	2.00 (2, 2)	0.030
MC (O)	1.64 (1, 2)	0.030
MC (N)	2.09 (1, 3)	0.090
MC (NA-N)	3.00 (3, 3)	0.010
MC (NTN-a)	2.98 (3, 3)	0.010
MC (NTN-b)	3.00 (3, 3)	0.010
MC (N-Wi)	2.04 (2, 2)	0.030
MC (N:O)	3.00 (3, 3)	0.010
MC (SYR-I)	2.37 (2, 3)	0.050
MC (SYR-II)	1.00 (1, 1)	1.000
MC (E)	2.00 (2, 2)	0.010

*HPD Cis= 最高后验密度置信区 *HPD Cis=Highest posterior density confidence intervals

2.5 ShX14 分离物的多重 RT-PCR 验证

ShX14 分离物的多重 RT-PCR 扩增到 3 个特异片段，分别约为 1000、600 和 400bp（图 3），与 PVYNTN-NW 株系（SYR-I 型）的 1076、633 和 441bp 的 3 个特异条带大小相一致，而阴性（健康马铃薯叶片）和空白对照均未扩增到相应的片段（Ali et al., 2010）。该结果进一步确定 ShX14 分离物为 PVY^{NTN-NW} 株系（SYR-I 型）。

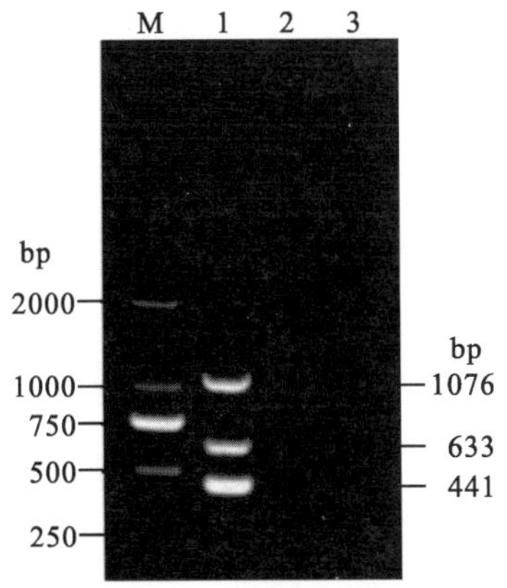

图 3 ShX14 分离物多重 RT-PCR 扩增结果

M：DNA 分子量标准（D2000）；1：ShX14 分离物的多重 RT-PCR 扩增产物；2 ～ 3：阴性对照和空白对照

Fig. 3 Results of the multiplex RT-PCR assay of PVY ShX14 isolate

M: Marker DNA (D2000); 1: Multiplex RT-PCR products of PVY ShX14 isolate; 2–3: Negative control 1 (healthy plant) and negative control 2 (water), respectively

3 讨论

3.1 ShX14 分离物基因组特征

对于 ShX14 分离物的 CP 基因（GenBank 登录号为 KC506395），笔者前期研究发现其为 PVY^{O} 型（高芳銮等，2013），这与其血清学检测结果相一致，但进一步分析扩增获得的 P1、VPg 基因片段却为 PVY^{N} 型，初步表明该分离物可能为 N×O 重组分离物。当将该分离物全基因组分为 R1、R2 和 R33 个区段（1 ～ 5673、5674 ～ 8382、8383 ～ 9183）分别与常见的 PVY 分离物 BLAST 比对发现，R1 区段与 $PVY^{N:O}$、PVY^{NTN}（a）、$PVY^{NTN\text{-}NW}$（SYR-I 型）的核苷酸序列高度同源，平均序列一致性分别为 98%、98% 和 98.5%；R2 区段与 PVY^{NTN}（a）、$PVY^{NTN\text{-}NW}$（SYR-I 型）核苷酸序列高度同源（平均一致性分别为 98.91% 和 99%），而与 $PVY^{N:O}$ 的核苷酸序列差异较大，序列一致性仅为 84%；R3 区段与 $PVY^{N:O}$、$PVY^{NTN\text{-}NW}$（SYR-I 型）核苷酸序列高度同源（平均一致性均为 99%），而与 PVY^{NTN}（a）的核苷酸序列差异较大，序列一致性仅为 91.91%。综合 3 个区段比对结果可知，该分离物与常见的重组株系 $PVY^{N\text{-}Wi}$ 和 PVY^{NTN} 在基因组结构差异较大，而与 $PVY^{NTN\text{-}NW}$（SYR-I 型）的较为接近，且核苷酸、氨基酸序列一致性均超过 98%（表 2），表明其与 $PVY^{NTN\text{-}NW}$ 株系（SYR-I 型）的分离物序列高度同源。

Simplot 分析结果显示 ShX14 分离物基因组内存在明显的基因重组现象，即 PVY^{N} 和 PVY^{O} 在不同的基因区段发生重组（图 1A）。GARD 检测到 3 个显著的重组位点（2318、5674 和 8385nt），这些重组位点的平均模型支持率均超过 93.64%，表明这些位点为 ShX14 分离物重组位点的置信度非常高。重组分析得到 3 个位点分别位于 HC-Pro/P3、VPg、NIb/CP 基因内，与 $PVY^{NTN\text{-}NW}$ 株系（SYR-I 型）的重组位点基本一致。综合序列比对及重组分析结果，表明 ShX14 分离物为 $PVY^{NTN\text{-}NW}$ 株系（SYR-I 型）的重组分离物。

3.2 PVY 系统发育及其分离物与株系关联分析

在系统发育关系上，ShX14 分离物与 $PVY^{NTN\text{-}NW}$ 株系（SYR-I 型）的亲缘关系最近，故而推断可能其归属于 $PVY^{NTN\text{-}NW}$ 株系（SYR-I 型），并得到 PVY 分离物与株系关联性及多重 RT-PCR 结果的验证。由于 *Potyvirus* 的病毒存在株系分化，为了获得更为精确、可靠的结果，Adams 等（2005）建议用全基因组序列作分子标记研究该属病毒的系统进化。为此，本研究基于全基因组序列采用 ML 法重建了 PVY 的系统发育关系，结果显示同一种株系的 PVY 分离物以较高的置信值优先相聚成簇，株系内差异小于株系间差异，表明 PVY 具有显著的株系特异性，并真实地反映出 PVY 株系的分子进化关系。

进化生物学认为，在进化模型确定的情况下，ML 法是与进化事实吻合度最好的一种算法（黄原，2012）。在 PVY 的分子进化中，Karasev 等（2011）研究认为 $PVY^{N\text{-}Wi}$ 与 $PVY^{N:O}$ 两个株系独立进化，起源不同，而 Singh 等（2008）则认为两者有着共同的进化起源。导致两种相悖结论的可能原因有二：首先，Karasev 等（2011）使用的是建树相对准确、计算速度快的 NJ 法，但它是一种基于距离的算法，将序列上的所有位点等同对待，适用于进化距离小、信息位点少的短序列，故而 NJ 树多作为系统发育分析的初始树；其次，通过比较基因组结构，不难发现 $PVY^{N\text{-}Wi}$ 株系与 $PVY^{N:O}$ 株系在 P1 基因区段上差异较大，前者的 P1 基因为 N×O 重组型，而后者则是 PVYN 型。而 Singh 等（2008）通过 PVY 分离物的传统生物学特性结合分子生物学特征，表明两者进化起源相同。本研究的系统发育分析结果显示，$PVY^{N\text{-}Wi}$ 株系与 $PVY^{N:O}$ 株系以较高的置信值（77%）相聚成簇，表明两者有着共同的进化起源，进一步支持了 Singh 等（2008）的分类结果。

在 PVY 的系统发育分析中，参考分离物的选择往往带有主观性，系统发育分析的结果主要根据系统发育树节点分枝上的置信值（或后验概率）进行粗略判断，所选分离物是否具有株系代表性多未进行统计检验，而参考分离物又是影响系统发育分析结果的重要因素之一。本文采用系统发育与性状关联分析，通过 AI 和 PS 两个统计参数反映所选的分离物与所属株系的关联强度（表 3），尽管个别株系统计检验不显著，但大部分所选的 PVY 分离物与所属株系之间存在较强的关联性，分析结果显示 ShX14 分离物与 $PVY^{NTN\text{-}NW}$ 株系（SYR-I 型）存在较强的关联性（P=0.05），表明其归属于 $PVY^{NTN\text{-}NW}$ 株系（SYR-I 型）。因此，本研究所选的分离物，可以作为今后 PVY 系统发育分析及株系分子鉴定的标准参考。

4 结论

克隆、测序获得 PVY 榆林分离物 ShX14 的全基因组序列，通过序列比对、重组检测、系统发育等综合分析揭示其基因组结构特征，并确定其归属于 $PVY^{NTN\text{-}NW}$ 株系（SYR-I 型），为后续深入开展该分离物相关功能研究等奠定了基础。

参考文献

[1] 高芳銮 , 沈建国 , 史凤阳 , 等 . 马铃薯 Y 病毒 pipo 基因的分子变异及结构特征分析 [J]. 遗传 , 2013, 035(009):1125-1134.

[2] 高芳銮 , 沈建国 , 史凤阳 , 等 . 中国马铃薯 Y 病毒的检测鉴定及 CP 基因的分子变异 [J]. 中国农业科学 , 2013, 46(015):3125-3133.

[3] 黄原 . 分子系统发生学 [J]. 遗传 , 2012(11):1455-1455.

[4] 罗红香 , 李余湘 , 刘会忠 , 等 . 贵州黔南烟草马铃薯 Y 病毒 (PVY) 株系的血清学鉴定 [J]. 中国烟草科学 ,2012, 000(005):60-62.

[5] ADAMS M J, ANTONIW J F, FAUQUET C M. Molecular criteria for genus and species discrimination within the family *Potyviridae*[J]. Archives of Virology, 2005, 150: 459-479.

[6] CHIKH ALI M, MAOKA T, NATSUAKI T, et al. $PVY^{NTN\text{-}NW}$, a novel recombinant strain of Potato virus Y predominating in potato fields in Syria[J]. Plant Pathology,2010, 59: 31-41.

[7] CHIKH ALI M, MAOKA T, NATSUAKI K T, et al. The simultaneous differentiation of Potato virus Y strains including the newly described strain PVY(NTN-NW) by multiplex PCR assay[J]. Journal of virological methods, 2010, 165: 15-20.

[8] CHUNG Y W, MILLER W A, ATKINS J F, et al. An Overlapping Essential Gene in the Potyviridae[J]. Proceedings of the National Academy of Sciences of the United States of America, 2008, 105: 5897-5902.

[9] DRUMMOND A J, SUCHARD M A, XIE D, et al. Bayesian phylogenetics with BEAUti and the BEAST. 1.7[J]. Molecular biology and evolution,2012, 29: 1969-1973.

[10] GLAIS L, TRIBODET M, KERLAN C. Genomic variability in Potato potyvirus Y (PVY): evidence that PVY(N)W and PVY(NTN) variants are single to multiple recombinants between PVY(O) and PVY(N) isolates[J]. Archives of Virology, 2002, 147: 363.

[11] HU X, KARASEV A V, BROWN C J, et al. Sequence characteristics of potato virus Y recombinants[J]. Journal of General Virology, 2009a, 90: 3033-3041.

[12] HU X, NIE X, HE C, et al. Differential pathogenicity of two different recombinant PVY^{NTN} isolates in Physalis floridana is likely determined by the coat protein gene[J]. Virology journal, 2011, 8: 207-207.

[13] HU X, HE C, XIAO Y,et al. Molecular characterization and detection of recombinant isolates of potato virus Y from China[J]. Archives of Virology, 2009b, 154: 1303-1312.

[14] JIANYI Y, AMBRISH R, YANG Z. BioLiP: a semi-manually curated database for biologically relevant ligand-protein interactions[J]. Nucleic acids research, 2013, 41: 1096-1103.

[15] KARASEV A V, HU X, BROWN C J, et al. Genetic Diversity of the Ordinary Strain of Potato virus Y (PVY) and Origin of

Recombinant PVY Strains[J]. Phytopathology, 2011, 101: 778.

[16] KOSAKOVSKY POND S L, POSADA D, GRAVENOR M B, et al. GARD: a genetic algorithm for recombination detection[J]. Bioinformatics, 2006, 22: 3096.

[17] LOLE K S, BOLLINGER R C, PARANJAPE R S, et al. Full-length human immunodeficiency virus type 1 genomes from subtype C-infected seroconverters in India, with evidence of intersubtype recombination[J]. Journal of virology, 1999, 73: 152-160.

[18] NIE X, SINGH R P, SINGH M. Molecular and pathological characterization of N:O isolates of the Potato virus Y from Manitoba Canada[J]. Canadian Journal of Plant Pathology, 2004, 26: 573-583.

[19] NOLTE P, WHITWORTH J L, THORNTON M K,et al. Effect of Seedborne Potato virus Y on Performance of Russet Burbank Russet Norkotah, and Shepody Potato[J]. Plant disease, 2004, 88: 248-252.

[20] NYLANDER J A A. MrModeltest v2.3. Program distributed by the author.Evolutionary Biology Centre Uppsala University. https://www.abc.se/~nylander/mrmodeltest2/mrmodeltest2.html, 2004.

[21] OGAWA T, TOMITAKA Y, NAKAGAWA A, et al. Genetic structure of a population of Potato virus Y inducing potato tuber necrotic ringspot disease in Japan; comparison with North American and European populations[J]. Virus Research, 2008, 131: 199-212.

[22] PARKER J, RAMBAUT A, PYBUS O G. Correlating viral phenotypes with phylogeny: Accounting for phylogenetic uncertainty[J]. Infection, genetics and evolution : journal of molecular epidemiology and evolutionary genetics in infectious diseases, 2008, 8: 239-246.

[23] QUENOUILLE J, VASSILAKOS N, MOURY B. Potato virus Y: a major crop pathogen that has provided major insights into the evolution of viral pathogenicity[J]. Molecular Plant Pathology, 2013, 14: 439-452.

[24] RAHMAN M, AKANDA A. Performance of seed potato produced from sprout cutting, stem cutting and conventional tuber against PVY and PLRV[J]. Bangladesh Journal of Agricultural Research, 2009, 34: 609-622.

[25] SINGH R P, VALKONEN J P T, GRAY S M, et al. Discussion paper: The naming of Potato virus Y strains infecting potato[J]. Archives of Virology, 2008, 153: 1-13.

[26] TAMURA K, PETERSON D, PETERSON N, et al. MEGA5: molecular evolutionary genetics analysis using maximum likelihood, evolutionary distance, and maximum parsimony methods[J]. Molecular Biology & Evolution, 2011, 28: 2731.

[27] TIAN Y P, LIU J L, ZHANG C L, et al. Genetic diversity of Potato virus Y infecting tobacco crops in China[J]. Phytopathology, 2011, 101: 377.

[28] WANG B, JIA J L, WANG X Q, et al. Molecular characterization of two recombinant potato virus Y isolates from China[J]. Archives of Virology, 2012, 157: 401-403.

[29] WHITWORTH J L, NOLTE P, MCINTOSH C, et al. Effect of Potato virus Y on Yield of Three Potato Cultivars Grown Under Different Nitrogen Levels[J]. Plant Disease, 2006, 90(1): 73-76.

Molecular characterization and detection of a recombinant isolate of bamboo mosaic virus from China

Wenwu Lin[1], Fangluan Gao[1], Wenting Yang[1], Chaowei Yu[1], Jie Zhang[1], Lingli Chen[1], Zujian Wu[1], Yau-Heiu Hsu[2], Lianhui Xie[1]

(1 Fujian Key Lab of Plant Virology, Institute of Plant Virology, Fujian Agriculture and Forestry University, Fuzhou 350002, Fujian; 2 Graduate Institute of Biotechnology, National Chung Hsing University, Taichung 40227, Taiwan)

Abstract:The complete genome sequences of three isolates of bamboo mosaic virus (BaMV) from mainland China were determined and compared to those of BaMV isolates from Taiwan. Sequence analysis showed that isolate BaMV-JXYBZ1 from Fuzhou shares 98% nucleotide sequence identity with BaMV-YTHSL14 from nucleotides 2586 to 6306, and more than 94% nucleotide sequenceidentity with BaMV-MUZHUBZ2 in other regions. Recombination and phylogenetic analyses indicate that BaMV-JXYBZ1 is a recombinant with one recombination breakpoint. To our knowledge, this is the first report of aBaMV recombinant worldwide.

Bamboo, valued for both its ornamental and edible qualities and rich forest resources, is widely distributed intropical and subtropical areas. A virus inducing mosaic symptoms on bamboo, bamboo mosaic virus (BaMV),was first described in Brazil, and this virus presents asignificant threat to the bamboo industry(Lin et al., 1977). Subsequently, the disease caused by BaMV was discovered in Taiwan, California and Florida, where it hascaused economic losses (Lin et al., 1979; Lin et al., 1995; Elliott and Zettler, 1996). BaMV is a member of the genus *Potexvirus* in the family *Alphaflexiviridae*(Lin et al., 1994). Virus particles have a flexuous rod-shaped morphology with a modal length of 490 nm and possess asingle-strand, positive-sense RNA genome of approximately 6.4kb containing five conserved open reading frames (ORFs), a 5′ methyl cap and a 3′ poly(A) tail (Lin et al., 1977; Lin et al., 1994).ORF1 encodes a 155-kDa replication-related protein consisting of a capping enzyme domain, a helicase-like domain (HLD), and an RNA-dependent RNA polymerase domain (RdRp) from the N-terminus to the C-terminus (Li et al., 2001a; Li et al., 2001b; Huang et al., 2005). ORF2, ORF3 and ORF4 encode movement proteins of 28, 13, and 6kDa, respectively, named triple gene block proteins (Lin et al., 1994).The 25kDa coat protein is encoded by ORF5 and is associated with virion encapsidation, replication, and cell-to-cell and long-distance virus movement (Lan et al., 2010). Although the structure and function of the BaMV genome has been deeply researched, only five complete genomes of wild BaMV isolates from Taiwan island (BaMV-V, BaMV-O, BaMV-Pu, BaMV-Au and BaMV-Bo) have been deposited in GenBank, under accession numbers L77962, NC_001642, AB636266,AB636267 and AB543679, respectively, and no recombinant isolate has yet been reported worldwide.

Three BaMV isolates (BaMV-JXYBZ1, BaMV-MUZHUBZ2 and BaMV-YTHSL14) were collected from bamboo plants of different species located in Fuzhou, Fujian province, China (Table S1). The symptoms

Archive of Virology. 2016, 161:1091-1094

Received 11 October 2015; Accepted 25 January 2016; Published online 29 February 2016 on bamboo leaves caused by these three isolates are shown in Fig.S1. Total RNA was extracted using an RNeasy Plant Mini Kit (Qiagen, Hilden, Germany), and cDNA was synthesized using an oligo dT(18) primer with a GoScript^{TM}Reverse Transcription System (Promega Corporation, Madison, USA) according to the manufacturer's instructions.

Based on the preliminary BaMV sequences downloaded from GenBank, primer sets (Table S2) were designed to allow the complete genomes of these three BaMV isolates to be amplified and sequenced by the Sanger method, and the 5′ RACE method was used as described previously for sequencing the ends of the genome (Lin et al., 1994). Twelve independent clones of each segment were sequenced for each isolate to exclude the possibility of mixed infection by divergent BaMV isolates. Moreover, a long distance-PCRtechnique was used to confirm the complete genome sequences using the primer pair 1F/5R. The final sequences were as sembled using DNAMAN version 8 (Lynnon, Quebec, Canada). Nucleotide and amino acid sequenceidentities were searched using the BLAST software package(http://www.ncbi.nlm.nih.gov/blast). To further characterize these BaMV isolates, five additional complete genome sequences were retrieved from GenBank to serveas reference sequences. Multiple sequence alignments were performed with MUSCLE, implemented in MEGA5 (Tamura et al., 2011). Phylogenetic analysis was performed by the maximum likelihood(ML) method implemented in MEGA5 (Tamura et al., 2011). The *GTR+G+I* model determined by Mr Model Test (Nylander, 2004) was used to reconstruct a phylogenetic tree. ML topology was evaluated with 1000 bootstrap replicates. Recombination events were detected using RDP4 (Martin et al., 2010), and the recombination points were confirmed and plotted with Simplot 3.5.1 (Lole et al., 1999). The putative recombination sites were numbered based on visual observation. Additionally, the confidence levels of the recombination breakpoints were evaluated by phylogenetic analysis using non-recombinant segments. The complete genomes of three isolates obtained in this study were deposited in the GenBank databases with the accession numbers KT591184-KT591186. To biologically characterize these three BaMV isolates, infectious cDNA clones were constructed, and *Nicotiana benthamiana* was challenged to confirm their infectivity (Fig. S2).

The complete sequence of BaMV-JXYBZ1 is 6364 nucleotides (nt) in length, excluding the poly (A) tail. The 5′and 3′ untranslated regions (UTR) consist of 93 nt and140 nt, respectively. Like other BaMV isolates, the complete genome of BaMV-JXYBZ1 contains five ORFs of constant length for their conserved functions. The genome structure of BaMV-YTHSL14 and BaMV-MUZHUBZ2 was identical to that of BaMV-JXYBZ1. All isolates contained conserved functional motifs and amino acids inspecific regions, such as Arg-16 and Arg-21 in theN-terminal region of the TGBp1 (Lin et al., 2004), Cys-109 and Cys-112 at the C-terminal tail of TGBp2 (Tseng et al., 2009), a glycine-rich motifin the N-terminal region of CP (Lan et al., 2010), and a conserved hexanucleotide sequence (ACc/uUAA) and AAUAAA motifin the 3′ UTR of BaMV genomic RNA (Chiu et al., 2002; Chen et al., 2005). When compared to the available genome sequences of other isolates of BaMV, the nucleotide sequence identities range from 82% to 99% (Table 1a).

Sequence analysis showed that isolate BaMV-JXYBZ1 shares 95% overall nucleotide sequence identity with BaMV-YTHSL14 and 92% with BaMV-MUZHUBZ2 at the genomic level. At the gene level, the C-terminal region of ORF1 and complete ORF2-5 of BaMV-JXYBZ1 share higher nucleotide and amino acid sequence similarity with YTHSL14 than with BaMV-MUZHUBZ2 (Table 1b), in agreement with complete genome analysis described above. In the N-terminal region of ORF1, BaMV-JXYBZ1 shares 94% nucleotide sequence identity and 98% amino acid sequence identity with BaMV-MUZHUBZ2 but shares only 90% nucleotide sequence identity and 96% amino acid sequence identity with BaMV-YTHSL14. This indicates that BaMV-JXYBZ1 might be a recombinant, with BaMV-YTHSL14 and BaMV-MUZHUBZ2 aspotential parents.

Detailed genomic information for allisolates is presented in Table 1.

Table 1 (a) Percentage of nucleotide sequence identity of the complete genome sequences of BaMV isolates. (b) Percentage of sequences identity of the genomic regions of the BaMV-JXYBZ1 isolate (amino acids in parentheses) to other BaMV isolates for which complete genome sequences are available

Virus isolate	BaMV-Au	BaMV-Bo	BaMV-Pu	BaMV- V	BaMV-O	BaMV-JXYBZ1	BaMV- MUZHUBZ2	BaMV- YTHSL14
(a)								
BaMV-Au	-	99	95	93	90	82	82	82
BaMV-Bo	99	-	95	93	90	82	82	82
BaMV-Pu	95	95	-	93	90	82	82	82
BaMV-V	93	93	93	-	90	83	83	82
BaMV-O	90	90	90	90	-	82	82	82
BaMV-JXYBZ1	82	82	82	83	82	-	92	95
BaMV- MUZHUBZ2	82	82	82	83	82	92	-	90
BaMV-YTHSL14	82	82	82	82	82	95	90	-

Genome position	BaMV-Au	BaMV-Bo	BaMV-Pu	BaMV-V	BaMV-O	BaMV-MUZHUBZ2	BaMV-YTHSL14
(b)							
5^{0}-UTR	95/-	95/-	95/-	94/-	88/-	98/-	97/-
ORF1	81/91	82/92	81/92	82/92	81/92	93/98	93/97
ORF2	82/93	82/93	82/94	83/94	82/92	89/96	98/100
ORF3	86/91	86/91	87/92	89/93	86/91	92/97	98/99
ORF4	76/79	76/79	79/83	79/79	77/81	89/85	98/96
ORF5	82/93	82/93	81/91	81/92	82/90	90/93	98/98
3^{0}-UTR	96/-	96/-	95/-	96/-	96/-	99/-	99/-

Phylogenetic analysis indicated that all isolates can be divided into two groups with high bootstrap support (Fig. S3). The BaMV isolate from Taiwan was in group 1, and isolates from mainland China were clustered in group 2. Comparing all BaMV isolates, recombination analyses identified one likely recombination event in the BaMV-JXYBZ1 genome(Fig. 1A), and BaMV-MUZHUBZ2 and BaMV-YTHSL14 were identified as its parents at a high level of significance by the RDP4 suite (RDP, $P = 4.56\times10^{-9}$; BOOTSCAN, $P\leqslant 1.13\times10^{-17}$; MAXCHI, $P\leqslant 1.30\times10^{-21}$; CHIMAERA,$P\leqslant 1.27\times 10^{-23}$; SISCAN, $P\leqslant 4.39 \times 10^{-17}$and 3SEQ, $P\leqslant 6.03 \times 10^{-29}$). In the similarity and bootscan plots (Fig. 1B, C), the putative recombinant BaMVJXYBZ1 showed a high degree of sequence similarity andphylo genetic clustering with different groups in two regions. For the first 2567 nucleotides, BaMV-JXYBZ1 has higher nucleotide sequence similarity to BaMV-MUZHUBZ2 (94% identity), whereas for the positions 2568 to 6306, it closely resembles the BaMV-YTHSL14 isolate, with 98% nucleotide sequence identity. Further, the phylogenies inferred fromthe non-recombinant fragments suggests that the phylogenetic in congruence is significant (Fig. 1D), supporting one breakpoint at the C-terminus of the helicase-like domain of the replicase in the complete genome of BaMV-JXYBZ1. Taken together, these analyses suggest that BaMV-JXYBZ1 is a novel recombinant from China. To our knowledge, this is thefirst report of a BaMV recombinant worldwide. The results of this study demonstrate the complexity of BaMV infection in wild bamboo. Additional surveillance for recombinant BaMV may be needed, as well as a survey of its effect on bamboo quality in cultivation areas.

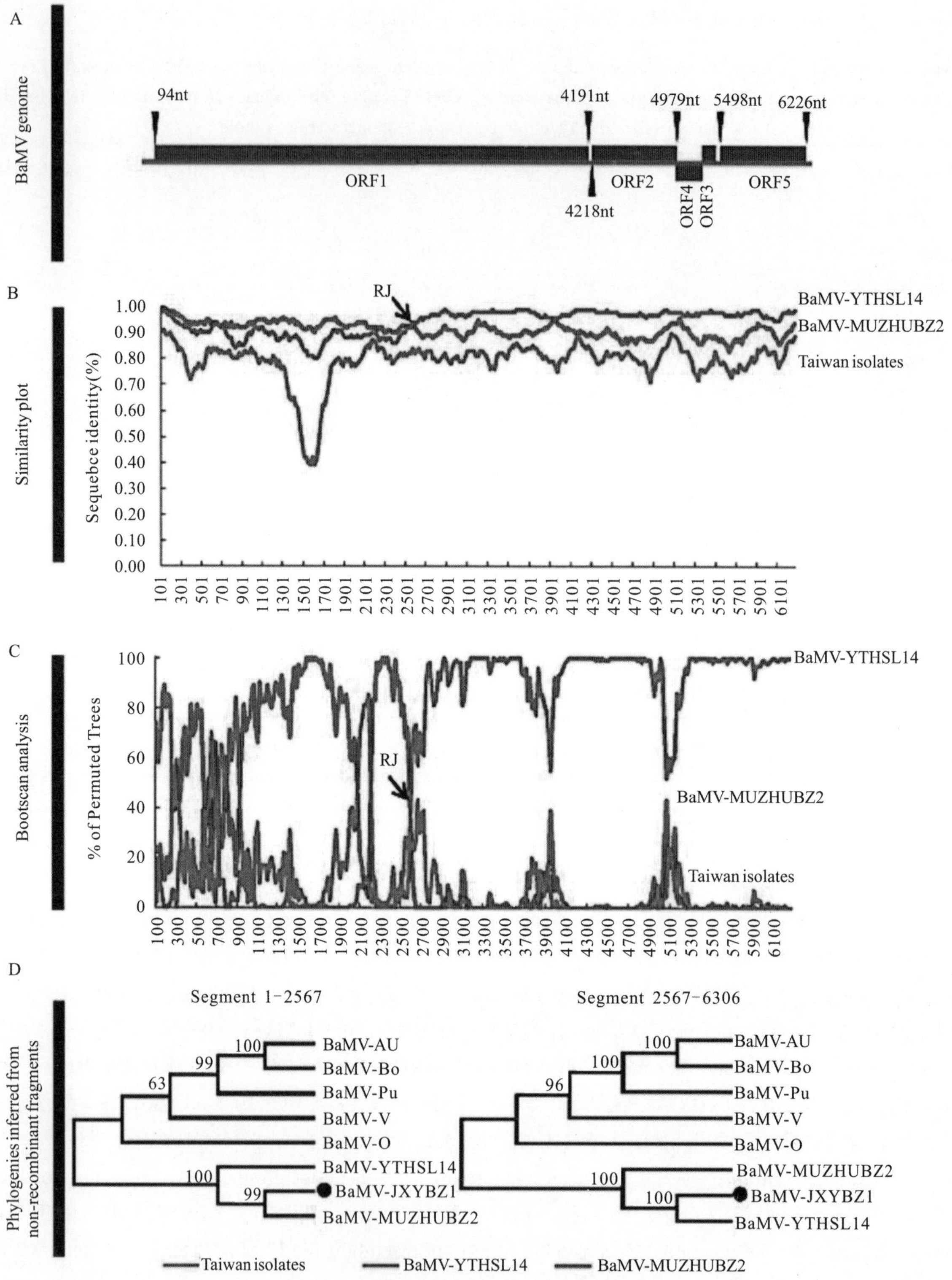

Fig. 1 Recombination analysis of BaMV genomes

(A) Schematic diagram of the BaMV genome. (B) Similarity plot. (C) Bootscan analysis. (D) Individual phylogenies reconstructed from non-recombinant fragments identified using MEGA5. A window size of 200bp and a step size of 20bp were used for the similarity plot and bootscan analysis. The recombinant isolate is indicated by a black dot

Acknowledgements

This study was funded by National Basic Research Program of China (973 Program) (Grant No.

2014CB138402), Natural Science Foundation of China (NSFC) (Grant No. 31171821) and National Key Technology Support Program (Grant No.2012BAD19B03).

References

[1] CHEN I H, CHOU W J, LEE P Y, et al. The AAUAAA motif of bamboo mosaic virus RNA is involved in minus-strand RNA synthesis and plus-strand RNA polyadenylation[J]. Journal of virology, 2005, 79: 14555-14561.

[2] CHIU W W, HSU Y H, TSAI C H. Specificity analysis of the conserved hexanucleotides for the replication of bamboo mosaic potexvirus RNA[J]. Virus Research, 2002, 83: 159-167.

[3] ELLIOTT M, ZETTLER F. Bamboo mosaic virus detected in ornamental bamboo species in Florida[J]. Selected Proceedings of the Florida State Horticultural, 1996, 24.

[4] HUANG Y L, HSU Y H, HAN Y T, et al. mRNA guanylation catalyzed by the S-adenosylmethionine-dependent guanylyltransferase of bamboo mosaic virus[J]. Journal of Biological Chemistry, 2005, 280: 13153-13162.

[5] LAN P, YEH W B, TSAI C W, et al. A unique glycine-rich motif at the N-terminal region of bamboo mosaic virus coat protein is required for symptom expression[J]. Mol Plant Microbe Interact, 2010, 23: 903-914.

[6] LI Y I, CHEN Y J, HSU Y H, et al. Characterization of the AdoMet-dependent guanylyltransferase activity that is associated with the N terminus of bamboo mosaic virus replicase[J]. Journal of Virology, 2001, 75: 782-788.

[7] LI Y I, SHIH T W, HSU Y H, et al. The helicase-like domain of plant potexvirus replicase participates in formation of RNA. 5′ cap structure by exhibiting RNA. 5′-triphosphatase activity[J]. Journal of Virology, 2001b, 75: 12114-12120.

[8] LIN M K, CHANG B Y, LIAO J T, et al. Arg-16 and Arg-21 in the N-terminal region of the triple-gene-block protein 1 of bamboo mosaic virus are essential for virus movement[J]. Journal of General Virology, 2004, 85: 251-259.

[9] LIN M, KITAJIMA E, CUPERTINO F, et al. Partial purification and some properties of bamboo mosaic virus[J]. Phytopathology, 1977, 67: 1439-1443.

[10] LIN N S, LIN B Y, LO N W, et al. Nucleotide sequence of the genomic RNA of bamboo mosaic potexvirus[J]. Journal of General Virology, 1994, 75: 2513-2518.

[11] LIN N, CHEN M, KIANG T, et al. Preliminary studies on bamboo mosaic virus disease in Taiwan[J]. Taiwan For Res Inst, 1979, 317: 1-10.

[12] LIN N, LIN B, YEH T, et al. First report of bamboo mosaic virus and its associated satellite RNA on bamboo in the U S[J]. Plant Disease, 1995, 79.

[13] LOLE K S, BOLLINGER R C, PARANJAPE R S, Full-length human immunodeficiency virus type 1 genomes from subtype C-infected seroconverters in India, with evidence of intersubtype recombination[J]. Journal of Virology, 1999, 73: 152-160.

[14] MARTIN D P, LEMEY P, LOTT M, et al. RDP3: a flexible and fast computer program for analyzing recombination[J]. Bioinformatics, 2010, 26: 2462-2463.

[15] NYLANDER J. MrModeltest v2. 2. Program distributed by the author Evolutionary Biology Centre. http://www.abc.se/nylander/mrmodeltest2/mrmodeltest2.html, 2004.

[16] TAMURA K, PETERSON D, PETERSON N, et al. MEGA5: molecular evolutionary genetics analysis using maximum likelihood, evolutionary distance, and maximum parsimony methods[J]. Molecular Biology & Evolution, 2011, 28: 2731-2739.

[17] TSENG Y H, HSU H T, CHOU Y L, et al. The two conserved cysteine residues of the triple gene block protein 2 are critical for both cell-to-cell and systemic movement of bamboo mosaic virus[J]. Molecular Plant-Microbe Interactions, 2009, 22: 1379-1388.

Adaptive evolution and demographic history contribute to the divergent population genetic structure of *Potato virus Y* between China and Japan

Fangluan Gao, Wenchao Zou, Lianhui Xie, Jiasui Zhan

(Fujian Key Laboratory of Plant Virology, Institute of Plant Virology, Fujian Agriculture and Forestry University, Fuzhou,China)

Abstract: *Potato virus Y* (PVY) is an important plant pathogen causing considerable economic loss to potato production. Knowledge of the population genetic structure and evolutionary biology of the pathogen, particularly at a transnational scale, is limited but vital in developing sustainable management schemes. In this study, the population genetic structure and molecular evolution of PVY were studied using 127 first protein (P1) and 137 coat protein (CP) sequences generated from isolates collected from potato in China and Japan. High genetic differentiation was found between the populations from the two countries, with higher nucleotide diversity in Japan than China in both genes and a K_{ST} value of 0.216 in the concatenated sequences of the two genes. Sequences from the two countries clustered together according to their geographic origin. Further analyses showed that spatial genetic structure in the PVY populations was likely caused by demographic dynamics of the pathogen and natural selection generated by habitat heterogeneity. Purifying selection was detected at the majority of polymorphic sites although some clade-specific codons were under positive selection. In past decades, PVY has undergone a population expansion in China, whereas in Japan, the population size of the pathogen has remained relatively constant.

Keyword: Bayesian skyline plots; demographic history; natural selection; phylogenetic analysis; *Potato virus Y*

1 Introduction

Genetic drift, gene flow, and natural selection are three main evolutionary forces shaping the spatial population genetic structure of species (Zhan and McDonald, 2004). Under constrained gene flow, stochastic changes in allele frequencies among geographic populations can result in random fixation of neutral alleles, leading to nonadaptive differentiation (Wright, 1938). On the other hand, divergent selection for various ecological or physiological characters among genetically isolated populations may lead to adaptive population subdivision (Koskella and Vos, 2015; Yang et al., 2016). In plant pathology, understanding how populations are spatially structured is important to project the evolutionary trajectories of pathogens and formulate approaches for sustainable plant disease management (Zhan el al., 2014). For example, many soil-borne plant pathogens are highly spatially structured (Gilbert, 2002) and regional *R* gene deployment may be an appropriate method for an effective and durable control of plant disease caused by the pathogens. On the other hand, recombination can facilitate the reshufflings of genomes and *R* gene pyramids may be less efficient in managing sexual pathogens usually characterized by higher genetic variation distributed at a fine scale (McDonald and Linde, 2002).

Plant pathogens can vary greatly in spatial population genetic structure (Barrett et al., 2008; Zhan et

Evolutionary Applications.2017, 10(4):379-390.
Received 10 August 2016; Accepted 10 January 2017

al., 2015), depending on the biotic and abiotic factors they associated with such as host genetics, pathogen biology, physical environments, and the ways of human intervention during and post agricultural production. These factors can affect individually and interactively on the extent of genetic drift, gene flow, and selection, therefore influencing the generation and maintenance of spatial population structure (Burdon and Thrall, 2008; Bergholz et al., 2011; Zhan and McDonald, 2013). Natural selection is expected to play a central role in the spatial population genetic structure of plant pathogens in agricultural ecosystems (Thrall et al., 2011), primarily due to variation in host genetics, fungicide applications, climatic conditions, and agricultural practices among regions. Directional selection for virulence, fungicide resistance, and other ecological characters related to particular biogeographic environments can drive the rapid accumulation of adaptive genetic differentiation in plant pathogen populations (Achtman and Wagner, 2008). At the same time though, variation in these same factors among regions can also strongly influence the demographic dynamics of plant pathogens, generating nonadaptive genetic differentiation in the pathogen populations. However, despite recognition of its importance, an in-depth understanding of how patterns of spatial population genetic structure are generated and maintained and the main evolutionary mechanisms responsible for these patterns are still limited for many plant pathogens, but critical for sustainable plant disease management (Zhan et al., 2015; He et al., 2016).

Potato virus Y (PVY) is a member of the genus *Potyvirus* in the family *Potyviridae*. Its genome has a single-stranded positive-sense RNA of ~9.7kb, encoding a polyprotein that is cleaved into 10 mature functional proteins (King et al., 2011). Additionally, a short polypeptide (PIPO) is expressed within the P3 cistron by frame shifting (Chung et al., 2008). Among the 11 functional proteins encoded, the first protein (P1) and the coat protein (CP) are thought to play an important role in the adaptation of potyviruses to host species (Valli et al., 2007). Furthermore, the CP gene has been frequently used in strain identification, species classification, and phylogenetic analysis of potyviruses (Cuevas et al., 2012).

PVY is one of the most destructive pathogens affecting potato (*Solanum tuberosum* L.), the third largest food crop in the world (Birch et al., 2012) and widely grown in many Asian countries including China and Japan. It can cause 40%–70% yield reduction in potato production (Nolte et al., 2004) and significantly reduce the quality of seed tubers. As the largest producer in the world, China accounts for 26.3% and 22.2% of the global potato acreages and yields, respectively (Wang et al., 2011), and potato production in the country is expected to substantially increase in coming decades due to government support and dietary shifts (Kearney, 2010). PVY is one of the main factors constraining further development of Chinese potato industry (Wang et al., 2011).

Knowledge of the population genetics and evolutionary biology of PVY, particularly at a transnational scale, is relatively limited compared to other important plant pathogens such as *Magnaporthe* (Tredway et al., 2005), *Pyrenophthora* (Gurung et al., 2013), *Verticillium* (Short et al., 2015), and *Phytophthora* (Tian et al., 2016). Many factors such as technology and resource availability may partially contribute to this shortage. In addition, importation of living pathogens even for scientific reasons is strictly forbidden in many countries including China due to quarantine regulations. As a consequence, analysis of biotrophic pathogens with no or few morphological characters used to be very challenging for many researchers. With the advance of PCR-based sequencing technology and sharing of sequence data in many public domains, empirical analysis of the population genetic structure and evolutionary biology of biotrophic plant pathogens such as PVY at an international scale has become possible. This approach has been used by several laboratories to infer the evolution of PVY (Ogawa et al., 2008; Cuevas et al., 2012), usually involving bona fide geographical populations. Ogawa and colleagues (Ogawa et al., 2008; Ogawa et al., 2012) compared the population structure of PVY in Japan, Europe, and North America and found some unique evolutionary patterns associated with the pathogen in Japan. For example, there were only three nonrecombinant subpopulations in Japan (PVY^{O}, PVY^{N}, and PVY^{NTN}), whereas six subpopulations existed worldwide. Furthermore, unlike results from other regions where the subpopulations all have the same age, it is believed that the Japanese PVY^{O} subpopulation is older than the PVY^{N} and PVY^{NTN} subpopulations.

The objectives of this study were to (i) compare the population genetic structure of PVY in China and Japan and (ii) determine the main evolutionary and demographic mechanisms responsible for the observed population genetic structure.

2 Materials and methods

2.1 Viral sequences

Eighty-five PVY isolates, confirmed by enzyme-linked immunosorbent assay (ELISA) with a broad-spectrum PVY antibody (Agdia, Elkhart, USA), were randomly collected from potato (*Solanumtu berosum*) across a range of geographical locations in China including Fujian, Hunan, and Hebei provinces in 2011 and 2012. Total RNAs were extracted using an RNA simple Kit according to the manufacturer's instructions (TianGen, Beijing, China). Full-length cDNAs were synthesized by RT-PCR using Oligo$(dt)_{18}$ and TransScript® First-Strand cDNA Synthesis SuperMix (TransGen, Beijing, China) and amplified with two pairs of degenerate primers as described previously (Gao et al., 2014). PCR amplifications of cDNAs were performed in a total volume of 50μl composed of 5.0μL of TransTaq™ 10× HiFi Buffer II, 4.0 μL of dNTPs (2.5 mM), 2.0μL of forward primer (10μmol/L), 2.0μL of reverse primer (10μmol/L), 34.5μL of ddH_2O, 0.5μL of TransTaq™ HiFi Polymerase (5U/μL), and 2.0μL of template cDNA. The PCR program was initially denatured at 94°C for 5min; followed by 30 cycles of 94°C for 30s, 53°C (P1 gene), or 55°C (CP gene) for 30s and 72°C for 1min; ended with an extension at 72°C for 10min. PCR products were separated on 1% agarose gels by electrophoresis, visualized using a UV transilluminator and cleaned using a TIANgel Maxi Purification Kit (TianGen).

Purified PCR fragments were ligated into T-tailed pEASY-T5 Zero vector (TransGen) and transformed into competent *Escherichia coli* strain Trans1-T1 (TransGen). Recombinant plasmids were extracted and sequenced in both directions by Nanjing GenScript Biological Technology Co., Ltd. (GenScript, Nanjing, China). Due to high mutation rate in RNA viruses, at least four cDNA clones derived from two separate PCR reactions were sequenced for each PVY isolate to ensure consensus. Only the sequence identical in at least three clones was used for further analyses to eliminate potential heterogeneity introduced by Taq polymerase. All sequences generated in this study were deposited in GenBank databases. In addition, 24 sequences of P1 gene and 26 sequences of CP gene from other parts of China (including Guizhou, Shandong, Heilongjiang, Liaoning, and Shaanxi provinces) and 34 sequences each of CP and P1 genes from Japan were retrieved from GenBank (Table S1). Consequently, the combined sequences from China were generated from PVY samples collected between 2005 and 2012 and the Japanese sequences were generated from samples collected between 1995 and 2012. As host-driven adaptation could affect the diversification of viral isolates, only PVY sequences derived from potato were retrieved and included in the analysis of population genetic spatial structure.

2.2 Genetic diversity and population genetic differentiation

Nucleotide sequences of the P1 and CP genes were aligned using the MUSCLE algorithm (Edgar, 2004) implemented in MEGA5 (Tamura et al., 2011). A nucleotide identity matrix was generated using BioEdit software (Hall, 1999) after all gaps were removed. Haplotype diversity (H_d) and nucleotide diversity (π) were estimated using DnaSP 5.0 (Librado and Rozas, 2009). Pairwise F_{ST}, a measure of genetic differentiation among populations, was computed in Arlequin 3.5 (Excoffier and Lischer, 2010). Genetic differentiation among populations was also evaluated by K_{ST} and S_{nn} using DnaSP 5.10 (Hudson et al., 1992; Hudson, 2000; Librado and Rozas, 2009). The hypothesis of deviation from null population differentiation was tested by 1000 permutations of the original data.

2.3 Recombination and phylogeography analyses

Putative recombination joints (RJ) and parental sequences were identified using seven methods (RDP, GENECONV, BOOTSCAN, MAXCHI, CHIMAERA, SiSCAN, and 3SEQ) implemented in the RDP4 suites (Martin et al., 2010). The probability of a putative recombination event was corrected by a Bonferroni procedure with a cutoff of $p<0.01$. To avoid false identification, only events supported by at least four of the seven methods were considered to be recombinants. Recombinants were removed from the subsequent reconstruction of phylogenetic trees.

Phylogenetic trees were reconstructed by the Bayesian inference (BI) implemented in MrBayes 3.2.5 (Ronquist et al., 2012) and maximum likelihood (ML) implemented in MEGA5 (Tamura et al., 2011) using the $GTR + G + I$ nucleotide substitution model determined by MrModeltest (Nylander, 2008). BI was run in 2,000,000 generations of Markov chains that were sampled every 100 generations to establish convergence of all parameters. The effective sample size (ESS) of parameters was checked by Tracer 1.6 to ensure values above 200 as advised by the programmers with the first 25% of sampled trees burn-in. Topology robustness of ML trees was assessed by 1,000 bootstraps. ML-BPs and BI-PPs were plotted on Bayesian 50% majority-rule consensus trees using FigTree 1.4.2 and Illustrator CS5 (Adobe).

The effect of geographic origin on PVY populations was evaluated by phylogeny–trait association analysis, and BaTS 2.0 (Parker, Rambaut, and Pybus, 2008) was used to compute an association index (AI), parsimony score (PS), and maximum monophyletic clade (MC). Low AI and PS scores and high MC scores suggest a strong PVY–geography association. The topology of reconstructed trees was tested using BaTS by comparing it with the randomized trees generated from 10,000 reshufflings of tip characters. Topology robustness was determined in BaTS by comparing it with the null distribution of trees obtained from 10,000 bootstraps of tip characters.

2.4 Natural selection

HyPhy 2.10b (Kosakovsky Pond,et al., 2005) and PAML 4.7 (Yang, 2007) were used to identify nucleotide sites in P1 and CP cistrons that were likely to be involved in PVY adaptation. Three codon-based approaches, that is, IFEL (internal branches fixed-effects likelihood), REL (random-effects likelihood), and MEME (mixed effects model of evolution) (Kosakovsky Pond et al., 2006, 2011; Murrell et al., 2012), were included in the HyPhy package. Only sites simultaneously identified by IFEL and MEME with $p<0.05$ and >0.95 posterior probability identified by REL were considered to be under selection. In addition, the ratio of nonsynonymous (dN) to synonymous (dS) substitution (ω = dN/dS) was calculated for each gene using CODEML algorithm implemented in PAML 4.7 (Yang, 2007). Three different nested models (M3 *vs.* M0, M2a *vs.* M1a, and M8 *vs.* M7) were compared and likelihood-ratio tests (LRTs) were applied to select the one that best fitted the data. When the LRT was significant (p-value < 0.01), the Bayes empirical Bayes (BEB) method (Yang, et al., 2005) was used to identify amino acid residues that are likely to have evolved under positive selection based on a posterior probability threshold of 0.95.

2.5 Population historic dynamics

Two different approaches were used to explore the demographic history of PVY populations in the two countries. First, Tajima's D and Fu's F_S statistics implemented in Arlequin 3.5 (Excoffier and Lischer, 2010) were used to determine the neutrality of the P1 and CP genes by 1,000 permutations of the original sequences. Tajima's D test identifies evolutionary events such as population expansion, bottlenecks, and selection by comparing the estimated number of segregating sites with the mean pairwise difference among sequences (Tajima, 1989). Fu's F_S is sensitive to population demographic expansion and usually displays a negative value (Fu, 1997) in expanding populations. Following these calculations, Bayesian skyline plots (BSP) was generated to explore demographical history using BEAST 1.8.2 (Drummond et al., 2012). Sampling times of the sequences were used to calibrate the molecular clock. Date-randomization tests (DRTs) were performed in R 3.3.1 using the Tip Dating Beast package (Rieux and Khatchikian, 2016) to determine the temporal signal in data sets. A data set was considered to have an adequate spread in sampling time if its average rate did not fall within the 95% confidence intervals (CIs) generated from 20 replicates of randomized dates (Ramsden et al., 2009; Duchêne et al., 2015). A Bayes factors test indicated that the relaxed uncorrelated exponential model was a better fit to the sequence data than the relaxed uncorrelated lognormal model and was chosen to estimate the molecular clock of P1 and CP genes. The MCMC was run for 5×10^8 generations to ensure convergence of all parameters. Convergence and ESS (>200) of the parameters were checked using Tracer 1.6.

3 Results

3.1 Sequence variation in P1 and CP genes

Seventy-three P1 and 78 CP genes were sequenced in this study and were deposited in GenBank (Table S1). Two P1 sequences had a nucleotide T insertion at position 33 (accession number: KF722821) and a nucleotide (A) insertion at position 814 (accession number: KF771018), respectively. Moreover, two P1 sequences (accession numbers: KX451346 and KX451344) contained premature termination codons (PTCs) at position 208–210 and 619–621. A single CP sequence (accession number: KC296822) had an 11-nucleotide deletion at the beginning of the gene. These nonfunctional sequences were removed from further analyses of the population genetic structure. In addition, 34 complete P1 and CP sequences originated from Japan were retrieved from GenBank (Table S1). Consequently, a total of 127 complete P1 and 137 complete CP sequences were included in the population genetic analysis of the virus.

The average nucleotide identities in the P1 and CP genes among the viral isolates from China were 89.9% (72.3%—100%) and 96.3% (87.0 %—100%), while those from Japan were 89.1% (71.5%—100%) and 95.7 % (88.7%—100%), respectively. All typical motifs of potyviruses were detected in the deduced amino acid sequences. However, the conserved N25 of CP protein in the reference sequences (O-139, N-605, C1-SON41, Adgen, and Chile3) was replaced by S (isolates NTNHIR3 and T13) or T (isolate ONGOB6) in the Japanese isolates and the E68 residue was replaced by G (isolates XQ03) or K (isolate Laiwu3) in Chinese isolates (Fig. S1).

One hundred and seventeen haplotypes were identified in the 127 complete sequences of P1 gene with an overall haplotype diversity of 0.998 and nucleotide diversity of 0.119 when the sequences from China and Japan were considered together (Table 1). The most common haplotype was detected four times in P1 sequences and 10 times in CP sequences. No identical haplotypes were detected in the two countries in both genes. When the sequences were considered according to individual geographic location, higher nucleotide but lower haplotype diversity was found in Japan than in China. In the CP genes, 121 haplotypes were identified in the 137 complete sequences, with an overall haplotype diversity of 0.984 and nucleotide diversity of 0.053 when sequences from the two countries were pooled. Similar to the P1 gene, higher nucleotide diversity but lower haplotype diversity was found in the CP gene from Japan than from China when they were considered separately.

3.2 Genetic differentiation between sequences from China and Japan

K_{ST}, S_{nn}, and F_{ST} were 0.141, 0.972, and 0.294 in the P1 gene and 0.289, 0.949, and 0.520 in the CP gene (Table 2), respectively. All of these indexes were significantly higher than the theoretical expectation of no population differentiation. K_{ST} and F_{ST} values were higher in the CP gene than in the P1 gene.

Table 1 Sample sizes and genetic variation of the P1 and CP genes in the *Potato virus Y* populations sampled from China and Japan

Gene	Country	Sample size	Haplotypes	Haplotype diversity	Nucleotide diversity
P1	China	93	90	0.999	0.100
	Japan	34	27	0.980	0.109
	All	127	117	0.998	0.119
CP	China	103	99	0.999	0.036
	Japan	34	22	0.914	0.043
	All	137	121	0.994	0.053

Table 2 Statistical tests for population differentiation between P1 and CP genes in the *Potato virus Y* populations sampled from China and Japan

Gene	K_{ST}	*P-value*	S_{nn}	*P-value*	F_{ST}	*P-value*
P1	0.141	0.000***	0.972	0.000***	0.294	0.000***
CP	0.289	0.000***	0.949	0.000***	0.520	0.000***
Concatenated	0.216	0.000***	0.977	0.000***	0.390	0.000***

* 0.01 < P < 0.05.; **0.001 < p < 0.01.; ***P < 0.001.

3.3 Recombination analyses

Forty-two P1 sequences, all originating from China, were identified with a high level of confidence as recombinants by all seven methods implemented in the RDP package (p-values ≤ 1.27×10^{-2}, Table S2). All of these sequences were derived from three recombination events. Similarly, three recombination breakpoints were detected in eight CP sequences from China and four CP sequences from Japan by at least six of the seven approaches (p-values ≤ 1.69×10^{-2}, Table S2).

3.4 Phylogenetic analysis

After removal of putative recombinants, 85 complete P1 sequences and 125 complete CP sequences were included in the phylogenetic analysis. Both P1 and CP sequences fell into two distinct clades (N clade and O clade) with high posterior probabilities (PPs≥0.99) and maximum likelihood (ML) bootstrap (BPs≥70%) supports (Fig. 2). In the N clade of P1 tree, all 41 Chinese isolates were clustered into subclade 1 and all but three (NTNHIR3, NTNKGAM2, and NTNTK1) of the 27 Japanese sequences were grouped into subclade 2. Similarly, all 10 Chinese isolates were clustered into subclade 3 and all seven Japanese isolates were grouped into subclade 4. In the N clade of the CP tree, all nine Chinese sequences were clustered into subclade 1, whereas all 23 Japanese sequences were grouped into subclade 2. In contrast, O clade was composed of a mixture of Chinese and Japanese sequences. A similar pattern was found when only the P1 and CP sequences from GenBank (i.e., excluding the sequences generated in this study) were used to construct a phylogenetic tree (data not shown). Distinct differences in the population genetic structure of PVY between China and Japan were also found by phylogeny–geography association analysis. Significant MC ($p < 0.01$), AI ($p < 0.001$), and PS ($p < 0.001$) were detected in both P1 and CP genes when sequences from the two countries were compared (Table 3). However, in general, a phylogeny–geography association was not found within countries (MC: $p > 0.05$; Table 3) with the exception of PVY isolates from Fujian and Heibei in China and Kyushu and Honshu in Japan.

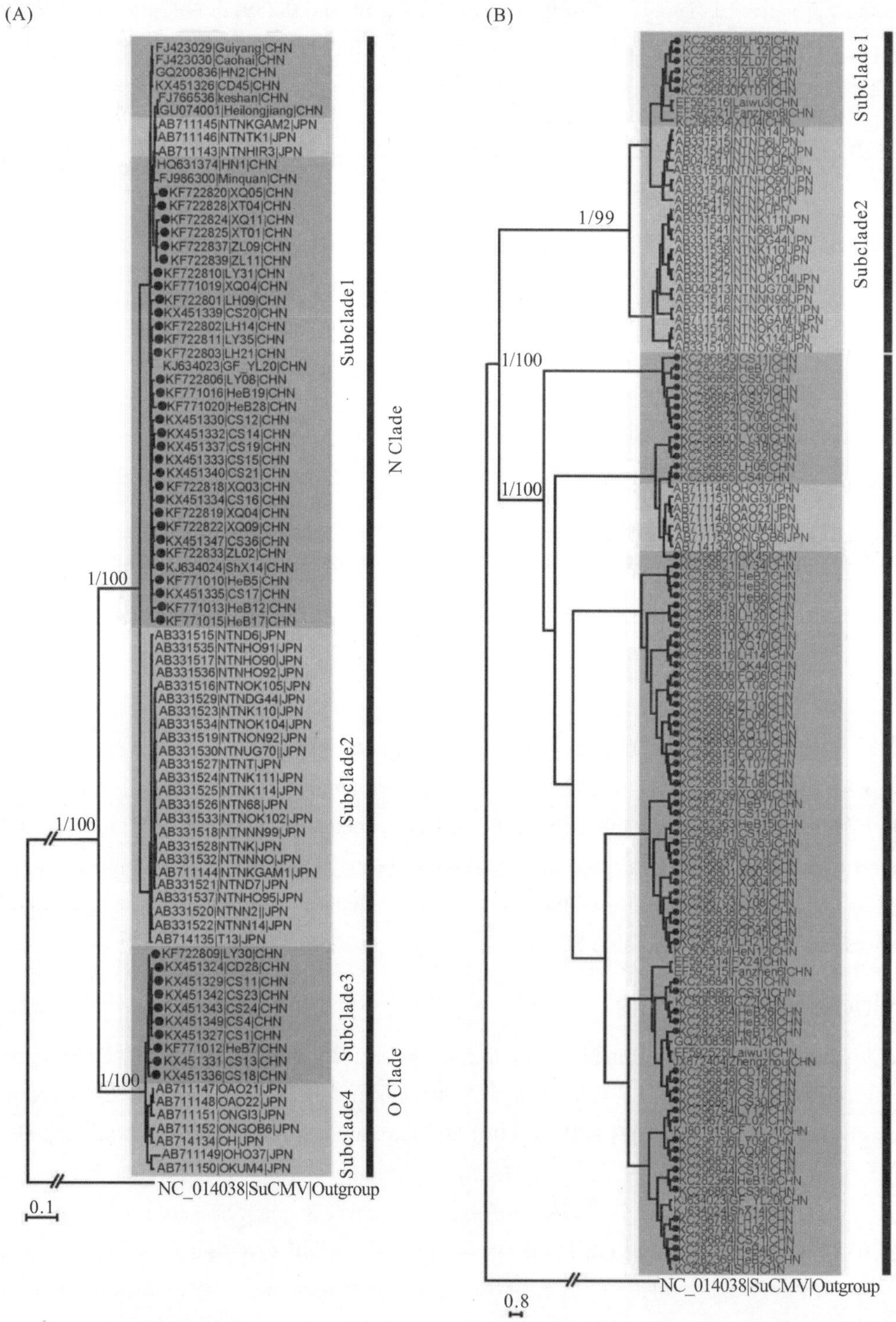

Fig. 1 Bayesian phylogenetic trees of *Potato virus Y* (PVY) isolates based on P1 (A) and CP (B) coding regions

For three key nodes in a tree, the Bayesian posterior probabilities and maximum likelihood bootstrap percentage are indicated above the branches (Bayesian posterior/bootstrap). PVY isolates from this study are indicated by black dots. The distance unit is substitutions/site.

Table 3 Tests of geography–phylogeny association for P1 and CP genes in PVY isolates originated from China and Japan

Gene	Clade	Statistic	Isolates	Observed Mean (95% HPD)	Null Mean (95% HPD)	Significance
P1		*AI*		0.06 (0.02, 0.19)	6.29 (5.51, 6.97)	< 0.001***
		PS		4.15 (4.00, 5.00)	37.63 (35.15, 39.74)	< 0.001***
	N	*MC* (China)	41	30.71 (28.00, 31.00)	3.38 (2.55, 4.41)	0.010**
		MC (Japan)	27	23.42 (17.00, 24.00)	3.26 (2.43, 4.62)	0.010**
	O	*MC* (China)	10	9.99 (9.98, 10.00)	1.31 (1.01, 2.01)	0.010**
		MC (Japan)	7	5.39 (3.00, 7.00)	1.18 (1.00, 1.99)	0.010**
CP		*AI*		0.27 (0.02, 0.52)	5.56 (4.80, 6.22)	< 0.001***
		PS		5.81 (5.00, 6.00)	33.34 (30.66, 35.45)	< 0.001***
	N	*MC* (China)	9	9.00 (9.00, 9.00)	1.18 (1.00, 1.81)	0.010**
		MC (Japan)	23	22.83 (22.80, 23.00)	1.80 (1.37, 2.19)	0.010**
	O	*MC* (China)	86	74.29 (55.00, 82.00)	6.31 (4.86, 8.49)	0.010**
		MC (Japan)	7	3.76 (3.00, 6.00)	1.14 (1.00, 1.69)	0.010**

AI, association index; PS, parsimony score; MC, maximum monophyletic clade; HPD, highest probability density interval; n/a: no data available because of insufficient sample size ($n< 2$).Significance thresholds: *$0.01 < p< 0.05$; ** $0.001 < p< 0.01$; ***$p< 0.001$.

3.5 Detecting natural selection

Purifying selection was detected in the majority of polymorphic sites by PAML packages (Fig. 2). Evidence of positive selection was detected in the 1st codon of CP sequences in the O clade using the PAML approach (PP > 0.99, Table 4). No positive selection was detected in the N clade. In P1 sequences, no positive selection was detected in either the N or O clade by PAML (Table 4). However, positive selection in the P1 gene was detected in the 3rd and 5th codons of the N clade and the 3rd, 78th, and 241st codons of the O clade by the REL approach implemented in the HyPhy package.

The REL approach also detected positive selection in the 1st, 9th, 11th, 15th, and 58th codons of CP sequences in the O clade and in the 1st codon of the N clade. Similarly, positive selection was found in the 1st codon both in the O and N clades of CP sequences by the MEME approach ($p< 0.05$; Table 4). Positive selection was also found in the 1st and 138th codons of CP sequences in the O clade by IFEL (<0.05; Table 4). However, only the 1st codon in the N-terminal region of the CP protein, one of the cleavage sites in the PVY genome, was found to be under positive selection by both PAML and three approaches implemented in the HyPhy package. Further analysis showed that Glu (87.50%) dominated at the NIb/CP cleavage site in the N clade of the CP protein, whereas Ala (40.86%) and Glu (56.99%) were found to have a high frequency at the cleavage site in the O clade of the CP protein.

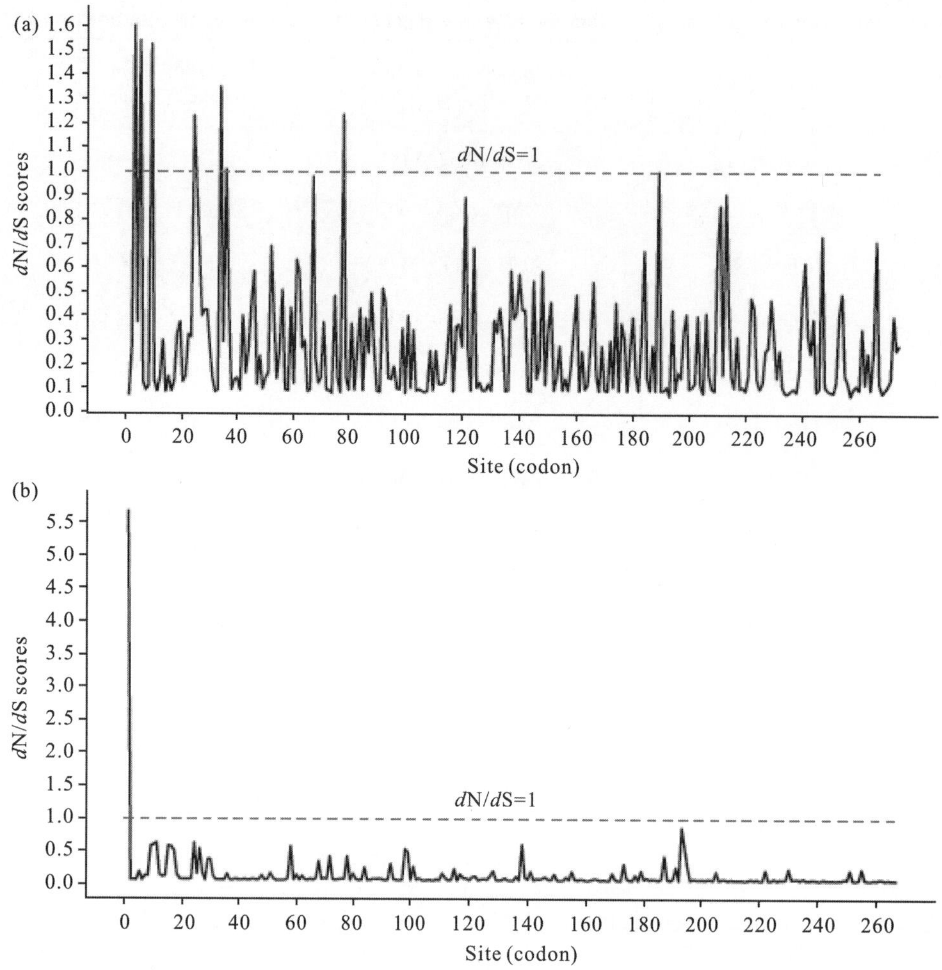

Fig. 2 Sliding window plot of *d*N/*d*S ratios for P1 (a) and CP (b) genes

Sites under neutral (*d*N/*d*S = 1) are indicated in red dotted line. The window size is 20 codons, and the offset between windows is one codon

3.6 Demographic history

Significantly negative Fu's F_S was detected in both P1 and CP sequences from China (Table 5); Tajima's D was positive for the P1 sequences and negative for the CP sequences from China, but none of them were significant. All Tajima's D and Fu's F_S in the sequences from Japan were positive but not significant (Table 5). All data included in the demographic analysis passed the DRTs that showed no overlaps between the original estimate of evolutionary rate and 95% CIs generated from 20 replicates of date randomization (Fig. S1). Coalescence-based BSP revealed an explicit demographic history for the PVY populations of China and Japan (Fig. 3)—showing that the PVY from China experienced a population expansion prior to a period of stability, whereas the population from Japan remained relatively constant throughout the past decades.

Table 4 Putative codons in the P1 and CP sequences detected under positive selection using the PAML package and three approaches implemented in the HyPhy Software

Gene	Clade	PAML		HyPhy					
		Model M8[a]		IFEL		REL		MEME	
		Codon	PP[b]	Codon	*P* value	Codon	PP[b]	Codon	*P* value
P1	N Clade					3	0.999*		
						5	1.000**		
	O Clade					3	0.970*		
						78	0.984*		
						241	0.972*		
CP	N Clade					1	0.994**	1	0.029*
	O Clade	1	0.999*	1	0.023*	1	0.991**	1	0.007**
				138	0.042*	9	0.981*		
						11	0.959*		
						15	0.977*		
						58	0.982*		

A For each PVY clade, model M8 performed better than model M7 in an LRT. Model M2a also performed better than models M0, M1a, or M3 in LRTs and provided results similar to M8.b PP, Posterior probability that individual codon positions belong to the positively selected category using the Bayes Empirical Bayes (BEB) method implemented in PAML and the random-effects likelihood (REL) approach implemented in the HyPhy, respectively.*Posterior probability >95% or $p< 0.05$.**Posterior probability >99 % or $0.001 <p< 0.01$

Table 5 Neutrality tests for P1 and CP sequences of PVY originated from China and Japan by Tajima's D and Fu's F_S

Population	P1		CP	
	Tajima's *D*	Fu's F_S	Tajima's *D*	Fu's F_S
China	0.649[ns]	-10.605**	-1.285[ns]	-23.966***
Japan	0.771[ns]	2.050[ns]	0.702[ns]	2.387[ns]

*$0.01 < p < 0.05$.; **$0.001 < p < 0.01$.; ***$p < 0.001$.

4 Discussion

Consistent with previous results (Ogawa et al., 2008, 2012; Tian et al., 2011), high genetic diversities (Table 1) were found in both P1 and CP genes in the current study. Each of these two genes encodes a protein with ecological functions that are important to the survival and adaptation of PVY in local biotic and abiotic environments (Quenouille et al., 2013). In addition to the high mutation rate in RNA viruses (Domingo et al., 2012), high genetic variation found in both current and previous studies is possibly attributed to the high recombination rate in PVY populations (Gibbs and Ohshima, 2010). In recent decades, recombinant PVYs were found to be prevalent in the global population of the pathogen sampled from potato production areas (Quenouille et al., 2013). In this study, 52.69% and 11.76% of concatenated sequences in the Chinese and Japanese populations show a mixed genomic structure of PVY^N and PVY^O (Table S2). These results re-enforce widespread concern related to the propensity for rapid adaptation of PVY to changing biotic and abiotic environments. Such adaptation may cause problems for the deployment of major gene-mediated host resistance in potatoes, tomato, and other crops (Karasev and Gray, 2013).

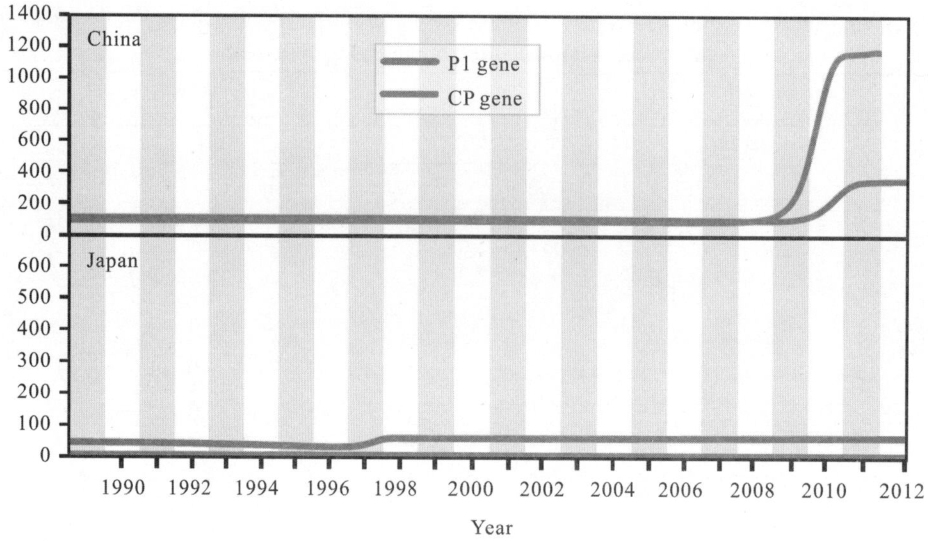

Fig. 3 Population dynamics of genetic diversity in *Potato virus Y* (PVY)

Bayesian skyline plots of the P1 and CP segments for PVY in China (top) and Japan (bottom). The y-axes represent a measure of relative genetic diversity, and x- axis is measured in calendar years

Our study also reveals a significant difference in the molecular population genetic structure between the PVY sequences originating from potato hosts in China and Japan—a pattern that is supported by comparative analyses of genetic variation (Table 1), population differentiation (Table 2), and phylogeny–geography association (Fig. 1; Table 3) in the two genes (each represented by ~130 sequences). It has been documented that measurements of the diversity of genetic variation are highly sensitive to sample sizes (Bashalkhanov et al., 2009) and populations with larger sample sizes tend to have higher observed genetic diversity. Despite smaller sample sizes included in the Japanese than Chinese populations, both P1 and CP sequences originating from Japan show higher nucleotide diversity than those from China, suggesting that differences in nucleotide diversity between the two viral populations are unlikely to be the result of sampling error. However, the higher haplotype diversity in Chinese sequences may be associated with the somewhat larger sample size as compared to Japanese sequences (93 vs. 34 isolates). Indeed, when we used a bootstrapping approach described previously (Zhan et al., 2003) to standardize sample sizes with the smaller population (Japan in this case) and recalculated haplotype diversity, the difference in genetic variation between the two populations disappeared (Table S3). The bootstrap was conducted using the Resampling Stats add-in package for Excel (Blank et al., 2001) with 100 replicates. For each bootstrap replication, haplotype diversity was recorded from a random sample of 34 sequences (the actual size of Japanese population). The mean and variance of the diversity were calculated and used for a t-test.

Three indices were used to test population differentiation between Chinese and Japanese PVY sequences. K_{ST} and F_{ST} measure the relative proportion of total genetic diversity attributable to among-population differences. They range from 0.00 to 1.00. A value of one for K_{ST} or F_{ST} indicates that populations investigated are completely isolated, while a value of zero indicates the populations studied are identical (Hudson et al., 1992). K_{ST} or F_{ST} values between 0.15 and 0.25 indicate high population differentiation, and values greater than 0.25 indicate very high genetic differentiation among populations (Balloux and Lugon-Moulin, 2002). S_{nn} measures how often the nearest sequences are found in the same populations. Its value would be close to one when populations are highly structured and near 0.5 when populations are identical (Hudson, 2000). Therefore, K_{ST} and F_{ST} are approximately equal to S_{nn}—0.5. All three indices are higher than 0.15 in the current study (Table 2), suggesting a high differentiation of PVY sequences between the two populations. K_{ST} is the most powerful index for detecting spatial population structure for recombination and high mutation populations (Hudson et al., 1992), which is the case in PVY (Blanchard et al., 2008).

Phylogenetic analyses provide additional support for differentiation between the Chinese and Japanese populations of PVY (Fig. 1). Sequences tend to cluster according to their geographic origin, consistent with previous results derived from more and wider geographic locations (Cuevas et al., 2012). Although some clades such as the O clade of the CP tree are composed of sequences from both locations, the hypothesis of random association between PVY sequences from China and Japan and their geographic origins is rejected by all three statistics (AS, MC and PS) (Table 3). Phylogenetic trees are reconstructed under the assumption of constant evolutionary rates and many evolutionary processes may affect the correct reconstruction of phylogenetic topology (Revell et al., 2008). The polyphyletic nature of these different clades could be generated by some sequences experiencing a complex evolutionary history such as recombination. For example, in the P1 tree, three Japanese sequences (NTNHIR3, NTNTK1, and NTNKGAM2) were clustered together with Chinese sequences in a subclade (Fig. 1). It is likely they are recombinants that were not detected by the techniques used here. Our analyses suggest that the CP sequences of these three isolates are also most likely derived from recombination (Table S2).

Both stochastic and deterministic events could lead to the observed pattern of spatial structure (Caruso et al., 2011). Stochastic events caused by genetic drift may lead to nonadaptive population differentiation between the two countries. In nature, the extent of population differentiation generated by random genetic drift is expected to increase over geographic distance. Although geographic distance between some locations (ex. 2,771.10 km between Guizhou and Heilongjiang) within China is greater than that between China and Japan (ex. 819.41km between Fujian in China and Okinawa in Japan), PVY sequences were clustered according to country origin (Fig. 1; Table 3). Because our data (non-neutrality) are not appropriate fortesting the stochastic event, the contribution of genetic drift could not be ruled out. However, we believe that deterministic events are the main factor responsible for the observed pattern of spatial population genetic structure in the pathogen.

The hypothesized contribution of deterministic events to the observed pattern of spatial population genetic structure is supported by neutrality tests in the deduced amino acid sequences and in the sequence–geography association analysis. The finding that most codons in the P1 and CP sequences were under purifying selection suggests that most mutations in the PVY genomes are harmful and consequently eliminated by natural selection. In this case, the selective agents may be habitat differences between the two countries such as differences in the potato cultivars grown and climatic conditions. The main potato cultivars grown in Japan include Irish Cobbler, May Queen, Kitaakari, Touya, Dejima, Nishiyutaka, Toyoshiro, Konafubuki, Northern Ruby, and Shadow Queen (Kawakami et al., 2015). None of these cultivars are used in China. In China, potatoes are grown over a much wider geographic area, ranging from a subtropical climatic zone in the south (e.g., Fujian) to a temperate continental climate zone in Northern China (e.g., Heilongjiang) using more diverse cultivars (Jansky et al., 2009), while in Japan, potatoes are mainly grown in the north (Kawakami et al., 2015).

Interestingly, the 1st codon translated to the cleavage site in the CP protein was detected to be under positive selection by both PAML and HyPhy with high confidence levels ($PP > 0.99$ or $p< 0.05$, Table 4) and the signal of positive selection increased when more sequences were included in the analysis (data not shown). Positive selection in cleavage sites has also been found in other viruses such as the human immunodeficiency virus (HIV, Banke et al., 2009). Successful cleavages to form functional cis-elements are crucial for survival and reproduction of PVY (Tena Fernandez et al., 2013). This process is catalyzed by proteases that are under constant change by mutation (Yu et al., 1995). Positive selection in the CP cleavage site could serve as a reversal mechanism compensatory to mutations in the protease in PVY and other viruses (Banke et al., 2009).

Differences in spatial structure between the Chinese and Japanese sequences may also result from differences in the demographic dynamics of the PVY populations. Sudden increases or decreases in population size associated with demographic events can affect the generation, maintenance, and distribution of genetic variation, not only directly through genetic drift and mutation, but also indirectly through impacts on the efficiency of natural selection to remove or amplify mutations as well as on migration and recombination (Wang and Whitlock, 2003). Indeed, when we performed demographic analyses, we found that PVY populations in China were small but had undergone recent expansion, possibly associated with increasing potato production in the current years (Wang et al., 2011), while in Japan, they were large but stably maintained (Fig. 3). This resultis consistent with the potato cultivation of the two countries in the past decades. Potato acreage in Japan maintained stable overthe last several decades (Kawakami et al., 2015) while increased in China from ~2.3 million hectares in 1980s to ~5.4 million hectares recently (http:// faostat.fao.org). The temporal scale of samples can impact the estimate of demographic dynamics (Drummond, Pybus, and Rambaut, 2003). In our study, the temporal scale (1995–2012) in the PVY sequences from Japan was 10 years longer than that (2005–2012) from China. However, we do not believe that this difference would have affected our conclusions because all data passed the DRTs (Ramsden et al., 2009; Duchêne et al., 2015) with a high level of confidence.

In summary, our study represents one of a few attempts to understand patterns and causes of spatial population genetic structure across political borders in PVY, a destructive pathogen of potato and many other *Solanaceous* crops. The finding of two subpopulations indicates distinct gene pools exist in PVY from China and Japan. This suggests that strict quarantine regulation is needed to prevent the movement of novel alleles or allelic combination of PVY between the two countries when trading plant materials that are hosts for this pathogen (e.g., potato and tobacco). However, sequences included in this analysis were relatively limited both in sample sizes and sites. In particular, the temporal scales might be relatively short for analysis of demographic events. Further study with larger, multiple-location, and longer temporal scale samples may be required to confirm the results and generalize the findings.

Acknowledgements

This work was supported by the China Agriculture Research System (Grant No. CARS-10-P11), P. R. China, granted to Jiasui Zhan. We thank Drs. Jeremy J. Burdon and Peter Thrall in CSIRO, Australia, for proof-reading the manuscript and Dr. Zhenguo Du in FAFU and Dr. Lin Zhang in NNU for comments and suggestions to the manuscript.

References

[1] ACHTMAN M, WAGNER M. Microbial diversity and the genetic nature of microbial species[J]. Nature Reviews Microbiology, 2008, 6: 431-440.

[2] BALLOUX F, LUGON-MOULIN N. The estimation of population differentiation with microsatellite markers[J]. Molecular Ecology, 2010, 11: 155-165.

[3] BANKE S, LILLEMARK M R, GERSTOFT J, et al. Positive selection pressure introduces secondary mutations at Gag cleavage sites in Human immunodeficiency virus Type 1 harboring major protease resistance mutations[J]. Journal of Virology, 2009, 83: 8916-8924.

[4] BARRETT L G, THRALL P H, BURDON J J, et al. Life history determines genetic structure and evolutionary potential of host-parasite interactions[J]. Trends in Ecology & Evolution, 2008, 23: 678-685.

[5] BASHALKHANOV S, PANDEY M, RAJORA O P. A simple method for estimating genetic diversity in large populations from finite sample sizes[J]. BMC Genetics, 2009, 10: 1-10.

[6] BERGHOLZ P W, NOAR J D, BUCKLEY D H. Environmental patterns are imposed on the population structure of *Escherichia coli* after fecal deposition[J]. Applied and Environmental Microbiology, 2011, 77: 211-219.

[7] BIRCH P R J, BRYAN G, FENTON B, et al. Crops that feed the world 8: Potato: are the trends of increased global production sustainable?[J]. Food Security, 2012, 4: 477-508.

[8] BLANCHARD A, ROLLAND M, LACROIX C, et al. *Potato virus Y*: A century of evolution. Current Topics in Virology, 2008, 7: 21-32.

[9] BLANK S, SEITER C, BRUCE P. Resampling stats in excel version 4. Arlington VA. 2001.

[10] BURDON J J, THRALL P H. Pathogen evolution across the agroecological interface: Implications for disease management[J]. Evolutionary Applications, 2010, 1: 57-65.

[11] CARUSO T, CHAN Y, LACAP D C, et al. Stochastic and deterministic processes interact in the assembly of desert microbial communities on a global scale[J]. Isme Journal, 2011, 5, 1406-1413.

[12] CHUNG B Y, MILLER W A, ATKINS J F, et al. An overlapping essential gene in the *Potyviridae*[J]. Proceedings of the National Academy of Sciences of the United States of America, 2008, 105: 5897-5902.

[13] CUEVAS J, DELAUNAY A, RUPAR M, et al. Molecular evolution and phylogeography of *Potato virus Y* based on the CP gene[J]. Journal of General Virology, 2012, 93: 2496-2501.

[14] DOMINGO E, SHELDON J, PERALES C. Viral quasispecies evolution[J]. Microbiology & Molecular Biology Reviews Mmbr, 2012, 76: 159-216.

[15] DRUMMOND A J, PYBUS O G, RAMBAUT A. Inference of viral evolutionary rates from molecular sequences[J]. Advances in Parasitology, 2003, 54: 331-488.

[16] DRUMMOND A J, SUCHARD M A, XIE D, et al. Bayesian phylogenetics with BEAUti and the BEAST. 1.7[J]. Molecular Biology & Evolution, 2012, 29: 1969-1973.

[17] DUCHÊNE S, DUCHÊNE D, HOLMES E C, et al. The performance of the date-randomization test in phylogenetic analyses of time-structured virus data[J]. Molecular Biology and Evolution, 2015, 32: 1895-1906.

[18] EDGAR R C. MUSCLE: Multiple sequence alignment with high accuracy and high throughput[J]. Nucleic Acids Research, 2004, 32: 1792-1797.

[19] EXCOFFIER L, LISCHER H E. Arlequin suite ver 3.5: A new series of programs to perform population genetics analyses under Linux and Windows[J]. Molecular Ecology Resources, 2010, 10: 564-567.

[20] FU Y X. Statistical tests of neutrality of mutations against population growth, hitchhiking and background selection[J]. Genetics, 1997, 147: 915-925.

[21] GAO F, CHANG F, SHEN J, et al. Complete genome analysis of a novel recombinant isolate of *Potato virus Y* from China[J]. Archives of virology, 2014, 159: 3439-3442.

[22] GIBBS A, OHSHIMA K. Potyviruses and the digital revolution[J]. Annual Review of Phytopathology, 2010, 48: 205-223.

[23] GILBERT G S. Evolutionary ecology of plant disease in natural ecosystems[J]. Annual Review of Phytopathology, 2002, 40: 13-43.

[24] GURUNG S, SHORT D P, ADHIKARI T B. Global population structure and migration patterns suggest significant population differentiation among isolates of Pyrenophoratritici-repentis[J]. Fungal Genetics & Biology, 2013, 52: 32-41.

[25] HALL T A.BioEdit: A user-friendly biological sequence alignment editor and analysis program for Windows 95/98/NT[J]. Nuclc Acids Symposium Series, 1999, 41: 95-98.

[26] HE D C, ZHAN J, CHENG Z B, et al. Viruliferous rate of small brown planthopper is a good indicator of rice stripe disease epidemics[J]. Scientific reports,2016, 6: 21376.

[27] HUDSON R R. A new statistic for detecting genetic differentiation[J]. Genetics, 2000, 155: 2011-2014.

[28] HUDSON R R, BOOS D D, KAPLAN N L. A statistical test for detecting geographic subdivision[J]. Molecular Biology & Evolution, 1992, 9: 138-151.

[29] JANSKY S H, JIN L P, XIE K Y, et al. Potato production and breeding in China[J]. Potato Research, 2009, 52: 57-65.

[30] Karasev A V, Gray S M. Continuous and emerging challenges of *Potato virus Y* in potato[J]. Annual Review of Phytopathology, 2013, 51: 571-586.

[31] KAWAKAMI T, OOHORI H, TAJIMA K. Seed potato production system in Japan, starting from foundation seed of potato[J]. Breeding Science, 2015, 65: 17-25.

[32] KEARNEY J. Food consumption trends and drivers[J]. Philosophical Transactions of The Royal Society B Biological Sciences, 2010, 365: 2793-2807.

[33] KING A M Q, LEFKOWITZ E, ADAMS M J, et al. Virus Taxonomy Ninth Report of the International Committee on Taxonomy of Viruses[J]. Amsterdam: Elsevier Academic Press. 2011.

[34] KOSAKOVSKY POND S L, FROST S DW, GROSSMAN Z, et al. Adaptation to different human populations by HIV- 1 revealed by codon- based analyses[J]. PLoS Computational Biology, 2006, 2: e62.

[35] KOSAKOVSKY POND S L, FROST S DW, MUSE S V. HyPhy: Hypothesis testing using phylogenies[J]. Bioinformatics, 2005, 21: 676-679.

[36] KOSAKOVSKY POND S L, MURRELL B, FOURMENT M, et al. A random effects branchsite model for detecting episodic diversifying selection[J]. Molecular Biology & Evolution, 2011, 28: 3033-3043.

[37] KOSKELLA B, VOS M. Adaptation in natural microbial populations[J]. Annual Review of Ecology, Evolution, and Systematics, 2015, 46: 503-522.

[38] LIBRADO P, ROZAS J. DnaSPv5: A software for comprehensive analysis of DNA polymorphism data[J]. Bioinformatics, 2009, 25: 1451-1452.

[39] MARTIN D P, LEMEY P, LOTT M, et al. RDP3: A flexible and fast computer program for analyzing recombination[J]. Bioinformatics, 2010, 26: 2462-2463.

[40] MCDONALD B A, LINDE C. Pathogen population genetics, evolutionary potential, and durable resistance[J]. Euphytica, 2002, 40: 349-379.

[41] MURRELL B, WERTHEIM J O, MOOLA S, et al. Detecting individual sites subject to episodic diversifying selection[J]. PLOS Genetics, 2012, 8: e1002764.

[42] NOLTE P, WHITWORTH J L, THORNTON M K, et al. Effect of seedborne *Potato virus Y* on performance of Russet Burbank Russet Norkotah, and Shepody potato[J]. Plant disease, 2004, 88: 248-252.

[43] NYLANDER J A A. MrModeltest v2.3. Program distributed by the author. http://www.abc.se/nylander/mrmodeltest2/mrmodeltest2.html, 2004.

[44] OGAWA T, NAKAGAWA A, HATAYA T, et al. The genetic structure of populations of *Potato virus Y* in Japan; based on the analysis of 20 full genomic sequences[J]. Journal of Phytopathology, 2012, 160: 661-673.

[45] OGAWA T, TOMITAKA Y, NAKAGAWA A, et al. Genetic structure of a population of *Potato virus Y* inducing potato tuber necrotic ringspot disease in Japan; comparison with North American and European populations[J]. Virus Research, 2008, 131: 199-212.

[46] PARKER J, RAMBAUT A, PYBUS O G. Correlating viral phenotypes with phylogeny: Accounting for phylogenetic uncertainty[J]. Infection Genetics & Evolution, 2008, 8: 239-246.

[47] QUENOUILLE J, VASSILAKOS N, MOURY B. *Potato virus Y*: A major crop pathogen that has provided major insights into the evolution of viral pathogenicity[J]. Molecular Plant Pathology, 2013, 14: 439-452.

[48] RAMSDEN C, HOLMES E C, CHARLESTON M A. Hantavirus evolution in relation to its rodent and insectivore hosts: No evidence for codivergence[J]. Molecular Biology & Evolution, 2009, 26: 143-153.

[49] REVELL L J, HARMON L J, COLLAR D C. Phylogenetic signal, evolutionary process, and rate[J]. Systematic Biology, 2008, 57: 591-601.

[50] RIEUX A, KHATCHIKIAN C E. TipDatingBeast: An R package to assist the implementation of phylogenetic tip- dating tests usingbeast[J]. Molecular Ecology Resources, 2017.

[51] RONQUIST F, TESLENKO M, VAN DER MARK P, et al. MrBayes 3.2: Efficient Bayesian phylogenetic inference and model choice across a large model space[J]. Systematic Biology, 2012, 61: 539-542.

[52] SHORT D P, GURUNG S, GLADIEUX P, et al.Globally invading populations of the fungal plant pathogen Verticilliumdahliae are dominated by multiple divergent lineages[J]. Environ Microbiol, 2015, 17: 2824-2840.

[53] TAJIMA F. Statistical method for testing the neutral mutation hypothesis by DNA polymorphism[J]. Genetics, 1989, 123: 585-595.

[54] TAMURA K, PETERSON D, PETERSON N, et al. MEGA5: Molecular evolutionary genetics analysis using maximum likelihood, evolutionary distance, and maximum parsimony methods[J]. Molecular Biology & Evolution, 2011, 28: 2731-2739.

[55] TENA FERNANDEZ F, GONZALEZ I, DOBLAS P, et al. The influence of cis-acting P1 protein and translational elements on the expression of *Potato virus Y* helper- component proteinase HCPro in heterologous systems and its suppression of silencing activity[J]. Molecular Plant Pathology, 2013, 14: 530-541.

[56] THRALL P H, OAKESHOTT J G, FITT G, et al. Evolution in agriculture: The application of evolutionary approaches to the management of biotic interactions in agro- ecosystems[J]. Evolutionary applications,2011, 4: 200-215.

[57] TIAN Y P, LIU J L, ZHANG C L, et al. Genetic diversity of *Potato virus Y* infecting tobacco crops in China[J]. Phytopathology,2011, 101: 377-387.

[58] TIAN Y E, YIN J L, SUN J P, et al. Population genetic analysis of Phytophthorainfestans in northwestern China[J]. Plant Pathology,2016, 65: 17-25.

[59] TREDWAY L P, STEVENSON K L, BURPEE L L. Genetic structure of Magnaporthegrisea populations associated with St. Augustinegrass and tall fescue in Georgia[J]. Phytopathology,2005, 95: 463-471.

[60] VALLI A, LÓPEZ-MOYA J J, GARCÍA J A. Recombination and gene duplication in the evolutionary diversification of P1 proteins in the family Potyviridae[J]. Journal of General Virology,2007, 88: 1016-1028.

[61] WANG B, MA Y, ZHANG Z, et al. Potato viruses in China[J]. Crop Protection, 2011, 30: 1117-1123.

[62] WANG J, WHITLOCK M C. Estimating effective population size and migration rates from genetic samples over space and time[J]. Genetics, 2003, 163: 429-446.

[63] WRIGHT S. Size of population and breeding structure in relation to evolution[J]. science, 1938, 87, 425-431.

[64] YANG Z. PAML. 4: Phylogenetic analysis by maximum likelihood[J]. Molecular Biology and Evolution, 2007, 24: 1586-1591.

[65] YANG Z, WONG W S, NIELSEN R. Bayes empirical Bayes inference of amino acid sites under positive selection[J]. Molecular Biology & Evolution, 2005, 22: 1107-1118.

[66] YANG L N, ZHU W, WU E J, et al. Trade-offs and evolution of thermal adaptation in the Irish potato famine pathogen Phytophthorainfestans[J]. Molecular Ecology, 2016, 25: 4047-4058.

[67] YU S F, BENTON P, BOVEE M, et al. Defective RNA replication by poliovirus mutants deficient in 2A protease cleavage activity[J]. Journal of Virology, 1995, 69: 247-252.

[68] ZHAN J, MCDONALD B A. The interaction among evolutionary forces in the pathogenic fungus Mycosphaerellagraminicola[J]. Fungal Genetics & Biology, 2004, 41: 590-599.

[69] ZHAN J S, MCDONALD B A. Experimental measures of pathogen competition and relative fitness[J]. Annual Review of

Phytopathology, 2013, 51:131-153.

[70] ZHAN J, PETTWAY R E, MCDONALD B A. The global genetic structure of the wheat pathogen *Mycosphaerellagraminicola* is characterized by high nuclear diversity, low mitochondrial diversity, regular recombination, and gene flow[J]. Fungal Genetics & Biology, 2003, 38: 286-297.

[71] ZHAN J S, THRALL P H, BURDON J J. Achieving sustainable plant disease management through evolutionary principles[J]. Trends in Plant Science,2014, 19: 570-575.

[72] ZHAN J S, THRALL P H, PAPAIX J, et al. Playing on a Pathogen's Weakness: Using Evolution to Guide Sustainable Plant DiseaseControl Strategies[J]. Annual Review of Phytopathology, 2015, 53: 19-43.

Identification and characterization of bamboo mosaic virus isolates from a naturally occurring coinfection in *Bambusa xiashanensis*

Wenwu Lin[1,2], Lu Wang[1,2], Wenkai Yan[1,2], Lingli Chen[1,2], Huihuang Chen[1,2], Wenting Yang[1,2], Maohui Guo[1,2], Zujian Wu[1,2], Liang Yang[1,2], Lianhui Xie[1,2]

(1 State Key Laboratory of Ecological Pest Control for Fujian and Taiwan Crops, Fujian Agriculture and Forestry University, Fuzhou 350002, China;

2 Fujian Key Laboratory of Plant Virology, Virology, Institute of Plant Fujian Agriculture and Forestry University, Fuzhou 350002, China)

Abstract: Bamboo mosaic virus (BaMV) is a well-characterized virus and a model of virus-host interaction in plants. Here, we identified naturally occurring BaMV isolates from Fujian Province, China and furthermore describe a naturally occurring BaMV coinfection in bamboo (*Bambusa xiashanensis*) plants. Two different types of BaMV were identified, represented by isolates BaMV-XSNZHA7 (X7) and BaMV-XSNZHA10 (X10). The phylogenetic relationships between X7- and X10-like isolates and published BaMV isolates were determined based on genomic RNA and amino acid sequences. Three clusters were identified, indicating that BaMV is highly diverse. The in planta viral replication kinetics were determined for X7 and X10 in single infections and in an X7/X10 coinfection. The peak viral load during coinfection was significantly greater than that during single infection with either virus and contained a slightly higher proportion of X10 virus than X7, suggesting that X10-like viruses may have a fitness advantage when compared to X7-like viruses.

Bamboomosaic virus (BaMV; classified in genus *Potexvirus*, family *Flexiviridae*) was originally described as the etiological agent of mosaic symptoms occurring on two species of bamboo plants, *Bambusa multiplex* Raeusch and *B. vulgaris* Schreder, in Brazil in 1977 (Lin et al., 1977). Subsequently, BaMV epidemics have been reported in Taiwan, California, Florida and mainland China (Lin et al., 1979; Lin et al., 1995; Elliott and Zettler, 1996; Lin et al., 2016). BaMV has been intensively studied for more than 20 years and is an important model system for plant-virus interactions (Liou et al., 2015). The BaMV genome consists of a single, positive-sense RNA (~6400nt) that contains five open reading frames (ORFs) (Lin et al., 1994). ORF1 encodes a 155-kDa replication-related protein with three functional domains: (1) an N-terminal mRNA capping enzyme, (2) a central RNA helicase, and (3) a C-terminal RNA-dependent RNA polymerase (RdRp) (Li et al., 2001a; Li et al., 2001b; Huang et al., 2005). ORFs 2–4 encode the triple gene block (TGB) proteins TGBp1,TGBp2, and TGBp3, which are essential for virus movement (Beck et al., 1991). The 25-kDa coat protein encoded by ORF5 is associated with virion encapsidation, replication, and cell-to-cell and long-distance viral movement (Lan et al., 2010). BaMV is the only known potexvirus that is associated with satellite RNAs (satBaMV) (Lin and Hsu, 1994). Although the structure and function of the BaMV genome has been well-characterized, there is limited publically available molecular sequence data. Thus, the genetic variability of BaMV populations is not well understood.

A more complete knowledge about the genetic variability of plant viruses is critical to understand plant

Archive of Virology. 2017, 162: 1335–1339
Received 10 August 2016; Accepted 20 November 2016; Published online 3 January 2017

virus evolution, virus-plant interactions, and for developing sustainable methods of controlling plant viral diseases. Genetic variability is critical for survival, and RNA plant viruses have high mutation rates creating dynamic quasispecies pools that allow rapid evolution to occur in response to changing environments.

Many studies have suggested that a single host can be coinfected with multiple viruses or divergent strains of a virus, which would alter the structure of the population and influence viral evolution. For example, coinfection with multiple viral strains is a prerequisite for viral recombination, which is an important source of variation for DNA viruses such as *Tomato yellow leaf curl virus* (TYLCV) (García-Andrés et al., 2006) and RNA viruses such as *Pepino mosaic virus* (PepMV) and *Soybean dwarf virus* (SDV) (Gómez et al., 2009; Schneider et al., 2011). Within the host, coinfection creates an environment characterized by competition and asymmetrical antagonism between viruses; even less virulent viral strains may have an advantage and accumulate to a greater extent than in a single infection (Gómez et al., 2009). Here, to provide additional information about the genetic diversity of BaMV, we cloned and characterized eight new isolates from different bamboo hosts using PCR and direct sequencing (Table 1). We also examined the in planta dynamics and fitness of two of the isolated BaMV strains alone and during coinfection.

Table 1 Collection data regarding the sequenced *Bamboo mosaic virus* (BaMV) isolates included in this study (isolates in bold were newly obtained)

Isolate	Host	Location	Contributor	GenBank #	With or without satBaMV? (Yes or No)	Nucleotide identity of complete genome when compared to BaMV-O isolate (%)	References
BaMV-Au	*Arundinaria usawai*	Taiwan	Y. Hisamoto	AB636267	Unknown	90.3	Unpublished
BaMV-Bo	*Bambusa oldhamii*	Taiwan	Y. Hisamoto	AB543679	Unknown	90.3	Unpublished
BaMV-Pu	*Pseudosasa usawai*	Taiwan	Y. Hisamoto	AB636266	Unknown	90.5	Unpublished
BaMV-O	*Bambusa oldhamii*	Taiwan	N. S Lin	NC_001642	No	100	（Lin et al., 1994）
BaMV-V	*Bambusa vulgaris*	Taiwan	C.-C Yang	L77962	Yes	90.1	Unpublished
BaMV-BSTLZHA6	*Dendrocalamus brandisii*	Fujian	W. Lin	KX648525	Yes	80.5	This study
BaMV-XSNZHA10	*Bambusa xiashanensis*	Fujian	W. Lin	KX648532	Yes	80.5	This study
BaMV-LLZHA5	*Bambusa albo-lineata*	Fujian	W. Lin	KX648530	Yes	80.6	This study
BaMV- CHGZHA4	*Bambusa pervariabilis*	Fujian	W. Lin	KX648526	Yes	80.6	This study
BaMV-HCLTZHA15	*Bambusa vulgaris*	Fujian	W. Lin	KX648527	Yes	80.6	This study
BaMV-LGHQZHA1	*Neosinocalamus affinis*	Fujian	W. Lin	KX648529	Yes	82.2	This study
BaMV-XSNZHA7	*Bambusa xiashanensis*	Fujian	W. Lin	KX648531	Yes	82.1	This study
BaMV-HQZSL5	*Dendrocalamus tsiangii*	Fujian	W. Lin	KX648528	Yes	82.2	This study
BaMV- MUZHUBZ2	*Bambusa rutila*	Fujian	W. Lin	KT591185	No	81.9	（Lin et al., 2016）
BaMV-JXYBZ1	*Phyllostachys aureosulcata*	Fujian	W. Lin	KT591184	Yes	81.8	（Lin et al., 2016）
BaMV-YTHSL14	*Bambusa rigida*	Fujian	W. Lin	KT591186	Yes	82.1	（Lin et al., 2016）

Bamboo samples with mosaic symptoms belonging to eight bamboo species were collected in Zhangzhou Bamboo Cultivation Garden, Fujian Province, China, in June 2015. Total RNA was extracted with the RNeasy Plant Mini Kit (Qiagen, Hilden, Germany) and cDNA was synthesized using oligo dT(18) primers with the GoScript^{TM} Reverse Transcription system (Promega, Madison, USA), according to the manufacturer's instructions. The complete genomes of BaMV and satBaMV in the infected samples were sequenced as described previously including the use of 5′-rapid amplification of cDNA ends (RACE) (Lin et al., 2016). Specific primer pairs were designed, based on BaMV sequences downloaded from GenBank, to amplify two large overlapping segments from samples that were coinfected with two viral strains (Fig. S1; Table S1). At least 10 independent clones of each viral genome segment were sequenced and the complete genomes were assembled based on the overlapping region (286bp). The accuracy of genome assembly was confirmed using long-distance PCR. Multiple sequence alignment was performed for the isolated viral genome segments and eight publically available BaMV sequences (GenBank) using the MUSCLE program in the MEGA5 platform (Tamura et al., 2011). Phylogenetic analysis was performed with the maximum-likelihood (ML) method implemented in MEGA5 (Tamura et al., 2011). MrModelTest was used to determine the *GTR+G+I* model that was used to reconstruct a phylogenetic tree (Nylander, 2004). ML topology was evaluated using 1000 bootstrap replicates.

Based on their divergent genome sequences, isolates X7 and X10 were identified as two viral strains isolated from the same bamboo sample (*B. xiashanensis*). To proceed further, both isolates were used to generate agroinfectious clones (to be described elsewhere) that were introduced into *Agrobacterium tumefaciens* GV3101 using the Gene-PulserXcell (Bio-Rad) system with default settings. A 5-mL culture of *A. tumefaciens* was grown overnight at 28℃ in LB-broth medium containing 25mg/L kanamycin and 25mg/L rifampicin. *A. tumefaciens* cells were harvested after overnight culture and resuspended in infiltration medium (10mM $MgCl_2$, 10mM MES, and 200mM acetosyringone), adjusted to an optical density of 1.0 at 600nm and incubated at room temperature for 2–3h. For coinfection experiments, separate cultures were mixed at a 1:1 ratio and used to infect the host plant. The culture was then infiltrated into the underside of leaves of *N. benthamiana* plants (at a 2- to 3-leaf stage) in similar-size spots by the use of a 1-mL syringe without a needle. The BaMV-S isolate (pCamBaMV-S) was used as a positive control and *A. tumefaciens* harboring the empty plasmid pCambiaTunos was used as the negative control (Cotton et al., 2009; Liou et al., 2014). Infected plants were maintained in a greenhouse at 28℃ under 16-h light/8-h dark conditions for up to 28 dpai (days post agroinoculation). All of the tissues above the infected leaves from each plant were snap frozen in liquid nitrogen and then ground into a homogenized powder. Total RNA was extracted from four biological replicates at 7, 14, 21, and 28 dpai. Reverse transcription was performed as described above. The fitness of each BaMV isolate was estimated by measuring the accumulated viral load in single-infection and coinfection systems using real-time quantitative PCR (qPCR) performed on a CFX96 real time PCR system (Bio-Rad). A highly conserved fragment of ORF1 (137bp) was selected for qPCR amplification (Fig. S1A). The *N. benthamiana* actin gene was used as an internal control to normalize the level of viral gene expression. The relative accumulation of viral genomic RNA for each isolate was compared to BaMV-S at 28 dpai for each time point collected. The primer sets are provided in the supporting information (Table S1).

Based on the sequence alignments from eight complete BaMV genome sequences, we identified two different viral strains in the primary samples collected. Of these, six sequences were isolated from plants infected with a single BaMV strain and two were from a plant coinfected with two different BaMV strains. Despite being collected from different bamboo species, and in some cases different genera, the isolates BaMV-BSTLZHA1, BaMVCHGZHA4, BaMV-HCLTZHA15, BaMV-LLZHA5 and BaMV-XSNZHA10 belong to the same strain, which we designated X10-like. The isolates BaMV-HQZSL5, BaMV-LGHQZHA1 and BaMV-XSNZHA7 are members of another strain, designated X7-like. The isolates within the X7- and X10-like strains have highly similar nucleotide sequence identities (> 99%). Therefore, a specific restriction endonuclease cleavage assay using SalI and XhoI was developed to identify infections with X7- and X10-like isolates in single and coinfection settings (Fig. S1A, Fig.S2; Table S1). SatBaMVs were detected with all newly identified BaMV isolates and shared similar genomic structure with the known isolates in GenBank (Fig. S1B).

The BaMV isolates (published and newly isolated) included in this study fell into three different phylogenetic clusters (Fig. 1). The published isolates from Taiwan clustered together and were distinct from the mainland Chinese isolates. The X10-like isolates clustered into a new phylogenetic sublineage, which differed from the other sequenced isolates at approximately 19% of the nucleotide positions. The X7-like isolates clustered with other isolates collected from Fuzhou, China, but only shared approximately 90% nucleotide identity with the other Fuzhou isolates (Lin et al., 2016). Phylogenetic trees based on the amino acid sequences of ORF1 (Replicase), ORF2 (TGBp1) and ORF5 (CP) showed very similar topologies (Fig. S3).

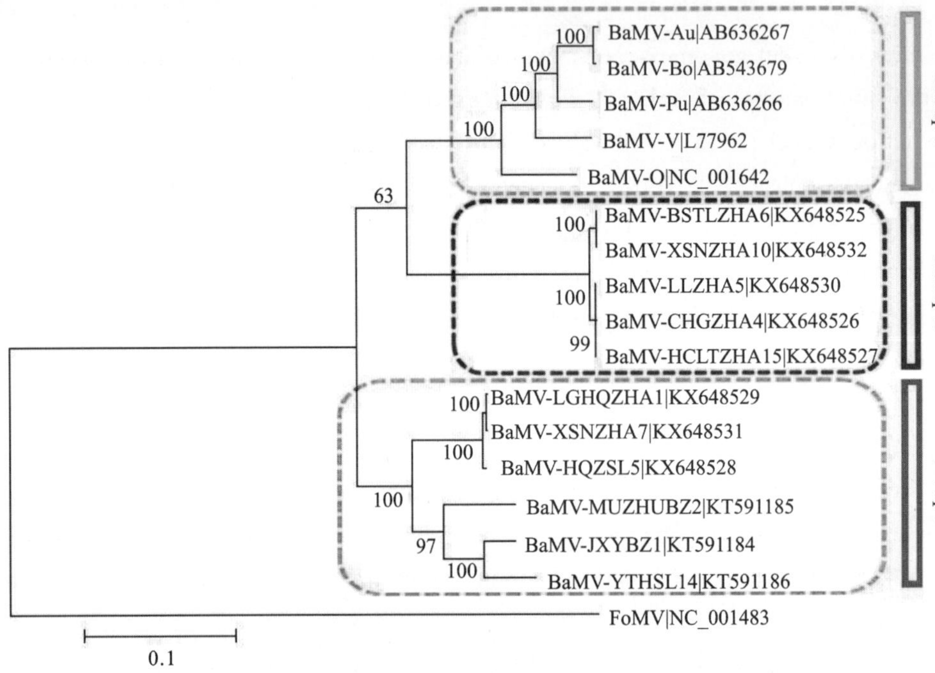

Fig. 1 Maximum-likelihood phylogenetic tree constructed from the complete genomes of 16 BaMV isolates.
The tree topology was evaluated based on 1,000 bootstrap replicates. For each node, bootstrap percentages are given on the branches. A *Foxtail mosaic virus* isolate (NC_001483) served as the outgroup

In planta viral accumulation differed significantly depending on the viral isolate and the type of infection (Fig. 2). In single infections, BaMV-S had the highest fitness level based on the accumulation of viral RNA at 7,14 and 21dpai. Regarding the comparison between X7 and X10, X10 accumulated higher levels of viral RNA than X7 during the course of infection, although there was no statistically significant difference at 21dpai. Of note, X10 continued to increase from 21 to 28 dpai, while the positive control and X7 both decreased. At 28 dpai, the relative accumulation of X10 was even higher than that of BaMV-S and at least twice as high than that of X7 on average. It is believed that plants are generally more susceptible to viral disease in early than in late phases of growth (Develey-Rivière and Galiana, 2007). This may reflect an increase in resistance over time, with plants gradually increasing their ability to control viral infection and colonization. In this study, 21dpai seemed to be a turning point in infection with BaMV-S and X7 viruses, i.e. after this time the plants had an increased ability to resist them. These results suggest that X10 may have a fitness advantage compared with the other isolates, especially in the late phase of single viral infection. In coinfections with X7 and X10 (MIX[X7/X10]), there was a marked increase in viral accumulation from 14 to 21dpai compared to the single infections and positive control. At its peak at 21dpai, the accumulation of viral RNA in the MIX(X7/X10) infection was approximately two-fold greater than that for the positive control, BaMV-S. These results contrast with a previous study in the PepMV model which indicated that asymmetrical antagonism was observed during mixed infections (Gómez et al., 2009).

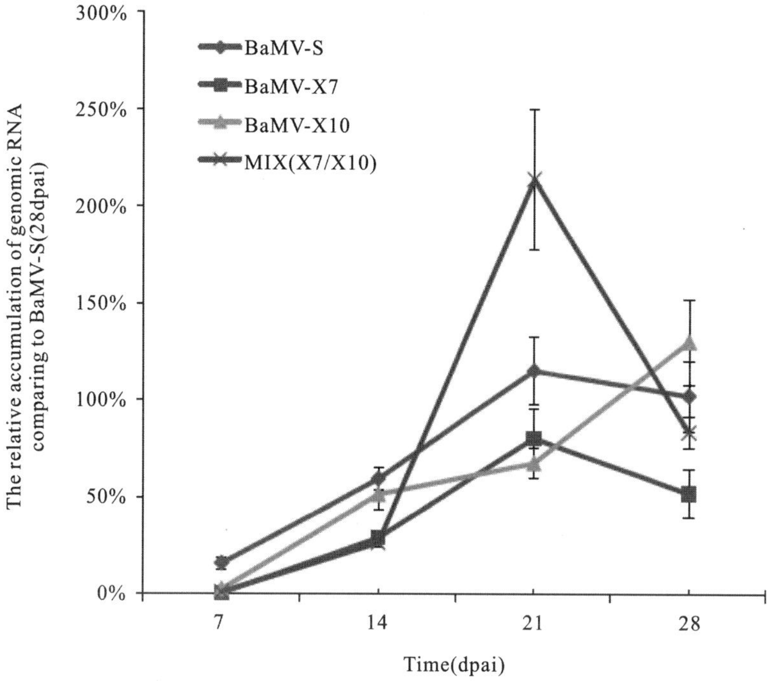

Fig. 2 Viral RNA accumulation of BaMV isolates X7 and X10 in single and mixed infections

The fitness of BaMV-XSNZHA7 (X7) and BaMV-XSNZHA10 (X10) was estimated by measuring the accumulation of viral genomic RNA in infected *N. benthamiana* plants using quantitative real-time PCR at 7, 14, 21, and 28 dpai (days post-agroinoculation)

An assay based on restriction endonuclease cleavage was used to estimate the relative proportions of X7 and X10 during coinfection (Fig. S4). The endonuclease assay was developed because specific primer sets for a PCR based assay were not adequate to distinguish between these two viral strains. The optimal cycle number to amplify the genomic fragments for digestion was empirically determined and defined as the point at which PCR products of the desired size (~2202 bp) were obtained without obvious heteroduplexes that migrated slower (Fig. S4A). One microgram of DNA was subjected to cleavage with *Sal*I and *Xho*I overnight. These experiments were repeated at least three times. Relative quantification of gel bands was performed using the GeneTools system (Syngene, UK) and Microsoft Excel 2010. In general, X10 had a higher viral load (more accumulation) than X7 at all of the time points evaluated (Fig. S4B), suggesting that the X10-like viruses are fitter than the X7-like viruses. This result is consistent with that of the qPCR assay with X7- and X10-infected hosts alone. The mosaic symptoms caused by each single infection and the coinfection at 21dpai are shown in Fig. S5. BaMV-S can cause severe mosaic symptoms on systemic leaves of *N. benthamiana*, compared to the mild mosaic symptoms caused by X10 and MIX[X7/X10] and the minimal symptoms caused by X7. This suggests that X7 is a mild BaMV isolate with lower pathogenicity than X10. The severity of symptoms is dependent not only on virus load, but also on the specific sequences of the viral genome and the interaction between the virus and the host.

Here, we identified and characterized two different BaMV strains (X7 and X10) and eight new BaMV isolates, as well as a naturally occurring mixed infection with two BaMV strains. While there was no evidence that recombination events occurred between X7 and X10 in this study, previous studies suggest that recombination can occur (Martin et al., 2010). We believe that coinfection within the same plant, and possibly the same cell, is likely to lead to a recombination event. The results of this study provide a theoretical basis for the formation of a recombinant BaMV isolate. Further examination of the genetic variability of BaMV will be critical to understand viral evolution and may provide insights into novel molecular virus-plant interactions.

Acknowledgements

This work was supported by Grants from the Natural Science Foundation of Fujian Province of China (2014J06011), the Doctoral Fund of Ministry of Education of China (20123515120005), the FAFU Science Fund for Distinguished Young Scholars (Grant No. xjq201402), and the China Scholarship Council (201608350081).

References

[1] BECK D L, GUILFORD P J, VOOT D M, et al. Triple gene block proteins of white clover mosaic potexvirus are required for transport[J]. Virology, 1991, 183: 695-702.

[2] COTTON S, GRANGEON R, THIVIERGE K, et al. Turnip mosaic virus RNA replication complex vesicles are mobile, align with microfilaments, and are each derived from a single viral genome[J]. Journal of Virology, 2010, 83: 10460-10471.

[3] DEVELEY-RIVIÈRE M P,GALIANA E. Resistance to pathogens and host developmental stage: a multifaceted relationship within the plant kingdom[J]. New Phytologist, 2007, 175: 405-416.

[4] ELLIOTT M, ZETTLER F. Bamboo mosaic virus detected in ornamental bamboo species in Florida[J]. Proceedings of the Florida State Horticultural Society, 24.

[5] GARCÍA ANDRÉS S, MONCI F, NAVAS CASTILLO J,et al. Begomovirus genetic diversity in the native plant reservoir *Solanum nigrum*: evidence for the presence of a new virus species of recombinant nature[J]. Virology, 2006, 350: 433-442.

[6] GÓMEZ P, SEMPERE R, ELENA S F, et al. Mixed infections of Pepino mosaic virus strains modulate the evolutionary dynamics of this emergent virus[J]. Journal of Virology, 2009, 83: 12378-12387.

[7] HUANG Y L, HSU Y H, HAN Y T, et al. mRNA guanylation catalyzed by the S-adenosylmethionine-dependent guanylyltransferase of bamboo mosaic virus[J]. Journal of Biological Chemistry, 2005, 280: 13153-13162.

[8] LAN P, YEH W B, TSAI C W,et al. A unique glycine-rich motif at the N-terminal region of bamboo mosaic virus coat protein is required for symptom expression[J]. Mol Plant Microbe Interact, 2010, 23: 903-914.

[9] LI Y I, CHEN Y J, HSU Y H,et al. Characterization of the AdoMet-dependent guanylyltransferase activity that is associated with the N terminus of bamboo mosaic virus replicase[J]. Journal of Virology, 2001a, 75: 782-788.

[10] LI Y I, SHIH T W, HSU Y H, et al. The helicase-like domain of plant potexvirus replicase participates in formation of RNA. 5′ cap structure by exhibiting RNA. 5′-triphosphatase activity[J]. Journal of Virology, 2001b, 75: 12114-12120.

[11] LIN M, KITAJIMA E, CUPERTINO F, et al. Partial purification and some properties of bamboo mosaic virus[J]. Phytopathology, 1977, 67: 1439-1443.

[12] LIN N S, HSU Y H. A satellite RNA associated with bamboo mosaic potexvirus[J]. Virology, 1994, 202: 707-714.

[13] LIN N S, LIN B Y, LO N W, et al. Nucleotide sequence of the genomic RNA of bamboo mosaic potexvirus[J]. Journal of General Virology, 1994, 75: 2513-2518.

[14] LIN N, CHEN M, CHIANG T, et al. Preliminary studies on bamboo mosaic disease in Taiwan. Bulletin. Shih Yen Pao Kao. 1979.

[15] LIN N, LIN B, YEH T, et al. First report of bamboo mosaic virus and its associated satellite RNA on bamboo in the US[J]. Plant Disease, 1995, 79.

[16] LIN W, GAO F, YANG W, et al. Molecular characterization and detection of a recombinant isolate of bamboo mosaic virus from China[J]. Archives of Virology, 2016, 161: 1091-1094.

[17] LIOU M R, HU C C, CHOU Y L, et al. Viral elements and host cellular proteins in intercellular movement of bamboo mosaic virus[J]. Curr Opin Virol, 2015, 12: 99-108.

[18] LIOU M R, HUANG Y W, HU C C, et al. A dual gene-silencing vector system for monocot and dicot plants[J]. Plant Biotechnol J, 2014,12: 330-343.

[19] MARTIN D P, LEMEY P, LOTT M, et al. RDP3: a flexible and fast computer program for analyzing recombination[J]. Bioinformatics, 2010, 26: 2462-2463.

[20] NYLANDER J. MrModeltest v2. Program distributed by the author[J]. Evolutionary Biology Centre Uppsala University 2004.

[21] SCHNEIDER W L, DAMSTEEGT V D, STONE A L, et al. Molecular analysis of soybean dwarf virus isolates in the eastern United States confirms the presence of both D and Y strains and provides evidence of mixed infections and recombination[J]. Virology, 2011, 412: 46-54.

[22] TAMURA K, PETERSON D, PETERSON N, et al. MEGA5: molecular evolutionary genetics analysis using maximum likelihood, evolutionary distance, and maximum parsimony methods[J]. Narnia, 2011, 28: 2731-2739.

III 检测鉴定与流行调控

这部分论文通过症状观察、生物学鉴定、分子检测、血清学检测、电镜观察等手段监测和鉴定了我国几种植物病毒病的发生情况和致病性特征。另外，从田间观测数据和生态学角度总结了植物病毒及其他病原物引起的病害发生与流行的特征，从可持续性植物病害防控的角度作了深入探讨和建议。

我国水稻条纹病毒致病性的分化与差异分析

程文金，邓慧颖，谢荔岩，林奇英，吴祖建，谢联辉

(福建农林大学植物病毒研究所，福建省植物病毒学重点实验室 福建 福州 350002)

摘要：以云南灰飞虱（*Laodelphax striatellus*）为介体昆虫，以4个水稻品种为寄主，通过人工接种对采自云南、江苏、河南、山东、安徽等省的水稻条纹病毒（*Rice stripe virus*, RSV）23个分离物进行传毒试验。致病性评价结果表明，23个分离物可划分为5个等级，即强致病型、次强致病型、中致病型、弱致病型、极弱致病型。致病性分化表现在地理差异、毒株差异及水稻品种差异三方面。不同分离物在寄主发病率、发病时间、症状表现以及对灰飞虱传毒效率等方面都存在差异。

关键词：水稻条纹病毒；致病性；灰飞虱

中图分类号：S435.111.4+9　**文献标识码**：A　**文章编号**：1671-5470(2009)06-0561-06

Pathogenicity differentiation and difference analysis of *rice stripe virus* of China

Wenjin Cheng, Huiying Deng, LiyanXie, Qiying Lin, Zujian Wu, LianhuiXie

(Key Lab of Plant Virology of Fujian Province, Institute of Plant Virology, Fujian Agriculture And Forestry University, Fuzhou, Fujian 350002, China)

Abstract: Twenty-three isolates of Rice stripe virus (RSV) from Yunnan, Jiangsu, Shandong, Henan, Anhui were inoculated to four rice varieties by *Landelphax striatellus*. According to the valuations of pathogenicity, the isolates could be divided into five grades, namely strong virus (sv), semi-strongvirus (ss), middle strongvirus (ms), weak virus (wv) and least weak virus (lw). Pathogenicity differentiation was shown to correlate with geographic distribution, strain differences and rice varieties. There were differences among isolates in the morbidity, time to exhibit disease, symptom and influences to *Landelphax striatellus's* transmission.

Key words: *Rice stripe virus*; pathogenicity; *Landelphax striatellus*

由水稻条纹病毒（*Rice stripe virus*, RSV）引起的水稻条纹叶枯病在我国16个省（市）的水稻种植区发生或流行，造成巨大损失（谢式半钰，1969；林奇英和谢联辉，1990）。本世纪以来，该病在江苏、上海、河南、山东、浙江、安徽、辽宁等省份相继暴发，以江苏一带病情最重（程兆榜等，2002；王桂云等，2005；高苓昌等，2006；弓利英等，2006；桑海旭等，2006；张国鸣等，2006；张华中和王忠义，2008）。

对RSV致病性分化的研究在国内外已有大量报道。学者根据不同标准来划分RSV, Ishiietal将其划分为展叶型和卷叶型2个株系（Ishiim, 1967）；Kisimoto将其划分为黄化型和白化型2个株系（Kisimotor, 1972）；Hayashietal将其划分为鸿巢、P、N3个株系（Hayashi et al., 1989）。林含新等的研究也表明，来

福建农林大学学报（自然科学版）. 2009, 38(6): 6-10

收稿日期：2009-03-24；修回日期：2009-06-18

基金项目：农业部公益性行业（农业）科研专项资助项目（nyhyzx07-051）和国家自然科学基金资助项目（30671357）

自我国不同地区的 7 个 RSV 分离物存在明显的致病性差异（林含新等 , 2002）。对病害暴发区和常发区灰飞虱携带的 22 个分离物进行致病性测定，结果表明，参试分离物可划分为强致病型（HV）、次强致病型（SH）、中致病型（MV）、次弱致病型（SL）、弱致病型（LV）5 个类型。他们主要针对寄主的发病率及症状的差异对 SL 和 LV2 种类型进行分析。本研究以发病率的差异来划分我国各省 23 个 RSV 分离物的致病性，并分析了不同分离物在发病时间、症状表现、对灰飞虱传毒影响等方面的差异，为病害防治提供参考。

1 材料与方法

1.1 介体昆虫及毒源

介体灰飞虱采于云南田间，将无带毒的个体后代饲养于水稻品种台中 1 号上。水稻条纹病毒 JHZ06 分离物于 2006 年 7 月采于江苏洪泽田间；其他分离物于 2007 年 7 ～ 8 月分别采自云南、河南、江苏、安徽、山东等省。

1.2 供试水稻品种

水稻品种武育粳 3 号、合系 39、花优 63 等由福建农林大学植物病毒研究所保存；水稻品种日本晴由福建农林大学遗传所提供。不同品种水稻经清水浸泡发芽后，长至芽长 1cm 左右，新叶未抽出时用于传毒。

1.3 传毒及发病率统计

1 ～ 2 龄无毒灰飞虱于病株上饲毒 48h。过循回期后，集团传毒接种于水稻品种武育粳 3 号合系 39、日本晴、花优 63，苗虫比为 1 ∶ 2 ～ 1 ∶ 3。单虫单苗传毒于取食 48h 后立即接种于水稻品种日本晴及花优 63，单只灰飞虱取食单棵水稻苗，每个分离物接种 30 棵水稻苗。供试水稻苗 25℃接种 48h 后移至 30℃显症。每天记录发病情况，接种 40d 后累计结果。

1.4 致病性评价

按株发病率将致病性划分为 5 个等级，即强致病型（>20%）、次强致病型（10.0% ～ 20.0%）、中致病型（5.0% ～ 10.0%）、弱致病型（0.0% ～ 5.0%）和极弱致病型（0.0%）。

2 结果与分析

2.1 RSV 致病性测定及其分化特点

根据集团传毒在武育粳 3 号及合系 39 上的发病率，将 21 个 RSV 分离物（HKF07、ABB07 因植株生长弱小未用于该次试验）的致病性分为 5 个等级，即强致病型、次强致病型、中致病型、弱致病型和极弱致病型（表 1）。21 个分离物在致病上存在明显的分化现象，即存在多个差异的致病性等级。多数分离物分布在次强致病型、中致病型、弱致病型 3 个中间等级中。致病性的分化特点主要表现在如下 3 个方面。

（1)21 个分离物在地理上形成以江苏地区为中心并向外辐射的致病群结构。强致病型分离物共 3 个，均为江苏分离物。次强致病型分离物共 5 个，山东的 3 个分离物全部划分在该等级中，其他 2 个分离物分别来自云南、江苏。中致病型的分离物有 5 个，2 个分布在江苏，1 个分布在河南，2 个分布在云南。弱致病型的分离物有 7 个，5 个分布在云南，2 个分布在江苏。江苏分离物 JYC07-2 为极弱致病型。

（2）在同一地区，不同分离物致病性的表现不同。例如：云南曲靖的 2 个分离物分别为次强和弱致病型；江苏 9 个分离物在 5 个等级中均有分布。江苏分离物属于混合致病群，云南分离物多为弱致病型。

（3）致病性分化还体现在同一分离物在不同水稻品种的发病率上。次强致病型的分离物 YQJ07-2、SJN07 在合系 39 上的发病率表现为弱致病型；弱致病型分离物 YBS07 与 JNJ07 在合系 39 上的发病率表现为次强致病型。这可能与病毒的准种结构有关。

表 1 我国水稻条纹病毒 21 个分离物在武育粳 3 号及合系 39 上的致病性

Table1 Pathogenicities of 21 isolates of rice stripe virus in China on rice varieties Wuyujing3 and Hexi39

致病性	数量	分离物及来源	致病性 / %	
			武育粳 3 号	合系 39
强致病型	3	JHZ06(江苏洪泽)	50.0	27.3
		JXH07(江苏兴化)	27.3	18.2
		JYC07(江苏盐城)	22.7	31.8
次强致病型	5	SLY07(山东临沂)	18.2	9.1
		YQJ07-2(云南曲靖)	18.2	4.5
		SHZ07(山东菏泽)	18.2	18.2
		JNJ07-2(江苏南京)	18.2	18.2
		SJN07(山东济宁)	13.6	4.5
中致病型	5	YLQ07(云南禄劝)	9.1	0.0
		HZZ07(河南郑州)	9.1	0.0
		JNJ07-3(江苏南京)	9.1	0.0
		JYC07-3(江苏盐城)	9.1	9.1
		YWD07(云南武定)	9.1	13.6
弱致病型	7	YCX07(云南楚雄)	4.5	0.0
		YBS07(云南保山)	4.5	13.6
		JXH07-2(江苏兴化)	4.5	4.5
		YBS07-2(云南保山)	4.5	0.0
		YBS07-3(云南保山)	4.5	0.0
		JNJ07(江苏南京)	4.5	18.2
		YQJ07(云南禄劝)	0.0	9.1
极弱致病型	1	JYC07-2(江苏盐城)	0.0	0.0

为进一步验证致病性分化特点，另选取表型症状明显、植株完整的 12 个 RSV 代表分离物（HKF07、ABB07 分别采于河南开封及安徽蚌埠田间）以集团传毒与单虫单苗传毒方式将其分别接种于水稻品种日本晴与花优 63。发病率统计结果（表 2）表明，12 个分离物同样表现出明显的致病性分化现象；江苏分离物仍属于混合致病群，以 JHZ06 的致病性最强；江苏周边省市的分离物属于强、次强致病型；云南有 2 个分离物属于强致病型，有 3 个分离物属于中、弱致病型；无极弱致病型分离物。YQJ07、YBS07 分离物在日本晴上的发病率高，表明以弱致病性分离物为多的云南种群中隐藏着强致病性分离物。

2.2 不同分离物间的差异分析

以不同分离物和水稻品种为因素对发病率作两因素无重复方差分析（表 1），结果表明，发病率在不同分离物间存在极显著差异，而在水稻品种武育粳 3 号、合系 39 间的差异不显著（表 3）。

通过单虫单苗传毒和集团传毒 2 种方式比较 JHZ、JYC07、SLY07、SJN07、ABB07、YLQ07、YWD07、YCX07、YBS07、YQJ07 等 10 个分离物（同期试验时，HKF07 及 JXH07 分离物因供试灰飞虱若虫大量死亡不计入统计）。结果表明：不同分离物表现症状所需的时间存在差异。不同时期接种的水稻幼苗发病时间也存在差异。饲毒后前 4d 接种幼苗，JYC07 分离物的发病时间为 33d，其他分离物均超过 40d 仍未见症状出现。饲毒后 4 ~ 6d、6 ~ 8d、8 ~ 10d 3 个时期接种苗，发病时间最短的是 JYC07 及 YBS07 两个分离物，均为 2d；发病时间最长的是 YWD07 分离物，为 22d（表 4）。对不同分离物在各个时期接种苗的发病时间进行方差分析，分离物因素共 10 个水平；不同时期接种的水稻品种各设为一个水平，将集团传毒与单虫单苗传毒分开统计，即水稻品种因素共 8 个水平。发病时间无重复两因素方差分析结果表明，发病时间在不同分离物间以及水稻品种间的差异分别达到显著和极显著水平（表 5）。

表 2 我国水稻条纹病毒 12 个分离物在日本晴及花优 63 上的致病性鉴定

Table2 Pathogenicity identification of 12 isolates of *rice stripe virus* in China on rice varieties of Ribenqing and Huayou63

致病性	数量	分离物	日本晴 J	日本晴 D	花优 63J	花优 63D
强致病型	4	JHZ06	54.5%	10.0	13.6%	6.7
		SJN07	36.4%	10.0	4.5%	6.7
		YQJ07	31.8%	0.0	13.6%	6.7
		YBS07	27.3%	10.0	4.5%	3.3
次强致病型	4	JYC07	13.6%	6.7	22.7%	10.0
		SLY07	13.6%	3.3	4.5%	6.7
		HKF07	13.6%	10.0	4.5%	0.0
		ABB07	13.6%	0.0	9.1%	0.0
中致病型	2	YLQ07	9.1%	6.7	4.5%	3.3
		YCX07	9.1%	0	4.5%	0.0
弱致病型	2	YWD07	4.5%	0.0	0.0%	0.0
		JXH07	4.5%	13.3	0.0%	10.0

日本晴 J— 集团传毒在日本晴上的发病率；日本晴 D— 单虫单苗传毒在日本晴上的发病率；花优 63J— 集团传毒在花优 63 上的发病率；花优 63D— 单虫单苗传毒在花优 63 上的发病率；致病性划分以日本 J 为依据。

表 3 RSV 不同分离物发病率方差分析

Table 3 Variance analysis of morbidity of different RSV isolates

变异来源	平方和	自由度	均方	F
分离物	0.238407	15	0.015894	4.52**
水稻品种	0.012641	1	0.012641	3.60
误差	0.052731	15	0.003516	
总和	0.303786	31		

水稻品种因素取武育粳 3 号及合系 39 两个水平；分离物因素共 16 个水平 (JYC07、JNJ07、JXH07 等 3 个水平发病率取相应分离物平均值)。

表 4 10 个分离物的发病时间

Table 4 Time of the disease incidence of ten isolates

分离物	JHZ06	JYC07	SLY07	SJN07	ABB07	YLQ07	YWD07	YCX07	YBS07	YQJ07
发病时间 / d	4	2	4	4	4	6	22	9	2	6

表 5 RSV 10 个分离物发病时间方差分析

Table 5 Variance analysis of disease incidence time of ten isolates

变异来源	平方和	自由度	均方	F
分离物	0.308064	9	0.034229	2.77*
水稻品种	0.355037	7	0.050720	4.11**
误差	0.777567	63	0.012342	
总和	1.440668	79		

对水稻品种武育粳 3 号、合系 39 的症状观察结果表明，不同分离物在 2 个水稻品种上的症状表现主要分为 2 种类型。第 1 种类型为云南类型，即云南分离物侵染后，寄主症状以展叶、条纹状为主，无致死趋势。第 2 种是云南外类型，即云南外分离物侵染后，寄主症状的表现以下垂、枯黄为主，甚至死亡。云南类型症状明显轻于云南外类型。

以接种苗发病为标准鉴定灰飞虱能否传毒，则带毒率表现为灰飞虱群体中至少能传毒 1 次的个体比例，而多次传毒率为带毒灰飞虱在多次接种试验中表现出的传毒比率。带毒率与多次传毒率即为灰飞虱传毒效率的评价指标。通过单虫单苗传毒方式对 JHZ、JYC07、SLY07、SJN07、ABB07、YLQ07、

YWD07、YCX07、YBS07、YQJ07 十个分离物对应灰飞虱带毒率及多次传毒率的分析显示，致病强的分离物对应灰飞虱的带毒率及多次传毒率都比较高（表 6）。YCX07、YWD07、ABB07 分离物的带毒率为 0.0%；YQJ07 分离物带毒率为 6.7%；YBS07、YLQ07、SLY07、SJN07、JYC07、JHZ 分离物带毒率为 10.0%。饲毒后 4 ～ 6d、6 ～ 8d、8 ～ 10d 连续 3 次接种，带毒灰飞虱能进行 2 次传毒的分离物有 YBS07、SJN07、JYC07、JHZ06 等 4 个分离物，2 次传毒率分别为 33.3%、66.7%、66.7%、66.7%。能进行 3 次传毒的分离物有 SJN07、JYC07、JHZ06，3 次传毒率分别为 66.7%、33.3%、33.3%。对灰飞虱传毒效率影响的差异直接导致了不同分离物对水稻品种的发病率差异。

表 6　不同分离物对灰飞虱的带毒率及传毒效率

Table 6　Toxic rate and transmission efficiency of *Landelphax striatellus* by different isolates

分离物	带毒率	2 次传毒率	3 次传毒率	分离物	带毒率	2 次传毒率	3 次传毒率
YCX07	0.0	0.0	0.0	SLY07	10.0	0.0	0.0
YBS07	10.0	33.3	0.0	SJN07	10.0	66.7	66.7
YQJ07	6.7	0.0	0.0	JYC07	10.0	66.7	33.3
YLQ07	10.0	0.0	0.0	JHZ06	10.0	66.7	33.3
YWD07	0.0	0.0	0.0	AB 07	0.0	0.0	0.0

注：带毒率 = (接种幼苗发病的灰飞虱数 / 接种总灰飞虱数) ×100%; 2 次 (或 3 次) 传毒率 = [能传毒 2 次 (或 3 次) 的灰飞虱数 / 带毒灰飞虱总数] ×10%

3　讨论

3.1　RSV 致病性的分析方法

RSV 致病性评价指标主要有寄主植物的发病率及症状严重度（高东明和李爱民，1991；林含新等，2002；程兆榜等，2008）。由于寄主发病还受到灰飞虱的传毒特性、生长环境、土壤营养及人为活动等因素影响。因而，致病型评价是一个系统分析的结果，除了以发病率或症状表现划分致病型等级外，还要进行其他方面的描述。对 RSV 致病性分析包括以下几方面：一是寄主发病率，并以此作为致病型等级划分的指标；二是发病时间，即最快观测到寄主症状表现所需的时间；三是症状表现，以发病初期寄主症状表现为主，后期生长为辅；四是病毒对灰飞虱传毒的影响，分析相应灰飞虱的带毒率、传毒效率等特性。供试水稻应选择多个品种，并以单次测定中最易感病品种（如武育粳 3 号或日本晴）为准，以放大各分离物的数据差异，便于比较。

3.2　致病性分化与病害流行的关系

致病性测定结果表明，在病害最严重的江苏省的 RSV 自然种群分布有强致病型分离物，并属于一个混合致病群。在江苏省周边病害流行的山东省、河南省分布有次强及中致病型分离物。在病害偶发的云南省分布着以弱致病型为主的分离物。说明我国 RSV 自然种群存在准种结构，即 RNA 病毒种群不是以单一的类型组成，而是以许多相关类型共存的形式存在（Holland et al., 1992；Domingo and Holland, 1997）。病毒准种的遗传多样性受控于病毒与寄主的相互作用，其高度多样性能使病毒迅速适应新的生境，造成更大的危害（Schneiderw, 2000）。RSV 自然种群致病性分化在地理上的差异可能与病毒所处的环境气候有关。云南四季如春，温差小，作物耕作在季节上变化小；云南以外的气候则冬寒夏炎，温差大，作物耕作在季节上变化大。而且，南方与北方间存在明显差异的耕作制度，这使得云南外 RSV 自然种群的寄主比云南的自然种群更频繁，即病毒与寄主间的互作更多样、多变。因此，云南外的 RSV 自然种群具有相对较高的多样性，相应产生较强的致病性，导致病害的大范围暴发。另外，同一分离物在不同水稻品种上表现不同发病率即致病型的现象可能与 RSV 单个分离物的准种结构有关，即在单个 RSV 病株上存在不同致病型的变种。而同一分离物的不同变种在不同寄主上的分离与定植可能存在差异，从而表现出不同的致病型。这进一步加大了病害流行的复杂性与难以预测性。

3.3 致病性评价与病害防治

对 23 个 RSV 分离物致病性的测定与评价结果表明，地理来源不同的 RSV 分离物存在不同程度的致病性差异；划分的 5 个等级大致上能反映我国主要 RSV 病害区的致病性结构与分布。江苏的 RSV 自然种群属于混合致病群，存在丰富的致病类型。因而，病害防治的重点应该是通过耕作改制，切断、压低介体灰飞虱，并及时清除病株，降低病毒的致病型多样性等减弱准种结构性。对发病严重以及灰飞虱肆虐的局部地区更要采取有效的生态、理化措施，阻断病害的扩散。江苏周边如山东、河南、安徽、上海等省市的 RSV 群体是病害流行的延伸区。对这些地区病害防治的重点应该是清除灰飞虱，阻止灰飞虱向其他地区迁飞；农药施用则以二代灰飞虱防治为主，抑制第二发病高峰。云南的 RSV 自然种群以弱致病型为主，但其隐藏的强致病型分离物不容忽视。特别应该注意病害严重地区（如保山、曲靖）的病情及其介体昆虫发生动态，及时采取有效措施加以防患。对尚未发生病害的周边地区亦应加强病害及其介体的监控力度。还可参考园林介壳虫综合治理信息咨询专家系统（罗佳 2008）防治灰飞虱，利用计算机软件建立起灰飞虱监测及防治的综合管理系统，以达到防治病害的目的。

参考文献

[1] 程兆榜，杨荣明，周益军，等．江苏稻区水稻条纹叶枯病发生新规律 [J]. 江苏农业科学，2002: 39-41.

[2] 程兆榜，任春梅，周益军，等．水稻条纹病毒不同地区分离物的致病性研究 [J]. 植物病理学报，2008 38: 126-131.

[3] 高东明，李爱民．江、浙两省主要粳、糯稻品种对条纹叶枯病的抗性测定 [J]. 浙江农业科学，1991: 96-98.

[4] 高苓昌，宋克勤，张洪瑞，等．黄淮稻区水稻条纹叶枯病的发病特点与综合防治 [J]. 山东农业科学，2006: 66-67.

[5] 弓利英，夏立，马丽，等．沿黄稻区水稻条纹叶枯病的发生及防治技术 [J]. 河南农业科学，2006, 35: 64-65.

[6] 林含新，魏太云，吴祖建，等．我国水稻条纹病毒 7 个分离物的致病性和化学特性比较 [J]. 福建农林大学学报（自然科学版），2002,31: 164-167.

[7] 林奇英，谢联辉．水稻条纹叶枯病的研究：Ⅱ病害的分布和损失 [J]. 福建农学院学报，1990: 421-425.

[8] 罗佳，唐乐尘，龙国伟．园林介壳虫综合治理信息咨询专家系统 [J]. 福建林学院学报，2008(04): 374-380.

[9] 桑海旭，王井士，边应权，等．辽宁省水稻条纹叶枯病严重发生的原因及防治对策 [J]. 垦殖与稻作，2006: 52-55.

[10] 王桂云，汪祖国，姚其林，等．水稻条纹叶枯病发生及综合防治技术初探 [J]. 亚热带农业研究，2005, 1: 36-38.

[11] 谢式半钰，邱人璋．台湾水稻新病毒病——条纹叶枯病 [J]. 植物保护会刊，1969, 11（4）: 175.

[12] 张国鸣，王华弟，戴德江．浙江省水稻条纹叶枯病发生发展态势与防控对策措施 [J]. 中国植保导刊，2006, 26: 20-21.

[13] 张华中，王忠义．2008. 来安县水稻条纹叶枯病发生原因及治理对策．安徽农学通报，14: 89-90.

[14] DOMINGO E, HOLLAND J. RNA virus mutations and fitness for survival[J]. Annual Review of Microbiology, 1997, 51: 151-178.

[15] HAYASHI T, USUGI T, NAKANO M, et al. On the strains of Rice stripe virus (1): An attempt to detect strains by difference of molecular size of disease-specific proteins[J]. Kyushu Plant Protection Research, 1989, 35(1): 1-2.

[16] HOLLAND J, JD DE LA TORRE J C, STEINHAUER D. RNA virus populations as quasispecies[J]. Curr Top Microbiol Immunol, 1992, 176: 1-20.

[17] Ishiim O. On strains of rice stripe virus(in Japanese)[J]. Annu Phyto. Soc. Japan, 1967, 32(1): 80-83.

[18] Kisimotor. Ecology of rice stripe disease(in Japanese)[J]. Hereditas, 1972, (12): 34-40.

[19] SCHNEIDERW L R J. Genetic diversity in RNA virus quasi species is controlled by host-virus interactions[J]. Journal of Virology, 2001, 75: 6566-6571.

[20] SCHNEIDERW L R J. Evolutionarily related Sindbis-like plant viruses maintain different levels of population diversity in a commonhost[J]. Journal of Virology, 2000,74 (7):3130-3134.

First report of the occurrence of *Sweet potato leaf curl virus* in tall morningglory (*Ipomoea purpurea*) in China

Caixia Yang, Zujian Wu, Lianhui Xie

(Institute of Plant Virology, Fujian Agriculture and Forestry University, Fuzhou, 350002, China)

Natural occurrence of *Sweet potato leaf curl virus* (SPLCV) has been reported in *Ipomoea batatas*(sweet potato, Convolvulaceae) or *I. indica* (Convolvulaceae) in several countries including the United States, Sicily, and China (Lotrakul et al., 1998; Briddon et al., 2006; Luan et al., 2007). In September of 2007, while collecting samples showing begomovirus-like symptoms in the Chinese province of Fujian, we observed tall morningglory (*I. purpurea*(L.) Roth, also known as *Pharbitis purpurea* (L.) Voigt), plants with slightly yellow mosaic and crinkled leaves. Total DNA was extracted from leaves of these plants and tested by rolling circle amplification (Haible et al., 2006). Amplification products were digested by the restriction enzyme *Bam*HI for 30min. Restriction products (2.8kb) were then cloned into pMD18T vector (Takara Biotechnology, China) and sequenced. Comparison of complete DNA sequences by Clustal V analysis revealed that these samples were infected by the same virus, and an isolate denoted F-p1 was selected for further sequence analysis. F-p1 was 2,828 nucleotides, with the typical genomic organization of begomoviral DNA-A (GenBank Accession No. FJ515896). F-p1 was compared with the DNA sequences available in the NCBI database using BLAST. The whole DNA sequence showed the highest nucleotide sequence identity (92.1%) with an isolate of SPLCV (GenBank Accession No. FJ176701) from Jiangsu Province of China. The result confirmed that the samples from the symptomatic tall morningglory were infected by SPLCV. To our knowledge, this is the first report of the natural occurrence of SPLCV in *I. purpurea*, a common weed species in China.

References

[1] BRIDDON R, BULL S, BEDFORD I. Occurrence of Sweet potato leaf curl virus in Sicily[J]. Plant Pathology, 2010, 55.

[2] HAIBLE D, KOBER S, JESKE H. Rolling circle amplification revolutionizes diagnosis and genomics of geminiviruses[J]. Journal of Virological Methods, 2006, 135(1): 9-16.

[3] LOTRAKUL P, VALVERDE R A A, CLARK C A, et al. Detection of a geminivirus infecting sweet potato in the United States[J]. Plant Disease, 1998, 82(11): 1253-1257.

[4] LUAN Y S, ZHANG J, LIU D M, et al. Molecular characterization of sweet potato leaf curl virus isolate from China (SPLCV-CN) and its phylogenetic relationship with other members of the Geminiviridae[J]. Virus Genes, 2007, 35(2): 379-385.

Plant Disease. 2009, 93(7): 764

棉花皱缩花叶病的初步研究

章松柏[1,2]，张长青[1]，吴祖建[2]，谢联辉[2]

(1 长江大学农学院，湖北荆州 434025；

2 农药生物化学教育部重点实验室，福建农林大学，福建福州 350002)

摘要：2007 年湖北荆州地区一些棉田棉花叶片上发生一种新的棉花病害，暂定名为棉花皱缩花叶病。通过症状描述、电镜观察和生物学接种对其进行了初步分析。结果显示：该病害症状疑似病毒病，但与已经报道的棉花病毒病症状都不相同。电镜观察显示，病株叶片汁液中存在 4 种杆状病毒颗粒，大小分别约是，18nm×300nm、18nm×500nm、18nm×800nm、18nm×1 100nm。生物学接种结果表明，该病不能经种子、棉蚜、烟粉虱、汁液摩擦 4 种方式传播。

关键词：棉花皱缩花叶病；电镜观察；生物学接种

中图分类号：S432 .1　**文献标识码**：A　**文章编号**：1004-3268(2010)03-0048-03

Preliminary study on cotton crimple mosaic disease

Songbai Zhang[1,2], Changqing Zhang[1], Zujian Wu[2],Lianhui Xie[2]

(1 College of Agriculture, Yangtze University, Jingzhou 434025, China;

2 Key Laboratory of Pesticide and Biochemistry, State Education Commission, Fujian Agriculture and Forestry University, Fuzhou 350002, China)

Abstract:In 2007, a virus-like disease occurred in some cotton fields of Jingzhou, Hubei Province, named cotton crimple mosaic disease temporarily. By symptom observation, electron microscope observation and biological inoculation experiment, the disease was preliminarily analyzed. The results showed that the disease symptoms were similar to viral disease though different from the cotton viral disease known so far. Four different kinds of bacilliform viruses were observed through electron microscope from the sap of diseased cotton leaf and their sizes were 18nm×300nm, 18nm×500nm, 18nm×800nm, 18nm×1100nm. No disease transmission was found using , cotton aphid, tobacco whitefly and sap inoculation.

Keywords:Cotton crumple mosaic disease; Electron microscope observation; Biological inoculation

棉花作为一种重要的经济作物，提供了世界上绝大多数的天然纤维。我国是世界上最大的棉花生产和消费国之一，棉花生产在整个国家经济中具有十分重要的战略地位（张桂寅 , 2005）。因此，保障棉花生产，增加棉农收入，十分重要。但棉花病害问题严重制约着棉花的生产和发展，如全球范围分布的棉花枯萎病和黄萎病，每年给棉花产业造成数十亿美元的经济损失（王红梅 , 2005）；局部地区分布的棉花曲叶病（cotton leafcurldisease, CLCuD），在巴基斯坦、印度等国已造成严重危害（Briddon and

中国学术期刊电子出版社 . 2010, (3): 48-50

收稿日期：2009-11-06

基金项目：长江大学校基金项目 (2006Z2070)；农业部农业公益性行业科研专项 (nyhyzx07-051)

Markham, 2000; 郭荣 , 2005; 青玲和周雪平 , 2005）；其他棉花病害也不同程度制约着棉花生产的可持续发展。因此，必须了解这些病害发生规律，密切监视各种棉花病害的流行发展动态，以便抑制主要病害的发生，控制次要病害的发展，保障棉花生产。

2007 年在湖北省荆州地区进行棉花病虫害调查时，发现一种新的、疑似病毒病的棉花病害，但与已经报道的棉花病毒病皆不相同，经相关专家初步鉴定为棉花病毒病，暂命名为棉花皱缩花叶病。以下是研究初报，以供参考。

1 材料和方法

1.1 材料

棉花病株采自湖北省荆州市太湖农场棉花田，病叶于 -70℃冰箱中保存，活株种植于防虫温室中，病株种子脱绒处理后室温保持于干燥环境中。感病品种杂交抗虫棉太 D5 号（鄂杂棉 10 号）购自荆州市农资公司；烟粉虱、棉蚜分别采自无病棉田棉花植株上，并在网室中种植的太 D5 号棉花上繁殖备用。

1.2 症状观察

于发病田间仔细观察发病症状，对照健康植株观察叶片颜色、叶片有无变形，叶脉是否明脉，茎秆表面和内部有无变化，植株有无矮化现象，病田有无发病中心，植株分枝情况等。

1.3 电镜观察

采用汁液负染法（夏更寿和郭志平，2007）：取样品组织加入 PBS 缓冲液（pH 7.0）充分研磨，10000g 离心 1min，取上清液备用。用铺有 Formver 膜的电镜铜网沾取上述病毒汁液吸附 10min，然后用滤纸吸干，加 2% 磷钨酸（pH 7.0）或 2 醋酸铀（pH 4.0）1 滴，染色 5min，再吸干，待干燥后，于投射电镜下观察有无病毒粒体。

1.4 病毒粗提

按照吴兴泉的方法粗提棉花体内的病毒（吴兴泉，2002）。棉花叶片 50~60g 加 100mL PB 缓冲液（0.1mol/L，pH 8.0，含 1% 巯基乙醇，10% 乙醇）充分研磨，3 层纱布过滤；取过滤液 8 000g，4℃离心 20min；取上清液加 1% Triton-X100，4℃磁力搅拌 1h; 5 500g，4℃离心 20min；取上清加入 0.2mol/L NaCl，4% PEG6000，4℃搅拌 1h，室温静置 1h; 10 000g，4℃离心 30min，取沉淀悬浮于 3mL PB 缓冲液（0.05mol/L，pH 8.0，含 1% Triton-X100）中，转入另一离心管中，并用 2mL 上述缓冲液洗涤 1 次; 8 000g，4℃离心 10min，上清液即为病毒粗提液，于 -70℃冰箱中保存备用。

1.5 生物学试验

生物学接种试验在防虫温室中进行。具体操作如下:（1）种传试验，播种在病株上采集的棉花种子，观察有无发病;（2）棉蚜传播试验 , 参考吴云峰等方法（吴云峰，1999），在病毒粗提液中加入少量的蔗糖，让棉蚜透过薄膜吸取病毒粗提液 1 h，转移棉蚜至三叶期棉花幼苗上，1h 后用药剂杀死棉蚜;（3）烟粉虱传播试验，用密网箱收集烟粉虱后，罩在病株枝条上，吸食 8d 后，转移至三叶期棉花幼苗上，集团接种，每隔 6h 拨动棉花 1 次，确保烟粉虱接触每一株幼苗，5d 后用药剂杀死烟粉虱;（4）汁液摩擦传播试验，再用磷酸缓冲液稀释 10 倍后加入石英砂，参考谢联辉等方法（谢联辉和林奇英，2004），选用病症明显的嫩叶剪碎，加入适量的磷酸缓冲液（0.05mol/L，pH 8.0）研碎过滤，获取病株汁液 2mL，摩擦接种 50 株棉花幼苗。

2 结果与分析

2.1 病害症状

该病发生在荆州市太湖农场部分棉花田上，只在杂交抗虫棉太 D5 号（鄂杂棉 10 号）棉花品种上发现。发病棉花田块有明显的发病中心，发病率为 5% ～ 10%，棉花病株挂桃少甚至不挂桃，棉桃也小；棉花病株矮化现象明显，高度不及正常植株的 2/3，主枝生长受到抑制，分枝较多；病叶较正常叶片嫩绿，脉明，皱缩，多带状，少数叶片倒舌状，晚分枝枝条上叶片有时呈线形，叶片不能展开；茎秆早期点状物增多，后期则逐渐粗糙形成粗皮（图 1）。

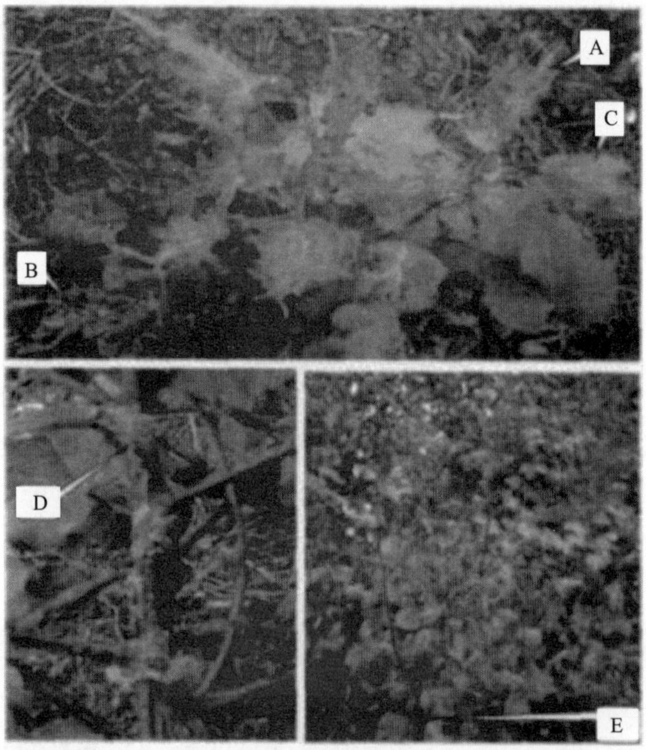

图 1　棉花皱缩花叶病的症状

A. 病叶明脉、皱缩、花叶，多带状叶；B. 部分病叶倒舌状；C. 部分新生病叶线形；D. 病株茎秆粗皮；E. 植株矮化，主茎秆不明显，分枝增多

2.2 电镜观察结果

电镜观察显示病株叶片汁液中存在 4 种杆状病毒颗粒，大小分别约是 18nm×300nm、18nm×500nm、18nm×800nm、18nm×1100nm（图 2）。

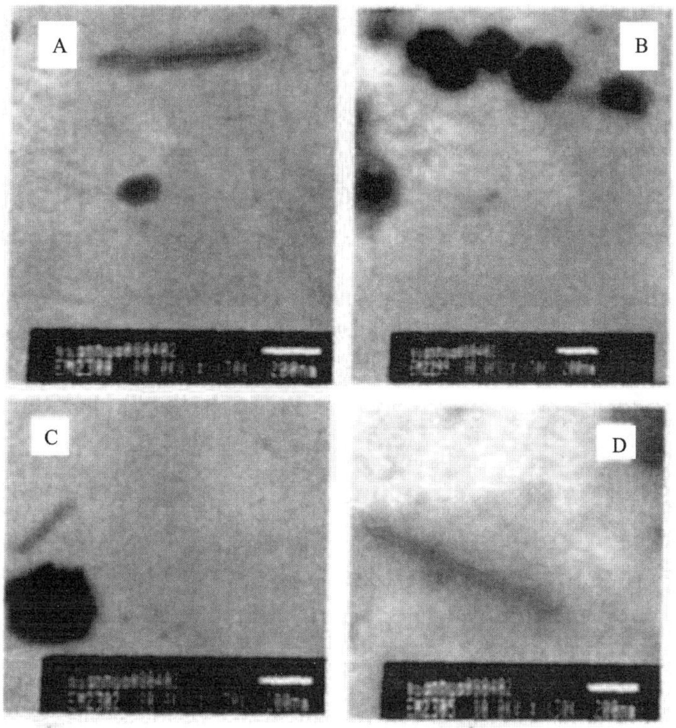

图 2 棉花皱缩花叶病叶片汁液电镜观察结果

A. 病毒粒体（18nm×500nm）；B. 病毒粒体（18nm×1100 nm）；C. 病毒粒体（18nm×300nm）；D. 病毒粒体（18nm×800nm）

2.3 生物学试验结果

生物学接种试验分别于2007年和2008年在防虫温室中进行。种传试验播种300粒种子，成活164株；蚜传接种试验接种 70 株，成活 63 株；汁液摩擦接种 150 株，成活 132 株；烟粉虱传毒接种试验接种棉花 80 株，成活 65 株。但 4 种方法接种的病株无一发病。如果这种病害是由病毒引起的，则说明这 4 种传毒方式不能传播该病害。

3 讨论

在诊断棉花皱缩花叶病的过程中，有些专家认为这种症状是农药 2, 4-D 引起的。但 2, 4-D 引起的病害往往在棉田中较大面积发生，无发病中心，而从棉花皱缩花叶病的症状上来看，是典型的病毒病症状。因此，初步认定该病害是一种植物病毒病。植物病毒病的鉴定需要做很多工作，其中关键是科赫法则。但在生物学接种试验中，4 种传毒方式皆不能够传播这种病害，因此，不能够完成科赫法则所要求的程序，也就无法确定该病害是由本研究电镜观察到的 4 种病毒引起的。值得注意的是，病汁液中存在长达 1100nm 的病毒粒体，这在病毒中比较少见（洪健等 , 1999, 2001）。其他几种病毒粒体是粗提过程中该病毒的断裂所致，还是不同的病毒粒体，或是多分体病毒，需要进一步研究。

许多植物病害的流行都是在人们忽视它或不认识的情况下开始积累的，如棉花枯黄萎病、棉花曲叶病的流行和扩散。我国是世界上最大的棉花生产国和消费国之一，棉花总产量和总消费量均占世界的 1/4。棉花安全生产非常重要，在做好栽培育种的同时，必须加强棉花病虫害的监控和管理。尽管目前我国还没有棉花病毒病发生的报道，但必须保持警惕，及时了解邻国巴基斯坦和印度棉花曲叶病的发生动态，同时投入人力、物力于棉花新发病害（如棉花皱缩花叶病）的研究，力争将新发病害拒绝于国门外，或消灭在萌芽状态。

参考文献

[1] 郭荣 . 对棉花生产构成严重威胁的病害 —— 棉花曲叶病毒病 [J]. 中国植保导刊 , 2005(02):46-47.
[2] 洪健 , 陈集双 , 周雪平 , 等 . 植物病毒的电镜诊断 [J]. 电子显微学报 , 1999, 18 (3): 274-289.
[3] 洪健 , 陈集双 , 周雪平 , 等 . 植物病毒的电镜诊断 (续篇)—— 新增植物病毒科和属及其形态学和细胞病理学特征 [J]. 电子显微学报 , 2001, 20 (6): 772-779.
[4] 青玲 , 周雪平 . 棉花曲叶病的研究进展 [J]. 植物病理学报 , 2005, 35 (3): 193-200.
[5] 王红梅 . 棉花抗黄萎病遗传及分子标记研究 [D]. 武汉：华中农业大学 ,2005.
[6] 吴云峰 . 植物病毒学原理与方法 [M]. 西安：西安地图出版社 , 1999, 72-98.
[7] 吴兴泉 . 福建马铃薯病毒的分子鉴定与检测技术 [D]. 福州：福建农林大学 , 2002.
[8] 夏更寿 , 郭志平 . 马铃薯病毒试管苗保存技术及病毒侵染力的研究 [J]. 扬州大学学报 (农业与生命科学版), 2007,28(4):103-105.
[9] 谢联辉 , 林奇英 . 植物病毒学 [M]. 2 版 . 北京：中国农业出版社 .2004.
[10] 张桂寅 . 棉花黄萎病抗性表现及其基因表达的研究 [D]. 保定：河北农业大学 , 2005.
[11] BRIDDON R W, MARKHAM P G. Cotton leaf curl disease[J]. Virus Research, 2000, 71: 151-159.

水稻黑条矮缩病的发生和病毒检测

章松柏[1,2]，李大勇[3]，肖冬来[2]，张长青[1]，吴祖建[2]，谢联辉[2]

（1 长江大学农学院，湖北荆州 434025；2 福建农林大学农药生物化学教育部重点实验室，福州 350002；
3 湖北省荆州市植保站，湖北荆州 434000）

摘要： 2009 年湖北荆州部分水稻田发生严重的水稻矮缩减产的现象。经过症状观察、10% SDS -PAGE、RT -PCR 等方法，诊断水稻发病是由水稻黑条矮缩病毒引起的。

关键词： 水稻黑条矮缩病毒；10% SDS-PAGE；RT-PCR

中图分类号： S432.1 **文献标识码：** A **文章编号：** 0439 － 8114（2010）03 － 0592-03

Occurance and Virus Molecular Detection of Rice Black Streaked Dwarf Disease

Songbai Zhang[1,2], Dayong Li[3], Donglai Xiao[2], Changqing Zhang[1], Zujian Wu[2], Lianhui Xie[2]

(1 College of Agriculture,Yangtze University, Jingzhou 434025, Hubei, China; 2 Key Laboratory of Pesticide and Biochemistry, State Education Ministry, Fujian Agriculture and Forestry University, Fuzhou 350002, China; 3 Plant Protecting Station of Jingzhou City, Jingzhou 434000, Hubei, China)

Abstact: There had not yet been reported that rice plant disease was caused by rice black streaked dwarf virus in Hubei. But in 2009, a rice dwarf disease, similar to rice black streaked dwarf disease, caused rice yield great loss in some counties of Jingzhou city. By symptoms observed, 10% SDS-PAGE and RT-PCR detection, the pathogen of rice dwarf disease was identified as rice black streaked dwarf virus.

Key words: rice black streaked dwarf virus; 10%SDS-PAGE; RT-PCR

水稻黑条矮缩病毒（*Rice black streaked dwarf virus*, RBSDV）是呼肠孤病毒科（*Reoviridae*）中的斐济病毒属（*Phytorevirus*）成员（Desselberger, 2002），可由灰飞虱、白背飞虱、白带飞虱以持久性方式传播，自然寄主包括玉米、水稻和小麦等禾本科作物（杨本荣和马巧月，1983; 张恒木 , 2001）。受 RBSDV 侵染的水稻叶色浓绿，植株矮缩，最大特征是叶脉上产生长度不同的蜡白色短条状突起或肿瘤，后期变为黑褐色，叶缘叶尖常扭曲，不结穗或很少结穗，以致产量损失严重。在我国，玉米粗缩病和水稻黑条矮缩病均是由 RBSDV 引起的病毒病害（方守国等 , 2000; 张恒木 , 2001; 陈声祥和张巧艳 , 2005）。20 世纪 50 年代该病毒在我国玉米和水稻上首次报道后，70 ～ 80 年代以来一直零星或局部发生，到 90 年代后在我国的许多地区流行，造成严重的损失。该病毒在南方（江苏、浙江、广东等省）水稻产区引起水稻黑条矮缩病发生，而在北方主要引起玉米粗缩病发生。湖北省地处南北过渡地带，玉米和水稻

湖北农业科学 . 2010, 49(3): 592-594
收稿日期：2009-11-30
基金项目：农业部农业公益性行业科研专项项目（nyhyzx07-051）；长江大学校基金项目（2006Z2070）

都有种植，且水稻种植面积大于玉米，传毒介体灰飞虱也存在，但只有玉米粗缩病流行的报道，而未见有水稻黑条矮缩病发生的报道（王朝辉，2004）。

20 世纪 70 年代湖北鲜有水稻矮缩病毒等水稻病毒病报道；90 年代以后，江浙一带、广东、福建等相继报道有多种水稻病毒病的大流行，而湖北水稻产区始终未有水稻病毒病发生的相关报道；2009 年，江汉平原的公安、松滋、石首等水稻产区晚稻大面积发生矮缩病，当地植保站联合湖南邻近县市植保站对病害进行了初步鉴定，认为可能是水稻霜霉病，并按此病防治方法指导农民生产，最终未见效果。应荆州市植保站邀请，在植物病毒学家谢联辉院士的指导下，笔者对发生在公安、松滋等县市的水稻矮缩病样品进行基因组核酸分析和 RT-PCR 扩增，确诊该病病原物为水稻黑条矮缩病毒。

1 材料与方法

1.1 材料

水稻矮缩病株于 2009 年 10 月 13 日采自湖北省公安县发病田间，活株剪掉上半部后种植于福建农林大学植物病毒所防虫网室内，病叶则保存于 –70℃超低温冰箱中；对照样品为已经鉴定的海南崖城水稻黑条矮缩病病株和健康水稻叶片。

1.2 症状观察

于发病田间仔细观察发病症状，注意叶片颜色、叶脉、茎秆部有无瘤状突起、抽穗和结穗情况等。

1.3 dsRNA 病毒基因组的提取

病毒基因组（dsRNA）的提取参考周雪平等（周雪平和李德葆，1995）病株中直接提取病毒基因组的方法，略有改进。步骤如下：采集具有典型症状的水稻病叶，液氮研磨成粉末。用 2 倍体积的 2×STE 及酚、氯仿抽提，低速离心。上清液用无水乙醇调至体积分数为 17%，注入用 17% 乙醇（V /V）的 1×STE 平衡过的纤维素（CF-11）柱中，然后用 80mL 含 17% 乙醇的 1×STE 洗脱，再用不含乙醇的 1×STE 洗脱，洗脱液用乙醇沉淀后真空抽干，即为 RBSDV-dsRNA。

1.4 SDS-聚丙烯酰胺凝胶电泳检测

提取的病毒 dsRNA 基因组加等体积 2× 凝胶加样缓冲液［100mmol/L Tris-HCl（pH 8.0），2% 巯基乙醇，4% SDS，0.2% 溴酚蓝，20% 甘油］煮沸 10min，取 10μL 加样于 10% SDS- 聚丙烯酰胺凝胶上，加 DNA Marker 和 RGDV dsRNA 基因组做对照；采用 TBE 缓冲体系，于 120V 电泳 10h，EB 染色或银染观察结果。

1.5 RT-PCR 检测

RT- PCR 检测用引物采用扩增 RBSDV S9ORF1 的引物

（S9F：5′-GGAATTCATGGCAGACCAAGAGCGGGGAG-3′，

S9R：5′-CGCGGATCCTCAAACGTCCAATTTCAAGG-3′）。

RT：对照样品组织中的总 RNA 采用总 RNA 分离试剂盒（天根生化科技有限公司）抽提，待检测样品直接使用已经分离得到的 dsRNA 病毒基因组。随后，薄壁 PCR 管加入 4μL 的总 RNA 或 dsRNA，1μL S9R 引物（20μmol /L），95℃变性处理 5min 后，迅速置冰浴中冷却 3min。上述处理的薄壁 PCR 管中加入 4μL 5× 反转录缓冲液，1μL 10mmol / LdNTP，10μL DEPC 处理水。加 0.5μL（20U）RNasin（Sigma），0.5μL（100U）M-MLV 反转录酶（Promega）。42℃条件下反应 40min，95℃ 5min。PCR：10μL 反转录混合物，5μL 10× 缓冲液，2μL 10mmol / LdNTP，3μL 25mmol/L MgCl2，1μL S9F（20μmol/L），1μL S9R

(20μmol / L)，0.25 μL（1.25U）*Ex-Taq*（TaKaRa），加去离子水至 50μL。95℃变性 5min，30 个循环的扩增（95℃ 30s，54℃ 50s，72℃ 1min），最后 72℃延伸 10min。产物经 1%琼脂糖凝胶电泳，EB 染色后 BioRad 凝胶成像系统观察记录。

2 结果与分析

2.1 病害症状

在湖北省荆州地区该病只在晚稻上发生。晚稻早期受 RBSDV 侵染后叶色浓绿，植株矮缩，分蘖增多，病株成丛生状，不结穗或很少结穗，以致产量损失严重；病株中下部茎秆上有乳白色、呈蜡点状纵向排列的瘤状突起；后期感病植株矮化不明显，但叶色深绿，茎秆表面也有少量蜡点状小瘤突；与报道不同的是，上部叶片叶尖未有明显卷曲，叶脉未见短条状瘤突；水稻品种的抗性不同，如奥两优 28 是高感品种，发病后呈现水稻黑条矮缩病所描述的典型症状。

2.2 SDS - PAGE 检测结果

从 5g 水稻病叶抽提 RBSDV dsRNA 基因组，然后进行 10% SDS - PAGE 检测，结果如图 1。因为 RBSDV dsRNA 基因组具有特异性的电泳图谱，所以通过 10% SDS-PAGE 可以直接检测病原的有无。从图 1 中可以看到 RBSDV 特异性的 dsRNA 基因组电泳图谱，证明该病原为 RBSDV。SDS-PAGE 还可以比较病原基因组 dsRNA 的带型特征，研究其基因组 dsRNA 多样性。但是该方法检测病毒需要大量的病叶，工作量较大，对于少量的病叶和单头介体及大批量的样品不适用（王朝辉等 , 2001；章松柏等 , 2005）。

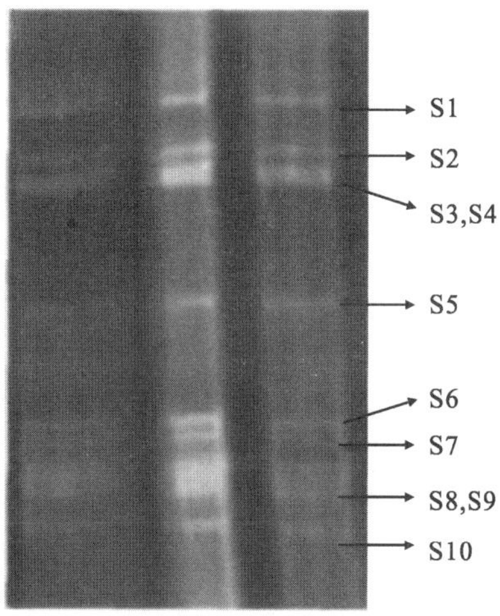

图 1 RBSDV dsRNA 基因组 10% SDS–PAGE 图谱

2.3 RT - PCR 检测结果

采用实验室保存的、用于扩增 RBSDV S9ORF1 的引物对（S9F / S9R），能稳定地从病株叶组织总 RNA 抽提物或病毒 dsRNA 基因组中经 RT - PCR 扩增出电泳呈单一明亮条带的 DNA 产物，产物分子量大小为 1000bp 左右，符合预期（1043bp），阴性对照健康植株无明显的扩增产物，阳性对照海南崖城样品有相同大小的条带（图 2）。说明湖北荆州地区各县市发生的水稻矮缩病为水稻黑条矮缩病毒引起。

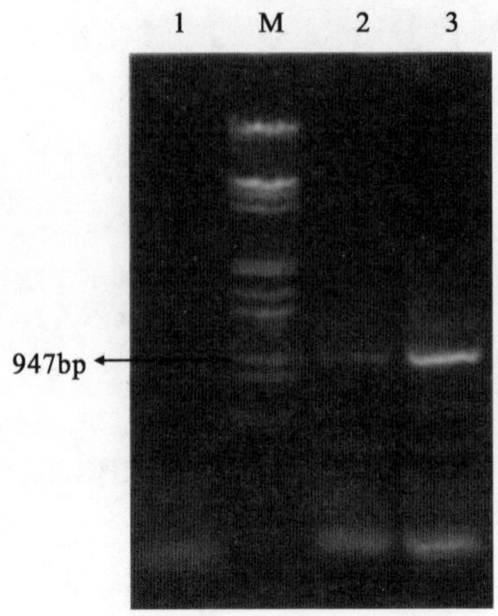

图 2 样品的 RT–PCR 检测

1. 健康水稻样品；M 核酸标准分子量；2. 海南崖城样品；3. 湖北公安样品

3 讨论

在病害诊断中，病原检测及鉴定至关重要。2009 年该病在荆州地区发病初期，被误诊为水稻霜霉病，并按此病防治方法指导农民用药，最终未见效果，给农民造成了很大的损失。采用 10% SDSPAGE 对提取的病毒 dsRNA 基因组进行核酸分析、用特异性引物（S9F / S9R）对病株样品进行了病原 RT-PCR 特异性检测，结果稳定可靠。病原鉴定后可以有针对性地进行防治指导，避免不必要的损失。下一步，可以利用该技术对昆虫的带毒率进行检测，用于病害的流行和预测预报。

玉米粗缩病在湖北早就有报道，并造成流行和较为严重的损失，但只在玉米上发生，而湖北发生玉米粗缩病的地区多有水稻，传毒介体灰飞虱也存在，却一直未有水稻黑条矮缩病发生的报道。并且 20 世纪 90 年代以来华东各稻区、华南各稻区、湖南和江西两省部分稻区相继报道有多种水稻病毒病的大流行，而湖北水稻产区始终未见水稻病毒病发生的相关报道，其中原因值得研究。近年来水稻黑条矮缩病在越南和我国海南、湖南、广西、广东、江西等地大面积发生，专家推测可能是带毒介体由南向北迁飞过程中传播而引起的（周国辉等 , 2008 ；徐海莲等 , 2009）。这次湖北省的公安、石首等部分水稻产区发病严重，这些发病的水稻产区都邻近湖南，因此，推测该病是由迁飞性介体沿湖南扩散传入。同时专家认为引起这些地区水稻矮缩病的病原是南方黑条矮缩病毒，传毒的介体主要是白背飞虱（周国辉等 , 2008）。至于 2009 年湖北省荆州地区部分稻区水稻黑条矮缩病病原是 RBSDV 还是 SRBSDV，以及主要传毒介体是灰飞虱还是白背飞虱，还有待于进一步研究。

参考文献

[1] 陈声祥 , 张巧艳 . 我国水稻黑条矮缩病和玉米粗缩病研究进展 [J]. 植物保护学报 , 2005, 32（1）: 97-108.

[2] 方守国 , 于嘉林 , 冯继东 , 等 . 我国玉米粗缩病株上发现的水稻黑条矮缩病毒 [J]. 农业生物技术学报 , 2000, 8（1）: 5.

[3] 王朝辉 . 水稻黑条矮缩病毒玉米分离物的分子特性及其侵染体系 [D]. 北京：中国农业大学 , 2004.

[4] 王朝辉 , 周益军 , 范永坚 , 等 . 应用 RT-PCR、斑点杂交法和 SDS-PAGE 检测水稻黑条矮缩病毒 [J]. 南京农业大学学报 , 2001, 24: 24-28.

[5] 徐海莲 , 肖水仙 , 刘银发 , 等 . 水稻黑条矮缩病的发生原因和防治对策 [J]. 安徽农学通报 , 2009, 15: 157.

[6] 杨本荣 , 马巧月 . 玉米粗缩病的病毒寄主范围研究 [J]. 植物病理学报 , 1983, 15（13）: 1-8.

[7] 张恒木 . 水稻黑条矮缩病毒分子生物学 [D]. 杭州：浙江大学 , 2001.

[8] 章松柏，吴祖建，段永平，等．水稻矮缩病毒的检测和介体传毒能力初步分析 [J]. 安徽农业科学，2005, 33: 2263-2264.
[9] 周国辉，温锦君，蔡德江，等．呼肠孤病毒科斐济病毒属一新种：南方水稻黑条矮缩病毒 [J]. 科学通报，2008, 53（1）：1-9.
[10] 周雪平，李德葆．双链 RNA 技术在植物病毒研究中的应用 [J]. 生物技术，1995: 1-4.
[11] VAN REGENMORTEL M H V，FAUQUET C M，BISHOP D H L. Taxonomy,classification and nomenclature of viruses:Seventh report of the international committee on taxonomy of viruses[R].New York, San Diege: Academic Press, 2000, 615-645.

Advances in the studies of Rice stripe virus

Donglai Xiao, Weiming Li, Taiyun Wei, Zujian Wu, Lianhui Xie

(Institute of Plant Virology, Fujian Agricultural and Forestry University, Fuzhou 350002, China)

Abstract: Rice stripe virus(RSV), the type member of the genus *Tenuivirus*, is one of the most economically important pathogens of rice and is repeatedly epidemic in China, Japan and Korea. The latest achievements of the studies on the biological functions of virus-encoded proteins, pathogenicity differentiation and genetic diversity of virus, virus-plant host interactions and management of virus were reviewed. The current problems encountered during studies and some approaches for further research were discussed.

Keywords: Rice stripe virus; achievements; biological functions of virus-encoded proteins

1 Introduction

The rice stripe disease was first described in Japan in 1897 (Kisimoto, 1965). Its causative agent, designated as Rice stripe virus (RSV), is a typical member of the genus *Tenuivirus* (Haenni et al., 2005), which is not assigned to any virus family yet. RSV is one of the most economically important pathogens of rice and is repeatedly epidemic in China, Japan and Korea (Lin et al., 1990). Symptoms caused by RSV in rice vary with the plant cultivars and developmental stage, and include two types, i.e. folded-leaf and unfolded-leaf with common features of chlorotic mottling (Lin et al., 1991; Ishii and Ono, 1966). RSV is transmitted by a small brown planthopper, *Laodelphax striatellus* Fallen (Hemiptera, Delphacidae), in circulative– propagative and transovarial manners (Falk and Tsai, 1998).

The rice stripe disease was first reported to occur in Jiangsu–Zhejiang–Shanghai (JZS) district in 1963 and was later discovered in 16 provinces of China (Lin et al., 1990). Since 2000, the disease has circulated widely in Jiangsu Province and has become more severe. For instance, in the years 2004 and 2005, the RSV-affected areas in Jiangsu Province reached 1.7 million hm^2, while serious incidence areas were 1 million hm^2 (Zhang et al., 2007a). Due to climatic changes as well as cultivation of susceptible cultivars, the rice stripe disease is not only becoming increasingly serious in rice annually (Zhang et al., 2007c), but also extending to wheat crops and caused great losses in wheat production in recent years (Cheng et al., 2007). Many efforts have been made in investigating the biology and molecular biology of RSV, as reviewed previously (Xie et al., 2001; Sun and Jiang, 2006; Zhang et al., 2007a). Herein, we highlight the advances in the studies of molecular biology and virus-host interaction of RSV.

2 Physical properties

RSV is a multicomponent RNA virus comprising four single-stranded RNA segments, designated as RNA1-4 in the decreasing size (Toriyama and Watanabe, 1989). The 8.9-kb RNA1 is completely negative sense

Frontiersof Agriculture in China. 2010, 4(3): 287–292

Received 13 April 2010; Accepted 25 May 2010.

and contains a single open reading frame (ORF) in the viral-complementary (vc) sense encoding the 336.8-kDa protein, the putative RNA-dependent RNA polymerase (RdRP) (Toriyama et al., 1994). The ambisense 3.5-kb RNA2 encodes a 22.8-kDa nonstructural protein (NS2) in the viral (v) sense and the 94-kDa NSvc2 protein in the viral complementary sense (Takahashi et al., 1993). In an expression strategy similar to the RNA2, the 2.7-kb RNA3 encodes a 23.9-kDa nonstructural protein (NS3) in the v sense and a 35.1-kDa coat protein (CP) in the vc sense (Kakutani et al., 1991), while the 2.1-kb RNA4 encodes two nonstructural proteins, including a 20.5-kDa disease-specific protein (SP) in the v sense and a 32.4-kDa NSvc4 in the vc sense (Kakutani et al., 1990; Zhu et al., 1992).

Each genomic RNA is encapsidated by multiple copies of the viral CP to form virion (Gingery, 1988), a filamentous particle, 500–2000nm, but with a diameter of only about 3–8nm (Koganezawa, 1977). The infection of RSV in plant cells induces the formation of four types of inclusion bodies, including electron-dense amorphous semi-electron-opaque inclusion bodies (dASO), ring-like structure, fibrillar amorphous semi-electron-opaque inclusion bodies (fASO) and filamentous electron-opaque inclusion bodies (FEO), which contain distinct sets of viral protein(s) (Liang et al., 2005b).

The genomic RNAs share highly conserved and complementary terminal sequences in 5′ ends (5′-ACACAAAGUCC-3′) and 3′ ends (5′-GACUUUGUGU-3′), with the exception of the varied sequence (5′-GACUAUGUGU-3′) in 3′ end of RNA1. Furthermore, 20 complementary pairing bases exist in 5′ and 3′ terminuses of each segment and form the characteristic panhandle structure typical of negative-strand viral RNAs, which may be the recognition site of RdRP, while the mismatch at the sixth base from the 3′ end is probably involved in the replication, transcription and/or the assembly of ssRNA1 into the filamentous particles (Takahashi et al., 1990; Barbier et al., 1992). The A/U rich sequences locate in the intergenic regions of RNA2, RNA3 and RNA4 and potentially form inverted repeats, which are supposed to be involved in the regulation of transcription termination in negative-sense RNA viruses (Zhang et al., 2007b). The transcription of RSV mRNAs was presumed to be achieved by cap– snatching mechanism, as evidenced by the fact that the 5′ terminus of mRNA, encoded by both vRNAs and vcRNAs, contains 10 to 23 host-derived bases, which may function as the capped RNA primers that initiate viral mRNA synthesis (Shimizu et al., 1996).

3 Biological functions of RSV-encoded protein

Basic knowledge on virus-encoded proteins is essential to understand this pathogen. The RdRP is presumed to be encoded by RNA1, as evidenced by the fact that a high level of RdRP activity associated with RSV filamentous particles in *vitro* is attributed to a minor polypeptide (Toriyama, 1986) with molecular weight (MW) of about 230kDa, indicating that only the largest genome segment, RNA1, is enough to express such a protein. Sequence analysis reaffirmed that the vc sequence of RNA1 does contain an ORF encoding a putative protein that shares high similarity with the RdRP of phleboviruses L protein (Toriyama et al., 1994). Interestingly, the ORF within the RNA1 encodes a protein of 2919 amino acids with an estimated MW of 337kDa, which differs in the MW (230kDa) of the polypeptide described above, suggesting that vcRNA1 encoded protein probably undergoes post translational cleavage. An ovarian tumour (OTU)-like domain was recently found within the N-terminal region (aa 1–500) of deduced RdRP and is absolutely conserved among all known RSV isolates (Zhang et al., 2007b). This finding indicates that the RSV polymerase should possess an endonuclease activity, allowing it to cleave host mRNAs 10–23nt from their 5′-capped ends and, subsequently, use the resulting capped leader to prime transcription of the viral genome (Shimizu et al., 1996).

The RdRP is responsible for viral RNA transcription and replication in initial infected cells, whereas movement protein (MP) is required to assist viral spread from cell to cell. Rice grassy stunt virus (RGSV) is one of the known members in the genus *Tenuvirus*. A small amount of RGSV NS2 was detected in cell-wall, organelle-enriched and crude membrane fractions (Chomchan et al., 2002) typical of the subcellular localization of virus MP, suggesting that RGSV NS2 as well as its counterpart RSV NS2 functions as the putative MP of tenuviruses. Furthermore, previous reports showed that RSV NS2 was associated with cell wall in virus-infected rice plants (Takahashi et al., 2002), and the transient expression of NS2-GFP led to formation of punctate structures in association with cell membrane in *Nicotiana benthamiana* leaves (Wei et al., unpublished data). These data together suggest the possible function of RSV NS2 as the candidate of RSV MP. Recently, NSvc4 was shown to have the ability to move between cells and complement the cell-to-cell movement of a movement- defective mutant of Potato virus X in *N. benthamiana* leaves, indicating that RSV NSvc4 is a putative MP of RSV (Xiong et al., 2008), though the direct evidence to show that this protein assists RSV for cell-to-cell movement in infected rice cells is unavailable. The transient expression of RSV NSvc4-GFP led to formation of punctate structures targeting to plasmodesmata (PD). Furthermore, the early secretory pathway and actomyosin motility system are required for the proper targeting of NSvc4 to PD *in planta*. (Wei et al., unpublished dada). The ribonucleo protein particles (RNPs) represent the minimal infectious unit for RSV, which contains at least the RdRPs, CPs, and genome length viral RNAs (Ramírez and Haenni, 1994; Falk and Tsai, 1998). RSV NSvc4 interacts with the CP by using in vitro binding experiments (Zhang et al., 2008a) and binds nucleic acid (Liang et al., 2005a), suggesting that NSvc4 associates with the viral ribonucleocapsid for transport from cell to cell. These data suggest that the NS2 and NSvc4 of RSV may coordinate the formation of PD-associated structures that facilitate the intercellular movement of RSV in infected plants.

Tenuivirus has phylogenetic similarity to the members of *Bunyaviridae* (Xie et al., 2001). RSV NSvc2 is analogous to the glycoproteins of bunyaviruses and topoviruses, the members of *Bunyaviridae*, which are involved in the receptor-mediated uptake of their particles (Liang et al., 2005b). Transient expression of seven proteins encoded by RSV fused with fluorescent proteins in *N. benthamiana* leaves revealed that only NSvc2 is associated with membranous structures, corresponding to two distinct transmembrane domains within the amino acid sequences of NSvc4 predicted by computer tools. The FEO-like inclusion bodies contain NSvc2 in gut lumen of epithelial cells of insect vector (Liang et al., 2005b), suggesting that NSvc2 may have a role in mediating the entry of RSV RNPs into insect vector cells in the form of FEO inclusion bodies.

Characterization of the gene silencing suppressor is one of recent progress in RSV research. The RSV NS3 was demonstrated to be able to strongly suppress local GFP silencing and prevent long-distance spread of silencing signals (Xiong et al., 2009). This result is consistent with the previous finding that the NS3 protein of the Rice hojablanca virus (RHBV), a tenuivirus, functions as a genesilencing suppressor (Bucher et al., 2003). However, it seems that the RSV NS3 is not a pathogenicity determinant because no any visible symptom or deformation appears in transgenic rice and tobacco plants expressing this protein (Xiong et al., 2009). Like Cucumber mosaic virus (CMV) 2b (Brigneti et al., 1998), a well-known gene-silencing suppressor, the RSV NS3 accumulates predominantly in nuclei leading by a nuclear localization signal (NLS), which is essential for silencing suppression (Xiong et al., 2009). The functional mechanism of the RSV NS3 is unknown yet. However, a counterpart of the RSV NS3, the NSs of Rift valley fever virus (RVFV, genus *Phlebovirus*), has been demonstrated to block interfer on production by inhibiting host gene transcription (Billecocq et al., 2004). The RSV NS3 and RVFV NSs have the same sublocalization in nucleus (Billecocq et al., 2004; Xiong et al., 2009). It will be interesting to further investigate if these two proteins share similar functional manner.

The SP is the major non-structural protein often accumulating to a very high level in a form of cytoplasmic inclusion body in RSV-infected rice and L. *striatellus* cells (Liang et al., 2005b). This protein could be detected in chloroplast, cytoplasm and nuclear of infected rice (Liu et al., 2000) and in ovum, surface of chorion, the mid-gut lumen and the columnar cells of the viruliferous females *L. striatellus* (Wu et al., 2001). Interestingly, the total amount of SP in chloroplasts correlated very well with the severity of chlorotic symptom in rice (Lin et al., 1998; Liu et al., 2000), suggesting the potential role of this protein in pathogenesis. In addition, SP was presumed to have function in vector transmission during the early period of RSV-planthopper recognition (Gray and Banerjee, 1999) but lacking of direct evidence.

As described above, the infection of RSV induces various inclusion bodies containing different virus-encoded proteins in diseased plant or viruliferous insect vector. Because these inclusion bodies may be generated by the aggregation of virus-encoded proteins and host proteins, some of them may represent the viral replication sites at the later stage of viral infection or the aggresomes of dissociated proteins. The time course of observation of these inclusion bodies formation would help us to characterize their biogenesis in the process of viral infection. The application of protoplast or the cultures of small brown planthopper vector cells in monolayers would help us to understand the earlier biogenesis of these inclusion bodies in near future.

4 Pathogenicity differentiation and genetic diversity of RSV

Pathogenicity is usually varied among RSV isolates from different regions. According to the difference in molecular weight of SP, RSV was classified as three different strains, Hongchao, P and N (Hayashi et al., 1989). Based on the average infection rates to different varieties, RSV isolates from China were divided into 5 pathotypes, including High virulence (HV), Slight high (SH), Moderate virulence (MV), Slight low (SL) and Low virulence (LV) (Cheng et al., 2008).

The genetic diversity and population structure of RSV in China were estimated based on the nucleotide sequences of isolates collected from different geographical regions during 1997–2004. RSV isolates could be divided into two or three subtypes. The population from eastern China is composed only of subtype I/IB isolates. In contrast, the population from Yunnan province (southwest China) is composed mainly of subtype II isolates, but also contains a small proportion of subtype I/IB isolates and subtype IA isolates (Wei et al., 2009). However, the frequent gene flow of RSV occurs within the subpopulations from different districts in eastern China or Yunnan province. The most recent outbreak of RSV in eastern China was not due to the invasion of new RSV subtype(s) (Wei et al., 2009). Extensive reassortment events were found, but significant recombination event was rarely detected in RSV natural population.

5 RSV-plant host interaction

Virus-host interaction is always one of the hot topics in the field of virology. The interaction between RSV and both plant species and insect vector is largely unknown but has some recent progress specific in RSV-plant interaction. Investigation of the transcriptional profiles of RSV-infected rice variety WuYun3 revealed that host genes involved in pathways of phosphate, flavonoid and brassinoliele synthesis are remarkably induced, but genes involved in gibberellin synthesis pathway are repressed, indicating that these biological pathways are regulated under the RSV infection (Zhang et al., 2008b). Analysis of transcriptional profiles of RSV-infected rice variety Nipponbare in Japan indicated that RSV modifies the rice cellular conditions for the enhancement of activity of virus propagation, accumulation of substrates for production of progenies, and suppression of host defense systems (Satoh et al., 2010). The host endogenous hormone may play critical role(s) during the RSV-host interaction, as showing that the synthesis of endogenous IAA is obviously enhanced in the early stage of RSV infection but is repressed after appearance of typical symptoms (Yang et al., 2008), and exogenous application of abscisic acid (ABA) is able to improve the host resistance to RSV (Ding et al., 2008). As a parasite, virus encoded a limited number of proteins; thereby, some unknown yet host factors are recruited and utilized to establish successful infection. Similar to TMV, two host factors, a J protein and a member of the hsp20 family, have been demonstrated to interact with the RSV NSvc4, the putative MP of virus (Lu et al., 2009). However, the biological function of the interaction is unknown.

6 Management of RSV

Based on the properties of rice virus diseases, it is important to skillfully use the criteria of resisting, averting, eradicating and curing according to the results of forecasting and admonishing (Xie et al., 1994). Transplanting or seeding date seriously affects the incidence of rice stripe disease, so it is essential to delay the date of cultivation to avoid the peak of migration of the vector planthopper, which is earlier multiplied in wheat fields (Bae and Kim, 1994). Since the large number of viruliferous planthoppers is one of the main factors causing disease epidemic, spraying pesticide is also an effective way to control this disease.

Comparably, screening the resistant genes as well as constructing virus-resistant varieties is a more economic and environment friendly method, and many positive steps have been taken towards this aim. By using techniques of plant genetic engineering, the transgenic japonica variety of rice harboring RSV *cp* gene was generated, conferring high level of resistance to RSV (Hayakawa et al., 1992); while two ribozymes were designed to cleave RSV genome RNA4 specifically *in vitro* (Liu et al., 1996), showing an alternative strategy against this virus. As the much valuable role of Quantitative trait locus (QTL) in assisting selective breeding, QTL analysis has been used widely to investigate host genes related to RSV resistance. The *Stv-a*, *Stv-b* and *Stv-b^i* are three rice genes well known for their essential role in RSV resistance. Both *Stv-a* and *Stv-b* exist in Japanese upland varieties, and *Stv-a* locates on chromosome 6 (Washio et al., 1968). The *Stv-b^i*, an allele of *Stv-b*, locates on chromosome 11 of indica-type varieties, and the resistance mechanism is different from other resistance genes producing HR (Hayano-Saito et al., 1998; Hayano-Saito et al., 2000). In addition, numbers of QTLs on chromosomes 1, 2, 3, 5, 7 and 8 (Ding et al., 2004; Maeda et al., 2006; Sun and Jiang, 2006; Wu et al., 2009) have been reported to be related to RSV resistance. Interestingly, the QTL on chromosome 11 is also associated with the resistance to brown planthopper (BPH) and small brown planthopper (SBPH) resistance (Ding et al., 2004; Sun and Jiang, 2006; Duan et al., 2007). A novel strategy, Vector-Insect-Symbiont Technology (VIST), was proposed recently (Wen et al., 2003), aiming to use one of the endosymbiont *Wolbachia* in *L. striatellus* to control RSV. This method needs further investigation, but is promising in application. Some countermeasures such as the introduction of RSV-resistant varieties, including Yangjing 93, Zhengdao 99, Yandao 8, Xudao 3, Yangliangyou 6 and Fengyouxiangzhan, and the abandonment of rice varieties, including Wuyujing 3 and Wujing, which are highly susceptible to RSV, have been adopted. As a result, in Shanghai and Jiangsu, the virus carrier rates of *L. striatellus* in 2005 to 2009 which were 1.3%, 23.2%, 23.1%, 17.2% and 11.2% respectively, were decreasing. This means that the epidemic potential of rice stripe disease has been significantly reduced.

7 Perspectives

Numerous approaches have been taken towards a better understanding of RSV molecular biology from the virus and host perspectives, leading to fundamental knowledge on aspects of RSV genomes, movement, and virus-host interactions as outlined above. However, due to the lack of a reverse genetic system, the investigation of RSV is greatly impeded, and the knowledge in areas of RNA synthesis, assembly, systemic movement, and transmission is still largely unknown. To greatly improve the ability to manage RSV and protect the rice production, we are looking forward to more significant progress in all of these areas of RSV in the near future, e.g., the cultures of small brown planthopper vector cells in monolayers would allow us to investigate the earlier RSV replication cycle, including entry, replication, assembly, intracellular trafficking and release; the serial passage of RSV in rice plants without insect vector transmission would select the transmission-defective virus isolate, which allows us to identify the transmission-related virus-encoded proteins; the application of artificial MicroRNAs and RNA inhibition constructs derived from viral or host replication-related genes would confer efficient RSV resistance in plants; the screening of rice mutant lines would select effective inherent RSV-resistant genes. As the typical member of the genus *Tenuivirus*, RSV has several unique characteristics such as genome containing four negative-sense RNA segments, exploiting ambisense gene expression strategy, multiplication in plant and insect vector cells, as well as using various mutation mechanisms, including deletion, insertion, recombination and reassortment. Thus, the studies of RSV would help us to extend our understanding of plant virology.

Acknowledgements

This work was supported by the Major Project of Chinese National Programs for Fundamental Research and Development (No. 2010CB126203), the National Natural Science Foundation of China (Grant Nos. 30770090, 30970135) and the Special Social Commonweal Research Programs of the State (Agriculture), China (No. nyhyzx07-051).

References

[1] BAE S D, KIM D K. Occurrence of small brown planthopper *Laodelphax striatellus* Fallen and incidence of rice virus disease bydifferent seeding date in dry seeded rice[J]. Korean Journal of Applied Entomology,1994, 33:173-177.

[2] BARBIER P, TAKAHASHI M, NAKAMURA I, et al. Solubilization and promoter analysis of RNA polymerase from Rice stripe virus[J]. Journal of virology,1992, 6610: 6171-6174.

[3] BILLECOCQ A, SPIEGEL M, VIALAT P, et al. NSs protein of Rift Valley fever virus blocks interferon production by inhibiting host gene transcription[J]. Journal of virology, 2004, 7818:9798-9806.

[4] BRIGNETI G, VOINNET O, LI W X, et al. Viral pathogenicity determinants are suppressors of transgene silencing in *Nicotiana benthamiana*[J]. The EMBO journal,1998, 1722:6739-6746.

[5] BUCHER E, SIJEN T, DE HAAN P, et al. Negative-strand to spoviruses and tenuiviruses carry a gene for a suppressor of gene silencing at analogous genomic positions[J]. Journal of virology, 2003, 772: 1329-1336.

[6] CHENG Z B, DENG J H, REN C M, et al. Occurrence of Rice stripe viruson wheat in Jiangsu provinceand the identification of its pathogen with RT-PCR. Journal of Triticeae Crops, 2007,27: 1138-1142.

[7] CHENG Z B, REN C M, ZHOU Y J,et al.Pathogenecity of Rice stripe tenuivirus isolates from different areas[J]. Acta

Phytopathologica Sinica, 2008, 382: 126-131.

[8] CHOMCHAN P, MIRANDA G J, SHIRAKO Y. Detection of rice grassystunt tenuivirus nonstructural proteins p2, p5 and p6 from infected rice plants and from viruliferous brown planthoppers[J]. Archives of Virology, 2002, 14712: 2291-2300.

[9] DING X L, JIANG L, LIU S J, et al. QTL analysis for rice stripe disease resistance gene using recombinant inbred lines RILs derived from crossing of Kinmaze and DV85[J]. Acta Genetica Sinica, 2004, 313: 287-292.

[10] DING X L, XIE L Y, LIN Q Y, et al.Callose deposition in resistant and susceptible rice varieties under Rice stripe virus stress[J]. Acta Phytophylacica Sinica,2008, 351: 19-22.

[11] DUAN C X, WAN J M, ZHAI H Q, et al. Quantitative trait loci mapping of resistance to *Laodelphax striatellus* (Homoptera: Delphacidae) in rice using recombinant inbred lines[J]. Journal of Economic Entomology, 2007 1004: 1450-1455.

[12] FALK B W, TSAI J H. Biology and molecular biology of viruses inthe genus Tenuivirus[J]. Annual Review of Phytopathology, 1998, 361: 139-163.

[13] GINGERY R E. The Rice stripe virus group. In: Milne R G, ed. Theplant virus 4: The filamentous plant viruses[J]. New York: Plenum Press, 1988, 297-329.

[14] GRAY S M, BANERJEE N. Mechanisms of arthropod transmission of plant and animal viruses[J]. Microbiology and molecular biology reviews : MMBR,1999, 631: 128-148.

[15] HAENNI A L, DE MIRANDA J R, FALK B W, et al. Tenuivirus. In: Fauquet CM, Mayo MA, Maniloff J, Desselberger U, Ball LA, eds. Virus Taxonomy:Eighth Report of the International Committee on Taxonomy of Viruses[J]. San Diego: Elsevier Academic Press, 1999, 717-723.

[16] HAYAKAWA T, ZHU Y, ITOH K, et al. Genetically engineered rice resistant to Rice stripe virus, an insect-transmitted virus[J]. Proceedings of the National Academy of Sciences of the United States of America, 1992, 8920:9865-9869.

[17] HAYANO-SAITO Y, SAITO K, NAKAMURA S, et al. Fine physical mapping of the rice stripe resistance gene locus *Stvb-i*[J]. Theor Appl Genet, 2000, 101(1-2): 59-63.

[18] HAYANO-SAITO Y, TSUJI T, FUJII K, et al. Localization of the rice stripe disease resistance gene Stvb-i, by graphical genotyping and linkage analyses with molecular markers[J]. TAG Theoretical and Applied Genetics,1998, 968: 1044-1049.

[19] HAYASHI T, USUGI T, NAKANO M, et al. On the strains of Rice stripe virus1: An attempt to detect strains by difference of molecular size of disease-specific proteins[J]. Kyushu Plant Protection Research, 1989, 35: 1-2.

[20] ISHII M, ONO K. Strains of Rice stripe virus[J]. Ann Phytopathological Soc Jpn, 1966, 32:83.

[21] KAKUTANI T, HAYANO Y, HAYASHI T,et al.Ambisense segment 4 of Rice stripe virus: possible evolutionary relationship with phleboviruses and uukuviruses *Bunyaviridae*[J]. Journal of General Virology, 1990, 717:1427-1432.

[22] KAKUTANI T, HAYANO Y, HAYASHI T, et al.Ambisense segment 3 of Rice stripe virus: the first instance of a virus containing two ambisensesegments[J]. Journal of General Virology, 1991, 722: 465-468.

[23] KISIMOTO R. On the transovarial passage of the Rice stripe virusthrough the small brown planthopper *Laodelphax striatelllus* Fallen[J]. Conference on relationships between Arthropods and plant-pathogenic viruses, 1965, Tokyo: 73-90.

[24] KOGANEZAWA H. Purification and properties of Rice stripe virus[J]. The Phytopathological Society of Japan, 1975, 10: 151-154.

[25] LIANG D L, MA X Q, QU Z C, et al. Nucleic acid binding property of the gene products of Rice stripe virus[J]. Virus Genes, 2005a, 312:203-209.

[26] LIANG D L, QU Z C, MA X Q, et al. Detection and localization of Rice stripe virus gene products in vivo[J]. Virus Genes,2005b, 312: 211-221.

[27] LIN Q T, LIN H X, WU Z J, et al. Accumulations of coat protein and disease specific protein of Rice stripe virusin its host[J]. Journal of Fujian Agricultural University, 1998, 273:257-260.

[28] LIN Q Y, XIE L H, XIE L Y, et al. Studies on ricestripe II Symptoms and transmission of the disease[J]. Journal of Fujian Agriculture and Forestry University Natural Science, 1991, 211: 24-28.

[29] LIN Q Y, XIE L H, ZHOU Z J, et al. Studies on ricestripe: I Distribution of and losses caused by the disease[J]. Journal of Fujian Agriculture and Forestry University Natural Science Edition,1990, 194: 421-425.

[30] LIU L H, WU Z J, LIN Q Y, et al.Cytopathological observation of rice stripe[J]. Acta Phytopathologica Sinica, 2000, 304: 306-311.

[31] LIU L, CHEN S X, QIU B S, et al. Effect of ribozymes cleaving conserved and coding region of Rice stripe virus RNA[J]. Acta Phy-topathologica Sinica, 1996,112: 157-163.

[32] LU L M, DU Z G, QIN M L, et al. A putative movement protein of Rice stripe virus, interacts with a type I DnaJ protein and a small Hsp of rice[J]. Virus Genes, 2009, 382: 320-327.

[33] MAEDA H, MATSUSHITA K, IIDA S, et al. Characterization of two QTLs controlling resistance to Rice stripe virus detected in a Japanese upland rice line Kanto 72[J]. Japanese Journal of Breeding, 2006, 564: 359-364.

[34] RAMÍREZ B C, HAENNI A L. Molecular biology of tenuiviruses, are markable group of plant viruses[J]. Journal of General Virology, 1994, 753: 467-475.

[35] SATOH K, KONDOH H, SASAYA T, et al. Selective modification of rice *Oryza sativa* gene expression by Rice stripe virusinfection[J]. Journal of General Virology, 2010, 911: 294-305.

[36] SHIMIZU T, TORIYAMA S, TAKAHASHI M, et al. Non-viral sequences at the 5′ termini of mRNAs derived from virus-sense and virus-complementary sequences of the ambisense RNA segments of rice stripe tenuivirus[J]. Journal of General Virology, 1996, 773: 541-546.

[37] SUN D Z, JIANG L. Research on the inheritance and breeding of Rice stripe resistance[J]. Chinese Agricultural Science Bulletin, 2006, 2212:318-322.

[38] TAKAHASHI M, ISHIKAWA K, MATSUDA I, et al. Immuno-electron microscopic localization of 22.8K protein of Rice stripe virusin infected plant cells[J]. Annals of the Phytopathological Society of Japan, 2002, 682: 212.

[39] TAKAHASHI M, TORIYAMA S, HAMAMATSU C, et al. Nucleotide sequence and possible ambisense coding strategy of Rice stripe virus RNA segment 2[J]. Journal of General Virology, 1993, 744: 769-773.

[40] TAKAHASHI M, TORIYAMA S, KIKUCHI Y, et al. Complementarity between the 5′- and 3′-terminal sequences of Rice stripe virusRNAs[J]. Journal of General Virology, 1990, 7112: 2817-2821.

[41] TORIYAMA S. An RNA-dependent RNA polymerase associated with the filamentous nucleoproteins of Rice stripe virus[J]. Journal of General Virology, 1986, 677: 1247-1255.

[42] TORIYAMA S, TAKAHASHI M, SANO Y, et al. Nucleotide sequence of RNA. 1, the largest genomic segment of Rice stripe virus, the prototype of the tenuiviruses[J]. Journal of General Virology, 1994, 7512:3569-3579.

[43] TORIYAMA S, WATANABE Y. Characterization of single- and double-stranded RNAs in particles of Rice stripe virus[J]. Journal of General Virology, 1989, 703: 505-511.

[44] WASHIO O, EZUKA A, SAKURAI Y, et al. Testing method for, genetics of and breeding for resistance to rice stripe disease[J]. Bulletin of the Chugoku National Agricultural Experiment Station Series A,1968,16: 39-179.

[45] WEI T Y, YANG J G, LIAO F L, et al. Genetic diversity and population structure of Rice stripe virusin China[J]. Journal of General Virology, 2009, 904: 1025-1034.

[46] WEN J G, HU Y Q, YAN J, et al. Infection of *Wolbachia pipientis* in the small brown planthopper *Laodelphax striatellus*[J]. Journal of Shanghai Jiaotong University Agricultural Science, 2003, 21Suppl: 35-37.

[47] WU A Z, ZHAO Y, QU Z C, et al. Subcellular localization of the stripe disease-specific protein encoded by Rice stripe virus RSV in its vector, the small brown planthopper *Laodelphax striatellus*[J]. Chinese Science Bulletin, 2001, 4621: 1819-1822.

[48] WU S J, ZHONG H, ZHOU Y, et al. Identification of Q TLs for the resistance to Rice stripe virusin the indica rice variety Dular[J]. Euphytica, 2009, 1653:557-565.

[49] XIE L H, LIN Q Y, WU Z J, et al. Diagnosis, monitoring and control strategies of rice virus diseases in China[J]. Journal of Fujian Agricultural University Natural Science, 1994, 233:280-285.

[50] XIE L H, WEI T Y, LIN H X, et al. Advances inmolecular biology of Rice stripe virus[J]. Journal of Fujian Agricultural University, 2001, 303: 269-279.

[51] XIONG R Y, WU J X, ZHOU Y J, et al. Identification of amovement protein of rice stripe tenuivirus[J]. Journal of Virology, 2008, 8224: 12304-12311.

[52] XIONG R Y, WU J X, ZHOU Y J, et al. Characterization and subcellular localization of an RNA silencing suppressor encoded by Rice stripe tenuivirus[J]. Virology, 2009,3871: 29-40.

[53] YANG J G, WANG W T, DING X L, et al. Auxin regulation in the interaction between Rice stripe virusand rice[J]. Chinese Journal of Agricultural Biotechnology, 2009, 164: 628-634.

[54] ZHANG H M, SUN H R, WANG H D, et al. Advances in thestudies of molecular biology of Rice stripe virus[J]. Acta Phytophylacica Sinica, 2007a, 344: 436-440.

[55] ZHANG H M, YANG J, SUN H R, et al. Genomic analysis of Rice stripe virus Zhejiang isolate shows the presence of an OTU-like domain in the RNA1 protein and a novel sequence motif conserved within the intergenic regions of ambisense segments of tenuiviruses[J]. Archives of Virology, 2007b, 15210: 1917-1923.

[56] ZHANG K Y, XIONG R Y, WU J X, et al. Detection of the proteins encoded by *Rice stripe virus* in *Laodelphax striatellus* Fallén and interactions *in vitro* between CP and the four proteins[J].Scientia Agricultura Sinica, 2008a, 4112: 4063-4068.

[57] ZHANG S X, LI L, WANG X F, et al. Molecular variation ofthe population of Rice stripe virus and the resistance of rice varieties in northern China[J]. Plant Protection, 2007c, 335: 45-50.

[58] ZHANG X T, XIE L Y, LIN Q Y, et al. Transcriptional profiling in rice seedlings infected by Rice stripe virus[J]. Acta Laser Biology Sinica, 2008b, 175: 620-629.

[59] ZHU Y F, HAYAKAWA T, TORIYAMA S. Complete nucleotide sequence of RNA.4 of Rice stripe virus isolate T, and comparison with another isolate and with maize stripe virus[J]. Journal of General Virology, 1992, 735: 1309-1312.

湖北发生的水稻矮缩病是南方水稻黑条矮缩病毒引起的

章松柏[1,3]，罗汉刚[2]，张求东[2]，张长青[3]，吴祖建[1]，谢联辉[1]

（1 福建农林大学植物病毒研究所，福建福州 350002；2 湖北省植物保护总站，湖北武汉 430070；3 长江大学农学院，湖北荆州 434025）

摘要：近年来由水稻黑条矮缩病毒和南方水稻黑条矮缩病毒引起的水稻黑条矮缩病在华南、华东等地区暴发流行，给水稻生产带来严重损失，而湖北一直未有水稻黑条矮缩病流行的报道。然而，2009 年和 2010 年在湖北省长江沿线的荆州、孝感、咸宁、黄石、鄂州、黄冈等水稻种植区普遍发生水稻矮缩减产的现象。经过症状观察、dsRNA 电泳图谱分析、RT-PCR 等方法，诊断该病是由南方水稻黑条矮缩病毒引起，说明该病毒已经随介体昆虫白背飞虱从我国南方传播至中部水稻种植区。

关键词：水稻黑条矮缩病毒；南方水稻黑条矮缩病毒；基因组电泳；反转录聚合酶链式反应；病毒鉴定

中图分类号：S432.1；S435.111.4+9 **文献标识码：**A **文章编号：**1001-7216(2011)02-0223-04

A dwarf disease on rice in hubei province, china is caused by southern *rice black-streaked dwarf virus*

Songbai Zhang[1,3], Hangang Luo[2], Qiudong Zhang[2], Changqing Zhang[3], Zujian Wu[1], Lianhui Xie[1]

(1 Institute of Plant Virology, Fujian Agriculture and Forestry University, Fuzhou 350002, China; 2 Hubei Plant Protection Station, Wuhan 430070, China; 3 College of Agriculture, Yangtze University, Jingzhou 434025, China)

Abstract: Recently, the epidemic of rice black-streaked dwarf disease caused by rice black-streaked dwarf virus or southern rice black-streaked dwarf virus has caused great yield losses in rice production in South China and East China. However, the occurrence of this disease in Hubei Province, China has not been reported. In 2009 and 2010, a rice dwarf disease, similar to rice black-streaked dwarf disease, caused great rice yield losses in some counties of Hubei Province along the Yangtze River. By symptom observation, dsRNA profiling and RT-PCR, the pathogen of the rice dwarf disease was identified as southern rice black-streaked dwarf virus (SRBSDV). This indicates that SRBSDV has been spread from the south to the central rice-growing area of China by the insect vector *Sogatella furcifera*.

Keywords:rice black-streaked dwarf virus; southern rice black-streaked dwarf virus; genome electrophoresis; reverse transcription polymerase chain reaction;virus identification

2009 年湖北省的荆州、咸宁、孝感等市部分水稻产区晚稻上大面积发生水稻矮缩病，初步鉴定为水稻黑条矮缩病（章松柏等 , 2010）。2010 年该病又进一步扩展至湖北黄冈、黄石、鄂州等市，发

中国水稻科学 . 2011 , 25(2):223-226

收稿日期：2010-04-26；修改稿收到日期：2010-08-14

基金项目：国家转基因生物新品种培育科技重大专项资助项目；长江大学基金资助项目

病程度更为严重，面积更广。虽然该病发病症状与水稻黑条矮缩病相似，但有些症状如病株地上茎节部产生高位分枝及倒生须根，却与周国辉等（周国辉等, 2004）报道的一致。周国辉等于 2001 年在广东省阳西县首次发现一种症状类似于水稻黑条矮缩病的水稻新病毒病，并于 2008 年鉴定该病毒为呼肠孤病毒科（*Reoviridae*）斐济病毒属（*Fijivirus*）第 2 组的一个新种，暂命名为南方水稻黑条矮缩病毒（southern rice black-streaked dwarf virus, SRBSDV）（Zhou et al., 2008）。与此同时，张恒木等也认为该病毒是一个新的斐济病毒属的病毒，并将其命名为水稻黑条矮缩病毒 -2（rice black-streaked dwarf virus-2, RBSDV-2）（Zhang et al., 2008）。之所以命名为南方水稻黑条矮缩病毒或水稻黑条矮缩病毒 -2 是因为该病毒与水稻黑条矮缩病毒在传播方式、寄主范围、症状、粒子形态、基因组电泳图谱、基因组结构及已知片段（S7 ～ S10）功能等方面都极其接近（周国辉等, 2004; Zhou et al., 2008; Zhang et al., 2008; Büchen, 2010）。但与水稻黑条矮缩病毒主要传播介体为灰飞虱不同的是，南方水稻黑条矮缩病毒主要由白背飞虱传播。白背飞虱在我国中部与北部地带不能越冬（沈君辉等, 2003 ; 秦厚国等, 2003），所以近几年报道的由该病毒引起的水稻矮缩病都局限在华南地区（周国辉等, 2004; Zhou et al., 2008）。而水稻黑条矮缩病毒引起的病害流行范围较大，在南方（江苏、浙江、广东等省）水稻产区引起水稻黑条矮缩病发生，而在北方主要引起玉米粗缩病（方守国等, 2000; 张恒木等, 2001; 陈声祥等, 2005）。有意思的是，由水稻黑条矮缩病毒引起的玉米粗缩病在湖北早就有报道，并曾流行和造成较为严重的损失（方守国等, 2000），但只是在玉米上发生危害，而湖北发生玉米粗缩病的地区多有水稻种植，传毒介体灰飞虱也存在，却一直未有水稻黑条矮缩病发生的报道。RBSDV 和 SRBSDV 引起的水稻矮缩病尽管在病害症状、流行规律上有很多相似点，但还是有很大差别，防控措施也不尽相同（周国辉等, 2004; Zhou et al., 2008），因此，及时鉴定水稻矮缩病病原对病害防控是必要的。应湖北省植物保护总站邀请，作者对发生在荆州、孝感、咸宁、黄石、黄冈等市的水稻矮缩病样品进行 dsRNA 基因组核酸分析和 RT-PCR 扩增和序列分析，确诊引起上述地区发病的病原为南方水稻黑条矮缩病毒，以下是具体研究报告。

1 材料与方法

1.1 材料

水稻矮缩病 22 个矮化样品于 2009 ～ 2010 年分别采自湖北省发病田间，采集地点包括孝感市郊区、荆州市公安县、石首市、监利县，咸宁市嘉鱼县、崇阳县、通城县，黄冈市黄州区、黄梅县、浠水县，鄂州市阳新县，黄石市大冶市等 12 个县市。活株剪掉上半部后种植于福建农林大学植物病毒研究所防虫网室内，病株地上部分则保存于 −70℃超低温冰箱中。对照样品为已经鉴定的福建福州水稻黑条矮缩病病株、福建连城水稻南方黑条矮缩病病株以及实验室保存的水稻锯齿叶矮缩病毒（rice ragged stunt virus, RRSV）和水稻矮缩病毒（rice dwarf virus, RDV）dsRNA 基因组核酸。灰飞虱和白背飞虱样品于 2009 年 10 月采自湖北省荆州市公安县发病田间，无水乙醇脱水处理后保存于 −70℃超低温冰箱中。

1.2 症状观察

于发病田间仔细观察发病症状，注意叶片颜色，叶脉、茎秆部有无瘤状突起，抽穗和结穗情况等。

1.3 病毒 dsRNA 基因组的提取和电泳

基于传统提取方法的原理，参照 dsRNA 传统提取法（周雪平等, 1995），用 2mL 离心管代替针筒或层析柱，进行病毒 dsRNA 基因组的快速提取，简要步骤如下：采集具有典型症状的水稻病叶或病茎 0.5g 左右，液氮研磨成粉末。用 2 倍体积的 2×STE 及酚、氯仿抽提，低速离心。上清液用无水乙醇调至含量为 17%，注入用含 17% 乙醇的 1 ×STE 平衡过的纤维素（CF11）中，然后用含 17% 乙醇的 1 ×STE 反复洗脱，直至纤维素 CF-11 变白为止。再用不含乙醇的 1×STE 洗脱，洗脱液用异丙醇、乙醇沉淀后真空抽干，即为 dsRNA。提取的病毒 dsRNA 基因组各片段用 1% 的琼脂糖凝胶电泳加以分离，用溴化乙锭染色并观察结果。

1.4 引物设计和 RT–PCR 检测

根据已经发表的湖北 RBSDV 分离物 S10 片段（AF227205）和湖北 SRBSDV 分离物的 S10 片段（HM585270）序列及序列差异设计特异性引物 PJF、PPJR 和 PNJR。PJF 根据 S10 片段保守区域设计，为扩展两种病毒特异性片段的通用正向引物。PPJR、PNJR 根据变异区域设计，前者用于扩增 RBSDV 特异性片段的反向引物，后者为扩增 RBSDV 特异性片段的反向引物。引物具体序列等信息见表 1。

表 1 RBSDV 和 SRBSDV 的检测引物

Table 1 Primers used for detecting rice blacked-streak dwarf virus (RBSDV) and southern rice blacked-streak dwarf virus (SRBSDV)

引物名称 Primer name	位置 Position	方向 Direction	序列 Sequence (5′-3′)	检测的病毒及片段大小 The virus detected and fragment size
PJF	592 -610 bp	+	5′-TGTCGTGAAGT TCCTGCTC-3′	RBSDV, 1045bp
PPJR	1637 -1616 bp	−	5′-GAAGAAACGTTGGCGGAAAGT-3′	
PJF	592 -610 bp	+	5′-TGTCGTGAAGTTCCTGCTC-3′	SRBSDV, 412bp
PNJR	1004 -986 bp	−	5′-GAATTGCCATCGACTCCTT-3′	

采自湖北公安发病田间的灰飞虱和白背飞虱样品的总 RNA 采用总 RNA 提取试剂盒（天根生化科技有限公司）抽提，待检测的 23 个湖北样品和对照样品（福建福州、福建连城）直接使用已经分离得到的 dsRNA 病毒基因组。随后，PCR 管中加入 3μL 的总 RNA 或 dsRNA，1μL PJF 引物（10μmol/L），8.5μL DEPC 处理水，95℃下变性处理 5min 后，迅速冰浴冷却 5min。随后按要求依次加入 RNasin、反转录缓冲液、dNTP 和 M-MuLV 反转录酶（Fermentas），42℃下反应 60 min，70℃下 10min，–20℃下保存备用。PCR 体系按照 TaKaRa 的 Ex-Taq 酶使用说明进行。反应程序如下：94℃下变性 4min，30 个循环的扩增（95℃下 30s，52℃下 45s，72℃下 1min），最后 72℃下延伸 10min，产物经 1% 琼脂糖凝胶电泳，用溴化乙锭染色后在 BioRad 凝胶成像系统下观察记录。

1.5 检测片段克隆和序列分析

RT-PCR 扩增所得的湖北公安样品检测片段连接 pMD18-T（TaKaRa），转化克隆验证后委托金思特生物科技有限公司测序。运用 DNA Star 5.01 软件（DNA STAR Inc.,USA）对所获序列进行加工整理和同源性分析。

2 结果与分析

2.1 病害症状

2009 年和 2010 年于湖北省的荆州、咸宁、孝感、黄冈、黄石、鄂州等市发病水稻种植区田间调查发现，该病只在中晚稻上发生。中稻发病较轻，发病面积和损失较小；晚稻发病较普遍，部分田块或品种损失较大。晚稻早期受病毒侵染后叶色浓绿，植株矮缩，分蘖增多，病株呈丛生状，不结穗或很少结穗，以致产量损失严重；病株中下部茎秆上有乳白色、呈纵向排列的蜡点状瘤状突起；后期感病植株矮化不明显，但叶色深绿，茎秆表面也有少量蜡点状小瘤突；与水稻黑条矮缩病毒引起的水稻矮缩病不同的是，叶脉未见短条状瘤状突起；病株地上茎节部产生高位分枝及倒生须根；未发现抗病的品种，但感病程度不同，如高度感病的品种有宜优 207、鄂糯 7 号、珍糯等，较为感病的有金优 117、晚籼 98、天优 8 号、天优 998、扬两优 6 号、金优 928、丰两优 299、武优 308、华优 332、三香优 714、金优

725、丰两优 4 号等。

2.2 dsRNA 基因组琼脂糖凝胶电泳

每个地方样品取 0.5g 水稻病叶或病茎，按照改进后的方法抽提病毒基因组 dsRNA，然后进行电泳检测（图 1）。因为 dsRNA 病毒基因组具有特异性的电泳图谱，所以通过琼脂糖凝胶电泳可以直接检测病原的有无和初步判断病原的种类（章松柏等 , 2005 ；章松柏等 , 2009）。从图 1 中可以看到 3 种水稻 dsRNA 病毒（RDV、RBSDV 或 SRBSDV、RRSV）基因组特异性电泳图谱，但无法判断引起湖北水稻矮缩病的病原是 RBSDV 还是 SRBSDV，因为这两个病毒的 dsRNA 病毒基因组电泳图谱几乎一致。此外，图 1A 显示该病毒在茎和叶中的含量有很大的差异，茎中的含量远大于在叶片中的含量。

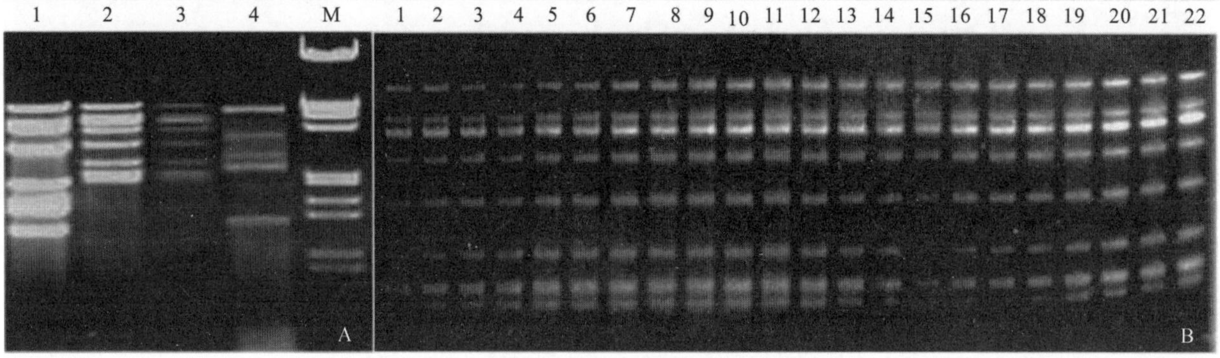

图 1　水稻病毒 dsRNA 基因组电泳分析

Fig. 1　Amplification of dsRNA of rice viruses

A: M - 标准分子量；1- 水稻矮缩病毒；2- 湖北公安样品（茎）；3 - 湖北公安样品（叶）；4 - 水稻锯齿叶矮缩病毒；B:1- 福建福州 RBSDV 样品；2- 福建连城 SRBSDV 样品；3 ～ 22- 湖北各地样品（2010 年采集）

A:M DNA Marker;1, Rice dwarf virus; 2, Samples from Gongan, Hubei (stems); 3, Samples from Gongan, Hubei (leaves); 4, Rice ragged stunt virus.B: 1, RBSDV samples from Fuzhou, Fujian; 2, SRBSDV samples from Liancheng, Fujian; 3-22, Samples collected from Hubei in 2010

2.3 RT-PCR 检测和序列分析

采用表 1 中扩增 RBSDV 或 SRBSDV S10 特异性片段的第 2 对引物（PJF/PPJR、PJF/PNJR），都能稳定地从昆虫总 RNA 抽提物或病毒 dsRNA 基因组中扩增出呈单一明亮条带的 DNA 电泳产物，产物分子量大小分别为 1000bp、400bp 左右，与预期相符（图 2）。因为对照样品（福州 RBSDV 和 SRBSDV 样品）都已经得到测序验证，所以可以判断 2009 年和 2010 年湖北长江沿线或长江以南各县市发生的水稻矮缩病为南方水稻黑条矮缩病毒引起，白背飞虱和灰飞虱都可以携带该病毒。对湖北公安样品检测片段进行测序和序列分析（表 2），结果显示检测片段与 SRBSDV 各分离物相应片段的同源性高达 99% 以上，与 RBSDV 湖北分离物相应片段的同源性只有 76.5%，而引起湖北玉米粗缩病的病原 RBSDV 湖北分离物与表 2 中 SRBSDV 各分离物的相应片段同源性也只有 76.5% 左右。这进一步说明 2009 年和 2010 年引起湖北水稻矮缩病的病原是 SRBSDV，而不是已报道的引起湖北玉米粗缩病的病原 RBSDV。

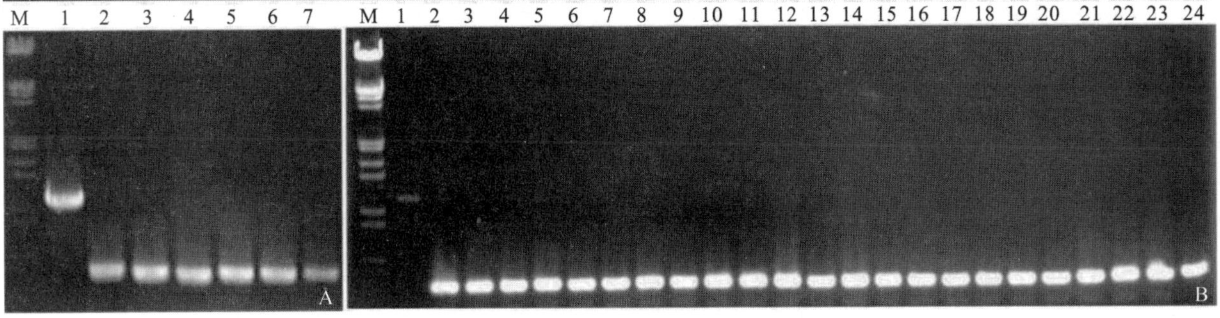

图 2 水稻样品的 RT-PCR 检测

Fig. 2 RT-PCR detection of rice samples.

A:M. 标准分子量；1. 福建福州 RBSDV 样品；2. 福建福州 SRBSDV 样品；3. 福建连城 SRBSDV 样品；4. 湖北公安样品；5. 湖北孝感样品；6. 白背飞虱；7. 灰飞虱（均为 2009 年样品）。B:M. 标准分子量；1. 福建福州 RBSDV 样品；2. 福建连城 SRBSDV 样品；3 ～ 22. 湖北各地样品（2010 年采集）；23. 湖北公安样品（2009 年采集）；24. 湖北孝感样品（2009 年采集）

A:MDNA marker; 1, RBSDV samples from Fuzhou, Fujian; 2, SRBSDV samples from Fuzhou, Fujian; 3, SRBSDV samples from Liancheng, Fujian; 4, Samples from Gongan, Hubei; 5, Samples from Xiaogan, Hubei; 6, *Sogatella furcifera*; 7, *Laodelphax striatellus*. All samples were collected in 2009. B:M DNA marker; 1, RBSDV samples from Fuzhou, Fujian; 2, SRBSDV samples from Fuzhou, Fujian; 3–22, Samples collected from Hubei in 2010; 23, Samples collected from Gongan, Hubei in 2009; 24, Samples collected from Xiaogan, Hubei in 2009

表 2 湖北公安样品的检测片段与 SRBSDV 和 RBSDV 相应片段的同源性比较

Table 2 Nucleotide identity between fragments amplified from S10 of the virus isolates from Gongan, Hubei and corresponding regions of SRBSDV or RBSDV genomes (%)

病毒分离物 Isolate of virus	湖北公安样品 Samples from Gongan, Hubei	RBSDV 湖北分离物 RBSDV isolate from Hubei
RBSDV 湖北分离物 RBSDV isolate from Hubei Province	76.5	
RBSDV-2 浙江分离物 RBSDV-2 isolate from Zhejiang Province	99.8	76 .5
SRBS DV 福建分离物 SRBSDV isolate from Fujian Province	99.3	76 .5
SRBS DV 海南分离物 SRBSDV isolate from Hainan Province	99.5	76 .8

3 讨论

在病害诊断中，病原检测及鉴定至关重要。2009 年该病在荆州地区发病初期，被误诊为水稻霜霉病，并按霜霉病防治方法指导农民生产，最终未见效果，给农民造成了很大的损失。由于 RBSDV、MRDV 和 SRBSDV 等斐济病毒属第 2 组成员 dsRNA 基因组电泳图谱极其相似，不能根据 dsRNA 图谱加以鉴定，所以需进一步通过序列分析等方法来鉴定。本研究根据 SRBSDV 基因组片段 S10 与 RBSDV 相应片段核苷酸序列差异设计引物（PJF/PPJR、PJF/PNJR），对 22 个湖北水稻矮化样品、2 个湖北介体昆虫样品以及已经鉴定的福建水稻矮化样品进行了病原 RT-PCR 检测，结果稳定可靠，能特异性地检测 RBSDV 样品（检测片段大小为 1000bp 左右），也能特异性地扩增 SRBSDV 样品（检测片段大小为 420bp 左右），所获得的扩增产物经克隆测序后，能方便地进行分子鉴定。SRBSDV 是新近确立的斐济病毒属一个暂定新种（Zhou et al., 2008; Zhang et al., 2008）。自 2001 年周国辉等首次发现以来，该病毒在华南各主要稻区已经普遍存在。近两年在海南、湖南、广西、广东、江西、湖南、福建、湖北、浙江等地大面积暴发流行的水稻黑条矮缩病，专家推测可能都是 SRBSDV 引起的。海南、福建、湖北等 3 省发生的水稻黑条矮缩病已经得到本实验室的验证，证明是由南方水稻黑条矮缩病毒引起，广西、江西、湖南、浙江等地经过会议或私下交流也予以确认，证明南方黑条矮缩病毒是病原（或病原之一）。尤其值得注意的是，2009 年湖北仅 2~3 个县市报道该病的发生，2010 年已经达到 12 个县市，发病程度也更为严重。说明 SRBSDV 引起的水稻矮缩病毒已经从华南地区扩散至华中、华东地区，对此，应予以特别重视。与 RBSDV 田间主要传播介体为灰飞虱不同的是，SRBSDV 在田间主要由白背飞虱传播（周国辉等 , 2004; Zhou et al., 2008）。白背飞虱是一种典型的迁飞性害虫，每年春季由南向北迁移，秋季由北向南回迁，分布范围几乎覆盖我国全境。上面提及的省份，除海南、福建、广东、广西等华南地区是白背飞虱越冬虫源地外（沈君辉等 , 2003 ；秦厚国等 , 2003），白背飞虱在其他省份多是不能够越冬的。因此，推测近两年这些省份水稻黑条矮缩病的暴发流行是带毒介体白背飞虱由南向北迁飞过程中传播 SRBSDV 而引起的。从目前水稻病毒病发展趋势来看，南方水稻黑条矮缩病毒引起的水稻黑条矮缩病成为水稻上最严重的病毒病之一，已经给水稻生产带来严重损失，鉴于它是新病毒，所以对它进行病害流行学等基础和应用研究势在必行。

参考文献

[1] 陈声祥 , 张巧艳 . 我国水稻黑条矮缩病和玉米粗缩病的研究进展 [J]. 植物保护学报 ,2005,32(1):97-108.

[2] 方守国 , 于嘉林 , 冯继东 , 等 . 我国玉米粗缩病株上发现的水稻黑条矮缩病毒 [J]. 农业生物技术学报 ,2000,8(1):5.

[3] 秦厚国 , 叶正襄 , 舒畅 , 等 . 白背飞虱种群治理理论与实践 [M]. 南昌：江西科学技术出版社 , 2003, 42-47.

[4] 沈君辉 , 尚金梅 , 刘光杰 . 中国的白背飞虱研究概况 [J]. 中国水稻科学 ,2003,17(增):7-22.

[5] 徐海莲 , 肖水仙 , 刘银发 , 等 . 水稻黑条矮缩病的发生原因和防治对策 [J]. 安徽农学通报 ,2009,15(13):157.

[6] 张恒木 , 雷娟利 , 陈剑平 . 浙江和河北发生的一种水稻、小麦、玉米矮缩病是水稻黑条矮缩病毒引起的 [J]. 中国病毒学 ,2001,16(3):246-251.

[7] 章松柏 , 李大勇 , 肖冬来 , 等 . 湖北水稻黑条矮缩病的发生和病毒检测 [J]. 湖北农业科学 ,2010,49(3):592-594.

[8] 章松柏 , 吴祖建 , 段永平 , 等 . 水稻矮缩病毒的检测和介体传毒能力初步分析 [J]. 安徽农业科学 ,2005,33(12):2263-2264, 2287.

[9] 章松柏 , 吴祖建 , 段永平 , 等 . 水稻矮缩病毒基因组遗传多样性的初步研究 [J]. 长江大学学报 : 自科版 ,2009,6(4):55-57.

[10] 周国辉 , 许东林 , 李华平 . 广东发生水稻黑条矮缩病病原分子鉴定 [C]// 中国植物病理学会学术年会 . 2004 年学术年会论文集 . 北京 : 中国农业科学技术出版社 , 210-212.

[11] 周雪平 , 李德葆 . 双链 RNA 技术在植物病毒研究中的应用 [J]. 生物技术 ,1995,5(1):1-4.

[12] BŮCHEN O, SMOND C. The Universal Virus Database: ICTVdB (ver.4)[DB/ OL]. 2006, NewYork:Columbia University.

[13] ZHANG H M, YANG J, CHEN J P, et al. A black-streaked dwarf disease on rice in China is caused by a novel Fiji-virus[J]. Archives of Virology, 2008, 153:1893-1898.

[14] ZHOU G H, WEN J J, CAI D J, et al. Southern rice black-streaked dwarf virus: A new proposed *Fiji-virus* species in the family Reoviridae[J]. Chinese Science Bulletin, 2008, 53(23):3677-3685.

单季稻小麦轮作区灰飞虱发生规律

程兆榜[1,2]，何敦春[2]，陈全战[3]，季英华[1]，任春梅[1]，魏利辉[1]，周益军[1]，范永坚[1]，谢联辉[2]

（1 江苏省农业科学院植物保护研究所 南京 210014；2 福建农林大学病毒研究所 福州 350002；3 南京晓庄学院 南京 211171）

摘要：为了明确单季稻 - 小麦轮作区灰飞虱 *Laodelphax striatellus* Fallén 的发生规律，采用灯诱法、盘拍法、盘刮法、黄盘诱集法对灰飞虱周年发生，特别是在寄主转移时的消长规律进行了系统研究。结果表明：灰飞虱主要在适宜藏匿的场所越冬，其中常规耕翻田主要在田边田头的禾本科枯死杂草或其他密生植物、稻套麦田主要在田中稻残桩中，灰飞虱的越冬效率为 50% ～ 60%。稻套麦田灰飞虱夏冬寄主间的转移效率约为 10%，是常规耕翻麦田的 20 倍，其冬后基数分别是机械浅旋耕和常规耕翻田的 35 倍和 77 倍。灰飞虱周年只有一个迁移扩散高峰期，与小麦收割时间基本吻合。移栽后水稻上灰飞虱数量锐减，大田前期灰飞虱主要来源于秧苗带入的未孵化卵块。通过比较发现稻套麦优化了灰飞虱的越冬环境，是近年来灰飞虱大发生及其传播的水稻病毒病大流行的主要原因。根据上述研究结果提出了压低灰飞虱越冬基数是灰飞虱水稻病毒病能够得以有效控制的前提这一新观点。

关键词：灰飞虱；寄主转移；越冬；稻套麦；水稻条纹叶枯病；水稻黑条矮缩病；控制策略

Factors affecting the occurrence of *Laodelphaxstriatellus* in a single rice-wheat rotation

Zhaobang Cheng[1,2], Dunchun He[2], Quanzhan Chen[3], Yinghua Ji[1], Chunmei Ren[1],Lihui Wei[1], Yijun Zhou[1], Yongjian Fan[1], Lianhui Xie[2]

(1 Institute of Plant Protection, Jiangsu Academy of Agriculture Sciences, Nanjing 210014, China; 2 Institute of Plant Virology, Fujian Agriculture and Forestry University, Fuzhou 350002, China; 3 Nanjing Xiaozhuang University, Nanjing 211171, China)

Abstract: We investigated seasonal population dynamics and between-host dispersion of the small brown planthopper *Laodelphax striatellus* Fallén (SBPH) in rice-wheat rotation fields, using light traps, yellow pan water traps and striking and scraping insects from wheat plants with collection trays. We found that in conventional cropping system wheat fields SBPH over wintered mainly on senescent gramineous grasses and /or other dense-growth plants while in the rice-wheat intercropping system they overwinter mainly on paddy stubble. On average, 50%–60% of SBPH population can successfully overwinter. In the rice-wheat intercropping system, without tillage the rate of transmitting SBPH from rice to wheat was about 10%; 20 times higher than with tillage. The number of SBPH successfully overwintering in the rice-wheat intercropping system with no tillage was 35

应用昆虫学报 . 2013, 50(3): 706-717
收稿日期：2012-03-23；接受日期：2012-05-31.
基金项目：国家自然科学基金 (30300230，31201506)；公益性行业 (农业) 科研专项 (201003031)；江苏省支撑计划 (BE2009385) 。

and 77 times higher than that with shallow tillage and conventional tillage, respectively. Massive migration in SBPH populations occurred only once a year, peaking at the time when wheat is harvested. The planthoppers migrated into rice seedling beds nearby. The population of adult SBPH decreased sharply after transplanting, with the new population establishing in the paddy field mainly emerging from unhatched eggs carried on the rice seedlings. We conclude that rice-wheat intercropping creates a favorable environment for the overwintering of SBPH and that this could be the main factor contributing to severe outbreaks of SBPH and rice stripe virus disease in recent years. Therefore, we propose the use of rice-wheat intercropping with a shallow tillage to control SBPH-transmitted rice virus diseases such as rice stripe virus disease and rice black-streaked dwarf virus disease, etc.

Key words: small brown planthopper; host transfer; overwinter; wheat interplanting with rice; rice stripe virus disease; rice black-streaked dwarf disease; control strategy

灰飞虱 *Laodelphax striatellus* Fallén（SBPH）属同翅目（Homoptera）飞虱科（Delphacidae）叶稻虱亚科（Delphaçinae），是水稻上常见的 3 种稻飞虱的一种，主要分布于东亚的亚热带季风气候和温带季风气候区，可在当地越冬为居留型昆虫（中国农业百科全书编辑委员会，1990）。受发生数量和种群遗传特性的限制，与白背飞虱 *Sogatellafurcifera Horváth* 和褐飞虱 *Nilaparvata lugens* Stål 不同，灰飞虱对水稻造成的直接为害一般较轻，偶有大发生导致水稻产量严重损失的报道，如 1977 ～ 1979 年在日本北海道、1985 ～ 1986 年在中国台湾（Okuyama and Kajino, 1980; Chen and Ko, 1986）。灰飞虱作为一种重要的传毒昆虫，在水稻上传播水稻条纹病毒（rice stripe tenuivirus）引发水稻条纹叶枯病、玉米和水稻上传播水稻黑条矮缩病毒（rice black streak dwarf fijivirus）引发玉米粗缩病和水稻黑条矮缩病、小麦上传播北方禾谷类花叶病毒（northern cereal mosaic cytorhabdovirus）引发小麦丛矮病等，从而造成农作物产量的重大损失，其中其传播的水稻条纹叶枯病和玉米粗缩病曾多次暴发流行（林奇英等，1990；苗洪芹等，2003）。2000 年以来，水稻条纹叶枯病在江苏等地暴发成灾并持续流行（程兆榜等，2002；朱金良等，2008），2007 ～ 2009 年在水稻条纹叶枯病危害得以基本控制之际水稻黑条矮缩病又突然在苏北和江苏沿海严重发生（季英华等，2009），这均与灰飞虱的发生相关联。

江苏为稻麦轮作区，是灰飞虱适宜生存地区之一。江苏灰飞虱发生规律已有一些报道（浦茂华，1963；程兆榜等，2002），目前已经明确灰飞虱在江苏的生活史和基本发生规律，明确了灰飞虱的越冬虫态、周年发生世代和数量、迁移时间、起飞行为、水稻栽培方式对灰飞虱数量的影响、抗药性与种群暴发关系、温度与种群发生关系、寄主适应性等（王瑞等，2008；朱金良等，2008；乔慧等，2009；王彦华等，2010；张爱民和刘向东，2010；张强翔等，2011），除此之外，刘向东等（2006）根据灰飞虱种群内外因互作关系对近年来江苏灰飞虱暴发成灾原因进行了分析，葛红等对江苏南通地区的灰飞虱寄主转移间的一般性规律进行了描述（葛红等，2010）。现有的研究从不同的侧面反映了灰飞虱的发生规律，但在江苏单季稻小麦轮作区灰飞虱种群消长及其与水稻条纹叶枯病流行危害程度密切相关的几个关键问题，如灰飞虱寄主转移规律、秧田和稻田中的消长规律及灰飞虱的越冬和越夏规律等方面还缺少系统研究。针对上述问题，笔者对多年来以江苏为代表地区的单季稻 - 小麦耕作方式下灰飞虱发生规律进行了研究，以期能对灰飞虱暴发成灾原因的理解及其所传播病毒病的防治有所帮助。

1 材料与方法

1.1 试验地点

实验在灰飞虱普遍发生和水稻条纹叶枯病流行的江苏稻麦轮作区进行，其中江苏东台、洪泽、姜堰、楚州、武进等地主要为传统的耕翻麦，江苏建湖为稻套麦。

1.2 灰飞虱的调查方法

根据不同的需要分别采用盘拍法、盘刮法、灯诱法、黄盆诱集法调查灰飞虱发生的数量、虫态，其中灰飞虱越冬场所调查、麦田灰飞虱发生动态、稻田灰飞虱消长规律采用盘拍法，灰飞虱周年迁移动态和水稻上灰飞虱侵入动态调查采用灯诱法和黄盆诱集法，秧田灰飞虱消长动态采用盘刮法。具体方法参考肖庆璞和程兆榜（2003），略作改进。

1.2.1 灯诱法

在田间设置 200W 黑光灯或白炽灯一只，每日收集灯下诱到的昆虫并进行分类，记载日灰飞虱虫量。设灯时间为 4 月 15 日至 10 月 15 日。

1.2.2 盘拍法

采用 40cm × 30cm 白瓷盘置于植株基部（苗期）或顶部（穗期）拍查，每处理定点调查 5 点，每点调查 300 株，或每处理调查 10 点，每点 1 ～ 15 盘，每盘代表面积 $0.05m^2$。

1.2.3 盘刮法

在秧苗顶端刮取。白磁盘 40cm × 30cm，每次每处理随机调查 20 盘，每盘刮取距离 1m，代表面积 $0.4m^2$。

1.2.4 黄盘诱集法

自水稻出苗始至水稻收割，在高于稻株顶端 20cm 以上设置一个 40cm × 30cm 的黄盘，黄盘的高度随植株的高度逐步调整。每日上午 6: 30 记录黄盘中诱集到的灰飞虱成虫数量。

2 结果与分析

2.1 灰飞虱的越冬及秋冬寄主转移

为了解灰飞虱越冬方式，2001 ～ 2002 年在江苏东台对常规耕翻麦田进行普查，结果发现灰飞虱从夏季生活寄主水稻向冬季越冬寄主转移时有一定的选择性，主要转移寄主有大小麦、游草、胡萝卜、野茭白、芹菜、芫荽、再生稻等（表 1），其中有的并不是灰飞虱的生活寄主（如芹菜、胡萝卜、芫荽等），这些寄主或是植株个体较小但种植密度大（如大小麦），或是生长连片（如游草）或是植株个体生长茂盛（如芫荽）或是有可藏避之处（如再生稻的稻残桩中），其共同特性是均适合灰飞虱藏匿。非转移寄主有油菜、大蒜等，这些寄主植株个体较大、分枝少、种植密度小，与上述转移寄主相反不适合藏匿。对东台丁溪和溱东两地同一寄主上的灰飞虱转移情况比较分析，丁溪的芹菜和胡萝卜比溱东高两个叶龄，植株普遍较大而两地沟渠边野生游草生育期相近，在 3 种植物上查见的灰飞虱量前者分别是后者的 11 倍、34 倍、1.3 倍，显示植株大小与灰飞虱的越冬数量存在一定的相关性，植株越大越适合藏匿灰飞虱的数量越多。同一田块中不同寄主灰飞虱越冬数量进行比较，耕翻田小麦上单位面积灰飞虱越冬密度远小于田埂、田头上的其他植物，这与小麦出苗最迟，苗最小也有一定关联。综上所述，灰飞虱越冬前的寄主转移以寻找临时栖息可藏匿的避难场所为主。

为进一步了解灰飞虱越冬之后的状况，2002 年 3 月在江苏姜堰对同类型的常规耕翻麦田进行调查。调查结果显示，与预期的灰飞虱主要在小麦上的情况相反，该类型麦田主要在田边、田头、沟渠中的禾本科枯草中查见灰飞虱（表 2）。近田埂的小麦苗和未被破坏的稻残桩上也有灰飞虱零星存在，而田中间的小麦苗上几乎查不到灰飞虱。由此说明水稻收割后农田耕翻，残存灰飞虱主要转移至田边、田头的夏季禾本科杂草里（冬季成为枯草）和稻残桩等场所越冬，这些枯死的植物虽不能为灰飞虱提供食源，但可作为灰飞虱良好的藏匿场所，这与冬前东台的调查结果相吻合。

表 1 2001—2002 年东台灰飞虱越冬情况调查

Table 1 Overwinter investigation of SBPH in Dongtai city, China from 2001 to 2002

调查日期（年/月/日）Date (year/month/day)	调查地点 Area	灰飞虱数量（头/667m²）Number of *laodelphax striatellus* per 667m²												
		小麦 Wheat	大麦 Barley	油菜 Rape	蚕豆 Broad bean	大蒜 Garlic	游草 Clubhead cutgrass herb	芹菜 Celery	胡萝卜 Carrot	雪里蕻 Potherb mustard	芫荽 coriander	野茭白 Wild zizaniar	白萝卜 Ternip	再生稻 Rationing rice
2001/11/20	丁溪 Dingxi	0	—	0	120	0	28800	13200	40800	0	—	—	—	—
	溱东 Qindong	0	0	0	0	0	21600	1200	1200	—	—	4800	0	123200
2001/12/10	丁溪 Dingxi	240	—	0	0	0	9600	—	13200	1200	0	—	—	—
	溱东 Qindong	0	180	0	0	0	—	—	—	—	—	—	0	—
2001/12/20	丁溪 Dingxi	640	—	0	0	0	3600	—	14400	2400	1200	—	—	—
	溱东 Qindong	120	1280	0	0	0	—	—	—	—	—	—	0	—
2001/12/30	丁溪 Dingxi	720	—	0	0	0	2400	—	2400	3600	1200	—	—	—
	溱东 Qindong	240	1520	0	0	0	—	—	—	—	—	—	0	—
2002/01/10	丁溪 Dingxi	1520	—	0	0	0	1200	—	52800	—	10800	—	—	—
	溱东 Qindong	240	1600	0	0	0	—	—	—	—	—	—	0	—

注：每次每种作物调查 50 盘，表中数据为平均值。其中溱东大麦前茬为水稻、小麦前茬为棉花，丁溪大、小麦前茬均为水稻。

Every plants was investigated 50 trays every time. Data in the table are average. The previous crop of barley was rice and that of wheat was cotton in Zhendong, and that of both barley and wheat was rice in Dingxi.

表 2 常规耕翻田灰飞虱越冬场所调查（姜堰，2002.03）

Table 2 Hibernant habitat investigation of SBPH in conventional ploughed land in Jiangyan city, China in 2002

场所 Habit	对象 Plants	调查盘数（盘）Trays	灰飞虱数量（头）Number of SBPH	平均密度（头/m²）Average density
田埂 Field ridge	枯草 Withered weeds	10	15	30
田槽 Field groove	枯草 Withered weeds	10	180	360
田头 Near boundary from the field	禾本科杂草 Gramineous weeds	10	10	20
田头 Near boundary from the field	油菜 Rape	20	2	2
田边 Near boundary in the field	小麦 Wheat	20	3	3
田边 Near boundary in the field	稻残桩 Withered paddy stubble	20	19	19
田中 Middle of field	小麦 Wheat	30	0	0

注：盘拍法调查，搪瓷盘 40cm × 30cm，每盘代表面积 0.05m²

Investigated with a tray of 40cm length and 30cm width which represented 0.05m²

近几年来稻套麦在江苏的种植面积越来越大，对此类麦田调查发现稻桩而不是麦苗成为灰飞虱主要越冬场所（表 3），平均每丛稻桩有灰飞虱 3 ～ 20 头。2005 年江苏建湖调查，稻套麦田灰飞虱越冬后数量分别是浅旋耕和常规耕翻田的 35 倍和 77 倍，稻套麦田越冬后灰飞虱最高虫量达每公顷 860 万头以上（表 4）。灰飞虱主要在稻残桩里而不是在麦苗上越冬，与上述在东台和姜堰耕翻田调查所得出的灰飞虱越冬时主要寻找一个适宜藏匿的处所而不一定是生活寄主的推论相吻合。

表 3　同一田块中稻桩与小麦上灰飞虱数量比较

Table 3　The SBPH number on paddy stubble and wheat in the same field

地点 Area	时间(年/月/日) Date(year/month /day)	灰飞虱数量（头/10m²）Number of SBPH per 10m² 稻桩 Paddy stubble	 麦苗 Wheat seedling
江苏建湖 Jianhu Jiangsu	2005 /03 /30	2690	25
江苏大丰 Dafeng Jiangsu	2005 /04 /04	1010	5
江苏姜堰 Jiangyan Jiangsu	2004 /03 /25	660	4

表 4　不同类型田块灰飞虱越冬虫量比较（江苏建湖，2005.03）

Table 4　Hibernant SBPH number in different farming lands in Jianhu, Jiangsu Province in 2005

田块类型 Farming type	调查田块 fields	虫量（头/667m²）SBPH number per 667m² 最低 Lowest	 最高 Highest	 平均 Average
传统耕翻麦田 Conventional plowed fields	5	500	4850	1872
机械浅旋耕麦田 Field of slight plowed with mechanism	16	1200	12000	4055
稻套麦田 Field of wheat interplanting with rice	30	26600	574750	144337

定点定田块调查发现，不同类型麦田灰飞虱的越冬效率相差不大，但其冬前寄主间转移效率差别很大。稻套麦田灰飞虱的秋冬转移率在 10% 左右，约为常规耕翻麦田的 20 倍（表 5）。由此可见稻套麦等轻型耕作方式在季节转换时对灰飞虱赖以生存的农田生境破坏较小，灰飞虱从夏季栖息地（稻田）向冬季栖息地（麦田）转移时损失较小。

2.2　灰飞虱周年迁移扩散及在水稻上的迁入动态

采用黑光灯诱集法调查灰飞虱周年迁移扩散动态，结果发现在江苏灰飞虱周年只有一个迁移高峰（图 1），主要集中在 5 月底至 6 月上中旬，时间上与江苏各地小麦成熟和收割时间基本同步。其中苏南（如武进）小麦成熟期早于苏北（如洪泽），相应灰飞虱迁移扩散时间上苏南亦早于苏北，由此说明稻麦轮作区灰飞虱集中迁移与小麦成熟相关。在水稻整个生长季节中，虽然灰飞虱在水稻上有所发生，但在灯下极少诱集到灰飞虱，这与笔者在田间调查时稻田灰飞虱成虫尤其是雌虫 95% 以上为非迁飞型的短翅型相吻合。

表 5　不同类型稻田灰飞虱稻－麦转移效率及越冬情况的调查（江苏建湖，2004 ～ 2005）
Table 5 Investigation of SBPH transfer from rice to wheat and overwinter in Jianhu, Jiangsu Province from 2004 to 2005

麦田类型 wheat field	灰飞虱虫量（万头 /667m²）SBPH number per 667m² (ten thousand head)				
	水稻收获前（10 月 21 日）Before rice harvest	水稻收获后（11 月 4 日）After rice harvest	冬前转移率 (%) Transfer efficiency (%)	翌年春天（3 月 31 日）Next spring	越冬存活率 (%) Survival rate of overwinter(%)
稻套麦田 1 No.1 Field of wheat interplanting with rice	197.5	18.6	9.4	11.1	59.7
稻套麦田 2 No.2 Field of wheat interplanting with rice	49.0	5.4	11.0	2.8	51.9
常规耕翻田 Conventional Ploughing field	62.4	0.32	0.5	0.18	56.2

注：冬前转移率（%）= 水稻收获后虫量 / 水稻收获前虫量 ×100；越冬存活率（%）= 翌年春天虫量 / 水稻收获后虫量 ×100。
Transfer efficiency（%）= number of SBPH after rice harvest / number of SBPH before rice harvest × 100.

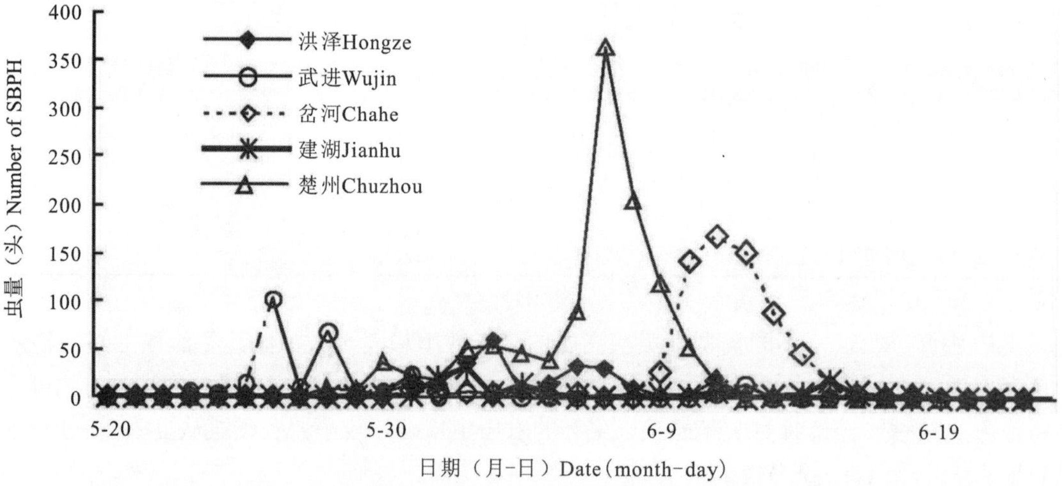

图 1　2002 年江苏 5 个地区灰飞虱迁移扩散动态（黑光灯诱集法）
Fig.1 Migration and dispersal dynamics of SBPH by light trap in 2002 in Hongze, Chahe, Chuzhou, Wujin and Jianhu, Jiangsu Province

有观点认为灰飞虱趋光性不强，采用黑光灯诱集的效果有限，灯下数据可能并不完全反映水稻上灰飞虱的迁入情况，为进一步探明水稻上灰飞虱的迁入规律，于水稻出苗开始在秧田和稻田设置黄盘进行诱集至水稻收割。结果表明，与灯下数据类似，常规水育移栽稻上水稻全生育期只有一个灰飞虱迁入高峰，时间为 5 月下旬至 6 月上中旬，不同地区灰飞虱迁入高峰出现的具体时间略有不同（图 2），但均与当地小麦收割时间基本同步。对江苏洪泽地区 2002 ～ 2007 年连续 6 年观测，年度间常规播期水稻灰飞虱迁入时间基本相同，但迁入量逐年升高（图 3），这与近年来小麦上灰飞虱的越冬虫量和一代灰飞虱发生量逐年升高的趋势基本一致。以上研究进一步说明水稻上灰飞虱迁入动态与当地当年小麦上灰飞虱的发生动态相关，佐证了灰飞虱在江苏是一种内源性昆虫的观点。

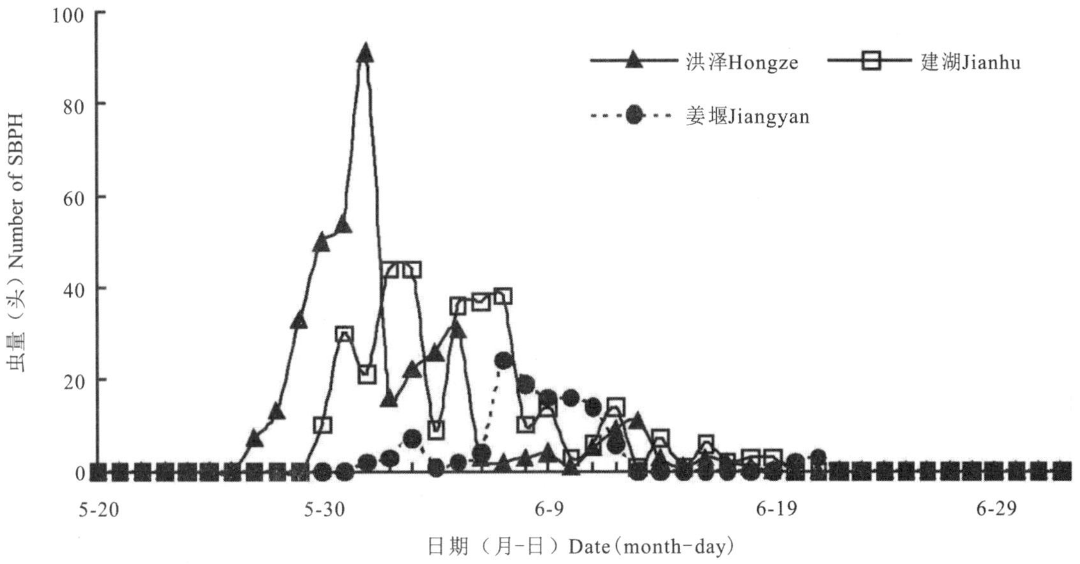

图 2 2002 年江苏 3 个地区水稻上灰飞虱迁入动态(黄盘诱集法)

Fig. 2 Incursion dynamics of SBPH on rice by yellow pan water trap in 2002 in Hongze, Jianhu and Jiangyan, Jiangsu Province

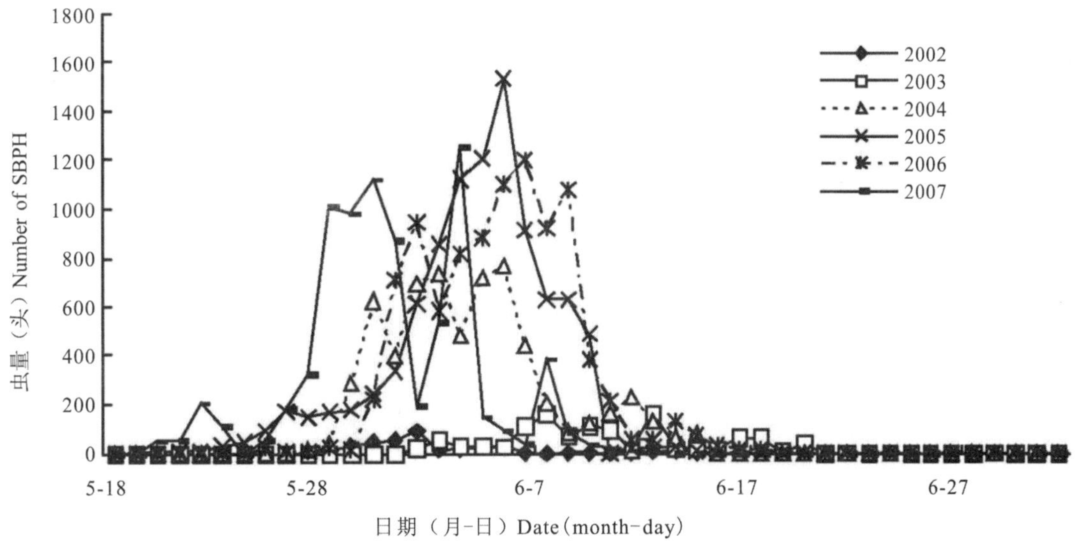

图 3 2002-2007 年江苏洪泽水稻上灰飞虱迁入动态(黄盘诱集法)

Fig. 3 Incursion dynamics of SBPH on rice by yellow pan water trap from 2002 to 2007 in Hongze, Jiangsu Province

2.3 水稻上灰飞虱的消长动态

2.3.1 秧田灰飞虱的发生动态

2002 年对江苏姜堰和洪泽自水稻出苗开始对常规播期的秧田灰飞虱逐日观察，结果表明：秧田灰飞虱从零到有，初查灰飞虱均为成虫，显示秧田灰飞虱由它处迁移而来。于 5 月下旬至 6 月中下旬时虫量急剧升高，在时间上与灯下诱集和黄盘诱集结果同步(图 4)，秧田末期可查见若虫。对洪泽地区 2002、2004、2006 年秧田灰飞虱的消长动态进行比较，年度之间秧田灰飞虱发生时间略有差异，但在发生量上差异巨大(图 5)，这与黄盘诱集结果一致。秧田末期(6 月 15 日左右)开始查见初孵若虫。

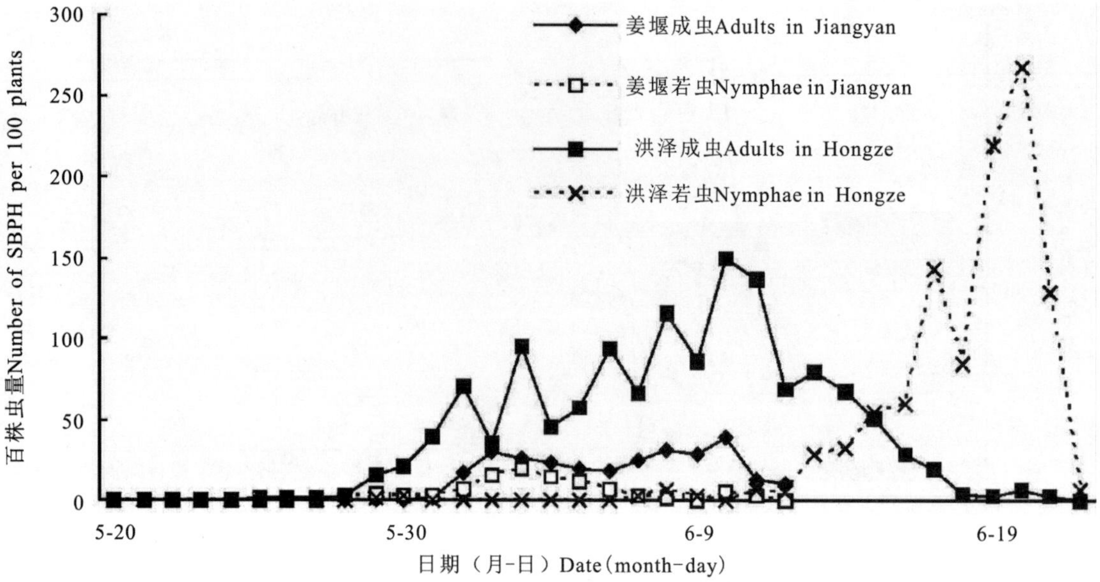

图 4 2002 年江苏秧田灰飞虱消长动态

Fig. 4 SBPH population dynamic in rice seedling bed in Jiangsu in 2002

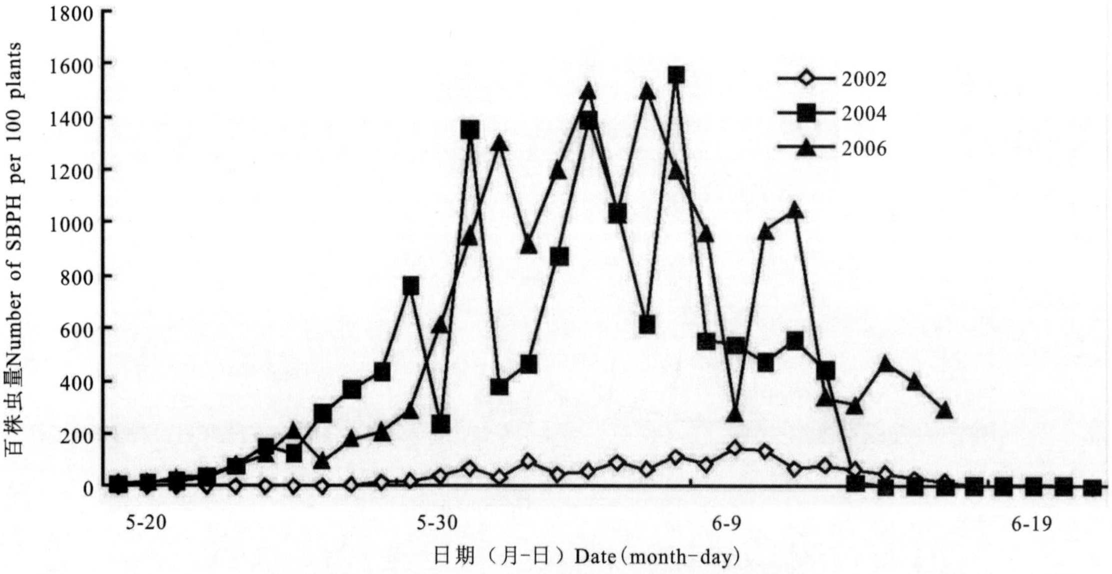

图 5 江苏洪泽不同年份秧田灰飞虱成虫消长动态

Fig. 5 SBPH adults dynamic in rice seedling bed in Jiangsu in 2002, 2004 and 2006

2.3.2 稻田灰飞虱的发生动态

对 2002 年江苏姜堰和洪泽大田灰飞虱发生动态逐日调查结果表明，在大田前期（6 月中下旬和 7 月）田间灰飞虱主要为若虫，而成虫量一直很低，可能与灰飞虱不耐高温从而影响了其若虫的成活率所致（图 6)。对江苏洪泽 2002、2004、2006 年 3 年稻田灰飞虱发生动态比较发现（图 7)，年度之间灰飞虱发生趋势一致，在移栽后的 10d 左右无若虫出现，移栽较早有部分成虫进入大田，整个水稻生长前期成虫发生量较低。年度之间发生量有一定的区别。

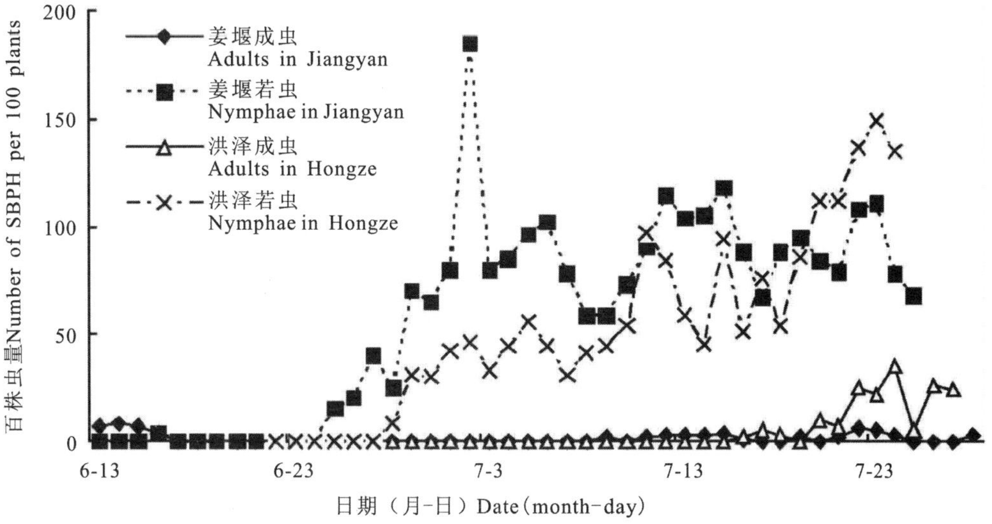

图6 2002年江苏稻田前期灰飞虱消长动态

Fig. 6 SBPH population dynamic in newly transplanted rice field in Jiangsu in 2002

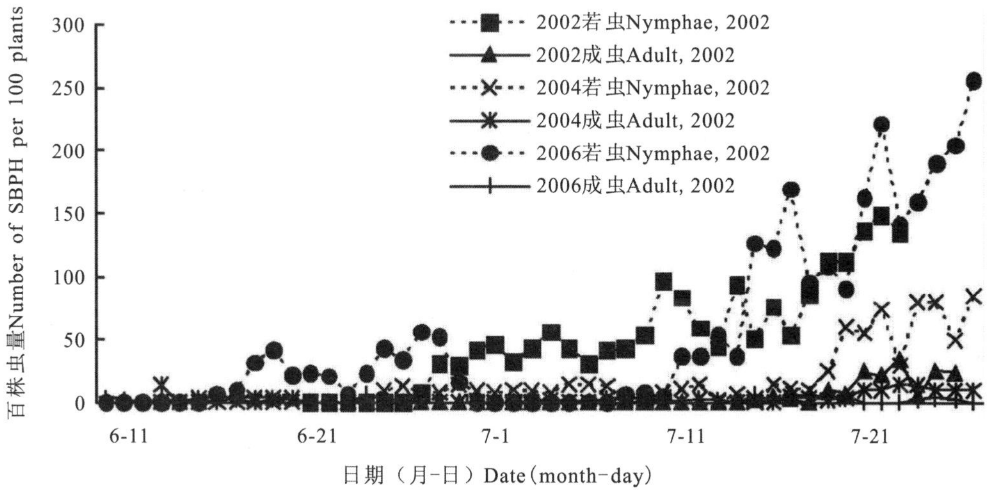

图7 江苏洪泽2002、2004、2006年大田前期灰飞虱发生动态

Fig. 7 SBPH population dynamic in newly transplanted rice field in Hongze, Jiangsu Province in 2002, 2004 and 2006

经统计，2002～2006年7～9月在8块系统调查田普查到的灰飞虱成虫2584头，仅有106头为长翅型，大田查见灰飞虱成虫95%以上为短翅型，无主动迁移扩散能力，这也部分解释了为什么7月份开始灯下未诱集到灰飞虱的原因。

3 讨论

3.1 江苏灰飞虱的越冬方式

据资料记载（浦茂华，1963），在江苏冬季灰飞虱在麦、紫云英、蚕豆、胡萝卜、野茭白和芫荽等植物上或于田埂、荒地、沟渠及路旁杂草丛中、土缝中以休眠或滞育方式越冬。本文对江苏多点的调查结果显示，灰飞虱在冬前主要转移至能够提供藏身之处的寄主植物上，该寄主植物可以是其生活寄主，也可以是已经枯死的禾本科植物或非生活寄主，只要适合藏匿即可。冬后气温尚未回升之前调查，灰飞虱主要在枯草丛、稻残桩等适合灰飞虱藏匿的场所查见，而并非是在寄主植物小麦上，也佐证了这一观点。植株大小与灰飞虱的转移数量多少存在一定的相关性，耕翻田中灰飞虱主要在田埂、田边而不是田中的小麦上，免耕的稻套麦田主要在田中的稻残桩上越冬等，进一步说明灰飞虱主要在藏匿场所越冬。这与灰飞虱越冬时处于滞育的生理状态无需在冬季取食相符（中国农业百科全书编辑委员会，1990），前人研究中的灰飞虱在土缝中或紫云英田越冬的观察结果（浦茂华，1963）也说明灰飞虱冬季仅在这些场所藏匿。

明确灰飞虱主要是以藏匿的方式越冬对于制定灰飞虱基数控制的策略制定具有重要指导作用。如果灰飞虱是以藏匿方式越冬，则破坏其藏匿的越冬场所将有可能大大减少其越冬基数从而可以有效控制其发生数量，而如果灰飞虱是在生活寄主上特别是在冬季种植的主要农作物如小麦上越冬，则难以实施以破坏越冬场所的形式来控制灰飞虱的越冬基数。本文研究中所展示的耕翻田灰飞虱数量远少于稻套麦田则很好地说明了这一点，耕翻田在耕翻过程中适宜灰飞虱藏匿的稻残桩被破坏，大大减少了灰飞虱的越冬场所，从而导致灰飞虱的越冬基数锐减，说明破坏灰飞虱的越冬场所可有效控制其越冬基数。

3.2 稻套麦与灰飞虱及其传播的水稻病毒病大发生的关系

利用灰飞虱主要在适合藏匿的场所越冬这一观点可以较好地解释近年来灰飞虱暴发的成因。在传统耕翻种植方式（据调查 2000 年前江苏等灰飞虱适生地和水稻条纹叶枯病适宜流行区地主要采用此耕作方式）下，灰飞虱主要在田边、田头、灌溉水槽的杂草上越冬。这些地点面积有限，约占农田总面积的 1/20，加上灰飞虱主要在这些场所的枯死杂草丛和土缝中越冬，其越冬基数较低。根据普查结果并综合考虑多种因素粗略估计每公顷灰飞虱越冬基数在 7500 ～ 30000 头。灰飞虱越冬基数低，与 2000 年以来采用这一耕作方式地区以及 2000 年前江苏地区灰飞虱和水稻条纹叶枯病轻发相连。近年来由于农村青壮年劳动力的外移，江苏小麦播种方式发生了很大变化，以减轻劳动强度为主要目标的轻型栽培技术应运而生，稻套麦、浅旋耕技术逐渐被农民接受和应用（陈洪礼等，2007）。稻套麦等免耕面积扩大后，灰飞虱越冬场所从田边田头的有限避难所转移到田中稻残桩这一广阔栖息地，灰飞虱越冬基数显著增大。据 2005 ～ 2007 年间江苏全省初步调查，免耕田中灰飞虱越冬数量一般在每公顷 300000 ～ 2250000 头，约是耕翻田的 40 ～ 100 倍。灰飞虱越冬基数的显著增大与江苏近年来麦田和秧田灰飞虱的发生量连年居高不下同步。

稻套麦田不仅留下稻残桩给灰飞虱提供广泛而优越的越冬场所，而且保护了灰飞虱种群在进行寄主间转移时免遭大的损失，从而优化了灰飞虱冬季的生存环境。据江苏省农林厅统计 2000 年以来江苏稻套麦种植面积持续增大，这可能是江苏近年来灰飞虱数量在 2001 ～ 2004 年连续攀高后并于 2005 ～ 2011 年连年居高不下的主要原因。灰飞虱是病毒的传播介体，免耕技术的推广和应用面积的扩大，扩展了适宜灰飞虱越冬的场所，导致灰飞虱超常发生可能是当前水稻条纹叶枯病、水稻黑条矮缩病大流行的根本原因之一。

3.3 利用移栽期控制灰飞虱 2 代若虫

灰飞虱可在本地越冬为内源性昆虫，近年来有研究认为江苏的灰飞虱也存在远距离迁移的可能（贺媛等，2012），但缺乏直接的证据。作为内源性昆虫，其周年繁殖的关键环节和薄弱环节是控制其发生危害的关键点。从稻麦轮作区灰飞虱的周年生活史可以看出，在传统育秧移栽栽培方式下灰飞虱种群增长主要有 3 个关键期：秋季由水稻田向小麦田等冬季作物或场所转移期、夏季由小麦向水稻转移期和由秧田向大田转移期，这 3 个时期涉及灰飞虱生活寄主和生存场所的重大改变，对灰飞虱种群数量影响很大。其中从小麦向秧田转移最为人所重视，一般认为这是灰飞虱危害及传播水稻病毒的关键时期（周益军，2010），从秧田向大田转移也很重要但常被放在一个相对次要的位置。本文研究结果显示秧田向大田转移期对移栽稻 - 小麦轮作区灰飞虱的种群消长影响巨大，水稻移栽后 7d 内几乎无灰飞虱成若虫，直至 7 ～ 10d 后才陆续出现 1 龄若虫。由此说明移栽这一农事操作在短期内几乎消灭或赶走了稻苗上的绝大部分灰飞虱成若虫，从2代灰飞虱卵历期23～26℃下为10～12d（蒲茂华，1963）判断，大田灰飞虱若虫来自于秧田后期成虫在秧苗上所产的未孵化卵块于大田孵化而来，而在秧田已经孵化和正在孵化的灰飞虱由于移栽这一农事操作被杀灭。如果适期推迟移栽期，一是可以避免一代灰飞虱成虫直接进入大田产卵，二是促使秧苗上的灰飞虱卵块在秧田孵化并在移栽时自然杀灭从而最大限度控制2代若虫的发生量，如此可能会达到不使用化学农药而自然控制水稻大田早期灰飞虱发生量的目的。因水稻条纹病毒可以经卵传播（Kisimoto and Yamada, 1991），水稻大田中的重要病毒源来自于灰飞虱2代若虫的带毒虫（结果另文发表），如何利用移栽期控制2代若虫的发生量，对于防治由灰飞虱传播的水稻条纹叶枯病等水稻病毒病害可能有积极意义，这已在江苏等地小面积应用取得良好的效果（结果另文发表），已有文献报道的推迟播种期可以减轻水稻条纹叶枯病的发生（朱金良，2008），也可能是推迟播种期自然推迟了移栽期而达到了适期移栽的目的。如何进行适期移栽并提高该项技术在防虫控病中的有效性值得进一步深入研究。

3.4 利用浅旋耕等技术压低灰飞虱越冬基数的源头控制与灰飞虱及其传播病毒病的治理

由灰飞虱传播的水稻条纹叶枯病是目前江苏等地的粳稻种植区水稻上最为重要的病害，水稻条纹叶枯病的防治上治虫防病、抗病品种、推迟播期、清洁田园等措施最为常用（刘水芳等，2007；王华弟等，2008；朱金良等，2008），这些措施在很大程度上控制了条纹叶枯病的危害，对水稻的保产增产起到了积极作用。然而江苏在该病防治上投入大量的人力、物力、财力后，病害流行潜力依然居高不下，2003～2011年江苏省农林厅每年均发布水稻条纹叶枯病大发生的预警，每年均需要实施紧急措施进行全民动员防治。一般病毒病的发生具有周期性、间隙性等特点（谢联辉，2008），像2000年以来水稻条纹叶枯病在江苏连续流行12年的例子历史上尚是首次，这可能是在病害防治上没有针对根源从而导致一直处于被动防治的状态。灰飞虱是该病的传毒介体，在此病的流行中起主要作用，近年来水稻条纹叶枯病的连年暴发态势与灰飞虱发生量连年攀高同步发生。不解决灰飞虱超常发生这一根本问题，始终存在一个巨大的传毒群体和病害流行潜力，以上所说的水稻条纹叶枯病各项防治措施欲发挥作用难度也大。秧田期灰飞虱的迁入期长达20d左右，且每天连续迁入，在灰飞虱暴发状态下用杀虫剂防治灰飞虱防不胜防，据实地调查，2004～2006年江苏许多农民对感水稻条纹叶枯病的水稻品种上灰飞虱的防治已达一日一次甚至一日2次施药的恐怖地步，如此局面下治虫防病策略的有效性、实用性和科学性越来越弱；在秧田株虫量达到10～20头的情况下，灰飞虱已成为一种水稻虫害而不仅仅是传毒介体，即使使用抗病品种同样需要防治灰飞虱，同时在抗条纹叶枯病品种上严重发生水稻黑条矮缩病也给使用抗病品种防治条纹叶枯病的效果大打折扣；推迟播期在许多地区受到气候条件的限制难以普遍实施，在有些情况下反而加重条纹叶枯病的发生，因灰飞虱发生量过大，推迟播期导致秧田期水稻条纹叶枯病严重发生的例子近年来在生产上屡屡出现（未发表数据）；田园周围灰飞虱与麦田和秧田中的相比数量有限，清洁田园的防病效果在病害大暴发的态势下其作用也有限，只在病害轻发之时起一定作用。

本文研究结果显示，灰飞虱越冬主要是寻找藏匿场所，2000年以来免耕技术如稻套麦应用面积的扩大，扩展了灰飞虱的越冬场所，优化了灰飞虱的越冬环境，与灰飞虱发生量的持续增高关系密切。耕翻田与稻套麦田相比灰飞虱发生数量差别显著，如能恢复农田耕翻种麦的耕作方式，可从源头上压低灰飞虱的越冬基数。结合前文分析，从源头控制灰飞虱的基数，应是目前控制灰飞虱传水稻病毒流行的根本出路，也是治虫防病、抗病品种、推迟播期、清洁田园等措施能够有效发挥作用的前提。类似的改进耕作栽培制度可有效控制水稻条纹叶枯病的流行在1966～1969年江苏苏南（浙江省农业科学院植物保护研究所病毒病研究组，1985）和1973～1976年在日本北海道（Hibino, 1996）的生产实践中得以证明。从目前农村形势来看，由于青壮劳动力大量转移，推广应用轻型栽培的发展趋势难以逆转，欲恢复到传统的深耕翻种麦方式几无可能。综合考虑，大力推广机械浅旋耕技术，既符合轻型栽培的发展趋势，又可压低灰飞虱的越冬基数，可作为水稻条纹叶枯病、水稻黑条矮缩病等灰飞虱传病毒病害防治上的首要措施，以从介体控制的角度充分降低灰飞虱传病毒病的流行潜力。

参考文献

[1] 陈洪礼，蔡建华，田文科，不同轻型简化稻作技术经济分析与应用前景评价 [J]. 中国稻米，2007, (5) : 41-42.

[2] 程兆榜，杨荣明，周益军，等．江苏稻区水稻条纹叶枯病新规律 [J]. 江苏农业科学 ,2002,(1) : 39-41.

[3] 葛红，季桦，徐莉，等．灰飞虱寄主转移规律及栽培技术对其种群数量的影响 [J]. 金陵科技学院学报 ,2010,26(2) : 69-71.

[4] 贺媛，朱宇波，侯洋旸，等．江浙麦区灰飞虱春季种群的发生消长和迁飞动态 [J]. 中国水稻科学 ,2012,26(1): 109-117.

[5] 季英华，任春梅，程兆榜，等．江苏省近年来新发生的一种水稻矮缩病害病原初步鉴定 [J]. 江苏农业学报 ,2009,25(6) : 1263-1267.

[6] 林奇英，谢联辉，周仲驹，等．水稻条纹叶枯病的研究Ⅱ．病害的分布和损失 [J]. 福建农学院学报 ,1990,19(4) : 421-425.

[7] 刘水芳，于福安，顾红艳，等．天津稻区水稻条纹叶枯病发生动态与综合防治 [J]. 中国农学通报 ,2007,23(5) : 302-305.

[8] 刘向东，翟保平，刘慈明．灰飞虱种群暴发成灾原因剖析 [J]. 昆虫知识 ,2006,43(2) : 141-146.

[9] 苗洪芹，陈巽祯，曹克强，等．玉米粗缩病的流行因素与预测模型 [J]. 河北农业大学学报 ,2003,26(2) : 60-64.

[10] 浦茂华．苏南灰稻虱的初步研究 [J]. 昆虫学报 ,1963,12(2) : 117-136.

[11] 乔慧，刘芳，罗举，等．不同植物上灰飞虱适合度的研究 [J]. 中国水稻科学 ,2009,23(1) : 71-78.

[12] 王华弟，陈剑平，祝增荣，等．水稻条纹叶枯病的为害损失及防治指标 [J]. 中国水稻科学 ,2008,22(2) : 203-207.

[13] 王瑞，沈慧梅，胡高，等．灰飞虱的起飞和扩散行为 [J]. 昆虫知识 ,2008,45(1) : 42-45.

[14] 王彦华，吴长兴，赵学平，等．灰飞虱对杀虫剂抗药性的研究进展 [J]. 植物保护 ,2010,36(4) : 29-35.

[15] 肖庆璞，程兆榜．灰飞虱的调查和观察方法 [J]. 江苏农业科学 (增), 2003: 8-9.

[16] 谢联辉．植物病原病毒学 [M]. 北京：中国农业出版社 , 2008:291-301.

[17] 张爱民，刘向东．灰飞虱的种群特性及其与温度的关系 [J]. 应用昆虫学报 , 2010, 47(2) : 326-330.

[18] 张强翔，任应党，林克剑，等．沿黄稻区灰飞虱越冬种群的时空分布及抽样技术研究 [J]. 应用昆虫学报 , 2011, 48(3) : 616-621.

[19] 浙江省农业科学院植物保护研究所病毒病研究组．水稻病毒病 [M]. 北京：农业出版社 , 1985:22-25.

[20] 中国农业百科全书编辑委员会．中国农业百科全书 : 昆虫卷 [M]. 北京：中国农业出版社 , 1990:1-598.

[21] 周益军．水稻条纹叶枯病 [M]. 南京：江苏科学技术出版社 , 2010:169-212.

[22] 朱金良，祝增荣，周瀛，等．水稻播种期对灰飞虱及其传播的条纹叶枯病发生流行的影响 [J]. 中国农业科学 , 2008,41(10) :3052-3059.

[23] CHEN C C, KO W F. Studies on the time of rice stripe virus infection and field experiments on disease control[J].Res.Bull.Taichung District Agric.Improv.Sta.(Taiwan),1986, 12: 51-59.

[24] HIBINO H. Biology and epidemiology of rice viruses[J]. Annual Review of Phytopathology, 1996,34(1) : 249-274.

[25] KISIMOTO R, YAMADA Y. Present status of controlling rice stripe virus//HadidiA,Khetarpal Ｒ K,Koganezawa(eds.). Plant Virus Disease Control[J].St.Paul.Minnesota: APS Press.1991, 470-481.

[26] OKUYAMA S，KAJINO Y. Studies on the control of the small brown planthopper transmitting rice stripe disease ID Disease occurrence and the rate of infective vectors. *Hokuno Northern Agriculture* (Japan),1980, 47(7): 10-22.

水稻锯齿叶矮缩病毒的检测及介体传毒特性

章松柏[1,2]，宋国威[1]，杨靓[1]，吴祖建[1]，谢联辉[1]

（1 福建农林大学植物病毒研究所 / 福建省植物病毒学重点实验室，福建福州 350002；2 长江大学农学院，湖北荆州 434025）

摘要：分别建立了水稻锯齿叶矮缩病毒（RRSV）病株的 dsRNA 基因组检测法和单头介体褐飞虱带毒的 RT-PCR 法，并结合生物学接种试验，对介体传毒特性进行了初步分析。结果显示：dsRNA 基因组鉴定法可以从 0.5g 病株样品中快速检测到 RRSV，RT-PCR 法可以灵敏地应用于褐飞虱带毒、传毒情况的检测；饲毒后褐飞虱成虫、若虫的带毒率分别为 75.0%、68.2%，传毒率分别为 50.0%、32.5%，说明褐飞虱种群传播 RRSV 的能力很强，是高度亲和的群体。

关键词：RT-PCR；dsRNA 基因组检测法；水稻锯齿叶矮缩病毒；褐飞虱

中图分类号：S432.1　**文献标识码：**A　**文章编号：**1671-5470(2013)03-0225-05

Determination of *Rice ragged stunt virus* and vector transmission characteristics

Songbai Zhang[1,2], Guowei Song[1], Liang Yang[1], Zujian Wu[1], Lianhui Xie[1]

(1 Key Laboratory of Plant Virology of Fujian Province / Institute of Plant Virology, Fujian Agriculture and Forestry University, Fuzhou, Fujian 350002, China; 2 College of Agriculture, Yangtze University, Jingzhou, Hubei 434025, China)

Abstract: RT-PCR and viral dsRNA identification assays were established to detect *Rice ragged stunt virus* (RRSV) in *Nilaparvata lugens* and rice plants，and vector transmission characteristics were analyzed using artificial inoculation. The results showed that viral dsRNA identification assay could detect virus from rice samples as less as 0.5g, and RT-PCR assay was suitable for detecting RRSV in individual brown planthopper. The viruliferous ratios of nymphs and adults of *N. lugens* inoculated by RRSV were 68.2% and 75.0%, respectively. 50.0% of nymphs and 32.5% of adults could transmit RRSV after virus acquisition. These results revealed that the brown planthopper population had high compatibility to RRSV with strong capacity to transmit the virus.

Keywords: RT-PCR; viral dsRNA identification assay; *Rice ragged stunt virus*; *Nilaparvata lugens*

水稻锯齿叶矮缩病是由水稻锯齿叶矮缩病毒（*Rice ragged stunt virus*, RRSV）引起的一种植物呼肠孤病毒病（Eishhro et al., 1979; Hibino, 1977），由介体昆虫褐飞虱（*Nilaparvatalugens*）持久性不经卵传播，并随着褐飞虱的长距离迁飞而扩散（Hibino, 1977; Senboku, 1978; Xie et al.,1980; Morinaka et al.,1981; Lu et al., 1999）。水稻锯齿叶矮缩病于 20 世纪 70 年代末至 80 年代末在亚洲一些国家先后暴发流行，造成水稻产量严重损失（Henog and Hardy, 2009）；2006 年该病又在越南大面积暴发（Du et al., 2007）；近几年，

福建农林大学学报 (自然科学版). 2013, 42(03): 225-229
收稿日期：2012-10-19
基金项目：国家 973 计划项目 (2010CB126200); 国家自然科学基金项目 (30970135)

我国福建、广东、广西等省部分地区发病严重（郑路平等，2008；黄立胜等，2011；刘红艳等，2012）。发病植株矮化，叶尖旋卷，叶缘缺刻呈锯齿状，叶鞘和叶片基部常有脉肿表现并导致水稻不育（谢联辉和林奇英，1996；Eishhro et al.,1979; Lu et al.,1999）。

由于水稻锯齿叶病毒病的很多症状与近几年流行的南方水稻黑条矮缩病（病原病毒为 *Southern rice black-streaked dwarf virus*，SRBSDV）（Zhou et al., 2008）相似，快速鉴定病毒显得至关重要。同时，RRSV 只能由褐飞虱等少数介体带毒传播，测定介体的带毒率及其对病毒的亲和性，有利于做好水稻病毒病的监测工作（林含新等，2000；吴建国等，2010）。目前，检测水稻病毒的方法多用 RT-PCR 法（雷娟利等，2011；刘红艳等，2012；Zhou et al., 2008），灵敏度虽高，但耗时长。为此，本研究拟建立水稻病株的 dsRNA 基因组核酸鉴定法和检测单头褐飞虱介体带毒的 RT-PCR 法，dsRNA 基因组核酸鉴定法未经过 RT-PCR 过程，耗时较短，对仪器和成本的要求也较低（章松柏等，2011）；并在此基础上，结合生物学接种试验，模拟自然条件，对介体传毒特性进行初步分析。

1 材料与方法

1.1 材料

水稻检测样品于 2010 年采集于江西省大余县和福建省龙岩市发病田间（采自同一田块），对照样品为已经鉴定的 RRSV 和 SRBSDV 福州分离物。用于生物学接种试验的褐飞虱为 2009 年 10 月采自福州水稻锯齿叶矮缩病田间并经人工饲养和分离纯化后的无毒群体。接种水稻品种为感病水稻品种 II 优航 2 号，购自福建省农嘉种业股份有限公司。

1.2 病毒 dsRNA 基因组检测法

参照 dsRNA 传统提取法（周雪平和李德葆，1995），用离心管代替针筒或层析柱，进行病毒 dsRNA 基因组的快速提取。简要步骤：采集具有典型症状的水稻病组织 0.5g，液氮研磨成粉末；用 2 倍体积的 2×STE 缓冲液及酚、氯仿抽提，低速离心；将上清液用无水乙醇调节至含量为 17%，注入经含 17% 乙醇的 1×STE 缓冲液平衡后的纤维素（CF-11）中，然后用含 17% 乙醇的 1×STE 缓冲液洗脱；再用不含乙醇的 1×STE 洗脱，洗脱液经异丙醇、乙醇沉淀后真空抽干，即为病毒的 dsRNA 基因组核酸。将提取的病毒 dsRNA 基因组核酸各片段用 1% 琼脂糖凝胶电泳分离，EB 染色观察结果。

1.3 生物学接种

1.3.1 褐飞虱饲毒

在自然条件下，感病植株能够吸引更多介体昆虫取食。因此，该环节模拟自然条件，随机选取一批雌性褐飞虱成虫在病株上产卵 2 ～ 3d，让卵在病株上孵化、生长发育、饲毒并度过循回期（29℃下循回期平均为 7.6d（谢联辉和林奇英，1996；Xie and Lin, 1980）。

1.3.2 单管单苗接种

在卵孵化后的第 9d 随机选取 80 头褐飞虱若虫进行单虫单管单苗接种，传毒 6d（变成成虫之前），然后将若虫低温保存；在卵孵化后的第 15d 随机选取 40 头成虫进行单虫单管单苗接种，传毒 8d，然后将传毒成虫低温保存。接种水稻苗均处于二叶一心期，接种苗每 24h 换 1 次，按编号种植于自然条件下的防虫网室内。

1.3.3 集团接种

在卵孵化后第 15、16、17、18d 分别选取 35、37、37、38 头成虫进行集团接种，接种 60 株水稻幼苗，不换苗，每天需赶虫 1 次，让介体尽可能接触每株幼苗，5d 后将幼苗种植于自然条件下的防虫网室内。

1.4 RT–PCR 法检测褐飞虱带毒率

取传毒后低温保存的成虫 24 头和若虫 22 头，用 RT-PCR 法检测褐飞虱是否带毒及其带毒率，以无毒的褐飞虱作为阴性对照，已经鉴定的 RRSV 福州分离物作为阳性对照。引物 P9F/P9（P9F:5′ATGAAGACTGCCTTTGCCAGA3′, P9R:5′CTACCCCGAGGCCTTCTGAGA3′），根据泰国分离物 S9 片段（Genbank 登录号：NC_003757）设计，扩增片段大小为 1017bp。单头褐飞虱总 RNA 提取参照 Trizol 试剂盒（北京天根生化科技有限公司）的说明，最后用 20μL 无核酸酶灭菌水溶解。反转录按照反转录试剂盒（Fermentas）说明操作。PCR 反应体系按照 TaKaRa 的 rTaq 酶使用说明进行扩增，反应程序为 94℃变性 4min，32 个循环（94℃ 30s，52℃ 30s，72℃ 45s），最后 72℃延伸 10min；产物经 1% 琼脂糖凝胶电泳、EB 染色后，用 BioRad 凝胶成像系统观察、记录。获得阳性褐飞虱样品后，选择其中 1 个样品，分别将其总 RNA 稀释 2、5、10、20、40 倍，然后取 5μL 稀释液为模板，按照上述 RT-PCR 法检测该方法的灵敏度。

2 结果与分析

2.1 水稻样品的检测结果

因为 dsRNA 病毒基因组具有特异性的电泳图谱，所以琼脂糖凝胶电泳可以直接检测病原的有无，初步判断病原的种类（章松柏等，2009，2011；刘红艳等，2012）。从图 1 可以得知，江西省大余县和福建省龙岩市同一发病田块存在 2 种 dsRNA 病毒［RRSV 和 RBSDV（*Rice black-streaked dwarf virus*）或 SRBSDV］。该方法优化了传统提取方法的步骤，降低了对样品质量的要求，可以在 3h 内从 0.5g 单个病株样品中提取质量较高的 dsRNA，能够直接应用于 dsRNA 病毒的检测。

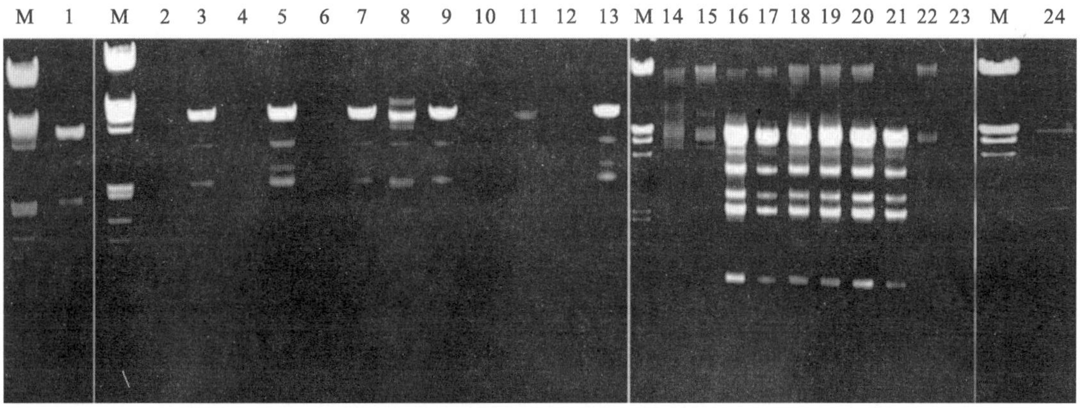

图 1 水稻病毒 dsRNA 基因组电泳图

Fig.1 Electrophoretogram showing dsRNA profiles of rice viruses

M: 标准分子质量；1 和 24: RRSV 和 SRBSDV 基因组核酸；2 ～ 13: 江西大余县水稻样品；14 ～ 23: 福建省龙岩市水稻样品．其中，3、5、7、9、13、16 ～ 21 为 RRSV，2、4、6、8、10 ～ 12、14 ～ 15、22 ～ 23 为 SRBSDV 或 RBSDV

2.2 生物学接种结果

由表 1、表 2 得出，能够传毒的褐飞虱成虫、若虫分别占供试成虫、若虫总数的 50.0%、32.5%；褐飞虱若虫接种于 480 株水稻后有 55 株发病，发病率为 11.5%，褐飞虱成虫接种于 320 株水稻后有 44 株发病，发病率为 13.8%；模拟自然条件下，褐飞虱集团接种的发病率为 5.0% ～ 8.3%（表 3）。

表 1　褐飞虱成虫传播 RRSV 结果 [1)]

Table 1　Serial daily transmission of RRSV by adults of *N. lugens*

传毒成虫	水稻发病情况							
	第 15 天	第 16 天	第 17 天	第 18 天	第 19 天	第 20 天	第 21 天	第 22 天
A2	○	○	○	○	●	○	○	○
A3	○	○	○	○	○	○	●	△
A6	●	○	●	○	○	●	●	●
A7	○	○	○	○	○	●	○	○
A15	○	●	○	●	○	○	●	●
A16	○	○	○	○	○	○	●	○
A17	○	○	○	●	○	●	○	●
A20	●	●	○	●	○	●	●	○
A21	○	●	○	○	○	●	●	○
A24	○	○	○	○	○	●	●	○
A25	○	●	○	○	○	○	○	○
A27	○	●	○	○	○	○	○	○
A29	○	○	○	○	○	●	●	●
A30	○	○	○	●	○	●	○	○
A33	●	○	●	○	○	○	○	○
A34	○	○	○	●	○	○	●	△
A35	○	○	○	○	○	○	●	●
A36	○	○	○	○	○	●	●	●
A37	○	○	○	○	○	●	○	○
A39	○	○	○	○	●	○	○	○

1) ○ 代表接种水稻未发病；● 代表接种水稻发病；△代表褐飞虱死亡

表 2　褐飞虱若虫传播 RRSV 结果 [1)]

Table 2　Serial daily transmission of RRSV by nymphs of *N. lugens*

传毒若虫	水稻发病情况						传毒若虫	水稻发病情况					
	第 9 天	第 10 天	第11天	第 12 天	第 13 天	第 14 天		第 9 天	第 10 天	第11天	第 12天	第 13天	第 14 天
N2	○	●	○	○	●	●	N36	○	○	●	○	○	▲
N7	●	○	●	○	○	○	N41	○	●	●	●	●	●
N11	●	●	●	○	○	○	N42	○	○	○	●	●	●
N12	○	○	○	●	○	●	N43	○	○	○	○	○	●
N15	○	○	○	○	○	●	N55	○	○	○	○	●	●
N17	○	○	●	○	○	●	N57	○	●	●	●	○	○
N20	○	○	●	○	○	○	N58	○	●	○	○	○	○
N21	○	○	●	○	○	○	N59	○	○	●	●	○	○
N25	○	●	○	●	○	●	N62	●	○	●	○	○	●
N30	●	○	●	○	○	○	N65	○	●	○	●	●	●
N32	○	●	○	●	○	○	N72	○	○	○	●	○	●
N34	○	○	○	●	○	○	N75	○	○	○	●	○	●
N35	●	○	○	○	○	●	N78	○	○	●	○	○	○

1) ○ 代表接种水稻未发病；● 代表接种水稻发病；▲ 代表若虫变为成虫

表 3 褐飞虱集团接种后的水稻发病率

Table 3　Morbidity of rice inoculated *N. lugens* group

批次	接种时间 /d	接种株数 / 株	介体数量 / 头	发病株数 / 株	发病率 / %
1	5	60	35	3	5.0
2	5	60	37	5	8.3
3	5	60	37	4	6.7
4	5	60	38	3	5.0

2.3　褐飞虱介体带毒率

结果显示，24 头成虫中有 18 头为阳性反应，该批成虫的带毒率为 75.0% 左右；22 头若虫中有 15

头为阳性反应，该批若虫的带毒率为 68.2% 左右（图 2）。灵敏度检测结果表明，样品总 RNA 稀释 20 倍后依然可以检测到目的条带，说明 RT-PCR 在检测单头褐飞虱携带 RRSV 上有较高的灵敏度。

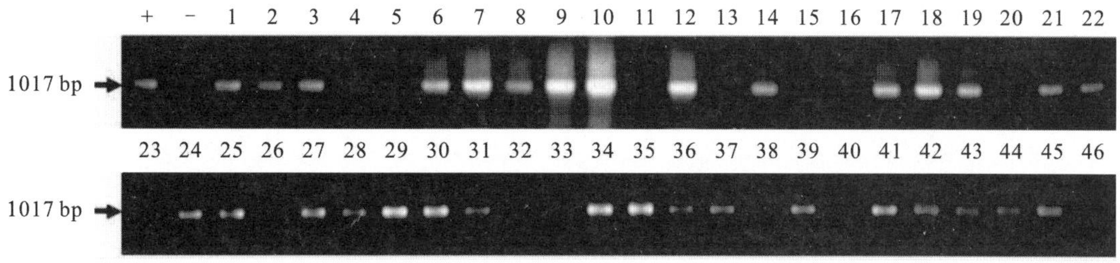

图 2　单头褐飞虱体内 RRSV 的 RT-PCR 检测

Fig.2　RRSV detection in single *N. lugens* by RT-PCR

+：阳性对照；－：阴性对照；1 ～ 22：褐飞虱若虫；23 ～ 46：褐飞虱成虫

3　讨论

本研究建立了检测 RRSV 的 dsRNA 基因组检测法和单头褐飞虱的 RT-PCR 法：dsRNA 基因组检测法可直接提取病毒的基因组进行检测，对样品量的要求较低（0.5g 水稻样品），同时可以批量处理水稻样品；RT-PCR 法能够非常灵敏地检测单头褐飞虱是否带毒，也可以批量处理样品，其结果可以反映介体昆虫的带毒率。尽管介体的带毒率一般高于传毒率，但根据试验可以推导两者之间的线性关系，建立水稻锯齿叶矮缩病预测模型，从而指导农业生产，减少病毒危害，提高经济效益。

结合 2.2 和 2.3，可以归纳福建福州褐飞虱群体的传毒特性。（1）种群获毒、传毒能力强。RT-PCR 检测结果显示，褐飞虱成虫、若虫的带毒率分别为 75.0%、68.2%，均比以往的报道（沈菊英等 , 1989; Hibino et al., 1977; Zhou and Ling, 1979; Xie et al., 1980; Lu et al., 1999）高。以往报道中饲毒天数一般为 2d 左右，与本试验“模拟自然条件下在病株上产卵、饲毒和度过循回期”不同，这可能是介体带毒率高的原因之一；同时，室内饲养的褐飞虱携带褐飞虱呼肠孤病毒（*Nilaparvata lugens reovirus*, NLRV）的比率较自然条件下高，获得 RRSV 后携带 RRSV 的比率比自然条件低（试验数据，未发表），推测褐飞虱体内的 NLRV 在某种程度上干扰了 RRSV 的入侵或复制，这也可能是介体带毒率高的另一个原因。结合生物学接种试验数据，显示带毒且能够传毒的成虫、若虫比率高达 66.7%、47.3%。由此推测，褐飞虱群体与 RRSV 高度亲和，具有较强的传毒能力。（2）循回期长。根据表 1、表 2 无法确定最短循回期，但最长循回期达 20d（编号 A3、A16、A35 的褐飞虱成虫）。（3）传毒持久。褐飞虱成虫、若虫通过循回期后，最多能够连续传毒 5d（编号 N41 的褐飞虱若虫），一般都有明显的传毒间歇现象，间歇期最短 1d，最长 4d（编号 N35 的褐飞虱若虫）。此外，模拟自然条件下，病株上饲毒的褐飞虱群体造成的水稻发病率为 5.0% ～ 8.3%，这与 2010 年福建省植物病毒学重点实验室调查的福建水稻锯齿叶矮缩病流行情况相吻合。

参考文献

[1] 黄立胜 , 李国君 , 卓晓光 , 等 . 广东雷州 2011 年警惕晚稻受到齿叶矮缩病毒威胁 [J]. 中国植保导刊 , 2011,31(9): 44-45.

[2] 雷娟利 , 吕永平 , 金登迪 , 等 . 应用 RT-PCR 方法检测水稻植株和介体昆虫体内的水稻齿叶矮缩病毒 [J]. 植物病理学报 , 2001,31(4): 306-309.

[3] 林含新 , 林奇田 , 魏太云 , 等 . 水稻品种对水稻条纹病毒及其介体灰飞虱的抗性鉴定 [J]. 福建农林大学学报 (自然版), 2000,29(4): 453-458.

[4] 刘红艳，章友爱，廖咏梅，等．广西南方水稻黑条矮缩病毒及水稻齿叶矮缩病毒的分子检测 [J]. 南方农业学报，2012, 43(7): 955-960.

[5] 沈菊英，彭宝珍，龚祖埙．水稻齿叶矮缩病毒在水稻病叶及传毒媒介昆虫组织内的形态 [J]. 上海农业学报，1989,5(2): 15-18.

[6] 吴建国，巴俊伟，李冠义，等．16 个水稻品种对水稻矮缩病毒抗性的鉴定 [J]. 福建农林大学学报（自然版），2010,39(1): 10-14.

[7] 谢联辉，林奇英．水稻病毒病 [M] // 方中达．中国农业百科全书：植物病理学卷．北京：农业出版社 ,427-430.

[8] 章松柏，罗汉刚，张求东，等．湖北发生的水稻矮缩病是南方水稻黑条矮缩病毒引起的 [J]. 中国水稻科学，2011,25(2): 223-226.

[9] 章松柏，吴祖建，段永平，等．水稻矮缩病毒基因组遗传多样性的初步研究 [J]. 长江大学学报：自科版，2009,6(4): 55-57.

[10] 郑璐平，谢荔岩，连玲丽，等．水稻齿叶矮缩病毒的研究进展 [J]. 中国农业科技导报，2008,10 (5): 8-12.

[11] 周雪平，李德葆．双链 RNA 技术在植物病毒研究中的应用 [J]. 生物技术 ,1995,5(1): 1-4.

[12] DU P V, CABUNAGAN R C, CABAUATAN P Q, et al. Yellowing syndrome of rice: etiology, current status, and future challenges[J]. Omonrice, 2007, 15: 94-101.

[13] EISHIIRO S, TOSHIHIRO S, KULCHAWEE K, et al. *Rice ragged stunt virus*, a new member of plant reovirus group[J]. Annals of the Phytopathological Society of Japan, 2009, 45: 436-443.

[14] HENOG K L, HARDY B. Planthoppers: new threats to the sustainability of intensive rice production systems in Asia[M]. Los Baňos: International Rice Research Institute, 2009, 357-368.

[15] HIBINO H. A virus disease of rice (*Kerdilhampa*) transmitted by brown planthopper *Nilaparvata lugens* Stal. in Indonesia[J]. contributions central research institute for agriculture, 1977, 35: 1-15.

[16] LU X B, PENG B Z, ZHOU G Y, et al. Localization of PS9 in *Rice Ragged Stunt Oryzavirus* and its role in virus transmission by brown planthopper[J]. Acta Biochimicaet Biophysica Sinica, 1999, 31: 180-184.

[17] MORINAKA T, CHETTANACHIT D, PUTTA M, et al. Nilaparvata bakeri transmission of rice ragged stunt virus[J]. International Rice Research Notes,1981, 6 (5): 12-13.

[18] SENBOKU T. Transmission of rice ragged stunt disease by *Nilaparvata lugens* in Japan[J]. International Rice Research Notes, 1977, 3 (2): 8.

[19] XIE L H, LIN Q Y. *Rice ragged stunt virus* disease, a new record of rice virus disease in China[J]. A Monthly Journal of Science, 1980, 25: 961-968.

[20] ZHOU G H, WEN J J, CAI D J, et al. *Southern rice black-streaked dwarf virus*: a new proposed *Fijivirus* species in the family Reoviridae[J]. Chinese Science Bulletin, 2008, 53 (23): 1-9.

[21] ZHOU L K, LING K C. Rice ragged stunt disease in China[J]. International Rice Research Notes, 1979, 4 (6): 10.

A new nepovirus identified in mulberry (*Morusalba* L.) in China

Quanyou Lu[1, 2, 3], Zujian Wu[1], Zhisong Xia[2, 3], Lianhui Xie[1]

(1 Institute of Plant Virology, Fujian Agriculture and Forestry University, Fuzhou 350002, Fujian, China; 2 Key Laboratory of Genetic Improvement of Silkworm and Mulberry, Ministry of Agriculture, Sericultural Research Institute, Chinese Academy of Agricultural Sciences, Zhenjiang 212018, Jiangsu, China; 3 Jiangsu University of Science and Technology, Zhenjiang 212003, Jiangsu, China)

Abstract: An isometric virus was identified in mulberry leaves showing symptoms of mulberry mosaic leaf roll (MMLR) disease. Its genome consists of two (+) ssRNAs. RNA1 and RNA2 have 7183 and 3742 nucleotides, excluding the 3′-terminal poly(A) tail. Based on phylogenetic analysis of the RNA1-encoded polyprotein and CP amino acid sequences, the properties of the the 3′-UTR of RNA1 and RNA2, and <75% identity in the CP amino acid sequence, this virus is proposed to be a new member of the genus *Nepovirus*, subgroup A. Since a causal relationship between this virus and MMLR has not been established, it is tentatively referred to as MMLR-associated virus.

Mulberry (*Morus alba* L.) is a deciduous perennial woody plant, the leaves of which are used for raising silkworms (*Bombyx mori*). In the past two decades, mulberry mosaic leaf roll (MMLR) disease, which had previously been called mulberry mosaic dwarf disease (Huang et al.,1992), has become a major constraint to sericultural production. MMLR was first described in an ancient book written in approximately 1624 and was named ‘long-sang’. The infectivity of the disease was demonstrated by graft transmission experiments (Xia and Lu, 2004). The characteristic symptoms are mosaic and upward rolling of leaves, reduction in leaf size, protuberance on leaf veins in some cases, and severe stunting.

The causal agent of MMLR remains ambiguous. Examination of diseased tissue by transmission electron microscopy in 1974 failed to provide evidence of a phytoplasma infection, but a filamentous virus particle of 11–13nm in diameter was found (The virology group et al.,1974). Thereafter, there were no further reports on MMLR etiology for about 30years until a viroid-like low-molecular-weight (LMW) RNA was isolated from MMLR-affected leaves by return polyacrylamide gel electrophoresis (R-PAGE) (Lu and Xia, 2006). Although this indicated that the LMW RNA is a possible causal agent of MMLR, all attempts to inoculate the LMW RNA to healthy mulberry for inducing disease symptoms were unsuccessful. In addition, only 20% of the samples with MMLR symptoms were found to contain LMW RNA by R-PAGE (Lu and Xia, 2010) or RT-PCR. Thus, the relationship between the LMW RNA and MMLR needs further studies.

Recently, an isometric virus was found in MMLR-affected leaves. From partially purified virus preparations, we randomly amplified and cloned cDNA fragments using random anchor primers in a RT-PCR protocol that uses minimal amounts of template and does not require prior knowledge of its sequence (Bohlander et al.,1992; Agindotan et al., 2010). Sequences similar to those encoding RNA-dependent RNA polymerases (RdRps) of RNA viruses were obtained. However, we failed to fulfill Koch’s postulates. Since the relationship

Archive of Virology. 2015, 160:851–855

Received 26 June 2014; Accepted 30 December 2014; Published online11 January 2015 between MMLR and the virus described here is unknown, it is tentatively referred to as mulberry mosaic leaf roll- associated virus (MMLRaV) because it was originally isolated from MMLR-affected mulberry trees. Herein, we report results of our further analysis of the genome of this mulberry virus.

Mulberry leaves (diseased and healthy) were collected from orchards at Sericultural Research Institute, Chinese Academy of Agricultural Sciences (Zhenjiang, Jiangsu province), in May 2010. The healthy leaves tested negative for MMLRaV in RT-PCR using MMLRaV-specific primers. Partially purified virus preparations were extracted from mulberry leaves as described (Xie et al.,1991). Extraction of viral RNA and total RNA of mulberry leaves were performed using a MiniBEST Viral RNA/DNA Extraction Kit Ver. 4.0 (TaKaRa) and a TRIzol kit (TaKaRa) according to the manufacturer's instructions.

Viral RNA was amplified using a reverse transcription (RT)-PCR procedure described by Agindotan et al., 2010. First, the 20-μl mixture used for the RT reaction was incubated at 30℃ for 10min and 42℃ for 60min using a random anchor primer (5′-TGGTAGCTCTTGAT- CANNNNNN-3′) (Bohlander et al.,1992). At the end of the reaction, 1 μL of RNase H (TaKaRa) was added, and the mixture was incubated at 37℃ for 30min and then at 70℃ for 10min. Second, the product of the RT reaction was used directly in PCR using a random anchor primer and the anchor primer 5′-AGAGTTGGTAGCTCTTGATC-3′ (Bohlander et al.,1992) to generate amplicons. PCR products were visualized as a smear of products (data not shown), and the fragments from 500 to 750bp were excised and recovered from the gel slice using a UNIQ-10 Column DNA Gel Extraction Kit (Sangon, Shanghai). The recovered PCR products were ligated into pUCm-T (Sangon, Shanghai) using T4 DNA ligase (TaKaRa) and cloned directly into the vector TOP10. Positive clones were selected by PCR using primers com- plementary to the M13 sites of the pUCm-T vector and subsequently sequenced (Sangon, Shanghai). A total of 200 recombinant clones were sequenced. These sequences were assembled using DNASTAR's SeqMan software. BLASTx analysis showed that a partial sequence 5695nt in length was significantly similar to RdRp sequences encoded by viruses of the subfamily *Comovirinae*. Another partial sequence 3315nt in size was significantly similar to the MP and CP of viruses of the genus *Nepovirus* in the subfamily *Comovirinae*. Based on the genome organization of nepoviruses, the 5695-nt and 3315-nt RNAs were accordingly named RNA1 and RNA2, respectively.

To obtain the exact sequences of the 5′end and the 3′ ends of both RNA1 and RNA2, rapid amplification of cDNA ends at the 5′ and 3′ -termini (5′ , 3′ -RACE) was carried out using total RNA extracted from the diseased mulberry as template and the commercially available cDNA amplification kit SMART^{TM} RACE (Clontech Laboratories, Inc.). The 5′ and 3′ -terminal sequences of each RNA strand were determined for seven clones. The results showed that, excluding the poly (A) tail, the full-length sequence of RNA1 is 7183nt long, and RNA2 is 3742nt long. The sequences were deposited in the GenBank database with the accession numbers KC904083 and KC904084, respectively.

The 5′ untranslated region (UTR) of RNA1 is 256nt in length. The first eight nucleotides of the extreme 5′ terminus of RNA1, UUGGAAAA, were identical to those of RNA2. The 3′ -UTR consists of 618nt, excluding the poly(A) tail. RNA1 contains a single ORF. Translation of this ORF (6309nt) leads to a polyprotein composed of 2102 aa with a molecular mass of 235.1kDa. The deduced amino acid sequence of the RNA1 ORF was compared with those of proteins deposited in the GenBank database. The N-terminal region of the putative ORF product contains motifs A (GXXXXGKS motif) to C of thentP- binding helicase (HEL) of superfamily 3 of positive-strand RNA viruses (Gorbalenya et al.,1990). Following the HEL, a motif (E-x_1 -T-x_3- N-x_4-R, x refers to any amino acid) with characteristics of a known viral genome-linked protein (VPg) was found in which the residue T (italicized) substituted for the Y reported previously (Mayo and Fritsch, 1994). The C-terminal region of the RNA1-encoded ORF contained eight conserved motifs that are characteristic of RdRp (Koonin, 1991), including the GDD motif typically found in many RdRps of plant, animal and bacterial viruses (Kamer and Argos, 1984). The amino acid sequence of this region shared identities ranging from 42%–60%, 44%–50% and 42%–46% with those of RdRps of viruses assigned to the genera *Nepovirus*, *Comovirus* and *Fabavirus* (family *Secoviridae*), respectively. In addition, the putative catalytic triad of H, E, and C conserved in the viral cysteine proteases (C-Pro) was found RNA1 is presented in Fig. 1 by analogy to that of other nepoviruses.

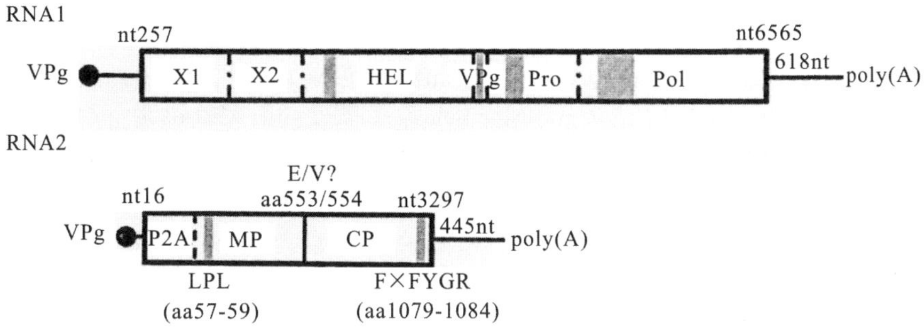

Fig.1 Schematic presentation of the genomic organization of RNA1and RNA2 of MMLRaV

The bold boxes indicate the open reading frame (ORF). The gray zones indicate conserved motifs located indifferent domains of the polyprotein encoded by RNA1 and RNA2. The numbers indicate the positions of nucleotides and amino acids calculated from the 50 terminus of the RNAs and the N-terminus ofthe polyprotein encoded by RNA2, respectively. The straight lines at the 30 termini indicate the UTRs, and the numbers indicate the UTR length. The question mark indicates that this cleavage site was predicted but not experimentally determined. The abbreviations are asfollows: HEL, helicase; VPg, viral genome-linked protein; C-pro, 3C-like proteinase; Pol, RNA-dependent RNA polymerase; MP, move-ment protein; CP, capsid protein

Phylogenetic analysis was performed to compare the deduced amino acid sequence of the RNA1-encoded polyprotein with that of different viruses from the subfamily *Comovirinae*. This analysis showed that MMLRaV clustered with viruses of the genus Nepovirus of the subfamily *Comovirinae* and was most closely related to melon mild mottle virus (MMMoV) (Tomitaka et al.,2001) (Fig.2A).

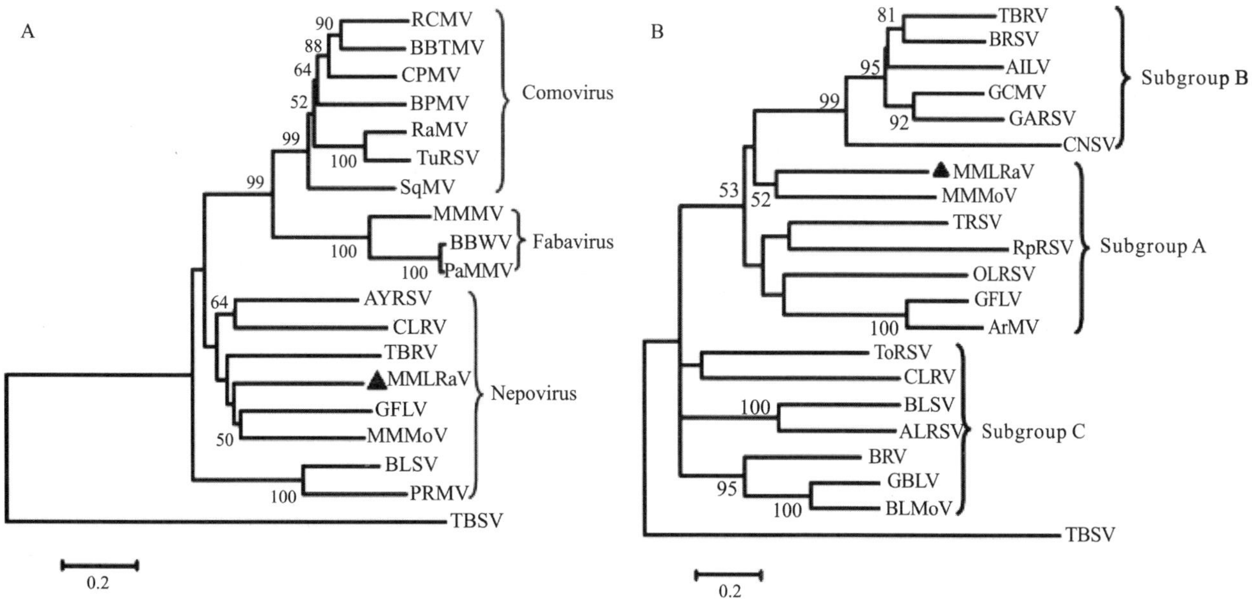

Fig.2 Phylogenetic tree based on the alignment of amino acidsequences of the RNA1-encoded polyprotein of MMLRaV and other viruses in the family *Secoviridae*

(A) and the RNA2-encoded capsid protein of nepoviruses (B). Multiple alignments were performed using CLUSTAL X 1.83 (Thompson et al.,1997). Phylogenetic trees were constructed using the NJ method in the program MEGA 3.1 (Kumar et al.,2004). Horizontal branch length indicates 0.2 replacements per site. TBSV (tomato bushy stunt virus),an unrelated virus, was selected as an outgroup. The virus abbreviations and GenBank accession numbers are as follows: A. MMMoV(BAJ16223); GFLV (ADJ10922); TBRV, tomato black ring virus (NP_958841); CLRV, cherry leaf roll virus (ACZ65483); AYRSV, artichoke yellow ring spot virus (CAJ33467); PRMV, peach rosette mosaic virus (AAB69867); BLSV, blueberry latent spherical virus(BAL04700); BBWV, broad bean wild virus 2 (ACK86674);MMMV, mikania micrantha mosaic virus (ACI22650); PaMMV, patchouli mild mosaic virus (NP_733967); CPMV, cowpea mosaicvirus (CAA25029); RCMV, red clover mottle virus (CAA46104); BBTMV, broad bean true mosaic virus (ADI60054); BPMV, beanpod mottle virus (NP_734070); SqMV, squash mosaic virus(NP_734012); RaMV, radish mosaic virus (AAY32935); TuRSV, turnip ringspot virus (ABS90367); TBSV (AAC32730); mMMLRaV(this study). B. MMMoV (BAJ16224); ArMV, arabis mosaic virus (ACF32435); TRSV (AF461164_1); RpRSV, raspberry ringspot virus(AAY63801); GFLV (ABC96691); CNSV, cycas necrotic stunt virus(NP_733975); GCMV (NP_734052); OLRSV, olive latent ringspotvirus (CAB90217); TBRV (CAA56792); BRSV, beet ringspot virus(NP_733980); AILV, artichoke Italian latent virus (CAA60707); GARSV, grapevine Anatolian ringspot virus (AAQ56596); ToRSV,tomato ringspot virus (NP_733973); GBLV, grapevine Bulgarian latent virus (CBO65362); CLRV (AAB27443); BLMoV, blueberry leaf mottle virus (AAA64608); BLSV (BAL04701); BRV, blackcurrant reversion virus (NP_733982); ALRSV, apricot latent ringspot virus (CAC05656); TBSV (BAF37070). mMMLRaV (this study)

The 5′ UTR of RNA2 was 15nt long. The nucleotides of the extreme 5′ terminus of RNA2 were UUGGAAAAUCU and in agreement with the consensus sequence (U-U/G-GAAAA-U/A-U/A-U/A) at the 5′end of nepovirus RNA2 (Fuchs et al.,1989). The 3′ -UTR consists of 445nt, excluding the poly(A) tail. The octanucleotide UUUCUUUU was present once in the 3′ -UTR of RNA2. The octanucleotide was found in the 3′ -UTR of the RNA2 of nepoviruses such as grapevine chrome mosaic virus (GCMV), grapevine fan leaf virus (GFLV) isolate-F13 and tobacco ringspot virus (TRSV) isolate S and in the 5′- UTR of GFLV-F13 and TRSV-S RNA2 (Serghini et al.,1990). RNA2 contains a single ORF (3282nt). Translation of the ORF leads to a polyprotein composed of 1093 aa corresponding to a molecular mass of 120.5kDa. The deduced amino acid sequence encoded by the RNA2 ORF shared the highest identity (32%) with polyprotein 2 of GFLV (ACR46367), a subgroup-A nepovirus. The RNA2-encoded polyprotein of subgroup-A and -B nepoviruses is cleaved into three mature proteins, P2A, MP, and CP (Sanfacon et al., 2012). Possible cleavage sites between MP and CP of MMLRaV were predicted using the NetPicoRNA 1.0 Server (Blom et al., 1996; von Bargen et al., 2012). A putative cleavage site at aa position ^{553}E/V^{554} of the RNA2-encoded polyprotein would result in a mature CP with a calculated molecular mass of 59kDa. This was consistent with the CP size estimated by Western blotting (Fig. 3), although further experimental evidence will beto be located between the VPg and RdRp domains. The deduced genomic organization of required to determine the exact position of the cleavagesite. The conserved nepovirus CP motif FXFYGR (Le Gall et al.,1995) was observed in the C-terminus of the MMLRaV CP. The deduced CP amino acid sequence of MMLRaV shared <28% identity with that of other nepoviruses. The conserved nepovirus MP motif LPL (Koonin et al.,1991) was found upstream of the putative MP domain. The deduced organization of MMLRaV RNA2 is presented in Fig.1, based on comparisons with other nepoviruses.

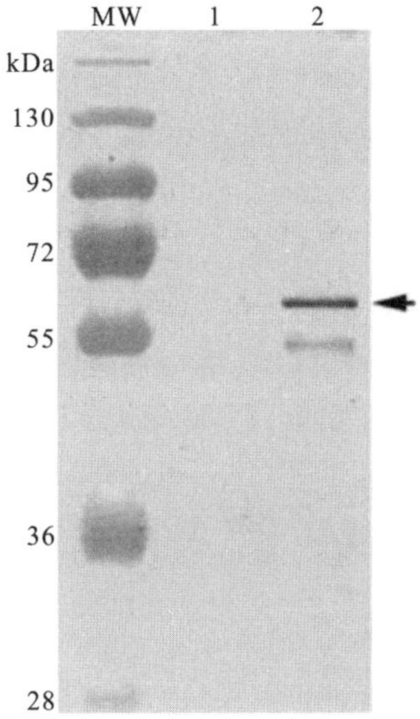

Fig.3 Western blot analysis of MMLRaV CP using a healthy mulberry sample as a negative control (1) and a partially purifiedvirus sample (2)

Pre-stained proteins (Fermentas, Lithuania) wereused as size markers (MW). The healthy mulberry sample was processed in the same ways as the sample shown in lane 2. Anantiserum produced using the prokaryotic expression product of thepartial CP gene of MMLRaV (a 38-kDa protein encoded by the 30 half of the MMLRaV CP gene) for injecting rabbits at Jiangsu University, China, was used as the primary antibody. HRP (horseradish peroxidase)-conjugated goat anti-rabbit IgG (Beyotime) was used as these condary antibody. The immuno-reaction was visualized with DAB(3,3′-diaminobenzidine) as substrate. The arrow indicates the position of the CP band. The additional faint band may correspond to adegradation product of the MMLRaV CP

Phylogenetic analysis based on the CP amino acid sequence suggests that MMLRaV is a member of the genus *Nepovirus*, subgroup A or B, and is closely related to MMMoV (Fig. 2B). However, the CP amino acid sequence identity between MMLRaV and MMMoV was remarkably low (27%).

The 3′ -UTR sequences of RNA1 and RNA2 were 89% identical in the last 376nt, excluding poly (A) tail, whereas the 5′ -UTR sequence identity and size between RNA1 and RNA2 of MMLRaV were remarkably different. The properties of the 3′ - and 5′ -UTRs of MMLRaV were similar to those of subgroup Anepoviruses, where the 3′ - and 5′ -UTRs between RNA1 and RNA2 are similar but not identical in sequence. This is different from subgroup B nepoviruses, where the 3′ -UTRs of RNA1 and RNA2 are identical, while the 5′ -UTRs just show sequence similarity in both RNAs (Sanfacon et al., 2012). Additionally, the full length of the MMLRaV RNA2 was 3,742nt, excluding the poly(A) tail, in agreement with subgroup A nepoviruses, which have an RNA2 of 3,700–4,000nt in length. These properties suggested that MMLRaV is a member of the genus *Nepovirus*, subgroup A.

Several decades ago, a spherical virus 22–25nm in diameter was isolated in Japan from mulberry trees showing both ringspot and filiform leaf symptoms. Based on virion size and substructure, the virus was assigned to the genus *Nepovirus* and named mulberry ringspot virus (MRSV) (Tsuchizaki et al.,1971). MMLRaV is similar to MRSV in particle morphology, but there are some differences between the two viruses: 1) Leaves of MRSV-infected mulberry trees mainly showed ringspot symptoms, which have never been observed on MMLR-affected mulberry. 2) MRSV was assigned to subgroup B on the basis of the relative molecular mass of RNA2 and the sedimentation coefficients of particle components, while MMLRaV was clustered into subgroup A based on the phylogenetic analysis of the putative amino acid sequence of CP and the identi ties between RNA1 and RNA2 in their 3′ - and 5′ -UTR sequences. 3) The particle size of MMLRaV, about 28–30nm in diameter (data not shown), is different from that of MRSV. Therefore, we hypothesize that MMLRaV is different from MRSV. In addition, only partial genomic sequences of four mulberry viruses are currently available in GenBank, namely mulberry mosaic virus (genus *Begomovirus*, family *Geminiviridae*), mulberry cryptic virus 1 (family *Partitiviridae*), mulberry endornavirus 1 (genus *Endornavirus*, family *Endornaviridae*) and mulberry vein banding virus (genus *Tospovirus*, family *Bunyaviridae*). Thus, MMLRaV is different from all four mulberry viruses that have been sequenced so far.

The species demarcation criteria in the genus Nepovirus are less than 80% identity in the deduced amino acid sequence of the conserved RdRp domain and less than 75% identity in the deduced amino acid sequence of CP (Sanfacon et al.,2012). The amino acid sequence deduced from the conserved RdRp domain of MMLRaV shared the highest identity (61%) with that of MMMoV, and the deduced CP amino acid sequence of MMLRaV shared identities of <28% with other nepoviruses. This suggests that MMLRaV belongs to a new species in the genus *Nepovirus*.

In conclusion, we have determined the complete genome sequence of MMLRaV, a new subgroup-A nepovirus first identified from mulberry in China.

Acknowledgements

This study was supported by funds from the National Basic Research Program of China (973 Program) (2014CB138402) and the Natural Science Foundation of Jiangsu province (No. BK2006084). The authors thank James L. Starr (TexasA&M University) for his editorial assistance in the preparation of this manuscript.

References

[1] AGINDOTAN B O, AHONSI M O, DOMIER L L,et al. Application of sequence-independent amplification SIA for the identification of RNA viruses in bioenergy crops[J]. Journal of Virological Methods, 2010, 169:119-128.

[2] BLOM N, HANSEN J, BLAAS D,et al. Cleavage site analysis in picornaviral polyproteins: discovering cellular targets by neural networks[J]. Protein Science, 2010, 5:2203-2216.

[3] BOHLANDER S K, ESPINOSA R, LE BEAU M M, et al. A method for the rapid sequence-independent amplification of microdissected chromosomal material[J]. Genomics, 1992, 13:1322-1324.

[4] FUCHS M, PINCK M, SERGHINI M A, et al. The nucleotide sequence of satellite RNA in grapevine fanleaf virus, strain F13[J]. Journal of General Virology, 1989, 70:955-962.

[5] GORBALENYA A E, KOONIN E V, WOLF Y I. A new superfamily of putative N TP-binding domains encoded by genomes of small DNA and RNA viruses[J]. FEBS Letters, 1990, 262:145-148.

[6] HUANG E T, TIAN L D, XIAO L Z, et al. The pratical mulberry protection (in Chinese)[J]. Sichuan Science and Technology Publishing House Chengdu. 1992.

[7] KAMER G, ARGOS P. Primary structural comparison of RNA dependent polymerases from plant, animal and bacterial viruses[J]. Nucleic Acids Research, 1984, 12:7269-7282.

[8] KOONIN E V. The phylogeny of RNA-dependent RNA polymerases of positive-strand RNA viruses[J]. Journal of General Virology, 1991, 72:2197-2206.

[9] KOONIN E V, MUSHEGIAN A R, RYABOV E V, et al. Diverse groups of plant RNA and DNA viruses share related movement proteins that may possess chaperone-like activity[J]. Journal of General Virology, 1991, 72:2895-2903.

[10] KUMAR S, TAMURA K, NEI M. MEGA3: integrated software for molecular evolutionary genetics analysis and sequence alignment[J]. Briefings in Bioinformatics, 2004, 5:150-163.

[11] LE GALL O, CANDRESSE T, DUNEZ J. A multiple alignment of the capsid protein sequences of nepoviruses and comoviruses

suggests a common structure[J]. Archives of Virology, 1995, 140:2041-2053.

[12] LU Q Y, XIA Z S. Preliminary report on the pathogen of mulberry mosaic dwarf disease[J]. Science of Sericulture, 2006, 32: 249-251.

[13] LU Q Y, XIA Z S. Determination and analysis of small viroid-like RNA isolated from leaf tissues of mulberry with mosaic dwarf disease[J]. Acta Sericol Sin, 36:1017-1021.

[14] MAYO M A, FRITSCH C. A possible consensus sequence for VPg of viruses in the family Comoviridae[J]. Febs Letters, 1994, 354:129-130.

[15] SANFACON H, IWANAMI T, KARASEV A V, et al. Family Secoviridae. In: King A M Q, Adams M J, Carstens E B,Lefkowitz E J (eds) Virus taxonomy: classification and nomenclature of viruses: ninth report of the international committee on taxonomy of viruses[J]. Elsevier/ Academic Press Amsterdam, pp, 2012,881-899.

[16] SERGHINI M A, FUCHS M, PINCK M, et al. RNA2 of grapevine fan leaf virus: sequence analysis and coat protein cistron location[J]. Journal of General Virology, 1990, 71:1433-1441.

[17] THOMPSON J D, GIBSON T J, PLEWNIAK F, et al. The ClustalX windows interface: flexible strategies for multiple sequence alignment aided by quality analysis tools [J]. Nucleic Acids Research, 25:4876-4882.

[18] TOMITAKA Y, USUGI T, YASUDA F, et al. A novel member of the genus Nepovirus isolated from Cucumismelo in Japan[J]. Phytopathology, 2011, 101:316-322.

[19] TSUCHIZAKI T, HIBINO H, SAITO Y. Mulberry ringspot virus isolated from mulberry showing ringspot symptom[J]. Japanese Journal of Phytopathology, 2009, 37:266-271.

[20] VON BARGEN S, LANGER J, ROBEL J, et al. Complete nucleotide sequence of Cherry leaf roll virus (CLRV), a subgroup C nepovirus[J]. Virus Research, 2012, 163:678-683.

[21] XIA Z S, LU Q Y. The graft-transmission test of the mul-berry mosaic dwarf disease in summer[J]. China Seric, 2004, 25:29-30.

[22] XIE L H, ZHOU Z J, LIN Q Y, et al. Study on rice stripe disease: III Pathogen of the disease[J]. Journal of Fujian Agricultural College, 1991, 20:144-149.

Playing on a Pathogen's Weakness: Using Evolution to Guide Sustainable Plant Disease Control Strategies

Jiasui Zhan[1,2], Peter H. Thrall[3], Julien Papaïx[4,5], Lianhui Xie[2], Jeremy J. Burdon[3]

(1 Key Laboratory for Biopesticide and Chemical Biology, Ministry of Education, Fujian Agriculture and Forestry University, Fuzhou, 350002, China; 2 Fujian Key Laboratory of Plant Virology, Institute of Plant Virology, Fujian Agriculture and Forestry University, Fuzhou, 350002, China; 3 CSIRO Agriculture Flagship, Canberra, ACT 2601, Australia; 4 INRA, Santé des Planteset Environnement, UR 1290 BIOGER-CPP, 78850 Thiverval-Grignon, France; 5 INRA, Mathématiques et Informatiques Appliquées, UR 341 MIAJ, 78352 Jouy-en-Josas, France)

Abstract: Wild plants and their associated pathogens are involved in ongoing interactions over millennia that have been modified by coevolutionary processes to limit the spatial extent and temporal duration of disease epidemics. These interactions are disrupted by modern agricultural practices and social activities, such as intensified monoculture using superior varieties and international trading of agricultural commodities. These activities, when supplemented with high resource inputs and the broad application of agrochemicals,create conditions uniquely conducive to widespread plant disease epidemics and rapid pathogen evolution. To be effective and durable, sustainable disease management requires a significant shift in emphasis to overtly include ecoevolutionary principles in the design of adaptive management programs aimed at minimizing the evolutionary potential of plant pathogens by reducing their genetic variation, stabilizing their evolutionary dynamics, and preventing dissemination of pathogen variants carrying new infectivity or resistance to agrochemicals.

Keywords:ecological disease management; spatiotemporal resistance gene deployment; trade-offs; evolutionary plant pathology; multilayer disease forecasting

1 Introduction

As the global human population increases and arable land declines, food security has become a major challenge for society. Much of the progress in securing food supplies in modern society has resulted from enhancing productivity through significant resource inputs (e.g., water and fertilizers), increased cultivation frequency (multiple cropping), and the use of genetically uniform plant varieties with superior yield and quality (monocultures). The latter agricultural practice in particular has narrowed the genetic base of all major crops, thereby increasing the potential for significant losses caused by plant diseases. In the long-term, such practices are unsustainable in terms of both disease management and resource conservation.

Despite growing interest in the use of adult plant and quantitatively based disease resistance (St. Clair, 2010; Zhang et al., 2012) expressed against all pathotypes of a pathogen, losses due to plant diseases are mainly managed through the use of major resistance (R) genes introduced into varieties with superior agronomic characters or by the application of agrochemicals. As new resistant varieties or agrochemicals become widely

Annual Review Phytopathology. 2015, 53:19–43.
Published online 4 May 2015

utilized, their efficacy in disease control is often countered by novel mutants that arise in local pathogen populations. Novel mutants increase in frequency through selection due to their high fitness and may spread to other cropping areas or even to new continents through natural dispersal (e.g., Ug99 spreading out of Africa;1) (Singh et al., 2011) or global trading (Brasier, 2008). Thus, despite the considerable investment required to develop new resistant varieties or agrochemicals, their effectiveness may be significantly reduced within only a few years of deployment (Fraaije et al., 2005; Delmotte et al., 2014). In essence, agricultural intensification and human activities disrupt the coevolutionary dynamics typically found between plants and pathogens in unmanaged ecosystems (Burdon and Thrall, 2014) and provide favorable environments for pathogens to rapidly develop high infectivity and aggressiveness.

Boom-and-bust cycles (Johnson, 1961) of disease epidemics result from complex interactions among hosts, pathogens,and environments that drive changes in population genetics composition and pathogen evolutionary trajectories. These interactions and their consequences for disease development can be modified by agricultural practices that aim to change the use of R genes and agrochemicals to create conditions favorable for hosts to maximize defense responses while minimizing infection. In this review, we first briefly elucidate the mechanisms that shape host-pathogen coevolution and their implications for the ecology and epidemiology of plant diseases in natural ecosystems. We then discuss the common practices and approaches used in modern agriculture to maximize productivity and how these may affect patterns of disease occurrence and epidemics as well as coevolutionary interactions between crop plant hosts and their associated pathogens. Finally, we discuss how to apply evolutionary principles to achieve sustainable disease management in agricultural ecosystems.

2 Disease occurrence and pathogen evolution in natural ecosystems

As disease-causing agents, plant pathogens can cause mortality and reduced fitness of individual plants, resulting in shifts in genetic composition and rapid declines of plant populations. Although there are records of major disease outbreaks [e.g., tree diseases such as chestnut blight (Anagnostakis, 1987), Dutch elm disease (Lanier et al., 1988), white pine blister rust (Kinloch Jr, 2003), and sudden oak death (Garbelotto et al., 2001)], disease epidemics in natural ecosystems are typically ephemeral and limited in spatial scale (Burdon et al., 2013). Indeed, all of the major tree disease epidemics noted above represent invasive pathogens that were transported from their center of origin to new environments where they encountered naïve host populations with no evolutionary history of interaction. In contrast, in situations involving hosts and pathogens that have long histories of association, host populations tend to be distributed patchily across landscapes and contain considerable within-and among-population diversity in genetically based disease resistance (Rauscher et al., 2010; Laine et al., 2011).

2.1 Red queen dynamics reflect the evolution of host resistance and pathogen infectivity

Over millennia, pathogens and plants have engaged in an evolutionary battle, with pathogens attempting to overcome plant defenses and plants attempting to resist loss due to pathogen attack. In natural ecosystems, these interactions occur against a backdrop of ecological and environmental heterogeneity in which pathogen impact is tempered by patchiness in the suitability of environments for disease development as well as by variable and small host populations. Two distinct hypotheses have been proposed to describe the coevolutionary trajectories that might result in such situations. Directional selection or arms-race dynamics envisage evolutionary change occurring as a consequence of novel resistance and infectivity alleles sequentially sweeping through populations. Alternatively, disruptive selection is based on a process of negative frequency dependence maintaining allele frequencies in a dynamically fluctuating state (Fisher, 1930; Brown and Tellier, 2011; Burdon et al., 2013). Practical evidence for evolutionary arms races is seen in many agricultural situations in which resistance breeding followed by sequential but widespread introduction of one or two major R genes has frequently been accompanied by boom-and-bust disease dynamics, as once-common pathotypes are rapidly replaced by variants able to overcome new R genes. In unmanaged systems, the major disease outbreaks that accompany many invasive diseases also show characteristics of such evolutionary trajectories, particularly during the earlier stages of invasion (Burdon et al., 2013). In contrast, longer-term natural host-pathogen interactions show more complex interactions in which negative frequency dependence is especially apparent (Thrall et al., 2012; Papaïx et al., 2014). Recent experimental studies of coevolution between *Pseudomonas fluorescens* and phage also support the view that, in antagonistic interactions, fluctuating selection is more likely to be maintained over time than arms-race dynamics (S., 1943; Gómez and Buckling, 2011; Hall et al., 2011).

2.1.1 Coevolutionary advantages of plant pathogens.

Evolutionary rates in species are negatively correlated with generation times (Andreasen and Baldwin, 2001; Thomas et al., 2010; Zhong et al., 2014). Species with shorter generation times tend to have higher evolutionary rates possibly because their genomes are copied more frequently, generating more DNA replication errors per unit time (Martin and Palumbi, 1993; Herrick, 2011). In host-pathogen interactions, pathogens are generally at an evolutionary advantage because of shorter generation times relative to their hosts. Short generation times not only ensure that pathogens generate more mutations in a fixed period of time but also enhance the opportunity for rapid adaptive change, as new mutations or gene combinations generated in each pathogen generation can quickly challenge host defense systems. Over a few pathogen generations, advantageous mutants can come to dominate a pathogen population, there by rendering host defense systems ineffective. Clearly, this advantage is magnified in cropping systems in which host defenses are relatively static and genetically homogeneous over a range of spatiotemporal scales.

Plant pathogens are also at an evolutionary advantage owing to their typically very large population sizes. Single disease lesions may generate large numbers of pathogen propagules, each of which can serve as an independent unit of reproduction. Population size contributes positively to the generation of new mutants and negatively to the loss of existing mutants (Zhan et al., 2014). Natural populations with large effective sizes tend to have greater genetic variation (Wright, 1931, 1938; Subramanian, 2013), as more mutants are expected and fewer are lost due to random genetic drift. Therefore, ultimately for any management strategy to successfully reduce pathogen evolutionary potential, it must directly or indirectly result in a significant reduction of effective pathogen population size.

2.1.2 Coevolutionary advantages of plants.

The obvious disadvantages that hosts face with respect to generation time and population size led to the idea that sexual recombination might provide effective evolutionary compensation by generating novel combinations of resistance factors (Red Queen hypothesis; (Hamilton, 1980; Hamilton et al., 1990)).In the short term,sexual reproduction provides hosts with a mechanism for the production of genetically unique progeny in each generation by shuffling existing defense related alleles.Sexual reproduction can also increase defense polymorphism in host populations by generating new alleles through intragenic recombination (Brunner et al., 2008; Sabat et al., 2008; He et al., 2010; Ferreira and Briones, 2012).In the long-term,sexual reproduction facilitates the formation of novel mutant alleles and allelic combinations favorable for host defense and enhances the removal of deleterious mutations that otherwise could accumulate in plants during asexual reproduction (Kondrashov, 1988; Hurst and Peck, 1996; Moran, 1996; Innocenti et al., 2011; Jaramillo et al., 2013).

Plant populations in nature are highly patchy with many diverse genotypes of the same species as well as phylogenetically closely and more distantly related species distributed in the community. Furthermore, population-level resistance, as well as other genetic and ecological traits, may vary greatly from year-to-year (Burdon and Thompson, 1995; Meyer et al., 2010). This spatiotemporal feature of the genetic structure of plant populations can have several negative effects on the evolution of corresponding pathogens. For example, pathogens need to invest significant resources in the production of sufficient offspring so as to track spatiotemporal, demographic, and genetic variation in their hosts. Moreover, high genetic variation and environmental heterogeneity among host populations mean that a large percentage of pathogen propagules may not lead to infection because they cannot find the right hosts at the right time and place. Simulation modeling studies suggest that further diversity and disruptive selection pressures may occur if host populations contain both quantitative and qualitative resistances, akin to a mixture of hard and soft selection pressures. Diversifying selection generated by such spatiotemporal diversity in a mosaic host population structure has the potential to impede the emergence of highly infective plant pathogens (Thrall and Burdon, 2002; Sapoukhina et al., 2009).

2.2 Trade-offs and fitness penalties can impede the emergence of super infective pathogens

Evolutionary theory hypothesizes the existence of trade-offs between pathogen infectivity or aggressiveness (Zhan and McDonald, 2013) and other life-history traits. In gene-for-gene interactions (Flor, 1955, 1956), the fitness cost by mutations from noninfectivity to infectivity has been documented in the pathology literature (Jenner et al., 2002; Abramovitch et al., 2006; Caffier et al., 2010; Zhan and McDonald, 2013). By gaining or accumulating infectivity, pathogens are able to infect more hosts but at the same time reduce their competitiveness when the corresponding resistance is absent; thus such costs are context-dependent. Isolates of the rust pathogen *Melampsora lini* carrying more infectivity factors against *Linum marginale* tend to produce fewer spores per pustule than less-infective strains (Thrall and Burdon, 2003), and the historic frequency of *Rhynchosporium secalis* on barley is negatively correlated with the complexity of its race structure (Zhan et al., 2012).

Trade-off theory is also relevant to plant pathosystems lacking gene-for-gene interactions due to a conflict between pathogen aggressiveness and transmission rates. Although higher aggressiveness resulting from greater host exploitation potentially increases plant-pathogen reproduction rates, it also shortens the time the pathogen is able to persist on an infected host and be transmitted (Doumayrou et al., 2013). This trade-off between reproduction and transmission often selects for pathogens with intermediate aggressiveness, although the optimum strategy depends on host and pathogen life history (e.g., host density, host range, transmission mode; (Lenski and May, 1994)). A classic example of this type of trade-off is seen in myxomatosis in rabbits. The myxoma virus was introduced into Australia to control European rabbit populations in 1950. Initially, the virus was extremely aggressive, leading to high rabbit mortality (~99%). However, a sharp drop in aggressiveness was observed after only a few pandemics (Mykytowycz, 1953), as the highly aggressive strain was killing the rabbit host too quickly, thereby reducing its chance of being transmitted.

In nature, plant pathogens compete with not only other genotypes of the same species but also other pathogen species. In these circumstances, it is not uncommon to find that pathogens with higher aggressiveness or more infectivity factors on a particular host show lower ability to compete on other hosts and poorer adaptation to stress and/or fluctuating environments than other pathogens that are less aggressive or carry fewer infectivity factors (Abang et al., 2006; Gandon and Nuismer, 2008; Sommerhalder et al., 2011; Brown et al., 2012). The average infectivity of plant pathogens in natural ecosystems can be further attenuated by trade-offs with other competing species or by selection for adaptation to different abiotic environments. For example, the infectivity of *Serratia marcescens*, a nonobligatory, opportunistic bacterial pathogen, decreases in environmental refuges through life-history trade-offs or random accumulation of mutations that impair infectivity traits under relaxed selection (Mikonranta et al., 2012).

3 Disease occurrence and evolution of plant pathogens in modern agricultural ecosystems

In modern society, agriculture has moved from small-scale traditional production methods typified by low inputs, single-cycle annual cropping, and spatiotemporal diversification to highly mechanized, industrial-style systems with high resource inputs, intensification, and reduced diversity (monocultures). These shifts impose crucial changes on patterns of host-pathogen dynamics and generally favor more widespread epidemics (i.e., larger pathogen population sizes) and thus rapid evolution of pathogen infectivity. In this section, we briefly describe how some farming practices adopted in modern agriculture and associated trends in global connectivity (e.g., trade patterns) may affect pathogen epidemiology and coevolutionary trajectories in cropping systems.

3.1 Monocultures (single species, single variety) impose strong selection on plant pathogens and drive their evolution

Most modern agriculture is characterized by situations in which crop varieties with closely related genetic backgrounds are planted across large areas. This practice is especially common in developed countries where labor costs are high and agriculture is highly mechanized. Examples of large-scale monocultures are seen in wheat production in the United States and Australia, and in soybean production in Brazil and North China. Growing crops with identical physiological characters and genetic architectures can result in greater yield per unit area because management procedures can be standardized and mechanized, competition from other species for nutrients can be more easily controlled, and better use can be made of available light and space. Standardized management protocols can result in less loss from inefficient planting and harvesting, whereas enhanced product uniformity is important for end-use processing and marketing.

Monocultures in agriculture are mainly composed of single varieties that have been bred for high yield and resistance to certain diseases. In the short term, the narrowed genetic base of crops in monoculture has made them more vulnerable to disease epidemics. Pathogen strains that can infect one individual in a crop can affect all and hence can quickly spread over an entire field or region because reduced environmental heterogeneity and genetic barriers do not exist to prevent or mitigate spread. Monocultures impose strong directional selection on pathogens, and their widespread use is considered to be responsible for rapid loss in the effectiveness of R genes and agrochemicals. Novel infectivity that emerges in pathogen populations as a result of single point mutations can quickly increase in frequency through strong directional selection and spread through natural or human-mediated gene flow even if the overall fitness of new pathogen variants is somewhat lower. This process of stepwise mutation, in which pathogen populations respond to the release of new resistant varieties, has been particularly well documented in the interaction between wheat and its stem (*Puccinia graminis tritici*; (Watson and De Sousa, 1983)) and stripe rust (*Puccinia striiformis*; (Wellings, 2007)) pathogens. The theoretical and empirical consequences of this escalating evolution of plant pathogens under monoculture have been well documented (Zhan et al., 2002; Marshall et al., 2009; Sommerhalder et al., 2011).

Somewhat ironically, the main source of disease resistance in crops is wild relatives in primary germplasm collections. Natural populations of wild relatives occurring in environments conducive to disease may contain many different R genes and alleles that individually confer resistance against a portion of the pathogen population (Dinoor, 1970; Laine et al., 2011). Under conditions suitable for disease development, most host individuals in these populations incur some disease, but the diversity of resistances (including a range of minor gene resistance) and their distribution in the population as a whole keep epidemic development in check (Thrall and Burdon, 2000). However, when these R genes are introduced into cultivated crops and deployed in agricultural ecosystems as genetically uniform stands growing under enhanced nutrient conditions, their effectiveness in controlling plant diseases does not persist in the long-term.

4 Production intensification minimizes the effect of trade-offs that restrict the emergence of super infectivity

Intensification has been a major driver of agricultural productivity, achieved mainly through planting varieties with high yield potential, increasing resource availability, and increasing the frequency of cropping. Such agricultural practices can have several impacts on disease epidemics and pathogen evolution. Maintaining or improving genetically based disease resistance is often seen as a distraction in efforts aimed specifically at high yield potential (Pink, 2002), and in some cases a single-minded focus on yield can result in changes in the vulnerability of the plant to previously unimportant pathogens. Thus, the use of Texas male-sterile cytoplasm in maize in the 1970s resulted in the accidental development of an entire crop that suffered major losses to *Helminosporium maydis*, a previously insignificant disease (Tatum, 1971; Ullstrup, 1972; Thomas et al., 2010). Increased nutrient and water availability can also make crops more vulnerable to pathogen attack. High nitrogen levels generally favor mildew diseases of a range of crops (Bainbridge, 1974; David et al., 2003; Keller et al., 2003; Chen et al., 2006), while irrigation may favor the development and spread of soilborne diseases (Cook and Papendick, 1972; Jefferson and Gossen, 2002), such as cotton wilt (El-Zik, 1985; Morrow and Krieg, 1990), especially if water is recycled on farm.

Similarly, increasing the frequency of cropping favors pathogen growth and reproduction by providing green tissue bridges across previous seasonal gaps, thereby minimizing disruptive selection previously forced on pathogens by reliance on off-season survival strategies, including shifts to saprophytic stages (Gubbins and Gilligan, 1997; David et al., 2003; Abang et al., 2006; Altizer et al., 2006; Chen et al., 2006; Sommerhalder et al., 2011). In nature, the spatial structure of local populations can reduce transmission and the evolution of pathogen infectivity (Messinger and Ostling, 2009), but growing crops over wide regions and agricultural intensification tend to minimize spatiotemporal heterogeneity, thereby potentially favoring increased pathogen infectivity and aggressiveness. In part, this is because continuous tissue availability associated with agricultural intensification not only supports large pathogen population sizes, which in turn minimize the effect of drift with regard to reducing pathogen genetic variation, but is also likely to diminish trade-offs between aggressiveness and transmission in pathogen populations as crop planting provides new tissues to support further growth and reproduction of the pathogen in the next season.

4.1 Global connectivity facilitates disease invasion and the spread of novel pathotypes, which increase the evolutionary potential of pathogen populations

In the current era of globalization, the world has become more interdependent than at any other time in human history. Increased international trade in agricultural products, exchange of plant materials for research and teaching, and hitchhiking of pathogens via various transportation modes have greatly increased the spread of many infectious plant diseases across major geographicboundaries. For example, a number of pathogens, including wheat stripe rust (*P. striiformis* f.sp.*tritici;*(Wellings, 2011; Morgounov et al., 2012)), apple scab (*Venturia inaequalis*; (Gladieux et al., 2008)), chestnut blight (*Cryphonectria parasitica*; (Freinkel, 2009)), and *Phytophthoraramorum*(Waage JK, 2008; Grünwald et al., 2012), have all spread at various times to various countries, where they have created major disease epidemics.

Vulnerability of many agricultural crops to exotic pathogens often occurs because breeding has proceeded for many years without challenge by the relevant pathogen. As a consequence, through a combination of genetic drift and negative selection associated with potential fitness costs, both major and minor gene resistance may be lost when diseases are absent. When host and pathogen are reunited, the pathogen essentially encounters naïve hosts and causes significant damage due to the lack of relevant coevolutionary responses (e.g., see (Anagnostakis, 1987)). In agricultural settings, there are many examples of severe crop loss caused by pathogens appearing in an area for the first time [e.g., the introductions of sugarcane rust (*Puccinia melanocephela*) into the Americas (Purdy et al., 1985), citrus canker (*Xanthomonas axonopodis* pv. citri) into the United States (Gottwald et al., 2001), and wheat stripe rust (*P. striiformis*) into Australia (O'Brien et al., 1980; Brown and Hovmøller, 2002)]. Similar phenomena have been associated with the introduction of pathogens into natural plant communities. For example, chestnut blight caused by the fungal pathogen *C. parasitica* causes fewer problems in Asia, the putative center of origin for the pathogen (Liu and Milgroom, 2007), than in North America, where the American chestnut has now ceased to be a significant part of the forest ecosystems that it once dominated (Freinkel, 2009).

In addition to the invasion of new diseases, globalization can promote the spread of new infectivity and agrochemical resistance among existing pathogen populations, for example the invasion of *Phytophthora infestans* mating type A2 into Europe in the 1980s (Hohl and Iselin, 1984; Grünwald and Flier, 2005; Judelson and Blanco, 2005). This caused the simultaneous breakdown of host R genes and rendering of agrochemicals ineffective across a large geographical area within a short time. Long-distance dispersal also enhances the evolutionary potential of pathogens by constantly introducing novel genetic variation from distant populations. For example, following the hitchhiking of novel pathotypes on the jet stream from Africa, the pathogenicity structure of *P. graminis* tritici populations in Australia changed rapidly, resulting in the rise of new dominant lineages (Watson, 1981). Long-distance dispersal can also contribute to shifting the evolutionary trajectory of pathogens toward higher aggressiveness as a consequence of escalating competition among less-related pathogen genotypes (Koskella et al., 2006; López-Villavicencio et al., 2007; López- Villavicencio et al., 2011; Zhan and McDonald, 2013).

The extent to which globalization impacts pathogen invasion and evolutionary potential is primarily determined by dispersal ability. Thus, soilborne or rain-splashed pathogens have lower potential for long-distance movement, and globalization can be the main driver for the increasing spread of these diseases. The horticultural trade in particular has been responsible for several major disease invasions (Brasier, 2008), including the appearance of *P. ramorum* in Europe (Waage JK, 2008; Grünwald et al., 2012). For airborne pathogens such as rusts and powdery mildews, spores can be carried by air currents for thousands of kilometers, leading to continental and intercontinental movement [e.g., wheat stem rust spreading along the *Puccinia* flyway throughout the Great Plains in the United States or barley powdery mildew dispersal in Europe (Brown and Rant, 2013)].Globalization may therefore not play such an important role in the spread and evolution of these pathogens.

4.2 Use of quantitative resistance may select for higher pathogen aggressiveness and pathogen diversity

Due to the rapidity with which major gene resistance is often rendered ineffective in cropping systems, interest in the use of quantitative resistance or a combination of quantitative and qualitative resistance to control plant pathogens has increased in recent decades. Quantitative host resistance reduces disease severity and slows down epidemic development but does not prevent it. The majority of quantitative resistance is pathotype nonspecific (Pink, 2002; Ballini et al., 2008; Marcel et al., 2008). In this type of host-pathogen interaction, all pathogen genotypes have the ability to cause infection, colonize host tissue, and reproduce, but vary in the extent to which disease develops. Competition among pathogen genotypes is expected to be low because all isolates can reproduce and host-imposed selection on pathogens to evolve at aggressiveness loci is expected to be weaker than for qualitative resistance,making it a more durable strategy for controlling plant diseases. Indeed, it has been documented both theoretically and experimentally that quantitative resistance can mitigate the evolution of plant pathogens because of decreased competition among pathogen genotypes (Zhan et al., 2002; Sommerhalder et al., 2011).

However, because quantitative resistance reduces the pathogen's within-host growth rate, it may also select for higher pathogen aggressiveness (Gandon and Michalakis, 2000), as has been demonstrated for the fungal pathogen *Zymoseptoria tritici*, which causes Septoria leaf blotch disease on wheat (Cowger and Mundt, 2002; Zhan et al., 2006; Yang et al., 2013), as well as other pathogens (Pariaud et al., 2009b; Pariaud et al., 2009a; Montarry et al., 2012). Moreover, co-infection may result in the increased aggressiveness of pathogens (Susi et al., 2015). Quantitative resistance may also favor greater within-host pathogen diversity than major gene resistance, thus increasing the evolutionary potential of the pathogen. In addition, due to positive associations between chemical resistance and aggressiveness (Angiolella et al., 2008; Yang et al., 2013) in both plant and animal pathogens, quantitative host resistance may also select for basal pathogen resistance to agrochemicals with multiple action modes. Quantitative resistance also selects for pathogen genotypes with recombinant backgrounds (Zhan et al., 2007), suggesting that this type of resistance might promote sexual recombination, thereby further increasing the pathogen's ability to evolve.

4.3 Use of agrochemicals may also select for pathogens with higher infectivity

During the growing season, when a crop is insufficiently well protected by genetic means to prevent epidemic development, agrochemicals may be the only alternative for efficient disease control. Undoubtedly, the invention and application of modern agrochemicals have contributed greatly to global food supply and security. For some crops,such as potatos in Western Europe,good harvests rely heavily on repeated application of up to 15 fungicide sprays in a single growing season. However, the continuing application of agrochemicals on a large geographic scale not only causes adverse environmental effects (e.g., impacts on beneficial arthropod predators and pollinators) but also speeds up changes in pathogen populations and increases their overall aggressiveness.

In many situations, the emergence of agrochemical resistance involves detoxification processes that range from reducing the influx of synthetic compounds into pathogen cells to enhancing their efflux through the action of efflux/influx pumps located in the cytoplasmic membrane (Truong-Bolduc et al., 2005; Chen et al., 2006; Gupta and Chattoo, 2008). These synthetic agrochemicals may share structural or functional characteristics with natural compounds that are produced by resistant hosts and that have lethal or inhibitory effects on the establishment, survival, and reproduction of pathogens (Akins, 2005). Some efflux pumps have the ability to transport a broad range of structurally unrelated compounds during pathogen infection and can affect both aggressiveness and agrochemical resistance in plant pathogens (Urban et al., 1999; Sun et al., 2006) For example, there is a significant positive correlation between aggressiveness and agrochemical resistance in the wheat pathogen *Z. tritici* (Yang et al., 2013).

4.4 Breeding for high yield and quality artificially removes undesired traits, which may be crucial for hosts to combat pathogens

Throughout their evolutionary history, plants have developed an array of constitutive and inducible arsenals, including a diversity of complex natural compounds (phytochemicals) such as terpenoids, phenolics, and alkaloids, which are known to reduce the susceptibility of plants to a range of insect, molluscan, and fungal parasites. Many of these chemical compounds or physical structures are critically important for the survival and reproduction of plants growing in the wild but also affect the quality and palatability of agricultural products or are even toxic to humans and/or animals. Such traits are often targets for removal in human-directed breeding programs or may be screened out unintentionally by plant breeders due to negative correlations between host resistance to pathogens and crop productivity when pathogens are absent (Staub and Grumet, 1993; Brown, 2002; Zeller et al., 2012). Indeed, fitness costs associated with both R-gene and minor-gene resistance have been reported on many occasions (Brown and Hovmøller, 2002; Burdon and Thrall, 2003; Brown and Rant, 2013). For example, transgenic *Arabidopsis thaliana* with a *Pseudomonas syringae* pv. *maculicola* resistance produced 9% fewer seeds per plant than isogenic lines lacking the R gene (Tian et al., 2003), whereas the presence of minor gene resistance for protection against *Peronospora parasitica* in *Brassica rapa* carried a 6% fitness cost in a disease-free environment (Mitchell- Olds and Bradley, 1996; Michelmore et al., 2013). However, such fitness costs are typically context-dependent, and there is a marked contrast between the costs seen in many controlled experimental studies (Brown and Tellier, 2011) and the costs seen under field conditions. Essentially, long-term selection by farmers and breeders has almost certainly selected traits that minimize any yield penalties.

5 Approaches for sustainable plant disease management

A wide range of approaches have been developed to control plant diseases in agricultural ecosystems, including host resistance, agrochemicals, biological agents, and crop management strategies. Undoubtedly, these have individually contributed significantly to short-term epidemic prevention and control but applied without broader consideration of their epidemiological and evolutionary implications may offer little hope of sustainable plant disease control. Perhaps the only exception to this involves localized engineering solutions that change the environment in a semi-permanent or permanent way. Thus, improved drainage through ditching, the installation of field drains, or the use of raised beds has a direct negative effect on the potential for many soilborne diseases to spread and increase (Hobbs et al., 2007).

Understanding of interactions between plants and communities of bacteria, fungi, and other microorganisms in the soil has been a major area of investigation for many years, as scientists have sought to limit soilborne diseases (Thurston, 1990). The advent of high-throughput molecular technologies has made possible a much fuller accounting of the soil microbial communities associated with particular crops (Van Elsas et al., 2008) and of how these communities may be affected by soil biogeochemistry, environmental factors, or even crop identity. However, a major challenge that remains is how such information may be used to design management approaches or develop crop varieties that foster beneficial soil communities and promote plant growth while suppressing pathogen activity (Mazzola, 2004). The application of coevolutionary principles offers a valuable way of tackling these complex community interactions (Kinkel et al., 2011).

Approaches to disease control that depend on resistant varieties and agrochemicals usually show high effectiveness when they are newly deployed. However, due to the high evolutionary potential of many plant pathogens (McDonald and Linde, 2002), novel genotypes can rapidly emerge via mutation or recombination. When this happens, particular disease control approaches can be rapidly rendered ineffective as the novel genotypes increase in frequency through natural selection and quickly spread to other locations to cause failure of control over large geographic scales [e.g., change in infectivity and agrochemical susceptibility in *Blumeria graminis* attacking barley in western Europe (Caffier et al., 1996; O'Hara and Brown, 1996; Blatter et al., 1998; Hovmøller et al., 2000; Bousset et al., 2002)]. As a consequence, sustainable disease management has two distinct but interdependent goals: an immediate epidemiological one of reducing the incidence, severity, and frequency of disease epidemics and a longer-term evolutionary one of reducing the rate of evolution of new pathotypes (Heikkilä, 2011; Zhan et al., 2014). Sustainable disease control could be achieved through an integrative disease management program that is able to minimize the evolutionary potential of plant pathogens by reducing their genetic variation, stabilize their evolutionary dynamics by diversifying selection, and restrict the migration of pathogens carrying new infectivity genes or agrochemical resistance.

To achieve sustainable disease management requires a shift in mindset away from a sole focus on yield and productivity to a broader integration of productivity with underlying ecological, economic, and environmental dimensions. Multidisciplinary collaboration is critical for the development of effective sustainable disease management strategies,and evolutionary plant pathologists and geneticists need to play a major role in designing management practices that maximize host plant defense while simultaneously minimizing opportunities for pathogens to evolve. Most importantly, the approaches used to control plant pathogens must be adaptive. Pathogens are often highly spatiotemporally variable with new genotypes constantly arising through mutation, recombination, and gene flow. However, current disease control methods are relatively static. To overcome the inexorable shift toward infectivity with respect to the current deployment of R genes, we must apply Red Queen principles by adaptively shifting through time and space the mix of resistances, agrochemicals, and agronomic practices designed to reduce pathogen loads and force the pathogen into a pattern of disruptive selection. At the same time, however, all control approaches must be practical and economically attractive if they are to be adopted; thus, prediction of optimal strategies must integrate aspects of sustainable farm profitability, disease epidemiology, and pathogen evolution.

5.1 Maximize Genetic Drift to Reduce Genetic Variation of Plant Pathogens and Mitigate Their Ability to Evolve

Genetic variation plays a critical role in the evolution of pathogen infectivity with the potential for a population to adapt to fluctuating environments depending upon the level of additive genetic variance in characters that are relevant to fitness (Fisher, 1930). Pathogen populations with higher genetic variation have increased mean fitness, resilience, and adaptability to changing environments and can evolve more rapidly to overcome control methods such as antibiotics, fungicides, and resistant varieties. They are therefore more difficult to control.

Genetic variation in pathogen populations is derived from interactions among mutation, gene flow, recombination, random genetic drift, and natural selection shaped by coevolution with their hosts. Mutation, recombination, and gene flow increase genetic variation either through generating new DNA sequences, rearranging existing DNA sequences, or introducing new genetic material from other populations. Genetic drift decreases genetic variation in pathogen populations by randomly purging mutants regardless of whether they are beneficial, neutral, or harmful. Natural selection can increase or decrease genetic variation, depending on the type of selection. Directional selection favoring some traits in pathogen populations may lead to a rapid decrease in genetic variation, whereas balancing or frequency-dependent selection tends to increase the amount of variation through selection for rare alleles or heterozygotes.Whereas mutation rate is an intrinsic characteristic of species and cannot be regulated, genetic drift (also gene flow, see next section) can be manipulated through changes in agricultural practices and cultivation systems (Thrall et al., 2011). Recombination may be difficult to influence for many pathogens, although note the success of barberry eradication programs for wheat stem rust control in the United States (130; also see the next section). Essentially, the particular combination of life-history characters possessed by a pathogen is an important determinant of the most effective strategies that can be invoked to effect evolutionary control (Barrett et al., 2008).

The extent of genetic drift in pathogens is determined by effective population size. In a random mating population, effective population size is usually close to half the number of individuals involved in the breeding system (Nunney, 1995). Relatively few pathogens fit such idealized models (clonal reproduction is common), and genetic drift in such populations is likely to be more severe. The negative effect of genetic drift on pathogen evolution can be maximized by reducing pathogen population size through changes in agricultural practices and cultivation systems, such as season-long fallows, field hygiene, intercropping, and crop rotation (Zhan et al., 2014). Reducing population size decreases the number of mutations occurring per unit time and decreases the probability that new alleles will be generated or that existing alleles will persist. Platz & Sheppard (Platz and Sheppard, 2007) ascribed low wheat stem rust occurring in 1973–1974 in the northern wheat growing areas of Australia (when a major epidemic was occurring farther south) to sustained small pathogen population sizes that reduced the probability of the arrival of new infectivity variants able to overcome the R genes in use at the time. Reducing population size also increases the probability of mating among close relatives (inbreeding), which further reduces the effective size of pathogen populations because the reproductive success of individuals in this case is no longer dependent on the independent alleles at each locus (Nunney, 1999). Because the chance of random fixation of a mutant is equal to its frequency, mutants at low frequency are generally at a disadvantage and may be lost to genetic drift.

Field hygiene, such as stubble burning or burial, and the use of clean seed not only reduce pathogen population size and evolutionary potential but also postpone the onset and intensity of epidemics through reduced inoculum availability. This is particularly relevant for monocyclic pathogens because epidemics of these diseases are largely determined by the density of primary inoculum. Seasonal fallows force annual epidemic cycles in foliar diseases so that local pathogen populations undergo regular cycles of expansion and contraction. At the end of each epidemic cycle, only a small fraction of the pathogen population survives to begin the next epidemic cycle, enhancing the effect of genetic drift in reducing and removing novel mutants (Zhan et al., 2001). Because generations with the smallest population size make a disproportional contribution to long-term estimates of effective population size (Lande and Barrowdough, 1987; Ellstrand and Elam, 1993), sudden crashes in pathogen populations resulting from agricultural practices, such as field hygiene, in one season can have a major impact on longer-term genetic variation and pathogen evolutionary potential, especially where this involves rare alleles that may be important in the emergence of new infectivity or agrochemical resistance.

Crop rotations create founder effects that occur when a small number of reproductive units from neighboring populations colonize the new crops. These genetic bottlenecks increase the chance of mating between related individuals, which further reduces effective pathogen population size. Using R genes, applications of agrochemicals and effective sanitation practices can reduce both epidemic severity, by hindering disease development, and the rate of evolution in the pathogen population, by decreasing its effective population size. However, continuous use of these approaches may facilitate the evolution of pathogen infectivity; thus, dynamic spatiotemporal variation of these approaches is required for sustainable disease management (see next section). Disease tolerance may be useful for controlling plant diseases in the short term but may have a longer-term negative impact on sustainable disease management because host tolerance allows plant pathogens to maintain large population sizes and enhances their ability to adapt to host defenses (Horns and Hood, 2012). Tolerant varieties could also impose higher disease burdens on vulnerable crops in surrounding areas; however, although this concept has been discussed extensively in the theoretical literature, there are few convincing examples of agricultural crop varieties that show evidence of significant disease tolerance. The practice of minimum tillage has significant physical benefits (reduced erosion, more effective water use, etc.) but has variable impacts on pathogen populations, with both increases and decreases reported (Bailey and Duczek, 1996; Bockus and Shroyer, 1998).

5.2 Minimize gene flow to reduce genetic variation of plant pathogens and prevent the spread of new diseases or infectivity genes

The impact of gene flow on the evolution of plant pathogens can be multidimensional. Gene flow is a critical factor promoting pathogen evolution via the spread of new infectivity and agrochemical resistance among pathogen populations across large geographical areas. Like mutation, gene flow directly serves as a source of genetic variation by introducing novel alleles and/or allele combinations from neighboring populations and indirectly by increasing effective pathogen population sizes. Gene flow links fragmented populations and reduces the chance that related individuals mate, thereby increasing effective population size and pathogen evolutionary ability (S., 1943; Slatkin and Voelm, 1991; Zhan et al., 2001).

A significant body of literature suggests that in animal host-parasite systems when genetic relatedness among pathogen strains is high, selection operates to increase cooperation among related pathogen genotypes. Conversely, increasing genetic separation increases the likelihood of superinfection and competitive exclusion (Zhan and McDonald, 2013). At the same time, theory predicts that competition among less-related genotypes will select for higher aggressiveness. If these competition mechanisms hold true for plant pathogens, then gene flow, by reducing genetic relatedness within pathogen populations, may facilitate the emergence of higher aggressiveness. Although a number of studies have attempted to measure competition among plant-pathogen strains (Zhan and McDonald, 2013), interpretation of these is often fraught with difficulties. However, infection experiments involving the systemic smut *Microbotryum lychnidis-dioicae* and its host *Silene latifolia* suggest that competition among pathogen genotypes increases as their genetic relatedness decreases(Koskellaet al., 2006; López-Villavicencio et al., 2007).

Susi and colleagues (Susi et al., 2015) demonstrated that co-infection by pathotypes of *Podosphaera plantaginis* attacking *Plantago lanceolata* resulted in higher aggressiveness (spore production) than occurred in situations involving single infections by individual pathotypes. Importantly, field surveys by these authors further showed a strong positive correlation between population levels of coinfection and the severity of epidemics observed during growing seasons. Given that co-infection levels in local populations of *P.lanceolata* were positively correlated with among-population connectivity, this argues strongly that a full analysis of R-gene packaging and deployment options in agricultural cropping systems must include specific consideration of spatial structure and its likely impact on pathogen evolution.

Quarantine is one of the most important procedures used to limit cross-continent or crosscountry gene flow in pathogens. Population genetics surveys indicate that the level of genetic variation of many pathogens, including *Z.tritici* (Zhan et al., 2003) and *R.secalis* (McDonald et al., 1999; Salamati et al., 2000), is lower in Australia than North America or Europe, possibly due to Australia's strict quarantine regulations. Quarantine can be effective in limiting invasion of new diseases caused by soilborne and seed-borne pathogens, or by pathogens having short-distance dispersal, but is less effective in preventing gene flow in airborne pathogens such as rusts and powdery mildews.

With the exception of special internal quarantine boundaries, however, quarantine does not regulate the movement of pathogens within political borders. Public and especially grower awareness of biosafety and biosecurity is crucial to mitigate gene flow in pathogens among regions within a country. Regional gene deployment may serve as a barrier to reduce gene flow for pathogens lacking long dispersal ability. Using clean seed and plant material, preventing the circulation of diseased plant products in markets, eradicating suspected new diseases and pathotypes, and maintaining comprehensive programs of farm and machinery cleanliness are also important tools for reducing movement and genetic exchange among pathogen populations. Governments can use a range of policy mechanisms to regulate gene flow, including the use of public information to encourage appropriate action and voluntary compliance. Incentives to growers to voluntarily report, control, and monitor disease outbreaks can be enhanced with the use of government financial instruments.

In practice, once a pathogen has become established in a new territory, it is common to relax quarantine restrictions. This practice may lead to the introduction of additional genetic variation, thereby increasing the pathogen's evolutionary potential and making disease control more difficult in the long-term. The genetic diversity of the pine pathogen *Sphaeropsis sapinea* is consistent with historical records of the frequency and quantity of seed and germplasm importation to and within the Southern Hemisphere. Diversity is high in South Africa, moderate in New Zealand, and low in Australia, corresponding to multiple seed importations in South Africa, moderate seed importations in New Zealand, and few seed importations in Australia (BURGESS and WINGFIELD, 2002). Relaxing quarantine also allows the movement of novel genes and genotypes among geographically isolated populations, possibly accelerating the breakdown of R genes. It should also be noted that appropriate introduction of seed and germplasm can diversify the resistance base.

5.3 Minimize sources of primary inoculum and new pathogenicity through eradication of alternate hosts

For pathogens that require alternate hosts to complete their life cycle, removing or reducing the density of wild hosts near cropping areas may be an effective approach to delay disease epidemics and retard pathogen evolution. Not only do alternate wild hosts serve as sources of primary inoculum, leading to early disease occurrence and epidemic development, but pathogens from these hosts also provide a source of novel infectivity genes as a consequence of selection. Alternate wild hosts are essential for sexual reproduction in those rust pathogens that have evolved a heteroecious lifestyle. Barberry (*Berberis vulgaris*) is the alternate host of wheat stem rust (*P. graminis tritici*). In the Great Plains region of the United States, aeciospores generated on barberry infect young wheat crops, acting not only as the primary inoculum to initiate new epidemics but also as a source of novel infectivity gene combinations. A concerted campaign of eradicating barberry plants over much of the twentieth century resulted in delayed onset of wheat stem rust epidemics of ~10d compared to the beginning of the program and significantly reduced aggressiveness and race complexity of *P. graminis* tritici in the Great Plains (Roelfs, 1982).

5.4 Utilize dynamic disease management programs to reduce selection pressure on pathogens by promoting diversifying selection

For the foreseeable future, the use of resistant varieties will continue to be the major approach to control plant diseases. However, the strong natural selection generated by current agricultural preferences for strategies involving disease-free or near-free clean crops (Fig. 1a) leads to a loss of effectiveness of the major gene-mediated resistance found in most varieties within five years of consecutive cultivation (Bayles et al., 2000; Cowger et al., 2000; Caffier and Laurens, 2005; Brun et al., 2010; Peressotti et al., 2010). Because it generally requires five to ten years to introgress an R gene into a commercial variety and the source of naturally occurring R genes is finite, the current approach to the use of R genes is clearly not sustainable in the long-term. To counter this, different R genes have been pyramided or combined with adult plant resistance genes (Brun et al., 2010). Deep knowledge of plant immune systems raises the possibility of increasing the durability of resistance through rational deployment of multiple nucleotide-binding leucine-rich repeat receptors that recognize distinctly different pathogen core receptors (Mikonranta et al., 2012; Dangl et al., 2013; Schornack et al., 2013). Whether such an approach is sufficient to present the pathogen with an insurmountable evolutionary barrier remains to be seen (Fig. 1a), although existing techniques that can determine the precise landing site for introgressed genes make it possible for multiple R genes with major effect to be introduced as a single cassette, thereby increasing breeding efficiency.

In a major dichotomy to the genetically uniform clean-crop approach that has promoted strong directional selection in the pathogen, considerable scientific interest has been shown in the use of varietal mixtures and spatial deployment constraints that together present pathogens with a diversifying selection environment (Fig. 1b). The practical application of varietal mixtures has generally been limited because of concerns associated with differences in the agronomy (especially time to maturity) and quality attributes of different lines. However, there have been significant successes in the use of three-way barley mixtures to reduce the incidence of mildew (Wolfe, 1985) and to control a range of diseases in rice, particularly blast (Inukai et al., 1994; Zhu et al., 2000). These and other examples have shown significant reductions in disease and increases in yield and quality (Zhu et al., 2000; Mundt, 2002) as well as stabilization in the population genetics dynamics of pathogens (Chin and Wolfe, 1984; Dileone and Mundt, 1994; Lannou and Mundt, 1997; Marshall et al., 2009).

Natural selection also operates in plant-pathogen systems lacking gene-for-gene interactions, resulting in erosion of resistance control by multiple genes of small effect (Cowger et al., 2000; Krenz et al., 2008; Poland et al., 2009), although this typically occurs at a slower rate than in R-gene systems. In multigenic systems, natural selection driven by the presence of partially resistant varieties favors an increase in the frequency of pathotypes with higher aggressiveness. Mutation and recombination among selected pathotypes create new genetic variation for the next cycle of selection. This type of recurrent selection increases the overall aggressiveness of the pathogen population over time (Burdon et al., 2014).

Crop rotation has been widely accepted as an economically and ecologically benign way of controlling plant disease. In many areas, this approach is not always practical because of climatic, skill, machinery, and other constraints, leading to the continuous cultivation of the same crop. In parts of Western Canada, some crops (e.g.,barley) have been continuously grown on the same land for decades. Furthermore, control through rotation among different species does not necessarily work for many soilborne pathogens that have long-lived resting structures such as sclerotia. Indeed, sclerotia of *Sclerotinia* can survive up to 20 years in the soil before germination. For such situations, creating a series of isogenic lines differing in R genes and using them in an R-gene rotation could help avoid disease problems. Thus, for both foliar diseases of barley (Turkington et al., 2005) and blackleg of rape (Marcroft SJ, 2012), sequentially planting different varieties of the same species on the same land substantially reduced disease severity as compared to continuous cultivation of the same varieties.

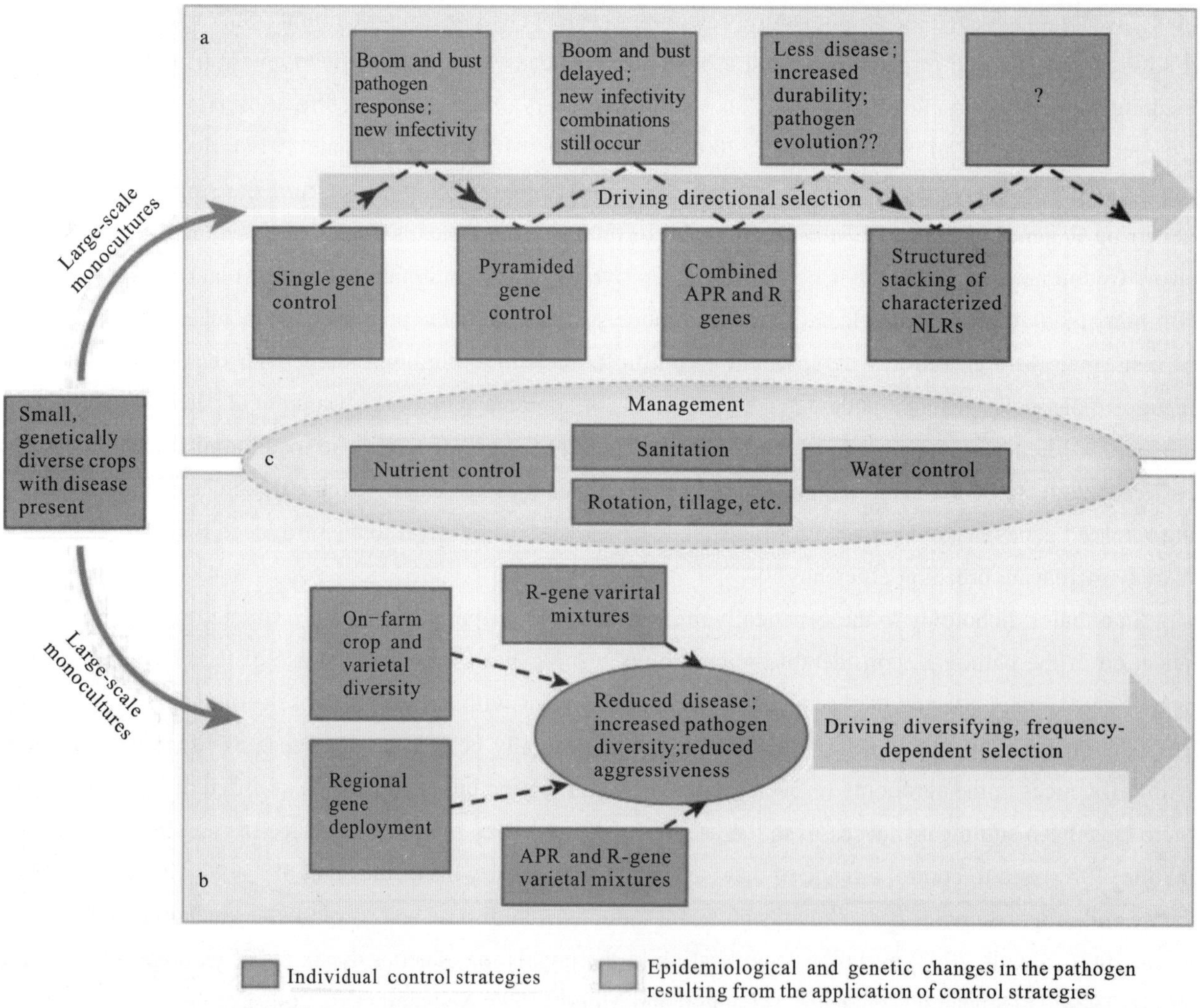

Fig. 1 A conceptual diagram describing the effects of genetics and deployment of resistance (R) genes and other agricultural practices on disease epidemics and pathogen evolution

Starting from the archetypal preindustrialized situation of small diverse crops in which disease was regularly present, two divergent approaches are possible: (a) Disease control through the use of R genes has progressed from individual use to multiple use, through to combination with adult plant resistance (APR) and the future possibility of specific stacking of well-characterized nucleotide-binding leucine-rich repeat receptors. This path has driven directional selection in the pathogen with changing epidemiologically consequences. (b) Alternative disease control strategies that promote frequency-dependent selection and less disease see the integration of a range of approaches that diversify protection mechanisms. (c) In both situations, more traditional management practices can contribute to significant levels of disease control. Abbreviation: NLRs, NOD (nucleotide oligomerization domain)-like receptors

R-gene rotation may also reduce pathogen evolutionary potential and help to maximize the effective useful life of R genes. Temporal dynamics in R-gene deployment generate a pattern of selection on pathogen populations that differs in impact from both pure stands and varietal mixtures. In pure stands, strong directional selection initially reduces pathogen genetic variation but continuous use of the same resistance allows pathogens to recover variation quickly. Alternatively, in varietal mixtures, disruptive selection may slow pathogen evolution but maintain high genetic variation and thus high evolutionary potential. R-gene rotation has the potential to combine the advantages of rapid short-term reduction in pathogen genetic variation that may occur in pure stands with subsequent prevention of the recovery of such variation via the disruptive selection that occurs through temporal rotation of different R genes. For example, remnant genetic variation surviving from the first R gene may be selected against by the second gene. In addition, R-gene rotation creates cycles of diversifying selection that prevent infectivity alleles from becoming dominant in the pathogen population (Fig. 2). Subsequently, genetic drift may wipe these infectivity alleles out of the pathogen population (Burdon et al., 2014; Zhan et al., 2014) when their frequency is low, thus reducing the probability that infectivity alleles accumulate in pathogen populations.

R-gene rotation combined with other deployment strategies, such as pyramiding, regional gene deployment (Pedersen and Leath, 1988; Hittalmani et al., 2000; Sapoukhina et al., 2009; Papaïx et al., 2011), and mixtures, could help to achieve more sustainable disease management and deserves greater theoretical and experimental attention. Indeed, although intercropping is often regarded as a disease management approach for smallholder agriculture in the developing world, increasingly such approaches are being shown to reduce disease in larger farm settings (Boudreau, 2013). Thus, for example, intercropping field peas with barley or wheat reduced Ascochyta blight by between 20% and 50% (Hauggaard-Nielsen et al., 2008; Schoeny et al., 2010; Schornack et al., 2013). Indeed, of more than 200 such studies assessed (Boudreau, 2013), 73% showed evidence of reduced disease.

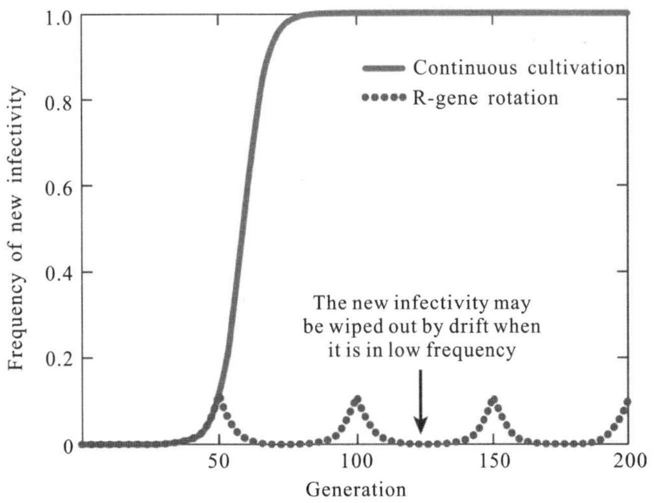

Fig. 2 Expected patterns of pathogen evolution (infectivity) to overcome major R genes under continuous cultivation and R-gene rotation, assuming no recombination, a selection coefficient of 0.2, and a recurrent mutation rate to new infectivity of 10^{-6}

Under the scheme of continuous cultivation, cultivars with major R genes are created, released, and continuously used until farmers decide to remove them from commercial use. Under the scheme of R-gene rotation, two cultivars with different major R genes are used alternatively at 50-generation intervals. Withdrawal of a cultivar with a major R gene from commercial production before its corresponding infectivity allele becomes dominant may favor negative selection to reduce the frequency of the new infectivity allele, and subsequently genetic drift can work to eliminate the infectivity allele from the pathogen population

5.5 Develop Multicomponent Disease Forecasting to Maximize Agrochemical Efficiency and Reduce Application

Although generating ecological and environmental concerns, agrochemicals play a major role in practical approaches to disease control in many crops (e.g., fungicides are essentially the only mechanism available to control mildew in grapevines). Currently, the timing and frequency of agrochemical applications mainly depend on affordability, preference, and experience, rather than efficiency. Multicomponent disease forecasting is essential for more effective use of agrochemicals (Gent et al., 2013), but it requires the collection of information relating to resistance and infectivity gene distributions,agrochemical resistance,and climatic conditions that must be coupled with early detection and disease monitoring technology. First, efforts should be made to gain, and provide to growers, information on the geographic distributions of R-gene cultivars currently available and the corresponding infectivity gene regarding target pathogens, the functional activity of agrochemicals, and the type and level of agrochemical resistance present in pathogen populations (first component). Particular attention should be paid to production areas with comparable host resistance and pathogen infectivity (second component). When environmental conditions in these areas are conducive to pathogens in a particular season, early diagnosis with molecular tools should be used to identify the pathotypes infecting hosts (third component) and the type and level of agrochemicals to be used (fourth component). Spatiotemporal variation in applications to avert directional selection (similar to R-gene deployment, as proposed in the previous section) is also critical for effective use of agrochemicals.

6 Concluding remarks

Plant diseases caused by pathogenic microbes have presented a major challenge to agricultural production over millennia. With the intensification of agriculture in modern society, plant crops are more vulnerable to pathogen attack and risk of disastrous epidemics. In this review, we discuss the use of evolutionary principles to effectively and durably control plant diseases with less resource input and reduced adverse impacts on the environment. To date, both genetic and agronomic approaches have been used to improve epidemiological outcomes in the short-term. However, too often a lack of consideration of the pathogen's ability to evolve in the medium-term has resulted in subsequent loss of disease control. Here, our discussion has focused on how to achieve the goal of sustainable plant disease management by regulating agricultural practices to minimize the evolution of pathogens. Principally, sustainable disease management may be achieved through the manipulation of agricultural practices to maximize the defense ability of plants, but a successful outcome requires due consideration of, and proactive response to, the ability of pathogens to evolve.

7 Summary points

1. The behaviors of pathogen populations in natural ecosystems and in agricultural ecosystems have much in common, especially when consideration is given to the evolutionary forces that drive changes in pathogen populations.

2. Agricultural intensification and monoculture coupled with globalization in modern society disrupt the coevolutionary dynamics of plant-pathogen interactions, as experienced in natural ecosystems, escalating the evolutionary rate of pathogens and increasing the risk of plant disease pandemics.

3. No one strategy for disease control necessarily works for all pathogens or for one pathogen in all environments. The efficacy of particular control strategies (including those based on evolutionary principles) is affected by environment and life-history interactions.

4. Disease management programs having high efficacy in short-term disease control are not necessarily the most effective with regard to minimizing pathogen evolutionary potential in the long-term. Evolutionary biologists need to play a larger role in designing plant disease management programs that can control both disease epidemics within growing seasons and pathogen evolution by generating environments conducive for host health and defense but with unfavorable conditions for pathogen survival, reproduction, and transmission.

5. Disease control programs that combine strategies to reduce pathogen population sizes, limit pathogen dispersal, and force pathogens to endure fluctuating, diversifying selection forces have the best prospects for reducing the emergence of new infectivity and new agrochemical resistance as well as for limiting the evolution of increased aggressiveness. Such an integrated approach would have good potential for long-term plant disease control.

8 Future issues

1. Integration of ecological and evolutionarily attractive disease control strategies with economic practicalities is a continuing issue; mechanisms that fail to increase on-farm profitability are unlikely to be adopted.

2. The increasing use of remote sensing and spatial positioning systems coupled with planting, harvesting, and spraying equipment will broaden control options by making more precise treatments a greater reality.

3. Simulation modeling of the interactive effects of particular disease control strategies on both epidemiological and evolutionary dynamics of pathogens will provide an increasingly important guide to future directions but must include the ecological realism defined by environment and life-history interactions.

4. Understanding interactions at the interface between agricultural and natural host pathogen systems will provide means of reducing disease occurrence in crops and will maximize opportunities to affect the evolutionary trajectories of pathogen population structure.

5. Increasing global information regarding the emergence of novel pathotypes can support efforts in preemptive breeding aimed at countering the increasingly rapid dispersal of pathogens around the world.

9 disclosure statement

The authors are not aware of any affiliations, memberships, funding, or financial holdings that might be perceived as affecting the objectivity of this review.

Acknowledgements

This project was supported by the National Natural Science Foundation of China grant numbers U1405213 and 31371901 and the state earmarked fund for Modern Agricultural Industry and Technology System grant number CARS-10.

References

[1] ABANG M M, BAUM M, CECCARELLI S, et al. Differential selection on Rhynchosporium secalis during parasitic and saprophytic

phases in the barley scald disease cycle[J]. Phytopathology, 2006, 96: 1214-1222.

[2] ABRAMOVITCH R B, ANDERSON J C, MARTIN G B. Bacterial elicitation and evasion of plant innate immunity[J]. Nature Reviews Molecular Cell Biology, 2006, 7: 601.

[3] AKINS R A. An update on antifungal targets and mechanisms of resistance in Candida albicans[J]. Medical Mycology, 2005, 43: 285-318.

[4] ALTIZER S, DOBSON A, HOSSEINI P, et al. Seasonality and the dynamics of infectious diseases[J]. Ecology Letters, 2010, 9: 467-484.

[5] ANAGNOSTAKIS S L. Chestnut blight: the classical problem of an introduced pathogen[J]. Mycologia, 1987, 79: 23-37.

[6] ANDREASEN K, BALDWIN B G. Unequal evolutionary rates between annual and perennial lineages of checker mallows (Sidalcea Malvaceae): evidence from 18S–26S rDNA internal and external transcribed spacers[J]. molecular biology and evolution, 2001, 18: 936-944.

[7] ANGIOLELLA L, STRINGARO A R, DE BERNARDIS F, et al. Increase of virulence and its phenotypic traits in drug-resistant strains of Candida albicans[J]. Antimicrobial Agents & Chemotherapy, 2008, 52: 927-936.

[8] BAILEY K, DUCZEK L. Managing cereal diseases under reduced tillage[J]. Canadian Journal of Plant Pathology, 1996, 18: 159-167.

[9] BAINBRIDGE A. Effect of nitrogen nutrition of the host on barley powdery mildew[J]. Plant Pathology, 1974, 23: 160-161.

[10] BALLINI E, MOREL J-B, DROC G, et al. A genome-wide meta-analysis of rice blast resistance genes and quantitative trait loci provides new insights into partial and complete resistance[J]. Molecular Plant-Microbe Interaction, 2008, 21: 859-868.

[11] BARRETT L G, THRALL P H, BURDON J J, et al. Life history determines genetic structure and evolutionary potential of host–parasite interactions[J]. Trends in Ecology & Evolution, 2008, 23: 678-685.

[12] BAYLES R, FLATH K, HOVMØLLER M, et al. Breakdown of the Yr17 resistance to yellow rust of wheat in northern Europe[J]. Agronomie, 2000, 20(7).

[13] BLATTER R, BROWN J, WOLFE M. Genetic control of the resistance of Erysiphe graminis f. sp. hordei to five triazole fungicides[J]. Plant Pathology, 1998, 47: 570-579.

[14] BOCKUS W, SHROYER J. The impact of reduced tillage on soilborne plant pathogens[J]. Annual Review of Phytopathology, 1998, 36: 485-500.

[15] BOUDREAU M A. Diseases in intercropping systems[J]. Annual Review of Phytopathology, 2013, 51: 499-519.

[16] BOUSSET L, HOVMØLLER M, CAFFIER V, et al. Observed and predicted changes over eight years in frequency of barley powdery mildew avirulent to spring barley in France and Denmark[J]. Plant Pathology, 2010, 51: 33-44.

[17] BRASIER C. The biosecurity threat to the UK and global environment from international trade in plants[J]. Plant Pathol, 2008, 57: 792-808.

[18] BROWN J, RANT J. Fitness costs and trade-offs of disease resistance and their consequences for breeding arable crops[J]. Plant Pathol, 2013, 62: 83-95.

[19] BROWN J K. Yield penalties of disease resistance in crops[J]. Current Opinion in Plant Biology, 2002, 5: 339-344.

[20] BROWN J K, HOVMØLLER M S. Aerial dispersal of pathogens on the global and continental scales and its impact on plant disease[J]. Science, 2002, 297: 537-541.

[21] BROWN J K, TELLIER A. Plant-parasite coevolution: bridging the gap between genetics and ecology[J]. Annual Review of Phytopathology, 2010, 49: 345-367.

[22] BROWN S P, CORNFORTH D M, MIDEO N. Evolution of virulence in opportunistic pathogens: generalism, plasticity, and control[J]. Trends in Microbiology, 2012, 20: 336-342.

[23] BRUN H, CHÈVRE A M, FITT B D, et al. Quantitative resistance increases the durability of qualitative resistance to Leptosphaeria maculans in Brassica napus[J]. New Phytologist, 2010, 185: 285-299.

[24] BRUNNER P C, STEFANATO F L, MCDONALD B A. Evolution of the CYP51 gene in Mycosphaerella graminicola: evidence for intragenic recombination and selective replacement[J]. Molecular Plant Pathology, 2010, 9: 305-316.

[25] BURDON J, THOMPSON J N. Changed patterns of resistance in a population of Linum marginale attacked by the rust pathogen Melampsora lini[J]. Journal of Ecology, 1995: 199-206.

[26] BURDON J, THRALL P. What have we learned from studies of wild plant-pathogen associations?—the dynamic interplay of time, space and life-history[J]. European Journal of Plant Pathology, 2014, 138: 417-429.

[27] BURDON J J, THRALL P H. The fitness costs to plants of resistance to pathogens[J]. Genome Biology, 2003, 4: 227.

[28] BURDON J J, THRALL P H, ERICSON L. Genes, communities & invasive species: understanding the ecological and evolutionary dynamics of host–pathogen interactions[J]. Current Opinion in Plant Biology, 2013, 16: 400-405.

[29] BURDON J J, BARRETT L G, REBETZKE G, et al. Guiding deployment of resistance in cereals using evolutionary principles[J]. Evolutionary Applications, 2014, 7: 609-624.

[30] BURGESS T, WINGFIELD M J. Quarantine is important in restricting the spread of exotic seed-borne tree pathogens in the southern hemisphere[J]. The International Forestry Review, 2002: 56-65.

[31] CAFFIER V, LAURENS F. Breakdown of P12, a major gene of resistance to apple powdery mildew, in a French experimental orchard[J]. Plant Pathology, 2005, 54: 116-124.

[32] CAFFIER V, HOFFSTADT T, LECONTE M, et al.Seasonal changes in pathotype complexity in French populations of barley powdery mildew[J]. Plant Pathology, 1996, 45: 454-468.

[33] CAFFIER V, DIDELOT F, PUMO B, et al. Aggressiveness of eight Venturia inaequalis isolates virulent or avirulent to the major resistance gene Rvi6 on a non-Rvi6 apple cultivar[J]. Plant Pathology, 2010, 59: 1072-1080.

[34] CHEN P R, BAE T, WILLIAMS W A,et al. An oxidation-sensing mechanism is used by the global regulator MgrA in Staphylococcus aureus[J]. Nature Chemical Biology, 2006, 2: 591.

[35] CHIN K, WOLFE M. Selection on Erysiphe graminis in pure and mixed stands of barley[J]. Plant Pathology, 2010, 33: 535-546.

[36] COOK R, PAPENDICK R. Influence of water potential of soils and plants on root disease[J]. Annual Review of Phytopathology, 1972, 10: 349-374.

[37] COWGER C, MUNDT C C. Aggressiveness of Mycosphaerella graminicola isolates from susceptible and partially resistant wheat cultivars[J]. Phytopathology, 2002, 92: 624-630.

[38] COWGER C, HOFFER M, MUNDT C. Specific adaptation by Mycosphaerella graminicola to a resistant wheat cultivar[J]. Plant Pathology, 2010, 49: 445-451.

[39] DANGL J L, HORVATH D M, STASKAWICZ B J. Pivoting the plant immune system from dissection to deployment[J]. Science, 2013, 341: 746-751.

[40] DAVID M, SWIADER J, WILLIAMS K, EASTBURN D. Nitrogen nutrition, but not potassium, affects powdery mildew development in Hiemalis begonia[J]. Journal of Plant Nutrition, 2003, 26: 159-176.

[41] DELMOTTE F, MESTRE P, SCHNEIDER C,et al. Rapid and multiregional adaptation to host partial resistance in a plant pathogenic oomycete: evidence from European populations of Plasmopara viticola, the causal agent of grapevine downy mildew[J]. Infection Genetics & Evolution Journal of Molecular Epidemiology & Evolutionary Genetics in Infectious Diseases, 2014, 27: 500-508.

[42] DILEONE J, MUNDT C. Effect of wheat cultivar mixtures on populations of Puccinia striiformis races[J]. Plant Pathology, 2010, 43: 917-930.

[43] DINOOR A. Sources of oat crown rust resistance in hexaploid and tetraploid wild oats in Israel[J]. Canadian Journal of Botany, 1970, 48: 153-161.

[44] DOUMAYROU J, AVELLAN A, FROISSART R, et al. An experimental test of the transmission-virulence trade-off hypothesis in a plant virus[J]. Evolution, 2013, 67: 477-486.

[45] EL-ZIK K M. Integrated control of Verticillium wilt of cotton[J]. Plant Disease, 1985, 69: 1025-1032.

[46] ELLSTRAND N C, ELAM D R. Population genetic consequences of small population size: implications for plant conservation[J]. Annual Review of Ecology & Systematics, 1993, 24: 217-242.

[47] FERREIRA R C, BRIONES M R. Phylogenetic evidence based on Trypanosoma cruzi nuclear gene sequences and information entropy suggest that inter-strain intragenic recombination is a basic mechanism underlying the allele diversity of hybrid strains[J]. Infection, Genetics and Evolution, 2012, 12: 1064-1071.

[48] FISHER R. The Genetical Theory of Natural Selection[M]. Oxford University Press, 1930, 272 p.

[49] FLOR H. Host-parasite interactions in flax rust-its genetics and other implications[J]. Phytopathology, 1955, 45: 680-685.

[50] FLOR H. The complementary genic systems in flax and flax rust[M]. In Advances in genetics (Elsevier), 1956, 29-54.

[51] FRAAIJE B, COOLS H, FOUNTAINE J, et al. Role of ascospores in further spread of QoI-resistant cytochrome b alleles (G143A) in field populations of Mycosphaerella graminicola[J]. Phytopathology, 2005, 95: 933-941.

[52] FREINKEL S. American chestnut: the life, death, and rebirth of a perfect tree[J]. Winterthur Portfolio, 2009.

[53] GANDON S, MICHALAKIS Y. Evolution of parasite virulence against qualitative or quantitative host resistance[J]. Proceedings of the Royal Society B: Biological Sciences, 2000, 267: 985-990.

[54] GANDON S, NUISMER S L. Interactions between genetic drift, gene flow, and selection mosaics drive parasite local adaptation[J]. American Naturalist, 2009, 173: 212-224.

[55] GARBELOTTO M, SVIHRA P, RIZZO D. New pests and diseases: Sudden oak death syndrome fells 3 oak species[J]. California Agriculture, 2001, 55: 9-19.

[56] GENT D H, MAHAFFEE W F, MCROBERTS N, et al.The use and role of predictive systems in disease management[J]. Annual Review of Phytopathology, 2013, 51: 267-289.

[57] GLADIEUX P, ZHANG X-G, AFOUFA-BASTIEN D, et al. On the origin and spread of the scab disease of apple: out of central Asia[J]. PLoS ONE, 2008, 3: e1455.

[58] GÓMEZ P, BUCKLING A. Bacteria-phage antagonistic coevolution in soil[J]. Science, 2011, 332: 106-109.

[59] GOTTWALD T R, HUGHES G, GRAHAM J H, et al. The citrus canker epidemic in Florida: the scientific basis of regulatory eradication policy for an invasive species[J]. Phytopathology, 2001, 91: 30-34.

[60] GRÜNWALD N J, FLIER W G. The biology of Phytophthora infestans at its center of origin[J]. Annual Review of Phytopathology, 2005, 43.

[61] GRÜNWALD N J, GARBELOTTO M, GOSS E M, et al. Emergence of the sudden oak death pathogen Phytophthora ramorum[J]. Trends in Microbiology, 2012, 20: 131-138.

[62] GUBBINS S, GILLIGAN C A. Persistence of host-parasite interactions in a disturbed environment[J]. Journal of Theoretical Biology, 1997, 188: 241-258.

[63] GUPTA A, CHATTOO B B. Functional analysis of a novel ABC transporter ABC4 from Magnaporthe grisea[J]. FEMS Microbiology Letters, 2008, 278: 22-28.

[64] HALL A R, SCANLAN P D, MORGAN A D, et al. Host–parasite coevolutionary arms races give way to fluctuating selection[J]. Ecology Letters, 2011, 14: 635-642.

[65] HAMILTON W D. Sex versus non-sex versus parasite[J]. Oikos, 1980: 282-290.

[66] HAMILTON W D, AXELROD R, TANESE R. Sexual reproduction as an adaptation to resist parasites (a review)[J]. Proceedings of the National Academy of Sciences of the United States of America,1990, 87: 3566-3573.

[67] HAUGGAARD-NIELSEN H, JØRNSGAARD B, KINANE J, et al. Grain legume–cereal intercropping: The practical application of diversity, competition and facilitation in arable and organic cropping systems[J]. Renewable Agriculture and Food Systems, 2008, 23: 3-12.

[68] HE C-Q, DING N-Z, HE M, et al. Intragenic recombination as a mechanism of genetic diversity in bluetongue virus[J]. Journal of

Virology, 2010, 84: 11487-11495.

[69] HEIKKILÄ J. Economics of biosecurity across levels of decision-making: a review[J]. Agronomy for Sustainable Development, 2011, 31: 119.

[70] HERRICK J. Genetic variation and DNA replication timing, or why is there late replicating DNA?[J]. Evolution, 2011, 65: 3031-3047.

[71] HITTALMANI S, PARCO A, MEW T, et al. Fine mapping and DNA marker-assisted pyramiding of the three major genes for blast resistance in rice[J]. Theoretical & Applied Genetics, 2000, 100: 1121-1128.

[72] HOBBS P R, SAYRE K, GUPTA R. The role of conservation agriculture in sustainable agriculture[J]. Philosophical Transactions of The Royal Society B Biological Sciences, 2008, 363: 543-555.

[73] HOHL H R, ISELIN K. Strains of Phytophthora infestans from Switzerland with A2 mating type behaviour[J]. Transactions of the British Mycological Society, 1984, 83: 529-530.

[74] HORNS F, HOOD M E. The evolution of disease resistance and tolerance in spatially structured populations[J]. Ecology & Evolution, 2012, 2: 1705-1711.

[75] HOVMØLLER M, CAFFIER V, JALLI M. The European barley powdery mildew virulence survey and disease nursery 1993-1999[J]. Agronomie, 2000, 20(2).

[76] HURST L D, PECK J R. Recent advances in understanding of the evolution and maintenance of sex[J]. Trends in Ecology & Evolution,1996, 11: 46-52.

[77] INNOCENTI P, MORROW E H, DOWLING D K. Experimental evidence supports a sex-specific selective sieve in mitochondrial genome evolution[J]. Science, 2011, 332: 845-848.

[78] INUKAI T, NELSON R, ZEIGLER R S, et al. Allelism of blast resistance genes in near-isogenic lines of rice[J]. Phytopathology, 1994, 84: 1278-1283.

[79] JARAMILLO N, DOMINGO E, MUÑOZ-EGEA M C, et al. Evidence of Muller's ratchet in herpes simplex virus type 1[J]. Journal of General Virology, 2013, 94: 366-375.

[80] JEFFERSON P G, GOSSEN B D. Irrigation increases Verticillium wilt incidence in a susceptible alfalfa cultivar[J]. Plant Disease, 2002, 86: 588-592.

[81] JENNER C E, WANG X, PONZ F, et al.A fitness cost for *Turnip mosaic virus* to overcome host resistance[J]. Virus Research, 2002, 86: 1-6.

[82] JOHNSON T. Man-guided evolution in plant rusts[J]. Science, 1961, 133: 357-362.

[83] JUDELSON H S, BLANCO F A. The spores of Phytophthora: weapons of the plant destroyer[J]. Nature Reviews Microbiology, 2005, 3: 47.

[84] KELLER M, ROGIERS S Y, SCHULTZ H R.Nitrogen and ultraviolet radiation modify grapevines' susceptibility to powdery mildew[J]. Vitis Geilweilerh of, 2003, 42: 87-94.

[85] KINKEL L L, BAKKER M G, SCHLATTER D C. A coevolutionary framework for managing disease-suppressive soils[J]. Annual Review of Phytopathology, 2011, 49: 47-67.

[86] KINLOCH JR B B. White pine blister rust in North America: past and prognosis[J]. Phytopathology, 2003, 93: 1044-1047.

[87] KONDRASHOV A S. Deleterious mutations and the evolution of sexual reproduction[J]. Nature,1988, 336: 435.

[88] KOSKELLA B, GIRAUD T, HOOD M. Pathogen relatedness affects the prevalence of within-host competition[J]. The American Naturalist, 2006, 168: 121-126.

[89]KRENZ J, SACKETT K, MUNDT C. Specificity of incomplete resistance to Mycosphaerella graminicola in wheat[J]. Phytopathology, 2008, 98: 555-561.

[90] LAINE A L, BURDON J J, DODDS P N, et al. Spatial variation in disease resistance: from molecules to metapopulations[J]. Journal of Ecology, 2011, 99: 96-112.

[91] LANDE R, BARROWDOUGH G. Effective population size, genetic variation, and their use in population[M]. Viable populations for conservation: 87, 1987.

[92] LANIER G, SCHUBERT D, MANION P. Dutch elm disease and elm yellows in central New York: out of the frying pan into the fire[J]. Plant Disease, 1988, 72: 189-194.

[93] LANNOU C, MUNDT C. Evolution of a pathogen population in host mixtures: rate of emergence of complex races[J]. Theoretical & Applied Genetics, 1997, 94: 991-999.

[94] LENSKI R E, MAY R M. The evolution of virulence in parasites and pathogens: reconciliation between two competing hypotheses[J]. Journal of Theoretical Biology, 1994, 169: 253-265.

[95] LIU Y-C, MILGROOM M G. High diversity of vegetative compatibility types in Cryphonectria parasitica in Japan and China[J]. Mycologia, 2007, 99: 279-284.

[96] LÓPEZ-VILLAVICENCIO M, JONOT O, COANTIC A, et al. Multiple infections by the anther smut pathogen are frequent and involve related strains[J]. PLoS Pathogens, 2007, 3: e176.

[97] LÓPEZ-VILLAVICENCIO M, COURJOL F, GIBSON A K, et al. Competition, cooperation among kin, and virulence in multiple infections[J]. Evolution, 2011, 65: 1357-1366.

[98] MARCEL T C, GORGUET B, TA M T, et al. Isolate specificity of quantitative trait loci for partial resistance of barley to Puccinia hordei confirmed in mapping populations and near-isogenic lines[J]. New Phytologist, 2008, 177: 743-755.

[99] MARCROFT S J, VD.W A, SALISBURY P A, et al. Rotation of canola (Brassicanapus) cultivars with different complements of blackleg resistance genes decreases disease severity[J]. Plant Pathology, 2012, 61:934–944.

[100] MARSHALL B, NEWTON A, ZHAN J. Evolution of pathogen aggressiveness under cultivar mixtures[J]. Plant Pathology, 2010, 58: 378-388.

[101] MARTIN A P, PALUMBI S R. Body size, metabolic rate, generation time, and the molecular clock[J]. Proceedings of the National Academy of Sciences, 1993, 90: 4087-4091.

[102] MAZZOLA M. Assessment and management of soil microbial community structure for disease suppression[J]. Annual Review of Phytopathology, 2004, 42: 35-59.

[103] MCDONALD B A, LINDE C. Pathogen population genetics, evolutionary potential, and durable resistance[J]. Annual Review of Phytopathology, 2002, 40: 349-379.

[104] MCDONALD B A, ZHAN J, BURDON J J. Genetic structure of Rhynchosporium secalis in Australia[J]. Phytopathology, 1999, 89: 639-645.

[105] MESSINGER S M, OSTLING A. The consequences of spatial structure for the evolution of pathogen transmission rate and virulence[J]. American Naturalist, 2009, 174: 441-454.

[106] MEYER S E, NELSON D L, CLEMENT S, et al. Ecological genetics of the Bromus tectorum (Poaceae)–Ustilago bullata (Ustilaginaceae) pathosystem: A role for frequency-dependent selection?[J]. American Journal of Botany, 2010, 97: 1304-1312.

[107] MICHELMORE R W, CHRISTOPOULOU M, CALDWELL K S. Impacts of resistance gene genetics, function, and evolution on a durable future[J]. Annual Review of Phytopathology, 2013, 51: 291-319.

[108] MIKONRANTA L, FRIMAN V-P, LAAKSO J. Life history trade-offs and relaxed selection can decrease bacterial virulence in environmental reservoirs[J]. PLoS One, 2012, 7: e43801.

[109] MITCHELL-OLDS T, BRADLEY D. Genetics of Brassica rapa. 3. Costs of disease resistance to three fungal pathogens[J]. Evolution, 1996, 50: 1859-1865.

[110] MONTARRY J, CARTIER E, JACQUEMOND M, et al. Virus adaptation to quantitative plant resistance: erosion or breakdown?[J]. Journal of Evolutionary Biology, 2012, 25: 2242-2252.

[111] MORAN N A. Accelerated evolution and Muller's rachet in endosymbiotic bacteria[J]. Proceedings of the National Academy of Sciences, 1996, 93: 2873-2878.

[112] MORGOUNOV A, TUFAN H A, SHARMA R, et al. Global incidence of wheat rusts and powdery mildew during 1969–2010 and durability of resistance of winter wheat variety Bezostaya 1[J]. European Journal of Plant Pathology, 2012, 132: 323-340.

[113] MORROW M, KRIEG D. Cotton management strategies for a short growing season environment: Water-nitrogen considerations[J]. Agronomy Journal, 1990, 82: 52-56.

[114] MUNDT C. Use of multiline cultivars and cultivar mixtures for disease management[J]. Annual Review of Phytopathology, 2002, 40: 381-410.

[115] MYKYTOWYCZ R. An attenuated strain of the myxomatosis virus recovered from the field[J]. Nature, 1953, 172: 448.

[116] NUNNEY L. Measuring the ratio of effective population size to adult numbers using genetic and ecological data[J]. Evolution, 1995, 49: 389-392.

[117] NUNNEY L. The effective size of a hierarchically structured population[J]. Evolution, 1999, 53: 1-10.

[118] O'HARA R, BROWN J. Immigration of the barley mildew pathogen into field plots of barley[J]. Plant Pathology, 2003, 45: 1071-1076.

[119] O'BRIEN L, BROWN J, YOUNG R, et al. Occurrence and distribution of wheat stripe rust in Victoria and susceptibility of commercial wheat cultivars[J]. Australasian Plant Pathology, 1980, 9: 14-14.

[120] PAPAÏX J, BURDON J J, LANNOU C, et al. Evolution of pathogen specialisation in a host metapopulation: joint effects of host and pathogen dispersal[J]. PLOS Computational Biology, 2014, 10: e1003633.

[121] PAPAÏX J, GOYEAU H, DU CHEYRON P, et al. Influence of cultivated landscape composition on variety resistance: an assessment based on wheat leaf rust epidemics[J]. New Phytologist, 2011, 191: 1095-1107.

[122] PARIAUD B, ROBERT C, GOYEAU H, et al. Aggressiveness components and adaptation to a host cultivar in wheat leaf rust[J]. Phytopathology, 2009a, 99: 869-878.

[123] PARIAUD B, RAVIGNÉ V, HALKETT F, et al. Aggressiveness and its role in the adaptation of plant pathogens[J]. Plant Pathology, 2009b, 58: 409-424.

[124] PEDERSEN W L, LEATH S. Pyramiding major genes for resistance to maintain residual effects[J]. Annual Review of Phytopathology, 1988, 26: 369-378.

[125] PERESSOTTI E, WIEDEMANN-MERDINOGLU S, DELMOTTE F, et al. Breakdown of resistance to grapevine downy mildew upon limited deployment of a resistant variety[J]. BMC Plant Biology, 2010, 10: 147.

[126] PINK D A. Strategies using genes for non-durable disease resistance[J]. Euphytica, 2002, 124: 227-236.

[127] PLATZ G, SHEPPARD J. Sustained genetic control of wheat rust diseases in north-eastern Australia[J]. Crop & Pasture Science, 2007, 58: 854-857.

[128] POLAND J A, BALINT-KURTI P J, WISSER R J, et al. Shades of gray: the world of quantitative disease resistance[J]. Trends in Plant Science, 2009, 14: 21-29.

[129] PURDY L, KRUPA S, DEAN J. Introduction of sugarcane rust into the Americas and its spread to Florida[J]. Plant Disease, 1985, 69: 689-693.

[130] RAUSCHER G, SIMKO I, MAYTON H, et al. Quantitative resistance to late blight from Solanum berthaultii cosegregates with R Pi-ber: insights in stability through isolates and environment[J]. Tag.theoretical & Applied Genetics.theoretische Und Angewandte Genetik, 2010, 121: 1553-1567.

[131] ROELFS A P. Effects of Barberry eradication[J]. Plant Disease, 1982, 66: 177.

[132] S, W. Isolation by distance[J]. Genetics, 1943, 28:114.

[133] SABAT A J, WLADYKA B, KOSOWSKA-SHICK K,et al. Polymorphism, genetic exchange and intragenic recombination of the aureolysin gene among Staphylococcus aureus strains[J]. BMC Microbiology, 2008, 8: 129.

[134] SALAMATI S, ZHAN J, BURDON J J, et al. The genetic structure of field populations of Rhynchosporium secalis from three continents suggests moderate gene flow and regular recombination[J]. Phytopathology, 2000, 90: 901-908.

[135] SAPOUKHINA N, DUREL C-E, LE CAM B. Spatial deployment of gene-for-gene resistance governs evolution and spread of pathogen populations[J]. Theoretical Ecology, 2009, 2: 229.

[136] SCHOENY A, JUMEL S, ROUAULT F, et al. Effect and underlying mechanisms of pea-cereal intercropping on the epidemic development of ascochyta blight[J]. European Journal of Plant Pathology, 2010, 126: 317-331.

[137] SCHORNACK S, MOSCOU M J, WARD E R, et al. Engineering plant disease resistance based on TAL effectors[J]. Annual Review of Phytopathology, 2013, 51: 383-406.

[138] SINGH R P, HODSON D P, HUERTA-ESPINO J, et al. The emergence of Ug99 races of the stem rust fungus is a threat to world wheat production[J]. Annual Review of Phytopathology, 2011, 49: 465-481.

[139] SLATKIN M, VOELM L. FST in a hierarchical island model[J]. Genetics, 1991, 127: 627-629.

[140] SOMMERHALDER R J, MCDONALD B A, MASCHER F, et al. Effect of hosts on competition among clones and evidence of differential selection between pathogenic and saprophytic phases in experimental populations of the wheat pathogen Phaeosphaeria nodorum[J]. BMC Evolutionary Biology, 2011, 11: 188.

[141] ST. CLAIR D A. Quantitative disease resistance and quantitative resistance loci in breeding[J]. Annual Review of Phytopathology, 2010, 48: 247-268.

[142] STAUB J E, GRUMET R. Selection for multiple disease resistance reduces cucumber yield potential[J]. Euphytica, 1993, 67: 205-213.

[143] SUBRAMANIAN S. Significance of population size on the fixation of nonsynonymous mutations in genes under varying levels of selection pressure[J]. Genetics, 2013, 193: 995-1002.

[144] SUN C B, SURESH A, DENG Y Z, et al. A multidrug resistance transporter in Magnaporthe is required for host penetration and for survival during oxidative stress[J]. Plant Cell, 2006, 18: 3686-3705.

[145] SUSI H, BARRÈS B, VALE P F, et al. Co-infection alters population dynamics of infectious disease[J]. Nature Communications, 2015, 6: 5975.

[146] TATUM L. The southern corn leaf blight epidemic[J]. Science, 1971, 171: 1113-1116.

[147] THOMAS J A, WELCH J J, LANFEAR R, et al. A generation time effect on the rate of molecular evolution in invertebrates[J]. Molecular Biology and Evolution, 2010, 27: 1173-1180.

[148] THRALL P, BURDON J. Effect of resistance variation in a natural plant host–pathogen metapopulation on disease dynamics[J]. Plant Pathology, 2010, 49: 767-773.

[149] THRALL P, BURDON J. Evolution of gene-for-gene systems in metapopulations: the effect of spatial scale of host and pathogen dispersal[J]. Plant Pathology, 2010, 51: 169-184.

[150] THRALL P H, BURDON J J. Evolution of virulence in a plant host-pathogen metapopulation[J]. Science, 2003, 299: 1735-1737.

[151] THRALL P H, LAINE A L, RAVENSDALE M, et al. Rapid genetic change underpins antagonistic coevolution in a natural host-pathogen metapopulation[J]. Ecology Letters, 2012, 15: 425-435.

[152] THRALL P H, OAKESHOTT J G, FITT G, et al. Evolution in agriculture: the application of evolutionary approaches to the management of biotic interactions in agro-ecosystems[J]. Evolutionary Applications, 2011, 4: 200-215.

[153] THURSTON H D. Plant disease management practices of traditional farmers[J]. Plant Disease, 1990, 74: 96-102.

[154] TIAN D, TRAW M, CHEN J, et al. Fitness costs of R-gene-mediated resistance in Arabidopsis thaliana[J]. Nature, 2003, 423: 74.

[155] TRUONG-BOLDUC Q, DUNMAN P, STRAHILEVITZ J, et al. MgrA is a multiple regulator of two new efflux pumps in Staphylococcus aureus[J]. Journal of Bacteriology, 2005, 187: 2395-2405.

[156] TURKINGTON T, CLAYTON G, HARKER K, et al. Cultivar rotation as a strategy to reduce leaf diseases under barley monoculture[J]. Canadian Journal of Plant Pathology, 2005, 27: 283-290.

[157] ULLSTRUP A. The impacts of the southern corn leaf blight epidemics of 1970-1971[J]. Annual Review of Phytopathology, 1972, 10: 37-50.

[158] URBAN M, BHARGAVA T, HAMER J E. An ATP-driven efflux pump is a novel pathogenicity factor in rice blast disease[J].

Embo Journal, 2014, 18: 512-521.

[159] VAN ELSAS J, SPEKSNIJDER A, VAN OVERBEEK L. A procedure for the metagenomics exploration of disease-suppressive soils[J]. Journal of Microbiological Methods, 2008, 75: 515-522.

[160] WAAGE J K, M J. Agricultural biosecurity[J]. Philosophical Transactions of the Royal Society of London, 2008, 363:863-876.

[161] WATSON I. Wheat and its rust parasites in Australia[J]. Wheat Science Today & Tomorrow, 1981: 129-147.

[162] WATSON I, DE SOUSA C. Long distance transport of spores of Puccinia graminis tritici in the southern hemisphere. Proceedings of the linnean society of new south wales, 1983, 311-321.

[163] WELLINGS C. Puccinia striiformis in Australia: a review of the incursion, evolution, and adaptation of stripe rust in the period 1979–2006[J]. Australian Journal of Agricultural Research, 2007, 58: 567-575.

[164] WELLINGS C R. Global status of stripe rust: a review of historical and current threats[J]. Euphytica, 2011, 179: 129-141.

[165] WOLFE M. The current status and prospects of multiline cultivars and variety mixtures for disease resistance[J]. Annu.rev. phytopathol, 1985, 23: 251-273.

[166] WRIGHT S. Evolution in Mendelian populations[J]. Genetics, 1931, 16: 97.

[167] WRIGHT S. Size of population and breeding structure in relation to evolution[J]. science, 1938, 87: 430-431.

[168] YANG L, GAO F, SHANG L, et al. Association between virulence and triazole tolerance in the phytopathogenic fungus Mycosphaerella graminicola[J]. PLoS One, 2013, 8: e59568.

[169] ZELLER S L, KALININA O, SCHMID B. Costs of resistance to fungal pathogens in genetically modified wheat[J]. Journal of Plant Ecology, 2013, 6: 92-100.

[170] ZHAN J, MCDONALD B A. Experimental measures of pathogen competition and relative fitness[J]. Annual Review of Phytopathology, 2013, 51: 131-153.

[171] ZHAN J, MUNDT C C, MCDONALD B A. Using restriction fragment length polymorphisms to assess temporal variation and estimate the number of ascospores that initiate epidemics in field populations of Mycosphaerella graminicola[J]. Phytopathology, 2001, 91: 1011-1017.

[172] ZHAN J, PETTWAY R E, MCDONALD B A. The global genetic structure of the wheat pathogen Mycosphaerella graminicola is characterized by high nuclear diversity, low mitochondrial diversity, regular recombination, and gene flow[J]. Fungal Genetics & Biology, 2003, 38: 286-297.

[173] ZHAN J, STEFANATO F, MCDONALD B A. Selection for increased cyproconazole tolerance in Mycosphaerella graminicola through local adaptation and in response to host resistance[J]. Molecular Plant Pathology, 2006, 7: 259-268.

[174] ZHAN J, MUNDT C C, MCDONALD B A. Sexual reproduction facilitates the adaptation of parasites to antagonistic host environments: evidence from empirical study in the wheat-Mycosphaerella graminicola system[J]. International Journal for Parasitology, 2007, 37: 861-870.

[175] ZHAN J, THRALL P H, BURDON J J. Achieving sustainable plant disease management through evolutionary principles[J]. Trends Plant Science, 2014, 19: 570-575.

[176] ZHAN J, MUNDT C C, HOFFER M, et al. Local adaptation and effect of host genotype on the rate of pathogen evolution: an experimental test in a plant pathosystem[J]. Journal of Evolutionary Biology, 2010, 15: 634-647.

[177] ZHAN J, YANG L, ZHU W, et al. Pathogen populations evolve to greater race complexity in agricultural systems–evidence from analysis of Rhynchosporium secalis virulence data[J]. PLoS One, 2012, 7: e38611.

[178] ZHANG H, WANG C, CHENG Y, et al. Histological and cytological characterization of adult plant resistance to wheat stripe rust[J]. Plant Cell Reports, 2012, 31: 2121-2137.

[179] ZHONG B, FONG R, COLLINS L J, et al. Two new fern chloroplasts and decelerated evolution linked to the long generation time in tree ferns[J]. Genome Biology Evolution, 2014, 6: 1166-1173.

[180] ZHU Y, CHEN H, FAN J, et al. Genetic diversity and disease control in rice[J]. Nature, 2000, 406: 718.

Viruliferous rate of small brown planthopper is a good indicator of rice stripe disease epidemics

Dunchun He[1], Jiasui Zhan[1], Zhaobang Cheng[2],Lianhui Xie[1]

(1 Fujian Key Lab of Plant Virology, Institute of Plant Virology, Fujian Agriculture and Forestry University, Fuzhou, 350002, China; 2 Institute of Plant Protection, Jiangsu Academy of Agricultural Sciences, Nanjing, 210014, China.)

Abstract: *Rice stripe virus* (RSV), its vector insect (small brown planthopper, SBPH) and climatic conditions in Jiangsu, China were monitored between 2002 and 2012 to determine key biotic and abiotic factors driving epidemics of the disease. Average disease severity, disease incidence and viruliferous rate of SBPH peaked in 2004 and then gradually decreased. Disease severity of RSV was positively correlated with viruliferous rate of the vector but not with the population density of the insect, suggesting that the proportion of vectors infected by the virus rather than the absolute number of vectors plays an important role in RSV epidemics and could be used for disease forecasting. The finding of a positive correlation of disease severity and viruliferous rate among years suggests that local infection is likely the main source of primary inoculum of RSV. Of the two main climatic factors, temperature plays a more important role than rainfall in RSV epidemics.

1 Introduction

RSV disease is one of the most important viral diseases in East Asia (Yasuo et al., 1968; Lijun et al., 2003; Sun et al., 2007), causing up to 50% grain loss (Toriyama, 1986; Hibino, 1996; Cho et al., 2013a; Li et al., 2015a) thereby threatening rice production and global food security. The disease was first reported in 1897 in Japan (Shinkai, 1985) and has caused several major epidemics in many countries including Japan, China, South Korea, North Korea and Ukraine. For example, in Japan, 13%—19% of rice fields were infected by RSV between 1963 and 1967, leading to annual grain losses of ~40,000 metric tons (Yasuo et al., 1968).

In China, RSV was first reported in 1963 (Zhu et al., 1964), spreading rapidly into many rice production areas of the country (> 20 provinces). It generally causes 10%—20% disease incidence, but more severe epidemics resulting in complete grain loss have been documented in many regions (Zhang et al., 2007; Wu et al., 2009; Xiong et al., 2009; Xiao et al., 2010; Li et al., 2015a). Interestingly, the disease was almost absent in Asia in the last decade of the 20th century but re-emerged after 2000 as the most important rice disease in China, particularly in Jiangsu Province, in Japan and South Korea. The resurgence disease induced greater damage compared to its pre-1990 outbreaks, causing 50%—100% grain loss in these regions (Zhang et al., 2013). In 2004 alone, ~1,570,000 hectares of paddy rice in China were infected by the disease, accounting for 79% of total rice production in the country (Zhou et al., 2012).

Like all other vector-transmitted diseases, the occurrence of RSV and the subsequent development of epidemics result from a complex interaction among five biotic and abiotic factors -the pathogen (RSV),

Scientific reports.2016, 6: 21376.

Received 30 November 2015; Accepted 21 January 2016; Published 22 February 2016 the vector (small brown planthopper, SBPH, *Laodelphax striatellus* Fallén), the host (rice), climate (e.g. temperature, wind, rainfall etc.) and human activity(Hibino, 1996; Abo and Sy, 1997; Wang et al., 2008; Wei et al., 2009; Otuka et al., 2010). Hibino (1996) argued that the main agronomic factors contributing to viral disease epidemics were the widespread use of susceptible varieties, the large-scale cultivation of winter wheat (an alternative host), the sowing and planting time of rice, and the time gap between early and late sowing in rice. Most rice virologists share his view (Kiritani, 1983; Kiritani et al., 1987; Bae and Kim, 1994; Wang et al., 2008; Zhu et al., 2009; Zhu et al., 2011). With regard to rice stripe disease, viral source, the density of small brown planthopper (SBPH), rice resistance and temperature were generally considered as the main factors driving the occurrence of the vector and through these viral epidemics (Wan et al., 2015).

Kishimoto and Yamada (1986) (Kisimoto, 1986) proposed the use of SBPH density and viruliferous rate in the overwintering generation to predict RSV epidemics and empirically divided epidemics of RSV into normal-, epidemic- and transitional-level. Viruliferous SBPH density, rice resistance, cultivation time, transplanting time, sowing time, climatic factors (Murakami and Kanda,1986;Yamamura, 1998; Yamamura and Yokozawa, 2002; Yamamura et al., 2006; Zhu et al., 2009; Shimizu et al., 2011; Zhu et al., 2011; Deng et al., 2013) alone or in combination of 2–3 of these factors have also been proposed as alternative predictors of RSV epidemics by other researchers. However, most previous models were built upon data generated from sprayed fields where disease occurrence and epidemics were under intensive human intervention and hence where some factors, for example, total SBPH density, rice variety, and other cultural practices can be manipulated, but others such as the viruliferous rate of the vector, or temperature cannot. To effectively predict and manage RSV disease, requires knowledge of the biotic and abiotic factors influencing its occurrence and epidemic patterns in both agro-ecosystems experiencing extensive human intervention and semi-natural ecosystems largely lacking direct human input (Kisimoto, 1993; Li et al., 2015b). In this case, establishing unsprayed fields resembling ecosystems with minimal disease controls as a counterpart to sprayed fields is important to fulfill the ultimate goal of better predicting and managing RSV disease.

Unlike studies of RSV epidemics in the 1960–1970s, in the current experiments co-incident with the current upsurge in RSV, we established a set of sprayed and unsprayed fields in 8 counties in Jiangsu Province where disease and vector were monitored. Disease data gathered from both sprayed and unsprayed fields over a 10-year period of epidemics were evaluated in parallel with vector and climatic data with an objective to develop a lower cost, more effective and ecologically friendly approach for managing RSV disease. The specific goals of the current study were to: 1) understand the temporal dynamics of RSV disease and its vector in one of the largest rice production areas in China; 2) determine the main factors responsible for the epidemics of RSV in the region; and 3) develop a mathematical model to predict the epidemics of RSV based on key factors determining its epidemics.

2 Results

2.1 Temporal dynamics in RSV severity, SBPH density and viruliferous rate

In unsprayedfields, rice stripe disease was observed only sporadically before 2001. The average disease severity rose quickly from ~12% in 2002 to peak at ~63% in 2004 before gradually decreased to ~2% in 2012 (Fig. 1, Table S1). Though varying in scale, similar patterns of temporal dynamics in disease severity were observed among the regions (8 counties, Fig. 1) and between sprayed and unsprayed fields. The percentage of rice acreage infected by RSV (disease incidence) showed a similar temporal trend as disease severity (Fig. 1). Total SBPH density and viruliferous rate varied greatly among years and regions (Fig. 2, Table S2). On the other hand, there was less spatiotemporal variation in viruliferous SBPH density (Fig. 2). Over the period of the study, viruliferous rate peaked at approximately the same time as disease severity but viruliferous and total SBPH density peaked in 2007. lagging three years behind disease severity.

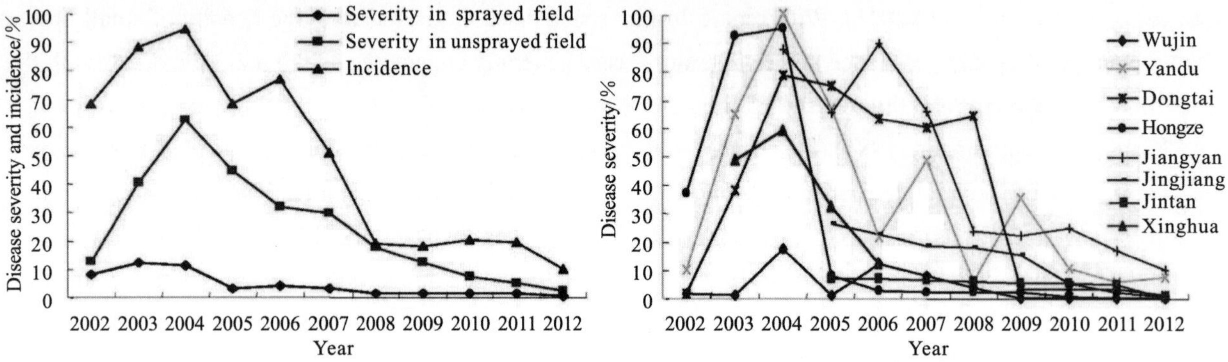

Fig. 1 Temporal dynamics of RSV epidemics between 2002 and 2012 in Jiangsu Province

1) Disease severity and incidence (% of acreage infected) across the eight counties; 2) Disease severity in unsprayed fields by county

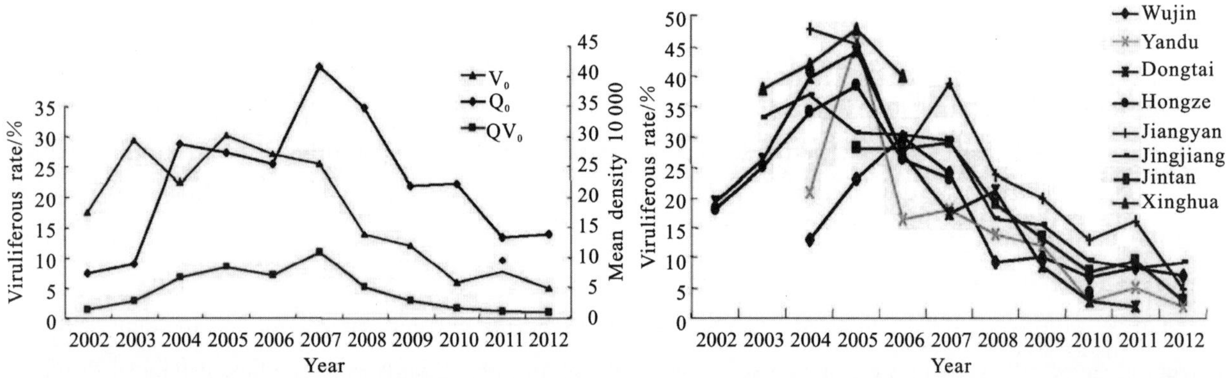

Fig. 2 Temporal dynamics of density and viruliferous characteristics of SBPH between 2002 and 2012 in Jiangsu Province

1) total SBPH density, viruliferous rate and viruliferous SBPH density at overwintering time across the eight counties; 2) viruliferous rate at overwintering time by county. Note: Q_0, V_0 and QV_0 are total SBPH density, viruliferous rate and viruliferous SBPH density at overwintering stage, respectively

2.2 Associations of rice stripe disease with SBPH traits

Rice stripe disease severity in both sprayed and unsprayed fields in any particular year was positively correlated with viruliferous rate in the preceding, current and following years at all epidemic stages but not with viruliferous or total SBPH density (Table 1). Though the overall patterns were similar across epidemic stages and field treatments, the association between disease severity and viruliferous rate in unsprayed fields was stronger than that in sprayed fields and that in the first generation was higher than that in overwintering vectors. The correlations between disease severity of the three years (preceding, current and following) and total SBPH density at the peak stage in those years were significantly negative but no associations were found between the two parameters at other epidemic stages. The relationship between disease severity and viruliferous rate of overwintering SBPH fitted a logistic model (Table 2) as expressed by $\ln (y \cdot 100) = 4.2-0.1861/x$, where y is the expected disease severity and x is the viruliferous rate of overwintering SBPH.

Table 1 Associations of RSV severity with density and viruliferous characters of SBPH in Jiangsu Province (Data summed across all sites)

		Severity in unsprayed fields			Severity in sprayed fields		
		Preceding year	Current year	Following year	Preceding year	Current year	Following year
Preceding year	Q_0	−0.076	−0.147	−0.117	0.098	0.067	0.133
	V_0	0.652**	0.528**	0.426**	0.474**	0.346*	−0.008
	Q_1	0.014	−0.107	−0.118	−0.012	−0.056	0.012
	V_1	0.814**	0.756**	0.684**	0.646**	0.510**	0.195
	Q_{max}	−0.447**	−0.622**	−0.723**	−0.465**	−0.568**	−0.556**
	QV_0	0.139	0.063	0.057	0.267	0.238	0.237
	QV_1	0.190	0.254	0.458**	0.151	0.197	0.196
Current year	Q_0	−0.015	−0.015	−0.147	0.164	0.098	0.067
	V_0	0.601**	0.652**	0.528**	0.645**	0.474**	0.346*
	Q_1	0.187	0.014	−0.107	0.180	−0.012	−0.056
	V_1	0.758**	0.814**	0.756**	0.775**	0.646**	0.510**
	Q_{max}	−0.356*	−0.447**	−0.622**	−0.332*	−0.465**	−0.568**
	QV_0	0.099	0.139	0.063	0.355*	0.267	0.238
	QV_1	0.246	0.190	0.254	0.278	0.151	0.197
Following year	Q_0	−0.043	−0.015	−0.076	0.169	0.164	0.098
	V_0	0.456**	0.601**	0.652**	0.640**	0.645**	0.474**
	Q_1	0.216	0.187	0.014	0.259	0.180	−0.012
	V_1	0.401*	0.758**	0.814**	0.824**	0.775**	0.646**
	Q_{max}	−0.331	−0.356*	−0.447**	−0.239	−0.332*	−0.465**
	QV_0	0.012	0.099	0.139	0.239	0.355*	0.267
	QV_1	0.291	0.246	0.190	0.479**	0.278	0.151

Q_0: overwintering SBPH density, V_0: viruliferous rate of overwintering SBPH,Q_1: first generation SBPH density, V_1: viruliferous rate of first generation SBPH, Qmax: SBPH density of peak stage, QV_0: viruliferous population density of overwintering SBPH ($Q_0\times V_0$), QV1: viruliferous population density of first generation SBPH ($Q_1\times V_1$). **significant at p = 0.01; *significant at p = 0.05

Table 2 The test statistic of S-curve estimation

Adjusted R^2	F	Sig.	T		Sig.	
			1/x	Constant	1/x	Constant
0.829	49.410	0.000	-7.029	17.807	0.000	0.000

2.3 Associations of rice stripe disease with climatic factors

Among relevant climate factors, mean temperature in May alone was significantly and negatively correlated with rice stripe disease severity in unsprayed fields (Table 3). In sprayed fields, disease severity was also significantly negatively correlated with mean temperatures in May and weakly so with May minimum temperatures. Temperatures in other months and rainfall showed no correlation with disease severity.

Table 3 Associations of rice stripe disease severity with monthly temperature and rainfall in SBPH overwintering and rice growing seasons (December to May)

Severity	Month	Mean T (C)	Min T (C)	Max temperature	Mean rainfall daily
Unsprayed field	Dec	0.026	0.066	0.044	-0.159
	Jan	-0.012	0.045	0.022	0.161
	Feb	0.084	0.017	0.020	-0.082
	Mar	-0.066	-0.045	0.015	-0.084
	Apr	-0.238	-0.030	0.007	0.051
	May	-0.389*	-0.228	-0.119	0.033
Sprayed field	Dec	0.048	–0.078	0.137	–0.065
	Jan	0.147	–0.046	0.180	0.181
	Feb	0.107	–0.121	0.069	0.01
	Mar	0.034	–0.138	0.168	0.030
	Apr	–0.122	–0.049	0.133	0.080
	May	–0.431**	–0.274	–0.035	0.077

**significant at p = 0.01; *significant at p = 0.05

2.4 Associations of disease severity among epidemic years

In unsprayed fields, rice stripe disease severity in any given year was positively and significantly correlated with that of preceding and current year levels in sprayed fields and the percentage of diseased fields in that year (Table 4). Furthermore, the mean disease severity in unsprayed fields in the eight counties was also positively correlated with that in sprayed fields (r_{10} = 0.760, p = 0.007) and the percentage of diseased fields (r_{10} = 0.850, p = 0.001).

Table 4 Associations of disease severity among epidemic years

	Severity in unsprayed field in preceding year	Severity in sprayed field in current year	Severity in unsprayed field in current year	Disease incidence in current year
Severity in sprayed field in preceding year	0.722**	0.844**	0.691**	0.618**
Severity in unsprayed field in preceding year		0.538**	0.735**	0.417**
Severity in sprayed field in current year			0.722**	0.778**
Severity in unsprayed field in current year				0.583**

**significant at p = 0.01

2.5 Associations among SBPH traits

Total SBPH density, viruliferous rate and viruliferous SBPH density in the first generation were positively correlated with those in the overwintering generation (Table 5). Though negative, the correlation between total SBPH density and viruliferous rate was not significant.

2.6 Associations between SBPH and climatic conditions

Most correlations between viruliferous rate and temperature were not significant both in the overwintering and first generation of SBPH. With a few exceptions, total SBPH density in the overwintering and first generation was positively correlated with maximum temperature and negatively correlated with minimum temperature in each month (Table 6).

Table 5 Associations between density and viruliferious characteristics of SBPH populations

	V_0	Q_1	V_1	Q_{max}	QV_0	QV_1
Q_0	−0.186	0.641**	−0.419*	0.113	0.871**	0.386*
V_0		−0.119	0.859**	−0.400*	0.283	0.272
Q_1			−0.108	0.235	0.555**	0.544**
V_1				−0.222	−0.013	0.423*
Q_{max}					−0.025	−0.127
QV_0						0.458**

Q_0:overwintering SBPH density, V_0: viruliferous rate of overwintering SBPH, Q_1: first generation SBPH density,V_1: viruliferous rate of first generation SBPH, Qmax: SBPH density of peak stage, QV_0: viruliferous population density of overwintering SBPH (Q0 × V0), QV_1: viruliferous population density of first generation SBPH($Q_1 \times V_1$). **significant at p = 0.01; *significant at p = 0.05

Table 6 Associations of local temperature and rainfall with density and viruliferous characteristics of the SBPH population

	Month	Mean T (C)	Min T (C)	Max temperature	Mean rainfall daily
V_0	Dec	0.048	0.235	−0.128	−0.097
	Jan	0.057	0.293	−0.168	0.375*
	Feb	0.032	0.213	−0.160	0.020
	Mar	0.163	0.164	−0.110	−0.042
Q_0	Dec	0.445**	−0.453**	0.537**	−0.184
	Jan	0.320*	−0.369*	0.464**	0.007
	Feb	0.226	−0.405**	0.575**	−0.045
	Mar	0.307*	−0.495**	0.563**	−0.131
QV_0	Dec	0.338*	−0.014	0.125	0.106
	Jan	0.423**	0.019	0.157	0.221
	Feb	0.313*	0.112	0.133	0.209
	Mar	0.320*	−0.172	0.316	0.078
V_1	Dec	−0.108	0.062	−0.006	0.024
	Jan	−0.147	0.054	−0.073	0.375*
	Feb	−0.021	0.062	−0.326	0.117
	Mar	−0.114	−0.074	−0.163	0.099
	Apr	−0.198	0.240	−0.036	0.218
	May	−0.441**	−0.200	−0.404*	0.291
Q_1	Dec	0.367*	−0.326*	0.426**	−0.302*
	Jan	0.147	−0.258	0.332*	−0.234
	Feb	0.1396	−0.371*	0.435**	−0.347*
	Mar	0.163	−0.355*	0.334*	−0.434**
	Apr	−0.089	−0.482**	0.353*	−0.317*
	May	0.207	−0.428**	0.385*	−0.212
QV_1	Dec	0.303	−0.337	0.394*	−0.128
	Jan	0.247	−0.135	0.228	0.126
	Feb	0.176	−0.137	0.157	−0.090
	Mar	0.225	−0.181	0.206	−0.219
	Apr	0.112	−0.053	0.336	−0.046
	May	0.065	−0.455**	0.121	0.025
Q_{max}	Dec	−0.064	0.009	−0.035	0.081
	Jan	−0.285	−0.059	−0.031	−0.089
	Feb	−0.086	−0.073	0.079	0.032
	Mar	0.016	−0.024	−0.085	0.041
	Apr	−0.187	−0.147	−0.109	−0.056
	May	0.222	0.033	0.119	0.198

Q_0: overwintering SBPH density, V_0: viruliferous rate of overwintering SBPH, Q_1: first generation SBPH density, V_1: viruliferous rate of first generation SBPH, Q_{max}: SBPH density of peak stage, QV_0:viruliferous population density of overwintering SBPH ($Q_0 \times V_0$), QV_1: viruliferous population density of first generation SBPH ($Q_1 \times V_1$). **significant at p = 0.01; *significant at p = 0.05

3 Discussion

Viruliferous population density has been widely used as a management indicator for insect-transmitted viral diseases (Froissart et al., 2010; Deng et al., 2013). However, our analysis shows that the RSV disease severity does not correlate with viruliferous SBPH density in either sprayed or unsprayed fields. Rather, it is positively correlated with viruliferous rate, consistent with previous studies in other rice viral diseases (Xie and Lin, 1980; Zhu et al., 2009). This result suggests that, the proportion, rather than the absolute size of the SBPH population infected by the virus, plays the decisive role in RSV epidemics. This counter-intuitive result is likely to result from a trade-off between the role of infected SBPH insects in disease epidemics and their own reproduction. While infection of insects by the virus is a pre-requirement for its transmission, and therefore disease epidemics, it also reduces the survival and reproductive ability of the insects themselves (Nasu, 1963; Fujita et al., 2013). This result also suggests that viruliferous rate could be a good indicator of RSV disease epidemics and management such that the equation $\ln(y\cdot 100) = 4.2-0.1861/x$ [where y is the expected disease severity and x is the viruliferous rate of overwintering SBPH] could be used for disease forecasting. Whether such a relationship is only relevant to plant viruses that propagate vertically within insect vectors and transmitted to their progenies ina persistent manner such as RSV (Koganezawa et al., 1975; Falk and Tsai, 1998), or is ubiquitous in all insect-transmitted plant viruses needs further study.

The observation that the proportion, rather than the absolute size of the SBPH population infected by the virus plays the decisive role in RSV disease epidemics is also supported by negative associations of total SBPH density with RSV severity and viruliferous rate. Viruliferous rates of SBPH in all epidemic stages were significantly and positively correlated with rice stripe disease severity in the preceding, current and following year in sprayed and unsprayed fields, which suggests that the viruliferous rate in a given year likely results from disease epidemics in the preceding year. The preceding epidemic is the primary viral source for new infection, as indicated by a positive correlation between disease severities among years. Under a constant viral source, increasing insect density reduces the chance of any individual insect becoming infected, leading to a negative correlation between total SBPH density and viruliferous rate. Furthermore, large numbers of uninfected insects competing with infected ones to feed on a newly established crop, reduce the chance of viruliferous insects transmitting the virus and hence causing disease epidemics.

The positive correlation between disease severities among years indicates that local infection is the main source of primary inoculum, as reported earlier (Kisimoto, 1989; Hoshizaki, 1997; Cheng et al., 2013; Otuka, 2013). In this scenario, it is expected that disease severity in the fields would change in a consistent direction. Unexpectedly, RSV severity in the areas studied showed a bell distribution, increasing from 2002 to 2004 and then declining. RNA viruses such as RSV are characterized by high mutation rates attributing to their lack of proofing mechanisms in genome duplication (Domingo and Holland, 1997; Roossinck, 1997; Elena and Sanjuán, 2007; Lauring and Andino, 2010; Sanjuán et al., 2010; Cho et al., 2013b). Genetic meltdown leading to changes in infectivity (of either insect or host plant) due to rapid accumulation of deleterious mutations (Domingo and Holland, 1997; Roossinck, 1997; Elena and Sanjuán, 2007; Lauring and Andino, 2010; Sanjuán et al., 2010; Cho et al., 2013b; Huang et al., 2015) could partially contribute to the switch of increasing epidemics to decreasing epidemics over the observed time scale. However, the observed changing epidemic pattern in the current study is most likely due to some form of human intervention. In this part of China SBPH overwinters mainly on wheat. Since the onset of the current epidemic cycle, local governments have launched an initiative to eradicate SBPH overwintering sites by reducing or stopping wheat cultivation in the region. Virus acquisition of SBPH depends on the probability of transovarial transmission or acquisition from diseased host plants. Viruliferous rates of SBPH will decrease gradually without any viral source supplement (Kisimoto, 1967; Li et al., 2015c).

With the exception of the first generation of SBPH density (Q_1), no associations were detected between rainfall and insect parameters. On the other hand, many correlations between temperature and insect parameters were significant. These results indicate that temperature plays a more important role than rainfall in the survival and reproduction of SPBH. The optimum temperature for survival and reproduction of SBPH is about 25°C and higher winter temperatures are apparently favorable for the overwintering of SBPH (Liu and Zhang, 2013), as supported by the positive correlation between mean temperature and overwintering SBPH density (Q_0) (Table 6). In addition, winter temperature also alters SBPH feeding behavior with warmer winters enhancing feeding activity as indicated by the positive correlation between temperature and viruliferous population density of overwintering SBPH (QV_0). However, the impact of temperature on SBPH survival, reproduction and feeding behavior is mainly restricted to the overwintering stage. In other phases of life cycle, the contribution of thermal fluctuation to the survival, reproduction and feeding behavior is negligible as indicated by a lack of correlation involving local temperature with SBPH density of peak stage (Q_{max}), Q_1, viruliferous rate (V_1) or viruliferous population density (QV_1) of first generation SBPH.

In this study, we collected data from both sprayed and unsprayed rice fields. Disease in unsprayed fields is less influenced by human activities, and therefore more closely resembles disease and epidemic occurrence in semi-natural ecosystems. Such areas are a potential source of primary inoculum for a new epidemic in the coming season. Comparing the temporal dynamics of RSV in sprayed and unsprayed fields is important as a means of untangling some of the differences between agricultural and more natural systems that drive the epidemiology of this disease. Such an approach could be of value if applied to studies of other plant viral diseases.

4 Materials and Methods

Rice stripe disease severity was monitored annually over the period 2002–2012 inclusively in a total of 4000 insecticide-sprayed (chemical used) paddy fields distributed across the eight major rice–growing counties of Jiangsu Province, China. Each county was represented by five towns and each town was represented by five villages. In each village, 20 paddy fields each at least 667m^2 in size were randomly selected for the experiment. An identical sampling protocol was used over the entire 10 year period with disease severity in each field being determined from 200 plants distributed at five points determined using a random spatial sampling approach 55. RSV severity at each sampling point was recorded using a five-grade system (< 5%, 5%–10%, 10%–20%, 20%–50% and > 50%) as described previously (Lin et al., 1979). The disease assessment took place at the same phenological stage (late tillering state) every year-a stage that experience had shown to provide a reliable estimate of epidemic development. An unsprayed (no insecticide application) field with a size of ~667m^2 was used as a control in each village with disease severity being assessed in the same way as for the sprayed fields. The mean disease severity (MDSev) in each county was estimated as:

$$MDSev=\sum[(DS\times A)_{F1}+\cdots+(DS\times A)_{F20}]/[A_{F1}+A_{F2}+\cdots+A_{F20}] \quad (1)$$

where DS is the mid-range figure of each disease severity grade (i.e. 2.5%, 7.5%, 15%, 35%, 75%), A is the area of the field, and Fn is the field number. The percentage of infected fields was determined by dividing the number of fields in each county infected by virus by the total number of fields. Total SBPH density and viruliferous rate of SBPH were recorded during the overwintering (December to March) and first generation phases (March to May) in wheat fields and at the peak phase (June) in the rice paddy fields. Total SBPH density was determined by the plant-flapping (on the roots in seedling stage or flag leaves in earing stage) approach using a porcelain plate (40cm × 30cm) as described previously (Li et al., 2013). The number of insects in each sampling point was counted from 20 porcelain plates and five sampling points each with 300 plants in a field were included in density analysis. SBPH viruliferous rate of each county in each year was determined by Dot-ELISA (Takahashi et al., 1991) using 98 insects randomly selected from the total collection. RSV monoclonal antibody was prepared by our laboratory using Goat-anti-Mouse IgG, chromomeric substrate and other reagents were purchased from the Sigma Company (No.398, Huaihai Road, Shanghai). The viruliferous rate across eight counties in each year was calculated as:

$$VRate= \sum(MD_i \times MVR_i)/TD_8 \; (i= 1, 2, 3,\cdots ,8) \quad (2)$$

where MD is the total SBPH density, MVR is the SBPH viruliferous rate of each county, TD is the total SBPH density across eight counties and i is the random order of these eight counties. Climate data including monthly mean, minimum and maximum temperature and daily mean rainfall from December to May for each year were obtained from local weather stations in the eight counties (Table S3). Disease severity, SBPH density and viruliferous rate in sprayed and unsprayed fields were tabulated over the epidemic years for each individual county as well as for the combined data from different counties. The association between and among disease severity, total SBPH density, viruliferous rate, viruliferous SBPH density and climatic parameters including strength, direction and quantity of the associations were analyzed using Spearman's correlation (Puth et al., 2015) taking the variables in each year and county as random parameters. In these analyses, all variables were ranked and correlation coefficients were calculated according to the order of the variable. Spearman correlation coefficient was calculated as:

$$\theta = \sum(R_i-R)(S_i- S)/[(R_i- R)^2 (S_i- S)^2]^{1/2} \quad (3)$$

where R_i and S_i are the rank of element i in the independent and dependent variables and R and S are the mean of independent and dependent variables, respectively.

In association analyses, it is necessary to guard against non-independent elements among the variables as they have the potential to affect the reliability of the extent and direction of estimates. However, by converting data to ranks (1,2,3,4, ⋯ , etc.) and estimating correlation coefficients on these ranked values the Spearman correlation is less dependent on the randomness of variable elements (Eisinga et al., 2013). Therefore, because severity, and possible driving forces such as SBPH density of RSV epidemics, may not be independent from year to year, we used this statistic to evaluate the associations between disease, vector and climatic data.

References

[1] ABO M E, SY A A. Rice virus diseases: epidemiology and management strategies[J]. Journal of Sustainable Agriculture,1997, 11:113-134.

[2] BAE S D, KIM D K. Occurrence of small brown planthopper (Laodelphax stiatellus Fallén) and incidence of rice virus disease by different seeding dates in dry seeded rice[J]. Korean Journal of Applied Entomology, 1994, 33: 173-177.

[3] CAI L, et al. Detecting Rice stripe virus (RSV) in the small brown planthopper (Laodelphax striatellus) with high specificity by RT-PCR[J]. Journal of Virological Methods, 2003, 112:115-120.

[4] CHENG Z B et al. Factors affecting the occurrence of Laodelphax striatellus in a single rice-wheat rotation[J]. Chinese Journal of Applied Entomology, 2013, 50: 706-717 (in Chinese).

[5] CHO S Y, et al. One-step multiplex reverse transcription-polymerase chain reaction for the simultaneous detection of three rice viruses[J]. Journal of Virological Methods, 2013, 193:674-678.

[6] CHO W K, LIAN S, KIM S M, et al. Current insights into research on Rice stripe virus[J]. Plant Pathology Journal, 2013, 29:223-233.

[7] DENG J H, LI S, HONG J, et al. Investigation on subcellular localization of Rice stripe virus in its vector small brown planthopper by electron microscopy[J]. Virology Journal, 2013, 10: 310-317.

[8] DOMINGO E,HOLLAND J J. RNA virus mutations and fitness for survival[J]. Annual Review of Microbiology, 1997, 51:151-178.

[9] EISINGA R, GROTENHUIS M, PELZER B. The reliability of a two-item scale: Pearson, Cronbach, or Spearman-Brown?[J]. International Journal of Public Health, 2013, 58: 637-642.

[10] ELENA S F, SANJUÁN R. Virus evolution: Insights from an experimental approach[J]. Annual Review of Ecology Evolution and Systematics, 2007, 38:27-52.

[11] FALK B W,TSAI J H. Biology and molecular biology of viruses in the genus Tenuivirus. Annual Review. Phytopathol, 1998, 36: 139-163.

[12] FROISSART,R, DOUMAYROU,J, VUILLAUME F, et al. The virulence-transmission trade-off in vector-borne plant viruses: a review of (non-) existing studies[J]. Philosophical Transactions of the Royal Society B: Biological Sciences, 2010, 365: 1907-1918.

[13] FUJITA D, KOHLI A, HORGAN F G. Rice resistance to planthoppers and leafhoppers[J]. Critical Reviews in Plant Sciences, 2013, 32:162-191.

[14] HIBINO H. Biology and epidemiology of rice viruses[J]. Annual Review of Phytopathology, 1996, 34:249-274.

[15] HOSHIZAKI S. Allozyme polymorphism and geographic variation in the small brown planthopper *Laodelphax striatellus* (Homoptera: Delphacidae)[J]. Biochemical Genetics, 1997, 35: 383-393.

[16] HUANG L, et al. Analysis of genetic variation and diversity of Rice stripe virus populations through high-throughput sequencing[J]. Frontiers in Plant Science, 2015, 6:176.

[17] KIRITANI K. Changes in cropping practices and the incidence of hopper-borne diseases of rice in Japan[J]. Plant virus epidemiology. The spread and control of insect-borne viruses, 1983, 239-247.

[18] KIRITANI K, NAKASUJI F,MIYAI S. Systems approaches for management of insect-borne rice disease[M]. Springer New York, 1987, 3:57-80.

[19] KISHIMOTO R. Genetic variability in the ability of a planthopper vector, *Laodelphax striatellus* (Fallén) to acquire the Rice stripe virus[J]. Virology, 1967, 32:144-152.

[20] KISIMOTO R, YAMADA Y A. Planthopper-rice virus epidemiology model: rice stripe and small brown planthopper *Laodelphaxstria-tellus* Fallén[J]. Plant Virus Epidemics Monitoring Modelling & Predicting Outbreaks, 1986: 327-344.

[21] KISHIMOTO R. Flexible diapause response to photoperiod of a laboratory selected line in the small brown planthopper *Laodelphaxstria-tellus* Fallén[J]. Applied Entomology & Zoology, 2008, 24: 157-159.

[22] KISHIMOTO R. Biology and monitoring of vectors in rice stripe epidemiology[J]. Extension Bull, 1993, 373:1-9.

[23] KOGANEZAWA H, DOI Y, YORA K. Purification of Rice stripe virus[J]. The Phytopathological Society of Japan,1975, 41:148-154.

[24] LAURING A S, ANDINO R. Quasispecies theory and the behavior of RNA viruses[J]. PLoS Pathogens, 2010, 6: e1001005.

[25] SANJUÁN R, NEBOT M R, CHIRICO N, et al. Viral mutation rates[J]. Journal of Virology, 2010, 84:9733-9748.

[26] LI A, et al. Identification and fine mapping of qRBSDV-6 MH, a major QTL for resistance to Rice black-streaked dwarf virus disease[J]. Molecular Breeding, 2013, 32:1-13.

[27] LI J, XIANG C Y, YANG J, et al. Interaction of HSP20 with a viral RdRp changes its sub-cellular localization and distribution

pattern in plants[J]. Scientific Reports, 2015, 5:14016.

[28] LI S, WANG X, XU J X, et al. A simplified method for simultaneous detection of Rice stripe virus and Rice black-streaked dwarf virus in insect vector[J]. Journal of Virological Methods, 2015, 211: 32-35.

[29] LI S, et al. Rice stripe virus affects the viability of its vector offspring by changing developmental gene expression in embryos[J]. Scientific Reports, 2015, 5:7883.

[30] LIN C S, POUSHINSKY G, MAUER M. An examination of five sampling methods under random and clustered disease distributions using simulation[J]. Canadian Journal of Plant Science, 1979, 59:121-130.

[31] LIU X D, ZHANG A M. High temperature determines the ups and downs of small brown planthopper *Laodelphax striatellus* population[J]. Insect Science, 2013, 20:385-392.

[32] MURAKAMI M, KANDA T. Occurrence of insect pests in rice stripe disease resistant cultivar[J]. Annual Report of the Kanto-Tosan Plant Protection Society, 1986, 33: 186-187.

[33] NASU S. Studies on some leafhoppers and planthoppers which transmit virus diseases of rice plant in Japan[J]. Bull. Kyushu AgricExpStn, 1963, 8: 153-349.

[34] OTUKA A, et al. The overseas mass migration of small brown planthopper, *Laodelphax striatellus*, and subsequent outbreak of rice stripe disease in western Jap[J]. Applied Entomology & Zoology, 2010, 45:259-266.

[35] OTUKA A. Migration of rice planthoppers and their vectored re-emerging and novel rice viruses in East Asia[J]. Frontiers in Microbiology, 2013, 4:309.

[36] PUTH M T, NEUHÄUSER M, RUXTON G D. Effective use of Spearman's and Kendall's correlation coefficients forassociation between two measured traits[J]. Animal Behaviour, 2015, 102: 77-84.

[37] ROOSSINCK M J. Mechanisms of plant virus evolution[J].Annual Review Phytopathol, 1997, 35:191-209.

[38] SHINKAI A. Present situation of rice stripe disease[J]. Plant Prot Jap, 1985, 11:503-507.

[39] SHIMIZU T, NAKAZONO-NAGAOKA E, UEHARA-ICHIKI T, et al. Targeting specific genes for RNA interference is crucial to the development of strong resistance to Rice stripe virus[J]. Plant Biotechnology Journal, 2011, 9:503-512.

[40] SUN D Z, et al. Quantitative trait loci for resistance to stripe disease in rice (Oryza sativa)[J]. Rice Science, 2007, 14:157-160.

[41] TAKAHASHI Y, OMURA T, SHOHARA K, et al. Comparison of four serological methods for practical detection of ten viruses of rice in plants and insects[J]. Plant Disease, 1991, 75:458-461.

[42] TORIYAMA S. Rice stripe virus: prototype of a new group of viruses that replicate in plants and insects[J]. Microbiological Sciences, 1986, 3: 347-351.

[43] WANG H D, et al. Recent Rice stripe virus epidemics in Zhejiang Province China, and experiments on sowing date, disease-yield loss relationships, and seedling susceptibility[J]. Plant Disease, 2008, 92: 1190-1196.

[44] WAN G, et al. Rice stripe virus counters reduced fecundity in its insect vector by modifying insect physiology, primary endosymbionts and feeding behavior[J]. Scientific Reports, 2015, 5:12527.

[45] WEI T Y, et al. Genetic diversity and population structure of Rice stripe virus in China[J]. Journal of General Virology, 2009, 90:1025-1034.

[46] WU S J, et al. Identification of QTLs for the resistance to Rice stripe virus in the indica rice variety[J]. Euphytica, 2009, 165: 557-565.

[47] XIAO D L, LI W M, WEI T Y, et al. Advances in the studies of Rice stripe virus[J]. Frontiers Of Agriculture In China, 2010, 4:287-292.

[48] XIE L H, LIN Q Y. Studies on the epidemic forecasting of rice transitory yellowing and rice dwarf diseases[J]. Fujian Journal of Agricultural Sciences, 1980, 9:32-43 (in Chinese).

[49] XIONG R, WU J, ZHOU Y, et al. Characterization and subcellular localization of an RNA silencing suppress orencoded by rice stripe Tenuivirus[J]. Virology, 2009, 387:29-40.

[50] YASUO S, ISHII M, YAMAGUCHI T. Studies on rice stripe disease[J]. I Epidemiological and ecological studies on rice stripe disease in the Kanto-Tosan district of central part of Japan. Review Plant Protection Research, 1968, 1: 96-104.

[51] YAMAMURA K. Stabilization effects of spatial aggregation of vectors in plant disease systems[J]. Researches on Population Ecology, 1998, 40, 227-238.

[52] YAMAMURA K, YOKOZAWA M. Prediction of a geographical shift in the epidemic of Rice stripe virus disease transmitted by small brown planthopper, (*Laodelphax striatellus* Fallén) (Hemiptera: Delphacidae), under global warming[J]. Applied Entomology & Zoology, 2002, 37, 181-190.

[53] YAMAMURA K, YOKOZAWA M, NISHIMORI M, et al. How to analyze long-term insect population dynamics under climate change: 50-year data of three insect pests in paddy fields[J]. Population Ecology, 2006, 48:31-48.

[54] ZHANG H M, SUN H R, WANG H D, et al. Advances in the studies of molecular biology of Rice stripe virus[J]. Acta Phytophylacica Sinica, 2007, 34:436-450.

[55] ZHANG H M, WANG H D, YANG J J, et al. Detection, occurrence, and survey of rice stripe and black-streaked dwarf diseases in Zhejiang Province, China[J]. Rice Science, 2013, 20:383-390.

[56] ZHU F M, XIAO Q P, WANG F M. Several new diseases occurring in rice south of the Yangtze River. Plant Protection, 1964, 2:100-102 (in Chinese).

[57] ZHOU Y J, LI S, CHENG Z B, et al. Research advances in rice stripe disease in China[J]. Jiangsu Journal of Agricultural Sciences, 2012, 28:1007-1015 (in Chinese).

[58] ZHU J L, et al. Effect of rice sowing date on occurrence of small brown planthopper and epidemics of planthopper-transmitted rice stripe viral disease[J]. Agricultural Sciences in China, 2009, 8:332-341.

[59] ZHU J L, et al. Effect of sowing/transplanting time on occurrence of main locally-overwintering insect pests and diseases[J]. Acta Agriculturae Zhejiang ensis, 2011,23:329-334 (in Chinese).

Problems, challenges and future of plant disease management: from an ecological point of view

Dunchun He[1], Jiasui Zhan[1, 2], Lianhui Xie[1, 2]

(1 Fujian Key Lab of Plant Virology, Institute of Plant Virology, Fujian Agriculture and Forestry University, Fuzhou 350002, P.R. China; 2 Key Lab for Biopesticide and Chemical Biology, Ministry of Education/Fujian Agriculture and Forestry University, Fuzhou 350002, P.R. China)

Abstract: Plant disease management faces ever-growing challenges due to: (i) increasing demands for total, safe and diverse foods to support the booming global population and its improving living standards; (ii) reducing production potential in agriculture due to competition for land in fertile areas and exhaustion of marginal arable lands; (iii) deteriorating ecology of agro-ecosystems and depletion of natural resources; and (iv) increased risk of disease epidemics resulting from agricultural intensification and monocultures. Future plant disease management should aim to strengthen food security for a stable society while simultaneously safeguarding the health of associated ecosystems and reducing dependency on natural resources. To achieve these multiple functionalities, sustainable plant disease management should place emphases on rational adaptation of resistance, avoidance, elimination and remediation strategies individually and collectively, guided by traits of specific host-pathogen associations using evolutionary ecology principles to create environmental (biotic and abiotic) conditions favorable for host growth and development while adverse to pathogen reproduction and evolution.

Keywords: disease resistance; avoidance; elimination and remediation; ecological plant disease management; evolutionary principle; food security; plant disease economy

1 Introduction

Plant disease has been a major factor influencing food production and human societal development over thousands of years (Palmgren et al., 2015). Throughout the early agricultural era, the occurrence of plant disease epidemics was seen as a punishment from the gods and overt plant disease management approaches were extremely limited. Given generally low yields and the general lack of significant food reserves, once disease epidemics occurred food shortages could easily develop resulting in disastrous effects on human society - such as the Irish Famine caused by potato late blight in the 1840s and the 1943 Bengal famine caused by rice brown spot (Bourke, 1964; Padmanabhan, 1973; Strange and Scott, 2005). Despite the contribution of scientific and technological advances to significant reductions in the frequency and intensity of epidemics in recent times, 20%—30% of actual production is still lost due to plant diseases per year (Oerke and Dehne, 2004; Oerke, 2005). These losses reflect incomplete knowledge relating to the causes and mechanisms behind epidemic development, a situation that unsurprisingly reflects a lack of adequate approaches to even efficiently manage them, let alone eliminate them. Furthermore, many plant disease management strategies together with

Journal of Integrative Agriculture. 2016, 15(4): 705–715
Received 29 October 2015; Accepted 1 February2016

many agronomic practices used in modern agriculture have also generated unintended problems including loss of biodiversity and other natural resources (Lucas, 2011; Gonthier et al., 2014), environmental deterioration (Enserink et al., 2013), and accelerated evolution in pathogens (Zhan et al.,2002; Sommerhalder et al., 2010; Zhan and McDonald, 2013).

Over agricultural history, plant disease management has experienced four major phases: (i) limited intervention in ancient farming systems; (ii) mechanical and temporal disease suppression approaches (rogueing, ploughing, rotations); (iii) widespread use of major gene resistances and pesticides before and following the first Green Revolution; and (iv) integrated pest management and ecological management emphasizing synergic effects on the economy, society and both the agricultural and natural environments. Ecological management of plant diseases is not a simple return to farming systems of ancient times. Rather, it aims to use evolutionary principles and thinking to maximize the regulatory functions of nature to create suitable environments for healthy hosts ensuring high and stable yield through the efficient use of natural and societal resources including high disease resistance to create environments adverse for the infection, reproduction, transmission and evolution of pathogens (Zhan and McDonald, 2013; Zhan et al., 2014, 2015). In addition, short-term and long-term economic and societal impacts should be evaluated for each plant disease management scheme. To achieve the goal of sustainable plant disease management, multidisciplinary collaboration involving natural and biological sciences such as plant pathology, breeding, agronomy, soil science, environmental science, economics and social science is needed.

2 The nature of plant disease epidemics and current situation of management

In natural systems, host plant and pathogen are constantly changing with pathogens evolving new pathogenicity to overcome host defense systems and plants evolving to reduce pathogen attack. These coevolutionary interactions occur within ecological settings in whichpathogen evolution and impact are tempered by environmental patchiness while host evolution is constrained by small population sizes and long generation times (Burdon and Thrall, 2009; Iranzo et al., 2015). In contrast, in modern agriculture systems, increasing requirements for high productivity and good quality of some specific crops and/or varieties forces the shift of agricultural practices to large-scale, intensive and specialized cultivation. In turn, this disrupts the co-evolutionary dynamics between host plants and pathogens as observed in natural systems, increasing the frequency and severity of disease epidemics and the spread of new diseases. However, largely due to a failure to place relationships among agricultural practices, disease epidemics and economic returns in an ecological and evolutionary context, plant disease management strategies adopted over the past 50–100 years rarely reflected these changes in risk and patterns of disease occurrence. Consequently, plant disease management can easily fall in the conumdrum where increasing efforts to control plant diseases actually promotes further disease problems.

Plant disease results from complex interactions among biotic and abiotic factors including hosts, pathogen and environments, to which should be added vectors for some diseases and human activities that modify the interaction intentionally or unintentionally through agricultural practices such as cropping systems, resistance gene deployment (Burdon et al., 2014) and application of pesticides (Fig. 1). In past decades, plant disease management and other agricultural practices have created ecological environments favorable for pathogen infection, reproduction, transmission and evolution as described in the following sections. This increases the negative impacts of plant disease on food security and human society.

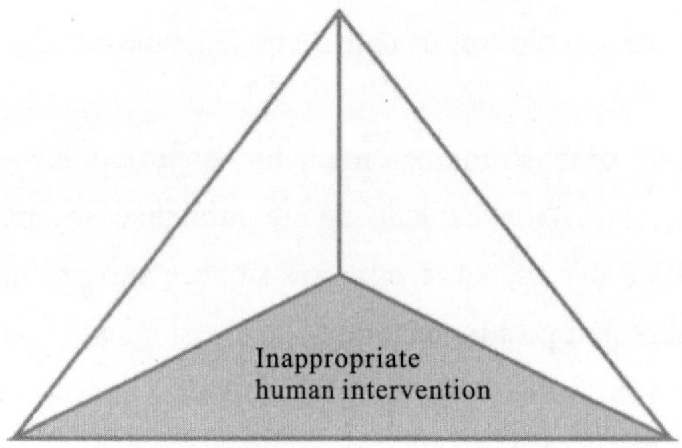

Fig. 1 The mechanism of plant disease epidemics in agricultural system

Plant disease epidemiology in modern agriculture is driven by disrupted eco-systems that create conditions favorable for pathogens but adverse to hosts due to inappropriate human intervention

2.1 Ecological environments adverse to host plants but favorable for pathogens

Healthy soils are the key to sustainable agriculture including plant disease management through their impact on pathogen density particularly of soil-borne diseases (Magdoff, 2001; Janvier et al., 2007), the structure of beneficial microbe communities and the availability of organic and inorganic nutrition for plant growth and development (Larkin, 2015; van Bruggen et al., 2016). Over past decades, water and air pollution resulting from industrial emission and agricultural wastes, and over-use of chemicals to nurse plants and manage pests and weeds has generated many near-irreversible changes reducing farmland quality through soil compaction, reduced organic material, mineral imbalances, and heavy metal and pesticide residue contamination (Kosalec et al.,2009; Lu et al., 2015; Tripathy et al., 2015). Furthermore, this deterioration in farmland quality may further reduce host plant immunity against pathogen infection.

Agricultural management strategies can have a major impact on soil quality (Bancal et al., 2008) with consequent effects on disease incidence. Thus most practices designed to improve soil quality by increasing beneficial microorganisms and the microbial biodiversity of farmland through activities such as organic matter supplementation also help to suppress the development of most diseases (Welbaum et al., 2004; Bonilla et al., 2012; Page et al.,2013). Crop rotation typically improves the physical and chemical properties of soil such as nutrition balance as well as the diversity of microbial communities (Ball et al., 2005). On the other hand, field management and production techniques such as continuous cropping and monocultures of single crops or varieties increase the risk of disease occurrence and epidemics by allowing pathogens to accumulate high inoculum loads. This is especially the case for soil-borne diseases but is also true of many foliar diseases. Such strategies also facilitate the breakdown of disease management strategies based on the use of limited numbers of resistance genes or pesticides due to the enhancement of selection pressures on pathogens due to reduced host diversity and the widespread use of pesticides with the same modes of action (Zhan et al., 2002; Sommerhalder et al., 2010).

2.2 Single and static management strategies increasing the intensity of plant disease outbreaks

Plant pathogens are difficult to control partly due to their rapid spatiotemporal dynamics and rapid evolution (Strange and Scott, 2005) associated with high genetic diversity and short generation times that together promote their ability to overcome the currently most effective disease control approaches based on major R gene resistance and industrial pesticides. Integrated pest management (IPM) approaches advocated in the last century were intended to manage plant diseases by assembling diverse approaches according to particular diseases, time and locations. However, the application of chemical pesticides has almost become the main and even only one approach of IPM strategy, particularly for crops lacking major resistance (Guedes et al., 2015). It has been reported that the rate of increased pesticide application has been far more than that of gained food production in recent decades (Popp et al., 2013), indicating the reduced efficiency and economic return of using pesticides to manage plant diseases. Usually, pesticides are used in a prescribed manner that standardizes type, time, frequency and dosage of application regardless of the particular crop's plant resistance status, environmental conditions and pathogen chemical sensitivity. Such a fixed and static strategy of pesticide application not only reduces management efficiency and increases costs, but also brings many unnecessary negative effects to environments and society such as toxicity to humans and livestock, and ecological degradation as discussed earlier.

2.3 The grey box of plant disease epidemic mechanisms

To achieve efficient and sustainable plant disease management, it is important to use an integrated systems thinking approach to understand the entire interaction between host and pathogen and its interplay with the broader environment (Burdon and Thrall, 2008). Thus while an understanding of the pathogenesis and epidemic principles of plant pathogens, and of the genetic, biological and physiological mechanisms of host plant defenses is important, so too is the knowledge of interactions with other microbial populations, and the ecological niche of the pathogen. More importantly, efforts should be made to comprehensively understand the effects of human activities such as agricultural practices (e.g., monoculture, rotation, etc.), international trading, and application of pesticides on the generation and evolution of new virulence, health of plant development and interactions among plants, pathogens, vectors and the environment (Xie et al., 2009). This is essential because it is commonly believed that many major disease outbreaks in history were mainly induced by human.

Correct and quick disease diagnosis, identification and forecasting are always fundamental (Miller et al., 2009). Technologies for diagnosis and identification have been well established and have been greatly assisted by the rise of molecular diagnostic kits. Despite this, their application in commercial production is still highly variable with some agri-companies and farmers still relying on experience, which can cause misdiagnoses and improper use of management approaches. In comparison with diagnosis and identification, disease forecasting requires a much deep understanding of pathogenesis and epidemic principles and the ecological and environmental interactions of plant pathogens with other biotic and abiotic factors. Because of this complexity, with a very few notable exceptions (for example, potato late blight in the north-eastern USA), accurate disease forecasting is still very limited.

2.4 Lack of models including externalities in the economic analysis of plant disease management

Externalities emerges when the effect of plant disease management on other parties is not reflected in the calculation of cost and profit. Externalities associated with plant disease management may be positive or negative and can be divided into short- and long-term ecological, social and economic components (biotic and abiotic, see Table 1). Negative externalities of plant disease management include environmental pollution, toxin production affecting humans or livestock, ecological damage, resource depletion, reduced disease management efficiency and costs associated with meeting minimum chemical residues on produce. Positive externalities include benefits to disease management in neighboring farms, reduced evolutionary potential of pathogens, and ensuring social stability and safety. Currently, these externalities are not included in economic analyses of plant disease management. Farmers are only responsible for the direct costs associated with pesticide application but not the costs associated with residue removal and ecology restoration, while those who apply ecologically-friendly approaches to disease management receive no additional benefit. Because farmers only pay the direct costs associated with plant disease management, they strongly select strategies that generate the best immediate economic returns while largely discounting potential negative impacts on the environment. To date, some highly effective disease management strategies have been used without sufficient regard to their long-term ecological impacts.

Table 1 Short- and long-term goals of plant disease management

Short-term benefit	Long-term benefit
High and stable yields	High efficiency and security (without residue, pollution and catastrophic effects)
Quality improvement	Sustainable production capacity
Low input	Reduced speed of pathogen evolution
High output	Social stability

Regulatory policy associated with industrial sewerage management provides a model for how externalities could be captured in assessing plant disease management strategies. The system levies the discharge of industrial wastes into surroundings to alleviate environmental pollution. Due to the transformation of externality to products, net profits of management strategies depend not only on the quality of the commodity but also the level of potential damage to environments. Taking pesticide and ecological management of plant diseases as examples, actual profits are substantially reduced for the former but increased for the latter when externalities are included in economic analysis (Fig. 2).

3 Challenges of plant disease management- rational management

Plant pathology faces ever-growing challenges. On one hand, societal demand for total, high quality and diverse food increases due to booming global population which is expected to reach 9 billion in 2050 (Godfray et al., 2010), and improving life standards. On the other hand, diminishing arable lands, and depleting natural resources reduce the potential for increasing agricultural productivity (Ray et al., 2013). Furthermore, monocultures, intensification and other high resource (fertilizer, water and pesticides) input agriculture practices aimed at maximum yield as the sole target, thereby facilitating the evolution and epidemics of plant diseases globally (Zhan et al., 2014, 2015). Into the future, much greater emphasis must be given to sustainable plant disease management strategies that ensure food security and societal development but also have less adverse impacts on environments and natural resources.

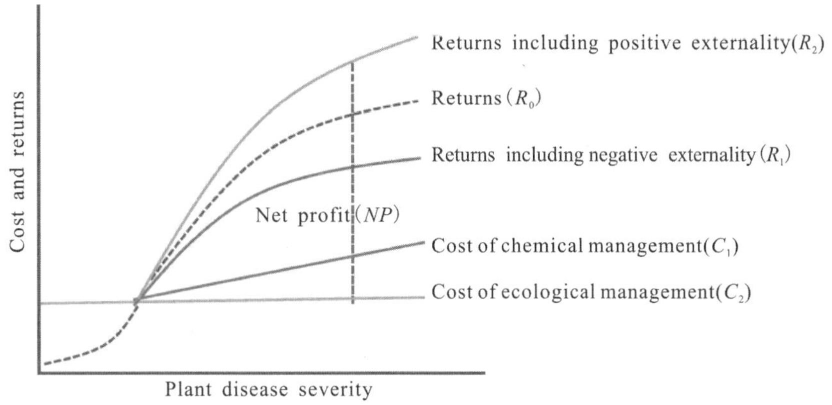

Fig. 2 Externality of plant disease management

If plant diseases are controlled by chemical application, management cost is positively associated with disease severity. On the other hand, if plant diseases are managed with ecological approaches aiming to improve plant health, cost is relatively stable with less negative impact to environments. According to Diminishing Marginal Returns principle, the profit (R_1-C_1) of chemical management is progressively reduced after peak disease, while the profit (R_2–C_2) of ecological management remains stable

To meet the challenge, plant disease management strategies, current agricultural practices and plant disease management strategies must change. Three components (society, economics and ecology) should be considered in future plant disease management strategies. Providing safe and adequate food for society is always the most important task of plant disease management. Plant disease management should strike to ensure food security and social stability by increasing crop productivity, reducing food contamination by microbial toxins, and guaranteeing the supply of diverse and reasonable priced foods. In regard to economic considerations, the ratio of input and return of plant disease management approaches should be better measured and evaluated including direct and indirect economic benefits and costs in a short- and long-term framework including externalities, opportunity costs, technical benefits, etc. In an ecological context, plant disease management should not only consider how to use ecological principles to reduce disease epidemics through changing agricultural practices but also how the strategies may impact on agricultural and ecological sustainability.

4 The ecological way to rational management of plant disease

4.1 Changes in the philosophy of plant disease management

To achieve rational and sustainable outcomes, the philosophy of plant disease control should shift the focus from managing pathogens (or insect vectors) to managing host plants and from the sole goal of high productivity to multiple goals of high yield, efficiency, good quality and safety.

4.2 Ecological management of plant disease

The key to sustainable plant disease management is to establish an agro-ecological system that is favorable to plant growth and development at the population level and adverse to pathogen evolution and epidemic development based on interactions among plants, pathogens, vectors and environments (Xie, 2003; Acosta-Leal et al., 2011). This management system includes two main components: multiple goals (high yield, efficiency, good quality, and safety) and dynamic and integrated approaches guided by a comprehensive understanding the evolutionary ecology of particular host-pathogen interactions. This integrated approach shows great promise in overcoming the problems and challenges associated with current strategies of plant disease management to optimize its economic, ecological and social benefits.

4.3 The core of ecological plant disease management

The core of ecological plant disease management is to manipulate the environments of host-pathogen interactions in the favor of hosts through the balanced application of RAER (resistance, avoidance, elimination and remedy) strategy (Fig. 3). Agricultural pathogens vary substantially in disease ecology, epidemic patterns, evolutionary potential and economic impact (Table 1) and RAER strategy should be applied according to specific conditions of the host-pathogen interactions involved (Xie and Lin, 1984; Xie et al., 1984, 1994). Some plant disease management approaches such as crop rotation may achieve the equivalent of resistance, avoidance, elimination and remedy effects simultaneously and could be applied widely to future agriculture despite some practical constraints (Pywell et al., 2015).

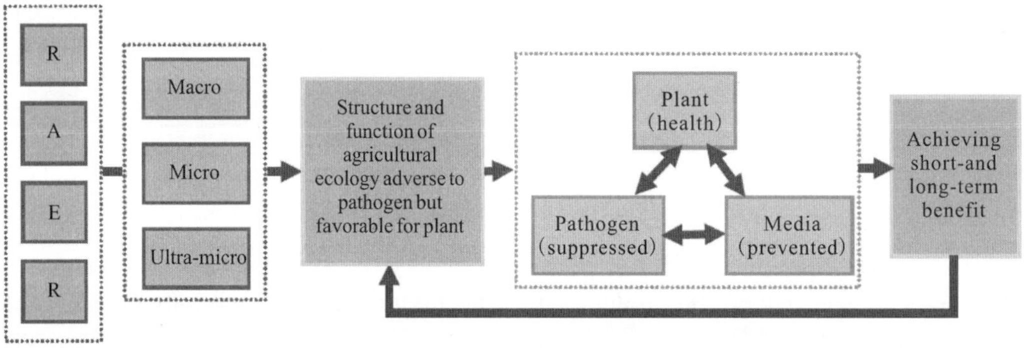

Fig. 3 Ecological management of plant disease

The ecological management of plant disease is to regulate host-pathogen interaction through the rational application of RAER (resistance, avoidance, elimination, and remediation) principle to create the environment favorable for host plant population but adverse to the spread and evolution of pathogen and/or vector for sustainability

4.3.1 Resistance

Host resistance is the most effective and convenient approach for plant disease management (Xie et al., 1983, 1994). Host resistance can be induced or constitutive, systematic or local and qualitatively or quantitative. Most resistances in crops are introduced from land-races or wild relatives through plant breeding (Manosalva et al.,2015; Palmgren et al., 2015). Qualitative (or major gene) resistance is highly effective but due to elevated evolution of plant pathogens under modern agricultural practices (Hall et al., 2010; Fraile et al., 2011; Thrall et al., 2012), many qualitative resistances lose their effectiveness only a few years after commercial release (Kiyosawa, 1982; Fry, 2008), particularly when they are used in large-scale monocultures. Compared with this qualitative resistance, quantitative resistance is less effective, reducing disease epidemics rather than preventing infection, but more durable due to the lower selection pressure it places on pathogens. On the other hand, induced resistance is thought to have an advantage over constitutive resistance primarily due to lower resource allocation when it is not needed (Anderson and O'Toole, 2008; Wei et al., 2015).

In addition to the genetics of host resistance, other elements termed the "ten principles of agricultural practices" (soil, nutrition, water, seed, population density, plant protection, field management, farming machine technology, light and air, Fig. 4) can also affect the resistance level of host plants (Savary et al., 2011; Szechyńska-Hebda et al., 2015). Changing any of these elements may modify the environment in a way that is either favorable or adverse to plant or pathogen (Table 2). Though managing plant diseases through a whole farming system approach (Plantegenest et al., 2007; Yuen and Mila, 2015) may still allow some disease development, may need more labor and other inputs particularly on establishment, and may need supplementary support from other strategies such as the application of pesticides, it has been used successfully to control rice blast (*Magnaporthe oryzae*) and tungro (*Rice tungro virus*) disease on a large scale (Xie et al., 1983; Xie and Lin, 1988; Xie, 2003).

Table 2 The RAER function of each plant disease management approach

Approach	Resistance(R)[1)]	Avoidance(A)[2)]	Elimination(E)[3)]	Remediation(R)[4)]
Single R gene	√			
R gene mixture	√	√		
R gene rotation	√	√		
R gene pyramid	√	√		
R gene regional deployment	√	√		
R induced	√			√
Tolerance	√	√		
Quarantine			√	
Isolation		√	√	
Hygiene	√	√	√	
Seed cleaning			√	
Cropping system adjustment	√	√	√	
Rotation	√	√	√	
Interplant	√	√	√	
Planting time adjustment		√	√	
Species diversity	√	√	√	
Field landscape	√	√	√	
Forecasting		√	√	
Pesticide		√		√
Bio-control agent				√
Physical treatment		√	√	√

1) Resistant to the infection of pathogen for host plant.

2) Avoid peak and key stage of pathogen reproduction.

3) Eliminate infectious sources.

4) Kill pathogen and/or vector.

Increasing host heterogeneity through intercropping or mixing crop varieties with different genetic and physiological properties such as type of resistance (quantitative versus quantitative) has been proved to be one of the most effective ecological approaches to manage plant diseases. This approach not only reduces disease epidemics, increases nutrition efficiency, productivity and yield stability in short term but also improves soil fertility and slows down pathogen evolution (Burdon and Thrall, 2008; Parnell et al., 2010; Pérez-Reche et al., 2010; Brooker et al., 2015; Papaïx et al.,2015; Tack et al., 2015), thereby extend the lifespan of resistant varieties (Table 2). For example, increasing host population heterogeneity by intercropping different rice varieties greatly reduced dependence on fungicide application to manage rice blast, while simultaneously increasing the quality and quantity of production significantly (Zhu et al., 2000, 2003). Varietal mixture has also been used successfully to control potato late and early blights (data not shown). In addition to mixture or intercropping technologies, other approaches in resistance gene deployment, such as R gene rotation and pyramiding, can also be used to assist in the ecological management of plant diseases (Coutts et al., 2011).

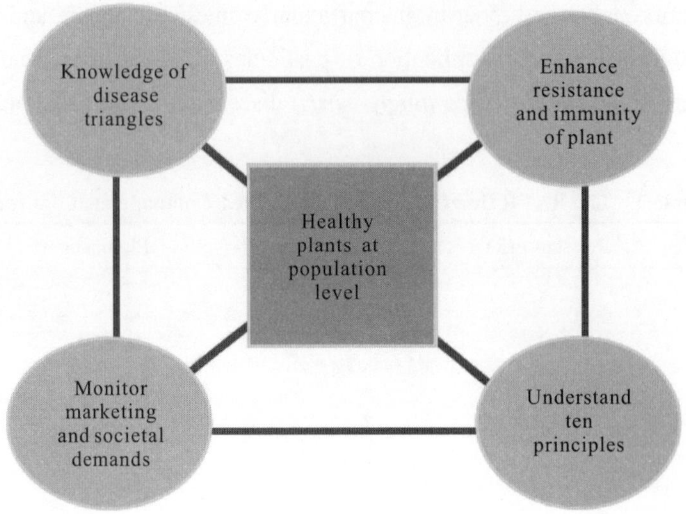

Fig. 4 Solution to improve healthy condition of plant population in agro-ecological system

Sustainable plant disease management that aims at improving plant health at population level can be achieved through a comprehensive understanding the mechanisms of plant disease epidemics, and multi-functions of agro-ecosystems, and interactions among biotic and abiotic conditions, marketing and societal demands. "Knowledge of disease triangles" means plant-pathogen-environment interaction and plant population-biotic factors-abiotic factors interaction. "Ten principles" mean soil, nutrition, water, seed, population density, plant protection, field management, farming machine technique, light and air

4.3.2 Disease avoidance

This approach aims to ensure a mismatch between critical periods of crop and pathogen development by changing the cultivation pattern of host plants spatially and temporally such as through variation in planting time, planting location or cultivation system (Table 2). It is a complex approach that requires a comprehensive understanding of host susceptibility through different phenological development stages, likely meteorological conditions, disease ecology, and pathogen and pathotype distributions. The efficiency of spatial avoidance including regional R gene deployment and varietal mixture is mainly determined by pathogen distribution and transmission modes. Spatial avoidance may be effective to manage plant disease caused by soil- or water-borne pathogens such as many diseases caused by nematodes and bacteria, but this approach is unlikely to be useful for the management of air-borne diseases which can be spread over a significant distance over a single course of epidemics.

Temporary avoidance approaches include changing crop planting times and crop rotation. The effectiveness of changing planting times to control plant diseases is heavily dependent on within-season climatic conditions particularly for polycyclic pathogens in which primary inoculum does not play a determinant role in disease epidemics. However, in rice virus disease, this approach shortens the exposure of plants to the pathogen during their most sensitive period by avoiding the peak stage of vectors transmission (Xie et al., 1983, 1984, 2001; Xie and Lin, 1988; Tiongco et al., 1990). A second class of temporary avoidance, i.e., crop rotation, is expected to be particularly effective in controlling plant diseases caused by soil-borne pathogens. Indeed, rotation has been shown to be very effective to control bacterial wilt of potato, banana, tobacco (Ong et al., 2007; Peeters et al.,2013) and black and root rot of sweet potato (Huang et al.,2014). For pathogens transmitted by insect vector such as many viruses, the key for disease avoidance is to understand the ecology-overwintering site, migration patterns, wind direction - as well as the reproductive biology of the insects (Acosta-Leal et al., 2011).

4.3.3 Elimination

Some methods (Table 2) that eliminate overwintering sites (places and hosts) of pathogens and their transmission vectors generate remarkable plant disease management outcomes with no or minimum adverse ecological impacts by eradicating or reducing inoculum sources. The key obstacle to using an elimination strategy to managing plant diseases is to identify the correct sources of primary inoculum. Misidentification of primary inoculum sources not only reduce management efficiency but also result in resource waste. If a disease exhibits epidemics continuously over many years despite heavy human intervention, it is necessary to re-check whether the crucial points of the disease cycle have been misidentified, to determine whether eradication at those points is actually feasible and to general rethink the strategies for management. A large number of agricultural practices have proved to be very effective in eliminating or reducing sources of pathogen inoculum by adapting farming systems to remove diseased plant tissues, volunteer host plants and secondary crops, etc. Foremost among these practices, crop rotation is a convenient method of disease elimination that can not only eliminate the pathogen (especially some of those that are soil-borne) and potential reservoir hosts, but may also improve soil quality (nutrition balance and physical structure) supporting healthier crop populations. The practice of ploughing soils after harvesting dramatically reduces the population density of the insect vector, *Nephotettixvirescens*, and therefore, the viral source of rice tungro disease (Hirao and Ho, 1987) by reducing the vector's overwintering sites. Similar to disease resistance and avoidance, a disease-elimination strategy should also be built on a proper understanding of the various interactions occurring among hosts, pathogens, vectors in an ecological and epidemiological context as well as with due consideration of the economic threshold of management (Fig. 2–4).

A successful case of applying an elimination strategy to control plant disease comes from wheat stem rust in China. Several major epidemics of the disease occurred in spring wheat in Northeast China and winter wheat in Fujian Province, southern China between 1948 and 1965 despite the wide use of major resistant varieties and chemical pesticides. Investigation found that *Pucciniagraminis* var. tritici, the causal agent of wheat stem rust, overwintered on cultivated winter wheat sown in August in Putian County, Fujian Province. Elimination of these *P. graministritici* overwintering sites by persuading Putian farmers to change their cropping systems from growing winter wheat to potatoes and broad beans has been marked by the occurrence of no major epidemics of wheat stem rust since then. Indeed, the disease almost disappeared in the China after the 1990s (Xie, 2003). Another successful example of application of an elimination strategy to control plant disease comes from rice stripe disease. The disease has been rampant in Jiangsu Province, the main rice production area of China for nearly10 years (2001–2010). Due to a lack of resistant varieties, the disease was mainly controlled by insecticide application to kill the insect vector, *Laodelphaxstriatellus*. However, after 2008, the strategy of managing Rice stripe virus was switched from sole insecticide application to the combined use of insecticides with primary inoculum source elimination achieved by abandoning a local common practice of wheat-rice rotations (this removed the overwintering sites of the vector). The disease was eventually brought under full control in the past few years.

4.3.4 Remedy

Spraying pesticides to kill pathogens and/or their insect vectors is an inseparable part of plant disease management when other approaches cannot achieve the required level of pathogen population density reduction and epidemic amelioration. However, the use of pesticides in an integrated disease management system is not to eradicate the disease completely but to control it to the most appropriate extent as guided by ecological and economical thresholds. During pesticide application, factors such as action modes and pathogen resistance should be considered (Siegwart et al., 2015). To increase their efficiency of application and reduce negative impacts on the environment, pesticides should be used in combination with disease forecasts and knowledge of the pathogen population genetic structure (Zhan et al., 2015) to determine the best time and frequency of application and to choose the type and utilization dosage of the pesticides (van den Berg et al., 2013).

Remedy successes could also be achieved by other approaches (Table 2) than synthetic fungicides (Xie, 2003), such as naturally occurring plant compounds with biological control activity - for example protein y3 that is extracted from edible fungi and other microbes (*Bacillus spp.*) (Wu et al.,2003; Luna et al., 2011; Chen et al., 2013; Kumar et al., 2014). To ensure effective use of such bio-pesticides a better understanding of their properties and application procedure is important as is information about relevant biological features and the transmission mode of pathogens. For example, adding viral therapeutic agents or biological control agents in 1–2 sprays at the rice seedling and turning green stage can not only reduce viruliferous insect population density, but also protect the plant from further infection (Xie et al.,1979). Combining pesticides with other biotic and abiotic approaches such as biological agents, soil pH adjustment and UV irradiation has proved to be very effective in long-term control of tomato and lettuce root rot (Tu, 2002; Lee, 2015).

5 The future of plant disease management

Sustainable plant disease management (Fig.3 and 4) requires a multi-dimensional consideration of the impacts of management approaches on economics, sociology and ecology by fully understanding the mechanisms of plant disease epidemics, the functioning of healthy agro-ecosystems and individual and collective roles of RAER approaches on disease management. This model of plant disease management seeks not only to increase agricultural productivity and improve food quality but also to protect the ecological environment and natural resources. To achieve this goal, future research in ecological plant disease management should focus on: (i) epidemic and evolutionary patterns of plant disease under changing environments and agricultural production philosophies; (ii) the role of ecological considerations in agricultural productivity and crop health; (iii) social-economic analysis of plant disease epidemics and management; and (iv) technology development for integrating management of major crop diseases with ecological principles.

Acknowledgements

This work was supported by the Fujian Technology Plan Project, China (2012N4001), the National Natural Science Foundation of China (U1405213) and the Ministry of Science and Technology of National 973 Program of China (2014CB160315).

References

[1] ACOSTA-LEAL R, DUFFY S, XIONG Z, et al. Advances in Plant Virus Evolution: Translating Evolutionary Insights into Better Disease Management[J]. Phytopathology, 2011, 101: 1136-1148.

[2] ANDERSON G G, O'TOOLE G A. Innate and induced resistance mechanisms of bacterial biofilms[J]. Current topics in microbiology and immunology, 2008, ed:85-105.

[3] BALL B C, BINGHAM I, REES R M, et al. The role of crop rotations in determining soil structure and crop growth conditions[J]. Canadian Journal of Soil Science, 2005, 85: 557-577.

[4] BANCAL M O, ROCHE R, BANCAL P. Late foliar diseases in wheat crops decrease nitrogen yield through N uptake rather than through variations in N remobilization[J]. Annals of Botany, 2008, 102: 579-590.

[5] BONILLA N, CAZORLA F M, MARTINEZ-ALONSO M, et al. Organic amendments and land management affect bacterial

community composition, diversity and biomass in avocado crop soils[J]. Plant Soil, 2012, 357: 215-226.

[6] BROOKER R W, BENNETT A E, CONG W F, et al. Improving intercropping: a synthesis of research in agronomy, plant physiology and ecology[J]. New Phytologist, 2015, 206: 107-117.

[7] BURDON J J, THRALL P H. Pathogen evolution across the agro-ecological interface: implications for disease management[J]. Evolutionary Applications, 2010, 1: 57-65.

[8] BURDON J J, THRALL P H. Coevolution of Plants and Their Pathogens in Natural Habitats[J]. Science, 2009, 324: 755-756.

[9] BURDON J J, BARRETT L G, REBETZKE G, et al. Guiding deployment of resistance in cereals using evolutionary principles[J]. Evolutionary Applications, 2014, 7: 609-624.

[10] CHEN Y, YAN F, CHAI Y, et al. Biocontrol of tomato wilt disease by *Bacillus subtilis* isolates from natural environments depends on conserved genes mediating biofilm formation[J]. Environmental Microbiology, 2013, 15: 848-864.

[11] COUTTS B A, KEHOE M A, JONES R AC. Minimising losses caused by *Zucchini yellow mosaic virus* in vegetable cucurbit crops in tropical, sub-tropical and Mediterranean environments through cultural methods and host resistance[J]. Virus Research, 2011, 159: 141-160.

[12] ENSERINK M, HINES P J, VIGNIERI S N, et al. The Pesticide Paradox[J]. Science, 2013, 341: 729-729.

[13] FRAILE A, PAGAN I, ANASTASIO G, et al. Rapid Genetic Diversification and High Fitness Penalties Associated with Pathogenicity Evolution in a Plant Virus[J]. Molecular Biology & Evolution, 2011, 28: 1425-1437.

[14] FRY W. *Phytophthora infestans*: the plant (and R gene) destroyer[J]. Molecular Plant Pathology, 2008, 9: 385-402.

[15] GODFRAY H CJ, BEDDINGTON J R, CRUTE I R, et al. Food Security: The Challenge of Feeding 9 Billion People[J]. Science, 2010, 327: 812-818.

[16] GONTHIER D J, ENNIS K K, FARINAS S, et al. Biodiversity conservation in agriculture requires a multi-scale approach[J]. Proceedings Biological Sciences, 2014, 281.

[17] GUEDES R NC., SMAGGHE G, STARK J D, et al. Pesticide-Induced Stress in Arthropod Pests for Optimized Integrated Pest Management Programs[J]. Annual Review of Entomology, 2015: 43-62.

[18] HALL C, WELCH J, KOWBEL D J, et al. Evolution and Diversity of a Fungal Self/Nonself Recognition Locus[J]. PLoS One, 2010.

[19] HIRAO J, HO K. Status of rice pests and their control measures in the double cropping area of the Muda irrigation scheme Malaysia[J]. Tropical Agriculture Research Series, 1987, 20: 107-115.

[20] HUANG L F, LUO Z X, FANG B P, et al. Advances in the researches on bacterial stem and root rot of sweet potato caused by *Dickeya dadantii*[J]. Acta Phytophylacica Sinica, 2014, 41: 18-122. (in Chinese)

[21] IRANZO J, LOBKOVSKY A E, WOLF Y I, et al. Immunity, suicide or both? Ecological determinants for the combined evolution of anti-pathogen defense systems[J]. BMC Evolutionary Biology, 2015, 15: 324.

[22] JANVIER C, VILLENEUVE F, ALABOUVETTE C, et al. Soil health through soil disease suppression: Which strategy from descriptors to indicators?[J]. Soil Biology & Biochemistry, 2007, 39: 1-23.

[23] KIYOSAWA S. Genetic and epidemiological modeling of breakdown of plant disease resistance[J]. Annual Review Phytopathol, 1982, 20: 93-117.

[24] KOSALEC I, CVEK J, TOMIC S. Contaminants of medicinal herbs and herbal products[J]. Archives of Industrial Hygiene & Toxicology, 2009, 60: 485-501.

[25] KUMAR G P, AHMED S KM H, DESAI S D, et al. In vitro screening for abiotic stress tolerance in potent biocontrol and plant growth promoting strains of pseudomonas and Bacillus spp[J]. International Journal of Bacteriology, 2014, 6: 1-6.

[26] LARKIN R P. Soil Health Paradigms and Implications for Disease Management[J]. Annual Review of Phytopathology, 2015:199-221.

[27] LEE S, GE C, BOHREROVA Z, et al. Enhancing plant productivity while suppressing biofilm growth in a windowfarm system

using beneficial bacteria and ultraviolet irradiation[J]. Canadian Journal of Microbiology, 2015, 61: 457-466.

[28] LU Y, SONG S, WANG R, et al. T Impacts of soil and water pollution on food safety and health risks in China[J]. Environment International, 2015, 77: 5-15.

[29] LUCAS J A. Advances in plant disease and pest management[J]. Journal of Agricultural Science, 2011, 149: 91-114.

[30] LUNA E, PASTOR V, ROBERT J, et al. Callose Deposition: A Multifaceted Plant Defense Response[J]. Molecular Plant Microbe In, 2011, 24: 183-193.

[31] MAGDOFF F. Concept, components, and strategies of soil health in agroecosystems[J]. Journal of Nematology, 2001, 33: 169-172.

[32] MANOSALVA P, MANOHAR M, VON REUSS S H,et al. Conserved nematode signalling molecules elicit plant defenses and pathogen resistance[J]. Nature Communications, 2015, 6(7).

[33] MILLER S A, BEED F D, HARMON C L. Plant Disease Diagnostic Capabilities and Networks[J]. Annual Review of Phytopathology, 2009:15-38.

[34] OERKE E C. Crop losses to pests[J]. Journal of Agricultural Science, 2006, 144: 31-43.

[35] OERKE E C, DEHNE H W. Safeguarding production-losses in major crops and the role of crop protection[J]. Crop Protection, 2004: 23: 275-285.

[36] ONG K L, FORTNUM B A, KLUEPFEL D A, et al. Winter cover crops reduce bacterial wilt of flue-cured tobacco[J]. Plant Health Progress, 2007, doi:10.1094/PHP-2007-0522-01-RS

[37] PADMANABHAN S Y. The great Bengal famine[J]. Annual Review Phytopathol, 1973, 11: 11-24.

[38] PARNELL S, GOTTWALD T R, VAN DEN BOSCH F, et al. Optimal strategies for the eradication of Asiatic citrus canker in heterogeneous host landscapes[J]. Phytopathology, 2009, 99:1370-1376.

[39] PAGE K, DANG Y, DALAL R. Impacts of conservation tillage on soil quality, including soil-borne crop diseases, with a focus on semi-arid grain cropping systems[J]. Australasian Plant Pathology, 2013, 42: 363-377.

[40] PALMGREN M G, EDENBRANDT A K, VEDEL S E, et al. Are we ready for back-to-nature crop breeding?[J]. Trends in Plant Science, 2015, 20: 155-164.

[41] PAPAIX J, BURDON J J, ZHAN J, et al. Crop pathogen emergence and evolution in agro-ecological landscapes[J]. Evolutionary Applications, 2015, 8: 385-402.

[42] PARNELL S, GOTTWALD T R, VAN DEN BOSCH F, et al. Optimal Strategies for the Eradication of Asiatic Citrus Canker in Heterogeneous Host Landscapes[J]. Phytopathology, 2009, 99: 1370-1376.

[43] PEETERS N, GUIDOT A, VAILLEAU F, et al. Ralstonia solanacearum, a widespread bacterial plant pathogen in the post-genomic era[J]. Molecular Plant Pathology, 2013, 14: 651-662.

[44] PEREZ-RECHE F J, TARASKIN S N, COSTA L DF, et al. Complexity and anisotropy in host morphology make populations less susceptible to epidemic outbreaks[J]. Journal of the Royal Society Interface, 2010, 7: 1083-1092.

[45] PLANTEGENEST M, LE MAY C, FABRE F. Landscape epidemiology of plant diseases[J]. Journal of the Royal Society Interface, 2007, 4: 963-972.

[46] POPP J, PETO K, NAGY J. Pesticide productivity and food security[J]. Agronomy for Sustainable Development, 2013, 33: 243-255.

[47] PYWELL R F, HEARD M S, WOODCOCK B A, et al. Wildlife-friendly farming increases crop yield: evidence for ecological intensification[J]. Proceedings of the Royal Society Biological Sciences, 2015: 282.

[48] RAY D K, MUELLER N D, WEST P C, et al. Yield Trends Are Insufficient to Double Global Crop Production by 2050[J]. PLOS ONE, 2013, 8(6):e66428.

[49] SAVARY S, MILA A, WILLOCQUET L, et al. Risk Factors for Crop Health Under Global Change and Agricultural Shifts: A Framework of Analyses Using Rice in Tropical and Subtropical Asia as a Model[J]. Phytopathology, 2011, 101: 696-709.

[50] SIEGWART M, GRAILLOT B, LOPEZ C B, et al. Resistance to bio-insecticides or how to enhance their sustainability: a review[J].

Frontiers in Plant Science, 2015, 6:381.

[51] SORNMERHALDER R J, MCDONALD B A, MASCHER F, et al. Sexual Recombinants Make a Significant Contribution to Epidemics Caused by the Wheat Pathogen Phaeosphaeria nodorum[J]. Phytopathology, 2010, 100: 855-862.

[52] STRANGE R N, SCOTT P R. Plant disease: A threat to global food security[J]. Annual Review of Phytopathology, 2005, 83-116.

[53] SZECHYNSKA-HEBDA M, WASEK I, GOLEBIOWSKA-PIKANIA G, et al. Photosynthesis-dependent physiological and genetic crosstalk between cold acclimation and cold-induced resistance to fungal pathogens in triticale (Triticosecale Wittm)[J]. Journal of Plant Physiology, 2015, 177: 30-43.

[54] TACK A JM., LAINE A L, BURDON J J, et al. Below-ground abiotic and biotic heterogeneity shapes above-ground infection outcomes and spatial divergence in a host-parasite interaction[J]. New Phytologist, 2015, 207: 1159-1169.

[55] THRALL P H, LAINE A L, RAVENSDALE M, et al. Rapid genetic change underpins antagonistic coevolution in a natural host-pathogen metapopulation[J]. Ecology Letters, 2012, 15: 425-435.

[56] TRIPATHY V, BASAK B B, VARGHESE T S, et al. Residues and contaminants in medicinal herbs-A review[J]. Phytochemistry Letters, 2015, 14: 67-78.

[57] TU J C. An integrated control of Pythium root rot of greenhouse tomato. Mededelingen Faculteit Landbouwkundige en Toegepaste Biologische Wetenschappen Universiteit Gent, 2002,67: 209-216.

[58] VAN BRUGGEN A HC, GAMLIEL A, FINCKH M R. Plant disease management in organic farming systems[J]. Pest Manage Science, 2016, 72: 30-44.

[59] VAN DEN BERG F, VAN DEN BOSCH F, PAVELEY N D. Optimal Fungicide Application Timings for Disease Control Are Also an Effective Anti-Resistance Strategy: A Case Study for *Zymoseptoria tritici* (*Mycosphaerella graminicola*) on Wheat[J]. Phytopathology, 2013, 103: 1209-1219.

[60] WEI T, WANG L, ZHOU X, et al. PopW activates PAMP-triggered immunity in controlling tomato bacterial spot disease[J]. Biochemical and Biophysical Research Communications, 2015, 463: 746-750.

[61] WELBAUM G E, STURZ A V, DONG Z M, et al. Managing soil microorganisms to improve productivity of agro-ecosystems[J]. Critical Reviews in Plant Sciences,2004, 23: 175-193.

[62] WU L, WU Z, LIN Q, et al. Purification and activities of an alkaline protein from mushroom Coprinus comatus[J]. Acta Microbiologica Sinica, 2003, 43: 793-798.

[63] XIE L H, CHEN Z X, LIN Q Y. The chemotherapy of rice virus disease. The Collected Papers of Virology, 1979:44-48. (in Chinese)

[64] XIE L H, LIN Q Y. Progress in the research of virus diseases of rice in China[J]. Entia Agricultura Sinica, 1984, 17: 58-65. (in Chinese)

[65] XIE L H, LIN Q Y. The occurence and control of rice virus disease in China[J]. In: The Progress of Integrated Control of Rice Disease and Insect in China. Chinese Plant Protection Station Science and Technology Press of Zhejiang Province Hangzhou. 1988, pp: 255-264. (in Chinese)

[66] XIE L H, LIN Q Y, WU Z J, et al. Diagnosis, detection and control of rice virus disease in China[J]. Journal of Fujian Agriculture University, 1994, 23: 280-285. (in Chinese)

[67] XIE L H, LIN Q Y, XIE L M, et al. The occurrence and control of rice bunchy disease[J]. Acta Phytopathologica Sinica, 1984, 14: 33-38. (in Chinese)

[68] XIE L H, LIN Q Y, XU X R. Plant Disease: Economics Pathology and Molecular Biology[J]. Science Press Beijing. 2009, pp:157-159. (in Chinese)

[69] XIE L H, LIN Q Y, ZHU Q L, et al. The occurrence and control of rice tungro disease in Fujian Province[J]. Journal of Fujian Agriculture College, 1983, 12: 275-284. (in Chinese)

[70] XIE L H, WEI T Y, LIN H X, et al. Advances of molecular biology of Rice stripe virus[J]. Journal of Fujian Agriculture and Forestry University, 2001, 30: 269-279. (in Chinese)

[71]XIE L H. Plant protection strategy of China in the 21 century[J]. Review of China Agricultural Science and Technology, 2003, 5: 5-7. (in Chinese)

[72] YUEN J, MILA A. Landscape-Scale Disease Risk Quantification and Prediction[J]. In Annual Review Of Phytopathology Vol 53, NK VanAlfen, 2015, ed:471-484.

[73] ZHAN J, MCDONALD B A. Experimental Measures of Pathogen Competition and Relative Fitness[J]. In Annual Review of Phytopathology Vol 51, NK VanAlfen, 2013, ed:131-153.

[74] ZHAN J, THRALL P H, BURDON J J. Achieving sustainable plant disease management through evolutionary principles[J]. Trends Plant Science, 2014, 19: 570-575.

[75] ZHAN J, MUNDT C C, HOFFER M E, et al. Local adaptation and effect of host genotype on the rate of pathogen evolution: an experimental test in a plant pathosystem[J]. Journal of Evolutionary Biology, 2010, 15: 634-647.

[76] ZHAN J, THRALL P H, BURDON J J. Achieving sustainable plant disease management through evolutionary principles. Trends Plant Science, 2014, 19:570-575.

[77] ZHAN J, THRALL P H, PAPAIX J, et al. Playing on a Pathogen' s Weakness: Using Evolution to Guide Sustainable Plant Disease Control Strategies[J]. Annual Review of Phytopathology, 2015, 53(1):19-43.

[78] ZHU Y Y, CHEN H R, FAN J H, et al. Genetic diversity and disease control in rice[J]. Nature, 2000, 406: 718-722.

[79] ZHU Y Y, CHEN H R, FAN J H, et al. The use of rice varietal diversity for rice blast control[J]. Scientia Agricultural Sinica, 2003, 36:521-527. (in Chinese)

Ⅳ 寄主抗性与天然产物

这部分论文围绕“抗、避、除、治”的植物病毒防控四字原则中的抗性和治疗这两个方面内容，深入研究植物自身的抗性基因对于病毒病的防控，同时探索利用自然界中可诱导植物抗性及对植物病毒侵染具有不同程度抑制作用的天然产物，这种抗病毒方式有较广阔的应用前景。在此基础上，通过对不同作物品种进行抗性评价，对兼具高产量、高品质和高抗性的优良品种筛选进行了初步探索。

金鸡菊（*Coreopsis drummondii*）的抗 TMV 活性物质

陈启建，欧阳明安，吴祖建，谢联辉，林奇英

（福建农林大学植物病毒研究所　福州　350002）

摘要：采用活性跟踪法从金鸡菊根中分离获得抗病毒活性物质，经质谱和核磁共振分析，鉴定该物质为 1- 苯基 -1,3,5- 三庚炔。采用半叶枯斑法、叶圆盘法测定了该物质对烟草花叶病毒的抑制效果，结果表明，0.2mg/mL 的该化合物对 TMV 表现出较好的体外抑制侵染和增殖活性，其对 TMV 侵染和复制的抑制率分别为 73.5% 和 84.3%。实时荧光定量 PCR 测定结果表明，该化合物对 TMV 外壳蛋白基因的表达有明显的抑制作用，0.2mg/mL 的该化合物对 TMV 外壳蛋白基因表达的抑制率为 79.8%。

关键词：金鸡菊；抗 TMV 活性；1- 苯基 -1,3,5- 三庚炔；活性物质分离

Anti-TMV Active Substances from *Coreopsis drummondii*

Qijian Chen, Ming'an Ouyang, Zujian Wu, Lianhui Xie, Qiying Lin

(Institute of Plant Virology, Fujian Agriculture and Forestry University, Fuzhou 350002, China)

Abstract: Production of plant secondary metabolites as a rich source of antimicrobial agents has been extensively applied in Chinese traditional medicine to treat viral diseases. Therefore, direct selection of antiphytoviral compounds from plants can be used to identify new potent antiviral agents. To screen antiviral compounds, tobacco mosaic virus (TMV) was chosen as a model target and the extracts with anti-TMV activity from 108 species of plants were analyzed. Among them, an extract from *Coreopsis drummondii*, predominantly suppressing the infection and replication of TMV was selected, and a constituent with anti-virus activity was isolated from it by bioassay. The compound was identified as 1-phenyl-1,3,5-triheptalkyne using MS and NMR. Its activity against TMV was investigated by local lesion and leaf discs assay. The results indicated that the compound showed a significant inhibitory activity against TMV *in vitro*, with 73.5% inhibitory rate against the infection of TMV and 84.3% against replication of TMV at a concentration of 0.2mg/mL, while its inhibitory effect on TMV-CP gene expression in tobacco leaf discs was examined by real-time quantitative PCR, which showed that the compound obviously suppressed the expression of TMV-CP gene, with 79.8% inhibitory rate at the concentration of 0.2mg/mL. Taken together, the anti-TMV activity of the compound contributes itself as a new potential candidate for development of antiviral drugs. Fig. 6, Tab 1, Ref 19.

Keywords: *Coreopsis drummondii*; anti-TMV activity; 1-phenyl-1,3,5-triheptalkyne; isolation of active substance

烟草花叶病毒（*Tobacco mosaic virus*, TMV）引起的病毒病不但危害严重，其防治也相当困难。

应用与环境生物学报 . 2009, 15(5): 621-625
收稿日期：2008-08-19
基金项目：国家自然科学基金项目（No. 30770089）和福建省教育厅项目（No.2006F5017）资助

TMV 具有寄主范围广、抗逆性强、在生产上造成的危害大等特点。TMV 的寄主范围很广，可侵染十字花科、茄科、菊科、藜科及苋科等多种植物，如烟草、番茄、茄子、辣椒、菠菜等；TMV 的危害不仅造成农作物产量的损失，而且还使农产品的品质大大降低，给农业生产造成了极大损失，据估计全世界每年因 TMV 而造成的经济损失就超过 1 亿美元（Wu et al., 1995）。由于植物病毒的绝对寄生，对其防治目前尚无理想的药剂（Deng et al., 2004）。已有研究表明一些植物的次生代谢产物对病毒有良好的抑制效果（Ouyang et al., 2007; Wu et al., 2007; Shen et al., 2008; Yao et al., 2001; Chen et al., 2008），因此，从植物中筛选抗病毒活性物质进而开发新型抗病毒剂具有广阔的前景。

我们在对 108 种植物浸提物抗病毒活性筛选中发现，金鸡菊（*Coreopsis drummondii*）浸提物对 TMV 具有较好的抑制作用。为进一步探究该植物中的抗病毒活性物质及其抗病毒活性，采用活性跟踪法对其中的抗病毒物质进行分离纯化，并以烟草 /TMV 体系对分离获得的单体物质进行了抗病毒活性测定，现将研究结果报道如下。

1　材料与方法

1.1　材料

供试植物：新鲜金鸡菊样品采自福建省福州金山，经鉴定后用自来水洗净其外表面，晾干后保存备用。

供试病毒：烟草花叶病毒普通株系保存于福建农林大学植物病毒研究所，其抗血清是由提纯的 TMV 经免疫家兔获得的，效价为 1: 20480。

供试烟草：TMV 枯斑寄主为心叶烟（*Nicotiana glutinosa*），系统侵染寄主为普通烟草（*N. tabacum*）K_{326} 品种。

仪器与试剂：Bruker AM-400 核磁共振仪，Agilent 6890 N/5973i 型气质联用仪，RI2000 型高效液相色谱仪，Biometra T3 Thermocycler PCR 仪，层析用 200 ～ 300 目硅胶（青岛海洋化工厂产品），薄层层析用 GF254 层析板（青岛海洋化工厂产品），石油醚（60℃～ 90℃）和丙酮为分析纯，95% 乙醇为化学纯，总 RNA 抽提试剂盒和荧光定量 PCR 试剂盒（SYBR Premix Ex Taq TM ）购自宝生物（大连）工程有限公司。

1.2　方法

1.2.1　抗病毒物质的提取与分离

提取：晾干后的金鸡菊根部于烘干箱中 45℃烘干后，经粉碎机粉碎并过 40 目筛。称取 2.8kg 的干粉，用 6 倍体积的 95% 的乙醇浸泡 4 次，每次 3d，过滤浸液，滤液减压浓缩，合并每次浓缩所得浸膏，即为乙醇提取物（227g）。

分离：样品的乙醇浸膏依次用石油醚、氯仿、正丁醇进行萃取。萃取时每次加入的溶剂体积大约是水体积的一半，直至加入的溶剂接近无色后更换另一种溶剂进行萃取。采用正丁醇萃取时，先于正丁醇中加入蒸馏水使之处于水饱和状态后再进行萃取。分别对所得的萃取物进行抗 TMV 活性测定，选择其中抗病毒活性最好的石油醚萃取物，采用柱层析对其进一步分离。将 200 ～ 300 目层析用硅胶置烘干箱中，于 110℃下烘干活化 1h，冷却后，用石油醚装柱。取石油醚萃取物 35g 用石油醚溶解后，加入一定量的硅胶混匀，待溶剂挥干后上柱。依次以石油醚、氯仿、丙酮为洗脱剂进行洗脱，流出液每 250mL 收集一份，分别对各收集组分进行薄层层析分析，合并含相同组分的部分，并测定其抗病毒活性。发现由第 20、21、22 部分合并成的流出物抗病毒活性最好。取该流出物 3.8g 再次进行硅胶层析分离，以不同配比的石油醚 — 丙酮为洗脱剂进行梯度洗脱，分别获得 1 个黄色和 1 个白色混晶 R1 和 R2。分别采用高效液相色谱对 R1 和 R2 进一步分离纯化，以石油醚 — 丙酮（9 : 1）为洗脱剂，流速为 4mL/min 进行分离，从 R1 获得化合物 1 和化合物 2；以氯仿 — 丙酮（9：1）为洗脱剂，流速为 4mL/min 进行分离，从 R2 中获得化合物 3、化合物 4 和化合物 5。

1.2.2　提取物结构鉴定

采用 Agilent 6890 N/5973i 气质联用仪进行化合物质谱分析。色谱柱为 HP-5 MS（30m×0.25mm×0.25μm）；载气为氦气，纯度 > 99.99%，流速 1mL /min；进样口温度为 280℃，分流比为 50：1；进样量为 1μL；炉温：初始 80℃，以 20℃ / min 上到 280℃；溶剂延迟 3min。质谱条件：电离方式为 EI，轰击能量 70 eV，离子源温度 250℃；传输线温度为 280℃；质谱扫描范围为 50 ～ 500amu。

核磁共振测定采用 Bruker AM-400 核磁共振仪。

1.2.3　体外抗 TMV 活性测定

取分离获得的单体 10mg 充分溶解于 1mL 的二甲亚砜（DMSO）中，配制成 10mg/mL 的母液备用。活性测定时取一定量的母液加蒸馏水配置成系列浓度，再分别加入提纯的 TMV，使病毒终浓度为 0.01mg/mL。混合 10min 后，采用半叶法摩擦接种枯斑寄主心叶烟，以含相同浓度的二甲亚砜和 TMV 的蒸馏水溶液为对照，每处理接种 5 个半叶，重复 2 次，待接种叶出现明显的枯斑后，记录枯斑数目，计算抑制率。

抑制率（%）= [（对照组平均枯斑数－处理组平均枯斑数）/ 对照组平均枯斑数]×100。

1.2.4　抑制 TMV 增殖作用的测定

采用叶盘法测定提取物对 TMV 增殖的抑制作用。选取生长良好的普通烟 K_{326}，叶片接种 TMV 6h 后，用打孔器从接种叶打取直径约为 15mm 的叶圆盘，将叶圆盘分别飘浮于用母液配制成不同浓度的提取物水溶液中，并分别以飘浮于蒸馏水中接种病毒的叶圆盘和不接种病毒的健康烟草叶圆盘为阳性对照和阴性对照，每处理 6 片叶圆盘。处理 48h 后分别将各叶圆盘用 1 : 20（w/V）的碳酸盐缓冲液包被，间接 ELISA 法检测其 D_{405nm} 值，试验重复 2 次，取其平均值。

1.2.5　实时荧光定量 PCR 测定对 TMV 外壳蛋白基因表达的影响

引物设计与合成：病毒外壳蛋白基因（*CP*）及管家基因（*Actin*）的引物见表 1，由宝生物（大连）工程有限公司设计并合成。

表 1　实时定量 PCR 引物

Table 1　Primers for real-time PCR

Gene	Primer sequence	Product size(bp)
CP	F: 5′-ATTAGACCCGCTAGTCACAGCAC-3′	84
	R: 5′-GTGGGGTTCGCCTGATTTT-3′	
Actin	F: 5′-CAAGGAAATCACCGCTTTGG-3′	106
	R: 5′-AAGGGATGCGAGGATGGA-3′	

RNA 的提取：烟草叶片中总 RNA 的提取采用 Trizol 法提取，提取的 RNA 经紫外分光光度计测定 $A_{260\ nm}$ /$A_{280\ nm}$ 值进行定量和纯度分析。

cDNA 合成：以总 RNA 为模板，分别用 TMV 外壳蛋白基因及管家基因的反向引物进行逆转录合成上述 2 个基因的 cDNA。cDNA 合成采用 20μL 逆转录反应体系：总 RNA 4μL，3′ 引物（10^{-6}mol/L）1μL，DEPC H_2O 7.5μL，5×buffer 4μL，RNA 抑制剂（40U/μL）0.5μL，dNTP（10^{-6}mol/L）2μL，AMV 逆转录酶（5U/μL）1μL，逆转录反应条件为 42℃ 60min，70℃ 15min，反应结束后加入 80μL 灭菌超纯水，混匀，–40℃低温保存备用。

PCR 反应体系与程序：实时荧光定量 PCR 反应体系：SYBR Premix Ex Taq TM 12.5μL，引物 F、R（10^{-6}mol/L）各 0.5μL，cDNA 2μL，PCR 级水 9.5μL，总反应体系为 25μL。反应程序：95℃ 10s（×1），95℃ 5s（×40），56℃ 25s（×40），在每个循环结束后检测荧光信号。

标准曲线的制作：将合成获得的 2 个基因的 cDNA 分别用含 Carrier tRNA 的 DEPC 水做 10 倍系列稀释，从 1×10、1×10^2、1×10^3、1×10^4 直至 1×10^5。采用荧光定量 PCR 试剂盒（SYBR Premix Ex Taq TM），在 T3 Thermocycler PCR 仪上进行 PCR 反应。随着 PCR 反应在每一个循环结束后测定吸光值，即可获得以 cDNA 稀释度的对数值为纵坐标，C_t 值（C_t 值是指每个反应体系内的荧光信号到达设定的阈值时所经历的循环数）为横坐标的标准曲线。

熔解曲线的测定：熔解曲线的测定于荧光 PCR 反应结束后进行，从 65℃到 90℃每隔 0.2℃测定其吸光值，即可得到以温度为横坐标、吸光值为纵坐标的熔解曲线。

对病毒外壳蛋白基因表达的影响测定：将活性物质用蒸馏水配制成质量体积浓度为 0.2mg/mL 溶液，以不含活性物质的蒸馏水为对照溶液，取双层滤纸分别于配制的溶液中浸泡数分钟，将吸饱溶液的滤纸分别铺放于不同的培养皿中。选取生长良好的普通烟品种 K_{326}，接种 TMV（0.02mg/mL）30min 后，从接种叶取直径约为 15mm 的圆盘，叶面朝上分别置于吸有各处理溶液的滤纸上，盖好皿盖，置于光照培养箱中，25℃下光照 / 黑暗（14h/10h）交替培养，并随时补充相应的溶液进行保湿。48h 后取出并用蒸馏水漂洗各圆盘，用滤纸吸干表面水分后，分别提取各处理叶圆盘中的总 RNA，对各处理中病毒 *CP* 基因表达情况进行实时荧光定量 PCR 检测，试验重复 3 次，结果以目标基因与内参基因（*Actin*）的 mRNA 拷贝数比值表示。

2 结果与分析

2.1 提取物结构分析

化合物 1 为白色针状结晶，经鉴定该化合物 1- 苯基 -1,3,5- 三庚炔（1-phenyl-1,3,5-triheptalkyne），分子量为 164，其化学式为 $C_{13}H_8$，化学结构式见图 1. EI-MS（*m/z*）164（M^+）；$IR\nu_{max}$（film）/cm^{-1}：2 219；^{1}H-NMRδ7.57（2H，dd, *J* = 8.0, 1.4 Hz），7.28（2H，br.d，*J* = 8.0 Hz），7.35（1H，m），1.85（3H，s）；^{13}C-NMRδ121.1（C-1′），133.5（C-2′），129.3（C-3′），130.5（C-4′），129.3（C-5′），133.5（C-6′），76.3（C-1），75.2（C-2），59.6（C-3），65.3（C-4）68.5（C-5），80.3（C-6），4.3（C-7）。化合物 1 为首次从该植物中分离获得。

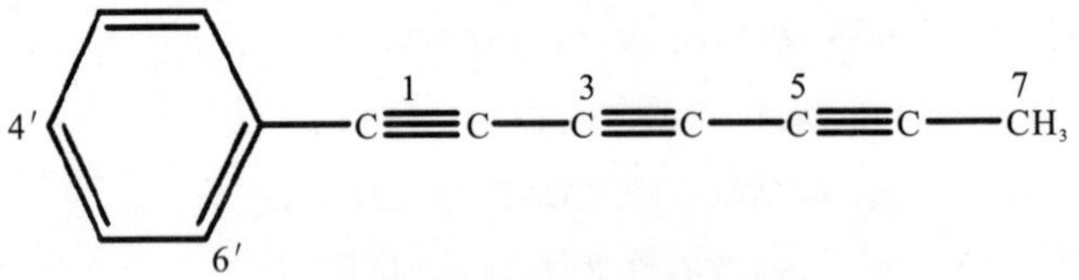

图 1 化合物 1 的化学结构式

Fig. 1 Chemical structure of compound 1 from *C. drummondii*

化合物 2 为淡黄色无定形粉末，结构鉴定结果表明，该化合物为（*1E*）-1-（3-hydroxy-butenyl）-1,2,4-trihydroxy-2,6,6-trimethyl-cyclohexane [反 -1-（3- 羧基 - 丁烯基）-1,2,4- 三羧基 -2,6,6- 三甲基 - 环己烷]，化学式为 $C_{13}H_{24}O_4$，分子量为 244。化合物 3 为白色片状结晶，经鉴定该化合物为β- 豆甾醇（β-stigmasterol），化学式为 $C_{29}H_{48}O$，分子量为 412。化合物 4 为白色片状结晶，经鉴定该化合物为豆甾醇（Stigmasterol），化学式为 $C_{29}H_{48}O$，分子量为 412，与化合物 3 为同分异构体。

2.2 体外抑制 TMV 活性

采用半叶枯斑法分别测定分离获得的化合物与提纯病毒混合后对 TMV 的抑制活性，当各化合物的质量浓度为 0.4mg/mL 时，化合物 1、2、3、4 和 5 对 TMV 的抑制率分别为 83.3%、42.6%、34.1%、31.2% 和 24.1%。结果表明，当质量浓度为 0.4mg/mL 时，只有化合物 1 表现出较好的抑制效果，其余 4 个化合物的抑制率均较低。

对抑制效果较好的化合物 1 进一步测定其在不同浓度下的体外抑制 TMV 效果，测定结果（图 2）表明，化合物 1 对 TMV 具有体外抑制作用，其抑制作用随浓度的增大而逐渐明显，具有浓度依赖性，其浓度在 0.05mg/mL 以上时对 TMV 有明显的抑制活性，大于 0.1mg/mL 浓度时其对 TMV 有良好的体外抑制作用。

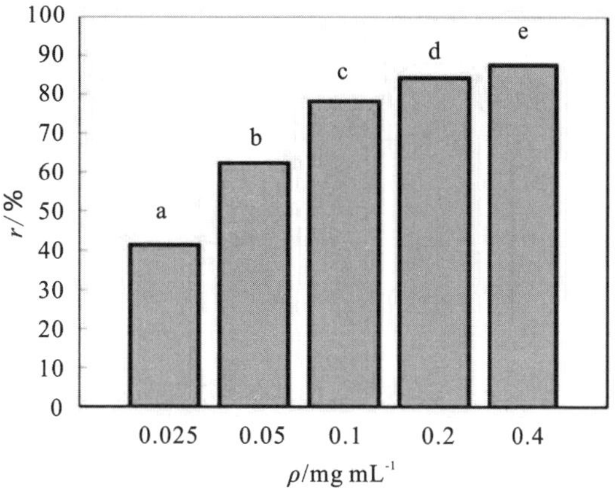

图 2 化合物 1 对 TMV 的体外抑制效果

Fig. 2 Inhibiting effect of compound 1 on TMV in vitro

柱上端不同字母表示在 0.05 水平差异显著，下同

Different letters in columns mean significant difference at P = 0.05 by Duncan's multiple range test. The same below

2.3 对 TMV 增殖的抑制作用

不同浓度的化合物 1 对烟草叶圆盘中 TMV 复制的抑制效果（图 3）显示，在离体条件下，化合物 1 对 TMV 复制的抑制作用随其浓度的增大而升高，质量浓度在 0.025mg/mL 时对烟草叶圆盘中 TMV 复制的抑制率仅为 41.3%，0.05mg/mL 浓度时对 TMV 复制的抑制率为 62.4%，0.1mg/mL 浓度时其抑制率达 78.2%，0.2mg/mL 浓度时其抑制率可达 84.3%，结果表明，在离体条件下化合物 1 对 TMV 的增殖具有明显的抑制作用，其抑制效果具有浓度依赖性，随浓度的提高而增大。

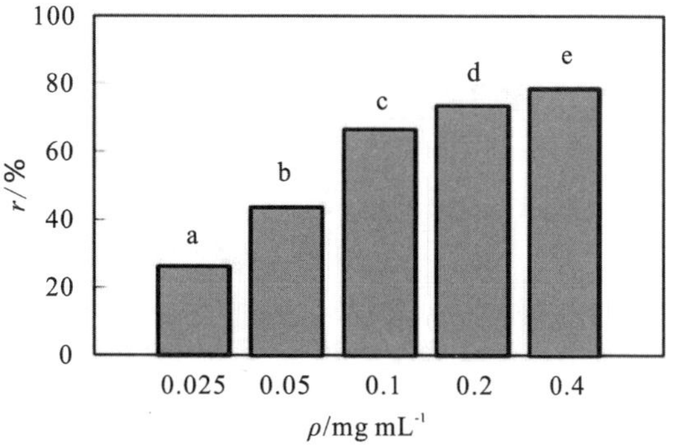

图 3 化合物 1 对 TMV 增殖的抑制效果

Fig. 3 Inhibiting effect of compound 1 on TMV replication in leaf discs

2.4 对 TMV CP 基因表达的抑制效果

2.4.1 标准曲线的制作

将逆转录合成的 cDNA 用 DEPC 水做 10 倍系列稀释，经实时荧光定量 PCR 仪测定的标准曲线（图 4）显示，*Actin* 的标准曲线方程为：$y_a = -0.2827x_a + 9.50$，其相关系数为 0.998；外壳蛋白基因的标准曲线方程为：$y_c = -0.251x_c + 8.22$，其相关系数为 0.991。其中 y 代表起始模板拷贝数以 10 为底的对数，x 代表 C_t 值。2 个基因标准曲线方程的相关系数均大于 0.99，说明标准曲线线性好，可用于通过未知样品的 C_t 值来计算未知样品的浓度。从图 4 还可以看出，在 10 ～ 10^5 范围内，起始模板拷贝数以 10 为底的对数和 C_t 值有很好的线性关系，表明可以在较宽的范围内对目的基因进行定量检测。

2.4.2 熔解曲线分析

利用熔解曲线可将特异和非特异产物区分开，图 5 为 TMV 外壳蛋白基因（*CP*）和烟草管家基因（*Actin*）扩增产物的熔解曲线。从图中可以看出，TMV 外壳蛋白基因扩增产物的特异峰出现在 77.8℃处（图 5-a），*Actin* 的扩增产物在 82.3℃处也出现特异峰（图 5-b），2 个特异峰峰值之前的曲线均比较平滑，未见其他显著的峰值，说明没有引物二聚体及非特异性扩增产物出现，表明本试验中对 2 个基因扩增所使用的引物均有强的特异性，且扩增条件合理，所获得的扩增产物均具有较强特异性。

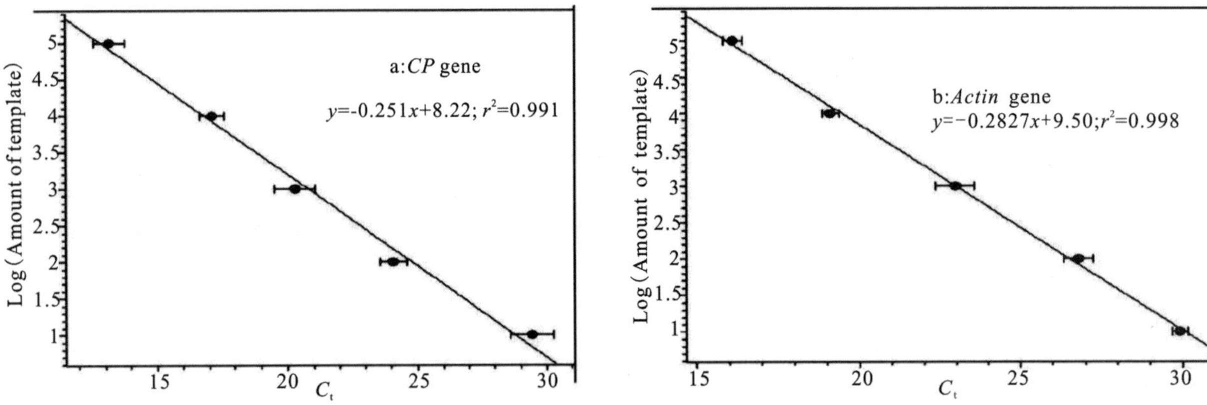

图 4 标准曲线

Fig. 4 Standard curves of CP gene and Actin gene

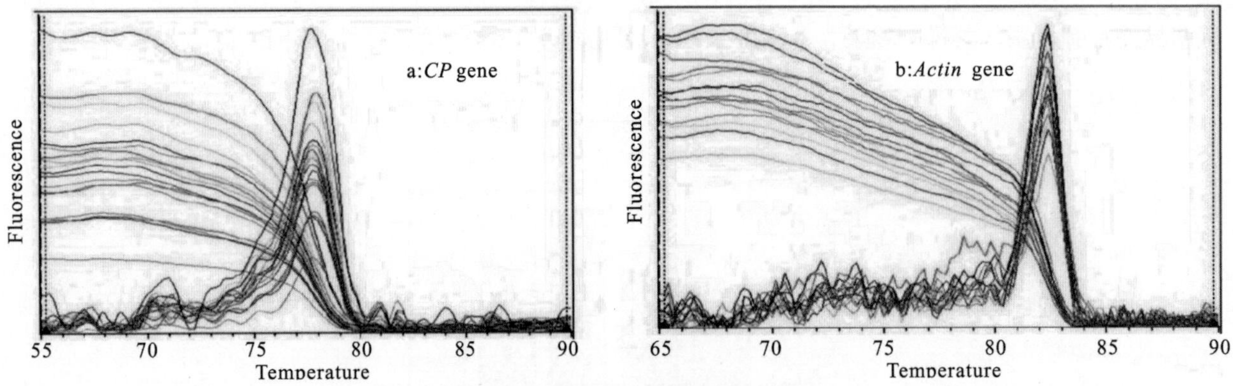

图 5 扩增产物的熔解曲线

Fig. 5 Dissociation curves of CP gene and Actin gene

2.4.3 对病毒外壳蛋白基因表达的抑制

采用实时荧光定量 PCR 对化合物 1 处理的病毒外壳蛋白基因表达情况进行了检测，在相对定量检测中，为减少样本间 RNA 提取效率、逆转录及扩增效率的差异对定量结果造成的误差，在测定目的基因的同时测定内源性管家基因 *Actin* 作为内参基因，结果以目标基因与内参基因（*Actin*）的 mRNA 拷贝数比值进行评价（图 6）。从图 6 中可看出，与对照相比，化合物 1 处理的病毒外壳蛋白基因的表达水平明显下降（$P < 0.05$）。对照组病毒 *CP* 基因表达水平（目的基因拷贝数与管家基因拷贝数的比值）为 37.28±8.88，处理组 *CP* 基因的表达水平为 7.52±1.82，约是对照组表达水平的 20.2%。结果表明，0.2mg/mL 的化合物处理对烟草中 TMV *CP* 基因的表达具有显著的抑制作用。

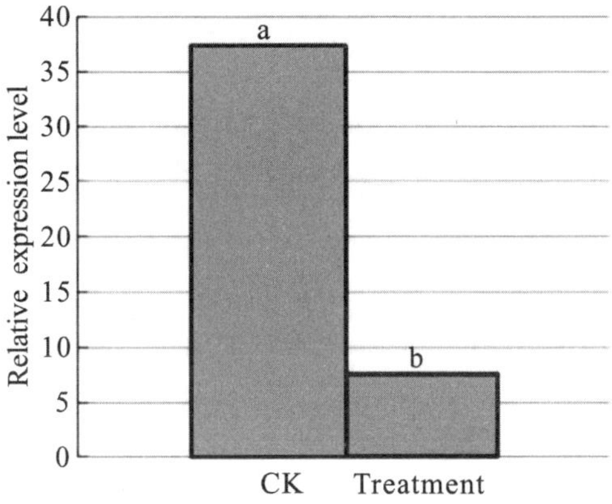

图 6　化合物 1 处理对 TMV CP 基因表达的影响

Fig. 6　Effect of compound 1 on TMV CP gene expression

3　讨论

烟草花叶病毒基因组为正义单链 RNA，可编码 4 种蛋白，其中相对分子质量为 126×10^3 和 183×10^3（126×10^3 的通读蛋白）的复制酶是其基因组 RNA 的翻译产物，17.5×10^3 的外壳蛋白和 30×10^3 的运动蛋白是 TMV 亚基因组 RNA 的翻译产物（Goregaoker and Culver, 2003; Reichel and Beachy, 2000）。外壳蛋白的主要功能是保护被其包被的 RNA 不受降解，也是 TMV 粒体在寄主体内长距离运动所必需的（Saito et al., 1987; Hilf and Dawson, 1993）。有研究表明，从植物中分离出的次生代谢物质双裂孕烷甾体 glaucogenin C 及其苷类化合物对 TMV 亚基因组具有选择性抑制作用，从而表现出对 TMV 复制的良好抑制效果（Li et al., 2007）。商文静等（2007）采用实时荧光定量 PCR 方法研究了壳寡糖诱导对烟草体内 TMV-CP 基因表达的影响，结果表明，经壳寡糖诱导的烟草表现出对 TMV 的系统获得抗性，其叶片中 *CP* 基因的表达量比未经处理的对照烟草显著降低（Shang et al., 2007）。因此，寄主植物体内 TMV-CP 基因的表达量可反应 TMV 在寄主体内的增殖动态。本试验采用荧光定量 PCR 方法测定了接种 TMV 的烟草叶盘经 1- 苯基 -1,3,5- 三庚炔处理后其中 *CP* 基因的相对表达量，可在分子水平上较为精确地评价该化合物对病毒增殖的抑制效果，为进一步研究奠定基础。

目前，植物病毒病的防治仍缺乏有效的防治药剂，一些植物源抗病毒剂，如由从油菜中提取的混合脂肪酸研制而成的耐病毒诱导剂 88-D（Sun and Lei, 1995），采用百合科和忍冬科的一些植物配制成的植毒灵（Huang et al.,1997），由锦葵科植物中提取的抗病毒物质多羟基双萘醛制成的抗病毒剂 WCT- Ⅱ（Zhang et al., 2005），虽然已在病毒病防治生产实践中发挥了重要的作用，但由于药剂品种较单一，且植物病毒具有较其他病原微生物易变异的特点，使病毒容易对这些药剂产生抗性。因此，若能获得一些新的抗病毒活性物质，进而研制成新的抗病毒剂，将其与其他抗病毒剂交替使用或混配使用，既可减缓病毒抗药性的产生，又可提高病毒病的防效，在农业生产上将具有广阔的应用前景。TMV 是正义单链 RNA 病毒的模式病毒，烟草 /TMV 体系是最常被用于抗植物病毒药物的筛选及其作用机制等研究的体系（Herms et al., 2002; Menard et al., 2004; Murphy and Carr, 2002）。本试验以烟草 /TMV 体系对金鸡菊提取物的抗病毒活性进行评价，采用活性跟踪法对其中的抗病毒成分进行分离，获得 4 种化合物单体，其中 1- 苯基 -1,3,5- 三庚炔为首次从金鸡菊中分离获得。其抗病毒活性测定结果表明，0.2mg/mL 的该化合物具有显著的抗 TMV 活性；实时荧光定量 PCR 测定结果表明该物质可明显抑制烟草体内 TMV 外壳蛋白基因的表达。这些研究结果表明，该物质可作为研制新型抗病毒剂的候选物质进一步加以探究，其抗病毒机制和理化性质等方面值得进一步研究。

参考文献

[1] CHEN Q J, OUYANG M A, XIE L H, et al. Isolation and detection of anti-TMV activity constituent from Parthenium hysterophorus[J]. Acta Laser Biology Sinica, 2008, 17(4): 544-548.

[2] DENG G, WAN B, HU H Z, et al. Biological activity of ningnanmycin on tobacco mosaic virus[J]. Chinese Journal of Applied and Environmental Biology, 2004, 10: 695-698.

[3] GOREGAOKER S P, CULVER J N. Oligomerization and activity of the helicase domain of the tobacco mosaic virus 126- and 183-kilodalton replicase proteins[J]. Journal of Virology, 2003, 77(6): 3549-3556.

[4] HERMS S, SEEHAUS, K, KOEHLE H, et al. A strobilurin fungicide enhances the resistance of tobacco against tobacco mosaic virus and Pseudomonas syringae Pv tabaci[J]. Plant Physiology, 2002, 130: 120-127.

[5] HILF M E, DAWSON W O. The tobamovirus capsid protein functions as a host-specific determinant of long-distance movement[J]. Virology, 1993, 193: 106-114.

[6] HUANG Z X, CHEN W J, CHENG L Z, et al. Preliminary studies on the control effect of chemical "Zhiduling" on tobacco mosaic virus[J]. Southwest China Journal of Agricultural Sciences, 1997, 10(2): 94-96.

[7] LI Y M, WANG L H, LI S L, et al. Seco-pregnane steroids target the subgenomic RNA of alphavirus-like RNA viruses[J]. Proceedings of the National Academy of Sciences of the United States of America, 2007, 104(19): 8083-8088.

[8] MENARD R, ALBAN S, DE RUFFRAY P, et al. β-1,3 glucan sulfate, but not β-1,3 glucan, induces the salicylic acid signaling pathway in tobacco and arabidopsis[J]. The Plant Cell Online, 2004, 16: 3020-3032.

[9] MURPHY A M, CARR J P. Salicylic acid has cell-specific effects on tobacco mosaic virus replication and cell-to-cell movement[J]. Plant Physiol, 2002, 128: 552-563.

[10] OUYANG M A, WEIN Y S, ZHANG Z K, et al. Inhibitory activity against tobacco mosaic virus (TMV) replication of pinoresinol and syringaresinol lignans and their glycosides from the root of Rhusjavanica var. roxburghian[J]. Journal of Agricultural & Food Chemistry, 2007, 55(16): 6460-6465.

[11] REICHEL C, BEACHY R N. Degradation of tobacco mosaic virus movement protein by the 26S proteasome[J]. Journal of Virology, 2000, 74(7): 3330-3337.

[12] SAITO T, MESHI T, TAKAMATSU N, et al. Coat protein gene sequence of tobacco mosaic virus encodes a host response determinant[J]. Proceedings of the National Academy of Sciences of the United States of America, 1987, 84: 6074-6077.

[13] SHANG W J, WU Y F, SHANG H S, et al. Inhibitory effect to TMV-CP gene expression in tobacco induced by chito-oligosaccharides[J]. Acta Phytopathological Sinica, 2007, 37(6): 637-641.

[14] SHEN J G, ZHANG Z K, WU Z J, et al. Antiphytoviral activity of bruceine-D from Brucea javanica seeds[J]. Pest Management Science, 2010, 64(2): 191-196.

[15] SUN F C, LEI X Y. The virus inhibitor 88-D induces production of PR proteins and resistance to TMV infection in tobacco.Acta Phytopathol Sin,1995, 25(4): 345-349.

[16] WU Y F, CAO R, WEI N S, et al. Screening and application of bio-virus pesticide. World Agric, 1995, (5): 35-36.

[17] WU Z J, OUYANG M A, WANG C Z, et al. Anti-tobacco mosaic virus (TMV) triterpenoid saponins from the leaves of Ilex

oblonga[J]. Journal of Agricultural & Food Chemistry, 2007, 55(5): 1712-1717.

[18] YAO Y C, AN T Y, GAO J, et al. Research of chemistry and bioactivity of active compounds antiphytovirus in Cynanehum komarovii[J]. Chinese Journal of Chemistry, 21(11): 1024-1028.

[19] ZHANG J X, WU Y F, FAN B. Antiviral physiopathology of WCT extracts of polyhydroxy dinaphthaldehyde[J]. Acta Phytopathol Sin, 35(6): 514-519.

YP3：食用菌榆黄蘑中新的植物病毒抑制物蛋白

吴丽萍[1]，曹郁生[1]，吴祖建[2]，林奇英[2]，谢联辉[2]

（1 南昌大学生命科学与食品工程学院，南昌 330047；2 福建农林大学植物病毒研究所，福州 350002）

摘要：食用菌中含有多种抗病毒蛋白，可用于植物保护。采用硫酸铵分级沉淀、DEAE- 纤维素离子交换层析和 Sephacryl™S-200 凝胶层析方法，从食用菌榆黄蘑新鲜子实体中提取到一单亚基蛋白，命名为 YP3。利用 SDS- 聚丙烯酰胺不连续凝胶电泳初步估计其分子量大约为 27.6 kDa。氨基酸组成分析表明该蛋白非常类似其他的抗植物病毒蛋白，并且几乎不含糖。其 N- 末端序列为 NRDVAACARFIDDFCDTLTP，在 GenBank 中没有找到同源序列。浓度为 0.24mg/L 时蛋白 YP3 对烟草花叶病毒（TMV 20mg/L）侵染心叶烟的抑制率为 50%。同时还发现 YP3 对供试的细菌和真菌没有抑制活性，对胃癌细胞株 MGC-803、肝癌细胞株 SMMC-7721、肺癌细胞株 SPC-A1 的细胞增殖具有一定的抑制作用，其 IC_{50} 大约为 20mg/L。

关键词：食用菌；榆黄蘑；蛋白；抗病毒活性

中图分类号：Q939.99; R978.7　**文献标识码：**A

YP3: a novel plant virus inhibitory protein from mushroom *Pleurotus citrinopileatus*

Liping Wu[1, 2], Yusheng Cao[1], Zujian Wu[2], Qiying Lin[2],Lianhu Xie[2]

(1 Academe of Life Sciences and Food Enginering, Nanchang University, Nanchang 330047, China; 2 Institute of Plant Virology, Fujian Agriculture and Forestry University, Fuzhou 350002, China)

Abstract: Many proteins from mushrooms can be used as potential antiviral agents, and in plant protection.An antiviral protein (YP3) was isolated from fruiting bodies of the fungus *Pleurotus citrinopileatus* with a proce dure involving ammonium sulfate precipitation, ion-exchange chromatography on DEAE-cellulose, and gel filtrationon Sephacryl™S-200 High Resolution. The molecular weight of protein was determined to be about 27.6kDa by Sodium dodecyl sulfate-polyacryl amide gel electrophoresis (SDS-PAGE). The amino acid analysis indicated that it was similar to some other antiviral proteins.The protein was almost not sacchariferous. The sequence of N-terminal of YP3 was NRDVAACARFIDDFCDTLTP, which had no homology with other sequences in GenBank. YP3 at 0.24mg/L achieved 50.0% inhibition of Tobacco mosaic virus (TMV 20mg/L) lesions in *Nicotiana glutinosa* leaves; YP3 is inactive to some bacteria and fungi tested; YP3 inhibited stomach cancer cell line MGC-803, liver cancer cell line SMMC-7721 and lung cancer cell line SPC-A1with an IC_{50} of about 20mg/L.

Keywords: mushroom; *Pleurotus citrinopileatus*; protein;antiviral activity

天然产物研究与开发 . 2009, 21: 371-376

Received 24 March 2008; Accepted 9 April 2008

1 Introduction

In the last decades and as an alternative to convention alchemical agents,a large number of phytochemicals have been recognized as a way to control infections caused by viruses (Abad et al., 2000).Plants and microorganisms have long been used as remedies, and many are now being collected and examined in an attempt to identify possible sources of antiviral agents.Since the first is olation of anantiviral protein (Pokeweed antiviral protein,PAP) from *Phytolacca Americana* (Kassanis and Leczkowski, 1948), many antiviral proteins had been purified from higher plants, including *Mirabilis jalapa* (Kubo et al.,1990), *Clerodendrum inerme* (Prasad et al.,1995; Praveen et al.,2001), *Silene schafta* (Alexandre et al.,1997), *Dianthus sinensis* L.(Choet al., 2000), *Phytolacca insularis* (Sang et al., 2000), *Cucumis figarei* (Fujiware et al., 2001), *Sambucus nigra* L. (Vandenbussche et al., 2004), *Beta vulgaris* Linn. (Iglesias et al., 2005), and *Celosia cristata* (Balasubrah et al., 2006). The proteins from different plants share different characterization and antiviral effect. Proteins like PAP were regarded as biopesticides and sprayed directly to crop. These were the gene sources of transgenic plants. Further more, many of them showed not only broad spectrum plant virus resistance, but also human virus and tumor inhibition.

Many mushrooms belonging to Pleurotaceae are edible and medicinally used. *Pleurotus citrinopileatus* (Pleurotaceae), a mushroom, is used for weak kidney. Protein YP46-46 isolated from it is an agglutinin and can inhibit *Tobacco mosaic virus* (TMV) and *Hepatitis B virus* (Fu et al., 2002). We purified and analyzed partial biochemical characteristics and activities of YP3, another protein from this mushroom.

2 Materials and Methods

2.1 Materials

Pleurotus citrinopileatus was presented by Juncao Reseach Institute of Fujian Agriculture and Forestry University. TMV, *Nicotiana glutinosa*, *Ralstonia solanaceance*, *Xanthomonas canpestrispv*. oryzae, *Altemaria brassicae*, *Colltotrichum higginsianum*, *Penicillium expansum*, *Fusarium oxysporum*. f.sp.cucumerinum and *Phytophthora capsici*, were stored at Institute of Plant Virology of Fujian Agriculture and Forestry University. The cell strain of MGC-803 was from DrCHEN Rui chuan(Xia men University, China) the cell line of SMMC-7721 and SPC-A1 from China Center for Type Culture Collection. DEAE-sepharose Fast Flow and Sephacryl^{TM}S-200 High Resolution were from Amersham Pharmacia Biotech, Uppsala, Sweden. Ultrafiltration membranes were from Millipore, Bedford, MA. Medium RPMI1640 was the product of Gibco BRL. MTT was from Fluka Co. Molecular weight markers were from Biochemistry Institute of Shanghai.Other chemicals and reagents were of analytical grade.

2.2 Methods

2.2.1 Isolation and molecular mass determination of protein YP3

Fresh fruiting body of *Pleurotus citrinopileatus* (150g) was stored at 4℃ for 4–5h after grinded in 300 mL distilled water and 1%(v/v)β-mercaptoethanol.The mixture was added with 40%–60% saturation $(NH_4)_2SO_4$ and centrifugated at 10000r/min for 10min. Precipitation dissolved in 0.01mol/L PBS (Na_2HPO_4-NaH_2PO_4, pH7.2) was applied to a DEAE-Sepharose Fast Flow column (1.6×10cm), which was eluted with the same buffer with a linear gradient of 0–0.5mol/L NaCl. The fractions contained TMV-resistant activity were collected, concentrated, desalted using 10kDa cutoff ultrafiltration membranes and applied to a Sephacryl^{TM}S-200 High Resolution column equilibrated with the same buffer (contain 0.15mol/L NaCl). Individual peaks were collected,concentrated. Purified protein YP3 is in one of the two OD_{280} peaks.

The protein was detected and its molecular weight determined by Sodium dodecyl sulfate-polyacrylamide gel electrophoresis (SDS-PAGE) and PAGE according to the method of Laemm li (Fuet al., 1970). Electrophoresis was performed on a 12% polyacrylaminde gel and stained with Coomassie blue R-250.The protein concentration was detected by the method of Bradford (Bradford et al., 1976).

2.2.2 N−terminal sequencing and concentration of sugar of protein detection

YP3 was transferred to PVDF film (Sigma) after SDS-PAGE and stained with Coomassie blue R-250.The N-terminal amino acid sequence of YP3 was determined at Hunan Normal College, China, by the automated Edman degradation method with an Applied Biosystems model 477, a protein sequencer and an on-line phenylthiohydantoin analyser. Carbohydrate-containing bands were detected with periodic acid/Schiff reagent (Sun et al., 1987) and phenol-vitriol (Li et al., 1998).

2.2.3 Amino acid composition analysis

Amino acid composition analysis was carried out in Microbiology Institute of Fujian Province.

2.2.4 Activity of YP3 on TMV detection

The half leaf method was used.TMV (20mg/L) was mixed with a sample of YP3 protein or with a similar volume of distilled water, and then inoculated into the right and left half of a *Nicotiana glutinosa* leaf, respectively. The number of local lesions was recorded, and inhibition (%) was calculated according to the formula. $I=[1-(I_s/I_w)]\times100$ Where I is inhibition, L_s is number of local lesions after treatment with sample and L_w is number of local lesions after treatment with water.

2.2.5 Antibacterial and fungus proof assay (Chen, 1990)

An inhibition zone assay was used with the following modified procedure: bacteria (100μL of one milliliter containing10^7–10^8 colony forming unit) were spread on to 7mL beef peptone plates. Aperture was formed on medium and added with sample and control. Plates were incubated at 37℃ for 24h. The diameters of inhibition zones were recorded in mm minus the aperture diameters. Inhibition on conidiophore bourgeoning was detected by the method of suspending drop. Growth speed assay was introduced to determinate inhibition of mycelia growth. An aperture (0.8cm diameter), on which a fungus colony was eugenic, was transplanted onto Potato Dextrose Agar PDA nutrient plates (25mg/L sample). The diameters of colony were recorded. Cytotoxicity of YP3 on tum or cell lines (Mosmann et al., 1983) MTT was used as the experiment reagent to examine YP3 inhibiting three kind softum or cells growth. The cells were inoculated in 96 wells plate and cultivated for 24h, and then treated with YP3 for sametime, and the control was cultivation medium. The 96 wells plate was washed by PBS and PBS containing 0.2g/LMTT was added to and it was cultivated at 37℃ for 4h, then DMSO (90% DMSO/10% 0.1mol/L Gly-NaOH pH9.8) was added after MTT was discarded. 30 min later OD_{570} was detected. Inhibition rate of YP3 was calculated according to the formula: Inhibition rate (%) = (1−OD_{570} of experiment/OD_{570} of control)×100IC_{50} is the concentration to reach an inhibition rate of 50%.

3 Results

3.1 Isolation and characteristics of YP3

Fresh fruiting body of *P. citrinopileatus* was grinded, and the mixture was added with 40%–60% saturation

$(NH_4)_2SO_4$ and centrifugated at 10000r/min for 10min.The ammonium sulfate precipitation was applied to ion-exchange chromatography, and the production revealed one main peak when detected the OD_{280} (Fig. 1), which displayed antiviral activity and was determined by SDS-PAGEto be several proteins (Fig. 3). At that while, the linear gradient of NaCl was about 0.1–0.3mol/L.These proteins were different in molecular weight and concentration. Gel filtration of this fraction gave a peak B containing nearly a protein with a molecular mass about 27.6kDa when analyzed by SDS-PAGE (Fig. 2, Fig. 3), and this displayed antiviral activity also. This protein was then designated YP3.Because β-mercapto ethanol was added in mixture when mushroom was grinded, YP3 was determined to contain only on subunit.

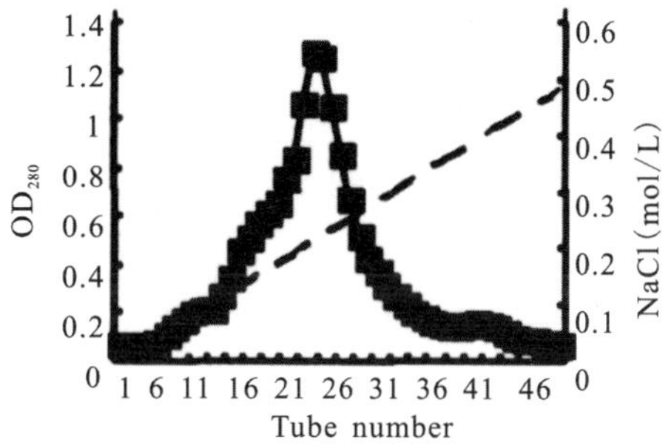

Fig. 1 Ion-exchange chromatography of *Pleurotus citrinopileatus* extract

Note: Elution profile of YP3 from a DEAE-Sepharose Fast Flow column.Fractions were eluted with a linear gradient of 0–0.5 mol/L sodium chloride(------). OD_{280nm} is indicated(■). Inhibitory effect of the fractions (-) was analyzed with a local lesion assay on tobacco leaves. Active fractions in tubes 13–26 were collected

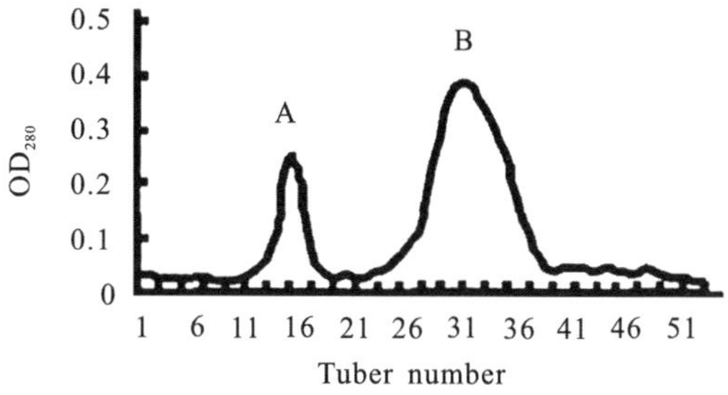

Fig. 2 Gel filtration of *Pleurotus citrinopileatus* extract

Note: Elution profile of YP3 from a Sephacryl™S-200 High Resolution column. Fractions containing antiviral activity from Fig. 1 were applied to a column previously equilibrated with the same buffer as Fig.1. Inhibitory effect of the fractions (peak A and B) was also analyzed as Fig.1 (B:The absorbance at 280 nm of YP3)

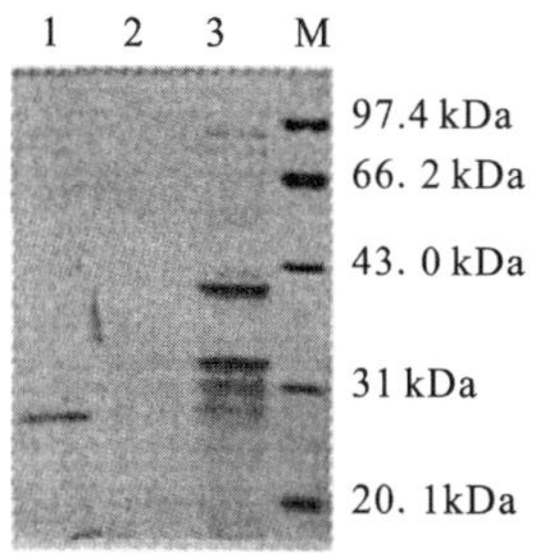

Fig. 3 Electrophorese analysis of purified fractions Note

Protein samples at each purification step electrophoresed on a 12% SDS-polyacrylamide gel, and stained with coomassie brilliant blue R-250. Lane1, fraction B from Sephacryl™S-200 High Resolution column; lane2, Blank; lane3, fractions from DEAE-Sepharose column; lane M, molecularm as smarkers Rabbit Phosphorylase b (97.4kDa), Bovine Serum Albumin (66.2kDa), Rabbit Actin (43kDa), Bovin Carbonic Anhydrase (31kDa), Trysin Inhibitor (20.1kDa), Hen Egg Lyso zyme(14.4kDa)

3.2 N–Terminal sequence of protein and the concentration of sugar

YP3 was transferred to PVDF film after SDS-PAGE and stained with Coomassie blue R-250.The N-terminal amino acid sequence of YP3 was determined by the automated Edman degradation method and an on-line phenylthiohydantoin analyser. The result showed that the 20 amino acid sequence of N-terminal was VYINKLTPPCGTMYYACEAV. We compared it with other proteins equence, and did not find any homologous sequence in GenBank, suggesting that it should be a news equence. The submission number was P83481 in Swiss-Prot (yp3, instead of YP3, was used as the name of protein when logging in Swiss-Prot).

When staining with periodic acid/Schiff reagent after SDS-PAGE, The protein showed no reaction, and after quantified by the method of phenol-vitriol, it was calculated that this protein possessed 0.2% of sugar, so YP3 had very low saccharinity and may not be aglycoprotein, like most function alproteins inplants.

3.3 Amino acid composition of protein

The amine acid composition of YP3 was shown in Table1. During the course of experiment, Trp was destroyed completely and so could not be calculated. Gln and Asn are transformed into Glu and Asp, respectively, thereby they are calculated incorporatively. We compared the amino acid composition of YP3 with that of other plant virus inhibiting proteins and RNase (because then ucleicacid of virus could be degraded by RNase when virus was uncovered by coatprotein). It was found that they contained abundant aminoacid of Asp, Lys, Ala, Glu, Ser and Thr (Chen, 1999).

3.4 Bioactivities of protein

The half leaf method was used to detect the inactivation activity of YP3 to TMV in vitro.The inhibition rate against TMV in its local lesion host, *Nicotiana glutinosa*, is 50.0% at a concentration of 0.24mg/L (Fig. 4A). Unlike its high inactivating activity to TMV, YP3 was not antibacterial and could not inhibit the growth of hyphae, nor the germination of spore, it can even accelerate the growth of *Colltotrichum higginsianum*. YP3 could inhibit the survival and multiplication of three tumor cell lines and the activity is not dose dependent (Fig. 4B).

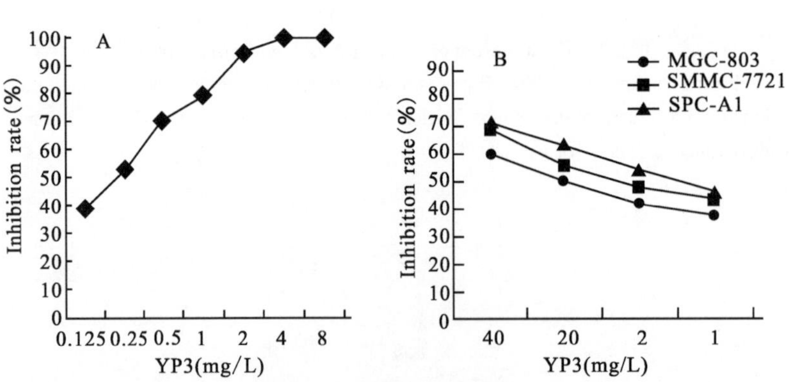

Fig.4 Bioactivities of YP3.

A. Effect of YP3 inactivating TMV The half leaf method was used and the host is *Nicotianag glutinosa*; B. Effect of YP3 inhibiting on cell multiplication. MTT was used as the experiment reagent to examine YP3 in habiting three kinds of tumor cells growth

4 Discussion

Till now many proteins have been isolated from mushroom and possess various activities. In this work, we purified an antiviral protein (YP3) from *Pleurotus citrinopileatus* by three steps of ammonium sulfate precipitation, ion-exchange chromatography on DEAE-cellulose and gel filtration on Sephacryl^{TM}S-200 High Resolution. Its molecular mass was about 27.6kDa. It may be a new sequence with low saccharinity and a novel antiviral and anticancer gene resource. YP3 had a remarkable inhibition on TMV with an inhibition rate up to 50.0% at a concentration of only 0.24mg/L, which was far lower than any other anti-TMV proteins from mushroom (Fu et al.2002; Kobayashi et al., 1987; Sun et al., 2001; Wu et al., 2003).We did not find any TMV that had been destroyed or parted in eyeshot (data not shown),when TMV was incubated with YP3 and observed through electron microscope by the method of negative dye and colloidal gold dye. The possible mechanism of YP3 inhibiting TMV should be that it competes with TMV for infection court. Such mechanism is unknown and need to be investigated further.

YP3 richly contained acidamino acid (Table1). When compared with YP3, other anti-TMV protein sand RNase all comprised abundant Asp, Lys and more Ala, Glu, Ser and Thr (Guo, 1999) (Table2). In addition, amino acid component in some kind of RNase, which can resist virus infection, was similar to that of inhibitor from higher plant, and they might have similar structural domain or structure near activity site (Table2). All of these will give us information to study the mechanism of YP3 inhibiting on TMV infection.

YP3 had no inhibiting activity to any bacteria and fungi tested, and it can accelerate them ycelia growth of *Colltotrichum higginsianum* indeed. The reason may be that YP3 is a lectin. A lot of research proved that most lectins can be mitogenic (Mohd and Rizwan, 2003), and lectins from edible may be important for mycelia development and morphologization (Wang and Bun, 1998).Whether it was the reason remained unknown, and more evidence was needed to prove it.

YP3 could inhibit the multiplication of tumor cells including MGC-803, SMMC-7721 and SPC-A1. It was similar to the protein PAP, which not only had inhibition to the multiplication of tumor cells, but also to some human's virus, such as human immunity virus.It may be a new anticancer medicine resource, but further study such as toxicity and pharma cological mechanism should be carried out.

Table 1 Amino acid composition of YP3

Amino acid	Content (mg/L)	Residues in one molecular
Asx	35.93	23
Thr	16.12	11
Ser	14.94	12
Glx	32.92	19
Ala	17.79	17
Cys	20.24	14
Val	22.22	16
Met	34.19	19
Ile	14.64	9
Leu	16.86	11
Tyr	18.77	9
Phe	35.18	18
Lys	16.82	10
His	4.18	2
Trp	ND	ND
Arg	5.9	3
Phe	ND	ND
Phe	35.18	18

Note: ND: not determined; Asx=Asp+Asn; Glx=Glu+Gln

Table 2 Amino acid component in different inhibitors and RNase

Amino Acid	Residues in one molecular						
	Inhibitor						
	PAP	PAP-II	PAP-S	FBP	AAVP	YP_3	RNase
Asx	31	26	34	28	16	23	15
Thr	17	18	14	15	10	11	10
Ser	18	16	15	15	8	12	15
Glx	26	25	29	19	12	19	12
Gly	16	21	18	18	11	16	3
Ala	15	19	16	18	12	17	12
Cys	3	4	2	0	0	14	8
Met	4	8	6	1	1	19	4
Ile	17	15	20	10	8	9	3
Leu	24	23	27	14	10	11	2
Tyr	19	17	16	10	5	9	9
Phe	10	11	7	11	6	18	3
Lys	18	20	24	7	4	10	10
His	3	4	2	2	2	2	4
Trp	11	16	10	5		ND	6
Arg	10	9	14	13	3	3	4
Pro	14	12	12	7	3	ND	4
MW(kDa)	29000	30000	30000	23000	15800	27400	14000

References

[1] ABADM J, GUERRA J A, BERMEJO P, et al. Search for antiviral activity in higher plant extracts[J]. Phytotherapy Research, 2000, 14: 604-607.

[2] ALEXANDRE M A V, GUZZO S D, HARAKAVA R, et al. Partial purification of a virus inhibitor from *Silene schafta* leaves[J]. Fitopatologia Brasileira, 1997, 22: 171-177.

[3] BALASUBRAHMANYAM A, BARANWAL V K, LODHA M L, et al. Purification and properties of growth stage-dependent antiviral proteins from the leaves of *Celosia cristata*[J]. Plant Science, 2000, 154: 13-21.

[4] BEGAM M, NARWAL S, ROY S, et al. An antiviral protein having deoxyribonuclease and ribonuclease activity from leaves of the post-flowering stage of *Celosia cristata*[J]. Biochemistry, 2006, 71 (Suppl.1): 44-48.

[5] BRADFORD M M. Arapid and sensitive method for the quantitation of microgram quantities of protein utilizing the principle of protein-dye binding[J]. Analytical Biochemistry, 1976, 72: 248-254.

[6] CHEN N C. Bioassays Technology of Pesticide[J]. Beijing: Beijing Agriculture University Press, 1990, 65.

[7] CHO H J, LEE S J, KIM S, et al. Isolationand characterization of cDNAs encoding ribosome in activating protein from *Dianthus sinensis* L[J]. Molecules and Cells, 2012, 10: 135-141.

[8] FU M J, LIN Q Y, WU Z J, et al. Purification of an antiviral protein in *Pleurotus citrino* and its activities against tobacco mosaic virus and hepatitis B virus[J]. Virologica Sinica, 2002, 17: 350-353.

[9] FUJIWARE M, KANAMORI T, OHKI S T, et al. Purification and partial characterization of Figaren, an R Nase-like novel antiviral

protein from *Cucumis figarei*[J]. Journal of General Plant Pathology, 2001, 67: 152-158.

[10] GUO B S. Advance in research on inhibiting matters to plant virus[J]. Journal of Agricultural University of Hebei, 1999, 22(3): 62-67.

[11] IGLESIAS R, PEREZ Y, DE TORRE C, et al. Molecular characterization and systemic induction of single-chain ribosome-inactivating proteins (RIPs) in sugar beet (*Beta vulgaris*) leaves[J]. Journal of Experimental Botany, 2005, 56: 1675-1684.

[12] KASSANIS B, KLECZKOWSKI A. The isolation and some proper ties of a virus –inhibiting protein from *Phytolacca esculenta*[J]. Journal of General Microbiology, 1948, 2: 143-153.

[13] KUBO S, IKEDA T, IMAIZUMI S, et al. A potent plant virus inhibitor found in *Mirabilis jalapa* L[J]. Japanese Journal of Phytopathology, 1990, 56: 481-487.

[14] KOBAYASHI N, HIRAMASTRU A, AKATUKA T. Purification and chemical properties of an inhibitor of plant virus infection from fruiting bodies of *Lentinula edodes*[J]. Agricultural and Biological Chemistry, 1987, 51: 883-890.

[15] LEAMMLI U K. Cleavage of structural proteins during the assembly of the head of bacteriophage 74[J]. Nature (Load), 1970, 227: 680-685.

[16] LI R L. Experiment of Biochemistry. Wuhan: Wuhan University Press, 1998, 9-10.

[17] MOHD T A, RIZWAN H K.Mitogeniclectins[J]. Medical Science Monitor, 2003, 9: 265-268.

[18] MOSMANN T. Rapid colorimetric assay for cellular growth and survival: Application toproliferaton and cytoxicity assays[J]. Immunol Methods, 1983, 65: 55-63.

[19] PRASAD V, SRIVASTAVA S, VARSHA H N. Two basic proteins isolated from *Clerodendrum inerme* Gaertn. are inducers of systemic antiviral resistance in susceptible plants[J]. Plant Science, 1995, 110: 73-82.

[20] PRAVEEN S, TRIPATHI S, VARMA A. Isolation and characterization of an inducer protein (Crip-31) from *Clerodendrum inerme* leaves responsible for induction of systemic resistance against viruses[J]. Plant Science, 2001, 161: 453-459.

[21] SANG S K, CHOI Y, MOON Y H, et al. Systemic induction of a *Phytolacca insularis* antiviral protein gene by mechanical wounding, jasmonic acid, and abscisic acid[J]. Plant Molecular Biology, 2000, 43: 439-450.

[22] SUN C. Generic methods of identifying activity and purification of lectin[J]. In: ZhangWei-jie. Biochemistry Technology of Complex Carbohydrate. Shanghai: Shanghai Science and Technology Press, 1987, 342-344.

[23] SUN H, WU Z J, XIE L H, et al. Purification and characterization of AAVP, a protein inhibitor of T MV infection, from the edible fungus *Agrocybe aegerita*[J]. Sheng Wu Hua Xue Yu Sheng Wu Wu LI XueBao, 2001, 33: 351-354.

[24] VANDENBUSSCHE F, DESMYTER S, CIANI M, et al. Analysis of the in plant an antiviral activity of elderberry ribosome-inactivating proteins[J]. European Journal of Biochemistry, 2010, 271: 1508-1515.

[25] WANG H X, BUN T N, OOI V. Lectins from mushrooms[J]. Mycological Research, 1998, 102: 897-906

[26] WU L P, WU Z J, LIN Q Y, et al. Purification and activities of an alkaline protein from mushroom *Coprinus comatus*[J]. Acta Microbiologica Sinica, 2003, 43: 793-798.

毛头鬼伞多糖诱导烟草体内水杨酸的积累

吴艳兵[1,2]，谢荔岩[1]，谢联辉[1]，林奇英[1]

(1 福建农林大学植物病毒研究所，福建 福州 350002; 2 河南科技学院资源与环境学院，河南新乡 453003)

摘要： 以烟草感病品种普通烟 K_{326} 为材料，以毛头鬼伞多糖（CCP）为诱导因子，采用高效液相色谱法（HPLC）测定了 CCP 对烟草体内水杨酸（SA）含量的影响。结果表明，CCP 可以诱导烟草感病品种体内 SA 含量增高，SA 含量在接种烟草花叶病毒（TMV）后 84h 达到最大值，为 458.927ng • g^{-1}。

关键词： 高效液相色谱法；毛头鬼伞多糖；诱导抗性；烟草花叶病毒；水杨酸

中图分类号： S435.72　**文献标识码：** A　**文章编号：** 1671-5470（2009）01-0006-05

Accumulation of salicylic acid in tobacco treated with polysaccharide from *Coprinuscomatus*

YanbingWu[1, 2], LiyanXie[1], LianhuiXie[1],QiyinLin[1]

(1 Institute of Plant Virology, Fujian Agriculture and Forestry University, Fuzhou, Fujian 350002, China; 2 Department of Resource and Environment, Henan College of Science and Technology, Xinxiang, Henan 453003, China)

Abstract:Using to bacco K_{326} (*Nicotianatabacum* cv., K_{326}) as experimental material, and polysaccharide from *Coprinus comatus*(CCP) as elicitor to investigate the effects of CCP onendogenousSAcontents by HPLC (high performance liquid chromotography).The results showed that the endogenous SA contents in leaves of tobacco K_{326} treated with CCP increased, and the peak value in SA content curve was 458.927ng • g^{-1} at 84 hafter inoculation in leaves of tobacco K_{326}.

Keywords:high performance liquid chromotography (HPLC); *Coprinus comatus* polysaccharide (CCP); induced resistance; *Tobacco mosaic virus* (TMV); salicylicacid(SA)

毛头鬼伞（*Coprinus comatus*）又名鸡腿菇，因其形似鸡腿而得名，是一种具有较高开发价值的珍稀食药同源食用菌。研究证实，毛头鬼伞具有降血糖、降血脂、提高免疫活性和抗肿瘤等生物学功能（刘艳芳和张劲松，2003）。有关毛头鬼伞及其他食用菌对植物病毒的抑制作用已有报道（吴丽萍等，2003；吴丽萍等，2004；张超等，2005；吴艳兵等，2007a；吴艳兵等，2007b）。水杨酸（salicylicacid, SA）是诱发系统抗性的信号物质，对植物抗病基因介导的抗性的产生起重要作用。Malamy et al 认为 SA 可以作为游离的信号物质传递，内源 SA 可作为诱导植物防卫反应的信号分子（Malamy et al., 1992）。许多研究表明，在系统获得抗性基因表达，以及系统获得抗性之前，内源 SA 先积累，而许多诱抗因子诱导植

福建农林大学学报（自然科学版）. 2009, 38 (1) : 6-10
收稿日期：2008-01-11
基金项目：福建省教育厅项目 (K04037)；福建省科技厅重点项目（99 -Z-193）

物系统获得抗性都需要积累 SA。关于毛头鬼伞多糖（CCP）对植株内源 SA 含量的影响迄今尚未见报道。本研究以自行制备的多糖为诱导剂，以感病品种普通烟 K_{326} 为试材，测定多糖诱导后烟草叶片中内源 SA 的积累情况，以期为 CCP 诱导烟草 TMV（*Tobacco mosaic virus*）的抗性研究提供依据。

1 材料与方法

1.1 材料

1.1.1 供试植物、毒源

烟草花叶病毒毒源由福建农林大学植物病毒研究所保存。测定系统侵染寄主为普通烟 K_{326}（*Nicotiana tabacum* cv., K_{326}）。

1.1.2 毛头鬼伞子实体多糖

新鲜毛头鬼伞子实体清洗干净后，用匀浆机破碎，加入一定体积的无水乙醇，水浴中回流提取 2h，以去除油脂，然后离心分离，残渣用于多糖提取。沸水浴重复提取 3 次。合并各次提取液，浓缩，加无水乙醇沉淀，冷冻干燥，得到粗多糖；然后将粗多糖用水溶解，用 Sevag（氯仿：正丁醇 =4 ： 1）剧烈振荡 20min，离心除去中间层沉淀，重复此过程若干次，直到上清液无蛋白质被测出，浓缩、冷冻干燥后得到多糖 CCP。

1.2 方法

1.2.1 接种方法和烟草幼苗处理

1g 病叶加 10mL 磷酸缓冲液（pH 7）研磨成匀浆，然后离心，取上清液进行摩擦接种。选取长势一致的烟草感病品种 K_{326} 幼苗植株，分别进行以下 4 种处理：（1）0.1mg • mL^{-1} CCP+TMV（喷 CCP 3 d 后，接种 TMV）；（2）清水 +TMV（喷清水 3d 后，接种 TMV）；（3）0.1mg • mL^{-1} CCP（喷 CCP 3d 后，不接种 TMV）；（4）清水（喷清水 3d 后，不接种 TMV）。于接种后 12、36、60、84、108h 取各处理第 4 片叶测定 SA 含量。

1.2.2 SA 含量的测定

（1）烟草叶片中 SA 的提取：取新鲜烟草叶片 0.5g，加液氮研磨后加 5mL 95%（体积分数）甲醇，经超声波提取 20min，在转速为 10000r • min^{-1} 的条件下离心 10min，上清液减压浓缩蒸干；沉淀经 1mL 2mol • L^{-1}NaOH 溶液重新悬浮，70℃加热 2h，用 250μL 36.9%（质量分数）的 HCl 在 70℃条件下酸化 1h，待冷却至 4℃后减压浓缩蒸干，流动相重新溶解后经孔径为 0.22mm 微孔滤膜过滤，用于 HPLC 分析。

（2）测定波长的选择：精确称取水杨酸对照样品，加冰醋酸 + 甲醇（体积比 1 ： 10）溶液，制成 10mg • mL^{-1} 的水杨酸溶液。采用分光光度法在 200 ～ 350nm 的波长范围内扫描紫外吸收图谱（图 1），水杨酸对照样品溶液在（302±2）nm 和（235±2）nm 的波长处有最大吸收，因在 302nm 处样品干扰峰少，故选定 302nm 波长来测定水杨酸的波长。

（3）水杨酸检测色谱条件：色谱柱用 C18 柱，流动相为甲醇 + 水 + 冰醋酸（体积比为 8 ： 4 ： 1），检测器为 DAD，检测波长为 302nm，流动相流速为 0.8mL • min^{-1}，水杨酸保留时间为 4.45min。

（4）标准曲线建立：精确称取水杨酸对照样品约 10.00mg，置 100mL 量瓶中，加冰醋酸 + 甲醇（体积比为 1 ： 10）溶液，使其溶解并稀释至刻度，摇匀。精确吸取上述溶液 1mL，置 100mL 量瓶中加冰醋酸 + 甲醇（体积比为 1 ： 10）溶液，使其溶解并稀释至刻度，摇匀。分别取 2、4、8、12、16、20μL 上述溶液依次注入液相色谱仪，浓度分别为 0.1、0.2、0.4、0.6、0.8、1.0mg • mL^{-1}，测定、记录色谱图，并进行回归处理。

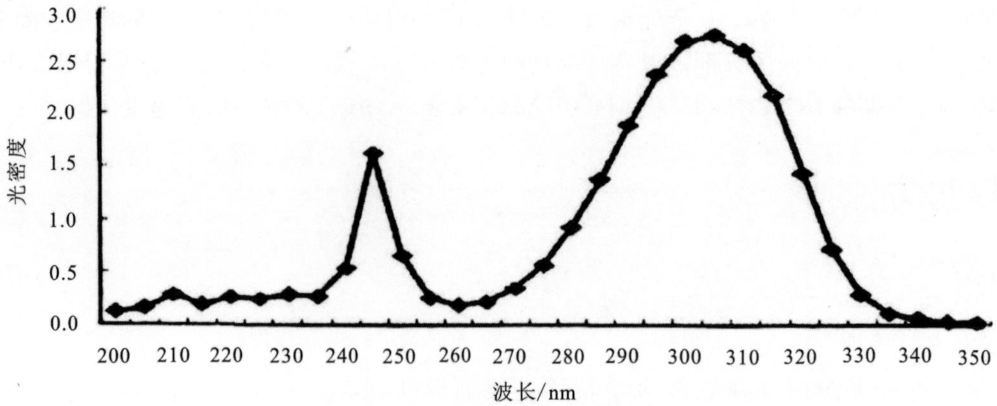

图 1　水杨酸波长紫外扫描图谱
Fig.1 Ultraviolet spectrum of SA

2　结果与分析

2.1　标准曲线制作

将不同浓度的标准溶液在上述液相色谱条件下分别进行测定，并以浓度为横坐标，峰面积为纵坐标绘制标准曲线。结果显示 SA 在 0~1.0mg·mL^{-1} 范围内有很好的线性关系，回归方程为 y=34.46634x+0.373364，相关系数 r 为 0.99972（图 2）。

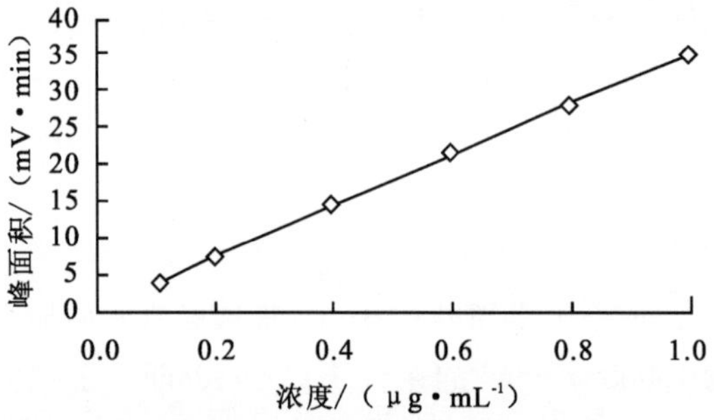

图 2　SA 的标准曲线
Fig. 2　The standard curve of SA

2.2　烟草叶片中 SA 的 HPLC 检测

按照上述检测条件和样品处理方法对 SA 进行检测，得到 SA 色谱图（图 3）。从图 3a 可以看出，保留时间为 4.45min 时 SA 色谱图出现峰值。

2.3　HPLC 检测方法的添加回收率

在样品中添加不同浓度的标准 SA，按照上述条件测定 SA 含量，结果如表 1 所示。从表 1 可看出，此方法的 SA 的添加回收率为 85.5%±3.5%，符合检测的要求，说明该测定方法比较准确。

2.4　CCP 对感病品种普通烟 K_{326} 叶片中 SA 含量的影响

分别于接种 TMV 后 12、36、60、84、108h 测定了 4 种处理（水、CCP、水 +TMV 和 CCP+TMV）叶片中的 SA 含量。从图 4 可以看出：CCP 和 TMV 均能诱导烟草体内 SA 含量的升高，CCP 处理组的烟草叶片 SA 含量在接种 TMV 后 12h 就明显提高，其诱导 SA 量与 TMV 本身诱导的 SA 量相当；但在接种 TMV 12h 后，其诱导 SA 量增加没有 TMV 本身显著。CCP 诱导的 SA 含量在接种 TMV 后 84h 达到最大值，为 458.927ng • g^{-1}；CCP+TMV 处理的普通烟 K_{326} 病叶中 SA 含量也高于水 +TMV 处理的病叶中 SA 含量，CCP+TMV 处理后 SA 含量最高可达 913.570ng • g^{-1}。这进一步说明除 TMV 本身诱导 SA 含量增加外，CCP 也可诱导 SA 含量的增加。

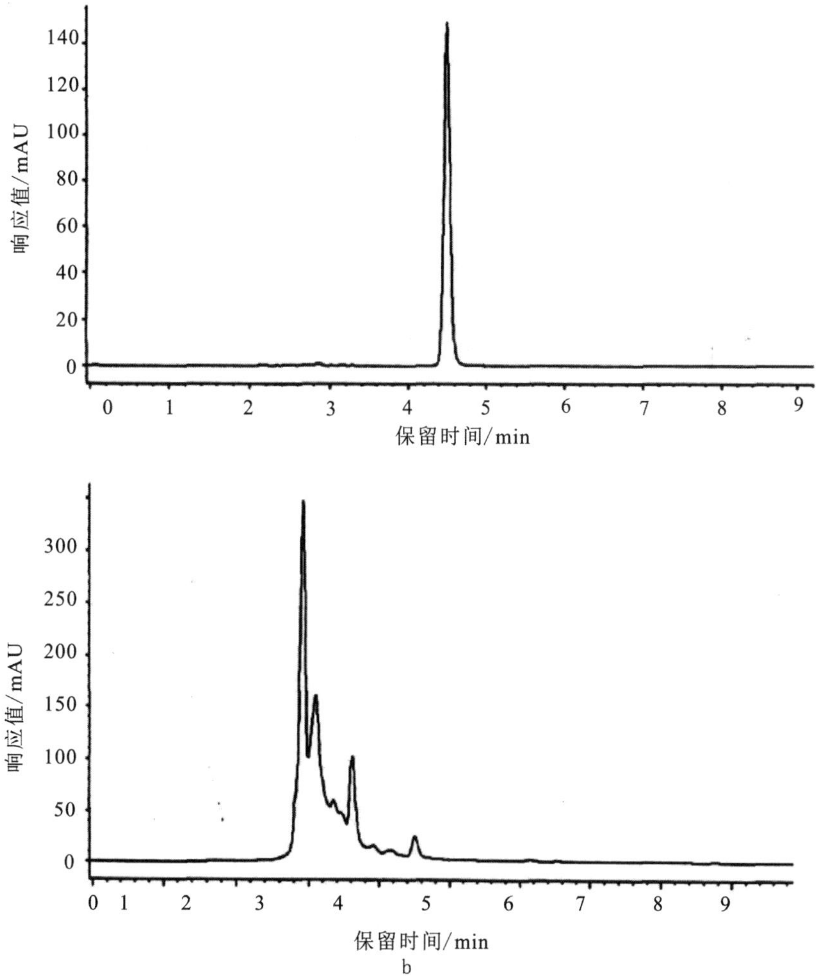

a. 标准SA；b.烟草叶片SA

图 3 标准 SA HPLC 色谱图

Fig. 3 The HPLC chromatogram of SA

表 1 HPLC 检测方法添加回收率

Table1 The recovery rates of SA by HPLC method

试验次序	SA 加入量 /μg	SA 测得量 /μg	回收率 /%	平均值 /%	标准偏差
1	0.5	0.395	79.0	85.5	3.5
2	0.5	0.427	85.4		
3	1.0	0.895	89.5		
4	1.0	0.864	86.4		
5	2.0	1.732	86.6		
6	2.0	1.718	85.9		

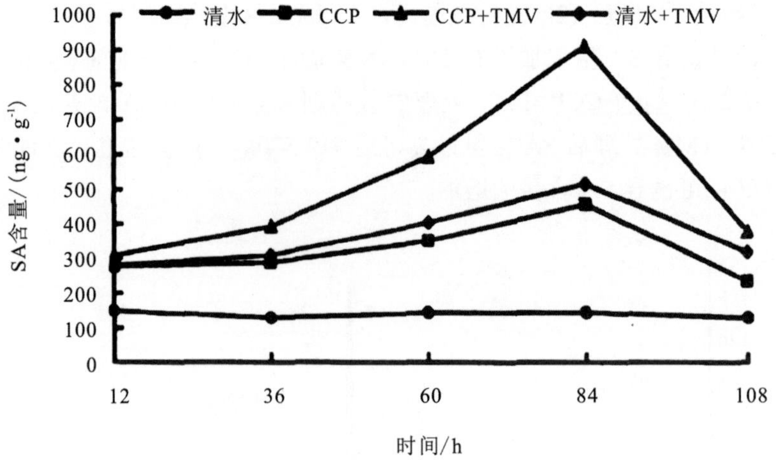

图 4　CCP 诱导处理普通烟 K_{326} 叶片中 SA 含量

Fig. 4　SA content in leaves of K_{326} treated with CCP

3　讨论

影响 SA 含量的因素很多，主要有提取溶液和流动相的 pH、温度、光照、抗感病品种、提取方法以及仪器本身的检测限和灵敏度。因此在测定 SA 的过程中，必须预先调整提取溶液和流动相的 pH，在接种 TMV 和烟草培育过程中要给予一定的光照（White, 1979; Enyedi et al., 1992）。Malamy 等发现在 22℃时抗病烟草植物接种 TMV 后，体内 SA 含量显著提高，并产生坏死反应（HR）和病程相关蛋白（PRPs）；但如果在 32℃的高温条件下，SA 含量就不会升高，HR 和 PRPs 也都受到抑制，此时施用外源SA 可以诱导PRPs的形成，但仍不产生HR（Malamy et al., 1992）。因此，温度的控制就显得尤为重要。

本试验采用的 HPLC 方法线性关系良好，提取 SA 的方法可以保证其与杂质很好地分离，检测方法的添加回收率符合检测要求，这些都说明此测定方法比较准确。CCP 处理的普通烟 K_{326} 健叶中 SA 含量高于清水处理的健叶中 SA 含量，CCP 处理后 SA 含量在 84h 达到最大值，为 458.927ng·g^{-1}；接种后，CCP+TMV 处理的普通烟 K_{326} 病叶中 SA 含量也高于水 +TMV 处理的病叶中 SA 含量，CCP+TMV 处理后 SA 含量最高可达 913.570ng·g^{-1}。在以前的试验中也发现，CCP 可以增强烟草的抗病性，表现为对 TMV 的侵染和扩展的抑制，以及保护酶系 POD、PPO、PAL、β-1，3- 葡聚糖酶、几丁质酶活性增强（吴艳兵等 , 2007a；吴艳兵等 , 2007b）。李宝聚等研究也表明随着接种病原菌时间的延长，葡聚六糖对葡萄霜霉病的防效下降（李宝聚等 , 2005）。余文英等研究表明外源 SA 能够降低膜脂过氧化作用，可诱导甘薯对薯瘟病产生抗性（余文英等 , 2008）。张晓燕等研究表明抗病毒 VA 可以诱导烟草产生 SA，且抗病品种 SA 的产生量明显高于感病品种，诱导烟草对 TMV 的抗性与 SA 的含量有密切关系（张晓燕等 , 2001）。本研究结果也表明 CCP 可以诱导普通烟叶片内 SA 含量升高，但 CCP 诱导烟草产生的抗性与 SA 含量的关系还有待今后更深入的研究。

参考文献

[1]李宝聚 , 范海延 , 孙艳秋 , 等 . 葡聚六糖诱导黄瓜体内水杨酸的积累及其与抗霜霉病关系的初步研究 [J]. 园艺学报 , 2005, 32 (1): 115-117.

[2] 刘艳芳 , 张劲松 . 鸡腿菇药理活性概述 [J]. 食用菌学报 , 2003, 10 (2): 60-63.

[3] 吴丽萍 , 吴祖建 , 林奇英 , 等 . 毛头鬼伞（*Coprinus comatus*）中一种碱性蛋白的纯化及其活性 [J]. 微生物学报 , 2003,43 (6): 793-798.

[4] 吴丽萍 , 吴祖建 , 林奇英 , 等 . 一种食用菌提取物 y3 对烟草花叶病毒的钝化作用及其机制 [J]. 中国病毒学 , 2004, 19 (1): 54-57.

[5] 吴艳兵，谢荔岩，谢联辉，等．毛头鬼伞多糖对烟草酶活性和同工酶谱的影响 [J]. 微生物学杂志，2007a, 27(5): 29-33.
[6] 吴艳兵，谢荔岩，谢联辉，等．毛头鬼伞多糖抗烟草花叶病毒（TMV）活性研究初报 [J]. 中国农学通报，2007b, 23 (5): 338-341.
[7]余文英，王伟英，邱永祥，等．水杨酸对甘薯抗薯瘟病和抗氧化酶系统的影响 [J]. 福建农林大学学报：自然科学版，2008, 37 (1): 23-26.
[8] 张超，操海群，陈莉，等．食用菌多糖对植物病毒抑制作用的初步研究 [J]. 安徽农业大学学报，2005, 32 (1): 15-18.
[9] 张晓燕，商振清，李兴红，等．抗病毒剂 VA 诱导烟草对 TMV 的抗性与水杨酸含量的关系 [J]. 河北林果研究，2001, 16 (4): 307-310.
[10] ENYEDI A J, YALPANI N, SILVERMAN P, et al. Localization, conjugation, and function of salicylic acid in tobacco during the hypersensitive reaction to tobacco mosaic virus[J]. Proceedings of the National Academy of Sciences of the United States of America, 1992, 89 (6): 2480-2484.
[11] MALAMY J, HENNIG J, KLESSIG D F. Temperature-dependent induction of salicylic acid and its conjugates during the resistance response to tobacco mosaic virus Infection[J]. The Plant Cell, 1992, 4 (3): 359-366.
[12] WHITE R F. Acetylsalicylic acid (aspirin) induces resistance to tobacco mosaic virus in tobacco[J]. Virology, 1979, 99 (2): 410-412.

Identification of two marine fungi and evaluation of their antivirus and antitumor activities

Shuo Shen[1, 2], Wei Li[1], Ming'an Ouyang[1, 2], Zujian Wu[2], Qiying Lin[2], Lianhui Xie[2]

(1 Key Laboratory of Bio-pesticide and Chemistry-Biology, Ministry of Education,Fujian Agriculture and Forestry University, Fuzhou 350002, China;

2 Institute of Plant Virology, Fujian Agriculture and Forestry University, Fuzhou 350002, China)

Abstract: To identify two marine fungi and evaluate the inhibitory effects of their crude extracts on *Tobacco mosaic virus* and two tumor cell lines.Crude extracts was obtained by extracting with MeOH and evaporated in vacuo. The extracts was water-soluble fraction which was dissolved in water, and the other fraction was water insoluble. The fungi were identified by morphology and Internal Transcribed Spcer (ITS) rDNA molecular methods. The inhibitory effect on *Tobacco mosaic virus* was evaluated by indirect enzyme linked immuno- sorbent assay, and the anti-tumor activity was tested by methyl thiazolyl tetrazoliummethod.The fungi were identified as *Penicilliumoxalicum* and *Neosartoryafischeri* There crude extracts inhibited *Tobacco mosaic virus* and two tumor cell lines. The active fraction named 0312F_1 inhibited *Tobacco mosaic virus* and tumor cell lines and was water-soluble. The fraction named 1008F_1 inhibited *Tobacco mosaic virus* and was insoluble inwater, whereas the fraction inhibited tumor cell lines was water-soluble.The active fraction named 0312F_1 inhibited *Tobacco mosaic virus* was different from that named 1008F_1inhibited *Tobacco mosaic virus*. The active fraction named 0312F_1 inhibited tumor cell lines was the same as that named 1008F_1. Furthermore, the inhibitory activity of water-soluble fraction named 0312F_1 against BEL-7404 cell line was much higher than that against SGC-7901 cell lines, whereas the inhibitory activity of active fraction named 1008F_1 against SGC-7901 cell line was muchhigher.

Keywords: marine fungi; inhibitory activity; *Tobacco mosaic virus* (TMV); anti-tumor; identification

1 Introduction

The oceans are unique resources that provide a diverse array of natural products (Cragg and Newman, 1999; Schwartsmann et al., 2000). The marine environment has more than a million described species (Nuijen et al., 2000). Microorganisms are increasingly exploited as a source of new pharmaceuticals (Nicoletti et al., 2008). Due to their huge numbers, their activities have a global impact on Earth. This is particularly relevant when considering marine microorganisms (Pedrós-Alió and Simó, 2002). The chemical and biological diversity of the marine environment is immeasurable and therefore is an extraordinary resource for the discovery of new drugs (Simmons et al., 2005). Over the past few years, about 3000 new compounds from various marine sources have been reported and some have entered clinical trials (Carte, 1996; Schwartsmann et al., 2001). In this paper, we describe the isolation, biological activities of 50 strains isolated from marine organisms. Finally, we identified 2 marine fungi with both higher inhibitory activities against cancer cells proliferation and TMV replication under both morphology and molecular methods.

Acta Microbiologica Sinica. 2009, 49(9): 1240-1246.

Received 21 December 2008; Accepted 19 June 2009.

2 Materials and methods

2.1 Meterials

2.1.1 Culture media

PDA solid and liquid media, Gause's medium.

2.1.2 Cancer cell lines

SGC-7901 and BEL-7404 tumor cell lines were provided by Hepatobiliary Surgery Institute of Fujian Province, Union Hospital affiliated with Fujian Medical University.

2.1.3 PCR reaction primers

ITS1: 5′-TCCGTAGGTGAACCTGCGG-3′,ITS4: 5′-TCCTCCGCTTATTGATATGC-3′(White et al., 1990; Gardes and Bruns, 1993).

2.1.4 Reagents

All PCR reactions related reagents were bought from TaKaRa company. DNA mag-extracted kitwas bought from TOYOBO company. Other reagents were all analytic reagents.

2.2 Sample collection, isolation and cultivation of fungi and actinomycetes

In September, 2006, 14 samples were collected at Changle beach in Fujian, Fuzhou, China. The samples included seasediments, seaweeds, crabs, sea anemones, several types of snails, medusa and mussels, et al.

The strains were isolated by the following procedures (Atlas and Parks, 1993; Imhoff and Stöhr, 2003), then, the plates were incubated at 28℃ for 3 to 20d. After that, the isolates were grown on PDA solid and liquid media. After different periods of incubation, the broths were harvested (Val et al., 2001).

2.3 Cancer cell lines culture conditions

Cancer cell lines were maintained in RPMI 1640 medium supplemented with 10% fetal bovine serum and 100 IUml of penicillin-streptomycin in an atmosphere of 5% CO_2 at 37℃ .

2.4 Extraction procedure

The fermentation broth was extracted using the methods described in our previous work (Wu et al., 2007a). The other fraction was water insoluble fraction. The extracts were dissolved in DMSO, and diluted to different concentrations.

2.5 Bioactivity assays

The inhibitory activity against TMV replication was performed using the methods described in our previous work (Wu et al., 2007b).

The inhibitory activity against cancer cells proliferation was performed using MTT assay (Twentyman et al., 1989). 5′-Fluoruoracial at a concentration of 50μg/mL was as the positive control. Experiments were conducted in triplicate.

2.6 IC_{50} measurement

IC_{50} values were determined by the inhibitory activities against cancer cell proliferation and TMV replication. The 50% inhibitory concentration for each extract was calculated from concentration-effect-curves after linear regression analysis.

2.7 Morphology identification of fungi with bioactivity

Fungi were identified to the genus level based on microscopic morphology. The results were according to Handbook of fungal identification (Wei, 1979).

2.8 Molecular identification of fungi with bioactivity

The fungi genomic DNA were extracted using DNA mag-extracted kit. The sequences of the ITS regions were amplified by PCR reactions using primers of ITS1 and ITS4. PCR reactions were carried out in a final volume of 25μL (1μL genomic DNA template, 12.75μL ddH_2O,5μLTakara PCRbuffer, 5μLdNTPs (2.5mol/L), 0.5μL of each primer (10μmol/L), and 0.25μL Qiagen HotStarTaq enzyme). After an initial denaturation for 10min at 95℃, the reaction was run for 36 cycles with the following parameters: denaturation for 1min at 94℃ , annealing for 1min at 50℃ and extension for 1 min at 72℃ . Final extension followed for 10min at 72℃ . Subsequently, the PCR products were sequenced and compared with the sequence reported in GeneBank. Sequence homology trees were constructed by DNAMAN 6.0.40, and phylogenetic trees were constructed under maximum likelihood (ML) method.

2.9 Statistical analysis

Statistical analysis was performed using SPSS 13.0. One-way ANOVA was used to analyze statistical comparisons between groups. Differences with P-values less than 0.05 were considered to be statistically significant.

3 RESULTS

3.1 Bioactivities of the fermentation broths of all isolates

The extracts from the fungi $0312F_1$ and $1008F_1$ showed both anti-TMV replication and anti-tumor activities (higher than 50%) (Table 1). To isolate the active fractions of these two fungi, the fermention broth was extracted with MeOH.

Table 1 The taxa with anti-TMV replication, anti-tumor activity

Strains No.	Inhibition rate[a](%)	Inhibition rate[b](%)	Inhibition rate[c](%)	Strains No.	Inhibition rate[a](%)	Inhibition rate[b](%)	Inhibition rate[c](%)
$0101F_2$	30.27	36.92	12.79	$0702F_2$	—	48.25	46.73
$0102F_1$	34.02	—	1.07	$0701F_3$	10.16	45.67	36.36
$0103F_1$	—	56.00	65.17	$0702F_3$	46.70	35.13	20.73
$0301F_1$	—	20.85	35.22	$0801F_1$	18.95	41.22	41.69
$0302F_1$	21.88	3.38	41.88	$1001F_1$	66.56	21.82	26.35
$0303F_1$	—	54.24	13.09	$1002F_1$	60.96	15.18	68.97
$0304F_1$	—	11.03	1.81	$1003F_1$	44.69	21.90	42.49
$0305F_1$	19.23	32.38	58.72	$1004F_1$	32.57	47.98	41.57
$0306F_1$	—	17.07	4.53	$1005F_1$	47.02	20.54	10.65
$0307F_1$	27.23	16.84	14.76	$1006F_1$	—	18.88	7.36
$0308F_1$	39.20	21.39	1.76	$1007F_1$	57.96	29.22	36.36
$0309F_1$	42.14	—	59.29	$1008F_1$	65.89	63.60	56.84
$0311F_1$	—	32.05	20.39	$1009F_1$	40.14	13.52	60.25
$0312F_1$	65.55	52.24	84.61	$1010F_1$	59.63	35.88	19.89
$0313F_1$	58.12	33.93	23.77	$1011F_1$	—	26.81	6.23
$0317F_1$	31.69	63.77	49.94	$1001F_2$	41.25	1.53	31.41
$0318F_1$	—	25.76	13.36	$1002F_2$	20.97	58.75	63.58

续表

Strains No.	Inhibition rate^a(%)	Inhibition rate^b(%)	Inhibition rate^c(%)	Strains No.	Inhibition rate^a(%)	Inhibition rate^b(%)	Inhibition rate^c(%)
$0301F_2$	64.46	34.96	21.89	$1101F_1$	51.95	34.59	1.83
$0302F_2$	47.98	46.62	16.14	$1102F_1$	30.05	19.67	13.27
$0303F_2$	10.71	73.85	68.56	$1104F_1$	—	30.39	12.07
$0301F_3$	23.58	31.12	45.53	$1105F_1$	25.65	35.70	21.49
$0302F_3$	66.49	35.10	56.03	$1301F_2$	13.66	—	31.20
$0402F_1$	26.96	2.40	52.99	$1302F_2$	42.12	35.19	—
$0403F_1$	27.68	1.97	16.88	$1401F_3$	19.02	43.79	21.17
$0502F_1$	46.98	31.56	38.59	$1001B_1$	18.77	18.41	30.17
$0701F_1$	55.07	38.81	—	$1101B_1$	28.98	6.20	13.25

Fungi fermentation broth diluted in 10^{-1}.^{a}: anti-TMV replication rates of fermentation broth of strains. ^{b}: anti-tumor activity rates of fermentation broth of strains against SGC-7901 cell line. ^{c}: anti- tumor activity rates of fermentation broth of strains against BEL-7404 cell line.

3.2 Bioactivities of the MeOH extracts of two fungi

The anti-TMV replication and anti-tumor fractions of stain $0312F_1$were both water soluble, the anti-TMV replication fraction of stain $1008F_1$was insoluble in water, whereas the anti-tumor fraction was also water soluble (Table 2).

3.3 Morphology identification of two active fungi

Themorphologic character of strain $0312F_1$ was similar to *Penicillium sp.* based on microscopic morphology after growing on the solid medium for about 7d at 28℃ (Fig. 1).

The morphologic character of strain $1008F_1$ was similar to *Aspergillus sp.* based on microscopic morphology after growing on the solid medium for about 7d at 28℃ (Fig. 2).

Table 2 The IC_{50} values of two fungi with bioactivity (*$p < 0.05$)

No.	IC_{50}^a (mg/mL±SD)	IC_{50}^b (mg/mL±SD)	IC_{50}^c (mg/mL±SD)
A1	$0.554^* \pm 0.23$	$2.921^* \pm 0.24$	$0.107^* \pm 0.08$
C1	1.807 ± 0.17	244.119 ± 12.36	1.604 ± 0.54
A2	2.136 ± 0.92	$0.020^* \pm 0.01$	$0.198^* \pm 0.06$
C2	$0.871^* \pm 0.42$	1.130 ± 0.35	0.908 ± 0.10

Extracts concentration diluted in 2mg/mL,1mg/mL,0.5mg/mL,0.25mg/mL and 0.125mg/mL.

A1: water-soluble fraction of MeOH extracts of strain 0312F1. A2: water-soluble fraction of MeOH extracts of strain 1008F1. C1: water insoluble fraction of MeOH extracts of strain 0312F1. C2: water insoluble fraction of MeOH extracts strain 1008F1.a: IC50 values of extracts of two fungi of anti-TMV replication rates, b: IC50 values of extracts of two fungi of anti-tumor activity rates against SGC-7901 cell line,c: IC50 values of extracts of two fungi of anti-tumor activityrates against BEL-7404 cell line. One-way ANOVA was used to analyze statistical comparisons between groups A1 and C1, A2 and C2, separately.

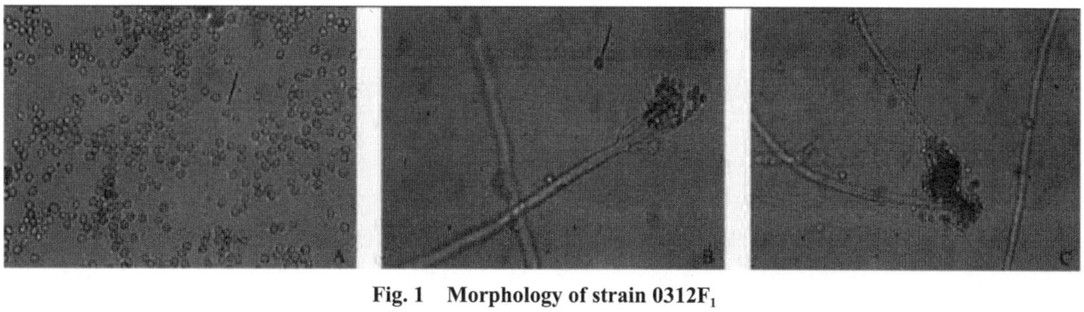

Fig. 1 Morphology of strain $0312F_1$

A: Conidia of strain $0312F_1$ (400×); (B) Conidiophores of strain $0312F_1$ (400×). (C) Conidia and conidiophores of $0312F_1$ (400×)

Fig.2 Morphology of strain $1008F_1$

(A) Conidiophores of strain $1008F_1$ (400×); (B) Conidia of strain $1008F_1$ (400×); (C) Conidia and conidiophores of $1008F_1$(400×)

3.4 Molecular identification of two active fungi

Comparison study supported a strong relationship between strain $0312F_1$ (GenBank accession number: EU926977) and members of *Penicillium sp.*, and particularly revealed the highest homology(100%)with EF103455.1 (*Penicillium oxalicum*). Homology relationship of closely related microorganisms is shown (Fig. 3).

Comparison study supported a strong relationship between strain $1008F_1$ (GenBank accession number: EU926976) and members of *Neosartorya sp.* (sexual phase of *Aspergillus sp.*) and particularly revealed the highest homology (100%) with AF176661.1 (*Neosartorya fischeri*). Homology relationship of closely related microorganisms is shown (Fig. 4).

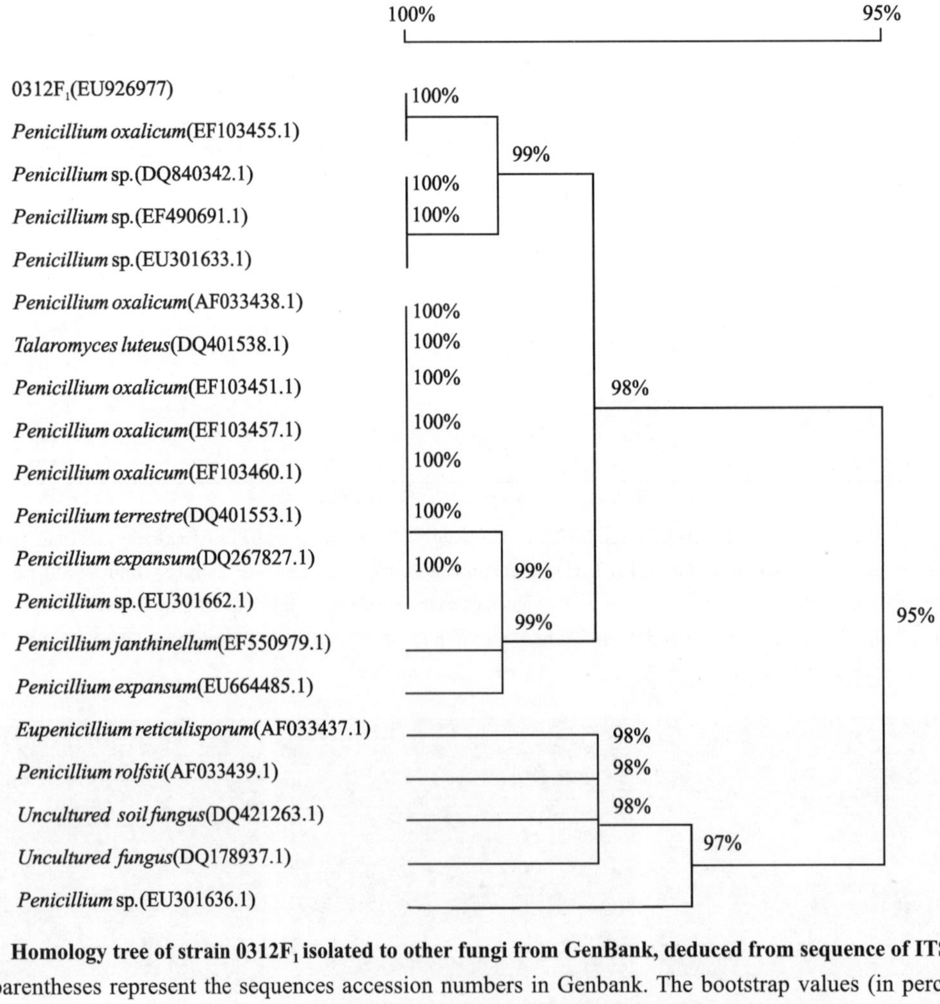

Fig.3 Homology tree of strain $0312F_1$ isolated to other fungi from GenBank, deduced from sequence of ITS rDNA

Numbers in parentheses represent the sequences accession numbers in Genbank. The bootstrap values (in percent) calculated illustrated homology relationship between strain $0312F_1$ and other strains. Bar, 95%–100% homology persentage

4 Discussions

There have been many researches about Isolation of *Penicillium sp.* and *Aspergillus sp.* frommarine organisms in china particularly from Mangrove (Liu et al., 2007; Meng et al., 2007). So strains in *Penicillium sp.* and *Aspergillus sp.* are universal. But most of them are associated with diverse bioactivity, for example, antitumor activity, antimicrobial activity, and activity associated producing related enzymes (Li et al., 2000; Li et al., 2001; Yang, 2002; Zhang et al., 2002; Wang and Yue, 2005; Chen and Zhang, 2006; Tan et al., 2006; Han et al., 2007; Zhou et al., 2007). Strains associated with anti-phytovirus activity are less reported compared to these.

The fermentation broths of most of the strains showed anti-tumor activities against two cell lines (Table 1), but the inhibition rates against SGC-7901 cell line were lower than those against BEL-7404 cell line. It might be a result of tumor cell lines specificity.

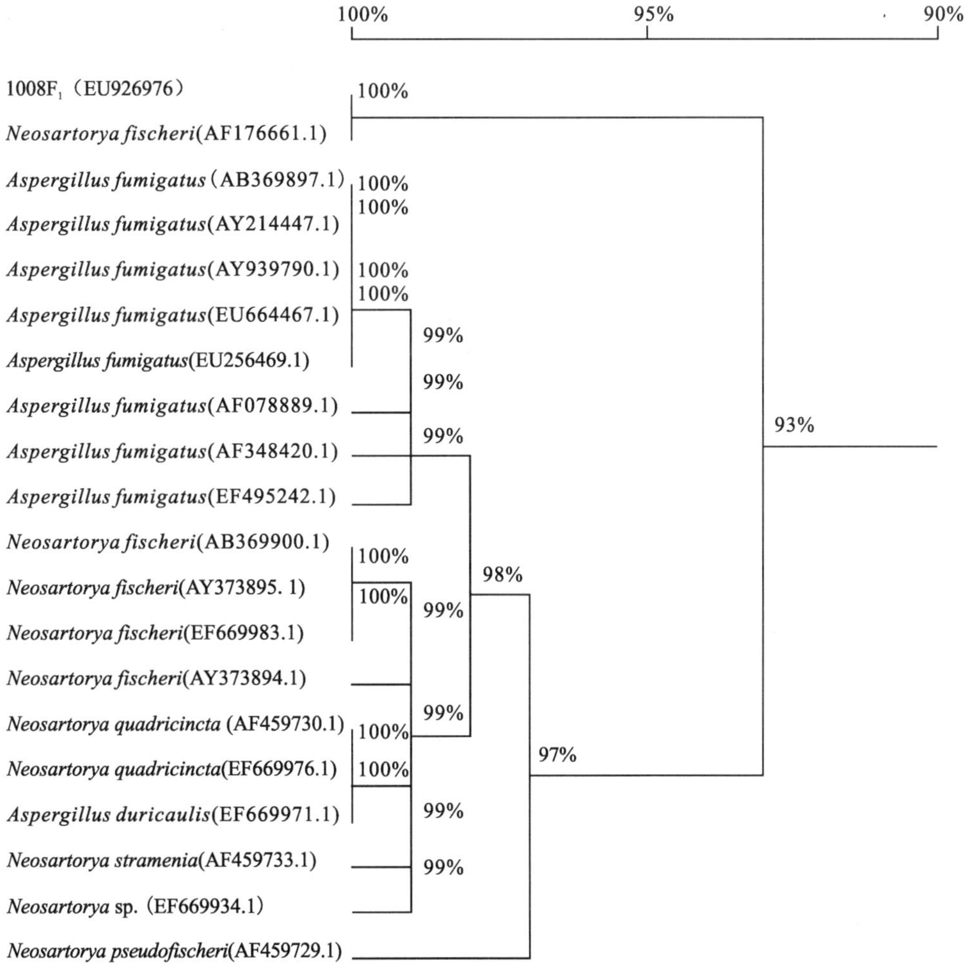

Fig.4 Homology tree of strain 1008F$_1$ isolated to other fungi from GenBank, deduced from sequence of ITS rDNA

Numbers in parentheses represent the sequences accession numbers in Genbank. The bootstrap values (in percent) calculated illustrated homology relationship between strain 1008F$_1$ and otherstrains. Bar, 90%–100% homology persentage

Since the 1990s, marine-derived fungi have been recognized as a rich source of novel bioactive metabolites (Zhang et al., 2007b). The IC$_{50}$ values (Table 2) of the crude extracts were in a range of 0.020mg/mL to 244.119mg/mL, from which the active fractions were identified. A lot of anti-tumor and anti-microbial compounds of alkaloids have been isolated from strains in *Aspergillus sp.*, suggesting that active fractions might consist of alkaloids to a certain extent (Zhang et al., 2007a; Zhao et al., 2007). Our further research will emphasize isolation of the active compounds from the extracts from fermentation broth of the two fungi.

References

[1] ATLAS R M, PARK L C. Handbook of Microbiological Media[M]. Boca Raton: CRC Press Inc, corporate Blvd. 2000.

[2] CARTE B K. Biomedical potential of marine natural products[J]. Bioscience, 1996, 46: 271-286.

[3] CHEN B, ZHANG S. Antimicrobial Characteristics of Marine Fungi Aspergillus sp.MF134. Journal of Huaqiao University, 2006, 27: 307-309.

[4] CRAGG G M, NEWMAN D J. Discovery and development of antineoplastic agents from natural sources[M]. Taylor & Francis, 1999, 17: 153-163.

[5] GARDES M, BRUNS T D. ITS primers with enhanced specificity for basidiomycetes- application to the identification of mycorrhizae and rusts[J]. Molecular Ecology, 1993, 2: 113-118.

[6] HAN X, XU X, CUI C. Diketopiperazines produced by marine-derived Aspergillus fumigatus H1-04 and their antitumor activities[J]. Chinese Journal of Medicinal Chemistry, 2007, 17: 155-159.

[7] IMHOFF J F, STÖHR R. Sponge-associated bacteria: general overview and special aspects of bacteria associated with Halichondria panicea[J]. Progress in Molecular & Subcellular Biology, 2003, 35-57.

[8] LI S, ZHONG Y, DAI X, et al. Study of Anti-Fungi Action of Marine Mold I Isolation of Marine Mold M182 Having Broad Spectrum Antimicrobiol Activities[J]. Marine Science Bulleten, 2000, 19: 29-33

[9] LI S, WANG J, YANG J, et al. Isolation, identification and characterization of a fungus strain producing antifungal antibiotic[J]. Mycosystema, 2001, 20: 362–367.

[10] LIU A, WU X, XU T. Research advances in endophytic fungi of mangrove[J]. Chinese Journal of Applied Ecology, 2007,18(4): 912- 918.

[11] MENG Q H, ZHANG H, ZHANG J. Isolation and Identification of Marine Fungi from Lüju River in Bohai-Sea[J]. Journal of Tianjin Normal University, 2007, 27: 28-31.

[12] NICOLETTI R, BUOMMINO E, DE FILIPPIS A, et al. Bioprospecting for antagonistic Penicillium strains as a resource of new antitumor compounds[J]. World Journal of Microbiology&Biotechnology, 2008, 2: 189-195.

[13] NUIJEN B, BOUMA M, MANADA C, et al. Pharmaceutical development of anticancer agents derived from marine sources[J]. Anti-cancer drugs, 2000, 11: 793-811.

[14] PEDRÓS-ALIÓ C, SIMÓ R. Studying marine microorganisms from space[J]. International Microbiology, 2002, 5: 195-200.

[15] SCHWARTSMANN G, BRONDANI A, BERLINCK R,et al. Marine organisms and other novel natural sources of new cancer drugs[J]. Annals of Oncology Official Journal of the European Society for Medical Oncology, 2000, 11: 235-243.

[16] SCHWARTSMANN G, DA ROCHA A B, BERLINCK R G, et al. Marine organisms as a source of new anticancer agents[J]. Lancet Oncology, 2001, 2: 221-225.

[17] SIMMONS T L, ANDRIANASOLO E, MCPHAIL K, et al. Marine natural products as anticancer drugs[J]. Molecular Cancer Therapeutics, 2005, 4: 333-342.

[18] TAN Z Q, ZHANG R Y, YANG J S, et al. Antagonistic Action Comparison of Marine Penicillium sp. T03 to Four Plant Pathogens[J]. Chinese Journal of Biological Control, 2006, 10: 112-114

[19] TWENTYMAN P R, FOX N E, REES J K. Chemosensitivity testing of fresh leukaemia cells using the M TT colorimetric assay[J]. British Journal of Haematology, 1989, 71: 19-24.

[20] VAL A, PLATAS G, BASILIO A, et al. Screening of antimicrobial activities in red, green and brown macroalgae from Gran Canaria (Canary Islands Spain)[J]. International Microbiology, 2001, 4: 35-40.

[21] WANG T, YUE S. Morphological observations and studies on cellulase properties of the oceanic penicillium sp. FS010441. Shandong Nongye Kexue (China), 2005.

[22] WEI J. Handbook of fungal identification[J]. Shanghai Science and Technology Press, 1979, 1-802.

[23] WHITE T, BRUNS T, LEE S, et al. Amplification and direct sequencing of fungal ribosomal RNA genes for phylogenetics //PCR protocols: A guide to methods and applications[J]. Academic Press San Diego CA, 1990.

[24] WU Z-J, OUYANG M-A, WANG C-Z, et al. Six new triterpenoid saponins from the leaves of Ilex oblonga and their inhibitory activities against TMV replication[J]. Chemical & Pharmaceutical Bulletin, 2007a, 55: 422-427.

[25] WU Z-J, OUYANG M-A, WANG C-Z, et al. Anti-tobacco mosaic virus (TMV) triterpenoid saponins from the leaves of Ilex oblonga[J]. Journal of Agricultural & Food Chemistry, 2007b, 55: 1712-1717.

[26] YANG J. Antimicrobial activity of secondary metabolites of marine Penicillum spp[J]. Chinese Journal of Tropical Crops, 2002, 23: 73-76.

[27] ZHANG G, WANG T, ZHAG Z, et al. Screening and characterization of the psychrophilic amylase-producing fungus[J]. Marine Sciences, 2002, 26: 3-5.

[28] ZHANG M, FANG Y, ZHU T, et al. Study on Indole-Quinazolines Alkaloids from Marine-Derived Fungus Aspergillus sydowi PFW-13 and their Anti-Tumor Activities[J]. Journal of Chinese Pharmaceutical Sciences, 2007a, 42(24): 1848-1851.

[29] ZHANG Y, WANG S, LI X M, et al. New sphingolipids with a previously unreported 9-methyl-C20-sphingosine moiety from a marine algous endophytic fungus Aspergillus niger EN-13[J]. Lipids, 2007b, 42: 759-764.

[30] ZHAO W-Y, ZHU Q-S, GU Q-J. Studies on the Chemical Constituents of Secondary Metabolites of Marine-derived Aspergillus fumigatus (Ⅱ) [J]. Journal of Qingdao University of Science and Technology (Natural Science Edition), 2007, 28: 199-201

[31] ZHOU M-H, ZHAO X-W, ZHOU L, et al. Screening of amylase excreting marine Aspergillus sp. and preliminary studies on the characteristics of the amylases[J]. Journal of Marine Sciences, 2007.25: 59-65

16 个水稻品种对水稻矮缩病毒抗性的鉴定

吴建国，巴俊伟，李冠义，林奇英，吴祖建，谢联辉

（福建农林大学植物病毒研究所，福建省植物病毒学重点实验室，福建福州 350002）

摘要：16 个水稻品种对水稻矮缩病毒 (*Rice dwarf virus*，RDV) 的抗性鉴定结果表明：不同水稻品种对 RDV 的抗性存在明显差异。根据各品种的发病率划分其抗性等级。免疫品种 1 个，占供试品种的 6.25%；中抗品种 6 个，占 37.50%；中感品种 3 个，占 18.75%；高感品种 6 个，占 37.50%。感病品种潜育期较短，为 11 ～ 15d；抗病品种潜育期较长，为 15 ～ 20d。水稻品种的发病率与叶蝉带毒率密切相关，相关系数为 99.99%。协优 46 兼具抗虫性和抗病性；宜香 2292 仅具抗病性；冈优 734 为免疫品种，是水稻矮缩病抗性育种的良好材料。

关键词：水稻矮缩病毒；品种抗性；叶蝉；抗病性；抗虫性

中图分类号：S4　**文献标识码：**A　**文章编号：**1671-5470（2010）01-0010-05

Identification of the resistance of 16 rice varieties to *Rice dwarf virus*

Jianguo Wu, Junwei Ba, Guanyi Li, Qiying Lin, Zujian Wu, Lianhui Xie

(Institute of Plant Virology, Fujian Agriculture and Forestry University, Key Laboratory of Plant Virology of Fujian Province, Fuzhou, Fujian 350002, China)

Abstract: Susceptibilities of 16 bred or extended rice varieties to *Rice dwarf virus* (RDV) were investigated using pathogens is rate as an indicator of resistance grade. The results showed that significant variance existed among the varieties. Among the 16 varieties studied, 1 variety was immune to RDV, accounting for 6.25%; 6 varieties were middle-resistant, accounting for 37.5%; 3 varieties were middle–susceptible, accounting for 18.75%; 6 varieties were disease-sensitive, accounting for 37.5%. The latent period (11–15d) of susceptible variety, was significantly shorter than that of resistance variety (15–20d). Further investigations revealed that the infection rate of rice variety correlated with the ration of leaf hopper (*Nephotrttix cincticeps*) carrying virus, and the correlation coefficient was 99.99%. The results of identification were as follows: both of RDV resistance and insect resistance were high in Xieyou 46 (*Oryza sativa Lsspindicacv* Xieyou 46); Yixiang 2292 (*Oryza sativa Lsspindicacv* Yixiang 2292) had resistance to RDV; Ganyou 734(*Oryzasativa Lsspindicacv* Ganyou 734) was an immune variety, which was proved to be an ideal material for RDV resistance breeding.

Keywords: *Rice dwarf virus* (RDV); variety resistance; leafhopper (*Nephotrttix cincticeps*); disease-resistant character; insect-resistant character

水稻矮缩病毒（*Rice dwarf virus*, RDV），属于呼肠孤病毒科（*Reoviridae*）植物呼肠孤病毒属

福建农林大学学报 (自然科学版). 2009, (01): 10-14

收稿日期：2009-04-20

基金项目：国家自然科学基金资助项目（30770090）；教育部博士点专项科研基金资助项目（20050389009）；福建省科技厅资助项目（K03005）

(*Phytoreovirus*)，同属成员还有水稻瘤矮病毒（*Rice gall dwarf virus*, RGDV）和三叶草伤瘤病毒（*Wound tumor virus*, WTV）（Boccardo et al.,1984; Fauquet et al.,2005）。RDV 全基因组由 12 条 dsRNA 组成，完整的 RDV 病毒粒体为球形正二十面体结构，直径为 70nm 左右（Zheng et al., 2000; Nakagawa et al., 2003）。在自然条件下，RDV 由黑尾叶蝉（*Nephotettix cincticeps*）或电光叶蝉（*Recilima dorsal*）传播，而且能在叶蝉体内复制，经卵传给子代（Boccardo et al.,1984）。该病毒广泛分布于日本、朝鲜、菲律宾、尼泊尔等东南亚水稻种植区，造成多种水稻品种的矮缩和减产，在我国南方水稻产区普遍发生，造成重大经济损失（李毅等 , 2001）。

水稻品种对水稻病毒抗性方面的研究国内外已有报道。1963 年日本中国农业实验场发现巴基斯坦籼稻 Modan 与农林 8 号回交 5 次的 B5F7 系统表现高抗，并开始了对抗水稻条纹叶枯病育种的系统研究（陈涛等 , 2006）。邢祖颐等（刑祖颐等 , 1985）对水稻条纹叶枯病抗源筛选和抗性育种方面进行研究，筛选并鉴定出塔杜康（Todukan）、特特普（Tetep）、窄叶青 8 号等抗性品种，同时选育出中作 180、中作 9 号及中作 59 号等抗性品种。张仲凯等选用转化 RDV 编码非结构蛋白基因（S6）、编码小核心蛋白基因（S7）和编码外壳蛋白基因（S8）的转基因水稻株系，通过田间试验测定其对 RDV 的抗性，结果表明，RDV-S6 株系的抗性较转 RDV-S7 和 RDV-S8 株系强（张仲凯等 , 2006）。Zheng et al. 将水稻矮缩病毒外壳蛋白基因转入粳稻“中华 8 号”中，获得了抗 RDV 的转基因水稻。Shimizu et al. 针对 RDV 的 S12（Pns12）和 S4（Pns4）分别构建 RNAi 载体转化水稻，发现针对 Pns12 的 RNAi 转基因水稻表现出对 RDV 高度抗性，而 Pns4 则表现出弱抗性，并延迟了 RDV 症状的出现。

而不同水稻品种对水稻矮缩病毒的抗性研究尚未见报道。鉴于此，本试验旨在根据当前水稻矮缩病发生现状及其特点，利用当前农业生产上大面积推广的优良粳稻和籼稻品种，筛选抗 RDV 的水稻品种，以期找到抗性较好的优良水稻品种，为水稻矮缩病的防治提供依据。

1 材料与方法

1.1 供试材料

1.1.1 毒源

水稻矮缩病毒由福建省植物病毒学重点实验室分离纯化，并保存在水稻感病品种台中 1 号上。

1.1.2 昆虫

为福建农林大学植物病毒研究所试验田的黑尾叶蝉（*Nephotrttix cincticeps*）和本研究所人工饲育的无毒叶蝉种群。

1.1.3 水稻品种

武育粳 3 号、楚粳 27、楚粳 26、合系 39、楚恢 15、楚粳 24、沈农 606、冈优 182、川丰 2 号、T 优 5570、协优 46、威优 77、冈优 734 均为本实验室保存；宜香 2292 由四川省宜宾市农业科学研究所林纲老师惠赠；岫 136-12 和岫 87-15 由云南省保山市农业科学研究所钏兴宽老师惠赠。

1.1.4 试剂与仪器

主要试剂有碳酸盐、PBST 和 PBST-PVP 缓冲液、RDV 抗血清（本实验室制备），以及羊抗兔抗体和底物缓冲液（对一硝基苯基磷酸二钠、PNPP，均为 Sigma 产品）。主要仪器有酶联仪（BioRad）等。

1.2 方法

1.2.1 间接 ELISA

参考谢联辉等（谢联辉等 , 2004）方法。在进行水稻田间抗病性比较试验前，对试验田间（田间 RDV 发病率在 60% 左右）的叶蝉进行 ELISA 带毒率检测，结果表明，阳性、阴性和阴性对照（CK）的

D_{405nm} 值分别为 0.17 ～ 0.19、0.07 ～ 0.09 和 0.09 ～ 0.10。以检测样品 D_{405nm} 与阴性对照样品 D_{405nm} 的比值≥ 2 记为阳性。检测结果表明田间叶蝉带毒率为 14.23%。

1.2.2 高带毒率叶蝉的筛选与饲育

取长期饲养的无毒黑尾叶蝉转入防虫网内病株症状明显的水稻矮缩病病株上群体饲毒 3 ～ 5d 后，将叶蝉转入试管内，纱布封口，供水稻品种的抗性鉴定试验（单虫单苗测定）；或将无毒黑尾叶蝉转入防虫网，在 25 ～ 30℃长期饲育于水稻矮缩病病株上。水稻品种抗性鉴定试验的供试黑尾叶蝉虫龄均一致。

1.2.3 品种抗病性测定

品种抗病性测定试验在本所隔虫区内完成。采取集团接种法，即将长期饲育的带毒黑尾叶蝉置于防虫箱内接种。网箱的空间为 60cm×60cm×90cm，接种苗龄为二叶一心期，每苗有带毒叶蝉 1 ～ 2 头，让其传毒取食 5 ～ 7d，每天用竹竿在网笼内赶虫 2 ～ 3 次，尽量保证传毒的相对一致。每次各水稻品种测定 30 ～ 50 株，重复 3 次，然后将接种后的稻苗移栽于水泥池内，罩上防虫网，观察发病情况并记录数据及统计发病率。按照全国水稻病毒病科研协作组统一制定的抗性分级标准，以对照品种岫 13612 的发病率为 100%，统计各品种的相对发病率。依据发病率划分抗性等级，共 5 级。0 级：免疫（immunity, IM），不发病；1 级：高抗（highly resistant, HR），发病率为 0.10% ～ 5.00%。2 级：中抗（moderately resistant, MR），发病率为 5.10% ～ 30.00%。3 级：中感（moderately susceptible, MS），发病率为 30.10% ～ 60.00%。4 级：高感（highly susceptible, HS），发病率为 60.10% 以上（谢联辉等 , 1982）。

选择从以上试验获得的抗感差异显著的水稻品种进行进一步鉴定。抗性鉴定采用长期饲育在水稻矮缩病病株上的 2 ～ 3 龄若虫，以单虫单苗法进行品种抗性接种鉴定，每一水稻品种接种 30 ～ 50 株，重复 3 次，苗龄在二叶一心期，传毒 2 ～ 3d。将接种的幼苗移栽到水泥池中，罩上防虫网，观察并记录发病情况。

1.2.4 水稻品种抗虫性测定

选择抗感差异显著的水稻品种，将虫龄一致的叶蝉进行单虫单苗接种试验，每一品种接种 40 ～ 60 株，重复 3 次，并记录接种后 1 ～ 7d 的叶蝉死亡情况（以叶蝉死亡率衡量水稻品种抗虫性）。

2 结果与分析

2.1 不同水稻品种抗病性差异显著性

16 个水稻品种与水稻矮缩病毒互作的差异显著性测定结果（表 1）表明：冈优 734 为免疫品种，占供试品种的 6.25%；冈优 182、川丰 2 号、T 优 5570、宜香 2292、协优 46 和威优 77 品种表现为中抗，占 37.5%；楚恢 15、楚粳 24、沈农 606 表现为中感，占 18.75%；岫 136-12（CK）、岫 87-15、武育粳 3 号、楚粳 27、楚粳 26、合系 39 为高感品种，占 37.5%。从各品种的发病潜育期看，感病品种潜育期较抗病品种短，为 11 ～ 15d；抗病品种潜育期较长，为 15 ～ 20d。

对各水稻品种发病率的差异显著性分析表明：（1）不同水稻品种间的抗病性存在明显差异，岫 136-12（CK）和岫 87-15 与合系 39、楚恢 15 和楚粳 24 等均存在显著差异，与冈优 734、冈优 182、T 优 5570、宜香 2292 和协优 46 等均存在极显著差异；（2）同为粳稻品种的岫 136-12（CK）、岫 87-15、武育粳 3 号、楚粳 27、楚粳 26、合系 39、楚恢 15 和楚粳 24 等，前 6 个品种表现为高感，后 2 个品种表现为中感，显著性分析表明前 5 个品种与后 2 个品种的抗病性存在显著差异；（3）籼稻品种均比粳稻品种抗病，本试验中表现免疫和中抗的品种大多为籼稻，而表现为高感和中感的品种中除沈农 606 外均为粳稻品种。

表 1 自然条件下水稻品种对水稻矮缩病毒的抗性鉴定及其新复极差法测定

Table 1 Determination of the resistances of rice variety to RDV and its variance analysis in nature condition

品种	发病率 / %					潜育期 /d	差异显著性 (α=0.05)	抗性评价
	1st	2nd	3rd	平均	平均相对发病率 /% ± 标准差			
岫 136-12*(CK)	42.13	41.09	50.77	44.66	100.00 ±0.00	11.00	a	HS
岫 87-15*	43.11	39.93	47.87	43.64	97.93 ±4.07	12.00	a	HS
武育粳 3 号 *	43.85	45.00	38.65	42.5	96.57 ±17.92	13.33	ab	HS
楚粳 27*	42.13	37.29	26.56	35.23	81.06 ±32.16	15.33	abc	HS
楚粳 26*	40.69	42.13	22.38	35.07	81.02 ±25.29	14.34	abc	HS
合系 39*	34.76	34.76	15.00	28.17	65.55 ±31.20	13.67	bcd	HS
楚恢 15*	30.00	9.63	37.76	25.79	57.53 ±8.04	12.00	cde	MS
楚粳 24*	20.70	26.78	29.6	25.69	56.34 ±28.54	11.33	cde	MS
沈农 606	0.00	18.44	25.48	14.64	31.69 ±27.57	15.00	def	MS
冈优 182	9.10	12.92	15.79	12.60	28.04 ±5.59	18.33	ef	MR
川丰 2 号	0.00	0.00	18.44	6.15	12.11 ±20.96	19.00	f	MR
T 优 5570	0.00	0.00	16.22	5.41	11.18 ±9.89	17.00	f	MR
宜香 2292	7.92	0.00	7.49	5.14	10.65 ±18.44	15.00	f	MR
协优 46	0.00	0.00	10.47	3.49	6.87 ±11.90	18.33	f	MR
威优 77	0.00	0.00	10.47	3.49	6.87 ±11.91	20.00	f	MR
冈优 734	0.00	0.00	0.00	0.00	0.00 ±0.00	19.00	f	IM

* 表示粳稻品种；以岫 136-12 * 为对照，相对发病率为 100%

通过田间比较试验，选取性状较好、抗性差异显著的 4 个品种做进一步测定，结果（表 2）再次证实：岫 136-12、岫 87-15 两粳稻品种均表现高感，协优 46、宜香 2292 两籼稻品种均表现中抗；粳稻与籼稻在 $\alpha_{0.05}$ 和 $\alpha_{0.01}$ 水平上抗病性均存在显著差异。

表 2 人工接种条件下水稻品种对水稻矮缩病毒的抗性鉴定及其新复极差法测定

Table2 Determination of the resistances of rice variety to RDV and its variance analysis in inoculation condition

品种	发病率 /%					差异显著性		抗性评价
	1st	2nd	3rd	平均	平均相对发病率 /% ± 标准差	α=0.05	α=0.01	
岫 136-12*	90.00	80.37	66.53	78.97	100.00 ±0.00	a	A	HS
岫 87-15*	80.19	75.46	71.56	75.74	96.85 ±9.57	a	A	HS
协优 46	18.44	15.00	14.18	15.87	20.15 ±1.35	b	B	MR
宜香 2292	8.53	12.25	19.73	13.50	18.21 ±10.38	b	B	MR

* 表示粳稻品种

2.2 水稻品种抗虫性差异显著性

选取岫 136-12、协优 46、宜香 2292 三个水稻品种进行室内抗虫性的鉴定，差异显著性分析结果（表 3）表明：协优 46 在 0.05 和 0.01 显著水平上表现为差异显著；而岫 136-12、宜香 2292 之间没有明显差异。表 2 和表 3 结果表明，协优 46 的抗病性可能是由其品种抗虫性决定的，由于抗虫导致其发病率低而表现出对水稻矮缩病的中度抗性；宜香 2292 水稻品种的抗病性可能为实质性抗病性，表现出对 RDV 的中度抗性；岫 136-12 的感病性可能为实质性感病性，表现为对 RDV 的高感。

表 3 不同水稻品种抗虫性显著性分析

Table 3 Significances of difference of anti-insect character of different rice varieties

品种	死亡率 /%					差异显著性	
	1st	2nd	3rd	平均	平均相对死亡率 /% ± 标准差	α = 0.05	α = 0.01
协优 46	75.71	61.25	95.00	75.71	326.51 ±123.72	a	A
岫 136-12	16.36	27.69	32.14	16.36	100.00 ±26.44	b	B
宜香 2292	6.25	13.33	28.33	6.25	58.16 ±0.00	b	B

采取计数叶蝉死亡率来分析水稻品种抗虫性；对照组的岫 136-12（相对死亡率为 100%）黑尾叶蝉死亡率越高表明该品种抗虫性越好

2.3 叶蝉带毒率与水稻发病率之间的关系

由表 4 可知，两高感品种在田间黑尾叶蝉带毒率较低（12% ～ 14%）和相对选择压力（虫口密度）较大的情况下，品种发病率明显低于通过高毒虫室内接种的品种发病率，在 $\alpha_{0.05}$ 和 $\alpha_{0.01}$ 水平上都是显著的，按照鉴定结果以秧苗发病率推算循回饲毒的叶蝉带毒率为 80% ～ 90%。对于感病品种叶蝉带毒率的高低直接影响到品种的发病情况，分析结果表明叶蝉带毒率与水稻发病率呈正相关，相关系数 r=0.9999。

表 4 叶蝉带毒率与水稻发病率之间的相关性分析

Table 4 Relation between the ration of *Nephotrttix cincticeps* carrying virus and incidence of RDV

区组	品种	发病率 /%					差异显著性		相关系数 /r
		1st	2nd	3rd	平均	平均相对发病率 /% ± 标准差	α = 0.05	α = 0.01	
A	岫 87-15	97.15	93.75	90.00	93.63	100.13 ±5.85	a	A	0.9999
	岫 136-12	100.0	97.29	84.21	93.83	100.00 ±0.00	a	A	
B	岫 87-15	46.67	39.13	41.18	42.33	35.26 ±4.51	b	B	
	岫 136-12	43.18	45.00	37.21	41.79	34.54 ±3.77	b	B	

以岫 136-12 为对照（相对发病率为 100%）；区组 A 为室内鉴定数据；区组 B 为田间品比数据；以田间叶蝉带毒率 14.23% 和饲毒叶蝉带毒率 80% 计算相关系数

3 小结与讨论

目前，对水稻病毒病的防治仍以选育、推广抗病品种为主。本研究对当前选育和推广的一些品种的抗性鉴定结果表明，16 个水稻品种对 RDV 的抗性存在明显差异，从高感、中感、中抗到免疫均有。并将其中对水稻矮缩病毒表现中抗且农艺性状好的品种作进一步鉴定，为抗病育种及其抗性机制的深入研究提供抗源材料。本试验测定了 16 个水稻品种对 RDV 的抗性和 3 个品种对介体叶蝉的抗性，得到的结果与林含新等经生物学接种鉴定的 38 个水稻品种对水稻条纹病毒（*Rice stripe virus*, RSV）的抗性和 12 个品种对 RSV 介体灰飞虱的抗性（林含新等 , 2000），以及孙黛珍等利用强迫饲毒、集团接种等方法研究、分析了 8 个抗 RSV 水稻品种对 RSV 和介体灰飞虱的抗性（孙黛珍等 , 2006）进行比较，结果显示不同水稻品种的抗性特征存在差异，水稻品种的抗虫性与其抗病性是紧密相关的，且籼稻品种一般较粳稻品种更具抗病性和抗虫性。这些结果说明许多品种的抗 RDV 基因和抗介体叶蝉基因可能是连锁的。近年来，由于水稻条纹病毒、水稻黑条矮缩病毒（*Rice black streaked dwarf virus*, RBSDV）和水稻锯齿叶矮缩病毒（*Rice ragged stunt virus*, RRSV）在我国一些地区暴发流行，引起了植物病理育种学家的重视，他们将筛选和选育抗病毒水稻品种作为一种主要的防病和育种目标。实践证明，使用具有稳定抗性的优质水稻品种作为抗源，改良高产水稻品种对水稻病毒病的抗性是有效的。然而，随着病毒的变异，品种抗病性可能丧失。因此，加强对现有水稻资源的抗性筛选，获得抗性材料，并对一些新的抗病基因的遗传规律进行研究，将有助于我国水稻的抗水稻矮缩病育种研究。

参考文献

[1] 陈涛 , 张亚东 , 朱镇 , 等 . 水稻条纹叶枯病抗性遗传和育种研究进展 [J]. 江苏农业科学 , 2006,2:1-4.

[2] 李毅 , 陈章良 . 水稻病毒的分子生物学 [M]. 北京：科学出版社 ,2001.

[3] 林含新 , 林奇田 , 魏太云 , 等 . 水稻品种对水稻条纹病毒及其介体灰飞虱的抗性鉴定 [J]. 福建农林大学学报 (自然版), 2000,29(4):453-458.

[4] 孙黛珍 , 江玲 , 张迎信 , 等 . 8 个水稻品种的条纹叶枯病抗性特征 [J]. 中国水稻科学 , 2006,20(2):219-222.

[5] 谢联辉 , 林奇英 . 水稻品种对病毒病的抗性研究 [J]. 福建农林大学学报 (自然科学版),1982,2:15-18.

[6] 谢联辉 , 林奇英 . 植物病毒学 [M]. 2 版 , 北京：中国农业出版社，2004.

[7] 邢祖颐 , 何家奇 , 刘志武 , 等 . 籼粳稻杂交育种的研究 II 抗条纹叶枯病育种 [J]. 作物学报 , 1985,11(1):1-6.

[8] 张仲凯 , 董家红 , 李展 , 等 . 抗水稻矮缩病毒转基因水稻的农艺性状分析 [J]. 西南农业学报 , 2006,19(1):159-161.

[9] BOCCARDO G, MILNER G. Plant reovirusgroup[J]. Cmi/aab Description of Plant Viruses, 1984: 294.

[10] FAUQUET C M, MAYOM A, MANILOFF J, et al. Virus Taxonomy[J]. New York: Elsevier Academic Press.2005.

[11] NAKAGAWA A, MIYAZAKI N, TAKA J, et al. The atom icstructure of rice dwarf virus reveals the self-assembly mechanism of component proteins[J]. Structure, 2003, 11: 1227-1238.

[12] SHIMIZU T, YOSHIIM WEI T, HIROCHIKA H, et al. Silencing by RNAi of the gene for Pns12, a viroplasm matrix protein of rice dwarf virus, results in strong resistance of transgenic rice plants to the virus[J]. Plant Biotechnology Journal, 2010, 7: 24-32.

[13] ZHENG H, YU L, WEI C, et al. Assembly of double-shelled, virus-like particles in transgenic rice plants expressing two major structural proteins of Rice dwarf virus[J]. Journal of Virology, 2000, 74: 9808-9810.

[14] ZHENG H, LI Y, YU Z H, et al. Recovery of transgenic rice plants expression the rice dwarf virus outer coat protein gene (S8)[J]. Theoretical & Applied Genetics, 1997, 94 (3-4): 522-527.

河南省主要推广品种对小麦黄花叶病毒抗性的评价

孙炳剑[1,2]，李洪连[3]，杨新志[4]，谢联辉[1]，陈剑平[2]

（1 福建农林大学植物病毒研究所，福州 350002；2 浙江省植物有害生物防控国家重点实验室培育基地，农业部植物保护与生物技术重点实验室，浙江省植物病毒重点实验室，浙江省农业科学院病毒学与生物技术研究所，杭州 310021；3 河南农业大学植物保护学院，郑州 450002；4 河南省西平县植保植检站，西平 463900）

摘 要： 为了评价河南省主要推广品种对小麦黄花叶病毒（*Wheat yellow mosaic virus*，WYMV）的抗性，于 2006 ～ 2010 年在河南省西平县病圃进行了田间抗性鉴定试验和室内间接 ELISA 检测，并分析了病害严重程度对产量的影响。结果表明，在供试的 62 个品种中，仅有新麦 208 表现为免疫；豫麦 70-36、泛麦 5 号、阜麦 936、山东 95519、豫麦 70、高优 503、豫麦 9676、郑麦 366 和陕麦 229 等 9 个品种表现为抗病，占供试品种的 14.5%；濮优 938、兰考矮早 8、新原 958、花培 2 号、温优 1 号、豫麦 18、郑麦 9023、豫麦 47、豫农 201、偃展 4110、豫麦 36、百农 878 和豫麦 49-198 等 13 个品种表现为中抗，占供试品种的 21.0%；另外 39 个品种表现为感病，占供试品种的 62.9%。对 48 个品种进行了产量与病害严重度分析，发现随着病害的严重度增加，小麦的穗数、千粒重以及产量都有明显下降，严重度为 1 级时，平均减产 9.6%；严重度达到 2 级和 3 级时，平均减产分别为 30.3% 和 33.5%。

关键词：小麦黄花叶病毒；小麦品种；抗性；产量损失

Evaluation of commercial wheat cultivars for resistance to *Wheat yellow mosaic virus* in Henan

Bingjian Sun[1,2], Honglian Li[3], Xinzhi Yang[4], Lianhui Xie[1], Jianping Chen[2]

(1 Institute of Plant Virology, Fujian Agriculture and Forestry University, Fuzhou 350002, Fujian Province, China; 2 State Key Laboratory Breeding Base for Zhejiang Sustainable Pest and Disease Control, MOA Key Laboratory of Plant Protection and Biology, Zhejiang Provincial Key Laboratory of Plant Virology, Institute of Virology and Biotechnology, Zhejiang Academy of Agricultural Sciences, Hangzhou 310021, Zhejiang Province, China; 3 College of Plant Protection, Henan Agricultural University, Zhengzhou 450002, Henan Province, China; 4 Xiping Protection and Quarantine Station, Xiping 463900, Henan Province, China)

Abstract: To evaluate resistance of winter wheat cultivars currently grown in Henan Province, China to Wheat yellow mosaic virus (WYMV), the resistance of 62 commercial wheat cultivars was tested in a disease nursery at Xi ping, Henan Province in the years 2006 to 2010. Plants were also tested using an indirect ELISA in the laboratory and the influence of disease severity on wheat yield and grain quality was in vest igated. Of the cultivars tested, only Xinmai 208 appeared to be immune to WYMV and nine cultivars (14% of those tested) were resistant: Yumai 70-36, Fanmai 5, Fumai 936, Shandong 95519, Yumai 70,

植物保护学报 (Acta Phytophylacica Sinica). 2011, (2): 102-108
基金项目：国家小麦现代化产业体系专项 (cytx-03)
收稿日期：2010-09-25

Gaoyou 503, Yumai 9676, Zhengmai 366 and Shaanmai 229. Thirteen cultivars (21% of those tested) were moderately resistant (Puyou 938, Lankaoaizao 8, Xinyuan 958, Huapei 2, Wenyou 1, Yumai 18, Zhengmai 9023, Yumai 47, Yunong 201, Yanzhan 4110, Yumai 36, Bainong 878 and Yumai 49-198) and the remaining 39 cultivars (63% of those tested) were susceptible. In studies of 48 cultivars, Increasing disease severity decreased grain yields, spike number sand 1000-grain weights 9.6% yield was lost when disease severity was grade 1, and 30.3% and 33.5% yield was lost when the disease severity was grade 2 and 3 respectively.

Keywords: *Wheat yellow mosaic virus*; winter wheat cultivars; resistance; yield loss

河南省是我国重要的冬小麦产区之一，每年小麦播种面积超过 60 万 hm^2，总产量约占全国的 1/4。自 20 世纪 70 年代以来，小麦黄花叶病在河南南部发生，并逐渐向北蔓延，已成为河南南部和中部地区小麦重要病害。该病害由小麦黄花叶病毒（*Wheat yellow masaic virus*, WYMV）引起，经土壤中的禾谷多黏菌 *Polymyxa graminis* 传播，在江苏、河南、安徽、山东、陕西、湖北和四川等省均有分布（Chen et al.,2000），全国发病总面积在 70 万 hm^2 以上，约占小麦栽培面积的 20% ～ 30%，每年损失粮食约 40 万吨（陈剑平 , 2005）。由于该病毒存在于禾谷多黏菌休眠孢子体内，而禾谷多黏菌休眠孢子堆壁厚，具有很强的抗逆性，很难用化学方法防治，因此，种植抗病品种是防治该病的最有效措施（Kühne, 2009）。为了有效地控制病害，国内学者进行了小麦品种和种质资源的抗病性鉴定工作，筛选出大量抗病种质材料（阮义理等 , 1990；岳绪国等 , 2001），并选育出一些抗病品种，如扬辐 9311、宁麦 9 号（刘伟华等 , 2004）。同时，也鉴定了部分国外品种对我国小麦黄花叶病的抗性，但是由于农艺性状、产量、品质、适应性等方面的不足，一些抗病小麦品种不适宜在我国直接推广种植（曹延杰等 , 2010）。20 世纪 80 年代以来，河南省先后审定了 100 多个小麦品种（Yang et al.,2002），但是，由于在育种和品种推广过程中对抗性亲本利用和品种抗性评价重视不够，感病品种在病区大面积种植，促进了病害的蔓延和流行，并造成了严重的经济损失。因此，对目前在河南推广应用的小麦品种进行抗病性评价，对于河南省小麦黄花叶病区品种的合理布局以及病害的有效控制具有重要的指导作用。2006 ～ 2010 年，作者等在河南省西平县田间病圃进行小麦推广品种抗黄花叶病毒鉴定，并分析了不同病害严重度对产量的影响。

1 材料与方法

1.1 材料

供试小麦品种包括豫麦 18、豫麦 25、豫麦 34、豫麦 36、豫麦 41、豫麦 47、豫麦 49、豫麦 49-198、豫麦 49-986、豫麦 58、豫麦 60、豫麦 70、豫麦 70-36、豫麦 949、豫麦 9676、豫农 201、豫优 1 号、郑麦 004、郑麦 98、郑麦 366、郑麦 9023、郑麦 9094、郑麦 9405、郑农 16、新麦 9-998、新麦 208、新麦 11、新麦 19、新麦 18、新原 958、周麦 18、周麦 19、周麦 16、周麦 12、山东 95519、泛麦 5 号、阜麦 936、花培 2 号、西农 979、陕麦 229、濮优 938、温优 1 号、高优 503、矮抗 58、兰考矮早 8、科优 1 号、科麦 10 号、百农 878、科麦 13、偃展 4110、中育 6 号、百农 9904、科麦 2 号、太空 6 号、小偃 803、淮麦 16、高优 505、漯麦 4、温麦 19、藁麦 8901、濮麦 9 号、开麦 18 等 62 个黄淮麦区推广的品种。试验所用种子由河南省农业科学院、河南农业大学、国家小麦工程技术中心（河南）、新乡市农业科学院和温县农科所等多家育种单位提供。

ELISA 检测所用到的 1 抗为小麦黄花叶病毒抗兔血清，浙江省农业科学院植物病毒实验室自制；2 抗为羊抗兔 IgG- 碱性磷酸酶，美国 Sigma 公司产品。酶标仪的型号为 BIO-RAD680。

1.2 试验设计

试验于 2006 ～ 2010 年在河南省西平县师灵镇师灵村病圃进行。试验地多年以小麦 – 玉米的种植模式，土质为砂姜黑土，田块平整，水肥条件良好，小麦黄花叶病发病均匀且严重。2006 年度和 2007

年度的试验设计：每品种重复3次，小区面积10m^2，各处理小区随机排列，四周设感病品种新麦18为保护行，播种时间比常规播种提前5～7天。2008年度和2009年度的试验品种与前两年度一致，主要是验证品种的抗病性，每品种播1行，每行播种50粒种子，每隔9行播种1行感病对照品种，重复3次。病虫害防治和其他田间栽培管理同一般大田。

1.3 调查和评价方法

3月10日前后，在小麦返青期进行发病率和病情严重度调查。前两个年度试验调查采取五点取样，每处理每点调查20株，共100株；后两个年度，每处理调查全部小麦植株。小麦黄花叶病严重度分级标准，参照刘伟华等（刘伟华等, 2004）方法并作适当改进。0级，无症状；1级，新叶出现褪绿条纹或黄花叶症状；2级，多数叶片出现褪绿条纹或黄花叶症状，有时会出现新叶扭曲，植株矮化不明显；3级，全株呈现严重花叶症状，老叶上出现坏死斑，植株明显矮化，部分分蘖死亡或全株死亡。抗性分级标准见表1。

表1 小麦品种对小麦黄花叶病抗性标准

Table 1 Grading standard of resistance for wheat cultivars to wheat yellow mosaic disease

抗性 resistance	严重度 Disease degree	病株率（%）Incidence	ELISA
免疫 Immune（I）	0	0	—
抗病 Resistant（R）	0	0<10	+
中抗 Moderately resistant(MR）	1	≥ 10	+
感病 Susceptible（S）	>1	≥ 10	+

1.4 ELISA 检测

在病害发生盛期取样，每品种取10片有代表性的叶片，样品加20倍重量的包被液（0.05mol/L碳酸盐缓冲溶液，pH 9.6）研磨，12000r/min离心10min，存放于4℃备用。以粗提的WYMV制剂为阳性对照，健康叶片汁液为阴性对照，ELISA检测参照Yang等（2002）的方法，样品的OD405值大于阴性对照2倍即为阳性（+），小于阴性对照2倍为阴性（−）。

1.5 产量测定

在2007～2008年度，收获期调查48个供试品种的每平方米穗数、千粒重（9次重复计算平均值）、每平方米产量，每处理调查3次重复。

1.6 数据统计分析

运用Excel和SPSS统计分析软件，对产量、穗数、千粒重等数据进行均值、标准差等计算。按照病害的严重度将供试的48个品种划分成4组，进行均值及95%置信限计算，并采用DMRT法进行方差分析和多重比较。

2 结果与分析

2.1 不同品种田间病圃发病情况调查与抗性评价

田间鉴定结果表明，不同年份发病情况存在一定的变化，品种间存在明显差异，供试的62个品种中，仅有新麦208田间未见症状，ELISA检测结果呈阴性，表现为免疫；豫麦70-36、泛麦5号、阜麦 936、山东 95519、豫麦 70、高优 503、豫麦 9676、郑麦 366、陕麦 229 等 9 个品种田间常年发病率低于 10%，表现为稳定的抗病，占供试品种 14.5%，以上 10 个品种可以作为抗病品种在河南省小麦黄花叶病区推广；濮优 938、兰考矮早 8、新原 958、花培 2 号、温优 1 号、豫麦 18、郑麦 9023、豫麦 47、豫农 201、偃展 4110、豫麦 36、百农 878、豫麦 49-198 等 13 个品种表现为中抗，占供试品种的 21.0%；其余 39 个品种表现为感病，占供试品种的 62.9%，这些感病品种应避免在病区推广（表 2）。

表 2 河南小麦品种对小麦黄花叶病的反应（2006—2010）

Table 2 Responses of some commercial wheat cultivars to Wheat yellow mosaic virus in Henan（2006–2010）

品种 Cultivar	病株率 Incidence（%）	严重度 Disease degree	ELISA	抗性评价 Resistance	品种 Cultivar	病株率 Incidence（%）	严重度 Disease degree	ELISA	抗性评价 Resistance
新麦 208 Xinmai208	0	0	−	I	豫麦 49 Yumai49	24.4-100.0	1~2	+	S
豫麦 70-36 Yumai70-36	0	0	+	R	郑麦 9405 Zhengmai9405	35.7-71.4	1~2	+	S
泛麦 5 号 Fanmai5	0	0	+	R	豫麦 34 Yumai34	37.5-100.0	1~2	+	S
阜麦 936 Fumai936	0	0	+	R	科麦 10 号 Kemai10	37.5-100.0	1~2	+	S
山东 95519 Shandong95519	0	0	+	R	豫优 1 号 Yuyou1	46.6-94.0	1~2	+	S
豫麦 70 Yumai70	0~4.4	0~1	+	R	豫麦 58 Yumai58	59.2-100.0	1~2	+	S
高优 503 Gaoyou503	0~6.5	0~1	+	R	郑麦 9094 Zhengmai9094	66.7-100.0	1~2	+	S
豫麦 9676 Yumai9676	0~7.0	0~1	+	R	豫麦 25 Yumai25	74.4-100.0	1~2	+	S
郑麦 366 Zhengmai366	0~8.6	0~1	+	R	郑农 16 Zhengnong16	75.2-100.0	1~2	+	S
陕麦 229 Shanmai229	0~9.8	0~1	+	R	太空 6 号 Taikong6	82.3-100.0	1~2	+	S
濮优 938 Puyou938	0~11.7	0~1	+	MR	科麦 13 Kemai13	90.0-100.0	1~2	+	S
兰考矮早 8 Lankaoaizao8	0~13.3	0~1	+	MR	豫麦 41 Yumai41	95.6-100.0	1~2	+	S
新原 958 Xinyuan958	0~13.5	0~1	+	MR	高优 505 Gaoyou505	29.2-96.7	2	+	S
花培 2 号 Huapei2	0~14.8	0~1	+	MR	郑麦 98 Zhengmai98	63.9-91.7	2	+	S
温优 1 号 Wenyou1	0~20.6	0~1	+	MR	郑麦 004 Zhengmai004	59.8-100.0	2	+	S
豫麦 18 Yumai18	0~29.2	0~1	+	MR	西农 979 Xinong979	63.3-94.4	2	+	S
郑麦 9023 Zhengmai9023	0~33.5	0~1	+	MR	豫麦 49-986 Yumai49-986	72.2~100.0	2	+	S

续表

品种 Cultivar	病株率 Incidence （%）	严重度 Disease degree	ELISA	抗性评价 Resistance	品种 Cultivar	病株率 Incidence （%）	严重度 Disease degree	ELISA	抗性评价 Resistance
豫麦 47 Yumai47	0~59.2	1	+	MR	矮抗 58 Aikang58	100.0	2	+	S
豫农 201 Yunong201	0~50.5	1	+	MR	温麦 19 Wenmai19	45.9~100.0	2~3	+	S
偃展 4110 Yanzhan4110	0~89.0	1	+	MR	新麦 11 Xinmai11	48.7~86.3	2~3	+	S
豫麦 36 Yumai36	11.1~50.6	1	+	MR	豫麦 949 Yumai949	66.9~98.9	2~3	+	S
百农 878 Bainong878	25.9~81.0	1	+	MR	科优 1 号 Keyou1	47.7~100.0	2~3	+	S
豫麦 49-198 Yumai49-198	66.7~84.1	1	+	MR	科麦 2 号 Kemai2	48.0~100.0	2~3	+	S
漯麦 4 号 Luomai4	54.4~90.1	2~3	+	S	中育 6 号 Zhongyu6	90.0~100.0	2~3	+	S
新麦 9-998 Xinmai9-998	57.4~100.0	2~3	+	S	周麦 19 Zhoumai19	98.9~100.0	2~3	+	S
藁麦 8901 Gaomai8901	55.5~96.7	2~3	+	S	百农 9904 Bainong9904	100.0	2~3	+	S
周麦 18 Zhoumai18	68.3~100.0	2~3	+	S	濮麦 9 号 Pumai9	73.3~100.0	3	+	S
周麦 12 Zhoumai12	69.4~100.0	2~3	+	S	新麦 19 Xinmai19	53.2~100.0	3	+	S
周麦 16 Zhoumai16	80.4~97.9	2~3	+	S	新麦 18 Xinmai18	72.8~100.0	3	+	S
豫麦 60 Yumai60	71.3~100.0	2~3	+	S	开麦 18 Kaimai18	86.7~100.0	3	+	S
淮麦 16 Huaimai16	88.4~100.0	2~3	+	S	小堰 803 Xiaoyan803	100.0	3	+	S

注：数据是 2007 ～ 2010 年的调查结果。I 免疫；R 抗性；MR 中抗；S 感病。“－” ELISA 检测阴性；“+” ELISA 检测阳性。
Note: Data were from experiments done during 2007–2010. I: Immune; R: Resistance; MR: Moderately resistant; S: Susceptible. “–” shows a negative reaction; “+” represents a positive reaction.

2.2 产量测定

产量测定结果表明，严重度为 0 级的 10 个品种，穗数均大于 510 个 /m^2，千粒重大于 37g，产量大于 600g/m^2，其中泛麦 5 号、豫麦 70、高优 503 和阜麦 936 等 4 个品种产量大于 700g/m^2，泛麦 5 号的穗数和产量最高，分别为 762.5 个 /m^2 和 926.8g/m^2；严重度为 1 级、穗数大于 500 个 /m^2、产量大于 600g/m^2 的品种分别是郑麦 9023、百农 878、豫麦 58、豫麦 36 和新原 958；严重度为 2 级的品种有 13 个，除郑农 16 和郑麦 004 的产量大于 560g/m^2，其余品种的产量均低于 550g/m^2，豫麦 34、豫麦 25、郑麦 98、豫麦 49-986 和豫麦 41 的产量最低，只有 379.6 ～ 441.5g/m^2；严重度为 3 级的品种有 15 个，除中育 6 号、淮麦 16、温麦 19、科优 1 号、百农 9904 产量高于 500g/m^2 外，其余 10 个品种的产量均低于 500g/m^2（表 3）。

表 3 河南西平小麦黄花叶病毒病对 48 个供试品种的穗数、千粒重以及产量的影响(2008)

Table 3 Effects of Wheat yellow mosaic virus infection on spike numbers, 1000-grain weight and Yield of 48 wheat cultivars tested at Xiping, Henan (2008)

品种 Cultivar	严重度 Disease degree	穗数 Spike number (head/m^2)	千粒重 1000-grain weight(g)	产量 Yield (g/m^2)	品种 Cultivar	严重度 Disease degree	穗数 Spike number (head/m^2)	千粒重 1000-grain weight(g)	产量 Yield (g/m^2)
泛麦 5 号 Fanmai5	0	762.4±19.2	38.7±0.9	926.8±25.9	郑麦 9023 Zhengmai9023	1	596.9±13.9	44.4±1.8	796.4±49.5
豫麦 70 Yumai70	0	584.4±33.7	47.6±2.9	767.7±49.0	百农 878 Bainong878	1	556.1±67.1	45.4±1.9	769.9±20.3
高优 503 Gaoyou503	0	593.7±10.0	40.7±1.8	740.3±76.3	豫麦 58 Yumai58	1	633.6±40.7	43.3±1.0	691.6±75.6
阜麦 936 Fumai936	0	544.1±49.4	40.4±1.1	704.7±85.3	豫麦 36 Yumai36	1	585.6±59.1	43.3±1.5	666.1±28.1
花培 2 号 Huapei2	0	541.2±1.7	44.2±0.5	698.8±3.1	新原 958 Xinyuan958	1	584.9±32.5	42.6±1.8	660.4±23.3
郑麦 366 Zhengmai366	0	632.1±49.3	37.0±1.6	687.4±51.5	豫麦 47 Yumai47	1	528.0±37.5	40.7±2.0	595.0±18.1
新麦 208 Xinmai208	0	631.2±52.0	40.1±1.9	674.9±9.0	豫农 201 Yunong201	1	526.8±17.5	46.8±2.0	585.4±10.7
濮优 938 Puyou938	0	513.6±62.4	41.7±2.8	664.4±70.6	豫农 9676 Yunong9676	1	408.0±40.7	46.0±6.5	559.0±67.4
豫麦 70-36 Yumai70-36	0	512.4±61.8	41.2±2.3	634.7±41.5	郑麦 9405 Zhengmai9405	1	401.6±27.6	45.2±1.0	552.8±57.8
山东 95519 Shandong95519	0	548.1±36.0	42.7±2.1	604.9±75.1	偃展 4110 Yanzhan4110	1	570.0±76.4	45.0±1.8	543.6±22.2
郑农 16 Zhengnong16	2	375.2±36.4	43.5±2.7	612.1±70.7	淮麦 16 Huaimai16	3	398.4±41.6	41.7±2.0	539.7±53.6
郑麦 004 Zhengmai004	2	447.2±86.7	36.2±2.2	560.6±80.4	温麦 19 Wenmai19	3	374.4±14.6	39.2±0.6	553.4±22.7
科麦 13 Kemai13	2	557.7±37.3	36.8±2.8	546.2±53.0	科优 1 号 Keyou1	3	404.9±20.5	34.5±2.1	543.0±21.1
太空 6 号 Taikong6	2	535.2±84.0	43.8±2.1	542.1±18.3	百农 9904 Bainong9904	3	486.5±31.8	36.1±2.1	508.7±39.5
豫麦 60 Yumai60	2	388.8±43.4	41.6±1.7	537.6±52.6	豫农 949 Yunong949	3	575.3±67.3	44.2±1.2	496.7±50.5
豫优 1 号 Yuyou1	2	472.8±42.2	37.8±1.8	537.4±78.7	新麦 18 Xinmai18	3	478.4±72.0	32.1±2.0	477.2±89.6
周麦 16 Zhoumai16	2	430.4±16.3	40.0±0.3	528.1±61.5	新麦 19 Xinmai19	3	465.6±48.7	39.0±1.1	471.1±91.0
科麦 10 号 Kemai10	2	474.0±1.3	37.2±2.5	523.2±1.9	新麦 9-998 Xinmai9-998	3	357.6±50.3	38.7±1.9	465.2±59.0
豫麦 34 Yumai34	2	387.2±57.9	43.1±3.4	441.5±97.7	开麦 18 Kaimai18	3	424.8±40.7	38.2±1.1	432.2±44.1
豫麦 25 Yumai25	2	420.0±20.4	42.1±0.5	417.8±96.5	濮麦 9 号 Pumai9	3	449.6±39.8	30.8±1.7	420.2±30.1
郑麦 98 Zhengmai98	2	267.2±33.1	36.6±0.9	416.4±46.9	周麦 19 Zhoumai19	3	394.4±39.1	34.0±3.8	415.2±78.9
豫麦 49-986 Yumai49-986	2	447.6±22.1	42.5±0.8	399.0±32.2	小堰 803 Xiaoyan803	3	332.4±42.4	31.7±2.1	409.2±48.2
豫麦 41 Yumai41	2	378.5±37.5	40.6±2.1	379.6±83.6	漯麦 4 号 Luomai4	3	427.2±67.2	38.2±1.6	400.0±21.9
中育 6 号 Zhongyu6	3	403.2±39.4	42.7±0.5	554.9±63.3	科麦 2 号 Kemai2	3	314.4±45.1	39.9±1.6	397.4±75.6

2.3 病害严重度对穗数、千粒重及产量的影响

结果显示，随着病害严重度的增加，穗数、千粒重以及产量都有明显下降。严重度为 0 级品种的平均穗数为 586.3 个 /m^2，高于其他 3 组，并且严重度 0 级和 1 级两组品种间差异不显著，但是与 2 级和 3 级两组品种间差异达到显著水平（$P < 0.05$），2 级和 3 级两组品种的穗数分别只有 429.4 个 /m^2 和 419.1 个 /m^2；严重度为 0 级和 1 级两组品种的平均千粒重均大于 41g，两组之间差异不显著，但是严重度为 1 级的品种，与 2 级和 3 级的两组品种之间差异均达到显著水平（$P < 0.05$），严重度为 3 级的品种平均千粒重只有 37.4g；严重度为 0 级的品种，平均产量为 710.5g/m^2，显著大于其他 3 组（$P < 0.05$），1 级的品种平均产量为 642.0g/m^2，与 2 级和 3 级的两组品种差异达到显著水平（$P < 0.05$），严重度为 2 级的品种减产率达到 30.3%，3 级的品种平均减产率为 33.5%（表 4）。

表 4 不同病害严重度对供试小麦品种穗数、千粒重以及产量的影响

Table 4 Effects of disease severity on spike numbers, 1000-grain weights and grain yields of wheat cultivars tested

严重度 Disease degree	品种数量 Number	穗数 (95% 置信限) Spike number (95%LC) (head/m^2)	千粒重 (95% 置信限) 1000-grain weight (95% LC)(g)	产量 (95% 置信限) Yield(95%LC) (g/m^2)	产量损失 (95% 置信限) Yield loss(95%LC) (%)
0	10	586.3(532.4~640.3) a	41.4(39.3~43.5) b	710.5(646.5~774.4) a	-
1	10	539.2(483.7~594.6) a	44.3(43.0~45.6) a	642.0(577.4~706.7) b	－ 9.6(－ 0.5~ － 18.7)
2	13	429.4(384.0~474.7) b	40.1(38.4~41.9) b	495.5(450.7~540.3) c	－ 30.3(－ 24.0~ － 36.6)
3	15	419.1(382.5~455.8) b	37.4(35.1~39.7) c	472.3(440.1~504.5) c	－ 33.5(－ 29.0~ － 38.1)

注：数据为相同严重度品种的平均值。数据后不同英文字母表示差异显著（$P<0.05$,DMRT 法）。

Note: The data are means of the same disease severity. Means followed by different letters with in the same column were significantly different at $P<0.05$ by DMRT.

3 讨论

植物病毒病防治主要依靠选育推广抗病品种、化学防治昆虫介体和农业措施等。对于禾谷多黏菌传播的小麦病毒病而言，由于其介体休眠孢子堆壁厚且散布在土壤中，化学防治难以奏效，因而抗病品种仍是目前最有效的防治手段。从上世纪中后期以来，转基因技术已经应用于多种植物的育种工作，国内通过转病毒复制酶和外壳蛋白基因，获得一些抗小麦黄花叶病毒的材料（徐慧君等 , 2001 ；董槿等 , 2002 ；吴宏亚等 , 2006 ；刘永伟等 , 2007），但是由于转基因品种抗性的稳定性、农艺性状和安全性等原因，这些品种目前仍无法应用于农业生产。因此对通过常规技术育成的小麦品种进行抗病性评价，对于小麦黄花叶病毒病的防控具有重要意义。本研究通过 4 年田间病圃鉴定，筛选出新麦 208、豫麦 70-36、泛麦 5 号、阜麦 936、山东 95519、豫麦 70、高优 503、豫麦 9676、郑麦 366、陕麦 229、濮优 938 等 10 个表现稳定抗病的品种，可以在河南及周边小麦黄花叶病区推广种植。同时，在选择抗病品种时，需结合当地环境条件和主要病虫害发生情况，如郑麦 366，后期湿度大时，易感叶枯病、白粉病，不适宜在南部多雨地区种植。另外，河南省近年来大面积种植的品种，如 : 矮抗 58、西农 979、太空 6 号、周麦 12、新麦 19 和新麦 18 等 , 缺乏对小麦黄花叶病的抗性，在驻马店等地严重发病，应避免在老病区推广种植，在疑似病区引种时也应该慎重。此外，通过中抗品种和高抗品种轮作，或者以合理比例间作，丰富品种的多样性，降低病毒的选择压力，可以减缓新的病毒致病株系的出现，延长抗病品种的使用年限。本研究改进了病害田间调查的病情严重度分级标准和抗性评价标准，在 1 级标准中特别指出“新叶出现褪绿条纹和黄化症状”，以取代过去文献中描述的“轻度花叶”或“部分轻度花叶”（岳绪国等 , 2001 ；阮义理等 , 2001 ；刘伟华等 , 2004 ；陈爱大等 , 2009），使得症状描述更加直观。在病害发生盛期，如果仅有老叶的轻度花叶，新叶正常，很可能是其他生物或非生物因子所致。在 3

级标准中增加“部分分蘖死亡或全株死亡”，因为分蘖的死亡导致有效成穗率降低，直接引起产量降低，同时还可使症状描述更加完善。与烟草花叶病、玉米矮花叶病、玉米粗缩病、水稻黑条矮缩（陈声祥和张巧艳，2005）等其他作物病毒病不同，小麦黄花叶病的发病率并不能完全代表病害的危害情况。本研究中，小麦黄花叶病即使发病率达到 100%，如果病害严重度比较轻，病株随着气温的升高而很快隐症，生长后期的生物量和产量降低并不显著。即严重度达到 1 时，平均减产率仅为 9.1%，而百农 878、豫麦 58 等品种的发病率均超过 80%，但减产并不显著。

参考文献

[1] 曹廷杰，赵虹，王西成，等．河南省半冬性小麦品种主要农艺性状的演变规律 [J]. 麦类作物学报，2010, 30 (3): 439-442.

[2] 陈爱大，冷苏凤，杨红福，等．不同地区小麦梭条花叶病病毒致病力的差异 [J]. 麦类作物学报，2009, 29 (1): 157-159.

[3] 陈剑平．中国禾谷多黏菌传麦类病毒研究现状与进展 [J]. 自然科学进展，2005, 15 (5): 524 -533.

[4] 陈声祥，张巧艳．我国水稻黑条矮缩病和玉米粗缩病研究进展 [J]. 植物保护学报，2005, 32 (1): 97-103.

[5] 董槿，何震天，韩成贵，等．抗小麦黄花叶病毒转基因小麦的获得及病毒诱导的基因沉默 [J]. 科学通报，2002, 47 (10): 763-767.

[6] 刘伟华，何震天，耿波，等．小麦对黄花叶病的抗性鉴定及典型品种的遗传分析 [J]. 植物病理学报，2004, 34 (6): 542-547.

[7] 刘永伟，徐兆师，杜丽璞，等．病毒复制酶基因 Nib8 和 ERF 转录因子 W17 基因枪法共转化小麦 [J]. 作物学报，2007, 33 (9): 1548-1552.

[8] 阮义理，林美琛，陈剑平．小麦品种资源对小麦梭条斑花叶病的抗性 [J]. 植物保护学报 (2 期), 1990, 17(2):101-104.

[9]吴宏亚，张伯桥，高德荣，等．转 WYMV-Nib8 基因抗黄花叶病小麦的鉴定及优良株系的选育 [J]. 麦类作物学报，2006, 26 (6): 11-14.

[10] 徐惠君，庞俊兰，叶兴国，等．基因枪介导法向小麦导入黄花叶病毒复制酶基因的研究 [J]. 作物学报，2001, 27 (6): 688-695.

[11] 岳绪国，景德道，陈爱大．小麦抗梭条花叶病品种的田间筛选及抗性遗传研究初报 [J]. 麦类作物学报，2001, 21 (3): 22-25.

[12] CHEN J, CHEN J P, DUO J, et al. Sequence diversity in the coat protein coding region of wheat yellow mosaic by movirus isolates from China[J]. Journal of Phytopathology, 2010, 148 (9/10): 515-521.

[13] HAN C G, LI D W W, XING Y M M, et al. Wheat yellow mosaic virus widely occurring in wheat (*Triticum aestivum*) in China[J]. Plant Disease, 2000, 84 (6): 627-630.

[14] KŮHNE T. Soil-borne viruses affecting cereals-Known for long[J]. Virus Research, 2009, 141 (2): 174-183.

[15] YANG J P, CHEN J P, CHENG Y, et al. Responses of some American, European and Japanese wheat cultivars to soil-borne wheat Viruses in China[J]. Agricultural Sciences in China, 2002, 1 (10): 1141-1150.

Viral-inducible argonaute 18 confers broad-spectrum virus resistance in rice by sequestering a host microRNA

Jianguo Wu[1, 2], Zhirui Yang[1], Yu Wang[1], Lijia Zheng[1, 2], Ruiqiang Ye[3], Yinghua Ji[4], Shanshan Zhao[1], Shaoyi Ji[1], Ruofei Liu[1], Le Xu[3], Hong Zheng[1], Yijun Zhou[4], Xin Zhang[5], Xiaofeng Cao[6], Lianhui Xie[2], Zujian Wu[2], Yijun Qi[3], Yi Li[1]

(1 State Key Laboratory of Protein and Plant Gene Research, College of Life Sciences, Peking University, Beijing, China; 2 Fujian Province Key Laboratory of Plant Virology, Institute of Plant Virology, Fujian Agriculture and Forestry University, Fuzhou, China; 3 Center for Plant Biology, Tsinghua-Peking Center for Life Sciences, College of Life Sciences, Tsinghua University, Beijing, China; 4 Institute of Plant Protection, Jiangsu Academy of Agricultural Sciences, Nanjing, China; 5 Institute of Crop Science, Chinese Academy of Agricultural Sciences, Beijing, China; 6 State Key Laboratory of Plant Genomics and National Center for Plant Gene Research, Institute of Genetics and Developmental Biology, Beijing, China)

Abstract: Viral pathogens are a major threat to rice production worldwide. Although RNA interference (RNAi) is known to mediate antiviral immunity in plant and animal models, the mechanism of antiviral RNAi in rice and other economically important crops is poorly understood. Here, we report that rice resistance to evolutionarily diverse viruses requires Argonaute 18 (AGO18). Genetic studies reveal that the antiviral function of AGO18 depends on its activity to sequester microRNA168 (miR168) to alleviate repression of rice AGO1 essential for antiviral RNAi. Expression of miR168-resistant *AGO1a* in *ago18* background rescues or increases rice antiviral activity. Notably, stable transgenic expression of AGO18 confers broad-spectrum virus resistance in rice. Our findings uncover a novel cooperative antiviral activity of two distinct AGO proteins and suggest a new strategy for the control of viral diseases in rice.

1 Introduction

Small RNAs including small interfering RNAs (siRNAs) and microRNAs (miRNAs) are important regulators of gene expression via RNA interference (RNAi) pathways in many organisms (Baulcombe, 2004; Wang et al., 2011; Hauptmann and Meister, 2013; Meister, 2013). The RNAi machinery contains many proteins involved in the biogenesis and function of specific types of small RNAs (Chen, 2009; Wang et al., 2011; Hauptmann and Meister, 2013; Meister, 2013). Among these are two categories that act as the core of the RNA silencing machinery: DICER (DICER-LIKE or DCL in plants) and ARGONAUTE (AGO). DICER and DCL proteins mainly function to process double-stranded or folded stem-loop precursor RNAs into small RNA duplexes. AGO proteins incorporate one strand of an siRNA duplex to form an effector complex called RNA-induced silencing complex (RISC) to regulate the stability/translatability of target RNAs or to mediate methylation of target DNA sequences (Baulcombe, 2004; Wang et al., 2011; Meister, 2013; Pfaff et al., 2013; Li et al., 2013a; Rogers and Chen, 2013).

Elife.2015, 4: e05733.
Received 22 November 2014; Accepted 13 February 2015;Published17 February 2015

In plants, the RNAi machinery has undergone extensive amplification and functional specialization during evolution. For instance, the model plant *Arabidopsis thaliana* encodes four DCLs that function indistinct and yet overlapping RNAi pathways to control diverse biological processes ranging from development, response to abiotic stresses, to defense against pathogens (Deleris et al., 2006; Chapman and Carrington, 2007; Garcia-Ruiz et al., 2010; Vazquez et al., 2010). *Arabidopsis* encodes 10 AGOs whose functions are not all understood. Well-studied AGOs include AGO1 that mediates mRNA cleavage is critical for development, AGO4 that directs DNA methylation, and AGO2 that functions in DNA double strand break repair (Baumberger and Baulcombe, 2005; Mallory and Vaucheret, 2010; Ye et al., 2012; Wei et al., 2012). AGO1 is particularly notable in that its homeostasis is controlled at the transcriptional, post-transcriptional, and post-translational levels. At the post-transcriptional level, the *AGO1* mRNA is a target of miR168. Therefore, miR168-guided cleavage of *AGO1* mRNA by AGO1 protein exerts auto-regulation. Moreover, *AGO1* is co-expressed with miR168 and AGO1 protein can stabilize miR168 post-transcriptionally (Vaucheret et al., 2006; Vaucheret, 2008; Mallory and Vaucheret, 2010). At the post-translational level, the accumulation of AGO1 can be reduced by F-box proteins in a proteasome-independent manner through the autophagy pathway (Derrien et al., 2012; Rogers and Chen, 2013). The auto-regulation of AGO1 indicates that its level within a cell can be dynamic and this dynamics may significantly impact the biological activities of a plant.

RNA-mediated immunity against viruses operates in plants, fungi, invertebrates, and mammals to specifically destroy viral RNAs through the cellular RNA silencing machinery (Li et al., 2013b; Maillard et al., 2013). In plants, it is well known that AGO1 is a major effector of antiviral RNAi; AGO1 associates with virus-derived siRNAs (vsiRNAs) and mediates the degradation of viral RNAs. Furthermore, AGO2 and AGO7 are induced during viral infection, and both proteins can bind viral siRNAs. The antiviral function of AGO2 and AGO7 requires their slicing activity (Qu et al., 2008; Wang et al., 2011). AGO2 is repressed by AGO1-associated miR403, and AGO1 and AGO2 appear to exert antiviral functions in a non-redundant and cooperative manner. Specifically, AGO1 functions in the first layer of antiviral RNAi; when AGO1's antiviral function is inhibited, a second layer is activated involving AGO2 (Harvey et al., 2011; Jaubert et al., 2011; Scholthof et al., 2011; Wang et al., 2011; Carbonell et al., 2012; Xia et al., 2014). AGO2 also recruits miR393* to regulate plant immunity against bacterial infection (Zhang et al., 2011). As a counter-defense strategy, some plant viruses have evolved silencing suppressors to target AGO1 (Burgya′n and Havelda, 2011). Moreover, infection of many viruses can elevate the miR168 level to down-regulate AGO1, thereby nullifying this layer of host defense (Va′ rallyay et al., 2010). Thus, regulation of AGO1 by both host and viral factors plays a critical role in determining host responses to viral infection. Whether a host has positive regulators to check the viral counter-defense activities is not understood. How different AGOs have evolved to regulate plant responses to pathogen infection also remains an outstanding question (Ding and Voinnet, 2007; Ding, 2010; Garcia-Ruiz et al., 2010).

Rice (*Oryza sativa*) is one of the most important food crops as well as an experimental model plant for monocotyledonous plants (monocots). However, rice production and consequently food sustainability is under the constant threat of emerging and reemerging viral diseases. Rice viral pathogens are genetically diverse and many highly pathogenic viruses such as *Rice stripe Tenuivirus* (RSV, with a genome comprising 4 negative-stranded RNAs) and *Rice dwarf Phytoreovirus* (RDV, with a genome comprising 12 double-stranded RNAs) are transmitted persistently and solely by arthropod vectors (Hibino, 1996; Ren et al., 2010; Du et al., 2011). Because of the global circulation of these vectors and lack of virus resistance germplasms, the incidence and severity of rice viral diseases in many rice-growing regions are unpredictable. Infection by multiple viruses is also a common and severe challenge for other important crops. Therefore, developing new and effective strategies to control infection by multiple viruses for a crop, especially the prevalent food crops in the monocot group such as rice, maize, and wheat that have traditionally been under-investigated, is of paramount importance for human food sustainability. Successful development of such strategies requires a knowledge base of the mechanisms of viral infection and host defense responses.

The rice genome encodes five DCLs and 19 AGOs. How different AGOs regulate antiviral RNAi in rice is not known. In this study, we report a critical role of RNAi in antiviral defense in rice under natural infection conditions, when the plants were inoculated, as in the field, by viruliferous insect vectors brown planthopper (*Laodelphax striatellus*) and leafhopper (*Nephotettix cincticeps*) that transmit RSV and RDV, respectively. Intriguingly, we found that AGO18, a member of a new AGO clade that is conserved in monocots, is specifically induced by the infection of two taxonomically different viruses and is required for the antiviral function of AGO1. Loss-of-function *ago18* mutation abolishes, whereas over-expression of AGO18 increases, the AGO1 antiviral activity. We further demonstrated that AGO18 competes with AGO1 for binding miR168, resulting in elevated levels of AGO1 in the infected plants to enable antiviral defense. Expression of an miR168-resistant *AGO1a* variant in the ago18 background rescues or increases rice antiviral activity. The antiviral function of AGO18 requires its small RNA-binding, not slicing activity. Our findings reveal a novel mechanism AGO1 homeostasis regulation by AGO18 for antiviral defense and have significant implications in understanding the evolutionary amplification of RNA silencing mechanisms and in developing novel antiviral strategies.

2 Results

2.1 A role for AGO1 in rice antiviral immunity under natural infection conditions

Our understanding of RNAi as an antiviral defense mechanism in plants mainly comes from viral infection systems involving model plants *Arabidopsis* and *Nicotiana benthamiana* by mechanical inoculation with virions or in vitro viral RNA transcripts or by agro-infiltration with viral cDNAs. To investigate the importance of RNAi in antiviral immunity in economically important crops under natural infection conditions, we used the rice-RSV patho-system in which rice is infected by RSV through the transmission of insect vector brown planthopper.

Given the established role of AGO1 in antiviral defense in *Arabidopsis*, we tested whether AGO1 functions similarly in rice by inoculating an *ago1* RNAi line (Wu et al., 2009) by viruliferous brown planthopper carrying RSV to recapitulate the natural infection conditions. As shown in Fig. 1A, the *ago1* RNAi line, in which the expression of *AGO1s* is diminished (Fig. 1, Fig. S1A), was much more susceptible to RSV infection and showed more severe symptoms. Northern blots showed a remarkable increase in the accumulation of RSV genomic RNAs in the *ago1* RNAi line (Fig.1B). Since *AGO1a* and *1b* are expressed at much higher levels than *AGO1c* and *1d* in both mock-inoculated and RSV-infected rice plants (Fig. 1, Fig. 1B,C) (Kapoor et al., 2008), we presumed that AGO1a and 1b are more important for antiviral defense. Therefore, AGO1a and 1b were characterized further in subsequent studies. We immunoprecipitated AGO1a and 1b complexes from RSV-infected or mock-treated rice plants using specific antibodies. Northern blot analyses showed that RSV vsiRNAs were readily detected in the AGO1a and AGO1b complexes (Fig. 1C), suggesting that AGO1 is an effector of vsiRNA. Taking together, these data indicate a role for AGO1s in antiviral immunity under natural infection conditions in rice.

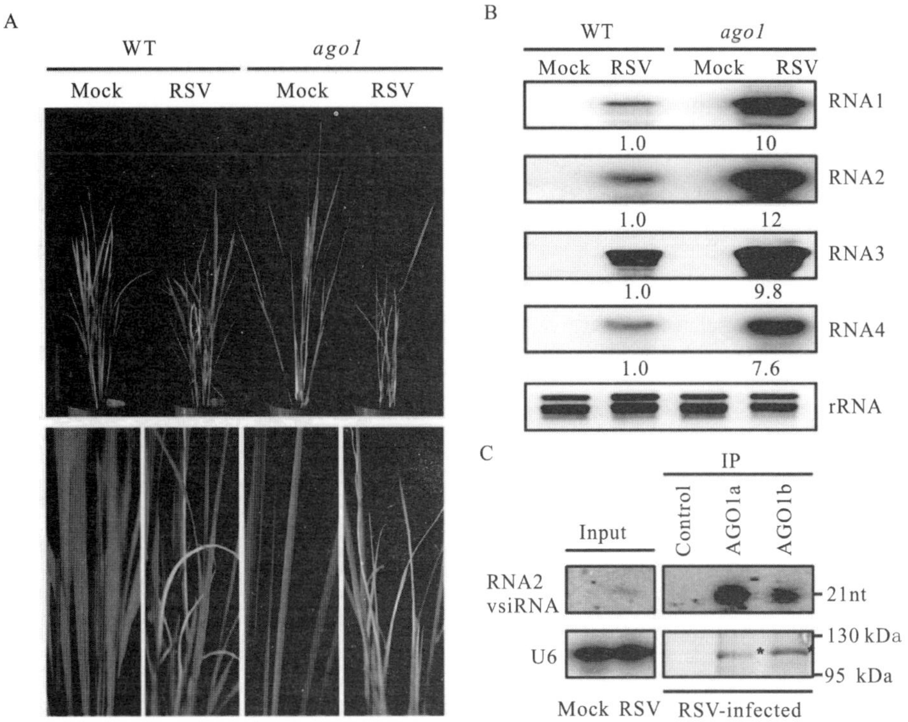

Fig. 1 AGO1 participates in antiviral immunity in rice

(**A**) Symptoms of WT and *ago1* RNAi rice plants infected with RSV, pictures were taken at 6 weeks post-inoculation. Scale bars, 15cm. (**B**) Detection of RSV genomic RNA segments in WT and *ago1* RNAi rice plants by Northern blot. The blots were hybridized with radiolabeled riboprobes specific for each RNA segment. rRNAs were stained with ethidium bromide and served as loading controls. The RNA signals were quantified and normalized to rRNAs, and the relative values were calculated by comparison with those in RSV-infected WT (arbitrarily set to 1.0). (**C**) Detection of vsiRNAs associated with AGO1a and AGO1b immunoprecipitates. The silver-stained gel shows that comparable amounts of AGO1 complexes were used for RNA preparation. The asterisks indicate the positions of AGO proteins. The position of a RNA size marker, electrophoresed in parallel, is shown to the right of the blots. U6 was also probed and served as a loading control

2.2 AGO18 is specifically induced by viral infection and required for antiviral immunity

We have previously shown that rice *AGO1* as well as *AGO18*, a member of a new AGO clade that is conserved in monocots (Fig. 2, Fig. S1), is highly induced by viral infection (Du et al., 2011). Consistent with mRNA expression patterns, AGO18 protein was hardly detectable in mock-inoculated rice plants but accumulated to a high level in RSV-infected plants (Fig. 2A). To further test whether RSV infection induces the transcription activity of *AGO18* promoter, we generated transgenic rice plants that express *β*-glucuronidase (GUS) under the control of *AGO18* promoter. RSV infection led to high levels of *GUS* expression in such plants as compared to the background levels in mock-inoculated plants (Fig. 2B, C). These data demonstrate that RSV infection, but not insect vector feeding, induced AGO18 expression at the transcriptional level.

To investigate how RSV infection triggers the expression of *AGO18*, we examined the possibility that viral proteins produced during infection induce AGO18 expression. To do this, we generated transgenic rice plants that over-express Myc-tagged RSV P2, CP, pC4, and P4 proteins (Xiong et al., 2008), respectively (Fig. 2, Fig.S2). We found that AGO18 was induced only in the transgenic lines that over-expressed CP, but not in those that over-expressed three other viral proteins (Fig. 2, Fig.S2). This suggests that RSV CP is an effector to induce AGO18 expression during viral infection.

The induction of AGO18 expression by RSV infection prompted us to investigate its function in antiviral defense. We obtained a *Tos17* retrotransposon insertional mutant *ago18* (NF6013) (Fig. 2, Fig.S3A) from the

Tos17 database and isolated homozygous plants (Fig. 2, Fig.S3B). The *ago18* mutant did not exhibit notable phenotypic differences from the WT plants in growth and development but were much more sensitive to RSV infection than the WT plants (Fig. 2D). Viral replication and infection rates also increased in *ago18* mutant plants (Fig. 2E,F;S1A). These results suggest that AGO18 may be dispensable for normal growth and development of rice, but is required for defense against RSV infection. Importantly, AGO1 is present in the ago18 mutants. Therefore, these data indicate that the antiviral function of AGO1 depends on the presence of AGO18. We further over-expressed *AGO18* under the control of the ACTIN promoter in the WT rice background. The antiviral activity of AGO18 in the three independent transgenic lines was directly correlated with the expression levels of AGO18 (Fig. 2G–J).

AGO18 was not only induced by and functioned against RSV, but also was induced by and functioned against RDV, a double-stranded RNA *Phytoreovirus* (Fig. 2; S4). This indicates that AGO18 is effective in positively regulating defense against a broad spectrum of viruses with diverse genome structures.

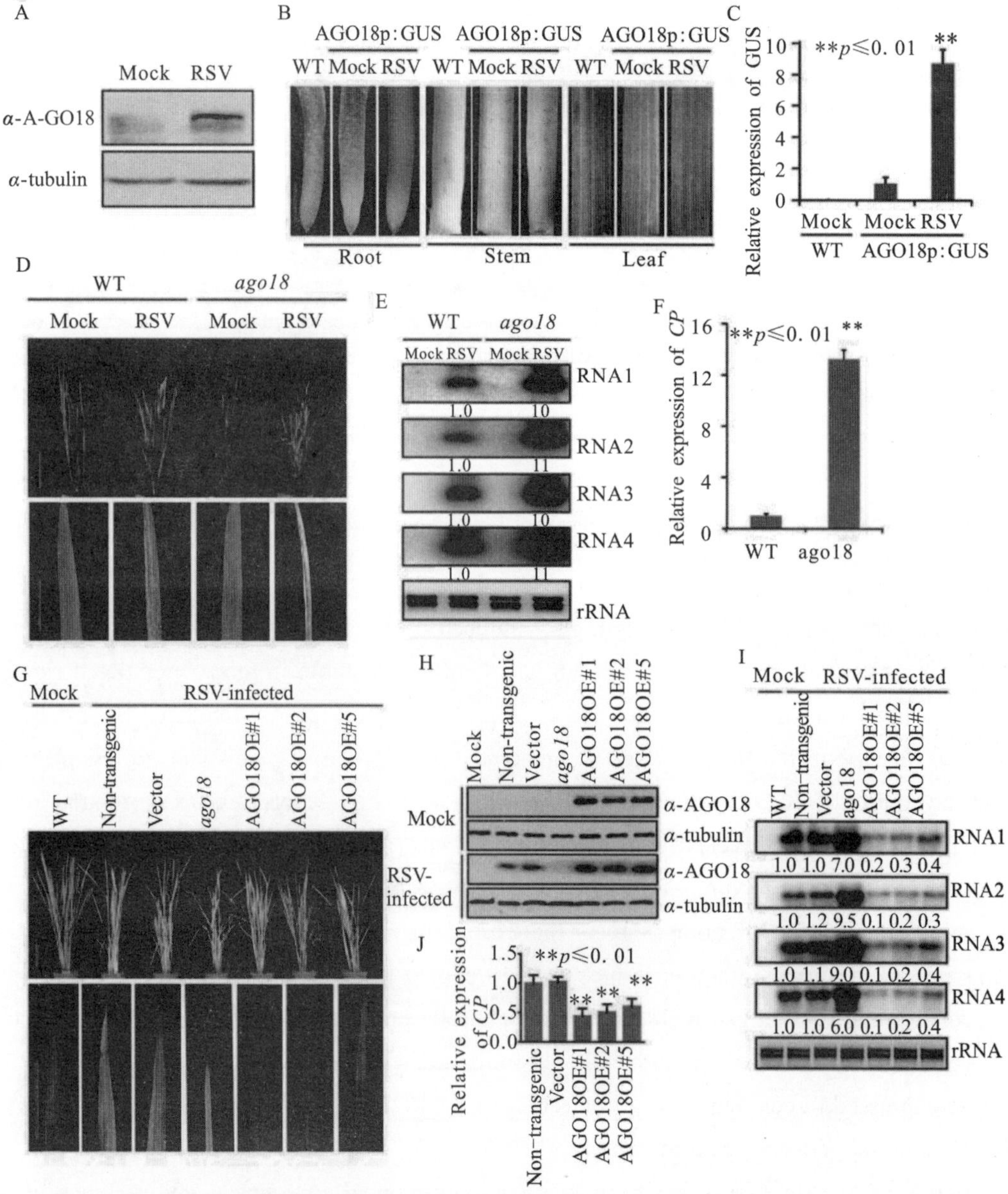

Fig. 2 AGO18 is induced by viral infection and confers antiviral immunity in rice

(**A**) Detection of AGO18 in mock- or RSV-inoculated rice plants by Western blot. Tubulin was probed and served as a loading control.

(**B**) Representative GUS staining images of mock- or RSV-inoculated *AGO18p:GUS* transgenic plants. Scale bars, 5 mm. (**C**) qRT-PCR analysis of the transcript level of *GUS* in the indicated plants. The average (± standard deviation) values from three biological repeats of qRT-PCR are shown. (**D**) Symptoms of wild-type (WT) and ago18 rice plants infected with RSV, pictures were taken at 6 weeks post-inoculation. Scale bars, 15cm. (**E**) Detection of RSV genomic RNA segments in WT and ago18 rice plants by Northern blot. The blots were hybridized with radiolabeled riboprobes specific for each RNA segment. rRNAs were stained with ethidium bromide and served as loading controls. The RNA signals were quantified and normalized to rRNAs, and the relative values were calculated by comparison with those in RSV-infected WT (arbitrarily set to 1.0). (**F**) Detection of RSV *CP* in the RSV-infected WT and *ago18* by qRT-PCR. The expression levels were normalized using the signal from *OsEF-1a*. The average (± standard deviation) values from three biological repeats of qRT-PCR are shown. (**G**) Symptoms of RSV-infected WT (Non-transgenic), *ago18* and transgenic lines overexpressing AGO18, pictures were taken at 6 weeks post-inoculation. Scale bars, 15cm (upper panel) and 5cm (lower panel). (**H**) Detection of AGO18 in mock- or RSV-inoculated plants as indicated. Tubulin was probed and served as a loading control. (**I**) Detection of RSV genomic RNA segments in the indicated plants by Northern blot. The blots were hybridized with radiolabeled riboprobes specific for each RNA segment. rRNAs were stained with ethidium bromide and served as loading controls. The RNA signals were quantified and normalized to rRNAs, and the relative values were calculated by comparison with those in RSV-infected WT (Non-transgenic) (arbitrarily set to 1.0). (**J**) Detection of RSV *CP* in the indicated RSV-infected plants by qRT-PCR. The expression levels were normalized using the signal from *OsEF-1a*. The average (± standard deviation) values from three biological repeats of qRT-PCR are shown

2.3 AGO18 unlikely functions as an effector of vsiRNAs

Given that AGO1 is present in the *ago18* mutant and AGO18 is present in ago1 RNAi plant and that both *ago18* mutant and *ago1* RNAi rice plants were very susceptible to viral infection, AGO1 and AGO18 evidently depend on each other for their antiviral activities. A key question is how this mutual dependence operates.

To investigate how AGO18 acts in antiviral defense, we first tested whether this protein may function as an effector of vsiRNAs, like AGO1. We immunoprecipitated AGO18 complexes from RSV-infected WT rice plants using AGO18 antibodies (Fig. 3A). Northern blots using RSV RNA2 probes showed much weaker signals for vsiRNAs in AGO18 (Fig. 3A), compared to those in AGO1a and AGO1b (Fig. 1C). We further profiled small RNAs in AGO18 complexes prepared from RSV-infected plants using deep sequencing. For comparison, small RNAs in total extracts, AGO1a, and AGO1b complexes were also profiled in parallel.We found that vsiRNAs accounted for about 4.48% of the total small RNAs in RSV-infected plants and 17.37%–9.77% of those in the AGO1a and AGO1b complexes (Fig. 3B; S1B), indicating that vsiRNAs are highly enriched in the AGO1 complexes. In contrast, vsiRNAs accounted for only about 2.52% of the AGO18-bound small RNAs (Fig. 3B; S1B). These data suggest that AGO18 is not a major effector of vsiRNAs. Together with the finding that AGO18 is required for the antiviral activity of AGO1 (see above), this observation suggests that AGO18 very likely uses a different strategy to regulate antiviral defense response in rice.

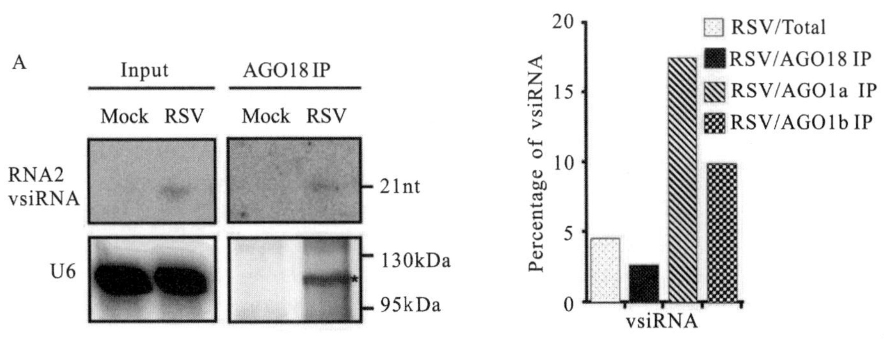

Fig. 3 AGO18 unlikely functions as an effector of vsiRNAs

(**A**) Detection of vsiRNAs in total extracts (Input) or AGO18 immunoprecipitates prepared from mock- or RSV-inoculated plants by

Northern blot. The silver-stained gel shows the quality of purified AGO18 complexes. The asterisk indicates the position of AGO18. The position of an RNA size marker is shown on the right of the blot. U6 is probed and served as a loading control. **(B)** Percentage of deep sequencing reads matching vsiRNAs in total reads obtained from total extracts, AGO1a, AGO1b, and AGO18-associated small RNAs. Samples for deep sequencing were prepared from RSV-infected rice plants

2.4 AGO18 competes with AGO1 for miR168 to up-regulate AGO1 upon viral infection

RSV infection increases the accumulation of miR168 as well as AGO18 (Du et al., 2011) (Fig. 2A–C and Fig. 4A), suggesting that the increased miR168 is not efficiently loaded into AGO1 and hence not effective in targeting AGO1. Intriguingly, from the small RNA deep sequencing analyses, we found that AGO18 recruited a large amount of miR168 in RSV-infected rice plants compared with AGO1a or AGO1b (Fig. 4A; S1C). Given that miR168 plays a critical role in AGO1 homeostasis in plants (Mallory and Vaucheret, 2010) and that the antiviral function of AGO1 requires the presence of AGO18, our observations suggest that AGO18 up-regulates AGO1 by competitively binding miR168 to enable antiviral defense.

Immunoprecipitation (IP)-northern blot analyses further confirmed that RSV infection of WT rice plants increased the association of miR168 with AGO18 but decreased its association with AGO1a and 1b, whereas several control miRNAs (including miR166 and miR156) showed no obvious changes (Fig. 4B). Consistent with these results, in RSV-infected ago18 mutants, where AGO18 is absent, more miR168 was now loaded into AGO1a and 1b (Fig. 4C), which was correlated with the reduced expression of AGO1 at mRNA (Fig. 4D) and protein (Fig. 4E) levels. Also in the AGO18OE#1 transgenic rice line, in which AGO18 was overexpressed, *AGO1a* and *1b* both increased at the mRNA level (Fig. 4F), likely as a result of more miR168 being loaded into AGO18. Consistent with our previous finding (Du et al., 2011), *AGO2* was induced by RSV infection in wild-type rice plants (Fig. 4F). Intriguingly, knockout or overexpression of AGO18 did not have an obvious effect on such induction, indicating that the induction of *AGO2* by RSV infection is independent of AGO18 (Fig. 4F). Thus, AGO18 specifically up-regulates AGO1 in antiviral defense.

To directly test whether AGO18 could indeed compete with AGO1 for binding miR168, we transiently expressed a miR168 precursor together with Flag-AGO1a or Flag-AGO1b in the presence or absence of Myc-AGO18 in *N. benthamiana* leaves. MiR444 was used as a control, because it is a species-specific miRNA in monocots that can be loaded by AGO1 (Wu et al., 2009), but not by AGO18 (Supplementary file 1C). Results from IP-northern experiments showed that when AGO18 was co-expressed with AGO1a or AGO1b, there was a nearly five-fold reduction in miR168 loading into AGO1a and AGO1b and concurrent increase in loading into AGO18 (Fig. 5). The control miR444 was specifically loaded into AGO1a and 1b and co-expression of AGO18 did not decrease its loading into AGO1a and 1b (Fig. 5). These data demonstrate that AGO18 can effectively compete with AGO1 for binding miR168.

2.5 Small RNA-binding but not slicing activity of AGO18 is required for its antiviral function

The above results show that AGO18 is unlikely an effector of vsiRNAs but has the novel activity of competing with AGO1 for binding miR168. To further test this, we analyzed the functions of AGO18 domains in antiviral defense. AGO family proteins contain four characteristic domains: an N-terminal domain and conserved PAZ, MID, and PIWI domains (Tolia and Joshua-Tor, 2007; Vaucheret, 2008). The PAZ domain binds to the 3′ end of small RNAs (Ma et al., 2004), whereas the MID domain recognizes the 5′ end (Kidner and Martienssen, 2005; Mi et al., 2008; Montgomery et al., 2008; Frank et al., 2010). The PIWI domain adopts an RNaseH-like structure and exhibits endonuclease (slicing) activity when an Asp-Asp-His (DDH) catalytic triad is present (Song et al., 2004; Rivas et al., 2005).

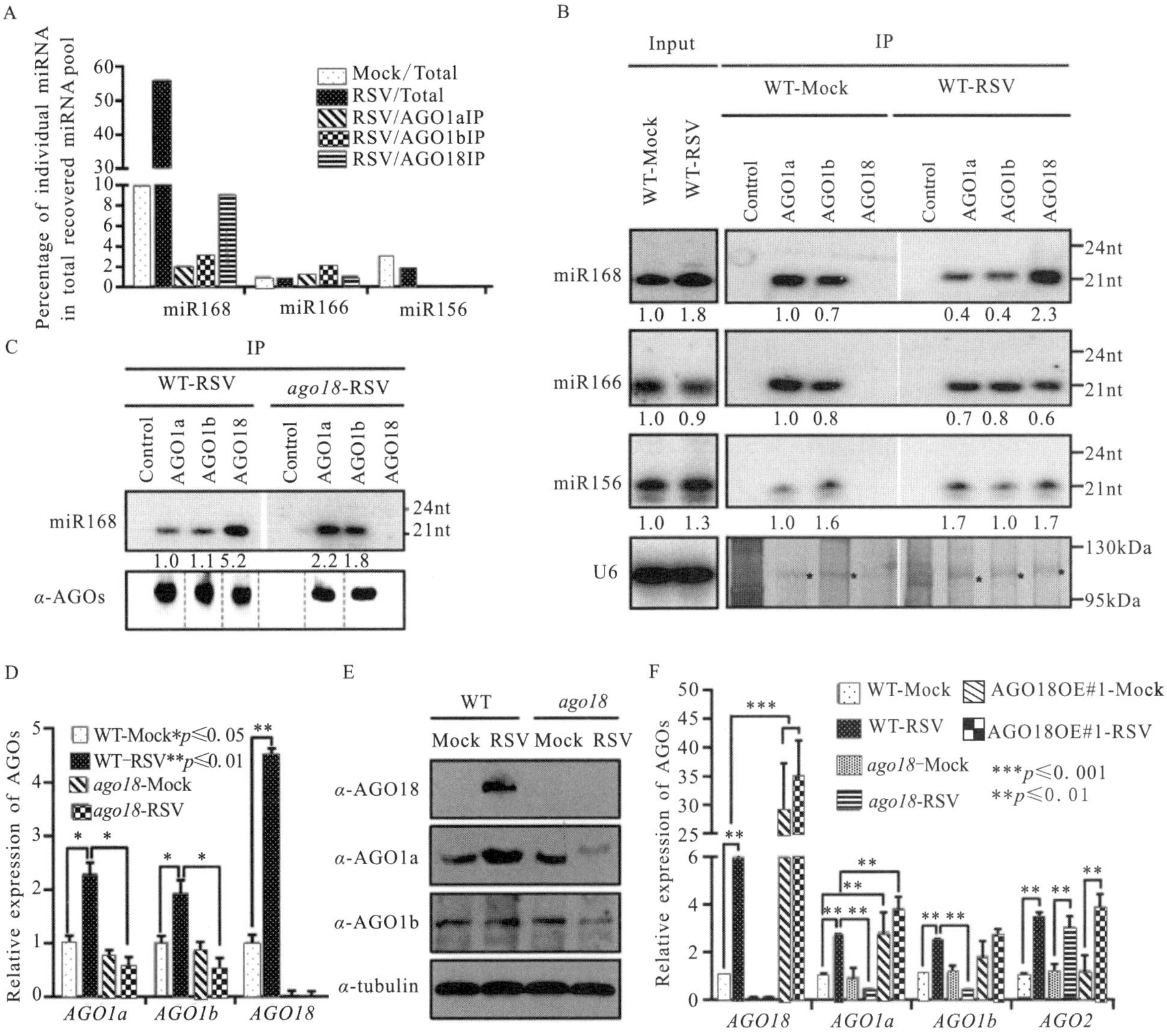

Fig. 4 AGO18 competes with AGO1 for miR168 to up-regulate AGO1 upon viral infection

(**A**) Percentage of deep sequencing reads matching the indicated miRNAs in total reads obtained from total extracts, AGO1a, AGO1b, and AGO18-associated small RNAs. Samples for deep sequencing were prepared from mock- or RSV-inoculated rice plants. (**B**) Detection of the indicated miRNAs in total extract (Input), AGO1a, AGO1b, and AGO18 complexes by Northern blot. The blots were stripped and reprobed for multiple times. The silver-stained gel shows that comparable amounts of different AGO complexes were used for RNA preparation. The asterisks indicate the positions of AGO proteins. The positions of RNA size markers are shown on the right of the blots. The RNA signals were quantified, and the relative values were calculated by comparison with those in total extracts or AGO1a complex prepared from mock-inoculated WT (arbitrarily set to 1.0). (**C**) Northern blot analysis showing miR168 from AGO1a, AGO1b, and AGO18 complexes in RSV-infected WT and *ago18* plants (upper panel). Western blot gel shows that comparable amounts of different AGO complexes were used for RNA preparation (lower panel). (**D**) qRT-PCR analysis of the levels of *AGO1a*, *AGO1b*, and *AGO18* in WT rice and *ago18* mutants with or without RSV infection. The expression levels were normalized using the signal from *OsEF-1a*. The average (± standard deviation) values from three biological repeats of qRT-PCR are shown. (**E**) Western blot showing AGO1a, AGO1b, and AGO18 protein levels in WT and *ago18* with or without RSV infection. Tubulin was probed and served as a loading control. (**F**) qRT-PCR analysis of the levels of *AGO1a*, *AGO1b*, *AGO2*, and AGO18 in the indicated plants. The expression levels were normalized using the signal from *OsEF-1a*. The average (± standard deviation) values from three biological repeats of qRT-PCR are shown

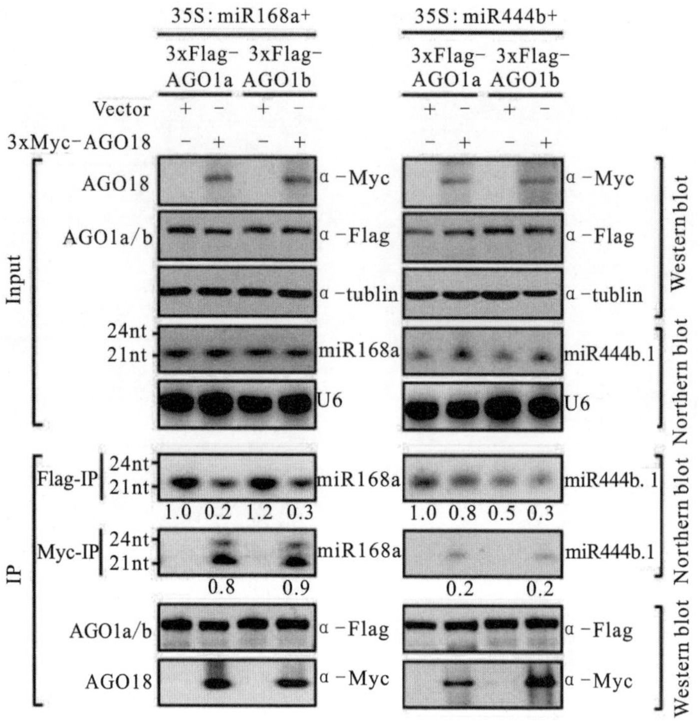

Fig. 5 AGO18 competes with AGO1 for miR168 in vitro

Specific AGO18–miR168 interaction was confirmed by in vitro assays in *N. benthamiana*. Constructs with the indicated combinations were introduced into *N. benthamiana* leaves for transient expression by agro-infiltration. Northern blots were conducted with total RNA (Input) and small RNAs recovered from immunoprecipitated AGO complexes (IP). Western blot analyses were done with the crude extract and aliquots of the IP products using anti-Flag or anti-Myc antibodies. The positions of RNA size markers are shown on the left of the blots. U6 was probed and served as a loading control

Our analysis indicates that AGO18 also contains all of the four AGO signature domains and the DDH catalytic triad (Fig. 6A). We investigated whether the small RNA binding and slicing activities of AGO18 are required for its role in antiviral defense by transgenic complementation experiments. We generated two AGO18 mutants, $AGO18^{Y537A/F538A}$ (YF/AA) and $AGO18^{D833A}$ (D833A). YF/AA contains alanine substitutions at two conserved residues (Y537F538) that are known to be required for small RNA binding (Ma et al., 2004; Guang et al., 2008; Ye et al., 2012), whereas D833A contains an alanine substitution at the DDH catalytic triad (Song et al., 2004; Rivas et al., 2005; Wee et al., 2012). We transgenically expressed WT AGO18, YF/AA, and D833A mutants under the control of the native *AGO18* promoter in the ago18 mutant background. We then inoculated these plants with RSV by insect vector transmission. Western blots confirmed the expression of WT and mutant AGO18 in the infected transgenic plants (Fig. 6B). Based on the disease symptoms (Fig. 6C) and accumulation of viral genomic RNAs (Fig. 6D), both WT AGO18 and D833A could mostly complement the *ago18* mutant for resistance to RSV infection, whereas the YF/AA mutant could not. These results suggest that AGO18 exerts its antiviral function mainly through small RNA binding, rather than slicing. We also measured the expression levels of *AGO1* in these transgenic lines that were infected by RSV. We found that *AGO1* expression was elevated by viral infection in the transgenic plants that express WT AGO18 or D833A mutant but not in those that express YF/AA (Fig. 6E), further suggesting that the small RNA binding but not slicing activity of AGO18 is required for its role in up-regulating AGO1.

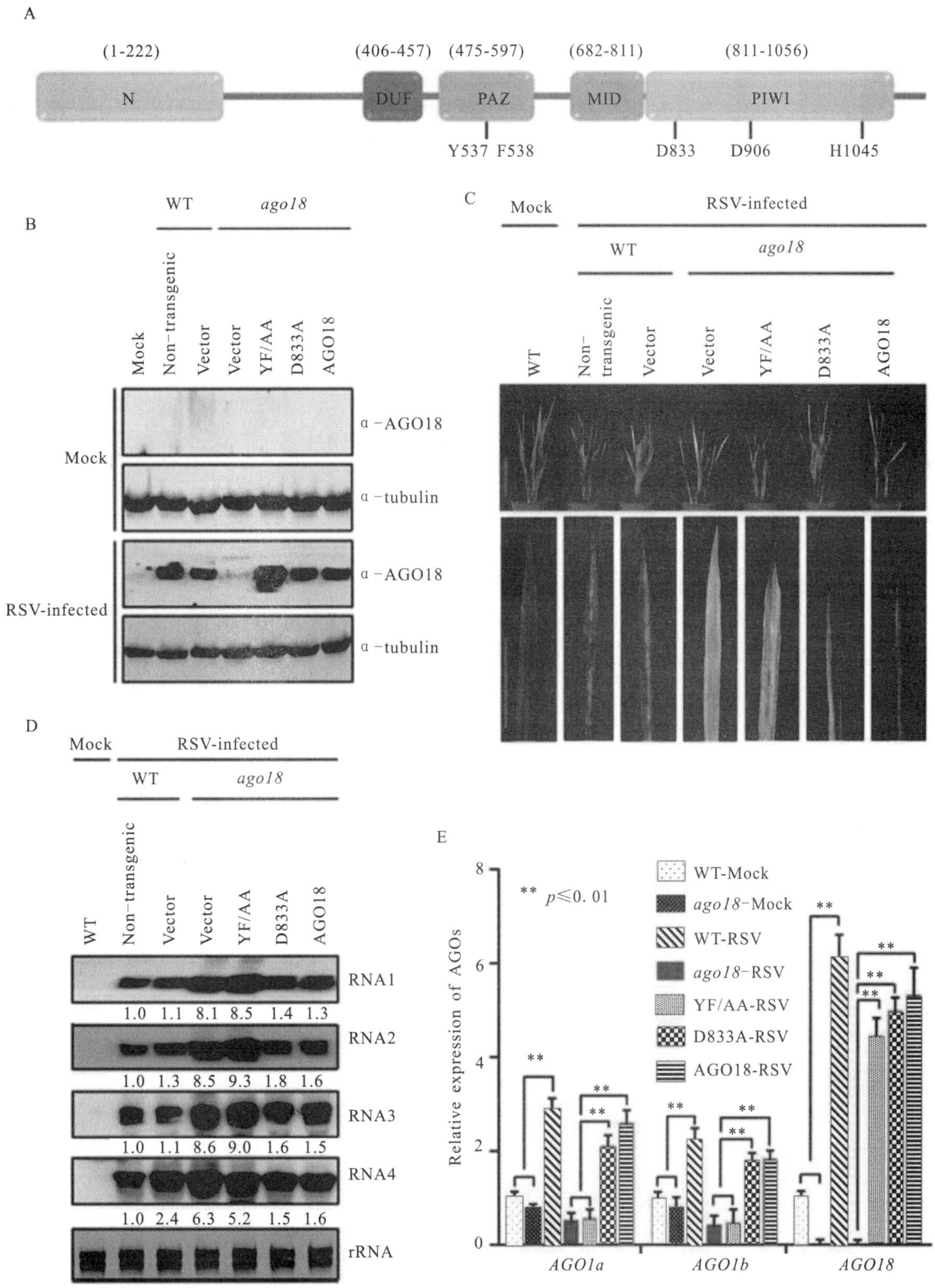

Fig. 6 Small RNA-binding but not slicing activity of AGO18 is required for its antiviral function

(A) Domain structure of AGO18 protein. AGO18 consists of a variable N-terminal domain and conserved C-terminal PAZ, MID, and PIWI domains. The residues Y537 and F538 required for small RNA binding, and D833, D906, and H1045 required for slicing are indicated. (B) Detection of AGO18 protein in mock- or RSV-inoculated WT (Non-trangenic) plants as well as ago18 mutants complemented with AGO18 and its derivatives by Western blot. Tubulin was probed and served as a loading control. (C) Symptoms of mock- or RSV-inoculated WT (Non-trangenic) plants as well as *ago18* mutants complemented with AGO18 or its derivatives, pictures were taken at 6weeks post-inoculation. Scale bars, 15cm (upper panel) and 5cm (lower panel). (D) Detection of RSV genomic RNA segments in mock- or RSV-inoculated WT (Non-trangenic) plants as well as *ago18* mutants complemented with AGO18 or its derivatives by Northern blot. The blots were hybridized with radiolabeled riboprobes specific for each RNA segment. rRNAs were stained with ethidium bromide and served as loading controls. The RNA signals were quantified and normalized to rRNAs, and the relative values were calculated by comparison with those in RSV-infected WT (arbitrarily set to 1.0). (E) qRT-PCR analysis of the levels of *AGO1a*, *AGO1b*, and *AGO18* in RSV-infected WT plants as well as ago18 mutants complemented with AGO18 or its derivatives. The expression levels were normalized using the signal from *OsEF-1a*. The average (± standard deviation) values from three biological repeats of qRT-PCR are shown

2.6 Expression of miR168-resistant *AGO1a* rescues the deficiency of *ago18* for viral resistance

The above studies demonstrated that AGO18 regulates AGO1 homeostasis by sequestering miR168 during viral infection. To further test this, we generated transgenic rice plants expressing wild type (AGO1a lines) and miR168-resistant *AGO1a* (AGO1a-Res lines) in the ago18 background, under the control of the native *AGO1a* promoter (Fig. 7A). The steady-state AGO1a/AGO1a-Res mRNA levels of both AGO1a and AGO1a-Res transgenes accumulated to higher levels relative to the *AGO1a* levels in wild-type (WT) non-transgenic and ago18 plants (Fig. 7B). It is noteworthy that *AGO1a* mRNA level was reduced by RSV infection in the AGO1a transgenic line, whereas AGO1a-Res transgenic lines had no significant reduction in *AGO1a-Res* mRNA levels upon RSV infection (Fig. 7B), indicating that AGO1a is subject to regulation by miR168/AGO18. Although the transcript levels of the miR168-resistant transgene *AGO1a* markedly increased, none of these lines showed notable phenotypic differences from WT or *ago18* plants in growth and development. These transgenic lines, however, were much more resistant to RSV infection than *ago18* and AGO1a plants (Fig. 7B,C). Viral replication significantly decreased in the AGO1a-Res plants (Fig. 7D). Thus, AGO1 over-expression could rescue the deficiency of *ago18* for viral resistance. These data provided further compelling evidence that AGO18 functions through sequestering miR168 to up-regulate AGO1 against virus infections.

Based on the above results, we conclude that AGO18 regulates AGO1 homeostasis by specifically loading miR168 during viral infection. Sequestering of miR168 by AGO18 leads to increased accumulation of AGO1 to embark on an effective antiviral defense response.

3 Discussion

AGO proteins are at the heart of RNAi machinery. Building upon the conserved core components of the RNA silencing machinery such as DCLs and AGOs, plants have evolved extensive variants of these components, especially AGOs. This evolutionary amplification of the machinery components implies amplification/diversification in regulatory mechanisms/functions. While the roles of some AGOs in specific functions of RNA silencing in gene regulation and innate immunity have been well studied, the roles of most AGOs remain unknown or poorly understood. Besides AGO1, only the slicing function of AGO2 appears to be also important for antiviral defense in *Arabidopsis* (Harvey et al., 2011; Jaubert et al., 2011; Scholth of et al., 2011; Wang et al., 2011; Carbonell et al., 2012; Xia et al., 2014). Current studies on the roles of several *Arabidopsis* AGOs in antiviral defense are dictated by the target RNA-slicing paradigm.

In the arms race between host defense and viral counter-defense, plant viruses have evolved various strategies, including encoding suppressors of RNA silencing, to interfere with different steps of the host RNA silencing defense pathways (Ding and Voinnet, 2007; Ding, 2010). In *Arabidopsis* and *N. benthamiana*, infection by many viruses leads to increased levels of *AGO1* mRNA readying the plants to combat viruses, but the viruses also simultaneously increase the expression levels of miR168 to down-regulate AGO1 to defeat host defense (Va'rallyay et al., 2010). RSV infection of rice also leads to an increase in miR168 (Fig. 4A,B) (Du et al., 2011). Thus, elevated miR168 expression appears to be a broad mechanism of viral counter-defense.

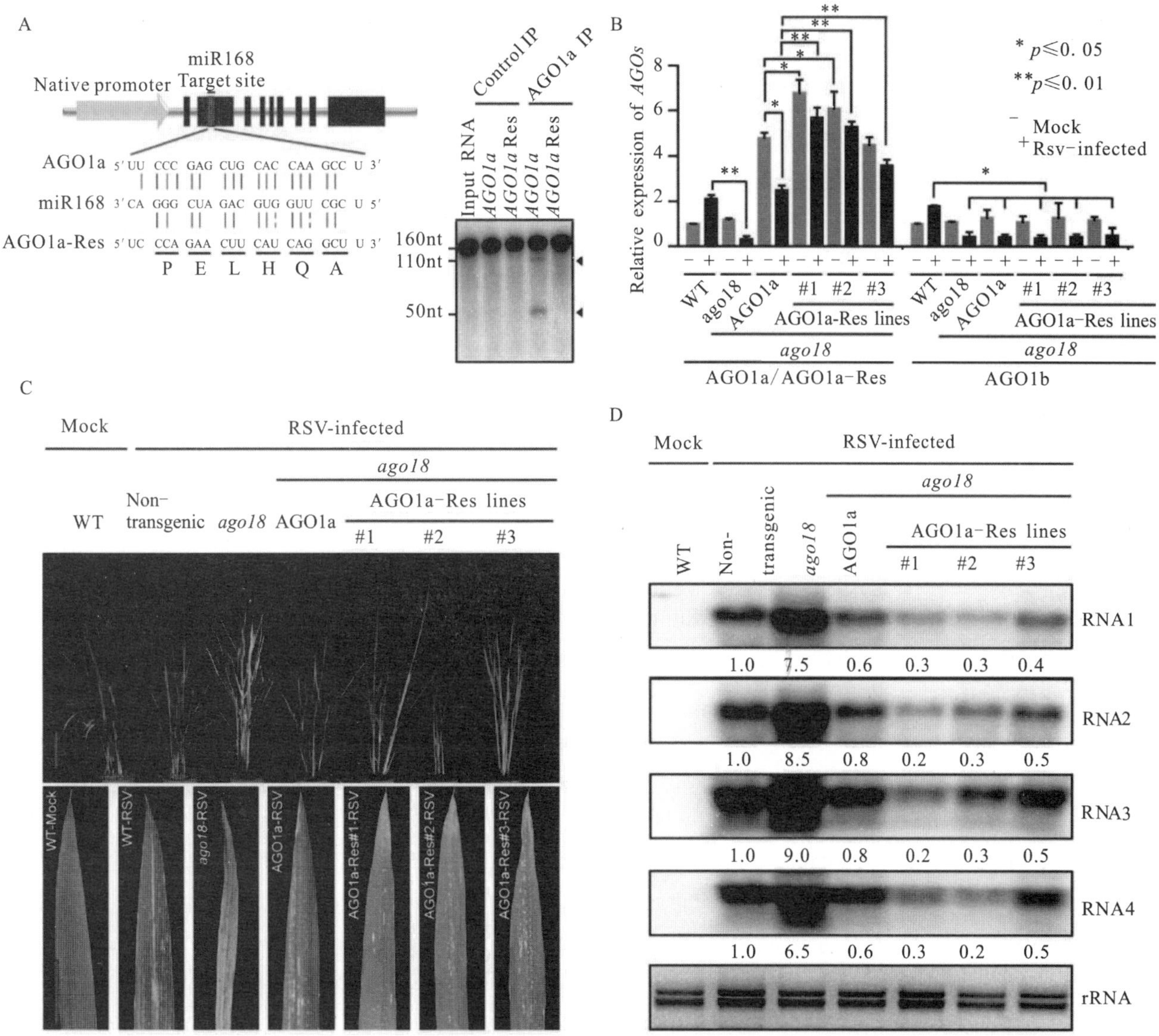

Fig. 7 Transgenic expression of miR168-resistant AGO1a rescued the deficiency of *ago18* for viral resistance

(A) Schematic drawing of *AGO1a* featuring the target site of miR168. In AGO1a that is resistant to miR168 cleavage (*AGO1a-Res*), seven synonymous nucleotide substitutions were introduced into the miR168 target site. The resistance of *AGO1a-Res* to miR168-directed cleavage was examined by in vitro cleavage assay using purified AGO1a complex. Positions of the cleavage products are indicated by arrows. (B) qRT-PCR analysis of *AGO1a* and *AGO1b* expression levels in mock (−) or RSV-inoculated (+) WT, *ago18*, and transgenic plants expressing *AGO1a* or *AGO1a-Res* in the *ago18* mutant background. The expression levels were normalized using the signal from OsEF-1a. The average (± standard deviation) values from three biological repeats of qRT-PCR are shown. (C) Symptoms of mock- or RSV-inoculated WT (Non-trangenic), ago18, and transgenic plants expressing AGO1a or AGO1a-Res in the ago18 mutant background. Scale bars, 15cm. (D) Detection of RSV genomic RNA segments in the indicated plants by Northern blot. The blots were hybridized with radiolabeled riboprobes specific for each RNA segment. rRNAs were stained with ethidium bromide and served as loading controls. The RNA signals were quantified and normalized to rRNAs, and the relative values were calculated by comparison with those in RSV-infected WT (Non-trangenic) (arbitrarily set to 1.0)

In this study, we discovered a novel mechanism of positive regulation of AGO1 activity by AGO18 in antiviral RNAi. We showed that AGO18 does not have antiviral function by itself, but its presence is required for the antiviral function of AGO1. AGO18 accomplishes this not via its slicing function but through its competition with AGO1 for binding miR168. This binding inhibits miR168-mediated auto-regulation of AGO1, thereby boosting the accumulation levels of AGO1 to enable antiviral function. Thus, AGO18 has evolved as a novel and dedicated positive regulator of the plant surveillance/ defense system. Our findings suggest that AGO18-bound miR168 is not competent in cleaving *AGO1* mRNA in RSV-infected rice plants, albeit AGO18 contains conserved DDH motif for slicing activity. Several possible mechanisms can be considered. First, the catalytic activity of AGO18 may not be as potent as that of AGO1 due to some sequence/structure differences between these two AGOs. Second, the slicing activity of AGO18 may be repressed by a yet-to-be-identified protein that specifically interacts with AGO18. Third, the cellular compartmentation of AGO18/miR168 complex may be distinct from that of AGO1/miR168 and prevent its access to *AGO1* mRNA. Finally, AGO18 binding of miR168 may subsequently induce the degradation of miR168. It will be highly interesting to test these possibilities in future studies.

There were previous examples of competitive binding of other miRNAs between other AGOs and AGO1 in *Arabidopsis*. miR168 can be incorporated into AGO10 to decrease the translation efficiency of *AGO1* mRNA in *Arabidopsis* (Mallory et al., 2009). The *Arabidopsis* miR166/165 are significantly enriched in AGO10-bound miRNAs, preventing them from being loaded into AGO1 to fulfill their normal roles in development (Zhu et al., 2011; Ji et al., 2011; Manavella et al., 2011). MiR390 is associated with AGO7 to mediated trans-acting siRNA biogenesis (Montgomery et al., 2008), through cooperative activity of AGO1. In addition, miR408 associates with both AGO1 and AGO2 redundantly to regulate Plantacyanin mRNA levels (Maunoury and Vaucheret, 2011). This competition, however, does not affect AGO1 homeostasis. In contrast to these negative regulations of AGO1, binding of miR168 by AGO18 in infected rice plants boosts AGO1 accumulation. Thus, among mechanisms of AGO-regulation of AGO1 homeostasis or activity, the up-regulation of AGO1 via AGO18 binding of miR168 represents a novel type of mechanism. Whether other AGOs function similarly in *Arabidopsis*, rice, and other plants to up-regulate AGO1 or another AGO to impact developmental processes or defense responses is an outstanding question to be addressed in future studies. It is also noteworthy that, in addition to miR168, several miRNAs including miR528, miR159a, and miR159b were also recruited by AGO18 in RSV-infected rice plants (Supplementary file 1C). Thus, for AGO18, we cannot rule out the possibility that AGO18 plays regulatory roles in other capacities or has its own independent biological functions.

It is important to emphasize that most plant viruses are transmitted to plants by insect vectors under natural infection conditions in the field, and yet the vast majority of studies so far on the mechanisms of RNA silencing-mediated antiviral responses employed infection methods such as mechanical inoculation with in vitro viral RNA transcripts or virions and infiltration with agrobacteria carrying engineered viral DNAs. These studies often used viral delivery methods with much higher levels of inoculum than natural field conditions. How some of such manipulations would alter host responses remains an outstanding question. Here, we used insect vectors that carry the natural forms of viruses to inoculate plants, best reflecting what happens under field infection conditions with regard to the behavior of viruses and host responses. This is not only important to dissect the natural infection and defense mechanisms, but also important for developing effective technologies to combat viral infection under natural infection conditions. Recent studies indicated that natural infection in early stage of virus infection may trigger different host-defense reaction (Garcia et al., 2014).

AGO18 is of special interest because it has evolved as a conserved clade in monocots that encompass many important crop plants for foods and biofuels. Whether it plays a similar role in antiviral defense in other monocots warrants further investigations. From a broader perspective, we expect that further studies on the extensive RNA silencing machinery components evolved in crop plants may lead to new discoveries about their functions in shaping plant diversity with regard to phenotypes and innate immunity mechanisms. Finally, sequestering miR168 by AGO18 to regulate AGO1 homeostasis represents an evolutionary novelty. It raises the question of whether small RNA competition-based regulations between other types of proteins have also evolved to regulate different biological processes in plants and other organisms. This mechanism functions against infection by two evolutionarily distinct viruses, and likely has broader significance in resistance against more viruses in a wide range of monocots. We propose that engineering rice and other cereals to over-express AGO18 may provide a new strategy for the control of diverse viral pathogens.

4 Materials and methods

4.1 Plant growth and virus inoculation

Plant growth and virus inoculation were essentially carried out as described (Du et al., 2011). Briefly, rice (*O. sativa* spp. japonica) seedlings were grown in a greenhouse at 28°C–30°C and 60% ± 5% relative humidity under natural sunlight for 4weeks. The viruliferous (RSV and RDV-carrying) insects of *L. striatellus* and *N. cincticeps*, as well as virus-free *L. striatellus* and *N. cincticeps* (mock) were used for inoculation. After feeding 3d, the insects were removed, and the rice seedlings were returned to the greenhouse to grow under the greenhouse conditions above. 3weeks post-inoculation when the newly developed leaves started to exhibit viral symptoms, the whole seedlings were harvested. For each sample, at least 15–20 rice seedlings were pooled for RNA extraction.

4.2 Non-preference test

Non-preference tests were performed for all rice seedlings of the different genetic backgrounds with the two viral transmission insect vectors brown planthopper (*L. striatellus*) and leafhopper (*N. cincticeps*). Details of the procedures were described previously (Hiroshi et al., 1994).

4.3 Histochemical GUS staining

Plants were infiltrated with 50mM sodium phosphate (pH 7.0), 10mM EDTA, and 0.5mg/ml X-gluc (Apollo Scientific, UK), followed by incubation at 37°C in the dark overnight and then destained in 70% ethanol before photographing.

4.4 Generation of antibodies against rice AGOs

Synthetic peptides AGO1aN (KKKTEPRNAGEC), AGO1bN (KKRTGSGSTGEC), and AGO18N (YHGDGERGYGRC) were used to raise rabbit polyclonal antibodies against AGO1a, AGO1b, and AGO18, respectively, essentially as described (Wu et al., 2009;Mi et al., 2008). The antisera were affinity purified and used for immunoprecipitation (IP) (1:50 dilution).

4.5 Constructs and transgenic lines

Gateway system (Invitrogen, Carlsbad, CA) was used to make binary constructs. Several destination vectors were created for transient expression in *N. benthamiana* and stable rice transformation. The binary gateway vector pMDC32 (Karimi et al., 2007) was modified to obtain rice AGO18 promoter (p32:pAGO18) or AGO1a promoter (p32:pAGO1a). Most cDNA and miRNA genes were cloned into pENTR/D vectors and *pENTR/D-Flag-AGO18D833A*, *YF/AA* and *AGO1a:miR168 Res* clones were prepared by using the QuikChange site-directed mutagenesis kit (Stratagene, La Jolla, CA). All these clones were confirmed by sequencing and transferred to the appropriate destination vectors by recombination using the Gateway LR Clonase II Enzyme mix (Invitrogen). *pCam2300*: *Actin1*::*OCS* and *pCam2300*:*35S*::*OCS* vectors (Wu et al., 2009) were used to generate *pCam2300*: *Actin1*:: *Flag- AGO18*, *pCam2300*:*35S*::*Myc-AGO18*, *Flag-AGO1a*, *Flag-AGO1b*, *miR168a*, and *miR444b*, respectively. All PCR primers are listed in Supplementary file 1D.

4.6 Transient expression in the *N. benthamiana*

Assays of transient expression in leaves of *N. benthamiana* were performed as described (Voinnet et al., 2003). A detailed protocol is available upon request.

4.7 Western blotting

Protein samples were boiled with same volume of 2×protein loading buffer at 95°C for 5min and separated by SDS-PAGE gel. Proteins were then transferred to PVDF membranes and detected with antibodies against AGO18, AGO1a, AGO1b, MYC (11667203001, Roche, Switzerland), FLAG (F1804, Sigma–Aldrich, St. Louis, MO), and tubulin (T5168, Sigma–Aldrich).

4.8 Purification of rice AGO-containing complexes and associated small RNAs

Rice AGO-containing complexes were immunopurified from RSV-inoculated rice plants as previously described (Qi et al., 2005; Wu et al., 2010). The quality of purification was examined by SDS-PAGE followed by silver staining or IP-western blotting, and the bands of expected sizes were confirmed as AGO proteins by mass spectrometry. RNAs were isolated from total cell extracts and from the purified AGO complexes by Trizol reagent (Invitrogen), resolved on a 15% denaturing PAGE gel, and visualized by SYBR-gold (Invitrogen) staining. Gel slices within the range of 18–28 nucleotides were excised, and the RNAs were eluted and purified for cloning.

4.9 Small RNA cloning and sequencing

Small RNA cloning for Illumina sequencing was carried out essentially as described previously (Mi et al., 2008; Wu et al., 2010). A detailed protocol is available upon request.

4.10 Small RNA Northern blotting

Northern blot analysis with total sRNAs or from purified AGO complexes was performed as described before (Qi et al., 2005). ^{32}P-end labeled oligonucleotide probes complementary to sRNAs were used for Northern blots. The sequences of the probes are listed in *Supplementary file 1D*.

4.11 Quantitative RT-PCR analysis

Total RNAs were extracted from rice plants with Trizol (Invitrogen). After removal of contaminating DNAs by digestion with RNase-free DNaseI (Promega, Madison, Wisconsin, USA), the RNAs were reverse transcribed by SuperScript III Reverse Transcriptase (Invitrogen) using oligo (dT). The cDNAswere then used as templates for quantitative PCR and RT-PCR. Quantitative PCR was performed using SYBR Green Real-time PCR Master Mix (Toyobo, Osaka, Japan). The rice *OsEF-1a* gene was detected in parallel and used as the internal control. All the other primers used are listed in *Supplementary file 1D*.

4.12 Bioinformatics analysis of small RNA data sets

The adaptor sequences in Illumina 1G sequencing reads were removed by using 'vectorstrip' in the EMBOSS package. The sRNA reads with length of 19–27nt were mapped to the rice nuclear, chloroplastic, and mitochondrial genomes (http://rice.plantbiology.msu.edu/, version 6.0). The sRNAs with perfect genomic matches were used for further analysis. Rice miRNA annotations were from miRBase (http://microrna.sanger.ac.uk/sequences, Release14) and our previous publication (Wu et al., 2009; Du et al., 2011). Statistical analysis of the sRNA data sets was done by using in-house-developed Perl scripts (*Supplementary file 2*).

4.13 miRNA cleavage assay

miRNA cleavage activity assay was performed essentially as described (Qi et al., 2005; Qi and Mi, 2010), using immunopurified rice AGO1a complex and in vitro-transcribed AGO1a or AGO1a-Res transcripts.

4.14 Accession numbers

Small RNA data sets generated in this study are deposited in the NCBI sequence read archive (SRA) (http://www.ncbi.nlm.nih.gov/sra) under accession number PRJNA273330.

Acknowledgements

We thank the National Institute of Biological Sciences (NIBS) Mass Spectrometry Center, Beijing, for sequencing AGO18 protein, NIBS Antibody Center and Integrated R&D Services—WuXi AppTec for generating the antisera used in this study. This work was supported by grants from the National Basic Research Program 973 (2014CB138400, 2011CB100703 and 2013CBA01403), Natural Science Foundation of China (31225015, 31421001, 31030005, 31420103904, and 31272018), Transgenic Research Program (2014ZX08010-001), and Doctoral Fund of Ministry of Education of China (20113515120004). JGW was supported in part by the Postdoctoral Fellowship of Peking-Tsinghua Center for Life Sciences.

References

[1] BAULCOMBE D. RNA silencing in plants[J]. Nature, 2004, 431: 356-363.

[2] BAUMBERGER N, BAULCOMBE D C. *Arabidopsis* ARGONAUTE1 is an RNA Slicer that selectively recruits microRNAs and short interfering RNAs[J]. Proceedings of the National Academy of Sciences of the United States of America, 2005,102: 11928-11933.

[3] BURGYA´N J, HAVELDA Z. Viral suppressors of RNA silencing[J]. Trends in Plant Science, 2011, 16: 265-272.

[4] CARBONELL A, FAHLGREN N, GARCIA-RUIZ H,et al. Functional analysis of three *Arabidopsis* ARGONAUTES using slicer-defective mutants[J]. The Plant Cell, 2012, 24: 3613-3629.

[5] CHAPMAN E J, CARRINGTON J C. Specialization and evolution of endogenous small RNA pathways[J]. Nature Reviews Genetics, 2007, 8: 884-896.

[6] CHEN X. Small RNAs and their roles in plant development[J]. Annual Review of Cell and Developmental Biology, 2009, 25: 21-44.

[7] DELERIS A, GALLEGO-BARTOLOME J, BAO J, et al. Hierarchical action and inhibition of plant Dicer-like proteins in antiviral defense[J]. Science, 2006, 313: 68-71.

[8] DERRIEN B, BAUMBERGER N, SCHEPETILNIKOV M,et al. Degradation of the antiviral component ARGONAUTE1 by the autophagy pathway[J]. Proceedings of the National Academy of Sciences of the United States of America, 2012, 109: 15942-15946.

[9] DING S W. RNA-based antiviral immunity[J]. Nature Reviews Immunology, 2010, 10: 632-644.

[10] DING S W, VOINNET O. Antiviral immunity directed by small RNAs[J]. Cell, 2007, 130: 413-426.

[11] DU P, WU J, ZHANG J, et al. Viral infection induces expression of novel phased microRNAs from conserved cellular microRNA precursors[J]. PLoS Pathogens, 2011, 7: e1002176.

[12] FRANK F, SONENBERG N, NAGAR B. Structural basis for 5′-nucleotide base-specific recognition of guide RNA by human AGO2[J]. Nature, 2010, 465: 818-822.

[13] GARCIA-RUIZ H, TAKEDA A, CHAPMAN E J, et al. *Arabidopsis* RNA-dependent RNA polymerases and dicer-like proteins in antiviral defense and small interfering RNA biogenesis during Turnip Mosaic virus infection[J]. Plant Cell, 2010, 22: 481-496.

[14] GARCIA D, GARCIA S, VOINNET O. Nonsense-mediated decay serves as a general viral restriction mechanism in plants[J]. Cell Host & Microbe, 2014, 16: 391-402.

[15] GUANG S, BOCHNER A F, PAVELEC D M,et al. An Argonaute transports siRNAs from the cytoplasm to the nucleus[J]. Science, 2009, 321: 537-541.

[16] HARVEY J J, LEWSEY M G, PATEL K,et al. An antiviral defense role of Ago2 in plants[J]. PLoS ONE, 2011, 6: e14639.

[17] HAUPTMANN J, MEISTER G. Argonaute regulation: two roads to the same destination[J]. Developmental Cell, 2013, 25: 553-554.

[18] HIBINO H. Biology and epidemiology of rice viruses[J]. Annual Review of Phytopathology, 1996, 34: 249-274.

[19] HIROSHI N, ISHIKAWA K, SHIMURA E. The resistance to rice stripe virus and small brown planthopper in rice variety Ir50[J]. Breeding Science, 2010, 44: 13-98.

[20] JAUBERT M, BHATTACHARJEE S, MELLO A F, et al. ARGONAUTE2 mediates RNA-silencing antiviral defenses against Potato virus X in *Arabidopsis*[J]. Plant Physiology, 2011, 156: 1556-1564.

[21] JI L, LIU X, YAN J,et al. ARGONAUTE10 and ARGONAUTE1 regulate the termination of floral stem cells through two microRNAs in *Arabidopsis*[J]. PLoS Genetics, 2011, 7: e1001358.

[22] KAPOOR M, ARORA R, LAMA T, et al. Genome-wide identification, organization and phylogenetic analysis of Dicer-like, argonaute and RNA-dependent RNA polymerase gene families and their expression analysis during reproductive development and stress in rice[J]. BMC Genomics, 2008, 9: 451.

[23] KARIMI M, DEPICKER A, HILSON P. Recombinational cloning with plant gateway vectors[J]. Plant Physiology, 2007, 145: 1144-1154.

[24] KIDNER C A, MARTIENSSEN R A. The developmental role of microRNA in plants[J]. Current Opinion in Plant Biology, 2005, 8: 38-44.

[25] LI S, LIU L, ZHUANG X, et al. MicroRNAs inhibit the translation of target MRNAs on the endoplasmic reticulum in *Arabidopsis*[J]. Cell, 2013a, 153: 562-574.

[26] LI Y, LU J, HAN Y,et al. RNA interference functions as an antiviral immunity mechanism in mammals[J]. Science, 2013b, 342: 231-234.

[27] MA J B, YE K, PATEL D J. Structural basis for overhang-specific small interfering RNA recognition by the PAZ domain[J]. Nature, 2004, 429: 318-322.

[28] MAILLARD P V, CIAUDO C, MARCHAIS A, et al. Antiviral RNA interference in mammalian cells[J]. Science, 2013, 342: 235-238.

[29] MALLORY A C, HINZE A, TUCKER M R, et al. Redundant and specific roles of the ARGONAUTE proteins AGO1 and ZLL in development and small RNA-directed gene silencing[J]. PLoS Genetics, 2009, 5: e1000646.

[30] MALLORY A, VAUCHERET H. Form, function, and regulation of ARGONAUTE proteins[J]. Plant Cell, 2010, 22: 3879-3889.

[31] MANAVELLA P A, WEIGEL D, WU L. Argonaute10 as a miRNA Locker[J]. Cell, 2011, 145: 173-174.

[32] MAUNOURY N, VAUCHERET H. AGO1 and AGO2 act redundantly in Mir408-mediated plantacyanin regulation[J]. PLoS ONE,

2011, 6: e28729.

[33] MEISTER G. Argonaute proteins: functional insights and emerging roles[J]. Nature Reviews Genetics, 2013, 14: 447-459.

[34] MI S, CAI T, HU Y, et al. Sorting of small RNAS into *Arabidopsis* argonaute complexes is directed by the 5′ terminal nucleotide[J]. Cell, 2008, 133: 116-127.

[35] MONTGOMERY T A, HOWELL M D, CUPERUS J T,et al. Specificity of ARGONAUTE7-miR390 interaction and dual functionality in TAS3 trans-acting siRNA formation[J]. Cell, 2008, 133: 128-141.

[36] PFAFF J, HENNIG J, HERZOG F, et al. Structural features of Argonaute-Gw182 protein interactions[J]. Proceedings of the National Academy of Sciences of the United States of America, 2013, 110: E3770-E3779.

[37] QI Y, DENLI A M, HANNON G J. Biochemical specialization within *Arabidopsis* RNA silencing pathways[J]. Molecular Cell, 2005, 19: 421-428.

[38] QI Y, MI S. Purification of *Arabidopsis* argonaute complexes and associated small RNAs[J]. Methods in Molecular Biology, 2010, 592: 243-254.

[39] QU F, YE X, MORRIS T J. *Arabidopsis* DRB4, AGO1, AGO7, and RDR6 participate in a D CL4-initiated antiviral RNA silencing pathway negatively regulated by DCL1[J]. Proceedings of the National Academy of Sciences of the United States of America, 2008, 105: 14732-14737.

[40] REN B, GUO Y, GAO F, et al. Multiple functions of Rice dwarf phytoreovirus Pns10 in suppressing systemic RNA silencing[J]. Journal of Virology, 2010, 84: 12914-12923.

[41] RIVAS F V, TOLIA N H, SONG J J, et al. Purified Argonaute2 and an siRNA form recombinant human[J]. Nature Structural & Molecular Biology.2005, 12: 340-349.

[42] ROGERS K, CHEN X. Biogenesis, turnover, and mode of action of plant microRNAs[J]. Plant Cell, 2013, 25: 2383-2399.

[43] SCHOLTHOF H B, ALVARADO V Y, VEGA-ARREGUIN J C, et al. Identification of an ARGONAUTE for antiviral RNA silencing in *Nicotiana benthamiana*[J]. Plant physiology, 2011, 156: 1548-1555.

[44] SONG J J, SMITH S K, HANNON G J, et al. Crystal structure of Argonaute and its implications for RISC slicer activity[J]. Science, 2004, 305: 1434-1437.

[45] TOLIA N H, JOSHUA-TOR L. Slicer and the argonautes[J]. Nature Chemical Biology, 2007, 3: 36-43.

[46] V′ARALLYAY ′E, VA′LO′CZI A, A′GYI A, et al. Plant virus-mediated induction of Mir168 is associated with repression of ARGONAUTE1 accumulation[J]. Embo Journal, 2014, 29: 3507-3519.

[47] VAUCHERET H. Plant argonautes[J]. Trends in Plant Science, 2008, 13: 350-358.

[48] VAUCHERET H, MALLORY A C, BARTEL D P. AGO1 homeostasis entails coexpression of M IR168 and AGO1 and preferential stabilization of MIR168 by AGO1[J]. Molecular Cell, 2006, 22: 129-136.

[49] VAZQUEZ F, LEGRAND S, WINDELS D. The biosynthetic pathways and biological scopes of plant small RNAs[J]. Trends in Plant Science, 2010, 15: 337-345.

[50] VOINNET O, RIVAS S, MESTRE P, et al. An enhanced transient expression system in plants based on suppression of gene silencing by the P19 protein of tomato bushy stunt virus[J]. The Plant Journal, 2003, 33: 949-956.

[51] WANG X B, JOVEL J, UDOMPORN P, et al. The 21-nucleotide, but not 22-nucleotide, viral secondary small interfering RNAs direct potent antiviral defense by two cooperative argonautes in *Arabidopsis* thaliana[J]. The Plant Cell,2011, 23: 1625-1638.

[52] WEE L M, FLORES-JASSO C F, SALOMON W E, et al. Argonaute divides its RNA guide into domains with distinct functions and RNA-binding properties[J]. Cell, 2012, 151: 1055-1067.

[53] WEI W, BA Z, GAO M, et al. A role for small RNAs in DNA double-strand break repair[J]. Cell, 2012, 149: 101-112.

[54] WU L, ZHANG Q, ZHOU H, et al. Rice microRNA effector complexes and targets[J]. Plant Cell, 2009, 21: 3421-3435.

[55] WU L, ZHOU H, ZHANG Q, et al. DNA methylation mediated by a microRNA pathway[J]. Molecular Cell, 2010,38: 465-475.

[56] XIA Z, PENG J, LI Y, et al. Characterization of small interfering RNAs derived from sugarcane mosaic virus in infected maize

plants by deep sequencing[J]. PLoS ONE, 2014, 9: e97013.

[57] XIONG R, WU J, ZHOU Y, et al. Identification of a Movement Protein of the Tenuivirus Rice Stripe Virus[J]. Journal of Virology, 2008, 82: 12304-12311.

[58] YE R, WANG W, IKI T, et al. Cytoplasmic assembly and selective nuclear import of *Arabidopsis* Argonaute4/siRNA complexes[J]. Molecular Cell, 2012, 46: 859-870.

[59] ZHANG X, ZHAO H, GAO S, et al. *Arabidopsis* Argonaute 2 regulates innate immunity via miRNA393(*)-mediated silencing of a Golgi-localized SNARE gene MEMB12[J]. Molecular Cell, 2011, 42: 356-366.

[60] ZHU H, HU F, WANG R, et al. *Arabidopsis* Argonaute10 specifically sequesters miR166/165 to regulate shoot apical meristem development[J]. Cell, 2011, 145: 242-256.

Host Pah1p phosphatidate phosphatase limits viral replication by regulating phospholipid synthesis

Zhenlu Zhang[1,2], Guijuan He[1,2], Gil-Soo Han[3], Jiantao Zhang[2], Nicholas Catanzaro[4], Arturo Diaz[5], Zujian Wu[1], George M. Carman[3], Lianhui Xie[1], Xiaofeng Wang[2]

(1 Fujian Province Key Laboratory of Plant Virology, Institute of Plant Virology, Fujian Agriculture and Forestry University, Fuzhou, Fujian, P. R. China; 2 Department of Plant Pathology, Physiology, and Weed Science, Virginia Tech, Blacksburg, VA, United States of America; 3 Department of Food Science and the Rutgers Center for Lipid Research, New Jersey Institute for Food, Nutrition, and Health, Rutgers University, New Brunswick, NJ, United States of America; 4 Department of Biomedical Sciences and Pathobiology, Virginia Maryland College of Veterinary Medicine, Virginia Tech, Blacksburg, VA, United States of America; 5 Department of Biology, La Sierra University, Riverside, VA, United States of America)

Abstract: Replication of positive-strand RNA viruses [(+) RNA viruses] takes place in membrane-bound viral replication complexes (VRCs). Formation of VRCs requires virus-mediated manipulation of cellular lipid synthesis. Here, we report significantly enhanced *Brome mosaic virus* (BMV) replication and much improved cell growth in yeast cells lacking *PAH1* (*pah1Δ*), the sole yeast ortholog of human *LIPIN* genes. *PAH1* encodes Pah1p (phosphatidic acid phosphohydrolase), which converts phosphatidate (PA) to diacylglycerol that is subsequently used for the synthesis of the storage lipid triacylglycerol. Inactivation of Pah1p leads to altered lipid composition, including high levels of PA, total phospholipids, ergosterol ester, and free fatty acids, as well as expansion of the nuclear membrane. In *pah1Δ* cells, BMV replication protein 1a and double-stranded RNA localized to the extended nuclear membrane, there was a significant increase in the number of VRCs formed, and BMV genomic replication increased by 2-fold compared to wild-type cells. In another yeast mutant that lacks both *PAH1* and *DGK1* (encodes diacylglycerol kinase converting diacylglycerol to PA), which has a normal nuclear membrane but maintains similar lipid compositional changes as in *pah1Δ* cells, BMV replicated as efficiently as in *pah1Δ* cells, suggesting that the altered lipid composition was responsible for the enhanced BMV replication. We further showed that increased levels of total phospholipids play an important role because the enhanced BMV replication required active synthesis of phosphatidylcholine, the major membrane phospholipid. Moreover, overexpression of a phosphatidylcholine synthesis gene (CHO_2) promoted BMV replication. Conversely, overexpression of *PAH1* or plant *PAH1* orthologs inhibited BMV replication in yeast or *Nicotiana benthamiana* plants. Competing with its host for limited resources, BMV inhibited host growth, which was markedly alleviated in pah1Δ cells. Our work suggests that Pah1p promotes storage lipid synthesis and thus represses phospholipid synthesis, which in turn restricts both viral replication and cell growth during viral infection.

1 Introduction

Positive-strand RNA viruses [(+) RNA viruses] are the largest of all virus classes and cause numerous important diseases in humans, animals, and plants. All of the well-studied (+)RNA viruses have been shown to

PLoS Pathogens. 2018, 14(4):e1006988.
Received 13 November 2017;Accepted 24 March 2018

remodel host intracellular membranes to build viral replication complexes (VRCs) for genomic replication (Paul et al.,2013; Laliberte et al.,2014; Wang et al.,2015; Hyodo et al.,2016). Because cellular lipids are the major building blocks of membranes, their metabolism and/or composition are crucial for virus-induced membrane rearrangements (Hyodo et al.,2016; Chukkapalli et al.,2012; Belov et al.,2012).

Brome mosaic virus (BMV) is the type member of the family *Bromoviridae* and a representative member of the alphavirus-like superfamily (Wang et al.). BMV induces spherular VRCs at the perinuclear endoplasmic reticulum (nER) membrane in the yeast *Saccharomyces cerevisiae* and in barley cells (Giovanni et al.,Restrepo-Hartwig et al.,1999; Schwartz et al.,2002; Diaz et al.,2014). BMV has three capped genomic RNAs and a subgenomic mRNA, RNA4. For viral replication, RNA1- and RNA2-encoded replication proteins 1a and 2a polymerase ($2a^{pol}$) are necessary and sufficient in barley and *Nicotiana benthamiana* (Wang et al.,Annamalai et al.,2005; Gopinath et al.,2005) as well as in yeast (Schwartz et al., 2002). With a central RNA-dependent RNA polymerase (RdRp) domain, $2a^{pol}$ serves as the replicase. In addition, the N-terminus of $2a^{pol}$ interacts with the C-terminal domain of 1a (Kan et al., 1992; O'Reilly et al.,1997; Chen et al., 2000). 1a has an N-terminal RNA capping domain that adds a cap to the 5′ end of viral RNAs (Ahola et al., 1999; Kong et al., 1999; Ahola et al., 2000)and a C-terminal ATPase/helicase-like domain that is required for translocating viral genomic RNAs into VRCs (Wang et al., 2005). 1a localizes to the nER membrane, which is the nuclear membrane or nuclear envelop, where it invaginates the outer nER membrane into the ER lumen to form spherules that have an overall negative membrane curvature (Restrepo-Hartwig et al., 1999; Wang et al., 2005). Spherules become VRCs when $2a^{pol}$ and viral genomic RNAs are recruited by 1a during viral replication (Schwartz et al., 2002). Several properties of 1a are required for this process, including its membrane association domain, an amphipathic α-helix (1a amino acids 392–407) (Liu et al., 2009), and its ability to self-interact (Diaz et al., 2012).

Lipids play crucial roles in BMV replication, similar to other (+) RNA viruses (Chukkapalli et al., 2012; Belov et al., 2012). In yeast, an ~30% increase of accumulated total fatty acids (FAs) per cell was induced by the expression of 1a along with the formation of spherules (Lee et al., 2003). A mild decrease in unsaturated FAs (UFAs) inhibited BMV RNA replication more than 20-fold (Lee et al., 2001; Lee et al., 2003). It was further shown that the decreased UFAs particularly affected the membranes surrounding VRCs, indicating that the lipid environment of VRC membranes is different from the rest of the nER membrane (Lee et al., 2001; Lee et al., 2003). BMV replication also requires host *ACB1*-encoded acyl-Coenzyme A (acyl-CoA) binding protein, which binds to long-chain fatty acyl-CoAs and is important in maintaining lipid homeostasis. In the *ACB1* deletion mutant, BMV RNA replication is inhibited by more than 10-fold and spherules are smaller in size but greater in number than those in wild-type (wt) cells (Zhang et al., 2012). Enhanced accumulation of phosphatidylcholine (PC) is also associated with BMV replication sites (Zhang et al.,2016). In addition, cellular PC synthesis enzyme Cho2p (phosphatidylethanolamine (PE) methyltransferase) (Fig 1A) is recruited to BMV replication sites by 1a via a specific 1a-Cho2p interaction, suggesting an enhanced PC synthesis at the viral replication sites. As expected, deletion of *CHO2* significantly inhibits BMV replication, raising the possibility of controlling the viral replication by blocking the 1a-mediated Cho2p recruitment (Zhang et al.,2016).

Phosphatidate (PA) is a common precursor for both phospholipids and storage lipids. PA is produced *de novo* from glycerol-3-phosphate (Athenstaedt et al.,1997; Athenstaedt et al.,1999) and can be converted to CDP-diacylglycerol (CDP-DAG) (Rattray et al., 1975; Caman et al., 1989; Paltaul et al., 1992), which is subsequently used to produce phospholipids, including PC, PE, phosphatidylinositol (PI), and phosphatidylserine (PS) (Fig. 1A). PA can also be converted to diacylglycerol (DAG) by PAH1-encoded Pah1p, which is an Mg^{2+}-dependent phosphatidate phosphatase, and further to TAG, the major storage lipid (Tauchi-Sato et al., 2002). In yeast, PA also regulates the expression of lipid synthesis genes by sequestering a transcription repressor, Opi1p (overproduction of inositol1), at the nER membrane(Loewen et al.,2004). When PA levels are low, Opi1p is released from the nuclear membranes and translocated to the nucleus to repress transcriptions of many genes involved in phospholipid synthesis, including *CHO2* and *OPI3* (Fig. 1A) (Leowen et al., 2004; Carman et al., 2011).

Pah1p is highly regulated given its important roles in directing PA for the synthesis of storage lipids and thus, away from phospholipid synthesis (Carman et al., 2006; Carman et al., 2009; Fernandez-Murray et al., 2016) Primarily localized in the cytosol as a hyperphosphorylated inactive form, Pah1p is dephosphorylated by a phosphatase complex that is composed of the catalytic subunit Nem1p (nuclear envelop morphology1) and the regulatory partner Spo7p (sporulation7) (Santos-Rosa et al., 2005; Han et al., 2007; Chio et al., 2012). The Nem1p-Spo7p complex also recruits Pah1p to ER membranes where the active Pah1p is associated with membranes via an insertion of an amphipathic α-helix (Karanasios et al., 2010; Choi et al., 2011). Both Nem1p and Spo7p are required for protein phosphatase activity and absence of either subunit inactivates the protein phosphatase activity of the complex, and thus, Pah1p PA phosphatase activity (Han et al., 2007). In *PAH1* deletion mutant (*pah1Δ*) cells, total phospholipid levels increase by ~2-fold while TAG levels decrease significantly and in addition, the nER membrane expanded compared to that of wt cells (Santos-Rosa et al., 2005; Fakas et al., 2011). It has been shown that in pah1Δ cells, the nER membrane always expands at the site close to nucleolus, the site of ribosomes biogenesis, and the chromosome DNA-occupied area remain the same as that in wt cells (Campbell et al., 2006). Pah1p shares structural and functional similarities to human lipins (lipin1, 2, and 3) as well as to AtPah1p and AtPah2p of *Arabidopsis thaliana* in that *LIPINs* or *AtPAHs* can complement phenotypical defects in yeast *pah1Δ* cells, including the decreased TAG and expanded nER membrane (Grimsey et al., 2008; Nakamura et al., 2009; Mietkiewska et al., 2011).

In a previous large-scale screening of a yeast deletion array, it was found that deleting *NEM1* or *SPO7* significantly enhanced BMV replication (Kushner et al., 2003). In addition, deletion of *PAH1* facilitates robust RNA replication of tomato bushy stunt virus (TBSV). TBSV normally replicates in peroxisomes but assemble their VRCs at expanded ER membranes in *pah1Δ* cells. In addition, TBSV VRCs in *pah1Δ* cells are more active than those in wtcells (Femandez et al., 2017; Chuang et al.,2014).

2 Results

2.1 There is a direct correlation between the inactivation of Pah1p and enhanced BMV RNA replication in yeast

In a previous genome-wide screen of yeast deletion mutants in which BMV RNA replication was measured by the expression of a *Renilla* luciferase reporter, there was a dramatic increase in BMV replication in yeast strains that had either *NEM1* or *SPO7* deleted (Kushner et al.,2003). The *pah1Δ* mutant, however, was missing from the library when the screen was performed. Since both Nem1p and Spo7p are required for Pah1p activation, deleting *NEM1*, *SPO7* or *PAH1* causes similar phenotypes in yeast (Siniossoglou et al.,1998; Santos-Rosa et al.,2005). To validate results of the screen and to determine the possible role of Pah1p in BMV replication, we tested BMV replication in *nem1Δ, spo7Δ and pah1Δ* single mutants, as well as the double mutants *pah1Δ nem1Δ and nem1Δ spo7Δ* by performing Northern hybridization with viral RNA strand-specific probes. As shown in Fig. 1B, in the *nem1Δ* or *spo7Δ* mutants, both negative- and positive-strand RNA accumulation increased by approximately 2-fold compared to that in wt cells. However, no further increase of BMV RNA replication was observed when both *NEM1* and *SPO7* were deleted agreeing well with the notion that each is necessary to activate Pah1p. Providing further support that Pah1p restricts BMV replication, deleting *PAH1* enhanced BMV positive- and negative-strand RNA3 accumulation by about 3-fold compared to that in wt cells (Fig. 1B). It should be noted, however, that cells in which both *PAH1* and *NEM1* were deleted consistently supported the highest levels of BMV genomic replication; thus, we used this double mutant in the majority of experiments described below (Fig. 1B). These data indicated that a lack of, or inactivation of, Pah1p promoted BMV replication in yeast, most likely through the increased production of PA and thus, increased total phospholipids and the expanded nER membrane.

To strengthen the notion that increased PA levels in *pah1Δ* cells is a major contributor to the enhanced BMV replication, we tested whether BMV replication was affected by overexpressing *DGK1*. *DGK1* encodes DAG kinase, which converts DAG to PA in yeast (Fig.1A). Similar to deleting *PAH1*, overexpressing *DGK1* leads to a decrease in TAG accumulation and an increase in PA levels, resulting in an expanded nER membrane in yeast cells (Han et al., 2008). As expected, overexpression of DGK1 also enhanced BMV RNA replication to levels comparable to that in *pah1*Δ cells (Fig.1C). To confirm that Dgk1p enzymatic activity was required for the effect, we used a Dgk1p mutant, *D177A*, which lacks DAG kinase activity and whose overexpression does not extend the nuclear membrane (Han et al., 2008). Indeed, overexpression of *D177A* did not promote BMV RNA replication (Fig.1C), consistent with the notion that redirecting lipid synthesis from TAG to phospholipids could enhance BMV replication.

In contrast to the above deletion mutants, overexpression of *PAH1* inhibited BMV replication ~2-fold (Fig.1D). Similar inhibition in BMV replication was also observed in yeast cells overexpressing *NEM1* or *SPO7*. These effects were comparable to that of *SKI8* (superkiller8), a well-known antiviral gene (Wickner, 1996; Kushner et al., 2003) (Fig.1D). Taken together, our results indicate that there is a positive correlation between the inactivation or disruption of Pah1p function and enhanced BMV replication levels in yeast, indicating that Pah1p is a limiting factor for BMV replication.

2.2 BMV 1a localizes to the extended nuclear membrane in cells lacking *PAH1*

A dramatically extended nER membrane is present in cells lacking *NEM1, SPO7,* and/or *PAH1* (Santos-Rosa et al., 2005;Siniossoglou et al., 1998) or when *DGK1* is overexpressed (Han et al., 2008). Since BMV 1a invaginates the outer nER membrane into the lumen to form spherules, the extended nER membrane in these mutant cells may provide an expanded surface area for VRC formation and thus, promote BMV replication (Schwartz et al., 2002).

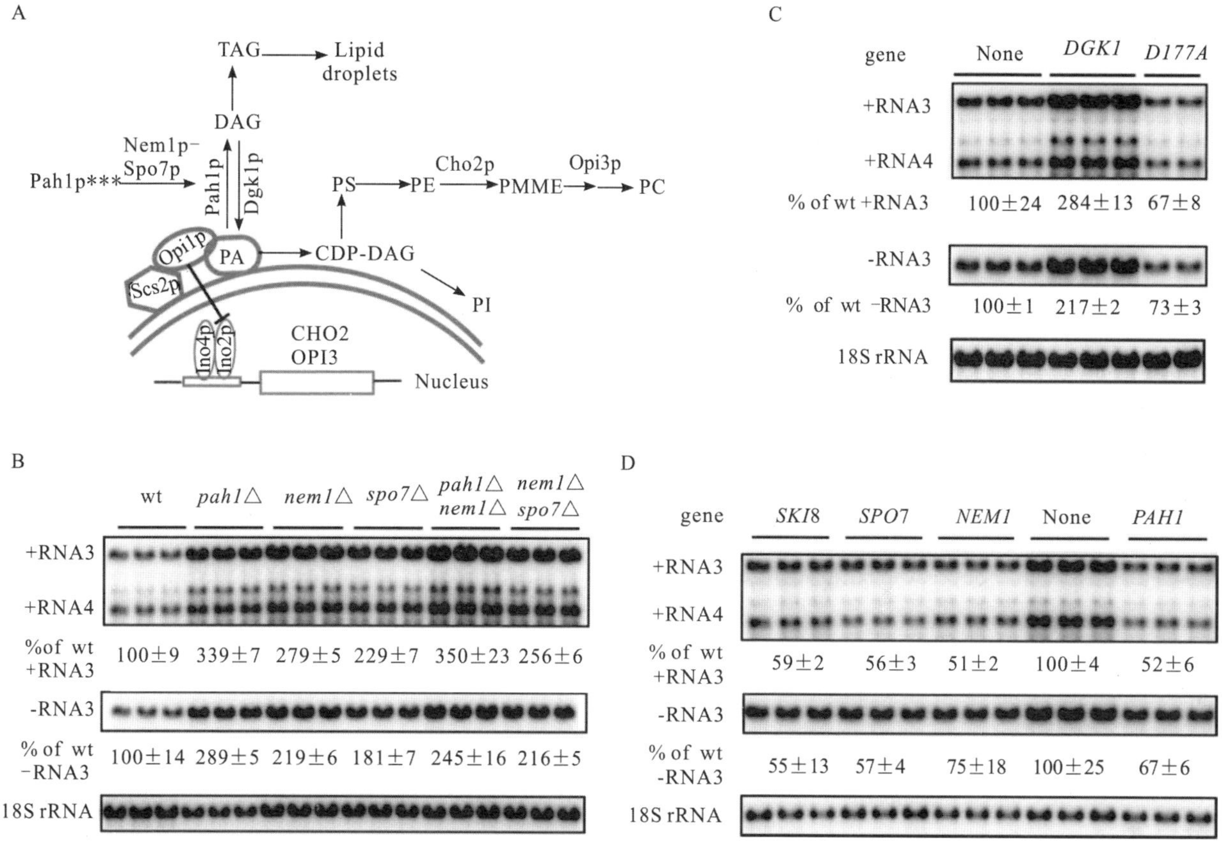

Fig.1 The inactivation or disruption of Pahlp promotes BMV genomic replication

(A) Diagram of lipid metabolism in yeast. Key enzymes are shown. PA serves as a substrate for phospholipids and TAG. PA and Scs2p bind to and sequester Opilp, keeping it from reaching to the nucleus, where Opilp interacts with Ino2p and represses transcription of *CHO2*, *OPI3* and other genes involved in phospholipid synthesis. Pahlp*** represents the hyperphosphorylated inactive Pahlp. (B) Accumulated BMV RNAs in wt and mutant cells with *PAH1* deleted or Pahlp inactivated. Positive- and negative-strand viral RNAs weredetected by using BMV RNA strand-specific probes. 18S rRNA was included as a control to eliminate loading variations. All experiments shown in the figureand in subsequent figures have been repeated multiple times and a representative figure is shown. (C) BMV replication in wt cells overexpressing wt or adefective mutant *D177A* of Dgklp. (D) BMV replication in wt cells overexpressing *PAH1*, *NEM1*, *SPO7*, or *SKI8*. *SKI8*, a well-known antiviral gene, serves as apositive control.

Here, we report that disruption ofPAH1 promotes BMV replication and results in the formation of VRCs that are 2-fold more abundant in number compared to those in wt cells, suggesting that a group of (+) RNA viruses could take advantage of the inactivation of Pahlp to promote their replication. We further demonstrate that the enhanced BMV replication phenotype is not due to the extended nER membrane but due to the increase in total phospholipid levels. In addition, we show that deleting PAH1 also alleviates BMV-inhibited yeast cell growth. We conclude that Pahlp, by targeting lipid flux away from phospholipid synthesis, constrains both viral replication and cell growth during BMV replication

We first examined whether the extended nER membrane was present in *pah1Δ nem1Δ* cells in the absence of BMV components using epifluorescence microscopy and transmission electron microscopy (TEM). ER membranes, which were identified using a GFP-tagged ER resident protein Scs2p (suppressor of choline sensitivity2, GFP-Scs2p), were observed as two-ring structures in wt cells (Fig. 2A). The larger outer ring is the peripheral ER membrane, which is underneath the plasma membrane in yeast. The smaller inner ring indicates the nER membrane, which surrounds the DAPI-stained, round-shaped nucleus. Like the misshapen nER membrane in *nem1Δ, spo7Δ,* and *pah1Δ* mutants, the nER membrane was extended in the *pah1Δ nem1Δ* mutant (Fig.2A). Agreeing well with previous report, the extended nuclear membrane was away from the DAPI-stained chromosome DNA area (Fig.2A) and has been shown to be close to the nucleolus (Campbell et al., 2006). Consistent with what was observed by epifluorescence microscopy, the strikingly proliferated nER membrane was also confirmed in *pah1Δ nem1Δ* cells using TEM (Fig.2B). To further characterize the extended nuclear membrane in *pah1Δ nem1Δ* cells, we measured the perimeter of nER membranes in wt and *pah1Δnem1Δ* cells. While the nuclear membrane perimeter in wt cells was ~5.9μm, it increased to approximately 9.2μm in the mutant, a 55% increase that was statistically significant (Fig.2C).

The expression of 1a, without other BMV components, induces spherule formation in the nER membrane of yeast (Schwartz et al., 2002). We tested whether the localization of 1a could be affected in *pah1Δ nem1Δ* cells. To visualize 1a, we first used a mCherry-tagged 1a, which primarily localized to the nER and partially localized to the peripheral ER in wt cells (Fig. 2D, upper panels) (Li et al., 2016). In mutant cells, 1a-mCherry dominantly co-localized with GFP-Scs2p at the extended nER membrane (Fig.2D). To further confirm that 1a was associated with the nER membrane, we used a GFP tagged nuclear pore complex component, Nup49p (Nuclear Pore 49, GFP Nup49p)(Wente et al., 1992). His6-tagged 1a, when expressed alone, co-localized with the GFP-Nup49-plabeled nER membrane in both wt and *pah1Δ nem1Δ* cells as determined by immunofluorescence microscopy (Fig.2E).

We next checked the accumulation and localization of BMV replication proteins during BMV replication. Both 1a and 2a^{pol} accumulated at higher levels in *pah1Δ nem1Δ* cells compared to those in wt cells based on Western blotting using anti-1a or 2a^{pol} antibodies (Fig.2F). The increased levels of 1a was consistent with the localization of 1a-mCherry and 1a-His6 along the expanded nER membrane (Fig.2D, E). To determine the site of BMV replication in *pah1Δ nem1Δ* cells, we tested the distribution of double-stranded RNA (dsRNA) using a dsRNA-specific monoclonal antibody J2. As a replication intermediate, dsRNA is considered a hallmark of viral VRCs (Cheng et al., 2015) and the J2 antibody has been commonly used to confirm localization of viral replication sites (Cheng et al., 2015; Cao et al., 2015). In wt cells, dsRNA signal co-localized nicely with that of 1a, as determined by immunofluorescence microscopy (Fig.2G). Moreover, both signals showed a half-ring structure surrounding the nucleus. However, at least two alterations were noticed in the majority of *pah1Δ nem1Δ* cells (Fig.2G): 1) Both dsRNA and 1a signals were not detected as a half-ring but localized at the extended nER membrane, and 2) Both signals extended away from the nucleus in many cells.

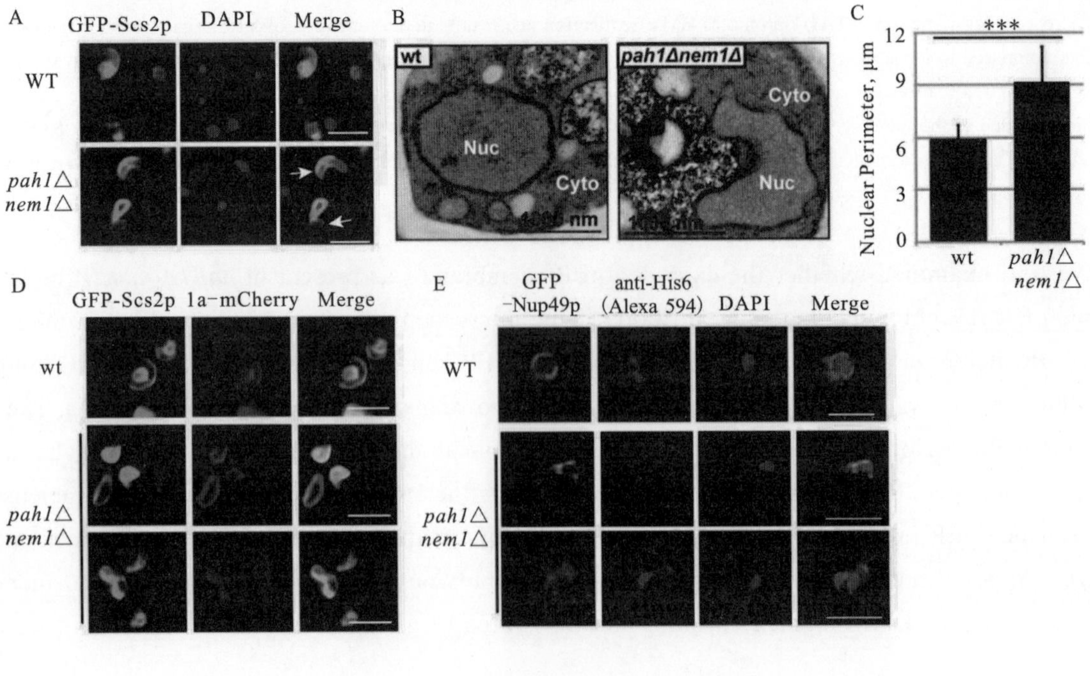

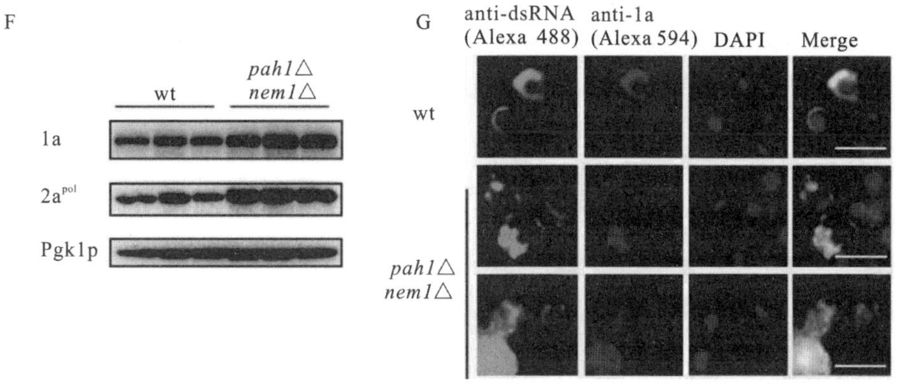

Fig.2 BMV 1a and double-stranded RNA localize to the extended nuclear membrane in *pah1Δ nem1Δ* cells

(A) Epifluorescence microscopic images of the extended nuclear membrane observed in *pah1Δ nem1Δ* cells. GFP tagged Scs2p, an ER membrane protein, represents ER membranes. Nuclei were stained with DAPI. White arrows indicate the extended nER, which is away from DAPI-stained chromosome DNA (blue). (Scale bar, 5μm) (B) Morphology of the nuclear membrane in wt and *pah1Δ nem1Δ* cells under transmission electron microscope. (C) The perimeters of nuclei in wt and *pah1Δ nem1Δ* cells. ***, p <0.001 (ANOVA single factor test). (D) Epifluorescence microscopic images showing the localization of 1a in wt and *pah1Δ nem1Δ* cells. The localization of 1a is indicated by mCherry, which is fused to the C-terminus of 1a. The yellow color in merged images represents the co-localization of 1a and Scs2p signals. (Scale bar, 5μm) (E) Immunofluorescence microscopic images showing the localization of 1a-His6 in wt and *pah1Δ nem1Δ* cells. Localization of 1a-His6 was detected using a polyclonal anti-His6 antibody and followed by a secondary anti-rabbit antibody conjugated to Alexa Fluor 594. GFP tagged Nup49p, a component of nuclear pore complexes, indicates the nuclear membrane. Nuclei were stained with DAPI. (Scale bar, 5μm) (F) Accumulated BMV 1a and $2a^{pol}$ in wt and *pah1Δ nem1Δ* cells. Total proteins were extracted from the same numbers of BMV replicating-yeast cells and analyzed by Western blotting using antibodies specific to 1a and $2a^{pol}$. Pgk1p serves as a loading control. (G) Immunofluorescence microscopic images of the co-localization of 1a and dsRNA signals in wt and *pah1Δ nem1Δ* cells. BMV 1a was detected with anti-1a antiserum followed by a secondary anti-rabbit antibody conjugated to Alexa Fluor 594. dsRNA was detected by a dsRNA-specific monoclonal antibody (J2) and a secondary anti-mouse antibody conjugated to Alexa Fluor 488. The yellow color in merged images represents the co-localization of 1a and dsRNA signals. Note 1a and dsRNA appear as a half ring in wt but not in mutant cells. Nuclei were stained with DAPI. (Scale bar, 5μm)

2.3 Substantially increased numbers of viral replication complexes are formed in *pah1Δ nem1Δ* cells compared to wild-type cells

To determine whether VRC assembly was affected in the *pah1Δ nem1Δ* mutant, we checked the morphology of spherular VRCs using TEM in both wt and mutant cells during BMV replication. In wt cells, viral spherular VRCs were found in the lumen of the nER membrane. In wt cells of the RS453 background, the average number of spherular VRCs per cell section was approximately 40 (40 ± 3) with an average diameter of ~53nm (53 ± 17nm, Fig.3A). In BMV replicating *pah1Δ nem1Δ* cells, an extended nER membrane was clearly observed (Fig.3B,C), similar to what was seen in mutant cells without BMV components (Fig. 2C). We found VRCs that were 24% smaller in diameter (40±10nm [mean ± SD], Fig.3B–E) but about 2.4-fold more abundant in number (97 ± 50, Fig.3B – E) compared to those in wt cells. These spherular VRCs were generated from membranes connected to the nER membrane. The increased numbers of VRCs is consistent with higher accumulation of both BMV 1a and $2a^{pol}$ (Fig.2F). To confirm that these smaller VRCs were active in viral replication, we performed immunogold electron microscopy analysis (IEM) using the J2 antibody (Cao et al.,2015). About 65% of the gold particles were associated with viral VRCs in BMV-replicating wt cells (65% ± 18%, n = 127) (Fig. 4A). A similar ratio was observed (64% ± 9, n = 297) in *pah1Δ nem1Δ* cells (Fig.4B). We have similarly detected BMV 1a in VRCs in wt and *pah1Δ nem1Δ* cells with similar ratios, 71% ± 8% (n = 110) and 75% (n = 206), respectively (Fig. S1). Given the fact that spherular VRCs are the site of RNA synthesis and that there was an increase in the accumulation of both positive- and negative-strand RNA in the mutant cells (Fig.1B), these results suggest that the smaller spherular VRCs in mutant cells support efficient viral RNA synthesis.

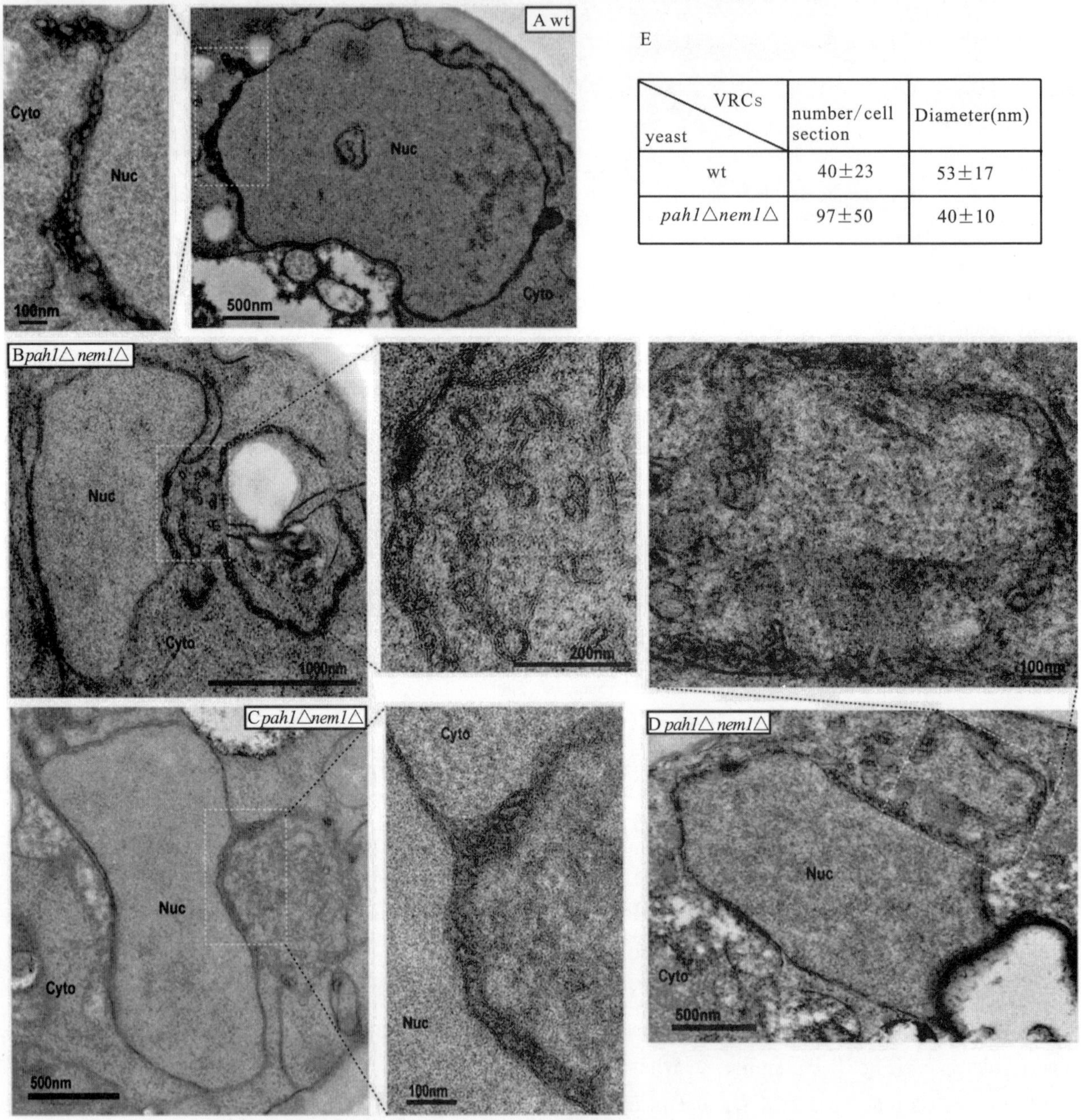

E

VRCs / yeast	number/cell section	Diameter(nm)
wt	40±23	53±17
pah1△nem1△	97±50	40±10

Fig.3 Number of spherular VRCs is substantially increased in *pah1Δ nem1Δ* cells

Electron micrographs of spherular VRCs formed in wt (A) and *pah1Δ nem1Δ* cells (B-D) are shown. Micrographs at a higher magnification of boxed areas are also shown. Note spherular VRCs are in membranes extended from the nER membrane. (E) Average number of VRCs per cell section and diameter of VRCs in wt and *pah1Δ nem1Δ* cells. Nuc, nucleus; Cyto, cytoplasm.In addition to smaller spherular VRCs, we also observed more dramatic membrane rearrangements in *pah1Δ nem1Δ* cells replicating BMV, usually multiple layers of bilayer membrane surrounding the nucleus (Fig. S2). These layers of membrane are likely generated during BMV replication because such structures have not been previously reported and were not observed in the absence of BMV replication (Fig. 2B). However, the nature of and the relationship of the layers to viral replication is currently unclear and is under further investigation

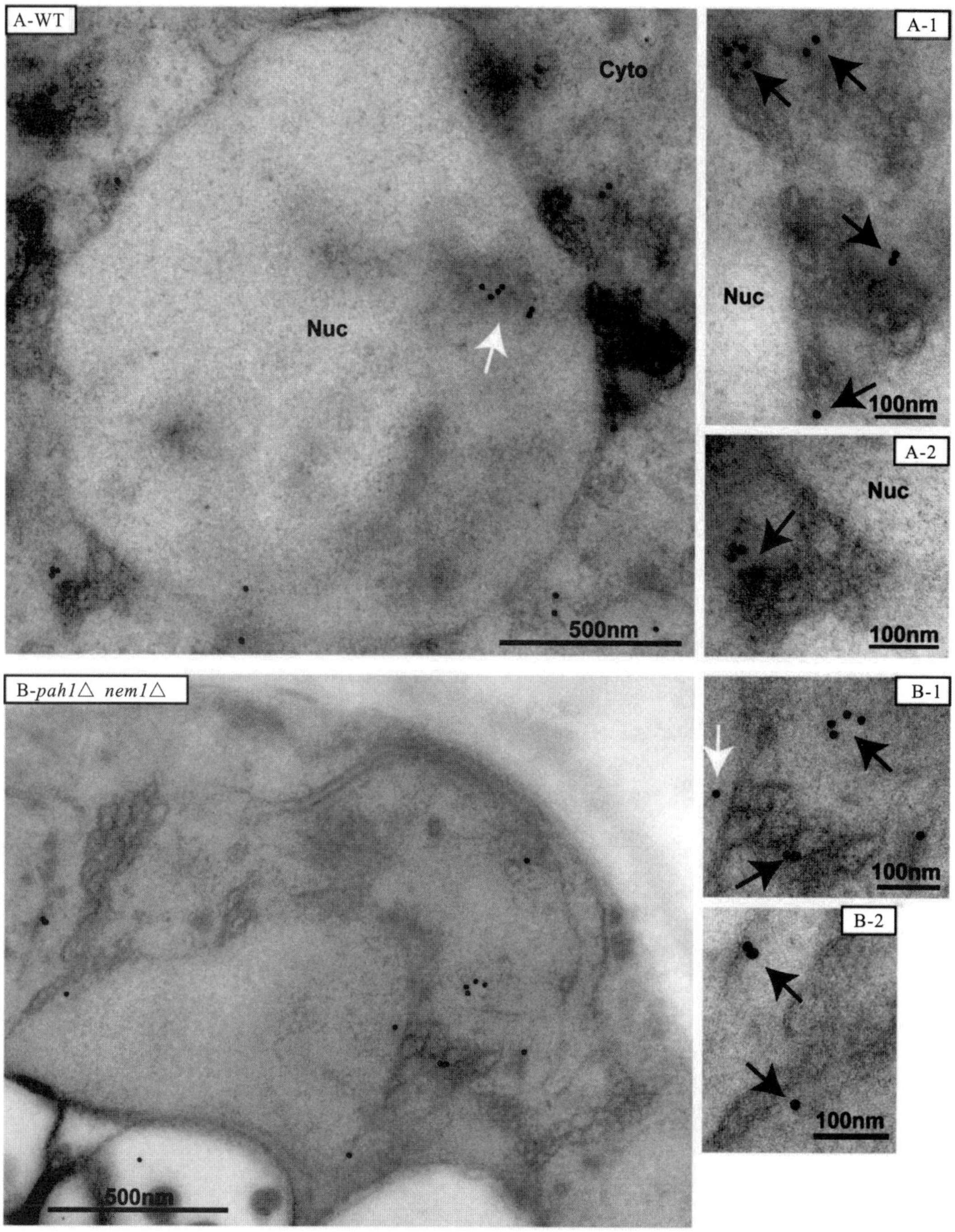

C

Yeast Strain	% of membrane-associated dsRNA	# of gold particles counted
WT	65±18	127
pah1△ *nem1*△	64±9	297

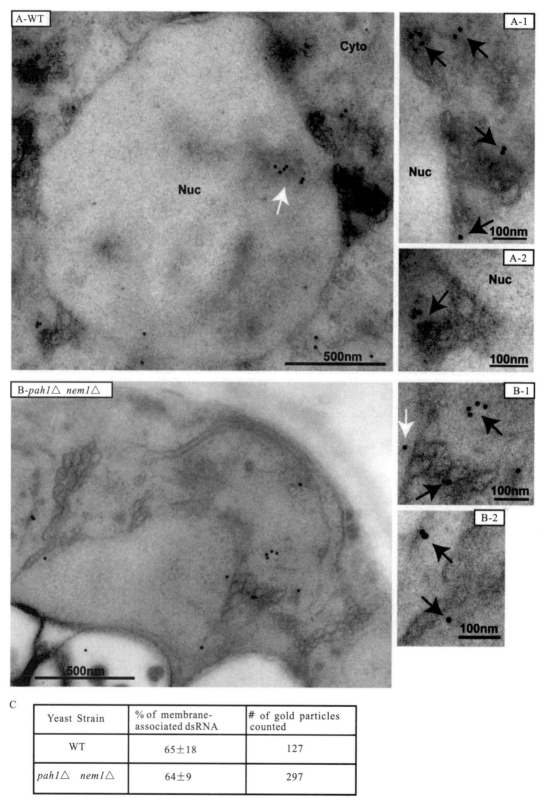

Fig.4 BMV replication sites are localized at the expanded nuclear ER membranes in *pah1*Δ *nem1*Δ cells

Immunogold labeling of dsRNA in wt (A) and *pah1*Δ *nem1*Δ (B) cells during BMV replication. Images at a higher magnification (A-1, A-2, B-1, and B-2) are also shown. Black arrows indicate the gold particles that were associated with membranes or spherular VRCs. White arrows indicate the gold particles that were not associated with membranes. (C) Number of total particles counted and the percentage of particles that were localized to the nER membrane and spherular structures among total counted particles in wt and *pah1*Δ *nem1*Δ cells. Particles within 20nm of the nER membrane or spherular VRCs, the distance spanned by primary and secondary antibodies, were counted as positive (Hayat,et al.,1991)

2.4 Total phospholipid levels increase in cells lacking *PAH1* in the presence of BMV replication

As reported previously, levels of total phospholipids, ergosterol esters (ErgE) and free FAs increased at the expense of TAG in *pah1Δ* mutant cells (Han et al., 2008). To confirm that similar altered lipid composition was still present in *pah1Δ nem1Δ* cells in the presence of BMV replication, we measured lipids of wt and mutant cells grown in the presence of [2 - ^{14}C] acetate to radiolabel neutral lipids and phospholipids (Fig. 5A). The mol percentages of each measured lipid was reported in Fig. 5B, C. The mol percentage of both DAG ($p < 0.01$) and TAG ($p < 0.001$) decreased significantly while total phospholipid levels increased ($p < 0.05$) in *pah1Δ nem1Δ* mutant cells compared to those in wt cells (Fig. 5B). Moreover, there was a significant decrease in ergosterol levels but a substantial increase in ErgE and free FAs levels in the presence of BMV (Fig. 5B). The similar compositional changes of all aforementioned lipids, in the absence of BMV, have been previously reported (Han et al.,2008), indicating that BMV did not alter the trend of lipid compositional changes in mutant cells.

The phospholipid composition was also altered in *pah1Δ nem1Δ* cells during BMV replication. Levels of PA ($p < 0.01$) and PE ($p < 0.01$) increased while there was a decrease in PS levels ($p < 0.01$) in the mutant compared to wt (Fig. 5C). However, there were no statistically significant changes in PC or PI levels (Fig. 5C). Thus, our data agrees with a previous report (Han et al.,2008), which showed that total phospholipid levels, PA in particular, increase upon deletion of *PAH1*, even in the presence of BMV replication.

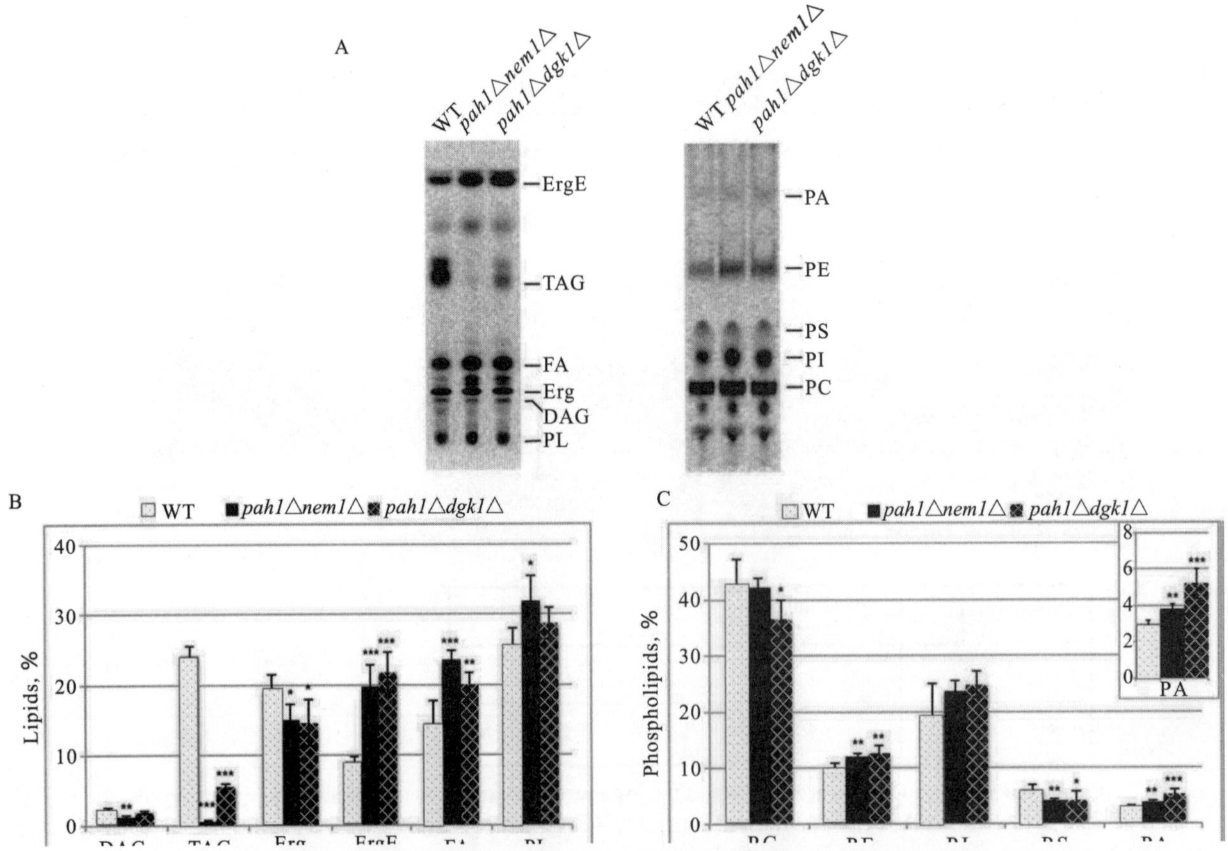

Fig.5 Increased total phospholipid levels in yeast cells lacking *PAH1* in the presence of BMV replication

WT, *pah1*Δ *nem1*Δ and *pah1*Δ *dgk1*Δ cells with BMV components were grown at 30ÊC in SC-Ura-Leu medium in the presence of galactose as the carbon source and [2-14C] acetate (1μCi/ml). Lipids were extracted, separated by the one-dimensional thin-layer chromatography system for phospholipids or neutral lipids, visualized by phosphoimaging and analyzed by Image Quant software. (A) Chromatograms of neutral lipid composition and total phospholipids (*left*), and phospholipid composition (*right*). The chromatograms shown in the panel are representative of three independent experiments. (B) and (C) The mol percentages shown for the individual neutral lipids and phospholipids were normalized to the total 14C-labeled chloroform fraction, which also contained the unidentified neutral lipids and phospholipids shown in (A). Each data point represents the average of three experiments ± S.D. (*error bars*). *, $P<0.05$; ***, $P<0.01$; ***, $p<0.001$ (based on single factor ANOVA test)

2.5 Extension of the nuclear membrane is not the major contributor to the increase in BMV genomic replication in cells lacking *PAH1*

In *pah1Δ* cells, several alterations may account for the enhanced BMV replication: 1) Since BMV assembles its VRCs at the nER membrane, the extended nER membrane will provide a larger surface area for the formation of BMV VRCs; 2) Since phospholipids are major components of membranes, the increased total phospholipid levels may provide building materials to form more VRCs. To determine which or both of these are the major contributor(s), we took advantage of the *pah1Δ dgk1Δ* mutant, in which both *PAH1* and *DGK1* are deleted. It was reported that the mutant has similar lipid compositional changes as those in the *pah1Δ* mutant but the nER membrane is normal (Han et al., 2008). We first checked the morphology of GFP-Nup49p-tracked nER membrane and confirmed that the nER membrane was indeed round shaped (Fig. 6A) and that the size of nuclei in *pah1Δ dgk1Δ* cells was similar to that of wt cells (Fig. 6B). The average perimeter of the nER membrane in *pah1Δ dgk1Δ* cells were 6.53μm (n = 136), a 10% increase over that of wt cells at 5.9μm (Fig. 6B). However, this increase is not statistically significant. In addition, we confirmed that 1a-His6 co-localized with GFP-Nup49p in the nER membrane (Fig. 6C). Consistent with the localization of 1a-His6, 1a and dsRNA were all localized at the round-shaped nER membrane during BMV replication in *pah1Δ dgk1Δ* cells (Fig. 6D). In addition, as seen in wt cells, both 1a and dsRNA localized as a half-ring in *pah1Δ dgk1Δ* cells. Surprisingly, 1a and $2a^{pol}$ still accumulated at much higher levels compared to wt cells, even the nER membrane was not extended (Fig. 6E). In addition, BMV replication increased up to ~2.5-fold, similar to that in the *pah1Δ nem1Δ* mutant (Fig. 6F). We also observed smaller but many more abundant spherular VRCs in *pah1Δ dgk1Δ* cells during BMV replication compared to those in wt cells (Fig. 7A, B). The average size of spherular VRCs was 42±9nm and the number of VRCs was about 79±44 per cell section (Fig. 7C). Lipid analysis indicated that the *pah1Δ dgk1Δ* and *pah1Δ nem1Δ* mutants shared similar trends in lipid compositional changes, including decreased DAG, TAG, and Erg but increased ErgE, free FAs, and total phospholipids (Fig. 5B). These data indicate that the altered lipid composition, but not the extended nuclear membrane, is responsible for the enhanced BMV genomic replication in cells with disrupted Pah1p activity.

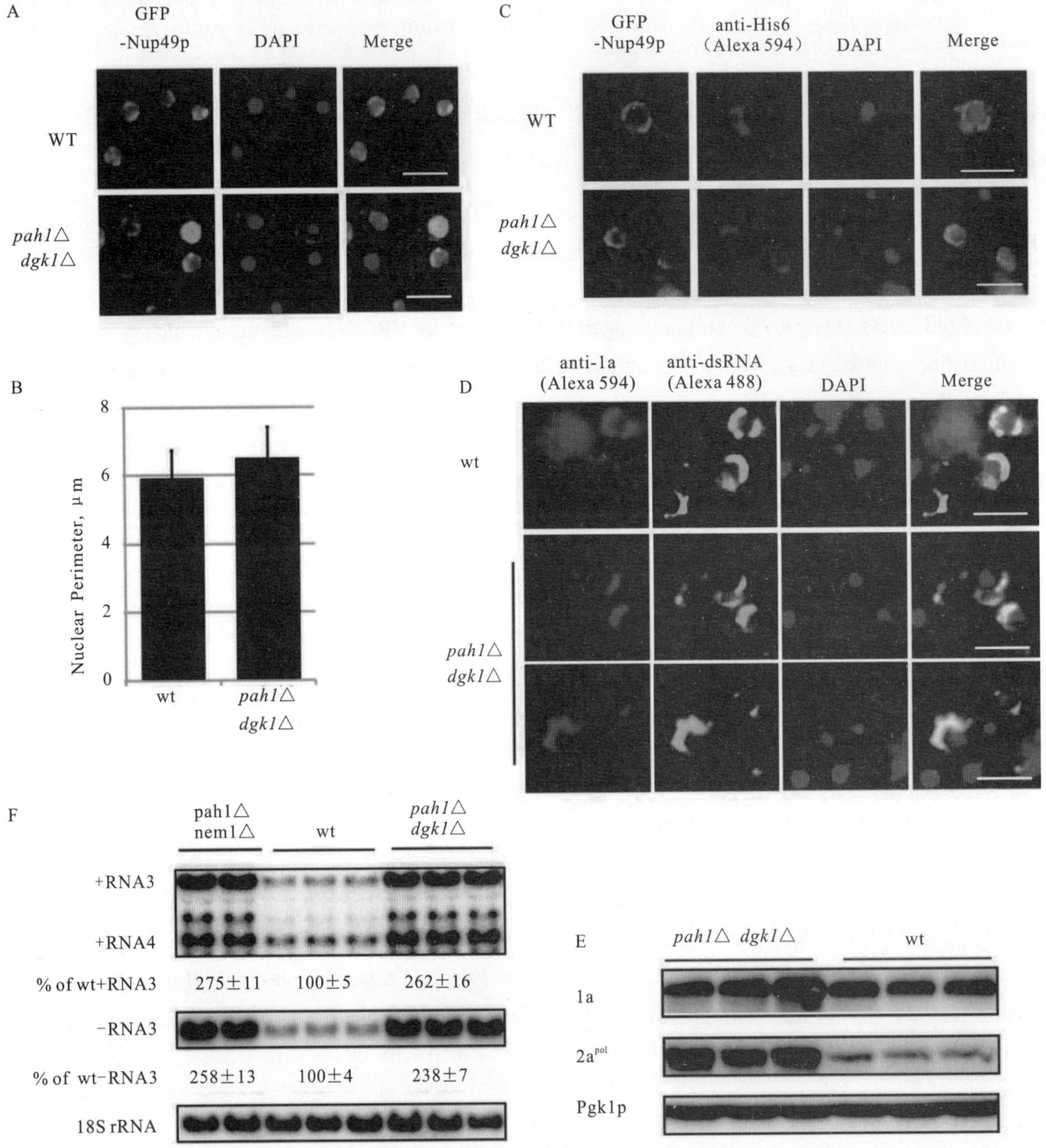

Fig.6 The extended nuclear membrane is not responsible for the enhanced BMV replication in cells that lack *PAH1*

(A) Epifluorescence microscopic images of the round-shaped nuclear membrane observed in wt and *pah1Δ dgk1Δ* cells. GFP-Nup49p was used as a nuclear membrane marker. Nuclei were stained with DAPI. (Scale bar, 5μm) (B) Nuclei perimeter measurements in wt and *pah1Δ dgk1Δ* cells. (C) BMV 1a localization in wt and *pah1Δ dgk1Δ* cells. GFP-Nup49p was used as a nuclear membrane marker. (Scale bar, 5μm) Note the wt cell (upper line) is the same one in Fig 2E (upper line). (D) Immunofluorescence microscopic images showing localization of dsRNA and 1a in wt and *pah1Δ dgk1Δ* cells. Note 1a and dsRNA appear as a half-ring structure in both wt and mutant cells. (Scale bar, 5μm) (E) Accumulated BMV 1a and $2a^{pol}$ in wt and *pah1Δ dgk1Δ* cells. Protein extraction and Western blotting were done as in Fig 2. (F) BMV replication in wt, *pah1Δ nem1Δ*, and *pah1Δ dgk1Δ* cells. Viral RNAs and 18S rRNA were detected as in Fig. 1

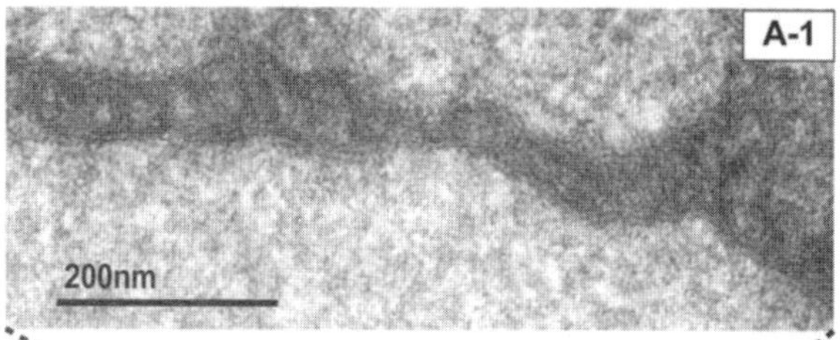

C

Yeast Strain	*pah1*△*dgk1*△
#VRCs/ cell section	79.3±44.4
Diameter (nm)	41.6±8.5

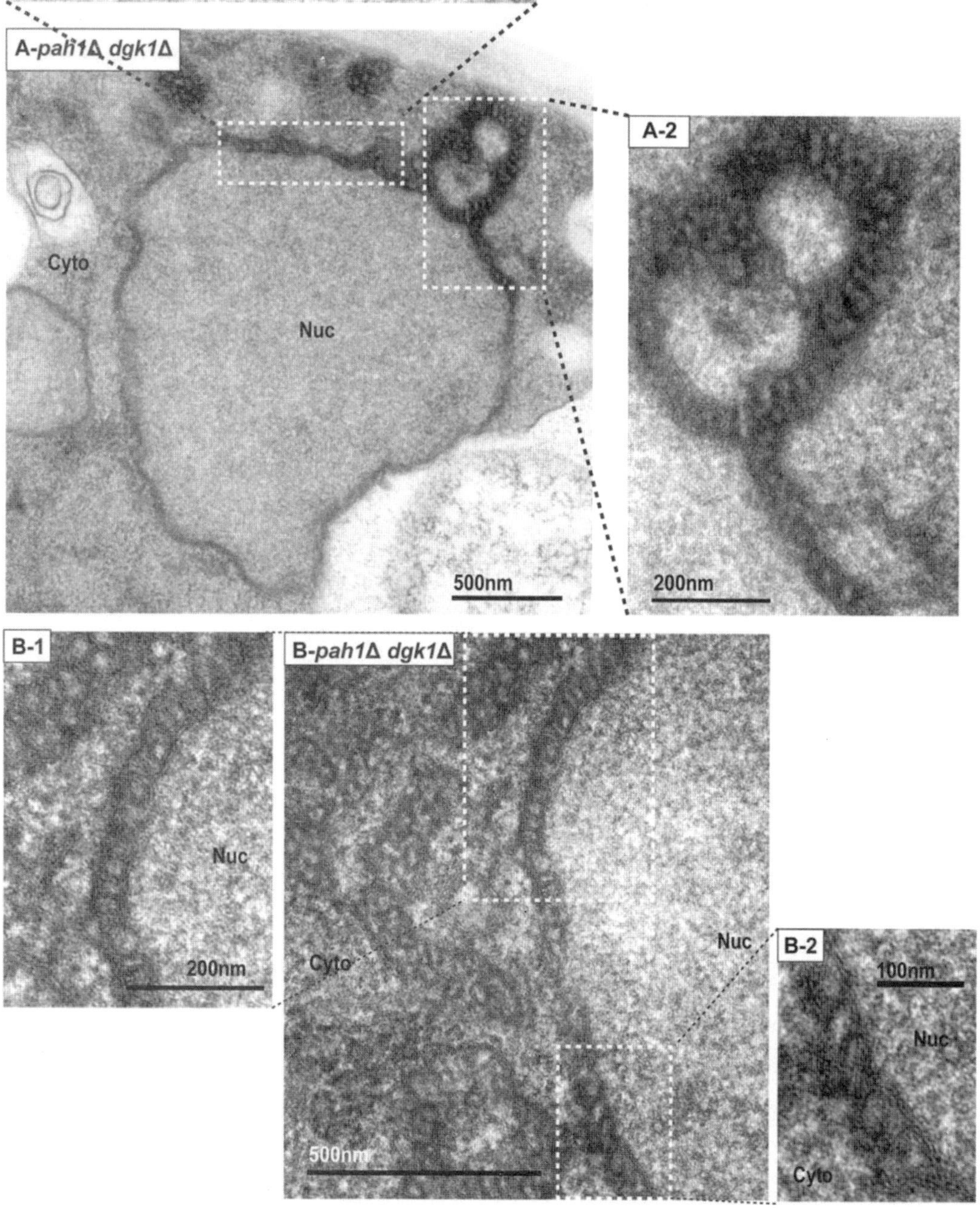

Fig.7 BMV replication complexes are associated with the perinuclear ER membrane in *pah1Δ dgk1Δ* cells

(A) and (B) Representative TEM images of spherular VRCs formed in *pah1Δ dgk1Δ* cells. Micrographs at a higher magnification of boxed areas (A-1, A-2, B-1, and B-2) are also shown. (C) Number and diameter of VRCs in *pah1Δ dgk1Δ* cells are shown. Nuc, nucleus; Cyto, cytoplasm

2.6 The contribution of enhanced phospholipids in the promoted BMV replication in cells lacking *PAH1*

We further tested whether increased total phospholipid levels, among lipid compositional changes, could play an important role in promoting VRC formation and BMV replication because phospholipids are major membrane components and both ErgE and free FAs are not present in membranes. We have previously shown that a pool of PC is synthesized in the site of viral replication by recruiting host Cho2p (Zhang et al.,2016), which is involved in converting PE to PC (Fig. 1A) (Henry et al.,2012). We first deleted *CHO2* and found that BMV replication was hardly detectable in the *cho2Δ* mutant in the RS453 background (Fig. 8A). When *CHO2* and *PAH1* were simultaneously deleted, positive-strand RNA3 accumulation increased by 23% but negative-strand RNA3 levels decreased by 33% compared to those in wt cells. However, comparing to that in the *pah1Δ nem1Δ* mutant, BMV replication significantly reduced (Fig.8A). Of note, 1a and $2a^{pol}$ proteins increased in the *pah1Δ cho2Δ* mutant background compared to those in wt cells (Fig.8B).

As a result of the increase in PA levels in the *pah1Δ* mutant, transcription of phospholipid synthesis genes increases due to the sequestration of the transcription repressor Opi1p (Fig. 1A) (Carman et al.,2011). To simulate those conditions, we tested whether enhanced *CHO2* expression would promote BMV replication. To achieve different levels of overexpression in wt cells, *CHO2* was expressed from a high-copy-number plasmid under its endogenous promoter (p426-*CHO2*) or from the strong GAL1 promoter (p3G-*CHO2*), respectively (Fig. 8C). An increase of 40% or 90% of negative-strand RNA3 over that in wt cells was associated with different levels of overexpressed Cho2p (Fig. 8C). An approximate 70% increase in positive-strand RNA was also noticed when *CHO2* was overexpressed (Fig. 8C). However, these increases in positive- and negative-strand RNA synthesis was not as significant as that in *pah1Δ* cells, suggesting other phospholipids besides PC contribute to the enhanced BMV replication phenotype in *pah1Δ* cells (Fig. 1B).

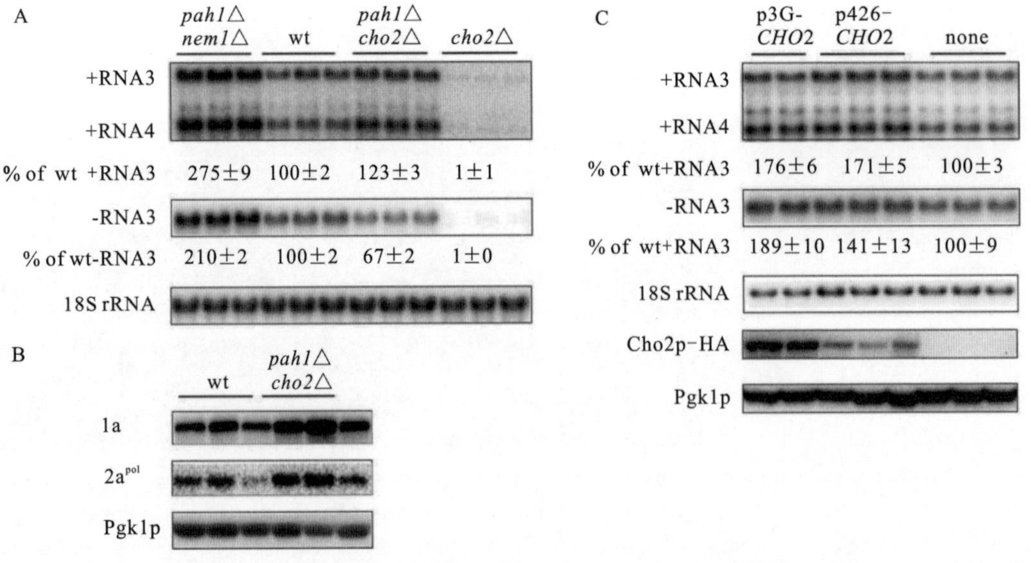

Fig.8 The active synthesis of phosphotidylcholine is required for the enhanced BMV replication in *pah1Δ* cells

(A) BMV replication in wt, *cho2Δ*, *pah1Δ cho2Δ* and *pah1Δ nem1Δ* cells. (B) BMV 1a and $2a^{pol}$ accumulation in wt and *pah1Δ cho2Δ* cells. Protein extraction and Western blotting were done as in Fig 2F. (C) Over-expression of *CHO2* promotes BMV replication in wt cells. Low or high levels of Cho2p-HA was expressed from p426-*CHO2* (a high-copy-number plasmid) or p3G-*CHO2* (a low-copy-number plasmid) under the control of the *CHO2* or *GAL1* promoter, respectively. The bottom panel shows accumulated Cho2p that was expressed from different vectors. Pgk1p serves as a loading control

2.7 Deleting *PAH1* improves yeast cell growth during BMV replication

Phospholipids are the major components of cellular membranes (Henry et al., 2012) and are utilized by various viruses for infection (Chukkapalli et al., 2012; Xu et al., 2015; Zhang et al., 2016; Belov et al., 2016; Altan-Bonnet et al., 2017). Viruses compete with their hosts for limited resources and, as a direct result, viral infections usually affect cell growth. We measured cell growth and calculated doubling times (based on growth during the exponential stage) of wt, *pah1Δ nem1Δ*, and *pah1Δ dgk1Δ* cells in the absence or presence of BMV in the galactose medium, which is to induce BMV replication (Fig. 9A). BMV replication substantially slowed down the growth of wt cells. In wt cells, the doubling time increased from ~ 4 hours/generation in the absence of BMV to ~ 9 hours/generation in the presence of BMV, an approximately 2-fold increase (Fig. 9B). In addition, the cell density of the culture expressing BMV components never reached to that of cells without BMV. Deleting *PAH1* profoundly improved the growth of cells with BMV replication (Fig. 9). The doubling times of *pah1Δ nem1Δ* and *pah1Δ dgk1Δ* mutants in the presence of BMV replication were approximately 4.5 and 7.2h/generation, respectively. It should be noted that these mutant cells grew at the same rate as wt cells in the absence of BMV components (Fig. 9), indicating that the growth differences between wt cells and the above mutants are directly related to BMV replication.

2.8 Expression of plant PAH1 orthologs inhibits BMV genomic replication in yeast and *Nicotiana benthamiana* plants

The PAP enzyme is present in yeast, plants and humans (Csaki et al., 2013). The two *PAH1* orthologs in *Arabidopsis thaliana* (*AtPAH1* and *AtPAH2*) could complement the phenotypical defects in *pah1Δ* cells (Nakamura et al., 2009; Grimsey et al., 2008; Mietkiewska et al., 2011) even though *Arabidopsis* and yeast Pah proteins share only ~14% identity at the protein level (Fig.10A). We have additionally identified five putative *PAH* genes in the genome of *Nicotiana benthamiana* based on the sequence similarity to *Arabidopsis AtPAH1* and *AtPAH2*: *NbPAH1A* (Niben101Scf01009g01015.1), *NbPAH1B* (Niben101Scf05306g01007.1), *NbPAH1C* (Niben101Scf07223g03002.1), *NbPAH2A* (Niben101Scf05628g01019.1), and *NbPAH2B* (Niben101Scf08200g05005.1). They can be classified into two clades, *NbPAH1A*, *1B*, and *1C* as one clade and *NbPAH2A* and *2B* as the other one, based on their sequence similarity to *AtPAH1* and *AtPAH2* and among themselves (Fig. 10A). To test the role of plant *PAHs* in BMV genomic replication, we expressed *NbPAH1A*, *NbPAH2A*, *AtPAH1* or *AtPAH2* in yeast cells to test whether their expression could inhibit BMV replication in a similar manner to that of yeast *PAH1* (Fig. 1D). All genes were expressed from a high-copy-number plasmid under the control of the *GAL1* promoter (Mietkiewska et al., 2011). Like yeast *PAH1*, the expression of plant orthologs inhibited BMV by ~40%–50% (Fig. 10B).

We next tested how BMV genome replication was affected when plant *PAH1* orthologs were highly expressed in *N.benthamiana*, which is a systemic host for BMV (Mise et al., 1993; Annamalai et al., 2005; Gopinath et al., 2005) and serves as a universal host for plant viruses (Goodin et al., 2008). AtPAH2, *NbPAH1A* and *NbPAH2A* were expressed from an enhanced cauliflower mosaic virus (CaMV) 35S promoter by agroinfiltration. BMV genome replication was inhibited by 40%–50% based on the accumulation of positive-strand RNA3 when *AtPAH2* or *NbPAH2A* was expressed. However, the expression of *NbPAH1A* only inhibited BMV replication by ~25%, suggesting that different plant *PAH1* orthologs may play different roles in plants (Fig. 10C).

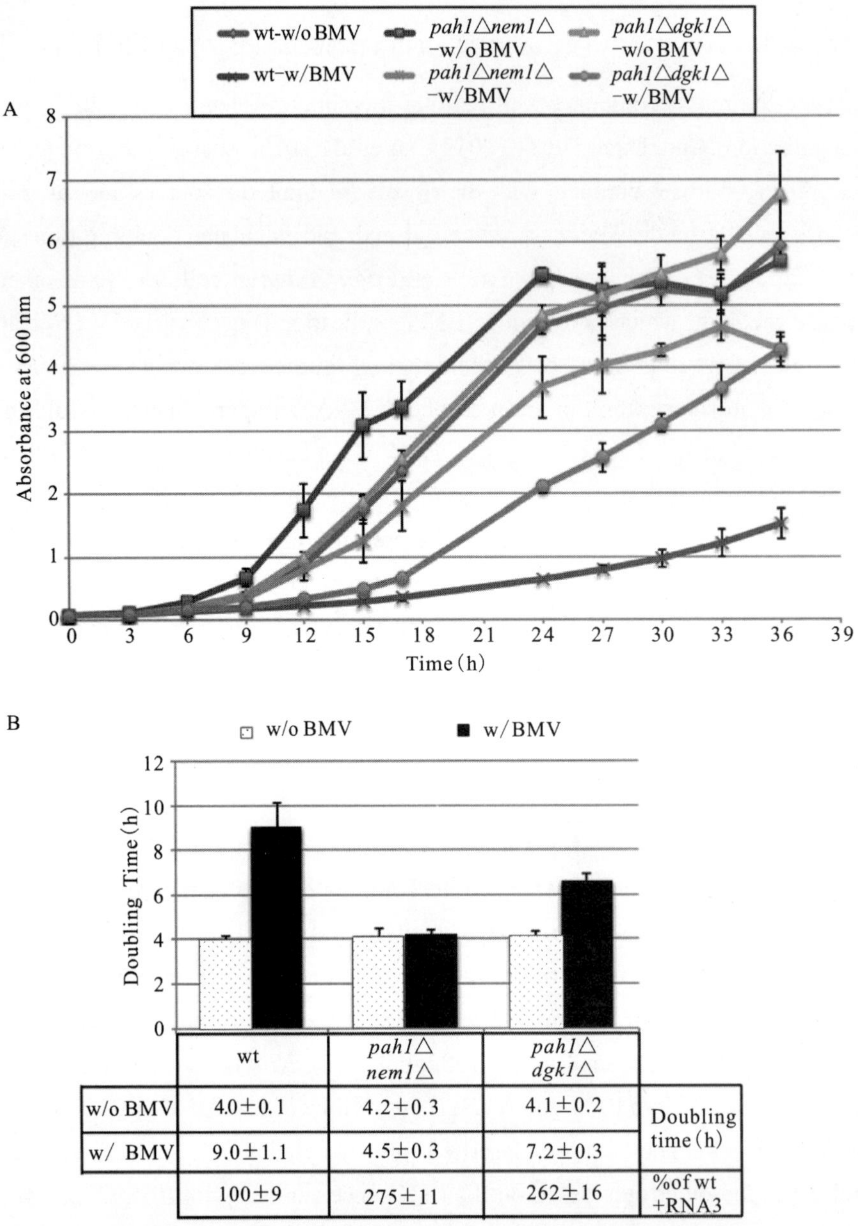

	wt	*pah1*△ *nem1*△	*pah1*△ *dgk1*△	
w/o BMV	4.0±0.1	4.2±0.3	4.1±0.2	Doubling time (h)
w/ BMV	9.0±1.1	4.5±0.3	7.2±0.3	
	100±9	275±11	262±16	%of wt +RNA3

Fig.9 Deleting *PAH1* alleviates BMV-repressed host cell growth

Cells of wt or mutants were grown in media using galactose as the carbon source in the absence or presence of BMV replication. (A) Growth curves of wt, *pah1*Δ *nem1*Δ, and *pah1*Δ *dgk1*Δ cells in theabsence or presence of BMV replication in 36h. (B) Doubling time of yeast strains in the absence or presence of BMV replication during exponential phase. Doubling time was calculated using the following equation: Doubling time = [hours cells grown * Ln(2)]/Ln[(A600nm at the end / A600nm at the start)]

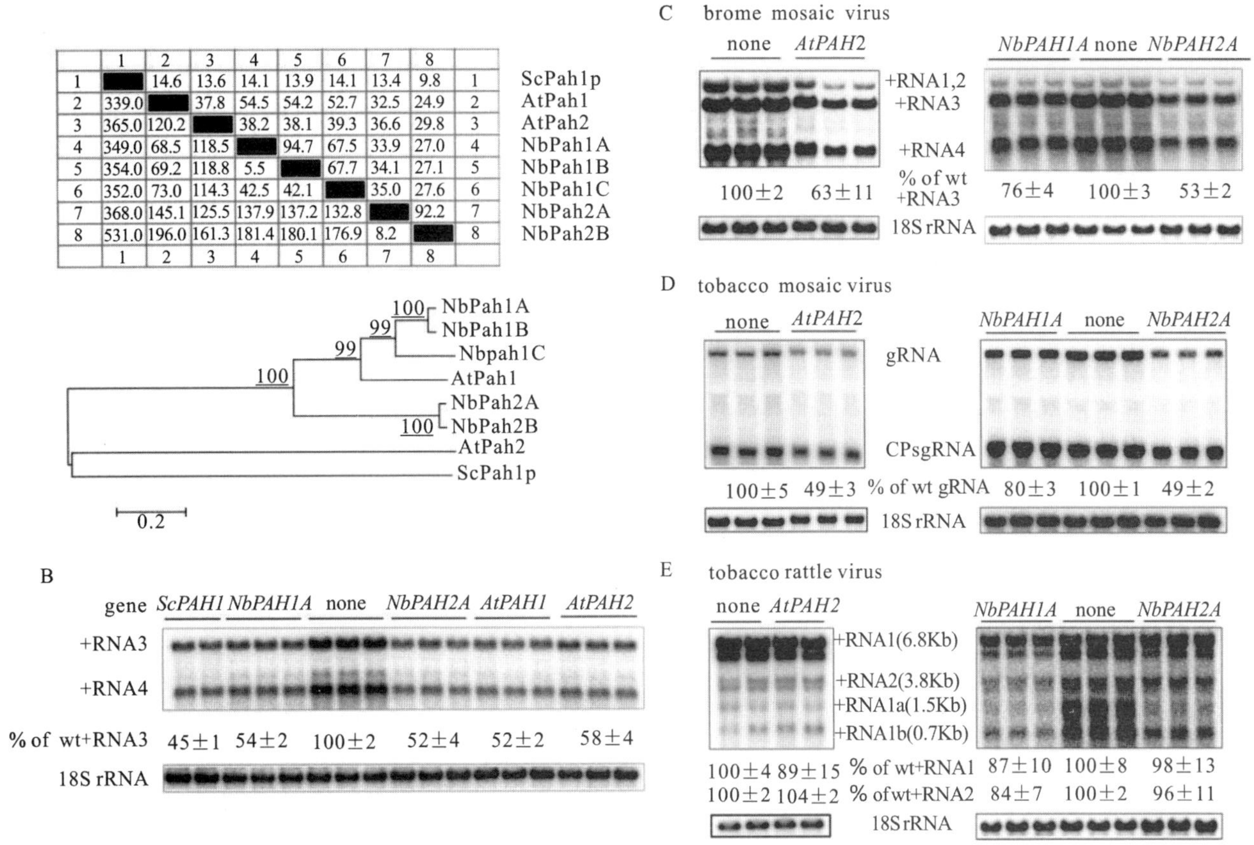

	1	2	3	4	5	6	7	8		
1		14.6	13.6	14.1	13.9	14.1	13.4	9.8	1	ScPah1p
2	339.0		37.8	54.5	54.2	52.7	32.5	24.9	2	AtPah1
3	365.0	120.2		38.2	38.1	39.3	36.6	29.8	3	AtPah2
4	349.0	68.5	118.5		94.7	67.5	33.9	27.0	4	NbPah1A
5	354.0	69.2	118.8	5.5		67.7	34.1	27.1	5	NbPah1B
6	352.0	73.0	114.3	42.5	42.1		35.0	27.6	6	NbPah1C
7	368.0	145.1	125.5	137.9	137.2	132.8		92.2	7	NbPah2A
8	531.0	196.0	161.3	181.4	180.1	176.9	8.2		8	NbPah2B
	1	2	3	4	5	6	7	8		

Fig.10 Expression of plant *PAH1* homologs inhibits BMV replication in yeast and *Nicotiana benthamiana*

(A) Homology analysis and phylogenetic tree of Pah1 proteins from yeast (ScPah1p), Arabidopsis (AtPah1 and AtPah2), and *N. benthamiana* (NbPah1A, 1B, 1C, 2A, and 2B). (B) BMV replication in wt yeast cells expressing *ScPAH1*, *AtPAH1*, *AtPAH2*, *NbPAH1A* or *NbPAH2A*. Positive-strand viral RNAs as well as 18S rRNA were detected as in Fig 1. Genome replications of BMV (C), TMV (D), or TRV (E) in *N. benthamiana* leaves expressing *AtPAH2*, *NbPAH1A* or *NbPAH2A*. BMV, TMV, and TRV were launched by agroinfiltration in *N. benthamiana* leaves 2d after agroinfiltration to express the plant *PAH1* orthologs. Virus-infected leaves were harvested 3d after agroinfiltration. RNA was extracted and viral positive-strand RNA was detected by using virus strand-specific probes as in Fig. 1

To determine whether the inhibition of viral replication by plant *PAH1* orthologs was specific to replication of BMV or if it was a general effect on other plant (+) RNA viruses, we also tested tobacco mosaic virus (TMV) and tobacco rattle virus (TRV). TMV replicates in association with ER membranes while TRV replicates on mitochondrial membranes (Harrison et al., 1976; Mas et al., 1999). We previously included both viruses and showed that a dominant negative mutant of *AtSNF7-2* (sucrose nonfermenting7) did not affect the replication of TMV and TRV but specifically affected BMV (Diaz et al.,2015). Here we found that TMV genome replication was inhibited by ~50% when *AtPAH2* or *NbPAH2A* was expressed in *N. benthamiana*. However, TMV replication was only slightly inhibited by the *NbPAH1A* expression (Fig. 10D), which similarly inhibited BMV replication at a lesser degree compared to *AtPAH2* and NbPAH2A (Fig. 10C). On the contrary, based on two genomic RNAs, RNA1 and 2, TRV genome replication was not significantly affected by any of plant orthologs in *N. benthamiana* (Fig. 10E). However, the accumulation of its subgenomic RNA1a levels were lower than that in untreated plants (Fig. 10E). It is unclear why the accumulation of different TRV RNAs was differently affected by the overexpression of *NbPAHs*.

3 Discussion

We report here that the host enzyme phosphatidic acid phosphohydrolase restricts BMV RNA replication by limiting the phospholipid synthesis and that BMV takes advantage of altered lipid composition, including the increased total phospholipids, in yeast cells lacking *PAH1* to assemble many more VRCs and substantially promote its genomic replication. Although deleting *PAH1* leads to several phenotypes that could facilitate BMV replication, our data suggest that the increased levels of total phospholipids but not the proliferated nER membrane is the primary contributor (Fig. 5, Fig.6). The rise in levels of total phospholipids in cells lacking *PAH1*, possibly with other altered lipids, also significantly improved cell growth during viral replication. It has been reported that in *pah1Δ* cells TBSV assembles its VRCs and replicates robustly in the extended ER membranes (Femandez et al.,2017), but the improved replicase activity is primarily responsible for the enhanced TBSV replication (Chuang et al.,2014). Thus, our work complements and expands the current understanding of Pah1p's role in balancing cellular phospholipid and storage lipid synthesis as well as in replication of various viruses.

3.1 Enhanced levels of total phospholipids, not the extended nuclear membrane, are primarily responsible for the increased BMV replication in cells that lack *PAH1*

Although both BMV and TBSV replicate at much higher levels in *pah1Δ* cells compared to wt cells, the mechanisms by which each virus takes advantage of *PAH1* deletion to improve its replication are different. Under normal conditions TBSV forms spherular VRCs in peroxisomes, however, it preferentially assembles VRCs in association with extended ER membranes in *pah1Δ nem1Δ* cells (Chuang et al., 2014; Femandez et al., 2017). It is not specified whether more VRCs are formed in *pah1Δ nem1Δ* cells (Femandez et al., 2017), but it is clear that TBSV VRCs isolated from the mutant cells are more efficient in supporting viral RNA synthesis *in vitro* than those from wt cells (Chuang et al., 2014). Our TEM data showed that in *pah1Δ nem1Δ* cells, viral spherules were present in the extended nER membrane and were about 2.4-fold more abundant in number than those in wt cells (Fig. 3B–D). Despite the fact that the nuclear membrane was normal in *pah1Δ dgk1Δ* cells (Fig. 6, Fig. 7) (Han et al., 2008), BMV replication levels and the number of VRCs were similar in the *pah1Δ dgk1Δ* and *pah1Δ nem1Δ* mutants (Fig. 6F, Fig. 7), indicating that the expanded nER membrane is not the major factor in promoting the VRC formation and viral genomic replication.

Two lines of evidence support that increased total phospholipid levels played an important role in the enhanced BMV replication in *pah1Δ* cells: (1) Reducing PC synthesis by deleting *CHO2* in the *pah1Δ* mutant diminished the substantially enhanced BMV replication (Fig. 8A); (2) Overexpressing *CHO2* in wt cells enhanced BMV replication (Fig. 8C). However, the enhancement in BMV genomic replication was not as significant as that in *pah1Δ* cells because total phospholipids, not just PC, increased in *pah1Δ* cells.

There are increases in three major lipids in cells lacking *PAH1*, either in the absence (Han et al., 2008) or presence of BMV (Fig. 5B), total phospholipids, ErgE, and free FAs. We focused on the roles of increased total phospholipids in this work because ErgE is not present in membranes and its involvement in genomic replication of any viruses has not been reported. Although it was reported previously that the expression of BMV 1a enhanced the accumulation of total FAs by 33% per yeast cell (Lee et al., 2003), it should point out that what we measured here was free FAs, which were not incorporated into phospholipids and not present in membranes. How are free FAs involved in BMV replication is not clear. Nevertheless, it merits further investigation whether ErgE or free FAs is involved in BMV replication. Although phospholipids are major building blocks of membranes and our work suggested that increased total phospholipids play an important role in the enhanced BMV replication (Fig. 8), the possible contribution of other significantly altered lipids, such as

ErgE and free FAs, cannot be totally ruled out.

3.2 Deleting *PAH1* facilitates both BMV replication and host cell growth

Lipid-containing cellular membranes are the sites where (+) RNA virus replication invariably takes place, although different viruses exploit specific organelle membranes and require different lipid microenvironment for their replication (Chukkapalli et al., 2012; Belov et al., 2012; Wang et al., 2015; Hyodo et al., 2016). For example, TBSV requires a PE-enriched microenvironment(Hu et al., 2015; Xu et al., 2016). Phosphatidylinostol-4-phosphate (PI4P) is produced in VRCs by hepatitis C virus (HCV)-recruited phosphatidylinositol-4-kinase III α (PI4KIIIα)(Berger et al., 2009; Reiss et al., 2011) or Coxsackievirus B3-engaged PI4KIIIβ(Hsu et al., 2010) either for the assembly or proper function of the VRCs, respectively. Our prior work showed that PC content is enriched at the viral replication sites of a number of (+) RNA viruses, including BMV, HCV, and poliovirus (Zhang et al., 2016).

It has recently been reported that *pah1Δ* cells are susceptible to abiotic stresses and have a short chronological life span (Park et al., 2015). Our data showed that deleting *PAH1* substantially improved host cell growth during viral replication. Yeast mutants that lack *PAH1*, either by itself or in combination with a second mutation, divided at a faster rate than wt cells during BMV replication (Fig. 9). It should be noted that the mutant cells divided at a similar rate to wt cells in the absence of BMV replication (Fig. 9). Both *pah1Δ nem1Δ* and *pah1Δ dgk1Δ* mutant grew faster than wt, suggesting that both of the increased viral replication and host cell growth were likely due to changes in lipid composition, possibly to the increase in total phospholipid levels. These results are consistent with the notion that BMV competes for the limited intracellular phospholipid resources with host cells, and that increasing total phospholipid levels could satisfy the requirements for both viral replication and host cell growth. As such, Pah1p serves as a limiting factor for BMV and possibly other (+) RNA viruses by directing lipid synthesis away from phospholipid synthesis, via converting PA to storage lipids. It is also possible that Pah1p promotes storage lipid synthesis at the onset of viral replication as a host reaction to stress response imposed by viral infection and in turn, limits cell growth.

3.3 Possible roles of PA in regulating nuclear membrane morphology and viral replication

In eukaryotic cells, besides serving as a key intermediate in lipid synthesis, PA is involved in multiple biological processes as a signaling molecule, such as cell growth and proliferation, secretion, endocytosis, and vesicular trafficking in mammalian cells (Waggoner et al., 1999; Sciorra et al., 2002; Testerink et al., 2002; Wang et al., 2006) as well as responses to biotic and abiotic stress and seed germination in plants (Van der et al., 2000; Yamaguchi et al., 2005; Testerink et al., 2005; Wang et al., 2006; Bargmann et al., 2006; Kirik et al., 2009).

Increased PA may be involved in viral replication through several nonexclusive mechanisms. One is through the extension of the nER membrane, which could provide more room for VRC assembly, such as the substantially enhanced replication of TBSV and related viruses (Chuang et al., 2014). Another possible mechanism is that higher PA levels may recruit PA-dependent effectors. As a signaling lipid, PA executes its function by binding to effector proteins and recruiting them to a specific membrane (Wang et al., 2006). Because such binding is dependent on the concentration of PA in the bilayer, higher levels of PA in the nER membrane may recruit its effectors more efficiently (Wang et al., 2006). Some of these effectors may play positive, yet unclear, roles in (+) RNA virus replication. This is supported by the enhanced BMV replication in the *pah1Δ dgk1Δ* mutant (Fig. 6), which has wtnER membrane but enhanced levels of total phospholipids including a high level of PA (Han et al., 2008). A third possible option is that the enhanced accumulation of PA and other phospholipids may affect protein conformation and stability. This is supported by increased accumulations of 1a and $2a^{pol}$ in both *pah1Δ nem1Δ* and *pah1Δ dgk1Δ* mutants (Fig. 2F, Fig. 6E). It is also possible that PA may play a direct role in the formation of VRCs because the incorporation of PA, a cone-shaped lipid, promotes the formation of negative curvature (Kooijman et al., 2005; McMahon et al., 2005; Burger et al., 2000). BMV spherules are formed by invaginating the outer nER membrane away from the cytoplasm, thus inducing a negative curvature. Higher PA levels may facilitate the formation of viral spherules, which may explain significantly increased numbers of viral VRCs formed in *pah1Δ nem1Δ* (Fig.3) and *pah1Δ dgk1Δ* cells (Fig.7).

Besides *de novo* synthesis, PA can be produced by phospholipase D (PLD)-catalyzed removal of the choline head group from PC. PLD-generated PA plays an important role in supporting replication of plant (+) RNA viruses (Hyodo et al., 2015). The replication protein p27 of red clover necrotic mosaic virus (RCNMV) in the *Tombusviridae* family binds to PA directly. Knocking down the expression or inactivation of PLD severely inhibited RCNMV replication (Hyodo et al., 2015). It should be noted that inhibition of PLD activity by addition of *n*-butanol in tobacco protoplasts also inhibited BMV replication, indicating an important role of PA in the replication of a group of plant (+) RNA viruses (Hyodo et al., 2015). Our data in yeast agree with the important role of PA in BMV replication in plants, although the sources of the increased PA are different.

In summary, our work suggests that altered lipid composition, likely through the enhanced total phospholipids, is the major factor not only for promoting BMV genomic replication but also for alleviating the virus-repressed cell growth in cells lacking Pah1p. Our data complement and extend prior findings on the role of PA in lipid metabolism and virus infections.

4 Materials and methods

4.1 Yeast strains and growth condition

All yeast strains used in this study are listed in Table 1 and were derived from the strain RS453 (*MATa ade2-1, his3-11, 15leu2-3, ura3-52, 112trp1-1*). The *spo7Δ* mutant was generated by replacing *SPO7* with a *HIS3MAX6* cassette. The *pah1Δ nem1Δ* mutant was made by replacing *NEM1* with a *KanMAX4* cassette in the *pah1::TRP1* background. In the majority of experiments presented, the *pah1Δ nem1Δ* mutant was used.

Yeast cells were grown at 30°C in synthetic complete (SC) medium containing 2% galactose as the carbon source. Histidine, leucine, uracil, or combinations of them were omitted from the medium depending on the selection markers of plasmids (Zhang et al., 2016). After two passages (24–48h) in SC medium, cells were harvested when the absorbance at 600nm (A_{600nm}) reached between 0.4–1.0 (Zhang et al., 2016).

Table 1 Yeast strains used in this study

Strain	Genotype	Ref./source
RS453	*MATa ade2-1, his3-11, ura3-52, 15 leu2-3, 112 trp1-1*	[54]
nem1Δ	RS453 *nem1::HIS3*	[53]
pah1Δ	RS453 *pah1::TRP1*	[54]
spo7Δ	RS453 *spo7::HIS3*	This study
cho2Δ	RS453 *pah1::TRP1 cho2::KanMX4*+YCplac33-URA3-*PAH1*	[54]
nem1Δ spo7Δ	RS453 *nem1::HIS3 spo7::HIS3*	[53]
pah1Δ cho2Δ	RS453 *pah1::TRP1 cho2::KanMX4*	[54]
pah1Δ nem1Δ	RS453 *pah1::TRP1 nem1::KanMX4*	This study
pah1Δ dgk1Δ	RS453 *pah1::TRP1 dgk1::HIS3*	[54]

4.2 Plasmids and antibodies

The plasmids used in this study are shown in Table 2 To launch BMV replication in yeast, plasmids pB12VG1 and pB3VG128 were used in the experiments as described before (Zhang et al., 2016). In the pB12VG1 plasmid, 1a is controlled by the *GAL1* promoter while $2a^{pol}$ is under the control of the *GAL10* promoter. RNA3 is under the control of the copper-inducible *CUP1* promoter but no copper was purposely included in the medium. *DGK1*, *DGK1-D177A* are overexpressed from a low-copy number plasmid YCplac33 under the control of the *GAL1* promoter and tagged with HA. The plasmid pB1YT3-mCherry was used to express mCherry-tagged BMV 1a. Rabbit anti-1a antiserum (a gift from Dr. Paul Ahlquist at University of Wisconsin Madison), mouse anti-His (Genescript, 6G2A9), mouse anti-dsRNA J2 antibody (English and Scientific Consulting, Hungary), and rabbit anti-HA (Thermo Fisher Scientific, 71–5500) were used at 1:100 dilution for Immunofluorescence microscopy and 1:10,000 or 1:3,000 for Western blotting. For Western blotting, we also used mouse anti-BMV $2a^{pol}$ at 1:3,000 dilution, and mouse anti-Pgk1p (Invitrogen, 459250) at 1:10,000 dilution.

4.3 RNA extraction and Northern blotting

Total RNA was extracted using a hot phenol method (Kohrer et al., 1991). Equal amounts of total RNA were used for Northern blotting analysis. P^{32}-labled probes specific to BMV positive- or negative-strand RNA or 18S rRNA were used in the hybridization. Radioactive signals were scanned using a Typhoon FLA 7000 phosphoimager and the intensity of radioactive signals were quantified by using Image Quant TL (GE healthcare). The 18S rRNA signal was used to normalize BMV RNA signals to eliminate loading variations (Zhang et al., 2016).

4.4 Western blotting

Two A_{600nm} units of yeast cells were harvested and total proteins were extracted as described previously (Lee et al.,2001). Equal volumes of total proteins were separated by 10% sodium dodecyl sulfate polyacrylamide gel electrophoresis (SDS-PAGE) and transferred to polyvinylidene difluoride (PVDF) membrane. Rabbit anti-BMV 1a (1:10,000 dilution), mouse anti-BMV $2a^{pol}$ (1:3,000 dilution), rabbit anti-HA (1:5,000 dilution), and mouse anti-Pgk1p (1:10,000 dilution) were used to detect 1a, $2a^{pol}$, HA, and Pgk1p (Zhang et al.,2016). Pgk1p was used as a loading control. Horseradish peroxidase (HRP)-conjugated anti-rabbit or anti-mouse antibodies (Thermo Fisher Scientific 32460 or 32430, 1:5,000 dilution) together with Supersignal West Femto substrate (Thermo Fisher Scientific, 34096) were used for signal detection.

Table 2 Plasmids used in this study

Plasmid	Description	Ref./ source
pB12VG1	BMV 1a and $2a^{pol}$ are driven by *GAL1* or *GAL10* promoter respectively in a CEN/LEU vector	[27]
pB3VG128-U	BMV RNA3 is under control of *CUP1* promoter in a CEN/URA vector	[27]
pB3VG128-H	BMV RNA3 is under control of *CUP1* promoter in a CEN/HIS vector	[27]
pB1YT3	BMV 1a is under control of *GAL1* promoter in a CEN/URA vector	[27]
pB1YT3-mCherry-L	1a-mCherry is under control of *GAL1* promoter in a CEN/LEU vector	[56]
pUN100-GFP-*NUP49*	GFP-Nup49p is constructed in pUN100, a CEN/LEU vector	[88]
p3G-*DGK1*-HA	*DGK1* is under control of *GAL1* promoter in a CEN/URA vector	This study

续表

Plasmid	Description	Ref./ source
p3G-*DGK1-D177A*- HA	*dgk1-D177A* is under control of *GAL1* promoter in a CEN/URA vector	This study
pBG1805-*SPO7*	*SPO7* is under control of *GAL1* promoter in the pBG1805, 2 μ /URA vector	[89]
pBG1805-*NEM1*	*NEM1* is under control of *GAL1* promoter in the pBG1805, 2 μ /URA vector	[89]
pBG1805-*PAH1*	*PAH1* is under the control of *GAL1* promoter in the pBG1805, 2 μ /URA vector	[89]
pBG1805-*SKI8*	*SKI8* is under control of *GAL1* promoter in the pBG1805, 2 μ /URA vector	[89]
p3G-*CHO2*-HA	*CHO2* is under control of *GAL1* promoter in a CEN/URA vector	[27]
p426-*CHO2*-HA	*CHO2* is under control of *CHO2* endogenous promoter in a 2 μ /URA vector	This study
YCplac33-*PAH1*	*PAH1* is under control of its endogenous promoter in a CEN/URA vector	[54]
p3G-*PAH1*-HA	*PAH1* is under control of *GAL1* promoter in a CEN/URA vector	This study
pYes2-*AtPAH1*	*AtPAH1* is under control of *GAL1* promoter in the pYES2.1/NT 2 μ /URA vector	[49]
pYes2-*AtPAH2*	*AtPAH2* is under control of *GAL1* promoter in the pYES2.1/NT 2 μ /URA vector	[49]
pYes2-*NbPAH1A*	*NbPAH1A* is under control of *GAL1* promoter in the pYES2.1/NT 2 μ /URA vector	This study
pYes2-*NbPAH2A*	*NbPAH2A* is under control of *GAL1* promoter in the pYES2.1/NT 2 μ /URA vector	This study
pPWHT- *NbPAH1A*	*NbPAH1A* is under control of an enhanced CaMV 35S promoter	This study
pPWHT- *NbPAH2A*	*NbPAH2A* is under control of an enhanced CaMV 35S promoter	This study
pAG2p-*AtPAH2*	*AtPAH2* is under control of an enhanced CaMV 35S promoter	This study

4.5 Electron microscopy

Samples were prepared as described previously (Zhang et al., 2016). Briefly, 10 A_{600nm} units of cells were fixed with 4% paraformaldehyde and 2% glutaraldehyde for 1h followed by secondary fixation in 1% osmium tetroxide for another 1h. After dehydration through an ethanol gradient, yeast cells were embedded in Spurr's resin (Electron Microscopy Sciences) for overnight. The sample sections were stained with uranyl acetate and lead citrate and observed under a JEOL JEM 1400 TEM at the Virginia-Maryland College of Veterinary Medicine.

For immunogold labeling, 4% paraformaldehyde and 0.5% glutaraldehyde were used to fix 10 A600nm units of cells for 1 hour and followed by 0.1% osmium tetroxide secondary fixation for another 15minutes. After dehydration through an ethanol gradient, yeast cells were embedded in LR White resin (Electron Microscopy Sciences) for overnight. Embedded samples were sectioned and nickel grids were used to hold the samples. After treated with blocking solution (AURION) for 30minutes, grids were incubated with primary antibody diluted in incubation buffer (PBS, pH 7.4, 0.15% AURION BSA-c and 15mM NaN3) and secondary antibody conjugated with colloidal gold particles (10nm or 15nm particles were conjugated to anti-mouse or anti-rabbit secondary antibody, AURION) diluted in incubation buffer. The primary antibodies were rabbit anti-1a antiserum (1:50), mouse-anti dsRNA monoclonal antibody J2 (1:50). Secondary antibodies were diluted at 1:20. Sections were counterstained with uranyl acetate (10minutes) and lead citrate (3 minutes) and observed under a JEOL JEM 1400 TEM at 80KV at the Virginia-Maryland College of Veterinary Medicine.

4.6 Immunofluorescence microscopy

Yeast cells were harvested and fixed with 4% formaldehyde for 30minutes. To prepare spheroplasts, the cell wall was removed by lyticase. After permeabilization with 0.1% Triton X-100 for 15min, the spheroplasts were incubated with primary antibodies (1:100 dilution) overnight at 4°C followed by incubation with secondary antibodies (1:100 dilution) for 1h at 37°C. Finally, the nucleus was stained with DAPI (Vector laboratories) for 10min. Samples were observed using a Zeiss epifluorescence microscope (Observer.Z1) at the Fralin microscopy facility, VT.

4.7 Measurement of yeast nuclear membrane perimeters

Measurements were performed with ImageJ software. Briefly, the scale was set based on the scale bar in images. The color threshold was adjusted to allow the spot to fit the nucleus perfectly and adding the target spot to the ROI (Region of Interest) manager by using the wand (tracing) tool. The perimeter was measured by performing the "measure" in the ROI manager tool.

4.8 Radiolabeling and analysis of lipids

The steady-state labeling of lipids with [2-^{14}C] acetate was performed as described previously (Moriock et al.,1988). Briefly, equal number of cells (2.5×10^5cells/ml) were inoculated into SC-Ura-Leu with galactose as carbon source along with the [2-^{14}C] acetate. The cells were grown to exponential phase (A_{600nm} = ~0.5) and harvested. Lipids were extracted (Bligh et al., 1959) from the radiolabeled cells, and then separated by one-dimensional TLC for neutral lipids (Henderson et al., 1992) or phospholipids (Vaden et al., 2005). The resolved lipids were visualized by phosphorimaging and quantified by Image Quant software using a standard curve of [2-^{14}C] acetate. The identity of radiolabeled lipids was confirmed by comparison with the migration of authentic standards visualized by staining with iodine vapor. The mol percentage of each neutral lipid or phospholipid was normalized to the total ^{14}C-labeled chloroform fraction. Single factor ANOVA was used for statistical analysis of lipid differences between wt and mutants.

4.9 BMV replication assay in *Nicotiana benthamiana*

Replication analysis of BMV, TRV, and TMV(Lindbo et al.,2007) in *N. benthamiana* was performed as previously reported (Liu et al.,2002). *Arabidopsis thaliana AtPAH2*, *N. benthamiana NbPAH1A* and *NbPAH2A* were expressed in *N. benthamiana* leaves by agroinfiltration following a protocol described before (Bendahmane et al., 2000). The *AtPAH2* was cloned into pAG2p vector between an enhanced CaMV 35S promoter and a terminator (Xu et al., 2015). The *NbPAH1A* and *NbPAH2A* were cloned into pPWHT vector between an enhanced CaMV 35S promoter and a terminator through gateway cloning. *N. benthamiana* leaves were first infiltrated with Agrobacteria (GV3101) harboring pAG2p *AtPAH2*, pPWHT-*NbPAH1A* or pPWHT-*NbPAH2A* plasmid. Two days later, the same leaves were infiltrated with the mixed Agrobacteria cultures harboring plasmids that launch BMV RNA 1, 2, and 3, or TRV1 and 2, or TMV. The infiltrated leaves were harvested 3d post viral inoculation. Total RNA was extracted following the hot phenol method. Viral RNA accumulation was analyzed by Northern blotting with BMV-, TRV-, or TMV-specific probes as described before (Xu et al., 2015).

Acknowledgements

We thank Haijie Liu, Jianhui Li, Elizabeth Barton, Nicholas Todd, and Nancy Kalaj for general assistance, Dr. Symeon Siniossoglou at University of Cambridge (UK) for providing yeast strains, Drs. Elzbieta Mietkiewska and Randall J. Weselake at University of Alberta (Canada), and Dr. Valérie Doye from Institute Jacques Monod (France) for providing plasmids, Drs. LiKa Liu and William Prinz from National Institute of Diabetes and Digestive and Kidney Diseases for help in lipid work, Drs. Tero Ahola and Janet Webster for critical reading of the manuscript. We appreciate the help from Ms. Kathy Lowe at Virginia-Maryland College of Veterinary Medicine (Virginia Tech) and Ms. Shannon Modla at the Delaware Biotechnology Institute for electron microscopy work, and Dr. Kristi DeCourcy at the Fralin Life Science Institute (Virginia Tech) for fluorescence microscopy work.

References

[1] AHOLA T, AHLQUIST P. Putative RNA capping activities encoded by brome mosaic virus: methylation and covalent binding of guanylate by replicase protein 1a[J]. Journal of Virology, 1999, 73 (12): 10061-10069.

[2] AHOLA T, DEN BOON J A, AHLQUIST P. Helicase and capping enzyme active site mutations in brome mosaic nvirus protein 1a cause defectsin template recruitment, negative-strand RNA synthesis, and viral RNA capping[J]. Journal of Virology, 2000, 74 (19): 8803-8811.

[3] ALTAN-BONNET N. Lipid Tales of Viral Replication and Transmission[J]. Trends in Cell Biology, 2017, 27 (3): 201-213.

[4] ANNAMALAI P, RAO A L. Replication-independent expression of genome components and capsid protein of brome mosaic virus in planta: a functional role for viral replicase in RNA packaging[J]. Virology, 2005, 338 (1): 96-111.

[5] ATHENSTAEDT K, DAUM G. Biosynthesis of phosphatidic acid in lipid particles and endoplasmic reticulum of *Saccharomyces cerevisiae*[J]. Journal of Bacteriology, 1997, 179 (24): 7611-7616.

[6] ATHENSTAEDT K, WEYS S, PALTAUF F, et al. Redundant systems of phosphatidic acid biosynthesis via acylation of glycerol-3-phosphate or dihydroxyacetone phosphate in the yeast *Saccharomyces cerevisiae*[J]. Journal of Bacteriology, 1999, 181 (5): 1458-1463.

[7] BARGMANN B O, LAXALT A M, RIET B T, et al. LePLDbeta1 activation and relocalization in suspension-cultured tomato cells treated with xylanase[J]. Plant Journal, 2010, 45 (3): 358-368.

[8] BELGAREH N, DOYE V. Dynamics of nuclear pore distribution in nucleoporin mutant yeast cells[J]. The Journal of Cell Biology, 1997, 136 (4): 747-759.

[9] BELOV G A. Dynamic lipid landscape of picornavirus replication organelles[J]. Current Opinion in Virology, 2016, 19: 1-6.

[10] BELOV G A, VAN KUPPEVELD F J. (+) RNA viruses rewire cellular pathways to build replication organelles[J]. Current Opinion in Virology, 2012, 2 (6): 740-747.

[11] BENDAHMANE A, QUERCI M, KANYUKA K, et al. Agrobacterium transient expression system as a tool for the isolation of disease resistance genes: application to the Rx2 locus in potato[J]. The Plant Journal, 2000, 21 (1): 73-81.

[12] BERGER K L, COOPER J D, HEATON N S, et al. Roles for endocytic trafficking and phosphatidylinositol 4-kinase III alpha in hepatitis C virus replication[J]. Proceedings of the National Academy of Sciences of the United States of America, 2009, 106 (18): 7577-7582.

[13] BLIGH E G, DYER W J. A rapid method of total lipid extraction and purification[J]. Canadian Journal of Biochemistry and Physiology, 1959, 37 (8): 911-7.

[14] BURGER K N. Greasing membrane fusion and fission machineries[J]. Traffic, 2010, 1 (8): 605-613.

[15] CAMPBELL J L, LORENZ A, WITKIN K L, et al. Yeast nuclear envelope subdomains with distinct abilities to resist membrane expansion[J]. Molecular Biology of the Cell, 2006, 17 (4): 1768-1778.

[16] CAO X, JIN X, ZHANG X, et al. Morphogenesis of Endoplasmic Reticulum Membrane-Invaginated Vesicles during Beet Black Scorch Virus Infection: Role of Auxiliary Replication Protein and New Implications of Three-Dimensional Architecture[J]. Journal of virology, 2015, 89 (12): 6184-6195.

[17] CARMAN G M, HAN G S.Phosphatidicacid phosphatase, a key enzymein the regulation of lipid synthesis[J]. Journal of Biological Chemistry, 2009, 284 (5): 2593-2597.

[18] CARMAN G M, HENRY S A. Phospholipid biosynthesis in yeast[J]. Annual Review of Biochemistry, 1989, 58: 635-669.

[19] CARMAN G M, HAN G S. Regulation of phospholipid synthesis in the yeast *Saccharomyces cerevisiae*[J]. Annual Review of Biochemistry, 2011, 80: 859-883.

[20] CARMAN G M, HAN G S. Roles of phosphatidate phosphatase enzymes in lipid metabolism[J]. Trends in Biochemical Sciences, 2006, 31 (12): 694-699.

[21] CHEN J, AHLQUIST P. Brome mosaic virus polymerase-like protein 2a is directed to the endoplasmic reticulum by helicase-like viral protein 1a[J]. Journal of Virology. 2000, 74 (9): 4310-4318.

[22] CHENG X, DENG P, CUI H, et al. Visualizing double-stranded RNA distribution and dynamics in living cells by dsRNA binding-dependent fluorescence complementation[J]. Virology, 2015, 485: 439-451.

[23] CHOI H S, SU W M, HAN G S, et al. Pho85p-Pho80p phosphorylation of yeast Pah1p phosphatidate phosphatase regulates its activity, location, abundance, and function in lipid metabolism[J]. Journal of Biological Chemistry, 2012, 287 (14): 11290-11301.

[24] CHOI H S, SU W M, MORGAN J M, et al. Phosphorylation of phosphatidate phosphatase regulates its membrane association and physiological functions in Saccharomyces cerevisiae: identification of SER (602), THR (723), AND SER(744) as the sites phosphorylated by CDC28 (CDK1)-encoded cyclin-dependent kinase[J]. Journal of Biological Chemistry, 2010, 286 (2): 1486-1498.

[25] CHUANG C, BARAJAS D, QIN J, et al. Inactivation of the host lipin gene accelerates RNA virus replication through viral exploitation of the expanded endoplasmic reticulum membrane[J]. PLoS Pathogens, 2014, 10 (2): e1003944.

[26] CHUKKAPALLI V, HEATON N S, RANDALL G. Lipids at the interface of virus-host interactions[J]. Current Opinion in Microbiology, 2012, 15 (4): 512-518.

[27] CSAKI L S, DWYER J R, FONG L G, et al. Lipins, lipinopathies, and the modulation of cellular lipid storage and signaling[J]. Progress in Lipid Research, 2013, 52 (3): 305-316.

[28] DIAZ A, GALLEI A, AHLQUIST P. Bromovirus RNA replication compartment formation requires concerted action of 1a's self-interacting RNA capping and helicase domains[J]. Journal of Virology, 2012, 86 (2): 821-834.

[29] DIAZ A, WANG X. Bromovirus-induced remodeling of host membranes during viral RNA replication[J]. Current Opinion in Virology, 2014, 9: 104-110.

[30] DIAZ A, ZHANG J, OLLWERTHER A, et al. Host ESCRT proteins are required for bromovirus RNA replication compartment assembly and function[J]. PLOS Pathogens, 2015, 11 (3): e1004742.

[31] FAKAS S, QIU Y, DIXON J L, et al. Phosphatidate phosphatase activity plays key role in protection against fatty acid-induced toxicity in yeast[J]. Journal of Biological Chemistry, 2011, 286 (33): 29074-29085.

[32] FERNANDEZDE CASTRO I, FERNANDEZ J J, et al. Three-dimensional imaging of the intracellular assembly of a functional viral RNA replicasecomplex[J]. Journal of Cell Science, 2016, 130 (1): 260-268.

[33] FERNANDEZ-MURRAY J P, MCMASTER C R. Lipid synthesis and membrane contact sites: a crossroads for cellular physiology[J]. Journal of lipid research, 2016, 57 (10): 1789-1805.

[34] GELPERIN D M, WHITE M A, WILKINSON M L, et al. Biochemical and genetic analysis of the yeast proteome with a movable ORF collection[J]. Genes, Development, 2005, 19 (23): 2816-2826.

[35] GIOVANNI P, MARTELLI MARCRELLO R. Virus-host relationships: symptomatological and ultrastructural aspects. In: RIB F, editor. The plant viruses.

[36] GOODIN M M, ZAITLIN D, NAIDU R A, et al. *Nicotiana benthamiana*: its history and future as a model for plant-pathogen interactions[J]. Molecular Plant-Microbe Interactions, 2008, 21 (8): 1015-1026.

[37] GOPINATH K, DRAGNEA B, KAO C. Interaction between Brome mosaic virus proteins and RNAs: effects on RNA replication, protein expression, and RNA stability[J]. Journal of Virology, 2005, 79 (22): 14222-14234.

[38] GRIMSEY N, HAN G S, O'HARA L, et al. Temporal and spatial regulation of the phosphatidate phosphatases lipin 1 and 2[J]. Journal of Biological Chemistry, 2008, 283 (43): 29166-29174.

[39] HAN G S, O'HARA L, CARMAN G M, et al. An unconventional diacylglycerol kinase that regulates phospholipid synthesis and nuclear membrane growth[J]. Journal of Biological Chemistry, 2008, 283 (29): 20433-20442.

[40] HAN G S, SINIOSSOGLOU S, CARMAN G M. The cellular functions of the yeast lipin homolog P AH1p are dependent on its phosphatidate phosphatase activity[J]. Journal of Biological Chemistry, 2007, 282 (51): 37026-37035.

[41] HARRISON B D, KUBO S, ROBINSON D J, et al. Multiplication Cycle of Tobacco Rattle Virus in Tobacco Mesophyll Protoplasts[J]. Journal of General Virology, 1976, 33 (Nov): 237-248.

[42] HAYAT M. Colloidal gold: principles, methods, and applications[J]. Hayat M, editor. San Diego: Academic Press.

[43] HENDERSON R, TOCHER D. Thin-layer chromatography[J]. Hamilton R, Hamilton S, editors. New York: IRL Press.

[44] HENRY S A, KOHLWEIN S D, CARMAN G M. Metabolism and regulation of glycerolipids in the yeast Saccharomyces cerevisiae[J]. Genetics, 2012, 190 (2): 317-349.

[45] HSU N Y, ILNYTSKA O, BELOV G, et al. Viral reorganization of the secretory pathway generates distinct organelles for RNA replication[J]. Cell, 2010, 141 (5): 799-811.

[46] HYODO K, OKUNO T. Pathogenesismediated by proviral host factors involved in translation and replication of plant positive-strand RNA viruses[J]. Current Opinion in Virology, 2016, 17: 11-18.

[47] HYODO K, TANIGUCHI T, MANABE Y, et al. Phosphatidic acid produced by phospholipase D promotes RNA replication of a plant RNA virus[J]. Plos Pathogens, 2015, 11 (5): e1004909.

[48] KAO C C, AHLQUIST P. Identification of the domains required for direct interaction of the helicase-like and polymerase-like RNA replication proteins of brome mosaic virus[J]. Journal of Virology, 1992, 66 (12): 7293-7302.

[49] KARANASIOS E, HAN G S, XU Z, et al. A phosphorylation-regulated amphipathic helix controls the membrane translocation and function of the yeast phosphatidate phosphatase[J]. Proceedings of the National Academy of Sciences of the United States of America, 2010, 107 (41): 17539-17544.

[50] KIRIK A, MUDGETT M B. SOBER1 phospholipase activity suppresses phosphatidic acid accumulation and plant immunity in response to bacterial effector AvrBsT[J]. Proceedings of the National Academy of Sciences of the United States of America, 2009, 106 (48): 20532-20537.

[51] KOHRER K, DOMDEY H. Preparation of high molecular weight RNA Methods Enzymol[J]. 1991, 194:398-405.

[52] KONG F, SIVAKUMARAN K, KAO C. The N-terminal half of the brome mosaic virus 1a protein has RNA capping-associated activities: specificity for GTP and S-adenosylmethionine[J]. Virology, 1999, 259 (1): 200-210.

[53] KOOIJMAN E E, CHUPIN V, FULLER N L, et al. Spontaneous curvature of phosphatidic acid and lysophosphatidic acid[J]. Biochemistry, 2005, 44 (6): 2097-2102.

[54] KUSHNER D B, LINDENBACH B D, GRDZELISHVILI V Z, et al. Systematic, genomewide identification of host genes affecting replication of a positive-strand RNA virus[J]. Proceedings of the National Academy of Sciences, 2003, 100 (26): 15764-15769.

[55] LALIBERTE J F, ZHENG H. Viral Manipulation of Plant Host Membranes[J]. Annual Review of Virology, 2014, 1 (1): 237-259.

[56] LEE W M, AHLQUIST P. Membrane Synthesis Specific Lipid Requirements, and Localized Lipid Composition Changes Associated with a Positive-Strand RNA Virus RNA Replication Protein[J]. Journal of Virology, 2003, 77 (23): 12819-12828.

[57] LEE W M, ISHIKAWA M, AHLQUIST P. Mutation of host delta9 fatty acid desaturase inhibits brome mosaic virus RNA replication between template recognition and RNA synthesis[J]. Journal of Virology, 2001, 75 (5): 2097-2106.

[58] LI J, FUCHS S, ZHANG J, et al. An unrecognized function for COPII components in recruiting the viral replication protein BMV. 1a to the perinuclear ER[J]. Journal of Cell Science, 2016, 129 (19): 3597-3608.

[59] LINDBO J A. TRBO: a high-efficiency tobacco mosaic virus RNA-based overexpression vector[J]. Plant Physiology, 2007, 145 (4): 1232-1240.

[60] LIU L, WESTLER W M, DEN BOON J A, et al. An amphipathic alpha-helix controls multiple roles of brome mosaic virus protein 1a in RNA replication complex assembly and function[J]. PLoS Pathogens, 2009, 5 (3): e1000351.

[61] LIU Y, SCHIFF M, DINESH-KUMAR S P. Virus-induced gene silencing in tomato[J]. The Plant Journal, 2002, 31 (6): 777-786.

[62] LOEWEN C J, GASPAR M L, JESCH S A, et al. Phospholipid metabolism regulated by a transcription factor sensing phosphatidic

acid[J]. Science, 2004, 304 (5677): 1644-1647.

[63] MAS P, BEACHY R N. Replication of tobacco mosaic virus on endoplasmic reticulum and role of the cytoskeleton and virus movement protein in intracellular distribution of viral RNA[J]. Journal of Cell Biology, 1999, 147 (5): 945-958.

[64] MCMAHON H T, GALLOP J L. Membrane curvature and mechanisms of dynamic cell membrane remodelling[J]. Nature, 2005, 438 (7068): 590-596.

[65] MIETKIEWSKA E, SILOTO R M, DEWALD J, et al. Lipins from plants are phosphatidate phosphatases that restore lipid synthesis in a pah1Delta mutant strain of *Saccharomyces cerevisiae*[J]. FEBS Journal, 2011, 278 (5): 764-775.

[66] MISE K, ALLISON R F, JANDA M, et al. Bromovirus movement protein genes play a crucial role in host specificity[J]. Journal of Virology, 1993, 67 (5): 2815-2823.

[67] MORLOCK K R, LIN Y P, CARMAN G M. Regulation of phosphatidate phosphatase activity by inositol in *Saccharomyces cerevisiae*[J]. Journal of Bacteriology, 1988, 170 (8): 3561-3566.

[68] NAKAMURA Y, KOIZUMI R, SHUI G, et al. Arabidopsis lipins mediate eukaryotic pathway of lipid metabolismand cope critically with phosphate starvation[J]. Proceedings of the National Academy of Sciences, 2009, 106 (49): 20978-20983.

[69] O'REILLY E K, PAUL J D, KAO C C. Analysis of the interaction of viral RNA replication proteins by using the yeast two-hybrid assay[J]. Journal of Virology, 1997, 71 (10): 7526-7532.

[70] PALTAUF F, KOHLWEIN S, HENRY S. Regulation and compartmentalization of lipid synthesis in yeast[J]. Cold Spring Harbor Monograph Archive, 1992, 415 (21B): 415-500.

[71] PARK Y, HAN G S, MILEYKOVSKAYA E, et al. Altered Lipid Synthesis by Lack of Yeast Pah1 Phosphatidate Phosphatase Reduces Chronological Life Span[J]. Journal of Biological Chemistry, 2015, 290 (42): 25382-25394.

[72] PAUL D, BARTENSCHLAGER R. Architecture and biogenesis of plus-strand RNA virus replication factories[J]. World Journal of Virology, 2013, 2 (2): 32-48.

[73] RATTRAY J B, SCHIBECI A, KIDBY D K. Lipids of yeasts[J]. Bacteriological Reviews, 1975, 39 (3): 197-231.

[74] REISS S, REBHAN I, BACKES P, et al. Recruitment and activation of a lipid kinase by hepatitis C virus NS5A is essential for integrity of the membranous replication compartment[J]. Cell host & microbe, 2011, 9 (1): 32-45.

[75] RESTREPO-HARTWIG M A, AHLQUIST P. Brome mosaic virus helicase- and polymerase-like proteins colocalize on the endoplasmic reticulum at sites of viral RNA synthesis[J]. Journal of Virology, 1996, 70 (12): 8908-8916.

[76] RESTREPO-HARTWIG M, AHLQUIST P. Brome mosaic virus RNA replication proteins 1a and 2a colocalize and 1a independently localizes on the yeast endoplasmic reticulum[J]. Journal of virology,1999, 73 (12): 10303-10309.

[77] SANTOS-ROSA H, LEUNG J, GRIMSEY N, et al. The yeast lipin Smp2 couples phospholipid biosynthesis to nuclear membrane growth[J]. EMBO Journal, 2014, 24 (11): 1931-1941.

[78] SCHWARTZ M, CHEN J, JANDA M, et al. A positive-strand RNA virus replication complex parallels form and function of retrovirus capsids[J]. Molecular Cell, 2002, 9 (3): 505-514.

[79] SCIORRA V A, MORRIS A J. Roles for lipid phosphate phosphatases in regulation of cellular signaling[J]. Biochim Biophys Acta, 2002, 1582 (1-3): 45-51.

[80] SINIOSSOGLOU S, SANTOS-ROSA H, RAPPSILBER J, et al. A novel complex of membrane proteins required for formation of a spherical nucleus[J]. EMBO Journal, 2014, 17 (22): 6449-6464.

[81] TAUCHI-SATO K, OZEKI S, HOUJOU T, et al. The surface of lipid droplets is a phospholipid monolayer with a unique Fatty Acid composition[J]. Journal of Biological Chemistry, 2002, 277 (46): 44507-44512.

[82] TESTERINK C, MUNNIK T. Phosphatidic acid: a multifunctional stress signaling lipid in plants[J]. Trends in Plant Science, 2005, 10 (8): 368-375.

[83] VADEN D L, GOHIL V M, GU Z, et al. Separation of yeast phospholipidsusing one-dimensional thin-layer chromatography[J]. Analytical Biochemistry, 2005, 338 (1): 162-164.

[84] VANDER LUIT A H, PIATTI T, VAN DOORN A, et al. Elicitation of suspension-cultured tomato cells triggers the formation of phosphatidic acid and diacylglycerol pyrophosphate[J]. Plant Physiology, 2000, 123 (4): 1507-1516.

[85] WAGGONER D W, XU J, SINGH I, et al. Structural organization of mammalian lipid phosphate phosphatases: implications for signal transduction[J]. Biochimicaet Biophysica Acta (BBA) Molecular and Cell Biology of Lipids, 1999, 1439 (2): 299-316.

[86] WANG A. Dissecting the molecular network of virus-plant interactions: the complex roles of host factors[J]. Annual Review of Phytopathology, 2015, 53: 45-66.

[87] WANG X, AHLQUIST P. Brome mosaic virus[J]. Encyclopedia of Virology (Third Edition), 2008:381-386.

[88] WANG X, DEVAIAH S P, ZHANG W,et al. Signaling functions of phosphatidic acid[J]. Progress in Lipid Research, 2006, 45 (3): 250-278.

[89] WANG X, LEE W M, WATANABE T, et al. Brome mosaic virus 1a nucleoside triphosphatase/ helicase domain plays crucial roles in recruiting RNA replication templates[J]. Journal of Virology, 2005, 79 (21): 13747-13758.

[90] WENTE S R, ROUT M P, BLOBEL G. A new family of yeast nuclear pore complex proteins[J]. The Journal of Cell Biology, 1992, 119 (4): 705-723.

[91] WICKNER R B. Double-stranded RNA viruses of Saccharomyces cerevisiae[J]. Microbiological Reviews, 1996, 60 (1): 250-265.

[92] XU K, NAGY P D. Enrichment of Phosphatidylethanolamine in Viral Replication Compartments via Coopting the Endosomal Rab5 Small GTPase by a Positive-Strand RNA Virus[J]. PLoS Biology, 2016, 14 (10): e2000128.

[93] XU K, NAGY P D. RNA virus replication depends on enrichment of phosphatidylethanolamine at replication sites in subcellular membranes[J]. Proceedings of the National Academy of Sciences of the United States of America, 2015, 112 (14): E1782-1791.

[94] YAMAGUCHI T, MINAMI E, UEKI J. Elicitor-induced activation of phospholipases plays an important role for the induction of defense responses in suspension-cultured rice cells[J]. Plant Cell Physiology, 2005, 46 (4): 579-587.

[95] ZHANG J, DIAZ A, MAO L, et al. Host acyl coenzyme A binding protein regulates replication complex assembly and activity of a positive-strand RNA virus[J]. Journal of Virology, 2012, 86 (9): 5110-5121.

[96] ZHANG J, ZHANG Z, CHUKKAPALLI V, et al. Positive-strand RNA viruses stimulate host phosphatidylcholine synthesis at viral replication sites[J]. Proceedings of the National Academy of Sciences of the United States of America, 2016, 113 (8): E1064-1073.

[97] ZHENG Z, ZOU J. The initial step of the glycerolipid pathway: identification of glycerol 3-phosphate/ dihydroxyacetone phosphate dual substrate acyltransferases in Saccharomyces cerevisiae[J]. Journal of Biological Chemistry, 2001, 276 (45): 41710-41716.

附 录

一、教材与专著

[1] 林传光，曾士迈，褚菊澂，李学书，谢联辉 . 1961. 植物免疫学 . 北京：农业出版社 .

[2] 梁训生，谢联辉主编 . 1994. 植物病毒学 . 北京：农业出版社 .

[3] 谢联辉主编 . 1997. 水稻病害 . 北京：中国农业出版社 .

[4] 谢联辉主编 . 1997. 赵修复文选，福州：福建科学技术出版社 .

[5] 谢联辉，林奇英，吴祖建 . 1999. 植物病毒名称及其归属 . 北京：中国农业出版社 .

[6] 谢联辉 . 2001. 水稻病毒：病理学与分子生物学 . 福州：福建科学技术出版社 .

[7] 谢联辉，林奇英主编 . 2004. 植物病毒学：第二版 . 北京：中国农业出版社 .

[8] 谢联辉，林奇英主编 . 2011. 植物病毒学：第三版 . 北京：中国农业出版社 .

[9] 谢联辉主编 . 2006. 普通植物病理学：第一版 . 北京：科学出版社 .

[10] 谢联辉主编 . 2013. 普通植物病理学：第二版 . 北京：科学出版社 .

[11] 谢联辉主编 . 2008. 植物病原病毒学 . 北京：中国农业出版社 .

[12] 谢联辉，林奇英，吴祖建 . 2009. 植物病毒：病理学与分子生物学 . 北京：科学出版社 .

[13] 谢联辉，林奇英，徐学荣 . 2009. 植物病害：经济学、病理学与分子生物学 . 北京：科学出版社 .

[14] 谢联辉，林奇英，吴祖建 . 2009. 天然产物：纯化、性质与功能 . 北京：科学出版社 .

[15] 谢联辉，尤民生，侯有明等 . 2011. 生物入侵——问题与对策 . 北京：科学出版社 .

[16] 徐学荣，谢联辉 . 2011. 植保经济学 . 福州：福建科学技术出版社 .

[17] 谢联辉，林奇英，魏太云，吴祖建 . 2016. 水稻病毒 . 北京：科学出版社 .

二、参编图书

[1] 谢联辉，林奇英 . 1964. 水稻的“三病”及其防治 . 农业科学技术常识：第一册 . 福建函授广播学校编印，11-21.

[2] 谢联辉 . 1981. 水稻病毒病测报方法 // 农业部作物病虫测报总站 . 农作物主要病虫测报办法 . 北京：农业出版社，27-39.

[3] 谢联辉，林奇英 . 1988. 我国水稻病毒病的发生和防治 . 曾昭慧 . 中国水稻病虫综合防治进展 . 杭州：浙江科学技术出版社，255-264.

[4] 谢联辉 . 1991. 水稻病毒病 // 北京农业大学，农业植物病理学 . 北京：农业出版社，31-45.

[5] 张学博，谢联辉 . 1991. 甘蔗病害 // 金善宝 . 中国农业百科全书 • 农作物卷 . 北京：农业出版社，182-183.

[6] 谢联辉，林奇英 . 1996. 水稻病毒病 // 方中达 . 中国农业百科全书 • 植物病理学卷 . 北京：农业出版社，427-430.

[7] 徐学荣，张巨勇，谢联辉，2005. 可持续发展通道与预警 // 汪同三、张守一、王崇举 . 21 世纪数量经济学 . 重庆：重庆出版社，69-80.

[8] 孙恢鸿，沈瑛，许志刚，谢联辉，林含新，2005. 水稻品种抗病性及其利用 // 李振歧、商鸿生：中国农作物抗病性及其利用 . 北京：中国农业出版社，263-327.

[9] 陈启建，谢联辉 . 2008. 植物病毒疫苗的研究与实践 // 邱德文 . 植物免疫与植物疫苗——研究与实践 . 北京：科学出版社，19-32.

[10] 谢联辉，2009. 水稻病毒病（含植原体病害）// 季良 . 植物病毒病防治与检疫：上册 . 北京：中国农业出版社，107-146.

三、论文

[1] 谢联辉，林德槛 . 1977. 引起水稻瘟兜的一种细菌性病害 —— 蕙头瘟 . 福建农业科技，(2):37-39.

[2] 谢联辉，林奇英 . 1979. 水稻黄矮病的流行预测和验证 . 植物保护，5（5）: 33-37.

[3] 谢联辉，陈昭炫，林奇英 . 1979. 水稻簇矮病研究 I . 簇矮病 —— 水稻上的一种新的病毒病 . 植物病理学报，9（2）: 93-100.

[4] 谢联辉，陈昭炫，林奇英 . 1979. 水稻病毒病化学治疗试验初报 . 病毒学集刊 , 44-48.

[5] 谢联辉，林奇英 . 1980. 锯齿叶矮缩病在我国水稻上的发现 . 植物病理学报，10（1）: 59-64.

[6] 谢联辉 . 1980. 水稻病毒病流行预测研究的几个问题 . 福建农学院学报，9（1）: 43-50.

[7] 谢联辉，林奇英 . 1980. 水稻黄叶病和矮缩病流行预测研究 . 福建农学院学报，9（2）: 32-43.

[8] Xie LH, Lin QY. 1980. Studies on bunchy stunt disease of rice, a new virus disease of rice plant. A Monthly Journal of Science，25（09）,785-789.

[9] Xie LH, Lin QY. 1980. Rice ragged stunt disease, a new record of rice virus disease in China. A Monthly Journal of Science，25（11）:960-963.

[10] 谢联辉，林奇英，黄金星 . 1981. 水稻矮缩病的两个新特征 . 植物保护，7（6）: 14.

[11] Xie LH,Lin QY. 1981. Rice bunchy stunt, a new virus disease. The paper presented at the 1981 International Rice Research Conference, Los Bonus, Philippins, 1-7.

[12] Xie LH, Lin QY, Guo JR. 1981. A new insect vector ofrice dwarf virus. International Rice Research Newsletter，6（5）:14.

[13] 谢联辉，林奇英 . 1982. 水稻品种对病毒病的抗性研究 . 福建农学院学报，11（2）: 15-18.

[14] 谢联辉，林奇英 . 1982. 水稻东格鲁病（球状病毒）在我国的发生 . 福建农学院学报，11（3）: 15-23.

[15] 谢联辉，林奇英，朱其亮 . 1982. 水稻簇矮病研究 II . 病害的分布、损失、寄主和越冬 . 植物病理学报，12（4）:16-20.

[16] Xie LH, Lin QY. 1982. Properties and concentrations of rice bunchy stunt virus. International Rice Research Newsletter，7（2）: 6-7.

[17] 谢联辉，林奇英 . 1983. 水稻簇矮病研究III . 病毒的体外抗性及其在寄主体内的分布 . 植物病理学报，13（3）: 15-19.

[18] 谢联辉，林奇英 . 1983. 福建水稻病毒病的诊断鉴定及其综合治理意见 . 福建农业科技，（5）: 26-27.

[19] 谢联辉，林奇英，朱其亮，赖桂炳，陈南周，黄茂进，陈时明 . 1983. 福建水稻东格鲁病发生和防治研究 . 福建农学院学报，12（4）: 275-284.

[20] 谢联辉 . 1984. 近年来我国新发现的水稻病毒病 . 植物保护，10（3）: 2-3.

[21] 谢联辉，林奇英，刘万年 . 1984. 福建甘薯丛枝病的病原体研究 . 福建农学院学报，13（1）: 85-88.

[22] 谢联辉，林奇英，谢黎明，赖桂炳 , 1984. 水稻簇矮病研究IV . 病害的发生发展和防治试验 . 植物病理学报，14（1）:33-38.

[23] 林奇英，谢联辉，王桦 . 1984. 水稻品种对东格鲁病及其介体昆虫的抗性研究 . 福建农业科技，

(4) :34-35.

[24] 林奇英，谢联辉，郭景荣 . 1984. 光照和食料对黑尾叶蝉生长繁殖及其传播水稻东格鲁病能力的影响 . 福建农学院学报，13 (3) :193-199.

[25] 谢联辉，林奇英，王少峰 . 1984. 水稻锯齿叶矮缩病毒抗血清的制备及其应用 . 植物病理学报，14 (3) :147-151.

[26] 谢联辉，林奇英 . 1984. 我国水稻病毒病研究的进展 . 中国农业科学，(6) :58-65.

[27] 林奇英，谢联辉，陈宇航，谢莉妍，郭景荣 . 1984. 水稻齿矮病毒寄主范围的研究 . 植物病理学报，14 (4) :247-248[研究简报].

[28] 林奇英，谢联辉 . 1985. 水稻黄萎病的病原体研究 . 福建农学院学报，14 (2) : 103-108.

[29] 林奇英，谢联辉，谢莉妍 . 1985. 水稻黄萎病的发生及其防治 . 福建农业科技，(4) :12-13.

[30] 谢联辉，林奇英，曾鸿棋，汤坤元 . 1985. 福建烟草病毒病病原鉴定初报 . 福建农学院学报，14 (2) :116[研究简报].

[31] Xie LH. 1985. Virus diseases of rice in China. International Symposium on Virus Diseases of Rice and Leguminous Crops in the Tropics October 1-5, 1985. Ministry of Agriculture, Forestry and Fisheries, Japan.

[32] Xie LH. 1986. Research on rice virus diseases in China. Tropical Agriculture Research Series, 19: 45-50.

[33] 林奇英，谢联辉 . 1986. 福建番茄病毒病的病原鉴定 . 武夷科学，6:275-278.

[34] 林奇英，谢联辉，谢莉妍，陶卉 . 1986. 烟草扁茎簇叶病的病原体 . 中国农业科学，(3) :92 [研究通讯].

[35] 林奇英，谢莉妍，谢联辉，黄白清 . 1986. 甘薯丛枝病的化疗试验 . 福建农业科技，(3) :25.

[36] 谢联辉，林奇英 . 1987. 热带水稻和豆科作物病毒病国际讨论会简介 . 病毒学杂志，(1) :85-88.

[37] 谢联辉，林奇英，黄如娟 . 1987. 水仙病毒病原鉴定初报 . 云南农业大学学报，2:113[研究简报].

[38] 谢联辉，林奇英，段永平 . 1987. 我国水稻病毒病的回顾与前瞻 . 病虫测报，(1) :41.

[39] 林奇英，谢联辉，黄如娟，谢莉妍 . 1987. 烟草品种对病毒病的抗性鉴定 . 中国烟草，(3):16-17.

[40] 周仲驹，林奇英，谢联辉，王桦 . 1987. 甘蔗褪绿线条病研究 I . 病名、症状、病情和传播 . 福建农学院学报，16 (2) :111-116.

[41] 周仲驹，林奇英，谢联辉，彭时尧 . 1987. 我国甘蔗白叶病的发生及其病原体的电镜观察 . 福建农学院学报，16 (2) :165-168.

[42] 周仲驹，谢联辉，林奇英，蔡小汀，王桦 . 1987. 福建蔗区甘蔗斐济病毒的鉴定 . 病毒学报，3 (3) : 302-304.

[43] 范永坚，周仲驹，林奇英，谢联辉，难波成任，山下修一，土居养二 . 1987. 中国几种水稻病毒病超薄切片的电镜观察 . 日本植物病理学会报，53 (3) :24-25[摘要].

[44] 陈宇航，周仲驹，林奇英，谢联辉 . 1988. 甘蔗花叶病毒株系研究初报 . 福建农学院学报，17 (1) :44-48.

[45] 谢联辉，林奇英，段永平 . 1988. 烟草花叶病的有效激抗剂的筛选 . 福建农学院学报，17 (4) :371-372.

[46] 周仲驹，林奇英，谢联辉 . 1988. 甘蔗病毒病及其类似病害的研究现状及其进展 . 四川甘蔗，(2) :28-34.

[47] Xie LH, Lin QY, Zhou ZJ, Song XG, Huang LJ. 1988. The pathogen of rice grassy stunt and its strains in China. 5th International Congress of Plant Pathology Abstracts of Papers, Kyoto, Japan, 383.

[48] 林奇英，唐乐尘，谢联辉 . 1989. 水稻暂黄病流行预测与通径分析 . 福建农学院学报，18 (1) :

37-41.

[49]林奇英，谢莉妍，谢联辉 . 1989. 百合扁茎簇叶病的病原体观察 . 植物病理学报，19（2）:78[研究简报].

[50]周仲驹，黄如娟，林奇英，谢联辉，陈宇航 . 1989. 甘蔗花叶病的发生及甘蔗品种的抗性 . 福建农学院学报，18（4）:520-525.

[51]彭时尧，周仲驹，林奇英，谢联辉 . 1989. 甘蔗叶片感染甘蔗花叶病毒后 ATPase 活性定位和超微结构变化 . 植物病理学报，19（2）:69-73.

[52]谢联辉，林奇英 . 1990. 中国水仙病毒病的病原学研究 . 中国农业科学，23（2）: 89-90[研究通讯].

[53]谢联辉，郑祥洋，林奇英 . 1990. 水仙潜隐病毒病病原鉴定 . 云南农业大学学报，5（1）: 17-20.

[54]郑祥洋，林奇英，谢联辉 . 1990. 水仙上分离出的烟草脆裂病毒的鉴定 . 福建农学院学报，19（1）: 58-63.

[55]林奇英，谢联辉，周仲驹，谢莉妍，吴祖建 . 1990. 水稻条纹叶枯病研究Ⅰ . 病害的分布和损失 . 福建农学院学报，19（4）: 421-425.

[56]周仲驹，林奇英 , 谢联辉 . 1990. 甘蔗花叶病毒株系研究现状 . 四川甘蔗，（3）: 1-7.

[57]Lin X, Wang YY, Zhang WZ, Xu JY, Lin QY, Xie LH, 1990. Amino acid component in protein of *Rice dwarf virus (RDV)* and laser Raman spectrum. ICLLS' 90, 452-454.

[58]林奇英，谢联辉，谢莉妍，周仲驹，宋秀高 . 1991. 水稻条纹叶枯病研究Ⅱ . 病害的症状和传播 . 福建农学院学报，20（1）: 24-28.

[59]谢联辉，周仲驹，林奇英，宋秀高，谢莉妍 . 1991. 水稻条纹叶枯病研究Ⅲ . 病害的病原性质 . 福建农学院学报，20（2）:144-149.

[60]谢联辉，郑祥洋，林奇英 . 1991. 水仙病毒血清学研究Ⅰ . 水仙黄条病毒抗血清的制备及其应用 . 中国病毒学 ，6（4）: 344-348.

[61]林奇英，谢联辉，谢莉妍 . 1991. 水稻簇矮病毒的提纯及其性质，中国农业科学，24（4）: 52-57.

[62]王明霞，张谷曼，谢联辉 . 1991. 福建长汀小米椒病毒病的病原鉴定 . 福建农学院学报，20（1）:34-40.

[63]周仲驹，施木田，林奇英，谢联辉 . 1991. 甘蔗花叶病在钾镁不同施用水平下对甘蔗产质的影响 . 植物保护学报 ，18（3）:288 [研究简报].

[64]周仲驹，林奇英，谢联辉，彭时尧 . 1991. 甘蔗褪绿线条病的研究Ⅱ . 病原形态及其所致甘蔗叶片的超微结构变化 . 福建农学院学报，20（3）: 276-280.

[65]唐乐尘，林奇英，谢联辉，吴祖建 . 1991. 植物病理学文献计算机检索系统研究 . 福建农学院学报，20（3）: 291-296.

[66]施木田，周仲驹，林奇英，谢联辉 . 1991. 受甘蔗花叶病毒侵染后甘蔗叶片及其叶绿体中 ATPase 活性的变化 . 福建农学院学报，20（3）: 357-360.

[67]周仲驹，林奇英，谢联辉 . 1992. 水稻东格鲁杆状病毒在我国的发生 . 植物病理学报，22（1）: 15-18.

[68]周仲驹，林奇英，谢联辉，彭时尧 . 1992. 水稻条纹叶枯病研究Ⅳ . 病叶细胞的病理变化 . 福建农学院学报，21（2）: 157-162.

[69]周仲驹，谢联辉，林奇英，陈启建 . 1992. 香蕉束顶病的病原研究 . 福建省科协首届青年学术年会 —— 中国科协首届青年学术年会卫星会议论文集 . 福州：福建科学技术出版社，727-731.

［70］Zhou ZJ, Xie LH, Lin QY. 1992. Epidemiology of *Banana bunchy top virus* in China's continent.5th International Plant Virus Epidemiology Symposium, Virus, Vectors and the Environment, Valenzano（BARI）, Italy, 27-31 July, 149-150.

［71］Xie LH, Lin QY, Wu ZJ. 1992. Diagnosis, monitoring and control strategy of rice virus diseases in China's continent. 5th International Plant Virus Epidemiology Symposium, Virus, Vectors and the Environment, Valenzano（BARI）, Italy, 27-31 July, 235-236.

［72］Zhou ZJ, Xie LH. 1992. Status of banana diseases in China. Fruits, 47（6）: 715-721.

［73］谢联辉 . 1993. 面向生产实际，开展病害研究 . 中国科学院院刊，8（1）: 61-62.

［74］胡翠凤，谢联辉，林奇英 . 1993. 激抗剂协调处理对烟草花叶病的防治效应 . 福建农学院学报，22（2）:183-187.

［75］谢联辉 . 1993. 水稻病毒与检疫问题 . 植物检疫，7（4）: 305.

［76］谢联辉，林奇英，谢莉妍，赖桂炳 . 1993. 水稻簇矮病研究Ⅴ . 病害的年际变化 . 植物病理学报，23（3）:253-258.

［77］周仲驹，林奇英，谢联辉，陈启建，郑国璋，吴黄泉 . 1993. 香蕉束顶病的研究Ⅰ. 病害的发生，流行与分布 . 福建农学院学报，22（3）: 305-310.

［78］周仲驹，陈启建，林奇英，谢联辉 . 1993. 香蕉束顶病的研究Ⅱ . 病害的症状、传播及其特性 . 福建农学院学报，22（4）:428-432.

［79］林奇英，谢联辉，谢莉妍 . 1993. 水稻簇矮病研究Ⅵ . 水稻种质对病毒的抗性评价 . 植物病理学报，23（4）:305-308.

［80］林奇英，谢联辉，谢莉妍，吴祖建，周仲驹 . 1993. 中菲两种水稻病毒病的比较研究Ⅱ . 水稻草状矮化病的病原学 . 农业科学集刊，1:203-206.

［81］林奇英，谢联辉，谢莉妍，王明锦 . 1993. 中菲两种水稻病毒病的比较研究Ⅲ . 水稻草状矮化病毒的株系 . 农业科学集刊，1:207-210.

［82］林奇英，谢联辉，谢莉妍，林金嫩 . 1993. 中菲两种水稻病毒病的比较研究Ⅳ . 水稻东格鲁病毒的株系 . 农业科学集刊，1: 211-214.

［83］吴祖建，林奇英，谢联辉，谢莉妍，宋秀高 . 1993. 中菲两种水稻病毒病的比较研究Ⅴ . 水稻种质对病毒及其介体的抗性 . 福建农业大学学报，23（1）:58-62.

［84］Lin QY, Xie LH, Chen ZX. 1993. Rice bunchy stunt virus, A new member of *Phytoreovirus* group. 6th International Congress of Plant Pathology Abstrcts of Papers, Montreal, Quebec, Canada（July 28-August 6, 1993）, 303.

［85］Xie LH, Lin QY, Zheng XY, Wu ZJ. 1993. The pathogen identification of narcissus virus diseases in China. 6th International Congress of Plant Pathology Abstrcts of Papers, Montreal, Quebec, Canada（July 28-August 6, 1993）, 303.

［86］Zhou ZJ, Xie LH, Lin QY, Chen QJ. 1993. A study on banana virus diseases in China. 6th International Congress of Plant Pathology Abstrcts of Papers, Montreal, Quebec, Canada（July 28-August 6, 1993）, 303.

［87］Zhou ZJ, Xie LH, Lin QY. 1993. An introduction of studies on banana bunchy top virus in Fujian, current banana research and development in China. Collected papers submitted to the 3rd Meeting of the INIBAP/ASPNET Regional Advisory Comittee and China and INIBAP/ASPNET RAC Meeting, September 6-9, SCAU, Guangzhou, China, 14-18.

［88］林奇英，谢联辉，谢莉妍，吴祖建，1994. 烟草带毒种子及其脱毒处理，福建烟草，（2）: 27-29.

[89]林奇英，谢联辉，谢莉妍，吴祖建，周仲驹 . 1994. 中菲两种水稻病毒病的比较研究Ⅰ . 水稻东格鲁病的病原学 . 中国农业科学，27(2) :1-6.

[90]林奇英，谢联辉，谢荔石，林星，王由义 . 1994. 水稻簇矮病研究Ⅶ . 病毒的光谱特性 . 植物病理学报，24(1) :5-9.

[91]谢联辉，林奇英，吴祖建，周仲驹，段永平 . 1994. 中国水稻病毒病的诊断，监测和防治对策 . 福建农业大学学报，23(3) :280-285.

[92]谢联辉，林奇英，谢莉妍，段永平，周仲驹，胡翠凤 . 1994. 福建烟草病毒种群及其发生频率的研究 . 中国烟草学报，2(1) :25-32.

[93]吴祖建，林奇英，谢联辉 . 1994. 农杆菌介导的病毒侵染方法在禾本科植物转化上研究进展 . 福建农业大学学报，23(4) : 411-415.

[94]Zhou ZJ, Peng SY, Xie LH. 1994. Cytochemical localization of ATPase and ultrastructural changes in the infected sugarcane leaves by mosaic virus. Current Trends in Sugarcane Pathology (ed. G. P. Rao et al.), 289-296.

[95]周仲驹，林奇英，谢联辉，陈启建 . 1995. 香蕉束顶病的研究 III. 传毒介体香蕉交脉蚜的发生规律 . 福建农业大学学报，24(1) : 32-38.

[96]徐平东，谢联辉 . 1995. 黄瓜花叶病毒分子生物学研究进展 . 山东大学学报(自然科学版)，29(增刊) : 30-36.

[97]吴祖建，林奇英，李本金，张丽丽，谢联辉 . 1995. 水稻品种对黄叶病的抗性鉴定 . 上海农学院学报，13(增刊) : 58-64.

[98]吴祖建，林奇英，林奇田，肖银玉，谢联辉 . 1995. 水稻矮缩病毒的提纯和抗血清制备 . 福建省科协第二届青年学术年会 - 中国科协第二届青年学术年会卫星会议论文集 . 福州：福建科学技术出版社 , 600-604.

[99]谢莉妍，吴祖建，林奇英，谢联辉 . 1995. 植物呼肠孤病毒的基因组结构和功能 . 福建省科协第二届青年学术年会 - 中国科协第二届青年学术年会卫星会议论文集 . 福州：福建科学技术出版社，605-608.

[100]张广志，谢联辉 . 1995. 香蕉束顶病毒的提纯 . 福建省科协第二届青年学术年会 - 中国科协第二届青年学术年会卫星会议论文集 . 福州：福建科学技术出版社，609-612.

[101]林含新，吴祖建，林奇英，谢联辉 . 1995. 应用 F(ab)$_2$-ELISA 和单克隆抗体检测水稻条纹病毒 . 福建省科协第二届青年学术年会 - 中国科协第二届青年学术年会卫星会议论文集 . 福州：福建科学技术出版社，613-616.

[102]鲁国东，黄大年，陶全洲，杨炜，谢联辉 . 1995. 以 Cosmid 质粒为载体的稻瘟病菌转化体系的建立及病菌基因文库的构建 . 中国水稻科学，9(3) : 156-160.

[103]吴祖建，谢联辉，大村敏博，石川浩一，日比启野行 . 1995. 中国福建省产イネ萎缩ウイルス(RDV-F)と日本产普通系(RDV-O)の性状の比较 . 日本植物病理学会报，61(3) : 272 [摘要].

[104]周仲驹，林奇英，谢联辉，徐平东 . 1996. 香蕉束顶病毒株系的研究 . 植物病理学报，25(1) : 63-68.

[105]周仲驹，林奇英，谢联辉，陈启建，吴祖建，黄国穗，蒋家富，郑国璋 . 1996. 香蕉束顶病的研究 IV. 病害的防治 . 福建农业大学学报，25(1) : 44-49.

[106]林含新，谢联辉 . 1996. RFLP 在植物类菌原体鉴定和分类中的应用 . 微生物学通报，23(2) : 98-101.

[107]杨文定，吴祖建，王苏燕，刘伟平，叶寅，谢联辉，田波 . 1996. 表达反义核酶 RNA 的转基因水稻对矮缩病毒复制和症状的抑制作用 . 中国病毒学，11(3) : 277-283.

[108]林含新，林奇英，谢联辉.1996.水仙病毒病及其研究进展.植物检疫，10(4)：227-229.

[109]徐平东，张广志，周仲驹，李梅，沈春奇，庄西卿，林奇英，谢联辉.1996.香蕉束顶病毒的提纯和血清学研究.热带作物学报，17(2)：42-46.

[110]王宗华，陈昭炫，谢联辉.1996.福建菌物资源研究与与利用现状，问题及对策.福建农业大学学报，25(1)：446-449

[111]Xie LH, Lin QY, Xie LY, Chen ZX. 1996. *Rice bunchy stunt virus*: a new member of *Phytoreoviruses*. Journal of Fujian Agricultural University, 25(3)：312-319.

[112]李尉民，Roger Hull，张成良，谢联辉.1997. RT-PCR检测南方菜豆花叶病毒.中国进出口动植检，30(1)：28-30.

[113]鲁国东，王宗华，谢联辉.1997.稻瘟病菌分子遗传学研究进展.福建农业大学学报，26(1)：56-63.

[114]徐平东，李梅，林奇英，谢联辉.1997.应用A蛋白夹心酶联免疫吸附法鉴定黄瓜花叶病毒血清组.福建农业大学学报，26(1)：64-69.

[115]徐平东，李梅，林奇英，谢联辉.1997.西番莲死顶病病原病毒鉴定.热带作物学报，18(2)：77-84.

[116]周仲驹，黄志宏，郑国璋，林奇英，谢联辉.1997.香蕉束顶病的研究V.病株的空间分布型及其抽样.福建农业大学学报，26(2)：177-181.

[117]鲁国东，黄大年，谢联辉.1997.稻瘟病菌的电击转化.福建农业大学学报，26(3)：298-302.

[118]王宗华，郑学勤，谢联辉，张学博，陈守仁，黄俊生.1997.稻瘟病菌生理小种RAPD分析及其与马唐瘟的差异.热带作物学报，18(2)：92-97.

[119]周仲驹，杨建设，陈启建，林奇英，谢联辉，刘国坤.1997.香蕉束顶病无公害抑制剂的筛选研究.全国青年农业科学学术年报(A卷)，北京：中国农业科技出版社，304-308.

[120]徐平东，谢联辉.1997.黄化丝状病毒属(Closterovirus)病毒及其分子生物学研究进展.中国病毒学，12(3)：193-202.

[121]林含新，林奇英，谢联辉.1997.水稻条纹病毒分子生物学研究进展.中国病毒学，12(3)：203-209.

[122]徐平东，李梅，林奇英，谢联辉.1997.黄瓜花叶病毒两亚组分离物寄主反应和血清学性质比较研究.植物病理学报，27(4)：353-360.

[123]王宗华，鲁国东，谢联辉，单卫星，李振岐.1998.对植物病原真菌群体遗传研究范畴及其意义的认识.植物病理学报，28(1)：5-9.

[124]周仲驹，徐平东，陈启建，林奇英，谢联辉.1998.香蕉束顶病毒的寄主及其在病害流行中的作用.植物病理学报，28(1)：67-71.

[125]王宗华，王宝华，鲁国东，谢联辉.1998.稻瘟病菌侵入前发育的生物学及分子调控机制.植物病理学研究.北京：中国农业科技出版社，65-69.

[126]周仲驹，谢联辉，林奇英，王桦.1998.我国甘蔗病毒及类似病害的发生、诊断和防治对策.植物病理学研究.北京：中国农业科技出版社，40-43.

[127]徐平东，谢联辉.1998.黄瓜花叶病毒亚组研究进展.福建农业大学学报，27(1)：82-91

[128]陈启建，周仲驹，林奇英，谢联辉.1998.甘蔗花叶病毒的提纯及抗血清的制备.甘蔗，5(1)：19-21.

[129]徐平东，沈春奇，林奇英，谢联辉.1998.黄瓜花叶病毒亚组I和II分离物的形态和理化性质研究.植物病害研究与防治，北京：中国农业科技出版社，201-203.

［130］周叶方，胡方平，叶谊，谢联辉．1998. 水稻细菌性条斑菌胞外产物的性状．福建农业大学学报，27（2）：185-190.

［131］胡方平，方敦煌，Young John，谢联辉．1998. 中国猕猴桃细菌性花腐病菌的鉴定．植物病理学报，28（2）：175-181.

［132］李尉民，Roger Hull，张成良，谢联辉．1998. 南方菜豆花叶病毒（SBMV）两典型株系特异cDNA 和 RNA 探针的制备及应用．植物病理学报，28（3）：243-248.

［133］林奇田，林含新，吴祖建，林奇英，谢联辉．1998. 水稻条纹病毒外壳蛋白和病害特异蛋白在寄主体内的积累．福建农业大学学报，27（3）：322-326.

［134］王宗华，鲁国东，赵志颖，王宝华，张学博，谢联辉，王艳丽，袁筱萍，沈瑛．1998. 福建稻瘟病菌群体遗传结构及其变异规律．中国农业科学，31（5）：7-12.

［135］鲁国东，王宗华，郑学勤，谢联辉．1998. cDNA 文库和 PCR 技术相结合的方法克隆目的基因．农业生物技术学报，6（3）：257-262.

［136］鲁国东，郑学勤，陈守才，谢联辉．1998. 稻瘟病菌 3- 磷酸甘油醛脱氢酶基因（gpd）的克隆及序列分析．热带作物学报，19（4）：83-89.

［137］林丽明，吴祖建，谢荔岩，林奇英，谢联辉．1998. 水稻草矮病毒与品种抗性的互作．福建农业大学学报，27（4）：444-448.

［138］Lu GD, Wang ZH, Xie LH, Zheng XQ. 1998. Rapid cloning full-length cDNA of the glyceraldehyde-3-phosphate dehydrogenase gene（gpd）from *Magnaporthe grisea*. Journal of Zhejiang Agricultural University 24（5）：468-474.

［139］Wang ZH, Lu GD, Wang BH, Zhao ZY, Xie LH, SY, Zhu LH. 1998. The homology and genetic lineage relationship of *Magnaporthe grisea* defined by POR6 and MGR586. Journal of Zhejiang Agricultural University 24（5）：481-486.

［140］王海河，谢联辉．1999. 植物病毒 RNA 间重组的研究现状．福建农业大学学报，28（1）：47-53.

［141］王海河，周仲驹，谢联辉．1999. 花椰菜花叶病毒（CaMV）的基因表达调控．微生物学杂志，19（1）：34-40.

［142］方敦煌，胡方平，谢联辉．1999. 福建省建宁县中华猕猴桃细菌性花腐病的初步调查研究．福建农业大学学报，28（1）：54-58.

［143］张春嵋，吴祖建，林奇英，谢联辉．1999. 植物抗病基因的研究进展．生物技术通报，15（1）：22-27.

［144］张春嵋，王建生，邵碧英，吴祖建，林奇英，谢联辉．1999. 鳖病原细菌的分离鉴定及胞外产物的初步分析．福建农业大学学报，28（1）：90-95.

［145］徐平东，李梅，林奇英，谢联辉．1999. 侵染西番莲属（Passiflora）植物的五个黄瓜花叶病毒分离物的特性比较．中国病毒学，14（1）：73-79.

［146］刘利华，林奇英，谢华安，谢联辉．1999. 病程相关蛋白与植物抗病性．福建农业学报，2（4）：53-56.

［147］徐平东，周仲驹，林奇英，谢联辉．1999. 黄瓜花叶病毒亚组 I 和 II 分离物外壳蛋白基因的序列分析与比较．病毒学报，15（2）：164-171.

［148］林丽明，吴祖建，谢荔岩，林奇英，谢联辉．1999. 水稻草矮病毒特异蛋白抗血清的制备及其应用．植物病理学报，29（2）：126-131.

［149］张春嵋，吴祖建，林奇英，谢联辉．1999. 纤细病毒属病毒的分子生物学研究进展．福建农业大学学报，28（4）：445-451.

[150] Lin HX, Lin QT, Wu ZJ, Lin QY and Xie LH. 1999. Purification and serology of disease- specific protein of rice stripe virus. Virologica Sinica 14（3）: 222-229.

[151] Lin HX, Wei TY, Wu ZJ, Lin QY and Xie LH Lin QY, Xie LH. 1999. Molecular variability in coat protein and disease-specific protein genes among seven isolates of *Rice stripe virus* in China. Abstracts for the XIth international Congress of Virology. 1999. 8. Australia. Sydney：235-236.

[152] Lin HX, Lin QT, Wu ZJ, Lin QY and Xie LH. 1999. Characterization of proteins and nucleic acid of *Rice stripe virus*. Virologca sinica 14（4）: 333-352.

[153] 郑耀通，林奇英，谢联辉. 2000. 水体环境的植物病毒及其生态效应. 中国病毒学，15（1）: 1-7.

[154] 魏太云，林含新，吴祖建，林奇英，谢联辉. 2000. 应用 PCR-RFLP 及 PCR-SSCP 技术研究我国水稻条纹病毒 RNA4 基因间隔区的变异. 农业生物技术学报，8（1）: 41-44.

[155] Guo YH, Lin SF, Yang XQ, Xie LH. 2000. 对虾白斑病的流行病学. 中山大学学报（自然科学版），39（增刊）: 190-194. 郭银汉，林诗发，杨小强，张诚，谢联辉. 2000. 对虾病毒性白斑病的流行病学. 中山大学学报（自然科学版），（S1）:190-194.

[156] 鲁国东，王宝华，赵志颖，郑学勤，谢联辉，王宗华. 2000. 福建稻瘟菌群体遗传多样性 RAPD 分析. 福建农业大学学报，29（1）: 54-59.

[157] 郭银汉，林诗发，杨小强，谢联辉. 2000. 福州地区对虾暴发性白斑病的病原鉴定. 福建农业大学学报，29（1）: 90-94.

[158] 魏太云，林含新，吴祖建，林奇英，谢联辉. 2000. 水稻条纹病毒两个分离物 RNA4 基因间隔区的序列比较. 中国病毒学，15（2）: 156-162.

[159] 张春嵋，吴祖建，林奇英，谢联辉. 2000. 水稻草矮病毒核衣壳蛋白基因克隆及在大肠杆菌中的表达. 中国病毒学，15（2）: 200-203.

[160] 王海河，蒋继宏，吴祖建，林奇英，谢联辉. 2000. 黄瓜花叶病毒 M 株系 RNA3 的变异分析及全长克隆的构建. 农业生物技术学报，8（2）: 180-185.

[161] 魏太云，林含新，吴祖建，林奇英，谢联辉. 2000. PCR-SSCP 技术在植物病毒学上的应用. 福建农业大学学报，29（2）: 181-186.

[162] 林尤剑，Rundell P A，谢联辉，Powell C A. 2000. 感染柑桔速衰病毒的墨西哥酸橙病株中病程相关蛋白的检测. 福建农业大学学报，29（2）: 187-192.

[163] 王宝华，鲁国东，张学博，谢联辉，王宗华，袁筱萍，沈英. 2000. 福建省稻瘟菌的育性及其交配型. 福建农业大学学报，29（2）: 193-196.

[164] 林尤剑，谢联辉，Rundell P A，Powell C A. 2000. 应用改进的多克隆抗体 Western blot 技术研究柑桔速衰病毒蛋白（英文）. 植物病理学报，30（3）: 250-256.

[165] 李利君，周仲驹，谢联辉. 2000. 利用斑点杂交法和 RT-PCR 技术检测甘蔗花叶病毒，福建农业大学学报，29（3）: 342-345.

[166] 郭银汉，林诗发，杨小强，张诚，谢联辉. 2000. 福州地区对虾白斑病病毒的超微结构. 中国病毒学，15（3）: 277-284.

[167] 于群，魏太云，林含新，吴祖建，林奇英，谢联辉. 2000. 水稻条纹病毒北京双桥（RSV-SQ）分离物 RNA4 片段序列分析. 农业生物技术学报，8（3）: 225-228.

[168] 刘利华，吴祖建，林奇英，谢联辉. 2000. 水稻条纹叶枯病细胞病理变化的观察. 植物病理学报，30（4）: 306-311.

[169] 张春嵋，林奇英，谢联辉. 2000. 水稻草矮病毒血清学和分子检测方法的比较. 中国病毒学，15（4）: 361-366.

[170] 林含新，林奇田，魏太云，吴祖建，林奇英，谢联辉. 2000. 水稻品种对水稻条纹病毒及其

介体灰飞虱的抗性鉴定 . 福建农业大学学报，29（4）：453-458.

［171］吴刚，吴祖建，谢联辉 . 2000. 水稻东格鲁病研究进展 . 福建农业大学学报，29（4）：459-464.

［172］Wang HH, Xie LH, Lin QY, Xu PD. 2000. Complete nucleotide sequences of RNA3 from *Cucumber mosaic virus*（CMV）isolates PE and XB and their transcription *in vitro*. The 1st Asian Conference on Plant Pathology. Beijing: China Agricultural Scientech Press, 104.

［173］Zhang CM, Lin QY, Xie LH. 2000. Construction of plant expression vector containing nucleocapsid protein genes of *Rice grassy stunt virus* and transformation of rice. The 1st Asian Conference on Plant Pathology. Beijing: China Agricultural Scientech Press, 124.

［174］Lin YJ, Rundell P A, Xie LH, Powell C A. 2000. In situ immunoassay for detection of *Citrus tristeza virus*. Plant Disease 84（9）：937-940

［175］Han SC, Wu ZJ, Yang HY, Wang R, Yie Y, Xie LH, Tien Po. 2000. Ribozyme-mediated resistance to *Rice dwarf virus* and the transgene silencing in the progeny of transgenic rice plants. Transgenic Research 9（2）：195-203.

［176］林含新，魏太云，吴祖建，林奇英，谢联辉 . 2001. 我国水稻条纹病毒一个强致病性分离物的 RNA4 序列测定与分析 . 微生物学报，41（1）：25-30.

［177］林含新，魏太云，吴祖建，林奇英，谢联辉 . 2001. 水稻条纹病毒外壳蛋白基因和病害特异性蛋白基因的克隆和序列分析 . 福建农业大学学报，30（1）：53-58.

［178］林尤剑，谢联辉，Powell C A. 2001. 橘蚜传播柑橘衰退病毒的研究进展 . 福建农业大学学报，30（1）：59-66.

［179］王海河，林奇英，谢联辉，吴祖建 . 2001. 黄瓜花叶病毒三个毒株对烟草细胞内防御酶系统及细胞膜通透性的影响 . 植物病理学报，31（1）：43-49.

［180］李利君，周仲驹，谢联辉 . 2001. 甘蔗花叶病毒 3′ 末端基因的克隆及外壳蛋白序列分析比较 . 中国病毒学，16（1）：45-50.

［181］张春嵋，谢荔岩，林奇英，谢联辉 . 2001. 水稻草矮病毒 *NS6* 基因在大肠杆菌中的表达及植物表达载体的构建 . 病毒学报，17（1）：90-92.

［182］孙慧，吴祖建，谢联辉，林奇英 . 2001. 杨树菇（*Agrocybe aegetita*）中一种抑制 TMV 侵染的蛋白质纯化及部分特征 . 生物化学与生物物理学报，33（3）：351- 354.

［183］林含新，魏太云，吴祖建，林奇英，谢联辉 . 2001. 应用 PCR-SSCP 技术快速检测我国水稻条纹病毒的分子变异 . 中国病毒学，16（2）：166-169.

［184］魏太云，林含新，吴祖建，林奇英，谢联辉 . 2001. 水稻条纹病毒 RNA4 基因间隔区的分子变异 . 病毒学报，17（2）：144-149, 203.

［185］王海河，谢联辉，林奇英 . 2001. 黄瓜花叶病毒西番莲分离物 RNA3 的 cDNA 全长克隆及序列分析 . 福建农业大学学报，30（2）：191-198

［186］谢联辉，魏太云，林含新，吴祖建，林奇英 . 2001. 水稻条纹病毒的分子生物学 . 福建业大学学报 , 30（3）:269-279.

［187］邵碧英 , 吴祖建，林奇英，谢联辉 . 2001. 烟草花叶病毒弱毒株的筛选及其交互保护作用 . 福建农业大学学报 , 30（3）:297-303.

［188］王海河 , 谢联辉 , 林奇英 . 2001. 黄瓜花叶病毒香蕉株系（CMV-Xb）RNA3 cDNA 的克隆和序列分析 . 中国病毒学 ,16（3）:217-221.

［189］王盛，吴祖建，林奇英，谢联辉 . 2001. 甘薯羽状斑驳病毒研究进展 . 福建农业大学学报，30（增刊）：2-9.

［190］林芩，吴祖建，林奇英，谢联辉 . 2001. 甘薯脱毒研究进展 . 福建农业大学学报，30（增刊）：10-14.

［191］邵碧英，吴祖建，林奇英，谢联辉 . 2001. 烟草花叶病毒弱毒株的致弱机理及交互保护作用机理 . 福建农业大学学报（增刊），30：19-28.

［192］郑杰，吴祖建，周仲驹，林奇英，谢联辉 . 2001. 香蕉束顶病毒研究进展 . 福建农业大学学报，30（增刊）：32-38.

［193］张铮，吴祖建，谢联辉 . 2001. 植物细胞程序性死亡 . 福建农业大学学报，30（增刊）：45-53.

［194］林芩，吴祖建，林奇英，谢联辉 . 2001. 甘薯分生组织培养配方的筛选 . 福建农业大学学报，30（增刊）：81-83.

［195］明艳林，吴祖建，谢联辉 . 2001. 水稻条纹病毒 CP、SP 进入叶绿体与褪绿症状的关系 . 福建农业大学学报，30（增刊）：147（简报）.

［196］魏太云，林含新，吴祖建，林奇英，谢联辉 . 2001. 寄主植物与昆虫介体中水稻条纹病毒的检测 . 福建农业大学学报，30（增刊）：165-170.

［197］吴兴泉，吴祖建，谢联辉，林奇英 . 2001. 核酸斑点杂交检测马铃薯 X 病毒 . 福建农业大学学报，30（增刊）：191-193.

［198］林毅，吴祖建，林奇英，谢联辉 . 2001. 核糖体失活蛋白及其对植物病毒病的控制 . 福建农业大学学报，30（增刊）：222-227.

［199］欧阳迪莎，谢联辉，施祖美，吴祖建 . 2001. 植物病害与持续农业 . 福建农业大学学报（社会科学版），4（增刊）：5-9.

［200］张春嵋，吴祖建，林丽明，林奇英，谢联辉 . 2001. 水稻草状矮化病毒沙县分离株基因组第六片断的序列分析 . 植物病理学报，31（4）：301-305.

［201］魏太云，林含新，谢联辉 . 2002. PCR-SSCP 分析条件的优化 . 福建农业大学学报（自然科学版），31（1）：22-25.

［202］沈建国，翟梅枝，林奇英，谢联辉 . 2002. 我国植物源农药研究进展 . 福建农业大学学报（自然科学版），31（1）. 26-31.

［203］付鸣佳，吴祖建，林奇英，谢联辉 . 2002. 美洲商陆抗病毒蛋白研究进展 . 生物技术通讯，13（1）：66-71.

［204］林琳，何志勇，杨冠珍，谢联辉，吴松刚，吴祥甫 . 2002. 人胎盘 TRAIL 基因的克隆和在大肠杆菌中的表达 . 药物生物技术，9（1）：12-15.

［205］邵碧英，吴祖建，林奇英，谢联辉 . 2002. 烟草花叶病毒强、弱毒株对烟草植株的影响 . 中国烟草科学，（1）：43-46.

［206］林琳，谢必峰，杨冠珍，施巧琴，林奇英，谢联辉，吴松刚，吴祥甫 . 2002. 扩展青霉 PF898 碱性脂肪 cDNA 的克隆及序列分析 . 中国生物化学与分子生物学报，18（1）：32-37.

［207］徐学荣，吴祖建，林奇英，谢联辉，2002，可持续植物保护及其徐学荣，吴祖建，林奇英，谢联辉，2002，可持续植物保护及其综合评价 . 农业现代化研究，（04）:314-317.

［208］徐学荣，吴祖建，林奇英，谢联辉，2002，绿色食品生产和消费的数量经济分析 . 经济数学，（03）:65-69.

［209］徐学荣，吴祖建，林奇英，谢联辉，2002，绿色食品需求函数的一种求解方法 . 运筹与管理，（05）:14-18.

［210］徐学荣，李宏宇，林奇英，谢联辉，2002，作物混合种植布局模型研究 . 农业技术经济，（02）：8-10.

［211］翟梅枝，沈建国，林奇英，谢联辉 . 2002. 中药生物碱成分的毛细管电泳分析 . 西北林学院学

报，17（1）：55-59.

［212］林含新，魏太云，吴祖建，林奇英，谢联辉 . 2002. 我国水稻条纹病毒 7 个分离物的致病性和化学特性比较 . 福建农林大学学报（自然科学版），31（2）：164-167.

［213］孙慧，吴祖建，林奇英，谢联辉 . 2002 小分子植物病毒抑制物质研究进展 . 福建农林大学学报（自然科学版），31（3）：311-316.

［214］吴兴泉，吴祖建，谢联辉，林奇英 . 2002. 马铃薯 S 病毒外壳蛋白基因的克隆与原核表达 . 中国病毒学，17（3）：248-251.

［215］付鸣佳，吴祖建，林奇英，谢联辉 . 2002. 榆黄蘑中一种抗病毒蛋白的纯化及其抗 TMV 和 HBV 的活性 . 中国病毒学，17（4）：350-353.

［216］王盛，吴祖建，林奇英，谢联辉，2002. 珊瑚藻藻红蛋白分离纯化技术及光谱学特性，福建农林大学学报（自然科学版），31（4）：495-499.

［217］付鸣佳，林健清，吴祖建，林奇英，谢联辉 . 2003. 杏鲍菇抗烟草花叶病毒蛋白的筛选 . 微生物学报，43（1）：29-34.

［218］魏太云，林含新，谢联辉 . 2003. 酵母双杂交系统在植物病毒学上的应用 . 福建农林大学学报（自然科学版），32（1）：50-54.

［219］徐学荣，吴祖建，林奇英，谢联辉 . 2003. 不同类型土壤作物混合种植布局优化模型 . 农业系统科学与综合研究，19（1）：63-65.

［220］魏太云，王辉，林含新，吴祖建，林奇英，谢联辉 . 2003. 我国水稻条纹病毒 RNA3 片断序列分析 —— 纤细病毒属重配的又一证据 . 生物化学与生物物理学报，35（1）：97-102.

［221］刘国坤，谢联辉，林奇英，吴祖建 . 2003. 介体线虫传播植物病毒专化性的研究进展 . 福建农林大学学报（自然科学版），32（1）：55-61.

［222］付鸣佳，吴祖建，林奇英，谢联辉，2003，金针菇中一种抗病毒蛋白的纯化及其抗烟草花叶病毒特性 . 福建农林大学学报（自然科学版），32（1）:84-88.

［223］林丽明，吴祖建，林奇英，谢联辉 . 2003. 水稻草矮病毒基因组 vRNA3NS3 基因的克隆、序列分析 . 农业生物技术学报，11（2）：187-191.

［224］徐学荣，吴祖建，张巨勇，谢联辉 . 2003. 可持续发展通道及预警研究 . 数学的实践与认识，33（2）：31-35.

［225］刘国坤，谢联辉，林奇英，吴祖建，陈启建 . 2003. 15 种植物的单宁提取物对烟草花叶病毒（TMV）的抑制作用 . 植物病理学报，33（3）：279-283.

［226］刘国坤，吴祖建，谢联辉，林奇英，陈启建 . 2003. 植物单宁对烟草花叶病毒的抑制活性 . 福建农林大学学报（自然科学版），32（3），292-295.

［227］陈启建，刘国坤，吴祖建，林奇英，谢联辉 . 2003. 三叶鬼针草中黄酮甙对烟草花叶病毒的抑制作用 . 福建农林大学学报（自然科学版），32（2）：191-184.

［228］翟梅枝，李晓明，林奇英，谢联辉 . 2003. 核桃叶抑菌成分的提取及其抑菌活性 . 西北林学院学报，18（4）：89-91.

［229］魏太云，林含新，吴祖建，林奇英，谢联辉 . 2003. 我国水稻条纹病毒种群遗传结构初步分析 . 植物病理学报，33（3）：284-285.

［230］魏太云，林含新，吴祖建，林奇英，谢联辉 . 2003. 水稻条纹病毒 NS2 基因遗传多样性分析 . 中国生物化学与分子生物学报，19（5）：600-605.

［231］魏太云，林含新，吴祖建，林奇英，谢联辉 . 2003. 水稻条纹病毒 RNA4 基因间隔区序列分析 —— 混合侵染及基因组重组证据 . 微生物学报，43（5）：577-585.

［232］翟梅枝，杨秀萍，林奇英，谢联辉，刘路 . 2003. 核桃叶提取物对杨毒蛾生物活性的研究 . 西

北林学院学报，18（2）：65-67.

［233］顾晓军 . 谢联辉 . 2003. 21 世纪我国农药发展的若干思考 . 世界科技研究与发展 . 25（2）：13-20.

［234］徐学荣，俞明，蔡艺，谢联辉 . 2003. 福建生态省建设的评价指标体系初探 . 农业系统科学与综合研究，19（2）：89-92.

［235］徐学荣，林奇英，施祖美，谢联辉 . 2003 科学组织农药使用 确保生态环境安全 . 农业现代化研究，24（增刊）：127-129.

［236］林琳，施巧琴，郭小玲，吴松刚，吴祥甫，谢联辉 . 2003. 扩展青霉碱性脂肪酶的纯化及 N-端氨基酸序列分析 . 厦门大学学报（自然科学版），30（5）：600-604.

［237］邵碧英，吴祖建，林奇英，谢联辉 . 2003. 烟草花叶病毒及其弱毒株基因组的 cDNA 克隆和序列分析 . 植物病理学报，33（4）：296-301.

［238］林丽明，张春嵋，谢荔岩，吴祖建，谢联辉 . 2003. 农杆菌介导的水稻草矮病毒 *NS6* 基因的转化，福建农林大学学报（自然科学版），32（3）288-291.

［239］邵碧英，吴祖建，林奇英，谢联辉 . 2003. 烟草花叶病毒复制酶介导抗性的研究进展 . 生物技术通讯，14（5）：416-418.

［240］徐学荣，姜培红，林奇英，施祖美，谢联辉 . 2003. 整合农药企业与资源利用效率问题的博弈分析 . 运筹与管理，12（5）：81-84.

［241］谢联辉 . 2003. 21 世纪我国植物保护问题的若干思考 . 中国农业科技导报，27（5）：5-7.

［242］刘振宇，林奇英，谢联辉 . 2003. 环境相容性农药发展的必然性和可能途径 . 世界科技研究与发展，（5）：11-16.

［243］林毅，陈国强，吴祖建，林奇英，谢联辉 . 2003. 绞股蓝抗 TMV 蛋白的分离及编码基因的序列分析 . 农业生物技术学报，11（4）：365-369.

［244］林毅，林奇英，谢联辉 . 2003. 绞股蓝核糖体失活蛋白的分离、克隆与表达 . 分子植物育种，1（5/6）：759-761.

［245］林毅，吴祖建，谢联辉，林奇英 . 2003. 抗病虫基因新资源：绞股蓝核糖体失活蛋白基因 . 分子植物育种，1（5/6）：763-765.

［246］欧阳迪莎，施祖美，吴祖建，林卿，徐学荣，谢联辉 . 2003. 植物病害与粮食安全 . 农业环境与发展，6：24-26.

［247］林毅，陈国强，吴祖建，谢联辉，林奇英 . 2003. 利用核糖体失活蛋白控制植物病虫害 . 云南农业大学学报，18（4）：52-56.

［248］林毅，陈国强，吴祖建，谢联辉，林奇英 . 2003. 绞股蓝核糖体失活蛋白的信号肽和上游非编码区 . 云南农业大学学报，18（4）：63-66.

［249］魏太云，林含新，谢联辉 . 2003. 植物病毒分子群体遗传学研究进展 . 福建农林大学学报（自然科学版），32（4）：453-457.

［250］魏太云，林含新，吴祖建，林奇英，谢联辉，2003，水稻条纹病毒中国分离物和日本分离物 RNA2 节段序列比较（英文）. 中国病毒学，（04）:73-78.

［251］魏太云，王辉，林含新，吴祖建，林奇英，谢联辉 . 2003. 我国水稻条纹病毒 RNA3 片段序列分析 —— 纤细病毒属重配的又一证据 . 生物化学与生物物理学报 ,（01）:97-103. 35（1）,97-103.

［252］吴丽萍，吴祖建，林奇英，谢联辉 . 2003. 毛头鬼伞（*Coprinus comatus*）中一种碱性蛋白的纯化及其活性 . 微生物学报，43（6）：793-798.

［253］何红，蔡学清，关雄，胡方平，谢联辉，2003，内生菌 BS-2 菌株的抗菌蛋白及其防病作用 . 植物病理学报，33（4）:373-378.

［254］吴兴泉，陈士华，吴祖建，林奇英，谢联辉，2003，马铃薯 X 病毒 CP 基因的原核表达及特

异性抗血清的制备 . 郑州工程学院学报，(02) :25-28.

[255]吴兴泉，陈士华，吴祖建，林奇英，谢联辉，2003，分子生物学技术在马铃薯病毒检测中的应用 . 中国马铃薯，17(3) :175-179.

[256]吴兴泉，陈士华，吴祖建，林奇英，谢联辉,2003，马铃薯Y病毒Pl基因的克隆与序列分析 . 中国病毒学，(04) :68-72.

[257]吴兴泉，陈士华，吴祖建，林奇英，谢联辉 . 2003 马铃薯A病毒CP基因的克隆与序列分析 . 植物保护，29(5)：25-28.

[258]孙慧，吴祖建，林奇英，谢联辉 . 2003. 小分子植物病毒抑制物质研究进展 . 福建农林大学学报(自然科学版)，32(3)：311-316.

[259]姜培红，徐学荣，谢联辉 . 2003. 县级植保站绩效综合评价 . 福建农林大学学报(哲学社会科学版)，6(增刊)：85-88.

[260]程兆榜，杨荣明，周益军，范永坚，谢联辉 . 2003. 关于水稻条纹叶枯病防治策略的思考 . 江苏农业科学，(增刊)：3-5.

[261]魏太云，林含新，吴祖建，林奇英，谢联辉 . 2003. Comparison of the RNA2 Segments Between Chinese Isolates and Japanese Isolates of Rice Stripe Virus. 中国病毒学，18(4)：381-386.

[262]王盛，钟伏弟，吴祖建，谢联辉，林奇英 . 2004. 抗病虫基因新资源：海洋绿藻孔石莼凝集素基因 . 分子植物育种，2(1)：153-155.

[263]王盛，钟伏弟，吴祖建，林奇英，谢联辉 . 2004. 一种新的藻红蛋白的亚基组成分析 . 福建农林大学学报(自然科学版)，33(1)：68-71.

[264]刘振宇，吴祖建，林奇英，谢联辉 . 2004. 羊栖菜多酚氧化酶特性 . 福建农林大学学报(自然科学版)，33(1)：56-59.

[265]林丽明，吴祖建，林奇英，谢联辉 . 2004. 农杆菌介导获得转水稻草矮病毒 *NS3* 基因水稻植株 . 福建农林大学学报(自然科学版)，33(1)：60-63.

[266]林毅，陈国强，吴祖建，林奇英，谢联辉 . 2004. 快速获得葫芦科核糖体失活蛋白新基因 . 农业生物技术学报，12(1)：8-12.

[267]陈宁，吴祖建，林奇英，谢联辉 . 2004. 灰树花中一种抗烟草花叶病毒蛋白质的纯化及其性质 . 生物化学与生物物理进展，31(3)：283-286.

[268]魏太云，林含新，吴祖建，林奇英，谢联辉 . 2004. 水稻条纹病毒中国分离物和日本分离物RNA1片断序列比较 . 植物病理学报，34(2)：141-145.

[269]金凤媚，林丽明，吴祖建，林奇英 . 2004. 转RGSV-SP基因水稻植株的再生 . 中国病毒学，19(2)：146-148.

[270]徐学荣，林奇英，谢联辉 . 2004. 绿色食品生产经营中的风险及其管理 . 农业系统科学与综合研究，20(2)：103-106.

[271]徐学荣，欧阳迪莎，林奇英，施祖美，谢联辉 . 2004. 农产品的价格和需求对无公害植保技术使用的影响 . 农业系统科学与综合研究，20(1)：16-19.

[272]王盛，钟伏弟，吴祖建，林奇英，谢联辉 . 2004. R-藻红蛋白免疫荧光探针标记方法的探索 . 福建农林大学学报(自然科学版)，33(2)：206-209.

[273]欧阳迪莎，徐学荣，林卿，谢联辉 . 2004. 优化有害生物管理，提升我国农产品竞争力 . 中国农业科技导报，6(3)：54-56.

[274]魏太云，林含新，吴祖建，林奇英，谢联辉 . 2004. 中国水稻条纹病毒两个亚种群代表性分离物全基因组核苷酸序列分析 . 中国农业科学，37(6)：846-850.

[275]林丽明，吴祖建，金凤媚，谢荔岩，谢联辉 . 2004. 水稻草矮病毒在水稻原生质体中的表达 .

微生物学报，44（4）：530-532.

［276］林丽明，谢联辉，林奇英，2004，水稻草状矮化病毒基因组 RNA1-3 的分子生物学．分子植物育种，2（3）:449-450.

［277］欧阳迪莎，徐学荣，林卿，谢联辉．2004. 农作物有害生物化学防治的外部性思考．农业现代化研究，25（增刊）：78-80.

［278］王盛，钟伏弟，吴祖建，林奇英，谢联辉．2004. 珊瑚藻 R- 藻红蛋白 *repA* 和 *repB* 基因全长 cDNA 克隆与序列分析．中国生物化学与分子生物学报，20（4）：428-433.

［279］郑耀通，林奇英，谢联辉．2004. TMV 在不同水体与温度条件下的灭活动力学．中国病毒学，19（4）：315-319.

［280］郑耀通，林奇英，谢联辉．2004. PV1、B. fp 在不同水样及温度条件下的灭活动力学．应用与环境生物学报．10（6）：794-797.

［281］耀通，林奇英，谢联辉．2004. 天然砂与修饰砂对病毒的吸附与去除．中国病毒学，19（2）：163-167.

［282］郑耀通，林奇英，谢联辉．2004. 闽江流域福州过境段水体病毒污染调查分析．中国环境监测，（05）:39-43.

［283］郑耀通，林奇英，谢联辉．2004. 氯与金属离子协同杀灭水中微生物的效果观察．中国消毒学杂志，（03）:34-37.

［284］郑耀通，林奇英，谢联辉．2004. 闽江流域福州区段肠道病毒污染程度的监测与预测．安全与环境学报，（05）:24-28.

［285］郑耀通，林奇英，谢联辉．2004. 废水活性污泥处理过程去除 TMV 效果研究．环境科学学报，（04）:625-632.

［286］郑耀通，林奇英，谢联辉．2004. 污水稳定塘菌 - 藻生态系统去除与灭活植物病毒 TMV 研究．环境科学学报，（06）:1128-1134.

［287］郑耀通，林奇英，谢联辉．2004. 水体病毒浓缩条件的优化．中国病毒学，（01）:63-67.

［288］祝雯，林志铿，吴祖建，林奇英，谢联辉．2004. 河蚬中活性蛋白 CFp-a 的分离纯化及其活性．中国水产科学，11（4）：349-353.

［289］祝雯，林志铿，吴祖建，林奇英，谢联辉，2004，河蚬糖蛋白对人肝癌细胞凋亡的影响．中国公共卫生，（06）:40-41.

［290］林毅，吴祖建，谢联辉，林奇英．2004. 绞股蓝 RIP 基因双子叶植物表达载体的构建及其对烟草叶盘的转化．江西农业大学学报，26（4）：589-592.

［291］林毅，陈国强，吴祖建，谢联辉，林奇英．2004. C- 末端缺失和完整的绞股蓝核糖体失活蛋白在大肠杆菌中的表达．江西农业大学学报，26（4）：593-595.

［292］翟梅枝，高芳銮，沈建国，林奇英，谢联辉．2004. 抗 TMV 的植物筛选及提取条件对抗病毒物质活性的影响．西北农林科技大学学报，32（7）：45-49.

［293］刘国坤，陈启建，吴祖建，林奇英，谢联辉．2004. 13 种植物提取物对烟草花叶病毒的活性．福建农林大学学报（自然科学版），33（3）：295-299.

［294］刘国坤，陈启建，吴祖建，林奇英，谢联辉，2004，几种植物提取物对 4 种植物病原真菌的抑制作用．福建农林大学学报，33（3）:295-299.

［295］陈启建，刘国坤，吴祖建，谢联辉，林奇英．2004. 26 种植物提取物抗烟草花叶病毒的活性．福建农林大学学报（自然科学版），33（3）：300-303.

［296］吴兴泉，陈士华，魏广彪，吴祖建，谢联辉．2004. 福建马铃薯 A 病毒的分子鉴定及检测技术．农业生物技术学报，12（1）：90-95.

［297］周莉娟，郑伟文，谢联辉．2004. *gfp/luxAB* 双标记载体在抗线虫菌株 BC2000 中的转化及表达检测．农业生物技术学报，12（5）：573-577.

［298］吴丽萍，吴祖建，林奇英，谢联辉，2004，一种食用菌提取物 y3 对烟草花叶病毒的钝化作用及其机制．中国病毒学，19（1）：54-57.

［299］王盛，钟伏弟，吴祖建，林奇英，谢联辉．2004. 珊瑚藻藻红蛋白 α 亚基脱辅基蛋白基因克隆与序列分析．农业生物技术学报，12（6）：733-734.

［300］沈建国，谢荔岩，翟梅枝，林奇英，谢联辉．2004. 杨梅叶提取物抗烟草花叶病毒活性及其化学成分初步研究．福建农林大学学报（自然科学版），33（4）：441-443.

［301］Wang S, Zhong FD, Zhang YJ, Wu ZJ, Lin QY, Xie LH. 2004. Molecular characterization of a new lectin from the marine alga *ulva pertusa*. Acta Biochimica et Biophysica Sincia 36（2）:111-117.

［302］Chen N, Wu ZJ, Lin QY, Xie LH. 2004. Purification and partial characterization of a protein inhibitor of tobacco mosaic virus infection from the maitake（*Grifola frondosa*）. Progress in Biochemistry and Biophysics 31（3）,283-286.

［303］Wang S, Zhong FD, Wu ZJ, Lin QY, Xie LH. 2004. Cloning and Sequeccing the γSubunit of R-Phycoerythrin from *Corallina officinalis*. Acta Batanica Sinica, 46（10）: 1135-1140.

［304］Lin LM, Wu ZJ, Xie LH, Lin QY. 2004. Gene cloning and expression of the *NS3* gene of *Rice grassy stunt virus* and its antiserum preparation. Chinese Journal of Agriculture Biotechnology 1（1）:49-54.

［305］杨小山，欧阳迪莎，徐学荣，吴祖建，金德凌．2005. 可持续植保对消除绿色壁垒的可行性分析及对策．福建农林大学学报（社会哲学版），8（1）：65-68.

［306］沈建国，谢荔岩，张正坤，谢联辉，林奇英．2005. 一种植物提取物对 CMV、PVY^N 及其昆虫介体的作用．中国农学通报，21（5）：341-343.

［307］付鸣佳，谢荔岩，吴祖建，林奇英，谢联辉．2005. 抗病毒蛋白抑制植物病毒的应用前景．生命科学研究，9（1）：1-5.

［308］林毅，谢荔岩，陈国强，吴祖建，谢联辉，林奇英，2005. 绞股蓝核糖体失活蛋白家族编码基因的 5 个 cDNA 及其下游非编码区．植物学通报，22（2）：163-168.

［309］付鸣佳，吴祖建，林奇英，谢联辉．2005. 金针菇中蛋白质含量的变化和其中一个蛋白质的生物活性．应用与环境生物学报，11（1）：40-44.

［310］刘振宇，谢荔岩，吴祖建，林奇英，谢联辉．2005. 孔石莼质体蓝素氨基酸序列分析和分子进化．分子植物育种，3（2）：203-208.

［311］刘振宇，谢荔岩，吴祖建，林奇英，谢联辉．2005. 孔石莼（*Ulva pertusa*）中一种抗 TMV 活性蛋白的纯化及其特性．植物病理学报，35（3）：256-261.

［312］陈启建，刘国坤，吴祖建，谢联辉，林奇英．2005. 大蒜精油对烟草花叶病毒的抑制作用．福建农林大学学报（自然科学版），34（1）：30-33.

［313］连玲丽，吴祖建，段永平，谢联辉．2005. 线虫寄生菌巴斯德杆菌的生物多样性研究进展．福建农林大学学报（自然科学版），34（1）：37-42.

［314］欧阳迪莎，何敦春，王庆，林卿，谢联辉．2005. 农业保险与可持续植保．福建农林大学学报（哲学社会科学版），8（2）：26-29.

［315］林白雪，黄志强，谢联辉．2005. 海洋细菌活性物质的研究进展．微生物学报，45（4）：657-660.

［316］吴兴泉，陈士华，魏广彪，吴祖建，谢联辉．2005. 福建马铃薯 S 病毒的分子鉴定及发生情况．植物保护学报，32（2）：133-137.

［317］李凡，杨金广，谭冠林，吴祖建，林奇英，陈海如，谢联辉．2005. 云南水稻条纹病毒病害

特异性蛋白基因及基因间隔趋序列分析 . 中国食用菌，24（增刊）：14-18.

［318］王盛，钟伏弟，吴祖建，林奇英，谢联辉 . 2005. 珊瑚藻藻红蛋白 β 亚基脱辅基蛋白基因克隆与序列分析 . 福建农林大学学报（自然科学版）. 34（3）：334-338.

［319］欧阳迪莎，何敦春，杨小山，林卿，谢联辉 . 2005. 植物病害管理中的政府行为 . 中国农业科技导报，7（3）：38-41.

［320］范国成，吴祖建，黄明年，练君，梁栋，林奇英，谢联辉 . 2005. 水稻瘤矮病毒基因组 S9 片断的基因结构特征 . 中国病毒学，20（5）：539-542.

［321］谢联辉，林奇英，徐学荣 . 2005. 植病经济与病害生态治理 . 中国农业大学学报，10（4）：39-42.

［322］范国成，吴祖建，林奇英，谢联辉 . 2005. 水稻瘤矮病毒基因组 S8 片断全序列测定及其结构分析 . 农业生物技术学报，13（5）：679-683.

［323］陈来，吴祖建，傅国胜，林奇英，谢联辉 . 2005. 灰飞虱胚胎组织细胞的分离和原代培养技术 . 昆虫学报，48（3）：455-459.

［324］张居念，林河通，谢联辉，林奇英 . 2005. 龙眼焦腐病菌及其生物学特性 . 福建农林大学学报（自然科学版），34（4）：425-429.

［325］周丽娟，郑伟文，谢联辉 . 2005. 线虫拮抗菌 BC2000 的分子鉴定及其 GFP 标记菌的生物学特性 . 福建农林大学学报（自然科学版），34（4）：430-433.

［326］李凡，杨金广，吴祖建，林奇英，陈海如，谢联辉 . 2005. 水稻条纹病毒病害特异性蛋白基因克隆及其与纤细病毒属成员的亲缘关系分析 . 植物病理学报，35（增刊）：135-136.

［327］林娇芬，林河通，谢联辉，林奇英，陈绍军，赵云峰 . 2005. 柿叶的化学成分、药理作用、临床应用及开发利用 . 食品与发酵工业，31（7）：90-96.

［328］章松柏，吴祖建，段永平，谢联辉，林奇英 . 2005. 单引物法同时克隆 RDV 基因组片段 S11、S12 及其序列分析 . 贵州农业科学，33（ 6）：27-29.

［329］章松柏，吴祖建，段永平，谢联辉，林奇英 . 2005. 水稻矮缩病毒的检测和介体传毒能力初步分析 . 安徽农业科学，33（12）：2263-2264,2287.

［330］章松柏，吴祖建，段永平，谢联辉. 林奇英. 2005. 一种实用的双链 RNA 病毒基因组克隆方法. 长江大学学报：自然科学版，2（2）：71-73.

［331］张晓婷，吴祖建，林奇英，谢联辉 . 2005. 双链 RNA 技术与植物病毒研究 . 云南农业大学学报，（04）：455-458.

［332］王林萍，林金科，庄佩芬，林奇英，谢联辉 . 2005. 常规茶与有机茶比较的经济分析 . 福建农林大学学报（哲学社会科学版，（03）：60-62.

［333］王林萍，林奇英，谢联辉 . 2005. 农药企业的社会责任探析 . 中国农业科技导报，7（6）:56-60.

［334］徐学荣，王林萍，谢联辉 . 2005. 农户植保行为及其影响因素的分析方法 . 乡镇经济，（12）:50-53.

［335］沈建国，谢荔岩，吴祖建，谢联辉，林奇英 . 2006. 药用植物提取物抗烟草花叶病毒活性的研究 . 中草药，37（2）：259-261.

［336］刘振宇，吴祖建，林奇英，谢联辉 . 2006. 孔石莼质体蓝素的柱色谱纯化及其对其 N- 端氨基酸序列的分析测定 . 色谱，24（3）：275-278.

［337］刘振宇，谢荔岩，吴祖建，林奇英，谢联辉 . 2006. 海藻蛋白质提取物对香蕉炭疽病的抑制作用 . 福建农林大学学报（自然科学版），35（1）：21-23.

［338］刘国坤，陈启建，吴祖建，林奇英，谢联辉 . 2006. 丹皮酚对烟草花叶病毒的抑制作用 . 福建农林大学学报（自然科学版），35（1）：17-20.

［339］何敦春，王林萍，欧阳迪莎，谢联辉 . 2006. 休闲经济与海峡西岸经济区建设 . 福建农林大学

学报（社会科学版），9（1）：21-25.

［340］李凡，杨金广，吴祖建，林奇英，陈海如，谢联辉. 2006. 水稻条纹病毒云南分离物CP基因克隆及序列比较分析. 云南农业大学学报，21（1）：48-51.

［341］沈硕，谢荔岩，林奇英，谢联辉. 2006. 组织培养技术在植物病理方面的应用研究进展. 中国农学通报，22（增刊）：150-155.

［342］杨彩霞，贾素平，刘舟，林奇英，谢联辉，吴祖建. 2006. 从福建省杂草赛葵上检测到粉虱传双生病毒. 中国农学通报，22（增刊）：156-159.

［343］陈启建，刘国坤，吴祖建，谢联辉，林奇英. 2006. 大蒜挥发油抗病毒花叶病毒机理. 福建农业学报，21（1）：24-27.

［344］李凡，林奇英，陈海如，谢联辉. 2006. 幽影病毒属病毒的研究现状与展望. 微生物学报，46（6）：1033-1037.

［345］刘伟，谢联辉. 2006. 芽孢杆菌对感染蔓割病甘薯活性氧代谢的效应. 福建农林大学学报（自然科学版），35（6）：569-572.

［346］李凡，林奇英，陈海如，谢联辉. 2006. 幽影病毒引起的几种主要植物病害. 微生物学通报，33（3）：151-556.

［347］吴兴泉，谭晓荣，陈士华，谢联辉. 2006. 马铃薯卷叶病毒福建分离物的基因克隆与序列分析. 河南农业大学学报，40（4）：391-393.

［348］吴兴泉，陈士华，谢联辉. 2006. 马铃薯X病毒的分子鉴定与检测技术. 河南农业科学，（02）：72-75.

［349］张居念，林河通，谢联辉，林奇英，王宗华. 2006. 龙眼果实潜伏性病原真菌的初步研究. 热带作物学报，27（4）：78-82.

［350］张福山，徐学荣，林奇英，谢联辉. 2006. 植物保护对粮食安全的影响分析. 中国农学通报，22（12）：505-510.

［351］王林萍，林奇英，谢联辉. 2006. 化工企业的社会责任探讨. 商业时代，28：84-86.

［352］侯长红，林光美，施祖美，谢联辉. 2006. 植病经济的内涵与研究方法评述. 福建农林大学学报（哲学社会科学版），9（4）：33-36.

［353］李凌绪，翟梅枝，林奇英，谢联辉. 2006. 海藻乙醇提取物抗真菌活性. 福建农林大学学报（自然科学版），35（4）：342-345.

［354］黄志强，林白雪，谢联辉. 2006. 产碱性蛋白酶海洋细菌的筛选与鉴定. 福建农林大学学报（自然科学版），35（4）：416-420.

［355］王林萍，徐学荣，林奇英，谢联辉. 2006. 论农药企业的社会责任. 科技和产业，6（2）：17-20.

［356］沈建国，张正坤，吴祖建，谢联辉，林奇英. 2006. 臭椿抗烟草花叶病毒活性物质的提取及其初步分离. 中国生物防治，23（4）：348-352.

［357］Lin YJ, Phyllis A. Rundell, Xie LH, and Charles A. Powell. 2006. Prereaction of *Citrus tristeza virus*（CTV）specific antibodies and labeled secondary antibodies increases speed of direct tissue blot immunoassay for CTV. Plant Disease 90:675-679.

［358］Liu B, Wu SJ, Song Q, Zhang XB, Xie LH. 2006. Two novel bacteriophages of thermophilic bacteria isolated from deep-sea hydrothermal fields. Current Microbiology 53:163-166.

［359］Liu B, Li HB, Wu SJ, Song Q, Zhang XB, Xie LH. 2006. A simple and rapid method for the dirrerentiation and identification of thermophilic bacteria. Canadian journal of microbiology 52: 753-758.

［360］Huang HN,Hua YY,Bao GR,Xie LH. 2006. The quantification of monacolin K in some red yeast rice from Fujian province and the comparison of the other product. Chemical & pharmaceutical bulletin 54（5）:

687-689.

[361]Kang CY, Tadashi Miayata, Wu G, Xie LH. 2006. Effects of enzyme inhibitors on acetylcholinesterase and detoxification enzymes in *Propylaea japonica* and *Lipaphis erysimi*. Proceedings of 5th International Workshop on Management of the Diamondback Moth and Other Crucifer Insect Pest, Beijing.

[362]吴艳兵，谢荔岩，谢联辉，林奇英 . 2007. 毛头鬼伞（*Coprinus comatus*）多糖的理化性质及体外抗氧化活性 . 激光生物学报，16（4）：438-442.

[363]吴艳兵，谢荔岩，谢联辉，林奇英，林诗发 . 2007. 毛头鬼伞多糖抗烟草花叶病毒（TMV）活性研究初报 . 中国农学通报，23（5）:338-341.

[364]吴艳兵，谢荔岩，谢联辉，林奇英 . 2007. 毛头鬼伞多糖对烟草酶活性和同工酶谱的影响 . 微生物学杂志，（05）:29-33.

[365]谢东扬，祝雯，吴祖建，林奇英，谢联辉 . 2007. 灵芝金属硫蛋白基因的克隆及序列分析 . 中国农学通报，23（5）：87-90.

[366]王林萍，徐学荣，林奇英，谢联辉 . 2007. 农药企业社会责任认知度调查分析 . 商业时代，15：62-64.

[367]张福山，徐学荣，林奇英，谢联辉 . 2007. 培育植保生态文化 促进可持续农业发展 . 福建农林大学学报（哲学社会科学版），10（2）：64-66,114.

[368]沈建国，张正坤，吴祖建，谢联辉，林奇英 . 2007. 臭椿和鸦胆子抗烟草花叶病毒作用研究 . 中国中药杂志，32（1）：27-29.

[369]谢东扬，祝雯，吴祖建，林奇英，谢联辉 . 2007. 灵芝中一种新的脱氧核糖核酸的纯化及特征 . 福建农林大学学报：自然科学版 ,36（5）:486-490.

[370]谢东扬，祝雯，吴祖建，林奇英，谢联辉，2007. 灵芝金属硫蛋白基因的克隆及序列分析 . 中国农学通报，23（5）:87-90.

[371]路炳声，黄志强，林白雪，谢联辉 . 2007. 海洋氧化短杆菌 15E 产碱性蛋白酶的发酵条件 . 福建农林大学学报（自然科学版）,36（6）: 591-595.

[372]杨小山，徐学荣，谢联辉，林奇英 . 2007. 农药管理能力和水平的综合评价指标体系与评价方法 . 福建农林大学学报（哲学社会科学版），10（5）：57-60.

[373]鹿连明，秦梅玲，谢荔岩，林奇英，吴祖建，谢联辉 . 2007. 利用酵母双杂交系统研究水稻条纹病毒三个功能蛋白的互作 . 美国农业科学与技术，1（1）：5-11.

[374]Zhang ZK, Ouyang MA, Wu ZJ, Lin QY, and Xie LH. 2007. Structure-activity relationship of triterpenes and triterpenoid glycosides against *Tobacco mosaic virus*. Planta medica 73:1457-1463.

[375]Wu G, Tadashi Miyata, Kang CY, Xie LH. 2007. Insecticide toxicity and synergism by enzyme inhibitors in 18 species of pest insects and natural enemies in crucifer vegetable crops. Pest Management Science 63: 500-510.

[376]Zhang YH, Tadashi Miyata, Wu ZJ, Wu G, Xie LH. 2007. Hydrolysis of acetylthiocholine iodide and reactivation of phoxim-inhibited acetylcholinesterase by pralidoxime chloride, obidoxime chloride and trimedoxime. Archives of toxicology 81: 785-792.

[377]丁新伦，谢荔岩，林奇英，吴祖建，谢联辉 . 2008. 水稻条纹病毒胁迫下抗、感病水稻品种胼胝质的沉积 . 植物保护学报，35（1）：19-22.

[378]连玲丽，谢荔岩，林奇英，谢联辉 . 2008. 芽孢杆菌三种抗菌素基因的杂交检测 . 激光生物学报，17（1）：81-85.

[379]程兆榜，任春梅，周益军，范永坚，谢联辉 . 2008. 水稻条纹病毒不同地区分离物的致病性研究 . 植物病理学报，38（2）：126-131.

[380]林白雪，黄志强，谢联辉. 2008. 海洋氧化短杆菌15E碱性蛋白酶的酶学性质. 福建农林大学学报(自然科学版)，37(2)：158-161.

[381]林董，郑璐平，谢荔岩，吴祖建，林奇英，谢联辉. 2008. GFP与水稻条纹病毒病害特异蛋白的融合基因在sf9昆虫细胞中的表达. 植物病理学报，38(3)：271-276.

[382]林董，何柳，谢荔岩，吴祖建，林奇英，谢联辉. 2008. RSV编码的4种蛋白在"AcMNPV-sf9昆虫细胞"体系中的重组表达. 福建农林大学学报(自然科学版)，37(3)：269-274.

[383]陈启建，欧阳明安，谢联辉，林奇英，2008. 银胶菊(*Parthenium hysterophorus*)中抗TMV活性成分的分离及活性测定. 激光生物学报，17(4)：544-548.

[384]丁新伦，张孟倩，谢荔岩，林奇英，吴祖建，谢联辉. 2008. 实时荧光定量PCR检测RSV胁迫下抗病、感病水稻中与脱落酸相关基因的差异表达. 激光生物学报，17(4). 464-469.

[385]鹿连明，林丽明，谢荔岩，林奇英，吴祖建，谢联辉. 2008. 水稻条纹病毒CP与叶绿体Rubisco SSU引导肽融合基因的构建及其原核表达. 农业生物技术学报，16(3). 530-536.

[386]鹿连明，秦梅玲，王萍，兰汉红，牛晓庆，谢荔岩，吴祖建，谢联辉. 2008. 利用免疫共沉淀技术研究RSV CP、SP和NSvc4三个蛋白的互作情况. 农业生物技术学报，16(5):891-897.

[387]鹿连明，秦梅玲，牛晓庆，兰汉红，王萍，谢荔岩，吴祖建，谢联辉，2008. 两个水稻品种(系)酵母双杂交cDNA文库的构建和比较分析. 激光生物学报，(05):656-662.

[388]吴艳兵，颜振敏，谢荔岩，林奇英，谢联辉. 2008. 天然抗烟草花叶病毒大分子物质研究进展. 微生物学通报. 35(7)：1096-1101.

[389]吴艳兵，谢荔岩，谢联辉，林奇英. 2008. 毛头鬼伞多糖CCP60a对TMV外壳蛋白的影响. 植物资源与环境学报，(03)：63-66.

[390]杨金广，方振兴，张孟倩，徐飞，王文婷，谢荔岩，林奇英，吴祖建，谢联辉. 2008. 应用Real-Time RT-PCR鉴定2个水稻品种(品系)对水稻条纹病毒的抗性差异. 华南农业大学学报，29(3). 25-28.

[391]杨金广，王文婷，丁新伦，郭利娟，方振兴，谢荔岩，林奇英，吴祖建，谢联辉. 2008. 水稻条纹病毒与水稻互作中的生长素调控. 农业生物技术学报，16(4)：628-634.

[392]张晓婷，谢荔岩，林奇英，吴祖建，谢联辉. 2008. Pathway Tools可视化分析水稻基因表达谱. 激光生物学报，17(3). 371-377.

[393]张晓婷，谢荔岩，林奇英，吴祖建，谢联辉. 2008. 水稻条纹病毒胁迫下的水稻全基因组表达谱. 激光生物学报，17(5):620-629.

[394]张正坤，沈建国，谢荔岩，谢联辉，林奇英. 2008. 鸦胆子素D对烟草抗烟草花叶病毒的诱导抗性和保护作用. 科技导报，26(8). 31-36.

[395]张正坤，吴祖建，沈建国，谢联辉，林奇英. 2008. 烟草花叶病毒运动蛋白的表达及特异性抗体制备. 福建农林大学学报(自然科学版)，37(3). 265-268.

[396]祝雯，谢东扬，林奇英，谢联辉，吴祖建. 2008. 杨树菇中一种脱氧核糖核酸酶的纯化及其性质. 中国生物制品学杂志，21(10)：869-872.

[397]吴丽萍，吴祖建，林奇英，谢联辉. 2008. 毛头鬼伞(*Coprinus comatus*)中一种抗病毒蛋白y^3特性和氨基酸序列分析. 中国生物化学与分子生物学报，24(7)：597-603.

[398]郑璐平，谢荔岩，姚锦爱，钟伏弟，林奇英，吴祖建，谢联辉. 2008. 孔石莼凝集素蛋白基因的克隆与表达. 激光生物学报，17(6)：762-767.

[399]郑璐平，谢荔岩，林奇英，谢联辉. 2008. 病毒诱导基因沉默的研究进展. 福建农林大学学报(自然科学版)，37(6)：636-640.

[400]郑璐平，谢荔岩，连玲丽，谢联辉. 2008. 水稻齿叶矮缩病毒的研究进展. 中国农业科技导

报，10（5）：8-12.

［401］徐学荣，张福山，谢联辉 . 2008. 植物保护的风险及其管理 . 农业系统科学与综合研究，24（2）:148-152.

［402］Liu F, Tadashi Miyata, Li CW, Wu ZJ, Wu G, Zhao SX, Xie LH. 2008. Effects of temperature on fitness costs, insecticide susceptibility and heat shock protein 70 in insecticide-resistant and susceptible plutella xylostella. Pesticide Biochemistry Physiology 91: 45-52.

［403］Yang JG, Dang YG, Li GY, Guo LJ, Wang WT, Tan QW, Lin QY, Wu ZJ, Xie LH. 2008. Anti-viral activity of *Ailanthus altissima* crude extract on *Rice stripe virus* in rice suspension cells. Phytoparasitica 36（4）: 405-408.

［404］Shen JG, Zhang ZK, Wu ZJ, Ouyang MA, Xie LH, and Lin QY. 2008. Antiphytoviral activity of bruceine-D from *Brucea javanica* seeds. Pest management science 64:191-196.

［405］Yang CX, Cui GJ, Zhang J, Weng XF, Xie LH and Wu ZJ. 2008. Molecular characterization of a distinct begomovirus species isolated from Emilia Sonchifolia. Journal of Plant Pathology 90（3）:475-478.

［406］Yang C, Jia S, Liu Z, Cui G, Xie L and Wu Z. 2008. Mixed Infection of Two Begomoviruses in *Malvastrum coromandelianum* in Fujian, China. Journal of Phytopathology 156, 553–555.

［407］沈硕，李玮，欧阳明安，吴祖建，林奇英，谢联辉 . 2009. 两株海洋真菌的鉴定及其次级代谢产物抑制烟草花叶病毒及抗肿瘤活性（英文）. 微生物学报，49（9）:1240-1246.

［408］程文金，吴祖建，谢联辉 . 2009. 水稻条纹病毒楚雄分离物一个重组 RNA 序列分析 . 中国农学通报，25（8）: 352-355.

［409］高芳銮，范国成，沈建国，谢荔岩，林奇英，吴祖建，谢联辉 . 2009. 水稻瘤矮病毒 P8 蛋白的结构分析及其表达 . 中国生物工程杂志，（8）: 51-56.

［410］许曼琳，段永平，吴祖建，谢荔岩，谢联辉，林奇英 . 2009. 芽孢杆菌两菌株对香蕉炭疽病菌的抑制作用及其机制 . 云南农业大学学报，（4）: 522-527.

［411］吴丽萍，曹郁生，吴祖建，林奇英，谢联辉 . 2009. YP3: 食用菌榆黄蘑中新的植物病毒抑制物蛋白（英文）. 天然产物研究与开发，（3）: 371-376.

［412］吴艳兵，谢荔岩，谢联辉，林奇英 . 2009. 毛头鬼伞多糖诱导烟草体内水杨酸的积累 . 福建农林大学学报（自然科学版），38（1）: 6-10.

［413］陈启建，欧阳明安，吴祖建，谢联辉，林奇英 . 2009. 金鸡菊（*Coreopsis*，*drummondii*）的抗 TMV 活性物质 . 应用与环境生物学报，15（5），621-625.

［414］程文金，邓慧颖，谢荔岩，林奇英，吴祖建，谢联辉 . 2009. 我国水稻条纹病毒致病性的分化与差异分析 . 福建农林大学学报（自然科学版）. 38（6），561-566.

［415］吴建国，蔡丽君，胡梅群，谢荔岩，林奇英，吴祖建，谢联辉 . 2009. 水稻瘤矮病毒 P3、P7、P8、Pn9、Pn10、Pn11、Pn12 的酵母双杂交载体的构建及自激活效应检测 . 热带作物学报，30（9）: 1364-1368.

［416］章松柏，吴祖建，段永平，王盛，林奇英，谢联辉 . 2009. 水稻矮缩病毒基因组遗传多样性的初步研究 . 长江大学学报（农学版）.（4），37-39.

［417］陈路劼，刘斌，林白雪，何柳，张宁，吴祖建，谢联辉 . 2009. 降解纤维素嗜热菌的分离及纤维素酶性质分析 . 福建农林大学学报（自然科学版）. 39（1），67-72.

［418］Wu G, Lin YW, Miyata T, Jiang SR, Xie LH. 2009. Positive correlation of methamidophos resistance between lipaphis erysimi and diaeretilla rape and effects of methamidophosingested by host insect on the parasitoid. Insect Science 16（2）,165-173.

［419］Wei TY, Yang JG, Liao FL, Gao FL, Lu LM, Zhang XT, Li F, Wu ZJ, Lin QY, Xie LH and Lin HX.

2009. Genetic diversity and population structure of rice stripe virus in China. Journal of General Virology 90, 1025–1034.

[420] Lu LM, Du ZG, Qin ML, Wang P, Lan HH, Niu XQ, Jia DS, Xie LY, Lin QY, Xie LH and Wu ZJ. 2009. Pc4, a putative movement protein of *Rice stripe virus*, interacts with a type I DnaJ protein and a small Hsp of rice. Virus Genes 38:320–327.

[421] Shen S, Li W, Ouyang MA, Wu ZJ, Lin QY, and Xie LH. 2009. Identification of two marine fungi and evaluation of their antivirus and antitumor activities. Acta Microbiologica Sinica 49（9）, 1240-1246.

[422] Yang JG, Wang WT, Ding XL, Guo LJ, Fang ZX, Xie LY, Lin QY, Wu ZJ, Xie LH. 2009. Auxin regulation in the interaction between *Rice stripe virus* and rice. Chinese Journal of Agricultural Biotechnology 6 : 27-33.

[423] Yang CX, Wu ZJ, and Xie LH. 2009. First report of the occurrence of sweet potato leaf curl virus in tall morningglory（*Ipomoea purpurea*）in China. Plant Disease 3（7）: 764.

[424] 沈硕，李玮，欧阳明安，谢联辉 . 2010. 两株海洋真菌的鉴定及其代谢产物的抑菌活性 . 中国生物防治，26（1）：62-68.

[425] 张宁宁，林白雪，何柳，余能富，刘斌，谢联辉，2010, 降解半纤维素嗜热菌的筛选及其酶学性质 . 福建农林大学学报（自然科学版），39（05）：528-533.

[426] 吴建国，巴俊伟，李冠义，林奇英，吴祖建，谢联辉 . 2010. 16 个水稻品种对水稻矮缩病毒抗性的鉴定 . 福建农林大学学报（自然科学版），39（1），10-14.

[427] 吴建国，王萍，谢荔岩，林奇英，吴祖建，谢联辉 . 2010. 水稻矮缩病毒对 3 种内源激素含量及代谢相关基因转录水平的影响 . 植物病理学报，40（2），151-158.

[428] 肖冬来，邓慧颖，谢荔岩，吴祖建，谢联辉 . 2010. 酵母双杂交系统筛选与水稻黑条矮缩病毒 P6 互作的水稻蛋白 . 热带作物学报，31（3）：435-439.

[429] 肖冬来，邓慧颖，谢荔岩，吴祖建，谢联辉 . 2010. 水稻条纹病毒胁迫下灰飞虱基因的差异表达 . 昆虫学报，（8）：914-920.

[430] 肖冬来，贾东升，吴建国，杜振国，谢荔岩，吴祖建，谢联辉 . 2010. 水稻条纹病毒 NS3 蛋白与水稻 3- 磷酸甘油醛脱氢酶（GAPDH）. 中国水稻科学，24（5）: 493-496.

[431] 张福山，徐学荣，林奇英，谢联辉 . 2010. 植物保护对中国粮食生产影响的经济分析 . 中国农学通报，26（3），320-326.

[432] 章松柏，李大勇，肖冬来，张长青，吴祖建，谢联辉 . 2010. 水稻黑条矮缩病的发生和病毒检测 . 湖北农业科学，49（3），592-594.

[433] 章松柏，张长青，吴祖建，谢联辉 . 2010. 棉花皱缩花叶病的初步研究 . 河南农业科学，（3），48-50.

[434] 刘振宇，吴祖建，谢荔岩，林奇英，谢联辉 . 2010. 孔石莼（Ulva pertusa）质体蓝素基因的克隆及基因特征分析 . 激光生物学报，20（1）：61-66.

[435] Shen S, Ding XA, Ouyang MA, Wu ZJ, Xie LH. 2010. A new phenolic glycoside and cytotoxic constituents from *Celosia argentea*. Journal of Asian Natural Products Research, 12（9）,821-827.

[436] Liu B, Wu S J and Xie L H. 2010. Complete genome sequence and proteomic analysis of a thermophilic bacteriophage BV1. Acta Oceanologica Sinica，29（3）：84-89.

[437] Xiao D L, Li W M, Wei T Y, Wu Z J and Xie L H. 2010. Advances in the studies of *Rice stripe virus*. Frontiers of Agriculture in China, 4（3）：287-292.

[438] Fan GC, Gao FL ,Wei TY, Huang MY, Xie LY, Wu ZJ, Lin QY, Xie LH. 2010. Expression of *Rice gall dwarf virus* outer coat protein gene（S8）in insect cells. Virologica Sinica, 25（6）: 401-408.

[439] Wu JG, Du ZG, Wang CZ, Cai LJ, Hu MQ, Lin QY, Wu ZJ, Li Y, Xie LH. 2010. Identification of Pns6, a putative movement protein of RRSV, as a silencing suppressor. Virology Journal 7:335 doi:10. 1186/1743-422X-7-335.

[440] Zhuang J, Cai G, Lin Q, Wu Z, Xie L. 2010. A bacteriophage-related chimeric marine virus infecting abalone. PLoS ONE, 5（11）: e13850. Published online 2010 November 5. doi: 10. 1371/journal. pone. 0013850.

[441] Wu ZJ, Wu JG, Scott Adkins, Xie LH, Li WM. 2010. Rice ragged stunt virus segment S6-encoded nonstructural protein Pns6 complements cell-to-cell movement of Tobacco mosaic virus-based chimeric virus. Virus Research, 152（1-2）：176-179.

[442] 罗金水，吴祖建，谢联辉 . 2011. 齿兰环斑病毒 CP 基因的原核表达及其产物抗血清制备 . 中国农学通报，27（19）:115-120.

[443] 罗金水，吴祖建，谢联辉 . 2011. 齿兰环斑病毒 CP 基因的原核表达及其产物抗血清制备 . 中国农学通报，27（19）:115-120.

[444] 孙炳剑，羊健，孙丽英，程兆榜，谢礼，姜鸿明，郑建强，赵倩，谢联辉，陈剑平 . 2011. 禾谷多黏菌传小麦病毒病的分布及变化动态 . 麦类作物学报，31（5）：969-973.

[445] 范国成，高芳銮，黄美英，谢荔岩，吴祖建，林奇英，谢联辉 . 2011. 水稻瘤矮病毒 S3 和 S8 基因共表达杆状病毒转移载体构建及重组病毒的鉴定 . 福建农林大学学报（自然科学版），40（2）：151-155.

[446] 连玲丽，谢荔岩，吴祖建，谢联辉，段永平 . 2011. 枯草芽孢杆菌 SB1 的抑菌活性及其对番茄青枯病的防治作用 . 植物病理学报，41（2）：219-224.

[447] 林白雪，刘斌，张宁宁，谢联辉 . 2011. 降解纤维素嗜热真菌的筛选与分子鉴定 . 福建农林大学学报（自然科学版），40（2）：182-186.

[448] 沈硕，李玮，欧阳明安，谢联辉 . 2011. 2 株海洋真菌的鉴定及其代谢产物的抑菌活性 . 中国生物防治，2010，26（1）：62-68.

[449] 吴建国，王春政，杜振国，蔡丽君，胡梅群，吴祖建，李毅，谢联辉 . 2011. 水稻瘤矮病毒 S12 编码第 2 个 RNA 沉默抑制子 . 中国科学：生命科学，41（1）：61-69.

[450] 章松柏，罗汉刚，张求东，张长青，吴祖建，谢联辉 . 2011. 湖北发生的水稻矮缩病是南方水稻黑条矮缩病毒引起的 . 中国水稻科学，25（2）：223-226.

[451] 孙炳剑，李洪连，杨新志，谢联辉，陈剑平 . 2011. 河南省主要推广品种对小麦黄花叶病毒抗性的评价 . 植物保护学报，38（2）：102-108.

[452] 肖冬来，邓慧颖，谢荔岩，吴祖建，谢联辉 . 2011. 灰飞虱酵母双杂交 cDNA 文库的构建及分析 . 植物保护，37（1）：19-23.

[453] 程兆榜，於春，任春梅，周益军，范永坚，谢联辉 . 水稻条纹病毒外壳蛋白叶绿体离体跨膜运输研究 . 南方农业学报，2011，42（12）：1476-1480.

[454] 张晓婷，谢荔岩，林奇英，吴祖建，谢联辉 . 水稻条纹病毒胁迫下的水稻蛋白质组学 . 植物病理学报，2011，41（3）：253-261.

[455] Yang CX, Luo JS, Zheng LM, Wu ZJ,Xie LH. 2011. Mixed infection of papaya leaf curl china virus and siegesbeckia yellow vein virus in siegesbeckia orientalis in China. Journal of General Plant Pahtology, 93（4）,81-81.

[456] Jiang DM, Li SF,Fu FH,Wu ZJ, Xie LH. 2011. First reported occurrence of coleus blumei viroid 3 from coleus blumei in china. Journal of General Plant Pahtology, 93（4），82-82.

[457] Zhuang HM, Wang KF, Miyata T, Wu ZJ, Wu G, Xie LH. 2011. Identification and expression

of caspase-1 gene under heat stress in insecticide-susceptible and resistant Plutella xylostella（Lepidoptera: Plutellidae）. Molecular Biology Reports, 38（4）, 2529-2539.

[458] Ye XJ, Ng TB, Wu ZJ, Xie LH, Fang EF, Wong JH, Pan WL, Wing SS, Zhang YB. 2011. Protein from red cabbage（Brassica oleracea）seeds with antifungal, antibacterial, and anticancer activities. Journal of Agricultural and Food Chemistry, 59（18）:10232-10238.

[459] Jiang DM, Wu ZJ, Xie LH, Teruo Sano, Li SF. 2011. Sap-direct RT-PCR for the rapid detection of coleus blumei viroids of the genus *Coleviroid* from natural host plants. Journal of Virological Methods, 174（1-2）:123-127.

[460] Ji X, Qian D, Wei C, Ye G, Zhang Z, Wu Z, Xie L, Li Y. 2011. Movement protein Pns6 of *Rice dwarf phytoreovirus* has both ATPase and RNA binding activities. PloS One, 6（9）: e24986.

[461] Yuan Z, Chen H, Chen Q, Omura T, Xie L, Wu Z, Wei T. 2011. The early secretory pathway and an actin-myosin VIII motility system are required for plasmodesmatal localization of the NSvc4 protein of *Rice stripe virus*. Virus Research 159:62-68.

[462] Du Z, Xiao D, Wu J, Jia D, Yuan Z, Liu Y, Hu L, Han Z, Wei T, Lin Q, Wu Z, Xie L. 2011. p2 of *Rice stripe virus*（RSV）interacts with OsSGS3 and is a silencing suppressor. Molecular Plant Pathology 12:808-814.

[463] Liu Y, Jia D, Chen H, Chen Q, Xie L, Wu Z, Wei T. 2011. The P7-1 protein of *Southern rice black-streaked dwarf virus*, a fijivirus, induces the formation of tubular structures in insect cells. Archives of Virology 156:1729–1736.

[464] 刘伟，林志伟，陈美霞，魏日凤，谢联辉 . 2012. 枯草芽孢杆菌绿色荧光蛋白高效表达载体的构建 . 热带作物学报，33（3）:467-471.

[465] 兆榜，任春梅，周益军，季英华，范永坚，谢联辉 . 灰飞虱来源的水稻条纹病毒外壳蛋白基因遗传多样性 . 植物病理学报，2012，42（6）：585-593.

[466] Liu B, Zhang NN, Zhao C, Lin BX, Xie LH, Huang YF. 2012. Characterzation of a recombinant thermostable xylanase from hot spring thermophilic geobacillus sp TC-W7. Journal of Microbiology and Biotechnology, 22（10）,1388-1394.

[467] Chen Q, Chen HY, Mao QZ, Liu QF, Shi MZ T, Uehara-Ichiki T, Wu ZJ, Xie LH, Omura T, Wei TY. 2012. Tubular structure induced by a plant virus facilitates viral spread in its vector insect. Plos Pathogens, 8（11）:e1003032.

[468] Jiang DM, Teruo Sano, Masaharu Tsuji, Hiroyuki Araki, Kyota Sagawa, Charith Raj Adkar Purushothama, Zhang ZX, Guo R, Xie LH, Wu ZJ, Wang HQ, Li SF. 2012. Comprehensive diversity analysis of viroids infecting grapevine in China and Japan. Virus Research, 169（1）: 237-245.

[469] Jia D, Guo N, Chen H, Akita F, Xie L, Omura T, Wei T. 2012. Assembly of the viroplasm by viral non-structural protein Pns10 is essential for persistent infection of Rice ragged stunt virus in its insect vector. Journal of General Virology, 93（Pt 10）:2299-309.

[470] Jia D, Chen H, Zheng A, Chen Q, Liu Q, Xie L, Wu Z, Wei T. 2012. Development of an insect vector cell culture and RNA interference system to investigate the functional role of fijivirus replication protein. Journal of Virology, 86（10）:5800-7. doi: 10. 1128/JVI. 07121-11. Epub 2012 Mar 7.

[471] 郑璐平，林辰，高芳銮，张超，谢荔岩，吴祖建，谢联辉 . 2013. 拟南芥柯浩体蛋白（Atcoilin）的功能预测及亚细胞定位 . 农业生物技术学报，21（11）:1270-1278.

[472] 高芳銮，沈建国，史凤阳，方治国，谢联辉，詹家绥 . 2013. 中国马铃薯 Y 病毒的检测鉴定及 CP 基因的分子变异 . 中国农业科学，46（15）:3125-3133.

［473］高芳銮，沈建国，史凤阳，常飞，谢联辉，詹家绥 . 2013. 马铃薯 Y 病毒 pipo 基因的分子变异及结构特征分析 . 遗传，35（9）:1125-1134.

［474］郑璐平，毛倩卓，林辰，谢荔岩，吴祖建，谢联辉 . 2013. 一个可指示核仁定位和柯浩体定位信号 Marker 的构建 . 中国细胞生物学学报 ,35（07）:1002-1007.

［475］程兆榜，何敦春，陈全战，季英华，任春梅，魏利辉，周益军，范永坚，谢联辉 . 2013. 单季稻小麦轮作区灰飞虱发生规律 . 应用昆虫学报，50（3）:706-717.

［476］贾东升，任堂雨，陈红燕，魏太云，谢联辉 . 2013. 白背飞虱体内 RNA 干扰技术体系的建立 . 福建农林大学学报（自然科学版），42（6）: 579-583.

［477］袁正杰，贾东升，吴祖建，魏太云，谢联辉 . 2013. NSvc4 和 CP 蛋白与水稻条纹病毒的致病相关 . 中国农业科学 . 46（1）: 45-53.

［478］章松柏，宋国威，杨靓，吴祖建，谢联辉 . 2013. 水稻锯齿叶矮缩病毒的检测及介体传毒特性 . 福建农林大学学报（自然科学版）. 4（3）: 225-229.

［479］郑璐平，林辰，吴祖建，谢联辉 . 2013. 水稻条纹病毒编码的 NS2 蛋白的亚细胞定位分析 . 福建农林大学学报（自然科学版）. 42（6）: 574-578.

［480］Sun BJ, Sun LY, Tugume AK, Adams MJ, Yang J, Xie LH, Chen JP. 2013. Selection pressure and founder effects constrain genetic variation in differentiated populations of soilborne *Bymovirus Wheat yellow mosaic virus (Potyviridae)* in China. Phytopathology, 103（9）,949-959.

［481］Jiang DM, Li SF, Fu FH, Wu ZJ, Xie LH. 2013. First Report of Coleus blumei viroid 5 from Coleus blumei in India and Indonesia. Plant Disease, 97（04）,561-561.

［482］Jiang DM, Hou WY, Teruo Sano, Kang N, Qin L, Wu ZJ, Li SF, Xie LH. 2013. detection and identification of viroids in the genus using a universal probe. Journal of Virological Methods 187（2）: 321-326.

［483］Zhang SB, Du ZG, Yang L, Yuan ZJ, Wu KC, Li GP, Wu ZJ, Xie LH. 2013. Identification and characterization of the interaction between viroplasm-associated proteins from two different plant-infecting reoviruses and eEF-1A of rice. Archives of Viology, 158. 10: 2031-2039.

［484］Yang CX, Zheng LM, Wu ZJ, Xie LH. 2013. Papaya leaf curl Guangdong virus and *Ageratum yellow vein virus* associated with leaf curl disease of tobacco in China. Journal of Phytopathol, 161（3）,201-204.

［485］Jiang DM, Li SF,Fu FH,Wu ZJ, Xie LH. 2013. First Report of Coleus blumei viroid 5 from Coleus blumei in India and Indonesia. Plant Disease 97（04）, 561-561.

［486］郑璐平，林辰，谢荔岩，吴祖建，谢联辉 . 2014. 水稻条纹病毒 NS2 和 NS3 基因共干扰转基因水稻的培育及抗病性分析 . 病毒学报，30（06):661-667.

［487］贾东升，马元元，杜雪，陈红燕，谢联辉，魏太云 . 2014. 水稻黑条矮缩病毒在灰飞虱消化系统的侵染和扩散过程 . 植物病理学报 . 44（2）: 188-194.

［488］郑立敏，刘华敏，陈红燕，贾东升，谢联辉，魏太云 . 2014. 干扰水稻瘤矮病毒（RGDV）非结构蛋白（Pns12）的表达抑制病毒在介体昆虫培养细胞内的复制 . 农业生物技术学报 . 22（11）: 1321-1328.

［489］许曼琳，杨燕燕，杨金广，谢宏峰，迟玉成，谢联辉 . 2014. 蓖麻枯萎病菌高效生防菌株 LX1 的筛选与鉴定 . 中国生物防治学报，30（2）:271-275.

［490］Zheng LM, Mao QZ, Xie LH, Wei TY. 2014. Infection route of rice grassy stunt virus a tenuivirus in the body of its brown planthopper vector Nilaparvata lugens（Hemiptera: Delphacidae）after ingestion of virus. Virus Research 188（8）: 170-173.

［491］Zhu HT, Zhuang J, Feng HL, Liang RF, Wang JY, Xie LH, Zhu P. 2014. Cryo-EM structure of isomeric molluscan hemocyanin triggered by viral infection. PLoS One, 9（6）:e98766.

[492] Yang L, Du ZG, Gao F, Wu KC, Xie LH, Li Y, Wu ZJ, Wu JG. 2014. Transcriptome profiling confirmed correlations between symptoms and transcriptional changes in RDV infected rice and revealed nucleolus as a possible target of RDV manipulation. Virology Journal, 11: 81.

[493] Jiang DM, Gao R, Qin L, Wu ZJ, Xie LH, Hou WY, Li SF，2014. Infectious cDNA clones of four viroids in Coleus blumei and molecular characterization of their progeny. Virus Research, 180, 97-101.

[494] Gao FL, Chang F, Shen JG, Shi FY, Xie LH, Zhan JS. 2014. Complete genome analysis of a novel recombinant isolate of *Potato virus Y* from China. Archives of Virology, 159（12）,3439-3442.

[495] Jia DS, Mao QZ, Chen HY, Wang AM, Liu YY,Wang HT, Xie LH, Wei TY. 2014. Virus-induced tubule:a vehicle for rapid spread of virions through basal lamina from midgut epithelium in the insect vector. Journal of Virology, 88（18）, 10488-10500.

[496] Xu ML, Gao FL, Yang JG, Wu JX, Xie LH, Chi YC. 2014. Complete genome sequence of *Peanut stripe virus* isolated in China. Journal of Phytopathol, 162（11-12）,829-832.

[497] 陈倩，张玲华，黄海宁，魏太云，谢联辉 . 2015. 水稻矮缩病毒非结构蛋白 Pns6 在病毒复制中的功能 . 中国科技论文，10（24）：2840-2846.

[498] 陈倩 , 张玲华 , 黄海宁 , 魏太云 , 谢联辉 . 2015. 干扰水稻矮缩病毒（RDV）非结构蛋白 Pns11 的表达可抑制病毒在介体黑尾叶蝉内的复制 . 农业生物技术学报，23（11）：1401-1409.

[499] 王辉 , KHERAPawan, 李双铃 , 任艳 , 袁美 , 庄伟建 ,VARSHNEYRajeevK, 郭宝珠 , 谢联辉 . 2015. SSR 分子标记在花生杂种鉴定中的应用 . 福建农林大学学报（自然科学版）, 44（4）:350-354.

[500] 卢全有 , 吴祖建 , 夏志松 , 谢联辉 . 2015. 桑花叶卷叶病相关病毒外壳蛋白基因的原核表达与免疫检测 . 蚕业科学 ,41（2）:218-225.

[501] 吴元兴 , 徐学荣 , 谢联辉 . 2015. 建立健全省域生态补偿机制的研究 —— 以福建省为例 . 福建论坛（人文社会科学版）,（4）:142-147.

[502] 高芳銮 , 常飞 , 沈建国 , 谢联辉 , 詹家绥 . 2015. PVY~（NTN-NW）榆林分离物的全基因组序列测定与分析 . 中国农业科学 , 48（2）:270-279.

[503] Zhuang J, Christopher J. Coates, Zhu HT, Zhu P, Wu ZJ, Xie LH. 2015. Identification of candidate antimicrobial peptides derived from abalone hemocyanin. Developmental & Comparative Immunology, 49（1）: 96-102.

[504] Chen Q, Chen H, Jia D, Mao Q, Xie L, Wei T. 2015. Nonstructural protein Pns12 of rice dwarf virus is a principal regulator for viral replication and infection in its insect vector. Virus research, 210: 54-61.

[505] Chen Q, Wang H, Ren T, Xie L, Wei T. 2015. Interaction between non-structural protein Pns10 of rice dwarf virus and cytoplasmic actin of leafhoppers is correlated with insect vector specificity. Journal of General Virology, 96（Pt 4）: 933-938.

[506] Zheng, L Du Z, Lin C, Mao Q, Wu K, Wu J, Xie L. 2015. *Rice stripe tenuivirus* p2 may recruit or manipulate nucleolar functions through an interaction with fibrillarin to promote virus systemic movement. Molecular plant pathology, 16（9）: 921-930.

[507] Wu JG, Yang ZR, Wang Y, Zheng LJ, Ye RQ, Ji YH, Zhao SS, Ji SY, Liu RF, Xu L, Zheng H, Zhou YJ, Zhang X, Cao XF, Xie LH, Wu ZJ, Qi YJ, Li Y. 2015. Viral-inducible argonaute18 confers broad-spectrum virus resistance in rice by sequestering a host microRNA. Elife4:e05733.

[508] Zhan JS, Thrall PH, Papaïx J, Xie LH, Burdon JJ. 2015. Playing on a pathogen's weakness: using evolution to guide sustainable plant disease control strategies. Annual Review of Phytopathology, 53:19-43.

[509] Zheng LM, Chen HY, Liu HM, Xie LH, Wei TY. 2015. Assembly of viroplasms by viral

nonstructural protein Pns9 is essential for persistent infection of rice gall dwarf virus in its insect vector. Virus Research, 196（22）: 162-169.

［510］Lu QY, Wu ZJ, Xia ZS, Xie LH. 2015. A new nepovirus identified in mulberry（Morus alba L.）in China. Archives of Virology, 160（3）:851-855.

［511］Zhuang J, Coates CJ, Zhu H, Zhu P, Wu ZJ, Xie LH. 2015. Identification of candidate antimicrobial peptides derived from abalone hemocyanin. Developmental And Comparative Immunology, 49（1）:96-102.

［512］Chen Q, Zhang L, Chen H, Xie L, Wei T. 2015. Nonstructural protein Pns4 of rice dwarf virus is essential for viral infection in its insect vector. Virology Journal, 12:211.

［513］Lu QY, Wu ZJ, Xia ZS, Xie LH. 2015. Complete genome sequence of a novel monopartite geminivirus identified in mulberry（Morus alba L.）. Archives of Virology, 160（8）, 2135-2138.

［514］Xu ML, YANG J G, WU J X, CHI Y C, XIE L H. 2015. First report of aspergillus niger causing root rot of peanut in China. Plant Disease, 99（2）,284-285.

［515］He DC, Zhan J, Cheng ZB, Xie LH. 2016. Viruliferous rate of small brown planthopper is a good indicator of rice stripe disease epidemics. Scientific Reports, 6, 21376.

［516］He DC, Zhan JS, Xie LH. 2016. Problems, challenges and future of plant disease management: from an ecological point of view. Journal of Integrative Agriculture, 15（4）: 705-715.

［517］Jun Zhuang, Christopher J. Coates, Qianzhuo Mao, Zujian Wu, Lianhui Xie. 2016. The antagonistic effect of *Banana bunchy top virus* multifunctional protein B4 against Fusarium oxysporum. Molecular plant pathology, 17（5）:669-679.

［518］Lin W, Gao F, Yang W, Yu C, Zhang J, Chen L, Wu Z, Hsu Y-H and Xie L. 2016. Molecular characterization and detection of a recombinant isolate of *Bamboo mosaic virus* from China. Archives of virology, 161（4）: 1091-1094.

［519］Zheng LP, Yao JA, Gao FL, Chen L, Zhang C, LL, Xie LY, Wu ZJ, Xie LH. 2016. The subcellular localization and functional analysis of fibrillarin2, a nucleolar protein in *nicotiana benthamiana*. Biomed Research International, 283-287.

［520］Wang H, Khera P, Huang B Y, Yuan M, Katam R, Zhuang W J, Harris-Shultz K, MoorE K M, Culbreath A K, Zhang X Y, Varshney R K, Xie L H, Guo B Z，2016. Analysis of genetic diversity and population structure of peanut cultivars and breeding lines from China, India and the US using simple sequence repeat markers. Journal of Integrative Plant Biology 58（5）,452-465.

［521］Gao B, Zhang JM, Wang YP, Chen F, Zheng CH, Xie LH. 2017. Genomic characterization of travel-associated dengue viruses isolated at entry-exit ports in Fujian province, China, 2013-2015. Japanese journal of infectious diseases doi: 10. 7883/yoken. JJID. 2016. 577.

［522］Gao B, Fang YL, Zhang JQ, Wu RQ, Xu BH, Xie LH. 2017. A DNA barcoding based study to identify main mosquito species in Taiwan and its difference from those in Mainland China. Comb Chem High Throughput Screen, 20（2）:147-152.

［523］Gao FL, Zou WC, Xie LH, Zhan J. 2017. Adaptive evolution and demographic history contribute to the divergent population genetic structure of Potato virus Y between China and Japan. Evolutionary Applications, DOI: 10. 1111/eva. 12459.

［524］Wu JG, Yang RX, Yang ZR, Yao SZ, Zhao SS, Wang Y, Li PC, i SongXWe, Jin L, Zhou T, Lan Y, Xie LH, Zhou XP, Chu CC, Qi YJ, Cao XF and Li Y. 2017. ROS accumulation and antiviral defence control by microRNA528 in rice. Nature Plants, 3, 16203.

［525］Zheng LJ, Zhang C, Shi CN, Wang Y, Zhou T, Sun F, Wang H, Zhao SS, Qin Q, Qiao R, Ding ZM,

Wei CH, Xie LH, Wu JG, Li Y. 2017. *Rice stripe virus* NS3 protein regulates primary miRNA processing through association with the miRNA biogenesis factor OsDRB1 and facilitates virus infection in rice. PLoS Pathogens, 13（10）: e1006662.

[526] Lin W, Wang L, Yan W, Chen L, Chen H, Yang W, Guo M, Wu Z, Yang L and Xie L. 2017. Identification and characterization of *Bamboo mosaic virus* isolates from a naturally occurring coinfection in *Bambusa xiashanensis*. Archives of virology, 162（5）: 1335-1339.

[527] Lin W, Yan W, Yang W, Yu C, Chen H, Zhang W, Wu Z, Yang L and Xie L. 2017. Characterisation of siRNAs derived from new isolates of *bamboo mosaic virus* and their associated satellites in infected ma bamboo（*Dendrocalamus latiflorus*）. Archives of virology, 162（2）: 505-510.

[528] Qin L, Han P P, Chen L Y, Walk T C, Li Y S, Hu X J, Xie L H, Liao H, Liao X，2017. Genome-wide identification and expression analysis of NRAMP family genes in soybean（*Glyccine Max L.*）. Frontiers in Plant Science, 1436.

[529] Xu ML, Xie HF, Wu JX, Xie LH, Yang JG, Chi YC. 2017. Translation initiation factor eIF4E and eIFiso4E are both required for *Peanut stripe virus* infection in peanut（*Arachis hypogaeaL.*）. Frontiers in Microbiology, 8:679.

[530] 高博，郑晖，李枢，谢联辉 . 2018. 登革热病毒（DENV）包膜蛋白特异性人源抗体筛选与鉴定 . 中国人兽共患病学报，34（10）:920-926.

[531] Gao B, Zhang JM, Xie LH. 2018. Structure analysis of effective chemical compounds against dengue virues isolated from *Isatis tinctoria*. Canadian Journal of Infectious Diseases & Medical Microbiology , 1-11.

[532] Zhang ZL, He GJ, Han GS, Zhang JT, Catanzaro N, Diaz A, WU ZJ, Carman GM, Xie LH, Wang XF. 2018. Host Pah1p phosphatidate phosphatase limits viral replication by regulating phospholipid synthesis. Plos Pathogens, e1006988.

[533] Zhang ZL, He GJ, Han GS, Zhang JT, Catanzaro N, Diaz A, Wu ZJ, Carman GM, He GJ, Zhang ZL, Sathanantham P, Zhang X, Wu ZJ, Xie LH, Whang XF. 2019. An engineered mutant of a host phospholipid synthesis gene inhibits viral replication without compromising host fitness. Journal Of Biological Chemistry. 294（38）,13973-13982.

[534] Zhuang J, Lin WW, Coates CJ, Shang PX, Wei TY,Wu ZJ, Xie LH. 2019. Cleavage of the Babuvirus movement protein B4 into functional peptides capable of host factor conjugation is required for virulence. Virologica Sinica, 34（3）,295-305.

[535] Lin WW, Liu YY, Molho M, Zhang SJ, Wang LS, Xie LH，Nagy PD. 2019. Co-oping the fermentation pathway for tombusvirus replication:Compartmentalization of cellular metabolic pathways for rapid ATP generation. Plos Pathogens, e1008092.

[536] Qin L, Walk T C, Han P P,Chen L Y, Zhang S, Li Y S, Hu X J, Xie L H, Yang Y, Liu J P, Lu X, Yu C B, Tian J, Shaff J E, Kochian L V, Liao X, Liao H. 2019. Adaption of roots to nitrogen deficiency revealed by 3D quantification and proteomic anlaysis. Plant Physiology, 179（1）,329-347.

[537] Zheng RR, Zhan JS, Liu LX, Ma YL, Wang ZS, Xie LH, He DC. 2019. Factors and minimal subsidy associated with tea farmers' willingness to adopt ecological pest management. Sustainability, 11（22）.

[538] 郑蓉蓉，刘路星，马妍丽，王自帅，陈少游，何敦春，谢联辉 . 2020. 基于 Logistic-ISM 模型的茶农采纳病虫生态调控技术的影响因素及层次结构分析 . 茶叶科学，40（5）：696-706.

[539] He DC, Burdon JJ, Xie LH, Zhan J. 2021. Triple bottom-line consideration of sustainable plant disease management: From economic, sociological and ecological perspectives. Journal of Integrative Agriculture，20（10）, 2581-2591.

四、翻译与审校论著

［1］谢联辉编译，1963. 论植物的免疫性 . 植物免疫学补充教材，福建农学院 . 1-97.

［2］谢联辉、林奇英等译 / 校，1978. 植物病理专辑，福建农学院译丛 .（1）：1-66.

［3］谢联辉、林奇英等译 / 校，1980. 植物病毒病，福建农学院译丛 .（1）:1-66.

［4］谢联辉、林奇英等译 / 校，1982. 植物病毒专辑，福建农学院译丛 .（3）:1-112.

［5］林奇英译，谢联辉校，1982. 用人工方法获得抗病突变体的遗传学分析（B. B. XBOXTOBA）植物病理学译丛（四）农业出版社，76-82.

［6］谢联辉译，胡文绣校，1983. 侵染环境和鉴定方法对抗病基因表现的作用（Э. Э. Гещел）.

［7］植物病理学译丛（五），农业出版社 . 12-16.

［8］谢联辉译 . 1986. 热带水稻和豆科植物国际讨论会论文摘要（筑波，TARC,Japan. 1985. 1-39）. 农业科学信息 . 2（2）:17-28.

［9］谢联辉、林奇英（译 / 校）：有关植病俄、英专业论文摘译，150 篇（条），分别刊于植物病理学文摘（1966、1978-1987）、农业科学信息（1985、1987）、国外农学——甘蔗（1985）等 .

五、博士后及其出站报告

［1］蒋继宏（谢联辉，林奇英）. 2000. 植物中抗病毒活性物质的分离及其作用机制研究 . 福建农业大学 .

［2］张巨勇（谢联辉，林奇英）. 2001. IPM 采用的经济学分析 [专著：有害生物综合治理（IPM）的经济学分析 . 北京：中国农业出版社，2004]. 福建农林大学 .

［3］李凡（谢联辉，林奇英）. 2006. 烟草丛顶病的病原物及其分子生物学 . 福建农林大学 .

［4］林河通（谢联辉，林奇英）. 2008. 真菌侵染所致龙眼果实采后病害的研究 . 福建农林大学 .

［5］吴刚（谢联辉，林奇英）. 2008. 菜田害虫和天敌抗药性进化的生态毒理学机制 . 福建农林大学 .

［6］魏远竹（谢联辉，林奇英）. 2010. 林业有害生物生态防治的经济学分析——以福建林业为例 . 福建农林大学 .

［7］王联德（林文雄，谢联辉）. 2010. 粉虱生物防治技术的若干问题研究 . 福建农林大学 .

［8］文才艺（谢联辉）. 2012. 抗植物病毒微生物代谢活性物质的筛选 . 福建农林大学 .

［9］吴昌标（谢联辉）. 2013. 柔嫩艾美耳球虫（*Eimeria tenella*）生物学特性及其生物防治 . 福建农林大学 .

［10］李熠（吴祖建）. 2014. 丝状子囊 nrDNA 进化研究 . 福建农林大学 .

［11］陈勇（魏太云）. 2015. 昆虫细胞自噬调控介体电光叶蝉传播水稻瘤矮病毒的机制 . 福建农林大学 .

［12］张洁（吴祖建）. 2017. 水稻病毒在水稻原生质体内的复制和表达 . 福建农林大学 .

［13］杨靓（吴祖建）. 2019. 水稻草矮病毒侵染导致水稻无效分蘖失控的分子机制 . 福建农林大学 .

［14］张金刚（吴建国）. 2020. 非保守 miRNA 在水稻和病毒互作中的功能 . 福建农林大学 .

六、博士及其学位论文

［1］周仲驹（谢联辉）. 1994. 香蕉束顶病的生物学、病原学、流行学和防治研究，福建农业大学 .

［2］吴祖建（谢联辉）. 1996. 水稻病毒病诊断、监测和防治系统的研究，福建农业大学 .

［3］鲁国东（谢联辉）. 1996. 稻瘟病菌转化体系的建立及三磷酸甘油醛脱氢酶基因的克隆，福建农业大学 .

［4］徐平东（谢联辉）. 1997. 中国黄瓜花叶病毒的亚组及其性质研究，福建农业大学 .

[5]李尉民（谢联辉）. 1997. 南方菜豆花叶病毒株系特异性检测、互补运动研究和菜豆株系 5′ 端的克隆，福建农业大学.

[6]王宗华（谢联辉）. 1997. 福建稻瘟菌的群体遗传规律，福建农业大学.

[7]胡方平（谢联辉）. 1997. 非荧光植物假单胞菌的分类和鉴定，福建农业大学.

[8]林含新（谢联辉）. 1999. 水稻条纹病毒的病原性质、致病性分化及分子变异①，福建农业大学.

[9]张春媚（林奇英，谢联辉）. 1999. 水稻草状矮化病毒基因组 RNA4-6 的分子生物学，福建农业大学.

[10]刘利华（林奇英，谢联辉）. 1999. 三种水稻病毒病的细胞病理学.

[11]王海河（谢联辉，林奇英）. 2000. 黄瓜花叶病毒三个株系引起烟草的病生理、细胞病理和 RNA3 的克隆，福建农业大学.

[12]孙慧（林奇英，谢联辉）. 2000. 抗植物病毒（TMV）活性的大型真菌的筛选和抗病毒蛋白的纯化及其性质，福建农业大学.

[13]郭银汉（谢联辉）. 2001. 福州地区对虾病毒病的病原诊断及流行病学，福建农林大学.

[14]林琳（谢联辉）. 2001. 扩展青霉 PF898 碱性脂肪酶基因的克隆与表达，福建农林大学.

[15]邵碧英（林奇英，谢联辉）. 2001. 烟草花叶病毒弱毒疫苗的研制及其分子生物学，福建农林大学.

[16]林丽明（谢联辉，林奇英）. 2002. 水稻草矮病毒基因组 RNA1-3 的分子生物学，福建农林大学.

[17]郑耀通（林奇英，谢联辉）. 2002. 闽江流域福州区段水体环境病毒污染、存活规律与灭活处理（专著：环境病毒学. 北京：化学工业出版社，2006），福建农林大学.

[18]吴兴泉（谢联辉，林奇英）. 2002. 福建马铃薯病毒的分子鉴定与检测技术，福建农林大学.

[19]付鸣佳（林奇英，谢联辉）. 2002. 食用菌抗病毒蛋白特性、基因克隆，福建农林大学.

[20]王盛（林奇英，谢联辉）. 2002. 珊瑚藻藻红蛋白分离纯化及相关特性，福建农林大学.

[21]魏太云（谢联辉，林奇英）. 2003. 水稻条纹病毒的基因组结构及其分子群体遗传，福建农林大学.

[22]翟梅枝（林奇英，谢联辉）. 2003. 植物次生物质的抗病活性及构效分析，福建农林大学.

[23]刘国坤（谢联辉，林奇英）. 2003. 植物源小分子物质对烟草花叶病毒及四种植物病原真菌的抑制作用，福建农林大学.

[24]林毅（林奇英，谢联辉）. 2003. 绞股蓝核糖体失活蛋白的分离、克隆与表达，福建农林大学.

[25]何红（胡方平，谢联辉）. 2003. 辣椒内生枯草芽孢杆菌（*Bacillus subtilis*）防病促生作用的研究，福建农林大学.

[26]徐学荣（谢联辉，林奇英）. 2004. 植保生态经济系统的分析与优化 [专著：植保经济学，福州：福建科学技术出版社，2012.]，福建农林大学.

[27]吴丽萍（林奇英，谢联辉）. 2004. 两种食用菌活性蛋白的分离纯化及其抗病特性，福建农林大学.

[28]刘振宇（林奇英，谢联辉）. 2004. 海藻中两种与植物抗病相关铜结合蛋白的分离、特性和基因特征，福建农林大学.

[29]祝雯（谢联辉，林奇英）. 2004. 河蚬 (*Corbicula fluminea*) 活性糖蛋白和多糖的分离纯化及其抗病特性，福建农林大学.

[30]欧阳迪莎（谢联辉，林卿）. 2005. 可持续农业中的植物病害管理，福建农林大学.

[31]洪荣标（谢联辉，林奇英）. 2005. 滨海湿地入侵植物的生态经济和生态安全管理，福建农林

① 全国百篇优秀博士学位论文

大学 .
[32]周莉娟（谢联辉）. 2005. GTP 在根结线虫（*Meloidogyne incogita*）拮抗菌 *Alcaligenes faecalis* 研究中的应用，福建农林大学 .
[33]沈建国（林奇英，谢联辉）. 2005. 两种药用植物对植物病毒及三种介体昆虫的生物活性，福建农林大学 .
[34]刘伟（谢联辉）. 2006. 芽孢杆菌（*Bacillus spp.* ）拮抗菌株的筛选及 TasA 基因研究，福建农林大学 .
[35]刘斌（谢联辉，章晓波）. 2006. 高温噬菌体分子特征及热稳定麦芽糖基淀粉酶的性质，福建农林大学 .
[36]黄宏南（谢联辉，林奇英）. 2006. 福建红曲的活性物质及其医疗保健效应，福建农林大学 .
[37]郑冬梅（赵士熙，谢联辉）. 2006. 中国生物农药产业发展研究 [专著：中国生物农药产业发展研究 . 北京：海洋出版社，2006]，福建农林大学 .
[38]林董（谢联辉，林奇英）. 2007. RSV 五个基因及 GFP-CP、GFP-SP 融合基因在 sf9 昆虫细胞中的表达，福建农林大学 .
[39]王林萍（谢联辉，林奇英）. 2007. 农药企业社会责任体系之构建 . [专著：企业社会责任体系的构建及实践——基于农药企业的分析 . 北京：科学出版社，2009.]，福建农林大学 .
[40]张福山（林奇英，谢联辉）. 2007. 植物保护对中国粮食生产安全影响的研究 [专著：植物保护与中国粮食生产安全研究 . 福州：福建科学技术出版社，2010]，福建农林大学 .
[41]连玲丽（谢联辉，林奇英）. 2007. 芽孢杆菌的生防菌株筛选及其抑病机理，福建农林大学 .
[42]方敦煌（谢联辉，林奇英）. 2007. 防治烟草赤星病根际芽孢杆菌的筛选及其抗菌物质研究，福建农林大学 .
[43]吴艳兵（林奇英，谢联辉）. 2007. 毛头鬼伞（*Coprinus comatus*）多糖的分离纯化及其抗烟草花叶病毒（TMV）作用机制，福建农林大学 .
[44]陈启建（林奇英，谢联辉）. 2007. 金鸡菊（*Coreopsis drummondii*）和小白菊（*Parthenium hysterophorus*）抗烟草花叶病毒活性研究，福建农林大学 .
[45]张晓婷（谢联辉，林奇英）. 2008. 水稻感染水稻条纹病毒后的基因转录谱和蛋白质表达谱，福建农林大学 .
[46]丁新伦（谢联辉，林奇英）. 2008. 脱落酸在水稻条纹病毒与寄主水稻互作的研究，福建农林大学 .
[47]杨金广（谢联辉，林奇英）. 2008. 水稻条纹病毒与寄主水稻互作中的生长素调控，福建农林大学 .
[48]鹿连明（谢联辉，林奇英）. 2008. 利用酵母双杂交系统研究水稻条纹病毒与寄主水稻间的互作，福建农林大学 .
[49]范国成（谢联辉，林奇英）. 2008. 水稻瘤矮病毒病毒样颗粒组装及 Pns12 蛋白的亚细胞定位，福建农林大学 .
[50]杨小山（谢联辉，林奇英）. 2008. 农药环境经济管理的主体行为及政策构思，福建农林大学 .
[51]张正坤（林奇英，谢联辉）. 2008. 鸦胆子活性物质抗烟草花叶病毒的作用机理及构效关系，福建农林大学 .
[52]程兆榜（谢联辉，范永坚）. 2009. 水稻条纹叶枯病生态控制原理及应用，福建农林大学 .
[53]程文金（谢联辉，吴祖建）. 2009. 水稻条纹病毒致病性分化与分子变异，福建农林大学 .
[54]杨彩霞（谢联辉）. 2009. 福建省六种双生病毒的分子鉴定及 RaMoV NSP 互作蛋白的筛选，福建农林大学 .

[55]胡志坚（林奇英，谢联辉）. 2009. 微囊藻毒素毒性及其致癌机制，福建农林大学.

[56]沈硕（谢联辉，欧阳明安）. 2009. 海洋真菌的鉴定及其次级代谢产物的研究，福建农林大学.

[57]吴建国（谢联辉）. 2010. 水稻矮缩病毒非结构蛋白 Pns11 的核仁定位及其与核仁蛋白 Fibrillarin 的互作，福建农林大学.

[58]林白雪（谢联辉）. 2010. 厦门近海温泉微生物多样性及热稳定纤维素酶的研究，福建农林大学.

[59]肖东来（谢联辉）. 2010. 水稻条纹病毒 NS2、NS3 蛋白及寄主间的互作，福建农林大学.

[60]杜振国（谢联辉，吴祖建）. 2011. 三种水稻病毒基因沉默抑制子的鉴定和研究，福建农林大学.

[61]孙炳剑（谢联辉，陈剑平）. 2011. 禾谷多黏菌传小麦病毒分布、遗传变异及品种抗性分析，福建农林大学.

[62]袁正杰（谢联辉，魏太云）. 2012. 水稻条纹病毒编码的 NSvc4 蛋白致病机理研究，福建农林大学.

[63]章松柏（谢联辉，吴祖建）. 2013. 南方水稻黑条矮缩病的监测及其病原病毒 Pns6 与寄主水稻的互作，福建农林大学.

[64]姜冬梅（谢联辉，李世访）. 2013. 锦紫苏类病毒的遗传变异及其与寄主之间的关系，福建农林大学.

[65]郑璐平（谢联辉，吴祖建）. 2013. 水稻条纹病毒基因 NS2 致病作用机理分析，福建农林大学.

[66]贾东升（谢联辉，魏太云）. 2013. SRBSDV 和 RRSV 在介体飞虱体内的侵染机理，福建农林大学.

[67]郑立敏（谢联辉，魏太云）. 2014. 水稻瘤矮病毒在介体昆虫内的增值及其诱导的细胞凋亡机制，福建农林大学.

[68]许曼琳（谢联辉，吴祖建）. 2014. 花生翻译起始因子 eIF4E 在 PStV 侵染过程中的功能分析，福建农林大学.

[69]高芳銮（谢联辉，詹家绥）. 2014. 中国马铃薯 Y 病毒群体遗传结构及其分子进化机制，福建农林大学.

[70]夏晓翠（魏太云，周冰峰）. 2014. 中华蜜蜂囊状幼虫病病毒在寄主体内的复制机制及其基于 RNAi 的防控策略，福建农林大学.

[71]王辉（谢联辉，郭宝珠）. 2015. 花生遗传图谱的构建以及番茄斑萎病毒和叶斑病抗性 QTL 的定位，福建农林大学.

[72]庄军（谢联辉）. 2015. BBTV 抗菌蛋白与血蓝蛋白抗菌肽的抑菌活性研究，福建农林大学.

[73]陈倩（谢联辉，魏太云）. 2015. 水稻矮缩病毒在介体叶蝉内的增值机理，福建农林大学.

[74]高博（谢联辉）. 2015. 天然产物对登革热病毒抑制作用研究，福建农林大学.

[75]邓萍（吴祖建，王爱民）. 2015. 芜菁花叶病毒 CI 蛋白参与病毒复制和运动及其与寄主的相互作用研究，福建农林大学.

[76]兰汉红（魏太云）. 2015. 介体昆虫 RNA 干扰途径调控水稻病毒的侵染，福建农林大学.

[77]何敦春（谢联辉，程兆榜，王林萍）. 2016. 水稻条纹叶枯病流行风险及管理，福建农林大学.

[78]刘小娟（吴祖建，杜振国）. 2016. 水稻条纹病毒和水稻草状矮化病毒“抓帽”机制比较研究，福建农林大学.

[79]张超（吴祖建，吴建国）. 2016. 基于 Small RNA 的水稻锯齿叶矮缩病毒致病机制研究，福建农林大学.

[80]郑立佳（谢联辉，李毅，吴建国）. 2017. 水稻条纹病毒调控 miRNA 加工的机制研究，福建农

林大学 .

［81］毛倩卓（魏太云）. 2017. 水稻瘤矮病毒经介体电光叶蝉水平和垂直传播的机制，福建农林大学 .

［82］熊桂红（吴祖建）. 2017. 水稻草状矮化病毒与寄主互作蛋白的鉴定与研究，福建农林大学 .

［83］吴元兴（谢联辉，徐学荣）. 2018. 福建省水稻病虫害专业化防治发展研究，福建农林大学 .

［84］隋雪莲（吴祖建、凌开树、陈启建）. 2018. 美国三种新兴蔬菜病毒病的特性研究及其检测方法的建立，福建农林大学 .

［85］张振鲁（谢联辉，王晓峰，吴祖建）. 2018. 酵母 Lipin 通过调控磷脂合成抑制病毒复制及寄主细胞生长，福建农林大学 .

［86］Muhammad Arif（吴祖建）. 2018. 福建省 begomoviruses 的检测和分子生物学研究，福建农林大学 .

［87］Saiful Islam（吴祖建）. 2018. 中巴蜻蜓分布和遗传多样性及其相关单链 DNA 病毒鉴定，福建农林大学 .

［88］Waqar Islam（吴祖建）. 2018. 巴基斯坦 *begomovirus* 和烟粉虱的遗传多样性及地理分布，福建农林大学 .

［89］林文武（谢联辉，Peter D. Nagy，吴祖建）. 2019. TBSV 招募寄主乙醇发酵相关酶促进病毒 RNA 合成的分子机制，福建农林大学 .

［90］丁作美（吴祖建，吴建国）. 2019. 水稻支链氨基酸转氨酶（OsBCATs）在水稻锯齿叶矮缩病毒侵染过程中的功能，福建农林大学 .

［91］王海涛（魏太云）. 2019. 水稻黄矮病毒侵染介体黑尾叶蝉中枢神经系统的机制，福建农林大学 .

［92］吴维（魏太云）. 2019. 黑尾叶蝉卵黄原蛋白 - 共生菌 - 水稻矮缩病毒三者互作介导的经卵传播机制，福建农林大学 .

［93］何桂娟（谢联辉，王晓峰，吴祖建）. 2020. 改造寄主磷脂合成酶 CHO2 基因抑制雀麦花叶病毒复制，福建农林大学 .

［94］郑蓉蓉（谢联辉，徐学荣，何敦春）. 2020. 茶树病虫生态防控的效益评价，福建农林大学 .

［95］俞超维（吴祖建）. 2021. 中国百香果病毒病调查鉴定及抗病相关基因的解析，福建农林大学 .